3rd International Conference on Mechanics and Mechatronics Research (ICMMR 2016)

MATEC Web of Conferences Volume 77 (2016)

Chongqing, China
15-17 June 2016

Editors:

D. Zhang
G. Cui
B. Zi
H. Ding

ISBN: 978-1-5108-3046-2

Printed from e-media with permission by:

Curran Associates, Inc.
57 Morehouse Lane
Red Hook, NY 12571

Some format issues inherent in the e-media version may also appear in this print version.

Printed by Curran Associates, Inc. (2016)

For additional information, please contact EDP Sciences – Web of Conferences at the address below.

EDP Sciences – Web of Conferences
17, Avenue du Hoggar
Parc d'Activité de Courtabœuf
BP 112
F-91944 Les Ulis Cedex A
France

Phone: +33 (0) 1 69 18 75 75
Fax: +33 (0) 1 69 28 84 91

contact-edps@webofconferences.org

Additional copies of this publication are available from:

Curran Associates, Inc.
57 Morehouse Lane
Red Hook, NY 12571 USA
Phone: 845-758-0400
Fax: 845-758-2633
Email: curran@proceedings.com
Web: www.proceedings.com

TABLE OF CONTENTS

CONFERENCE COMMITTEES

International Advisory Committee

Prof. Wang Jun, The Chinese University of Hong Kong, Hong Kong

Conference Committee Chair

Prof. Dan Zhang, York University, Canada

Program Committee Chairs

Prof. Bin CHEN, Vice director, Xi'an Jiaotong University, China
Prof. Sheng Guo, Beijing Jiaotong University, China
Prof. Ge Chen, DongHua University, China
Prof. Yuefa Fang, Beijing Jiaotong University, China
Prof. Bin Zi, Hefei University of Technology, China
Prof. Guohua Cui, Hebei University of Engineering, China
Prof. Huafeng Ding, China University of Geosciences, China
Prof. Yansong Wang, Shanghai University of Engineering and Technology, China

Technical Committee Chairs

Prof. Wei-chen Lee, National Taiwan University of Science and Technology, Taiwan
Prof. Ye Sun, Michigan Technological University, USA
Prof. Zhongchao Tan, University of Waterloo, Canada
Prof. Jie Zhang, University of Texas at Dallas, USA
Dr. Xiuling Wang, Purdue University Calumet, 46323, Hammond, USA
Dr. Uğur Arabacı, Gazi University, Turkey
Dr. Behrouz Takabi, Texas A&M University, College Station, Texas, USA
Prof. Dr. MELLAL Mohamed Arezki,LMSS, M"Hamed Bougara University, Algeria
Prof. Balhassn S. M. Ali, Bursa Orhangazi University, Turkey
Dr. Muhammad Naufal Bin Mansor, Universiti Malaysia Perlis (UniMAP), Malaysia
Prof. Dezhen Feng, University of Jinan, China
Dr. Zhang Yanyan, Harbin Institute of Technology, China
Prof. Atilla Altintas, 1Karabuk University, Turkey
Dr. Zhou Jiehua, Nanchang Hangkong University, China
Prof. CHENG Qi-you, China Helicopter Research and Development Institute, China
Dr. Alexander Popov, Lobachevsky State University of Nizhny Novgorod, Russia
Dr. YANG Xiaole, Beijing institute of space Mechanics & Electricity, China

Preface

2016 3rd International Conference on Mechanics and Mechatronics Research (ICMMR 2016) will be held during June 15-17, 2016, in Chongqing China. This conference covered recent trends and progresses made in the field of mechanics and mechatronics research. The technical objective of this conference is to address key issues associated with science and technology in this rapidly evolving field of research and to promote contact between basic researchers and technological needs for real industrial applications. This event is intended as a vehicle for the dissemination of research results on latest advances made in the area of mechanics and mechatronics research, which gave an excellent opportunity for the scientists and researchers around the world to have a common platform to exchange their findings and to discuss the developments and make a base for starting collaborations in the national and international level. It was also aimed at promoting contacts amongst researchers and research groups for the creation of multinational thematic and research networks, as well as promoting contacts for future collaborative joint projects within some of the international funding programs.

The conference program offered invited, oral and poster presentations. The activities of the conference included invited speeches and author presentations in sessions. The invited talks were focused on the potential application of mechanics and mechatronics research in various fields of application.

ICMMR 2016 conference proceedings are addressing diverse topics related to recent trends in mechanics and mechatronics research. We hope the volume will offer the readers a good opportunity to access the state-of -the art developments and future directions of researches on mechanics and mechatronics research.

We're truly grateful to all the members of the conference committee, the reviewers who spared their valuable time, for their advice which have certainly helped to improve the quality, accuracy, relevance and the sincere efforts to maintain the quality of each paper selected for this conference program and volume for publication. And thank every author and participant who has contributed to the success of the conference.

Prof. Dan Zhang, York University, Canada

ICMMR 2016 Committees,
June, 2016

Publication Chairs

Dr. Bin Wei, University of Ontario Institute of Technology, Canada
Dr. Jian Yang, Chongqing University, China

Statement of Peer review

In submitting conference proceedings to MATEC Web of Conferences, I certify to the Publisher that I adhere to the **Policy on Publishing Integrity** of the journal in order to safeguard good scientific practice in publishing.

1. All articles have been subjected to peer review administered by the proceedings editors.
2. Reviews have been conducted by expert referees, who have been requested to provide unbiased and constructive comments aimed, whenever possible, at improving the work.
3. Proceedings editors have taken all reasonable steps to ensure the quality of the materials they publish and their decision to accept or reject a paper for publication has been based only on the merits of the work and the relevance to the journal.

Title, date and place of the conference

2016 3rd International Conference on Mechanics and Mechatronics Research (ICMMR 2016)

Chongqing, China during June 15-17, 2016

Proceedings editor(s):

Prof. Dan Zhang, York University, Canada

Prof. Bin Zi, Hefei University of Technology, China

Prof. Guohua Cui, Hebei University of Engineering, China

Prof. Huafeng Ding, China University of Geosciences, China

Date and editor's signature

MATEC Web of Conferences 77, 01001 (2016) DOI: 10.1051/matecconf/20167701001
ICMMR 2016

The Study on Re Effect Correction for Laminar Wing with High Lift

Jieke Yao, Wenliang Feng, Lingying Lv and Bin Chen

Chengdu Aircraft Industrial (group) CO.LTD, 610092, Chengdu, China

Abstract. In the past years, Reynolds number (Re) effect correction is carried out by varied Re experiment. The experiment can obtain double logarithm linear curve between the least drag and Re, and the least drag in flying Re is got by linear exploration. However, Re effect exerts a great influence on transition position for laminar wing with high lift. Therefore, the lift, drag and longitudinal moment can't be acquired by traditional Re effect correction. In this paper, the transition prediction simulations for laminar wing high lift are caried to obtain unconstrained transition position in different Re by adopting γ-Re_θ transition prediction model. An agreement with experiment for transition position is achieved. According to the result of transition perdition and unconstrained transition test, Re effect correction in different height for the data of wind tunnel test is done. This method makes up for the traditional method, which Re effect correction for battle-plane is done for only drag. In present paper, Re effect correction for the lift, drag and longitudinal moment is done to acquire more accurate experimental data to ensure everything in flight.

1 Introduction

Usually, the scaled-down model is adopted in the wind tunnel tests, which makes a greater difference between the wind tunnel tests and actual fight situation for the Reynolds number (Re). Therefore, correction based on the Re effect for wind tunnel test data must be done.

Study on the Re effect of battle-plane shows that flow is almost all turbulence in the near boundary layer of the wings when it flies in the atmosphere. This attributes to apply thinner wings for the battle-plane. For such wings, Re has a smaller effect on the lift and longitudinal moment characteristics, but it exerts a powerful influence on the frictional drag on the surface of wing. Consequently, the Re effect correction is carried out for only the drag including the least drag (C_{Dmin}) and drag due to lift in the correction of wind tunnel data.

In order to pursue the higher cruising efficiency, civil aircraft and Unmanned Aerial Vehicle (UAV) having high height and speed properties in the modern world , such as supercritical aerofoil and supercritical laminar compound aerofoil, usually use high lift-drag ratio aerofoil. These aerofoils retain the natural laminar region at least 50% chord length in the designed cruising state. The research presents that the drag for laminar aerofoil is less than half usual turbulent aerofoil in the designed state. However, the natural laminar region disappears and the drag of aerofoil is rising rapidly when aerofoil deviates the designed state. And at the same time, the experimental research also shows that macroscopically aerodynamic characteristics containing the lift, drag and longitudinal moment for the supercritical aerofoil is influenced by Re because its flow structure and pressure distribution are changed by Re [1]. Transition position of supercritical aerofoil is also affected by Re. Transition which means transform from laminar flow to turbulence flow is very important concept in the boundary layer theory [2]-[5].

The correction method for Re effect of battle-plane is acquired by varied Re experiment. According to experiment, linearized curve, which C_{Dmin} varies according to Re, is obtained. Therefore, C_{Dmin} for experiment Re is corrected to actual fight Re by extrapolation.

Numerical simulation based on γ-Re_θ transition predicting mode for the wing is carried out in this paper, and the unconstrained transition position of wing in different Re is got. Meantime, forced transition test is carried out in the wind tunnel test, and aerodynamic data in the different transition position of wing is obtained. On the basis of results for the transition position of numerical calculation and data of unconstrained transition experiment, data correction based on Re effect for the lift and drag of wind tunnel test is done [6]-[11].

2 The illustration of computation

The commercial software ANSYS CFX, which is usually used by aerospace industry, is adopted for present study solving the Navier-Stokes (NS) equation. CFX is based finite-volume method of finite element, adopts coupled solution of full implicit-scheme and multi-mesh, and uses interpolation of 24 points for hexahedral mesh. Computational model employs hexahedral mesh in present paper, account of computational mesh is about 10 million. The geometry with grids is shown in the Fig. 1.

No-slip boundary and opening boundary are respectively set up for solid boundary and far field boundary in this paper, and transition computation chooses Shear Stress Transport (SST) γ-Re_θ model. The aim for transition model is solving intermittency function γ ($0 \le \gamma \le 1$), which is chances of turbulence for the flow pattern of a point in space, and the aim for combination of transition model and SST turbulence model is controlling occurrence for the transition [12]-[14]. However, in some other similar works, even very recent studies, concerning on the Computational Fluid Dynamics (CFD) investigation of a fluid flow, they have employed other turbulence model such as the RNG-k-ε [15], standard k-ε model [16], bybrid URANS-LES [17].

Figure 1. The geometry with grids

3 Forced transition experiment

In order to investigate the difference of aerodynamic characteristic on different transition position of wing, forced transition wind tunnel test is carried out by pasting special tape on the different position for the surface of wing. The results of experiments show that the changes, which C_{Dmin} , angle of attack for 0 lift (α_0), slope of lift curve ($C_{L\alpha}$) and longitudinal stability derivatives ($C_m CL$) change with the transition position, basically presents linear. Re for the experiment is 1.23×10^6, 1.51×10^6 and 1.76×10^6 for Mach number (Ma) of 0.4, 0.5 and 0.6, respectively. The results are showed from Fig. 2 to Figure 6. From these figs, it is seen that aerodynamic data of different Ma presents certain difference in the same transition position, which mean that Ma exert an influence on the aerodynamic characteristic. Because different Ma reflects different experiment Re, Re has an effect on aerodynamic data. And the influence of Re is greater.

Forced transition experiment can't directly obtain data for the transition position of wing, and there are also more difficult to get transition position by unconstrained transition experiment. In this paper, some data, such as α_0 , $C_{L\alpha}$, C_{Dmin} , are obtained by unconstrained transition experiment of full-aircraft, and the transition position of wing is acquired by linear extrapolation in the Fig. 2, Fig. 3 and Fig. 4.

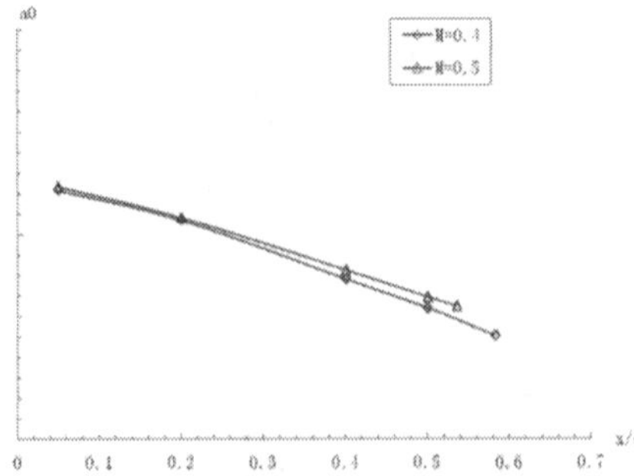

Figure 2. The relationship between α_0 and transition position

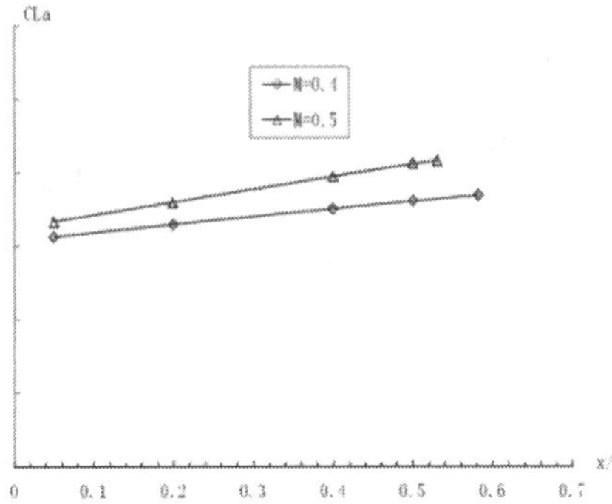

Figure 3. The relationship between $C_{L\alpha}$ and transition position

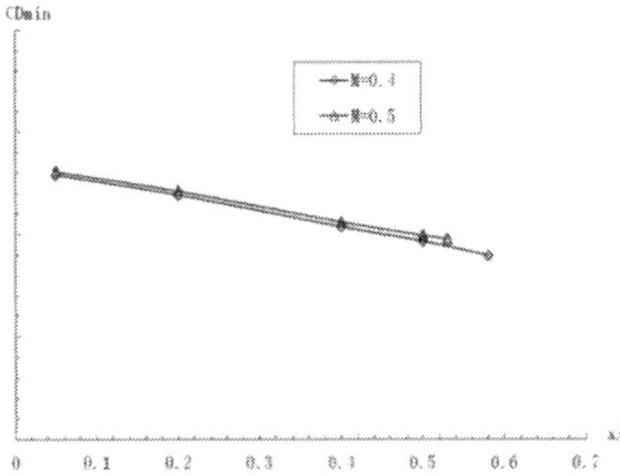

Figure 4. The relationship between C_{Dmin} and transition position

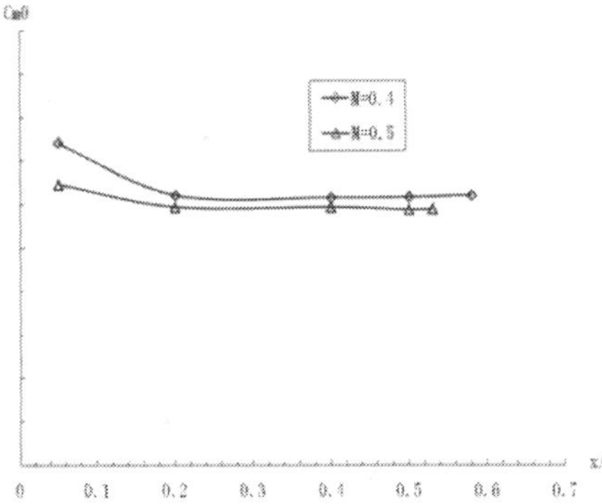

Figure 5. The relationship between moment for 0 lift (C_{m0}) and transition position

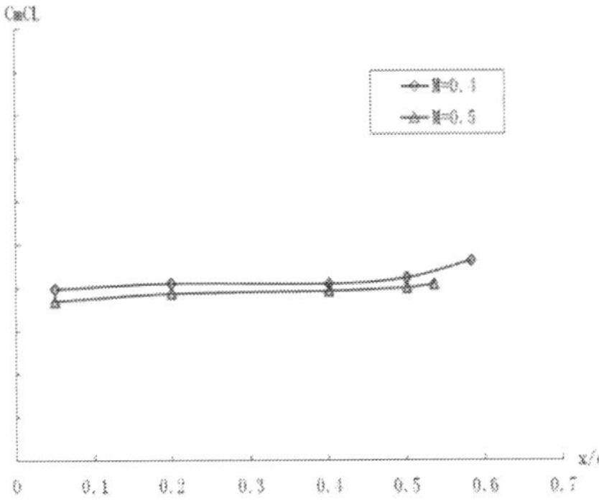

Figure 6. The relationship between $C_m CL$ and transition position

4 Supercritical characteristic of wing

In the initial design of wing, supercritical laminar compound airfoil was used to seek higher lift-drag ratio and critical Ma. In the later design, because the aircraft changed target and reduced flight speed, supercritical characteristic of wing wasn't used. Therefore, supercritical characteristic which is critical Ma wasn't fully verified in the wind tunnel test.

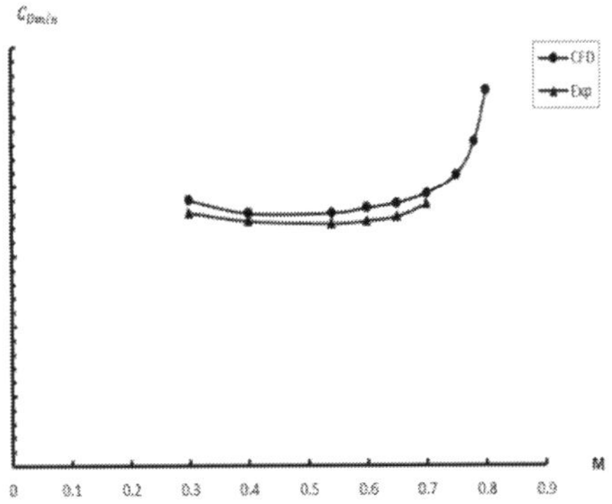

Figure 7. The comparison C_{Dmin} of wing between experiment and CFD

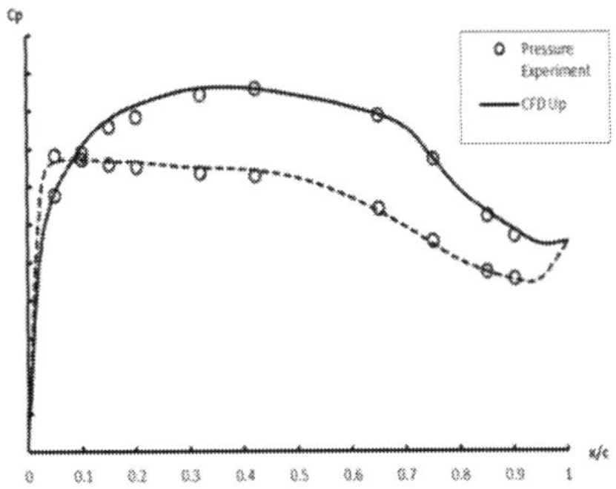

Figure 8. Pressure distribution of wing section (M=0.65, α=0°)

The comparison of C_{Dmin} for the wind tunnel test and CFD simulations is shown in Figure 6. From the result of the Fig. 7, the critical Ma ($\partial C_{Dmin}/\partial Ma$=0.18) is about 0.8. Fig. 8 and Fig. 9 show the comparison of pressure experiment and CFD simulation. That can be seen that the higher negative pressure doesn't present in the upper surface of the wing leading edge, pressure distribution is relatively flat in upper surface of middle wing. These can effectively weaken intensity of shock wave and increase the critical Ma. According to the above results, benign supercritical characteristic of wing can be presented.

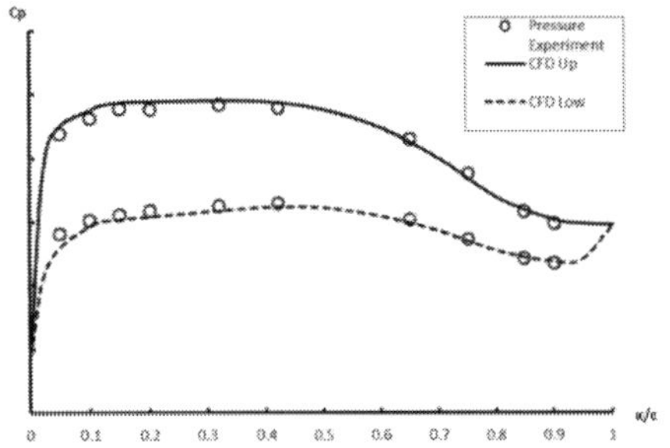

Figure 9. Pressure distribution of wing section (M=0.65, α=3°)

5 The comparison of computation and experiment

According to the result of forced transition for wing different position in the wind tunnel test, the experimental data for unconstrained transition position of wing is obtained by interpolation. Computational condition and experimental condition for unconstrained transition are identical.

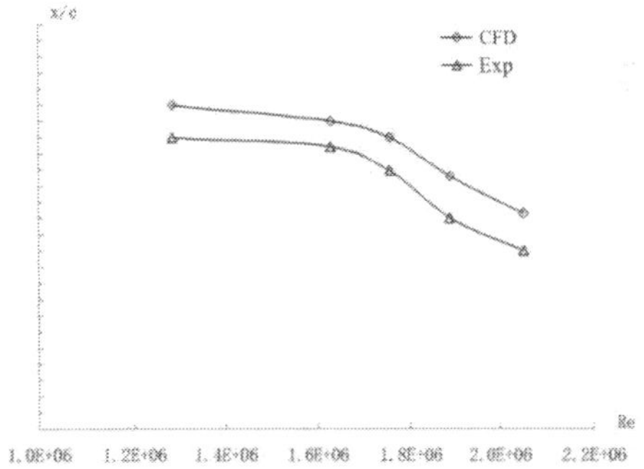

Figure 10. The comparison of transition position for CFD and experiment

Because Re change in a relatively smaller range in the wind tunnel test, the experimental Re is under 2.5×106. Fig. 10 gives the comparison of transition position for CFD simulation and the wind tunnel test. It can be seen that unconstrained transition position of wing for CFD simulation is back about 2% local chord than the tunnel test, varied trend, which transition position changes with Re is identical for CFD and experiment, and a great agreement is obtained.

Fig. 11, Fig. 12 and Fig. 13 show intermittent factor image on the upper surface of wing in different Re. It can be seen that intermittent factor is almost 0, it immediately increases when reaching the transition position.

Meantime, because of the influence of fuselage flow field, laminar range for inner section of wing is smaller than outer section of wing and the distribution of transition position along wingspan is no uniform.

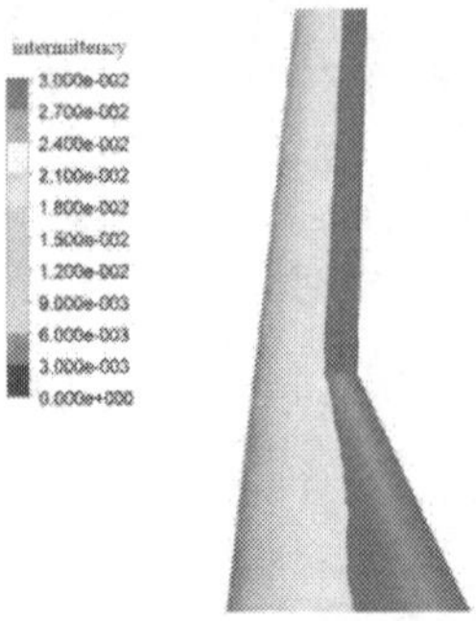

Figure 11. The distribution of intermittent factor (Re=2.0×10^6)

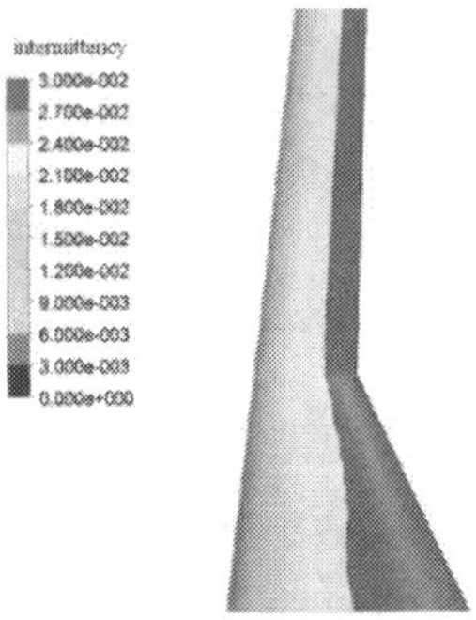

Figure 12. The distribution of intermittent factor (Re=5.0×10^6)

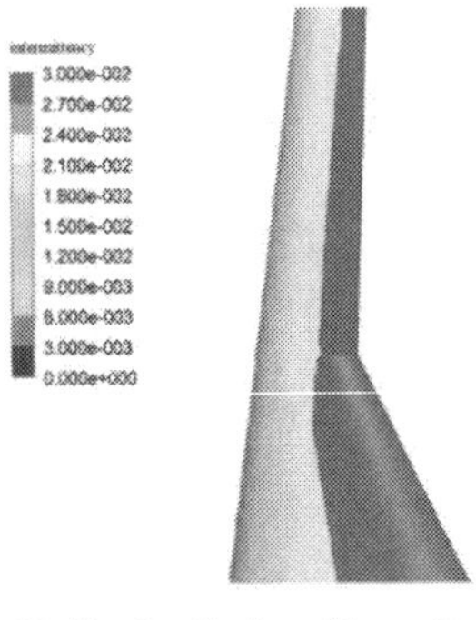

Figure 13. The distribution of intermittent factor (Re=8.0×10^6)

6 The correction method for Re effect

According to aircraft envelope, the range of flight Re is confirmed and numerical simulations for these Re are carried out to compute transition position in different height and speed. Because of deviation of 2% local chord for transition position of CFD and the wind tunnel test, the transition position for CFD is done certain correction.

Under the transition position of wing for different Re simulated by CFD and linear relation of transition position and C_{Dmin} , α_0 , $C_{L\alpha}$, C_{m0} and $C_m CL$, aerodynamic derivatives in corresponding Re are obtained by linear interpolation (form Fig. 2 to Fig. 6). The formulas of correction are as follows.

$$C_L^{Re} = C_L^{exp} + \Delta C_L^{Re} \tag{1}$$

$$\Delta C_L^{Re} = \Delta\alpha_0^{forced\ transition} + \Delta C_{L\alpha}^{forced\ transition} * \alpha \tag{2}$$

$$C_D^{Re} = C_D^{exp} + \Delta C_D^{Re} \tag{3}$$

$$\Delta C_D^{Re} = \Delta C_{Dmin}^{forced\ transition} \tag{4}$$

$$C_m^{Re} = C_m^{exp} + \Delta C_m^{Re} \tag{5}$$

$$\Delta C_m^{Re} = \Delta C_{m0}^{forced\ transition} + \Delta C_{mCL}^{forced\ transition} * C_L^{Re} \tag{6}$$

where $\Delta\alpha_0^{forced\ transition}$, $\Delta C_{L\alpha}^{forced\ transition}$, $\Delta C_{Dmin}^{forced\ transition}$, $\Delta C_{m0}^{forced\ transition}$, $\Delta C_{mCL}^{forced\ transition}$ are interpolated data in curve for forced transition experiment, respectively. According these experimental results, the derivatives for nonlinear range, such as $\Delta C_{L\alpha}^{forced\ transitin}$ and $\Delta C_{mCL}^{forced\ transition}$, are corrected.

Unconstrained transition position computed in different Ma and Re is shown in Fig. 14 From this fig, it can be seen that while Re is smaller than 7×10^6, transition position is hardly changed, however, when Re is greater than 7×10^6, transition position is forward sharply and laminar range is shorten.

Experimental speed simulated for wind tunnel test is only similar to the real flight, but aerodynamic data in different height can't be obtained. Different height for flight of aircraft indicates different Re. Therefore, Re effect correction for the data for the wind tunnel test must be done to obtain aerodynamic data in different height of flight.

Fig. 15 and Fig. 16 show polar and pitching moment curve corrected for Re effect in different height. Fig. 17 gives the comparison of data for wind tunnel test corrected by Re effect and aerodynamic data of flight test. The lift and drag of flight test are obtained by back-stepping for weight and thrust of engine for flight. From these results, an agreement is got for wind tunnel test data corrected by Re effect and flight test data, but certain differences are leaded to by thrust losing unable to obtain for flight data and data corrected.

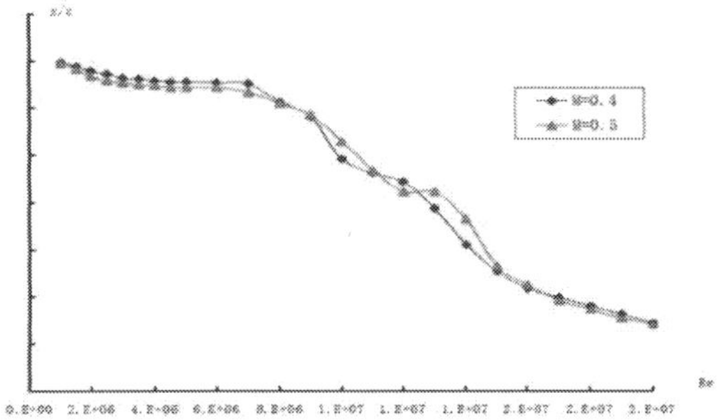

Figure 14. The relationship between transition position obtained by CFD and Re

DOI: 10.1051/matecconf/20167701001

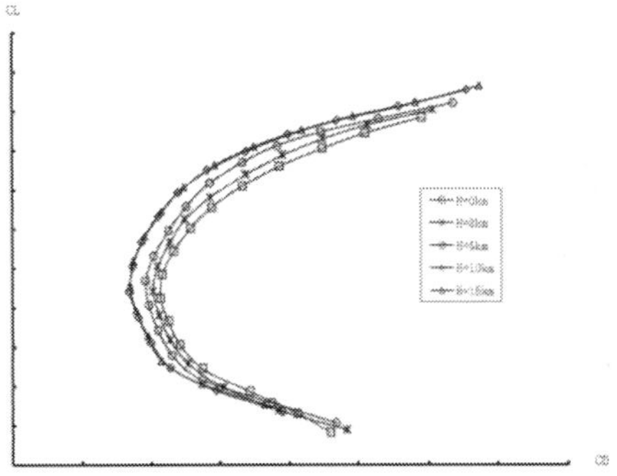

Figure 15. The polar of lift and drag corrected by Re effect

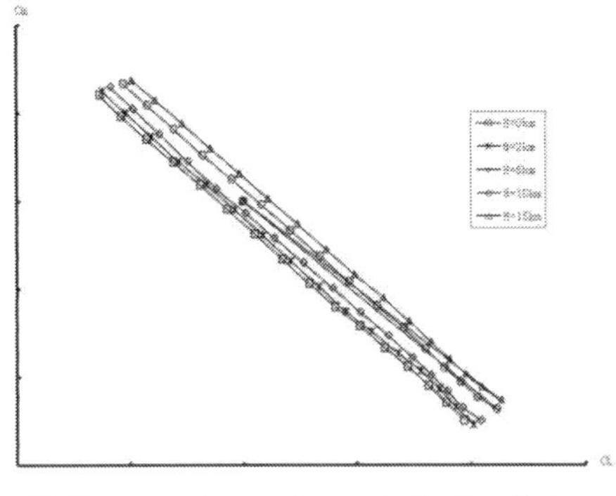

Figure 16. The curve of moment corrected by Re effect

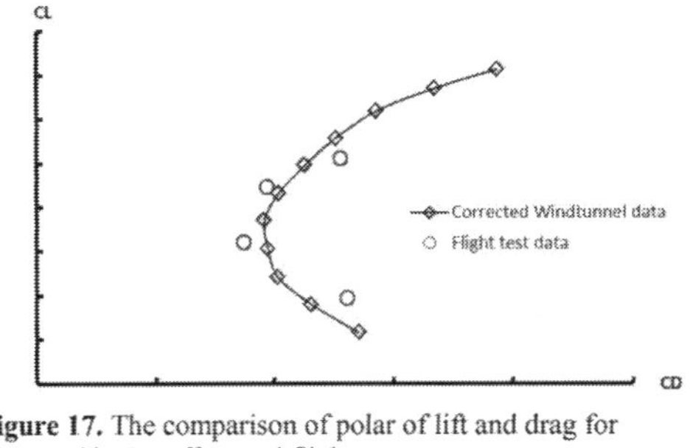

Figure 17. The comparison of polar of lift and drag for corrected by Re effect and flight test

7 Conclusion

According to the prediction for transition position and the method of Re effect correction for unconstrained transition experiment, more accurate and reliable data can be obtained. This corrected method has been applied in the aircraft of laminar wing with high lift and has been validated by experiment of flight. Present method may be used as reference to Re effect correction for data of wind tunnel test of the similar aircraft.

References

1. U. Frisch, B. Hasslacher, Y. Pomeau. "Lattice-gas automata for the Navier-Stokes equation". Physical Review Letters, 1986, vol. 56, pp. 1505-1508.
2. X. XU, D. W. Wei, D. H. Chen, Y. J. Wang. Analysis of flow over super-critical airfoil Reynolds number effects. CSTAM, 2012-B03-0140.
3. Y. C. Chen, J. T. Si, X. L. Han, J. L. Wei. Investigation of transition effect on the pressure distribution of super-critical wing. Acta Aerodynamica Sinica, **21** (4), pp. 470-475, (2003).
4. Q. Z. Yang, Z. Y. Zhang. Analysis of the boundary layer and aerodynamic characteristics of a super-critical laminar wing. Acta Aerodynamica Sinica, **25**(5), pp. 439-422, (2004).
5. Y. Chen, Z. H. Gao. Application of gamma-theta transition model to flows around airfoils. Acta Aerodynamica Sinica, **27**(4): 411-418, 2009.
6. D. B. DeGraff,D. R. Webster,J. K. Eaton. The effect of Reynolds number on boundary layer turbulence. Experimental Thermal and Fluid Science, **18**: 341-346, (1999).
7. S. Y,H. P. Modeling of flow transition using an intermittency transport equation. Journal of Fluids and Engineering, **122**(2): 273-284, (2000).
8. Y. B. Suzen,G. Xiong, P. G. Huang. Prediction of transition flows in a low-pressure turbine using an intermittency transport equation. AIAA-2000-2645.
9. Y. L. Zhang, G. X. Wang, D. H. Meng, Y. T. Wang. Calibration of γ-Reθ transition model. Acta Aerodynamica Sinica, **29**(3): 295-301, (2011).
10. R. B. Langtry,F. R. Menter. Transition modeling for general CFD applications in aeronautics. AIAA paper, 2005-522.
11. R. E. The role of laminar-turbulent transition in gas turbine engines. ASME Journal of Turbo machinery, **113**: 509-537, (1991).
12. R. B. Langtry, F. R. Menter, S. R. Likki, Y. B. Suzen, P. G. Iiuang, S. Volker. A correlation based transition model using local variables: part II-test case and industrial application. ASME-GT, 2004-53454.
13. F. R. Menter, R. B. Langtry, S. R. Likki, Y. B. Suzen, P. G. Iiuang, S. Volker. A correlation based transition model using local variables: part I-model formulation[R]. ASME-GT, 2004-53452.
14. R. B. Langtry. A correction-based transition model using local variables for unstructured parallelized CFD codes. Ph D Thesis, Stuttgart, (2006).
15. M. Miltner, M. Jordan. CFD simulation of straight and slightly swirling turblence free jets using difference RANS-turblence model[J]. Applied thermal engineering, **89** :1117-1126, (2015).
16. B. Takabi, H. Shokouhmand. Effect of Al203/water hybrid nanofluid on heat transfer and flow characteristics in turbulent regime[J]. International journal of modern pyhsics, **26**(4) : 1550047-1550050, (2015).
17. G. Vijayakumar, J. G. Brasseur, A. Lavely, B. Jayaraman, B. C. Craven. Interaction of atmospheric turbulence with blade boundary layer dynamics on a 5MW wind turbine using blade-boundary-layer-resolved CFD with hybird URANS-LES[C]. 34th wind energy symposium, California, USA, (2016).

MATEC Web of Conferences 77, 01002 (2016) DOI: 10.1051/matecconf/20167701002
ICMMR 2016

Research of Transformer Intelligent Evaluation and Diagnosis Method Based on DGA

Wang Feng, Bi jiangang, Zhang bowen, Yuan Shuai

China Electric Power Research Institute, Haidian District, Beijing, China

Abstract. Oil chromatography analysis is the foundation of transformer fault diagnosis, In this paper, based on knowledge base of the oil chromatographic analysis, an extended research combining the improved three ratio method, BP neural network, and case-based reasoning method for the transformer fault diagnosis was proposed, in the purpose of building a transformer Intelligent Evaluation Diagnosis model, to improve the accuracy of the transformer fault diagnosis, which is of a great significance for practice.

1 Introduction

With the development of society and economy, energy and environmental issues is becoming increasingly prominent. In order to meet the requirements of a sustainable future, the development and construction of a strong smart grid is an inevitable choice to promote China's economic development.

Safe operation of the power transformer is to ensure a reliable, safety power grid and the necessary conditions for stable operation. The main transformer on-line monitoring of transformer oil chromatogram is analysed to determine the latent fault transformer. Oil chromatographic analysis is one of the most effective methods of early detection and prevention of transformer failures. Because the neural network to outside input sample has a strong ability to identify and classify information processing characteristics of associative memory, making the application of artificial neural networks and fault diagnosis of transformer combined. Therefore, the neural network fault diagnosis method based on chromatographic analysis of transformer oil for transformer fault diagnosis provides a new way.

2 Transformer fault diagnosis research status

Power transformer fault, from the development process can be divided into two categories: sudden failure and latent failures. For sudden failure we can use some conservative measures for prevention, existing methods mainly for the purposes of latent failures. Latent failures can be divided into three categories: internal transformer partial discharge, local overheating and transformer insulation aging. Power transformer is closely related to the failure of the transformer condition, history log, and transformer models. Therefore, it is necessary to study the transformer fault, timely and accurate detection of early latent transformer fault detection and maintenance system to develop a reasonable, accurate service, improve the reliability of transformers [1].

Expert System has received a preliminary application, but the main problem is that the "bottleneck" problem of knowledge acquisition cannot effectively solve so many uncertainties that exist in fault diagnosis, these problems greatly affected the accuracy of fault diagnosis. Fuzzy diagnosis method using fuzzy matching principle troubleshooting, there are some false positives, generally rarely used. In a variety of intelligent algorithms, neural network model with good nonlinear approximation capability has been widely adopted, has been well developed in the transformer fault diagnosis [2].

In this paper, the traditional methods and BP intelligent algorithm combined with the introduction of Case-based inference and thought transformer fault diagnosis system, conducted in-depth research.

3 Establish transformer fault diagnosis model

The failure mechanism of transformers is often very complex, accompanied by a variety of fault generation, which can be used to characterize and determine the fault type of state amount is often more than one correspondence between the amount of state failure and more complicated. Failure mechanism is the basis for fault diagnosis master, research and state the amount of transformer failure relationship is very necessary. The actual amount of fault diagnosis system using state oil chromatogram, the core ground current, infrared, casing, top oil temperature, moisture.

3.1 Transformer condition data quality analysis

During the first transformer fault diagnosis before the collected data processing and analysis, the main purpose is to:

1) Random noise can be treated through treatment, to improve signal to noise ratio, in order to prevent false positives or false negatives fault made, the key is to improve the anti-jamming measures.

2) In addition to improving the SNR beyond, the need to collect data that can better reflect the information displayed, which can not simply set out the collected raw data, while outside to the inside to do some analysis and processing make it a reliable criterion for transformer fault diagnosis.

Quality analysis of the data included four areas: identification of invalid data; complete analysis; normative analysis; stable consistency analysis.

3.2 Transformer fault diagnosis process

Transformer fault diagnosis in two steps: to determine whether the fault and the fault type. In the absence of a transformer failure, want to run the state of the transformer to predict and forecast, after a failure of judgment transformer fault type, location, cause, degree, and maintenance decisions. Specific steps are as follows [3].

3.2.1 Establish fault sample library

Failure to create a sample database according to the device type, model, manufacturer, etc. and the accumulation of knowledge and provide training and validation samples for fault diagnosis algorithm.

3.2.2 Establish the diagnosis algorithm library

Diagnostic algorithm library provides various types of equipment through the package, all kinds of state amount of diagnostic algorithms. Diagnostic algorithm used in this paper include traditional DGA algorithm, advanced intelligent algorithm with BP algorithm combines Case-based inference.

3.2.3 Status diagnostics

Firstly whether the fault diagnosis, determine fault, respectively IEC, BP neural network and Case-based inference three methods to conduct a diagnosis, identified as overheating fault or electrical fault, and then use three methods for secondary diagnosis, determine the specific type of fault.

3.2.4 Diagnostic results

To determine what kind of fault to fault location, cause of the failure, malfunction, etc. rating judgment.

3.3 DGA principle

In the situation of failure, the transformer oil will be oxygen-decomposed and split-decomposed, and then create the hydrogen and low molecule gases. Unsaturation of the hydrocarbon gas will increase along with the pyrolysis energy density (temperature) growing. Different fault, due to the failure point energy is different at different temperatures and different insulation materials involved in its production of gas situation is different. The use of DGA for device internal fault judgment is based on this principle of gas production characteristics of insulating materials.

Part of the gas dissolved in transformer oil, part floating on the upper (free gas).The valuable gases are H_2、CH_4、C_2H_6、C_2H_2、C_2H_4、CO and CO_2. Table 1 gives an improved three-ratio method of encoding rules [4].

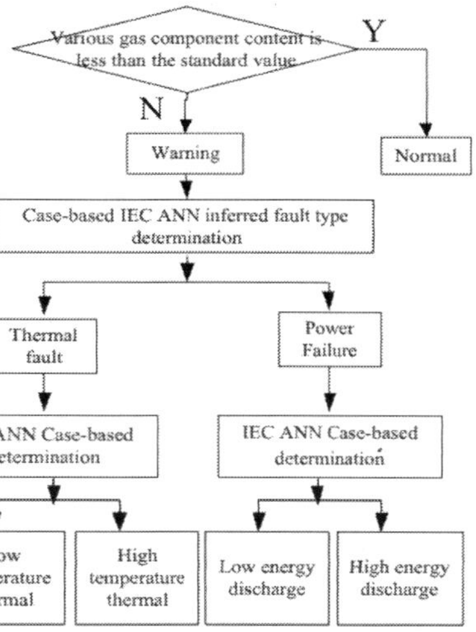

Figure 1. Integrated troubleshooting flowchart transformer

Table 1. Improved three-ratio encoding rules

The ratio range	Coding ratio range		
	C_2H_2/C_2H_4	CH_4/H_2	C_2H_4/C_2H_6
<0.1	0	1	0
0.1~1	1	0	0
1~3	1	2	1
⩾3	2	2	2

Improved three-ratio method (IEC) is based on the principle of a coding in table 1 to calculate the ratio of the gas, finding the corresponding fault code correlated to the type, you can achieve the purpose of fault diagnosis. The CO/CO_2 ratio needed to be analysed when the solid insulating material was involved in the fault. IEC accuracy rate is about 60 percent, three-ratio method in the early stages of failure is more effective diagnostic methods, But the presence of a fault coding incomplete ratio method, not judge or easy to misjudge, not fully reflect the fault condition and other shortcomings when

DOI: 10.1051/matecconf/20167701002

the combined effects of multiple faults. Based on IEC inadequate methods, we propose to combine it with intelligent algorithms [5].

3.4 Transformer fault diagnosis based on neural network

BP (Back Propagation) neural network is the earliest typical artificial neural network, that error back propagation algorithm. This paper uses a simple three-layer neural network includes an input layer, an hidden layer and an output layer. Specific configuration shown in Fig. 2:

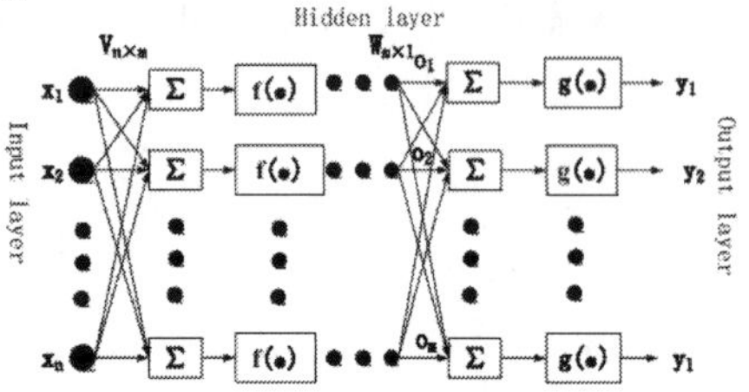

Figure 2. BP Neural network model structure diagram

Among them, the vector $x=(x_1,\ldots,x_n)^T$ 、 $o=(o_1,\ldots,o_m)^T$ 、 $y=(y_1,\ldots,y_l)^T$ 、 $d=(d_1,\ldots,d_l)^T$ 、 $v=(v_1,\ldots,v_m)^T$ and $w=(w_1,\ldots,w_m)^T$ respectively, the input vector (the content of each gas), hidden layer output vector, output vector (equipment failure type), the desired output power weight vector between the input layer and the hidden layer and the hidden layer and output layer between value vector [6-7].

We have established a three-layer BP neural network, set the input layer of the network are H_2、CH_4、C_2H_6、C_2H_2 、 C_2H_4 five gas concentration ， therefore, the number of nodes in the input layer is 5.Output layer are low energy discharge, high energy discharge, low heat, high temperature overheating , so the output node number is 4, the output of the corresponding type of fault output is 1, and the rest is zero.Hidden layer nodes generally use trial and error to determine,m=2n+1， n is the number of input layer nodes.Through experiments, we finalized a network of hidden nodes m is 10 , the error is relatively small networks. Fig. 3 shows the hidden layer node training error for different values of the network.

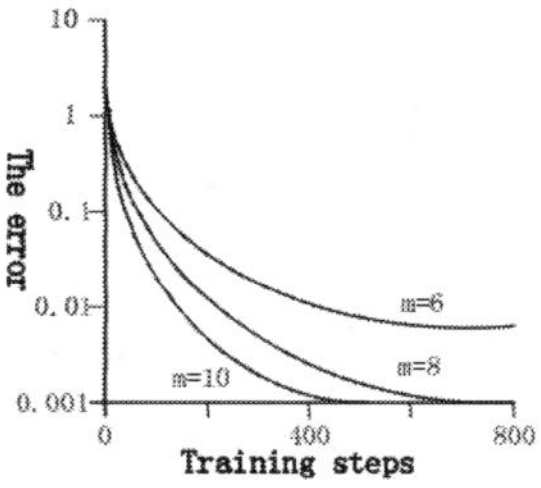

Figure 3. Hidden layer neurons error performance curve

In this paper, the excitation function is sigmoid function,network structure is 5×10×4 .Establishment of the system of five third-order nonlinear dynamic input four output model, shown in Figure 4.The connection weights of BP network of value v_{ij}、 w_{jk} and threshold θ_j are random value within [-1, 1]， The expected error is set to 0.001 ， learning rate η is 0.01 ， Maximum number of training step is 1000.

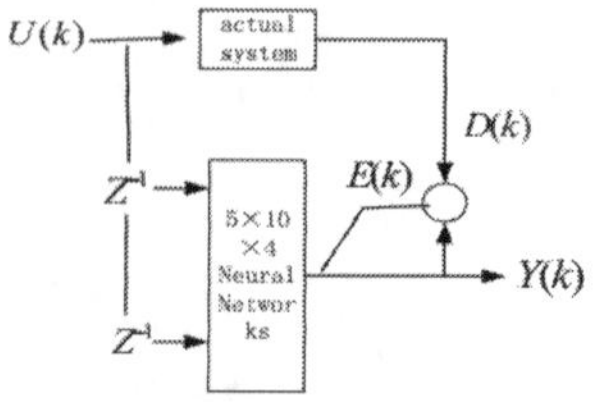

Figure 4. BP nonlinear dynamic model of neural networks

According to the network structure we have established a flow chart of BP neural network algorithm, can be seen from the calculation process flow diagram of the network.

a. Initialize weights and thresholds of the network, set between [-1, 1];

b. Training set to a given network, select the fault sample as the network training sample;

c. Layers obtained weights and thresholds, according to the weights and thresholds dressing formula to find the smallest error to make the network weights and thresholds;

d. Strike overall error, the error value calculated according to the formula error model;

e. Whether the termination condition; judgments based on the network settings to achieve maximum error of 0.001 or training times 1000;

f. If you meet the requirements to stop the train, otherwise go to step c to continue training.

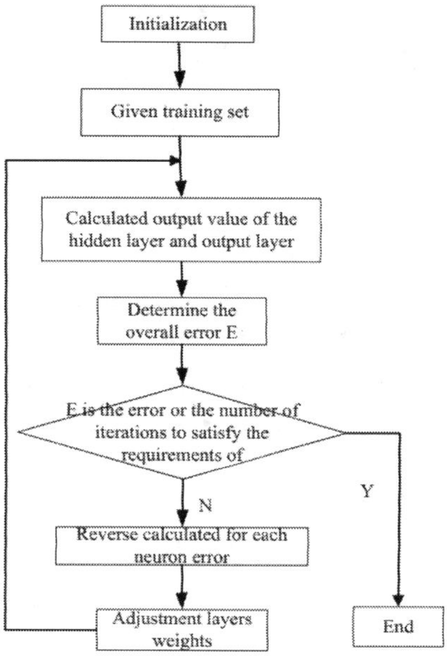

Figure 5. BP neural network algorithm flowchart

3.5 Transformer fault diagnosis based on case-based inference

Case-based analysis for the comparative analysis of similar equipment failure cases, infer accidents, breakdowns and corresponding treatment measures that may occur, which for transformer fault predictive analysis is very helpful.

Case-based analysis of a single point for the analysis of whether there is currently a fault, such as a recent one period of chromatographic data, the data repository with a typical case of an accident or malfunction of some equipment before fairly consistent, then it should be the police immediately, and fault type and treatment measures equipment failure once displayed to the user, in order to adopt the appropriate measures.

Multi-point case-based reasoning was used for analysis of fault thresholds. Consistent with a set of data trends and Case-based devices in a typical case library trends, then the device is currently in good health should also inform the user of the situation, and timely analysis of the two sets of data, propose appropriate implementation of countermeasures. Case-based multi-point analysis than just using a mathematical model to be more accurate and simple, it is a way of reasoning based on historical data.

Change of state of the transformer is nonlinear, especially for off-line data acquisition, affected long interval, the big objective factors such as changes in the environment, and cannot find a proper application of mathematical models only. Case-based analysis can just solve nonlinear prediction problem, but it is more difficult to obtain a sample of the data [8-9].

You can build a transformer fault samples Case-based library management system or through the production of electrical equipment on-line monitoring system. Taking of transformer fault samples have been collected, stored on your computer, we need to be divided into a few simple search space, thereby reducing the load on your computer to improve retrieval efficiency. Specific classification as shown below:

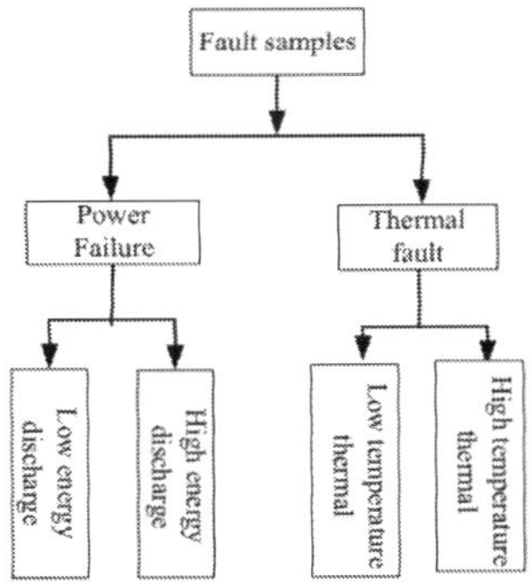

Figure 6. Case-based transformer fault diagnosis based on principles deduced

Selected according to the characteristics of the fault of the different features space retrieval, using the principles of fuzzy matching to find space where the sample, then the sample accurately find space located closest to the sample according to some examples of mathematical algorithms to find the type of fault samples [10].

4 Result Analysis

Transformer fault diagnosis and analysis of aspects, with the above methods, the introduction of the failure mechanism and diagnostic reasoning technology to form a comprehensive fault diagnosis method based on artificial neural networks.

DGA using neural networks and transformer fault diagnosis method based on Case-based inference test. Take the results of the test data based on condition monitoring, which includes low energy discharge, high energy discharge, low heat, high temperature overheat four fault types. Gas chromatographic analysis of the oil content of the individual components, the analysis results as input of neural network, trained neural network, a stable neural network model.

In terms of the type of fault diagnosis and analysis of substation equipment, with the diagnostic method of Fig. 1, the introduction of the failure mechanism and diagnostic reasoning technology to form a comprehensive fault diagnosis method based on artificial neural networks and Case-Based Reasoning transformer.

Table 2. Gas content of each component.

Groups	H2	CH4	C2H4	C2H6	C2H2
1	698.5	65.1	11.59	2.16	0
2	4.79	11.12	3.09	27.74	0.51
3	33.66	2.97	27.72	33.16	2.47
4	42.01	33.61	8.96	14.84	0.56
5	16.7	16.53	57.37	6.87	2.47
6	16.1	24.00	48.90	10.13	0.81
7	84.33	8.05	0.38	7.22	0
8	181	10.81	1	0	0
9	17.01	13.75	39.5	3.06	26.59
10	62.30	11.19	10.01	1.85	14.63
11	14	5	1	6	5
12	324	345	748	99.6	177
13	33.56	33.57	23.49	9.22	0.15
14	30.11	39.03	26.13	4.55	0.15
15	0.98	43.83	28.38	26.79	0

According to the models and methods, selected 15 groups of oil-immersed transformer DGA abnormal conditions were compared to verify the correctness of the algorithm:

 DOI: 10.1051/matecconf/20167701002

The above table, the question mark is an error of judgment or Missing items, analyze the accuracy of the algorithms for three-ratio method is 66.7%, neural network is 73.3%,integrated algorithm is 80%, which, combined with a comprehensive three-ratio method algorithm, Case-based neural network algorithms and knowledge base, and therefore its high diagnostic accuracy.

Various types of methods for the analysis of the results, the following conclusions:

Table 3. Diagnostic results.

Groups	Fault type	IEC	BP	Combined method
1	Partial Discharge	Local low energy density and discharge	Partial discharge or moisture	Partial discharge or moisture
2	High temperature thermal	Higher than the range of 700 ℃ thermal fault	?	Partial discharge or moisture
3	Low temperature thermal	150 ℃ lower than the thermal breakdown of	Low temperature thermal	High energy discharge / high temperature overheating
4	Low temperature thermal	General overheated wires	Low temperature thermal	Low temperature thermal
5	High temperature thermal	?	High temperature thermal	High energy discharge / high temperature overheating
6	High temperature thermal	High temperature thermal	High temperature thermal	High temperature thermal
7	Low energy discharge	?	Low energy discharge	Partial discharge or moisture
8	Partial Discharge	Arc discharge	?	Partial discharge or moisture
9	High energy discharge	High energy discharge	High energy discharge	Low energy discharge
10	High energy discharge	High energy discharge	Low energy discharge	High energy discharge
11	Low energy discharge	?	Low energy discharge	Low energy discharge or moisture
12	Low energy discharge and thermal	Low energy discharge and overheating	Low energy discharge	Low temperature thermal / temperature overheat
13	Low temperature thermal	Low temperature thermal	Low temperature thermal	Low temperature thermal
14	High temperature thermal	Higher than 700 thermal fault	High energy discharge / high temperature overheating	High energy discharge / high temperature overheating
15	Low temperature thermal	Low temperature thermal	Low temperature thermal	?

When the ratio method for diagnostic use, for the failure in the ratio range, the correct rate is relatively high, but cannot determine the ratio for the failure is not within the scope of the formation of the Missing.

1)Seen from the statistics, a comprehensive algorithm for low Missing two other single law for the type of fault is divided roughly, the correct rate is higher, but the simplistic division of the type of fault is not conducive to further determine the actual failure.

2)Ratio method cannot determine the effect of a variety of related faults, the diagnosis is often an error occurs.

3)Three-ratio method and neural network algorithms for high energy and low energy discharge discharge capacity is not enough to identify, often misjudge Missing phenomenon.

4)Synthesis algorithm can avoid inadequate three-ratio method and neural network algorithm, a more comprehensive diagnosis of the sample, according to the conclusions of the table shows, the higher its diagnostic accuracy.

Due to a variety of methods have advantages and disadvantages can be seen using a combination of methods to obtain higher accuracy, the results of fault diagnosis for better reference value. Therefore, this paper focuses on the opportunities DGA's Intelligent Fault Diagnosis Method has a good combination of practical significance.

4 Conclusion

This paper is the study of a comprehensive fault diagnosis method, according to the knowledge base and oil chromatogram troubleshooting cases library established transformer fault diagnosis method for assessing intelligence and neural networks combination of Case-based inference. Through practice tests proved the effectiveness of this integrated approach for latent internal transformer fault diagnosis, avoid misjudgement traditional three-ratio method, the miscarriage of justice or incomplete coding shortcomings, to a certain extent, improve the accuracy of fault diagnosis of transformer rate. But for the sudden failure, this method does not validate in future studies also need to sudden failure of the transformer were analyzed.

Acknowledgement

This work is in part supported by the National High-tech R&D Program of China (863 Program) (No.2015AA050204) and Science and Technology Program of State Grid Corporation of China (No.GY71-15-045).

References

1. People's Republic of China National Standard.GB7252-2001.Dissolved gases in transformer oil analysis and judgment guidelines Beijing: China Standard Press, (2002)
2. Zhengrong Wang, Xiaojian Deng. Transformer oil dissolved gas analysis online intelligent diagnostic systems. Power System Engineering[J], **14**（1）: 64~66, (2002)
3. Caixin Sun, Junfeng Guo. Gas analysis in the study of fuzzy clustering multi-mode fault diagnosis method of transformer oil dissolved. China Electrical Engineering[J], **21**（2）: 34~41, (2001)
4. Ferrito Sam J.A Comparative Study of Dissolved Gas Analysis Techniques: The Vacuum Extraction Method Versus the Direct Injection Method. IEEE Trans. On PD, **5**（1）: 220~225, (1990).
5. Inoue Y, Suganuma K, Kamba M, et al. Development of Oil-dissolved Hydrogen Gas Detector for Diagnosis of Transformers. IEEE Trans. On PD, 5（1）: 226~232, (1990)
6. Changchang Wang, Fuqi Li, Shengyou Gao. Online monitoring and fault diagnosis of electrical equipment[M], 12 (2011)
7. Jianmin Gu.State maintenance of electrical equipment and computer-aided analysis and forecast[J], 2（21）, 2, (2004)
8. Mitchell S. Lebold, Karl M. Reichard, Daniel Ferullo, David Boylan, Open System Architecture for Condition-based Maintenance: Overview and Training Material.
9. Penn State University / Applied Research Laboratory, The Boeing Company, and Machinery Information Management Open Standards Alliance (MIMOSA), Open System Architecture for Condition-based Maintenance: Primer.
10. Bengtsson M, Clsson E, Funk P, et a1, Technical design of condition based maintenance system—a case study using sound analysis and case-based reasoning. maintenance and reliability conference proceedings of the 8th Congress, Knoxville, USA, May 2-5, (2004)

MATEC Web of Conferences 77, 01003 (2016) DOI: 10.1051/matecconf/20167701003
ICMMR 2016

A Fast Kurtogram Demodulation Method in Rolling Bearing Fault Diagnosis

Li Li , Mian Li and Fafa Chen

College of Mechanical & Power Engineering, China Three Gorges University, Yichang City, Hubei Province, China

Abstract. Targeting at the problem of finding the best demodulation band when applying envelope analysis in rolling bearing fault diagnosis, this paper proposes a novel Fast Kurtogram Demodulation Method (FKDM) to solve the problem. FKDM is established based on the theory of spectrum kurtosis and the short-time Fourier Transform. It determines the best demodulation band firstly, which is also known as the central frequency and frequency resolution. Then, the fault signals can be demodulated in the obtained frequency band by using envelope demodulation algorithm. The FKDM method ensures the fault diagnosis correction by solving the problem of demodulation band selection. Applied FKDM in rolling bearing fault diagnosis and compared with conventional envelope analysis, the results demonstrate FKDM can achieve a better performance.

1. Introduction

Rolling bearing fault is difficult to diagnose. This is because of the high incidence of rolling bearing fault generation due to its complexity and poor working conditions. Also, the bearing signals are usually drowned by noises during operation, which makes the fault diagnosis difficult [1]. When a fault occurs in rolling bearing, it will interact with other rolling element surfaces, which makes the vibration signals to appear with the feature of amplitude modulation [2]. To diagnose this fault, the envelope analysis is one of the most commonly used methods, which mainly includes the Hilbert Transform and Teager energy operator demodulation [3]. Besides, wavelet transform and empirical mode decomposition have also been widely applied [4].

In the process of forming the envelope signals in envelope analysis method, the central frequency of demodulation band is selected based on prior experiences. However, this selection step is totally neglected in some cases, which posts a great influence on analysis results. In consideration of this problem, a novel Fast Kurtogram Demodulation Method (FKDM) is proposed.

The Fast Kurtogram is a method based on Spectral Kurtosis (SK) which is systematically defined by J. Antoni [5]. SK is used to measure the peak of square envelope. Moreover, SK is very sensitive to non-stationary patterns in a signal, and is able to indicate exactly the frequency at which those patterns occur. However, SK is only a theoretical concept and not so practical in application. By combining band-pass filter and short-time Fourier transform (STFT) with SK[6], Fast Kurtogram is introduced which is more practical and less complex while processing signals. With Fast Kurtogram, the best demodulation band can be located in advance, and signals can be filtered at specific frequency before obtaining the envelope. This method is known as FKDM.

2. The spectral kurtosis

In this section, Spectral kurtosis, the foundation of Fast Kurtogram, will be elaborated in detail.

2.1 Signal modeling

For rolling element bearings, many incipient faults produce a series of repetitive short transients, which in turn excite some structural resonances. They appear in forms of shocks which are not periodic and subjected to modulations by rotating relative frequency [7]. Hence, the signal modeling of rolling bearings $Y(t)$ can be presented as following:

$$Y(t) = X(t) + N(t) \tag{1}$$

where $X(t)$ is fault signal and $N(t)$ is additive noise. Furthermore, $X(t)$ can be presented as:

$$X(t) = \sum_k X_k h(t - \tau_k) \tag{2}$$

where $h(t)$ is the impulse response results from a single impact and where $\{X_k\}_{k\in Z}$ and $\{\tau_k\}_{k\in Z}$ are sequences of random variables which account for possible amplitudes and random occurrence of impact.

2.2 Wold-cramer decomposition

Wold-Cramer decomposition deals with fault signals expressed as a time-varying form, i.e.

$$Y(t)=\int_{-\infty}^{+\infty} e^{j2\pi ft}H(t,f)dX(f) \quad (3)$$

where $H(t,f)$ is the Fourier transform of impulse response, $h(t,s)$ and $dX(f)$ is the spectral process derived from $X(t)$. Also, $e^{j2\pi ft}$ $H(t,f)$ $dX(f)$ is the output at time t of an infinitely narrow-band filter central on frequency f.

2.3 Spectral moment and spectral kurtosis

The Wold-Cramer decomposition introduces $H(t,f)$ to describe fault signals. It will be accessed statistically by spectral moment, which will be defined as:

$$S_{2nY}(f)\triangleq E\{|H(t,f)dX(f)|^{2n}\}/df=E\{|H(t,f)^{2n}|\cdot S_{2nX} \quad (4)$$

$S_{2nY}(f)$ is also known as 2n-order instantaneous moment. In this way, the strength of the energy of $H(t,f)$ at frequency f can be measured.

Based on spectral moment, spectral cumulant i.e. combinations of several spectral moments of different orders can be achieved to characterize fault signals. Specifically, the fourth-order spectral cumulant is used:

$$C_{4Y}(f)=S_{4Y}(f)-2S_{2Y}^2(f), f\neq 0 \quad (5)$$

It can be shown that the larger the deviation of signals, the larger its fourth-order cumulant. Therefore, the energy-normalised fourth-order spectral cumulant will measure the peakiness of probability density function at frequency f. The definition of SK is given as:

$$K_Y(f)\triangleq\frac{C_{4Y}(f)}{S_{2Y}^2(f)}=\frac{S_{4Y}(f)}{S_{2Y}^2(f)}-2, f\neq 0 \quad (6)$$

SK can overcome the shortage of kurtosis as a global indicator which is not appropriate to measure fault signals masked by strong noise, and also can be applied locally in different frequency bands. Generally, SK can be seen as the kurtosis computed at the output of a perfect filter-bank at each frequency f.

3. Fast kurtogram demodulation method

In this section, the algorithm of Fast Kurtogram and its application to rolling bearing fault diagnosis will be illustrated.

3.1 STFT-based estimation of spectral kurtosis

$Y(n)$ is used to replace $Y(t)$ by changing the sampling period to 1. The STFT of fault signal at a given analysis window $w(n)$ of length N_w and a given step P can be defined as

$$Y_w(kP,f)\triangleq\sum_{n=-\infty}^{\infty} Y(n)w(n-kP)e^{-j2\pi nf} \quad (7)$$

Therefore, spectral moment can be written as

$$\hat{S}_{2nY}(f)\triangleq\left\langle |Y_w(kP,f)|^{2n}\right\rangle_k \quad (8)$$

It should be noted that the spectral moment are functions of window $w(n)$ and step P, so that the STFT-based estimator of spectral kurtosis can be defined as

$$\hat{K}_Y(f)=\frac{\hat{S}_{4Y}(f)}{\hat{S}_{2Y}^2(f)}-2, |f-\mathrm{mod}(1/2)|>\frac{1}{N_w} \quad (9)$$

3.2 Kurtogram

To address the question mentioned in section 2.3 of finding perfect filter-bank, a perfect band-pass filter should be designed to maximize the kurtosis of the envelope of the filtered signal. This problem is strictly equivalent to finding the central frequency f and the frequency resolution Δf which maximizes the STFT-based SK over all possible choices.

3.3 Fast Kurtogram[8]

The kurtogram is an appealing exploratory tool for rolling bearing fault signals. However, the complete exploration of (f , Δf) plane is a formidable task.

In this section, a fast algorithm for computing Kurtogram will be proposed, which is known as the Fast Kurtogram.

The algorithm is described in tree structure from binary to 1/3-bianry. Two quasi-analysis low-pass and high-pass filter $h_0(n)$ and $h_1(n)$ in the frequency bands[0;1/4] and [1/4;1/2] are constructed:

$$h_1(n)=h(n)e^{j3\pi n/4}=h(n)[cos(3\pi n/4)+j\sin(3\pi n/4)] \quad (10)$$

where $h(n)$ is a low-pass FIR filter with cut-off frequency $f_c=1/8+\varepsilon, \varepsilon>0$ .This process is iterated from level $k=0$, where $c_0(n)\equiv x(n)$ down to level

$k-1$. Noteworthy is the multiplication by $(-j)^n$ in order to turn the high-pass sequences into low-pass sequences. As a consequence, coefficient $c_k^i(n)$ can be interpreted as the complex envelope of signal $x(n)$ positioned on the central frequency. $f_i = (i+2^{-1})2^{-k-1}$

And with bandwidth (frequency resolution) $(\Delta f)_k = 2^{-k-1}$ The Kurtogram is finally estimated by computing the kurtosis of all sequences $c_k^i(n)$ $i = 0,\cdots,2^k-1$, $k = 0,\cdots,K-1$:

$$K_Y(f,\Delta f) = \frac{\left\langle \left|c_k^i(n)\right|^4 \right\rangle}{\left\langle \left|c_k^i(n)\right|^2 \right\rangle^2} - 2 \qquad (10)$$

The Fast Kurtogram representation at nodes $\{f_i;(\Delta f)_k\}$ of the (f , Δf) plane is illustrated in Fig.1(c), in which 2^K-1 kurtosis values are compounded.

After all the description from section 1 to 4, here we can put forward the specific process of rolling bearing fault diagnosis based on Fast Kurtogram i.e. FKDM. With Fast Kurtogram, best demodulation band can be determined by scientific theory which also overcomes the main shortage of conventional envelope analysis. The process has been illustrated in Fig. 1.

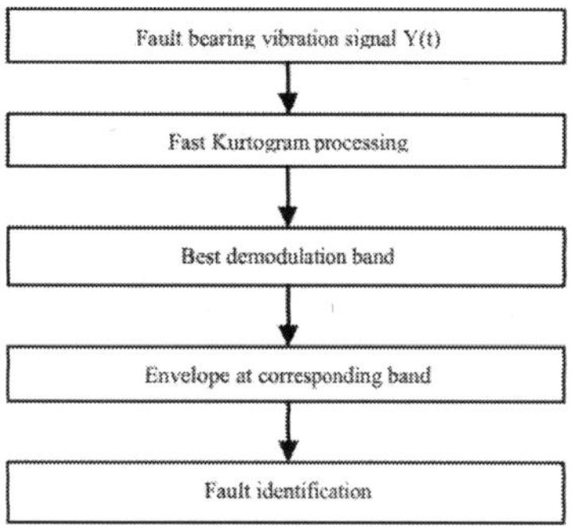

Figure 1. Flow chart of FKDM

4. Application of FKDM

In this section, FKDM will be applied to rolling bearing faults diagnosis. The results are presented and compared to conventional envelope demodulation method.

Rolling bearing fault signals are obtained from the Case Western Reserve University Bearing Data Center. Date was collected from bearing test stand which consists of a 2hp motor, a torque transducer, a dynamometer and control electronics. The rolling bearings selected are 6205-2, deep groove ball bearing. The size of the bearings is shown in Table 1. The sampling frequency is 12000Hz, with motor speed of 1797Rpm and rotating frequency fr=29.95Hz. Meanwhile, the fault frequency of the bearings can be calculated and the result is shown in Table 2.

Table 1. Size of Deep Groove Ball Bearing

Inside Diameter	Outside Diameter	Contact Angel	Ball Diameter	Number of Balls
25.00mm	52.00mm	0	8.20mm	9

Table 2. Fault Frequency of Different Fault Location in Rolling Elements

Fault location	Fault frequency
Inner race	fi=162Hz
Outer race	fo=107Hz
Ball	fb=141Hz

In the first case, ball fault signal is processed under both FKDM and envelope analysis. Result of envelope analysis is displayed in Fig. 2(a), and ball fault frequency of 141Hz can be found with certain magnitude and many unidentified frequency components appear in the envelope. However, the result of FKDM is different. As shown in Fig. 3(a), central frequency and frequency resolution (f , Δf)=(2875, 250) can be obtained, which is the best demodulation band. The corresponding envelope in Fig. 4(a) clearly reveals that the ball fault frequency of 141Hz is modulated by rotation speed of 29.95Hz.

In the second case, outer race fault signal is processed the same as above. Result of envelope analysis is displayed in Fig. 2(b), where rotation speed of 29.95Hz obtains the maximum amplitude of the whole envelope, which makes the outer race fault frequency not obvious. Meanwhile, FKDM not only reveals the best demodulation band (f , Δf)=(3375, 750) as shown in Fig. 3(b), but also clearly exhibits the outer race fault frequency of 107Hz and its harmonics in Fig. 4(b).

In the last case, inner race fault signal is processed the same way. Result of envelope analysis is displayed in Fig. 2(c), inner race fault frequency of 162Hz and its harmonics dominate the whole envelope. On the other hand, FKDM obtains the same result as shown in Fig.4(c).

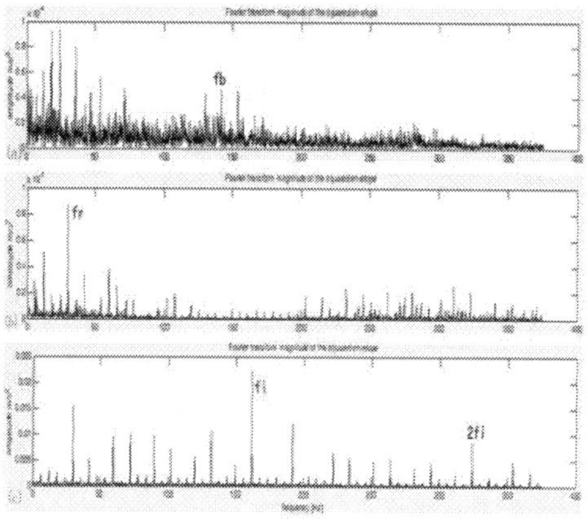

Figure 2. Envelope of (a) ball fault (b) outer race fault (c) inner race fault bearing signal with conventional envelope demodulation method

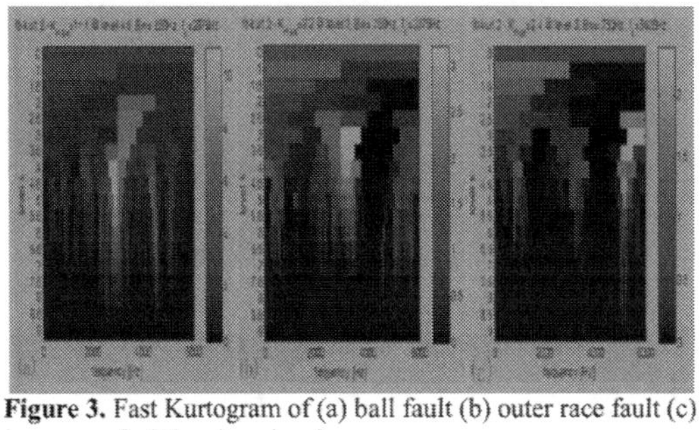

Figure 3. Fast Kurtogram of (a) ball fault (b) outer race fault (c) inner race fault bearing signal

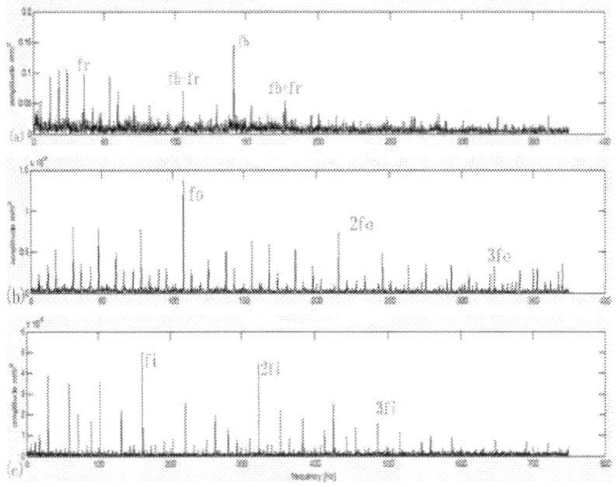

Figure 4. FKDM results of (a) ball fault (b) outer race fault (c) inner race fault bearing signal

5. Conclusion

As shown in section 4, in the diagnosis of outer race fault and ball fault rolling bearing, the FKDM is more advanced compared to conventional envelope analysis. Fault features are much more obvious when applying FKDM. In the case of inner race fault rolling bearing diagnosis, the two methods have both obtained certain fault feature, however, FKDM has obtained the best demodulation band, which would be useful in further research.

By delicately combining Fast Kurtogram with envelope analysis, FKDM not only keeps the advantage of envelope analysis for processing modulation signal, but also overcomes the shortage of finding optimum demodulation band. Moreover, the application result has also indicated that FKDM has a better performance.

Acknowledgement

This work described in this paper was supported by National Natural Science Foundation of China (No. 51175401) and 2015 Master Degree Paper Excellent Training Foundation of Three Gorges University (No.2015PY020)

References

1. Y. Lv, Q. L Zhu, and R. Yuan. Fault diagnosis of rolling bearing based on fast non-local means and envelope spectrum. Sensors. **15(1)**:1182-1198, (2015)
2. Y. Ming, J. Chen, G.M. Dong. Weak fault extraction of rolling bearing based on cyclic Wiener filter and envelope spectrum. Mechanical Systems and Signal Processing, **25**, 1773-1785, (2011)
3. M. Liang, I. Soltani Bozchalooi. An energy operator approach to joint application of amplitude and frequency demodulations for rolling bearing fault detection. Mechanical Systems and Signal Processing, **24**: 1473-1494, (2010).
4. D. Yu, J. Cheng, and Y. Yang, Application of EMD method and Hilbert spectrum to the fault diagnosis of roller bearings. Mech.Syst. Signal Process, **19(2)**, 259-270, (2005)
5. J. Antoni, R. B. Randall, The spectral kurtosis: a useful tool for characterizing none-stationary signals. Mechanical Systems and Signals Processing, **20(2)**:282-307, (2006)
6. J. Antoni, R. B. Randall, The spectral kurtosis: application to the vibratory surveillance and diagnostics of rotating machines. Mechanical Systems and Signal Processing, this issue, doi:10.1016/j.ymssp.2004.09.001
7. J.Aotoni, F.Bonnardot, A.Raad, M.ElBadaoui. Cyclostationary modeling of rotating machine vibration signals. Mechanical System and Signal Processing **18(6)** 1285-1314, (2004)
8. J.Antoni, Fast computation of the Kurtogram for the detection of transient faults. Mechanical Systems and Signals Processing, **21(1)**:108-124, (2007).

MATEC Web of Conferences 77, 01004 (2016) DOI: 10.1051/matecconf/20167701004
ICMMR 2016

Fault Diagnosis in Transformer Based on Weighted Degree of Grey Slope Incidence of Optimized Entropy

Anping Zhang, Sujie Geng

School of Economics and Management, Nanjing University of Science and Technology, Nanjing, Jiangsu Province, 210094, China

Abstract: Dissolved gas analysis (DGA) is an important method to find the hidden or incipient insulation faults of oil-immersed power transformer. However, code deficiency exists in the gas ratio methods specified by the IEC standard and complexity of fault diagnosis for power transformer. Hence a new model based on optimized weighted degree of grey slope incidence was put forward. Firstly, the entropy weight is used to determine objective weight of indices; then the model fault types are obtained by weighted degree of grey slope incidence. The combination of entropy weight with grey slope incidence analysis can fully utilize over all information of DGA and give full play to the superiority of grey slope incidence, which overcomes shortcomings of original grey slope incidence analysis. The experimental results also demonstrate that the improved method has higher accuracy compared with three-ratio method and general grey slope incidence analysis method. The diagnosis accuracy is 92.8%.

1 Introduction

Electric transformer is not only the most important and most expensive equipment, but also one of the most accident-prone equipments in electric system. In order to ensure safety and credibility of the transformer, it is important to find out the potential fault in the transformer as soon as possible and forecast the development trend of fault. It is an important instrumentality to monitor the running of electric transformer and ensure its safe and economical running. Its running reliability has direct relation to the safe running of electric system.

At present, dissolved gas analysis (DGA) is still the extensively using method in transformer fault diagnosis [1-2]. For example, IEC three-ratio method has not enough codes to effectively troubleshoot a transformer's operational condition in practical use. In recent years, many scholars both at home and abroad have done a lot of research in the field of transformer fault diagnosis. They also put forward a lot of methods, such as combining dissolved gas analysis with fuzzy logic, rough set theory, Bayesian networks, artificial neural networks, artificial immune, new radial basis function networks, genetic algorithm and support vector machine. These methods are relatively effective for transformer fault diagnosis, but each has advantages and disadvantages. Such as artificial neural networks with self-organizing, adaptive, fault-tolerant, self-learning and highly nonlinear approximation capability, is a strong tool to process nonlinear system. But the accuracy of the diagnosis depends from the completeness of samples and it is unable to process the incomplete sample [3-4]. Fuzzy set theory is good at treating uncertainty knowledge, but the determination of fuzzy membership function requires expert experience or repeated test [5]. Support vector machine has a powerful generalization ability, and overcomes shortcomings like excessively fits, slow convergence and easy to fall into local extreme value exiting in neural networks. But training time is long, and repeating training samples are large. In addition, the classification capability is not satisfactory under the incomplete information [6-7].

Transformer fault system is a grey system which partial information is known and partial information is unknown, the essence of transformer fault diagnosis would be the whitening process of grey system [8-11]. Therefore, we can take DGA as the characteristic vector, and use the grey incidence analysis method to diagnose transformer fault. Since the incidence coefficients of general grey slope incidence analysis method fluctuate easily, combining with the analysis of grey sequence entropy, this paper presents the model based on weighted degree of grey slope incidence of optimized entropy [12]. And then we have proved the effectiveness and practical value of this improved method according to specific fault examples.

2 The model of improved grey incidence analysis

2.1 The model of general grey slope incidence

The basic idea of grey incidence analysis is judging

whether a series of curves of sequences are close or not according to their correlation degree. Two curves are much closer, the correlation of sequences is bigger, contrarily, become more smaller. Therefore, the degree of grey incidence evaluates the correlation condition of two sequences on the whole. There is both a positive correlation and a negative correlation between them whose correlation degree can be reflected by the change in slope of sequences curves [13]. The definition of grey slope incidence can be summarized as follows.

Definition 1: Supposing $X_0 = [x_0(1), x_0(2), ..., x_0(n)]$ be a data sequence of a system's characteristic behavior, and $X_i = [x_i(1), x_i(2), ..., x_i(n)], i = 1, 2, \cdots, m$ relevant factor sequences. And

$$\begin{aligned}
&\Delta_0(k) = x_0(k) - x_0(k-1), k = 2, 3, \cdots, n \\
&\Delta_i(k) = x_i(k) - x_i(k-1), k = 2, 3, \cdots, n \\
&R_i = \max_k x_i(k) - \min_k x_i(k) \\
&R_0 = \max_k x_0(k) - \min_k x_0(k) \\
&\gamma_{0i}(k) = \mathrm{sgn}_k \frac{1}{1 + \left| \frac{|\Delta_i(k)|}{R_i} - \frac{|\Delta_0(k)|}{R_0} \right|}, \\
&k = 2, 3, \cdots, n, i = 1, 2, \cdots, m
\end{aligned} \tag{1}$$

where $\mathrm{sgn}_k = \begin{cases} 1, & if \Delta_0(k)\Delta_i(k) \geq 0 \\ -1, & if \Delta_0(k)\Delta_i(k) \prec 0 \end{cases}$ denotes a incidence sign function, the $\gamma_{0i}(k)$ represents the grey slope incidence coefficient. Then

$$\gamma_{0i} = \frac{1}{n-1} \sum_{k=2}^{n} \gamma_{0i}(k) \tag{2}$$

is referred to as the degree of grey slope incidence between X_0 and X_i.

2.2 The model based on weighted degree of grey slope incidence of optimized entropy

When calculating the grey slope incidence, we can't accurately reflect the relations between various sequences if we take the mean of slope incidence coefficient as grey slope incidence. Because the grey slope incidence coefficients of various points have a different influence on the data sequence of a system's characteristic behavior [14]. Accordingly, maybe it's better to add a weight to the model of calculating the grey slope incidence. The definition of weighted degree of grey slope incidence is as follows.

Definition 2: Given the data sequences:

$$\begin{aligned}
X_0 &= [x_0(1), x_0(2), ..., x_0(n)] \\
X_1 &= [x_1(1), x_1(2), ..., x_1(n)] \\
&\vdots \\
X_i &= [x_i(1), x_i(2), ..., x_i(n)] \\
&\vdots \\
X_m &= [x_m(1), x_m(2), ..., x_m(n)]
\end{aligned}$$

where X_0 is a data sequence of a system's characteristic behavior and X_i is relevant factor sequences. Then,

$$\gamma_{0i} = \sum_{k=2}^{n} w_k \gamma_{0i}(k) \tag{3}$$

where $\gamma_{0i}(k)$ represents the slope incidence coefficient between X_0 and X_i at point k, γ_{0i} is referred to as a weighted degree of grey slope incidence between X_0 and X_i, w_k is known as the weight coefficient, and $\sum_{k=2}^{n} w_k = 1 (0 \prec w_k \prec 1)$ is required.

Obviously, γ_{0i} meets four axioms of grey incidence, namely normalization, even symmetry, globality and proximity.

The key to the model of weighted degree of grey slope incidence may be how to fix the weight values which can reflect the effect of each point's slope on whole grey incidence. We hope to keep these effects balanced and objective, so entropy weight method is applied to weighting the slope incidence coefficient.

Based on the assumption that the different elements' relations are stable in the grey system, we will analyze the weighted slope incidence coefficient according to the grey entropy theory presented in paper [12]. First, this paper constructs a grey sequence based on the weighted slope incidence coefficient, and then it establishes an optimized model of the weight of slope incidence coefficients.

Definition 3: Let Map: $R_j \rightarrow P_j$

$$p_k = \frac{w_k \gamma_{0i}(k)}{\sum_{k=2}^{n} w_k \gamma_{0i}(k)}$$

be the distributed map of weighted slope incidence coefficient, map p_k is referred to as the density value of weighted slope incidence coefficient distribution.

The information contained in this sequence $\{p_k\}$ can't be completely certain, and then, for any k, there exits $p_k \geq 0$ and $\sum_{k=2}^{n} p_k = 1$. According to the definition taken from paper [12], the sequence $\{p_k\}$ is a grey sequence.

Definition 4: The given function

$$H_{\otimes}(R_i) = -\sum_{k=2}^{n} p_k \ln p_k \tag{4}$$

is the weighted grey incidence entropy of X_i.

The weighted grey incidence entropy $H_{\otimes}(R_i)$ share the same structure with Shannon entropy. They both have symmetry, non negativity, additive, convexity and extremum. However, Shannon entropy is probability entropy while grey entropy is non-probabilistic entropy, the grey properties of grey entropy depend on grey series' properties. However, according to the convexity and extremum of grey entropy function, this distinction does not affect the extremum properties of grey entropy.

As this sequence $\{p_k\}$ is grey sequence, on the basis of entropy increase principle of grey series presented in paper [12], the theorem of weighted slope incidence entropy increase can be given as follows.

Theorem 1: Assume that the distribution density sequence of weighted grey incidence coefficient is $P = \{p_k \mid \forall k, p_k \ge 0, \sum_{k=2}^{n} p_k = 1\}$, and $H_{\otimes}(R_i)$ is referred to as incidence entropy, then any change that tends to equalize p_k will increase incidence entropy $H_{\otimes}(R_i)$.

In the grey incidence analysis, the grey slope incidence coefficient reflects the influence of each point's slope on the data sequence of system's characteristic behavior. The influence ought to remain stable, so the incidence coefficient of slope at each point which is weighted should keep balance as much as possible [15-16]. Our goal is fixing the weight coefficient w_k so that the weighted slope incidence coefficient distribution density value p_k tends to balance. Therefore, on the basis of theorem 1, we exert a maximum constraint to the weighted incidence entropy $H_{\otimes}(R_i)$.

$$\begin{aligned} &\max H_{\otimes}(R_i) = -\sum_{k=2}^{n} p_k \ln p_k \\ &s.t. \sum_{k=2}^{n} w_k = 1, w_k \ge 0, k = 2, \cdots, n \end{aligned} \tag{5}$$

where

$$p_k = \frac{w_k \gamma_{0i}(k)}{\sum_{k=2}^{n} w_k \gamma_{0i}(k)}$$

Then, we formulate Lagrangian function

$$L(p_k, \lambda) = -\sum_{k=2}^{n} p_k \ln p_k + \lambda\left(\sum_{k=2}^{n} p_k - 1\right) \tag{6}$$

Based on the necessary condition for the existence of extremum, we obtain

$$\begin{cases} \dfrac{\partial L}{\partial p_k} = \ln p_k + 1 - \lambda = 0 \\ \dfrac{\partial L}{\partial \lambda} = \sum_{k=2}^{n} p_k - 1 = 0 \end{cases}$$

The solution is

$$p_k = \frac{1}{n-1} \tag{7}$$

That is to say, the distribution density values of slope incidence coefficient after weighted are equal.

As equation (3) implies

$$\begin{cases} \dfrac{w_k \gamma_{0i}(k)}{\sum_{k=2}^{n} w_k \gamma_{0i}(k)} = \dfrac{1}{n-1} \\ \sum_{k=2}^{n} w_k = 1 \end{cases} \tag{8}$$

We denote $\gamma_{0i}(k)$ by γ_m, where $m = k-1$, and $k = 2, \cdots, n$, then equation (8) can be expressed as

$$\begin{cases} \gamma_1 w_1 - \gamma_2 w_2 = 0 \\ \gamma_2 w_2 - \gamma_3 w_3 = 0 \\ \vdots \\ \gamma_{n-2} w_{n-2} - \gamma_{n-1} w_{n-1} = 0 \\ \sum_{m=1}^{n-1} w_m = 1 \end{cases} \tag{9}$$

which can be written on matrix forms

$$\Gamma \Omega = b \tag{10}$$

where

$$\Gamma = \begin{bmatrix} \gamma_1 & -\gamma_2 & & & \\ & \gamma_2 & -\gamma_3 & & \\ & & \ddots & & \\ & & & \gamma_{n-2} & -\gamma_{n-1} \\ 1 & 1 & 1 & 1 & 1 \end{bmatrix},$$

$$\Omega = \begin{bmatrix} w_1 \\ w_2 \\ \vdots \\ w_{n-2} \\ w_{n-1} \end{bmatrix}, \quad b = \begin{bmatrix} 0 \\ 0 \\ \vdots \\ 0 \\ 1 \end{bmatrix}$$

Then, we calculate the determinant of coefficient matrix Γ as follows:

$$|\Gamma| = \begin{vmatrix} \gamma_1 & -\gamma_2 & & & \\ & \gamma_2 & -\gamma_3 & & \\ & & \ddots & & \\ & & & \gamma_{n-2} & -\gamma_{n-1} \\ 1 & 1 & 1 & 1 & 1 \end{vmatrix} \tag{11}$$

For $m = 3,4,\cdots,n-1$, we add the m-1-th row of $|\Gamma|$ to the m-2-th row, the result is

$$|\Gamma| = \begin{vmatrix} \gamma_1 & & & & -\gamma_{n-1} \\ & \gamma_2 & & & -\gamma_{n-1} \\ & & \ddots & & -\gamma_{n-1} \\ & & & \gamma_{n-2} & -\gamma_{n-1} \\ 1 & 1 & 1 & 1 & 1 \end{vmatrix} \tag{12}$$

According to Equation (1), we know $\gamma_m \succ 0\ (m = 1,2,\cdots,n-2)$. Multiply the m-th column by $\frac{\gamma_{n-1}}{\gamma_m}$ and then add it to the n-1-th column, it is easy to get

$$|\Gamma| = \begin{vmatrix} \gamma_1 & & & & \\ & \gamma_2 & & & \\ & & \ddots & & \\ & & & \gamma_{n-2} & \\ 1 & 1 & 1 & 1 & 1+\sum_{m=1}^{n-2}\frac{\gamma_{n-1}}{\gamma_m} \end{vmatrix} \tag{13}$$

Hence $|\Gamma| = \prod_{m=1}^{n-2}\gamma_m(1+\sum_{m=1}^{n-2}\frac{\gamma_{n-1}}{\gamma_m}) = \prod_{m=1}^{n-2}\gamma_m(\sum_{m=1}^{n-1}\frac{\gamma_{n-1}}{\gamma_m}) \succ 0$, namely, the solution of the above linear equation is existent and unique.

As a consequence, combining with equation (10), we obtain the weight vector:

$$\Omega = \Gamma^{-1} b \tag{14}$$

3 Transformer fault diagnosis based on the model of weighted grey slope incidence

3.1 Establishment of normative fault mode vector for transformer

The basic principle of grey incidence analysis applied in transformer fault diagnosis is as follows. First, the state mode vector is composed of multiple parameters which are used to characterize the state of transformer, and we establish normative fault mode vector. Then, by comparing the incidence degree between the specific state under test and the normative mode vector, we may determine which state it belongs to so that we can diagnose the transformer's fault type.

Let the gases dissolved in the transformer oil which include hydrogen (H_2)、methane (CH_4)、ethylene (C_2H_4)、ethane (C_2H_6) and acetylene (C_2H_2) be the characteristic parameters, and we establish state mode vector as follows.

$$M = (M(1), M(2), M(3), M(4), M(5))$$

where, $M(k)$ denotes the component content of gas.

Based on a great deal of literature reading, relative data and information collecting, normative fault mode vector of transformer has been constructed by published typical chromatographic analysis data [16-17]. Then after extensive testing and constant adjustments, we establish the normative fault mode as shown in Table 1.

Table 1. Normative mode of transformer fault (μL/L)

Sequence	H_2	CH_4	C_2H_4	C_2H_6	C_2H_2
M_1	46.1	21.5	15.8	61.5	1.2
M_2	16	38.4	28	70	0
M_3	27.5	48.2	46	18.4	0
M_4	12.95	24.6	60.6	12.9	2.8
M_5	195.9	14.5	2.4	11.6	0
M_6	61.5	24.6	5.6	1.33	20.5
M_7	75.5	30.2	30.3	2.33	18.2
M_8	4.94	15.8	3.6	0.88	14.8
M_9	14.2	36.4	30.6	8.5	24.6

The normative fault mode consists of nine normative fault mode vectors, in which M_1 is the sequence for normal running, M_2 is the fault sequence for low temperature superheat (<300° C), M_3 is the fault sequence for medium temperature superheat (300~700° C), M_4 is the fault sequence for high temperature superheat (>700° C), M_5 is the sequence for partial discharge, M_6 is the sequence for low-energy discharge, M_7 is the sequence for high-energy discharge, M_8 is the sequence for low-energy discharge and superheat, M_9 is the sequence for high-energy discharge and superheat, so the normative fault mode vector matrix can be expressed as:

$$M=\begin{bmatrix} m_1 \\ m_2 \\ \vdots \\ m_9 \end{bmatrix}=\begin{bmatrix} m_1(1) & m_1(2) & m_1(3) & m_1(4) & m_1(5) \\ m_2(1) & m_2(2) & m_2(3) & m_2(4) & m_2(5) \\ \vdots & \vdots & \vdots & \vdots & \vdots \\ m_9(1) & m_9(2) & m_9(3) & m_9(4) & m_9(5) \end{bmatrix}$$

The fault mode vector under test is:

$$M_0=(m_0(1),m_0(2),m_0(3),m_0(4),m_0(5))$$

3.2 Case calculation

We have collected a large amount of data of oil chromatography during transformer practical operation from an electric power company in Yunnan. First, we establish the fault mode vector under test (M_0) for transformer, and according to equation (1), we calculate the slope incidence coefficient at each point between sequence M_0 and the sequence $M_i(i=1,2,\cdots,9)$ of normative fault mode, and build the matrix Γ. Second, we calculate the weight of each point's slope by equation (14). Then, we count the weighted degree of grey slope incidence between M_0 and $M_i(i=1,2,\cdots,9)$ based on equation (3).

According to the above model and method, we have done extensive experiments, and now several groups are randomly selected to prove the reliability of this method as shown in Table 2.

Table 2. Demonstration of fault diagnosis of transformer

No.	H_2	CH_4	C_2H_4	C_2H_6	C_2H_2	Actual results	Results of the improved method in this paper	Results of IEC three-ratio method
1	1565	93	47	34	0	M_5(partial discharge)	true	no coding
2	160	130	96	33	0	M_3(medium temperature superheat)	true	false
3	56	78	173	18	0	M_2(low temperature superheat)	false	true
4	345	112.3	51.5	27.5	58.8	M_6(low-energy discharge)	true	true
5	20.6	19.89	61.27	7.4	1.15	M_4(high temperature superheat)	true	no coding

In order to test the efficiency of the method based on the weighted grey slope incidence by entropy optimization, we randomly select 300 groups data of transformer oil chromatography from the collected data, calculate the accuracy of fault diagnosis, and then compare it with IEC three-ratio method and general grey slope incidence method. The comparison result is as shown in Table 3.

Table 3. Comparison of diagnosis result by three methods

diagnostic method	the modified method in this paper	IEC three-ratio method	the general grey slope-incidence method
Accuracy/%	92.8	77.5	89.7

4 Conclusions

Owing to the shortage of traditional transformer fault diagnosis method, the grey incidence method is put forward to diagnose transformer's fault in this paper, and we have optimized the model of the general grey slope-incidence by entropy weight method. The extensive specific examples have proved the stability of this improved method. The new method largely make up the drawback of IEC three-ratio method, is more reasonable than the traditional ones, and effectively improve the accuracy of fault diagnosis for transformer. The diagnosis accuracy achieves 92.8% when 300 groups of random data are selected and calculated. It reduces the error caused by system interference to make the analysis results more objective.

References

1. GB/T7252-2001. Guide Anal. Diagn. Gases Di. Trans. Oil.
2. C.X. Sun, W.G. Chen, J. Li, et al. B.J.: Sci. Press, (2003).
3. J. L. Guardado, J.L. Naredo, P. Moreno, et al. IEEE Trans. Power Del., **16** (2001)
4. X. M. Wang, W.S. Li, Y. Zhang, High Volt. Eng., **31** (2005)
5. J. Li, C.X. Sun, W.G. Chen, et al.. Proc. Chin. Soc. Electr. Eng., **23**(2003)
6. C.B. Wu. JLU, (2009)
7. Y. C. Xiao, X.M. Chen, H. G. Zhu, J.SJTU, **41**(2007)
8. J. L. Deng. HUST. Press, (2002)
9. H. P. Zheng, C. X. Sun, J. Li, et al. Proc. CSEE, **21**(2001)
10. S. F. Liu, Y. G. Dang, Z. G. Fang. B. J.: Sci. Press, (2004)
11. W. H. Wu. CSUST, (2009)
12. Q. S. Zhang, XJ. Guo, J.L. Deng. Syst. Eng.-Theory & Prac., **16**(1996)
13. L. Z. Cui, S.F. Liu, Z.P. Li, et al. Stat. & Inf. Forum, **25**(2010)
14. P. Purkait, S. Chakravorti. DEI, **9** (2002)
15. B. Song, P. Yu, Y.B. Luo, et al. Power Syst. Technol. Int. Conf. Power Con. K.M., (2002)
16. H. G. Deng, A. Luo, J. Cao, et al. Autom. Electr. Power Syst., **29**(2005)
17. M. Dong, Z. Yan, L. Yang, et al. Proc.-Chin. Soc. Electr. Eng., **26**(2006)

MATEC Web of Conferences 77, 01005 (2016)
ICMMR 2016 DOI: 10.1051/matecconf/20167701005

Study on Payload Effects on the Joint Motion Accuracy of Serial Mechanical Mechanisms

Dan Zhang[1], Bin Wei[2]

[1]*Department of Mechanical Engineering, York University, Toronto, Ontario, Canada*
[2]*Faculty of Engineering and Applied Science, University of Ontario Institute of Technology, Oshawa, Ontario, Canada*

Abstract. When the end-effector of the industrial serial mechanical manipulators grasps different payload masses, the output of joint motion will vary, which decreases end-effector positioning accuracy of the overall system. Based on the model reference adaptive control technique, the payload variation effect can be solved and therefore, to improve the positioning accuracy. This paper studies payload effects on the joint motion accuracy of serial mechanical mechanisms.

1 Introduction

Proportional–integral–derivative (PID) control is the widely used control method in many industries. For example, in many robotic arm used industries, PID control applies to each joint to control the whole robotic arm. By adjusting the PID gains of the PID controller, one can have desired output performance. In [1], a discrete PID controller was designed to employ it in nano scale systems. In [2], the PID controller with additional imposed nonlinear logic was reviewed for robot motion positioning control. In [3], the control strategies that implement planar micro assembly using groups of stress-engineered MEMS micro robots controlled through a single global control signal was presented. Model reference adaptive control is another control method that was proposed early by Landau [4] and it has been developed afterwards [5-8]. The reason that one needs to apply the adaptive control, especially the model reference adaptive control approach, is that traditional controllers cannot compensate the payload variations, i.e. when the end-effector grasps different payload masses, the joint output will vary under different payload masses, which will affect the positioning accuracy of the end-effector. Whereas for the model reference adaptive control, the above problem can be effectively resolved and payload variation effect can be compensated. For example, the MRAC (model reference adaptive control) proposed by R. Horowitz and later associated developments by other authors [9] contains an adaptive algorithm block and a position feedback loop which provides the difference between desired and actual position of the joints. This difference is acted upon by the integral portion of a seemingly PID control structure before feedback values of position and velocity are subtracted from it. The problem by using PID controller of not being able to compensate the payload variation is illustrated and the general theory of model reference control is presented in the following section.

2 Payload Variation Effect and Model Reference Control

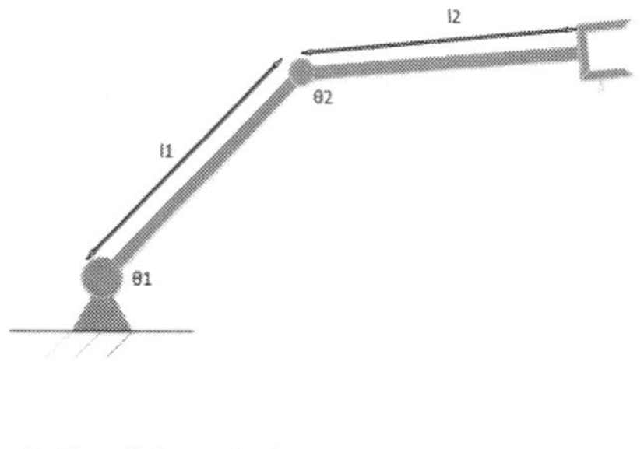

Figure 1. Two-link manipulator

In order to implement PID control of the two-DOF (degrees of freedom) link manipulator case, as shown in Fig. 1, the dynamic equation has to be derived. Here by using the Lagrange method, the torques applied to the joints are determined:

$$\tau_1 = \frac{d}{dt}\frac{\partial L}{\partial \dot{\theta}_1} - \frac{\partial L}{\partial \theta_1},\ \tau_2 = \frac{d}{dt}\frac{\partial L}{\partial \dot{\theta}_2} - \frac{\partial L}{\partial \theta_2} \quad (1)$$

The kinetic energy and potential energy for link 1 are expressed as:

$$K_1 = \frac{1}{2} m_1 (l_1 \dot{\theta}_1)^2,\ P_1 = m_1 g (l_1 \sin\theta_1) \quad (2)$$

For link 2, first write down the coordinates of the end of link 2, then differentiate them with respect to time in order to obtain the kinetic energy. Denote the Cartesian coordinates of the end of link 2 as (x2, y2).

$$x_2 = l_1 \cos\theta_1 + l_2 \cos(\theta_1 + \theta_2)$$

$$y_2 = l_1 \sin\theta_1 + l_2 \sin(\theta_1 + \theta_2) \tag{3}$$

Differentiate with respect to time results in:

$$\dot{x}_2 = -l_1 \dot{\theta}_1 \sin\theta_1 - l_2(\dot{\theta}_1 + \dot{\theta}_2)\sin(\theta_1 + \theta_2)$$

$$\dot{y}_2 = l_1 \dot{\theta}_1 \cos\theta_1 + l_2(\dot{\theta}_1 + \dot{\theta}_2)\cos(\theta_1 + \theta_2)$$

Therefore, the kinetic energy for link 2 are expressed as:

$$K_2 = \frac{1}{2}m_2 v_2^2 = \frac{1}{2}m_2 l_1^2 \dot{\theta}_1^2 + \frac{1}{2}m_2 l_2^2(\dot{\theta}_1 + \dot{\theta}_2)^2 + m_2 l_1 l_2 \cos\theta_2 \dot{\theta}_1(\dot{\theta}_1 + \dot{\theta}_2) \tag{4}$$

where $v_2^2 = \dot{x}_2^2 + \dot{y}_2^2$

The potential energy for link 2 are expressed as:

$$P_2 = m_2 g l_1 \sin\theta_1 + m_2 g l_2 \sin(\theta_1 + \theta_2) \tag{5}$$

The total kinetic and potential energy are expressed as:

$$P = P_1 + P_2 = (m_1 + m_2) g l_1 \sin\theta_1 + m_2 g l_2 \sin(\theta_1 + \theta_2)$$

$$K = K_1 + K_2 = \frac{1}{2}(m_1 + m_2) l_1^2 \dot{\theta}_1^2 + \frac{1}{2}m_2 l_2^2(\dot{\theta}_1 + \dot{\theta}_2)^2 + m_2 l_1 l_2 \cos\theta_2 \dot{\theta}_1(\dot{\theta}_1 + \dot{\theta}_2)$$

The Lagrange is obtained as:

$$L = K - P = \frac{1}{2}(m_1 + m_2) l_1^2 \dot{\theta}_1^2 + \frac{1}{2}m_2 l_2^2(\dot{\theta}_1 + \dot{\theta}_2)^2 + m_2 l_1 l_2 \cos\theta_2 \dot{\theta}_1(\dot{\theta}_1 + \dot{\theta}_2) - (m_1 + m_2) g l_1 \sin\theta_1 - m_2 g l_2 \sin(\theta_1 + \theta_2) \tag{6}$$

$$\tau_1 = \frac{d}{dt}\frac{\partial L}{\partial \dot{\theta}_1} - \frac{\partial L}{\partial \theta_1}$$

$$= ((m_1 + m_2) l_1^2 + m_2 l_2^2 + 2 m_2 l_1 l_2 \cos\theta_2)\ddot{\theta}_1 + (m_2 l_2^2 + m_2 l_1 l_2 \cos\theta_2)\ddot{\theta}_2$$

$$+(-2 m_2 l_1 l_2 \sin\theta_2)\dot{\theta}_1 \dot{\theta}_2 + (-m_2 l_1 l_2 \sin\theta_2)\dot{\theta}_2^2 + ((m_1 + m_2) l_1 \cos\theta_1 + m_2 l_2 \cos(\theta_1 + \theta_2))g$$

$$\tau_2 = \frac{d}{dt}\frac{\partial L}{\partial \dot{\theta}_2} - \frac{\partial L}{\partial \theta_2}$$

$$= (m_2 l_2^2 + m_2 l_1 l_2 \cos\theta_2)\ddot{\theta}_1 + (m_2 l_2^2)\ddot{\theta}_2 + (m_2 l_1 l_2 \sin\theta_2)\dot{\theta}_1^2 + m_2 l_2 \cos(\theta_1 + \theta_2) g$$

Put them in a matrix form, we can obtain the following:

$$\begin{bmatrix}\tau_1\\ \tau_2\end{bmatrix} = M\ddot{\theta} + N + Gg = \begin{bmatrix} m_{11} & m_{12}\\ m_{12} & m_{22}\end{bmatrix}\begin{bmatrix}\ddot{\theta}_1\\ \ddot{\theta}_2\end{bmatrix} + \begin{bmatrix}n_{11}\\ n_{21}\end{bmatrix} + \begin{bmatrix}g_{11}\\ g_{21}\end{bmatrix} g$$

where

$$m_{11} = (m_1 + m_2) l_1^2 + m_2 l_2^2 + 2 m_2 l_1 l_2 \cos\theta_2,$$

$$m_{12} = m_2 l_2^2 + m_2 l_1 l_2 \cos\theta_2, \quad m_{22} = m_2 l_2^2$$

$$n_{11} = 2(-m_2 l_1 l_2 \sin\theta_2)\dot{\theta}_1 \dot{\theta}_2 + (-m_2 l_1 l_2 \sin\theta_2)\dot{\theta}_2^2,$$

$$n_{21} = m_2 l_1 l_2 \sin\theta_2 \dot{\theta}_1^2$$

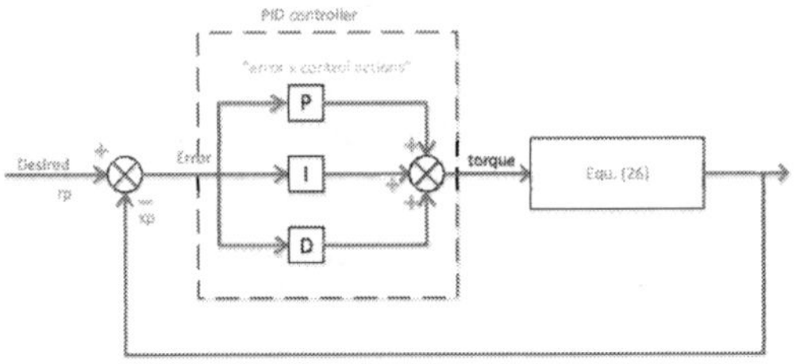

Figure 2. PID control of two-link manipulator

Now apply the PID controller, as shown in Fig. 2, the controller output is the torque, i.e.

$$K_p e + K_i \int e dt + K_d \dot{e} = \begin{bmatrix}\tau_1\\ \tau_2\end{bmatrix} \tag{7}$$

where error $e = r_p - x_p$

We know the two-link manipulator M and N matrices, the output from the manipulator (i.e. acceleration of the joints 1 and 2) can be determined as follows:

$$\begin{bmatrix}\tau_1\\ \tau_2\end{bmatrix} = M\ddot{\theta} + N + Gg = \begin{bmatrix} m_{11} & m_{12}\\ m_{12} & m_{22}\end{bmatrix}\begin{bmatrix}\ddot{\theta}_1\\ \ddot{\theta}_2\end{bmatrix} + \begin{bmatrix}n_{11}\\ n_{21}\end{bmatrix} + \begin{bmatrix}g_{11}\\ g_{21}\end{bmatrix} g$$

So

$$K_p e + K_i \int e dt + K_d \dot{e} = \begin{bmatrix}\tau_1\\ \tau_2\end{bmatrix} = M\ddot{\theta} + N + Gg$$

$$\Rightarrow \begin{bmatrix}\ddot{\theta}_1\\ \ddot{\theta}_2\end{bmatrix} = M^{-1}(K_p e + K_i \int e dt + K_d \dot{e} - N) \tag{8}$$

After deriving the acceleration of joints 1 and 2, take the time integral to obtain the velocity of joints 1 and 2 and take another integral to obtain the positon of joints 1 and 2.

$$\begin{bmatrix}\dot{\theta}_1\\ \dot{\theta}_2\end{bmatrix} = \int \begin{bmatrix}\ddot{\theta}_1\\ \ddot{\theta}_2\end{bmatrix} dt, \quad \begin{bmatrix}\theta_1\\ \theta_2\end{bmatrix} = \int \begin{bmatrix}\dot{\theta}_1\\ \dot{\theta}_2\end{bmatrix} dt \tag{9}$$

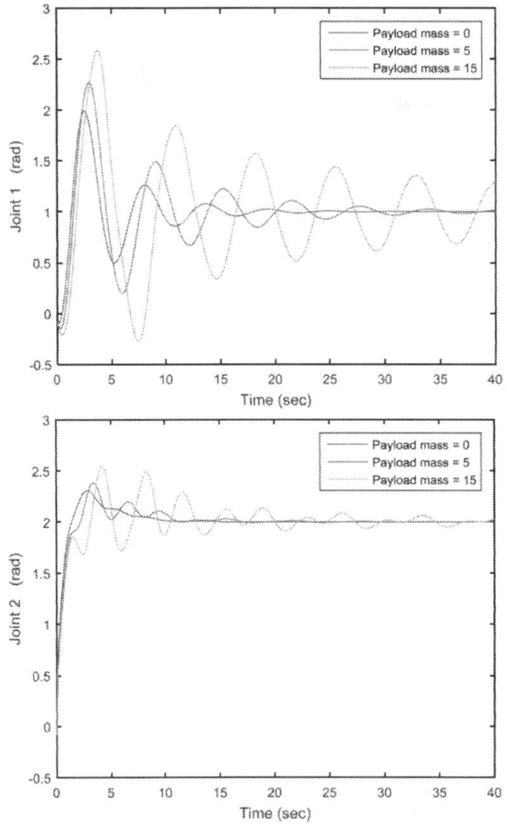

Figure 3. Joints 1 and 2 output

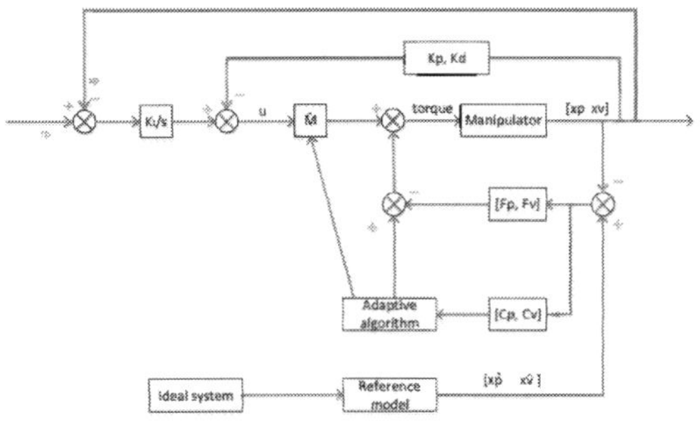

Figure 4. Model reference control approach

After applying different payload masses, the joints motion output are illustrated in Fig. 3. For the joint 1, when the payload is 0, joint one motion is quite steady, but when the payload increases to 5 and 15, one can see that joint 1 motion is not the same anymore, and also the joint output is going up and down as shown in Fig. 8. Same applies to joint 2. In order to address the above problem, model reference control is applied to compensate the payload variation effect. Fig. 4 shows a model reference control approach. Similarly with the PID control, the output from the controller can be determined as follows. For the model reference adaptive control approach,

$$ControllerOut = \tau = \hat{M}u + \hat{V} - F_p e - F_v \dot{e} \quad (10)$$

where $u = K_I \int (r_p - x_p) - K_p x_p - K_d x_v$

The manipulator dynamic equation is:

$$\tau = Ma + V + Gg \quad (11)$$

So the output from the manipulator (i.e. acceleration of joint) is:

$$\hat{M}u + \hat{V} - F_p e - F_v \dot{e} = \tau = Ma + V$$

$$\Rightarrow \begin{bmatrix} a_1 \\ a_2 \end{bmatrix} = \begin{bmatrix} \ddot{\theta}_1 \\ \ddot{\theta}_2 \end{bmatrix} = M^{-1}(\hat{M}u + \hat{V} - F_p e - F_v \dot{e} - V) \quad (12)$$

After deriving the acceleration of joint, take the time integral to obtain the velocity of joint and take another integral to obtain the positon of joint. The adaptive algorithm is derived as follows:

$$\int_0^T y^T(t)\tilde{M}u(t)dt = \int_0^T \begin{bmatrix} y_1 \\ y_2 \end{bmatrix}^T \begin{bmatrix} \tilde{m}_{11} & \tilde{m}_{12} \\ \tilde{m}_{12} & \tilde{m}_{22} \end{bmatrix} \begin{bmatrix} u_1 \\ u_2 \end{bmatrix} dt$$

$$= \int_0^T [y_1, y_2] \begin{bmatrix} \tilde{m}_{11} & \tilde{m}_{12} \\ \tilde{m}_{12} & \tilde{m}_{22} \end{bmatrix} \begin{bmatrix} u_1 \\ u_2 \end{bmatrix} dt$$

$$= \int_0^T \tilde{m}_{11} y_1 u_1 dt + \int_0^T \tilde{m}_{12}(y_1 u_2 + y_2 u_1) dt + \int_0^T \tilde{m}_{22} y_2 u_2 dt$$

Consider the first term in above equation, we need to find $\frac{d}{dt}\hat{m}_{11}(t) = \frac{d}{dt}\tilde{m}_{11}(t)$, so that $\int_0^T \tilde{m}_{11} y_1 u_1 dt \geq -\gamma^2$.

From

$$\int_0^T z(t)^T \dot{z}(t)dt = \frac{z(T)^T z(T)}{2} - \frac{z(0)^T z(0)}{2} \geq -\frac{z(0)^T z(0)}{2} = -\gamma_0^2$$

So by selecting $\frac{d}{dt}\hat{m}_{11}(t) = \frac{d}{dt}\tilde{m}_{11}(t) = k_{m11} y_1 u_1$

$$\Rightarrow y_1 u_1 = \frac{\dot{\tilde{m}}_{11}(t)}{k_{m11}} \quad (13)$$

Then

$$\int_0^T \tilde{m}_{11} y_1 u_1 dt = \int_0^T \tilde{m}_{11} \dot{\tilde{m}}_{11} \frac{1}{k_{m11}} dt = \frac{1}{k_{m11}} \int_0^T \tilde{m}_{11} \dot{\tilde{m}}_{11} dt \geq -\gamma^2$$

Using the same analysis on the other two terms, we obtain

$$\frac{d}{dt}\hat{m}_{12}(t) = \frac{d}{dt}\tilde{m}_{12}(t) = k_{m12}(y_1 u_2 + y_2 u_1)$$

$$\frac{d}{dt}\hat{m}_{22}(t) = \frac{d}{dt}\tilde{m}_{22}(t) = k_{m22} y_2 u_2$$

Derivation for M has finished, now using the same approach, we can obtain the adaptive algorithm for N as follows:

$$\frac{d}{dt}\hat{n}_{12}(t) = \frac{d}{dt}\tilde{n}_{12}(t) = k_{n12}(2y_1 x_{v1} x_{v2} - y_2 x_{v1}^2)$$

$$\frac{d}{dt}\hat{n}_{22}(t) = \frac{d}{dt}\tilde{n}_{22}(t) = k_{n22} y_1 x_{v2}^2$$

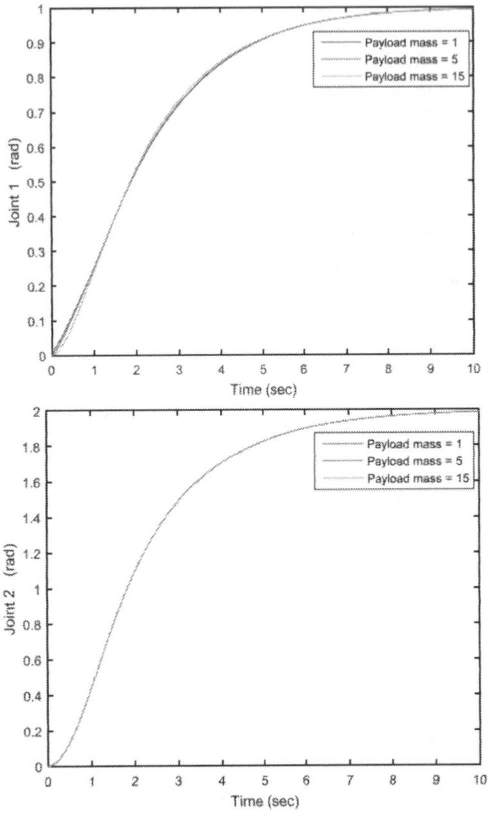

Figure 5. Joints 1 and 2 output

Fig. 5 shows the joints output that under different payload masses. By using the model reference adaptive control approach, three line coincide with each other under different payload masses, i.e. the payload mass variation effect has been compensated.

4 Conclusion

When the end-effector grasps different payload masses, the output of joint motion will vary, which will decrease the positioning accuracy of the end-effector. Based on the model reference adaptive control approach, the payload variation effect can be solved effectively and therefore, to improve the positioning accuracy. Future research will focus on hybrid control design by combining model reference adaptive control and PID control for serial robotic arms.

Acknowledgement

The authors would like to thank the financial support from the Natural Sciences and Engineering Research Council of Canada (NSERC) and Canada Research Chairs program.

References

1. Navid Razmjooy and Mehdi Ramezani, Using Quantum Gates to design a PID Controller for Nano robots. International Research Journal of Applied and Basic Sciences, Vol, 8 (12): 2354-2359, 2014.
2. Soukeyna Bouchebout, Aude Bolopion, Jean-Ochin Abrahamians, Stephane Regnier. An overview of multiple DoF magnetic actuated micro-robots. Journal of Micro-Nano Mechatronics, 7 (4), pp.97-113. 2012.
3. Bruce R Donald, Christopher G Levey, Igor Paprotny and Daniela Rus. Planning and control for micro assembly of structures composed of stress-engineered MEMS micro robots. The International Journal of Robotics Research, 32(2), pp. 218–246, 2013.
4. Landau, Y. D., Adaptive control: the model reference approach, Marcel Dekker, New York, 1979.
5. S. Dubowsky, D. Desforges, The application of model-referenced adaptive control to robotic manipulators, Journal of Dynamic Systems Measurement and Control, Vol. 101, (1979) 193-200.
6. Cao, C. and Hovakimyan, N., Design and Analysis of a Novel L1 Adaptive Control Architecture with Guaranteed Transient Performance, IEEE Transactions on Automatic Control, Vol. 53, No. 2, (2008) pp. 586-591.
7. Priyank Jain and Dr. M.J. Nigam. Design of a Model Reference Adaptive Controller Using Modified MIT Rule for a Second Order System. Advance in Electronic and Electric Engineering, Volume 3, Number 4 (2013) pp. 477-484.
8. Nhan T. Nguyen, Kalmanje Krishnakumar, Jovan Boskovic. An Optimal Control Modification to Model-Reference Adaptive Control for Fast Adaptation. AIAA Guidance, Navigation and Control Conference and Exhibit, August (2008) Honolulu, Hawaii. 1-19.
9. R. Horowitz, M. Tomizuka, An Adaptive Control Scheme for Mechanical Manipulators—Compensation of Nonlinearity and Decoupling Control, Journal of Dynamic Systems, Measurement, and Control, Volume 108, (1986) Issue 2, 1-9.

DOI: 10.1051/matecconf/20167701006

The Speed Control of Constant Tension Motor of Marine Crane

Xinyang Chen[1], Haiquan Chen[1], Wenhua Li[1], Shukai Shi[2], Yuqing Sun[1]

[1] *Dalian Maritime University, Maritime Engineering College, 116026 Dalian, China*
[2] *Dalian Maritime University, College of Public Management & the Humanities, 116026 Dalian, China*

Abstract. This article describes the working principle of the marine beacon crane hanging disc mechanical anti-sway device, and establish mathematical model on the rope controlling hanging disc of mechanical anti-sway device; Through matlab simulation analysis, this article obtains the relation curve between the velocity of traction rope of hanging disc and output frequency of the crane motor, combining rotary crane scaled model, this article carries out anti-sway experiment for the rotary crane to examine the crane's anti-sway effects.

1. Introduction

Marine crane is a kind of special crane in the transport operations, it needs to control the load with swing in the process of work, so it is particularly important to ensure the position precision when the Marine crane is running and transportation efficiency, as well as the operation safety. In order to reduce the weight of shaking, this article describes the marine crane hanging disc mechanical anti-sway device, through resistance the hanging objects' simple pendulum motion is equivalent to reduce pendulum length. When the marine crane is working, the traction rope of hanging disc must be state of tension in order to show damping effect. To solve the swing of object on the Marine crane, there are two main types of solving method, one is Mechanical anti-sway, another is intelligent anti-sway. In the aspect of mechanical anti-sway, KIMIAGHALAM B established linear kinetic model for Maryland locks on the Marine crane, applying model predictive control and optimal control method to control of object swing [1, 2]. In the aspect of intelligent anti-sway, the US Sandia National laboratory uses TSS POS/MV320 navigation system to measure ship orientation, use the combination of inertial measurement unit IMU, gyroscope and GPS to measure ship motion, and use position control strategy, all of which can better anti-sway of hanging objects [3, 4].

The article mainly to the establishment of the mathematical model of constant tension motor and calculate output frequency of the crane motor under different working condition.

2. Setting up the mathematical model

In this paper, a new type of rotary crane mechanical anti-sway device-hanging disc mechanical anti-sway device, the main purpose is decreasing rock caused by ship rocking conditions. Fig. 1 is crane disc anti-sway device model [5].

Crane mechanical rolling test system can be divided into three parts: the hoisting, derricking and slewing mechanism; the constant tension of electric drive system; hanging disc mechanical anti-sway device. Hanging disc mechanical anti-sway device includes: hanging disc, traction rope of hanging disc, anti-swing arm. The permanent magnetism synchronous servomotor of Siemens IFK7081 was adopted to be the constant tension motor, motor reducer's reduction ratio is 40:1. Hanging disc install above the hook, when the crane is working, traction rope of hanging disc must be synchronous movement with goods to keep the same degree of tension, at the same time, pressing the Anti-rolling button, the both sides of crane motor tighten at the same time to reduce the degree of swing.

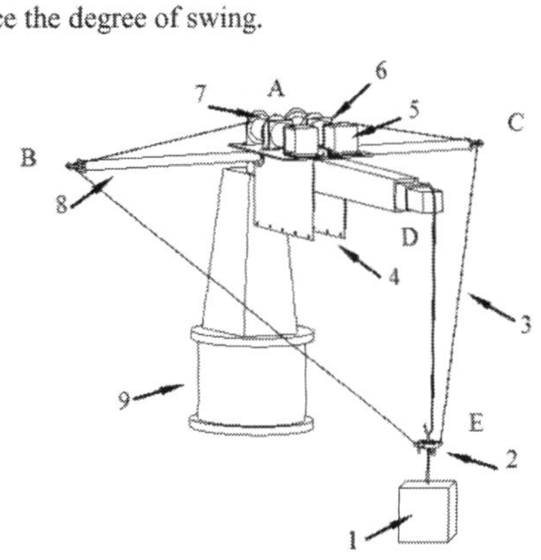

1 object 2 hanging disc 3 traction rope of hanging disc 4 base 5 motor 6 motor reducer 7 coiling block 8 anti-swing arm 9 Marine crane

Figure 1. Crane disc anti-sway device model

DOI: 10.1051/matecconf/20167701006

2.1 Model calculation of anti-swing arm

According to the shown in Fig. 1, anti-swing arm is very important in the anti-sway system to support the traction rope of hanging disc of normal work.According to the experiment, when the crane is working with 1000 kg load, the force of traction rope of hanging disc is 400N, anti-swing arm can be simplified as a cantilever beam to determine the size, material is Q235 steel. For mathematical modelling, the assumption of ideal machine is used. Therefore, the effects of notches on stress distribution could be neglected in the governing equations [6-8]. The force is shown in Fig. 2:

Figure 2. Force of anti-swing arm

The maximum bending moment and shear force of anti-swing arm:

$$M_{max} = Fl \tag{1}$$

$$Q_{max} = F \tag{2}$$

The maximum bending normal stress and maximum bending shear stress of anti-swing arm:

$$\sigma_{max} = \frac{M_{max}}{W_z} = \frac{Fl}{\frac{\pi D^3(1-\alpha^4)}{32}} \tag{3}$$

$$\tau_{max} = \frac{2Q_{max}}{A} = \frac{2F}{\pi D^2(1-\alpha^2)} \tag{4}$$

α stands for the ratio of anti-swing arm of inner and outer diameter;

D stands for the anti-swing arm of outer diameter;

l stands for the length of the anti-swing arm;

W_z stands for the bending section modulus.

$$\frac{\sigma_{max}}{\tau_{max}} = \frac{\frac{Fl}{\frac{\pi D^3(1-\alpha^4)}{32}}}{\frac{2F}{\pi D^2(1-\alpha^2)}} = \frac{16l}{D(1+\alpha^2)} \tag{5}$$

According to the formula, if is much larger than D, we can only consider bending normal stress.

According to the formula 3, bending normal stress of anti-swing arm is less than Allowable stress of Q235, Formula is as follows:

$$\sigma_{max} \cdot K \leq [\sigma] \tag{6}$$

K stands for the safety factor, K=1.2.

Through calculation, anti-swing arm of outer diameter is 40mm, anti-swing arm of inner diameter is 34mm.

The stress and strain of anti-swing arm need to be analyzed, we can get the displacement and the stiffness equation [9, 10]:

$$\begin{cases} \{f\} = [N]\{\delta\}_e \\ \{\varepsilon\} = [B]\{\delta\}_e \end{cases} \tag{7}$$

$\{f\}$ stands for force vector in the unit;

$\{\delta\}_e$ stands for displacement vector in the unit;

$[N]$ stands for deformation function matrix;

$\{\varepsilon\}$ stands for stain vector in the unit;

$[B]$ stands for strain matrix.

According to the constitutive equation,we can get the stress expression with the strain expression:

$$\{\sigma\} = [D][B]\{\delta\}_e \tag{8}$$

$\{\sigma\}$ stands for stress vector in the unit;

$[D]$ stands for elastic matrix related to the material.

According to the variational principle,wen can get balance equation, the equation shows the relationship between the nodal force and its displacement:

$$\{F\}_e = [k]_e\{\delta\}_e \tag{9}$$

$$[k]_e = \iiint [B]^T [D][B] dxdydz \tag{10}$$

By ANSYS simulation analysis, as shown in the figure below:

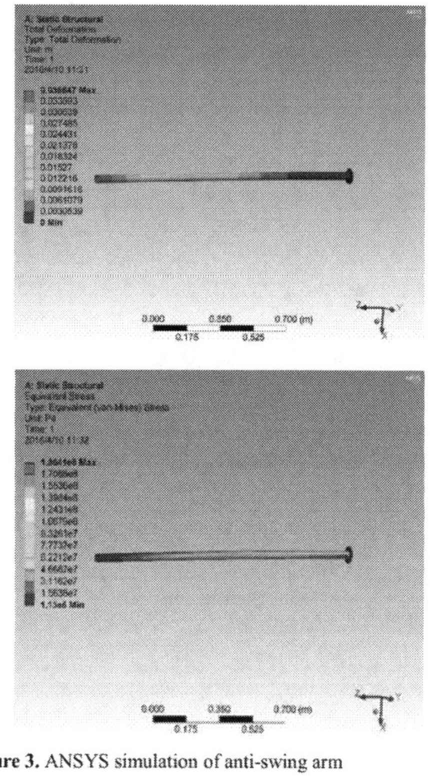

Figure 3. ANSYS simulation of anti-swing arm

According to Fig. 3, anti-swing arm's biggest variable is 3.6 cm, the maximum stress is 186.4 Mpa,, satisfying the requirements of Q235.

2.2 Condition 1: Crane lifting or falling

When the crane is lifting and falling stage, traction rope of hanging disc must be state of tension, the velocity is calculated as follows:

$$\cos\angle ADE=\frac{L_{AD}^{2}+L_{DE}^{2}-L_{AE}^{2}}{2*L_{DE}*L_{AD}} \qquad (11)$$

$$L_{BE}=\sqrt{L_{AB}^{2}+L_{AE}^{2}} \qquad (12)$$

$$L_{BE}(t)=\sqrt{L_{AB}^{2}+2*L_{AD}*L_{DE}(t)\cos\angle ADE-L_{AD}^{2}-L_{DE}(t)^{2}} \qquad (13)$$

2.3 Condition 2: Crane change-angle

When the crane is changing the angle, the velocity of traction rope of hanging disc is calculated as follows:

$$\cos(90-\xi\cdot t-\varphi)=\frac{L_{AD}^{2}+L_{DE}^{2}-L_{AE}^{2}}{2*L_{DE}*L_{AD}} \qquad (14)$$

$$L_{BE}=\sqrt{L_{AB}^{2}+L_{AE}^{2}} \qquad (15)$$

$$L_{BE}(t)=\sqrt{L_{AB}^{2}+2*L_{AD}*L_{DE}(t)\cos(90-\xi\cdot t-\varphi)-L_{AD}^{2}-L_{DE}(t)^{2}} \qquad (16)$$

Derivation for the time variable, we can draw down raction rope of hanging disc's speed stage of function. Traction rope of hanging disc is connected with oiling block, so we need to convert it into angular velocity:

$$\omega=\frac{\dot{L}_{BE}(t)}{R} \qquad (17)$$

As a result, the output of the motor frequency as follows:

$$\frac{\dot{L}_{BE}(t)*30}{R*\pi}Hz \qquad (18)$$

According to the formula, through matlab simulation analysis, this article obtains motor output frequency under different working condition.

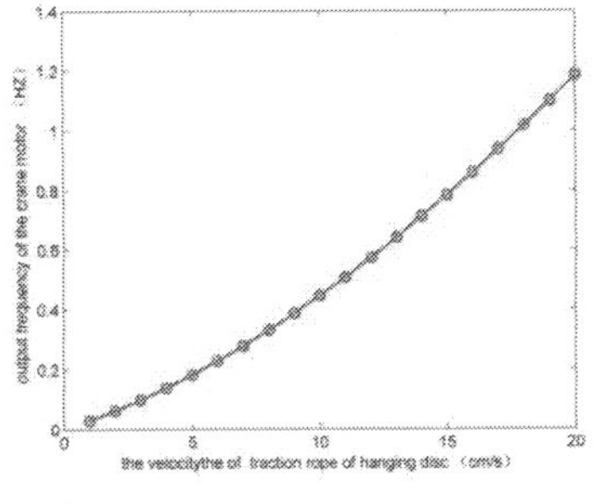

Figure 4. Motor output frequency of condition 1

The Fig. 4 shows that when the crane is lifting and falling stage, The motor output frequency approximation output direct ratio, The maximum output frequency motor is 1.1839HZ, the minimum output frequency is 0.0293HZ. At the same time, pressing the Anti-rolling button, the both sides of crane motor tighten at the same time to reduce the degree of swing.

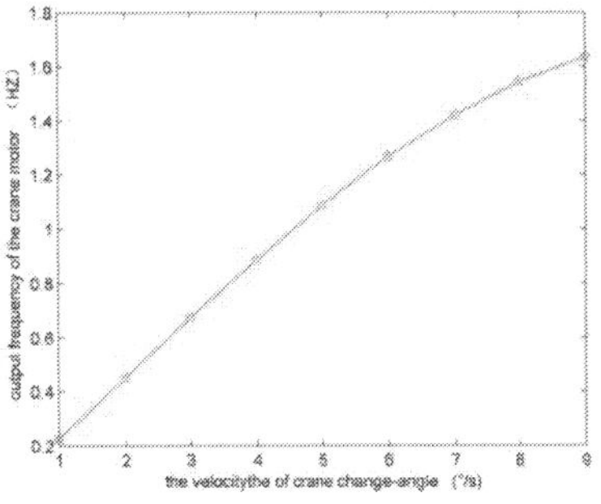

Figure 5. motor output frequency of Condition 2

The Fig. 5 shows that when the crane is changing the angle, the maximum output frequency motor is 1.6342HZ, the minimum output frequency is 0.2255HZ. At the same time, pressing the Anti-rolling button, the both sides of crane motor tighten at the same time to reduce the degree of swing.

3 Model test of hanging disc mechanical anti-sway device

According to the crane test, Experimental platform construction are all designed for no damage , not to damage the structure and function of existing equipment, test bench is shown in Fig. 6:

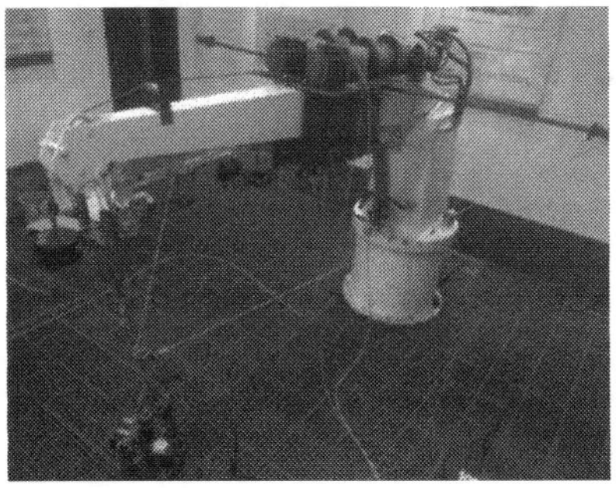

Figure 6. Experimental platform

Test rolling experiment steps:

1 Through PC corresponding control program, object is 100 kg, Rotary speed of crane without hanging disc mechanical anti-sway device is rotating, ship rocking' s angle of the shaking is 4° , Pendulum length is 1.5mm.

2 Through PC corresponding control program, object is 100 kg, Rotary speed of crane with hanging disc mechanical anti-sway device is rotating, ship rocking' s angle of the shaking is 4° . Anti-swing arm length is 1.5 m, traction rope of hanging disc length is 0.97 m.

By installing hanging disc mechanical anti-sway device, Hanging objects track is shown in Fig. 7.

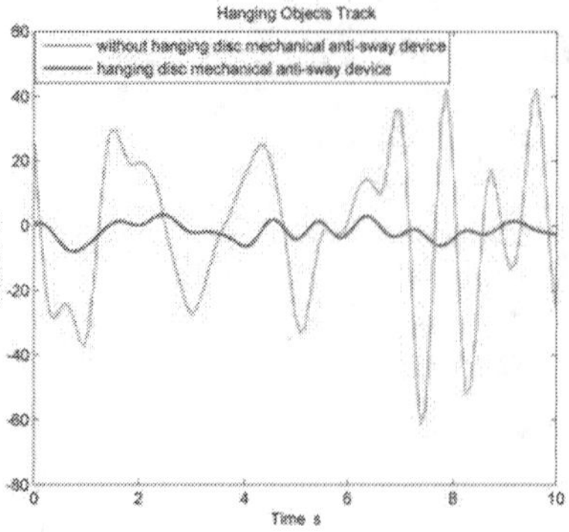

Figure 7. Hanging objects track

4. Conclusion

This paper establishes a mathematical model by the different conditions and shows simulation results of constant tension motor output frequency in order to the traction rope of hanging disc must be state of tension; hanging disc anti-sway device can effectively prevent the shake hanging objects, reducing the average swing around 60cm.

Acknowledgement

This paper is sponsored by the Fundamental Research Funds for the Central Universities (3132014332, 3132015025) and Natural Science Foundation of Liaoning Province (2015020132).

References

1. Kimiaghalam B, Homaifar A, Sayrrodsari B. An Applicaton of Model Predictive Control for a Shipboard Crane[C]// Proceedings of the 2001 American Control Conference. (2001)
2. Wen B, Homaifar A, Bikdash M, et al. Modeling and Optimal Control Design of Shipboard Crane[C]//Proceedings of the 1999 American Control Conference. (1999)
3. Robinett, R. D., Groom, K. N., Feddema, J. T., et al. Control System and Method for Payload Control in Mobile Platform Cranes: United States, US 6,496,765[P].2002-12-17.
4. Kimiaghalam B, Homaifar A, Sayrrodsari B. An applicaton of model predictive control for a shipboard crane[C]. Proceedings of the 2001 American Control Conference, Arlington, USA, June, 929-934, (2001)
5. Wang Sheng-hai1, YU Guang-chao, ZHANG Zhi-jiang, CHEN Hai-quan, SUN Yu-qing. Design Research of an Anti-sway Test Bed for Rotary Crane [J]. Ship Engineering, S1:113-116, (2015)
6. Parker G, Graziano M, Leban F, et al. Reducing Crane Payload Swing Using A Rider Block Tagline Control System[J]. OCEANSE, Aberdeen, June, 1-5, (2007)
7. Robinett, R. D., Groom, K. N., Feddema, J. T., & Parker, G. G.. Control system and method for payload control in mobile platform cranes[P]. United States Patent. No. US 6,496,765, December **17**, (2002)
8. Jang Jae Hoon , Kwon Sung-Ha , Jeung Eun Tae. Pendulation reduction on ship -mounted container crane via T-S fuzzy model[J]. J. Cent. South Univ., **19**:163-167, (2012)
9. R. M. T. Raja Ismail, Q. P. HaTrajectory. Tracking and Anti-sway Control of Three-dimensional Offshore Boom Cranes Using Second-order Sliding Modes[C]. IEEE International Conference on Automation Science and Engineering (CASE), Madison, WI, 17-20 Aug, 2013:996-1001.
10. Yongchun Fang, Pengcheng Wang, Ning Sun, et al. Dynamics Analysis and Nonlinear Control of an Offshore Boom Crane[J]. IEEE Transactions on Industrial Electronics, **61(1)**:414-427, (2014)

MATEC Web of Conferences 77, 01007 (2016) DOI: 10.1051/matecconf/20167701007
ICMMR 2016

Research on Dynamics and Stable Tracking Model of Six-axis Simulation Turntable

Zhou Jiehua[1] and Peng Xiafu[2]

[1]*School of Information Engineering, Nanchang Hangkong University, Nanchang 330063, China*
[2]*College of Information & Technology, Xiamen University, Xiamen, 361005, China*

Abstract. A new type of six-axis simulation turntable used to simulate ship swaying is proposed in the paper, and the dynamics and stable tracking model of six-axis simulation turntable are researched. Firstly, according to the design indexes of six-axis simulation/testing turntable, the dynamics of six-axis simulation turntable was established by Lagrange method. Then the dynamics was simulated by software Matlab, which can provide data and theoretical basis for dynamics compensation and driving motors selection of six-axis simulation turntable. Secondly, the stable tracking model of six-axis simulation turntable was established by geometric analysis method, and simulation and reverse checking computation results show that the stable tracking model is correct.

1 Introduction

Simulation turntable executing semi-physical simulation and test in the marine, aviation, aerospace and defense domain is one of key guarantee equipments, so it plays an extremely important role in the development process of related equipments [1-5]. Simulation turntable can realistically simulate the attitude of moving vehicle, and reproduce the dynamic characteristics in the runtime, which can help designer improve the related equipments [6-10]. Furthermore, simulation turntable can be used as a stable platform which can effectively isolate the motion of moving vehicle and ensure the attitude stabilization of equipments mounted on the turntable. Simulation turntable can be used for all types of moving vehicle as well as some handheld and ground-mounted applications, such as Hubble Space Telescope, motion simulation system, military system, tracking system and the handheld camera and so on [5].

When the six-axis simulation turntable is running, the dynamic factor and interaction between the frame will change, which resulting in decreased performance of anti-interference, speed stability, dynamic characteristics and control precision [11]. Therefore, if we want to achieve high-precision control for six-axis simulation turntable, it must be to compensate for its dynamic characteristics. Dynamics is the precondition achieving compensation of dynamic characteristics. In addition, dynamics is the precondition of driving motor selection in the design process of the six-axis simulation turntable. On the other hand, the ship will be swaying on the sea due to the action of sea wave, which causes the precision equipments on ship not work accurately [12]. So, it must be to research the compensation model of ship swaying for ensuring the normal work of the precision equipments on ship.

The above mentioned importance prompts the study of dynamics and stable tracking of six-axis simulation turntable. According to the process of research, the rest of this paper is organized as follows. In Section 2, we present the structure of six-axis simulation/testing turntable and establish its dynamics. Then the stable tracking model of six-axis simulation/testing turntable is researched in Section 3. Thereafter, the simulation and a short conclusion are drawn in Section 4.

2 Six-axis simulation turntable and dynamics

Six-axis simulation turntable is mainly composed of machinery body, motor servo control system, angle measurement system and computer control system. Machinery body is composed of upper body and lower body. The lower body is a three-axis turntable which simulates the rocking motion of a ship on the sea, and the upper body which simulates the stable and tracking motion of platform is also a three-axis turntable. They are both U-O-O type structure, which inner frame and middle frame are both O-type structure, and outer frame is U-type structure. The upper and lower body can be used alone or in combination. When they are used in combination, the upper body is mounted on the outer frame of the lower body, and the six-axis simulation turntable is shown in Fig. 1.

The Denavit-Hartenberg method is used to establish the following coordinates (shown in Fig. 2): (1) Fixed coordinate system (o_0-$x_0y_0z_0$): the intersection point between outer frame axis of lower body and ground surface is selected as the origin o_0, the direction of outer frame axis of lower body is selected as the axis-o_0z_0 and the upward direction is its positive direction, the axis-o_0x_0 is perpendicular to the axis-o_0z_0 and the outward direction is its positive direction, the aix-o_0y_0 meets the right-hand rule with the axis-o_0x_0 and axis-o_0z_0. (2) Frame coordinate system (o_i-$x_iy_iz_i$, i= 1, 2,...,6): i= 1, 2, 3 are used in outer frame, middle frame and inner frame of lower body respectively, i= 4, 5, 6 are used in outer frame, middle frame and inner frame of upper body respectively. When the i-th frame rotates, the frame coordinate system o_i-$x_iy_iz_i$ moves with it. The intersection point of three frames of lower body is selected as the origin o_i, where i= 1, 2, 3. The intersection point of three frames of upper body is selected as the origin o_i, where i= 4, 5, 6. The rotational axis direction of the i-th frame is the direction of axis-o_iz_i, where i= 1, 2,...,6. The normal direction of the i-th frame is the direction of axis-o_ix_i, where i= 1, 2, 4, 5. The middle frame axis of lower body and middle frame axis of upper body are the direction of axis-o_3x_3 and axis-o_6x_6 respectively, and the axis-o_ix_i, the axis-o_iy_i and the axis-o_iz_i all meet the right-hand rule.

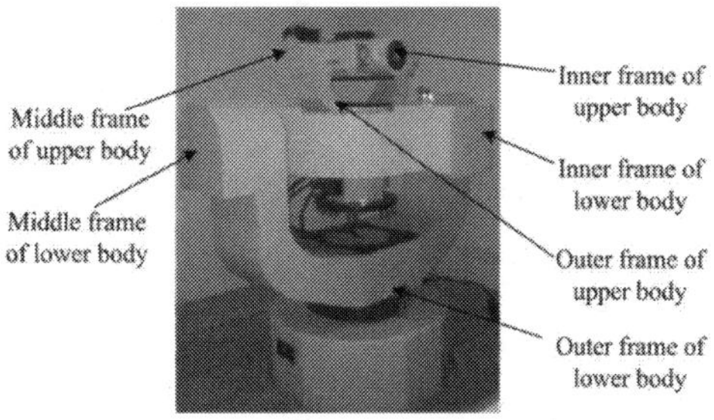

Figure 1. Machinery body of six-axis simulation turntable.

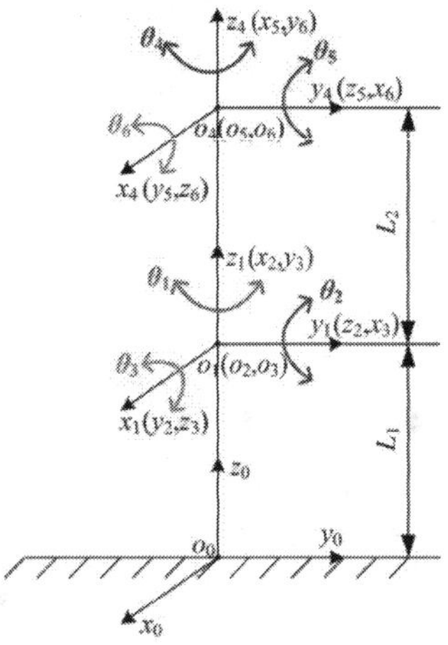

Figure 2. Coordinates of six-axis simulation turntable.

The link parameters of six-axis simulation turntable can be seen in Table 1, and the homogeneous transformation matrix between the coordinate system o_i-$x_iy_iz_i$ and coordinate system o_{i-1}-$x_{i-1}y_{i-1}z_{i-1}$ can be written as:

$$ {}^{i-1}_{i}\boldsymbol{T} = \begin{bmatrix} {}^{i-1}_{i}\boldsymbol{R} & {}^{i-1}_{i}\boldsymbol{P} \\ 0_{1\times 3} & 1 \end{bmatrix} \tag{1} $$

where

$$ {}^{i-1}_{i}\boldsymbol{R} = \begin{bmatrix} c\theta_i & -s\theta_i & 0 \\ s\theta_i c\alpha_{i-1} & c\theta_i c\alpha_{i-1} & -s\alpha_{i-1} \\ s\theta_i s\alpha_{i-1} & c\theta_i s\alpha_{i-1} & c\alpha_{i-1} \end{bmatrix} $$

$$ {}^{i-1}_{i}\boldsymbol{P} = \begin{bmatrix} a_{i-1} & -s\alpha_{i-1}d_i & c\alpha_{i-1}d_i \end{bmatrix}^{\mathrm{T}} $$

$s\theta_i$, $c\theta_i$, $s\alpha_{i-1}$ and $c\alpha_{i-1}$ are the abbreviation of $\sin(\theta_i)$, $\cos(\theta_i)$, $\sin(\alpha_{i-1})$ and $\cos(\alpha_{i-1})$ respectively.

Table 1. Link parameters of six-axis simulation turntable.

i	Link rotation angle α_{i-1}	Link length a_{i-1}	Offset d_i	Joint angle θ_i
1	0	0	L_1	θ_1
2	-π/2	0	0	θ_2-π/2
3	-π/2	0	0	θ_3-π/2
4	-π/2	0	L_2	θ_4-π/2
5	-π/2	0	0	θ_5-π/2
6	-π/2	0	0	θ_6-π/2

All middle frame gravity centers and inner frame gravity centers are coincident with their geometric centers, and all outer frame gravity centers are in the revolving axle of outer frame. The homogeneous coordinates of gravity center of each frame in the respective frame coordinate system are as follows: ${}^1\boldsymbol{C}_1$=[0, 0, b_1, 1]^{T}、${}^2\boldsymbol{C}_2$=[0, 0, 0, 1]^{T}、${}^3\boldsymbol{C}_3$=[0, 0, 0, 1]^{T}、${}^4\boldsymbol{C}_4$=[0, 0, b_2, 1]^{T}、${}^5\mathrm{C}_5$=[0, 0, 0, 1]^{T} and ${}^6\boldsymbol{C}_6$=[0, 0, 0, 1]^{T}. So, the homogeneous coordinates of gravity center of each frame in the fixed coordinate system can be calculated as follows:

$$ \begin{cases} {}^0\boldsymbol{C}_1 = \begin{bmatrix} 0 & 0 & b_1 + L_1 & 1 \end{bmatrix}^{\mathrm{T}} \\ {}^0\boldsymbol{C}_2 = \begin{bmatrix} 0 & 0 & L_1 & 1 \end{bmatrix}^{\mathrm{T}} \\ {}^0\boldsymbol{C}_3 = \begin{bmatrix} 0 & 0 & L_1 & 1 \end{bmatrix}^{\mathrm{T}} \\ {}^0\boldsymbol{C}_4 = \begin{bmatrix} (b_2+L_2)(s\theta_1 s\theta_3 + c\theta_1 s\theta_2 c\theta_3) \\ (b_2+L_2)(-c\theta_1 s\theta_3 + s\theta_1 s\theta_2 c\theta_3) \\ (b_2+L_2)c\theta_2 c\theta_3 + L_1 \\ 1 \end{bmatrix} \\ {}^0\boldsymbol{C}_5 = \begin{bmatrix} L_2(s\theta_1 s\theta_3 + c\theta_1 s\theta_2 c\theta_3) \\ L_2(-c\theta_1 s\theta_3 + s\theta_1 s\theta_2 c\theta_3) \\ L_2 c\theta_2 c\theta_3 + L_1 \\ 1 \end{bmatrix} \\ {}^0\boldsymbol{C}_6 = \begin{bmatrix} L_2(s\theta_1 s\theta_3 + c\theta_1 s\theta_2 c\theta_3) \\ L_2(-c\theta_1 s\theta_3 + s\theta_1 s\theta_2 c\theta_3) \\ L_2 c\theta_2 c\theta_3 + L_1 \\ 1 \end{bmatrix} \end{cases} \tag{2} $$

Setting the gravitational acceleration g=9.8m/s^2, and the weight vector $\boldsymbol{M}$=[m_1, m_2, m_3, m_4, m_5, m_6]. According to the equation (2), the total potential energy can be calculated and written as:

$$E_p = g[m_1 b_1 + (m_1 + m_2 + m_3 + m_4 + m_5 + m_6)L_1 + (m_4 b_2 + m_4 L_2 + m_5 L_2 + m_6 L_2)c\theta_2 c\theta_3] \quad (3)$$

By substituting the equation (1) into equation (2), the velocity vectors of gravity center of each frame in the fixed coordinate system can be calculated by differential method as follows:

$$\begin{cases} {}^0\boldsymbol{v}_{C_1} = {}^0\boldsymbol{v}_{C_2} = {}^0\boldsymbol{v}_{C_3} = \begin{bmatrix} 0 & 0 & 0 \end{bmatrix}^{\mathrm{T}} \\ {}^0\boldsymbol{v}_{C_4} = (b_2 + L_2)\boldsymbol{A}\begin{bmatrix} \dot{\theta}_1 & \dot{\theta}_2 & \dot{\theta}_3 \end{bmatrix}^{\mathrm{T}} \\ {}^0\boldsymbol{v}_{C_5} = {}^0\boldsymbol{v}_{C_6} = L_2\boldsymbol{A}\begin{bmatrix} \dot{\theta}_1 & \dot{\theta}_2 & \dot{\theta}_3 \end{bmatrix}^{\mathrm{T}} \end{cases} \quad (4)$$

where

$$\boldsymbol{A} = \begin{bmatrix} c\theta_1 s\theta_3 - s\theta_1 s\theta_2 c\theta_3 & c\theta_1 c\theta_2 c\theta_3 & s\theta_1 c\theta_3 - c\theta_1 s\theta_2 s\theta_3 \\ s\theta_1 s\theta_3 + c\theta_1 s\theta_2 c\theta_3 & s\theta_1 c\theta_2 c\theta_3 & -c\theta_1 c\theta_3 - s\theta_1 s\theta_2 s\theta_3 \\ 0 & -s\theta_2 c\theta_3 & -c\theta_2 s\theta_3 \end{bmatrix}$$

Because of ${}^{i+1}\boldsymbol{\omega}_{i+1} = {}^{i+1}_{i}R\,{}^{i}\boldsymbol{\omega}_i + \theta_{i+1}\,{}^{i+1}Z_{i+1}$, the angular velocity vector of each frame in the respective frame coordinate system can be displayed as:

$$\begin{cases} {}^1\boldsymbol{\omega}_1 = \begin{bmatrix} 0 & 0 & \dot{\theta}_1 \end{bmatrix}^{\mathrm{T}} \\ {}^2\boldsymbol{\omega}_2 = \begin{bmatrix} c\theta_2\dot{\theta}_1 & -s\theta_2\dot{\theta}_1 & \dot{\theta}_2 \end{bmatrix}^{\mathrm{T}} \\ {}^3\boldsymbol{\omega}_3 = \begin{bmatrix} s\theta_3 c\theta_2\dot{\theta}_1 + c\theta_3\dot{\theta}_2 & c\theta_3 c\theta_2\dot{\theta}_1 - s\theta_3\dot{\theta}_2 & -s\theta_2\dot{\theta}_1 + \dot{\theta}_3 \end{bmatrix}^{\mathrm{T}} \\ {}^4\boldsymbol{\omega}_4 = \begin{bmatrix} (s\theta_4 s\theta_3 c\theta_2 - c\theta_4 s\theta_2)\dot{\theta}_1 + s\theta_4 c\theta_3\dot{\theta}_2 + c\theta_4\dot{\theta}_3 \\ (c\theta_4 s\theta_3 c\theta_2 + s\theta_4 s\theta_2)\dot{\theta}_1 + c\theta_4 c\theta_3\dot{\theta}_2 - s\theta_4\dot{\theta}_3 \\ c\theta_3 c\theta_2\dot{\theta}_1 - s\theta_3\dot{\theta}_2 + \dot{\theta}_4 \end{bmatrix} \\ {}^5\boldsymbol{\omega}_5 = \begin{bmatrix} (s\theta_5 s\theta_4 s\theta_3 c\theta_2 - s\theta_5 c\theta_4 s\theta_2 + c\theta_5 c\theta_3 c\theta_2)\dot{\theta}_1 + \\ (s\theta_5 s\theta_4 c\theta_3 - c\theta_5 s\theta_3)\dot{\theta}_2 + s\theta_5 c\theta_4\dot{\theta}_3 + c\theta_5\dot{\theta}_4 \\ \\ (c\theta_5 s\theta_4 s\theta_3 c\theta_2 - c\theta_5 c\theta_4 s\theta_2 - s\theta_5 c\theta_3 c\theta_2)\dot{\theta}_1 + \\ (c\theta_5 s\theta_4 c\theta_3 + s\theta_5 s\theta_3)\dot{\theta}_2 + c\theta_5 c\theta_4\dot{\theta}_3 - s\theta_5\dot{\theta}_4 \\ \\ (c\theta_4 s\theta_3 c\theta_2 + s\theta_4 s\theta_2)\dot{\theta}_1 + c\theta_4 c\theta_3\dot{\theta}_2 - s\theta_4\dot{\theta}_3 + \dot{\theta}_5 \end{bmatrix} \\ {}^6\boldsymbol{\omega}_6 = \begin{bmatrix} (s\theta_6 s\theta_5 s\theta_4 s\theta_3 c\theta_2 - s\theta_6 s\theta_5 c\theta_4 s\theta_2 + s\theta_6 c\theta_5 c\theta_3 c\theta_2 + \\ c\theta_6 c\theta_4 s\theta_3 c\theta_2 + c\theta_6 s\theta_4 s\theta_2)\dot{\theta}_1 + (s\theta_6 s\theta_5 s\theta_4 c\theta_3 - \\ s\theta_6 c\theta_5 s\theta_3 + c\theta_6 c\theta_4 c\theta_3)\dot{\theta}_2 + (s\theta_6 s\theta_5 c\theta_4 - \\ c\theta_6 s\theta_4)\dot{\theta}_3 + (s\theta_6 c\theta_5 + c\theta_6)\dot{\theta}_4 \\ \\ (c\theta_6 s\theta_5 s\theta_4 s\theta_3 c\theta_2 - c\theta_6 s\theta_5 c\theta_4 s\theta_2 + c\theta_6 c\theta_5 c\theta_3 c\theta_2 - \\ s\theta_6 c\theta_4 s\theta_3 c\theta_2 - s\theta_6 s\theta_4 s\theta_2)\dot{\theta}_1 + (c\theta_6 s\theta_5 s\theta_4 c\theta_3 - \\ c\theta_6 c\theta_5 s\theta_3 - s\theta_6 c\theta_4 c\theta_3)\dot{\theta}_2 + (c\theta_6 s\theta_5 c\theta_4 + \\ s\theta_6 s\theta_4)\dot{\theta}_3 + (c\theta_6 c\theta_5 - s\theta_6)\dot{\theta}_4 \\ \\ (c\theta_5 s\theta_4 s\theta_3 c\theta_2 - c\theta_5 c\theta_4 s\theta_2 - s\theta_5 c\theta_3 c\theta_2)\dot{\theta}_1 + \\ (c\theta_5 s\theta_4 c\theta_3 + s\theta_5 s\theta_3)\dot{\theta}_2 + c\theta_5 c\theta_4\dot{\theta}_3 - s\theta_5\dot{\theta}_4 + \dot{\theta}_6 \end{bmatrix} \end{cases} \quad (5)$$

By substituting the equation (4) and equation (5) into $E_k = \sum_{i=1}^{6}\left[\frac{1}{2}m_i({}^0\boldsymbol{v}_{C_i})^{\mathrm{T}}\,{}^0\boldsymbol{v}_{C_i} + \frac{1}{2}({}^i\boldsymbol{\omega}_i)^{\mathrm{T}}\,{}^i\boldsymbol{I}_i\,{}^i\boldsymbol{\omega}_i\right]$, and the total kinetic energy of six-axis simulation turntable can be calculated, where ${}^i\boldsymbol{I}_i$ is the inertia matrix of the i-th frame in the respective frame coordinate system.

Let :

$$\begin{cases} \boldsymbol{\theta} = \begin{bmatrix} \theta_1 & \theta_2 & \theta_3 & \theta_4 & \theta_5 & \theta_6 \end{bmatrix}^{\mathrm{T}} \\ \dot{\boldsymbol{\theta}} = \begin{bmatrix} \dot{\theta}_1 & \dot{\theta}_2 & \dot{\theta}_3 & \dot{\theta}_4 & \dot{\theta}_5 & \dot{\theta}_6 \end{bmatrix}^{\mathrm{T}} \end{cases}$$

By substituting E_p and E_k into $\boldsymbol{\tau} = \frac{d}{dt}\frac{\partial E_k}{\partial \dot{\boldsymbol{\theta}}} - \frac{\partial E_k}{\partial \boldsymbol{\theta}} + \frac{\partial E_p}{\partial \boldsymbol{\theta}}$, and we can get the driving torque- $\boldsymbol{\tau}$ of each frame.

3 Stable tracking model

A tracking antenna is mounted on the plane of inner frame of upper body, and the task is that the antenna tracks the moving target in the sky. The steps to achieve the task as follows:

Step 1. Calculating the rotation angle-θ_4 where the outer frame of upper body rotates about its axis to achieve that the azimuth angle of antenna is equal to the azimuth angle of target in the fixed coordinate system.

Step 2. Calculating the rotation angle-θ_5 where the middle frame of upper body rotates about its axis to achieve that the inner frame axis is parallel to the ground surface.

Step 3. Calculating the rotation angle-θ_6 where the inner frame of upper body rotates about its axis to achieve that the antenna tracks to the target in the sky.

The unit reference vector of antenna in the frame coordinate system can be written as: $[1\ 0\ 0\ 0]^{\mathrm{T}}$. When three frames of lower body and outer frame of upper body rotate, the unit reference vector of antenna in the fixed coordinate system can be calculated as:

$$\begin{bmatrix} x_{01} \\ y_{01} \\ z_{01} \\ 0 \end{bmatrix} = {}^0_1T\,{}^1_2T\,{}^2_3T\,{}^3_4T \begin{bmatrix} 0 \\ 1 \\ 0 \\ 0 \end{bmatrix} \quad (6)$$

By substituting the equation (1) into equation (6), the unit reference vector of antenna in the fixed coordinate system can be written as:

$$\begin{cases} x_{01} = (c\theta_1 c\theta_2 c\theta_3 - s\theta_1 s\theta_3)s\theta_4 + c\theta_1 s\theta_2 c\theta_4 \\ y_{01} = (s\theta_1 c\theta_2 c\theta_3 + c\theta_1 s\theta_3)s\theta_4 + s\theta_1 s\theta_2 c\theta_4 \\ z_{01} = c\theta_2 c\theta_4 - s\theta_2 c\theta_3 s\theta_4 \end{cases} \quad (7)$$

The azimuth angle and elevation angle of target are A and E in the fixed coordinate system respectively. The azimuth angle of antenna is equal to the azimuth angle of target in the fixed coordinate system, which can get the following condition:

$$\tan A = x_{01} / y_{01} \tag{8}$$

By substituting the equation (7) into equation (8), the rotation angle-θ_4 can be calculated as:

$$\tan\theta_4 = \frac{s\theta_1 c\theta_2 c\theta_3 \tan A + c\theta_1 s\theta_3 \tan A - c\theta_1 c\theta_2 c\theta_3 + s\theta_1 s\theta_3}{s\theta_1 s\theta_2 \tan A - c\theta_1 s\theta_2} \tag{9}$$

The middle frame of upper body rotates about its axis to achieve that the inner frame axis is parallel to the ground surface after the azimuth angle of antenna is equal to the azimuth angle of target in the fixed coordinate system, the unit reference vector of axis- o_0z_0 in the frame coordinate system can be displayed as:

$$\begin{bmatrix} x_{20} \\ y_{20} \\ z_{20} \\ 0 \end{bmatrix} = {}^5_4T\,{}^4_3T\,{}^3_2T\,{}^2_1T\,{}^1_0T \begin{bmatrix} 0 \\ 0 \\ 1 \\ 0 \end{bmatrix} \tag{10}$$

By substituting the equation (1) into equation (10), x_{20} can be written as:

$$x_{20} = (s\theta_2 c\theta_3 c\theta_4 + c\theta_2 s\theta_4) s\theta_5 + s\theta_2 s\theta_3 c\theta_5 \tag{11}$$

If and only if x_{20}=0, the inner frame axis can be parallel to the ground surface. Therefore, according to the equation (11), the rotation angle-θ_5 can be calculated as:

$$\tan\theta_5 = (-s\theta_2 s\theta_3) / (s\theta_2 c\theta_3 c\theta_4 + c\theta_2 s\theta_4) \tag{12}$$

The difference value between the elevation angle of antenna and the elevation angle of target can be calculated as:

$$\sin\Delta E = z_{01} / 1 \tag{13}$$

And then the rotation angle-θ_6 can be calculated as:

$$\theta_6 = E - \Delta E = E - \arcsin(c\theta_2 c\theta_4 - s\theta_2 c\theta_3 s\theta_4) \tag{14}$$

4 Simulation and conclusion

Dynamics and stable tracking simulation studies were conducted on Matlab. The dynamics simulation was carried out with the angles θ_1=30°sin1.256t, θ_2=15°sin0.628t, θ_3=15°sin0.628t, θ_4=40°sin1.256t, θ_5=20°sin0.628t, θ_6=20°sin0.628t, the weights m_1=200kg, m_2=125kg, m_3=63kg, m_4=25kg, m_5=20kg, m_6=10kg and the moment of inertia of the i-th frame in the respective frame coordinate system J_1=217kgm^2, J_2=75kgm^2, J_3=24kgm^2, J_4=2.9kgm^2, J_5=2.9kgm^2, J_6=1.1kgm^2. According the dynamics, the driving torque of each frame can be calculated shown in Fig. 3, which can provide data and theoretical basis for dynamics compensation and driving motors selection of six-axis simulation turntable.

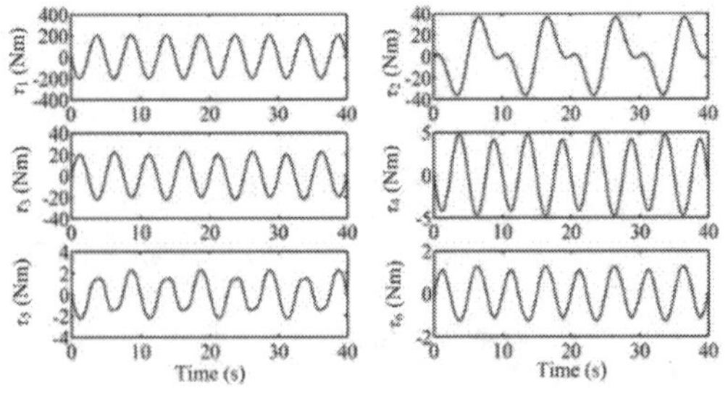

Figure 3. Driving torque of each frame.

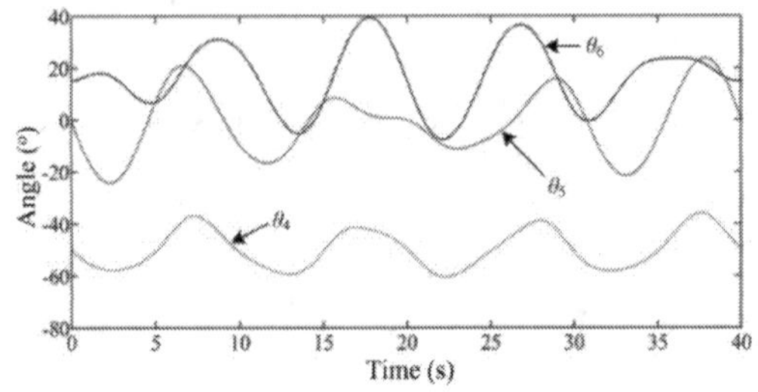

Figure 4. Compensatory angle of stable tracking model.

When the simulation turntable is running in the stable tracking mode, the rotation angle, azimuth angle and elevation angle of target are θ_1=10°sin0.628t, θ_2=15°sin0.785t, θ_3=20°sin0.628t, A=50° and E=25° respectively. According to the stable tracking model, the rotation angle θ_4, θ_5 and θ_6 shown in the figure 4 can be calculated, which achieves the tracking task. Furthermore, the angle θ_4, θ_5 and θ_6 being known conditions are substituted into the kinematics model, which calculates the azimuth angle A' and elevation angle E' of antenna. Calculating the error ΔA=A'–A and ΔE =E'–E shown in the figure 5, which shows the stable tracking model is correct.

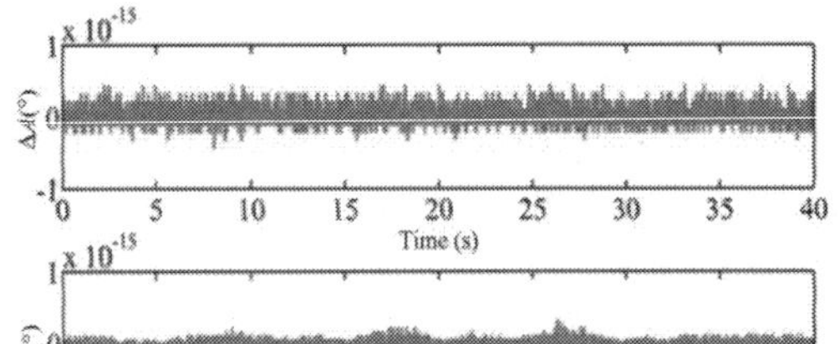

Figure 5. Calculation error of stable tracking model.

Acknowledgement

This work is supported by Foundation Nature Science Foundation of Jiangxi Province (20151BAB217022), Project of Jiangxi Provincial Department of Education (GJJ14549) and Doctor Scientific Research Foundation of Nanchang Hangkong University (EA201204428).

References

1. Wang Weiguo, Liu Tingxia, Li Yan, et al, Scondary planet tracking tactic arithmetic for three-axis turntable, Optics and Precision Engineering, **23**, 871-878 (2015)
2. Lin Cunhai, Cao Guangfeng, Decoupling and control of three-axis turret based on modified repetitive control system. Electro-optic Technology Application, **29**, 79-82 (2014)
3. Wu Yunjie, Tian Dapeng, Liu Youmin, Perfect tracking control for flight simulator, Control Theory & Applications, **28**, 414-420 (2011)
4. Zhao Yuhua, Yuan Feng, Ding Zhenliang, Error modeling and analysis of three-axis simulation turntable, Journal of Sichuan University (Engineering Science Edition), **42**, 227-231 (2010)
5. J. M. Hilkert, Inertially stabilized platform technology, IEEE Control Systems Magazine, **28**, 26-46 (2008)
6. Chen Songlin, Shan Meilin, Wang Libin, Disturbance observer-based robust perfect tracking control for flight simulator, Electric Machines and Control, **19**, 113-118 (2015)
7. Zhou Jiehua, Peng Xiafu, Zhong Xunyu, Six-aixs ship swaying simulation rotationtable and orientation stable tracking model, Ship Engineering, **33**, 53-57 (2008)
8. Wang Xiaojun, Li Dianpu, Zao Yan, et al, 3-axis steady-tracking with shipborne radar, Journal of Harbin Engineering University, **23**, 58-63 (2002)
9. Liu Chaojun, Liao Xiaozhong, Zhang Yuhe, Kinematics modeling and simulation of strapdown 3-axis stable tracking system, Journal of Beijing Institute of Technology, **20**, 313-316 (2000)
10. Fan Dapeng, Zhang Zhiyong, Fan Shixun, et al, Research of basic principles of E-O stabilization and tracking devices, Optics and Precision Engineering, **14**, 673-678 (2006)
11. Zhou Xinhong, Zhang Lizhong, Hong Jin, Vehicular laser stabilized sighting turntable dynamics modeling and simulation, Journal of Changchun University of Science and Technology (Natural Science Edition), **38**, 33-36 (2015)
12. J. H. Zhou, X. F. Peng, Research on horizontal stability model of six-axis simulation/testing turntable. *2015 International Conference on Electrical, Automation and Mechanical Engineering*, 680-683, (Phuket, 2015)

MATEC Web of Conferences 77, 01008 (2016) DOI: 10.1051/matecconf/20167701008
ICMMR 2016

Dynamic Simulation and Analysis on Elevating Process of Elevating Equilibrator

Wanpeng Ma , Jisheng Ma and Lijun Cao

Artillery Engineering Department, Ordance Engineering College, Shijiazhuang,China

Abstract. The elevating equilibrator is used to control the elevating angel of the director of a rocket launcher. Based on an elevating equilibrator, a method for co-simulation is introduced with dynamic simulation software ADAMS and hydraulic simulation software EASY5. A three-dimensional model of the elevating equilibrator is developed with Solidworks and is imported into ADAMS. Furthermore, the hydraulic system model is established by EASY5. The interaction of mechanical system and hydraulic system is analyzed. The analysis indicates that the modeling method for co-simulation is effective and practical.

1 Introduction

The elevating equilibrator is a relatively complex action mechanism and used to control the elevating angel of the director of a rocket launcher. As the mechanism is simulated, the mechanical system and hydraulic system should be analyzed together. In this paper, the virtual prototype of the elevating equilibrator is established with ADAMS and hydraulic system model is established with EASY5 respectively. Then co-simulation is achieved by dynamic data exchange between two models [1]. Because of a large load generated when elevating equilibrator working, the dynamic simulation and analysis on elevating process of elevating equilibrator is of great significance [2] .

2 Mechanical and hydraulic system modeling

2.1 Elevating equilibrator modeling

Figure 1. Elevating Equilibrator Model

The elevating equilibrator is constituted of hydraulic cylinders and energy accumulator. Significantly the hydraulic cylinder is composed of a fixed piston rod, a moving piston rod and sealing elements, and, the upward, downward and balance cavities are in the hydraulic cylinder. The accumulator includes a variable accumulator volume and an optional fluid tee-section that connects to adjacent components. The elevating equilibrator is shown in Fig. 1.

The elevating equilibrator model that is established and assembled in Solidworks software[3] is imported into ADAMS software and simplified on the basis of the dynamic relation. The dynamic simulation and analysis are in progress after the material properties defined and joints and load created. The virtual prototyping model is shown in Fig. 2.

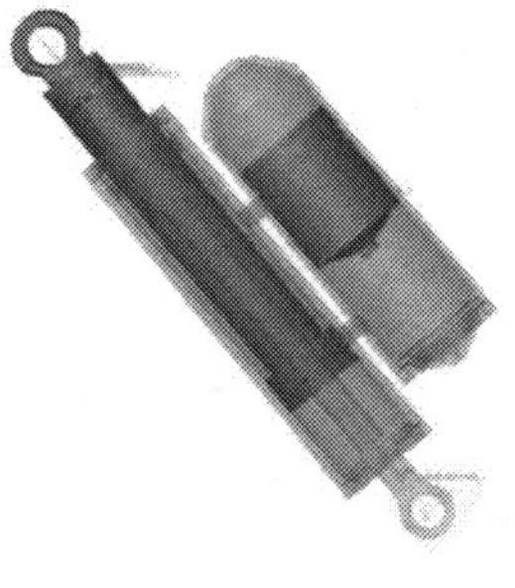

Figure 2. Simplified Model

2.2 Hydraulic system modeling

2.2.1 Hydraulic system circuit

Hydraulic oil flows into upward and downward cavity through the hydraulic system circuit, which controls the

 DOI: 10.1051/matecconf/20167701008

elevating process, and the location of machine is fixed by the hydraulic lock composed of two hydraulic control check valve. The circuit is established in EASY5 software, which includes the creating of hydraulic oil, pumps, sinks, cylinders and various control valves. Hydraulic system circuit is shown in Fig. 3.

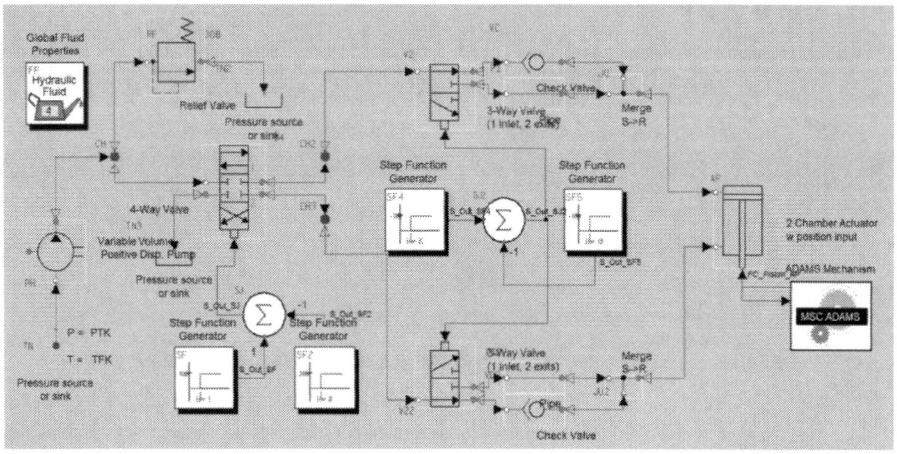

Figure 3. Hydraulic system modeling

In the hydraulic system model, the elevating process and the hydraulic lock are controlled by the 4-way valve and two 3-way valves respectively. The system pressure is limited for the relief valve, and energy is storage and reused by the energy accumulator.

2.2.2 Parameter setting

It is of significance to configure parameters of oil properties, power components, actuator components and various controllers. #10 aircraft hydraulic oil (SH0358-1995)[4] is chosen to be used in the circuit, and the fluid temperature(TC), ambient temperature and pressure are defined on basis of the working condition.

Force formula of the cylinder is:

$$F = P_1A_1 - P_2A_2 = P_1\frac{\pi D^2}{4} - P_2\frac{\pi(D^2 - d^2)}{4}$$

where F is the force of hydraulic cylinder, P_1P_2 stand for oil pressure of extend and retract chamber, and A_1A_2 stand for piston area of extend and retract chamber. The parameters of the cylinders and other components are designed, and the parameters are shown in Table 1.

Table 1. Parameters of Hydraulic Circuit

PC_RF(bar)	300
PFO_RF(bar)	320
TC_RF(s)	0.01
PCK_VC(bar)	1
APE_AP(cm2)	50
APR_AP(cm2)	40

3 Co-simulation of ADAMS and EASY5

New design variables are created in ADAMS software including an input variable (force) and two output variables (displacement, velocity) are shown in Table 2.

Table 2. State Variables Designed

State Variable	Type	Meaning
Force	Input	Force to Piston Rod
Displacement	Output	Displacement of Piston rod
velocity	Output	Velocity of Piston rod

And these variables are connected with ADAMS Mechanism Component (AD) in EASY5 software through the ADAMS/Control and EASY5/Extension interface. In Chosen Execution Mode, the Function Evaluation (no feed-through) is selected[5].

The ADAMS/View is started through the EASY5 Command Shell, and the control file of model (.cmd) and hydraulic control system file (.dll) are imported into virtual prototyping model. When interactive simulation was using, simulation process always failed because integral can not converge. Therefore the scripted simulation method is determined which controls integral steps by expression command[6]. Simulation integral step is 0.0005 before 13s and 0.005 before 15s, which is determined by repeated simulation test.

4 Results and analysis

4.1 Motion process analysis

Starting from 1s, the piston rod rises at a constant speed and drops from 8s at the same speed. When the 4-way valve is working, a shock will be engendered between hydraulic oil and both sides of piston rod. Because of that, at 1s and 8s, the velocity of piston rod will change rapidly and stabilize in a very short period of time. At 13s, the hydraulic circuit is locked by the hydraulic lock, and the motion of piston rod is stopped. The simulation results of the velocity and displacement are shown in Fig.4 and Fig.5

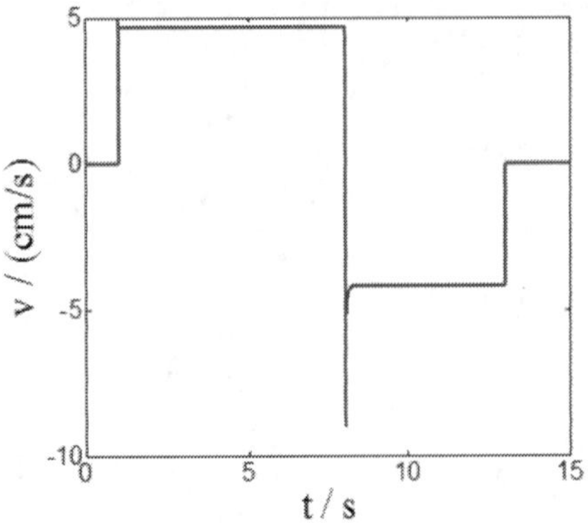

Figure 4. Velocity of Piston Rod

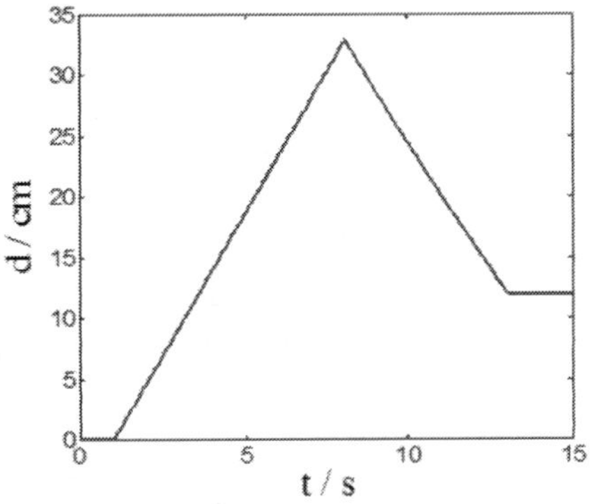

Figure 5. Displacement of Piston Rod

4.2 Force analysis of piston rod

In this simulation, the hydraulic oil pressure in the upward cavity and downward cavity are shown in Fig.6.

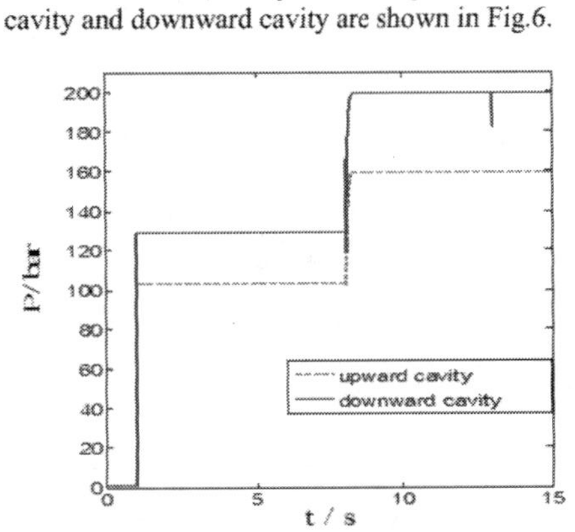

Figure 6. Changes of Pressure in Cylinder

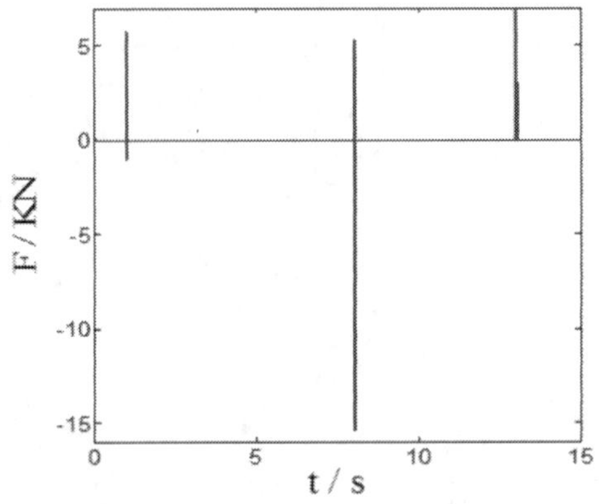

Figure 7. Shock Force of Oil Flow

When the direction of oil flow is changed by operate manual reversing valve, pressure of oil is changed rapidly at same time. At 13s, the hydraulic lock is closed up and pressure also is changed obviously. These changes in simulation results are in agreement with the actual working situation and these simulation values are credible and meaningful. The direction of oil flow changing, a shock between oil and piston rod is of significant impact for service life and dynamics performance of piston rod. The simulation results of shock force is shown in Fig.7, and at 1s, 8s and 13s the shock force is produced due to the change of the oil flow direction and the closure of the hydraulic lock. Seen from Fig.7, the maximum shock force occurs at the moment of change in the direction of oil flow at 8s, and this condition is in agreement with the actual working situation. Based on the above simulation results and analysis, the virtual prototyping model and hydraulic control system model are effective and practical, and the simulation results are credible and available for reference.

4.3 Flow analysis of relief valve and directional valve

Flow is a very important parameter for the hydraulic control system. At the level of balance machine complicated system, the flow parameters are difficult to be measured. Through the above analysis, the preliminary confirmation of co-simulation model established is consistent with the actual working conditions. So the flow analysis for the complex components can be finished by co-simulation.

The flow parameters of the relief valve and the 4-way valve are shown in Fig.. 8, 9, 10. From the figure, the flow can be analysed, when the mechanism is stopped, the relief valve is fully open, the flow is to a maximum value of 49.09 cm ^3/s. At the rising stage, the flow from port 1 of 4-way valve is positive, and the value of port 2 is negative. At the downward stage, the flow of port1 and port2 is just the opposite.

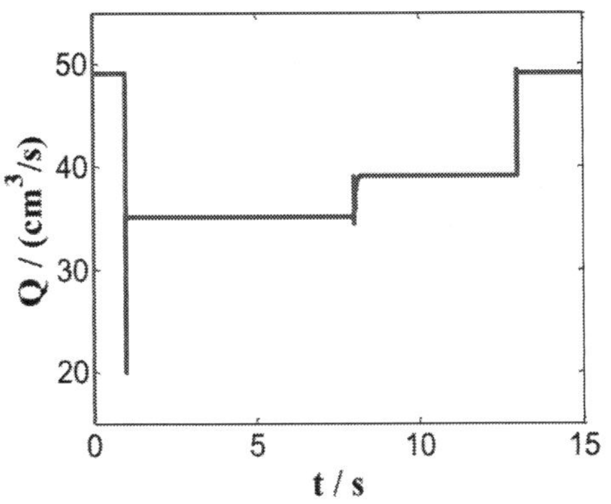

Figure 8. Flow of relief valve

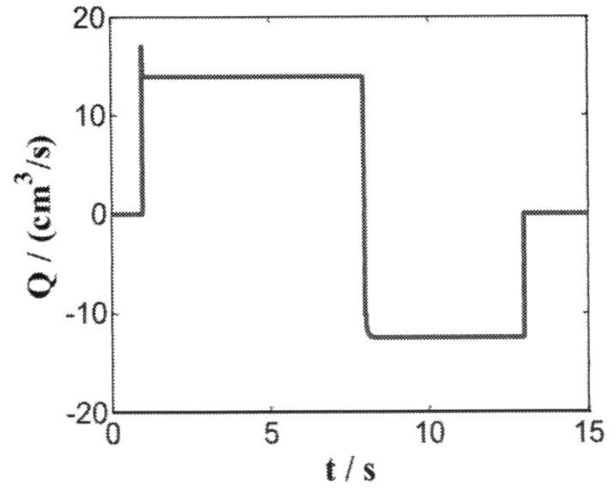

Figure 9. Flow of 4-way valve port1

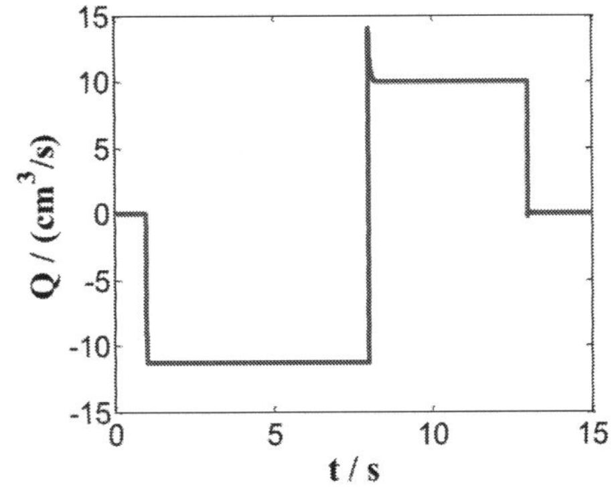

Figure 10. Flow of 4-way valve port2

Above the flow analysis of the relief valve and the 4-way valve is used to solve the problem, that it is difficult to be measured for the flow at the mechanism working. For it is difficult to place the sensors in the actual machine, the parameters of the flow is not easy achieved. It is feasible and effective to calculate the flow parameters through the simulation analysis. It provides the basis and reference for the further research of hydraulic system and related properties.

5 Conclusion

The virtual prototyping model of elevating equilibrator is established by Solidworks and imported into ADAMS. The hydraulic control system model is built in EASY5, and co-simulation is executed through dynamics data exchange. The models are in agreement with actual working situation and can provide reference for related research in the future. The direction of oil flow changing, the shock force to piston rod is of significant influence on the working performance and service life and should be considered on the design and calibration.

Using this model, the flow parameters of the relief valve and the valve are solved. It is difficult to add sensors in a complex mechanism when the actual mechanical is working. So the virtual prototyping simulation technology can be used to measure different parameters. This provides a new solution to solve practical engineering problems and can provide reference and basis for fault diagnosis and so on.

References

1. Sheng Longbo, MA Jisheng. Co-simulation of the Braking System of a Tracked Vehicle Based on ADAMS and EASY5[J]. Journal of Ordnance Engineering College (2012).
2. CHENG Xuejing, WANG Qing'an. Co-simulation of 40 k N Hydraulic Crane Based on ADAMS and EASY5 [J]. Machine Tool & Hydraulics (2015).
3. Lv Yingbo. *Solidworks 2016 Technical* Manual[M](2016).
4. Li Wenji. *Application of Aviation Hydraulic Oil*[M].68 (1995).
5. Zhou Jie, MA Jisheng. Dynamic Simulation Research on Mechanical-Hydraulic Coupling of a Certain Weapon Launch Tower's Pendulum System[J]. Machine Tool & Hydraulics (2014).
6. Zheng Jianrong. *ADAMS: Introduction and Improvement of Virtual Prototyping* Technology[M] (2002).

Study on Blade Fatigue Life of Rotating Power Machinery

Fu Xi, Kou Hai-jun and Zhang Jun-hong

State Key Laboratory of Engines, Tianjin University, Tianjin 300072, China

Abstract. The linear damage model (LDM) is widely applied in engineering calculation, but it does not consider the relationship between damage variable and load parameters. Therefore, the life prediction based on LDM is not satisfied for the aero-engine blades. Besides, it easily brings about error in predicting fatigue life by common nonlinear damage model which neglect the influence of torsional stress. Hence, a modified nonlinear continuum damage model (CDM) is put forward based on Chaboche nonlinear damage model in this research. And to determine the damage and fatigue life of TC4 material used in aero-engine blades, axial tension and compression fatigue test is conducted. Compared with LDM results, the fatigue life prediction results of the modified CDM in this work show a good agreement with the tests data. So the correctness of the modified model is verified. Finally, the fatigue life of a certain aero-engine high pressure compressor blade is predicted by the modified nonlinear continuum damage model.

1 Introduction

During aero-engine working, the compressor blades easily occur fatigue failure under complex loads, which is the main failure form for the blade [1-4]. Once the blade failure, a catastrophic accident will be caused in the engine and even the aircraft[5,6]. Therefore, the researches carried out for the blade fatigue failure are very significant for the online monitoring and fault diagnosis of aero-engine.

A suitable damage estimation model and fatigue life model are the key elements of further studying on life prediction of the blades. Among all the structural fatigue analysis methods, Palmgrem-Miner[7] linear damage accumulation model (LDM) is the most widely used.

$$D=\sum_{i=1}^{n}\frac{n_i}{N_i}=\sum r_i \tag{1}$$

where n_i is the cycles under a certain stress level; N_i is the failure cycles of structure under the same stress level; D is damage variable. The foundation for the establishment of LDM is independent of damage produced per cycle and loading parameters.

The fatigue damage of structure is a performance deterioration process under cyclic loading. Damage variable is closely associated with loading parameters, such as stress and strain[8]. Therefore, the life prediction based on Miner's rule is not satisfied. In order to overcome the deficiencies of LDM, researchers have proposed various models based on nonlinear continuum damage theory. However, most of the models are semi-empirical equations and lack of theoretical basis. In view of this problem, Kachanov[9] and Rabotnov[10] put forward the concepts of continuity degree and damage factor respectively. Marco and Starkey[11] firstly proposed a nonlinear load-dependent damage rule which allows correctly taking the effect of different loading sequences into consideration.

$$D=\sum_{i=1}^{n} r_i^{x_i} \tag{2}$$

where x_i is the variable related to loads. Some other nonlinear damage accumulation equations were also proposed in order to take into account of the influence of the multiple stress levels[12,13]. However, in those equations, load-related variables should be recalculated for different loading conditions. Although the predicted results from those equations agree with the experimental data well for some materials under certain conditions, it is complicated to calculate the corresponding coeffcients and difficult to apply in egineering.

In order to overcome the shortcoming, Chaboche [14] formulated a nonlinear continuum damage model (CDM) to describe structural performance deterioration based on fatigue fracture characteristic:

$$dD=f(D,\sigma)dn \tag{3}$$

This model mainly describes the inseparability between load parameters and damage variable. Jessen and Plumtree[15] conducted a stress-controlled fatigue test and developed a strain energy model to predict the fatigue life based on the Chaboche's work. And Cheng [16] proposed another nonlinear fatigue damage model by using the ductility exhaustion as the damage variable. Although the accuracy of fatigue damage prediction based on the above nonlinear damage models has been validated and widely accepted, those models are proposed to be mainly used in the situation without considering torsional stress.

In this work, a modified nonlinear model based on continuum damage theory is proposed. It suits for fatigue analysis of aero-engine compressor blades which bear bending stress and torsional stress caused by centrifugal and aerodynamic loads. And the correctness of the modified model is proved by basic conditions of continuum damage mechanics. In addition, fatigue tests are carried out to obtain parameters of mechanical property and fatigue characteristic for TC4 titanium alloy. The prediction results of fatigue life on the basis of the modified CDM have a good agreement with test results as well. Finally, the model is used to predict the fatigue life of a certain high pressure compressor blade.

2 Fatigue test of TC4 titanium alloy

TC4 titanium alloy is usually used to manufacture the aero-engine compressor blades due to its excellent mechanical property in high specific strength, high yield strength, high heat and corrosion resistance. In order to learn the fatigue properties of the TC4, the static tensile test is firstly carried out to obtain its mechanical parameters. Test results are listed in Table 1. The test specimen used rolled bar with 17-mm diameter (Fig. 1), 720°C annealing temperature, 1h heat preservation and air cooling.

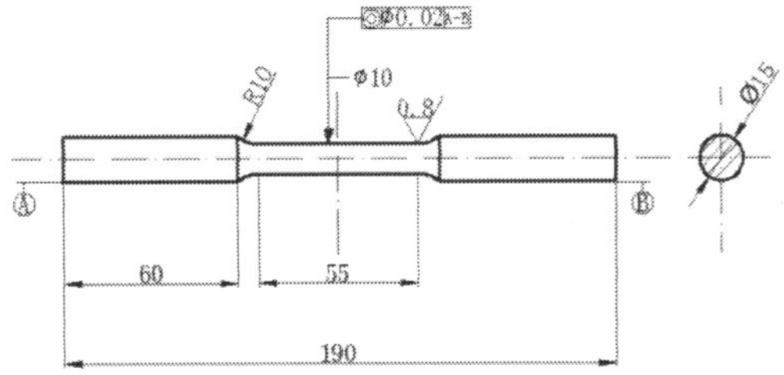

Figure 1. Static tensile test specimen.

Table 1. TC4 material properties.

Yield strength σ_s / MPa	Tensile strength σ_b / MPa	Elona -tion A / %	Reduction of area Z / %	Elasticity modulus E / GPa
975	1005	16	46	107

In order to study the cumulative damage of TC4 material, the axial tension and compression fatigue test are also carried out at ambient temperature. According to the national standard of China GB/T 3075-2008, the test specimen is designed as Fig. 2 and made from rolled bar with 17-mm diameter. This test is designed as a defect free fatigue test. So, the test specimen's surfaces are manually polished by metallographic sandpaper to make sure that the surface without notch and defects. The test frequency is set to 2 Hz, and stress ratio is R=1. The whole test results are shown in Table 2.

The fatigue limit is firstly preliminary judged through single point method which is between 300 ~ 350 Mpa. Then 4 stress amplitudes are inserted into them, they are 310, 320, 330 and 340 MPa respectively. Finally, according to test results, fatigue limit of TC4 alloy is determined to 340MPa. The S-N curve and cross-section of fracture specimen are listed in the Fig. 3 and 4 respectively.

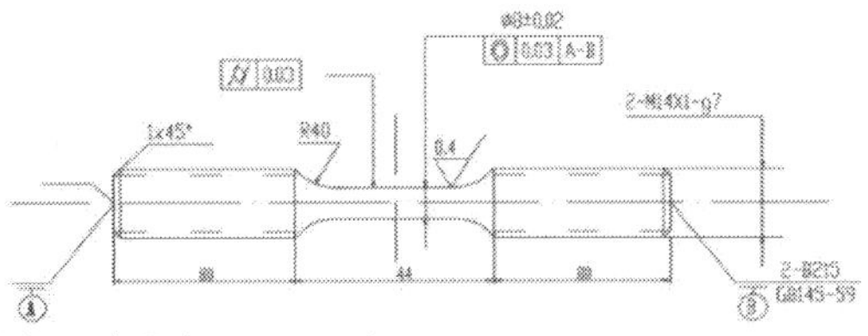

Figure 2. Fatigue test specimen.

Table 2. Axial tension and compression fatigue test results

No.	Maximum stress(MPa)	Cycle($\times 10^4$)	Results
1	550	11	Fractured
2	550	30	Fractured
3	500	32	Fractured
4	500	39	Fractured
5	450	98	Fractured
6	450	210	Fractured
7	400	51	Fractured
8	400	110	Fractured
9	400	130	Fractured
10	400	480	Fractured
11	350	440	Fractured
12	350	500	Fractured
13	340	910	Fractured
14	330	1000	Unfractured
15	320	820	Fractured
16	310	1000	Unfractured
17	300	1000	Unfractured

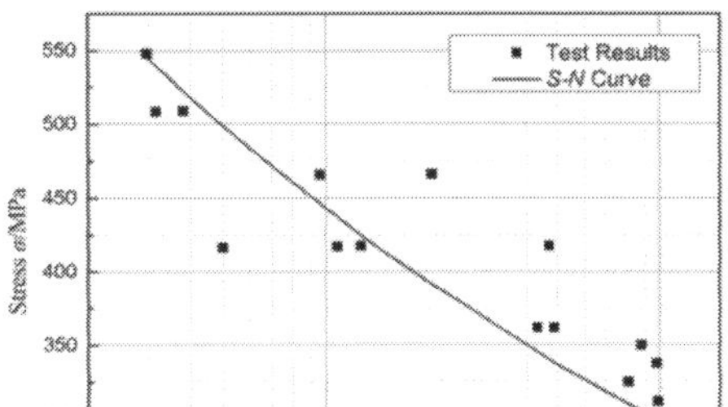

Figure 3. Test results and S-N curve of TC4.

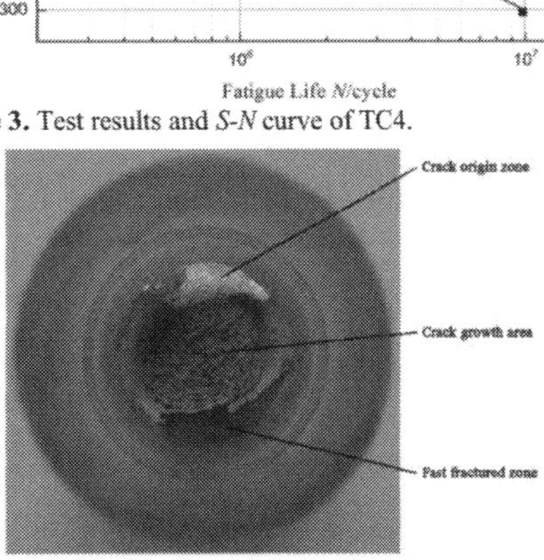

Figure 4. Section picture of fatigue test specimen.

Fig. 4 shows the cross-section of fracture specimen, it can be seen that the fatigue fractured section of specimen can be divided into three zones: the crack origin zone, the crack growth zone and the fast fractured zone. The crack origin zone is the earliest area of the crack, the section of which is smooth, crack propagates gradually along the normal direction of the cyclic load from surface to the inner layer, and the length is about 1/4 of the diameter. The section of crack growth area is rough, showing radial shape, this part occupies the vast majority of the cross-section area, in which crack growth rate is relatively stable. In the fast fractured zone, the section is wedge shaped, and the material is brittle fractured.

3 Nonlinear continuous damage model

For simple uniaxial fatigue problem, Chaboche proposed the following equation based on Eq.(3):

$$dD=\left[1-(1-D)^{1+\beta}\right]^{\alpha}\left[\frac{\sigma_a}{M(1-b\sigma_m)(1-D)}\right]^{\beta}dN \quad (4)$$

where β, M_0 and b are the material parameters; σ_m is mean stress; σ_a is stress amplitude; α is the parameter related with loads and damage:

$$\alpha=1-H\left\langle\frac{\sigma_{max}-\sigma_1(\sigma_m)}{\sigma_u-\sigma_{max}}\right\rangle \quad (5)$$

$$\sigma_1(\sigma_m)=\sigma_m+\sigma_{l_0}(1-b\sigma_m) \quad (6)$$

where σ_u is ultimate tensile strength; σ_{l0} is fatigue limit for a fully reversed loading conditon; H is material parameter, and $\langle x\rangle=\begin{cases}0, x\le 0\\ x, x>0\end{cases}$. Damage variable D is regulate 0 to represent the initial undamaged state and it is set to D=1 when the structures occurs fatigue failure.

The stress distribution of the compressor blade is very complex. The single axis fatigue model can not effectively supply a accurate prediction results of blade damage. So the model correction must be made. Considering the compressor blade working conditions and its structure characteristics, there are three kinds of vibration mode which include torsional, axial and tangential vibration. Duo to the complex aerodynamic loading, the torsional vibration of the blade will be promoted during the blade rotating. And the blade itself has the initial torsion angle, the effect of the torque on the damage must take into account[17].

According to the fourth strength theory:

$$\sigma_B=\left(\sigma_n^2+3\sigma_t^2\right)^{\frac{1}{2}} \quad (7)$$

where σ_B is total stress; σ_n is normal stress; σ_t is shear stress; Substituting formula (7) into (4), it can be obtained:

$$dD=\left[1-(1-D)^{1+\beta}\right]^{\alpha}\left[\frac{1-R}{2}\frac{\left(\sigma_B^2+3\sigma_t^2\right)^{\frac{1}{2}}}{M(1-b\sigma_m)(1-D)}\right]^{\beta}dn \quad (8)$$

Taking the whole process of material damage into account, on the basis of integral of upper formula from D=0 to D=1, the fatigue life expression under arbitrary level of load with loading parameters as independent variable can be obtained as:

$$N_f=\frac{1}{1-\alpha}\frac{1}{1+\beta}\left[\frac{2}{1-R}\frac{M(1-b\sigma_m)}{\left(\sigma_B^2+3\sigma_t^2\right)^{\frac{1}{2}}}\right]^{\beta} \quad (9)$$

In the course of above work, no damage occurs, any damage D is corresponding to the number of cycles n:

$$n=\frac{1}{1-\alpha}\frac{1}{1+\beta}\left[\frac{2}{1-R}\frac{M(1-b\sigma_m)}{\left(\sigma_B^2+3\sigma_t^2\right)^{\frac{1}{2}}}\right]^{\beta}\left[1-(1-D)^{1+\beta}\right]^{1-\alpha} \quad (10)$$

From formula (8), the cumulative damage equation under the constant loading condition is obtained:

$$D=1-\left\{1-\left[\frac{H(\sigma)}{M^{\beta}}(1+\beta)\left(\frac{1-R}{2}\frac{\left(\sigma_B^2+3\sigma_t^2\right)^{\frac{1}{2}}}{M(1-b\sigma_m)}\right)^{\beta}n\right]^{-H(\sigma)}\right\}^{-(1+\beta)} \quad (11)$$

Damage accumulation model needs to meet two basic conditions[18]:

1) Damage is not reversible, that is, in the loading cycle, the damage value must be monotonically increasing.

$$\frac{\partial D}{\partial n}>0 \quad (12)$$

2) In the case of a certain damage to the material, the relatively large load per cycle can cause more damage in the rest of the load cycle.

$$\frac{\partial^2 D}{\partial n\partial\sigma}>0 \quad (13)$$

By type (9) and (11), it can be drawn that

$$\frac{\partial D}{\partial n}=\frac{1}{(1+\beta)(1-\alpha)N_f}\left(\frac{n}{N_f}\right)^{\frac{\alpha}{1-\alpha}}\left[1-\left(\frac{n}{N_f}\right)^{\frac{1}{1-\alpha}}\right]^{\frac{\beta}{1+\beta}} \quad (14)$$

Thus, it can be obtained that $\frac{\partial D}{\partial n}>0$. With derivation of the upper formula, it can be proved that $\frac{\partial^2 D}{\partial n\partial\sigma}>0$.

Therefore, the modifed nonlinear continuum damage model proposed in this paper meets basic conditions of continuum damage mechanics. It indicates that this model is rational.

In order to further prove the accuracy of the modifed CDM proposed in this paper, a finite element (FE) model (Fig. 5a) is built to analyze the stress distribution. Then the damage evolution equation is adopted to calculate the fatigue life, the stress distribution is also shown in Figure

5b. Prediction life of specimen by making use of modifed CDM and LDM respectively, and the results are comparied with the test data, as shown in Fig. 6. It can be seen that the modified CDM is better than Miner's LDM in prediction accuracy of fatigue life, and it shows greater consistency with the test results.

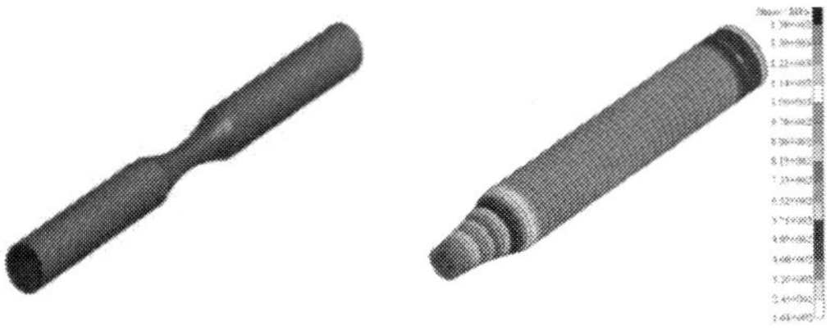

(a) Finite element model (b) Finite element results

Figure 5. FE model of test specimen.

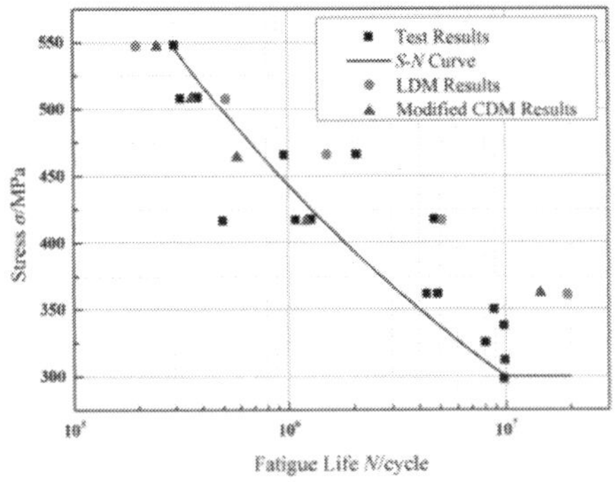

Figure 6. Comparison of life prediction results among modified CDM, LDM and test.

4 Fatigue life prediction of high pressure compressor blades

Bending and torsional stress of compressor blade caused by centrifugal and aerodynamic load is the main factors that lead to blade fatigue failure during aero-engine working. Hence, it is more reasonable to use the modified CDM proposed in this study to predict the fatigue life of compressor blade. The Fig. 7 shows assembled blade.

Figure 7. High pressure compressor blade

Table 3. Engine behavior.

Working conditions of high pressure compressor (RPM)					
Working conditions	Take-off	High-speed	Maximum-continuous	Cruise	Idle
Rotate speed	9561	9337	9172	8847	6966

Table 4. Spectrum of power allocation for aero-engine/1000h[17].

Working conditions	Take-off	High-speed	Maximum continuous	Cruise	Idle
Time scale / %	1.64	2.13	18.83	31.62	45.78

Based on the aero-engine test data, the whole blade working condition of the engine is simplified into five rated conditions, they are take-off, high-speed, maximum-continuous, cruise and idle (Table 3). And on this basis, the state power distribution spectrum is determined by the statistical data [19] of power distributing for aero-engine in 1000 flight hours as shown in Table 4.

Finally, fatigue failure life of the blade under the five typical working conditions is calculated respectively. Combined with state power distribution spectrum, the blade fatigue life under the whole working condition is predicted by modified CDM. The stress distribution of some typical wording conditions is shown in Fig. 8. It can be seen that the blade root of suction surface always receives the highest stress under no matter what working condition. According to damage accumulation theory, the blade root is determined as dangerous position. And finally, the predicting life is 333628 cycles.

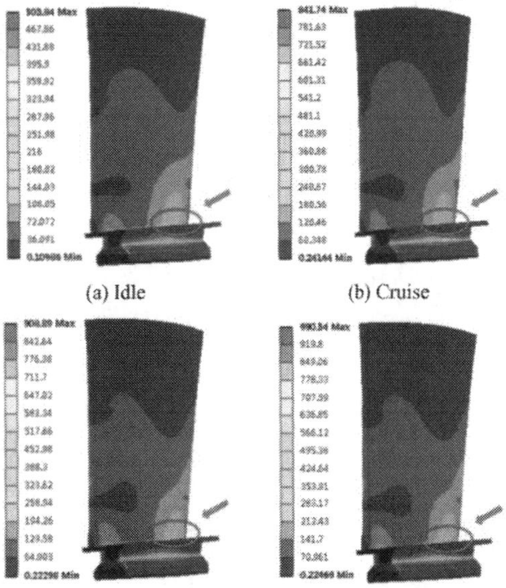

(a) Idle (b) Cruise

(c) Maximum-continuous (d) Cruise

Figure 8. Stress distribution of blade.

5 Conclusion

Based on the nonlinear continuum damage mechanics, a modified CDM is put forward, which takes into account the effect of torsional stress. The model has been proved to meet the basic conditions of continuum damage theory. Compared with LDM prediction results of fatigue life, the prediction result on the basis of the modified CDM proposed in this work shows a good agreement with the tests data. In addition, stress analysis and fatigue life predicion of a real compressor blade have been carried out. The blade root of suction surface always suffers the largest stress during the whole blade working condition. According to the damage accumulation theory, the fracture failure would occur in the region. And finally, the life prediction results of the blade is identified as 333628 cycles.

Acknowledgements

This work is supported in part by the National High-tech R&D Program of China (863 Program) 2014AA0415011 and the Specialized Research Fund for the Doctoral Program of Higher Education 20130032130005. The authors are grateful for these financial supports.

References

1. Bhaumik S, Sujata M, Venkataswamy M, et al, Failure of a low pressure turbine rotor blade of an aeroengine. Engineering Failure Analysis, **13**(2006)
2. Poursaeidi E, Aieneravaie M, Mohammadi M, Failure analysis of a second stage blade in a gas turbine engine. Engineering Failure Analysis, **15**(2008)
3. Poursaeidi E, Salavatian M, Failure analysis of generator rotor fan blades. Engineering Failure Analysis, **14**(2007)
4. Sreyas Krishnan S, Anish R, and Girish K E, Stress Analysis of a Rear Pressure Bulkhead of the Fuselage Structure and Fatigue Life Estimation. International Journal of Mechanical Engineering and Robotics Research, **2**(2013).
5. Sujeet S Narkhede and A Pateriya, Design and Analysis of Turbine Blade of Engine - A Review. International Journal of Mechanical Engineering and Robotics Research, **3**(2014).
6. Vaibhav R Pannase and A M Shende, Optimization Design, Modeling and Structural Analysis of Wind Mill Blade. International Journal of Mechanical Engineering and Robotics Research,**2**(2013).
7. Miner M. A, Cumulative damage in fatigue. Journal of applied mechanics, **12** (1945)
8. Fatemi A, Yang L, Cumulative fatigue damage and life prediction theories: a survey of the state of the art for homogeneous materials. Int. J. Fatigue, **20**(1998)
9. Kachanov L M, Time of the rupture process under creep conditions. Isv. Akad. Nauk. SSR. Otd Tekh. Nauk, **8**(1958)
10. Rabotnov Y N, On the equation of state of creep, *Proceedings of the Institution of Mechanical Engineers, Conference Proceedings*. SAGE Publications, **178**(1963)
11. Marco S M, Starkey W L, A concept of fatigue damage. Trans. ASME, **76**(1954)
12. Manson S. S, Halford G. R, Practical implementation of the double linear damage rule and damage curve approach for treating cumulative fatigue damage. Int. J. Fract. **17**(1981)
13. Rabotnov Y. N, Creep problems in structural members. North Holland, Amsterdam. (1969)
14. Chaboche J. L, Lesne P. M. A, Non-linear Continuous Fatigue Damage Model. Fatigue & fracture of engineering materials & structures, **11**(1988)
15. Jessen S. M, Plumtree A, Continuum damage mechanics applied to cyclic behaviour of a glass fibre composite pultrusion. Composites, **22** (1991)
16. Cheng G, Plumtree A. A, Fatigue damage acculiation moedl based on continuum damage mechanics and ductility exhaustion. Int. J. Fatigue. **20** (1998)
17. Wu Ying, *Fracture and fatigue*, Wuhan: China University of Geosciences Press, (2008)
18. Shang Deguang, Yao Weixing, Research on the cumulative model of single axis nonlinear continuous fatige damege. Journal of Aeronautical, **16**(1998)
19. Yu Zhufeng, *Design of structure and strength for nickel base single crystal turbine blade*, Beijing: Science Press, (2008)

The Research and Analysis on Failure Distribution Model of Diesel Engine Component Parts

Shaokun Wang, Yamei Zhou, Qiang Zhang, Shuai Yang, Guanfeng Wang, Xiang Li

School of mechanical engineering, Shandong University, Key Laboratory of High-efficiency and Clean Mechanical Manufacture (Shandong University), Ministry of Education, Jinan 250061, China

Abstract. Reliability research not only provides the direction of quality improvement for new product research and development, but also helps to get the product failure distribution model, so as to ensure the logistics organization to improve the service level. According to the reliability data from the database of research group, we analyzed the rule of failure distribution of high frequency fault component (fuel injection pump) and built a Weibull model. The parameters of the model are estimated and solved by using the uniform linear method and the least square method. After solving the model, the density function curves and the failure rate curves are drawn, and then the model of the demand forecasting of spare parts based on the failure distribution is obtained.

1 Introduction

At present, there is still a gap of the overall quality level of the engine between our country and the developed countries. Not only the average life expectancy is lower than abroad, and there will be various faults during operation. This phenomenon requires to improving the quality level and improving the quality of the engine. Reliability research not only provides the direction of quality improvement for new product research and development, but also helps to get the product failure distribution model, so as to ensure the logistics organization to improve the service level. In the reliability research, we need to analyze the failure law of products, in order to carry out reliability evaluation. The introduction of a life distribution model to fit the existing data is necessary.

Since Weibull distribution model was invented later, because the good life estimation results, it become a most widely used model in the field of reliability research quickly [1]. Abernethy summarizes a variety of commonly used parameters estimation model which is related to the Weibull distribution, method and application in product life distribution, and gives a lot of data analysis examples [2]. In terms of estimation and test method of life distribution, in addition to the traditional methods, people also established other parameter estimation methods, and introduced the research on genetic algorithm [3]. In order to study the warranty period of the new product, Mann assumed that the product failure data following the Weibull distribution, and proposed a method for estimating the reliability of the product to determine the warranty period of the new product [4].

2 Establishment and analysis of failure distribution model for fuel injection pump

Failure probability is a function of time t, which is a random variable.It is defined as:

$$F(t) = P\{T \leq t\} \quad t \geq 0 \tag{1}$$

F(t) is a failure distribution function, as known as the degree of unreliability. It is expressed the probability in the time t or before time t. The degree of reliability R(t)is the probability which is the product being completed in time t.

Probability density function of the Weibull distribution

$$f(t) = \frac{m}{\eta}\left(\frac{t-\gamma}{\eta}\right)^{m-1} e^{-\left(\frac{t-\gamma}{\eta}\right)^m} \tag{2}$$

where m is shape parameter, η is scale parameter, γ is position parameter.

Failure probability function

$$F(t) = \int_{\infty}^{t} f(x)dx = 1 - exp\left[-\left(\frac{t-\gamma}{\eta}\right)^m\right] \tag{3}$$

Failure rate function

$$\lambda(t) = \frac{m}{\eta}\left(\frac{t-\gamma}{\eta}\right)^{m-1} \tag{4}$$

Reliability function

$$R(t) = 1 - F(t) = exp[-(\frac{t-\gamma}{\eta})^m] \quad (5)$$

The failure distribution model of the fuel injection pump is established by using the uniform linear method to estimate the parameters.

For three-parameters Weibull model, t is product life, F(t) is the cumulative failure probability distribution function.

$$F(t) = \int_{\infty}^{t} f(x)dx = 1 - exp[-\frac{(t-\gamma)^m}{t_0}] \quad (6)$$

where $t_0 = \eta^m$. m is shape parameter m>0 ,γ is position parameter γ >0, $t_0$is scale parameter

The Weibull distribution linear processing

$$Y = lnln\left[\frac{1}{1-F(t)}\right] \quad (7)$$

Turn into$Y = bX + a$

where $X = \ln(t - \gamma)$

Two-parameters Weibull distribution

$$F(t) = \int_{\infty}^{t} f(x)dx = 1 - \exp[-\frac{(t)^m}{t_0}] \quad (8)$$

The sample data from small to large array, then get a set of data, t1<t2<......tn.

$$X_i = ln(t_i - \gamma)$$

When the number of samples is less than 20, the cumulative failure probability can be calculated by using the median rank,

$$F(t_i) = \frac{i-0.3}{n+0.4} \ (i = 1,2 \ldots, n) \quad (9)$$

Or mean rank,

$$F(t_i) = \frac{i}{n+1} \ (i = 1,2 \ldots, n) \quad (10)$$

When the number of samples is greater than 20, the cumulative failure probability is calculated by the related$F(t_i)$.

$$F(t_i) = \frac{i}{n+1} \ (i = 1,2 \ldots, n) \quad (11)$$

Mean rank formula is used here.

$$Y_i = \ln\ln\left[\frac{1}{1-F(t_i)}\right] = \ln\ln\left[\frac{1}{1-{}^{i}/_{n+1}}\right] \quad (12)$$

Establishing equation of straight line. Through the parameter estimation of a, b using least square method, obtained the parameters of Weibull distribution. The principle is to make the sun of squares of the deviation $(y_i - Y_i)$of the measured value and the calculated value minimum as the optimization criterion.

$$Q = \sum(y_i - Y_i)^2 = \sum(y_i - bx_i - a)^2 \quad (13)$$

$$Q = \sum(y_i - Y_i)^2 = \sum(y_i - bx_i - a)^2 \quad (14)$$

Using the least square method to obtain the optimal solution of the linear coefficient estimation.

$$\begin{cases} \frac{\partial Q}{\partial a} = -2\sum_{i=1}^{n}(y_i - bx_i - a)x_i = 0 \\ \frac{\partial Q}{\partial b} = -2\sum_{i=1}^{n}(y_i - bx_i - a) = 0 \end{cases} \quad (15)$$

The answer is:

$$\begin{cases} b = (\sum_{i=1}^{n} x_i y_i - n\overline{xy}) \Big/ (\sum_{i=1}^{n} x_i^2 - n\bar{x}^2) \\ a = \bar{y} - \bar{x}b \end{cases} \quad (16)$$

Using SPSS to calculate the parameters of the estimated value and obtain the failure distribution model, density function, failure rate function.

3 Establishment for parts demand model

On the basis of the statistical an analysing of the vehicle parts and determining the distribution of failure, forecasting the demand of spare parts is a practical method with strong feasibility. According to the sales data of the components of the same product, the life distribution model is obtained, which can calculate the failure probability and failure time of each component. Based on prediction sold parts demand quantity and demand cycle, not only can improve the efficiency of inventory stocking, and can be accurate to distribution, to streamline the cost saving resources, and service guarantee timeliness.

According to the failure rule, the failure rate function is obtained, and then the fault number is obtained from the failure rate and the sales volume. For mileage failure, according to the relationship between time and mileage, change the mileage to the time in the model, then the relationship between fault number and duration of using is obtained. Then calculate the failure of each time period, get the demand for spare parts.

Failure rate function

$$\lambda(t) = \frac{f(t)}{R(t)} = \frac{m(t-\gamma)^{m-1}}{t_0} \quad (17)$$

Preliminary fault number model

$$X = \int_0^T \lambda(t)N(t)dt \quad (18)$$

where N(t) indicates the number of remaining parts

If it is according to the statistics of the month, then, integration can be converted into summation formula.

$$X = \sum_1^k \lambda(t_i)N(t_i) \quad (19)$$

4 Case illustration and analysis

4.1. Solution and analysis of failure distribution model of injection pump

The data of this paper is to extract the fault data from the second half of 2012 from the database of the task group. By censoring, the fault mileage L divided into 6 groups.

Table1 Fault mileage and frequency

Group number	Interval lower limit /km	Interval upper limit /km	Group mid-value /km	Frequency /time	Frequency
1	250	21663	10956.5	19	0.365384615
2	21663	43076	32369.5	11	0.211538462
3	43076	64489	53782.5	9	0.173076923
4	64489	85902	75195.5	7	0.134615385
5	85902	107315	96608.5	4	0.076923077
6	107315	128740	118027.5	2	0.038461538

According to the previous section established the Weibull distribution model, the parameters were calculated as: b=0.790, a=-8.468

The correlation coefficient is: $R^2 = 0.968 \approx 1$ It is significantly correlated.

linear equation is $Y = 0.790X - 8.468$

Parameter

$$\begin{cases} m = 0.790 \\ t_0 = e^{8.468} = 4759.99 \end{cases}$$

Failure distribution model

$$F(t) = 1 - \exp[-\frac{t^{0.790}}{4759.99}] \quad (20)$$

Density function

$$f(t) = \frac{0.79}{4759.99} t^{-0.21} e^{-\frac{t^{0.79}}{4759.99}} \quad (21)$$

Failure rate function

$$\lambda(t) = \frac{f(t)}{R(t)} = \frac{m(t-\gamma)^{m-1}}{t_0} = \frac{0.79t^{-0.21}}{4759.99} \quad (22)$$

According to the model, the probability density function of fault distance is plotted, and the density function curve and the failure rate curve are plotted by Matlab.

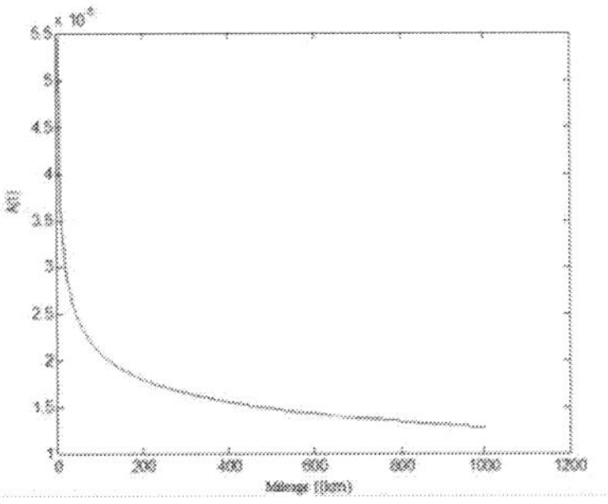

Figure 1 The failure rate curve.

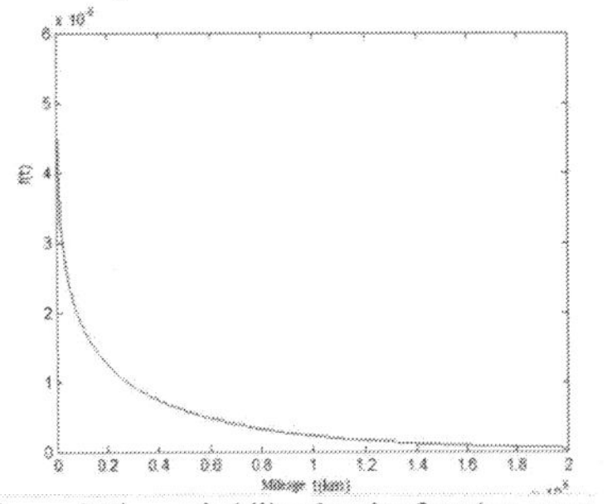

Figure 2 The probability density function curve.

It can be seen that the position parameter m<1, which belongs to the early fault period. According to Figure 1, failure rate curve is monotonically decreasing, which belongs to the early fault type.

The mean time between failure(MTBF) of the fuel injection pump is:

$$E(T) = MTBF = \int_0^{+\infty} R(t)dt = \int_0^{+\infty} tf(t)dt = \eta\Gamma(1+\frac{1}{m}) \quad (23)$$

$$\Gamma(x) = \int_0^{+\infty} t^{x-1}e^{-t}dt \quad (24)$$

Using MATLAB to solve this integral, then

$MTBF = \int_0^{+\infty} tf(t)dt = 31648$ (km)

Assuming that the average speed is 80km/h, convert the number of kilometers to time, then

MTBF=395.6 (h)

The mean time to repair(MTTR) can be obtained by observations as follows, assuming repair after 3 working days to repair, then

$$MTTR = \frac{1}{N_o}\sum_{i=1}^{n} tMi = 24(h) \quad (25)$$

The availability(A) is

$$A = \frac{MTBF}{MTBF + MTTR} = 0.942 \quad (26)$$

4.2 Parts demand model solution

DOI: 10.1051/matecconf/20167701010

Failure rate function of fuel injection pump

$$\lambda(t) = \frac{f(t)}{R(t)} = \frac{m(t-\gamma)^{m-1}}{t_0} = \frac{0.79t^{-0.21}}{4759.99} \qquad (27)$$

In order to facilitate the demonstration, it is assumed that: diesel engine is used on the bus, the daily mileage is 156km/d, 30 days per month, sales is 1000,then calculated the demand for each month respectively.

Set the lower limit is 0,the upper limit is 4680,put the failure rate function into it, using matlab to solve,after calculation, the failure number of the fuel injection pump in January is obtained.

X1=167

The result is the failure number within one month under the condition of theoretical state. In practice, many other factors need to be considered in order to make more accurate decision, such as environmental factors, seasonal factors and so on. In addition, according to this formula can calculate the failure number in each month. The data is shown in Fig. 3

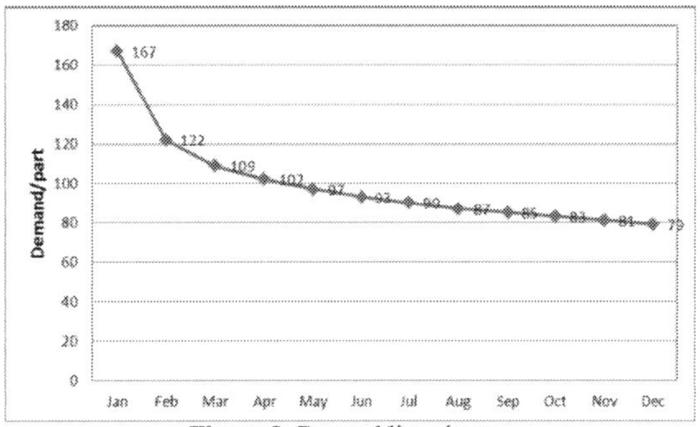

Figure 3. Demand line chart.

As can be seen from Fig. 3, the fuel injection pump which is in the early fault period, since the sales, the number of failures per month is decreased, the demand is also gradually reduced, according to the demand forecasting model can get the monthly demand theoretically. By analyzing the failure rule of the parts, the life and failure rate of the components are obtained.

When the parts to wear out failure, guide distribution in advance .

5 Conclusion

According to the fuel injection pump of diesel engine failure distribution, the Weibull model is established. Using the uniform linear method to estimate the parameters, the model is solved, and the failure distribution model of the injection pump is established. Using the matlab curve fitting, we can see that the curve of failure rate is monotonically decreasing, which belongs to the early fault type. Based on the failure distribution model, the life index of spare parts is calculated, and then the parts distribution demand model is established, and the theoretical optimal stocking quantity of each month is calculated by the example of the fuel injection pump. This model can ahead of guidance distribution, to enhance emergency response speed, improve service levels and make distribution cycle plan, monthly distribution. When the failure number is a large quantity, it is helpful in making flexible distribution measures and the cost reduction.

References

1. W Weibull. A statistical distribution function of wide applicability [J]. Journal of Applied Mechanics, 18(3):293-297. (1951).
2. Abernethy R B. The new weibull handbook(5th edition) [M]. North Palm Beach: Robert B: Abernethy, (2006).
3. Fang yuanhua, Fan zhendong, Hu changhua. Maximum Likelihood Optimization Estimation for Parameters of Reliability Life Distribution by GA [J]. AEROSPACE SHANGHA I, (2):50-53.(2006).
4. Mann N R. Warranty periods based on three ordered sample observations from a Weibull population [J]. IEEE Transactions on Reliability, 19(4):167-171.(1970).

MATEC Web of Conferences 77, 01011 (2016) DOI: 10.1051/matecconf/20167701011
ICMMR 2016

Performance Study of PID and Fuzzy Controllers for Position Control of 6 DOF arm Manipulator with Various Defuzzification Strategies

Ahmed Kharidege, Ding Jianbiao, Yajun Zhang

College of Mechanical and Electrical Engineering, Beijing University of Chemical Technology, Beijing 100029, China

Abstract. Today's arm manipulators are more and more demanding in terms of productivity. Conventional controllers are not always able to provide good and accurate results. To complete a position movement of the manipulator's end-effector, a set of joint angles of manipulator first required to be converted to the position coordinates by using the forward kinematics method, and each joint rotation is executed using a servomotor feedback control. The kinematic model has been validated using MATLAB® robotics toolbox. An end-effector based 6 degree of freedom (6-DOF) platform is proposed in this work which uses DC servomotor for actuation of the three revolute joints. PID controller is used as a reference benchmark. And FLC controller with different defuzzification strategies was employed. Results were compared in terms of time response criteria. Simulation results using MATLAB are demonstrated that PID has superior performance in terms of transient parameters. In Steady state response, both PID and FLC manage to converge to the desired output but in terms of overshot FLC is outperformed.

1 Introduction

Arm manipulators are the most commonly used robots in industrial applications. Arm manipulator consists of a series of links, connected by joints, which can be moved translationally or rotationally. Parts of all of the joints are actuated and extend from a base to an end-effector. Arm manipulator is analogous to the human hand, and the end-effector can be designed to perform any desired task [1]. To successfully fulfil the desired task, the kinematics and dynamics of a manipulator must be studied first, and then the controllers also need to be designed and implemented in the mechanical manipulator system.

Various controllers have been designed and applied in the robot manipulator. Different types of these controllers in terms of performance index will be evaluated. Proportional Integral Derivative (PID) controller may be the most widely used controller in the industrial and commercial applications for the early decades. Due to its simplicity of designing and implementation, so the first attempt is to apply PID control.

Due to high reliabilities, flexibilities and low costs, DC motors are widely used in industrial applications, robot manipulators and home appliances where speed and position control of motor are required. Numerous papers about DC motor fuzzy control system design. In [2] authors reviewed fuzzy-logic applications for robot systems and flexible links manipulators. Delibasi and Türker presented the position control of DC motor using FLC and PID control algorithms. The two controllers were designed based on LabVIEW program. Results showed that the fuzzy controller better than using PID controller for achieving the desired position. They featured the influence of fuzzy logic controller upon the performance of robot movement simulation [3]. Lin *et al.* compared PID and FLC for position control and observed that FLC performed better than PID [4]. Azevedo *et al.* have shown that FLC is less sensitive than PID to load variations [5]. Sreentha and Pradhanb designed fuzzy controller for the position control of single flexible link revolute. In this controller, time rate of change and error in joint angular displacement as input variables were used. Both theoretical and experimental results showed that the angular displacement at the base joint is highly oscillatory and settling time extremely large [6]. Mishra et.al, made a comparison between PID and FLC for servomotor control and described that PID parameters had to be tuned again under variations of plant parameters external disturbance or noise [7]. Young and Shih designed Fuzzy-PID controller of a six axis serial manipulator. Results compared with PID and two fuzzy controllers, Study shown that Fuzzy-PID has smaller steady state errors [8].

In this work, 6 DOF robot Manipulator proposed the robotic arm has five revolute joints include tool, wrist, elbow, shoulder, waist or base joints as shown in Figure 1. An important contribution is the use of servomotors for actuation of the revolute joints. The advantages of the servomotors is that they have a high accuracy and efficiency and reliability, low maintainability, and good dynamic control for variable speed applications .The rest of the paper are organized as follows: section 2 describes the kinematic model of the proposed platform, section 3 describes the DC servo motor design and value of transfer function. Different control schemes were introduced in section 4, simulations carried out in section 5 and the last section is the result analysis.

2 The Robot Model

In this work, 6 DOF robot Manipulator proposed the robotic arm has five revolute joints include tool, wrist, elbow, shoulder, waist or base joints as shown in Fig. 1. This robot is available for studies in machinery design and theory Laboratory in Beijing University of Chemical Technology, and used by students of Robotics.

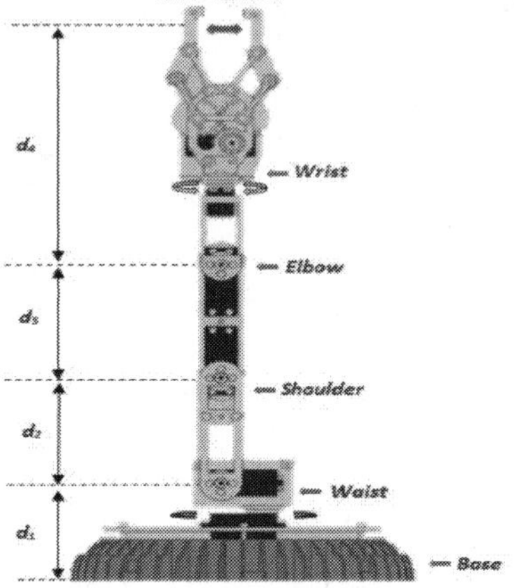

Figure 1. Arm Manipulator Model.

2.1 The kinematic model

Kinematic problem is associated with the motion of robots. It does not deal with the forces that cause the motion. It is further divided into two categories, a) Forward Kinematic, b) Inverse Kinematics, forward kinematic deals with the computation of end-effector coordinates knowing the achieved joint angles, the most commonly used methods for kinematic modelling are Denavit–Hartenberg (D-H) and successive screw displacement, In D-H formulation approach, first D-H parameters are defined. Having the knowledge of these parameters, the kinematic model can be described for any robot [9]. In contrast, inverse kinematic model computes the required joint angles for the given coordinates and is more complex than forward kinematic [10].

The relationship between the individual joints of the rehabilitation device and the position and orientation of the robot's end-effector is expressed concisely by the four D-H parameters given in Table.1. The four parameters a_i, αi, d_i, θ_i are generally known as the link length, link twist, link offset, and joint angle respectively [11].

Table 1.The link parameters of Manipulator (D-H parameters).

No	$\alpha_{i-1(°)}$	$a_{i-1(mm)}$	$d_{i-1(mm)}$	$\theta_{i-1(°)}$
1	90	0	0	θ_1
2	0	0	d_2	θ_2
3	0	0	d_3	θ_3
4	-90	0	d_4	θ_4
5	0	0	0	θ_5
6	0	0	0	End-effector

Table 2.The link lengths of Manipulator.

joint	waist	shoulder	elbow	wrist
symbol	d_1	d_2	d_3	d_4
Link length (MM)	93	80	81	172

Based on these parameters, the transformation matrix T_6^0 includes the overall rotation and translation of tool frame {6} with respect to base frame {0}. The transformation matrix is given by (1):

$$T_6^0=\begin{bmatrix} C_1(C_{234}C_5-S_{234}S_5) & C_1(-C_{234}S_5-S_1C_5) & -C_1(C_{23}S_4+S_{23}C_4) & C_1(C_{234}d_4+C_{23}d_3+C_2d_2) \\ S_1(C_{234}C_5-C_{234}S_5) & S_1(-C_{234}S_5+C_1C_5) & -S_1(C_{23}S_4+S_{23}C_4) & S_1(C_{234}d_4+C_{23}d_3+C_2d_2) \\ S_{234}C_5+C_{234}C_5 & -S_{234}S_5 & -S_4((S_{23})+C_4C_{23}) & S_{234}d_4+S_{23}d_3+S_2d_2 \\ 0 & 0 & 0 & 1 \end{bmatrix} \quad (1)$$

where:

$$C_i=\cos\theta_i,\ S_i=\sin\theta_i,\ S_{12}=S_1C_2+C_1S_2,\ C_{12}=C_1C_2-S_1S_2$$

$$C_{234}=C_2(C_3C_4-S_3S_4)-S_2(S_3C_4+C_3S_4),\ S_{234}=S_2(C_3C_4-S_3S_4)+C_2(S_3C_4+C_3S_4)$$

Based on the user-defined desired angles, the developed forward kinematic model computes the corresponding positions. These positions are calculated using MATLAB©, Fig. 2.shows the side and top view.

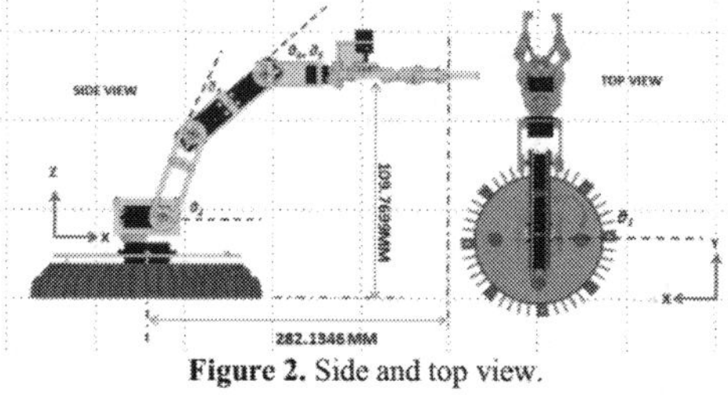

Figure 2. Side and top view.

2.2 Forward kinematic model validation

Going one step forward, for well visualizing, the derived forward kinematic model has been validated using MATLAB® environment, The Toolbox developed by Peter Corke provides functions for manipulating and converting between datatypes e.g (vectors, homogeneous transformations, roll-pitch-yaw and Euler angles) which are necessary to represent 3-Dimensional position and orientation [12]. Visual plot of a robot gives clear insight of the kinematic behaviour of a robot. Given various angle set as input to the developed forward kinematics model (1) and MATLAB toolbox, corresponding results have been compared and plotted.
Considering joint angles configuration [θ_1, θ_2, θ_3, θ_4, θ_5] as [0°,-60°,-30°,-30°, 0°], the position and orientation of the end-effector expressed in the base coordinates is:

$$T_6^0=\begin{bmatrix} 0 & 0 & 0 & 0 \\ 0.9999 & 0 & 0.0000 & 282.1346 \\ 0.9999 & 0 & 0.6250 & 109.7699 \\ 0 & 0 & 0 & 1.0000 \end{bmatrix} \quad (2)$$

Using the command 'fkine' in MATLAB Toolbox for Robotics, the same result has been obtained. The corresponding MATLAB plot for this joint configuration is illustrated in Figure 3.

Considering joint angles configuration [θ_1, θ_2, θ_3, θ_4, θ_5] as [90°,-60°,-30°,-30°,0°], the position and orientation of the end-effector expressed in the base coordinates is:

$$T_6^0 = \begin{bmatrix} 0 & 0 & 0 & 0 \\ 0.9999 & 0 & 0.0000 & 282.1346 \\ 0.9999 & 0 & 0.6250 & 109.7699 \\ 0 & 0 & 0 & 1.0000 \end{bmatrix} \quad (3)$$

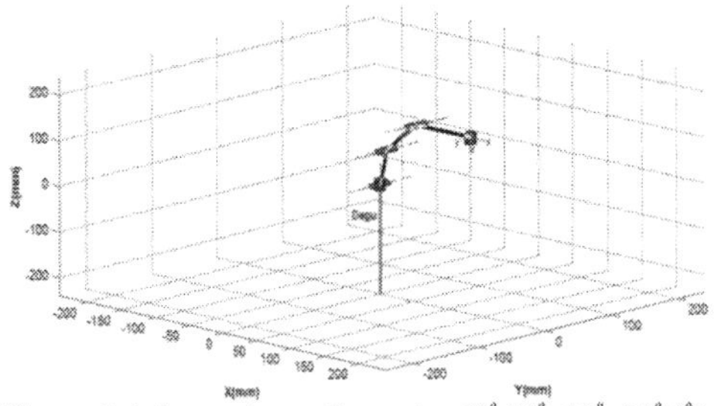

Figure 3. joint angle configuration [0°,60°,-30°,-30°,0°].

Using the command 'fkine' in MATLAB Toolbox for Robotics, the same result has been obtained. The corresponding MATLAB plot for this joint configuration is illustrated in Fig. 4.

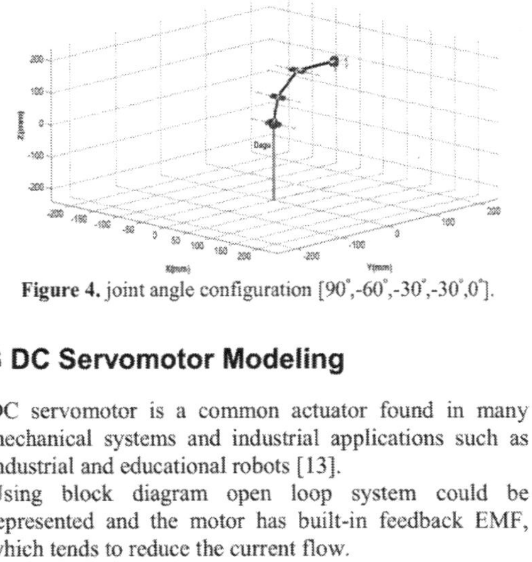

Figure 4. joint angle configuration [90°,-60°,-30°,-30°,0°].

3 DC Servomotor Modeling

DC servomotor is a common actuator found in many mechanical systems and industrial applications such as industrial and educational robots [13].
Using block diagram open loop system could be represented and the motor has built-in feedback EMF, which tends to reduce the current flow.

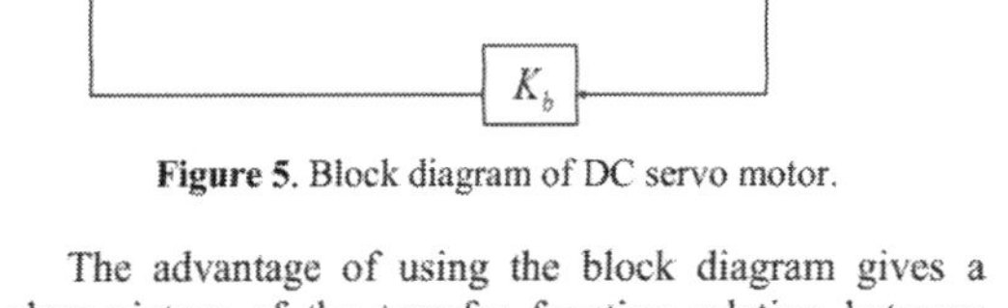

Figure 5. Block diagram of DC servo motor.

The advantage of using the block diagram gives a clear picture of the transfer function relation between each block of the system.

Using SIMULINK, the model of the motor was created. Table 3 shows DC servomotor parameters and values chosen for motor simulation.

Table 3. DC servo Motor Parameters.

Parameters	*Values*	*Unit*
Moment of inertia (B_m)	*0.00003*	*kg.m2*
Friction coefficient (J_m)	*0.196*	*N.ms*
Back EMF constant (K_b)	*0.04*	*v/ms-1*
Torque constant (K_t)	*0.7*	*Nm/A*
Electric resistance (R_a)	*2*	*Ohm*
Electric inductance (L_a)	*0.015*	*H*

The overall transfer function of the system is given below:

$$\frac{\theta_m(s)}{v_t(s)} = \frac{0.7}{0.000000045s^3 + 0.00036s^2 + 0.424s} \quad (4)$$

4 Controller Design

To be able to control a motion process, the precise position of object needs to be measurable and maintained. The designed system should respond to the applied input with a suitable overshoot a settling time and a zero steady state error as possible. To simplify the process of robot and control design and analysis, two suggested MATLAB function block models with its function block parameters for arm robot will be proposed.

4.1 PID Controller

PID controller [14] uses the Error $E_{(t)}$ between the reference input $R_{(t)}$and the output $U_{(t)}$ as input, and then generates a control signal $U_{(t)}$ for the controlled system. The transfer function of the PID controller has the following form:

$$G_{pid}(s) = K_p + K_i / s + K_d s \quad (5)$$

where K_p, K_i and K_d are the proportional, integral, and derivative gains respectively. Another useful equivalent form of the PID controller is in the form:

$$G_{pid}(s) = K_p(1 + 1/(T_i s) + T_d s \quad (6)$$

where $T_i = K_p / K_i$ and $T_D = K_d / K_p$ are known as integral and derivative time constant respectively.

For Tuning PID parameters there are some rules should be considered as:

- If the error $E_{(t)}$=$R_{(t)}$-$Y_{(t)}$ is positive high then the proportional gain K_p must be high, integral term K_i low and the derivative term K_d is low. Therefore this will speed the system output.
- If the current error is very small the PID parameters will have to be smaller value for proportional gain, larger value of integral time constant and larger value of derivative gain. So the speed of the system response will be small to reduce the overshoot of the output.

Parameters of PID controller were tuned using a Simulink block instead of conventional tuning methods. Initial values of parameters were defined as and the new parameter values were adjusted in just a few iterations as shown in Fig. 6. Thus the time and effort for tuning parameters decreased considerably.

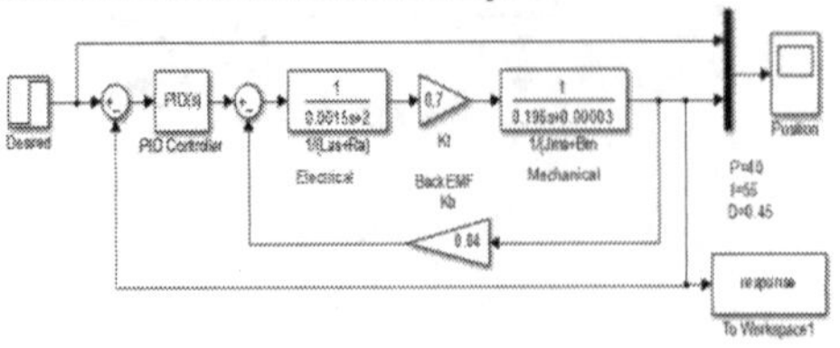

Figure 6. Block Diagram of Servomotor with PID Controller.

4.2 Fuzzy Logic Controller

In the design of fuzzy logic controllers there are four main components: fuzzifier, knowledge base, inference mechanism and Defuzzifier [15]. Fuzzifier converts a crisp input signal into a fuzzified signals identified by membership functions into fuzzy sets. The knowledge base consists of rule base and the data base. The inference mechanism evaluates which control rules are relevant at the current time and then decides what the input to the plant should be. Finally the defuzzification process converts the fuzzy output into crisp controlling signal.

Designing procedure of the fuzzy controller [16] is described as follows:

- Define the input and output of FLC. There are two inputs of FLC, the error $E_{(t)}$ and change of error $\Delta_{(t)}$ and One output is a control signal to the plant.
- Fuzzifying input and output variables. Each variable of fuzzy control inputs has five fuzzy sets ranging from negative high (NH) to positive high (PH).
- Design the inference mechanism rule to find the input-output relation. This paper uses Mamdani (Pro-Max) inference mechanism.
- Defuzzifying the output variable of fuzzy mechanism. Different Defuzzification methods were used and compared to obtain the control signal.

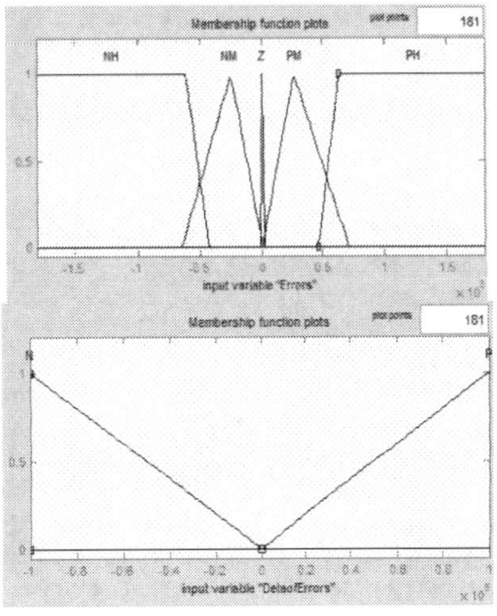

Figure 7. Membership Functions of $E_{(t)}$ and $\Delta E_{(t)}$.

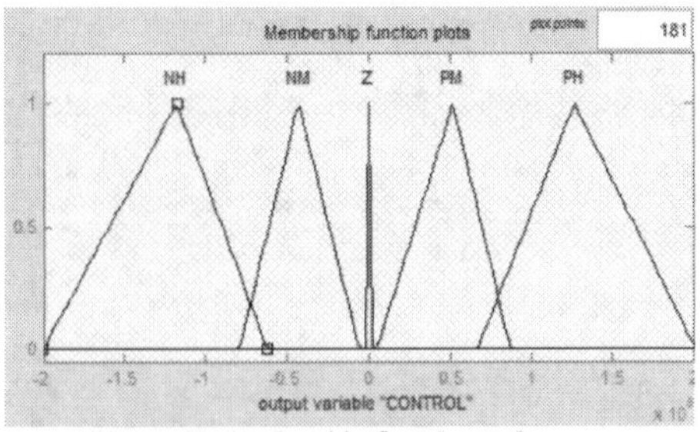

Figure 8. Membership functions of output.

The fuzzy membership functions for the two input parameters are shown in Fig. 7, and the membership function for the output is shown in Fig. 8. For error (E) as input and control as output, here NH means Negative High, NM means Negative Medium, Z means Zero, PM means Positive Medium, and PH means Positive High. For change of errors (Delta DE) as input, here N means Negative and P means Positive.

Table 4. Rule base.

No	Rule Description
1	If E is PH then control is PH
2	If E is NH then control is NH
3	If E is Z and DE is N then control is NM
4	If E is Z and DE is P then control is PM
5	If E is Z then control is Z
6	If E is NM then control is NM
7	If E is PM then control is PM

Rule base and membership functions were designed. Fig. 9 shows the Simulink model for FLC controller.

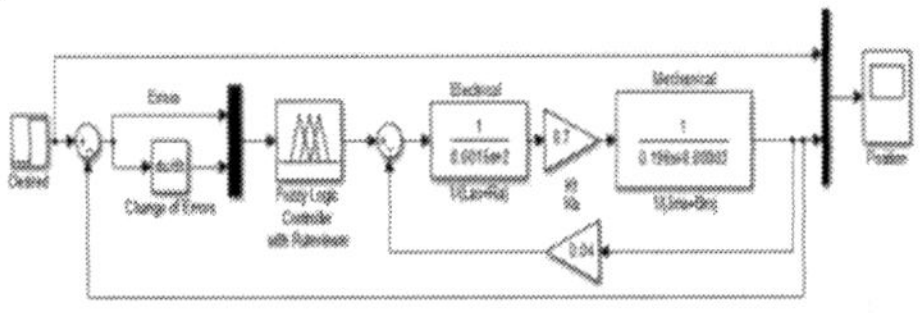

Figure 9. Block diagram of servomotor with fuzzy controller.

4.2.1 Defuzzification strategies

Defuzzification is a process with the aim to produce a non-fuzzy action; it transforms a union of fuzzy sets into a crisp value. The most common methods are as follows:

1- Centre of Gravity (COG). 2- Bisector of Area (BOA). 3- Mean of maximum (MOM). 4- Smallest of Maximum (SOM). 5- Largest of Maximum (LOM).

An idea of Centre of Gravity (COG) where the crisp control value u_{COG} is the abscissa of the centre of gravity of the fuzzy set, u_{COG} is calculated as follows:

$$u_{COG} = \frac{\sum_i \mu_c(x_i)x_i}{\sum_i \mu_c(x_i)} \tag{7}$$

where x_i is a point in the universe of the conclusion (i=1,2...) and $\mu_c(x_i)$ is the membership value of the resulting conclusion set. For continuous sets summations are replaced by integrals.

The (*BOA*) defuzzification method calculates the abscissa of the vertical line that divides the area of the resulting membership function into two equal areas.

$$\left| \sum_{i=1}^{j} \mu_c(x_i) - \sum_{i=j+1}^{i_{max}} \mu_c(x_i) \right|, i \langle j \langle i_{max} \qquad (8)$$

Here i_{max} is the index of the largest abscissa $x_{i_{max}}$, BOA is considered a computationally complex method.

Another approach to obtain the crisp value is to choose the point with the highest membership. There may be several points in the overall implied fuzzy set which have maximum membership value. Therefore it's a common practice to calculate the mean value of these points. This method is called (*MOM*) and the crisp value is calculated as follows:

$$u_{MOM} = \frac{\sum_{i \in I} x_i}{|I|}, I = \{ i \,|\, \mu_c(x_i) = \mu_{max} \} \qquad (9)$$

Here I is the (crisp) set of indices i where $\mu_c(x_i)$ reaches its maximum μ_{max}, and $|I|$ is its cardinality (the number of members).

One can also choose the leftmost point among the points which have maximum membership to the overall implied fuzzy set. This method is (*SOM*) defuzzification method. Crisp value is calculated as follows:

$$u_{SOM} = x_{\min(I)} \qquad (10)$$

Another possibility is to choose the right most point among the points which have maximum membership to the overall implied fuzzy set. This method is called (*LOM*) defuzzification method where crisp value is calculated as:

$$u_{LOM} = x_{\max(I)} \qquad (11)$$

These strategies proposed in the literature. Four of them will be adopted in this work.

5. Simulation Results

The simulation is carried out with the various Controllers. In order to assess the performance of the controller, PID and Fuzzy Controllers were designed and simulations are presented. In this work the first joint (Base θ_1) sought to move 90' in 8 second for the step time.

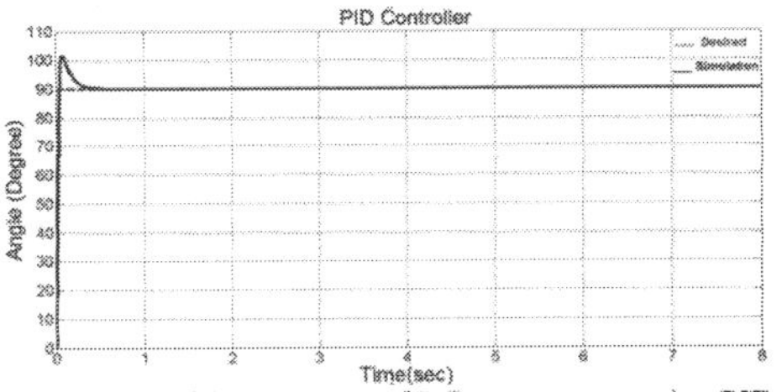

Figure 10. Position response of DC servo motor using PID Controller.

5.1 PID Controller

Using a PID controller, the position time response of the DC servomotor is shown in Fig. 10, where the dashed line refers to a desired signal, the solid line refers to a simulation result.

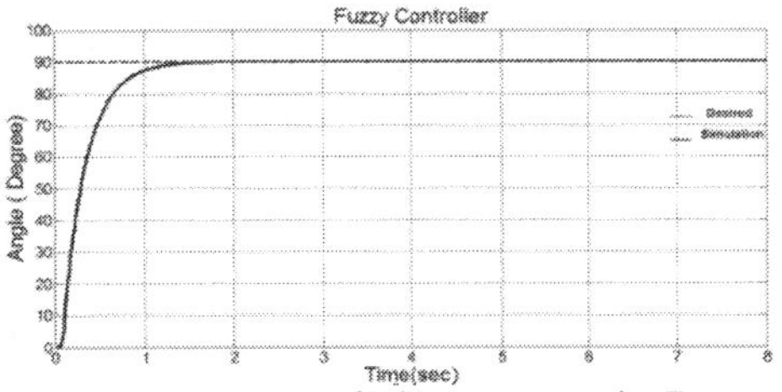

Figure 11. Response of DC servo motor using Fuzzy Controller.

5.2 Fuzzy controller

Using a fuzzy logic controller, the position time response of the DC servomotor is shown in Fig. 11, where the dashed line refers to a desired signal, the solid line refers to a simulation result.

6. Result Comparison

The model has been subjected to reference step inputs signal. Simulation results in Fig. 10.11 show the output response performance of PID and fuzzy Controller respectively. The results show that all Controllers can complete the desired motion of the Manipulator's Servomotor.

The time response parameters, including rise time (t_r), percent overshoot (Mp), settling time (t_s) and steady state error (E_{ss}) of the PID Controller and the fuzzy logic Controller (FLC) for the higher order system transfer function of Manipulator 's servo motor are obtained. PID controller produces faster rise time, settling time, whereas FLC producing lower steady state error with zero overshoot. The comparison between two controllers is summarized in Table (5).

Table 5. Comparison between PID and FLC Controllers.

Controller Type	*System Characteristics*			
	Rise time (t_r)	*Settling time (t_s)*	*Steady State Errors (E_{ss})*	*Overshoot (Mp%)*
PID	*0.04*	*2*	*1*	*10.72*
FLC	*0.2667*	*4*	*0.8*	*0*

Different defuzzification strategies were investigated and results are shown in Table 6.

Table 6. Comparison between various Defuzzification strategies of FLC Controller.

Defuzzification Methods	*System Characteristics*		
	Rise time (t_r)	*Steady State Errors (E_{ss})*	*Overshoot (Mp%)*
Center of Gravity	*0.4000*	*0.8*	*0*
Smallest of Maximum	*0.2667*	*0.86*	*0*
Mean of Maximum	*0.2667*	*0.85*	*0*
Bisector of Area	*0.2667*	*0.86*	*0*

The results obtained using the four Defuzzification strategies have been shown in Table 6. From this table, noticed that Bisector of Area, Mean of Maximum and Smallest of Maximum strategy are giving approximately the same results, whereas for the Centre of Gravity approach there is wide variations in the results that are obtained. Due to complex operations as fuzzification and particularly Defuzzification, implementing simplified Defuzzification strategy leads to optimize the system. That means recommended avoiding Centre of Gravity strategy.

7. Conclusion

One of the most problems with fuzzy controller is that the computing time is longer that for PID .The comparative analysis based on the simulation for Manipulator's servomotor is tabulated which shows the performance index of the two proposed controllers. This work presents design and performance comparison of PID and FLC strategies. Given step input signal to characterize the response, simulation results demonstrate that PID has superior performance in terms of transient parameters. In Steady state response, both PID and FLC manage to converge to the desired output but in terms of overshot FLC outperformed with zero value. Different defuzzification strategies were employed. Results shown that Bisector of Area, Mean of Maximum and Smallest of Maximum strategies are outperformed as compared with Centre of Gravity.

Acknowledgment

This work was financially supported by the Fundamental Research funds for the Central Universities (YS1403, ZY1521).

References

1. L. W. Tsai, Robot Analysis: The Mechanics of Serial and Parallel Manipulators, New York: Wiley, (1999).
2. M. Nazemizadeh1, M. Taheri and Sh Nazeri. "The application of fuzzy-logic method to control of robots : review study". International Journal of Mechanical Engineering and Robotics Research, Vol. **3**, No. 2, April 2014.
3. A. Delibasi, T. Turker, and G. Cansever. Real-Time DC motor position control by fuzzy logic and PID controllers using LabVIEW, IEEE International Conference on Mechatronics and Robotics (MechRob2004), September 13-15 Aachen, Germany.
4. Lin, P.H. Hwang, S. and Chou, J., Comparison on Fuzzy Logic and PID Controls for a DC Motor Position Controller. Conference Record of the 1994 IEEE Industry Applications Society Annual Meeting, 1930-1935, (1994).
5. Azevedo, H.R., Brandao, S.F.M. and Mota Alves, J.B., A Fuzzy Logic Controller for dc Motor Position Control. IEEE 2nd International Workshop on Emerging Technologies and Factory Automation. Design and Operations of Intelligent Factories. Workshop Proceedings, 18-27, (1993).
6. Sreenatha. A. G. Makarand. P., "Fuzzy Logic Controller for Position Control of Flexible Structures," Acta Astronaut journal, Vol. **50**,No.11. pp.665-671,(2002).
7. Mishra, M.K., Kothari, A.G., Kothari, D.P. and Ghosh, A., Development of a Fuzzy Logic Controller for Servo Systems. IEEE Region 10 International Conference on Global Connectivity in Energy, Computer, Communication and Control (TENCON '98), 204-207, (1998).
8. Yong and Shih, "Position Control of a Serial Manipulator Using Fuzzy-PID Controllers" International Journal of Automation and Smart Technology, doi: 10.5875/ausmt.v**5**i1.832, ISSN: 2223-9766.
9. J.J. Craig, (2004). Introduction to robotics: mechanics and control.
10. M.W. Spong, S. Hutchinson and M. Vidyasagar, (2006). Robot modeling and control, New York, John Wiley & Sons.
11. JJ Crage. Introduction to Robotics Mechanics and Control, 3rd Edition, Prentice Hall. (2005).
12. Corke, Peter (2011) Robotics, Vision and Control: Fundamental Algorithms in MATLAB, ISBN 978-3-642-20144-8, Springer Tracts in Advanced Robotics, 73. Springer, Berlin.
13. M. W. Spong, S. Hutchinson, and M. Vidyasagar, Robot Modeling and control: John Wiley & Sons New York, (2006).
14. R.P. Copeland and K.S. Rattan, "A Fuzzy Logic Supervisor for PID Control of Unknown Systems," IEEE International Symposium on Intelligent Control. pp. 22–26, Aug.(1994).
15. G. Feng, "A Survey on Analysis and Design of Model-Based Fuzzy Control Systems," IEEE Trans. on Fuzzy Sys, Vol. 14, No. 5, pp. 676–697, Oct. (2006).
16. C.C. Lee, "Fuzzy Logic in Control Systems: Fuzzy Logic Controller-Part I", IEEE Transactions on Systems, Man and Cybernetics. Vol. 20, No.2, pp. 404–418, March/April (1990).

Analytical kinematics and trajectory planning of large scale hexagonal modular mesh deployable antenna

Mariyam Sattar and Cheng Wei

School of Aeronautics Science & Engineering, Beihang University, Beijing, 100191, Haidian District, China

Abstract. This investigation aims to analyze the forward kinematics and deployment path of large scale hexagonal modular mesh deployable space antenna employing analytical procedure. At first a comprehensive review of technical contributions from 1987 to date in this class of antenna is completed and least targeted areas are extracted. Based on shortcomings a new design of modular antenna is proposed that has an aperture diameter of 13 meters and is composed of seven modules. Each module possesses an aperture of more than 6 meters and consists of 6 ribs arranged in radial pattern around the central beam. In order to ensure that the antenna will deploy successfully in space conditions, feasibility of design & its stability during deployment; analytical investigation into kinematics & antenna trajectory is carried out using loop closure equations and homogeneous coordinate transformation method. The concept of virtual link is introduced to develop formulations that will help to determine position, velocity and acceleration of each revolute joint in the rib with any small increment in linear motion of slider from top to bottom of central beam. The proposed solution techniques will help to replace time consuming and costly iterations for design, build more economical structure, reduce antenna manufacture time and reduce weight leading to reduced launch as well as experimentation costs before launch.

1 Introduction

Space deployable antenna is an essential aerospace equipment to land and space exploration activities [1], [2]. In past, extensive researches about antenna design have taken place ranging from highly accurate but heavy solid reflectors [3] to extremely light inflatable ones. In between these two extremes are the mesh antenna [4]. Hexagonal module antenna, with high stiffness, packaging efficiency & high accuracy of observation is one of its subcategories. The shape of this reflector is maintained by mesh surface supported by cable network and deployable truss. The integrity of support structure is the most important factor that needs to be confirmed before final launch. Current expedition is to research & apply some cost effective & efficient methods of analysis to minimize the risk of failure and confirm stability of design during launch or in-orbit operation.

2 Background Study

Nippon Telegraph and Telephone Corporation has been developing modular mesh deployable antenna since 1987 [5]-[7] for S-band mobile communication satellites. Since then methods for, basic module unit design, their integration & alignment, mesh & cable scrutiny, surface accuracy analysis, the deployment mechanism, in-orbit & ground modularized testing, analytical & simulation tools, have been developed to verify and maximize the performance of antenna. Initially, multi-beam, mechanical & measurement technologies were used to construct partially deployable hexa-link truss and tetra tri-gonal prism [8]. Later, a variety of basic modular units, such as square grid deployable [9]-[11], two types of basic modules with one being able to stow as cylinder & the other to thin plate form [12], [13], quadrilateral ribs with diagonal members for ETS VIII antenna [14]-[33], tri-foldable ribs [34], [35] and tri-foldable with bifurcated rib tips [36] for L-band application, have been proposed to integrate hexagonal reflector with aperture ranging from 1.7m to 30m in diameter. In [27], [29] six different rib schemes based on ETS VIII antenna units are discussed and fuzzy comprehensive evaluation method is applied [29] to determine the optimized configuration. Quest to increase aperture and reduce weight has led to design 13m, 14 modules ETS VIII antenna of around 340kg [15], [17] , 30m, 7 modular antenna composed of tri-foldable 4 sided ribs of 250kg [34], [35] and recently 15m aperture antenna of 58kg composed of tri-foldable bifurcated rib tips [36]. However, there is an increase in level of complexity of deployment mechanism with introduction of tri-foldable ribs in design. When modules are connected to each other displacement issue at connection points arise leading to shape and alignment errors. Harada *et. al*, [37] proposed an error approximation method to estimate variation of structural displacements at end points by deriving the covariance matrix for the displacement vector, while least square [24], correction by varying angles between ribs in the

 DOI: 10.1051/matecconf/20167701012

modules & equal envelop circle radius method [27] are employed to adjust alignment error. Meguro quantitatively assessed influence of alignment errors on surface accuracy and proposed to adjust tie cable length to improve it [10]. It is suggested [20] to use extremely stiff surface cables & relatively elastic ones as back & supporting cables to reduce deformation sensitivity & achieve high surface accuracy of antenna. Surface accuracy effected by thermal deformation in orbit is actively compensated by using shape measuring sensors [38]. Numerous numerical & analytical procedures are applied to refine parameters of modular antenna. Numerical procedures include MSC/NASTRAN [14], OOCD [12], [13], [32] & Origami/ETS [36] for design & structural analysis, CASA [9]-[11] for analysis of reaction forces & moments created by mesh & cable network , DADS [9]-[13] for deployment motion analysis, NEVADA/SINDA for thermal analysis [14], SPADE for flexibility analysis of truss [12], [13], [32], [39]-[41] & Flexible Multi-body dynamics simulation [14, 16, 42] to determine deployment driving force. Mitsugi analytically measured cable strains by integrating stiffness matrix of the support structure into nonlinear cable analysis equation to cope with elastic deformation [11]. In [43] strain induced in cable element during deployment are determined. Ando [39] used direct coordinate partitioning method to reduce DoF of joints contributing to design of deployable truss. Later, homogeneous coordinate transformation method is used for kinematics [30] & geometric modeling [24] of deployable truss with each module sharing the same geometric parameters. In [28] kinematics of rib unit is investigated by dividing it into three one DoF 4-link sub-structures. Misawa predicted [44] and verified [45] natural frequencies for stiffness design of deployed antenna. Later, in [31] multi-body dynamics theory is employed to determine rigid body dynamics of a rib. In [25], [26], [28] simulations on deployed antenna are worked in ANSYS to obtain free, harmonic & damped transient dynamic response in terms of natural frequencies & modes. Another targeted section is the influence of chord, beams & cables diameter [26, 28] and flexibility of linkages [31] on the dynamic properties. Stiffness & frequency can be enhanced by improving diameter of chord and diagonal beam. While flexibility that led to bending and fluctuation of velocity & acceleration can be controlled by using cell bodies of higher density, elastic modulus & cross-sectional area. Many ground experimental tests [9], [12], [13], [15]-[19], [21]-[23], [33], [40], [41], [43], [46], [47] are conducted to verify design criteria [9, 12], analysis method [12], [16], [19], [43], deployment reliability [13], [17], [18], [21] and in-orbit operation performance of antenna [23]. Ground deployment tests equipment mostly consisted of suspension wires, pulleys & counter weights, to overcome gravity [12], [13] however, in [46] a vertical support using a magnetically suspended slider system is proposed to create 0 G conditions. Meguro considered items like elastic deformation, contingency mode and mesh fold pattern for more realistic analysis of deployment [18]. Moreover, deployment testing on ground is more accurate if it is performed on a single module [40]. In ground deployment testing gravity interference highly affects the analysis results as in [9] unexpected loads arose in the hinge mechanisms. Doubtlessly, because of limited gravity compensation, gravity torque is induced that affect and change deployment characteristics. Despite the amount of extensive research into deployable structures, several high-profile failures have occurred. Failures in space are very expensive & extremely difficult to correct. The robustness of the concept has to be first proven by analytical models. Extraction from background study depicts that mostly simulations; ground and in-orbit experimental tests are completed to analyze the performance characteristics of modular antenna. Limited analytical procedures either on single rib of truss module or for one module is available to determine the kinematics while for dynamics and static topology optimization there is lack of any such technique. Here attempts are made to develop analytical methods that may assure stability and kinematic performance of large modular antenna prior to operation by checking the deployment trajectory. In the proceeding section, antenna design & problem description is conversed; formulation for kinematics and trajectory planning is obtained next, followed by conclusion and future recommendations.

3 Antenna details and problem description

Deployable reflector antenna under consideration consists of mesh surface, cable network and deployable truss structure. Mesh surface is made of gold-plated molybdenum or beryllium copper wire. Cable network consists of surface, tie and back cables. The mesh & cable network are attached to the rigid structure by stand-offs and tension cables. Rigid support is a deployable truss consisting of several basic modules. Each module is a hexagonal truncated pyramid composed of six basic ribs arranged in radial pattern around central beam. Basic rib is a deployable quadrilateral linkage with a diagonal member. Slider on central beam initiates and control deployment of ribs with the help of motor that releases the restrictive cables uniformly to avoid impact at the end of deployment. Deployable truss is designed to be spherical shape because all modules have almost the same design so, they are easy to handle and adjust the parabolic shape of mesh surface. In addition the number of modules can be conveniently varied to construct different antenna aperture sizes.

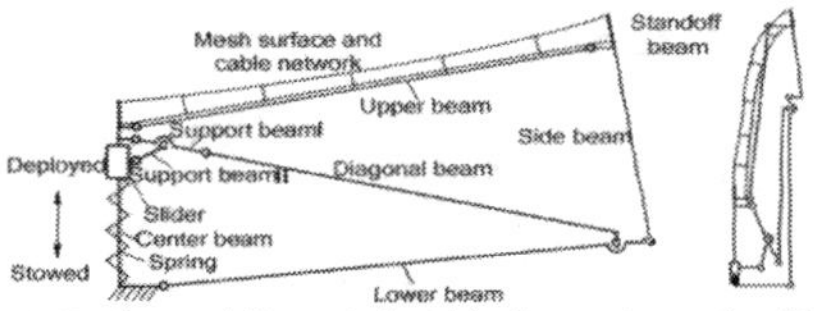

Fig. 1 Basic quadrilateral rib with diagonal member [14].

In present work, the antenna structure consists of 7 modules with each of it more than 6 m in diameter contributing to give an aperture diameter of 13 m. Basic rib shown in Fig. 1 is used as test specimen for initial development of analytical kinematics that will be

DOI: 10.1051/matecconf/20167701012

extended to one complete module then to seven modules truss. Analysis is based on following assumptions;

(1) Central beam is fixed and the reference configuration
(2) Rib is a planar & one DoF structure

3.1 Kinematics analysis

To design truss structure of 7 modular antenna and ensure that it will be deployed successfully in orbit forward kinematics analysis is carried out. Kinematics of basic rib is determined at the first step. As basic rib is a one DoF mechanism so, slider driven by spring moves up & down to complete synchronous deployment and retraction of six ribs in a module. In [28] basic rib is divided into three 1-DoF four link substructures (SS). SS-1 consists of slider, support beam (SB) I, central beam (CB) & SB-II. SS-II consists of SB-I, CB, diagonal and lower beam while third SS has upper, lower, side and CB. CB is assumed to be fixed to determine loop closure equations of all SS. Given the length of the links in the loop closure equations the corresponding angles of the revolute joints are calculated by the following equations;

From SS-I, the angular dimensions are calculate to be;

$$\theta_4 = \sin^{-1}\left[L_{MF}^2 + L_{EC}^2 - L_{DC}^2 \Big/ 2L_{MF}L_{EC}\right] \quad (1)$$

$$\theta_5 = \cos^{-1}\left[L_{FE} + L_{EC}\cos\theta_4 - L_{MD} \Big/ L_{DC}\right] \quad (2)$$

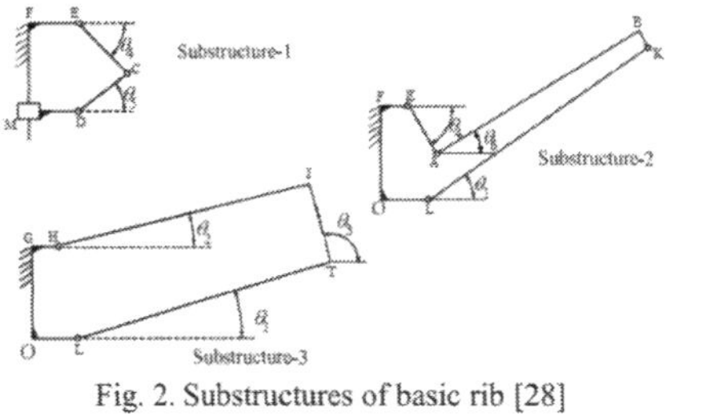

Fig. 2. Substructures of basic rib [28]

From SS-II;

$$\theta_6 = \left[\sin^{-1}\left(L_1 \Big/ \sqrt{P^2 + Q^2}\right) - Z\right] \quad (3)$$

where; L_1, P, Q & Z are given by;

$$L_1 = L_{LK}\cos\theta_1 + L_{OL} - L_{FE} - L_{EA}\cos\theta_4 \quad (3a)$$

$$P = L_{AB} \quad (3b)$$

$$Q = L_{BK} \quad (3c)$$

$$Z = \tan^{-1}\left(Q/P\right) \quad (3d)$$

Angle θ_1 is calculated to be;

$$\theta_1 = \left[\sin^{-1}\left(L_2 \Big/ \sqrt{R^2 + S^2}\right) - Y\right] \quad (4)$$

where; L_2 , R, S, Y are given by;

$$L_2 = L_{LK}^2 + L_a^2 + L_b^2 - L_{AB}^2 - L_{BK}^2 \quad (4a)$$

$$R = 2L_aL_{LK} \quad (4b)$$

$$S = 2L_bL_{LK} \quad (4c)$$

$$Y = \tan^{-1}(S/R) \quad (4d)$$

L_a & L_b will be calculated using following formula,

$$L_a = L_{OL} - L_{FE} - L_{EA}\cos\theta_4 \quad (4e)$$

$$L_b = L_{EA}\sin\theta_4 - L_{OF} \quad (4f)$$

Angles in SS-III are given by,

$$\theta_2 = T + \theta_3 \quad (5)$$

where; 'T' in the above equation is;

$$T = \cos^{-1}\left[L_{HJ}^2 + L_{JK}^2 - L_c^2 - L_d^2 \Big/ 2L_{HJ}L_{JK}\right] \quad (5a)$$

$$L_c = L_{OL} + L_{LK}\cos\theta_1 - L_{GH} \quad (5b)$$

$$L_d = L_{LK}\sin\theta_1 - L_{OG} \quad (5c)$$

For θ_3;

$$\theta_3 = \left[\sin^{-1}\left(L_3 \Big/ \sqrt{U^2 + V^2}\right) - X\right] \quad (6)$$

$$L_3 = L_{LK}\sin\theta_1 - L_{OG} \quad (6a)$$

$$U = L_{HJ} \times \sin T \quad (6b)$$

$$V = \left(L_{HJ} \times \cos T\right) - L_{JK} \quad (6c)$$

$$X = \tan^{-1} V/U \quad (6d)$$

The displacement of slider in SS-I is an input variable handling deployment and stowage by varying the six angular dimensions i.e. Θ_1, Θ_2, Θ_3, Θ_4, Θ_5 & Θ_6.

In order to determine the positions and hence the trajectory of connection point 'K' with motion of slider from top to bottom; virtual link between point 'M' and 'L' is considered. CB and link 'OL' will always maintain an angle of 90°. From geometry;

$$L_{ML} = \sqrt{L_{OM}^2 + L_{OL}^2} \quad (7)$$

where;

$$\alpha = \cos^{-1} L_{OL} / L_{ML} \quad (8)$$

So, the position of revolute at point 'K' with every increment in distance between point 'M' & 'F' is;

$$x_K = L_{OL} + L_{LK}\cos\theta_1 \quad (9)$$

$$x_K = L_{ML}\cos\alpha + L_{LK}\cos\theta_1 \quad (9a)$$

Similarly,

$$y_K = L_{ML}\sin\alpha + L_{LK}\sin\theta_1 \quad (10)$$

So, any change in length between points 'M' and 'F' will cause variation in distance between origin and point M. This will lead to change of length of link 'ML' and the angle it forms with the x-axis thus, changing the position 'K' of the revolute on lower beam with each millimeter variation in position of slider 'M'. As rib is one DoF mechanism so, position of any revolute on the rib with respect to translational motion of slider can be calculated in the same manner by introducing virtual link.

Now in order to extend the kinematics from basic rib to complete module the method using combination of homogeneous coordinates & transformation matrix is applied. A coordinate system is established at the bottom of central beam. X-axis is from origin to positive x-direction, y-axis is from bottom of center beam towards the top i.e. positive y-direction and z-axis points towards the reader. There are six ribs in each module. Consider the one under consideration as rib 1. Homogeneous coordinates of revolute at point 'K' on the bottom link of rib 1 are given by;

$$K_1 = [x_{k_1} \quad y_{k_1} \quad 0 \quad 1]^T \quad (11)$$

Six ribs are spaced at an equal angle of '$\pi/_3$' around the central beam. Transformation matrix to determine location of each rib in the module is actually the rotation by an angle 'β_n' around the y-axis and is given by;

$$Rot(y, \beta_n) = \begin{bmatrix} \cos\beta_n & 0 & \sin\beta_n & 0 \\ 0 & 1 & 0 & 0 \\ -\sin\beta_n & 0 & \cos\beta_n & 0 \\ 0 & 0 & 0 & 1 \end{bmatrix} \quad (12)$$

where, 'β_n' is given by '$^{n\pi}/_3$' ; & n = 0, 1, 2, 3, 4, 5 for one complete module. So, the homogeneous coordinates of any connection point in rib under consideration can be given by;

$$\begin{aligned} \mathrm{H}_i &= Rot(y, \beta_n) \times \mathrm{H}_1 \\ &= \begin{bmatrix} \cos\beta_n & 0 & \sin\beta_n & 0 \\ 0 & 1 & 0 & 0 \\ -\sin\beta_n & 0 & \cos\beta_n & 0 \\ 0 & 0 & 0 & 1 \end{bmatrix} \times \begin{bmatrix} x_{\mathrm{H}_1} \\ y_{\mathrm{H}_1} \\ 0 \\ 1 \end{bmatrix} \\ &= \begin{bmatrix} x_{\mathrm{H}_1} \times \cos\beta_n \\ y_{\mathrm{H}_1} \\ -x_{\mathrm{H}_1} \times \sin\beta_n \\ 1 \end{bmatrix} \end{aligned} \quad (13)$$

where 'H' will change from 'A to M' & 'i' will modify its value from '1 to 6', in fact the general formulation for 'i'is $i = n + 1$. So, the general model of deployment kinematic analysis giving the formulation to determine the position of any point on the structure has been developed. Provided the length of links this will help to analyze deployment process and trajectory of revolutes on the structure with change of position of slider. The nodal velocities and accelerations can be determined by taking the first and second derivative of nodal displacement equation with respect to time. The same procedure of transformation matrix will be extended to 7 modular antenna to determine kinematics and trajectory of connection points in rest of modules with respect to basic coordinate system.

4 Conclusion and future recommendations

Large deployable hexagonal modular mesh antenna is research object in this paper. A comprehensive review of technical contributions from 1987 to date in this class of antenna is completed and least targeted areas are extracted. Based on shortcomings a new design of a 7 modular deployable antenna of 13m aperture is proposed. Each module in design is more than 6m in diameter. In order to ensure that the antenna will deploy successfully in space conditions, analytical investigation into kinematics is carried out using loop closure equations and homogeneous coordinate transformation method. In order to ensure feasibility of design & its stability during deployment process formulations are developed to determine antenna trajectory.

In order to validate analytical kinematics, design of truss structure composed of 7 modules in Solid works is underway. The geometry will be imported to ADAMS software to work out deployment process and trajectory of overlapping points of modular structure that will help to authenticate analytical kinematics. The deployment sequence of the structure will be like the one depicted in Fig. 3.

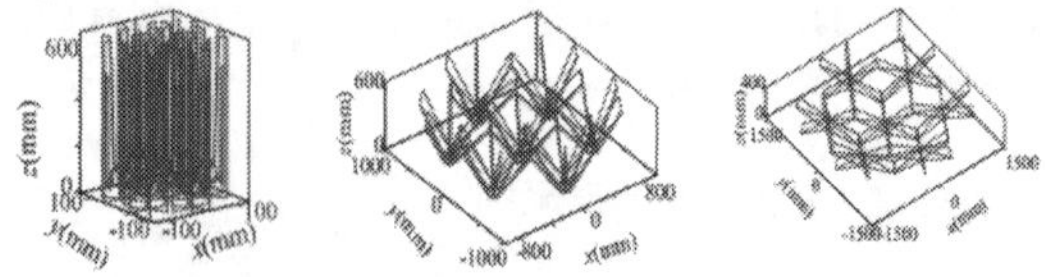

Fig. 3. Deployment sequence of deployable truss [28]

Previous researchers have used a module of 4.8m aperture dia. to design 13 m & 9.56 m aperture antenna by combining 14 & 7 modules respectively. The highest value of fundamental natural frequency achieved was 0.14 Hz [17] for 14 module antenna while for a 7 modular it is 0.25 Hz [48]. It is found that using 14 modules contribute to an increase in weight at the cost of stiffness and natural frequency. So, authors are working on 7 integrated modules to design 13m aperture antenna that will improve stiffness, reduce fabrication time, cost & complexity of design and will provide long distance and large scale communication with reduced number of modules. Its expected value of fundamental frequency will be between 0.14 Hz and 0.25 Hz. Analytical & simulation procedures will be attempted to verify these results. Analytical coupled dynamic model, incorporating the influence of thermal & 0G conditions, of deployable truss will be developed using Lagrange method to obtain response in terms of natural frequency. Positions & velocities obtained from kinematic model will be used to calculate potential & kinetic energies of the system to determine the Lagrangian which will further be solved for frequency response. The results will be validated by comparing the results with FE simulations run in ANSYS to get natural frequencies and the features of vibration modes. Deployable antenna is very large and their natural frequency is usually very low. Structures with frequencies falling in low frequency regime are more likely to observe resonance. So in order to avoid resonance deployable antenna is designed after setting some requirements in which main focus is to increase stiffness and reduce mass to acquire frequency response. Based on the natural frequency response of antenna improvements can be applied to the design to get optimized performance. Moreover the study of natural frequency response is important because it gives modal analysis to study the dynamics of vibrations, accuracy of design and stability of antenna.

Optimization can be done to replace time consuming iterations for design, build more economical structure, reduce antenna manufacture time and reduce weight leading to reduced launch & experimentation costs before launch. Authors intend to use Solid Isotropic Material with Penalization (SIMP) method to investigate the topology optimization of the structure. The objective function will be to maximize the value of stiffness and frequency response with least number of modules for desired mission application.

References

1. Angeletti P., M. Lisi, and G. Lucchi, *2[1st] International Communications Satellite Systems Conference and Exhibit, Yokohama, Japan*, (2003).
2. Mini F., et al., *The Second European Conference on Antennas and Propagation, EuCAP,* IET, (2007) .
3. Medzmariashvili E., et al., *Proceedings of Earth & Space*, (2006).
4. Gunnar T., *Deployable tensegrity structures for space applications.* Ph. D. Dissertation, (2002): p. 13-31.
5. Mitsugi J. and T. Yasaka, *Space Technology and* Science, (1990).
6. Mitsugi J. and T. Yasaka, *IAF, 42nd International* Astronautical *Congress, Montreal, Canada*, (1991).
7. Meguro A., et al., *20th International Symposium on* Space *Technology and Science, Gifu, Japan*, (1996).
8. Ebisui T., et al., *14th AIAA International* Communication *Satellite Systems Conference and Exhibit, Washington DC*, (1992).
9. Meguro A., *NASA Conference Publication*, (1993).
10. Meguro A., J. Mitsugi, and K. Ando, Electronics and Communications in Japan (Part I: Communications), (1994). **77**(8): p. 90-100.
11. Mitsugi J., Computers & structures, (1994). **51**(1): p. 47-56.
12. Mitsugi J., *Proc. Sixth European Space Mechanisms and* Tribology *Symposium* (1995), Technopark: Zurich, Switzerland.
13. Watanabe M., et al., Acta Astronautica, (1996). **39**(7): p. 497-505.
14. Nakamura K., et al., 18th ICSSC, AIAA-98-1229, (1998).
15. Meguro A., A. Tsujihata and N. Hamamoto, *Antennas and Propagation Society International* Symposium, (1999), *IEEE*.
16. Mitsugi J., et al., Acta Astronautica, (2000). **47**(1): p. 19-26.
17. Meguro A., et al., Acta Astronautica, (2000). **47**(2): p. 147-152.
18. Meguro A., et al., *42nd AIAA/ASME/ASCE/AHS/ASC* Structures, *Structural Dynamics and Material Conference and Exhibit*, (2000).
19. Miyasaka A., M. Hommat, and A. Tsujigata, *42th MAA/*ASME/*ASCE/AHS/ASC Structures, Structural Dynamics, and Materials Conference and Exhibit. Seattle, WA, United states: MAA*, (2001).
20. Harada S., A. Meguro, and M. Watanabe, *44th AIAA/ASME/ASCE/AHS/ASC Structures, Structural Dynamics, and Materials Conference*, (2003).
21. Yamada K., et al., *International Communications* Satellite *Systems Conference and Exhibit, Yokohama, Japan*, (2003).
22. Meguro A., H. Ishikawa and A. Tsujihata, J Spacecr Rockets,(2006). **43**(4): p. 780-787.
23. Meguro A., et al., Acta Astronautica, (2009). **65**(9): p. 1306-1316.
24. Tian D., R. Liu, and Z. Deng, *International Conference on Measuring Technology and* Mechatronics *Automation (ICMTMA)*, IEEE, (2010).
25. Tian D.K., et al., *Applied Mechanics and Materials*, Trans Tech Publ, (2010).
26. Tian D., et al. *9th World Congress on Intelligent* Control *and Automation (WCICA)*, IEEE, *(2011)*.
27. Fei C., Chinese Journal of Aeronautics, (2009).
28. Guo H., et al. *6th European Conference on Antennas and Propagation (EUCAP)*, IEEE, (2012).
29. Tian D., et al., *Applied Informatics and* Communication *(*2011), Springer. p. 178-185.
30. Liu R.Q., et al., *Advanced Materials Research*, Trans Tech Publ, (2012).
31. Wang S.M., S. Chen, and R. Yuan, *Applied Mechanics and Materials*, Trans Tech Publ, (2013).
32. Meguro A., S. Harada, and M. Watanabe, Acta Astronautica, (2003). **53**(11): p. 899-908.
33. Ishikawa H., A. Meguro, and M. Watanabe, Transactions of the Japan Society for Aeronautical and Space Sciences, (2004). **47**(157): p. 175-180.
34. Ozawa S., K. Shintate, and A. Tsujihata. *Proceedings of the 5th European Conference on Antennas and Propagation (EUCAP)*, IEEE, (2011).
35. Ozawa S., K. Shintate and A. Tsujihata. *Proceedings of the 5th European Conference on Antennas and* Propagation *(EUCAP)*, (2011).
36. Nakamura K., et al., *3rd AIAA Spacecraft Structures* Conference, (2016).
37. Harada S., M. Shimizu, and T. Mitsugi, AIAA-98, (1998).
38. Ueba M., et al. *15th AIAA International* Communications *Satellite Systems Conference, San Diego, CA*. (1994).
39. Ando K., J. Mitsugi and Y. Senbokuya, Computers & structures, (2000). **74**(1): p. 21-39.
40. Meguro A., H. Ishikawa, and A. Tsujihata, *Proceedings of the 43rd AIAA/ASME/ASCE/AHS/ASC Structure, Structural Dynamics and Material* Conference, (2002).
41. Tsujihata A., et al., *Large Deployable Reflector (LDR)* onboard *Engineering Test Satellite (ETS-VIII), (1998)*.
42. Mitsugi J. *Proceedings of 36th Structures, Structural* Dynamics, *and Material Conference, AIAA*, (1995).
43. Mitsugi J. and Y., *Proceedings of the 37th Structural Dynamics, and Materials Conference, AIAA*, (1996).
44. Misawa M. and J. Mizuno, *A Collection of Technical Papers: AIAA/ASME/ASCE/AHS/ASC Structures, Structural Dynamics and Materials Conference and Exhibit, AIAA/ASME/AHS Adaptive Structures Forum, AIAA Non-Deterministic Approaches Forum, AIAA Gossamer Spacecraft Forum: Seattle, Washington, 16-19 April, (*2001).
45. Misawa M. and A. Ogawa, *Analytical and* experimental *frequency verification of deployed satellite antenna*, (2003). **2**: p. 3.
46. Sugimoto T., et al.,*38th Structures, Structural* Dynamics, *and Materials Conference*, (1997).
47. Nagaoka N., et al., *44th AIAA / ASME / ASCE / AHS / ASC Structures, Structural Dynamics, and Materials Conference*, (2003).
48. Higuchi K., et al., *Proc. 50th AIAA / ASME / ASCE / AHS / ASC Structures, Structural Dynamics and Materials Conference, AIAA-2009-2609*, (2009).

Prediction and analysis of 3D hydrofoil and propeller under potential flow using panel method

Chen-Wei Chen, Ning Peng

Zhejiang University, Department of Naval Architectural and Ocean Engineering, Ocean College, 316021, Zhoushan, Zhejiang, China

Abstract. Potential flow over an airfoil plays an important historical role in the theory of airfoil. The governing equation for potential flow is Laplace's equation. One of Green's identities can be used to write a solution to Laplace's equation as a boundary integral. Using distributions of singularity solutions and determining their strength via the boundary conditions is the essence of panel method. This paper introduces a quick prediction method of three-dimensional hydrofoil and propeller performance based on panel method. The surface of hydrofoil and propeller is divided into numbers of quadrilateral panels. Combined sources with doublets singularities will be distributed on the corners of panels. Calculated blade pressure distributions of hydrofoil and propeller agree well with experimental data. Several sample calculations have been included using panel method.

1 Introduction

A singularity panel method based on Green's function integral equation is used to calculate the motion characters. The motion and general behavior of a fluid is governed by the fundamental laws of classical mechanics and thermodynamics and plays an important role in many fields. An introductory text on fluid mechanics, such as the study of Sabersky [1], surveys the basic concepts of fluid dynamics and the various mathematical models used to describe fluid flow under different restrictive assumptions. After the first successful application of numerical approach to the solution of flow around an arbitrary body by Hess and Smith [2] in the 1960's the singularity panel method has developed quickly.

Panel method was mainly applied to solving lifting flows for thick airplane, but now panel code can be also used in shipbuilding field [3] and even in submarine [4]. Now panel method code is suitable for marine propellers [5] and numerical resistance calculation of hulls [6]. The recent trend is the application of low-order methods [7], and many different panel methods are used now. This led to some comparisons [8] about different panel methods and other methods. Apparently, low-order panel methods are cheaper and faster to use. The panel method which is proposed by this paper is able to predict the three dimensional hydrofoil and propeller performance without performing a computational simulation. Generally, for performing the computational simulations, a high performance computer and a lot of computational time would be needed [9], [10]. However, using this method can reduce the computational cost.

In this paper, panel method is applied to the prediction of three-dimensional hydrofoil and propeller performance. Surface pressure distributions of hydrofoil and propeller will be calculated to compare with experimental data.

2 Solution of potential flow using panel method

Panel methods are numerical models based on simplified assumptions about inviscid and incompressible flow problem. In principle, if a problem can be solved by distributing unknown quantities on the boundary surface surrounding the body under simplified potential flow, characteristic coefficients of foil will be obtained. Under these assumptions, the vector velocity describing the flow field can be represented as the gradient of a scalar velocity potential. Therefore, the equation is as following:

$$\vec{Q} = \nabla\phi \quad (1)$$

A statement of conservation of mass in the flow field leads to Laplace's equation as the governing equation for the velocity potential. If the flow in the fluid region is considered to be incompressible and irrotational then the continuity equation reduces to Eq. (2).

$$\nabla^2\phi = 0 \quad (2)$$

$\nabla\Phi$ is indicated in the reference that is attached to the body. Apparently, the disturbance created by the motion must disappear far from the body as Eq. (3).

$$\lim_{r\to\infty}(\nabla\phi - v) = 0 \quad (3)$$

r is the distance from the body. We need to establish a general approach to the solution of incompressible potential flow problems. Eq. (2) is the governing equation and it can be solved by one of Green's identities like the following Eq. (4).

$$\int_S(\phi_1\nabla\phi_2 - \phi_2\nabla\phi_1)\cdot n dS = \int_V(\phi_1\nabla^2\phi_2 - \phi_2\nabla^2\phi_1)\cdot dV \quad (4)$$

This equation is one of Green's identities. The integral of surface is all the boundaries S. The body is enclosed in a volume V, and Φ_1, Φ_2 are two scalar functions of position. It is well known that any element which satisfies Laplace's equation can be written as an integral over the boundary surface S. In Eq. (4) S is the surface integral, and $S=S_B+S_W+S_\infty$ where S_B means body model boundaries, S_W means wake model boundaries and S_∞ means outer boundary.

One important observation is that Eq. (2) can be solved by distributing elementary solutions on the boundaries. Commonly singular solutions for panel methods includes vortex, source and doublet distributions. Now the solution is reduced to find the appropriate element strength over the known boundaries. When the potential is specified on the boundaries this mathematical problem is called the Dirichlet problem [11]. Another approach to the solution is to describe that the normal flow boundary condition on the solid boundaries is zero. This problem is named as the Neumann problem [11]. For a given boundary condition, the above solution can be used in specific condition respectively.

3 Two-dimensional foil performance in panel method

Based on the distribution of singularity elements, surface geometry, and the exact type of boundary conditions, computational methods can be constructed. What should be notified is by the combination of some types of singularities the panel method can be more accurate. Further improvement is obtained by changing spacing and density of grid, wake model, location of collocation points and singularity elements, method of enforcing the boundary conditions, and Kutta condition [12].

A step sequence (Fig. 1) is specified when a numerical solution of panel method is established. The panel code is developed following these steps.

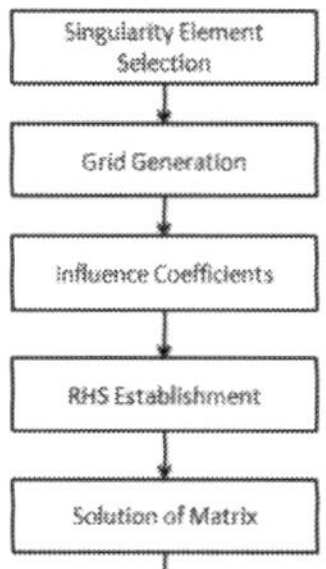

Figure 1. Steps toward constructing a numerical solution

In this paper we use the combination of source and doublet singularity elements on the surface of the foil. It means that every panel has a local source and doublet strength and either the source or the doublet strength values should be specified. Also, any doublet distribution can be replaced by a vortex distribution, and solutions based on the vortex can be used in this case too [13].

In the process of grid generation, the singularity elements' corner points and collocation points should be defined. Singularity elements are placed at corner points and the boundary conditions, such as zero normal flow on the solid surface, will be enforced at collocation points. Fig. 2 shows the distribution of collocation points and singularity elements on panels.

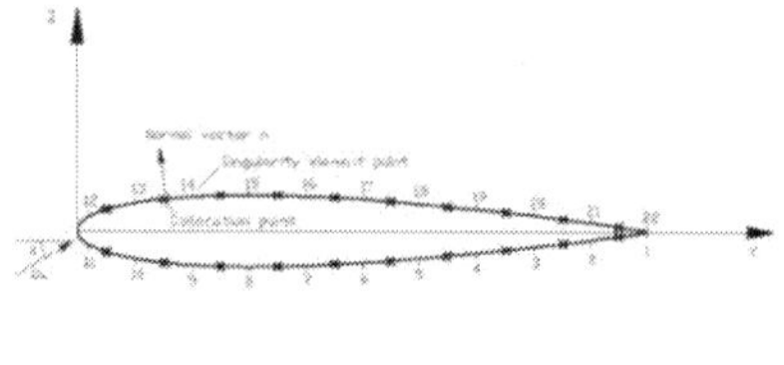

Figure 2. Distribution of singularity points and collocation points on foil

To represent a numerical solution the surface S is divided into N panels and the integration is applied to every panel, and the general solution to Eq. (2) can be obtained by several sources σ and doublets μ. Here we apply Dirichlet boundary condition to this case. Then we can obtain the integral equation :

$$\phi_{source}-\phi_{doublet}+\phi_\infty=constt \tag{5}$$

Here the equation is specified at each collocation point, so we obtain a linear algebraic equation for each point. As mentioned above, the steps of establishing such numerical solution are as follows.

The potential at point P can be included into two subroutines in MATLAB that calculate the potential at point (x, z) due to the doublet and source element j:

$$\Delta\phi_s=\phi_{source}(\sigma_j,x,z,x_j,z_j,x_{j+1},z_{j+1}) \tag{6}$$

$$\Delta\phi_d=\phi_{doublet}(\mu_j,x,z,x_j,z_j,x_{j+1},z_{j+1}) \tag{7}$$

As the $N+1$ panel corner points and N collocation points are generated, Dirichlet boundary condition will be used and thus the collocation points should be placed inside the body. Then the influence of the potential at collocation point i due to element j is obtainable.

$$b_{ij}=\phi_{source}(\sigma_j=1,x,z,x_j,z_j,x_{j+1},z_{j+1}) \tag{8}$$

$$c_{ij}=\phi_{doublet}(\mu_j=1,x,z,x_j,z_j,x_{j+1},z_{j+1}) \tag{9}$$

The influence coefficients will result in a $N\times N$ matrix, but the number of unknown elements are $N+1$ because of the wake doublet μ_w. Therefore additional equation is necessary by applying Kutta condition:

$$(\mu_1-\mu_N)+\mu_w=0 \tag{10}$$

In this case we replace μ_w with μ_N-μ_1. Therefore the order will be reduced to N. Only the first and the Nth columns should change because of the term $\pm c_{iw}$. So we can replace the doublet influence like this:

$$\begin{aligned} a_{ij}&=c_{ij} && j\neq 1,N \\ a_{i1}&=c_{i1}-c_{iW} && j=1 \\ a_{iN}&=c_{iN}+c_{iW} && j=N \end{aligned} \tag{11}$$

After the known matrix multiplication is moved to the right-hand side of this equation, we can obtain RHS vector.

Based on the theory of panel method above a MATLAB program is developed to calculate the strength of singularities, and then we can obtain the coefficient of pressure on the surface of the foil. The external potential Φ_u can be represented as the internal potential Φ_i plus the doublet strength μ. Then the local external tangential velocity component of each collocation point can be represented as Eq. (13) where l is the distance between the two adjacent collocation points. After that the pressure coefficient of jth panel can be calculated in Eq. (14).

$$\phi_u = \phi_i + \mu \tag{12}$$

$$Q_t = \frac{\partial \phi_u}{\partial l} \tag{13}$$

$$C_{pj} = 1 - \frac{Q_{tj}^2}{Q_\infty} \tag{14}$$

For the lack of wind-tunnel data, the calculated result is compared with the pressure distribution derived from Profili. Profili is one of the most famous tools for aeromodel designers and it offers a large amount of airfoils database and explicit coefficients of pressure.

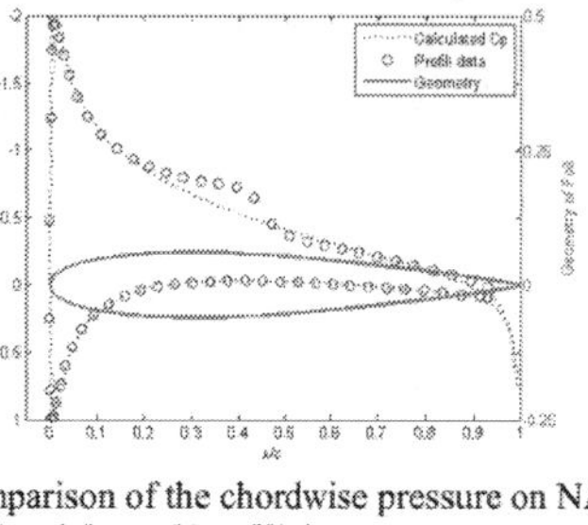

Figure 3. Comparison of the chordwise pressure on NACA0012 between calculated Cp and Profili data

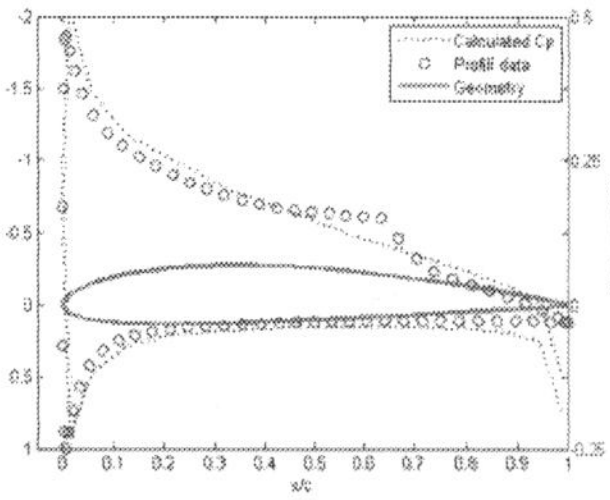

Figure 4. Comparison of the chordwise pressure on NACA2410 between calculated Cp and Profili data

Fig. 3 and Fig. 4 represent the result derived from the comparison between calculated coefficient of pressure and Profili data. The calculated result shows the coefficients of pressure of NACA0012 and NACA2410 at attack angle of 5 degrees. From Figure 4 we can see the coefficients of pressure on upper surface and lower surface specifically. Although it shows that the present panel method is suitable for calculating the coefficient of pressure on the surface of foil, still unconformity exsits when x/c is at 0.4 and 0.6 in Fig. 3 and Fig. 4 because of divided panel size and viscosity condition. As a whole this method is suitable for calculating the coefficient of pressure on the surface of foil.

A further step can be taken to do some secondary computations. After the C_p is obtained, coefficients of lift and drag are obtainable. The secondary computation can be illustrated by the case of NACA0012.

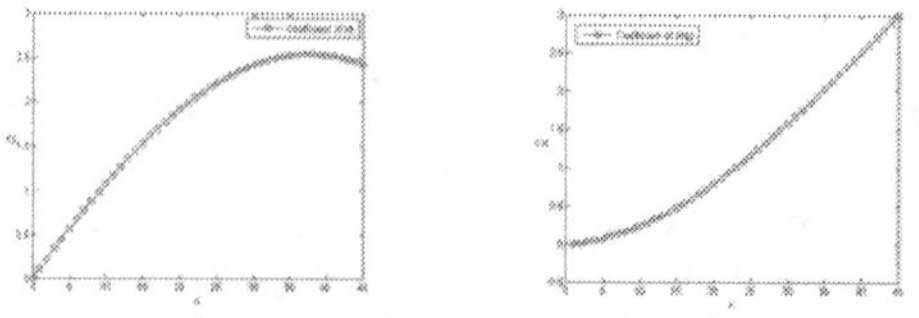

(a) (b)

Figure 5. Coefficients of lift and drag on NACA0012 at different attack angles

Fig. 5(a) represents the coefficients of lift on NACA0012 with the increase of attack angle, and it's obvious that the coefficients of lift on NACA0012 increases with the increase of attack angle. From Fig.8 we can see that the C_l reaches peak at 38 degrees, and then it decreases. Figure 5(b) shows the coefficients of drag on NACA0012 with increase of attack angle. Therefore, this program based on panel method provides a solution to potential flow around the surface of foil, and it needs less computational time and provide a quick estimation of performance of foil. But it is a sensitive program. The mismatching between calculated data and Profili data depends on the grid, the number of panels and other various parameters. These deficiencies can be improved in the future.

4 Prediction of hydrofoil and propeller performance

4.1 Geometry and wake models

Three-dimensional numerical solution based on panel method is similar to two-dimensional solution. In attempting to construct a potential-flow problem of three-dimensional case, only the wake and trailing conditions (Kutta condition) will need more additional attention [14]. In this case, the most difficult part is the modeling of geometry, and some solutions to arbitrary geometry need to be sought [15]. The geometries of hydrofoil and propeller are shown in Fig. 6.

The key point of this step is the definition of the propeller coordinate. The cartesian and cylindrical coordinates of a point on the camber surface with radial coordinate r and chordwise coordinate s can be expressed in terms of skew(θ), rake(x_m), chord (c), and camber(f) [16]. With this definition, the coordinate of a point on the camber surface of the key blade may be written as

$$\begin{cases} x_c = x_m + c(s-1/2)\sin\phi - f\cos\phi \\ \theta_c = \theta_m + c(s-1/2)\dfrac{\cos\phi}{r} + f\dfrac{\sin\phi}{r} \\ y_c = r\cos\theta_c \\ x_c = r\sin\theta_c \end{cases} \tag{15}$$

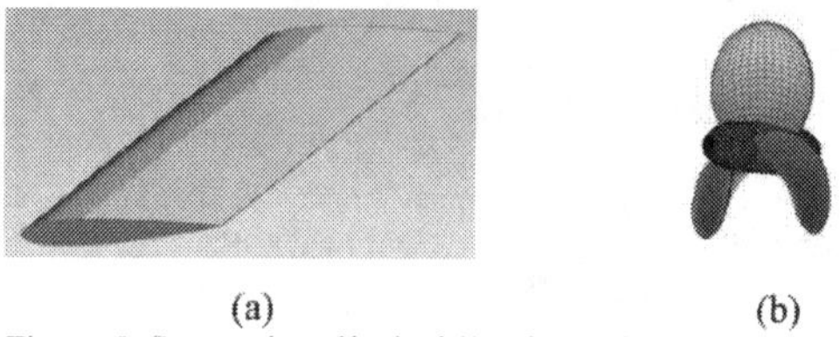

(a) (b)

Figure 6. Geometries of hydrofoil and propeller

Trailing vortex leaves the trailing edge of the blade and flows into the slipstream with the local veocity at that position. However, the wake surface is usually approximated by prescribed helical surface in order to avoid time consuming calculation [17]. In this paper, we will use the wake model based on QCM [18]. Namely, the wake surface leaves the trailing edge in the direction tangent to the mean camber surface in order to satisfy the Kutta condition. Then the pitch of wake changes linearly. In the down faraway stream, the wake becomes helical surface with the ultimate constant pitch.

4.2 Results and analysis

In this case, the basic panel singularity element has a constant strength source or doublet which is similar to two-dimensional case, and surface of panel is also planar. Following the previous equation, the Dirichlet boundary condition on a thick body can state that the perturbation potential inside is zero. Therefore this equation should be evaluated for every collocation point inside. l and k are the counters of panels. The influence coefficients C_k, C_l of the body and wake doublets and B_k of the sources can be calculated. Here the zero normal velocity boundary condition should be applied, and the equations on the collocation points of panel i:

$$\sum_{k=1}^{N} C_k^* \mu_k + \sum_{l=1}^{N_W} C_l^* \mu_l + \sum_{k=1}^{N} B_k^* \sigma_k = -\mathbf{Q}_\infty \cdot \mathbf{n}_i \tag{16}$$

In this three-dimensional panel method the shape of body is divided into panel elements. The grid is usually constructed of rectangular patches, and in this case the collocation points may be slightly under the surface. The counter k for every panel is assigned and the outward normal vector $\boldsymbol{n}_k$ is identified in Figure 7.

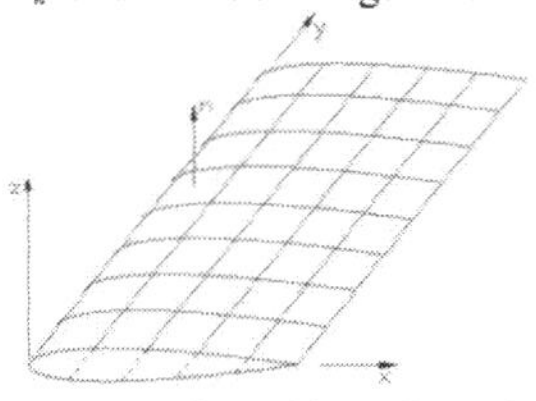

Figure 7. Panels on the surface of three-dimensional foil

Dirichlet boundary condition is used in this case. Therefore the doublet strengths are set as the unknowns. At this point, the potential at collocation point (under surface) of every panel is influenced by all the N other elements and then the first collocation point can be:

$$C_{11}\mu_1 + \cdots + C_{1l}\mu_l + \cdots + C_{1m}\mu_m + \cdots + C_{1N}\mu_N + \sum_{p=1}^{N_W} C_{1p}\mu_p + \sum_{k=1}^{N} B_{1k}\sigma_k = 0 \tag{17}$$

As two-dimensional panel method represented, the influence of wake panel at point 1, when single out from the term is

$$C_{1p}(\mu_l - \mu_m) \tag{18}$$

Here the p is the counter of wake panels. Therefore this second summation (Eq. (18)) can be substituted into the equation. In this case, the equation of the first point becomes:

$$C_{11}\mu_1 + \cdots + (C_{1l} + C_{1p})\mu_l + \cdots + (C_{1m} - C_{1p})\mu_m + \cdots + C_{1N}\mu_N + \sum_{k=1}^{N} B_{1k}\sigma_k = 0 \tag{19}$$

Therefore, this equation can be specified to:

$$\sum_{k=1}^{N} A_{1k}\mu_k = -\sum_{k=1}^{N} B_{1k}\sigma_k \tag{20}$$

Here $A_{lk}=C_{lk}$ if no wake flows from the panel and $A_{lk}=C_{lk}\pm C_{lp}$ if it is shedding a wake panel. Where the N linear equations for the N unknown μ_k and σ_k is already known. In this case, note that $a_{kk}=1/2$, except when the panel is at the trailing edge. Therefore, the right-hand-side matrix can be specified for the strengths of the sources are already known.

In this paper NACA0012 is selected as the root and tip geometry of the three-dimensional hydrofoil and the surface of the hydrofoil is divided into 72 panels. The influence of the wing's image can be added if needed. Here we set that the length of chord is 1 m, the length of semispan is 10 m, and the angle of attack is 5 degrees, and then the pressure coefficient distribution of the hydrofoil is obtainable.

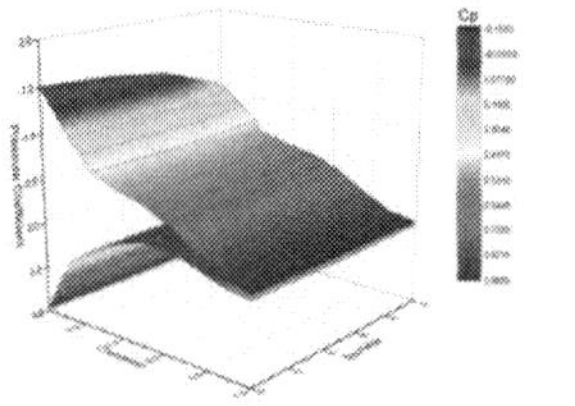

Figure 8. The pressure coefficient distribution of NACA0012 three-dimensional foil at the attack angle of 5 degrees

Fig. 8 illustrates the pressure coefficient distribution of upper surface and lower surface of NACA0012 three-dimensional foil. X-axis is chordwise axis, Y-axis is spanwise axis and Z-axis is pressure coefficient axis. Though the influence of every panel has been considerate, each section of the figure is similar to the two-dimensional case. With the result of pressure coefficient distribution, other performance indexes can be obtained by secondary computation: the coefficient of lift is equal to 0.4420, and coefficient of drag is 0.0244.

Table 1. Principal characteristics of DTRC Propeller 4119

r/R	P/D	Rake/D	Skew	Chord	f/C	t/C
0.20	1.105	0	0	0.32	0.014	0.20
0.25	1.103	0	0	0.34	0.019	0.18
0.30	1.102	0	0	0.36	0.023	0.16
0.40	1.098	0	0	0.40	0.023	0.11
0.50	1.093	0	0	0.44	0.021	0.09
0.60	1.087	0	0	0.46	0.020	0.07
0.70	1.083	0	0	0.46	0.020	0.05
0.80	1.081	0	0	0.43	0.019	0.04
0.90	1.078	0	0	0.36	0.018	0.03
0.95	1.077	0	0	0.28	0.016	0.03
1.00	1.075	0	0	0.03	0.000	0.00

In this paper we will calculate DTRC4119 propeller as example. Principal characteristics of propeller 4119 are shown in Table 1. The propeller has 3 blades with no skew and rake. The blade sections are designed with NACA 66 modified profiles and a camber line a=0.8. The advanced ratio is j=0.833. The comparison of chordwise pressure distributions at r/R=0.3 and r/R=0.7 between calculated data and experimental data is shown in Fig. 9.

Comparison of K_T, $10*K_Q$ and η between calculated data and experimental data is shown in Fig. 10.

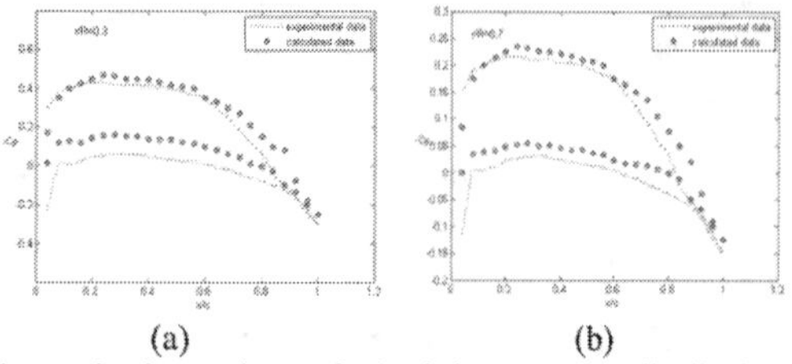

(a) (b)

Figure 9. Comparison of chordwise pressure distributions between calculated data and experimental data

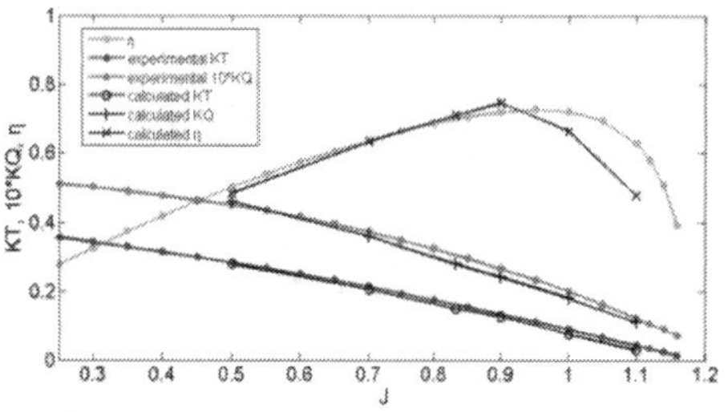

Figure 10. Comparison of K_T, $10*K_Q$ and η between calculated data and experimental data

5. Conclusion

In this paper, a potential-based panel method has been used to make prediction of hydrofoil and propeller performance. A panel method program was developed in MATLAB to calculate the pressure coefficients on the surface. This panel method program is certainly less time-consuming than CFD programs and even cheaper and easier to operate with more controlled variables. Fairly good agreements are obtained, which shows the present method is effective and accurate. However, some mismatching can be still found in some cases especially when the number of panels changes. Although many inputs can be set in panel code, the geometry of other creative shapes are still difficult to be modeled in panel code. Future work includes refinement of the wake model, generalization to the unsteady case, and interaction of propeller and rudder.

References

1. R. H. Sabersky, A. J. Acosta, E. G. Hauptmann, and E. M. Gates, Fluid Flow: A First Course in Fluid Mechanics, 4th ed., Englewood Cliffs, NJ: Prentice Hall, (1998).
2. Hess, J.L. and Smith, A.M.O., Calculation of Potential Flow about Arbitrary Three-Dimensional Bodies, Progress in Aeronautical Science, Pergamon Press , (1966).
3. X Lv, X Wu. Trim Optimization of Ship by a Potential-Based Panel Method, Advances in Mechanical Engineering, vol. 2013, no. 378140, pp. 528—534, Nov, (2013).
4. T. Gourlay., Edward D., A Havelock Source Panel Method for Near-surface Submarines, Journal of Marine Science & Application, 14: 215-224. (2015)
5. G. Luca, M. Roberto, Marine propellers performance and flow-field prediction by a free-wake panel method, Journal of Hydrodynamics, 26(5): 780-795, Apr., (2014).
6. MS. Tarafder, MT. Ali, Numerical prediction of wave-making resistance of pentamaran in unbounded water using a surface panel method, Procedia Engineering, 56(56): 287-296. (2013).
7. K. Sudhakar, GR. Shevare, Low-order panel method for internal flows, Journal of Aircraft, 28(4): 286-288, (2012).
8. J. Baltazar, D. Rijpkema, A Comparison of Panel Method and RANS Calculations for a Ducted Propeller System in Open-Water, Third International Symposium on Marine Propulsors smp'13, pp. 338—346, May. (2013).
9. B. Takabi, H. Shokouhmand, Effects of Al2O3-Cu/water hybrid nanofluid on heat transfer and flow characteristics in turbulent regime, International Journal of Modern Physics C, Vol. 26, No. 4, (2015).
10. B. Takabi, S. Salehi, Augmentation of the Heat Transfer Performance of a Sinusoidal Corrugated Enclosure by Employing Hybrid Nanofluid, Advances in Mechanical Engineering, Vol. 6, 147059 (2014).
11. I. Baydoun, Localisation Inverse Problem and Dirichlet-to-Neumann Operator for Absorbing Laplacian Transport, Journal of Modern Physics, 04(06): 772-779, (2013).
12. F. Mohebbi, M. Sellier. On the Kutta Condition in Potential Flow over Airfoil. Journal of Aerodynamics, (2014).
13. S. Lawton, C. Crawford, Investigation and Optimization of Blade Tip Winglets Using an Implicit Free Wake Vortex Method, Journal of Physics: Conference Series, 524(1), (2014).
14. G. H. Cottet, PD. Koumoutsakos, Vortex Methods: Theory and Practice, Measurement Science & Technology, 12(3):354, (2001).
15. T. Gourlay, Shallow Flow: A Program to Model Ship Hydrodynamics in Shallow Water, Asme International Conference on Ocean, (2014).
16. J. E. Kerwin, C S Lee. Prediction of steady and unsteady marine propeller perfomance by numerical lifting-surface theory[J]. Sname Transactions, (1978).
17. T. Hoshino, Hydrodynamic Analysis of Propellers in Steady Flow Using a Surface Panel Method[J]. Journal of the Society of Naval Architects of Japan, 1989(166):79-92. (1989).
18. T. Hoshino, N. Nakamura. Propeller design and analysis based on numerical lifting-surface calculation[J]. Design Criteria, (1988).

 DOI: 10.1051/matecconf/20167701014

The mechanical property analysis of the large supportive stage with super smooth surface

YueHua Li , BaoShan Zhao and Yong Yang

Tianjin Institute of Aerospace Mechanical and Electrical equipment, Tianjin 300045, P R China

Abstract. With the rapid development of satellite astronomy technique, In order to ensure the effective operation of the spacecraft in orbit, the spacecraft requires full ground physical simulation tests. So the construction of the ground simulation test system is an important guarantee for the development of space technology [1]. In the system of full physical simulation test for lunar orbit rendezvous and docking, the large supportive stage with super smooth surface is one of the most important large-scale precision test equipment, this stage can provide a high precision horizontal support surface in the range of 40m*30m for the test load, Each test load is usually supported by three circular air cushion, whose diameter is 200mm~400mm. High pressure gas is stored in the load equipment, and is released into the air through the air holes on the air cushion, A thin layer of air film is formed between the air cushion and the stage, the thickness of gas film is 10μm~15μm, The load equipment can freely move on the surface of a large stage with the gas film. There are strict requirements for the deformation of the system, Since the floating height is 10 μm and the test load is 3T. So the altitude difference of neighboring platforms is in range of 10μm. In order to guarantee the experiment effects, the paper analyses mechanical properties of the stage, distribute index of 10μm, and provide theoretical support for subsystem [2], [3].

1 Introduction

In order to ensure construction requirements of the GNC laboratory, and consider great difference of the Mechanics environment between ground and orbit, the microgravity simulation system has been developed. The spacecraft performance of Mechanics, machinery, navigation and control can be tested and verified [4].

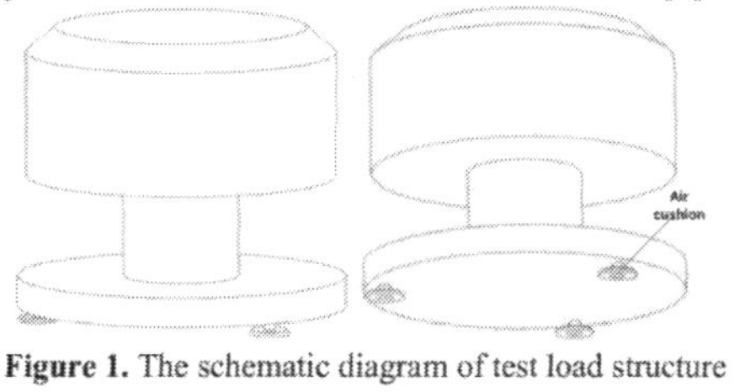

Figure 1. The schematic diagram of test load structure

A supportive surface of high-precision and high-levelness can be provided on the large supportive stage with super smooth surface [5]. Each test load is usually supported by three circular air cushion, whose diameter is 200mm ~ 400mm, The air cushions are distributed on the circumference of Diameter is 1~3m,As shown in Fig. 1. High pressure gas is stored in the load equipment , and is released into the air through the air holes on the air cushion, A thin layer of air film is formed between the air cushion and the stage, the thickness of gas film is 10μm ~15μm, The load equipment can freely move on the surface of a large stage with the gas film. As shown in Fig. 2. The weight of each test load is 3t, multiple loads can move freely on surface of the stage with the gas film Under the condition of no friction almost [6]. As shown in Fig. 3.

Figure 2. The side view of the large stage

Figure 3. The global graph of the large stage

The large stage is stitched together by 200 block "jinan green" Marble, whose size is 3m*2m*0.4m. Because a plane can be determined by three points, the Supporting method of single block platform is Three main supports and three auxiliary supports [7].as shown in Fig. 4. The main supports are 1,3,5, the auxiliary supports are 2,4,6.

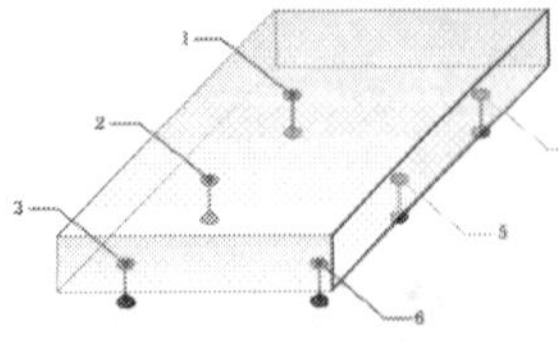

Figure 4. The diagram of supporting method

2 The subsystems of the large stage

The large is constituted by six subsystems. Respectively as (1)platform subsystem.(2)foundation subsystem. (3)assemblage subsystem. (4) testing subsystem (5)support and adjustment subsystem (6) integrated control subsystem.as shown in Fig. 5.

The functions of subsystems are as follows:

(1) The material of platform subsystem is "jinan green" marble. A high-levelness and High flatness surface can be provided by the platform, which is the controlled object by integrated control subsystem;
(2) The high grade cement is used to pour foundation subsystem, which is important to satisfy requirement of specifications.
(3) The assemblage subsystem includes method of assembling 200 platforms marble by using specialized car.
(4) The testing subsystem includes method and instrument of testing levelness and altitude difference.
(5) The support and adjustment subsystem includes some test devices, which is used to detect levelness and altitude difference.
(6) The integrated control subsystem includes driving equipment, control algorithm and software.

3 The index decomposition of altitude difference

The altitude difference of neighboring platforms is in range of 10μm when the weight of load is 3t.the altitude difference comes from two points as follows:

(1) The adjustment error of mechanical drive is defined as H_c.
(2) the elastic deformation of overall system is defined as H_k.

In the following, these two parts above would be analysed, as shown in Fig. 5, and the equation 1 is true.

$$H = H_C + H_K \tag{1}$$

3.1 The analysis of adjustment error

The adjustment error of mechanical drive caused by the two parts: one is the accuracy of integrated control subsystem, denoted by H_{cp}, the other is the Straightness of platform, denoted by H_{cl}, as shown in Fig. 6, and The equations 2 is true.

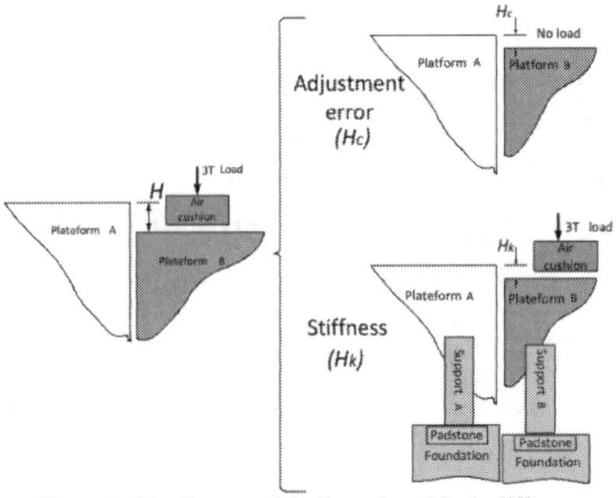

Figure 5. The diagram of decomposing altitude difference

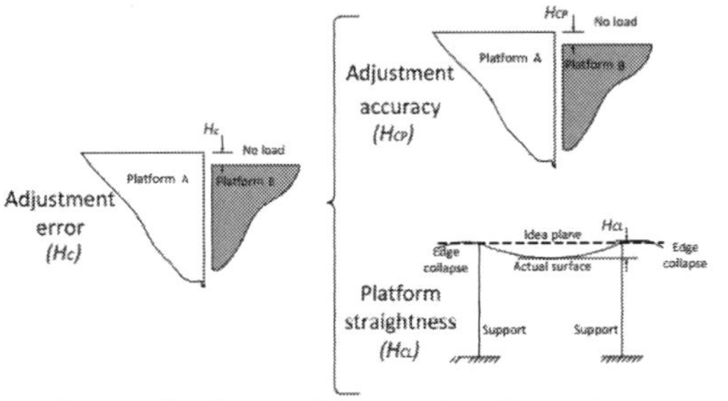

Figure 6. the diagram of decomposing adjustment error

$$H_C = H_{CP} + H_{CL} \tag{2}$$

3.1.1 The analysis of adjustment accuracy

The adjusting methods is that a platform is adjusted without load. so this adjustment accuracy is the accuracy without load.

Based on 3 points leveling method, the distance between support and the edge of platform is small. The speed of adjusting platform is slow, the Height adjustment of support is approximately equal to the lifting height of neighboring platforms. Basis of analysis and research, the adjustment accuracy of servo motor can reach 0.03 ° . With pienty of experiments and data analysis, the adjustment accuracy of altitude difference can reach 0.2μm. The following equation 3 is established.

$$H_{cp} = 0.2\mu m \tag{3}$$

3.1.2 The analysis of platform straightness

Considering the impact of machining、grinding、deformation and temperature. Because the impact of straightness, the altitude difference of neighbouring platforms is random. the straightness is got by statistical method usually.

In general, the overall profile is concave when the grinding of platform is finished. as shown in Fig. 7.

Because of the higher technical targets of altitude difference, the straightness of platform must be no greater than 4μm, and satisfy consistence of neighboring platforms.

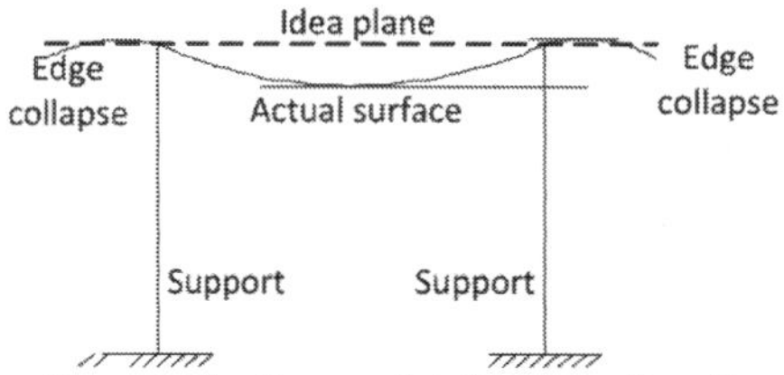

Figure 7. The diagram of platform overall profile

Considering the shape of the overall profile and the effects of temperature, the straightness of platform can reach 3.3μm by test in constant temperature room. The following equation 4 is established.

$$H_{CL} = 3.3\mu m \tag{4}$$

Through above analysis, the altitude difference of neighbouring platforms, importing data into equation 2, and the equation 5 is true.

$$H_C = H_{CP} + H_{CL} = 0.2 + 3.3 = 3.5\mu m \tag{5}$$

3.2 The analysis of stiffness

the stiffness of system has effects on altitude difference, include platform stiffness, support stiffness and foundation stiffness.as shown in Fig. 8, and the equation 6 is established.

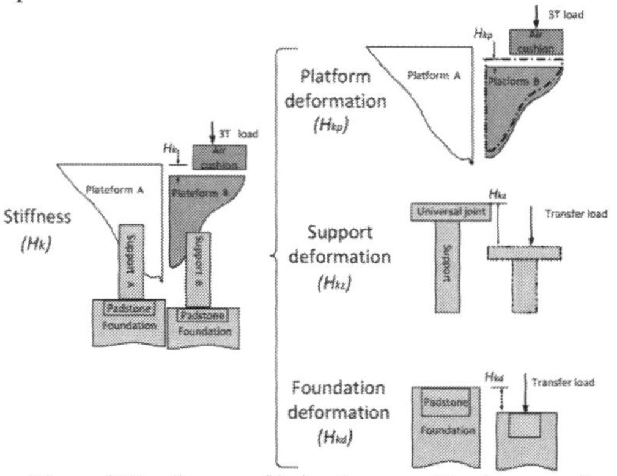

Figure 8. The diagram of index decomposition because of stiffness

$$H_K = H_{KP} + H_{KZ} + H_{KD} \tag{6}$$

3.2.1 The analysis of platform mechanical characteristics

The elasticity deformation of platform is different when the position of air bearing pad is different. As shown in figure 9, the worst condition can be got by calculation and analysis. In the worst condition, the 3t load can result in maximum deformation of platform. The deformation could be calculated by using finite element method [8], as shown in Fig. 10.

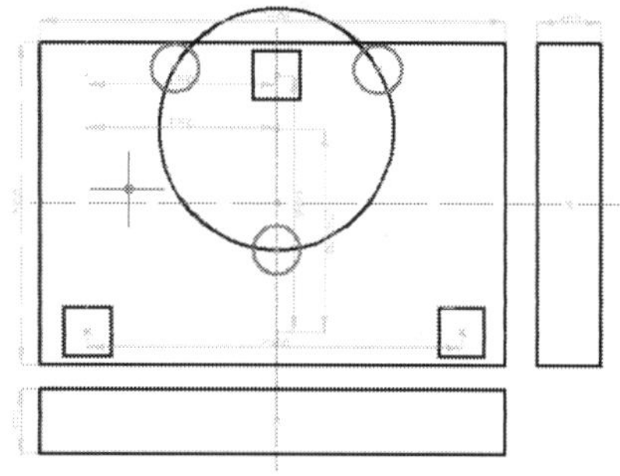

Figure 9. The diagram of the worst condition

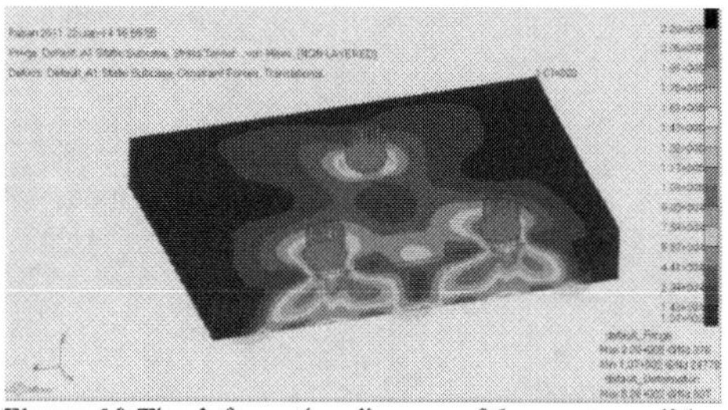

Figure 10. The deformation diagram of the worst condition

Through above analysis, the maximum deformation is 2.2μm when the load is 3t. The following equation 7 is established.

$$H_{KP} = 2.2\mu m \tag{7}$$

3.2.2 The mechanical characteristics analysis of foundation subsystem

In general, the weight of whole system and the 3t load would result in the deformation of foundation subsystem. The altitude difference is the most important, the paper would analyze foundation deformation, which is caused by 3t load. In the process of moving load, the maximum load variation of supporting mechanism can be got by a lot of calculation, this maximum value is the input condition of analyzing foundation deformation. The variation of support 5 is the maximum value,as shown in Fig. 9.

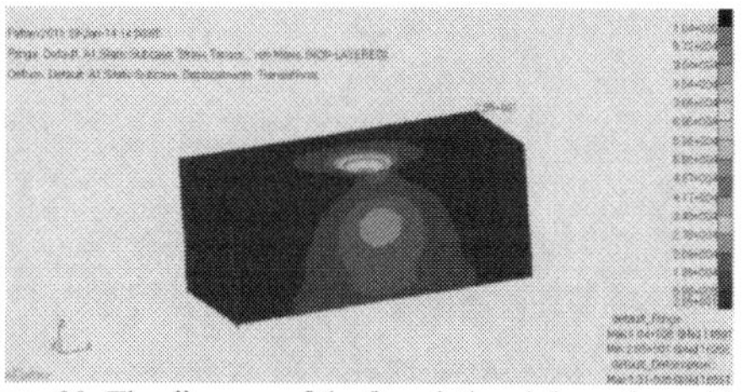

Figure 11. The diagram of the foundation deformation without steel plate

Using finite element method, the maximum variation value of support 5 is 12076.6N. The maximum elastic deformation of foundation subsystem is not more than

1.3μm without steel plate. The maximum elastic deformation is not more than 0.98μm with steel plate, and the maximum stress is not more than 0.146MPa, the safety factor is not less than 100. As shown in Fig. 11 and Fig. 12. The following equation 8 is established.

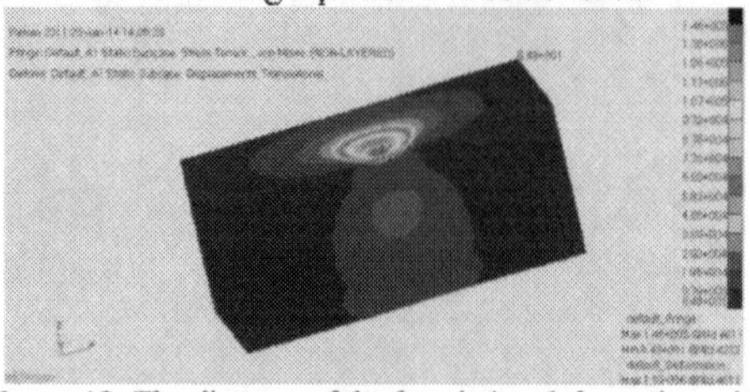

Figure 12. The diagram of the foundation deformation with steel plate(50mm).

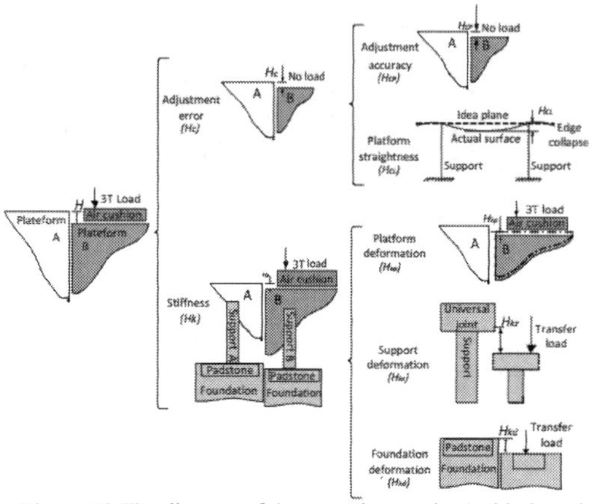

Figure 13. The diagram of decomposing mechanical index of altitude difference index

$$H_{KD} = 0.98\mu m \tag{8}$$

3.2.3 The mechanical characteristics analysis of support

there are more transmission structure, the elastic deformation of support is harder to get, so the index of support could be got by using residual quantity method, which is difference between the overall design index and deterministic subsystem index. The above index of subsystem will be summarized as follows:

(1) The index of adjustment accuracy

$$H_{CP} = 0.2\mu m$$

(2) The index of platform straightness

$$H_{CL} = 3.3\mu m$$

(3) The index of plateform elastic deformation

$$H_{KP} = 2.2\mu m$$

(4) The index of foundation elastic deformation

$$H_{KD} = 0.98\mu m$$

Using residual quantity method, the whole index of system is 10μm, the sum of above four parts is 6.68μm , so the index of support is 3.32μm. It is based on thinking about heavy load deformation, adjustable accuracy and adjustment range. The support can satisfy requirements by investigating. The following equation 9 is established.

$$H_{KZ} = 3.32\mu m \tag{9}$$

Through above analysis, the altitude difference because of stiffness is 6.5μm. Importing data into equation 6, and the equation 10 is true.

$$H_{K} = H_{KP} + H_{KZ} + H_{CD} = 2.2 + 3.3 + 0.98 = 6.48\mu m \tag{10}$$

Summarizing the figures of decomposing mechanical index as shown in Fig. 13.

4 The analysis of whole system mechanical characteristics

In summary, the whole model would be built by using Patran, the plate deformation of system would be analysed when the load is 3t. As shown in Fig. 14.

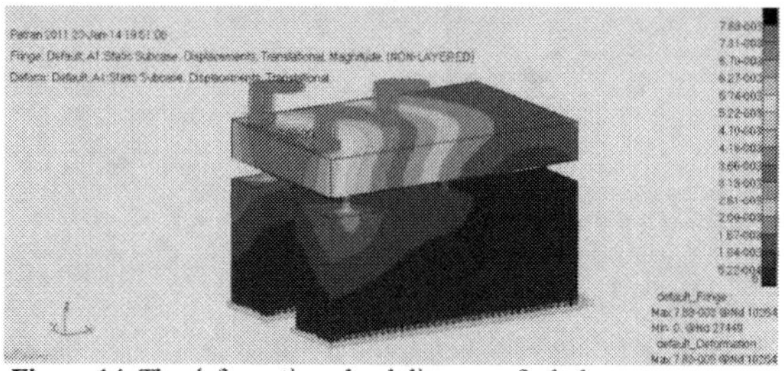

Figure 14. The deformation cloud diagram of whole system on 3t load

When the load is 3t, the levelness variation of plate is 0.5"<1", the deformation of platform meet index requirements. The maximum stress of foundation is 0.96MPa, the safety factor is greater than 200.

5 Summary

This paper analyzes and researches the mechanical indexs (altitude difference) of The large supportive stage with super smooth surface. Affecting factors are divided into and adjustment error and system stiffness. First, analyzing adjustment accuracy and platform straightness, which influence the adjustment error. Second, analyzing platform mechanical characteristics、support mechanical characteristics and foundation mechanical characteristics, which influence the system stiffness. Conclusions are as follows:

(1) the index of motor adjustment accuracy is 0.2μm;
(2) the index of platform straightness is 3.3μm;
(3) the index of platform deformation is 2.2μm;
(4) the index of foundation deformation is 0.98μm;
(5) the index of support deformation is 3.32μm;
(6) When the load is 3t, the deformation of whole system is 9.79μm, which is less than index requirement 10μm.

Above index assignment and mechanics calculation are meet the technical requirements. The theoretical support and design method are provided in this paper, which is based on objective and reasonable considerations.

References

1. White G C, Xu Y S. Active Vertical-Gravity Compensation System[J].IEEE Transactions on Instr- umentation and Measurement,1994,43(6):786-792.
2. J.S.Li, Mu X.G.Mu and J.J.Zhang. Physical Simulation for Satellite Control Systems[J]. Aerospace Control, 2004(4):37-43.
3. L.X.Xiao, Q.F.Jia and H.H.Deng: *Theoretical Mechanics* (Tianjin University press,1995).
4. N.M.Qi, W.H.Zhang, J.Z.Gao and M.Y.Huo, The Primary Discussion for the Ground Simulation System of Spatial Microgravity[J]. Aerospace Control, 2011:95-100.
5. Y.S.Yao, T.Mei. Simulation Method of Space Operation on the Ground-Buoyancy Mehod[J]. Chinese Journal of Mechanical Engineering, 2008:182-188.
6. D. Miller,et al. SPHERES: A tested for long duration satellite formation flying in micro-gravityconditions [A]. Proceedings of the AAS/AIAA Space Flight Mechanics Meeting[C]. Clearwater, Florida, January 23-26,2000:167-179.
7. C.M.Lv, L.Xi, G.H.Zhao, Y.H.Zhang. Experiment Microgravity System for Zero Gravity Airplane[J].Journal of Tsinghua University(Science and Technology), 2003:1064-1068.
8. P.Zeng: Finite Element Analysis and Applications(Tsinghua University press,2004).

MATEC Web of Conferences 77, 01015 (2016) DOI: 10.1051/matecconf/20167701015
ICMMR 2016

Controller Design and Analysis of Spacecraft Automatic Levelling and Equalizing Hoist Device based on Hanging Point Adjustment

Tang Laiying[1,2], Liu Guangtong[1,2], Guo Tao[1,2], Liu Zhe[1,2], Hu Ruiqin[1,2], Guo Dabao[1,2]

[1]*Beijing Institute of Spacecraft Environment Engineering, 100094 Beijing China*
[2]*Beijing Engineering Research Center of the Intelligent Assembly Technology and Equipment for Aerospace Product, 100094 Beijing China*

Abstract. Spacecraft Automatic Levelling and Equalizing Hoist Device (SALEHD) is a kind of hoisting device developed for eccentric spacecraft level-adjusting, based on hanging point adjustment by utilizing XY-workbench. To make the device automatically adjust the spacecraft to be levelling, the controller for SALEHD was designed in this paper. Through geometry and mechanics analysis for SALEHD and the spacecraft, the mathematical model of the controller is established. And then, the link of adaptive control and the link of variable structure control were added into the controller to adapt the unknown parameter and eliminate the interference of support vehicle. The stability of the controller was analysed, through constructing Lyapunov energy function. It was proved that the controller system is asymptotically stable, and converged to origin that is equilibrium point. So the controller can be applied in SALEHD availably and safely.

1 Introduction

Hoist device, as an important actuating unit in spacecraft hoisting, has been widely used in assembly and integration of spacecraft, such as propellant tank assembling, modules docking, spacecraft environment test, containing and transportation for spacecraft. So hoisting is an important process for quality control and safety guarantee in spacecraft assembly[1][2].

However, in spacecraft hoisting currently, the deviation of spacecraft centre of gravity (COG) from its geometric centre or theoretical centre is contributed to differences in assembly degree and operating condition. This would become hidden safety trouble such as inclinations, swing and clash, by special hoisting device of traditional fixed structure which could not self-adapt the eccentric spacecraft. So, self-adapted electromechanical hoist device needs to be researched to resolve these problems [3][4][5].

In this paper, adaptive controller based on hanging point adjustment by utilizing XY-workbench was designed, and stability of the controller was analysed, which is proved to be asymptotically stable and converged to equilibrium point.

2 Brief introduction of SALEHD

2.1 Composition of SALEHD[1][2]

SALEHD based on hanging point adjustment, mainly consists of mechanical part and electric part. Mechanical part consists of hanging beam, XY-workbench, ring and so on. Electric part consists of tilt sensor, tension sensors, electrical cabinet, LED displays, motors, power supply and control software. Compositions of SALEHD are shown in **Figure 1**.

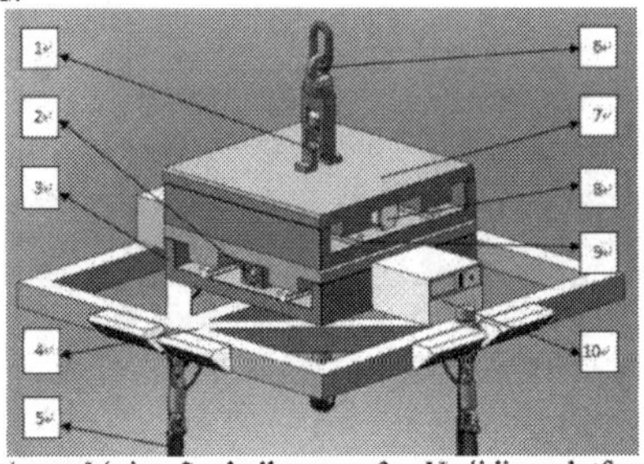

1—Universal joint; 2—ball screw; 3—Y sliding platform; 4—hanging beam; 5—sling; 6—ring; 7—X sliding platform; 8—motor; 9—slide; 10—electrical cabinet

Figure 1. Internal structure diagram of SALEHD

2.2 Workflow of SALEHD[1][2]

The specific level-adjusted workflow is described as follows:

(1) Spacecraft and SALEHD is connected by slings under SALEHD. Slings are adjusted to equal length, and preloaded.

(2) Point moving the crane to hoist spacecraft a little height, makes spacecraft generate a small inclination angle, and docking frame of spacecraft is suspended partially from docking frame of support vehicle.

(3) Horizontal degree of docking frame of spacecraft is measured by two-dimensional tilt sensor, after spacecraft is static.

(4) Spacecraft is landed back on the support vehicle.

(5) XY-workbench is moved to the specified location which is calculated by controller through the sensors.

(6) Repeat steps above, until the horizontal degree is within the allowable range.

3 Establishment of mathematical model

3.1 Simplification of boundary conditions[1][2]

(1) Spacecraft and support vehicle are in a horizontal state, before hoisting.

(2) Slings between spacecraft and SALEHD are equal length. And stress deformations of slings are ignored.

(3) Spacecraft and SALEHD are regarded as rigid bodies, whose stress deformations are ignored.

(4) Attitude angle between spacecraft and SALEHD are regarded as equal in static state, because slings are vertical and their lengths are equal.

3.2 Model building

According the simplification above, the system consisted by spacecraft and SALEHD is shown as Figure. 2.

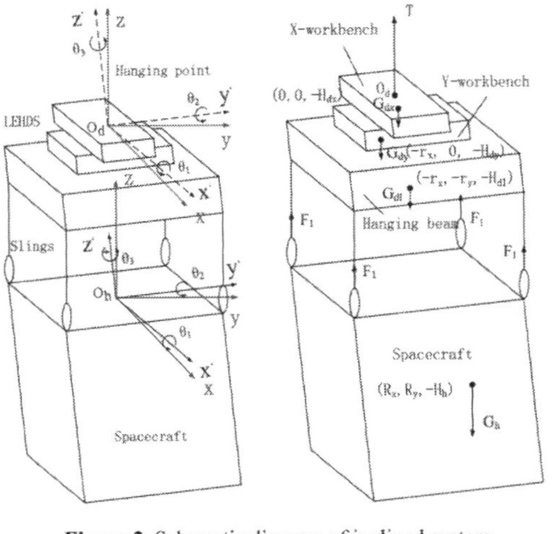

Figure 2. Schematic diagram of inclined system

3.2.1 Geometric model building

Taking hanging point O_d as the origin, body coordinate $\Sigma O_dX_d'Y_d'Z_d'$ is established. So, position vectors of COGs of X-workbench, Y-workbench and hanging beam can be expressed as Formula (1).

$$\begin{cases} \vec{V}'_{Gdx} = \left(0, 0, -H_{dx}\right)^T \\ \vec{V}'_{Gdy} = \left(-r_x, 0, -H_{dy}\right)^T \\ \vec{V}'_{Gdl} = \left(-r_x, -r_y, -H_{dl}\right)^T \end{cases} \tag{1}$$

Where

H_{dx}, H_{dy} , H_{dl} —respectively as distance from COGs of X-workbench, Y-workbench and hanging beam, to hanging point above SALEHD, m;

r_x, r_y —respectively as displacement of X-workbench and Y-workbench, m;

Taking spacecraft surface centre O_h as the origin, body coordinate $\Sigma O_hX_h'Y_h'Z_h'$ is established. So, position vectors of COGs of spacecraft can be expressed as Formula (2).

$$\vec{V}'_{Gh} = \left(R_x, R_y, -H_h\right)^T \tag{2}$$

Where

R_x, R_y, $-H_d$—respectively as position of COG of spacecraft, m;

3.2.2 Mechanical model building

1) Attitude and coordinate transformation for SALEHD

When SALEHD is tilted, tilted angle is less than 1°. So Rotation matrix could be simplified as Formula (3), approximated by infinitesimal angular displacement.

$$C_d = E - \boldsymbol{\Phi}^\times = \begin{pmatrix} 1 & -\theta_3 & \theta_2 \\ -\theta_3 & 1 & -\theta_1 \\ \theta_2 & -\theta_1 & 1 \end{pmatrix} = \begin{pmatrix} 1 & 0 & \theta_2 \\ 0 & 1 & -\theta_1 \\ -\theta_2 & \theta_1 & 1 \end{pmatrix} \tag{3}$$

Where

θ_1, θ_2, θ_3—respectively as turning angles around the axis X, Y, and Z of body coordinate of SALEHD. (Because of no moment in yaw direction, $\theta_3=0$)

SALEHD can be regarded as rotating around the hanging point above, so taking hanging point O_d as the origin, inertia coordinate ΣO_dXYZ is established. COG coordinate value of each part of SALEHD can be expressed as Formula (4).

$$\begin{cases} \vec{V}_{Gdx} = C_d\vec{V}'_{Gdx} = [-\theta_2H_{dx}, H_{dx}\theta_1, -H_{dx}]^T \\ \vec{V}_{Gdy} = C_d\vec{V}'_{Gdy} = [-r_x - \theta_2H_{dy}, \theta_1H_{dy}, \theta_2r_x - H_{dy}]^T \\ \vec{V}_{Gdl} = C_d\vec{V}'_{Gdl} = [-r_x - \theta_2H_{dl}, -r_y + \theta_1H_{dl}, \theta_2r_x - \theta_1r_y - H_{dl}]^T \end{cases} \tag{4}$$

2) Attitude and coordinate transformation for spacecraft.

When spacecraft is hoisted in static state, each sling is in vertical state, so COG coordinate value of spacecraft can be expressed in inertia coordinate ΣO_dXYZ as Formula (5)

$$\vec{V}_{Gh} = C(\vec{V}'_{Gh} - \vec{V}'_{hd1}) + (0,0,-L)^T + C\vec{V}'_{dd1} \tag{5}$$

Where

L—as length of sling between spacecraft and SALEHD, m;

Formula (1), (2), (3), (4) are taken into (5) to obtain (6)

$$\vec{V}_{Gh} = \begin{bmatrix} -r_x - \theta_2H_{dj} + R_x - \theta_2H_h \\ -r_y + \theta_1H_{dj} + R_y + \theta_1H_h \\ \theta_2r_x + \theta_1r_y - H_{dj} - L - \theta_2R_x + \theta_1R_y - H_h \end{bmatrix} \tag{6}$$

Where

H_{dj}—as vertical distance between hanging point below and above of SALEHD, m;

3) Moment balance

The force balance equation is established for the spacecraft as Formula (7).

$$G_h = F_1 + F_2 + F_3 + F_4 \tag{7}$$

The static torque balance equation is established for spacecraft in origin O_d of inertia coordinate ΣO_dXYZ.

$$\sum M = \vec{V}_{Gdx} \times \begin{bmatrix} 0 \\ 0 \\ -G_{dx} \end{bmatrix} + \vec{V}_{Gdx} \times \begin{bmatrix} 0 \\ 0 \\ -G_{dy} \end{bmatrix} + \vec{V}_{Gdl} \times \begin{bmatrix} 0 \\ 0 \\ -G_{dl} \end{bmatrix} + \vec{V}_{Gh} \times \begin{bmatrix} 0 \\ 0 \\ -G_{h} \end{bmatrix} = 0 \tag{8}$$

Formula (1), (7) are taken into (8) to obtain (9)

$$\begin{cases} \theta_1 = \dfrac{r_y G_{dl} + r_y G_h - R_y G_h}{H_{dx} G_{dx} + H_{dy} G_{dy} + H_{dl} G_{dl} + H_{dj} G_h + H_h G_h} \\ \theta_2 = \dfrac{R_x G_h - r_x G_{dy} - r_x G_{dl} - r_x G_h}{H_{dx} G_{dx} + H_{dy} G_{dy} + H_{dl} G_{dl} + H_{dj} G_h + H_h G_h} \end{cases} \tag{9}$$

When SALEHD is levelling, $\theta_1=\theta_2=0$, i.e.

$$\begin{cases} r_{yp} = \dfrac{R_y G_h}{G_{dl} + G_h} \\ r_{xp} = \dfrac{R_x G_h}{G_{dy} + G_{dl} + G_h} \end{cases} \tag{10}$$

Where

r_{xp}, r_{yp}—respectively as displacement of X-workbench and Y-workbench when SALEHD is leveling, m;

4 Design for variable structure adaptive controller

4.1 Control in known parameter and total suspended

From Formula (9) and (10), (11) can be obtained:

$$\begin{cases} \Delta r_y = r_{yp} - r_y = -\dfrac{H_{dx} G_{dx} + H_{dy} G_{dy} + H_{dl} G_{dl} + H_{dj} G_h + H_h G_h}{G_{dl} + G_h} \theta_1 \\ \Delta r_x = r_{xp} - r_x = \dfrac{H_{dx} G_{dx} + H_{dy} G_{dy} + H_{dl} G_{dl} + H_{dj} G_h + H_h G_h}{G_{dy} + G_{dl} + G_h} \theta_2 \end{cases} \tag{11}$$

Where

Δr_x, Δr_{yp}—respectively as difference from location to target position of X-workbench and Y-workbench, m;

In the condition of known parameter of spacecraft and SALEHD (such as G_h), Δr_x, Δr_{yp} can be calculated from Formula (11) and the tilt angles θ_1, θ_2. So motor can drive XY-workbench to equilibrium point by Δr_{xp}, Δr_{yp}, to make spacecraft levelling. System block diagram is shown in Figure.3.

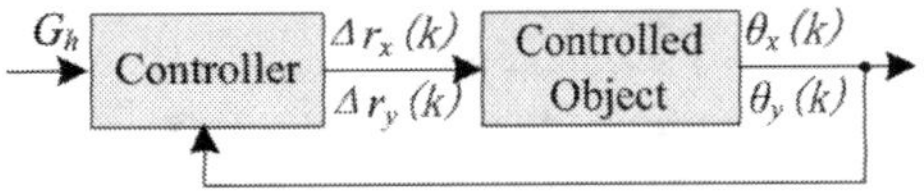

Figure 3. System block diagram

4.2 Adaptive control in unknown parameter and total suspended

In actual hoisting process, G_h, H_h, R_x, R_y are unknown, so they need to be estimated and identified. G_h can be approximated by tension sum of slings through Formula (7). R_x, R_y can be calculated by equilibrium position of XY-workbench through Formula (10). Only H_h can't be calculated and its influences decrease with the tilt angle decreasing, so it needs to be predicted, identified and corrected. Parameter identification module is added into control system, shown in Figure 4.

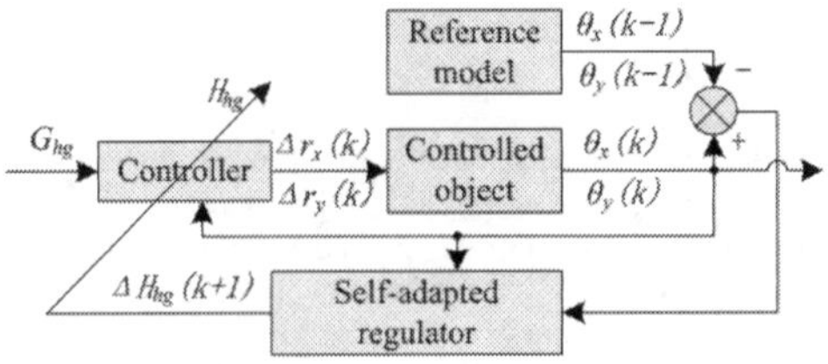

Figure 4. Block diagram of adaptive control system

Formula (7) and (11) are discretized as follows:

1) Discrete equation of controller

$$\begin{cases} \Delta r_x(k+1) = \dfrac{H_{dx} G_{dx} + H_{dy} G_{dy} + H_{dl} G_{dl} + H_{dj} G_{hg} + G_{hg} H_{hg}(k+1)}{G_{dy} + G_{dl} + G_{hg}} \theta_2(k) \\ \Delta r_y(k+1) = -\dfrac{H_{dx} G_{dx} + H_{dy} G_{dy} + H_{dl} G_{dl} + H_{dj} G_{hg} + G_{hg} H_{hg}(k+1)}{G_{dl} + G_{hg}} \theta_1(k) \end{cases}$$

$$G_{hg} = F_1 + F_2 + F_3 + F_4 \tag{12}$$

2) Discrete equation of controlled object

$$\begin{cases} \theta_1(k+1) = \dfrac{(G_{dl} + G_h)\sum_{i=0}^{k+1} \Delta r_y(i) - R_y G_h}{H_{dx} G_{dx} + H_{dy} G_{dy} + H_{dl} G_{dl} + H_{dj} G_h + H_h G_h} \\ \theta_2(k+1) = \dfrac{R_x G_h - (G_{dy} + G_{dl} + G_h)\sum_{i=0}^{k+1} \Delta r_x(i)}{H_{dx} G_{dx} + H_{dy} G_{dy} + H_{dl} G_{dl} + H_{dj} G_h + H_h G_h} \end{cases} \tag{13}$$

Comparing two output values θ, the relation between H_h and θ can be obtained, as follows:

$$\begin{cases} \dfrac{\theta_1(k+1)}{\theta_1(k)} = \dfrac{[H_h - H_{hg}(k+1)]G_h}{H_{dx} G_{dx} + H_{dy} G_{dy} + H_{dl} G_{dl} + H_{dj} G_h + H_h G_h} \\ \dfrac{\theta_2(k+1)}{\theta_2(k)} = \dfrac{[H_h - H_{hg}(k+1)]G_h}{H_{dx} G_{dx} + H_{dy} G_{dy} + H_{dl} G_{dl} + H_{dj} G_h + H_h G_h} \end{cases} \tag{14}$$

Discrete state equation of parameter regulator is obtained, through corrected value $H_{hg}(k+2)$ replacing true value H_h.

$$\begin{cases} H_{hg}(k+2) = \dfrac{H_{dx} G_{dx} + H_{dy} G_{dy} + H_{dl} G_{dl} + H_{dj} G_h + G_h H_{hg}(k+1)}{G_h[\theta_1(k) - \theta_1(k+1)]} \theta_1(k+1) \\ \quad + H_{hg}(k+1) \\ H_{hg}(k+2) = \dfrac{H_{dx} G_{dx} + H_{dy} G_{dy} + H_{dl} G_{dl} + H_{dj} G_h + G_h H_{hg}(k+1)}{G_h[\theta_2(k) - \theta_2(k+1)]} \theta_2(k+1) \\ \quad + H_{hg}(k+1) \end{cases} \tag{15}$$

And reference model is the K-1 time response model of the system, so discrete state equation of reference model can be described as follows:

$$\begin{cases} \theta_1(k+1) = \theta_1(k) \\ \theta_2(k+1) = \theta_2(k) \end{cases} \tag{16}$$

4.3 Variable structure adaptive control in unknown parameter and support vehicle

Before spacecraft leaving support vehicle, support vehicle makes a restrictive effect to prevent spacecraft from inclining further, compared with the total suspended state. The output value tends to a constant value, showing a certain saturation, which makes structure of the control system changed. So the influence of the support vehicle can be equivalent to a saturation nonlinearity series in output port of adaptive control system in unknown parameter and total suspended, as follows:

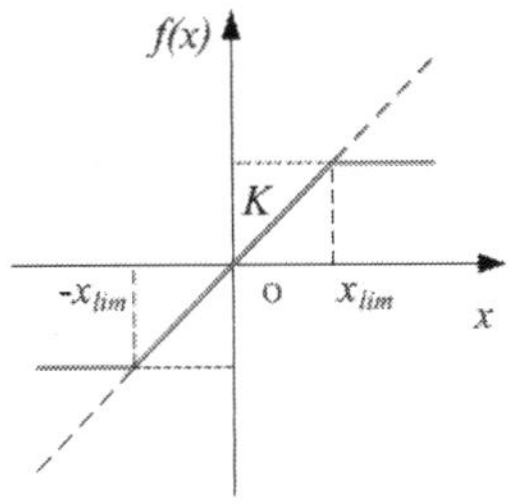

Figure 5. Chart of saturation nonlinearity

$$f(x)=\begin{cases}Kx;|x|\le x_{\lim}\\ Kx_{\lim};x>x_{\lim}\\ -Kx_{\lim};x<-x_{\lim}\end{cases}\quad K=1,x_{\lim}>0 \tag{17}$$

The system works in two different ranges. When the system working in the range of linearity, control mode is the adaptive control in total suspended. When the system working in the range of saturation nonlinearity, the output value tends to a constant value leading to the difference between reference model and actual output value is zero, which make the self-adapted system can't identify parameter or adjust parameter. So a judgment link is added in front of the self-adapted regulator, as follows.

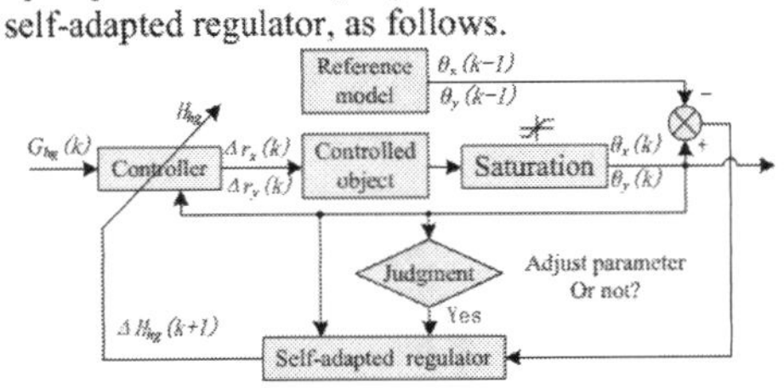

Figure 6. Block diagram of variable structure adaptive control system

$$H_{hg}(k+2)=\begin{cases}H_{hg}(k+1);|\theta_1(k)|\ge\theta_{\lim}-\delta,\text{且}|\theta_2(k)|\ge\theta_{\lim}-\delta\\ \dfrac{\theta_1(k+1)}{\theta_1(k)-\theta_1(k+1)}K+H_{hg}(k+1);|\theta_1(k)|<\theta_{\lim}-\delta\\ \dfrac{\theta_2(k+1)}{\theta_2(k)-\theta_2(k+1)}K+H_{hg}(k+1);|\theta_2(k)|<\theta_{\lim}-\delta\end{cases}$$

$$K=\frac{H_{dx}G_{dx}+H_{dy}G_{dy}+H_{dl}G_{dl}+H_{dj}G_{h}(k+1)+G_{h}(k+1)H_{hg}(k+1)}{G_{h}(k+1)} \tag{18}$$

Let: $x_1=\Delta r_x$; $x_2=\Delta r_y$; $x_3=\theta_1$; $x_4=\theta_2$; $r=G_{hg}$;

The final state equation of whole control system as follows:

$$\begin{cases}x_1(k+1)=\dfrac{H_{dx}G_{dx}+H_{dy}G_{dy}+H_{dl}G_{dl}+H_{dj}r(k+1)+r(k+1)H_{hg}(k+1)}{G_{dy}+G_{dl}+r(k+1)}x_4(k)\\ x_2(k+1)=-\dfrac{H_{dx}G_{dx}+H_{dy}G_{dy}+H_{dl}G_{dl}+H_{dj}r(k+1)+r(k+1)H_{hg}(k+1)}{G_{dl}+r(k+1)}x_3(k)\\ x_3(k+1)=f(\dfrac{(G_{dl}+G_h)\sum_{i=0}^{k+1}x_2(i)-R_yG_h}{H_{dx}G_{dx}+H_{dy}G_{dy}+H_{dl}G_{dl}+H_{dj}G_h+H_hG_h})\\ x_4(k+1)=f(\dfrac{R_xG_h-(G_{dy}+G_{dl}+G_h)\sum_{i=0}^{k+1}x_1(i)}{H_{dx}G_{dx}+H_{dy}G_{dy}+H_{dl}G_{dl}+H_{dj}G_h+H_hG_h})\end{cases}$$

$$r(k)=F_1(k)+F_2(k)+F_3(k)+F_4(k)$$

$$\begin{cases}y_1(k+1)=x_3(k+1)\\ y_2(k+1)=x_4(k+1)\end{cases} \tag{19}$$

Where

$$f(x)=\begin{cases}x;|x|\le x_{\lim}\\ x_{\lim};x>x_{\lim}\\ -x_{\lim};x<-x_{\lim}\end{cases}$$

$$H_{hg}(k+1)=\begin{cases}H_{hg}(k);|x_3(k-1)|\ge x_{\lim},\text{且}|x_4(k-1)|\ge x_{\lim}\\ \dfrac{x_3(k)}{x_3(k-1)-x_3(k)}K(k)+H_{hg}(k);|x_3(k-1)|<x_{\lim}\\ \dfrac{x_4(k)}{x_4(k-1)-x_4(k)}K(k)+H_{hg}(k);|x_4(k-1)|<x_{\lim}\end{cases}$$

$$K(k)=\frac{H_{dx}G_{dx}+H_{dy}G_{dy}+H_{dl}G_{dl}+H_{dj}r(k)+r(k)H_{hg}(k)}{r(k)}$$

5 Stability analyses

5.1 Stability analysis for Adaptive control in total suspended

Adaptive control system is nonlinear essentially, so the Lyapunov stability theorem is adopted to judge stability for the system. Positive definite quadratic function is introduced as Lyapunov energy function $V[x(k)]$ for discrete system, i.e.

$$V[\boldsymbol{x}(k)]=\boldsymbol{x}^T(k)P\boldsymbol{x}(k)$$

Where

$$\boldsymbol{x}(k)=[x_1(k),x_2(k),x_3(k),x_4(k)]^T$$

$$P=\begin{bmatrix}1&0&0&0\\0&1&0&0\\0&0&1&0\\0&0&0&1\end{bmatrix}$$

i.e.

$$V[\boldsymbol{x}(k)]=x_1^2(k)+x_2^2(k)+x_3^2(k)+x_4^2(k) \tag{20}$$

$V[x(k)]$ is positive definite, and the origin is balanced. If $\Delta V[x(k)]$ is negative definite, the system is asymptotically stable. i.e.

$$\begin{aligned}\Delta V[\boldsymbol{x}(k)]&=V[\boldsymbol{x}(k+1)]-V[\boldsymbol{x}(k)]\\&=[x_1^2(k+1)-x_1^2(k)]+[x_2^2(k+1)-x_2^2(k)]\\&+[x_3^2(k+1)-x_3^2(k)]+[x_4^2(k+1)-x_4^2(k)]\end{aligned} \tag{21}$$

Because the absolute value of the differences between true value and estimated value of COG height is far less than the true value. i.e.

$$|H_h-H_{hg}(k)|<|H_h| \tag{22}$$

Formula (22) is taken into (14) to obtain as follows

$$\begin{cases}\left|\dfrac{x_3(k+1)}{x_3(k)}\right|=\left|\dfrac{[H_h-H_{hg}(k)]G_h}{H_{dx}G_{dx}+H_{dy}G_{dy}+H_{dl}G_{dl}+H_{dj}G_h+H_hG_h}\right|\\ <\left|\dfrac{[H_h-H_{hg}(k)]G_h}{H_hG_h}\right|<1\\ \left|\dfrac{x_4(k+1)}{x_4(k)}\right|=\left|\dfrac{[H_h-H_{hg}(k)]G_h}{H_{dx}G_{dx}+H_{dy}G_{dy}+H_{dl}G_{dl}+H_{dj}G_h+H_hG_h}\right|\\ <\left|\dfrac{[H_h-H_{hg}(k)]G_h}{H_hG_h}\right|<1\end{cases} \tag{23}$$

i.e.

$$\begin{cases} x_3^2(k+1)-x_3^2(k)<0 \\ x_4^2(k+1)-x_4^2(k)<0 \end{cases} \tag{24}$$

Formula (19) is taken into (23) to obtain as follows

$$\begin{cases} \left|\frac{x_1(k+1)}{x_1(k)}\right| = \left|\frac{x_4(k+1)}{x_4(k)}\right| < 1 \\ \left|\frac{x_2(k+1)}{x_2(k)}\right| = \left|\frac{x_3(k+1)}{x_3(k)}\right| < 1 \end{cases} \tag{25}$$

i.e.

$$\begin{cases} x_1^2(k+1)-x_1^2(k)<0 \\ x_2^2(k+1)-x_2^2(k)<0 \end{cases} \tag{26}$$

Formula (24) and (26) are taken into (21) to obtain as follows.

$$\Delta V[\boldsymbol{x}(k)]<0$$

ΔV[x(k)] is negative definite, so the system is asymptotically stable and converged to the origin, that is the equilibrium point.

5.2 Stability analysis for variable structure adaptive control in support vehicle

Variable structure adaptive control system works in two different ranges. When system works in range of linearity, the stability of the system has been proved above. The stability of the system working in the range of saturation nonlinearity, only needs to be proved in this section.

Positive definite quadratic function is introduced as Lyapunov energy function *V[L(k)]* for discrete system, i.e.

$$V[\boldsymbol{L}(k)] = \boldsymbol{L}^T(k)\boldsymbol{P}\boldsymbol{L}(k)$$

Where

$$\boldsymbol{L}(k) = [L_x(k), L_y(k)]^T$$

$$\begin{cases} L_x(k) = r_{xp} - r_x(k) = r_{xp} - \sum_{i=0}^{k}\Delta r_x(k) \\ L_y(k) = r_{yp} - r_y(k) = r_{xp} - \sum_{i=0}^{k}\Delta r_x(k) \end{cases} \tag{27}$$

$$P = \begin{pmatrix} 1 & 0 \\ 0 & 1 \end{pmatrix}$$

i.e.

$$V[\boldsymbol{L}(k)] = L_x^2(k) + L_y^2(k) \tag{28}$$

Where

$L_x(k)$, $L_y(k)$—respectively as difference from location to equilibrium position of X-workbench or Y-workbench, m; i.e.

V[L(k)] is positive definite, and the origin is balanced. If *ΔV[L(k)]* is negative definite, the system is asymptotically stable. i.e.

$$\begin{aligned} \Delta V[\boldsymbol{L}(k)] &= V[\boldsymbol{L}(k+1)] - V[\boldsymbol{L}(k)] \\ &= L_x^2(k+1) + L_y^2(k+1) - [L_x^2(k) + L_y^2(k)] \\ &= [L_x^2(k+1) - L_x^2(k)] + [L_y^2(k+1) - L_y^2(k)] \end{aligned} \tag{29}$$

From Formula (11) and (19), (30) can be obtained as follows:

$$\begin{aligned} L_x(k+1) - L_x(k) &= -\Delta r_x(k+1) = -x_1(k+1) \\ &= -\frac{H_{dx}G_{dx} + H_{dy}G_{dy} + H_{dl}G_{dl} + H_{dj}r(k+1) + r(k+1)H_{hg}(k+1)}{G_{dy} + G_{dl} + r(k+1)} x_4(k) \\ &= -\frac{H_{dx}G_{dx} + H_{dy}G_{dy} + H_{dl}G_{dl} + H_{dj}r(k+1) + r(k+1)H_{hg}(k+1)}{G_{dy} + G_{dl} + r(k+1)} x_{\lim} \end{aligned} \tag{30}$$

Let

$$x_{\lim} = f[x_4'(k)]$$

$$x_4'(k) = \frac{R_x G_h - (G_{dy} + G_{dl} + G_h)\sum_{i=0}^{k+1} x_1(i)}{H_{dx}G_{dx} + H_{dy}G_{dy} + H_{dl}G_{dl} + H_{dj}G_h + H_h G_h} \tag{31}$$

From Formula (17), it can be obtained that x_{lim} and $x'_4(k)$ are both plus or both minus.

Formula (11) and (27) are taken into (31) to obtain as follows

$$x_4'(k) = \frac{(G_{dy} + G_{dl} + G_h)L_x(k)}{H_{dx}G_{dx} + H_{dy}G_{dy} + H_{dl}G_{dl} + H_{dj}G_h + H_h G_h} \tag{32}$$

So $x'_4(k)$ and $L_x(k)$ are same plus or same minus.

In summary, it can be obtained that $L_x(k+1)-L_x(k)$ and $Lx(k)$, one is plus and another is minus. Also, the k and $k+1$ time are both in the range of saturation, without crossing zero, so $L_x(k+1)$ and $L_x(k)$ are both plus or both minus.

If $L_x(k+1)>0$ *and* $L_x(k)>0$, then $L_x(k+1)-L_x(k)<0$, that is $L_x(k+1)<L_x(k)$, so $|L_x(k+1)|<|L_x(k)|$.

If $L_x(k+1)<0$ *and* $L_x(k)<0$, then $L_x(k+1)-L_x(k)>0$, that is $L_x(k+1)>L_x(k)$, so $|L_x(k+1)|<|L_x(k)|$.

In summary:

$$L_x^2(k+1) - L_x^2(k) < 0 \tag{33}$$

In the same way:

$$L_y^2(k+1) - L_y^2(k) < 0 \tag{34}$$

Formula (33) and (34) are taken into (28) to obtain as follows:

$$\Delta V[\boldsymbol{L}(k)] < 0 \tag{35}$$

ΔV[L(k)] is negative definite, so the system is asymptotically stable and converged to the origin that is the equilibrium point.

In summary, the system working in ranges of linearity and saturation nonlinearity both asymptotically stable, and converged to the same origin.

6 Conclusion

The variable structure adaptive controller designed in this paper, can adjust control structure by itself according to the actual support of the vehicle, and adapt different eccentric conditions by correcting parameter itself automatically. Its stability has been proved by virtual simulation also. At present, SALEHD with this controller, has been put into practical application in multiple spacecraft types and achieved certain economic effect.

References

1. Tang Laiying, Sun Gang, Chen Jinming, Fu Hao. Design and research of Spacecraft Level-adjusted Hoist equipment based on method of suspension centre

adjustment[J]. Spacecraft environment engineering, 2013(2): 219-224

2. Fu Hao, Sun Gang, Tang Laiying, Cui Junfeng. Design and Development of Spacecraft Automatic Leveling and Equalizing Hoist Device. International Journal of mechanical aerospace industrial mechatronic and manufacturing engineering. Vol :6, No 9, 2012
3. LIU Yu-gang. The technology improvement of assembling scheme for hoisting tool of a satellite [J]. Spacecraft environment engineering, 2004, 21(4): 48-53
4. ZHAO Pifu, QI Chun, CHEN Xianfeng. Study of hoisting technology of aerospace product [J]. Aerospace manufacturing technology, 2007,(3): 12-16
5. Yi Yu, Jianqing Yi, Chengdong Li, Dongbin Zhao, Jianhong Zhang, Control of a Rope-driven Self-leveling Device for Leveling Adjustment, 2009 American Control Conference Hyatt Regency Riverfront, St.Louis, MO, USA. June 10~12,2009

MATEC Web of Conferences 77, 01016 (2016) DOI: 10.1051/matecconf/20167701016
ICMMR 2016

A Coupled Helicopter Rotor/Fuselage Dynamics Model Using Finite Element Multi-body

Cheng Qi-you, Zhu Yan, Feng Zhi-zhuang, Chen Quan-long

Science and Technology on Rotorcraft Aeromechanics Laboratory, China Helicopter Research and Development Institute, Jingdezhen, 333001, China

Abstract. To develop a coupled rotor/flexible fuselage model for vibration reduction studies, the equation of coupled rotor-fuselage is set up based on the theory of multi-body dynamics, and the dynamic analysis model is established with the software MSC.ADMAS and MSC.NASTRAN. The frequencies and vibration acceleration responses of the system are calculated with the model of coupled rotor-fuselage, and the results are compared with those of uncoupled modeling method. Analysis results showed that compared with uncoupled model, the dynamic characteristic obtained by the model of coupled rotor-fuselage are some different. The intrinsic frequency of rotor is increased with the increase of rotational velocities. The results also show that the flying speed has obvious influence on the vibration acceleration responses of the fuselage. The vibration acceleration response in the vertical direction is much higher at the low speed and high speed flight conditions.

1 Introduction

The structural dynamic analysis of rotorcraft systems involves several interacting elastic components: several blades and rotors, fuselage, and possibly wings. It is convenient to think of such rotorcraft systems as multi-body elastic systems, i.e. a collection of mutually interacting elastic body.

As helicopter crew and passenger comfort has gained increased emphasis, vibration requirements have become more stringent [l-3]. Vibration reduction devices are usually used to meet these requirements. Even though there has been enormous progress with vibration suppression technology [4], cost and weight penalty has been excessive in part because of insufficient vibration prediction capability. To minimize the additional cost and weight, accurate vibration prediction is necessary at the early design stage.

A great deal of effort is devoted to improve the mathematical analysis of rotorcraft. Anubhav Datta [5] created a coupled rotor-fuselage analysis model using modal synthesis method. The model was used to analyze the vibration of helicopter. A coupled rotor-fuselage is set up based on finite element method and transfer matrix method [6]. Some author [7-9] developed the analysis model with Finite Element Multibody. However aerodynamic force is ignored in this model. Vellaichamy and Chopra [10] presented a coupled rotor-fuselage analysis using elastic blade and flexible fuselage modeling (stick model). Chiu [11] and Yeo [12] developed coupled rotor/flexible fuselage model for vibration reduction studies using 3-D fuselage. Euler and Navier-Stokes CFD codes have begun to be applied to the problem [13,14]. A recent rotor-body interactional aerodynamics study by Wachspress, Quackenbush and Boschitsch [15] shows the current state of the art. Interactional effects were studied using a full span constant vorticity contour free wake and a panel method to model the fuselage bodies and lifting surfaces. However, these analyses incorporated highly idealized aerodynamics such as uniform or linear inflow and quasi-steady aerodynamics so that vibration was substantially under predicted.

The present study is aimed at developing a coupled rotor/flexible fuselage model for vibration reduction studies. The equation of coupled rotor-fuselage is set up based on the theory of multi-body dynamics, and the dynamic analysis model is established with the software MSC.ADMAS and MSC.NASTRAN. The frequencies and vibration acceleration responses of the system are calculated with the model of coupled rotor-fuselage, and the results are compared with those of uncoupled modeling method. The influence of whirling speed on the intrinsic frequency and the influence of speed on vibration are also analyzed.

2 Theory

2.1 Coordinate systems

Several coordinate systems are used to describe blade sectional velocity or position vector. Fig.1 illustrates the ground-fixed, hub and fuselage coordinate systems. The definition of these coordinate systems is shown below:

1. The ground-fixed inertial coordinate system {Og , Xg, Yg, Zg }, which is fixed to the ground and does not move with the helicopter. Its unit vector is {ig, jg, kg}. And the gravity has the same direction of -kg.

2. The fuselage-fixed coordinate system {Of , Xf , Yf , Zf } ,with unit vector {if, jf, kf} fixed at the helicopter center of gravity. And six parameters (Xf, Yf, Zf, Φ Xf, Φ Yf, Φ Zf) are definite to describe the position of fuselage relate to the ground-fixed inertial coordinate system.

3. The hub-fixed coordinate system {OH, XH , YH , ZH }, with unit vector {iH, jH, kH} fixed at the hub. When there are several rotor system, the hub-fixed coordinate system can be definite as {OH i, XH i , YH i , ZH i } （i=1,2,···,IR）.

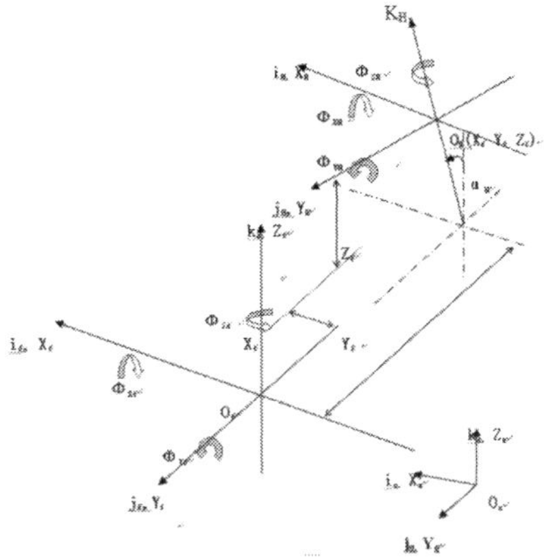

Figure 1. Coordinate systems

2.2 Mathematical Model

According to the definition of coordinate system, r_{no} and R_{np} are definite to describe the position of the origin point and random point on the blade to the ground-fixed inertial coordinate system. The Rnp can be expressed as:

$$R_{np} = r_{no} + A_{no}(r_{np} + u_{np}) \tag{1}$$

Where, A_{no} is the position matrix which the fuselage-fixed coordinate system relative to the ground-fixed inertial coordinate system. r_{np} is position vectors without deformation of point p. The u_{np} is the deformation of point p.

The position matrix of rotor-fixed coordinate system can be written as follows:

$$A_{no}=A_F A_H A_S A_P A_L A_T \tag{2}$$

Where A_f, A_H, A_S, A_P, A_L, A_T is position matrix of fuselage, hub, blade, flapping motion of blade, lagging motion of blade, and twist of blade, respectively.

The blade can be view as a several discrete elements, and the deformation of blade can be descript by the location of these elements. At the direction of small elastic deformation, the displacement of these elements can be transformed to the modal domain, using a truncated number of free vibration mode shapes. The deformation of point P can be written as follows:

$$u_{np} = \phi_i h \tag{3}$$

Where ϕ_i and h is mode shapes component and amplitude of node I, respectively.

We now substitute Eq.(3) into the Eqs.(1). We obtain the following relations:

$$R_{np} = r_{no} + A_{no}(r_{np} + \phi_i h) \tag{4}$$

It is definite that:

$$r_{no} = [x, y, z]^T$$
$$\Psi = [\psi, \theta, \phi]^T \tag{5}$$
$$h = [h_1, h_2, \cdots, h_m]^T$$

Where, x, y, z is coordinate of blade origin in the ground-fixed inertial coordinate system. ψ, θ, ϕ is Euler angle of the blade-fixed inertial coordinate system relative to the ground-fixed inertial coordinate system. h_m is vibration component of m-order amplitude.

The object of blade motion can be written as

$$\xi = [r_{no}^T, \Psi^T, h^T]^T \tag{6}$$

The velocity of node i can be obtained

$$v_i = [I, -A_{no}\left(\tilde{r}_{no} + \tilde{\phi}_i h\right)B, A_{no}\phi_i]\dot{\xi} \tag{7}$$

Where $\tilde{r}_{no}, \tilde{\phi}_i$ are position vector, and B is transform matrix.

From Eq.(7) , the kinetic energy and potential energy can be obtained

$$T = \frac{1}{2}\sum_{i=1}^{I} m_i v_i^T v_i = \frac{1}{2}\dot{\xi}^T M(\xi)\dot{\xi} \tag{8}$$

$$V = \frac{1}{2}\dot{\xi}^T K(\xi)\dot{\xi} \tag{9}$$

Where, The M is mass matrix. K is stiffness matrices. Taking Eq.(8) and Eq.(9) into the Lagrangian equation ,We obtain the following relations

$$M\ddot{\xi} + \dot{M}\dot{\xi} - \frac{1}{2}\left[\frac{\partial M}{\partial \xi}\dot{\xi}\right]^T \dot{\xi} + K\xi + f_g + D\dot{\xi} + \left[\frac{\partial \Omega}{\partial \xi}\right]\lambda = Q \tag{10}$$

Where, D is damping matrix, f_g is extended gravity, λ is Lagrange multiplier, Ω and Q are loads.

2.3 Model

The baseline rotor analysis is taken from MSC.ADAMS. The blade is assumed to be an elastic beam undergoing flap bending, lag bending, elastic twist, and axial deformation. The elastic rotor coupled equations include six hub degrees of motion. The rotor vibratory loads are transmitted to the fuselage through the hub and the effects of fuselage motion are included in the determination of blade loads.

The elastic fuselage is represented by a complete three dimensional finite element model, and a collection of elements (i.e. element library) is used to generate the structural dynamic model of the fuselage. The elements available are: beam, truss, non-structural mass, and a plate element. The concentrated non-structural masses of the helicopter such as transmission engine, fuel tanks, gearing, and payload are modeled using consistently derived finite element model for inertia properties. The non-structural mass element is capable of three translational displacements as well as three rotational degrees of freedom at each node. Due to the large number of degrees of freedom present in the finite element model of the fuselage, the equations are transformed to the modal domain, using a truncated number of free vibration mode shapes. The airframe modal data (eigenvalue, eigenvector, and generalized mass) are generated using NASTRAN and are used as an input to the coupled rotor/fuselage vibration analysis program. The couplings between rotor and fuselage are included in a consistent manner into the MSC.ADAMS.

3 Results and discussion

A civil helicopter is chosen for this study. The frequencies and vibration acceleration responses of the system are calculated with the model of coupled rotor-fuselage, and the results are compared with those of uncoupled modeling method.

Displayed in Table 1 is the frequency of rotor system with different rotational velocities. It is observed that the rotational velocity has obviously influence on the frequency. The frequency is increased with the rotational velocity. This is because the modulus of blades increases as rotational velocity increases. The calculate results are in accordance with the test results. However, the calculate results are slightly above the test results. This is because the restraint stiffness of calculate is slightly more than the test.

It is also observed that the results obtained with coupling method are slightly less than the one obtained with isolated rotor. This is because the restraint stiffness of rotor system with coupling method is slightly more than with isolated rotor.

Table 1. Rotor system frequency vs rotational velocities

rotational velocities (r/min)		290	328	386
Isolated Rotor	*calculate/Hz*	14.11	15.53	18.53
	experiment/Hz	13.63	15.24	18
Coupled method	*calculate /Hz*	13.27	14.58	17.3
comparison of two computation		5.82%	6.12%	6.64%

The points selected for vibration suppressions are located on rudder pedal (P1 and P2), flight-deck floor (P3 and P4), cockpit floor (P5), and main jackstay (P6 and P7).

The vibration acceleration responses without vibration-absorbing system of direction X, Y, and Z in the points selected for vibration suppression are shown in Figure 2.

The results show that the vibrations in the direction X and Y are relatively low. However the vibration acceleration response in the vertical direction is much higher than the other directions. It is means that the vibration suppression in the vertical direction is the most important. So the next discussion on the actuator optimization selection is focus on the vertical direction.

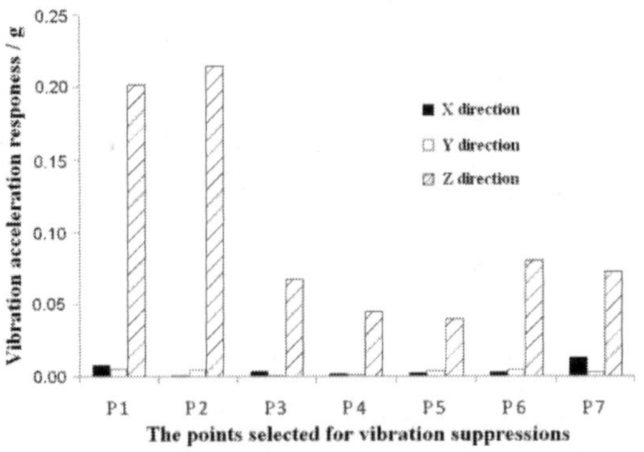

Figure 2. Vibration acceleration responses

The vibration acceleration responses of vertical direction in the key points of fuselage obtained with coupling method and uncoupled method are shown in Figure 3.

It is also observed that the results obtained with coupled method are slightly higher than the one obtained with uncoupled method. It is because the influence of rotor system on the infrequency of fuselage is ignored in the analysis of fuselage with the finite element method. However the frequencies of fuselage obtained with these two methods are different. If the frequency of fuselage obtained with coupling method is tend to the frequency of external load, the vibration acceleration responses will increase. Conversely, the vibration acceleration responses will decrease.

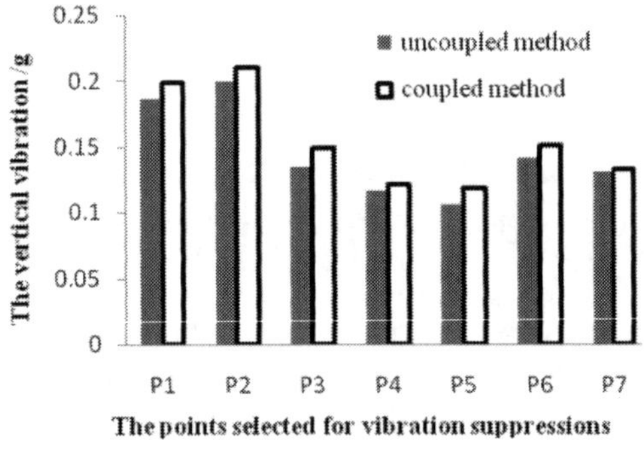

Figure 3. Vibration acceleration responses of computation

The influence of velocity on the vertical vibration in the key points of fuselage are shown in Figure 4. Three different speeds (75, 120, and 221 km/h) which respectively represent low speed, moderate speed, and high speed flight conditions.

The results show that there are some differences on the vibration at different flight speed. The vibration acceleration response in the vertical direction is much higher at the low speed and high speed flight conditions.

At the low speed conditions, the unequal spaced distribution of induced velocity is the main factor of serious vibration. And the vortex wake from the main rotor impinges directly on the tail boom that gives rise to fuselage vibration at the blade passage frequency.

Under the high speed flight condition, the vertical vibration is much higher. The airflow separation and contraction is the cause of vibration at the high speed flight condition.

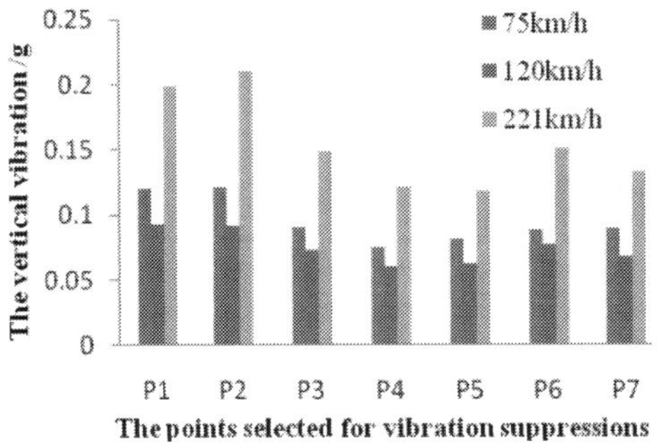

Figure 4. Influence of velocity on the vibration

4 Conclusions

The equation of coupled rotor-fuselage is set up based on the theory of multi-body dynamics, and the dynamic analysis model is established with the software MSC.ADMAS and MSC.NASTRAN. The frequencies and vibration acceleration responses of the system are calculated with the model of coupled rotor-fuselage, and the results are compared with those of uncoupled modeling method. Findings are listed below:

1) The calculate results are in accordance with the test results. However, the calculate results are slightly above the test results. This is because the restraint stiffness of calculate is slightly more than the test.

2) With the increase of rotational velocities, the modulus of blades is increased which make the intrinsic frequency of rotor increased.

3) The dynamic characteristics obtained by the model of coupled rotor-fuselage are some different from the results obtain by uncoupled model. If the frequency of fuselage obtained with coupling method is tend to the frequency of external load, the vibration acceleration responses will increase. Conversely, the vibration acceleration responses will decrease.

4) The flying speed has obvious influence on the vibration acceleration responses of the fuselage. The vibration acceleration response in the vertical direction is much higher at the low speed and high speed flight conditions. The unequal spaced distribution of induced velocity is the main factor of serious vibration at the low speed flight condition. However, the airflow separation and contraction is the cause of vibration at the high speed flight condition.

References

1. Crews, S. T., "Rotorcraft Vibration Criteria A New Perspective," *Proceedings of the 43rd Annual Forum of the American Helicopter Society*, St. Louis, MO, May 1987.
2. Chopra, I., "Status of Application of Smart Structures Technology to Rotorcraft Systems," Journal of the American Helicopter Society, Vol. 45, No. 4, October 2000, pp 228-252.
3. Roget, B., Chopra, I., "Wind Tunnel Testing of an Individual Blade Controller for a Dissimilar Rotor," Presented at the American Helicopter Society 60th Annual Forum, Baltimore, MD, June 7-10, 2004.
4. Reichert, G., "Helicopter Vibration Control-A Survey," *Vertica*, Vol. 5, No. 1, 1981.
5. Anubhav Datta, "FUNDAMENTAL UNDERSTANDING, PREDICTION AND VALIDATION OF ROTOR VIBRATORY LOADS IN STEADY LEVEL FLIGHT," Ph.D. dissertation, Dept. Elect. Eng., Harvard Univ., Cambridge, MA, 2004.
6. Hohenmeser K H, Yin S K., "The role of rotor impedance in the vibration analysis of rotorcraft," *Vertica*, Vol. 3, No. 2, pp. 187-204, 1979.
7. Bianchi F. *Helicopter Tail Rotor ANALYSIS EXPERIENCE IN AGUSTA WITH ADAMS*, Agusta, 1994.
8. Olivier A. Bauchau.,Carlo L. Bottasso., "T Modeling Rotorcraft Dynamics with Finite Element Multibody," *Mathematical and Computer Modeling*, pp. 1113-1137, 2001.
9. Altorio Bertogalli, Francesco Bisi., *Marco Lovera. MODELLING HELICOPTER ROTOR DYNAMICS WITH ADAMS*: 1995.
10. Vellaichamy, S., Chopra, I., "Effect of Modeling Techniques in the Coupled Rotor-Body Vibration Analysis," *Proceedings of the 34th Structures, Structural Dynamics and Materials Conference and Adaptive Structures Forum,* LaJolla, April 1993.
11. Chiu, T., Friedmann, P. P., "A Coupled Helicopter Rotor/Fuselage Aeroelastic Response Model for ACSR," *Proceedings of the 36th Structures, Structural Dynamics and Materials Conference and Adaptive Structures Forum,* New Orleans, April 1995.
12. Yeo, H., Chopra, I., "Coupled Rotor/Fuselage Vibration Analysis Using Detailed 3-D Airframe Models," Mathematical and Computer Modelling 33 (2001) pp. 1035-1054.
13. Zori, L. A. J., Mathur, S. R., Rajagopalan, R. G., "Three Dimensional Calculations of Rotor-Airframe Interaction in Forward Flight," american Helicopter Society 52nd Annual Forum Proceedings, Washington D.C., June 1992.
14. Bettschart, N., "Rotor Fuselage Interaction : Euler and Navier-Stokes Computations with an Actuator Disk," American Helicopter Society 55th Annual Forum Proceedings, Montreal, Canada, May 1999.
15. Wachspress, D. A., Quackenbush, T. R., Boschitsch, A. H., "Rotorcraft Interactional Aerodynamics with Fast Vortex/Fast Panel Methods," Journal of the American Helicopter Society, Vol. 48, No. 4, October 2003.

MATEC Web of Conferences 77, 01017 (2016) DOI: 10.1051/matecconf/20167701017
ICMMR 2016

Rotor Trajectories in Fluid-Film Bearings, Adjustment of Them and Energy Efficiency Parameters

Leonid Savin , Sergey Mayorov, Denis Shutin and Alexander Babin

Oryol State University, Department of Mechatronics and International Engineering, 302020 Naugorskoe shosse 29,Oryol, Russia

Abstract. The problem of rotor machines modification requires development of new approaches to the design of specific nodes. As bearings are considered the most important elements of rotor machines, such approaches can imply the application of active control. In the present paper using the example of fluid-film bearings the approach is considered to enchancing energy efficiency of rotor machines by means of generating optimal trajectories. Means are shown to analyze the dynamics of a rotor system and types of the obtained trajectories. The paper also covers the approach to setting the criteria of energy efficiency of rotor's trajectories. After the optimal trajectory is obtained, active bearings are proved feasible to control the displacement of a rotor by generating a force to match the actual trajectory to the calculated one, and by this to decrease the energy loss.

1 Introduction

Increasing competitiveness of a machine is closely linked to the necessity of increasing their performance and decreasing size and cost at the same time. In terms of transport or technological turbomachinery, e.g. compressors, pumps, expanders, this fact requires increasing the rotational frequencies of rotors [1]. Under the conditions of high revolution frequencies and dynamic loads on the most vital parts of the machine, it seems only perspective to set the rotor on the fluid-film bearings [1-5]. Here the provision on vibrational reliability is by all means one of the main operability criteria, as it defines both the stress acting in the rotor and the dynamic load on the supporting nodes, as well as the load transferred to the housing of the machine.

2 Nonlinear rotor dynamics problems

The following factors play a significant role in operation of a high-speed turbomachine and influence its performance: mass and stiffness distribution, damping properties of the rotor's elements, location of the bearings, seals and dampers; type, design and operational conditions of the bearings, seals and dampers; mass and inertia characteristics of the housing of a machine; overall operational conditions; etc. It has to be noted, that the presence of the fluid-film bearings makes the dynamic analysis a complex task due to the following facts: 1) these bearings are not standard elements, and for every design and specific conditions an adequate model has to be developed; 2) can trigger self-excited, parametric and chaotic oscillations [2-4]; 3) are dependent on the combination of operational and geometric parameters in terms of their dynamic characteristics [2-6].

It is due to that, that it is reasonable to search for the rational parameters of the rotor system during the stage of project calculations. Given the present requirements the problem is almost impossible to solve without using the modern CAD/CAE systems. In engineering practice of developing real machines there is basically no need in full modeling of the rotor system's dynamics. Instead, the following basic problems are solved [6-7]: 1) steady-state analysis; 2) determination of the natural frequencies and oscillations' shapes (modal analysis); 3) frequency response determination within the range of operational frequencies (harmonic analysis); 4) transient processes analysis, which allows modeling of the rotor's displacement caused by the action of external forces, e.g. imbalance, impulse loads (blade avulsion, various hit-like disturbances).

The motion of the rotor can be considered as of a rigid body or as of a linear elastic body. In the first approach generally one would have to solve a system of non-linear differential equations, in the second case however the task is complicated and the system now consists of differential equations in partial derivatives. Presently to solve the problems of second kind the methods of special discretization are used, e.g. finite element method, etc. [7-9]. So, when analyzing a rotor system, given the according boundary and initial conditions, the motion equations can be expressed as a system of differential equations as follows:

$$[M]\frac{d^2}{dt^2}\{Q\}+[B]\frac{d}{dt}\{Q\}+[K]\{Q\}=\{F\} \qquad (1)$$

where $\{Q\}$ – displacement vector; $[M]$, $[B]$, $[K]$ –mass, damping and stiffness matrixes of a system 'rotor – fluid-film bearing', which depend in a general case on displacement and time, and are defined by a number of chosen elements; $\{F\}$ – load vector.

The equation (1) can be considered as a base for all the previously mentioned types of analysis, based on which a specific task is formulated.

The boundary conditions of a model (1) are represented by external loads and reactions of the fluid film. External forces usually are known or can be statistically evaluated. Determination of the reactions of the fluid film requires the solution of the hydrodynamic equations given the cinematic parameters of the rotor [3, 6, 7]. However, as it is shown in [2 – 4, 6, 7, 10] given the high speeds of rotors' rotation and small values of operational eccentricity, a linear representation of the fluid film reactions can be used.

The dynamic model of a rotor-bearing node given the condition of applicability of a linear approach can be represented as an oscillator on a set of springs and dampers (figure 1a).

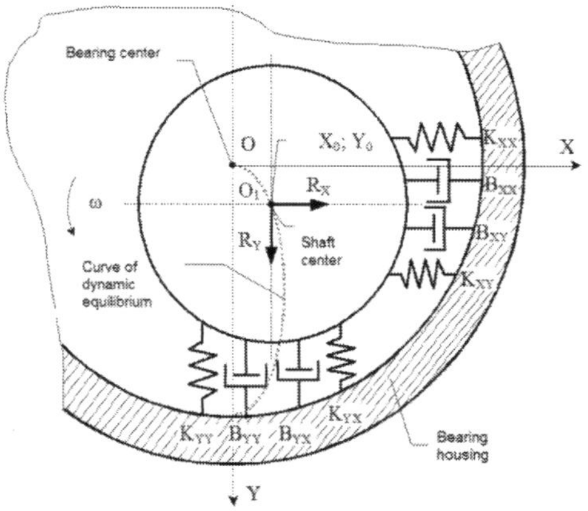

a)

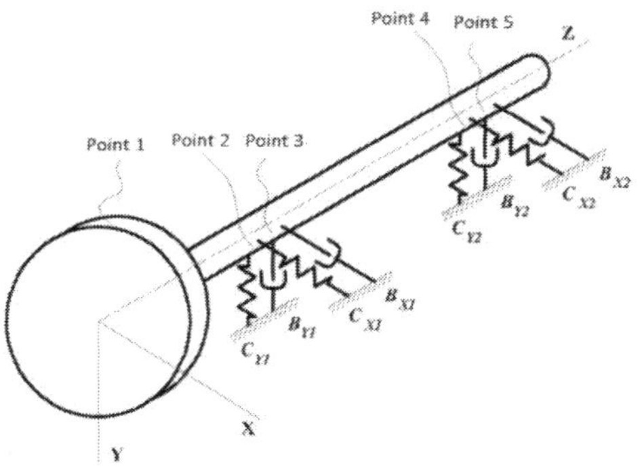

b)

Figure 1. Dynamic model of a bearing (a) and a rotor (b)

In this case the expressions for the reactions of the fluid film will have a following form, given the small change of coordinates and velocities:

$$R_X = R_{X0} - K_{XX}\Delta X - K_{XY}\Delta Y - B_{XX}\Delta\dot{X} - B_{XY}\Delta\dot{Y}; \quad (2)$$

$$R_Y = R_{Y0} - K_{YX}\Delta X - K_{YY}\Delta Y - B_{YX}\Delta\dot{X} - B_{YY}\Delta\dot{Y}, \quad (3)$$

where the stiffness and damping coefficients are as follows:

$$K_{XX} = -\left(\frac{\partial R_X}{\partial X}\right)_0; \quad K_{XY} = -\left(\frac{\partial R_X}{\partial Y}\right)_0; \quad K_{YX} = -\left(\frac{\partial R_Y}{\partial X}\right)_0;$$

$$K_{YY} = -\left(\frac{\partial R_Y}{\partial Y}\right)_0; \quad B_{XX} = -\left(\frac{\partial R_X}{\partial \dot{X}}\right)_0; \quad B_{XY} = -\left(\frac{\partial R_X}{\partial \dot{Y}}\right)_0;$$

$$B_{YX} = -\left(\frac{\partial R_Y}{\partial \dot{X}}\right)_0; \quad B_{YY} = -\left(\frac{\partial R_Y}{\partial \dot{Y}}\right)_0 \quad (4)$$

To solve the practical application problems of dynamic analysis it is convenient to consider the reaction forces of a bearing in the following form (figure 1b):

$$R_X = R_{X0} - C_X\Delta X - B_X\Delta\dot{X}; \quad (5)$$

$$R_Y = R_{Y0} - C_Y\Delta Y - B_Y\Delta\dot{Y}, \quad (6)$$

where the corresponding dynamic coefficients are:

$$C_X = K_{XX} + K_{XY}; \; B_X = B_{XX} + B_{XY}; \; C_Y = K_{YX} + K_{YY};$$

$$B_Y = B_{YX} + B_{YY}. \quad (7)$$

The dynamic coefficients calculation of the fluid film is an independent task. The determination of these coefficients for a specific bearing given the operational conditions in all cases but the simplest requires the development of a relatively complex models of a bearing itself, usually based on the Reynolds equation, and requires the application of numerical methods [3, 6, 11].

Methods of numerical calculation of the dynamic coefficients for various types of bearings under the conditions of turbulent non-isothermal flow of a compressible lubricant along with some calculation results are covered in high detail in [12].

So, in order to solve practical problems, a rotor on fluid-film bearings can be considered rigid deformable body supported by a system of strings and dampers. Of particular interest here is the determination of the behavior of some character points, namely: 1 – radial load point of application; 2 and 3, 4 and 5 – points where the rotor is supported in the corresponding directions.

3 Types of trajectories

Trajectory method, based on the analysis of unstable processes, is an effective means of studying the influence of non-linearity of the lubricant's film's reaction and allows modeling real dynamic behavior of a rotor. The method is based on a joint numerical integration of the hydrodynamic system of equations and the equations of rotor motion. By this one can trace the influence of some specific phenomena in the fluid-film (boiling and double-phase condition) on the rotor's dynamics.

The trajectories of a rotor are the locus which determines the location of the shaft's center, the motion of which is caused by some external forces and the reaction forces of the fluid film, in a specific moment in time. The trajectory is a phase portrait in generalized coordinates in the shaft's center's plain. This method allows studying the dynamics of a rotor with any values of eccentricity.

The type and sizes of a trajectory of a rotor are dependent on the type and character of the load and geometric and operational parameters of the system. Generally, the values and the directions of the external forces acting on the rotor change in time.

Depending on the character of the acting forces, the following types of rotor trajectories and stability conditions of a rotor system can be highlighted:

1) point-stable (focus) (figure 2e). Such trajectory occurs when a balanced rotor moves towards the equilibrium locus curve;

2) orbit-stable (stable limit cycle), when a rotor orbits along an elliptic trajectory (figure 2d). This type of trajectories is an attribute of an unbalance rotor, which performs forced oscillations caused by the centrifugal forces; the shape of the ellipse is defined by the ratio between the inertia and imbalance forces;

3) «odd shape» – unstable-limited condition, which is characterized by an open, evolving in a limited area of the radial gap trajectory of complex shape; evidence of the presence of self-excited oscillations caused by non-linear properties of the fluid film; rotor's workability persists;

4) closed curves of complex shape, which look like cardioid (figure 2a), various epicycloids (figure 2b and 2c), limaçon of Pascal (figure 2d), inclined '8' (figure 2g), etc. Here rotor is exposed to precession frequent self-oscillations and overlapping them synchronous oscillations due to imbalance (bi-harmonical oscillations);

5) unstable-unlimited condition where rotor operates in an unstable region, and the trajectory is an unfolding spiral seeking the borders of the radial gap (figure 2h).

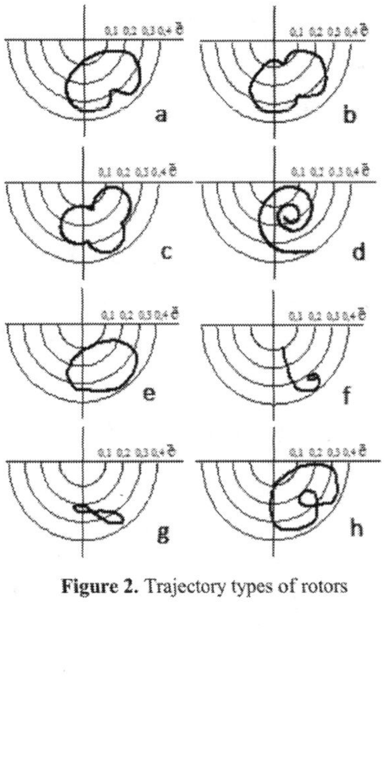

Figure 2. Trajectory types of rotors

So, the trajectory method allows evaluating the stability of the 'rotor–fluid-film bearing' system on the basis of the following assumptions:

1) the trajectory of the center of the shaft of a balanced rotor starting from some initial point ends with either a focus or a limit cycle;

2) if the motion of the rotor is unstable, the trajectory of the center of the shaft either moves monotonically from the initial position and eventually reaches the borders of the radial gap, or steadily fills some area not reaching the borders of the gap.

With that in mind, the conclusion on whether the system is stable or not is made considering the shape and the location of the trajectory.

4 Enegry Efficiency Parameters

As it has been previously said, the basic trend in rotor dynamics is parametric optimization, to which a big number of papers lately are dedicated. The main controversial problem of parametric optimization is the determination of the quality criteria for a specific rotor system. In the present study the criteria is suggested which is based on the effective use of the supplied energy.

The main function of any rotor system is the energy transfer from the drive to the execution device, and it is obvious that this energy characteristic can be expressed as kinetic energy of rotational motion:

$$T_{eff} = \frac{J\omega^2}{2} \tag{8}$$

where J –inertia momentum with respect to the rotation axis; ω – rotor's angular speed.

During the revolution of the rotor in fluid-film bearings the dissipation of energy will have two distinctive components: 1) dissipation of energy due to revolution, in other words, the action of viscous forces along the angular displacement A_{dis}^{rot} ; 2) dissipation of energy due to oscillations, in other words, the action of the forces on trajectories considered above of the rotor A_{dis}^{disp} ; and these values can be evaluated as follows:

$$A_{dis}^{rot} = \int_0^{\varphi^*} M_{fr} d\varphi \tag{9}$$

$$A_{dis}^{disp} = \oint_s \vec{R} \cdot d\vec{s} \tag{10}$$

where M_{fr} –friction momentum of a fluid-film bearing; $\vec{R}$ –reaction force vector in afluid-film bearing; $d\vec{s}$ – singular vector tangent to the trajectory (s) of shaft's center; φ^* – angle of a rotor displacement during the time on trajectory.

So, for every closed trajectory the following energy parameter can be introduced:

$$E_{eff} = \frac{A^{rot}_{dis} + A^{disp}_{dis}}{\int_0^{t^*} T_{eff}\,dt} \quad (11)$$

where t^* – time long enough to express the maximum trajectory duration.

It has to be noted, that in order to universalize the suggested criteria, it needs to be divided by the number of revolutions the rotor makes to cover the closed trajectory, so the following form is obtained:

$$E^*_{eff} = \frac{\left(A^{rot}_{dis} + A^{disp}_{dis}\right)}{\omega t^* \int_0^{t^*} T_{eff}\,dt} \quad (12)$$

The parametric optimization with a suggested criteria is indeed a laborious task in terms of calculation. One of the alternatives to it is active control of the fluid-film bearing's parameters. The variable parameters are here changed so that the trajectory with the lowest E^*_{eff} is obtained from a number of possible.

4 Rotor Trajectories Correction

The problem of optimal trajectory realization in fluid-film bearings is a natural case of a problem of trajectory determination. Such problem implies the necessity of a generated and adjustable force to act on the rotor, which would correct the rotor's location once the actual trajectory differs from the calculated. To implement this various types of modified fluid-film bearings can be used with various principles of rotor's location control. A general approach, however, remains and is in the necessity of influencing the pressure distribution in the fluid film so, that the reaction forces moved the rotor in a desired direction. The following ways of rotor's location control can be highlighted:

1) by means of adjusting the value and the configuration of the gap between a rotor and a bearing, e.g. tilting-pad bearings [13];

2) by means of adjusting the rheological properties of a lubricant in some areas of a bearing, e.g. magnetorheological lubrication [14];

3) by means of changing supply pressure to the area of friction, e.g. active hydrostatic bearings or hybrid bearings. In [15] a more detailed information and features of these bearings can be found. These bearings are relatively the simplest of all listed above. An active hybrid bearing is a traditional bearing with devices of supply pressure control. Modeling approach and the implementation of control in such bearings can be found in [16].

The modeling results of the described system are shown in the figure 3. Here a rotor system was modelled on hybrid bearings and rotor was exposed to complex load – during the operation the imbalance and radial force changed.

As it is clear from the calculation results, active hybrid bearings indeed allow implementation of the rotor's location control in fluid-film bearings. In the case above application of an active hybrid bearing resulted in decreasing the amplitude of rotor's oscillations, whereas in a traditional bearing at some point, due to the centrifugal forces and a constant radial force, the fractional frequency oscillations occurred. In the active hybrid bearing the oscillations remained synchronous.

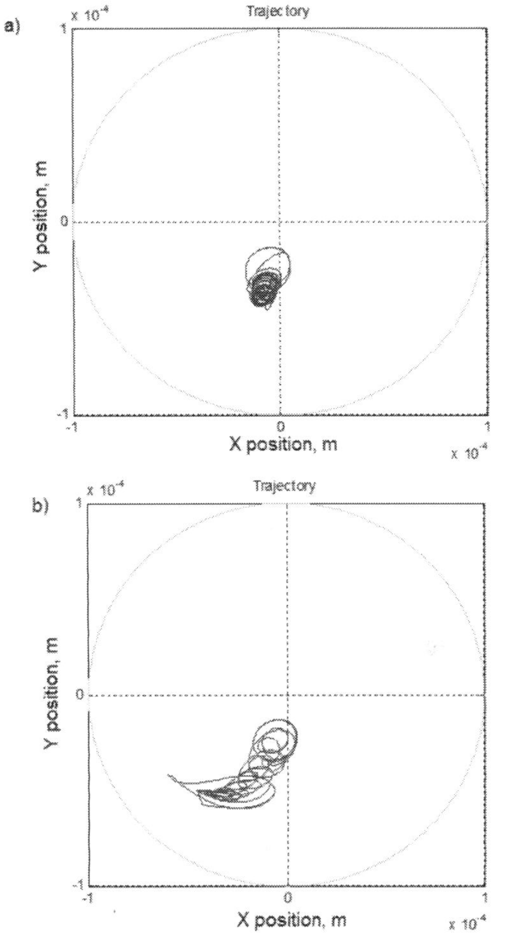

Figure 3. Trajectory types of rotors

5 Conclusion

The problem of rotor machines' modernization is in a close relation with development of new bearings. In this field the most perspective appear to be active fluid-film bearings. Given a relative simplicity in terms of design, they however allow actively adjusting rotor's displacement in a bearing in order to provide the most energy efficient operation. First of all, this is provided by means of lowering the friction losses, and here the determination of energy efficient trajectories is a serious calculation task. So, by combining the apparatus of optimal trajectories calculation with the principles of rotor's location control, it is now possible to develop bearings of new generation, which excel in terms of energy efficiency in comparison to today's solutions.

Acknowledgements

The research was implemented in the framework of the Russian Scientific Foundation project № 16-19-00186.

References

1. E. Logan, *Handbook of Turbomachinery* (Marcel Dekker, 1995)
2. Yu. Ravikovich, Constructions and *Design of Sliding Bearings for Aircraft Engines* (MAI 1995)
3. N. Arttemenko, *Hydrostatic Bearings of High-Speed Rotating Machinery* (Osnova 1992)
4. V. Lukanenko, *Oscillations of High-Speed Rotors on Hydrostatic Bearings and Methods of Reduction of Machines Vibration Activity* (Samara SC RAS 2001)
5. V. Maksimov, G. Batkis, *Tribology of Fliud-Film Bearings and Seals in High-Speed Machinery*. (FEN 1998)
6. M. Adams, *Rotating Machinery Vibration: From Analysis to Troubleshooting* (Marcel Dekker, 2001)
7. T. Yamamoto, Y. Ishida, *Linear and Nonlinear Rotordynamics. A Modern Treatment with Applications* (John Willey&Sons, 2001)
8. O. Zienkiewich, R. Taylor, *The Finite Element Method*, **1**, (Butterworth-Heinemann, 2000)
9. K. Bate, E. Wilson, *Numerical Methods of Analysis and FEM* (Stroyizdat, 1982)
10. O. Solomin, Dynamic Characteristics of Hybrid Bearings at the Conditions of Biphasic Lubricant State, *High School Proceedings – Mechanical Engineering*, **1**, 14–23 (2006)
11. A. Pugachev, M. Deckner. Shape Optimization of a Labyrinth Seal: Lleakage Minimization and Sensitivity of Rotordynamic Coefficients, *IMechE Ninth International Conference on Vibrations in RotatingMachinery*, 849–859
12. S.K. Guha, On the Steady-State Performance of Hydrodynamic Flexible Journal Bearings of Finite Width Lubricated By Ferro Fluids with Micro-Polar Effect, *International Journal of Mechanical Engineering and Robotics Research*, 1, 32-49 (2012)
13. D. Deckler, R. Veillette, M. Braun, F. Choy, Simulation and Control of an Active Tilting-Pad Journal Bearing, *STLE Tribology Trans*, **47**, 440-458 (2000)
14. J. Hesselbach, C. Abel-Keilhack, Active Hydrostatic Bearing with Magnetorheological Fluid, *Journal of Applied Physics*, **93**, 8441-8443, (2003)
15. L. Savin, D. Shutin, Control of Rotor Motion Using Active Fluid-Film Bearings, */Proceedings of the 9th IFToMM International Conference on Rotor Dynamics. Volume 21 of the series Mechanisms and Machine Science,* 1115-1122 (2015)
16. D. Shutin, R. Polyakov, L. Savin, A. Babin, Peculiarities of Reactions Control for Rotor Positioning in an Active Journal Hybrid Bearing, *International Journal of Mechanics*, **10**, 62-67 (2003)

 DOI: 10.1051/matecconf/20167701018

Design of an Electric Road-Rail Vehicle

Xu shun[1], Hou xiaobao[1], Bing Chenyang[1], Ma Zetao[2] and Cui Shumei[2]

[1]*Engineering and Technology Department, CRRC, Qingdao, China.*
[2]*Institute of Electromagnetic and Electronic Technology, Harbin Institute of Technology, Harbin, China.*

Abstract. In this paper, an electric road-rail vehicle and its control system are designed. The electric vehicle equipped a train couplers in front of it is used as a movable electric signal test bench for China Rail High-speed (CRH). Firstly, architecture of the vehicle is introduced. In the powertrain, hydraulic system is designed for the transition from road mode to rail mode, and vice versa. And in both road and rail modes, the vehicle is driven by an induction machine. Secondly, to improve the driving performance and satisfy different driving conditions, particular control strategy for the induction machine is put forward by using a load torque observer based on sliding mode control theory. The load observer based induction machine control system can improve the speed stability and robustness to the unexpected load disturbing. Finally, experiment is done and the results verify that the algorithm works well.

1 Introduction

Road-rail vehicle [1,2] is used as a train-trailer or for freight transport on rail. According to its different driving requirement, the vehicle should be able to run on the road as well as the rail. Because of the environmental pollution and the energy crisis, nowadays manufacturers and governments pay more and more attention to clean energy vehicles [3,4,5].As the development of the energy storage systems, long distance and heavy load condition is now not a problem anymore for electric vehicle. Due to the significant advantage of electric vehicle, in this paper, we propose a kind of electric road-rail vehicle which uses both hydraulic system and pure electric drive system. The electric road-rail vehicle is used as a test bed in workshop for automatic test of train couplers electric signals. The electrification leads to no pollution and noise, which would significantly improve the comfort of the operating environment.

In the paper, the architecture and control structure of the road-rail electric vehicle is presented and the control system is designed. The hydraulic system for the mode transition is introduced. Then the electric driving system and the control strategy is designed. To improve the driving performance, a load torque observer based on sliding mode control is design to realize robust speed control of the induction machine driving system. Simulation is done and the results show good performance of the control method.

2 Architecture of the proposed electric Road-rail vehicle

The electric road-rail vehicle is designed as a movable test bed for CRH2 train couplers. It is used to test the electrical signals of the train couplers for fault diagnosis. To efficiently fulfill the experiments, which means less operators and faster operation, the vehicle should be designed properly. Firstly, the vehicle should be driven as a normal vehicle on the road.Secondly, when testing the train couplers, the vehicle should be driven on the rail to couple with the under testing coupler on the train, which requires the speed of the vehicle should be robustly controlled. Furthermore, when running on the rail, the operators should be able to drive the vehicle remotely, which means the vehicle should be controlled by using a remote control.

Figure 1. The electric road-rail vehicle

As shown in Figure 1, the specific designed electric vehicle equipped a train coupler in front and the control cabinet in rear. The control cabinet includes a test control system and a driving system. The test control system is used for the coupler signals testing and the driving control system is used for driving control.

2.1 Architecture of the proposed electric road-rail vehicle

In the electric road-rail vehicle, both hydraulic system and electric driving system is required to satisfy different driving mode. The hydraulic system is designed to realize the mode transition. For example, when the vehicle is required to run on the rail, the guide wheels should be put down to engage the rail for direction guidance, and vice versa. The vehicle is propelled by an induction machine, which is driven by vector controlled inverter. The Architecture is shown in Fig. 3.

Figure 2. The test control system for train couplers

In the electric road-rail vehicle, the induction machine is used to propel the vehicle both in road mode and rail mode. PLC is used as the center controller and receives instruction from the driver. Guide wheels are used to guide the vehicle along the rail when the vehicle runs on the rail. When the guide wheels are put down, the rear of the vehicle is lifted up to disable the steering system.

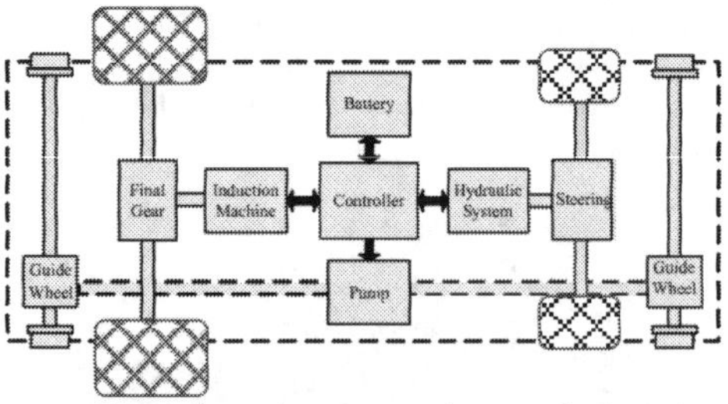

Figure 3. Architecture of the electric road-rail vehicle

Use a two-column format, and set the spacing between the columns at 8 mm. Insert "MATEC Web of Conferences" in even headers and the short form title of the conference in odd headers, except the first one. Please check with the organiser the exact short title of the conference. Do not add any page numbers. The articles will be quoted as follows:

2.2 Hydraulic system design of the road-rail vehicle

The hydraulic system is designed for the steering system and the guide wheels control system.

In the guide wheel control system, electromagnetic valves are used to realize the lifting and descending of the guide wheels. When the vehicle runs on the rail, only the guide wheels in rear and the tires in front brace the vehicle, while the guide wheels in front only guides the vehicle and the tires in rear is lifted away from the ground. This simplified designation guarantees that when the vehicle runs on the rail, the steering system is disabled, which means the vehicle is restricted along the rail direction reliably.

The highest pressure of the hydraulic system is 17Mpa which is enough for the action of the guide wheels in rear.

Besides, the hydraulic system makes sure that the vehicle is able to turn around in nearly 360 degree for free turning in the factory.

2.3 Induction machine drive system

In the electric road-rail vehicle, the induction machine is used. In different driving mode, the induction machine should be controlled in different ways[6,7,8].

When the vehicle runs on the road, the center controller receives torque command from the driver, which means the induction machine should be defined as torque controlled. In the torque control mode, the vehicle can be driven like it is a traditional vehicle. While when the vehicle runs on the rail, in some occasion, high performance of speed control is required for specific usage. For example, when the vehicle is required to be remotely controlled and located precisely. Thus, in the induction machine drive system, torque control mode and speed control mode is designed respectively to satisfy different driving acquirement.

In both torque control mode and speed control mode, the induction machine is controlled by a voltage fed inverter, applying vector control algorithm. While in torque control mode, the speed closed loop is canceled.

The diagrams of the two control strategies are shown in Fig. 4 and Fig. 5.

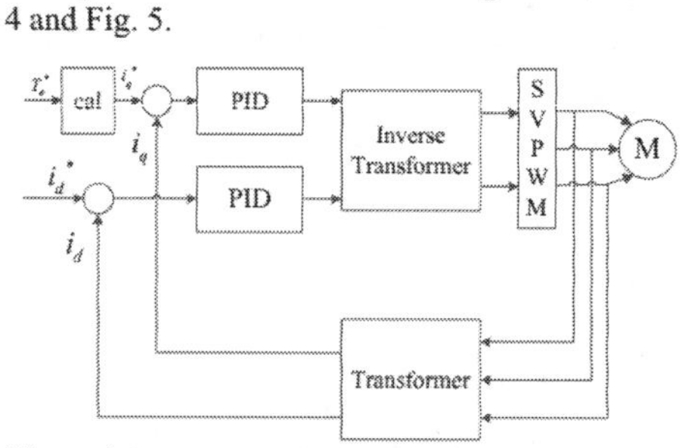

Figure 4. Torque control diagram of the induction machine

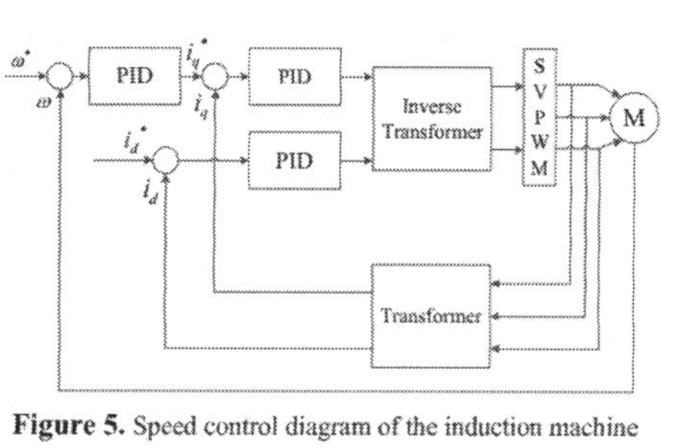

Figure 5. Speed control diagram of the induction machine

In the torque control mode, only current closed loop is required to realize torque control. While in speed control mode, the controller should include both current closed loop and speed closed loop. Due to the particular driving condition of the vehicle, field weakening is not an important point of the control strategy of the electric machine. Thus, no field weakening control strategy is discussed in this paper.

3 Load Torque Observer based on sliding mode theory

When the electric road-rail vehicle runs on the rail, high performance of speed control is required for particular application. To improve the robust performance of the speed control, load torque feed forward control strategy is used in the IM drive system.

3.1 Induction machine model

The induction machine model in the dq frame can be described as:

$$\begin{bmatrix} u_{ds} \\ u_{ds} \\ 0 \\ 0 \end{bmatrix} = \begin{bmatrix} R_s + pL_s & -\omega_s L_s & pL_m & -\omega_s L_m \\ \omega_s L_s & R_s + pL_s & \omega_s L_m & pL_m \\ pL_m & -\Delta\omega_r L_m & R_r + pL_r & -\Delta\omega_r L_r \\ -\Delta\omega_r L_m & pL_m & \Delta\omega_r L_r & R_r + pL_r \end{bmatrix} \begin{bmatrix} i_{ds} \\ i_{qs} \\ i_{dr} \\ i_{qr} \end{bmatrix} \quad (1)$$

Where u_{ds} and u_{ds} are the stator voltage in dq frame respectively and i_{ds} i_{qs} are the stator current respectively. R_s is the stator resistance and R_r is the rotor resistance. L_s is the stator inductance and L_r is the rotor induction. L_m is the mutual inductance.

Then, the electro-magnetic torque of the IM machine can be expressed as:

$$T_e = p_m L_m (i_{ds} i_{qr} - i_{qs} i_{dr}) \quad (2)$$

Where T_e is the electro-magnetic torque and p_m is the pole-pair number.

When implement vector control strategy, the inductance machine speed control system is shown in figure5. Current and speed information is used for double closed loop control to improve the responsibility and control performance of the driving system.

3.2 Load observer based on sliding mode theory

Sliding mode control has been applied in induction machine and PMSM machine control widely. For example, different kind of torque observers and sensorless control strategies use sliding mode control theory to improve the control performance of the machine [9-15].

The motion equation of the induction machine can be described as:

$$\frac{d\omega}{dt} = \frac{1}{J} T_e - \frac{1}{J} T_l \quad (3)$$

Where ω is the rotating speed, J is the equivalent inertia and T_l is the load torque.

To apply the sliding mode theory in a load torque observer, the state equation is built:

$$\frac{d\hat{\omega}}{dt} = \frac{1}{J} T_e - k\,\mathrm{sgn}(\hat{\omega} - \omega) \quad (4)$$

Where $\hat{\omega}$ is the estimated rotor speed, and sgn is the signal function.

Combining the equation (3) and equation (4), the speed error is now expressed as:

$$\frac{d\tilde{\omega}}{dt} = \frac{1}{J} T_l - k\,\mathrm{sgn}(\tilde{\omega}) \quad (5)$$

Where $\tilde{\omega}$ is the speed error between the estimated speed and the real rotor speed.

According to equation (5), when the speed error equals to zero, then the load torque can be expressed as:

$$T_l = Jk\,\mathrm{sgn}(\tilde{\omega}) \quad (6)$$

To stabilize the control loop, the Lyapunov stabilization condition is constructed as:

$$(\frac{1}{2} s^2)' = s\dot{s} \le 0 \quad (7)$$

Combining equation (6) and equation(7), the stabilization equation is now described as:

$$\tilde{\omega}\dot{\tilde{\omega}} = \frac{\tilde{\omega}}{J} T_l - k\tilde{\omega}\,\mathrm{sgn}(\tilde{\omega}) \le 0 \quad (8)$$

Thus, the relation below should be satisfied:

$$k \ge \left| \frac{T_l}{J} \right| \quad (9)$$

Furthermore, to acquire smooth load torque signal, a first order filter is used to filter the high frequency chattering of the observed signal:

$$G(s) = \frac{\omega_0}{s + \omega_0} \quad (10)$$

Where ω_0 is the cut-off frequency.

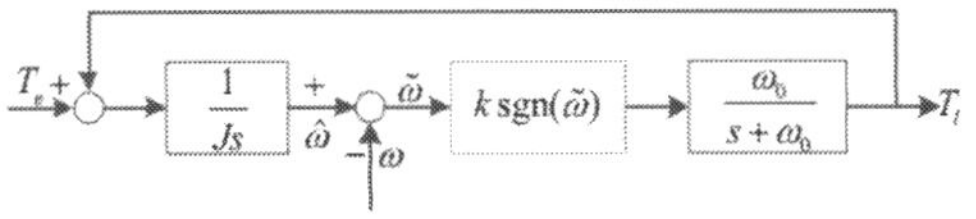

Figure 6 Load torque observer based on sliding mode theory

The diagram of the load torque observer based on sliding mode theory is shown in Figure 6.

3.3 Feed forward Torque Control Strategy

In the electric road-rail vehicle, the vector controlled strategy is investigated. To acquire high performance of speed control, the load torque feed forward control is implemented. The load torque is added into the control loop to improve the responsibility of the control system. When the driving conditions of the vehicle changed, the vehicle speed would change at the same time, the sliding mode observer then calculates the dynamic load torque and feeds to the current closed loop, which means the dynamic tuning process of the current closed loop is significantly cut short thus the responsibility and the robustness of the driving system is improved.

The complete control diagram of the induction machine is shown in fig7.The load torque observer

DOI: 10.1051/matecconf/20167701018

calculates the load torque and then the load torque is fed forward to the control loop to improve the robust performance.

The control strategy is implemented in a digital processor. As shown in figure8 and figure9, the real speed follows the target speed well. In figure9, disturbance is added at 50 second, the experiment results shows that the driving system has a little oscillation. This is because the sliding mode observer has chattering characteristic.

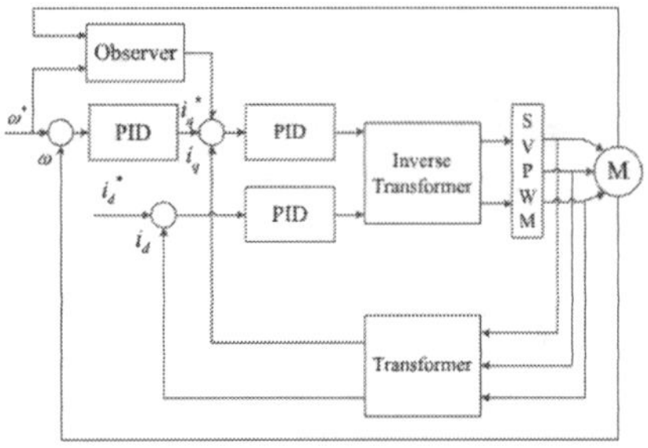

Figure 7. Feed forward speed control diagram of the induction machine

The control strategy is implemented in a digital processor. As shown in Figure 8 and Figure 9, the real speed follows the target speed well. In figure9, disturbance is added at 50 second, the experiment results shows that the driving system has a little oscillation. This is because the sliding mode observer has chattering characteristic.

4 Experimental results

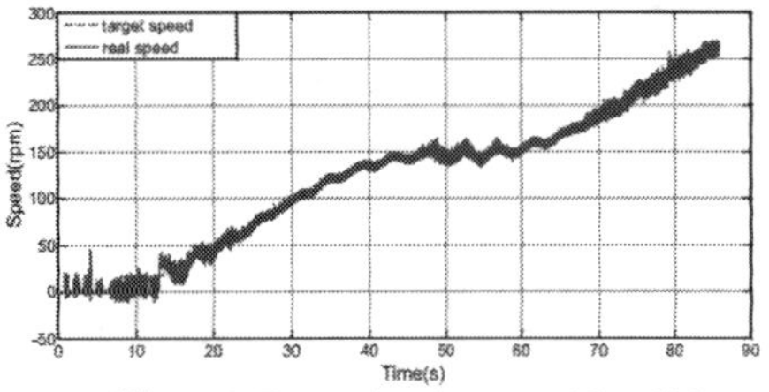

Figure 8. The speed performance of the vehicle

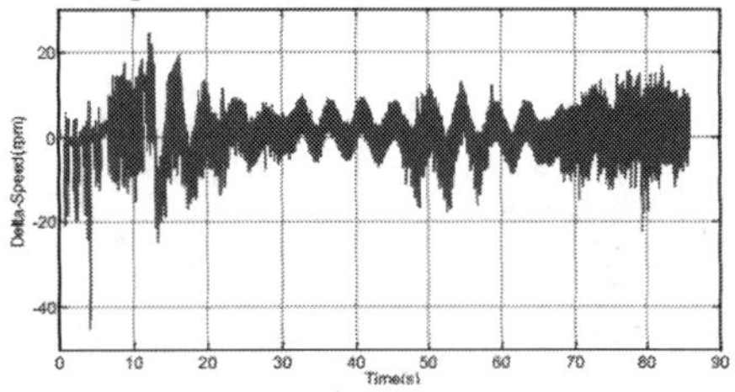

Figure 9. The speed error

5 Conclusion

In this paper, we design the control system for an electrified road-rail vehicle. In the vehicle, for different usage, the hydraulic system is designed to fulfill transition from road vehicle to rail vehicle, and vice versa. In the other hand, different induction machine control strategies are used for specific applications. That is, torque control is used for road driving and speed control is used for rail driving. A sliding mode torque observer is designed to improve the speed control performance of the vehicle. Simulation is done to verify the correctness of the algorithm.

References

1 Pei Lei, "Research on control system of electric road-rail towing vehicle", Harbin Institute of Technology, 2010.
2 Road to Rail. SRS
3 C. C. Chan. The State of the Art Electric and Hybrid Vehicles. Proceedings of The IEEE. 2002, 90(2):247~275
4 C. C. Chan. Electric Vehicles: Charge Forward. IEEE Power and Energy Magazine. 2004:2~12
5 9 Skudelny, Hans Christoph. Modern Directions of Research on Drive Systems for Electric Rail and Road Vehicles. 1980, 71(10):280~285.
6 Peter Vas. Vector control of AC Machines [M]. NewYork: Oxford University Press, 1990.
7 Peter Vas. Sensorless Vector and Direct Torque Control [M]. NewYork: Oxford University Press, 1998.
8 Vas P, Willems J L, Brown J E. The Application of Space-phasor Theory to the Analysis of Electrical Machines with Space Harmonics[J]. Archiv fur Elektrotechnik, 1987, 69:359-363.
9 Zhang Xiaoguang, Sun Li, Zhao Ke, "Sliding Mode Control of PMSM Based on a Novel Load Torque Sliding Mode Observer", Proceedings of the CSEE, 2012
10 Lai C K, Shyu K K. A novel motor drive design forincremental motion system via sliding-mode controlmethod[J]. IEEE Transaction on Industrial Electronics, 2005, 52(2): 499-507.
11 Roy A M C, Mohammad S I. Application of a sliding mode observer for position and speed estimation in switched reluctance motor drives[J]. IEEE Transactions onIndustry Application, 2001, 37(1): 51-58.
12 Tong Kewen, Zhang Xing. Sliding mode variable structure control of permanent magnet synchronous based on a novel reaching law [J]. Proceedings of the CSEE, 2008, 28(21): 102-106(in Chinese).
13 Roy A M C, Mohammad S I. Application of a sliding mode observer for position and speed estimation in switched reluctance motor drives [J]. IEEE Transactions on Industry Application, 2001, 37(1): 51-58.
14 Wai R. Total sliding-mode controller for PM synchronous servo motor drive using recurrent fuzzy neural network [J]. IEEE Transactions on Industrial Electronics, 2001, 48(5): 926-944.
15 Ha Q P, Nguyen Q H, Rye D C, et al. Fuzzy sliding-mode controllers with applications [J]. IEEE Transactions on Industrial Electronics, 2001, 48(1): 38-41.

Analysis and Optimization of Section-Thickness of the Manipulator for GIS Detection

Shuangyan Su[1,2,a], Yong Chang[2] and Yonghe Wei[1]

[1]*College of mechanical engineering, Shenyang Ligong University, 110159 Shenyang Liaoning, China*
[2]*The State Key Laboratory of Robotics, Shenyang Institute of Automation, Chinese Academy of Sciences, 110016 Shenyang Liaoning, China*

Abstract. In order to find out the influence of different section-thickness of the manipulator for GIS detection on the mass, strength and stiffness, the performance of the arm rod with different section thickness is analysed by MATLAB under the condition of a certain section height. Firstly, the influence of the parameters of each section-thickness on the mass, strength and stiffness is analysed by single variable analysis. Then the mathematical model is established, and the section-thickness of the arm is optimized using the method of nonlinear programming. The results are similar to the results of single variable analysis. Finally, the section thickness of the virtual prototype is optimized by the finite element analysis method. The correctness of MATLAB optimization results is verified. The optimal solution of the section-thickness of the GIS-detection-arm is obtained.

1 Introduction

The X-ray digital imaging detection system, which is the detection equipment in electricity substation, has been successfully applied to the detection of GIS [1-2] (gas insulated switchgear) and other equipment. At present, we mainly depend on the way of manual lifting X-ray detection equipment to detect GIS. Because of the large volume and weight of X-ray detection equipment, the layout of the equipment is slow, the labour intensity of workers is great, and the risk in manual detection is high. The efficiency is quite low. Aiming at the above-mentioned problems, the related groups development X-ray non-destructive detection mobile manipulator. The corresponding arm, installed on the mobile platform, lift X-ray detection device for GIS detection. As shown in Figure 1.

According to the requirement of better load capability and light weigh, the influence of different section-thickness on the mass, strength and stiffness of the main linkage of the manipulator, installed on the mobile platform, is analysed. This paper uses three methods to optimize the section-thickness of the arm rod.

The optimization design of the section-thickness of the arm linkage for GIS detection is carried out on the basis of the initial mechanism and structure design. The total weight of manipulator is taken as the objective function and the mechanical performance parameters as considered as constraint condition. The decrease of the manipulator weight is realized, meanwhile the premise that the mechanical properties are within the limits.

Figure1. GIS Detection

The research on the section shape and thickness has started from the end of 1980s. A parameter optimization method based on stiffness sensitivity analysis was proposed for the structural optimization of slender manipulator by Z. Luo and L. Liang et al in [3-4]. This work lays a theoretical foundation for the structural optimization for such manipulators. The structural and the control parameters of the whole arm system are optimized simultaneously using a genetic algorithm and the performance is compared with that of an arm with uniform links and an optimized control system by Y. Zhu, J. Qiu and T. Junji in [5]. The simulation result shows that faster regulation and less weight of the arm system can be achieved by simultaneous optimization. Stiffness model of space manipulator is established by H. Jia et al in [6]. Optimal solution checking formula is derived by Lagrange number multiplication in this paper. The corresponding mechanical properties are verified by changing the thickness of the truss by Y. Yang in [7].

[a] Corresponding author: 18766562369@163.com

According to the results, the analysis of the influence of the height of the cross section on the force is analysed.

For an attempt to resolve the problem in this paper, combined with the above mentioned methods, the three steps optimization principle for the optimization of section thickness of the GIS detection manipulator are proposed.

Step 1 Single variable analysis: Section-thickness value changed in turn, and its influence on the mass, strength and stiffness, get the proper ranges of each section-thickness value is analysed.

Step 2 Nonlinear programming method: Establish the objective function and the constraint conditions are defined firstly. Then the section-thickness is optimized by nonlinear programming method.

Step 3 Finite element analysis: Analyse the virtual prototype by the finite element analysis method is analysed. According to the results, the section-thickness is optimized by genetic algorithm.

Combined with three kinds of optimization results, the optimal solution of section thickness is obtained.

2 Establishment of Mechanical Model

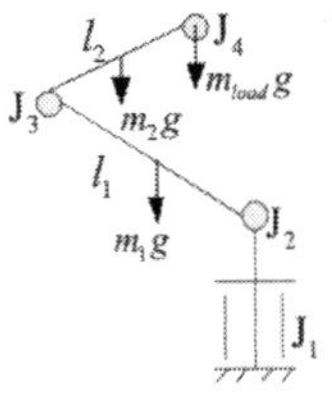

Figure 2. The Design Sketch of Manipulator Mechanism

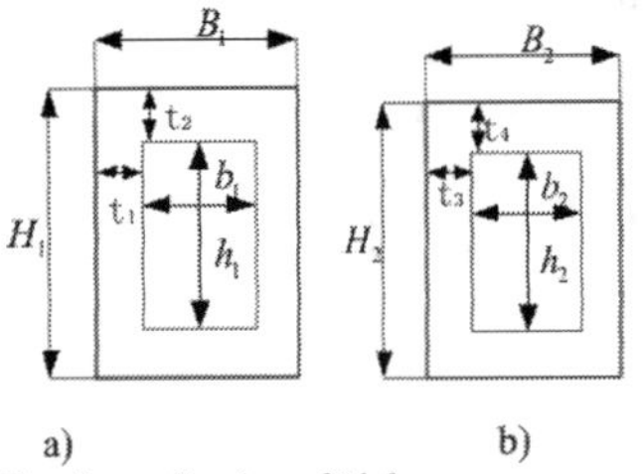

a) b)

Figure 3. The Cross-Section of Linkage

On the basis of the designed GIS detection manipulator, the mathematical model of the mechanics is established. The rod length of the two linkage are $l_1 = 1000\text{mm}$ and $l_2 = 780\text{mm}$, respectively. The load is about 30kg in end. The cross section-height and cross section-width of the first linkage are H_1=150mm and B_1=60mm, respectively. The cross section-height and cross section-width of the second linkage are H_2=120mm and B_2=45mm, respectively. We take Super-duralumin 7075 as linkage material. Considering the mass and the force of the linkage, the safety factor is obtained according to the initial calculation moment, and the continuous torque of the most dangerous joint moment is $450\ \text{N}\cdot\text{m}$.

Figure 2 is the design sketch of manipulator mechanism. According to the actual working process, it is the most dangerous condition when the manipulator is in the fully folded state.

Figure 3 a) and b) are the cross section of the linkage 1 and 2, respectively.

where $H_i(i=1,2)$ is the section-height of the arm rod 1 and 2, $B_i(i=1,2)$ is the section- width of the arm rod 1 and 2, respectively.

The masses of the two main linkage of the manipulator can be given by (1).

$$\begin{cases} m_1 = 0.81t_1 + 0.324t_2 - 0.0108t_1t_2 \\ m_2 = 0.505t_3 + 0.18954t_4 - 0.008t_3t_4 \end{cases} \quad (1)$$

where $m_i(i=1,2)$ is the mass of the arm rod, $t_i(i=1,3)$ is the side thickness, $t_i(i=2,4)$ is the thickness of top and down.

According to strength calculation formula of linkage $\sigma = \dfrac{M}{W_z}$, the normal stress of the two rod can be given by (2).

$$\begin{cases} \sigma_1 = \dfrac{2.7\times10^6}{150\times60^2-(150-2t_2)(60-2t_1)^2} \\ \sigma_2 = \dfrac{2.7\times10^6}{120\times45^2-(120-2t_4)(45-2t_3)^2} \end{cases} \quad (2)$$

where σ_i is the stress of arm rod, ω_i is the deformation of arm rod, M is the moment, and W_z is the unbending section coefficient.

According to the principle of superposition of deflection in [8], the deformations of the two linkages are obtained. The stiffness model is established by (3).

$$\begin{cases} \omega_1 = \dfrac{3.75\times10^7+1.736\times10^5m_1}{150\times60^3-(150-2t_2)(60-2t_1)^3} \\ \omega_2 = \dfrac{2.9\times10^4m_2}{120\times45^3-(120-2t_4)(45-2t_3)^3} \end{cases} \quad (3)$$

3 Single Variable Analysis

3.1. Relationship between section-thickness and mass

Firstly, the influence of the section-thickness parameter of linkage 1 on the mass is analysed by MATLAB. The relationship between thickness (t_1, t_2) and mass (m_1) by (1) is concluded, as shown in Figure 4.

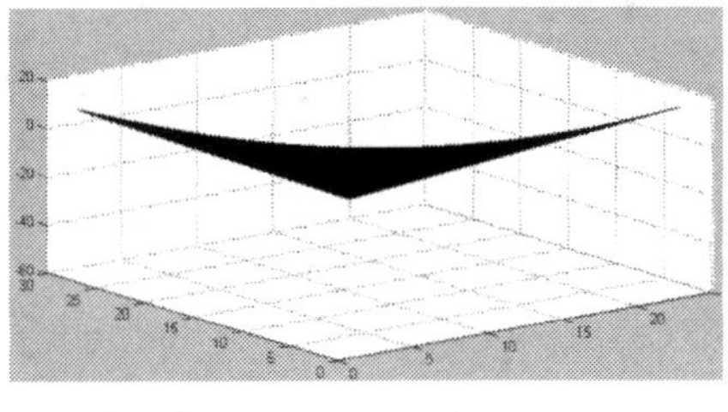

a) The relationship between t_1, t_2 and m_1

DOI: 10.1051/matecconf/20167701019

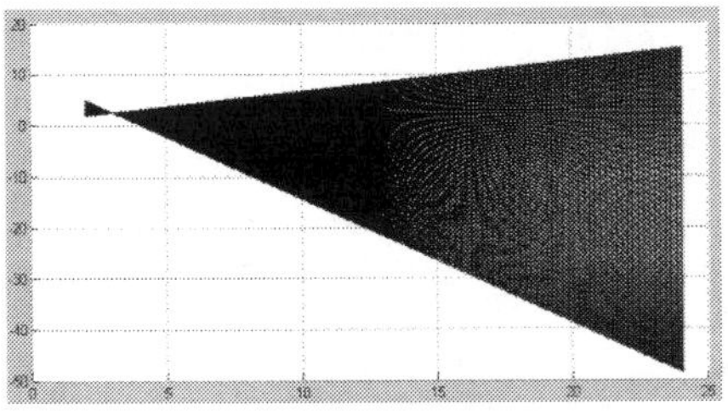

b) The relationship between t_1 and m_1

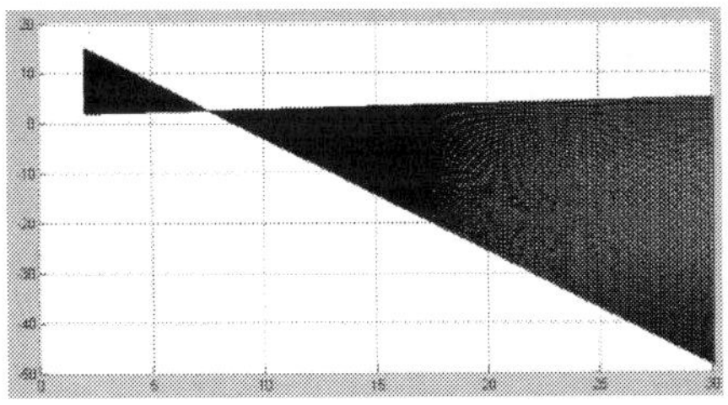

c) The relationship between t_2 and m_1

Figure 4. The Relationship between t_1, t_2 and m_1

From Figure 4, we can sum up that the mass variation range is not large when $t_1 \geq 4$ mm and the mass increases, but the range is not very large when $t_2 \geq 8$ mm.

3.2 Relationship between section-thickness and strength

The following we analyze the effect of section-thickness variation on strength is analysed and the relationship between thickness (t_1, t_2) and strength (σ_1) by (2) is conclued, as shown in Figure5.

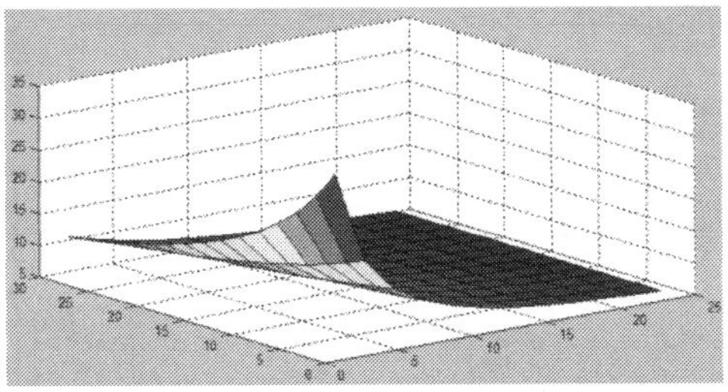

a) The relationship between thickness (t_1, t_2) and strength (σ_1)

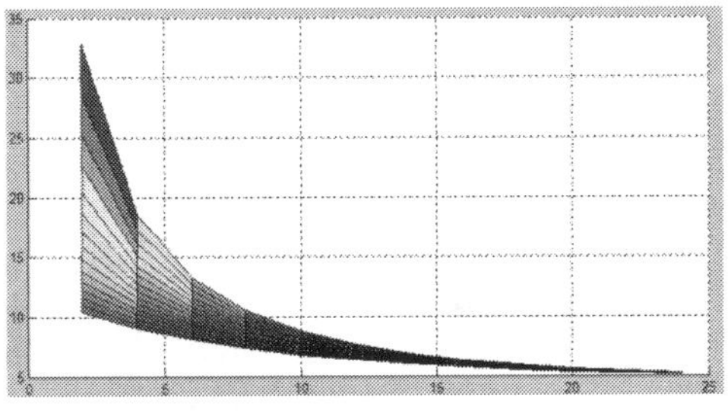

b) The relationship between thickness (t_1) and strength (σ_1)

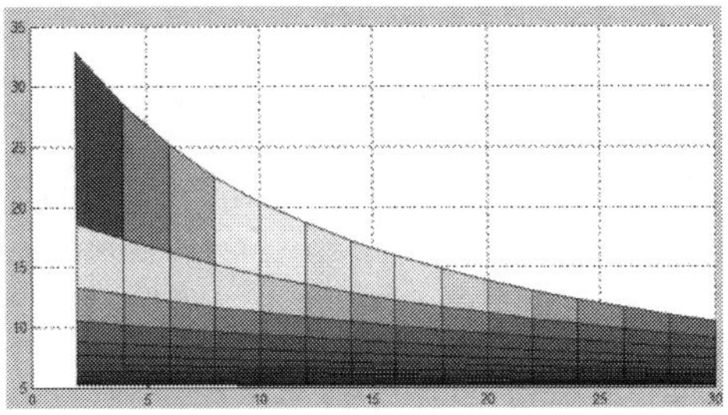

c) The relationship between thickness (t_2) and strength (σ_1)

Figure 5. the Relationship between Thickness (t_1, t_2) and Strength (σ_1)

Aluminum has good characteristics of stress. We can see from Figure 5 that when the thickness $t_1 \geq 2$ mm, $\sigma_1 \leq 33$ MPa and $t_2 \geq 2$ mm, $\sigma_1 \leq 33$ MPa. We can sum up that the thickness can be easily satisfied in terms of stress.

3.3 The relationship between section-thickness and stiffness

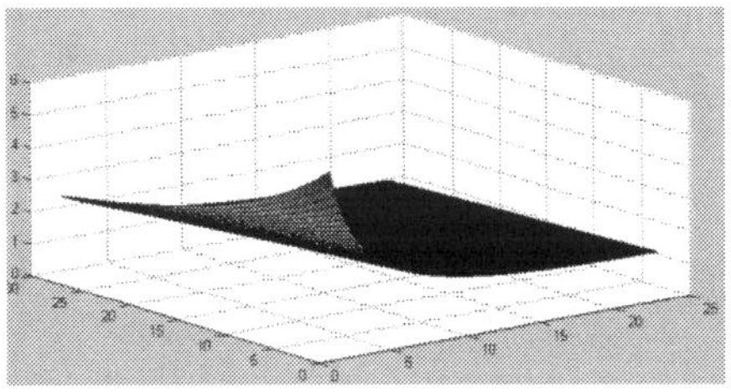

a) The relationship between thickness (t_1, t_2) and stiffness (ω_1)

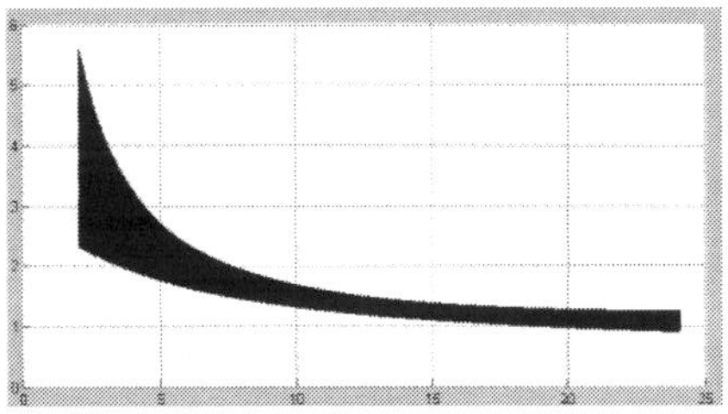

b) The relationship between thickness (t_1) and stiffness (ω_1)

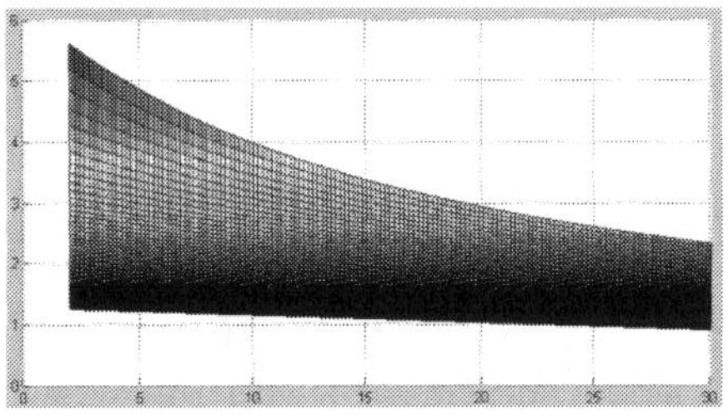

c) The relationship between thickness (t_2) and stiffness (ω_1)

Figure 6. The Relationship between Thickness (t_1, t_2) and Stiffness (ω_1)

DOI: 10.1051/matecconf/20167701019

The effect of section-thickness variation on stiffness is simulated and the relationship between thickness (t_1, t_2) and stiffness (ω_1) by (3) is described, as shown in Figure 6.

We can see from Figure 6 that when the thickness $t_1 \ge 5\,\text{mm}$, $\omega \le 5\,\text{mm}$ and $t_2 \ge 8\,\text{mm}$, $\omega \le 4\,\text{mm}$.

Combined with the above three kinds of analysis results, it can be concluded that the best section-thickness of the linkage 1 is $t_1 = 4\,\text{mm}$ and $t_2 = 8\,\text{mm}$.

Through the above analysis method, in the same way, it can be concluded that the optimum section-thickness of the linkage 2 is $t_3 = 3\,\text{mm}$ and $t_4 = 6\,\text{mm}$.

4 Nonlinear Programming

The effect of the thickness on the mass, strength and stiffness of the single variable is analysed, and the optimal solution is obtained.

According to the constraint conditions and objective function, the mathematical model of nonlinear programming is established [9-10]. The comprehensive analysis of the effects of two thickness variables on mass, strength and stiffness is carried out.

The nonlinear programming model is as follows:

$$\min\ m_i (i=1,2)$$

$$s.t.\begin{cases} \dfrac{M_1}{W_{z1}} = \sigma_1 \le [\sigma] \\ \omega_1 = \dfrac{M_1 l_1^3}{2EI} - \dfrac{m_1 g(l_1/2)}{6EI}\left(3l_1 - \dfrac{l_1}{2}\right) \le [\omega] \\ \dfrac{M_2}{W_{z2}} = \sigma_2 \le [\sigma] \\ \omega_2 = \dfrac{m_2 g(l_2/2)}{6EI}\left(3l_2 - \dfrac{l_2}{2}\right) + \dfrac{m_2 g l_2^2}{3EI} \le [\omega] \\ t_{i,\min} \le t_i \le t_{i,\max} (i=1,3) \\ t_{1,3} \le \dfrac{B_{1,2}}{2} \\ t_{i,\min} \le t_i \le t_{i,\max} (i=2,4) \\ t_{2,4} \le \dfrac{H_{1,2}}{2} \end{cases}$$

where E is elastic modulus of Aluminium alloy and I is moment of inertia of cross section.

Based on the above model, the section-thickness is optimized by the nonlinear programming method using *fmincon* function in MATLAB optimization toolbox. The results of t_1, t_2, t_3, t_4 are as follows:

8.6814 3.5183 5.9878 3.4042

The masses of the two linkages are as follows, respectively.

7.8420 3.5000

The optimization results difference between the nonlinear programming method and single variable analysis is very small.

5 Optimization of Finite Element Analysis

We model according to the designed parameters, as shown in Figure7.

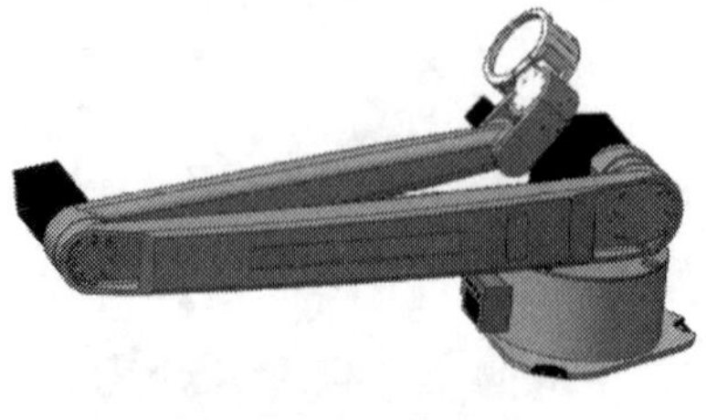

Figure 7. Model of Manipulator

The initial section-thickness values of the linkage 1 are t_1=7mm and t_2=12mm. The mass is 10kg. The results of finite element analysis of the rod 1 are as shown in Figure8:

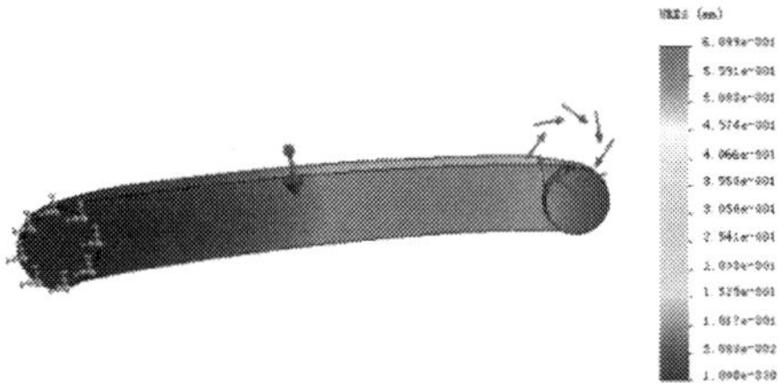

a) The analysis result of displacement

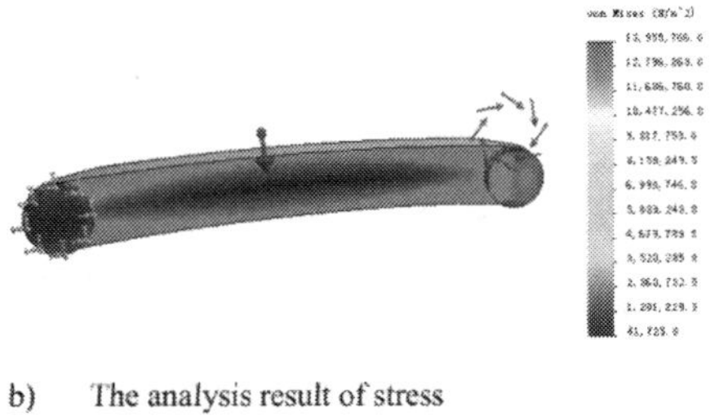

b) The analysis result of stress

Figure 8. The Results of Finite Element Analysis

Using finite element analysis [11] and genetic algorithm, we take the section thickness of t_1 and t_2 as design variables , take the stress and deformation as the constraint condition, and take mass as the optimization goal, to optimize the model. The optimization results of rod 1 and 2 are as shown in Table 1:

Table 1. The optimization results of rod 1 and 2

	Rod 1	Rod 2
$t_{1,3}$	8mm	3mm
$t_{2,4}$	4mm	3.5mm
Stress	21.393MPa	17.349MPa
Deformation	0.88929mm	0.00053mm
Mass	7.035kg	3.011kg

DOI: 10.1051/matecconf/20167701019

6 Conclusions

According to the structure of the typical example of GIS detection manipulator, the mathematical model is established. Analysis and optimization of the mass, stiffness and strength of the manipulator are carried out. The optimal solution is obtained through a single variable analysis. Then, the top and down section thickness and the side thickness of the linkage are optimized by the method of nonlinear programming. Finally, the demension of section is optimized by the finite element method. Through three different analysis optimization methods, the optimal solution of section thickness is obtained.

The optimization results show that the weight of the manipulator decreases obviously according to the decrease of the thickness, and the stress and deformation increases slightly. The work of this paper used as the preliminary research for designation of the manipulator in detail.

References

1. X. Zheng, J. Chen. New techniques for fault detection in GIS [J]. Guangdong electric power, 2001, 14 (5): 43-45.
2. http://baike.baidu.com/link?url=vt08j5CBKBdgz8hbpj8V5yqOF4IvHWKM4VeEdyzfxdLI1JvbRCsRLWmhTCJJrSJ8RWt9uPVxK9CTM6VXj8YweK
3. Z. Luo, L. Liang, Y. Chen, et al. Sensitivity analysis and parameter optimization of the stiffness of slender manipulator [J]. Journal of Northeastern University: Natural Science Edition, 2011, 32 (9): 1319-1323.
4. Shen J, Yoon D. A new scheme for efficient and direct shape optimization of complex structures represented by polygonal meshes[J]. International Journal for Numerical Methods in Engineering, 2003, 58(14):2201–2223.
5. Ye Z, Qiu J, Junji T. Simultaneous optimization of a two-link flexible robot arm[J]. Journal of Robotic Systems, 2001, 18(1):29–38.
6. H. Jia, Q. Yao, Q. Huang. Stiffness optimization of space manipulator based on mass distribution [J]. China Space Science and technology, 2008 (3): 45-52.
7. Y. Yang, X. Liang. Analysis of the influence of the height of the cross section of the arm on the strength of the rod [J]. construction machinery, 1994 (9): 24-27.
8. H. Liu. Advanced material mechanics [M]. higher education, 1985.
9. D. Xue. MATLAB solution to the problem of higher applied mathematics [M]. Tsinghua University press, 2008.
10. J. Chen. MATLAB collection [M]. Electronic Industry Press, 2011.
11. H. Zhang. ANSYS Workbench14.5 numerical simulation engineering case analysis [M]. Machinery Industry Press, 2013.

Estimation of the Appropriate Output of Diesel Generator for Microgrid System in Chuja Island

Seungmin Lee[1], Sang Heon Chae[2], Eel-Hwan Kim[2a], Ho Min Kim[1] and Ngoc-Thinh Quach[2]

[1]*Faculty of Wind Energy Engineering Graduate School, Jeju National University, S. Korea*
[2]*Deparetment of Electrical Engineering, Jeju National University, S. Korea*

Abstract. This paper proposes a method to estimate the appropriate minimum output of diesel generator in the microgrid system that composed of renewable energy, diesel generator and energy storage systems. While maximizing the output of the renewable energy in the microgrid, to maintain the power system stability is very important. To analyze the effectiveness of the proposed method, simulation is carried out by using the PSCAD/EMTDC program with past weather and power load data in Chuja Island located in the south area of Korean peninsula, respectively. In the simulation result, power quality of Chuja Island power system under variable renewable energy generation is stable. The proposed method is useful for operating the isolated microgrid system.

1 Introduction

In order to solve the energy supplying problem for isolated area, the use of microgrid has been promoted all over the world. The microgrid is an integration of wind turbine, photovoltaic (PV), energy storage system (ESS) and diesel generator [1]. The most important objectives of microgrid are minimization of operating diesel generator and grid stability [2]. Therefore the microgrid must be designed to meet these conditions. There are two types of microgrid. One is on-grid type of microgrid, which is connected to main grid. This microgrid is operated by renewable energy sources and conventional source from main grid. In this case, if the load of microgrid is higher than renewable energy sources, the microgrid will receive the demand power from main grid. If the load of microgrid is lower than renewable energy sources, the microgrid will transmit the redundant power to main grid [3]. Another is off-grid type of microgrid, which is isolated system and always operates stand-alone. Because this microgrid cannot share electric power with main grid, the diesel generator can be supported for grid stability by sustaining minimum operation. Therefore elaborate energy management system is essential to this microgrid. Recently, the government of South Korea has a plan named 'Green and energy self-reliance Island', the objective of the plan is to make off-grid type microgrid for isolated islands by using non-pollution energy. In South Korea, two islands named Gapa and Gasa were already installed to off-grid microgrid from this plan [4]. Jeju Special Self-governing Province will make another microgrid system in Chuja Island by 2019. From this point, this paper proposes the estimating method of operation range for diesel generator focused on Chuja Island to minimize air pollution and grid stability. The operation of microgrid system in Chuja Island will be verified by using PSCAD/EMTDC simulation program. PV and wind turbines output power will be calculated by using matlab program.

2 Overview of microgrid system in Chuja Island

Chuja Island is located in the north of Jeju Island in South Korea. Presently, the power grid on this island is only operated by seven diesel generators with the total capacity of 5.5 MW. However, the government has a plan to install microgrid system in Chuja Island by 2019 as illustrated in Fig. 1. In this plan, the capacity of components for microgrid system in Chuja Island is already determined. The components of Chuja Island microgrid consist of PV system with the total capacity of 1.6 MW, ESS with the total capacity of 12 MWh, wind turbines with the total capacity of 2.4 MW. The demand load of this island is estimated about 1.24 - 2.74 MW.

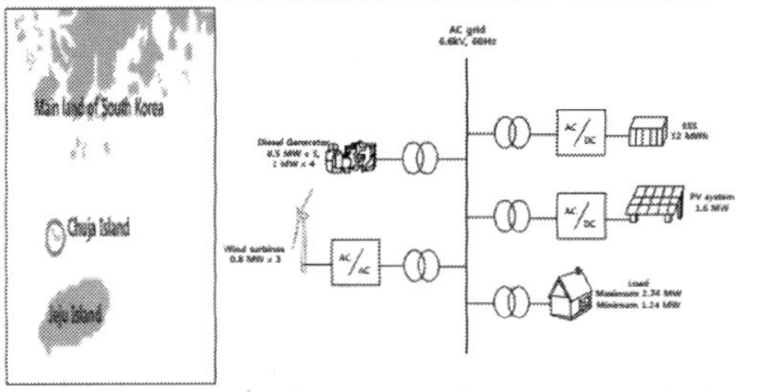

Figure 1. Configuration plan of microgrid system in Chuja Island

[a] Corresponding author: ehkim@jejunu.ac.kr

3 Modeling of microgrid system in Chuja Island

The microgrid system in Chuja Island is designed by PSCAD/EMTDC simulation program [5]. Line impedances of Chuja Island distribution grid will be applied by actual line model. The actual parameters of transformer are also considered for simulation. The distribution voltage for microgrid system in Chuja Island is 6.6 kV.

3.1. Energy storage system

In order to simulate proposed method, the ESS is modelled by shepherd nonlinear battery model referred to SAFT's model as shown in Fig. 2. The capacity of a basic lithium ion battery is 5.4 Ah and the nominal voltage is 4.2 V. Through series and parallel connection of the batteries, its capacity and system nominal voltage is calculated. The state of charge (SOC) of ESS is set from 0.2 to 0.8. Power conversion system (PCS) of ESS plays a role as a DC/AC inverter. The PCS is made from two level voltage source inverter. In this study, it will control for active power and reactive power based on current controller as illustrated in Fig. 3 [6].

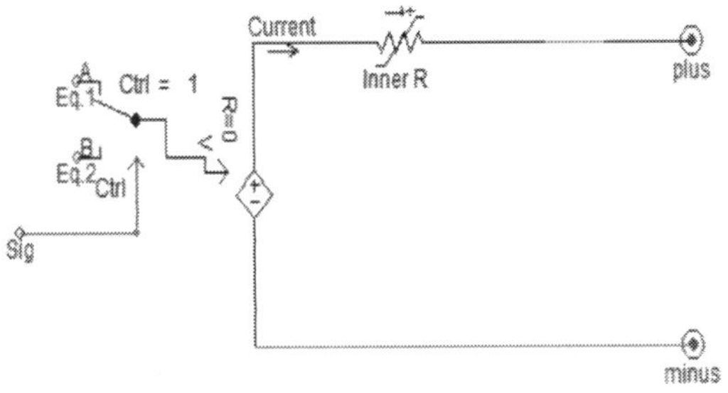

Figure 2. Battery equicalent model in PSCAD/EMTDC

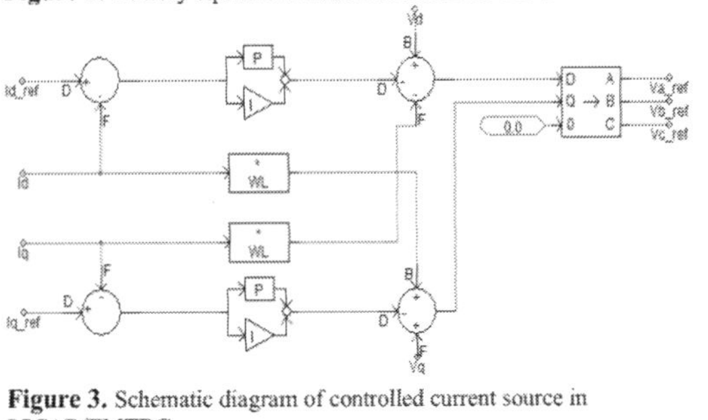

Figure 3. Schematic diagram of controlled current source in PSCAD/EMTDC

3.2 Wind turbine

The wind turbine is designed by controlled current source based on ENERCON's model with the each capacity of 0.8 MW as shown in Fig. 4. The generation power of wind turbine will be calculated from its power curve by actual wind speed data on Chuja Island. From this process, the input data of controlled current source in PSCAD/EMTDC will be gotten in advance by matlab program.

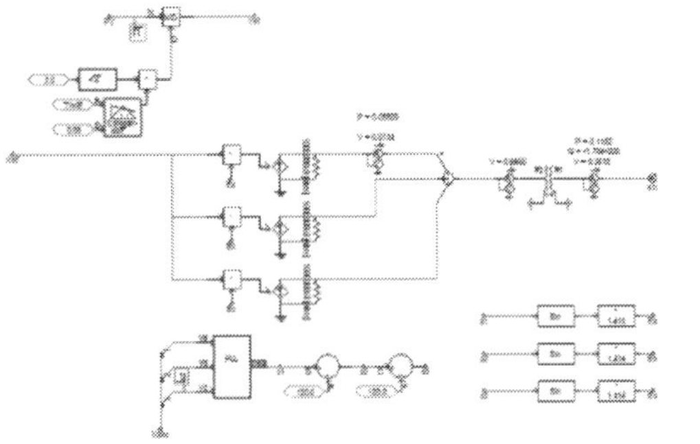

Figure 4. Equivalent current source model in PSCAD/EMTDC

3.3 PV system

The PV system is made from controlled current source similar with wind turbine simulation model [7]. Since it is based on DC source, its equivalent model will only generate active power. The generation power data of PV will be calculated from actual radiation data on Jeju Island nearby Chuja Island by matlab program. Gradient of PV panel is assumed at 30 degrees..

3.4 Diesel generator

The diesel generator is modelled from PSCAD/EMTDC library model as shown in Fig. 5. It uses actual parameter of thermal plant in Chuja Island. There are two type of diesel generator in Chuja Island. One is 0.5 MW. Another is 1 MW. The diesel generators consist of diesel engines and generators.

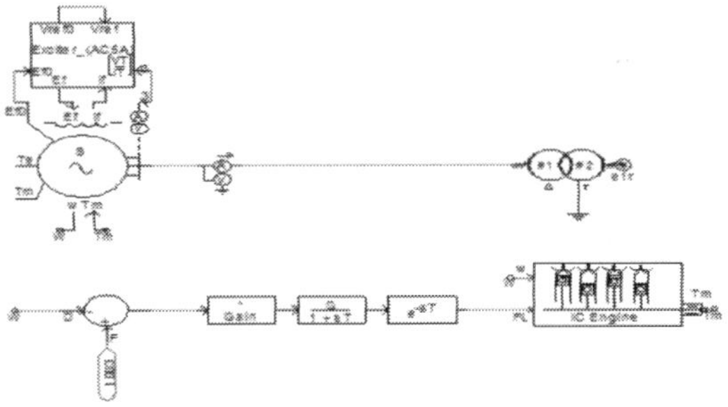

Figure 5. Diesel generator model in PSCAD/EMTDC

4 Estimation of appropriate output of diesel generator

Generally, one of the important points of microgrid is operation of diesel generator. It can support to the stability of microgrid more than any other sources because the diesel generator has the biggest moment of inertia. It also has robust characteristics for fluctuation of frequency and voltage. Although the diesel generator has many advantages for stability of microgrid, but the microgrid must operate with minimum diesel generator capacity in order to minimize air pollution from using fossil fuel.

Thus, this paper proposes the method of estimating

 DOI: 10.1051/matecconf/20167701020

appropriate output for diesel generator applied to microgrid for one day. This method is based on calculating renewable energy and consumption on demand load. The appropriate output for diesel generators will be calculated by using actual data of wind speed and solar radiation. Consequently, the wind and solar powers are calculated in advance. The diesel generation will be estimated in accordance with average load and renewable energy sources at specific date on Chuja Island.

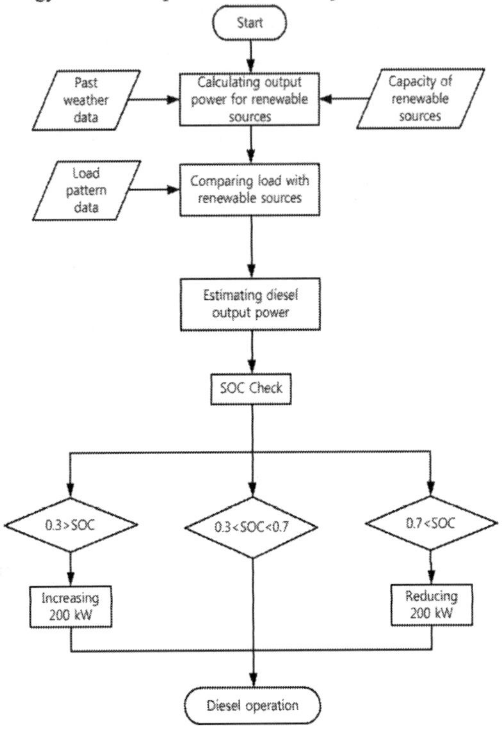

Figure 6. Flow chart of estimation for diesel generator output power

Because there is no radiation measuring device on Chuja Island, the solar radiation data on Jeju Island will be used for calculating the PV power on Chuja Island. The ESS plays a role as power balancing controller on grid. Fig. 6 shows a calculation process of appropriate output for diesel generator. In order to achieve a reliability, the past weather data at Dec. 13, 2015 will be used for calculating diesel appropriate output, because that date has highest fluctuation of output wind power in 2015. The appropriate output of the diesel generator is calculated as follows.

$$A_{DG} = A_{RES} - A_{DL} \tag{1}$$

where, A_{DG}: Diesel generation [kWh]

A_{RES} : Renewable energy [kWh]

A_{DL} : Power consumption [kWh]

Although the appropriate output of diesel generator will be calculated from renewable energy sources, it can be changed according to the SOC of ESS. The output power of diesel generator will depend on SOC of ESS because renewable energy sources have high intermittent characteristic and ESS capacity is limited. In this case, ESS will adjust the power balance between generation power and demand load. Therefore, it is essential to maintain the operation range of SOC. If the SOC is lower than 0.3, the diesel output must be increased to 200 kW. If the SOC is higher than 0.7, the diesel output must be decreased to 200 kW.

Table 1. Calculation results of microgrid on Chuja island at Dec. 13.

Item	Result
Load	2,477,565 kWh
Wind Turbine	1,319,417 kWh
PV System	185,511 kWh
Diesel Generator	972,636 kWh
Operation Capacity of Diesel Generation	648 kW

Table 1 shows the calculating results. From this point, the renewable energy sources and demand load are calculated for power demand per one day. The output power of diesel generator is estimated. Therefore, the diesel generator will be operated with a constant power of 648 kW. (**u**) for vectors. The order for brackets should be {[()]}, except where brackets have special significance.

5 Simulation results

The Fig. 7 shows the active power of microgrid system in Chuja Island. The ESS, PV and wind turbines are operating to response to the demand load. The diesel generator maintains the appropriate output as base generation for high stability of grid and reduction of CO2. In the duration time of 0:00 - 6:00, the output power of wind turbine is stable. The PV output power is almost zero. However, renewable energy sources increase in the duration time of 7:30 - 12:00. Therefore, the ESS charges for the demand load comparing with distributed resources. In duration time of 19:00 - 23:00, the output power fluctuation of wind turbine is highest. However the ESS charges power of wind turbine for grid stability.

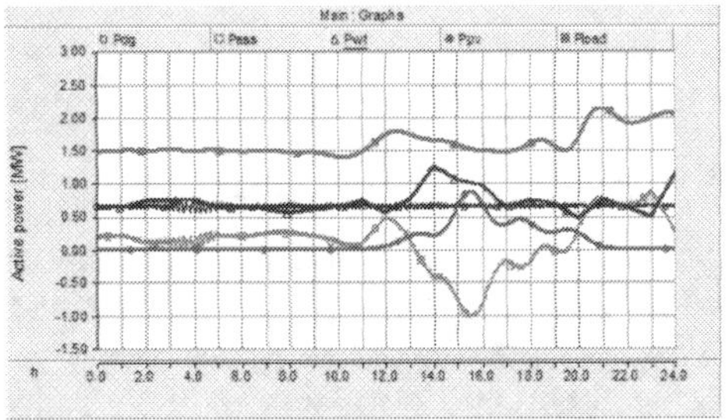

Figure 7. Active power of microgrid system in Chuja Island

The frequency of microgrid system in Chuja Island is illustrated in Fig. 8. The maximum frequency of this microgrid is 60.14 Hz, the minimum one is 59.89 Hz. It means that this microgrid system is achieved the frequency standard of South Korean government for 59.8 - 60.2 Hz [8]. Although the fluctuations of renewable energy sources are very high in the duration time of 07:00 - 17:00 and 19:00 - 23:00, the grid frequency maintain in stable state. The grid voltage of Chuja Island is seen in Fig. 9. The maximum voltage of this microgrid is 6.66 kV, the minimum one is 6.57 kV. The operation of this microgrid system is achieved the voltage standard of South Korean government for 6.0 - 6.9 kV [8]. The grid voltage stays in stable range.

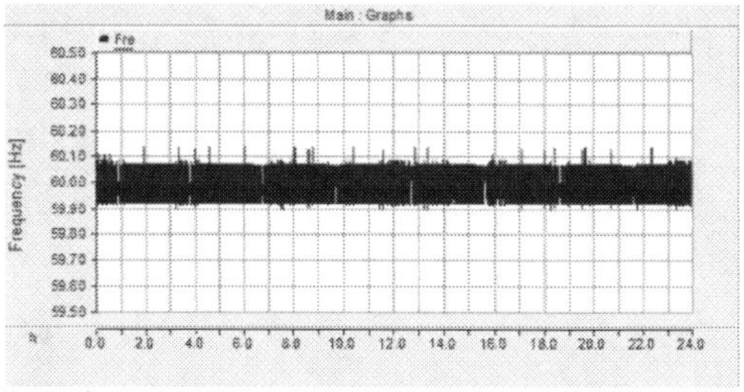

Figure 8. Frequency of microgrid system in Chuja Island

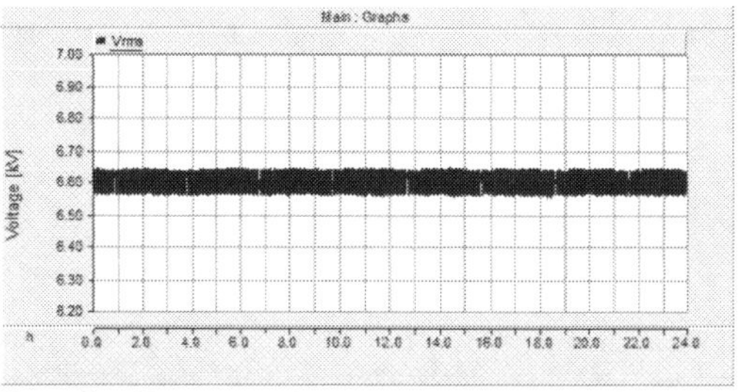

Figure 9. Grid voltage of microgrid system in Chuja Island

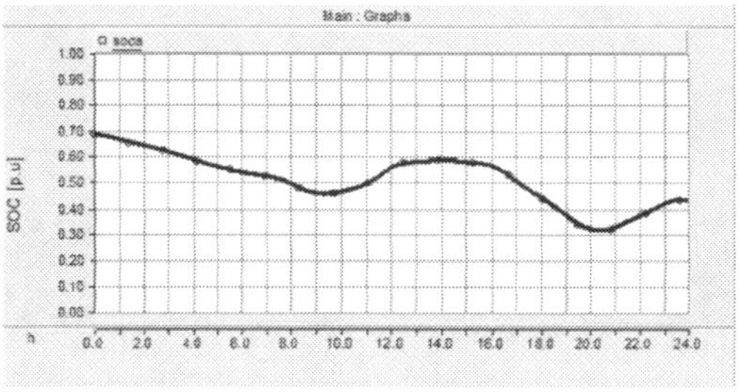

Figure 10. SOC of ESS in microgrid system in Chuja Island

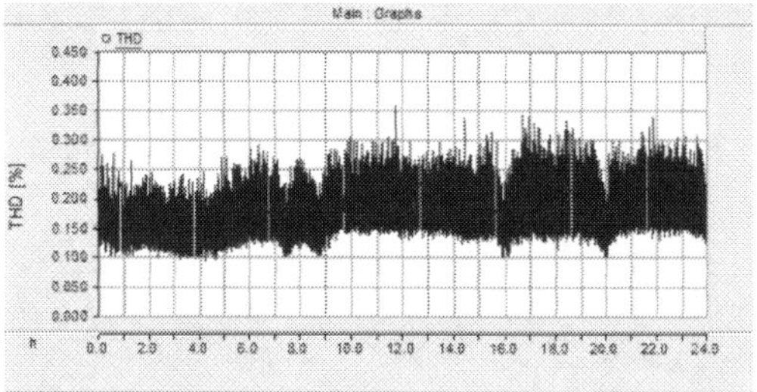

Figure 11. Total harmonics distortion of microgrid system in Chuja Island

The SOC of ESS on Chuja Island is shown in Fig. 10. In this study, The ESS operates in range of SOC from 0.2 to 0.8.

When the load is higher than the generation power, the ESS discharges the power for the demand load. Thus, the SOC is decreased. It is also increased in contrary case.

The voltage total harmonics distortion (VTHD) of this microgrid is shown in Fig. 11. the VTHD is lower than 0.5%. It means that the power quality is achieved the VTHD standard of South Korean government for 5%.

6 Conclusion

This paper proposes the method for estimating appropriate output of diesel generator in microgrid system in Chuja Island. With the proposed method, the microgrid system in Chuja Island can operate stably as the demand load. Although the renewable energy sources are variable, grid voltage and frequency are also stable and achieve the standard of Korean government. The VTHD is recorded less than 0.5%. It means that reliability of this microgrid system is quite reasonable in spite of inverter systems operation. The calculation result shows that the diesel generator operates at a constant power. Then, the ESS operates with range of SOC from 0.2 to 0.8 as setting point. In the future, by using the weather forecasting data, this method will be useful to estimate appropriate output of diesel generation in microgrids.

Acknowledgement

This work was supported by Graduate School of Specialized Wind Energy the Human Resources Development (NO. 20094020200020) and the Expansion of the Type Testing Site for Wind Turbines (NO. 2012T100201731) of the Korea Institute of Energy Technology Evaluation and Planning (KETEP) grant funded by Korea government Ministry of Trade, Industry and Energy.

References

1. CERT, Intergration of Distributed Energy Resource: the CERT Microgrid Concept, LBNL-50829, (2003).
2. Lasseter, T. H., Microgrids, IEEE Power Engineering Society Winter Meeting, pp. 146-149, (2001).
3. Seong Man Kim, Examples and Applications for Techniques of Microgrid, Jounal of the Electric World/Monthly Magazine Special, no. 4, pp. 37-44, (2014).
4. Policy Report, Carbon Free Island Jeju by 2030, Jeju Special Self-Gorvorning Province, May (2015).
5. Seung-Mo Song, Seung-Jin Oh, Jong-Hak Lee, Jae-Min Ahn, Kyung-Min Jin, Byung-Gi Kwon, Chang-Ho Choi, POSCO ICT., The Application and Verificaion of the 2 MVA BESS with Wind farm in Microgrid of Gapa, Jeju, KIPE, **19**, no. 4, pp. 303-311, (2014).
6. KERI, Development and Experiment Actual Site of Isolated Microgrid (2010101040050A), Report.

7. Dong-Wan Kim, A Study on the Improvement Scheme of Distributed Generation Facilites in Gapa Island Power Grid., Master Thesis, Jeju National University, (2014).
8. KEPCO, Standards of Transmission and Distribution.

MATEC Web of Conferences 77, 01021 (2016) DOI: 10.1051/matecconf/20167701021
ICMMR 2016

The Effects of Guide Pads on Bore Diameter Enlargement Magnitude in Deep Hole Drilling

Huang Zhang[1], Xingquan Shen[1], Yaoming Li[1], Fuhe Yang[1] and Paul Kwon[2]

[1] *North University of China, Taiyuan 030051, China*
[2] *Joincheer Mechanical & Electrical Company, Korea*

Abstract. Deep hole machining accuracy plays a crucial role in product's quality. The results of an investigation into the machining accuracy of deep drilling holes which highlights problems of bore diameter enlargement magnitude assessment are presented in this paper. Through the study of BTA deep drilling tool properties and mechanism, as well as the formation rule of deep hole surface, clarified the burnishing mechanism of guide pad in deep hole drilling, and obtained the directly relationship between surface formation rule and the guide pad. After drilling experimentrevealed the effect of guide pad on bore diameter enlargement magnitude.

1 Introduction

Deep-hole drilling is one of the most important processes for the production of a high-precision workpiece with high-quality holes [1]. The machining quality of deep-hole drilling has been studied for various machining conditions [2], [3]. Sakuma et al. [4] considered the behavior of the BTA solid boring system and postulated that the forming of a multi-corner hole is a certain kind of self-excited vibration. Deng et al. [5] used beam theory to investigate the hole straightness as affected by pilot bushing and intermediate support misalignments.

Literature investigation reveals that there are less rigorous theories to study the formation rule and its particularity of deep hole surface. And the guide mechanism of the guide paid as the most essential attributes of deep hole drilling is rarely studied. It's necessary to study the bore diameter enlargement magnitude, so as to improve the deep hole machining accuracy for strong support.

2 The theoretical research of the diameter enlargement magnitude

2.1 Properties and mechanics of BTA drilling tool

Deep hole drilling tools differ from conventional hole manufacturing tools due to a large length to diameter ratio, which have guide pads on the tool heads. In order to ensure the stability of drilling smoothly, the cutting forces are balanced by the reaction forces generated at the guide pads that are burnished on the inner hole surface, and consequently deep drilling tools are often referred to a 'self-piloting' tool, which allows high quality deep holes to be drilled. High pressure cutting fluids inject into the cutting region along the annular space between the drilling shaft and hole wall serving as coolant and lubricant at the same time, and flows over the cutting edge and returns with the cutting chips along the tool shaft inner hole, which is a kind of the inner ejection deep hole cutting machining method. Figure1 shows a typical deep drilling tool which has a single cutting edge. The outside corner of the cutting edge is in advance of the pads by a small distance L_L, so that they do not disrupt the cutting action.

Basedon the cutting force model of the deep holedrilling tools in Fig. 1, it can be seen intelligibly that the main cutting forces P_{MA},P_{MB} and P_{MD} occurring at deep hole drill each blade are in the same direction that alltowardto the first guide pad, andthey are balanced by the reaction forces P_{G1} and μP_{G1} generated at the first guide pad. The radial cutting force component P_{RD}, P_{RB} and P_{RA} are balanced by the reaction forces P_{G2} and μP_{G2} generated at the second guide pad. At the same time, all guide pads are under the effect of friction torque, ultimately to ensure that the guide pads have stabilizing and guiding effects.

2.2 The formation rule of deep hole surface

To determine the direct relationship between the guide and the law of machining surface formation, B. J. Griffiths [6] using a specially designed quick stop device revealed features of the mechanics of the deep drilling process. The generation of white layer formations, as shown in Fig. 2, is the result of plastic deformation that generated by the two guide pads burnished the machined hole surface continuous periodically, and the high temperature can exceed 500℃ in front of the pads, and combined with the rapid cooling effect caused by the pads revolving at high speed against the bore wall in the presence of the copious cutting fluid. Also it can be get that the hole diameter at the cut surface is less than at the burnished surface, which implying that considerable plastic flow has occurred.

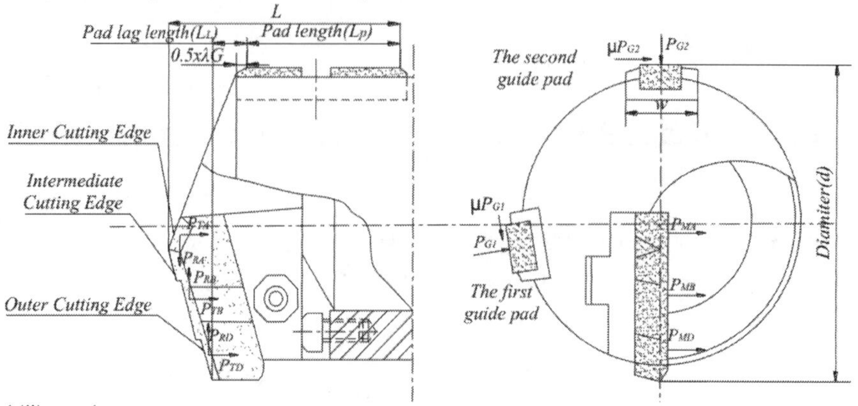

Figure 1. BTA drilling tool

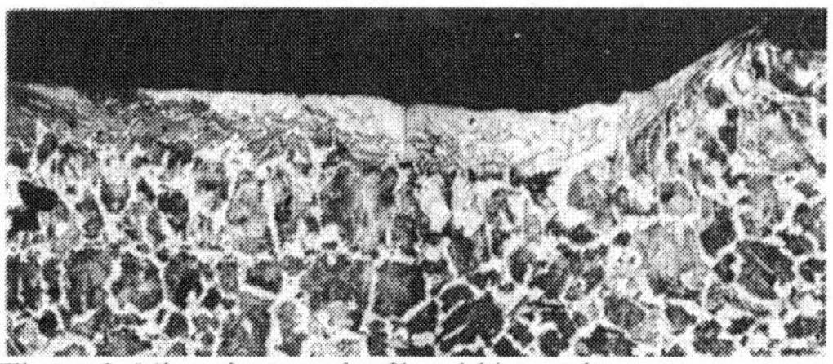

Figure 2. Microphotograph of burnishing region

The law of machining surface formation can be described as the ridge creation and modification occurs within the material burnished by guide pads in cycle, and the ridge truncation at the cutting edge land face, as shown in Fig. 3. Fig. 3(a) shows the hole form with the ridge burnished after the guide pads have passed, as the feed movement continue, the ridge has been truncated and area ABC removed in Fig. 3(b), when the tool has moved forward one feed rate, the area DEFG has been displaced by the guide pads, and in the forward direction, the new ridge HIG recreated and in the backward direction the plastic flow tends to fill in the previous indentation area KLEM.

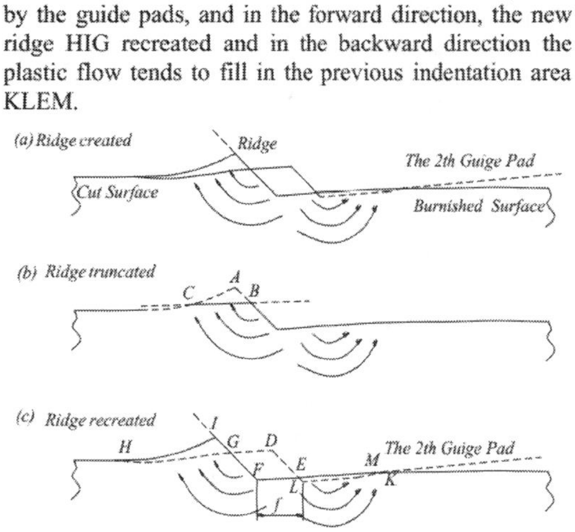

Figure 3. Schematic representation of deformation cycle

2.3 The affect factors of deep hole bore diameter enlargement magnitude

Comprehensive above research, the plastic deformation and elastic deformation are created simultaneously after the guide pads burnished, that lead to the enlargement of the machined hole. With the different workpiece materials, the plastic deformation part of the hole wall is differentand proportional to feed rate f. The deep hole bore diameter enlargement magnitude ΔD [7] caused by guide pads brunished can be represented astheequation (1):

$$\Delta D = \left(\frac{K_S \cdot D}{2HB \cdot w} - C + \frac{1}{4}\right) f \cdot \tan\lambda_G \qquad (1)$$

where K_S is the specific cutting force of workpiece material ($\times$ 9.8 N/mm^2), D is the hole diameter, HB is the brinell hardness of the workpiece, w is the width of the guide pad, C is the ratio of the feed rate to the actual length of the guide pad fore-end and hole wall contact surface, λ_G is the entrance angle of guide pad, as shown in Fig. 1. It is not difficult to draw that the thinnerhole wall, the bigger deformation, and the hole shape error will be produced.

In addition, The influencing factors of bore diameter enlargement also includes the following aspects:the cutting heat causes the material heat bilges cold shrink, the extension ofthe built-up edge on the outer edge cause the expansion of the aperturein the processing of drilling, and if the corner of outer cutting edge is blunt, it will generate elastic deformation and cutter relieving in the drilling region, that further cause the narrowing of the bore diameter.

3 Drilling experimental

To examine the effects of guide pads, a fundamental experiment of a BOTEK deep hole tools with diameter of

 DOI: 10.1051/matecconf/20167701021

30mm was made, and the correlative parameters are as follows: the width ω of guide pad is 7.9mm, the entrance angle $\lambda_G = 10°$, rotating rate n of workpiece is630 r/min, the workpiece materials are A2017-T4(HB=98) and A2017-0(HB=51), and assigning a value to C is 2, also the feed rate can be adjusted from 0 to $0.2mm \cdot r^{-1}$.

(a)
(b)

Figure 4. The effects of guide pad burnished, (a) and (b) respectively represent the guide paid and the machined hole

(a)
(b)

Figure 5. The wear state of guide pad, (a) and (b) respectively represent the first and second guide pad

The deep hole formed characteristics and wear state of guide pad are obtained by high depth of field microscope shown respectively in Fig. 4-Fig. 5, and comparatively analyzed the influence rules of deep hole bore diameter expanding quantity through adjusting feed rate. As shown in Figure 4, a matter worthy of reflection, the residual workpiece material after being squeezed is stored on the guide pad entrance, and at the machined hole exit with a redundant circular workpiece after the drilling is finishing. These are the characterization of the ridge creation after the guide paid passing. These results are very encouraging in that guide pads played an important role in the hole surface formation regularity, and enlarge influence to the machined surface simultaneously. In addition, Fig. 5 shows the wear states of first and second guide pad respectively. It can be seen that the wear state with coating peelingon the front end locale of guide pad is more serious than the back-end locale, and the wear stateon the first guide pad presented a dark brown is more serious than the second guide pad. That is mainly because the second guide pad is lagged behind the first guide pad about f/4. Also it can be come to the conclusion that the stress of the first guide paid is significantly bigger than the second guide paid which lead to the first guide paid is wear more serious through studying the cutting force model of the deep hole drilling tools in Fig. 1.

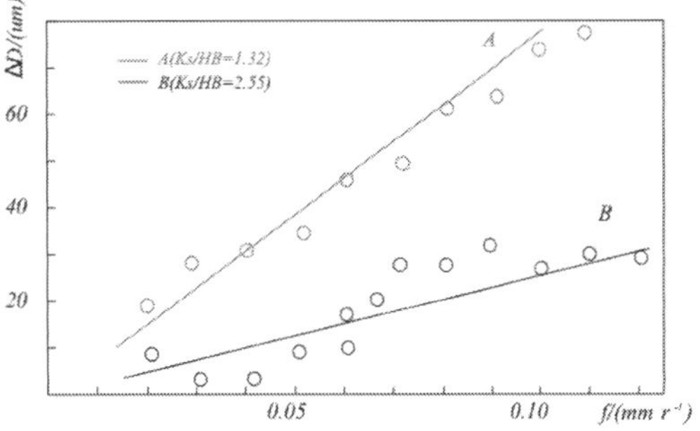

Figure 6. The changes of bore diameter enlargement magnitude along with feed rate

Fig. 6 shows the changes of bore diameter extensive magnitude along with feed rate under the condition of K_s/HB are 1.32 and 2.55 respectively. In the case of other factors unchanged, with the increase of the feed rate, the bore diameter extensive magnitude also increases accordingly. Combined with the Equation (1), because theK_S is associated with the strength and toughness of workpiece materials, while the HB value is related to the material heat treatment conditions, so in general, the bore diameter extensive magnitudeis remarkable for higher strength, toughness and without hardening material. It is not hard to analyze that no matter what kind of material, to increase feed rate will cause the proportional increase of bore diameter enlargement magnitude.

4 Conclusions

Through the description of the deep hole drilling process, revealed the formation of machining surface by BTA deep hole drilling tools, and put forward the root causes for the enlargement of the machined hole, that both the cutting stage and the burnishing stage, make the machined inner hole surface produce plastic deformation several times.

The influencing factors of bore diameter enlargement magnitude have been studied. Due to the bore diameter expanding and reducing factor is not unilaterally, it is conceivable that the actual diameter depends on the comprehensive effect of various factors, while the bore diameter enlargement magnitude will proportionally increase along with the increase feed rate. Ultimately, the bore diameter enlargement magnitude is directly related to the guide pad.

Combined with the mechanical analysis of BTA drilling model and the drilling experiments of different hardness materials, researched the wear state of guide pad and the deep hole formation characteristics, verified the burnishing effect of guide pads and the root causes of the machined hole enlargement.

Acknowledgements

This work was financially supported by the National Natural Science Foundation of China (51175482) and International S&T Cooperation Program of China (2013DFA70770).

References

[1] CS Deng, JH Chin. Roundness Errors in BTA Drilling and a Model of Waviness and Lobbing Caused by Resonant Forced Vibrations of Its Long Drill Shaft. Journal of manufacturing Science and Engineering, **126**: 524-533 (2004).
[2] Aized T, Amjad M. Quality improvement of deep-hole drilling process of AISI D2 [J]. The International Journal of Advanced Manufacturing Technology, **69**(9-12): 2493-2503 (2013).
[3] Matsuzaki K, Ryu T, Sueoka A, et al. Theoretical and experimental study on rifling mark generating phenomena in BTA deep hole drilling process (generating mechanism and countermeasure)[J]. International Journal of Machine Tools and Manufacture, **88**: 194-205 (2015).
[4] Sakuma K, Taguchi K, Katsuki A, et al. Self-guiding action of deep-hole-drilling tools [J]. CIRP Annals-Manufacturing Technology, **30**(1): 311-315(1981).
[5] Deng, C. S., Huang, J. C., and Chin, J. H., ''Effects of Support Misalignments in Deep-Hole Drill Shafts on Hole Straightness,'' Int. J. Mach. Tools Manuf., **41**(8): 1165–1188 (2001).
[6] Griffiths B J. Modelling complex force systems, Part 1: The cutting and pad forces in deep drilling [J]. Journal of engineering for industry, **115**(2): 169-176 (1993).
[7] Jun Wang. Modern deep processing technology [M]. Harbin: Harbin Institute of Technology Press, **2** (2005).

Implementation of multiaxial high cycle fatigue criterion in finite element analysis

Quoc Huy Vu, Dinh Quy Vu

Department of Aeronautical and Space Engineering, Hanoi University of Science and Technology, Vietnam

Abstract. Multiaxial high cycle fatigue analyses concern many mechanical components in several industries (aircraft, automobile, railway...). Many multiaxial fatigue criteria have been proposed in the literature for metals. This study deals with the integration of fatigue criteria in finite element code by using APDL (ANSYS Parametric Design Language) scripting language. Three criteria have been implemented including Crossland, Dang Van and Vu et al. [6] criteria. The predictions of Vu et al. criterion are in good accordance with the experimental data.

1 Introduction

Mechanical components in service are usually subjected to multiaxial complex loading and their required lifetimes are often in the range of high cycle fatigue (10^6 – 10^9 cycles). In order to determine the stress level allowable for the safety of the component, many multiaxial fatigue endurance criteria have been proposed in the literature for metals [1,2]. For application with real structure, the multiaxial fatigue criteria need to be implemented with finite element codes [3]. Obvious advantages of this approach: benefit the material properties data, 3D design of structure, meshing, setup of load and boundary conditions, stress analysis by finite element method and result post-processing. The finite element analysis programs allow using the stress analysis results for fatigue analysis. Many commercial CAE (Computer-Aided Engineering) softwares now provide the capability of fatigue analysis such as ANSYS, nCode, Msc Fatigue, Abaqus, Samcef... However, only limited softwares integrate the multiaxial fatigue criteria such as nCode (integration of Dang Van criterion), Samcef (integration of Crossland criterion).

2 Multiaxial high cycle fatigue criteria

From the point of view of numerical implementation, the most convenient are two categories: criteria based on stress invariants and criteria using critical plane approaches. Two classical criteria widely used are Crossland criterion [4] (invariant group) and Dang Van criterion [5] (critical plane group).

2.1 Crossland criterion

Crossland proposed a combination between the amplitude of the second invariant of the stress deviator ($J_{2,a}$) and the maximum hydrostatic stress ($\sigma_{H,\max}$):

$$\sqrt{J_{2,a}} + \alpha\sigma_{H,\max} \le \beta \qquad (1)$$

where α and β are material parameters. In case of out-of-phase bending, the value of $J_{2,a}$ can be defined as a function of macroscopic stress components:

$$\sqrt{J_{2,a}} = \frac{1}{\sqrt{2}}\sqrt{\left(\frac{\sigma_a^2}{3}+\tau_a^2\right)+\sqrt{\left(\frac{\sigma_a^2}{3}+\tau_a^2\right)^2-\frac{4}{3}\sigma_a^2\tau_a^2\sin^2\delta}} \qquad (2)$$

where δ is the phase shift angle between normal and shear stress.

2.2 Dang Van criterion

Dang Van proposed a combination of local shear stress $\tau(t)$ and the hydrostatic stress $\sigma_H(t)$ that looks for the maximal value in all grain orientations and in variable time. The simplified form of the Dang Van criterion can be formulated as:

$$\max_t\left[\tau(t) + \alpha.\sigma_H(t)\right] \le \beta \qquad (3)$$

where α and β are material parameters.

2.3 Vu et al. criterion

Vu et al. [6] proposed a multiaxial criterion for high cycle fatigue based on invariants of macroscopic stress tensor (denoted as V-H-N criterion). The presence of $J_{2,mean}$ quantity allows capturing accurately effects of phase shift and frequency on fatigue limit of material under multiaxial loading. The prediction capacity of the criterion is tested on 119 iso-frequency axial–torsion experiments and some other more complex loadings: biaxial loading and asyn-chronous loading and the results show that the criterion is in good accordance with the experimental data. Under proportional loading, the assessment of the criterion can be carried out from an analytical solution. Under others loading cases, the numerical implementation of the criterion is very simple and can be easily integrated in a damage model or in a finite element code. By introducing the quantity $J_{2,mean}$, the multiaxial fatigue endurance criterion is established as follows:

$$f = \sqrt{\gamma_1 J'_2(t)^2 + \gamma_2 J^2_{2,mean} + \gamma_3 I_f(I_{1,a}, I_{1,m})} \le \beta \quad (4)$$

where γ_1, γ_2, γ_3 and β are material parameters; $J'_2(t)$ and $J_{2,mean}$ capture shear stress effect and phase shift effect. $J'_2(t)$ is the second invariant of stress amplitude part defined from the deviator of the amplitude of the stress tensor $\underline{\underline{S}}^a(t)$ and $J_{2,mean}$ is the mean value of $J'_2(t)$ during period T:

$$J'_2(t) = \sqrt{\frac{1}{2}\underline{\underline{S}}^a(t) : \underline{\underline{S}}^a(t)} \quad (5)$$

$$J_{2,mean} = \frac{1}{T}\int_0^T J'_2(t)dt \quad (6)$$

$I_f(I_{1,a}, I_{1,m})$ is a function of $I_{1,a}$ and $I_{1,m}$ refeflecting resrespectively the effects of amplitude and mean value of the hydrostatic stress, in which $I_1(t) = tr(\sigma)$ is the first invariant of the stress tensor.

3 Implementation of multiaxial fatigue criteria in finite element code

To apply multiaxial fatigue criteria for real structures subjected to complex loading, we need integrate the fatigue criteria into a finite element code. In this study, in order to implement the criteria, a combination of APDL (ANSYS Parametric Design Language) with ANSYS Workbench is carried out. The APDL scripting language is used to build the multiaxial fatigue criteria (Crossland, Dang Van, V-H-N) while the ANSYS Workbench Mechanical defines all other parts of the structure simulation: geometry, mesh, loads, boundary conditions and results processing.

The APDL commands for multiaxial fatigue criteria perform following steps:

- to record values of stress tensor components in function of time increment during one period
- to calculate the value of hydrostatic stress
- to calculate the value of the equivalent stresses according to Crossland, Dang Van and V-H-N criteria
- to visualize the fatigue analysis results in the form of contour maps

4 Numerical example

To implement and compare the fatigue criteria mentioned above, a simple cylindrical specimen was chosen as a test example. The dimensions of specimen (all in mm) are shown in Fig. 1. The specimen is subjected to bending (P_b) – torsion (P_t) sinusoidal loading:

$$P_b = P_{ba}.\sin(2\pi ft) \qquad P_t = P_{ta}.\sin(2\pi ft - \delta) \quad (7)$$

where P_{ba} and P_{ta} are the bending and torsion load amplitude, δ the phase shift angle, *t* the time and *f* the frequency. The considered material is hard steel with bending limit fatigue f_{-1} = 313.9 MPa, torsion limit fatigue t_{-1} = 196.2 MPa and ultimate strength R_m = 680 MPa.

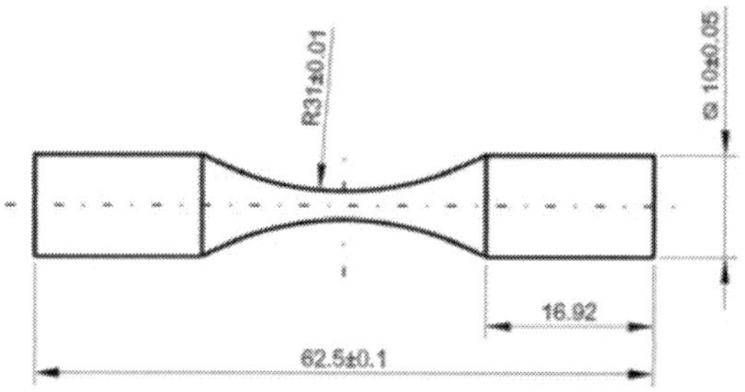

Figure 1. Specimen geometry.

To assess the predicting capacity of the fatigue criteria, comparison of the predicted fatigue limits against out-of-phase bending – torsion experimental results in the literature is carried out, an error index I is usually used:

$$I = \left(\frac{\sigma_{eq} - \beta}{\beta}\right).100\% \quad (8)$$

For the considered fatigue criteria and the assumed material properties, $\beta = t_{-1} = 196.2(MPa)$ and the equivalent stresses of the criteria have the values as follows:

- Crossland criterion:

$$\sigma_{eq,C} = \sqrt{J_{2,a}} + 0.143.\sigma_{H,max}$$

Dang Van criterion: $\sigma_{eq,DV} = \max[\tau(t) + 0.375.\sigma_H(t)]$

V-H-N criterion:

$$\sigma_{eq,V} = \sqrt{0.65 J'_2(t)^2 + 0.8636 J^2_{2,mean} + 18(I_{1,a} + I_{1,m})}$$

The error indexes I_C, I_{DV}, I_V refer respectively to the predicted results of Crossland, Dang Van and V-H-N criteria. Table 1 summarizes the prediction of criteria in some loading cases with different phase shift angle. The bending – torsion experimental results are collected from

[1]. As shown in Table 1, the predictions of Crossland and Dang Van criteria are dangerous under 90° out-of-phase loading case while the V-H-N criterion captures quite accurately the effect of phase shift including 90° case.

Table 1. Out-of-phase bending-torsion fatigue limit data of hard steel specimen.

P_{ba} (MPa)	P_{ta} (MPa)	δ (°)	I_C (%)	I_{DV} (%)	I_V (%)
138.1	167.1	0	-2.3	1.4	-1.99
140.4	169.9	30	-2.5	0.24	-1.01
145.7	176.3	60	-3.6	2.09	1.37
258.0	129.0	90	-17.8	-16.83	-3.19

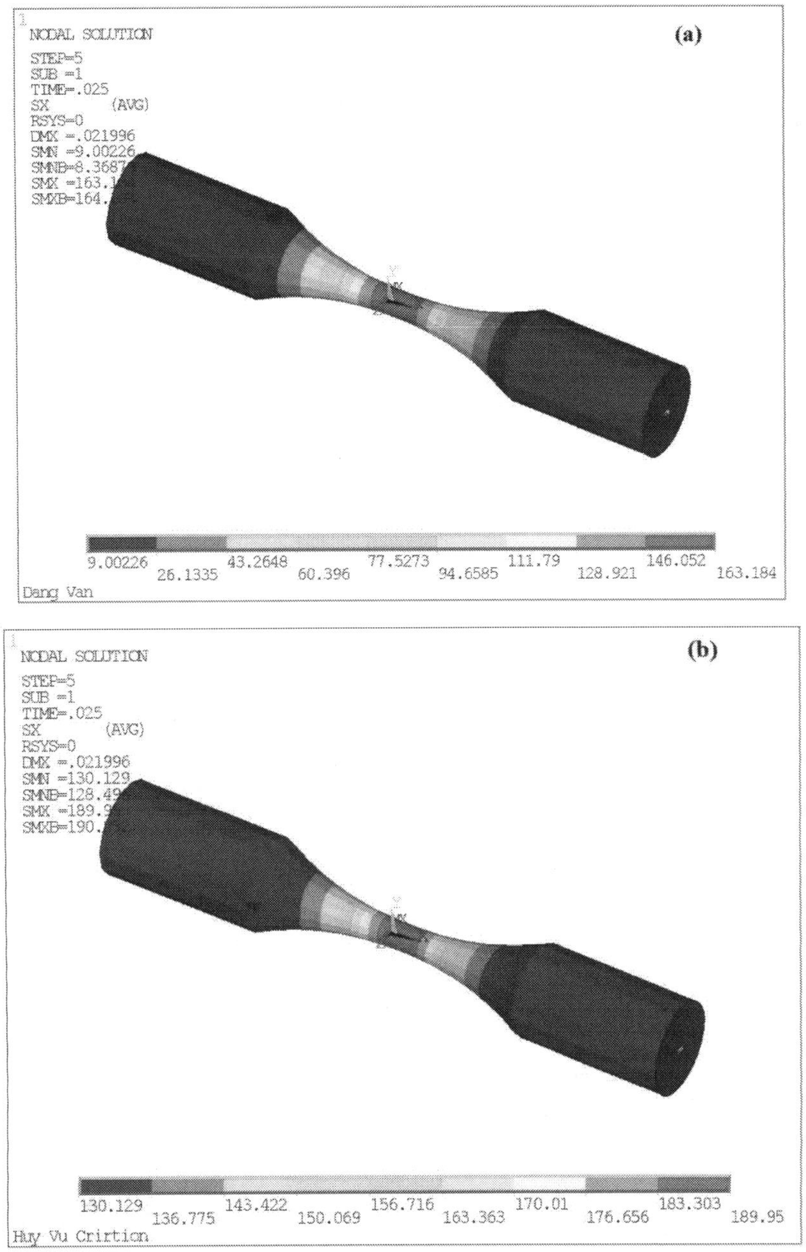

Figure 2. Field of equivalent stress (MPa) according to Dang Van (a) and V-H-N criteria (b).

The contours of equivalent stress (MPa) in the specimen according to Dang Van and V-H-N criteria are shown in Fig. 2, in which the variable SMX means maximal equivalent stress (MPa). While the contour shapes of two criteria are quite similar, the values of SMX show the difference between two criteria ($SMX_{Dang\ Van}$ = 163.14; $SMX_{V\text{-}H\text{-}N}$ = 189.95).

5 Conclusion

The numerical implementation of multiaxial fatigue criteria in finite element code is carried out in this paper. Three fatigue criteria have been tested including Crossland, Dang Van and V-H-N criteria. The combination of APDL script with ANSYS Workbench Mechanical shows a good performance to integrate multiaxial criteria into finite element code. From the computational point of view, the criterion V-H-N has the same order of simplicity as the Crossland criterion. The implementation of V-H-N criterion may expand for more complex analyses such as variable amplitude fatigue loading.

Acknowledgments

This research is funded by Vietnam National Foundation for Science and Technology Development (NAFOSTED) under grant number 107.02-2014.25.

References

1. I. V. Papadopoulos, P. Davoli, C. Gorla, M. Filippini and A. Bernasconi, Int. J. Fatigue, **19**(3), p. 219-235 (1997)
2. J. Papuga, Procedia Engineering, **66**, p. 587-597 (2013)
3. M. Mrzyglod, A. P. Zielinski, J. Theor. Appl. Mech., **44**(3), p. 691-712 (2006)
4. B. Crossland, *Proceedings of the international conference on fatigue of metals*, London, p. 138–149 (1956)
5. K. Dang Van, *Thèse de Doctorat ès Sciences.* Sci Tech de l'Armement, **47**(3), p. 647–722 (1973)
6. Q. H. Vu, D. Halm, Y. Nadot, Int. J. Fatigue, **32,** p. 1004-1014 (2010)

MATEC Web of Conferences 77, 01023 (2016) DOI: 10.1051/matecconf/20167701023
ICMMR 2016

A Wideband Stacked Patch Antenna with Printed Meandering Probe

Ruina Xing[1, 2], Jujin Li[1, 2] and Dan Sun[1, 3]

[1]*AVIC LEIHUA Electronic Technology Research Institute, 214063 Wuxi, Jiangsu, China*
[2]*Aviation Key Laboratory of Science and Technology on AISSS, 214063 Wuxi, Jiangsu, China*
[3]*School of Information Science and Technology, Fudan University, 200433 Shanghai, China*

Abstract. A broadband probe-fed stacked patch antenna is presented in this paper. The probe is composed of two printed feed-lines and three vias. The feed-lines help of compensating the inductances produced by the vias. Further, with the stacked patch method, the bandwidth of the antenna is enhanced. The simulated results show the antenna has a bandwidth over 40% (VSWR<2) and good radiation performance.

1 Introduction

With the increase of the application of microstrip patch antenna, its bandwidth enhancement has been one of the most attractive research fields. The most convenient method is to use electrically thick substrate. However, in the conventional edge- and probe-fed patch antennas, the method is greatly restricted because of the increasing spurious radiation of the feed-lines and the increasing inductance of the probe resulting from the thick substrate. Hence, in many broadband patch antennas with thick substrates, non-contact feeding method is employed, such as aperture- and proximity-coupling [1-5]. Howbeit, aperture-coupling will increase the back radiation, and the feed structure of proximity-coupling patch antenna will increase the spurious radiation. As compared with non-contact feeding method, direct feeding mechanism is easier to be achieved. In particular, probe-feeding has the potential that the inductance resulting from the probe can be compensated by modifying its structure. Accordingly, the limitation of the substrate thickness is overcome. In literature [6], a meandering probe fabricated by metallic plate was used, and a bandwidth of 24% was obtained. Further, the bandwidth of the antenna with the feeding approach was improved to 40% through combining the stacked patch technique [7]. However, the cross-section of the probe formed by the metallic plate is thick due to the requirement for the structure strength, which will influence the matching of antenna in high-frequency band. Furthermore, Furthermore, the metallic plate probe brings on the difficulty of fabrication in high-frequency band applications. For instance, it cannot be manufactured together with the patch by printed circuit technique and the antenna used the air substrate to install the metallic plate probe.

In this paper, a printed meandering probe, instead of the metallic plate probe, is designed to solve the aforementioned problem, and then a broadband stacked patch antenna is designed. With the method, the fabrication difficulty of the antenna is reduced, and its application range is improved to high-frequency band. The bandwidth of the antenna exceeds 40%, and its cross-polarization is less than −23dB.

2 Antenna configuration

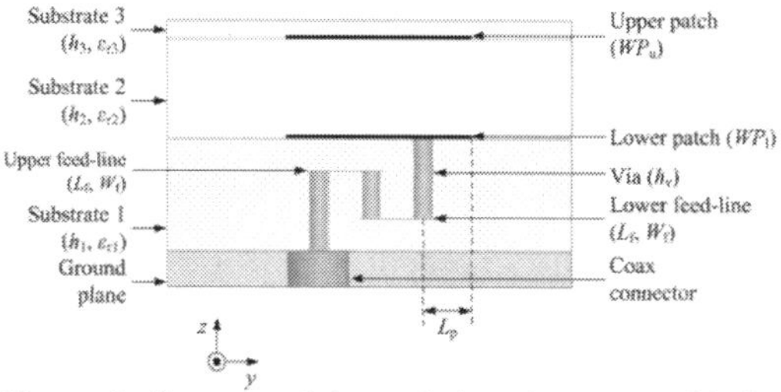

Figure 1. Geometry of the stacked patch antenna with the printed meandering probe.

The antenna fed by the printed meandering probe is shown in Figure 1. The antenna consists of three substrate layers, two stacked patches, a printed meandering probe, and a ground plane. The lower patch is on the upper surface of substrate 1, and the upper patch is located on the lower surface of substrate 3. The dimensions of the stacked patches are WP_l and WP_u, respectively. Two feed-lines are located on the different surfaces in substrate 1, and connected by a via centered with respect to the patch. Each feed-line is symmetrical with respect to the *x*-direction. Its length and width are L_f and W_f, respectively. The connection of the lower feed-line and the lower patch, as also the connection of the upper feed-line and coax connector, is the same as that of two feed-lines. The two vias have the height denote h_v. These vias and feed-lines constitute the printed meandering probe.

The distance between the feeding point of via and the edge of the lower patch is half of WP_l minus L_f. These vias and feed-lines constitute the printed meandering probe. To fabricate the patch circuits and the vias easily, PTFE with the permittivity of 2.2 is used as the material of substrates 1 and 3. Their heights are 3.175mm and 0.127mm, respectively. The material of substrates 2 is foam, whose height and permittivity are 3mm and 1.27, respectively.

3 Antenna design and results

The two feed-lines of the antenna partly neutralize the inductance produced by the vias, and the good impedance matching is achieved even if the electrically thick substrate is employed in the antenna. Accordingly, the bandwidth of the antenna is increased. To compensate the discontinuity produced by the vertical connection of the feed-lines and the vias, the bonding pads are added at their junctions. The antenna is simulated by HFSS v12. The parameters of the feeding structure and the patches are optimized by the particle swarm optimization (PSO) algorithm [8]. Their solution spacing is shown in Table 1.

Table 1. Range of the optimized parameters (unit: mm)

Parameters	WP_l	WP_u	h_v	W_f	L_f
Minimum	3	3	1.59	0.3	0.4
Maximum	12	12	3	2.5	$WP_l/2-0.8$

The numbers of the particles are ten, and the fitness function is defined as (1).

$$f(x)=\left(\left|\mathrm{VSWR}_{8\mathrm{GHz}}-1.5\right|+\left|\mathrm{VSWR}_{12\mathrm{GHz}}-1.5\right|\right)/2 \\ +\max\left(\mathrm{VSWR}_{9\mathrm{GHz}\text{-}11\mathrm{GHz}}\right)-1 \quad (1)$$

With the optimization of parameters, a designed instance is exhibited. Its designed parameters are as follows: WP_l=8.37mm, WP_u=8.4mm, W_f=2.16mm, L_f=1.735mm, h_v=2.26mm. The simulated bandwidth of the antenna is revealed in Figure 2. As can be seen, the operation band with VSWR below 2 is from 7.92GHz to 12.18GHz, and its relative bandwidth is 42.4%. Furthermore, in the frequency range of 8.17GHz to 12 .04GHz (38.3%), the VSWRs are all lower than 1.5, and the ripples of the VSWR curve are small across the entire impedance band. The bandwidth simulated result indicates that the antenna with the printed meandering probe have a good broadband response which is evenly matched with that of the antenna in [7], when retaining the advantage of easy fabrication.

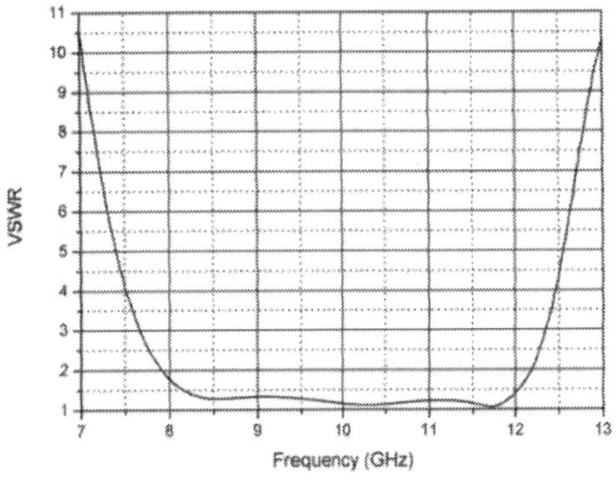

Figure 2. Simulated VSWR of the antenna.

Figure 3 shows the radiation patterns of the antenna. From the simulated results, it can be seen that the good polarization purity is achieved. The cross-polarization

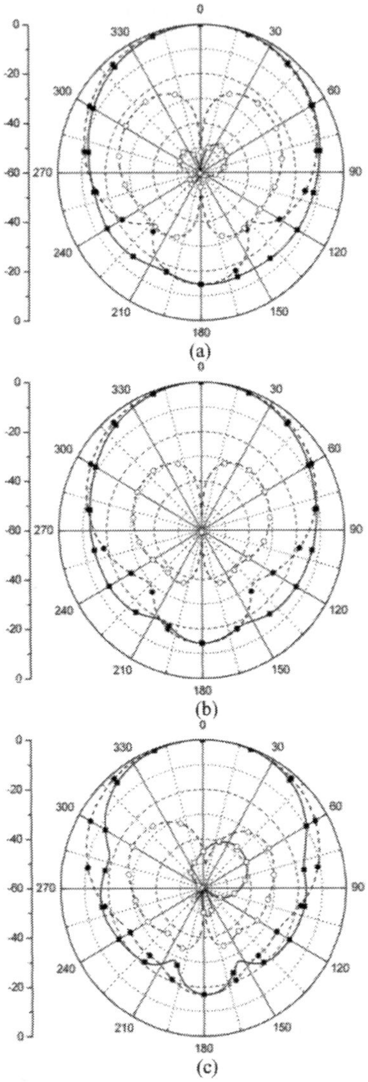

Figure 3. Simulated radiation patterns at different frequencies. (a) 8GHz; (b) 10GHz; (c) 12GHz; —■— Co-pol (E plane); -●- Co-pol (H-plane); —□— Cross-pol (E-plane); -○- Cross-pol (H-plane)

levels in E and H planes are less than −38dB and −23dB, respectively. And at the frequency of 10GHz, its values are lower than −58dB and −28dB, respectively. With the increase of frequency, the spurious radiation of the feed-lines increases. This results in a small asymmetry of the radiation pattern in E plane at 12GHz. In the meantime, the cross-polarization is deteriorated. Because the ground plane with the dimensions of 13.3mm×13.3mm is small, the front-to-back ratio is not high, whose value is only better than 14dB. But it can be improved simply by increasing the ground's sizes. On the whole, the antenna shows the pretty radiation performance in the operation band.

4 Feed-line and patch analyses

As is stated above, the matching performance of the antenna is influenced by the thickness of the cross-section of the feeding structure. If the feed-line is thick, its matching will be deteriorated rapidly. Figure 4 reveal VSWRs of the antenna with the different thickness of the feed-line. While its thickness is 0.018mm (typical value of the copper claddings on laminates), the matching performance is almost the same as that of the ideal condition (thickness=0mm). Also, this means the PTFE materials and the printed circuit technique will not influence the performance of the antenna. For ensuring the enough structure strength, the thickness of the feed-line composed of metallic plate is at least 0.1mm, and it reaches 1mm in most case. As can be seen in Figure 4, when its thickness is 0.1mm, the VSWRs of the antenna are increased, which means the matching performance is worsened. Furthermore, the degree of the deterioration will be serious when the thickness is more than 0.3mm. The phenomenon is caused by the mismatching of the resonant impedances. Figure 5 shows the impedance of the antenna with different thickness of the feed-line. Because the two staked patches produce two resonances, there are two peak values in the resistance curves seen from Figure 5(a). As the feed-line is thin (i.e. the curves of 0mm and 0.018mm), the resistances of the antenna are approach to 50-Ω in the operation band. In the meantime, the reactance of the antenna is close to zero. With the enhancement of the thickness, the resistances are increased. In particular, the resistances nearby the upper resonance are shot up. In addition, the range of reactance variation is increased extremely. These characteristics means the impedances of the antenna is mismatch 50-Ω coax connector.

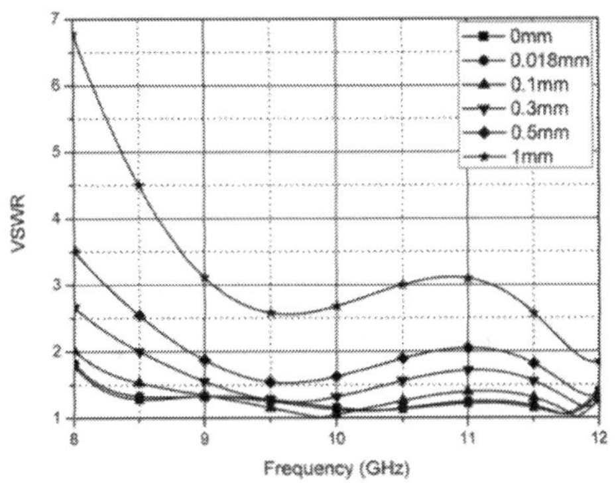

Figure 4. VSWR versus thickness of the feed-line.

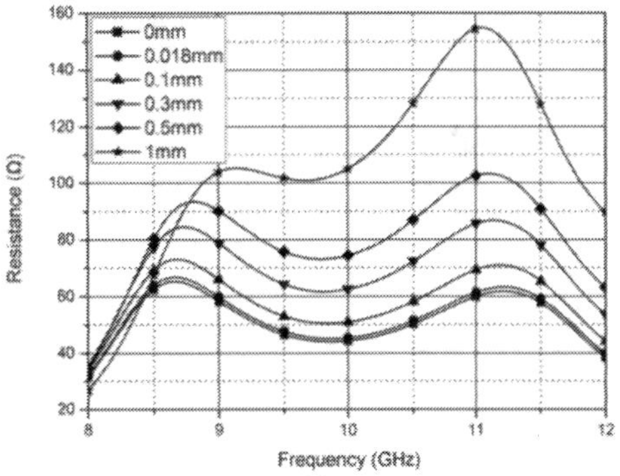

(a)

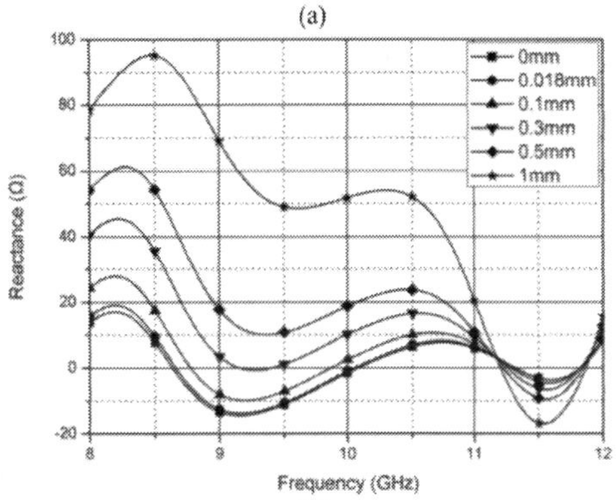

(b)

Figure 5. Input impedance versus thickness of the feed-line. (a) resistance; (b) reactance

The feeding structure influences not only the matching performance, but also the polarization purity. Figure 6 gives the maximum cross polarization of the antenna with different thickness of the feed-line. As compared with the effect of the feed-line on the matching

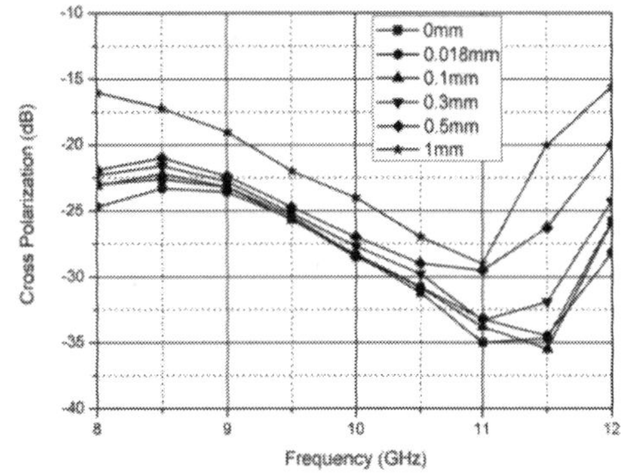

Figure 6. Maximum cross polarization level versus thickness of the feed-line.

performance, that on the cross polarization is significantly smaller with the thickness below 0.5mm. However, with the further increase of the thickness, the polarization purity degenerates appreciably. While its thickness equals 1mm, the cross polarization level is more than that of ideal condition by the maximum value of 14.7dB. The foregoing depictions indicate the feeding structure made up of the metallic plate is unfavorable for the application in high-frequency band.

The enlargement of bandwidth depends mainly on the stacked patch besides the feeding structure. The two patches affect the resonant resistance and frequency. Figure 7 demonstrates the impedance variations of different WP_l. As can be seen form Figure 7(a), the lower resonant resistance is improved with the increases of WP_l. Meanwhile, the frequency of the lower resonance is shifted downward to the lower-frequency band. As compared with the changes of the lower resonance, those of the upper resonance are relatively smaller. Its resistance and frequency are varied smoothly. With the further decrease relative to the optimum design, they remain stable basically. This means the lower patch dominates mostly the lower resonance.

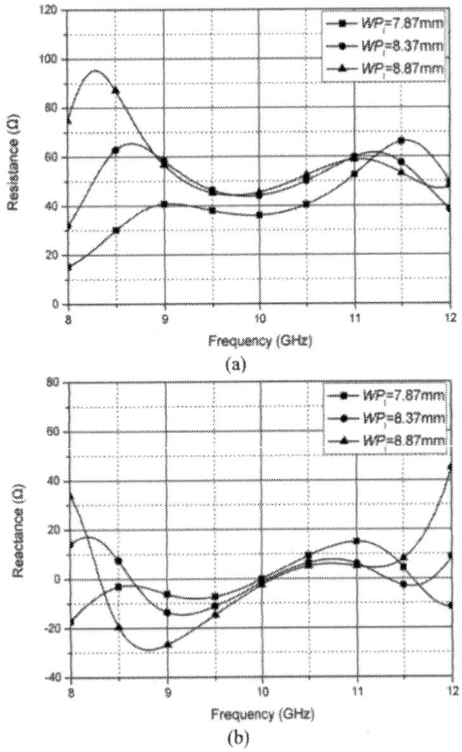

Figure 7. Input impedance versus WP_l. (a) resistance; (b) reactance

The effects on the impedance of different WP_u are shown in Figure 8. As WP_u increases, the lower resonant resistance is reduced, which is opposite to the trend with the changes of WP_l. This is because its extension means the lower patch size relative to the upper patch size decreases. It also can be seen that the increase of WP_u make the upper resonant frequency be shift downward to the lower-frequency band quite a bit, but the lower resonant frequency is almost unchanged. Also, the augment of WP_u results in increasing the upper resonant resistance. These behaviors indicate the upper resonance is influenced mainly by the upper patch. Actually, due to the coupling between the two patches, both of the two resonances are affected by the variation of each of them, which can be seen in Figures 7 and 8. The difference is the degree of their influences. On the whole, the large frequency interval and the small impedance alteration range between the two resonances contributes to improvement of the antenna bandwidth, and the balance of the two resonant impedances is prerequisite for achieving this goal.

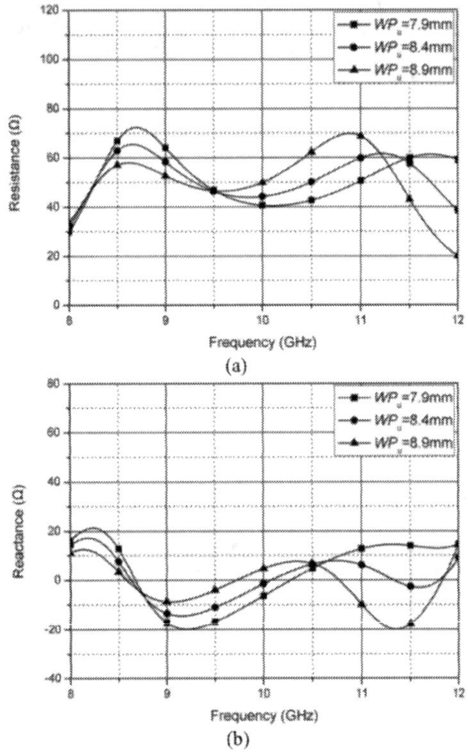

Figure 8. Input impedance versus WP_u. (a) resistance; (b) reactance

4 Conclusion

A wideband staked patch antenna fed by a printed meandering probe has been designed in this paper. The feeding structure helps in the broadband impedance matching of the antenna, and has the feature that easy of manufacture. The excellent simulated results of the antenna show the broadband over 40% (VSWR<2) and the cross-polarization below −23dB, which demonstrates that the method is effective. Furthermore, due to the simple and compact structure of the antenna, it is suitable for assembling an array.

References

1. S. D. Targonski, R. B. Waterhouse, D. M. Pozar, Design of wide-band aperture-stacked patch microstrip antennas, IEEE Trans. Antennas Propagat. **46**, 9, pp. 1245-1251 (1998)
2. Y. Qin, S. Gao, A. Sambell, E. Korolkiewicz, M. Elsdon, Broadband patch antenna with ring slot coupling, Electron. Lett. **40**, 1, pp. 5-6 (2004)
3. K. Oh, B. Kim, J. Choi, Design of dual and wideband aperture-stacked patch antenna with double-sided notches, Electron. Lett. **40**, 11, pp. 643-644 (2004)
4. I. Ang, B. L. Ooi, Broadband semi-circle-fed microstrip patch antennas, IET Microw. Antennas Propagat. **1**, 3, pp. 770-775 (2007)
5. D. Sun, L. You, A broadband impedance matching method for proximity-coupled microstrip antenna, IEEE Trans. Antennas Propagat. **58**, 4, pp. 1392-1397 (2010)
6. H. W. Lai, K. M. Luk, Wideband patch antenna with low cross-polarisation, Electron. Lett. **40**, 3, pp. 159-160 (2004)
7. H. W. Lai, K. M. Luk, Wideband Stacked patch antenna fed by meandering probe, Electron. Lett. **41**, 6, pp. 297-298 (2005)
8. S. Xu, Y. Rahmat-Samii, Boundary conditions in particle swarm optimization revisited, IEEE Trans. Antennas Propagat. 55, 3, pp. 760-765 (2007)

MATEC Web of Conferences 77, 01024 (2016) DOI: 10.1051/matecconf/20167701024
ICMMR 2016

Fault Identification Method of Ball Bearing Based on IAs and SVMs

Chao Chen, Zhaojun Yang,Heng Zhang, Binbin Xu ,Yifeng Ye and Yankun Wang

School of Mechanic Technology and Engineering, Jilin University, Changchun, 130025, China

Abstract. In order to effectively identify the bearing running condition, this paper proposed a new method which combines local mean decomposition (LMD) and support vector machine (SVM) together for ball bearing fault identification. Firstly, the gathered vibration signals were decomposed into a number of product functions (PFs) by LMD, with each PF corresponding to an instantaneous amplitude (IA) signal and instantaneous frequency (IF) signal. Then, introduce the concept of fault characteristic amplitude ratios which can be used to construct fault feature vectors; the extracted characteristic features were input into SVM to train and construct the fault identification model; the bearing running state identification was thereby realized. Cases of normal and fault were analyzed. Experimental results show that the proposed algorithm can diagnose the bearing failures reasonable and efficient.

1 Introduction

Bearing is the critical component of rotating machine which impacts the accuracy, reliability and lifetime of the machine. What's more, it's a common failure unit in rotating machines. Once it has any sort of faults, it would make the machine break down and lead to serious economic lose. Therefore, it is important to recognize the bearing running state [1].

Vibration measurements and signal analysis is a commonly used fault-detection technique employed in condition monitoring of ball bearings [2].The majority of the rotating machines use ball or rolling element bearings. Each bearing has one inner-race, one out-race and a set of balls or rolling elements [3].The ball bearings produce characteristic vibrations [4,5]. And when a defect develops in ball bearings, vibration characteristics and levels change. The vibrations caused by faults with distinctive characteristics reveal important information about the defect development within them that can be measured and compared with reference ones in order to perform the fault detection and diagnosis [6]. Hence, examining the vibration spectrum shows the information on the fault pattern. Normally, the process of ball bearing fault identification includes data acquisition, feature extraction and pattern recognition, and the later two are the priority [7].

The vibration signal of bearing has two typical characters. The first one is that bearing defect characters always perform as nonlinear and non-stationary which would affect the separation of the target signals, especially weaken the feature extraction of the fault information greatly [8]. Secondly compared with the complex noise the bearing defect characters are very weak which are usually submerged in the noise signals and hard to be picked out [9].

Targeting the first character, local mean decomposition (LMD) is proposed and applied to the fault feature extraction of ball bearings as LMD is successful for nonlinear and non-stationary signals LMD is a self-adaptive signal processing method to demodulate amplitude and frequency modulated signals [10]. By using LMD, vibration signal can be decomposed into a number of product functions (PFs), each of which is the product of an envelope signal and a purely frequency modulated signal from which physically meaningful instantaneous frequencies (IF) and instantaneous amplitudes (IA) can be obtained. Since the decomposition of the LMD is based on the local characteristics time scale of the data, LMD can be applied to nonlinear and non-stationary process perfectly [10,11].

As it is hard to obtain sufficient bearing fault samples in practice, in recent years, support vector machines (SVMs) have been widely applied for bearing fault diagnosis because SVMs specialize for small number of samples [12]. SVMs are supervised learning models with associated learning algorithms that analyze data, classify data and recognize fault patterns. Due to the noise in the vibration signal, the data need be preprocessing to get fault feature. Author uses Local mean decomposition (LMD) method to reduce the noise impact. And the fault feature vectors extracted by LMD were input into SVM to classify the fault pattern of ball bearing.

The rest of the article is organized as follows: a brief introduction to the LMD and SVMs are provided in Section 2 and 3 respectively. Section 4 presents the proposed method based on fault diagnosis method based on LMD and SVMs. Section 5 shows the detailed analysis of a case study using our proposed method. Section 6 draws a short conclusion.

2 LMD

LMD is a novel self-adaptive method based on partial characters of the signal which decomposes vibration signal into a linear combination of a set of mono-component signals. The LMD method is a fully data driven approach which involves progressively separating a FM component from an AM envelope signal and gets a series of PFs, instantaneous amplitude (IAs) and instantaneous frequency (IFs) [13]. The process of computing IA and IF by LMD is shown as Fig. 1.

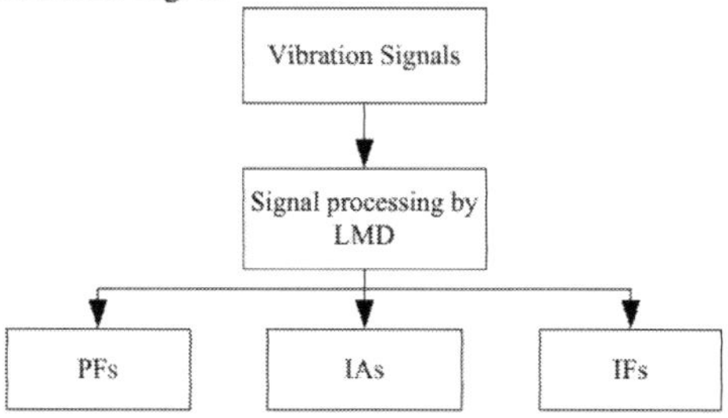

Figure 1. The sketch of computing IA and IF by LMD

3 SVMs

SVM is developed from the optimal separation plane under linearly separable condition. SVM uses a hypothesis space of linear functions in a high dimensional feature space to estimate decision surfaces directly rather than modeling a probability distribution across training data [14]. It uses support vector kernel to map the data from input space to a high dimensional feature space which facilitates the problem to be processed in linear form. The maximum-margin hyper-plane and margins are used to train trained samples from two classes [15]. The nearest samples are used to define the margin and are called support vectors. The basic principle of SVM is shown in the Fig. 2.

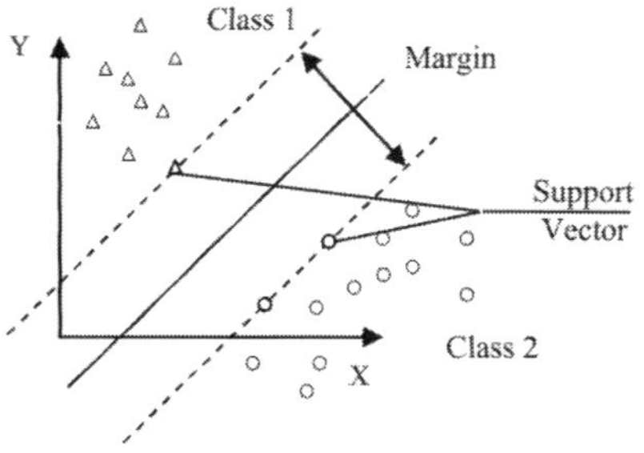

Figure 2. Basic principle of SVM

4 Fault diagnosis dombined LMD and SVM

While the ball bearing with different faults is operating, the corresponding special frequencies are produced in the vibration signals. Considering high-frequency band of vibration signal contains the key fault information of ball bearing [16], the fault characteristics could be extracted from the first numbers of PFs component. As PFs contain the information of the impulse and its severity level in each period, the fault characteristic parameters defined from PFs should reflect the working condition and fault pattern accurately. Namely, each PF obtained by LMD can be used to extract the fault feature of sample data. Yu [16] used Hilbert transformation to get the envelope spectrum of the envelope signal based on EMD. What's more, some characters of the envelope signal may be lost in Hilbert transformation. However, LMD gets the envelope spectrum directly and avoids information lost. Compared with support vector based on EMD, the support vector from LMD is more accuracy.

Based on the peak value could be found at corresponding fault characteristic frequencies in PFs when ball bearing has fault, three fault characteristics, namely fault instantaneous amplitude ratios (FIAR) γ ρ and η are designated as:

$$\gamma = \frac{A(f_o)}{A(f_i)},\ \rho = \frac{A(f_r)}{A(f_i)},\ \eta = \frac{A(f_c)}{A(f_i)} \quad (1)$$

where, f_i, f_o, f_b are ball bearing inner-race, out-race and ball fault characteristic frequencies respectively. $A(f_i), A(f_o)$, $A(f_b)$ and $A(f_c)$ are the corresponding instantaneous amplitude (IA) of the PFs at fault characteristic frequencies. The fault characteristic frequencies of the ball bearing with inner-race fault, out-race fault , ball fault and cage can be calculated as follows [17]:

$$f_i = \frac{1}{2}(1 + \frac{d}{D_m}\cos\alpha) f_n Z \quad (2)$$

$$f_o = \frac{1}{2}(1 - \frac{d}{D_m}\cos\alpha) f_n Z \quad (3)$$

$$f_b = \frac{1}{2}\frac{D_m}{d}(1 - (\frac{d}{D_m})^2 \cos^2\alpha) f_n Z \quad (4)$$

$$f_c = \frac{1}{2}(1 - \frac{d}{D_m}\cos\alpha) f_n \quad (5)$$

where d and Z are the diameter and number of the rolling elements, respectively; D_m is the pitch circle diameter; α is the contact angle of the roller bearing; f_n is the rotation frequency of the shaft.

The fault diagnosis method is given as following:

1)Sample N times at a certain frequency f_s under the condition that the ball bearing is normal, the ball bearing with inner-race faults, the ball bearing with out-race faults and the ball bearing with ball faults, respectively. And the 4N signals are taken as samples that are divided into two subsets, the training samples and test samples.

2) The original vibration signals are decomposed into some PFs and the first numbers of PFs that include the most dominant fault information are chosen to extract the feature.

3) For each PF, get the corresponding IA and IF. According to Eq. (1) calculate the characteristic amplitude ratios γ, ρ and η.

4) Design SVM classifiers. For SVM1, define the condition with inner-race fault as y=1 and the other 3 conditions as y=-1 thus the inner-race fault could be separated from other conditions by SVM1. Then, define the condition with outer-race fault as y=1 and the other conditions as y=-1 for SVM2, thus the outer-race fault could be separated from normal and ball fault condition by SVM2. Similarly, the ball fault could be separated from normal condition by SVM3.

5) The characteristic amplitude ratios of the training samples serve as the fault feature vectors to be input to the SVM and the classifiers are trained.

6) Actually the decision function $f(x)$ is determined only by the support vectors, so after they are obtained the fault characteristic vectors of test samples can be input to the trained SVM classifiers and then the working condition can be identified by the output of the SVM classifiers.

The flow chart of the identification method based on LMD and SVM are show in Fig. 3

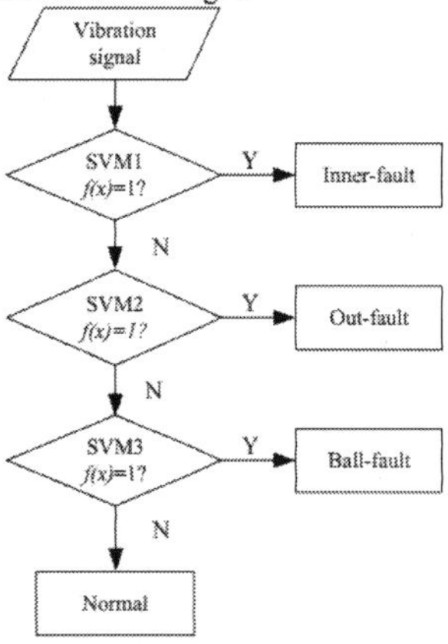

Figure 3. The flow chart of fault identification by SVM1, SVM2 and SVM3

5 Data acquisition and signal preprocessing

To evaluate the effectiveness of the proposed method based on LMD and SVMs for bearings, the test data were acquired by bearing accelerometer sensors under different operating loads and bearing conditions.The test bearing type is 7205C, which is a deep groove ball bearing. Accelerometer sensors were placed at the 12 o'clock position at the drive end of the motor housing. The test bearings support the motor shaft and the load was at the speed of 1800 rpm. The sampling frequency is 10kHz; and the sampling points are 1000. The vibration signals of ball bearing were selected as normal, out-race fault, inner-race fault and ball fault.

Eight vibration signals of ball bearings in each condition are obtained at random in which 2 groups are drawn out respectively as the training samples and the rest are taken as test data. Decompose the vibration signals of above four conditions, namely normal, inner-race fault, out-race fault and ball fault with LMD method into a number of PFs according to Ref. [13]. The one group results of LMD decomposition are shown in Fig. 4, respectively. The decomposed results have 3 PFs and 1 residue. The residue can be decomposed furthermore, but it makes no sense.

As main information of fault vibration signal of ball bearing is included in high-frequency bands, only first three PFs are chosen. According to Eqs. (2)–(5), the calculating results of fault characteristic frequencies of the ball bearing with inner-race fault, out-race fault ball fault and cage are 107.36 Hz,162.19 Hz, 141.17Hz and 18.02Hz, respectively. According to Eqs. (1), calculate the characteristic amplitude ratios γ_i , ρ_i and η_i in each PF envelope spectrum, which denotes the characteristic amplitude ratios of PF1, PF2 and PF3 .

Table 1. The identification results of proposed method for ball bearing

Test samples	Characteristic vector			SVM1		SVM2		SVM3		Fault pattern
	(1)	(2)	(3)	Distance	Re.	Distance	Re.	Distance	Re.	
Inner-race fault	0.7121	0.5524	0.4486	0.9120	1					Inner-race fault
Inner-race fault	0.5141	0.5554	0.6129	1.0531	1					Inner-race fault
Out-race fault	15.9789	28.0421	17.1438	-22.8354	-1	38.6380	1			Out-race fault
Out-race fault	4.5184	7.6925	6.6721	-4.3762	-1	9.2407	1			Out-race fault
Ball fault	1.3546	3.7431	2.3479	-2.0211	-1	-4.2254	-1	5.3312	1	Ball fault
Ball fault	1.3650	3.7159	1.8772	-1.8644	-1	-3.9307	-1	6.0255	1	Ball fault
Normal	0.6381	1.2295	1.7870	-0.9537	-1	-1.3854	-1	-2.7208	-1	Normal
Normal	0.7762	0.8196	2.0920	-0.7074	-1	-0.9113	-1	-1.3113	-1	Normal

Serve γ_i and ρ_i as the fault characteristic vectors to input to the SVM classifiers. By Ref. [14], design theclassifiers SVM1-3. After SVM classifiers have been trained, the fault characteristic vectors of test samples are input into the trained SVM classifiers and then the working condition is classified by the output of the SVM classifiers. In this progress, the inner-race fault is indentified by SVM1successfully. Finally, by SVM1-3, all the classification results are shown in Table 1 from which we can see that three SVM classifiers can identify the working conditions and fault patterns of ball bearing accurately.

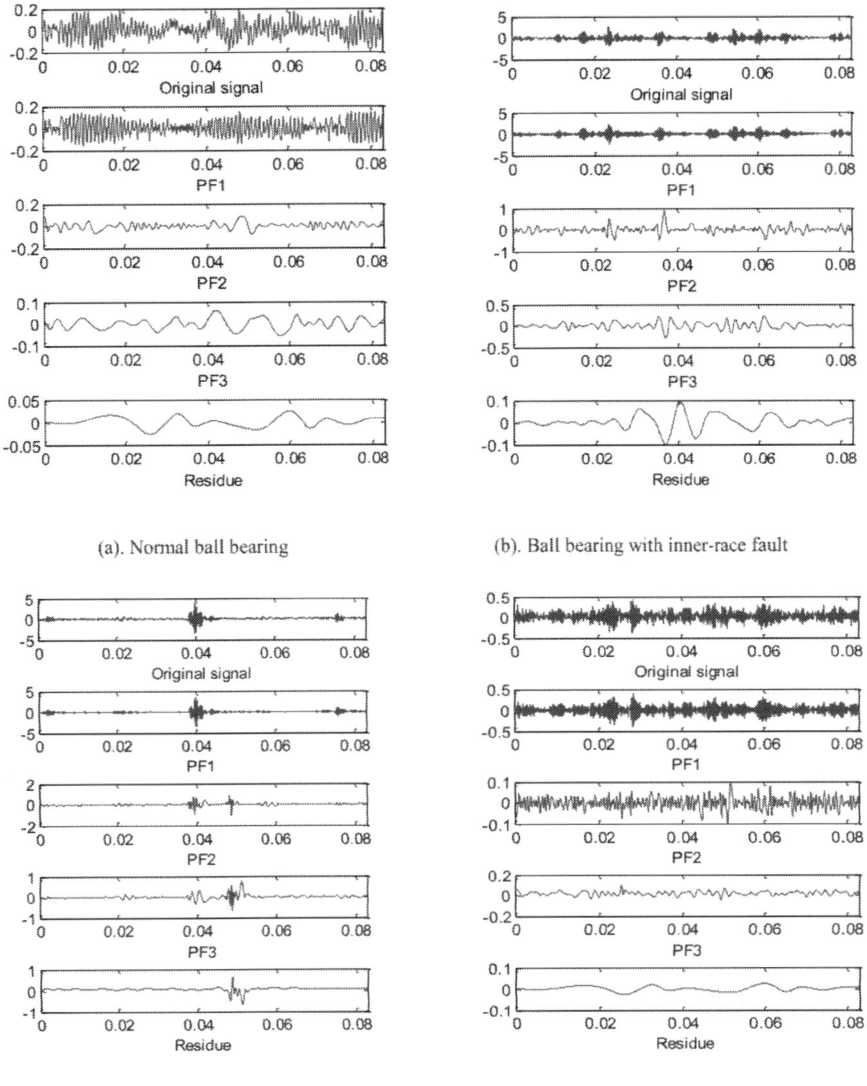

(a). Normal ball bearing (b). Ball bearing with inner-race fault

(c). Ball bearing with out-race fault (d). Ball bearing with ball fault

Figure 4. The original signal and LMD decomposed results of ball bearing

6 Conclusions

In this paper, original vibration signals of ball bearing are decomposed by LMD. Considering each PF corresponding to an IA signal and IF signal could reflect the intrinsic and real information of the analyzed signal, the characteristic amplitude ratios of fault characteristic frequencies in fault bearing and corresponding frequencies of normal one in each IA are proposed. And the fault characteristic extraction method based on PF envelope spectrum is proposed in this paper. In addition, unlike EMD needs Hilbert transform to get the IAs and IFs, SVM which gets the two signals directly avoids some information losing. It's more accuracy than EMD method. The paper shows the analysis results of four kinds of ball bearing vibration signals among which one is normal and the other three have inner-race, out-race and ball faults, respectively, that the fault diagnosis approach of ball bearing based on PF, IA and SVM can be applied to classify the working conditions and fault patterns of ball bearing effectively and accurately.

Acknowledgments

This work was financially supported by the Specialized Research Fund of the Application engineering of one thousand sets of CNC machines with Dometic CNC systems (Grant No. 2014ZX04001506l), the Postdoctoral fund project of Jilin province (Grant No. 801130020414)

References

1. W. Guo, P.W. Tse. A, Djordjevich, Faulty bearing signal recovery from large noise using a hybrid method based on spectral kurtosis and ensemble empirical mode decomposition, Meas. **45** (2012) 1308-1322
2. M. Zhao, J. Lin, X. Q. Xu, X. J. Li, Multi-fault detection of rolling element bearings under harsh working condition using IMF-based adaptive envelope order analysis, Sen **14**(2014) 20320-20346
3. W. T. Sui, S. Osman, W. Wang, An adaptive envelope spectrum technique for bearing fault detection, Meas. Sci. Tech, **25**(2014) 095004
4. J.P. Yang, S.X. Chen, Vibration predictions and verifications of disk drive spindle system with ball bearings. Comput Str. **80**(2002)1409-18.
5. R. Aini, H. Rahnejat, R.Gohar, Vibration modeling of rotating spindles supported by lubricated bearings. J Tri.**124**(2002)158-65.
6. P. Konar, P. Chattopadhyay. Bearing fault detection of induction motor using wavelet and Support VectorMachines (SVMs). Appl. Soft Com **11** (2011) 4203-4211
7. J. Yang, C.K. Peng, Y.S. Xu. Hierarchical entropy analysis for biological signals, J mput. Appl. Math **236** (2011) 728-742.
8. H. Xu, G. Chen. An intelligent fault identification method of rolling bearings based on LSSVM optimized by improved PSO. Mech. Sys. Sig. Pro **35** (2013) 167-175
9. B.T.Holm-Hansen, R.X. Gao, Vibration Analysis of a Sensor Integrated Ball Bearing. J Vib. Acou **122** (2000) 384-392
10. J. S. Smith, The local mean decomposition and its application to EEG perception data, J. R. Soc. Int **2** (2005) 443-454.
11. Y, Bu, J, Wu, J, Ma, et. al. The rolling bearing fault diagnosis based on LMD and LS-SVM. 26th CCDC 3797-3801
12. N. Cristianini, J. Shawe-Taylor. *Support Vector Machines and other Kernel Based Learning Methods*, Cambridge University Press, 2000
13. Y. Wang, Z. He, Y. Zi, A demodulation method based on improved local mean decomposition and its application in rub-impact fault diagnosis. Meas. Sci. Tech. **20**(2009)025704
14. J. Chenga, D. Yua, J.Tangb, Y.Yanga, Application of SVM and SVD technique based on EMD to the fault diagnosis of the rotating machinery. Shoc. Vib. **16**(2009) 89-98
15. J. Lee. *Advanced Electrical and Electronics Engineering*, Springer Science & Business Media, **2**(2011) 425
16. Y. Yang, D. Yu, J. Cheng. A fault diagnosis approach for roller bearing based on IMF envelope spectrum and SVM. Meas. **40** (2007) 943-950
17. S. Nandi, H. A. Toliyat, X. D. Li, A. Richard, condition monitoring and fault diagnosis of electrical motors-a review, IEEE T Energy Conver, **20**(2005) 719-729.

Early estimation of work contents for planning the one-of-a-kind production by the example of shipbuilding

Jan Niklas Sikorra, Axel Friedewald and Hermann Lödding

Hamburg University of Technology, Institute of Production Management and Technology, 21073 Hamburg, Germany

Abstract. A transparent, reliable early estimation of work contents is crucial for capacity planning in shipyards. Yet most of the shipyards rely on practical experience of specialized production planners, for the early capacity estimation, which are prone to errors, as a bill of material is not available in the early stages of production planning in the one-of-a-kind production. This is due to a yet unfinished construction of the ship. This paper will discuss a method for creating a computer-supported, transparent, more reliable way for the estimation of work content, at an early planning stage of a one-of-a-kind production, using a generated bill of material. The bill of material is generated automatically using parametrized templates and contains steel and outfitting parts. Therefore a tool has been developed that generates steel and outfitting parts from parametrized templates. From the bill-of-material, the work content in hours is estimated by linking the parts in the bill-of-material to a process. The operation time can then be calculated by dividing the work content by the capacity of the work system. The operation time can then be used for scheduling the one-of-a-kind production.

1 Introduction

Work content estimation in the early phase of planning the one-of-a-kind production of complex products is uncertain and prone to inconsistencies. Yet decisions made in the early phase of production planning in the one-of-a-kind production have a high impact on delivering on time. Furthermore as the price of the product is fixed, costs have to be estimated, which makes wrong estimations costly. Therefore generating better information for production planning in the one-of-a-kind production is a necessity to gain an advantage against competitors on the world market.

The uncertainty in the early phase of planning the one-of-a-kind production origins from a lack of design data, as design, production planning and work scheduling have to be done in parallel, to minimize the throughput time of a one-of-a-kind product.

The paper discusses an approach on how to estimate the work content in the early phase of planning the one-of-a-kind production by coping with the lack of design data, by the example of shipbuilding. It therefore describes a way of generating a bill of material (BOM) from past data, using parameterized templates. It also introduces a parameter based estimation for calculating the work content from the part data in the BOM. The work content can then be used for creating more realistic schedules and for estimating the capacity demand.

2 State of the Art

One-of-a-kind production means, that a product is only produced once. In the one-of-a-kind production of complex goods, like a building, a plant or a ship, design and planning are often done in parallel, which causes a lack of data in the early phase of planning a one-of-a-kind product [1–3].

In the early phase of planning the one-of-a-kind production and scheduling the work, little design information of the one-of-a-kind product is available, as the planning is done in parallel to the design, to reduce the throughput time of the whole project. Due to this a target conflict arises between having design data available to enable a certain planning and having a low throughput time for the whole project and low costs in total [4].

If all parts of a product, as well as the parts hierarchy were known and documented in the BOM and the processing times, lot sizes and set up times were known, the work content in hours could be calculated [5]. From the work content, the operation time in days can be derives, which is needed to schedule the production.

Yet without the design, the exact work steps are not known yet. Therefore calculating the work content has to base on vague estimations and old ships data. While there are the DIN 277 and DIN 276 for calculating costs of constructions, and a method to do so, such a method does not exist in shipbuilding [6]. There also are approaches of estimating cost for plants, whereas the parts are known

DOI: 10.1051/matecconf/20167701025

for the cost estimation the methods either work bottom up or on each level of the design hierarchy [7–9].

There is the necessity to guess a BOM at the early stage of planning a one-of-a kind production. Shipyards currently reuse old design data, but as the ships change tremendously, the approach of just reusing data does not deliver sufficient results [4].

First approaches to develop a tool to generate simple part data have been made in the research project GeneSim, in a different context. The idea in GeneSim was to fill gaps in simulation databases by generating parts. For this purpose a very simple software to create basic structural steel parts (profiles and plates) was developed [10]. The software is not yet able to create a sufficient BOM.

As a result the question is raised whether a BOM can be guessed at the early stage of planning. A sufficiently good BOM could on the one hand close the gap that the unfinished design creates in the early stage of planning the one-of-a-kind production, on the other hand it could create a new step in today's typical planning of the one-of-a-kind production which would lead to a higher transparency.

3 Planning workflow for work content estimation

Estimating a bill of material while planning the one-of-a-kind production delivers two advantages. The first advantage of estimating a BOM is a higher transparency in the planning process, as product uncertainties and process uncertainties are separated. This enables a profound controlling and identification of errors in the planning process is possible, which enables the planners to learn in a systematic way from their mistakes. The second advantage is, that a sufficiently good estimation of the BOM can improve the certainty of the planning.

This paper describes a new approach for planning the one-of-the-kind production (Figure 1).

It differs from traditional approaches by taking into account an estimated BOM. The BOM fills a knowledge gap in planning the one-of-a-kind production, by estimating important parts, which are not yet constructed. The approach as shown in figure 1 consists of eight steps. The first four steps deal with estimating parameters for the generation of important parts for the BOM. Steps five to seven deal with generating the BOM. Step eight describes the estimation of the work content. The work content can then be used for cost estimation and production scheduling.

The following paragraphs will describe each step of the ne planning workflow estimating a BOM in the one-of-a-kind production.

3.1 Product specifications

Already when negotiating a contract, first product specifications are set. Latest by closing the contract, all product specifications for the one-of-a-kind product are fixed. These product specification are the base for the planning approach described in this paper and should be used as early as they are available.

An example for a product specification in the shipbuilding industry would be the ice class. Knowing the ice class, product specifications can be used for calculating global and local parameters.

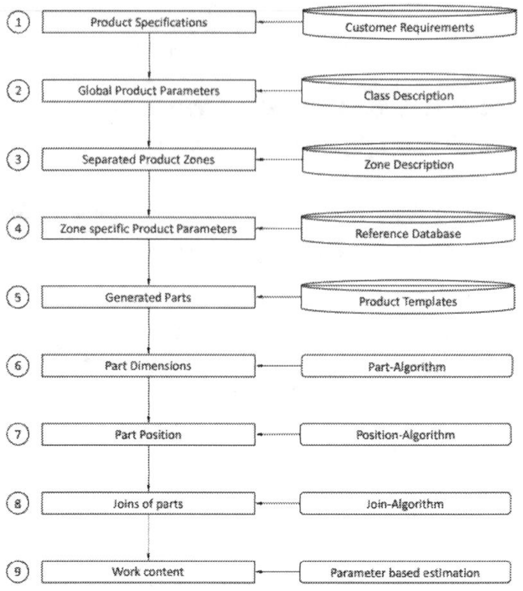

Figure 1. New approach for planning the one-of-a-kind production

3.2 Global product parameters

Global product parameters are derived from the ships specifications. To differentiate global and local parameters minimizes the work content for the planner, as global parameters are only derived once for a ship. The parameters are needed for the calculation of the BOM, as well as the local parameters. Global parameters can either be derived from guidelines, from former projects or calculations. For example the ship specification ice class would define a global parameter, the thickness of the hull.

3.3 Separated product zones

To create clusters with similarities in the design, the ship is divided into zones. While different zones might differ a lot, a zone is assumed to bare a lot of similarities. This allows the estimation of zone specific parameters which describe a zone relatively precisely. Therefore after global parameters are set, zones will be defined, which enclose an area with a specific function, for example the engine room of a ship, or the double bottom. If parameters were derived for areas with strongly differing functionalities, the variance of parameters in this zone might be too big and therefore the results of the estimation based on the parameters would be uncertain.

3.4 Zone specific product parameters

 DOI: 10.1051/matecconf/20167701025

Now that the one-of-a-kind product is divided into zones, zone specific parameters can be derived. Zone specific parameters are used to describe specifications that cannot be described on a global level with enough certainty. The parameters can once again be derived from directives, old projects or calculations. On a Ferry for example the stiffener spacing, width and height on the car deck can be derived from the expected load.

3.5 Generated parts

The calculating of the parts and their hierarchy starts at step five of figure 1. The goal of calculating parts and hierarchy is to create a BOM that reflects the new build one-of-a-kind product in the best way possible and can be used for work content estimation.

To minimize the planning efforts, the BOM is generated using templates. The templates can be parametrized using the global and local parameters derived in step one to four. The templates are structured hierarchic to reflect the structures of the designs, which's data they inhabit. The result is a hierarchic list of the parts, of the estimated design. For example a design panel (figure 2) can be created as a hierarchy consisting of the panel on the top level and below this plates and profiles.

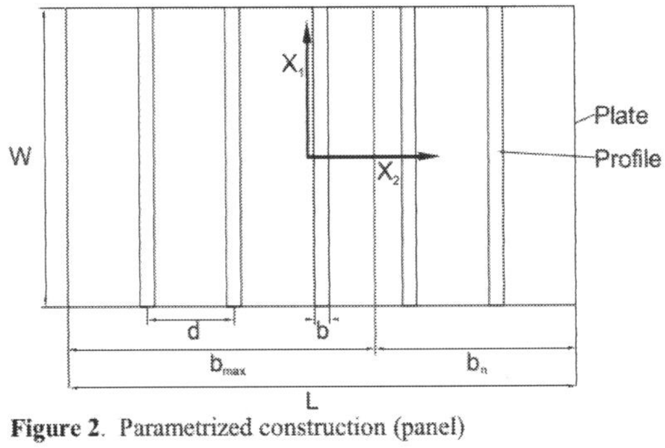

Figure 2. Parametrized construction (panel)

A complete BOM must inhabit all parts of the design. If a part occurs several times, all of the reoccurrences have to be calculated. Therefore the number of profiles ($n_{profiles}$) in the construction in figure 2 can be derived by rounding down the length of the construction (L) divided by the distance of the stiffeners (d) according to equation 1.

$$n_{profiles} = \left\lfloor \frac{L}{d} \right\rfloor \quad (1)$$

For a complete BOM, the reoccurrence of parts, which have a maximum size due to delivery or production constraints has to be calculated. An example for this is a plate field. In figure two, the maximum length of a plate is parameters is the parameter b_{max} in figure 2. The parameter b_{max} specifies the maximal length of a single steel plate. After a length of b_{max}, the plate has to be split. Therefore the number of single plates (n), a field of plates consists of is defined by the width (W) and length (L) of the field of plates and the maximum width (W_{max}) and maximum length (L_{max}) of the plate. Equation 2 shows the calculation of the number of plates by the example of length. The number of plates always is an integer, therefore the results of equations one and two have to be rounded up. Therefore n_L is the number of plates placed in the length direction.

$$n_L = \lceil L/b_{\max} \rceil \quad (2)$$

The same equation applies for calculations in width-direction. n_W therefore is the number of plates in width direction (equation 3).

$$n_W = \lceil W/W_{max} \rceil \quad (3)$$

The total number of plates in a field of plates calculates to n as stated in equation 4.

$$n = n_L \cdot n_W \quad (4)$$

Analog to the equations stated above, reoccurring parts can be calculated and included in the hierarchy calculated for the BOM.

3.6 Part dimensions

For the calculation of the work content, part dimensions are important, as the work content often varies with part specific parameters. An example would be a profile, which's length would determine the length of the joins, on which it would be welded onto a plate. Part dimensions can be set in three ways. The first possibility is that they are either are defined by global or local parameters, like a maximum width, or a default width. The second possibility is that the dimensions are defined by a constraint of the construction. For example the length of a profile would equal the width of the panel in figure 2. The third possibility is that they can be calculated. For example the width of the last plate in a field of plates would be calculated to b_n in figure 2. To calculate the length b_n, the length of the combined $n-1$ plates before the n-th plate are subtracted from the length of the complete construction L (equation 5).

$$b_n = L - b_{max} \cdot (n_L - 1) \quad (5)$$

3.7 Part position

The work content of a complex designs can depend on more parameters than the dimensions of the parts included. To calculate the work content for complex design, the position information for each part is important. For example in shipbuilding, the position information allows for the calculation of the joins of parts (step 8). This is due to three reason. The first reason is, that knowing the positon of a part, position specific parameters for calculating the work content can be introduced. The second way is a calculation of the joins in complex constructions. The intersections of the parts can differ from the length of the part. Therefore join estimation is more precise for calculating the work content than a length based parameter estimation. The position also allows part selection by defining a new sub design. The position of the part determines whether the part is in the sub design.

The calculation of the positions is divided into two steps. First, position for the templates at the bottom of the template hierarchy are calculated using a coordinate system that is placed in the center of a construction. In the second step, if templates are combined, the coordinate systems of the sub designs are transformed into the parent design.

In step one the position of each part in this coordinate system can be specified using a vector. Reoccurring parts are placed on a line. The calculation as shown in equation 6 will be explained by the example of a profile. Equation six shows the equation for the calculation of the position of reoccurring parts where i is the parts name and λ is the index of parts.

$$X_{i\lambda} = (\lambda - 1) \cdot \begin{pmatrix} x_1 \\ x_2 \\ x_3 \end{pmatrix} + \begin{pmatrix} c_1 \\ c_2 \\ c_3 \end{pmatrix} \quad (6)$$

The coordinates of the profiles in figure 2 can be calculated using equation 7. Therefore the coordinates of the j-th profile can be calculated to $X_{pr,j}$.

$$X_{pr,j} = (j-1)\begin{pmatrix} 0 \\ d \\ 0 \end{pmatrix} + \begin{pmatrix} 0 \\ -L/2 + d \\ 0 \end{pmatrix} \quad (7)$$

The calculation of the position of each part of an assembly can always be calculated using vectors.

If a complex construction is created, construction templates are combined using coordinate system transformations with Euler angles. Figure 3 shows an example of a construction created with the described method.

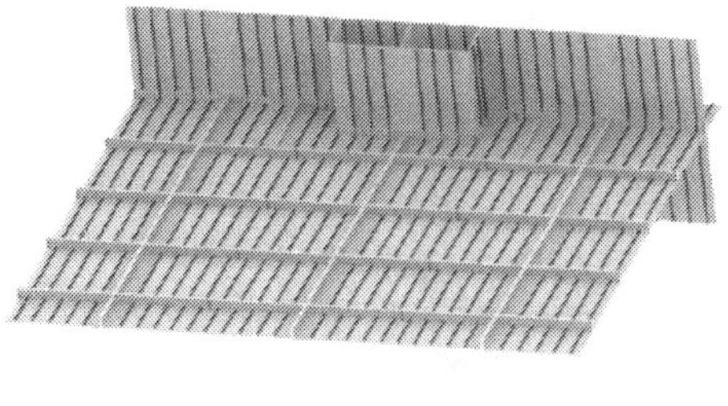

Figure 3. Generated ship's block

The advantage of the coordinate system transformation is that common assemblies can be reused and combined to create complex assemblies on a higher level in the hierarchy. Therefore the effort for describing the complex object is minimized.

3.8 Joins of parts

The example of shipbuilding includes a lot of steel construction. Main driver for work content when building a ship's steel hull is welding joins. Therefore a method has been developed to identify areas of contact of parts and calculate their edges and their positions. The edges are the joins of the parts. The algorithm for automatic join creation can also be used to calculate the joins of subassemblies to each other and therefore calculate the work content for different hierarchic levels of the manufacturing and assembly process in the production. This enables a faster and more reliable way of work content estimation. As the calculated joins can directly be used for parameter based work content estimation.

3.9 Work content

The work content (WC) in hours can be calculated using equation 7 (step eight of figure 1). The equation bases on work content estimation methods. Therefore the work content for a specific part can be calculated by a part specific parameter ($P_{Product}$), like its length, divided by its specific process speed ($v_{Process}$) in 1/min. The part specific parameter is introduced as the work content for the same part type might vary due to the parts volume for instance.

$$WC_p = \frac{P_{Product}}{V_{Process}} \cdot \frac{1}{60} \quad (8)$$

A basic table of possible parameters was created in the research conducted. For a welding the length of the join would be divided by the speed of welding. For an isolation, the work content of an isolation of a room would for example be calculated by the surface of the volume divided by the speed of isolating.

With the work content calculated, either costs for the work content of a one-of-a-kind product, in this case a ship, can be calculated, or the production can be scheduled using the estimated data.

4 Software prototype

The method described in this paper was implemented into a software prototype. It consists of two parts. The first part is the algorithm that enables the selection of templates and that calculates components, parts and their metadata. The second part displays the generated data and allows for its selection and manual manipulation. The software prototype is built to estimate a complete ship.

The old data is stored in templates, which can be created by hand. The templates created from old data are structured in the same hierarchy as the design. The template also specifies the parameters, the planner has to enter in order to calculate the parts, in our example the distance of stiffeners for example. Templates can be used nested, therefore one template can include another one and pass the parameters needed for the calculation to it. The software prototype then does the calculation for each of the nested templates. The nesting of the template prohibits reoccurring creation of the same assemblies. The software also contains a chooser for the templates. This tool allows the planner to visually choose a template for estimating a new assembly, which he thinks fits best for the future-to-build ship, which he is currently working on. The XML-Picker is shown in figure 4.

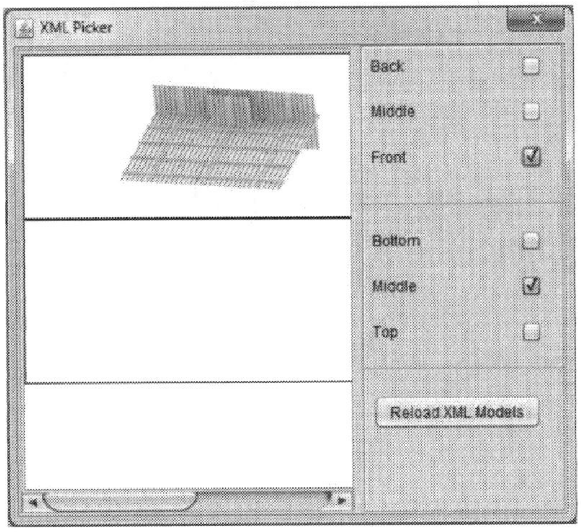

Figure 4. XML-Picker

The template is parsed by the software prototype, and the parts as well as the hierarchy are created. The parts and the hierarchy are than calculated by the first part of the software. The parts and their hierarchy are calculated using the mathematics described above. The parts calculated are then written to a database or output. The database can be used for manufacturing simulations or for work content estimations. Figure 5 shows the output of the parts in the software prototype.

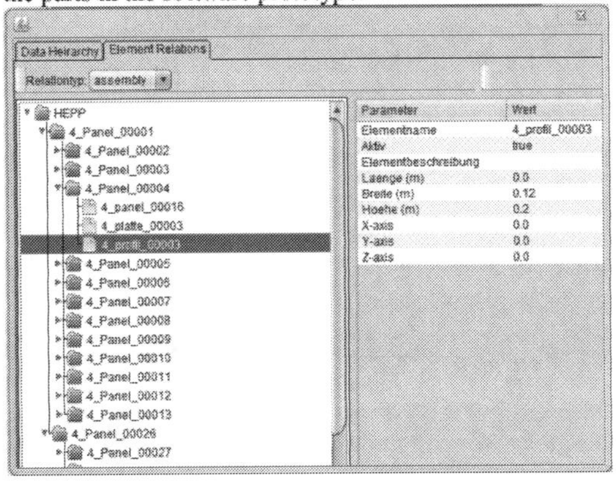

Figure 5. Part view in software prototype

5 Conclusion

The above paper explains on how to estimate the work content of a one-of-a-kind product in the early stage of planning. At first a logic on how to describe parts and assemblies in a way, that they can be parameterized, was developed.

In the second step, a software prototype was developed, to validate the logic developed. The software prototype can create parts of an assembly using parametrized templates.

In the future this approach will be tested in reality to allow to compare the estimations done using the tool to the estimations of the current estimation methodology.

Acknowledgments

The research, described in this paper was conducted in the research project HEPP, in cooperation with German shipyards, funded by the German Federal Ministry of Economic Affairs and Energy (BMWi) due to a decision of the German Bundestag.

References

1. F. Berner, B. Kochendörfer, R. Schach, Grundlagen der Baubetriebslehre 1. (2007)
2. U. Bracht, S. Wenzel, D. Geckler, Digitale Fabrik, 1. Aufl. (2009)
3. U. Stalleicken, Marktorientierte Produktion komplexer Investitionsgüter - Analysen, Strategien und Konzepte am Beispiel des deutschen Schiffbaus. (2000)
4. R. Lee Storch, C. P. Hammon, H. M. Bunch, R. C. Moore, Ship Production. (1995)
5. H.-P. Wiendahl, Fertigungsregelung: Logistische Beherrschung von Fertigungsabläufen auf Basis des Trichtermodells. (1997)
6. P. J. Fröhlich, Hochbaukosten-Flächen-Rauminhalte: DIN 276-DIN 277-DIN 18960; Kommentar und Erläuterungen. (2008)
7. B. Ebert, Technische Projekte, 1. Aufl. (2002)
8. C. Lühe, Modulare Kostenschätzung als Unterstützung der Anlagenplanung für die Angebots-und frühe Basic Engineering Phase. (2013)
9. P. Römisch, M. Weiß, Projektierungspraxis Verarbeitungsanlagen. (2014)
10. H. Lödding, A. Friedewald, L. Wagner, R. Wandt, Generisches Daten- und Modellmanagement für die schiffbauliche Produktionssimulation. (2011)

MATEC Web of Conferences 77, 01026 (2016) DOI: 10.1051/matecconf/20167701026
ICMMR 2016

The Sucker Rod Pump Parameters Optimization of Horizontal Wells with ASP Flooding Considering the Influence of Centralizers

Zhen-Hai Jiang

No.3 Oil Production Plant, Daqing Oilfield Company, China

Abstract. The centralizers can add the polished rod load, and affect the pumping system efficiency. A pumping system efficiency model has been built by considering the influence of centralizers on polished rod load. The influence of centralizers, stroke length, pumping speed and flooding period on sucker rod pumping system efficiency and production has been studied. The results indicate that the sucker rod pumping system efficiency and production rate decrease while the number of centralizer increases, but the influence of number of centralizer on sucker rod pumping system efficiency and production rate is not big while the influence of stroke length and pumping speed is significant. The sucker rod pumping system efficiency first increase obviously and then trend to be gentle with the stroke length and pumping speed increasing. The pumping system efficiency in main slug of ASP flooding is lowest. The optimization value of stroke length and pumping speed has been selected to maximize the pumping system efficiency. And hence get the reasonable pump depth, and system efficiency, it have reference significance to the reasonable production.

1 Introduction

The viscosity of production fluid is high which can increase the eccentric wear between sucker rod and tubing [1], [2], so it is necessary to install centralizers on sucker rods, but the centralizers will increase the friction of the sucker rod and tubing and liquid column in the up and down stroke so it can add the polished rod load, and affect the pumping system efficiency [3], [4]. When the wells oil production is stable, the flow of the formation into wellbore is balanced with the actual flow of sucker rod pump, the different swabbing parameter of the sucker rod pump correspond different bottom hole producing pressure and production.

This paper studies the influence of centralizer and sucker rod pump parameters on the efficiency of the system efficiency. With the highest system efficiency as the goal, based on the pumping system efficiency calculation model, pumping system efficiency calculation model and the influence of various parameters on the system efficiency, the influence of centralizers, stroke length, pumping speed and flooding period on sucker rod pumping system efficiency and production has been studied. It has reference significance to the reasonable production.

2 Pumping system efficiency calculation model

Pumping system efficiency calculation model considering the influence of gas in the pump, volume change of the oil, pump leakage and stroke loss on pumping system efficiency.

When ASP flooding oil well produced fluid flow in the wellbore, it Performance for an emulsion of oil- gas-water three phases.

$$\rho_{\mathrm{m}} = \frac{[(\rho_{\mathrm{o}} + R_{\mathrm{p}}\rho_{\mathrm{g}})(1-f_{\mathrm{w}}) + \rho_{\mathrm{w}} f_{\mathrm{w}}]}{\{(1-f_{\mathrm{w}})[B_{\mathrm{o}} + Zp_{\mathrm{sc}}T_{\mathrm{p}}(R_{\mathrm{p}} - R_{\mathrm{s}}) / p_{\mathrm{p}}T_{\mathrm{sc}}] + f_{\mathrm{w}}\}} \tag{1}$$

ASP flooding produced liquid rheological model for power rate, the Reynolds number use generalized Reynolds number Re when liquid flow through the suction valve of oil well pump.

$$\begin{aligned} \mathrm{Re}_{pl} &= 8^{1-n} \rho D^n v^{2-n} / \{k[(3n+1)/4n]^n\} \\ &= 8^{1-n} \rho D^n v^{2-n} / \{k[(6n+2)/n]^n\} \end{aligned} \tag{2}$$

The viscosity of ASP liquid through the suction valve hole is given by:

$$\mu = K\gamma^{n-1} \tag{3}$$

Fluid through the pump suction valve hole flow pressure drop Δp_{in} is given by:

$$\Delta p_{in} = \rho_m v^2 / (2\xi^2) = \rho_m A_p v_p^{\ 2} / (2A_v \xi^2) \tag{4}$$

The relation between the standardized valve flow coefficient and Re is given by:

$$\xi=4\times10^{-22}\,\mathrm{Re}^5-2\times10^{-17}\,\mathrm{Re}^4-6\times10^{-13}\,\mathrm{Re}^3+6.5\times10^{-9}\,\mathrm{Re}^2+4.78\times10^{-5}\,\mathrm{Re}+0.0106$$

Δp_{in} is submergence pressure that in order to ensure the filled degree of pump, When fluid flows through the pump suction valve, pump pressure for gas is given by:

$$p_i=p_s-\Delta p_{in} \quad (5)$$

Ratio of gas and liquid in the pump is:

$$R=[ZT_p p_{sc} R_p (1-f_w)(1/p_i-1/p_b)]/T_{sc}B_o \quad (6)$$

The influence of gas inside the pump on the pumping system efficiency is:

$$\eta_1=1.1/(1+R)-0.1 \quad (7)$$

The influence of oil volume change on the pumping system efficiency is [5]:

$$\eta_2=1/[B_o(1-f_w)+f_w] \quad (8)$$

The influence of pump leakage on the pumping system efficiency is approximately given by [6, 7]:

$$\eta_3=\eta_{31}\eta_{32}=0.8924 \quad (9)$$

where, η_{31} is space leakage between pump plunger and pump barrel, take 0.97; η_{32} is valve leakage, take 0.92.

The influence of stroke loss on the pumping system efficiency is [5]:

$$\eta_4=1-(D_p^2\gamma_1 L_f L/2.62\times10^{11}S)(\sum_{i=1}^{n}a_i/f_{ri}+1/f_t) \quad (10)$$

The computational formula for pumping system efficiency is:

$$\begin{aligned}\eta&=\eta_1\eta_2\eta_3\eta_4\\&=[\frac{1.1}{1+[ZT_p\rho_m gh_s R_p(1-f_w)(1/p_i-1/p_b)]/T_{sc}B_o}-0.1]\\&\times0.8924/[B_o(1-f_w)+f_w]\\&\times[1-(D_p^2\gamma_1 L_f L/2.62\times10^{11}S)(\sum_{i=1}^{n}a_i/f_{ri}+1/f_t)]\end{aligned} \quad (11)$$

3 System efficiency calculation model

System efficiency calculation model is given by:

$$\eta_s=p_e/p_{mi}=\frac{Q\cdot[h_1+(p_o-p_c)\times1000/\rho\cdot g]\cdot\rho\cdot g}{86400\cdot p_{mi}\times100\%} \quad (12)$$

p_{mi} is associated with the load of pumping unit. The actual input maximum power of electrical machinery is given by:

$$p_{mi}=0.81M_{max}n_p/9549 \quad (13)$$

The computational formula for the maximum torque N_{max} of crankshaft is:

$$\begin{aligned}N_{max}&=1800S\\&+0.202S[\rho_s g f_r L+\rho_1 g(f_p-f_r)L+I_u\\&+F_u+p_o-(\rho_s-\rho_1)gf_r L-I_d-F_d]\end{aligned} \quad (14)$$

I_u, I_d is the maximum inertial load of sucker rod string and liquid column on up and down stroke, respectively.

$$I_u=W_r(Sn_p^2/1790)(1+\lambda) \quad (15)$$

$$I_d=W_r(Sn_p^2/1790)(1-\lambda) \quad (16)$$

F_u, F_d is the maximum friction load on up and down stroke, respectively.

$$F_u=F_{rt}+F_{pb}+F_{tl} \quad (17)$$

$$\begin{aligned}F_d&=m\alpha1.5\%W_r+F_{pb}\\&+2\pi\mu L\left[(c^2-1)/\left[(c^2+1)\ln m-(c^2-1)\right]\right]v_{max}\\&+F_{cl}+F_v\end{aligned} \quad (18)$$

$$\mu=K\left[2(\pi Sn_p/60)/(d_t-d_r)\right]^{n-1} \quad (19)$$

F_{cl} is the friction between the centralizer and liquid column which occur in the down stroke and the direction of friction is upward, when there is a centralizer, fluid flow through the centralizer will generate additional pressure drop due to the centralizer flow section is small, then the friction between centralizer and the fluid column is given by:

$$F_{cl}=\Delta p_h A_c \quad (20)$$

$$\Delta p_h=0.25v_{max}^2\,A_c/A_p \quad (21)$$

F_{tl} is the friction between liquid column and the tubing, it occur in the up stroke so increases suspension point load. Calculating formula is:

$$F_{tl}=\Delta p_L(f_t-f_r) \quad (22)$$

F_{pb} is the friction force between piston and liner which exist all the time. The value can be calculated according to the following formula:

DOI: 10.1051/matecconf/20167701026

$$F_{pb} = 0.94 d_p / e - 140 \qquad (23)$$

F_v is the friction force of liquid through the floating valve, in the big production Wells with high viscosity, the resistance caused by the liquid through the floating valve is often the main reason for the bending of the sucker rod string, and the influence on the suspended load cannot be ignored. Calculation formula is given by:

$$F_v = \frac{1}{729}(Sn)^2 \rho_l \cdot f_p^3 / (\mu^2 \cdot f_o^2) \qquad (24)$$

4 System efficiency influence factors

The influence of centralizers, stroke length, pumping speed and flooding period on sucker rod pumping system efficiency show in Fig. 1- Fig. 4, respectively. The results indicate that the sucker rod pumping system efficiency decrease while the number of centralizer increases. The influence of stroke length and pumping speed on sucker rod pumping system efficiency is significant. The sucker rod pumping system efficiency increase obviously with the stroke length and pumping speed increasing, but while the stroke length is bigger than 5m and pumping speed more than 6 min-1 it increase gently. In different displacement stage, due to viscosity of produced liquid is different, by force in the wellbore is different, so the system efficiency is also different, the pumping system efficiency in main slug of ASP flooding is lowest.

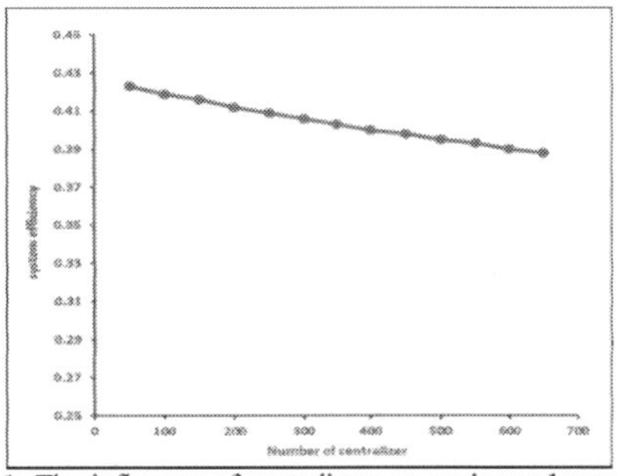

Figure 1. The influence of centralizers on sucker rod pumping system efficiency.

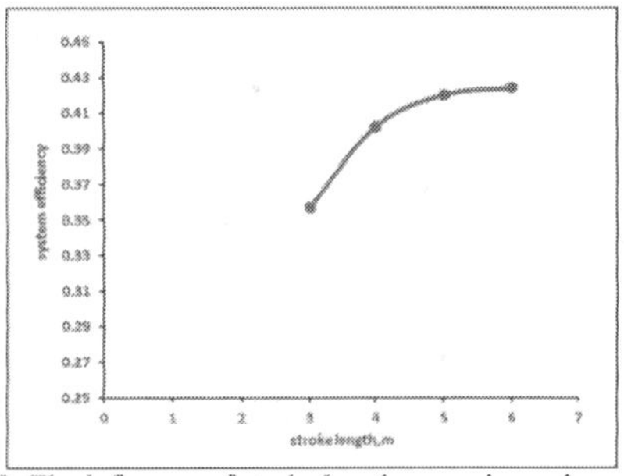

Figure 2. The influence of stroke length on sucker rod pumping system efficiency.

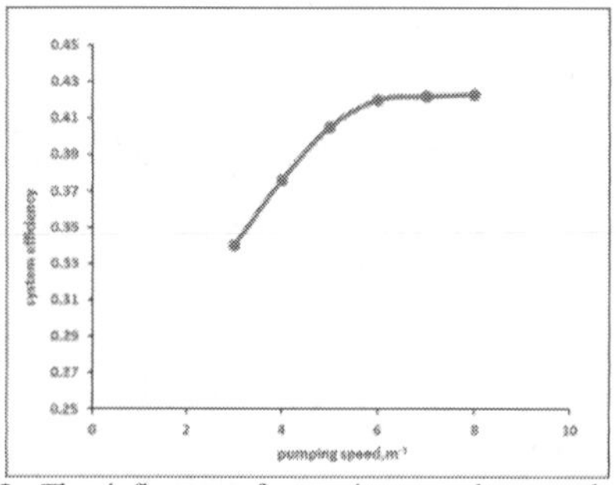

Figure 3. The influence of pumping speed on sucker rod pumping system efficiency.

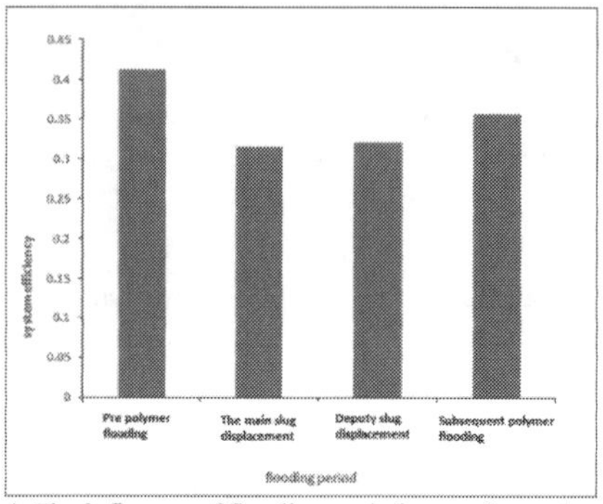

Figure 4. The influence of flooding period on sucker rod pumping system efficiency.

5 Determination of the reasonable pump depth and system efficiency of ASP flooding

The viscosity of production fluid is high which can make the condition complex of the wellbore. A low pump depth result in a low production of the oil well, on the contrary, the eccentric wear between sucker rod and tubing will be serious if pump depth is too high, so it is necessary for improve the well working condition, pumping system efficiency and production to get the reasonable pump depth.

According to the data of a production plant in Daqing, the optimization value of pump depth has been selected to maximize the pumping system efficiency. The biggest pump depth is 950m.

$$H_m > 0 \qquad (25)$$

$$H_f - (P_{wf} - P_c) / (\rho_{mix} g) + H_{pump} < 950 \qquad (26)$$

The inflow performance relationship curve of produced wells is different in different displacement stage, therefore the system efficiency of different displacement stage is different, need to calculate the change law of bottom hole flowing pressure, production, system efficiency, and pump depth under the different displacement stage with different submergence depth of ASP flood. The stroke range for 3-6 m, speed range is 3-8 times/min on the spot, get the optimization results of stroke , stroke times, bottom hole flowing pressure and pump depth under the different displacement stage with the maximum system efficiency of two kinds of pump diameter, as show in Table 1.

Table 1. Optimization results of pump depth and system efficiency under different flooding stages

Stage	Stroke[m]	Sumber of stroke[times/min]	Pump diameter[m]	Sump depth[m]	System efficiency[%]
Water flooding stage	5	3	0.083	787	45.9
	5	4	0.07	764	44.7
Pre polymer flooding	6	6	0.083	629	41.8
	6	7	0.07	616	39.8
Main slug	6	8	0.083	662	32.0
	6	8	0.07	638	26.3
Sub slug	6	8	0.083	632	33.7
	6	8	0.07	607.8	20.4
Subsequent polymer flooding	6	8	0.083	627	32.5
	6	8	0.07	608.6	21.0

6 Conclusion

(1) Asp flooding produced fluid viscosity is bigger, can produce a larger pressure drop in pump suction that affect the fluid level of pump and reduce the pump efficiency.

(2) The centralizer rubs with tubing and produced liquid can affect the suspension center load, and results in the decrease of system efficiency.

(3) when the submergence depth is too big, will cause the production is too high, bottom hole flowing pressure is too low, the depth of producing fluid level from well head is too big, result in pump depth more than deflection point. when the submergence depth is too small, will cause the production is too low, bottom hole flowing pressure is too high, producing fluid level reached the well head, we obtained the reasonable stroke, stroke times, bottom hole flowing pressure and pump depth under the different displacement stage with two kinds of pump diameter of ASP flooding.

R_p—produced gas-oil rate, m^3/m^3;
R_s—dissolved gas-oil ratio, m^3/m^3;
T_{sc}—standard state temperature, 293K;
T_p—flow temperature of Pump inlet, K;
P_p—pressure of Pump inlet, Pa;
D—diameter of suction valve, m;
ξ—valve flow coefficient;
A_p, A_v—the area of plunger and valve port, respectively, m^2;
v_p—the movement velocity of plunger, m/s.
p_s—submergence pressure, Pa.
L_f—working fluid level, m;
L—the total length of sucker rod, m;
f_t—the metallic cross-sectional area of tubing, m^2;
ai—the ratio between the sucker rod and the total length;
f_{ri}—the cross-sectional area of sucker rod, m^2;
D_p—diameter of pump, m;
γ_l—the specific gravity of liquid;
S—stroke, m.
P_e—available power, kW;
P_{mi}—the input power of electric motor, kW;
h_l—the well working fluid level, m;
P_o—the well tubing head pressure, MPa;
P_c—the well surface casing pressure, MPa.
n_p—pumping speed, min^{-1};
ρ_s—the density of rod material(steel);
f_r—the cross-sectional area of sucker rod, m^2;
c —the ratio of tubing inside diameter and sucker rod diameter, m;
A_c—the cross-section area of centralizer, m^2;
Δp_l—the frictional pressure drop of liquid in the annulus.

References

1. Wang Yan, Technology for Controlling Eccentric Wear of Sucker Rods and Tubing in Pumping Wells Lifting Fluids Containing Polymer, SPE 89927.
2. Hong-Sheng Han, Li-Ping Guo, Yu-wang Song, Effect of resistance in polymer flooding on sucker rob partial abrasion. Journal of Daqing Petroleum Institute, **27**(4):21~25, (2003).
3. O.Rivas, A. Newsky, M. Cedeno, et al, Sucker Rod Centralizers for Directional Wells, SPE21131
4. Rui-Xia Zhang, Zeng-Liang Li, Dong Jiang, Study of Sucker Rod Centralizer Disposition of Reciprocal Sucker Rod Pump, Oil Field Equipment, **37**(12):28-35, (2008).
5. Hong-Tao Dong, Tao Wu, Qu Liao, et al, Relationship between suitable submergence depth

with pump efficiency, Inner Mongolia Chemical, **18**: 36-39, (2010).
6. Qi-Biao Xia, Yong-Hui Liu, Xin-Fu Zhou, The frictional force calculation and the centralized distribution design of sucker rod string in directional wells, Drilling and Production Technology, **27**(1): 60-63, (2004).
7. Ming-Qiang Zhao, Chao-Dong Tan, Fu-Qing Yang, Study of sucker rod centralizer optimized design research, **11**: 51-53, (2012).

MATEC Web of Conferences **77**, 01027 (2016) DOI: 10.1051/matecconf/20167701027
ICMMR 2016

Gas Turbine Engine Optimization at Conceptual Designing

Andrey Y. Tkachenko, Venedikt S. Kuz'michev, Ilia N. Krupenich and Viktor N. Rybakov

Samara National Research University, Department of Theory of Aircraft Engines, 34, Moskovskoye Shosse, Samara, 443086, Russia

Abstract. This article is a survey of the basic aspects of a gas turbine engine's conceptual designing and the optimization methods for solving problems at this stage. In the beginning, the purpose of the conceptual design is considered. Then, the methods of solving some problems are subsequently described. Finally, article provides information about the computer-aided means for solution of these problems, which were developed.

1 Introduction

Conceptual design is the initial stage of the engine design. Working process parameters, geometrical parameters of the flow path, control laws are selected at this stage. They should ensure maximum fuel, ecological, weight, cost efficiency in compliance with all regulations and restrictions.

Most of the conceptual design tasks (see Fig. 1) are essentially problems of multicriteria conditional parametric and structural optimization.

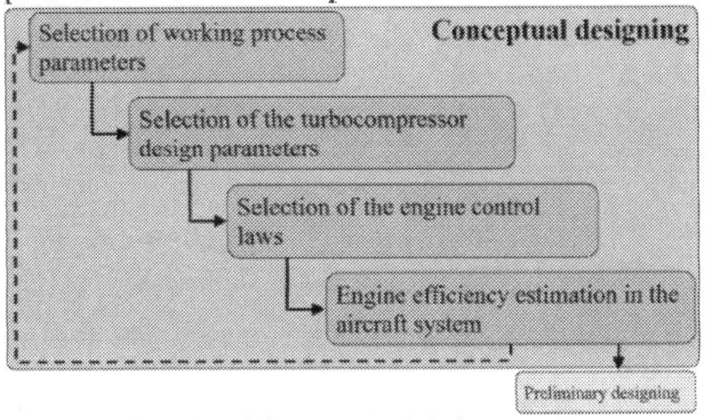

Figure 1. The tasks of the conceptual design.

Therefore, the crucial problem is the development of methods and models that can reliably, quickly and accurately find an adequate solution.

2 Optimization of working process parameters

The first task is the values selection of working process parameters that form a thermodynamic cycle [1]-[3]. Mathematical formulation of this problem is

$$X = arg\left(\min_{x_i} \max_{Z_j} Z_j(x_i, p) \middle| Q_j(x_i, p) \leq 0\right) \quad (1)$$

$$Q(x_i, p) = \Sigma\, q_k(x_i, p) \quad (2)$$

$$Z_j(x_i, p) = r_j\left(\frac{y_j(x_i, p)}{Y_j(p)} - 1\right) \quad (3)$$

$$y_j(x_i, p) = \left[C_{\text{T·KM}}, M_{\text{cy+T}}, a, \ldots\right] \quad (4)$$

Variable parameters are a total compressor pressure ratio, a combustor exit temperature, a bypass ratio, a fan pressure ratio and maybe some others.

In accordance with the system approach, the aircraft efficiency criteria should be used as performance indicators [4]-[6]. Some of which are a total mass of the power plant and fuel, a fuel cost per tonne-kilometre of transported cargo, an operating cost and so on.

The structure of the mathematical model is presented in Fig. 2.

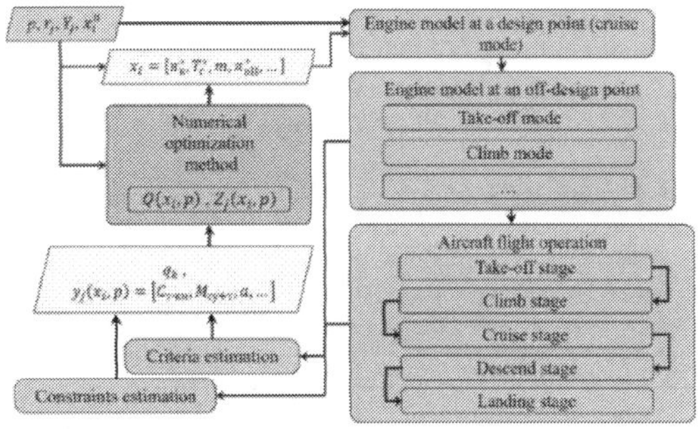

Figure 2. The structure of the mathematical model for optimization of the working process parameters.

After the working process parameters values were defined or refined for the current iteration, the design engine model is calculated. For turbofan engines, this calculation is usually performed for a cruise mode. Its results define the data for off design engine, such as, the flowpath areas at the main sections, the characteristic maps of the compressors and turbines.

Further, off-design engine model is calculated for various modes of operation. Thus, the diameter size of the engine, the powerplant weight and the engine performance can be defined.

As mentioned before, the important aspect is that the engine designing should be based on the aircraft performance estimation. Therefore, the next block performs the aircraft flight modelling.

Information about flight program, weight, geometric and aerodynamic characteristics of the airplane is used to obtain the required thrust level and, consequently, the operation mode of the engine at each point of the flight path.

The result of numerical integration of the aircraft dynamic equations is the flight duration and the total fuel consumption, which are used for criteria estimation.

The optimization module calculates values of the objective function and the termination criteria. For taking constrains into account the penalty method is well suited. If the termination condition is not achieved, the algorithm proposes new values of the variable parameters, and the calculation is reiterated.

Another important aspect is the multicriteria formulation of the problem. In this case, optimization task may be performed in two phases. In the first phase, the parameters are optimized in a single-criterion statement separately for each indicator, as shown in Fig. 3.

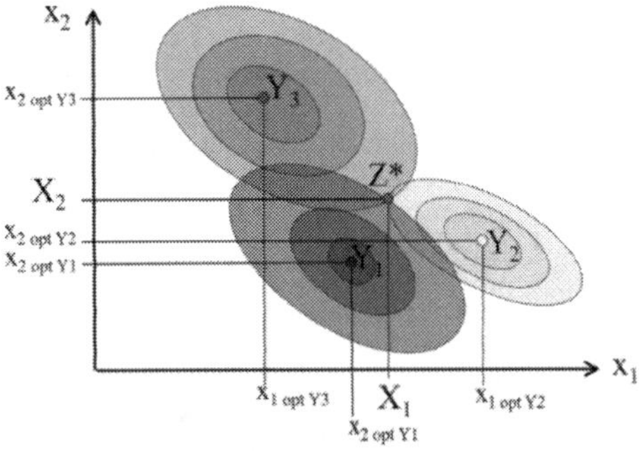

Figure 3. Optimization process phases.

Optimum indicator values are used to determine its normalized values in the next phase. So, in the second phase, the objective function is calculated as the maximum value from the set of normalized indicator values considering the constraints. Thus, the optimization problem in multi-criteria formulation applies the minimax principle. The Nelder-Mead method or differential evolution algorithm is commonly used for numerical optimization.

3 Optimization of an engine family based on a unified engine core

The modification of the base engine is common practice for designing of advanced engines with various applications. This approach provides high levels of reliability, service life of new engines and also reduces the time and cost of its development process.

The most complex part of engines is an engine core or a gas-generator. However, it is difficult to apply the same engine core in different engines, especially, if it was not originally designed for this purpose. Often, the engine core needs to be modified or scaled. The vivid example is the Rolls-Royce Trent engine family.

Therefore, a new method was developed, which allows to simultaneously optimize the working process parameters of the unified engine core and parameters of entire family of aircraft engines and energy power plants based on this engine core [7]. Mathematical formulation of this task and the model structure of this task are presented in Fig. 4.

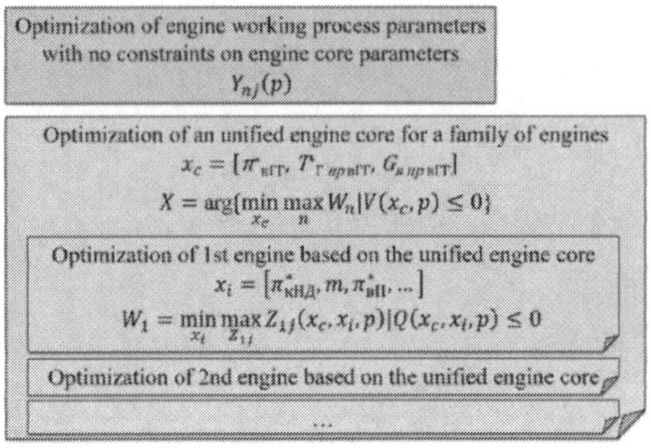

Figure 4. Model structure of optimization of an engine family based on a unified engine core.

This method differs from the previous one. In the first phase, the parameters are also optimized in a single-criterion statement separately for each engine of the family and for each indicator. The result is a set of optimal engines with different engine cores.

In the second phase, the outer optimization loop is used for engine core parameters selection. The inner subtasks run parallel optimizing remaining working process parameters for each engine of the family.

Thus, optimization of the unified engine core and engine family is a combination of external and internal problems of multicriteria optimization.

This method is the most versatile, because a parameters optimization of a sole engine is only, in fact, its particular case.

4 Optimization of the turbocompressor design parameters

The next basic task of a gas turbine engine conceptual design is optimization of the turbocompressor design parameters. It can be formulated as follows.

$$X = \alpha\rho\gamma\{\min_s \min_{x_{si}} F_s(x_{si}, p) | Q(x_{si}, p) \qquad (5)$$

$$F_s(x_{si}, p) = \sum_j Z_{sj}(x_{si}, p) \qquad (6)$$

$$Z_{sj}(x_{si}, p) = r_{sj}\left(\frac{y_{sj}(x_{si}, p)}{Y_{sj}(p)} - 1\right) \qquad (7)$$

Variable parameters are the geometrical parameters of blades and a flowpath, a number of blades in a blade row of turbines and compressors, as well as the parameters that define the scheme of the turbocompressor, such as a number of spools, types of compressors and turbines, a number of stages in the compressor and turbine spools and so on.

Optimization of the turbocompressor design parameters may be divided on two subtasks (see Fig. 5). The internal subtask is parametric optimization of geometrical parameters of blades and the number of blades in a blade row. The external subtask is structural optimization of the turbocompressor scheme.

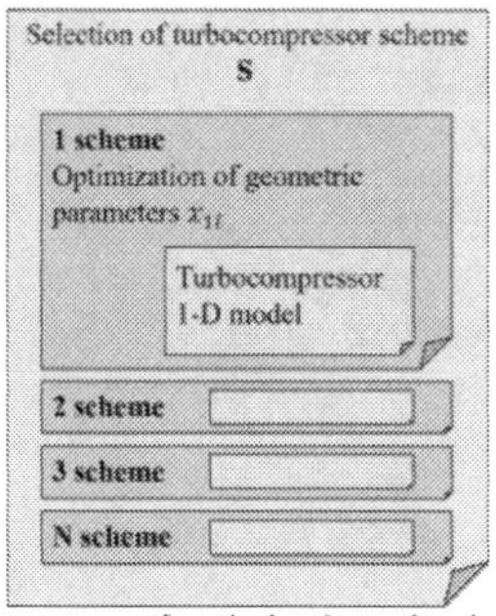

Figure 5. Task structure of optimization of turbocompressor design parameters.

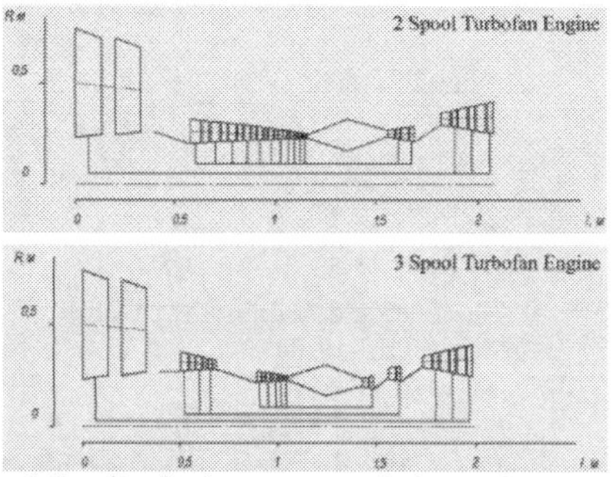

Figure 6. Results of turbocompressor design parameters optimization.

The performance estimation is performed using one-dimensional models of the turbocompressor units and weight assessment models. The one-dimensional model reflects the influence of variable parameters on the unit's efficiency and the total engine efficiency at design and off-design points.

If there are different types of indicators (float and integer), it is better to use the additive value of the normalized criteria as the objective function for multicriteria optimization.

Figure 6 shows the results of turbocompressor design parameters optimization for two engines. They have the same thermodynamic parameters, but different schemes.

5 Engine control optimization

The aim of this task is to derive the optimal control law for engine operation managing during the aircraft flight [8], [9]. For numerical modeling a continuous flight process defined by a system of differential equations have to be replaced by a discrete multi-step process (see Fig. 7). In this case, the control law may be represented as a tabular function of the engine control parameters at each discrete step of the process. Values of engine control parameters in particular step uniquely identify a system transitions from one state to another [10].

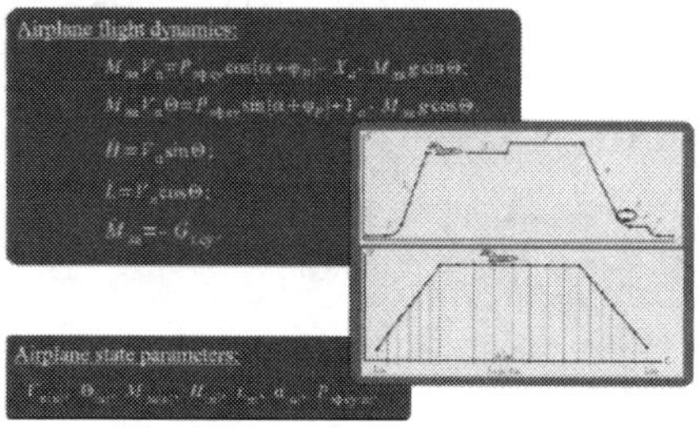

Figure 7. Aircraft flight process modelling.

Engine control optimization is a problem of finding a sequence of control parameters values u, which provide a minimum of the objective function at the end of the aircraft flight.

$$U = \alpha\rho\gamma\left\{\min_{x_i} \max_{z_j} Z_j(\{\boldsymbol{u}\},p) | \, T(\{\boldsymbol{u}\},p) \ \leq 0\right\} \quad (8)$$

$$U = \{\boldsymbol{u}\} *= \{u_1(L), u_2(L), \ldots u_n(L)\} * \quad (9)$$

$$T(\{\boldsymbol{u}\},p) = \Sigma\, t_k(\{\boldsymbol{u}\},p) + Q(\{\boldsymbol{u}\},p)\left(10 + \frac{n-i}{n}\right) \quad (10)$$

$$Q(\{\boldsymbol{u}\},p) = \Sigma\, q(\{\boldsymbol{u}\},p) \quad (11)$$

$$Z_j(\{\boldsymbol{u}\},p) = r_j\left(\frac{y_j(\{\boldsymbol{u}\},p)}{Y_j(p)} - 1\right) \quad (12)$$

The algorithm of this task is based on the numerical method of dynamic programming and the Bellman optimality principle. A system state at the end of a step depends only on the state at the beginning of this step and control parameters values at this step. Therefore, at the last step, it is easy to find the optimal control by means of a numerical method of parametric optimization. For the second last step, action is the same. However, changing the control at this step it is necessary to repeat control optimization at last step to obtain proper criteria values.

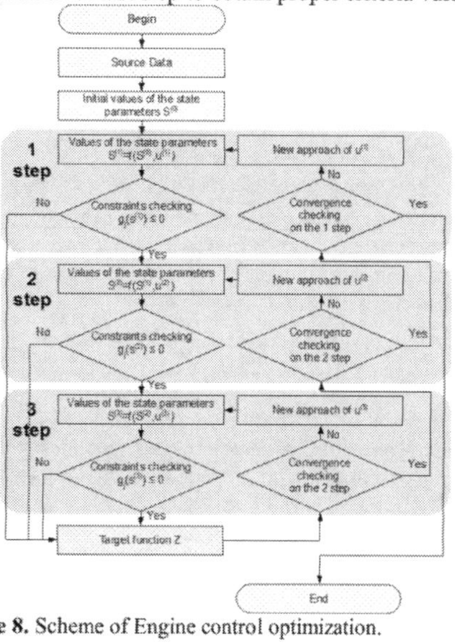

Figure 8. Scheme of Engine control optimization.

 DOI: 10.1051/matecconf/20167701027

Thus, this task can be represented as a sequence of nested optimization subtasks, so that control optimization at the first step is external to all others. For example, the control optimization algorithm of the three-step process is presented on the block diagram (see Fig. 8).

It should be mentioned that this algorithm is very time-expensive and demands high-performance computing facilities.

6 Conclusions

In this research we have investigated the methods and algorithms of solving such problems as:

• Optimization of the working process parameters at the design point of gas turbine engine;
• Optimization of the working process parameters of the unified engine core and the engine family based on it;
• Optimization of the turbocompressor design parameters;
• Engine control optimization.

The main purpose of this paper has been to develop the effective and reliable optimization methods which can be used for creating of computer-aided tools for aircraft gas turbine engines designing.

Algorithms of numerical methods, basic functions, modeling tools are implemented as a standalone software framework CARCASE. Computer-aided means for complex technical systems engineering can be created using this framework and domain specific models, such as CAE-system ASTRA for example (Fig. 9).

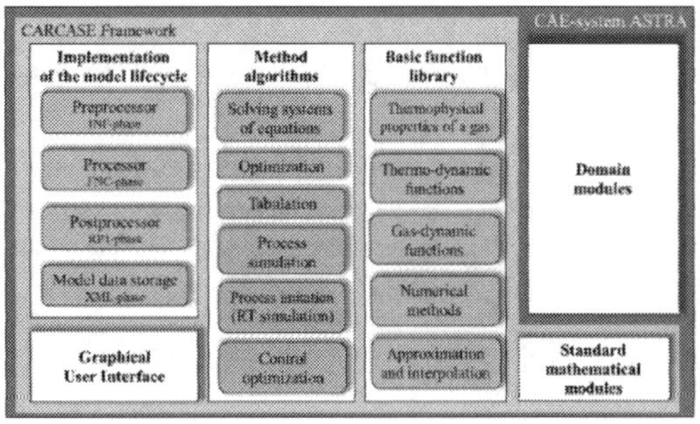

Figure 9. Structure of CAE-system ASTRA software.

CAE-system ASTRA is a computer software for solving of complex conceptual design problems, which are listed below:

• Computer modeling of engine working process;
• Thermodynamic analysis of a gas turbine engine cycle;
• Calculation at the design and off-design operation points;
• Evaluation of the impact of atmospheric and flight conditions;
• Performance prediction of a turbocompressor;
• Optimization of working process parameters;
• Optimization of turbocompressor design parameters;
• Optimization of engine control with aircraft flight modeling;
• Modeling and simulation of transient engine operation;
• Analysis of engine transient operation;
• Implementation of virtual test bench.

Acknowledgments

This work was supported by the Ministry of education and science of the Russian Federation in the framework of the implementation of the Program of increasing the competitiveness of Samara University among the world's leading scientific and educational centers for 2013-2020 years.

Nomenclature

$C_{т\cdot км}$	fuel consumption per ton of the payload per kilometer of distance, [kg/(tons·km)];
$M_{су+т}$	total mass of the power-plant and the fuel, consumed during the flight, [kg];
a	net cost of transportation, [1/(tons·km)];
π^*_{fII}	secondary flow pressure ratio;
m	bypass ratio;
$\pi^*_{кНД}$	low pressure compressor pressure ratio;
$\pi^*_{кГГ}$	engine core compressor pressure ratio;
π^*_{Σ}	overall pressure ratio;
$T^*_{Г.пр.вГГ}$	specific turbine inlet temperature, normalized by the parameters at the engine core inlet, [K];
$G_{В.пр.вГГ}$	specific air flow rate, normalized by the parameters at the engine core inlet, [kg/s];
L	flight range, [km];
p	set of source data;
x	variable parameters;
q	parametric constraint;
Q	total parametric constraint for engine core parameters;
V	total parametric constraint for engines parameters;
t	process constraint;
T	total process constraint;
y	efficiency criterion of gas turbine engines or aircraft;
Z	normalized value of efficiency criterion;
n	number of engine;
U	table of the control function values;
i,j,k	indexes.

References

1. P. Ahmadi, I. Dincer, M.A. Rosen, Thermodynamic modeling and multi-objective evolutionary-based optimization of a new multigeneration energy system, *Energy Conversion and Management*, **76**, (2013), pp. 282-300.
2. S. Campanari, Boncompagni, E. Macchi, Microturbines and trigeneration: Optimization strategies and multiple engine configuration effects, *Journal of Engineering for Gas Turbines and Power*, **126 (1)**, (2004), pp. 92-101.
3. S. Ramakrishnan, K.-Y. Teh, C.F. Edwards, Identification of optimal architecture for efficient simple-cycle gas turbine engines, *ASME International Mechanical Engineering Congress and Exposition, Proceedings*, **6**, (2010), pp. 539-548.
4. E. Van Driel, R. Slingerland, N. Antoine, A framework for aircraft power systems optimization,

Structural Dynamics and Materials Conference, **6**, (2006), pp. 3894-3904.
5. A. Guha, D. Boylan, P. Gallagher, Determination of optimum specific thrust for civil aero gas turbine engines: A multidisciplinary design synthesis and optimization, *Proceedings of the Institution of Mechanical Engineers, Part G: Journal of Aerospace Engineering*, **227(3)**, pp. 502-527.
6. K.E. Sheldon, C.M. Anderson-Cook, W.F. O'Brien, Using computer modeling to optimize cost and maintain performance constraints for the production of aircraft turbine engines, *Quality and Reliability Engineering International*, **20 (5)**, (2004), pp. 541-551.
7. V.S. Kuz'michev, V.V. Kulagin, A.Y. Tkachenko, I.N. Krupenich, V.N. Rybakov, Optimization of the turbofan engine with a given engine core: problem setting, *Vestnik of Samara State Aerospace University*, **5(36)**, vol. 1, (2012), pp. 165-169.
8. V.V.R. E Silva, W. Khatib, P.J. Fleming, Control system design for a gas turbine engine using evolutionary computing for multidisciplinary optimization, *Controle y Automacao*, **18 (4)**, (2007), pp. 471-478.
9. S. Kim, H. Thompson, P. Fleming, Active hierarchical fuzzy control for gas turbine altitude relighting using multi-objective optimization, *Proceedings of the ASME Turbo Expo*, **2**, (2006), pp. 853-861.
10. V.S. Kuz'michev, A.Y. Tkachenko, V.N. Rybakov, Aircraft flight simulation during the optimization of the working process parameters of the gas turbine engines, *Proceedings of the Samara Scientific Center of the Russian Academy of Sciences*, **vol.14, №1(2)**, (2012), pp. 491-494.

MATEC Web of Conferences 77, 01028 (2016) DOI: 10.1051/matecconf/20167701028
ICMMR 2016

Computer-Aided System of Virtual Testing of Gas Turbine Engines

Viktor N. Rybakov , Andrey Y. Tkachenko, Venedikt S. Kuz'michev and Ilia N. Krupenich

Samara National Research University, Department of Theory of Aircraft Engines, 34, Moskovskoye Shosse, Samara, 443086, Russia

Abstract. The article describes the concept of a virtual lab that includes subsystem of gas turbine engine simulation, subsystem of experiment planning, subsystem of measurement errors simulation, subsystem of simulator identification and others. The basis for virtual lab development is the computer-aided system of thermogasdynamic research and analysis "ASTRA". The features of gas turbine engine transient modes simulator are described. The principal difference between the simulators of transient and stationary modes of gas turbine engines is that the energy balance of the compressor and turbine becomes not applicable. The computer-aided system of virtual gas turbine engine testing was created using the developed transient modes simulator. This system solves the tasks of operational (throttling, speed, climatic, altitude) characteristics calculation, analysis of transient dynamics and selection of optimal control laws. Besides, the system of virtual gas turbine engine testing is a clear demonstration of gas turbine engine working process and the regularities of engine elements collaboration. The interface of the system of virtual gas turbine engine testing is described in the article and some screenshots of the interface elements are provided. The developed system of virtual gas turbine engine testing provides means for reducing the laboriousness of gas turbine engines testing. Besides, the implementation of this system in the learning process allows the diversification of lab works and therefore improve the quality of training.

1 Introduction

Modelling is the most popular way of complex systems (technical objects and control systems) examination during the designing, development and analyzing [1].

Simulation modelling is a method of experiment simulation: real-time experimental investigation of a virtual prototype of an engine that is implemented using a simulator instead of an engine itself.

One of the important ways to provide a high efficiency of the engine designing process is to develop a virtual model of its working process, suitable for use during all the stages of engine life cycle. The greater part of information about the engine behavior is obtained during the physical experimental investigations due to the lack of adequacy of modern simulators [2]. Physical experimental investigations of the gas turbine engines are quite time-consuming and expensive, so it is a good policy to combine them with a virtual investigations during the design and development stages.

Computer-aided system of termogasdynamic calculations and analysis of the gas turbine engines (ASTRA) [3], developed at the Samara State Aerospace University [4], [5] may be used as a basis for development of a virtual testing lab. Thermogasdynamic models of the gas turbine engines (GTEs), implemented using the ASTRA are suitable for calculating the performance data, optimizing the control programs of engines and other tasks of conceptual designing.

2 Computer-aided system of virtual testing of gas turbine engines

This paper suggests the concept of GTE virtual testing lab [6]-[8]. The principal elements of the lab are: measurement imitation subsystem, design of experiments subsystem, mathematical model identification subsystem, engine simulation subsystem (ASTRA), information subsystem, documentation subsystem and testing visualization subsystem (Fig. 1).

Design of experiments subsystem is intended to draw up a plan of virtual experiment.

Measurement imitation subsystem is intended to introduce deviations from the design values (imitating individual features of a certain engine generated during its production, e.g. manufacturing errors) and random measurement errors. This subsystem is crucial for methods of processing the results of the experiment.

Fig. 2 shows the scheme of deviations and measurement errors simulation.

Mathematical model identification subsystem. During the engine designing process the first-level mathematical models are mostly used. These models are the systems of nonlinear equations that interrelate the engine parameters with the characteristics of its elements Q and the external factors X (ambient conditions and the mode of operation) using the joint operation equations:

$$P = f(Q, X). \quad (1)$$

The essence of the identification process is to redefine the elements' characteristics Q using the measured results of the experimental investigations P. It should be mentioned that the number of unknown parameters A is usually much higher than the number of measured parameters P.

Documentation subsystem is intended to organize the results of experimental investigations and performance data.

Testing visualization subsystem provides the audio and visual imitation of the testing process (graphical representation of the measured parameters, imitation of the gas flow, noise imitation etc.).

Information subsystem is a collection of databases: of initial data, mathematical models and the testing results (Fig. 3).

The ASTRA CAE-system includes the numerical integration module that provides capability to assess the influence of various factors upon the behavior of an engine during the real-time simulation [9].

This feature provides capability to develop a virtual lab on the basis of ASTRA and to solve such problems as:
- Investigation of dependency of engine parameters upon the ambient conditions, mode of operation and flight condition;
- Investigation of transient modes of engine operation at various conditions and for various engine control program settings;
- Optimization of the control program for the processes of engine acceleration and deceleration;
- Visual representation of joint operation of the engine elements.

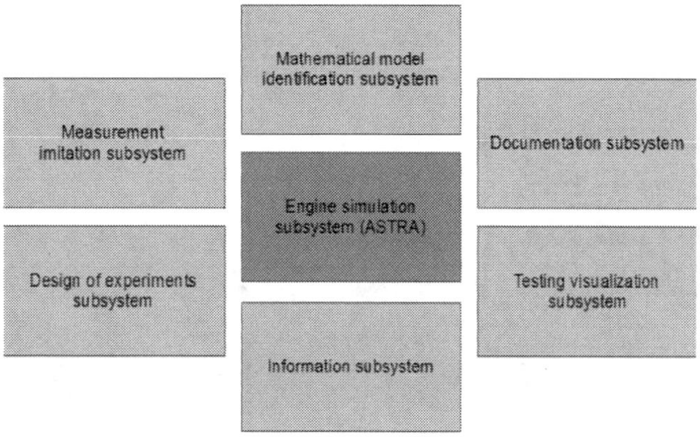

Figure 1. Virtual testing lab structure.

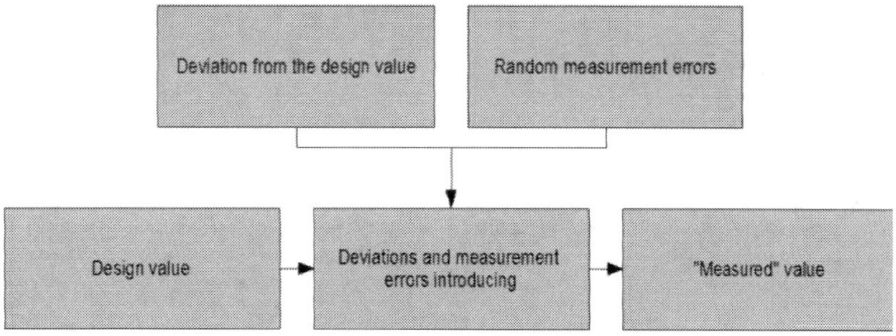

Figure 2. Scheme of deviations and measurement errors simulation.

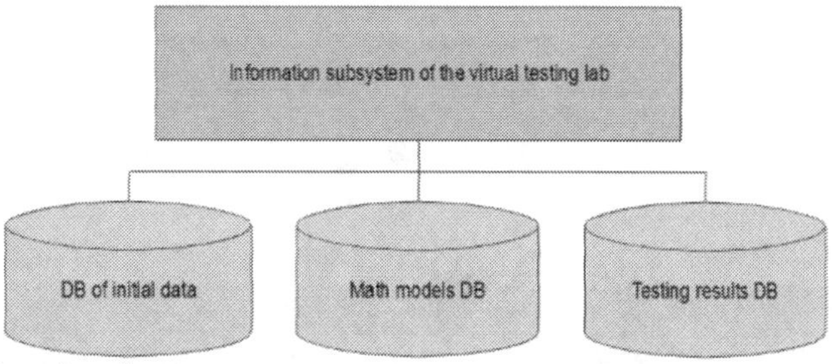

Figure 3. Scheme of the information subsystem

The power balance of turbine and compressor is not satisfied during the transients and it is the feature of this modes of operation. The difference between the turbine and compressor power determines the kinetic energy increment of the rotating mass of the rotor:

$$N_т \cdot \eta_m - N_к = \frac{dE}{dt}, \text{ where} \quad (2)$$

$E = I \cdot \omega^2/2$ – kinetic energy of the rotating mass of the rotor; I – rotor polar moment of inertia; $\omega = \pi n / 30$ – angular velocity.

Values of power produced by the turbine and consumed by the compressor at the given value of fuel flow-rate during the calculations of transient modes of operation are evaluated on the basis of combination of working process parameters and angular velocity of the rotor at the given moment of time.

These values, in turn, determine the value of angular acceleration. The angular velocity of rotor is evaluated by means of numerical integration of rotor motion equation. Other parameters are evaluated using the quasi-stationary concept (using the same equations as for the stationary modes of operation). Thus, the transient thermogasdynamic model includes the same modules as the steady-state thermogasdynamic model except for the modified module of the engine transmission.

This transient thermogasdynamic model may be used to imitate the engine behavior during the testing and thus provide the information required for the planning of physical experimental investigations.

This model was implemented as a virtual testing lab of low-sized gas turbine engine "ASTRA 5.0 T microGTU" using the object-oriented language (Java). Fig. 4 shows the interface of this virtual lab, running under various operating systems. The left panel may be used to change the values of external factors (airspeed, altitude, ambient temperature and pressure) and the parameters defining the mode of operation (angular velocity and the cross-section area of the nozzle exit section).

The middle panel provides the graphical representation of the main elements of the engine and its characteristic sections. The colour corresponds to the stagnation temperature along the flow-path. The diagrams below it represent the values stagnation temperature and pressure along the axis of engine.

The right panel shows the digital and pointer indicators of the main parameters of the engine.

Graphical representation of compressor characteristic with calculated operating points during the transients is located in the right-bottom corner of the window (Fig. 5).

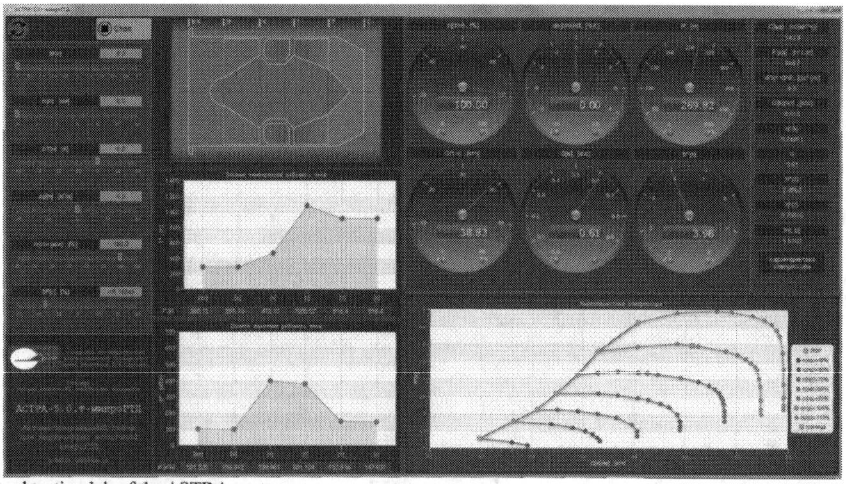

Figure 4. Virtual testing lab of the ASTRA system.

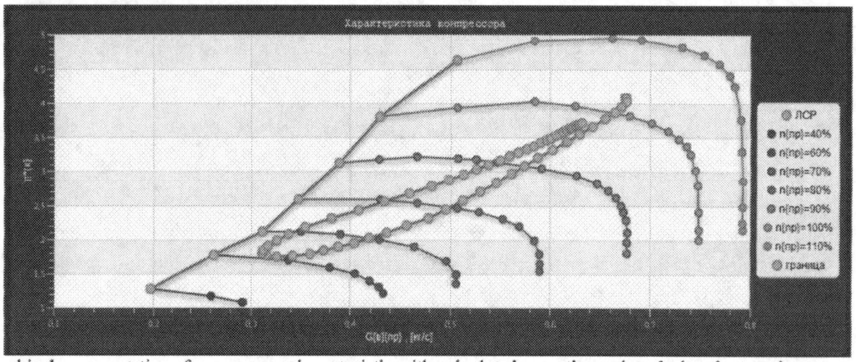

Figure 5. Graphical representation of compressor characteristic with calculated operating points during the transients.

3 Conclusion

A concept for creating the virtual lab using the real-time simulation modelling of the gas turbine engines is proposed.

Virtual testing lab intended for experimental investigation of a virtual prototype of an engine that is implemented using a simulator instead of an engine itself was developed at the Department of Aircraft Engines Theory of Samara National Research University using ASTRA – the computer-aided system of thermogasdynamic calculation and analysis.

Virtual experimental investigations may be used for mathematical model identification and providing the initial data for designing physical experiments decreasing time costs and expenses.

The work was financially supported by the Ministry of education and science of Russia in the framework of basic part of government assignment.

References

1. Popov, G.M., Goryachkin E. S., Baturin O.V. and Kolmakova, D.A., 2014, "Development of recommendations on building of the lightweight calculation mathematical models of the axial turbines of gas turbine engines", International Journal of Engineering and Technology, 6(5), pp. 2236-2243.
2. Kolmakova, D.A., Baturin, O.V. and Popov, G.M., 2014, "Knowledge lack impact assessment of the source date on numerical simulation results of operational process in axial flow turbine blade row", ARPN Journal of Engineering and Applied Sciences, 9(12), 2014, pp. 2880-2889.
3. Kuz'michev, V.S., Tkachenko, A.Y., Krupenich, I.N., Rybakov, V.N. "Composing a virtual model of gas turbine engine working process using the CAE system "ASTRA", Research Journal of Applied Sciences, 9(10), 2014, pp. 635-643.
4. Tkachenko, A.Yu., Kuzmichev, V.S., Kulagin, V.V., Krupenich, I.N., Rybakov, V.N. "Computer-aided system of thermogasdynamic calculations and analysis (ASTRA 4) of gas turbine engines and power plants", Problems and development prospects of propulsion engineering: materials of the international scientific and technical conference held on the 28-30th of June, 2011, Samara: SSAU, 2011. In 2 parts. Part 2. – pp. 80-82. (In Russ.).
5. Tkachenko, A.Yu., Krupenich, I.N. "Development of a virtual prototype of gas turbine engine at the stage of conceptual thermogasdynamic designusing the CAE-system «ASTRA»", Vestnik SSAU, 2012, №3(34), Part 2, pp. 333-342. (In Russ.).
6. Rybakov, V.N., Kuzmichev, V.S., Tkachenko, A.Yu. "Concept of construction of virtual lab for gas turbine engine testing", Vestnik SSAU, 2011, №3(27), Part 3, pp. 326-330. (In Russ.).
7. Rybakov, V.N., Kuzmichev, V.S., Kulagin, V.V., Krupenich, I.N., Fedorov, D.V. "Dataware of virtual lab for gas turbine engine testing", Vestnik SSAU, 2011, №3(27), Part 3, pp. 320-325. (In Russ.).
8. Kuzmichev, V.S., Rybakov, V.N. "Principles and methods of the virtual laboratory tests of gas turbine engines", Vestnik UGATU, 2012, Vol. 16, №2(47), pp. 199-202. (In Russ.).
9. Kuz'michev, V.S., Rybakov, V.N., Tkachenko, A.Y., Krupenich, I.N. "Optimization of working process parameters of gas turbine engines line on the basis of unified engine core", ARPN Journal of Engineering and Applied Sciences, 9 (10), 2014, pp. 1873-1878.

MATEC Web of Conferences 77, 01029 (2016) DOI: 10.1051/matecconf/20167701029
ICMMR 2016

Fault diagnosis for tilting-pad journal bearing based on SVD and LMD

Zhang Xiaotao and Li Weiguang

School of Mechanical and Automotive Engineering, South China University of Technology, Guangzhou, 510640, China

Abstract. Aiming at fault diagnosis for tilting-pad journal bearing with fluid support developed recently, a new method based on singular value decomposition (SVD) and local mean decomposition (LMD) is proposed. First, the phase space reconstruction of Hankel matrix and SVD method are used as pre-filter process unit to reduce the random noises in the original signal. Then the purified signal is decomposed by LMD into a series of production functions (PFs). Based on PFs, time frequency map and marginal spectrum can be obtained for fault diagnosis. Finally, this method is applied to numerical simulation and practical experiment data. The results show that the proposed method can effectively detect fault features of tilting-pad journal bearing.

1 Introduction

Tilting-pad journal bearing applied widely in high-speed rotary machinery has such advantages as strong carrying capacity, long operational life span and small friction factor, swing effect of tilting-pad further improving the stability of vibration [1]. The mechanical fulcrum that makes the stability of the operation system decline is one of the deficiencies of traditional tilting-pad journal bearing. Therefore, it is necessary to develop new type of tilting-pad journal bearing [2].

This paper developed one kind of three-pad tilting-pad journal bearing with fluid support. A large-scale test bench is established to evaluate its performance. Acceleration sensors are put near the screw bolt of footing on the test bench to acquire the vibration signal

The vibration signal from tilting-pad journal bearing contains large amount of information about the operation state, represented as non-stability and multi-component modulation signal. Moreover, this vibration signal contains large mutation and short-term shock component, which makes it hard to discriminate, particularly for the weak fault signal interfered by the surrounding equipment in the field. How to extract the fault feature from the interfered vibration signal is the key for the development of tilting-pad journal bearing.

There are many signal processing techniques that can extract the fault information from the vibration signal, windowed Fourier transform (WFT), wavelet transform and et al. Each of them has their own limitations. For WFT, once the window function is fixed, the size of the time-frequency window is unalterable, the time and frequency are the same for all the components that include different time scales [3]. Wavelet analysis could provide local features in both time and frequency domains and has the feature of multi-scale, which enables wavelet analysis to distinguish the abrupt components of the vibration signal [4]. Thus, wavelet analysis has been widely applied for fault diagnosis of rotary machinery [5]. However, wavelet analysis is essentially an adjustable window Fourier transformation. When a signal is decomposed by wavelet, only the rectangular time–frequency partition of time–frequency plane can be obtained and such tile partition would fail to guarantee instantaneous frequencies of each resulting component to have physical significance. Wavelet analysis has no self-adaptive feature in nature [6].

Empirical Mode Decomposition (EMD) is a self-adaptive signal processing method, which has been widely used for mechanical fault diagnosis and other fields [7]. However, there still exist many drawbacks in EMD such as the end effect, mode fixing, IMF stopping criterion, the problem of quick algorithm and envelope line, which are still underway.

The local mean decomposition (LMD) was proposed by Jonathan S. Smith [8] to decompose an amplitude and frequency modulated signal into a small set of product functions (PFs). The LMD method can be applied to analyse on wide variety of natural signals such as electrocardiograms, functional magnetic resonance imaging data, and earthquake data [9]. Here, we introduce the LMD method into tilting-pad journal bearing fault diagnosis.

SVD method has been widely applied in many fields during recent years, such as adaptive filter [10], noise reduction [11], etc. Therefore, in this paper SVD is used for noise reduction for vibration signal to get relatively purified signal before LMD decomposition.

This paper is organized as follows. Section 2 describes the theory of tilting-pad journal bearing with fluid support and tilting-pad journal bearing test bench. In section 3, the theory of SVD and LMD are introduced. In section 4, a simulation signal is processed to evaluate the proposed method. In section 5, the real vibration signal from the tilting-pad journal bearing test bench is tested. Finally, the conclusion is drawn in section 6.

2 Tilting-pad journal bearing

2.1. The principle

The structure of tilting-pad journal bearing is shown in Fig. 1. The whole bearing includes several pads. Each pad can swig on its mechanical pivot to adjust its position freely along with the axis movement to form wedge, adapting to the change of velocity and bearing load. The resultant force of oil-film pressure in each pad passes through the centre of axle journal of the rotor, not producing the tangential force that can cause instability. Theoretically tilting-pad journal bearing is fully stable. Therefore, it has been widely used in large rotating machinery, such as steam electric generating set, wind turbine and etc. [12]. However, its deficiencies in pad vibration and pivot friction damping will reduce working life and increase system instability.

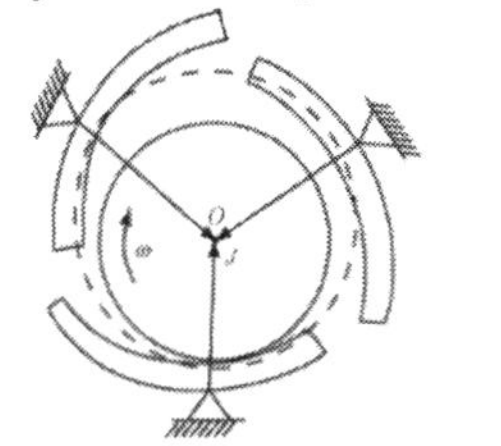

Figure 1. Schematic drawing of tilting-pad journal bearing with three mechanical pivots.

Three pads tilting-pad journal bearing with fluid support is developed in this paper, fluid support substituted for the mechanical pivot, being avoid of mechanical pivot friction and pad vibration. Tilting-pad journal bearing with fluid support has two layers of oil film, inner layer oil film and outer layer oil film. The inner oil film works as dynamic pressure oil film, the outer oil film works as static oil film or squeeze oil film.

2.2 Test bench

Figure 2. Tilting-pad journal bearing test bench. (a) Cast iron base. (b) Power servo motor. (c) Shaft coupling. (d) Front tilting-pad journal bearing. (e) Rotor. (f) Rear tilting-pad journal bearing. (g) Anti-thrust bearing. (h) Lubrication oil system.

A tilting-pad journal bearing test bench is built to evaluate the performance of the tilting-pad journal bearing as shown in Fig. 2.

The main parameters of the tilting-pad journal bearing test bench system are listed in table 1.

Table 1. Main parameters of the test bench.

Working bearing	Tilting-pad journal bearing, circular journal bearing
Journal diameter of rotor	Φ200h6 (mm)
Working velocity range	0～5000 (RPM)
Rotor quality	4975kg
Input oil temperature	30-45℃ adjustable
Oil feed pressure	0～2MPa
Lubrication oil	68# turbine oil, 0.06052Pa.s
Servo motor	3-phase AC motor, 200kw

3 Analysis methods

3.1 Principle of SVD

Based on phase space reconstruction theory, the acquired signal expressed as $x=(x_1,x_2,\ldots,x_N)$ is constructed to be Hankel matrix A

$$A=\begin{bmatrix} x_1 & x_2 & \ldots & x_q \\ x_2 & x_3 & \ldots & x_{q+1} \\ \ldots & \ldots & \ldots & \ldots \\ x_p & x_{p+1} & \ldots & x_N \end{bmatrix} \quad (1)$$

where A is $p\times q$ order matrix, N is the signal length, $N=p+q+1$.

Conduct SVD and we get

$$A=UDV^T \quad (2)$$

Where U and V^T are $p\times p$ and $q\times q$ orthogonal matrix respectively, D is $p\times q$ diagonal matrix, $D=diag(\lambda_1,\lambda_2,\ldots,\lambda_k,O)$, $k=\min(p,q)$, O is the zero matrix, λ_i ($i=1,2,\ldots,k$) are called the singular values of A. According to the theory of SVD and matrix optimal approximation under Frobenious normal, the previously bigger singular values reflect the useful signal, while the smaller singular values reflect the noise. If we set these smaller singular values to zero, the noise of acquired signal will be removed, then we reconstruct the signal with the rest singular values, we can get the denoised signal.

3.2 Principle of LMD

To extract fault information from a vibration signals, LMD decomposes the raw signal into a series of product functions (PFs). A PF, which is the product of an envelope signal and a frequency modulated signal, has actual physical significance. Based on the moving average method, LMD progressively smooth the signal. The details about LMD algorithm can be found in Refs.

[8, 9]. Based on the decomposition process, the original signal can be reconstructed by

$$x(t) = \sum_{i=1}^{k} PF_i(t) + U_k(t) \qquad (3)$$

That is, the raw signal $x(t)$ can be reconstructed by all the PFs and a monotonic function $U_k(t)$. LMD method can guarantee the information integrity of the raw signal and effectively extract useful and comprehensive fault information.

4 Simulation signal analysis

A modulation signal $x(t)$ is shown in Fig. 3 to verify the proposed method, data length 1000, Hankel matrix dimension 501×500.

$$x(t) = (1 + 1.5t)\cos(30\pi t) + 2\sin(70\pi t) + \eta(t)$$

where $\eta(t)$ reprents the additive Gaussian white noise,standard deviation 1.5.

Fig. 4 shows the singular values (SVs) distribution curve. According to singular values difference spectrum shown in Fig. 5, choose the second peak for noise reduction, the reconstructed signal is given in Fig. 6. The result of the reconstructed signal after LMD is shown in Fig. 7. From Fig. 7, it can be clearly seen that PF1 represents 35Hz sinusoidal frequency and PF2 represents another modulation signal, which shows that SVD method can remove noise effectively and be helpful for LMD result explicable.

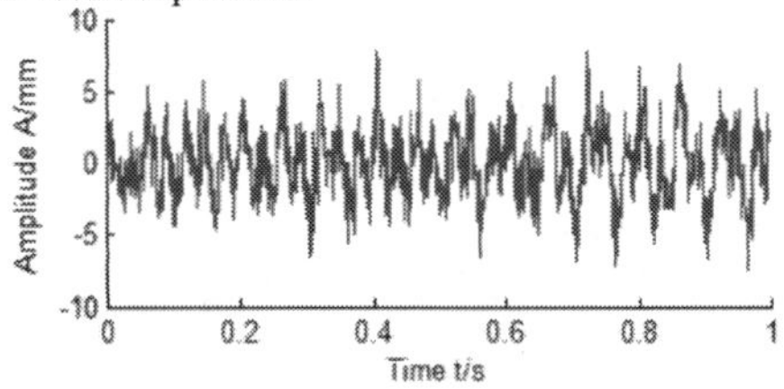

Figure 3. Simulation signal.

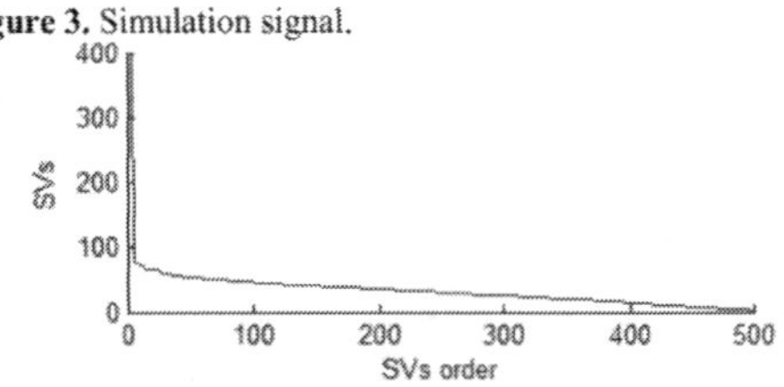

Figure 4. Singular values distribution curve of simulation signal.

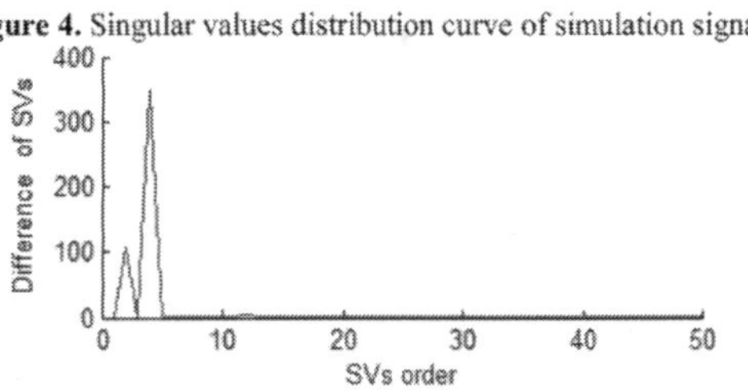

Figure 5. Difference spectrum of singular values.

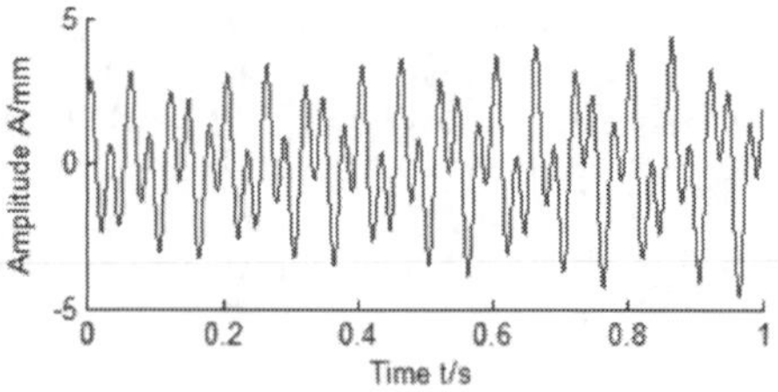

Figure 6. Reconstructed signal with the noise removed.

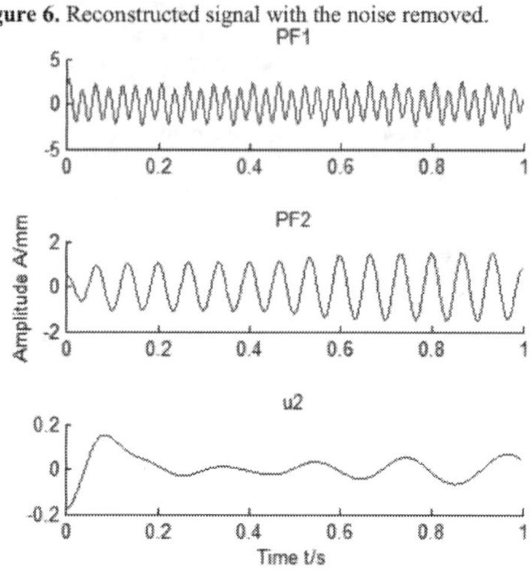

Figure 7. LMD result of the reconstructed signal.

5 Experiment research

5.1 Test bench

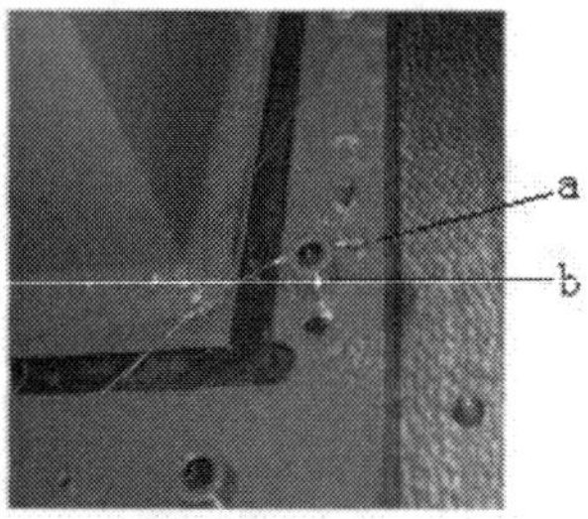

Figure 8. PCB accelerometers located along connecting bolts.(a) PCB accelerometers.(b) Connecting bolts.

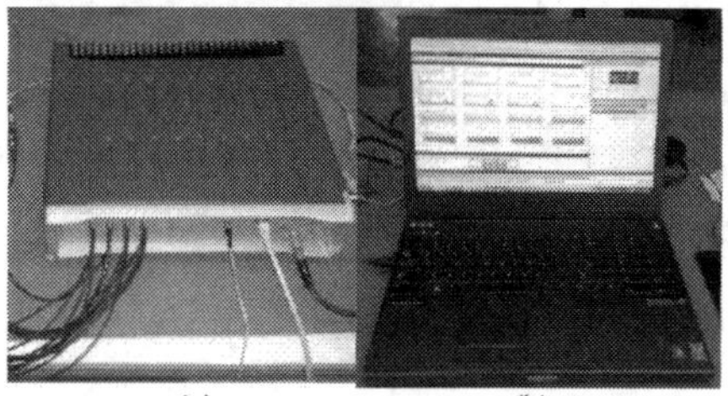

(a) (b)

Figure 9. LMS data acquisition system, (a) Hardware, SCADAS SCM202. (b) Software, Test.Lab 14A.

 DOI: 10.1051/matecconf/20167701029

Accelerometer is placed alongside connecting bolts as shown in Fig. 8. The model of accelerometer is PCB 333B32 single axis, sensibility 100mV/g, measurement range ±50g p-p. Data acquisition system is shown in Fig. 9.

5.2 Data analyses

The tilting-pad journal bearing test bench works at 3900r/min, collect signal with the data length 1024 by 2048Hz, shown in Fig. 10.

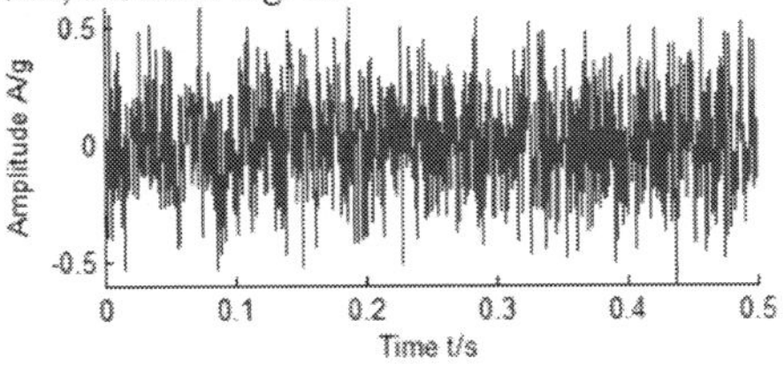

Figure 10. Collected signal from test bench.

As can be seen from Fig. 10, there is much noise which affects the following analysis. Thus, the method based on SVD is applied for noise reduction. Singular values distribution curve is shown in Fig. 11. According to the singular values difference spectrum shown in Fig. 12, we choose 13 orders for noise deduction, and then the reconstructed signal is shown in Fig. 13.

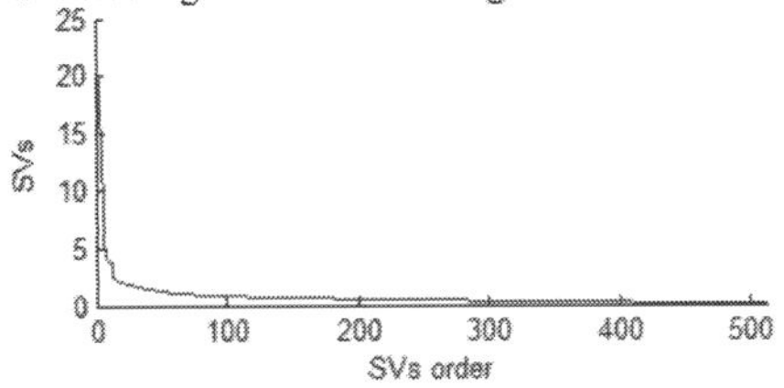

Figure 11. Distribution curve of singular values.

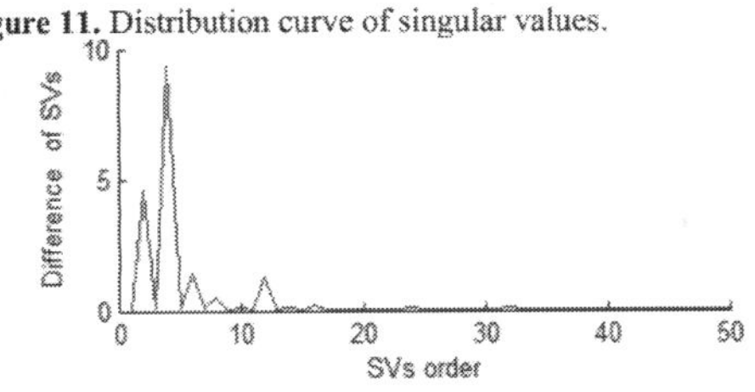

Figure 12. Difference spectrum of singular values.

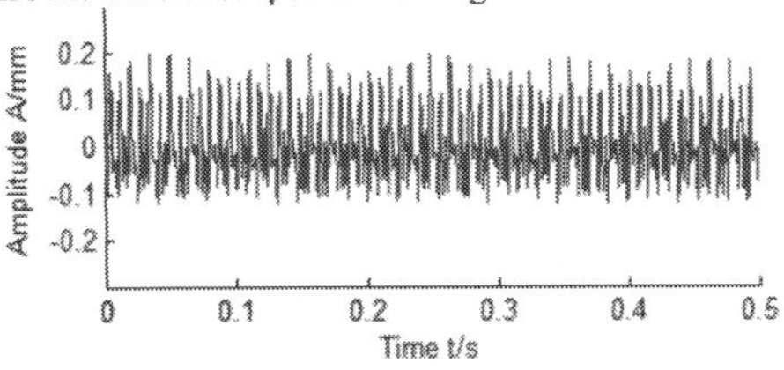

Figure 13. Restructured waveform.

From the comparison between Fig. 10 and Fig. 13, the noise is reduced effectively. Conduct LMD, the result is shown in Fig. 14.

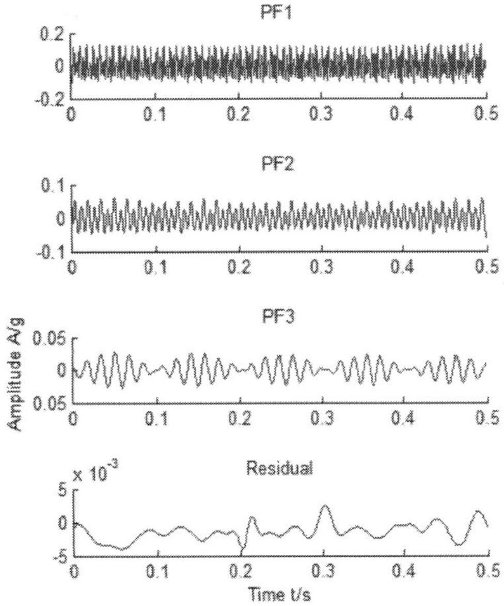

Figure 14. LMD Result of the signal after SVD denoising.

The useful PFs have strong correlation with the signal, while the useless PFs don't. Mutual information and correlation coefficient, which is used here, are common two methods to distinguish between these two kinds of PFs. The bigger the correlation between the PFs and original signal x is, the bigger the correlation coefficient ρ is. Correlation coefficient is defined as below:

$$\rho_{x,PF_i} = \frac{\sum_{k=1}^{N} x(k) PF_i(k)}{\left[\sum_{k=1}^{N} x^2(k) \sum_{k=1}^{N} PF_i^2(k)\right]^{1/2}} \tag{4}$$

The values of correlation coefficient between these PFs and signal before LMD are listed in Table 2, the value between the residual and the signal is also listed for comparison. From this table, we find the values of correlation coefficient between the PFs and the signal before LMD are far bigger than the value between the residual and the signal, which demonstrates that these PFs all are useful components.

Table 2. Correlation coefficient values.

Component	PF_1	PF_2	PF_3	Residual
Correlation coefficient	0.845	0.591	0.236	0.015

As can be seen from Fig. 14 the production functions (PFs) are non-stable, showing typical features of amplitude modulation and frequency modulation caused by the rubbing between rotor and stator [13].

Time-frequency map and marginal spectrum are shown in Fig. 15 and Fig. 16 respectively. As can be seen from Fig. 15, frequency varies dramatically with time. In Fig. 16, it is found there is wide range of spectrum around wording frequency (65Hz) and its frequency multiplication. Large amount of experiments shows that vibration produced by friction from mechanical equipment will cause nearly continuous spectrum in some bands. The continuous frequency band is the characteristic of friction fault [14].

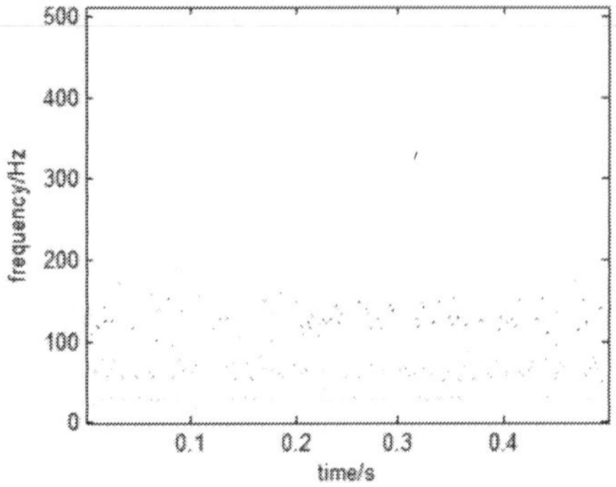

Figure 15. Time-frequency map.

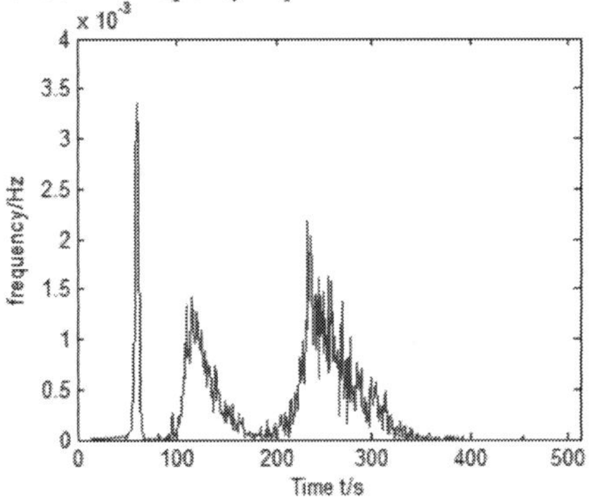

Figure 16. Marginal spectrum.

In order to verify this conclusion, stop the test bench and check the inner surface of the bottom pad of one tilting-pad journal bearing, find that there exists serious friction on it, as shown in Fig. 17.

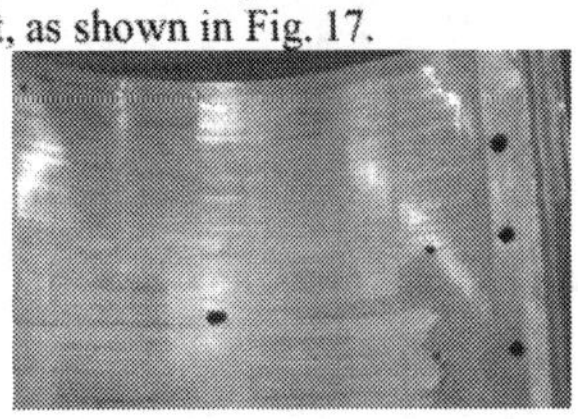

Figure 17. Inner surface of bottom pad of tilting-pad journal bearing.

6 Conclusion

Aiming at the fault diagnosis for the newly developed tilting-pad journal bearing with fluid support, this paper proposed the hybrid method based on SVD and LMD. Its efficiency has been evaluated on simulation analysis and experiment signal. The result demonstrates that the proposed method is feasible and effective for the fault diagnosis of the tilting-pad journal bearing with fluid support.

Tilting-pad journal bearing with fluid support is a new type of bearing, until now its theory foundation is still insufficient, which is an important project for future study. The reconstruction order of SVD has a great effect on the reconstructed signal and following signal analysis, so how to select effective reconstruction order may be another issue for investigation later.

References

1. L.Weiguang, Z. Xiaotao. Analysis on Static Performance of Tilting Pad Sliding Bearing. Lubrication Engineering **38(8)**, 6-12 (2014).
2. L. Suya. Bearing tester and testing technique. Bearing **8**, 55-59 (2011)
3. Y.B. Fu, C.K. Chui, C.L. Teo. Accurate two-dimensional cardiac strain calculation using adaptive windowed Fourier transform and Gabor wavelet transform. Int. J. Comput. Assisted Radiol. Surg. **8(1)**, 135–144(2013).
4. S. Mallat. A theory for multi-resolution decomposition, the wavelet representation. IEEE Trans. Pattern Anal. Mach. Intell. **11(7)**, 674–689(1989)
5. Z.K. Peng, F.L. Chu. Application of the wavelet transforms in machine condition monitoring and fault diagnostics: a review with bibliography. Mech. Syst. Signal Process. **18 (2)**, 199–221(1989).
6. S. Olhede, A.T. Walden. The Hilbert spectrum via wavelet projections. Proc. R. Soc. Lond **460**, 955–975(2004).
7. J.S. Cheng, D.J. Yu, Y. Yang. Energy operator demodulating approach based on EMD and its application in mechanical fault diagnosis. Chinese J. Mech. Engrg. **40(8)**, 115-118(2004).
8. Smith Jonathan S. The local mean decomposition and its application to EEG perception data. J. Roy. Soc. Int. **2,** 443-454(2005).
9. P. Cheolsoo, L. David, et al. The complex local mean decomposition,Neuro-computing. Neuro-computing, **74(6)** ,867-875(2011).
10. M. R. Bai, S.J. Elliott. Preconditioning multichannel adaptive filtering algorithms using EVD- and SVD-based signal prewhitening and system decoupling.J. of Sound and Vibration, **270**(4), 639–655(2004).
11. L. Lehtola, M. Karsikas, M. Koskinen, et al. Effects of noise and filtering on SVD-based morphological parameters of the T wave in the ECG. J. of Med. Eng. and Tech., **32(5)**, 400–407(2008).
12. Y. Lihua, Q. Shemiao, G. Haipeng, et al. The numerical analysis on the load capacity of aerodynamic tilting-pad journal bearings.Lubrication Engineering **32(4)**, 14-16(2007).
13. L. Luping, Z. Xinyuan, C. Jian. Typical features study on rubbing fault of steam turbine unit. J. of Vib. Mea. & Diag. **12(4)**, 281-285(2001).
14. H. Zhengjia, Z. Yanyang, M. Qingfeng, et al. *Fault diagnosis theory and application of non-stationary signal of mechanical equipment.* (Beijing:Higher Education Press, 2001).

MATEC Web of Conferences **77**, 01030 (2016) DOI: 10.1051/matecconf/20167701030
ICMMR 2016

Research on Single Drive and Multi-dimension Output Vibrator for Fuse Sand Filling Based on Decoupled 1T2R Parallel Mechanism

Qin Wei, Hang Lubin , Cheng Wushan and Huangfu Yabo

College of Mechanical Engineering, Shanghai University of Engineering Science, China

Abstract. In this paper a three-dimension vibrator for fuses' sand filling, including one-dimension vertically translation and two-dimension oscillating rotation (1T2R) is presented, which is used to improve the low efficiency in the process of traditional fuses' sand filling. The paper starts with a type synthesis of a new kind of 1T2R parallel mechanism (PM) which is constructed based on the POC set. After two of the driving joints in the PM are substituted by the spring-damp system, the design propose of single-input and multi-outputs can be realized. Then the kinematics analysis and simulation optimization of this mechanism was also carried out.

1 Introduction

Fuse is an important component of the smart grid user end system, some problems still remains in the production process of the fuse [1]. Since the fuse can improve the electrical stability and thermal stability of the electrical appliances [2], [3] and the density of quartz sand filled in the fuse box has a direct and significant influence on the breaking performance of the fuse [4]-[6].

Single-dimension sand filling vibration appliances [7], shown in Fig. 1. (a) (b), are widely used in domestic low voltage electrical enterprises. As is shown in Fig. 1.(c) The fuse sand filling unit consists of flexible sand filling nozzle 1, fuse 2, electromagnetic coil components 3, spring 4, base 5.

(a) (b) (c)
Figure 1. Mechanical fuses sand filling device

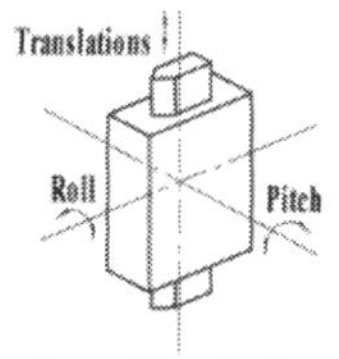

Figure 2. Expected motion of fuse in the process of sand filling

For the reasons that the traditional single-dimension vertical vibrator is difficult to meet the requirements of sand filling density. The paper proposed a novel three-dimension vibrator for fuses' sand filling, including one-dimension vertical translation and two-dimension rotations (shown in Fig. 2), based on POC set [8].

In comparison to the traditional fuse's sand filling vibrator, the two additional rotary vibration of the new designed PM exerted on the fuse can make the fuse get oscillated in roll and pitch directions, which can spread the quartz sands in the fuse uniformly, compact the fine-grained quartz sands and improve the density of fuse product.

The paper is organized as follows. The type synthesis of 1T2R PM is given in section 2. The kinematic analysis is carried out in section 3. The vibration simulation and geometry optimization of the 1T2R PM vibrator is carried out in section 4. Finally, conclusions are drawn in section 5.

2 Type synthesis of PM

The multi-dimension fuse vibration platform can improve the density of quartz sand filled in the fuse. The platform needs to achieve a motion of one translation and two rotations, shown in Fig. 2. The input-output characteristics of mechanism should meet the requirement of the decoupled judgment criteria, the POC set of the decoupled PM can be derived as $M_{P_a} = \begin{bmatrix} t^1 \\ r^2 \end{bmatrix}$.

2.1 The construction of decoupled PM

For the sake of reducing the control difficulty of the PM, the decoupled property should be combined with the mechanism in the design process.

According to the decoupled judgment criteria of the PM proposed by author of paper [9], the 1T2R PM shown in Fig. 3 is selected as the prototype to construct the multi-dimension vibrator because of its simplicity, and the structure synthesis and kinematics analysis are given in detail thereinafter, it consists of two basic kinematic chains (BKC). One of the loops is a BKC of planar five-bar loop with two driving translation joints P_1, P_2 and three revolute joints R_1, R_2, R_3, it can be denoted by $HSOC_1\{-P_1 \perp R_1 \| R_2 \| R_4 \perp P_2-\}$. The other loop is a BKC consist of a driving joints P_3, two spherical joints S_1, S_2 and a revolute joint R_2, it can be denoted by $SOC_2\{-P_3-S_1-S_2-R_2-\}$. The linkage S_2R_2 is the output moving platform, as is shown in Fig. 3.

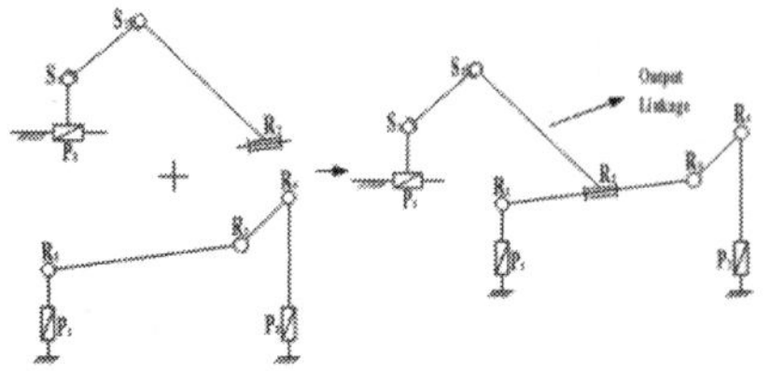

Figure 3. The construction of 1T2R PM

2.2 PM synthesis with the output motion characteristic matrix

1) POC matrix of each branch end

The POC matrix of the branch end of $HSOC_1$ and SOC_2 can be expressed as: $M_{b_1}=\begin{bmatrix} t^1(\| z) \\ r^2(\| \square(x,y)) \end{bmatrix}$, $M_{b_2}=\begin{bmatrix} t^3 \\ r^3 \end{bmatrix}$, respectively.

2) POC matrix of the moving platform of the PM

Substitute the POC matrix of the terminal branches of chain 1, 2 into formula $M_{P_a}=\bigcap_{i=1}^{v+1} M_{b_i}$ [10], the POC matrix of the moving platform is obtained.

$$M_{P_a}=\begin{bmatrix} t^1(\| z) \\ r^2(\| \square(x,y)) \end{bmatrix} \bigcap \begin{bmatrix} t^3 \\ r^3 \end{bmatrix} = \begin{bmatrix} t^1 \\ r^2 \end{bmatrix}$$

2.3 The DOF of the mechanism

The POC set of the moving platform is affected by the degrees of freedom (DOF) of the mechanism, so it's necessary to analyze whether the DOF of the PM can meet the demand of the POC set. According to the formula of DOF [11], we will obtain:

$$F=\sum_{i=1}^{m} f_i-\sum_{j=1}^{v} \xi_{L_j}=12-(6+3)=3 \quad (1)$$

Where m-- the number of kinematic pairs

f_i-- the DOF of the i th kinematic pair

v-- the number of the independent loops

ξ_{Li}-- the number of the independent loop of the j th basic loop.

3 Forward kinematics of 1T2R PM

The static and moving coordinate systems are established as shown in Fig. 4. The angle between the y-axis and the Y-axis is donated by α, The angle between the x-axis and the X-axis is donated by β, the angle between the X-axis and linkage R_3R_4 is denoted by γ, The longitudinal coordinate of the moving platform's center is denoted by z. the displacement of shifting pair $P_i(i=1,2,3)$ is denoted by $d_i(i=1,2,3)$, the distance of each two adjacent joints is denoted by $L_i(i=1,2...9)$.

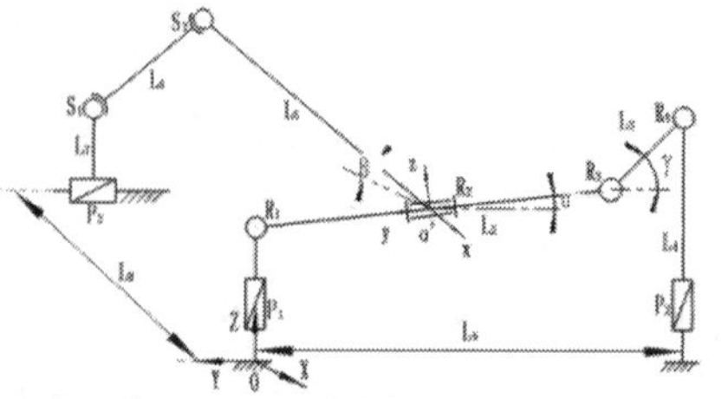

Figure 4. Coordinate system of 1T2R PM

3.1 The governing kinematics equation of the PM

Project the linkages in loop $-P_1 \perp R_1 \| R_3 \| R_4 \perp P_2-$ to the X-axis and the Y-axis respectively, and the projection equation of the loop will be obtained:

$$\begin{cases} L_2\cos\alpha + L_3\cos\gamma = L_9 \\ L_1 + d_1 + L_2\sin\alpha + L_3\sin\gamma = L_4 + d_2 \end{cases} \quad (3)$$

Since the coordinates of spherical joints S_1, S_2 must satisfy the geometry constraint equation below [12], [13]:

$$(X_{S_2}-X_{S_1})^2+(Y_{S_2}-Y_{S_1})^2+(Z_{S_2}-Z_{S_1})^2=L_5^2 \quad (4)$$

So, the simultaneous equations of the mechanism kinematics can be expressed as:

$$\begin{cases} L_2\cos\alpha + L_3\cos\gamma = L_9 \\ L_1 + d_1 + L_2\sin\alpha + L_3\sin\gamma = L_4 + d_2 \\ (X_{S_2}-X_{S_1})^2+(Y_{S_2}-Y_{S_1})^2+(Z_{S_2}-Z_{S_1})^2=L_5^2 \end{cases} \quad (5)$$

3.2 Forward position problem

The purpose of the forward position problem is to obtain the output parameters, i.e., the position parameters z, α, β of the moving platform's center, and the displacement d_1, d_2, d_3 of the shifting pairs P_1, P_2, P_3 are known.

$$z=L_1+d_1+L_2\sin\alpha=L_1+d_1+L_2\sin\left(2\arctan\frac{A\pm\sqrt{A^2+B^2-C^2}}{B-C}\right) \quad (6)$$

From Eq. (5), two linear equation in variables α, β are obtained, expressed in the following way:

$$A\sin\alpha + B\cos\alpha + C = 0 \quad (7)$$

$$D\sin\beta + E\cos\beta + F = 0 \quad (8)$$

From Eq. (7) and Eq. (8), we can obtain:

$$\alpha = 2\arctan\frac{A\pm\sqrt{A^2+B^2-C^2}}{B-C}$$

$$\beta = 2\arctan\frac{D\pm\sqrt{D^2+E^2-F^2}}{E-F}$$

where

$$
\begin{aligned}
A &= 2L_2(L_4 + d_2 - L_1 - d_1) \\
B &= 2L_2 L_9 \\
C &= L_2^2 - L_3^2 - (L_1 + d_1 - L_4 - d_2)^2 - (L_9 + d_1 + d_2)^2 \\
D &= (L_9 + 2d_3 + 3L_2\cos\alpha)L_5\sin\alpha + 2(L_1 + d_1 - L_7)L_5\cos\alpha \\
E &= -2L_5 L_8 \\
F &= -L_6^2 + L_8^2 + (\tfrac{1}{2}L_9 + d_3 + \tfrac{1}{2}L_2\cos\alpha)^2 + (L_1 + d_1 + L_2\sin\alpha - L_7)^2
\end{aligned}
$$

3.3 The reduction and selection of the driving pair

For the reasons that the number of the vibration generators have an important influence on fuse vibration's cost, we propose to reduce the number of the driving joints to one. Since the loop 1 has partly symmetry in the structure, the active driving joints in loop 1 are substituted by the force adaptable elastic element to realize the design purpose of single-input and multi-outputs, that is to say the driving pairs P_1, P_2 are constrained by the spring-damp system based on the mechanism decoupled principles, and the loop 1 becomes the force adaptable servo loop. Since only one driver is needed in this mechanism, the operation cost of the fuse vibration is reduced. The shifting pair P_3 is selected to be the vibration generator and the structure of the PM fuse vibrator is shown in Fig. 5.(a), and by the usage of the 24 potentiometers, which are shown in Fig. 5.(b), the vibration mode (working amplitude, pitch angle and roll angle) of the PM can be adjusted easily.

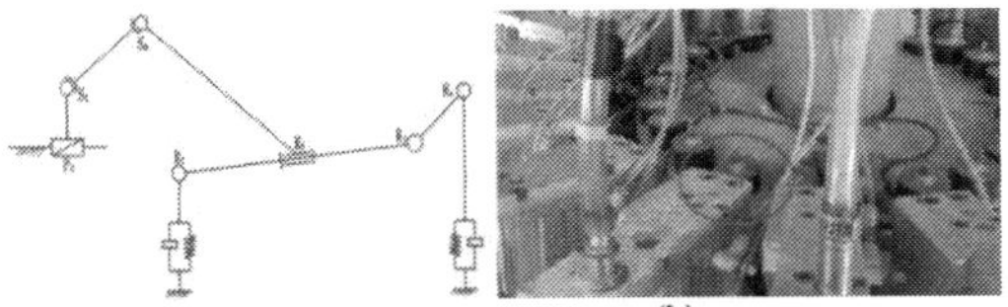

(a) (b)

Figure 5. The mechanism sketch of 1T2R PM vibrator constrained by the spring-damp system and the potentiometers

4 The simulation and optimization of the vibrator

Experiment shows that the fuse vibrator with a working amplitude of 1~2mm, a pitch angle of ±2°, a roll angle of ±1° can improve the fuse's sand filling density and the efficiency according to process parameters.

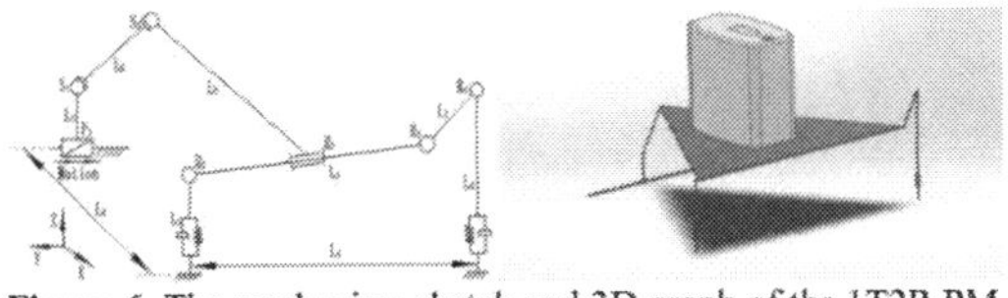

Figure 6. The mechanism sketch and 3D graph of the 1T2R PM

4.1 The initial setup of the PM vibrator's size

The simulation model of the PM is established in Solidworks, as shown in Fig. 6, and the size of the mechanism is set up according to the structure and the dynamic parameters of the vibrator [14]. The PM's structure parameters are set as: L_1=30mm, L_2=220mm, L_3=22mm, L_4=40.5mm, L_5=84mm, L_6=22mm, L_7=11mm, L_8=100mm. The stiffness of the springs between the shifting pairs P_1, P_2 is 5N/mm, and the damping coefficient is 0.035.

4.2 Dynamic simulation analysis

In order to test the motion performance of the 1T2R PM vibrator we designed, the motion simulation model of the vibrator is established in ADAMS, the kinematic pairs are constrained, and the fuse model is added onto the center of the moving platform. The final simulation model is shown as Fig. 7.

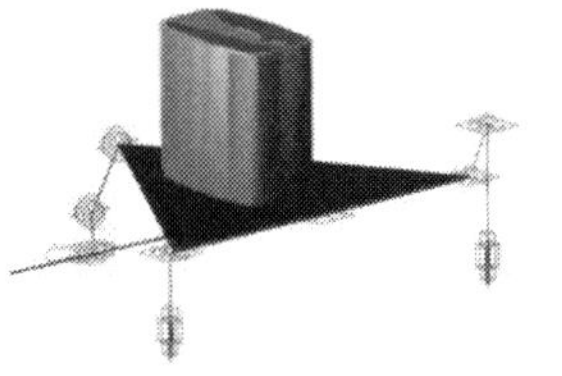

Figure 7. The structural diagram of the vibrator in ADAMS

According to the analysis of the type synthesis of the mechanism, the prismatic pair P_3 is chose to be the driving pair, and the driving force is added to it to generate the vibration, the expression is: d3=10sin(100πt) mm, 50Hz. The simulation results (the working amplitude, the pitch angle and the roll angle) of the moving platform are shown as Figure 8-10.

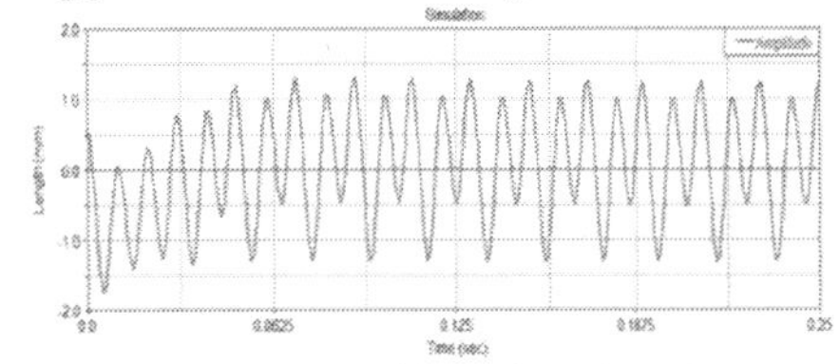

Figure 8. Simulation curve of fuse vibrator's working amplitude

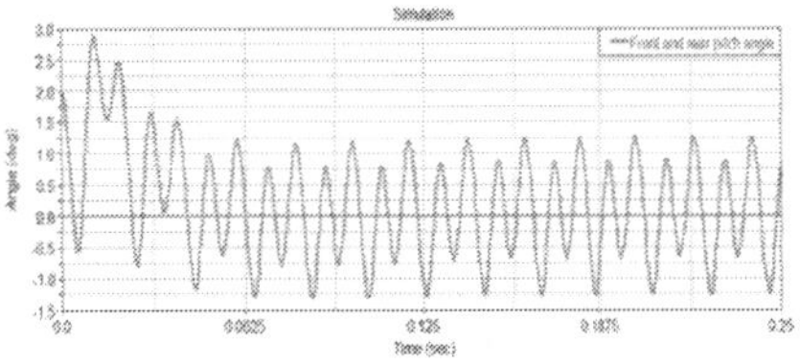

Figure 9. Simulation curve of fuse vibrator's pitch angle

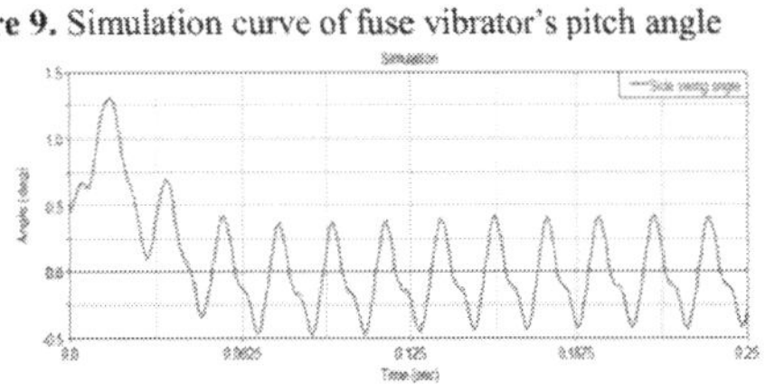

Figure 10. Simulation curve of fuse vibrator's roll angle

From the simulation curves above, we can see that the PM vibrator comes to a steady state after 0.0625s, the working amplitude and the pitch angle of the fuse vibration can meet the production index of mechanical sand filling, but the roll angle of the vibration is just ±0.43°, and can't meet the working demand of ±1°.

4.3 Model parameter optimization

The model parameter optimization must be carried out to make the roll angle of the vibration meet the production demand of mechanical sand filling as well.

By parametrical setting of each point in ADAMS, the influence of each design variables on the roll angle of the vibration can be analyzed, and we found that the Z coordinate of the spherical pair S1, the Z coordinate of the spherical pair S2 and the Z coordinate of revolute joint has a great influence on the roll angle of the vibration.

Then the coordinates of the three reference points are adjusted to obtain the roll angle of the vibration. By using the ADAMS to optimize the three design variables, the coordinates of the design variables and the corresponding roll angle are optimized. The optimization results are shown in Table I.

Table 1. The optimization of the roll angle

	Roll angle	S_{1Z}	S_{2Z}	R_{2Z}
Initial Value	±0.4398°	11.000	27.000	30.250
Optimization	±0.8503°	11.007	25.472	30.599

To sum up, the optimized size of the 1T2R PM fuse vibrator, which meet the demand of the mechanical sand filling working index can be set as: L_1=30mm; L_2=220mm; L_3=22mm; L_4=40.5mm; L_5=89.629mm; L_6=22.572mm; L_7=11.258mm; L_8=100mm.

5 Conclusion

(1) To improve the efficiency of the traditional single dimensional vertical sand filling vibrator, a three-dimension vibrator based on the decoupled PM for fuses' sand filling is presented.

(2) The decoupled 1T2R PM is established based on the POC set and the decoupled criterion.

(3) To reduce the number of the driving joints of the PM, a creative method of substituting the two prismatic pairs in loop 1 by spring-damp system is proposed. The mechanism can provide one vertical translation and two rotations output motion in the condition of one input motion.

(4) The motion simulation of the mechanism is carried out in ADAMS, and roll angle of the fuse vibration is optimized by using the principle of sensitive point.

Acknowledgement

The authors would like to acknowledge the financial support of the national science foundation of China (NSFC) with grant number of 51475050, Shanghai Science and Technology Committee with grant number of 12510501100, the Natural Science Foundation of Shanghai City with grant number of 14ZR1422700, and the Graduate student research Innovation Project of Shanghai University of Engineering Science (SUES) with grant number of E109031501012 and E109031501016.

References

1. X.M. Wang. Electric Power Technologic and Economics. **22**, 1-7 (2010).
2. M.A. Saqib, AD Stokes. IEEE Trans. Power. Del. **25**, 212-220 (2010).
3. A. Pleşca, INT J THERM SCI. **70**, 144-153 (2013)
4. X.Q. Chen, H.P. Fan, J.M. Wang. Low Voltage Apparatus. **01**,44-47 (1986).
5. W. Bussiere. J. Phys. D: Appl. Phys. **34**, 925 (2001)
6. Y.Y. Xiao, J.W. Zhuang, B Chen. J. Harbin Eng. Univ. **34**, 524-529 (2013).
7. Zhang Minliang, Cheng Wushan, Lin Haiou, Hang Lubin, Yang Huibin, Yue Min. A kind of intermittent rotary continuous circulating type fuse sand filling machine[P]. Shanghai: CN102915881A, 2013-02-06
8. T.L. Yang. *Theory and Application of Robot Mechanism Topology*, Beijing Science Press, (2012).
9. L.B. Hang, Y Wang, T.L. Yang. China Mechanical Engineering. **15**, 1035-1037,1041, (2004).
10. T.L. Yang, A.X. Liu, Q Jin, et al. J. Mech. Des. **131**, 021001-1-17, (2009).
11. T.L. Yang, DJ Sun. J. Mechanisms Robotics. **4**, 011001-1-17, (2012).
12. J J. Craig *Introduction to robotics: mechanics and control*, Upper Saddle River: Pearson Prentice Hall, (2005).
13. Y Yun, Y Li. J. Dyn. Sys. Meas. Control. **134**, 021001-1-9, (2012).
14. B Li, W Zhao, Z Deng. JMSY. **31**, 50-58, (2012).

MATEC Web of Conferences 77, 01031 (2016) DOI: 10.1051/matecconf/20167701031
ICMMR 2016

A Development and Application of the Fretting Fatigue Test Apparatus

Jun Wang[1,2], Hong Xu[2], Tiexiong Su[1], Huping Mao[1], Yi Zhang[1] and Yangang Zhang[1]

[1]*School of Materials Science and Engineering, North University of China, Taiyuan, Shanxi 030051, China*
[2]*School of Mechanical and Power Engineering, North University of China, Taiyuan, Shanxi 030051, China*

Abstract. Fretting fatigue occurs in contacting parts which are subjected to fluctuating loads and very small sliding movements at the same time. Fretting fatigue can significantly reduce the fatigue life of materials. In this investigation, a new device is introduced, which performed using a combination of a servo-hydraulic push-pull fatigue testing machine and a specially designed jig for appropriate positioning of pads. The axial and contact loads are measured by load cells. The functionality of the device is examined by making a comparison between fretting fatigue lives and plain fatigue lives of a number of ZL702A aluminum alloy specimens. The main advantages of the device are its simplicity and cheapness. It can be developed further for high and low temperature tests in future.

1 Introduction

Fretting is induced by vibrations or the application of bulk fatigue stresses to one or both of the contacting parts and widely occurs at the contact interface between two bodies when there is small amplitude oscillatory relative movement (typically 5-50 micrometers), such as dovetail assemblies in aircraft structures [1], [2], cables in overhead conductors [3] and mine hoist [4], riveted joints [5], prostheses [6], etc. At the edge of the contact there will be a stress singularity which results in early nucleation of cracks at the pad tip. The surface interface cracks are the result of partial slip, friction and etc. within the contact interface. The fretting results in enormous reduction of the endurance limit of a component by half or more, in comparison to the normal fatigue conditions [7].

By far, there is not a commercial universal machine available for fretting fatigue testing. To investigate the characteristics of fretting fatigue, various test apparatus has been developed for special purpose in different fields. Literature of fretting fatigue demonstrates that the fatigue load was imposed mostly by hydraulic actuator and occasionally by position control machine [7]; the contact pressure of fretting specimens is induced through the clamping pads attached to the proving ring, spring or other fretting fixtures. Geometries of the pads are flat, spherical, cylindrical, and bridge-type [3], [4], [8]-[10], but most specimens use flat fretting contact surface. Though there are diverse apparatus for fretting fatigue test, the design has to meet certain requirements to obtain efficient operation and results [11]: (i) the ability to produce a constant normal force between pad and specimen and the visibility of pad and specimen, (ii) a few microns relative movements, (iii) very high stiffness of the fixture, (iv) high precision of measurement.

The research background of the project was the fretting fatigue damage in engine aluminum blocks, and the aim is to provide accurate data to support theoretical design and the practical application of the alloy in vehicle engine. The cylinder blocks are the largest and most intricate single piece, which suffers high clamping forces from bolt assembly and oscillating dynamic forces from bearing load. Fretting wear occurs frequently on the aluminous cylinder block of vehicle engine [12]-[14].

Figure 1. A general view of the fretting fatigue test system.

2 Test apparatus

The fretting fatigue test system has been developed by the authors in this investigation. A general view of the system is depicted in Fig. 1. The system consists of three parts: the universal hydraulic fatigue testing device which produces the oscillatory axial displacement, the fretting fatigue test jig through which the pressure is transmitted by two pads to the specimen and the instrumentation. The instrumentation consists of load cells for measuring the contact and axial forces imposed on the specimens, the

 DOI: 10.1051/matecconf/20167701031

inverter for varying the frequency, and fatigue cycles counter.

As shown in Fig. 2, the schematic of fretting fatigue jig consists of two fixed mounts, two cross girders, two normal load cells, two load adjusting screws, two screw holders, two fretting pad holders and two fretting pads etc. The fixed mounts and cross girders constitutes the supporting platform of fretting fatigue jig, which completely tightened on the two main columns of the servo-hydraulic testing machine by the screws. The other parts of the jig are installed on the platform.

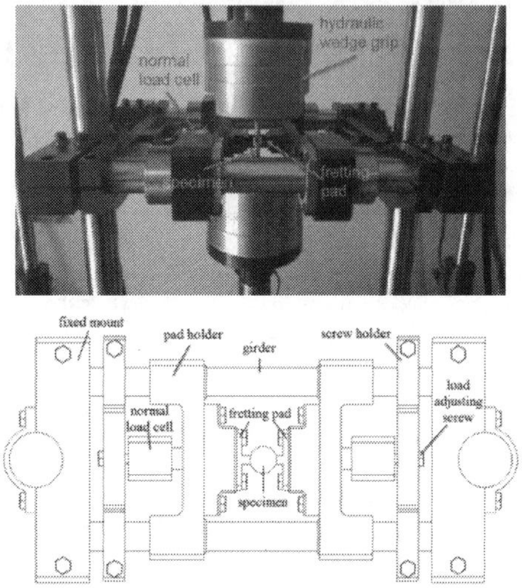

Figure 2. A schematic of the fretting fatigue jig.

The fretting pads fastened on the pad holders which can move along the cross girders always. The adjusting screw holders can move along the cross girders, to facilitate the installation of the specimen before the test. When the specimen installed by the hydraulic wedge grip of the fatigue testing machine, the screw holders and the pad holders can slide to make the fretting pad to press against a flat specimen just. Then the screw holders are fixed on the cross girders.

The normal load is induced by two adjusting screws which impelled the fretting pad clamped the specimen tightly and is real-time measured using two load cells. The load cell digital readings can ensure that the contact loads produced by adjusting screws in two opposite directions are equal and sufficiently accurate.

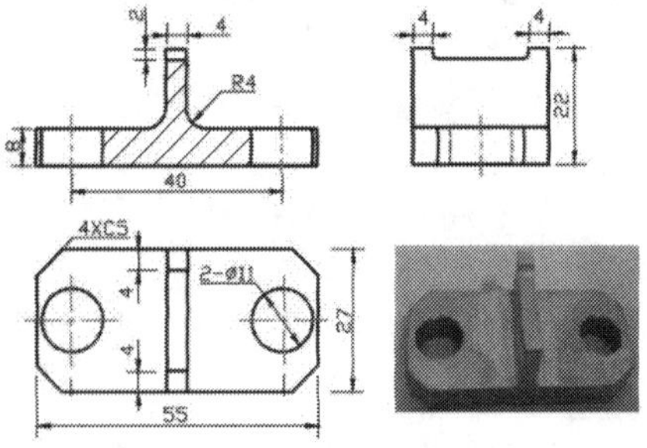

Figure 3. Dimensioned design drawings (dimensions are in mm) and entity of the fretting pad which used for transmitting the contact pressure.

Fig. 3 shows the geometry of the fretting pad. Each pad has two contact surfaces with the rectangular section 4×4mm, the total area of 32 mm2 (i.e. the contact area of the specimens), with not more than 0.32μm surface roughness which was obtained by using the low stress grind machining and the polishing processes.

To make the laboratory condition closer to actual working condition of engine blocks, the parallelism between the contact surfaces of the specimen and pad was maintained. The fretting fatigue system has the characteristics as below: (i) it can simulate flat-flat or round-flat contact states between specimens and pads by using pads with plane or cylindrical geometries, (ii) it produces a constant normal force of maximum 5kN between pads and specimens, (iii) it has the visibility of the pads and the specimens, (iv) its normal load be measured using compression load cells with the accuracy of ±1% FSO (full-scale output), which corresponds to ±0.05kN.

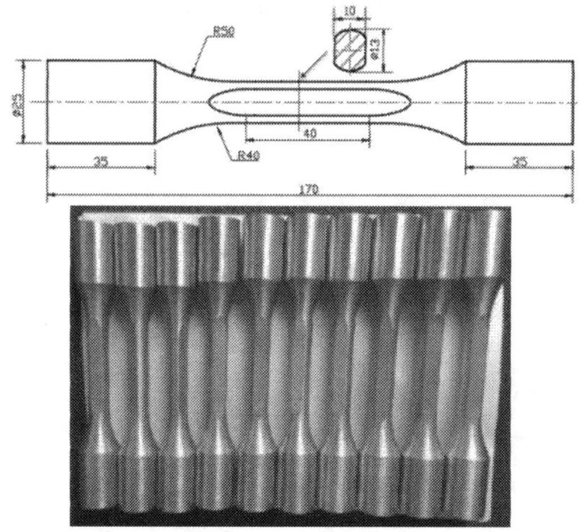

Figure 4. Dimensioned design drawings (dimensions are in mm) and entities of the fretting specimen.

3 Experiment results

3.1 Materials and specimens

ZL702A aluminum alloy was used in this investigation. The material's composition obtained using energy-dispersive X-ray apparatus is AL 91%, Si 7 %, Cu 1.55%, Mg 0.41% and Mn 0.22% (wt%). The flat specimen used in this work had a width of 13mm, a thickness of 10mm and a gauge length of 40mm. The specimen geometry is shown in Fig. 4. The specimens were processed by solid solution treatment and complete artificial aging. The gauge portions of all specimens were polished with SiC papers and cleaned with acetone. The surface average roughness of the specimen contact surface in the transverse direction is 0.32μm.

 DOI: 10.1051/matecconf/20167701031

3.2 Fretting fatigue tests

In this investigation, the experiments were conducted for stress ratio of R=0.1, frequency of 10 Hz at a constant contact force of 2kN, and alternating stress amplitudes of 86, 104, 121, 138 and 156MPa. A number of plain fatigue tests were also performed using the universal fatigue testing machine under the same conditions as stated above for comparison purposes. The results are given in Table I. As the results indicate, the plain and fretting fatigue lives both decrease with the alternating stress increasing. At the same alternating stress, the fretting fatigue life is much lower than the plain fatigue life, and the percent decrease of fatigue life due to fretting is found to be 59% at average. It is obtained from the experiments that the plain fatigue limit is 89 MPa, and the fretting fatigue limit is 42 MPa at a constant contact force of 2kN, which is only 47% of plain fatigue limit.

Table 1. Lives obtained from the experiments on fretting and plain fatigue.

Applied Stress σ_{max} /MPa	Fretting fatigue life N_f/cycles	Plain fatigue life N_p/cycles	Reduced proportion of life $(1-N_f/N_p)$%
86	842101	>10^7	-
104	73846	180588	59
121	17908	46945	62
138	15559	35592	56
156	811	2013	60

3.3 Morphology of fracture and wear Zone

The fracture and fretting wear surfaces of tested specimens were examined using a scanning electron microscope (SEM). The typical fracture morphologies of the plain and fretting fatigue specimens are illustrated in Fig. 5. The figure indicates that the fracture surface consists of two quite distinct regions: a fatigue zone created by crack propagation and an instant rupture zone which given rise to fracture of specimen when it is sufficiently weakened by the crack zone development. Fig. 5(a) shows that the plain fatigue crack initiated at the intersection of the plane and the cylinder of the specimens (shown by the arrow). The fatigue crack shows a fan-shaped radial propagation. Fig. 5(b) showed cracks of the fretting fatigue specimens originating from the edge of the contact region, expanding to the inner which is under the contact region of the specimen and then radially propagating throughout its thickness.

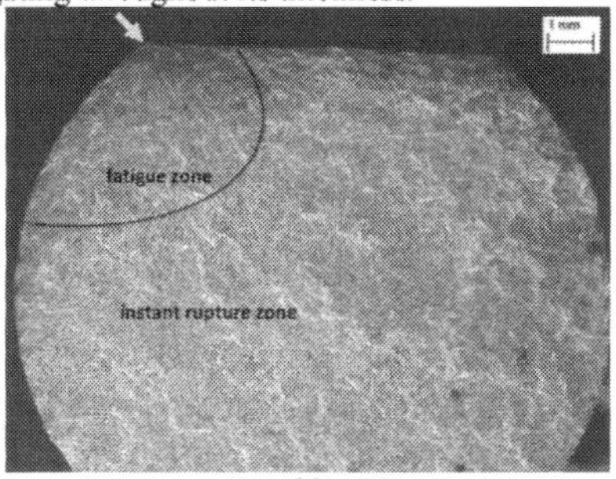

(a)

(b)

Figure 5. SEM micrographs of the fracture. (a) plain fatigue test specimen and (b) fretting fatigue test specimen.

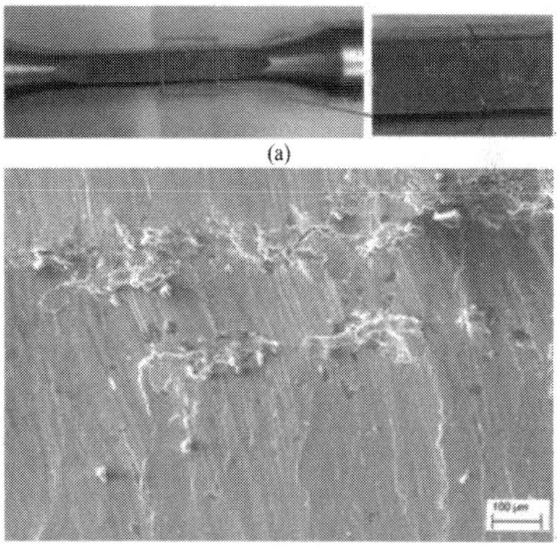

(a)

(b)

Figure 6. Fretting wear zone of fretting fatigue specimen. (a) Fretting scar and (b) SEM micrographs of the scar.

Fretting wear zone which are created in the contact area between pads and specimens is shown in Fig.6. Fretting fatigue specimens surface was worn which appears black by fretting and the fracture happened in the most severely worn area. The lamellar structure and plenty of debris presented on the fretting scars of fretting fatigue specimen.

4 Conclusion

Based on the universal hydraulic fatigue machine, a new fretting fatigue device typically is developed and manufactured by authors, which has enough accuracy and reliability and has enabled testing condition closer to actual working condition of engine blocks. The device is capable of exerting adjustable axial and contact loads. The frequency of the device can be varied from 1-50Hz. The functionality of the device is validated by fretting fatigue test results of a number of ZL702A aluminum alloy specimens. The device is well instrumented for varying and measuring axial and contact loads, fatigue cycles, test frequency.

Acknowledgment

This work supported by the National Nature Foundation of China (No. 51275489).

References

1. G.H. Majzoobi, M. Jaleh, Mater. Sci. Eng., **A 452-453**, 673-681, (2007).
2. D.S. Wei, Y.R. Wang, Eng. Fail. Anal., **25**, 144-155, (2012).
3. C.R.F. Azevedo, A.M.D. Henriques, A.R. Pulino Filho, J.L.A. Ferreira, J.A. Araújo, Eng. Fail. Anal., **16**, 136-151, (2009).
4. D.G. Wang, D.K. Zhang, Z.F. Zhang, S.R. Ge, Eng. Fail. Anal., **22**, 92-112, (2012).
5. L. Zhao, X.C. He, B.Y. Xing, Y. Lu, F.S. Gu, A. Ball, Mater. Des., **87**, 1010-1017, (2015).
6. K. Kim, J. Geringer, Wear, **296**, 497-503, (2012).
7. G.H. Majzoobi, R. Hojjati, M. Nematian, E. Zalnejad, A.R. Ahmadkhani, E. Hanifepoor, Transactions of the Indian Institute of Metals, **63**, 493-497, (2010).
8. G.Q. Wu, Z. Li, W. Sha, H.H. Li, L.J. Huang, Wear, **309**, 74-81, (2014).
9. M.P. Szolwinski, T.N. Farris, Wear, **221**, 24-36, (1998).
10. R. Sadeler, S. Atasoy, A. Arıcı, Y. Totik, J. Mater. Eng. Performance, **18**, 1280-1284, (2009).
11. N. Borms, D. De Schamphelaere, J. De Pauw, P. De Baets, W. De Waele, Sustainable Construction and Design, 370-377, (2011).
12. S. Cha, H. Chang, K.W. Lee, SAE Int. J. Mater. Manuf., **4**, 613-619, (2011).
13. Z.M. Wei, J.S. Tan, Z.M. Zhang, S.J. Wen, Z.Z. Chen, J.Y. Li, Vehicle Engine, **162**, 56-58, (2006).
14. J.S. Chen, C.Y. Wang, L.S. Bai, J. Mat. Eng., **18**, 21-23, (1990).

MATEC Web of Conferences 77, 01032 (2016) DOI: 10.1051/matecconf/20167701032
ICMMR 2016

Error Detection and Classification of Tangential Radial Single Machining of Non-Circular Gear

Jianqing Dai, Shengze Wang

School of mechanical engineering, Donghua University, Shanghai 201620, People's Republic of China

Abstract. The studies on non-circular gear cutting and using errors were generally based on theory of cylindrical gear errors, and the similarities and differences between cylindrical gear and non-circular gear were combined to study the change rule of non-circular gear contacting line and obtained corresponding conclusions. In this paper, from a different perspective, by starting from the non-circular gear engagement principle, on the basis of analyzing the concrete geometric error conditions, new type tooth profile algorithm routine is reasonably modified, and needed non-circular gear tooth profile error contrast figure is obtained. Furthermore, theoretical derivation is carried out by incorporating analytical method and contacting line incremental method, validity of computer simulation is verified, a reasonable classification method is proposed based on the characterization of the error characteristic code in the general case, Summed up the error genetic map which possess enormous significance to direct practice generation[1].

1 Introduction

Non-circular gear in terms of its essence can be considered as a new type of institution which integrates the advantages of both cam and gear. Non-circular gear mechanism not only inherited the numerous congenital advantage of the gear mechanism, but also has predictable characteristics of cam mechanism transmission rule; therefore the non-circular gear is a kind of gear mechanism can achieve predictable variable speed ratio transmission [2], [3]. After years of research and development, non-circular gear foundation design theory has been basically complete, but now the application scope of the non-circular gear and its inherent advantage doesn't match. The reason lies in the rather difficult manufacturing, the researches in manufacturing lack systematicness and completeness, especially that . the standard of non-circular gear machining error has not been set, which greatly restricted the speed that non-circular gear developed from theory to practical application [4].

Non-circular gear machining exist unique errors, respectively are the tool setting error [5], [6], the main reason of tool setting error is the human factor, and the error size is far greater than other sources of error. In the actual statistics tool setting error accounted for non-circular gear machining error causes up to about 70% [7], through the tool setting error analysis method can also be used in other areas of error, so the study on tool setting error has very important practical significance.

2 Error model

According to the principle of gear meshing, we can get the model of tangential and radial error. In tangential, radial and geometric eccentricity errors, Tangential and radial errors change the meshing position of pitch curve, which will change the shape of tooth profile curve, so we studied respectively with tangential error and radial error.

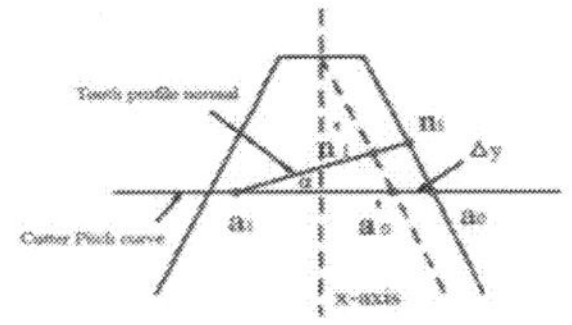

Figure 1. Tangential error model [1]

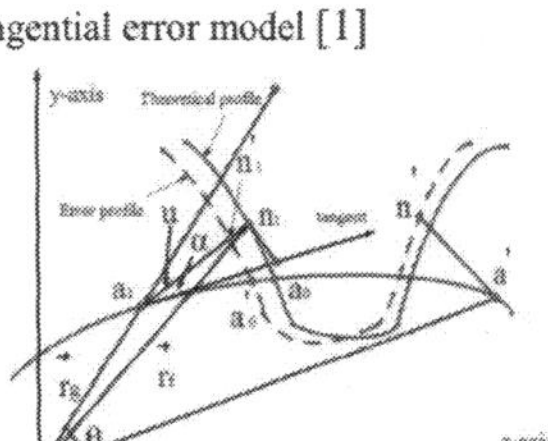

Figure 2. Tangential profile analytic graph of non-circular gear

As shown in Fig. 1[1], zero position of serrateknife just aligned the collinear with the zeroposition of workpiece at a time of zero, in actual production, because of the existence of tool setting error; the tooth cutter has Δy offset in the axial direction. The length of vector in Fig. 1 will cause $\overrightarrow{a_1n_1}$ to $\overrightarrow{a_1n_1'}$, the tooth profile normal in Fig. 3. will cause $\overrightarrow{a_1n_1}$ to $\overrightarrow{a_1'n_1}$.

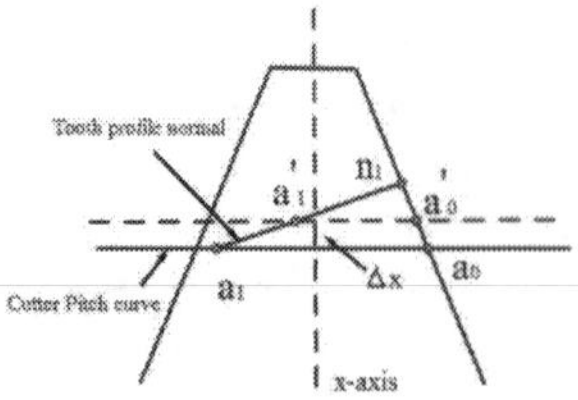

Figure 3. Radial error model

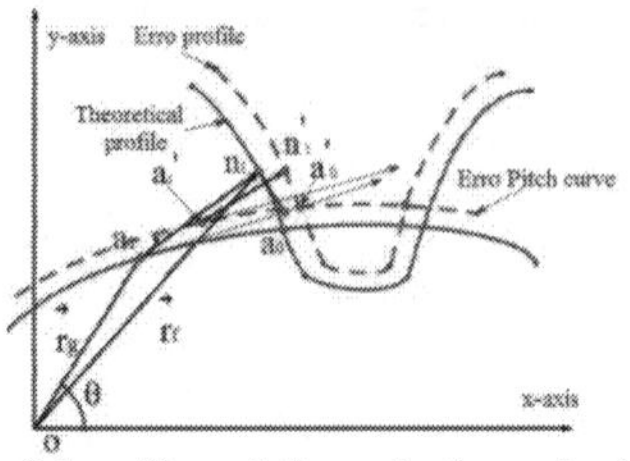

Figure 4. Radial profile analytic graph of non-circular gear

3 Computer simulation and graphic representation

We used beforehand programmed non-circular gear tooth profile algorithm program, the program's main algorithm idea is shown in Fig. 5 first of all, we carried out a generative simulation of the serrated knife according to the operation condition of the zero alignment, and we got the theory tooth profile diagram without the error, then we add a fixed parameter error in the x coordinate value of serrated knife small coordinate system to the subroutine to use the error offset in Fig. 1, he initial position of the x axis in the coordinate system is parallel to the direction of the y axis, in order to simulate the tool setting error condition. Then the program runs again to get the tooth profile pattern when the tool setting error appeared. Finally, two pictures were combined in the same coordinate system to form Fig. 1[1].

Fig. 6(a) and Fig. 6(c) shows the positive direction and negative direction of long axis of tooth profile comparison chart of cogging under the tool setting error, Fig. 6(b) and Fig. 6(d) shows the positive direction and negative direction of short axis of tooth profile comparison chart of cogging under the tool setting error. Small red circles shows the theory tooth profile without error in Fig. 6, while blue stars are tooth profile graphs that have the tool setting error.

From Fig. 6 we can see

(1) The left and right profile of the error profile is in the same direction as the theoretical profile.

(2) The formed tooth profile and the theory tooth profile after the excursion of tooth profile contrast diagrams are different, that is, error tooth profile isn't the simple equidistant line of theory tooth profile.

(3) At the bottom of the Error profile and theoretical profile of maintain level in normal direction, that is, the radial error is zero.

From Fig. 7 we can see

(4) The positive displacement in Fig. 7(a) is offset along the normal direction, the positive displacement in Fig. 7(b) is offset along the negative direction of the normal.

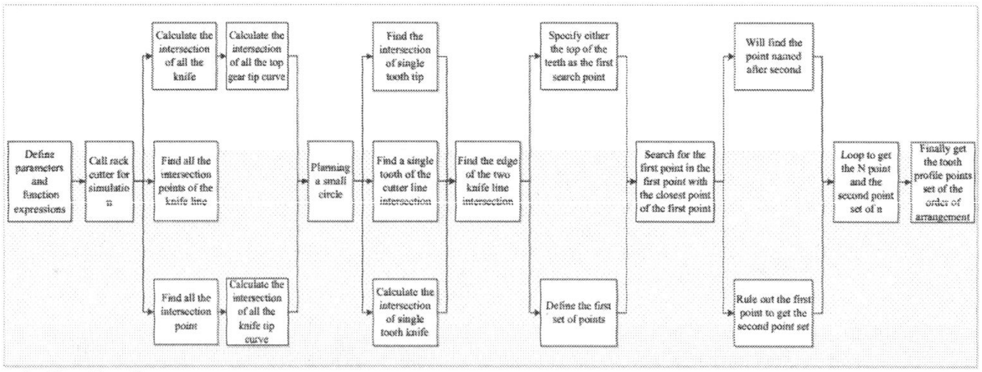

Figure 5. Algorithm flow

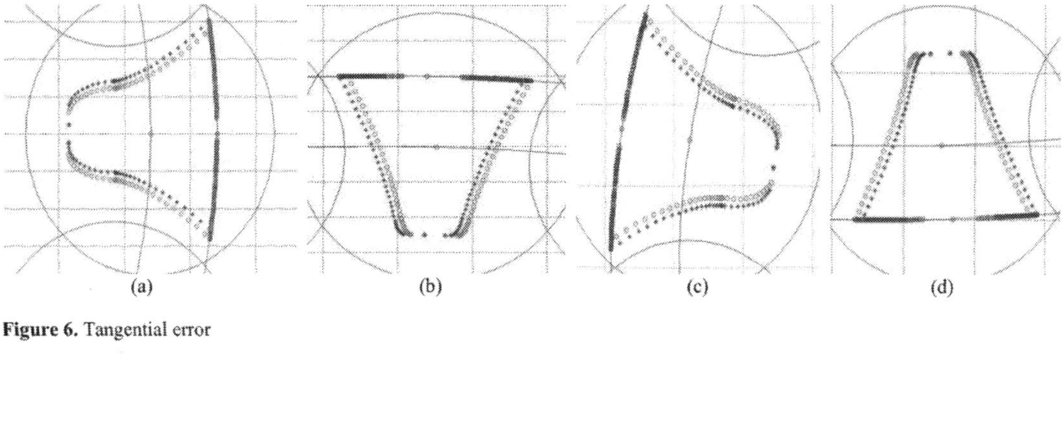

(a) (b) (c) (d)

Figure 6. Tangential error

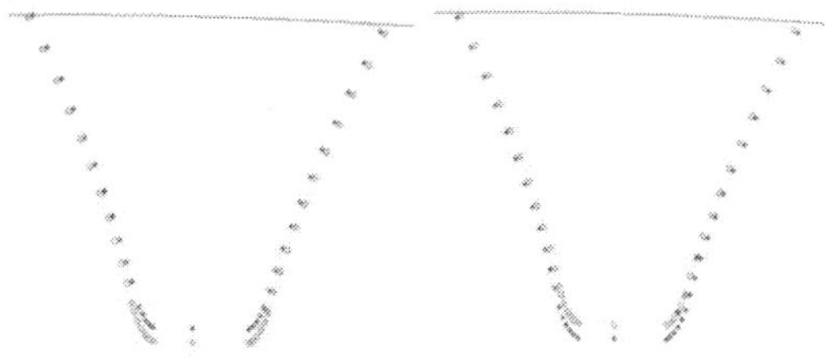

Figure7. Radial error

4 Theoretical verification and analysis

4.1 Theoretical verification [1]

In order to set the tooth profile normal distribution regulation of the non-circular beforehand for convenience, it's determined by the following [8].

$$\lambda - \mu = \text{constant} \tag{1}$$

From Fig. 1 we can see

$$a_1 n_1 = a_0 a_1 * \cos\alpha \tag{2}$$

$$S_1 = a_0 a_1 \tag{3}$$

The serrated knife excursions also comply with the above rule, but the arc length changes. In Fig. 2, vector equation of the tooth profile can be expressed by the following [9, 10].

$$\overline{r_f} = \overline{r_g} + \overline{an} \tag{4}$$

Right tooth profile equationis obtained when converting into rectangular coordinates.

$$\begin{cases} x_R = r_g \cos\theta + an\cos(\theta + u - \alpha) \\ y_R = r_g \sin\theta + an\sin(\theta + u - \alpha) \end{cases} \tag{5}$$

Left tooth profile equation

$$\begin{cases} x_L = r_g \cos\theta + a'n'\cos(\theta + u + \alpha - \pi) \\ y_L = r_g \sin\theta + a'n'\sin(\theta + u + \alpha - \pi) \end{cases} \tag{6}$$

The tooth profile normal length is acquired by the following

$$an = a_0 a \cos\alpha_u = S * \cos\alpha_u \tag{7}$$

$$an = a_0 a \cos\alpha_u = S * \cos\alpha_u \tag{8}$$

The analytical formula for theory tooth profile is above one offset Δy in Fig. 1 is as following after theory tooth profile.

$$a_0 a_0' = \Delta y \tag{9}$$

Simplified as error vector quantity

$$a_1 n_1' = a_1 n_1 - a_0 a_0' \cos\alpha_u \tag{10}$$

Formula is got after being converted into rectangular coordinate.

$$\begin{cases} x_n' = x_n \mp a_0 a_0' \cos\alpha_u tg\alpha_u \\ y_n' = y_n \mp a_0 a_0' \cos\alpha_u ctg\alpha_u \end{cases} \tag{11}$$

In the traditional meshing line increment method of cylindrical gear tooth profile error analysis method [9]:

$$\Delta F = \sum_{k=1}^{n} e_k \sin(K\varphi + \varphi_k) \tag{12}$$

e_k is eccentric error of gear, K is gear error frequency in a turn, φ is corner, φ_k is phase angle.

$$\Delta r_{re} = \frac{\Delta F_{left} + \Delta F_{right}}{2\sin\alpha} \tag{13}$$

$$\begin{cases} \Delta F_{runleft} = +e_{run} \sin\varphi \\ \Delta F_{runright} = -e_{run} \sin\varphi \end{cases} \tag{14}$$

Bring Eq. (14) into Eq. (13) to obtain radial direction error equation

$$\Delta r = \frac{e_{run} \sin\varphi - e_{run} \sin\varphi}{2\sin\alpha} \tag{15}$$

In accordance with the tangential error theory, it is also shown that the tooth profile of the Fig. 4 theory and the position of the generating error:

$$a_1 n_1 = a_0 a_1 * \cos\alpha \tag{16}$$

$$a_1' n_1' = a_0' a_1' * \cos\alpha \tag{17}$$

$$a_0' a_1' = a_0 a_1 - \frac{\Delta x(\sin\alpha + \cos\alpha)}{\sin\alpha} \tag{18}$$

According to the meshing principle, the instantaneous turning center of the gear cutter is a_1 point, The n_1 points on the tooth profile of the rack are corresponding to the n points on the conjugate teeth, When the gear cutter is up to Δx, a_1 points and a_0 were turned into graphics in the a_1' point and a_0'. The remaining part of the use of the formula in front of the $4 \sim 6$ that can be obtained n_1 and n_1' coordinates of the position.

4.2 Theoretical analysis

In Fig. 1[1], the Δy causes the a_0 to move to the a_0' position, causing the $\overline{a_1 n_1'} < \overline{a_1 n_1}$. In the Fig. 6 and Eq. (4)

DOI: 10.1051/matecconf/20167701032

an shorter makes the error profile of the left tooth profile to the counter clockwise direction, In the same way, the *an* of the right tooth profile makes the error profile to the same direction as the right tooth profile, This explains the reason for the graphical representation. By Eq. (15), we know due to the various curvature radiuses of points on the section curve, leading to the each included angles μ of each point radius vector and the tangent line are not identical, the vectorial direction of the offset is not exactly the same, so it can be proved the correctness of the graphical representation. The reason that radial error does not appear can be got by Eq. (15).

Promotion to general situation, we can get the tool clamping error is actual a specially case of variable tangential direction error. Tool clamping error is got when Δy is a constant value, when Δy is a variate value, in principle, we need to determine the relevant Δy value that each gear cutting angle of tooth profile should correspond with, and consider the variate value Δy as the serrated knife offset of corresponding angle.

In variable tangential direction error conditions, offset Δy changes follow the change of the cutting angle and comply with the sine function law $\Delta y = k\sin\theta$.

In the radial error contrast diagram, in the same way by formula $16 \sim 18$ and formula $4 \sim 6$, we can obtain that the radial error profile is not the line of the theoretical profile.

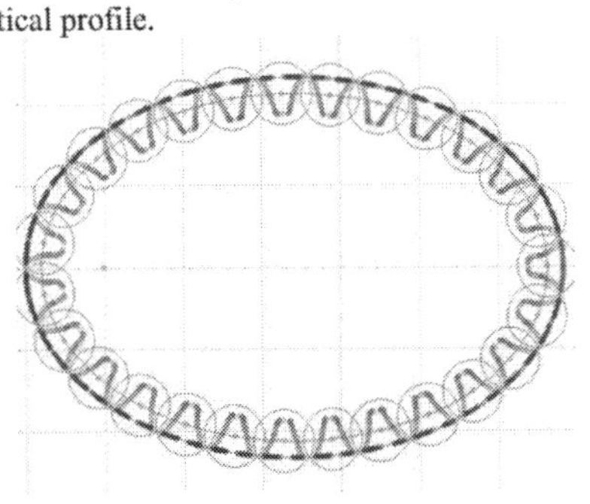

Figure 8. elliptic non-circular gear tooth profile

Figure 9. Solid model of non-circular gea

5 Experiment and classification

Using the preceding algorithm program we take the most common elliptic gear as an example. Fig. 8. for elliptic gear tooth profile, and we can process the non-circular gear solid model which is shown in Fig. 9. through the rapid prototyping system,Using the RoyalArm RA315 measuring arm (measurement accuracy of 0.05mm) which shows in Fig. 10. on the key points of non-circular gear contrate profile actual coordinates value and theoretical coordinate value difference to determine what kind of error [1].

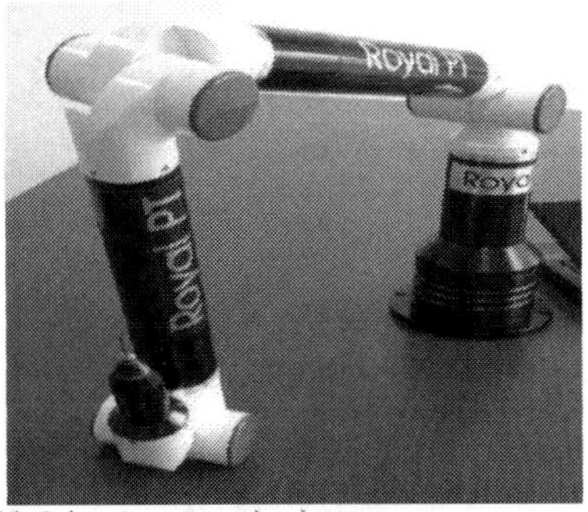

Figure 10. Joint arm measuring instrument

Table 1. Data sheet[1]

	Long axis normal direction		Short axis normal direction		Long axis negative direction		Short axis negative direction	
X- theoretical tooth top apex	35.2954	35.2954	16.7847	14.3307	-5.4062	-5.9078	14.3250	16.7847
Y- theoretical tooth top apex	-1.4003	1.4003	13.9980	13.9331	3.0883	0.3393	-13.9980	-13.9330
X- actual tooth top apex	35.30	35.30	16.70	14.45	-5.40	-5.90	14.40	16.90
Y- actual tooth top apex	1.30	1.45	13.95	14.00	3.00	0.25	-14.00	-13.95
X- theoretical tooth of root	33.0952		15.4782		-3.5371		15.4745	
Y- theoretical tooth of root	0		11.7433		1.3606		-11.7436	
X- actual tooth of root	33.10		15.40		-3.55		15.55	
Y- actual tooth of root	0.10		11.75		1.25		-11.70	

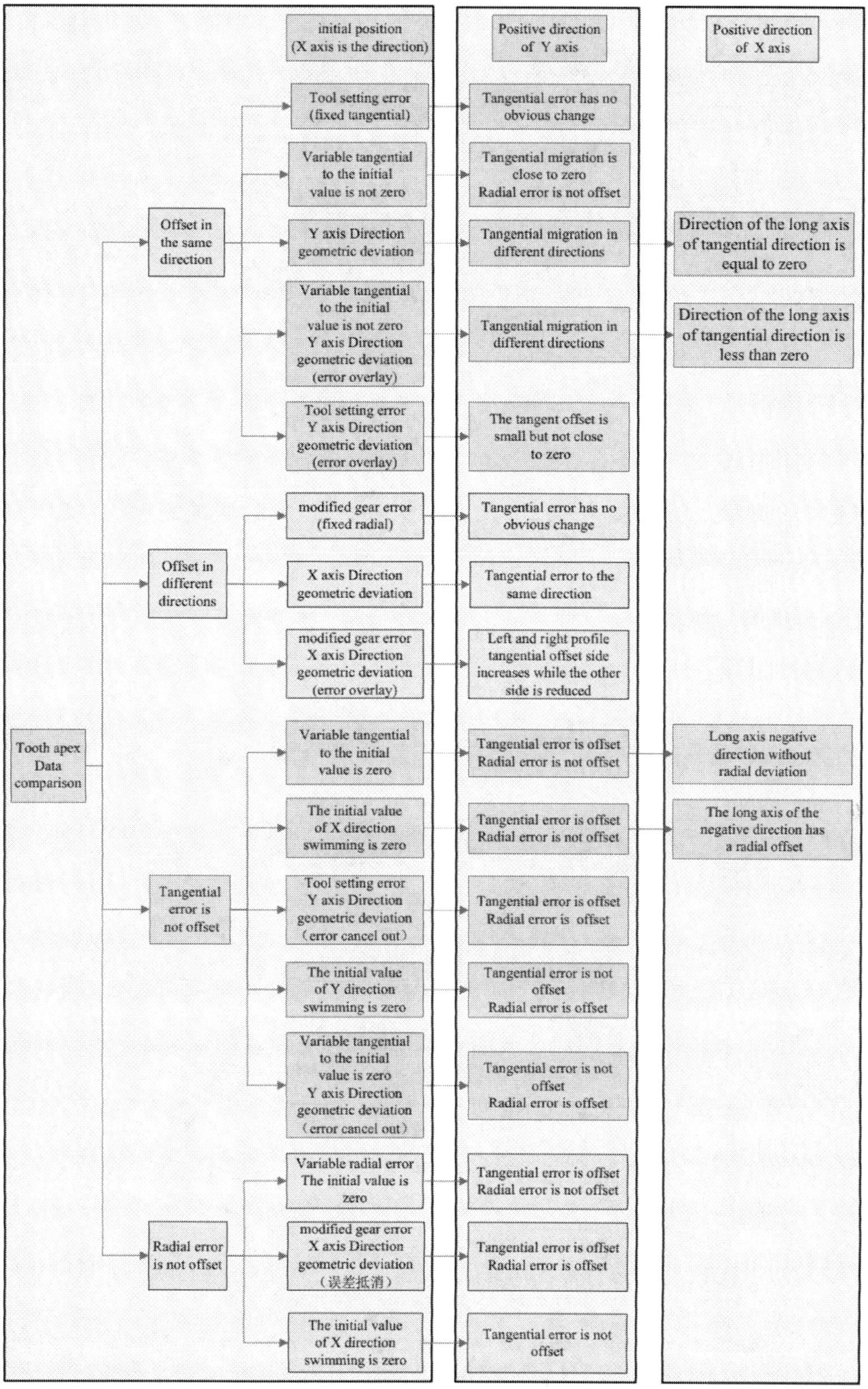

Figure 11. Characterization of characteristic code

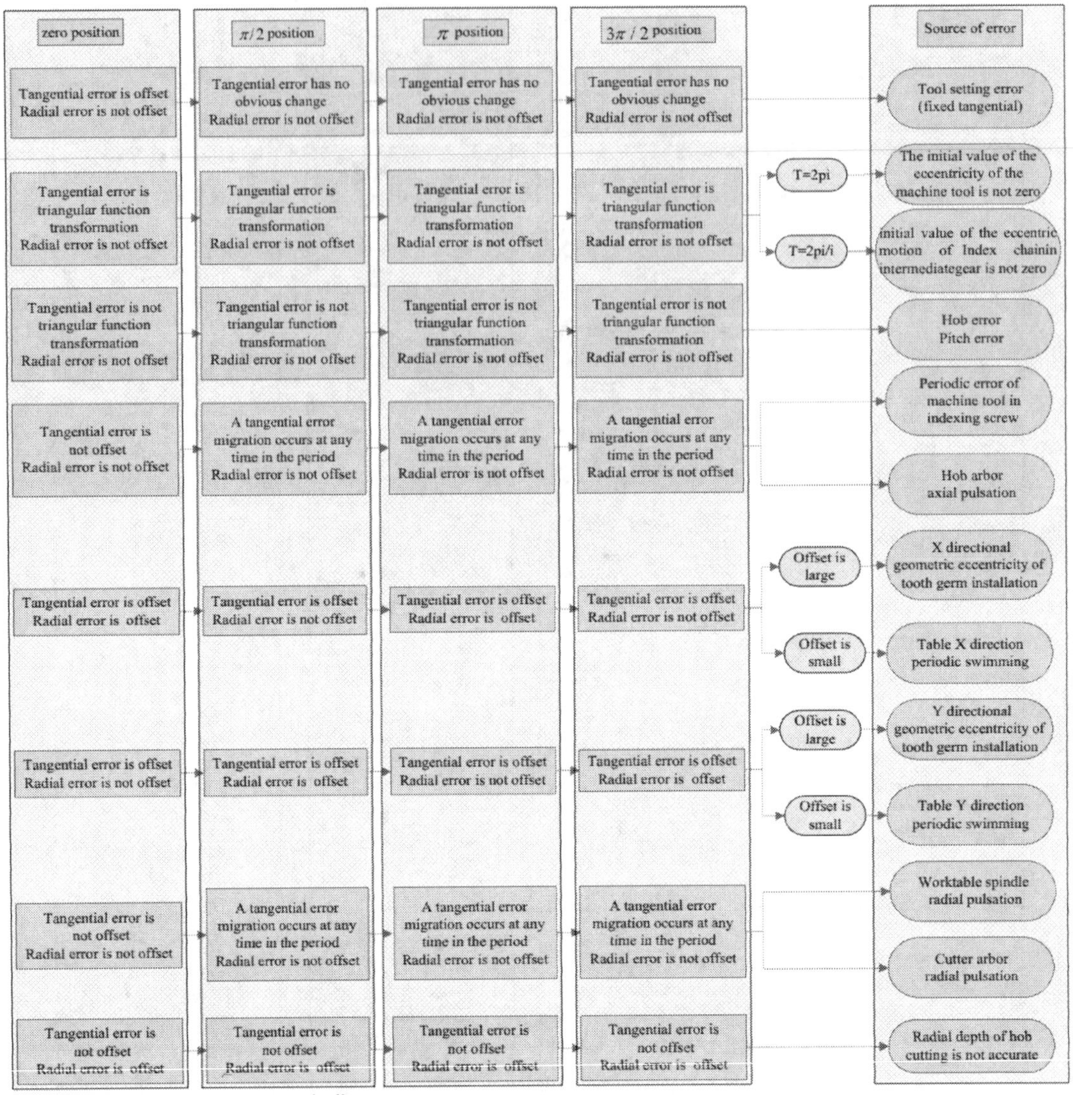

Figure 12. error factors of error genetic diagr

The experimental method is as follows [1].

(1) Non-circular gear processing must have a zero, that is, the initial processing bit. In the model above the left end elliptic focus as the origin of coordinates, the right level for the positive direction of axis, vertical axis direction, establish coordinate system.

(2) Because of the large number of non-circular gear tooth, large quantities of one of the measurement of each tooth space for all positions of the geometric point in the actual production is not acceptable, so we at the end of the ellipse long axis and short axis at both ends of each selected a tooth, coordinate measurement of tooth top apex and tooth root value.

(3) The actual coordinates the collection value and coordinate theory are compared to the values, the zero cogging data must be collected, the cases of zero cogging in long axis are the direction and the results are shown in Table I.

(4) According to the results of the comparison of the results in Fig. 11 can be classified to learn about the causes and sources of error. Because this article is only relates to cut to the field of error, Fig. 11 relates to the radial error and the geometric deviation theory proof in my other are published articles are presented in detail. This only references related to cut to the part of the conclusion of the error. In Table I at the coordinate data of zero tooth apex the conclusion can be obtained that the tooth profile is not offset, and the tangential direction is shifted in the same direction., From Fig. 11 we know this phenomenon may have five reasons: tool setting error, cut to the initial error is not zero, the direction of geometrical deviations, cut to the initial error is zero and the direction of the geometric deviation of these two factors superimposed, tool setting error and the orientation geometric deviation of these two factors superimposed. The data of the short axis normal direction

can be obtained in the radial direction, and there is no significant change in the tangential migration, from Fig. 11 we know the model belongs to the typical causes of tool setting error.

From this example we get the test method of the origin of the error and the characterization of the genetic code. In the actual production we are more concerned about how to quickly troubleshoot what are the process error factors that cause the error, so in Fig. 11 based combined with common cause of cut to error of process factors are summarized Fig. 12 non-circular gear error factors of the error genetic diagram. So in Fig. 11 based on combination of common cause of cutting error of process factors to get Fig. 12 non-circular gear error factors of the error genetic diagram. This will provide a powerful tool for the production of the cause of the problem quickly.

6 Innovation andweaknesses

Via the system analysis and the theoretical proof of the computer simulation results for non-circular gear tool setting error, we may obtain the conclusion that the tool setting error is a fixed variable tangential error, it's beneficial for the research complement of the non-circular gear error system and lay a solid foundation for the subsequent individual error of other conditions, and got the condition code of the tool setting error excursion, providing theory support for timely finding out the cause of the gear error when the factory inspection, thus improve the machining precision of the non-circular gear [1].

At the same time, we must realize that although all the errors mentioned above can be attributed to the error of tangential, radial and geometric errors, the error can be formed separately or in combination. But the source of the process error is different, and the combination of various errors is more complicated. So the Fig. 11 and Fig. 12 summarizes the graphical representation and the error factors are not one-to-one relationship, actual operation must be flexible use must not be applied mechanically, the geometric errors and the cycle of variation of the machine tool swimming factor in Fig. 11 and Fig. 12 can be guaranteed by $T = 2\pi$, but you need to get more data points with changes to determine the cause of the error when other $T \neq 2\pi$ Error period. In single error or simple two kinds of combination of errors when the method has accurate recognition ability, when the superposition of many error the method also has certain reference value, so it seems the method also has a lot of room for improvement.

References

1. Jianqing Dai,Shenzhe Wang, in:Analogue simulation and analysis of non-circular gear reentry error, ICMMR2016
2. F.L.Litvin. Generation of planar and helical elliptical gears by application of rack- cutter, hob, and shaper, 196:4321-4336. Comput. Methods Appl. Mech. Engrg, (2007).
3. Guoxing Sun, Chuanqiong Sun, Yongde Liu, in: The application prospect of non-circular gear on infinitely variable speed, 1: 1-4. Equipment maintenance technology, (2009).
4. Xueqiang Gao, in: Transmission technology overview of non-circular gear, 3:5-8. Mechanical drive, (2003).
5. Huiping Yan, Chibing Hu, Limei Liu, in: The research of straight tooth non-circular gear hobbing machining error analysis, 4: 18-20.Mechanical drive, (2007).
6. Fangyong Tian, Chibing Hu, Yancang Jiang, in: An automatic tool setting method of non-circular gear hobbing, 4: 41-43 [J].Mechanical research and application, (2010).
7. Liqun Wu, Hongxia Bi, in: The determination of axial special-shaped gear NC hobbing machining adjusted cutter point coordinates, 6:109-112.Coal mine machinery, (2008).
8. Yongping Liu, Peng Wang, Jia Li, in: The elliptical gear tooth profile design via analytical method, 2: 214-216. Machinery design and Manufacture, (2015).
9. Fusheng Li. Non-circular gear, (1975).
10. Mingtong Wang. Involute gear accuracy, (1983).

MATEC Web of Conferences 77, 01033 (2016) DOI: 10.1051/matecconf/20167701033
ICMMR 2016

A Comparative Study of Chain Clamping Fixture with Other Clamping Methods for Gate Valve Body: Cycle Time and Rigidity Study

Lokeswar Patnaik , Sunil Kumar, Deva Prasad S

Department of Mechanical Engineering, CVR College of Engineering, Hyd-501510, INDIA

Abstract. "Fixture rigidity influences the production rates". In this paper, a comparison and feasibility study is demonstrated to improve the productivity and existing fixture rigidity for machining on a gate valve body by comparing the existing technique *i.e.* top clamping fixture with other possible clamping methods including newly developed chain clamping fixture. The CAD models of the fixtures are prepared in CATIA V5. The existing fixture concept, clamping methods, cutting parameters and cycle times have been studied and compared with the values of chain clamping fixture obtained by trail out results. Along with the increased production rates, other parameters like tool life, noise and vibration levels and surface roughness values are also improved slightly.

1 Introduction

Gate valve body shown in Fig. 1 are machine elements which are used to control of flow of volatile, often toxic gases and liquids and keep them away from being emitted into atmosphere [1]. A gate valve body of cast steel has to undergo various machining processes and to accommodate rigidity while machining, fixtures are required. In this paper, traditional top clamping method to locate the gate valve is compared with other fixtures along with newly developed chain clamping fixture.

Figure 1. Valve body

B-0 (Front view) B90, B270 (Side view) B-180 (Back view)

Figure 2. B0, B90, B270 and B180 view for OP-10 orientation.

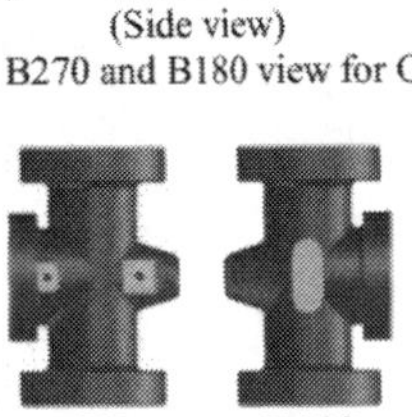

B0 B180

Figure 3. B0, B180 views of OP-20 orientation

Table 1. Operations wise tool requirement.

Operations	Tools required
B0 side	
7/8-9UNC tap	Ø19.5 Drill and 7/8"-9 UNC tap
Ø101.4 & Ø 97.1 Bore roughing & Finishing	Ø101.4 Finish Boring bar
Face milling Rough and finishing	Ø63 Face mill and Ø 20 End mill
3x45 degree chamfer	Chamfer Tool
B90 & B270 side	
Ø25.4 Hole	Ø25.4 Drill
Ø52.3 hole	Ø52 Rough boring bar Ø52.3 Finish Boring bar
Face milling	Ø63/Ø125 Face mill
B180 side	
Face milling	Ø63/Ø125 Face mill

The majors input to fixture design consists of product design data including CAD geometric model, the machining operation sequence, the quantity to be produced and the machines on which the various operations are to be performed [2]. The machining operations are done in a horizontal machining center (HMC). The B-axis machine tool combines the turning capabilities of a horizontal/vertical turning machine with milling and machining capabilities of a five-axis machining center. A B-axis machine can do the work of two because it supports the entire range of milling and turning operations possible, with the advantage of doing it in one setup [3]. OP-10 (Operation number 10 of process sheet of the component as per the production drawing) and OP-20 (Operation number 20) are orientations or set up of the component (Gate valve body) in which B-0, B-180 and B-270 are the views of the component in B-axis shown in Fig. 2 and Fig. 3 in which the machining have to be performed. The machining operations to be performed in operation number 10 (OP – 10) along with the tools to be used are shown in Table I

1.1 OP-10 Operational details

1.2 OP 20 Operational details

In B0 side of OP-20, the operations are 2 x 1/2"NPT Tap and spot facing whereas in B180 side face milling is the only operation.

2 Design evolutions

2.1 Existing technique- Top Clamping

Fig. 4 shows a top clamping fixture, this method employs concentrated clamping *i.e* clamping force will act on very small area on the component. Major drawbacks of top clamping technique are high tendency of component deformation, less rigidity than distributed load type and handling difficulties. This is popularly and widely used method but due to above mentioned demerits; its applications are limited to small components only where clamping force is significantly less. To avoid these limitations, alternative concepts were developed and evaluated.

Figure 4. Top clamping fixture.

2.2 Various concepts

2.2.1 Bottom Clamping

Advantages of bottom clamping concept shown in Fig. 5 are clamping against resting thus no component deformation, high clamping forces and easy to Handle. The disadvantages of this method are high over hang portion (Unsupported flange faces) in both the operations, it limits the maximum parameters of the machine that can be used for machining.

OP-10 OP-20

Figure 5. Bottom clamping

2.2.2 Concept 2: Strap Clamps

Fig. 6 shows the concept of strap clamping. In this arrangement two straps are used for clamping. The advantages of this set up are two strap Clamps against the resting V-Blocks *i.e*, eliminating component deformation and provides good clamping forces. Difficulies in orientation of component and increase in loading and unloading time are the major disadvantages of this concept.

OP-10 OP-20

Figure 6. Use of Two Strap Clamps

2.2.3 Chain Clamping with orientation by studs and additional clamping

In this concept, the component is clamped with two chain, one additional clamp and one suport pin as shown in Fig. 7. The orientation is easy, accessable and quick, also it is provided with top clamping for additional support as fall back option which is practically unnecessary as it will increase cycle time.

Figure 7. Chain clamping with additional clamp

2.2.4 Chain Clamping Fixture Final Concept

In OP-10, the component is located on v-blocks with resting and butting pads. It is orientated by two M12 Studs and clamping by two Chain Clamping Sets against resting. The clamping and orientation is shown in Fig. 8.

Figure 8. Chain Clamping Fixture concept OP-10

In OP-20, the component is oriented and located by two pins (Diamond pin type) diagonally which shown by color green and blue respectively. One central pin with top clamp is used for clamping the component. The fixture concept with component and without component is shown in Fig. 9.

Figure 9. Chain clamping fixture concept OP-20

3 Cycle time and cutting force calculations

The operation wise parameters are taken from Kennametal catalogue [4] and total time taken in OP-10 and OP-20 is found out which is shown in Table 2 and Table 3.

Table 2. OP-10 time study

OP- 10 CUTTING TIME					
SNo.	Operation	Tool	Cycle Time	RPM	Feed
1	Face milling-Rough	Ø63 FMC	8.08	1062	850
2	7/8-9UNC tapping	Ø19.5 Drill	4.00	978	97
3		Ø14 CHF tool	0.41	1431	100
4		7/8-9 UNC tap	1.47	200	430
5	Ø25.4 Hole	Ø25.4 Drill	13.06	778	77
6	Ø97.2 hole	Ø63 FMC	5.5	1110	1644
7	Ø101 hole – Roughing	Ø63 FMC	2.1	1110	1644
8	Ø101 hole – SF	Ø20 EM	0.45	1400	1200
9	Ø101.2 hole CHF	Ø14 CHF tool	1.06	900	600
10	Ø92 hole CHF	Ø14 CHF tool	0.50	900	600
11	Ø101.2 finish boring	Ø101.2 FBB	0.35	314	31
12	Ø52.1 hole-Rough	Ø52.1 RBB	6.02	616	61
13	Ø52.1 hole CHF	Ø14 CHF tool	0.14	900	600
14	Ø25.4 Hole CHF	Ø14 CHF tool	4.02	900	600
15	Ø186 OD CHF	Ø14 CHF tool	3.25	900	600
16	Ø52.4 Finish Boring	Ø52.4 FBB	8.3	848	42
17	Face milling-Finish	Ø63 FMC	7.13	1261	500
Total Time for OP-10			**66.23**		

Cutting time for OP-10 = 66.23 + 10 % cushion time = 72.5 min.

Non Cutting time for OP-10

Loading time = 10 min (maximum)
Work setting time = 10 min (maximum)
Unloading time = 6 min.
Total non-cutting time for OP-10 =26 min.
Total Cycle time for OP-10(cutting + Non-cutting)=98.5 min.

Table 3. OP-20 time study,

OP-20 FIXTURE CUTTING TIME					
S No.	Operation	Tool	Cycle Time	RPM	Feed
1	Face milling-Rough	Ø63 FMC	8.08	1110	888
2	1/2" NPT tapping	Ø18.3 Drill	0.96	1050	130
3		1/2"NPT tap	0.73	160	290
Total Time for OP-20			**9.77**		

Cutting time for OP-20 = 9.77 + 10 % cushion time = 10.8 min.
Non Cutting time for OP-10 includes
Loading time = 6 min.
Work setting time= 10 min.
Unloading time = 5 min.
Total non-cutting time for OP-20 = 21 min.

Total Cycle time for OP-20 (cutting + Non-cutting) =31.8 min.

The total time of both OP-10 and OP-20 operation is 130.3 min.

3.1 Cutting force calculations

For calculating the cutting forces, the operation in which max cutting forces will be induced was selected. The operations to be performed on the component are,

1. Face Milling of flanges (roughing and finishing),
2. Boring operation for ø53mm holes,
3. Drilling operation for φ25.4 mm holes.

Among the above the operations, major cutting forces will be induced during face milling (roughing) and then boring.

3.1.1 For side face rough milling operation (B-90 or 270)

Material of valve body is cast steel, cutting parameters and other details considered for cutting force calculations in this work are as follows:

Operation		: Milling
Cutter	(D)	: Ø 63 Face Mill cutter,
No of inserts	(Z)	: 4 inserts
Feed per tooth	(F_z)	: 0.15 mm to **0.2** mm
Depth of cut	(t)	: 1.5 - **2** mm
Feed per minute	(F_m)	: mm ($F_z \times Z \times n$)
Cutting speed	(V_c)	: 180 - **210** m/sec
Spindle speed	(n)	:RPM ($1000 \times V_c / \pi D$)
Width of cut	(b)	: 60 mm.
Cutting force	(F_c)	:Kgf ($(6120 \times P_c) / V_c$)

Clamping force (F) : KN ($1.5 \times F_c$)
Cutting power (P_c) :kW ($(t \times b \times F_m \times K_{fs})/(60 \times 10^5 \times \eta)$)
Where, η is Efficiency of the machine.
RPM (N) : Rev/min
Material removal rate (Q) :mm^3/sec ($b \times t \times F_m/60$)
Specific cutting force (K_{fs}) : 260 N/mm^2

PS: SI Units considered, in brackets formulae used in calculations.

Considering maximum values of above preferred parameters, Spindle speed, feed per min, metal removal rate, clamping force, cutting force and cutting Power are calculated, formulae given above in brackets at the parameter [5].

n = 1061.5 RPM *i.e.*, Taken as 1062 rpm
F_m = 850 mm
Q = 1700 mm^3/sec
P_c = 5.9 kW
F_c = 171.94 Kgf in Tangential direction.
= 1.686 kN
F =1.5 × 1.686 = 2.54 kN

3.2 Elements of Chain clamping fixture

For Valve body machining as provide in the component operational geometrical requirements, two setups were selected (OP-10 & OP-20) with two different fixtures with single part loadings.

3.2.1 OP-10 Fixture concept details

Fig. 8 is showing the fixture with component and without component. The component is located by V-block and orientated by two studs, which prevent the rotation of component in X-axis. It is clamped by the two set of chain. Maximum force applied by chain should be more than the clamping force. [6]

3.2.2 OP-20 fixture concept details

The location and orientation is done by cylindrical locating pin and diamond pin respectively. The clamping is done by top clamp with stud and washer as shown in figure 9.

4 Experimental details and results

4.1 Total Cycle time for OP-10

Total cycle time = 64 (Cutting time - CT) + 6 (Setting time - ST) = 70 min (With pallet changer)

In case of single pallet machine, cycle time will include loading (L) and unloading (UL) time also. Then, the total cycle time is,

Cycle time = 70 + 20 = 90 min.

4.2 Total Cycle time OP-20

Total cycle time = 10 (CT) + 5 (ST) = 15 min.
(With Pallet changer)

= 15 + 10 (L and UL) = 25 min.

Total cycle time = 115 min (for OP -10 and OP- 20)

The total cycle time for top clamping fixture is 145 min *i.e* 2.24 hr per component. Production rate with this cycle time and 300 working days in a year having 16 hr (two shifts in a day) of production in a day produces 1983 components per year

The total cycle time for chain clamping fixture is 115 min *i.e* 1.91 hr per component. Production rate with this cycle time having 300 working days per year and production time of 16 hr (two shifts in a day) yields 2513 components per year.

Therefore,

Increase in the Production rate of the component, P = 2513 – 1983 = 530 Components per Year

5 Conclusions

The distribution of pressure induced by clamping is uniform and noise reduction while machining is reduced significantly. By using chain clamping over the component the cycle time was reduced by *20% (from 145 min to 115 min)*for Gate valve body Component. The experimental results shown increase in Production rate by *530 numbers (26%)*components per year, from *1983 Numbers to 2513 Number per Year.* Other parameters like vibrations and noise levels were observed decreasing due to increased fixture rigidity.

References

1. PradipBhaskarPatil, V. R Gambhire, "Structural Analysis of Gate Valve Body using F.E.A". International Journal of Engineering Research and Technology, **3**, 6, June 2014.
2. Joe Cecil, Richard Mayer and Umesh Hari, "An integrated methodology of fixture design". Journal of intelligent manufacturing 7, 95- 106, (1996).
3. Information on http://www.mmsonline.com .
4. Information on "Kennametal catalogue".
5. CMTI, "Machine Tool Design Handbook", TMH 1982.
6. Information on www.machinetoolsonline.com

MATEC Web of Conferences 77, 01034 (2016) DOI: 10.1051/matecconf/20167701034
ICMMR 2016

Applied CATIA Secondary Development to Parametric Design of Active Magnetic Bearing

Zhang Li[1], Yu Meiyun[1] and Ren Yafeng[2]

[1]*Qingan group co., LTD., 710077 Xian, China*
[2]*Northwestern Polytechnical University, 710072 Xian, China*

Abstract. Based on the properties of active magnetic bearing, the electromagnetic parameters and structure parameters are analyzed, parametric design method is introduced to study the structure of active magnetic bearing. Through a program personalization process that is in accordance with active magnetic bearing is established. Personalization process aims to build the parametric model of active magnetic bearings and component library by use of CATIA secondary development. Component library is to build assembly model for a multiple degree of freedom magnetic bearing system. Parametric design is a method that provides the direction for its structural design.

1 Introduction

Active Magnetic Bearing (AMB) is a typical mechatronics system. Due to no contact, wear and lubrication, controlling its dynamic behavior and other series of characteristics [1], the limitation of traditional mechanical bearing has been breakthrough, it appeared that has been widely attention. The structure of an AMB is complex, the performance is synthetically influenced by structure parameters, electromagnetic parameters and control parameters [2], it also provides the reference for the structure design of other parts. Especially for small machinery, the miniaturization of structure is even more important. Therefore, AMB design has been in an important position.

Parametric design is a kind of intelligent design method, defined in advance a series of parameters and the component configuration is driven through parameterized drawing and parametric changing engine. When parametric model is established, it can be easily modified and optimized on structure by designers, and as a result, the design efficiency and design quality is greatly improved. At the same time, it also provides basis on the establishment of component library and assembly system.

2 Active magnetic bearing

2.1 Structure of radial AMB

The structure of radial AMB is similar with the motor, also divided into two parts: stator and rotor. The stator is composed of electromagnet and coil. Soft magnetic materials are used in the electromagnet. Silicon steel sheet is widely used in radial AMB the same as the rotor so that forms a magnetic circuit loop. Type of the stator slots includes trapezoidal groove, rectangular tank, circular groove. Difference type leads to different functions. For example the radial size that has a trapezoidal groove is small, the transition is smooth; for circular groove the radial size is big, the filling coefficient is high. So, the circular groove is the best choice [3], as shown in Fig. 1.

At the same time, In order to optimize the bearing geometry we distribute the space available in the bearing magnet optimally between iron and copper for maximum load capacity [4], Optimization can already be achieved.

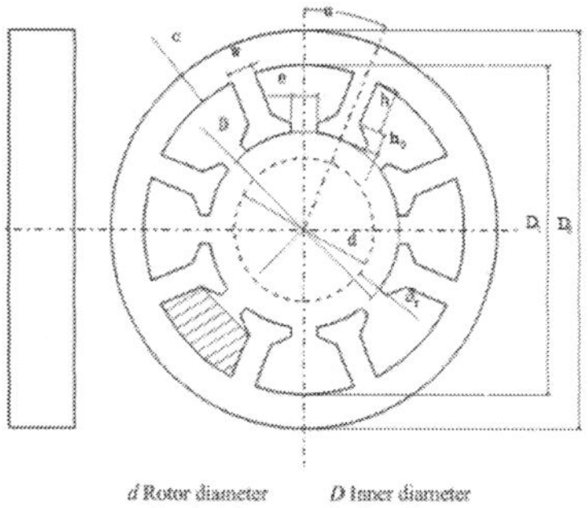

Figure 1. Typical geometry of the radial magnet bearing.

2.2 Calculating parameters

Seen from Fig. 1 the structure parameters of the radial magnetic bearing are rotor diameter *d*, nominal air gap σ_r, Inner diameter *D*, outer diameter D_0, leg width *w*, pole height *h* etc. Some parameters are mutual influence such as the distance between the poles e is about 10 ~ 20 times of air gap. The rotor parameters of motor determine the rotor diameter.

The value of nominal air gap associates with the rotor diameter *d*. From practical experience, when *d* is less than or equal to 100mm, σ_r is equal to between 0.2mm and 0.6mm; When *d* is greater than 100 mm, however, the air gap between 0.6mm and 1.2mm. Based on the geometric relationship we can get the inner diameter *D*, the bottom of the channel diameter D_1, stator diameter D_0 as shown in formula 1.

$$\begin{cases} D = d + 2\delta_r \\ D_1 = D + 2(h + h_0) \\ D_0 = D_1 + 2c \end{cases} \quad (1)$$

If the electromagnetic parameters of the maximum magnetic density B_{max} and coil current I_{max} are identified, we can deduce coil slot area A_{cu} such as formula 2, said the shadows in the Fig. 1.

$$\begin{cases} A_{cu} = \dfrac{\pi N_p d_w^2}{4\lambda} \\ d_w = \sqrt{\dfrac{I_{max}}{\pi \varepsilon_0}} \\ B_{max} = \mu_0 \dfrac{n I_{max}}{2\delta_r} \end{cases} \quad (2)$$

where λ is the filling coefficient, average value is between 0.5 and 0.7, d_w is the coil diameter. ε_0 is the coefficient of the sectional area and current. Type 3 is simultaneous deduced by type 2.

$$A_{cu} = \frac{B_{max}\delta_r}{2\lambda\mu_0\varepsilon_0} \quad (3)$$

From the formula 3 when σ_r and B_{max} is determined, the coil groove area will be also determined. Therefore, according to the geometric relation we get the formula 4, deducing the pole height *h*. Generally, magnetic pole height h_0 is between 0.1 to 0.3 times of *h*.

$$A_{cu} = \alpha h(D_1 - h) - hw \quad (4)$$

Leg width *w* is identified based on pole shoe arc length ($l = \pi D/N_p - e$), that *w* is equal to 0.5 to 0.8 times of *l*. N_p is the number of a column in active magnetic bearing. Static load capacity of an AMB F_{max} is shown in formula 5.

$$F_{max} = \frac{B_{max}^2 A_a \cos\alpha}{\mu_0} \quad (5)$$

The magnetic space A_a is shown in formula 6. K_s is the stator stack loading coefficient. *L* is bearing width (magnetically active). Table 1 shows some basic parameters of the radial magnetic bearing.

$$A_a = K_s wL \quad (6)$$

Table 1. Typical dimensions of radial AMB.

d mm	d_1 mm	D_0 mm	σ_r mm	$L_t - L$ mm	N_{max} r/min	F_0 N/ mm /T^2	N_p
40	24	98	0.3	11	900	3.9	8
60	36	116	0.3	12	600	5.9	8
80	49	138	0.4	16	450	7.8	8
100	74	164	0.6	21	360	9.8	1
120	89	191	0.8	25.5	300	11.8	16

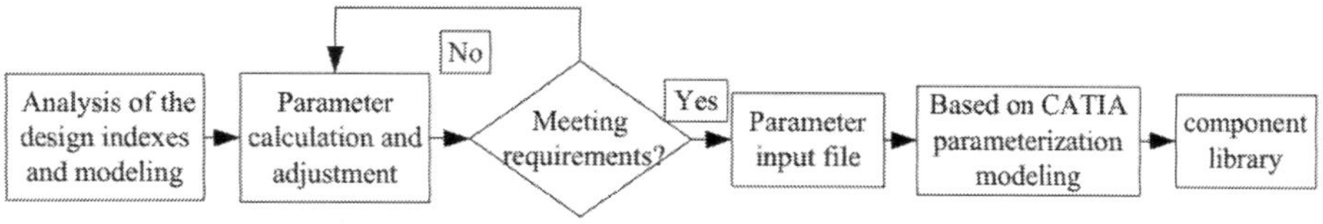

Figure 2. Flow charts of the parametric design on active magnetic bearing.

3 Parameterized design method of active magnetic bearing

3.1 Parametric technology

Parametric technology is one of the core technology of modern CAD/CAE system software. It is also a main research direction in the field of CAD software. In fact, parametric design is to realize the driving of the model by modifying the components parameters that govern the size or assembly relations or change the size of the part or some parts in advance [5].

3.2 Parametric design of AMB

Designers get a series of defined parameters through the analysis of the structure and the decomposition of magnetic bearing design indicators. Generally, parametric design process includes the analysis of design indexes and modeling, parameter calculation and adjustment, based on CATIA parameterization modeling, component library and so on, as shown in Fig. 2. The analysis of the design indexes and modeling, parameter calculation and adjustment are the core part of the design process, the latter is belong to the parametric design software implementation. And the assembly design of magnetic bearing system can be done by component library.

Analysis of the design indexes and modeling is based on the overall layout, the overall structure and so on. Design indexes include highest speed of rotor, maximum power, radial and axial maximum cutting force, selection of motor parameters, processing and manufacturing features, etc. Parameter calculation and adjustment is designed to estimate the interrelationship of electromagnetic parameters, structural parameters and optimize the parameters that do not meet the requirements of design and size. All of design parameters are outputted into standard format such as txt, xml file.

An interface of parameter data and the modeling that can quickly complete parametric modeling of magnetic bearing is designed. According to the modern database technology, we have established some component library to complete parameter assembly and the traceability of data information. Component library includes some parts library of a magnetic bearing system such as a rotor shaft library, a radial AMB library and an axial AMB library.

4 Realization of parametric design

For development of the parametric design in an AMB, CATIA software platform is introduced, Mainly based on the following consideration:

(1) Parametric design technology is very suitable for designing the shape of the same structure and only the change of size parameters on parts, and AMB just meet this characteristic.

(2) VB or VBA is used for secondary development of CATIA, CATIA provides powerful variable drive and parametric ability in order to achieve the parameters of the design goals, the process is that it provides the corresponding code drive to realize parameterized modeling by calling the underlying CATIA API interface function. Compared with the internal module development, this kind of way is more development depth.

(3) CATIA system is an open and wide application, particularly in the field of aviation and space flight.

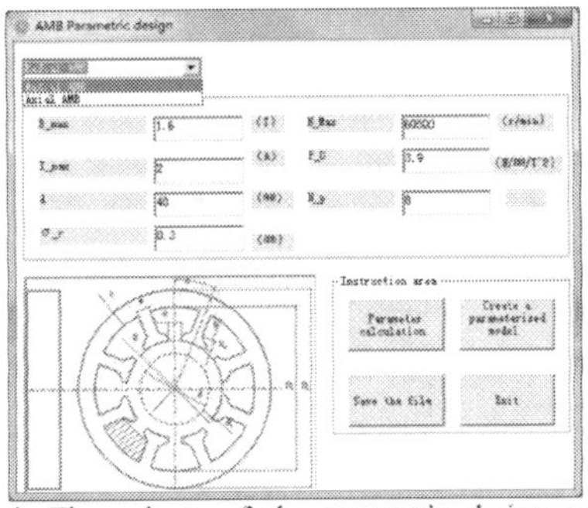

Figure 3. Flow charts of the parametric design on active magnetic bearing.

The development of this system uses Visual Basic 6.0 namely the Automation technology, use of COM technology and interface may realize to invoke the internal module of CATIA. Parametric design system framework and interface in VB is customized, as shown in Fig. 3.

Below is part of the code that is used to invoke internal module of CATIA, analyze and calculate the structure parameters of an AMB, finally complete the parameterized modeling, as shown in Fig. 4.

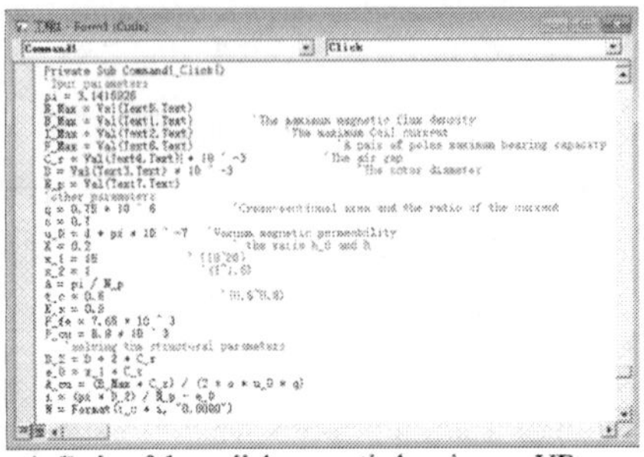

Figure 4. Code of the radial magnetic bearing on VB.

When users input or modify a set of design parameters, the parameterized model is directly generated, as shown in Fig. 5.

Figure 5. 3D model of the radial magnetic bearing.

5 Conclusion

The structure of active magnetic bearing is complex, its performance is influenced by structure parameters, electromagnetic parameters and control parameters. By adopting the method of parametric design, custom application driven magnetic bearing a parameterized model is built up, greatly improving the design efficiency, saving the cost. Assembly of magnetic bearing system integration is of great significance.

References

1. G. Schweitzer, H. Maslen. Magnetic bearing-Theory, design, and application to rational machinery, (2009).
2. A. Filatov, P. McMullen, L. Hawkins, E. Blumber. Magnetic Bearing Actuator Design for a Gas Expander Generator. Proceedings of the Ninth

 DOI: 10.1051/matecconf/20167701034

International Symposium on Magnetic Bearings, Lexington, KY, USA, (2004).

3. Yang jing, Gen Haipeng, Yu lie. Dynamic properties investigation about conical AMB. Proceeding of the eighth International Symposium on Magnetic Bearings, Mito, Japan, (2002).
4. C. Nataraj, E. Calver. Optional design of radial magnetic bearings. Proceeding of the sixth International Symposiuk on Magnetic Bearings, MIT, USA, (1998).
5. Xie Yuefeng, YU Xiongqing. Parametric design of aircraft configuration using API in CATIA. Computer Engineering and Design, **29**,14 (2008).

MATEC Web of Conferences 77, 01035 (2016) DOI: 10.1051/matecconf/20167701035
ICMMR 2016

Error Analysis and Compensation Method on the Mechanical Structure of the Hydraulic Control System

Luo Yanyan, Liu Jun, Zhang Li and Zhang Haining

Qingan group co., LTD., Xian, China

Abstract. Mechanical deformation of mechanical transmission part in hydraulic control system directly affects the loading accuracy of the system. For improving the mechanical properties of the system, The force analysis and motion analysis of mechanism are simulated based on the four-bar linkage structure (FLS), and kinematics simulation is designed by using Matlab program, then came to a system error bar graph. The system error was calculated accurately according to the results of the structural mechanics simulation made by Solidworks motion module. The structure of the system will be modified when systematic errors exceed the required limit values until it reach the required value.

1 Introduction

The electro-hydraulic servo method is widely used in hydraulic control system which has high efficiency, accurate control, easily design control law, etc. Especially, the accuracy of control is improved greatly in the case of rapid development of sensor technology. Because of the force deformation of system organization and tiny errors in system installation, will cause the change of hydraulic load torque, also it difficult to measure through the pressure sensor or displacement sensor in the system [1]. While the system error caused by the deformation is greater than the system error caused by sensor accuracy, the using of high precision sensor becomes meaningless.

In this essay, the mathematic model of FLS will be established, and the error caused by deformation will be analysed which is widely happened in many conditions of electro-hydraulic experiment. By using Matlab and Solidworks, the system error of FLS caused by mechanical deformation will be calculated. Finally, compensation algorithm was proposed.

2 Structure analysis of the hydraulic loading system

2.1 Analysis of force balance

The schematic diagram of the FLS is shown in Fig. 1. The OA lever represents the loaded parts, such as crane booms, mechanical arm, and aircraft door; the BC1D lever is the loading cylinder; point O, point C and point D are fixed point, and yet point BC1 is the hinge point of hydraulic cylinder interacting with CB lever; OA lever, CB lever, BC1D lever circles respectively point O, point C, point D; the swing Angle α of OA lever is decided by the elongation of actuator. While the specific load torque (known its output characteristic) force to the loaded parts (OA lever), hydraulic cylinder operated and output hydraulic pressure, and then FLS will transmit the hydraulic torque to cabin door [2], [3]. Take some reason such as site structure in to consideration, the swing angle α is the only angle feedback quantity which can be obtained.

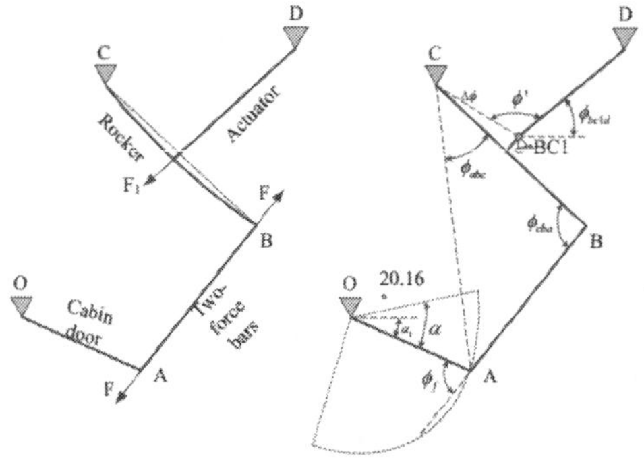

Figure 1. Schematic diagram of the FLS

Torque balance equation of system as follows.

$$\begin{cases} T_{oa} = F_{ab} \cdot l_{oa} \cdot \sin(\phi_f) \\ T_{cb} = F_{ab} \cdot l_{cb} \cdot \sin(\phi_{cba}) \\ T_{cb} = F_{bc1} \cdot l_{cbc1} \cdot \sin(\phi') \end{cases} \quad (1)$$

where T_{oa} is the torque of OA lever shaft with point O; T_{cb} is the torque of CB lever shaft with point C; F_{ab} is the stress suffered of two-force bars; F_{bcld} is the hydraulic pressure of actuator; ϕ_f is the angle between the two-force bars and the loaded parts; ϕ_{cba} is the angle between the two-force bars and the big rocker; ϕ' is the angle between actuator and the loaded parts [4].

2.2 Axis mathematical model

A coordinate system is established for the X axis in the horizontal direction, as shown in Fig. 1. The coordinates of point O, point C and point D are known in the design of an aircraft test bench structure, and the angle between the limit position of OA lever and horizontal direction is 20.16°. Coordinate of point A, point B, point BC1 as follows [5]:

$$\begin{cases} X_a = l_{oa} \cdot \cos(-\alpha + 20.16) \\ Y_a = l_{oa} \cdot \sin(-\alpha + 20.16) \end{cases} \tag{2}$$

$$\begin{cases} \phi_{ca} = \arctan(\dfrac{X_a - X_c}{Y_a - Y_c}) \\ \phi_{acb} = \arccos[(l_{bc}^2 + l_{ac}^2 - l_{ab}^2) / (2 \cdot l_{bc} \cdot l_{ac})] \\ \phi_{cb} = \phi_{ca} + \phi_{acb} \end{cases} \tag{3}$$

$$\begin{cases} X_b = X_c + l_{cb} \cdot \cos(\phi_{cb}) \\ Y_b = Y_c + l_{oa} \cdot \sin(\phi_{cb}) \end{cases} \tag{4}$$

$$\begin{cases} X_{bc1} = \dfrac{X_c + X_b}{2} + \sqrt{\left(\dfrac{l_{cb}}{2}\right)^2 + \Delta l^2} \cdot \cos(\phi_{cb} + 90) \\ Y_{bc1} = \dfrac{Y_c + Y_b}{2} + \sqrt{\left(\dfrac{l_{cb}}{2}\right)^2 + \Delta l^2} \cdot \sin(\phi_{cb} + 90) \end{cases} \tag{5}$$

where $X_{a/b/c}, Y_{a/b/c}$ are respective the horizontal / vertical coordinates of the point A, point B, point C ; ϕ_{ca} is the angle between the connection of point A and point C with the x-axis; ϕ_{cb} is the angle between the connection of point B and point C with the x-axis.

The trajectory of point A, point B, mid-point of CB lever and point BC1 are drawn by using Matlab software, as shown in Fig. 2.

2.3 Calculating the relative error

According to the axis mathematical model (equation (2), (3), (4), (5)) and torque balance equation (1), these parameters can be obtained ,such as coordinates of each point, angles, system of force, torque of each lever and force of each point in torque balance equation .The torque curve of OA could be calculated as follows:

$$T_{oa} = F_{bc1} \cdot \frac{l_{cbc1} \cdot \sin\phi'}{l_{cb} \cdot \sin\phi_{cba}} \cdot l_{oa} \cdot \sin\phi_f \tag{6}$$

After calculation, torque curve of OA lever was drawn with angle of OA lever increased from 0° to 122.8 ° (fully open), on the assumption that the system error is zero, output pressure of actuator is 50000 N, as shown in Fig. 3.

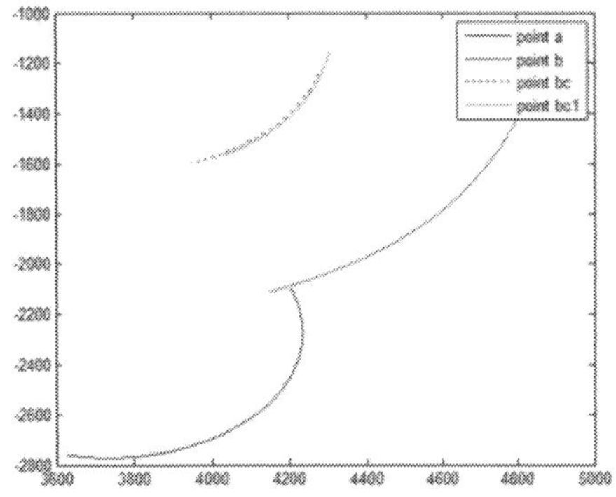

Figure 2. Trajectory of mechanism node.

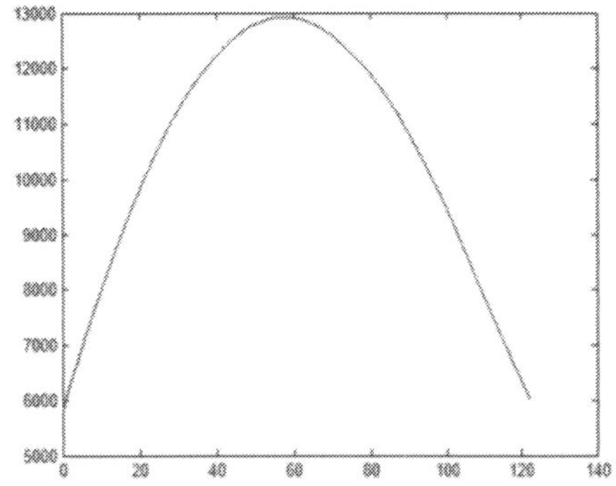

Figure 3. Torque curve of OA lever.

The system error was produced since the load mechanism (point C, point D) is skewing caused by force. The system error of point D was analysed firstly. Point D can offset from any direction to point D' under stress, while BC1D lever rotate to different directions. The coordinates of Point D' as shown in equation (7): Δl is the offset.

$$\begin{cases} X_d' = X_d + \Delta l \cdot \cos(\theta) \\ Y_d' = Y_d + \Delta l \cdot \sin(\theta) \end{cases} \tag{7}$$

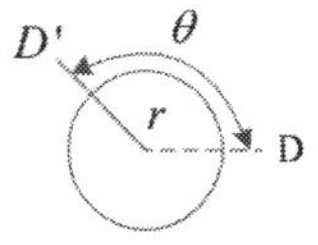

Figure 4. Course deviation of point D.

While the deviation of point D changes, we get a new parameter of the angle ϕ' by formula (8) 、 (9) and (10).

$$\phi_{cbc1} = -\phi_{cb} + \Delta\phi \tag{8}$$

$$\phi_{bc1d} = \sin^{-1}\left(\frac{Y_d - Y_{bc1}}{\sqrt{(Y_d - Y_{bc1})^2 + (X_d - X_{bc1})^2}}\right) \quad (9)$$

$$\phi' = \frac{\pi}{2} - \phi_{bc1d} + \frac{\pi}{2} + \phi_{cbc1} \quad (10)$$

By substitute the new ϕ' into formula 6, we can calculate a new torque of the OA lever.

Relative error of the system can be calculated with excursion of point D along x-axis, under the assumption that excursion of point D is 10 mm in the horizontal direction, as shown in equation (11).

$$\Delta w = \frac{T_f - T_0}{T_0} \times 100\% \quad (11)$$

where T_f is the torque of actual system; T_0 is the torque of theory system.

3 The simulation result

The results of relative error are shown in Fig. 5 after simulation and calculation.

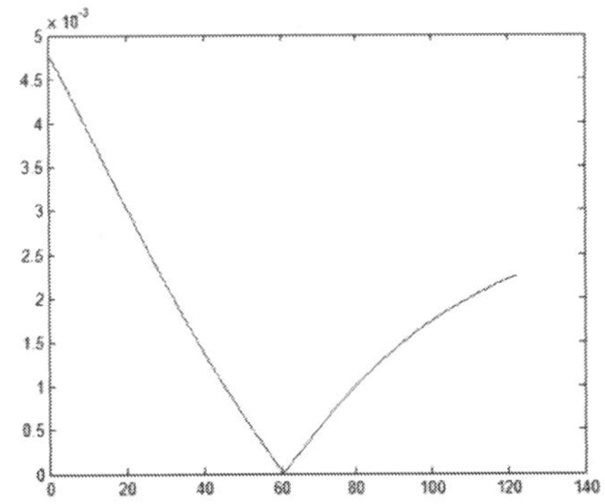

Figure 5. The relative error of system.

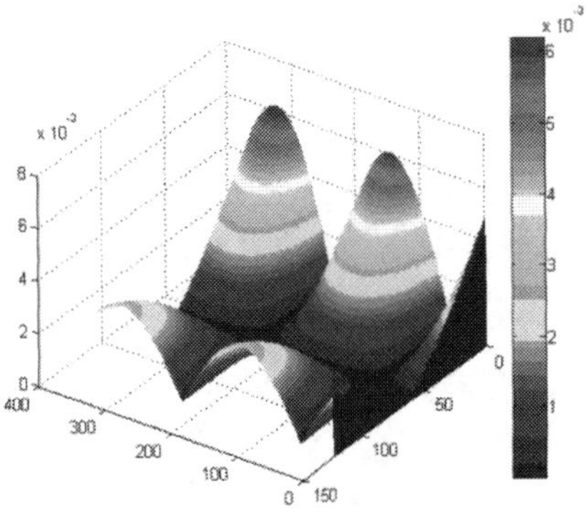

Figure 6. The system error while two hinged point moved simultaneously.

The system error was produced only consider the force deformation when point O and point C were skewing synchronously. The error matrix $W[\alpha \quad \theta]$ was established with the position deviation of point D extension to 360°. Where the α is swing angle of OA lever; the θ is swing angle of point D'; the values of error matrix is the relative error. As shown in Fig. 6.

After analysis, the relative error can be produced only in the offset of point D is 10 mm. The relative error is 0.62% while the swing angle of OA lever is 10.9° and the deviation angle of point D is 324°.

The matrix $W_1[\alpha \quad \theta \quad \theta_1]$ was established while the deviation was occurred synchronously between point C and point D, wherein the θ_1 is swing angle of point C'. The maximum relative error is 2.08% in the deviation of point C and point D is 10 mm. At this time, the swing angle of OA lever is 123°, the deviation angle of point D' is 145° and the deviation angle of point C' is 113°.

According to the above methods, the maximum relative error was calculated respective with the deviation value Δl range for 0.1, 0.5, 1, 5, 10, 50, 100. The semi-logarithmic graph of deviation-relative error was drawn using least-squares fitting method, as shown in Fig. 7.

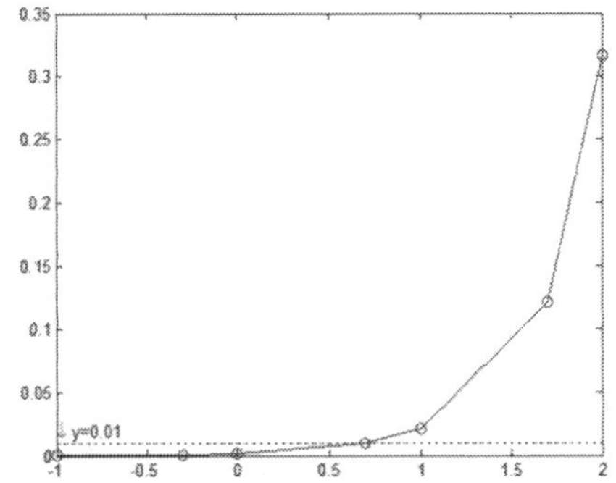

Figure 7. The curve between the maximum error and the force deformation.

In the above figure, the horizontal axis represents $\log(\Delta l)$ and the vertical axis represents the relative error. With the increase of the offset ratio, the relative error also will increase. Based on the curve data to make the relative error is less than 0.01, the deviation of loading point needs to be controlled within 5 mm. However, how to calculate the deviation of loading point in theory? Solidworks researched by Dassault company make it easy and solve the problem quickly.

4 Structural calculations of solidworks

Structural mechanics of hinge point of FLS were calculated by using the motion simulation module of Solidworks.

4.1 Establish 3D model of the structure

The precise mechanism model was drawn by using 3D modeling of Solidworks. The structure model of system take root at point D was drawn, as shown in Fig. 8. The single hanger pedestal welded on the 20# square steel tube.

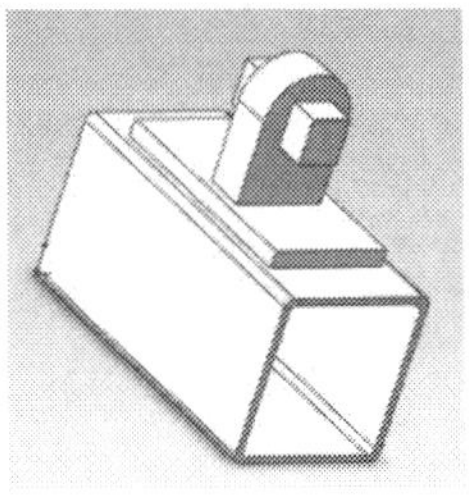

Figure 8. 3D model of hinge point (point D).

4.2 Choose material and load of the structure

After model was established, the setting of simulation can be started by clicking on the options of Solidworks motion. At the same time, the materials and load of the structure can be set up. The simulation results as shown in Fig. 9.

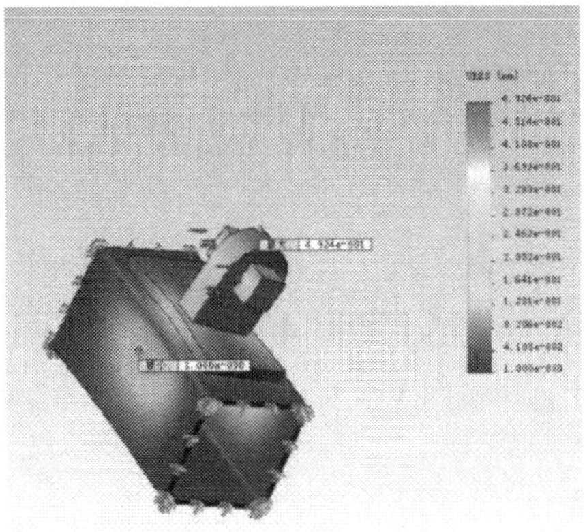

Figure 9. The simulation result of force deformation.

In the above figure, the deviation of the load point is 0.4 mm which meets the requirements of less than 5 mm. At this point, the relative error of the system is 0.08% with the output of error calculation program.

5 Conclusion

The deviation of load point's position caused by force deformation is inevitable for the design of hydraulic loading system. Therefore, the method provided in this paper can greatly improve the reliability of structure. It also can solve the problem of system control precision in the basic theory design and provide the data when needs to consider the accuracy deviation compensation in design control system.

References

1. Wang Zhanlin. Hydraulic Servo-control, Beijing University of Aeronautics & Astronautics Press, (1987).
2. Su Donghai. Liu Qinghe. Wu Shenglin, Dynamic Compensation and Simulation Analysis for Improving Control Performances of Electro-Hydraulic Load Simulator. China Mechanical Engineering. (1999).
3. Yuan Chaohui. Wang Lei, Compound Compensator Based on Cerebellar Model Articulation Controller for Surplus Torque in Passive Electric Loading System. Journal of Tongji University. (2004).
4. Jain R K. Majumder S. Dutta A, Multiple path generation by a flexible four-bar mechanism using ionic polymer metal composite. Journal of Intelligent Material Systems and Structures. (2012).
5. Kuo Y L.Cleghorn W L. Curvature and displacement based finite element analyses of flexible four-bar mechanisms. Journal of Vibration and Control, **17**(6), (2011).

DOI: 10.1051/matecconf/20167701036

The Stress Analysis and Tests on the Hinge Beam of the Diamond Synthesis Cubic Press

Ma Liang[1],Lu Beihua[1], Yuan Yuan[2] and Zhang Yinghong[1]

[1]*School of Mechanical Engineering, Guilin University of Electronic Technology, 541004 Guangxi, China*
[2]*Guilin Guiye Heavy Industries Co.,Ltd, 541199 Guangxi, China*

Abstract. To deal with the problem of the lug fractures of hinge beam caused by the fatigue and overload during the operation of the cubic press, the analysis methods of finite element are applied to the analysis of the internal stress distributions of the hinge beam. The simulation results show that the internal stress of the hinge beam mainly concentrates on the upper surface of the lug roots connecting the outer cylinder with the both lugs. According to the data of simulation and analysis as well as the actual fracture situations, considering the strain-test methods we have designed the schemes of testing the strain on the lugs of hinge beam. And the strain measurements of the lugs are completed by the repeated loading experiments. Comparing the data of simulation and analysis with the measured data has verified their consistency. It also confirms the model established by the simulation and analysis is reasonable and accurate at the same time.

1 Introduction

Strain gauge measurement [1] is the most common and effective mechanics testing method in the mechanical engineering testing technology. It has not only provided the scientific bases for analyzing the failure causes of the components and optimizing the structure, but also has a good guiding significance for the new product development and modification when we need to take the stress measurement and understand the stress distribution laws of the structures in the safety and reliability analysis of engineering structures.

The cubic press is the prime equipment for compositing the superhard materials in current China[2,3]. The hinge beam, as the key component of the hydraulic press, plays a decisive role in the safety and reliability of the whole hydraulic press system, and founding its internal stress distribution is helpful to improve the structure and performance of cubic press. At present, the analysis of hinge beam is so stagnant in the traditional software analysis based on a large number of assumptions and simplification that the analysis results can not accurately reflect the actual working stress of hinge beam. As it is shown in the Fig. 1 that it is likely to cause some unexpected accidents. Hence, the designed hinge beam of cubic press relying solely on the above analysis method can not accurately mirrors its true stress. Consequently, in order to analyze accurately and effectively the stress and the strain of hinge beam, it is significant for us to provide a set of the complete test schemes and the test data for verifying the theoretical analysis results.

This paper proposes the research of the stress analysis and tests on the hinge beam of the diamond synthesis cubic press. First, 3D design model of the hinge beam is established by ANSYS, a kind of the analysis software of finite element. And carrying out the static analysis and calculation of finite element, we has drawn the equivalent stress distribution nephogram and explored the stress distribution situations of hinge beam. Furthermore, based on the electrical stress-measurement technology [4], the actual tests of the stress distribution of hinge beam are carried out. And comparing the results of finite element analysis with the test data, the both are in the same line, which verifies better rationality and correctness of the established model in this paper.

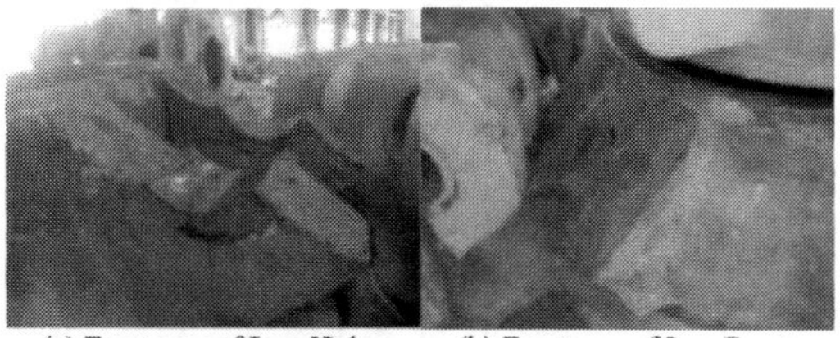

(a) Fractures of Lug Holes (b) Fractures of Lug Roots

Figure 1. The Fractures of the Hinge Beam

2 The framework and working principle of the cubic press

Main Consisting of the hinge beam, the working block and the piston, the hinged cubic press is an assembly body which is composed of 6 hinge beams through the

DOI: 10.1051/matecconf/20167701036

hinged constraints. The Fig. 2 shows the structure of the cubic press.

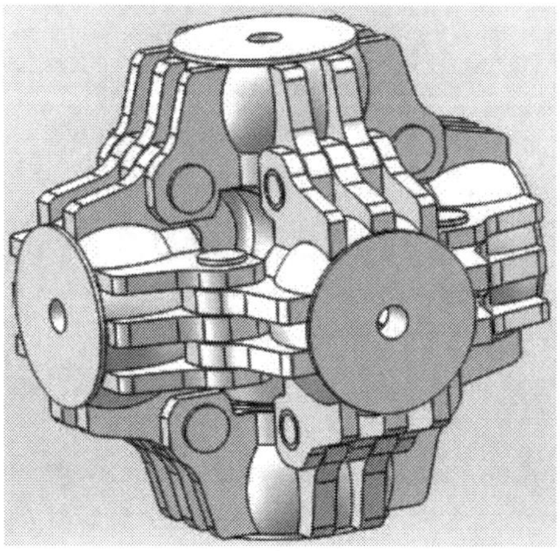

Figure 2. The hinged cubic press

The hinged cubic press is special equipment for producing the synthetic materials on the condition of the high temperature and high pressure [5]. In order to push the piston forward and generate the super-high pressure in the front of the piston, the hydraulic oil is pushed into the center of the main engine from the bottom of six working blocks. Furthermore, with the help of the cemented carbide anvil in the center of the main engine, a super-high pressure chamber with the sealed cube is formed around the raw materials. And then the high-pressure conditions on which the diamond and other superhard materials are growing are provided by the pressure the pyrophyllite transfers. At the same time, the high-temperature conditions on which the chambers make the synthetic materials grow are provided by the way of electric heating. After holding pressure and compositing the superhard materials for a period of time on the condition of the high temperature and high pressure, the hydraulic devices removed the anvils again and take out the pyrophyllite from the synthesis chambers, so that the device carry out a list of the operations of the pressure-relief. Finally, the overall operation of compositing the superhard materials is completed well. The working process of the whole equipments is completed cooperatively by the electric control system and the mechanical hydraulic system, and the working loads can be up to 100MPa.

3 The finite element analysis of hinge beam

3.1 Finite element model

Because the hinge beam is the main bearing component of the cubic press. In view of the actual working conditions, this paper employs ANSYS, a kind of the analysis software of finite element, to carry out the static analysis and calculation and to explore the stress distribution of the hinge beam [6]-[8]. To facilitate the structure discretization and element analysis and keep the higher calculation accuracy, we has simplified the model [9]: for one thing, the smooth holes take the place of the threaded holes; for another, the chamfers are omitted on the upper and the lower tables. And then we establish directly the 3D solid model by ANSYS.

The selected elements are related to the properties of the tetrahedral structure module Solid186 that is a 3D 20 nodes element, which is used to establish the mesh dividing model of hinge beam. Due to the structure symmetry of hinge beam, and also its internal loads are uniform and symmetric, as it is shown in the Figure 3 that we only need to draw the mesh model of the 1/4 hinge beam. The whole structures are made of the alloy structural steel of 35CrMo made in China.

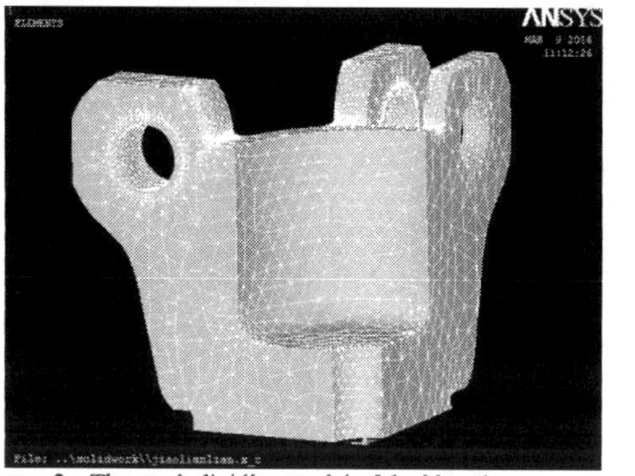

Figure 3. The mesh dividing model of the hinge beam

3.2 Adding constraints and loads

The hinged cubic press is the circulatory working equipment including three stages: loading, holding pressure and unloading on the condition of the high temperature and high pressure. Taking the test experiments into account this paper merely aims to simulate and analyze at the stage of holding pressure.

In order to ensure the consistency between the calculation results and the actual working situations, the method adding gradual the quantitative load to the hinge beam is applied to the whole analysis process when we use ANSYS to simulate and analyze. And assuming each lug to be in the uniform stress ,the surface constraints are exerted correspondingly over the bottom of hinge beam and the symmetric cross-section of a quarter of the model. In the same way, the surface loads are exerted correspondingly over the lugholes and the bottom of hinge beam.

3.3 Result analysis

In this paper, we establish the static analysis model of 1/4 hinge beam by PCG(Precondition Conjugate Gradient). In the light of the limited space of this article, as it is indicated in the Figure 4 that we merely draw the equivalent stress distribution nephogram of the 1/4 hinge beam on the condition of 100MPa. It is analyzed that the maximum tensile stress mainly concentrates on the upper surface of the lug roots connecting the outer cylinder with the both lugs. Its value is about 324.8MPa. Meanwhile, the stress value is only 5.32MPa in the middle of the two lug roots and below the two lug roots. In addition, considering the safety factor of 2~2.4 in the design of hinge beam and often normal-load operation of hinge beam in the working process, so it has enough strength. However, it can be seen in the Fig. 3 that the gradients of the stress focusing on the range of 5 millimeter between the lug roots and the upper surface are larger. According to the test data, it has preliminarily verified that the results obtaining from the finite element analysis are correct and consistent.

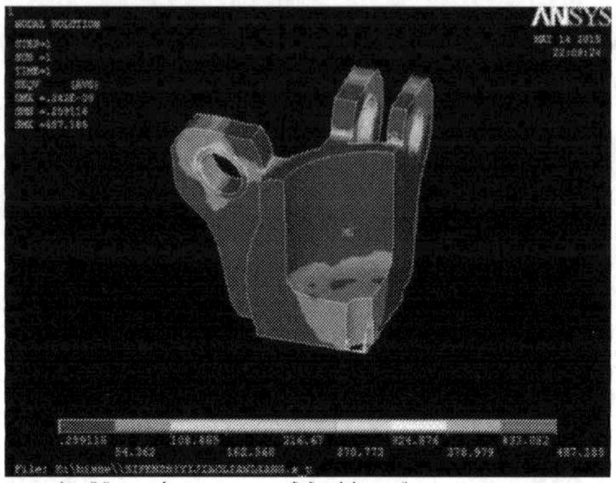

Figure 4. Von mises stress of the hinge beam

4 Experimental verification

4.1 Layout schemes of the strain gauges

According to the results of finite element analysis and the actual fracture position of the lugs, the two strain gauge groups defining the distance as 1 millimeter and consisting of 8 coaxial and multi-gate strain gauges in which the distance is designed as 3 millimeter are pasted in the lugs. So we can measure accurately the strain variation at those areas of the larger strain gradient. Additionally, in order to compare and make the test data more accurate, we choose respectively a lug to measure in the 4 directions of the two lugs and the three lugs.

4.2 Experimental tests and result analysis

We use the Wheatstone bridge circuit to measure the change of resistance strain gauge. Regarding 10MPa as the unit load increment, A successful and complete test is implemented by way of gradual loading from 10MPa to the maximum load 100MPa. Moreover, after each quantitative loading, the measurement of holding pressure for 3 minutes is carried out on the hinge beam so as to ensure the stability and reliability of the test data.

After the repeated tests of loading, holding pressure and unloading, we has gotten a set of experimental data. In consideration of the limited space of this article, as it is respectively shown in the Table 1 that we only list the related data of 8 test points on the condition of 100MPa about GF group of a hinge beam. It is concluded from Table1, that GF5 is the point of the maximum tensile stress 311.85MPa which is less than the yield limit of the selected material. Comparing the test data of the upper surface of the lug roots on the condition of 100MPa with the data of the finite element simulation, the relative error 4.18% between the test values and the simulation and calculation values verifies the fact that the both are consistent each other. It justifies the establishment and analysis of the finite element model is reasonable and accurate.

Table 1. The related data of 8 test points in GF group

Loads [Mpa]	Strain values[με]							
	GF1	GF2	GF3	GF4	GF5	GF6	GF7	GF8
10	67	82	150	178	203	162	151	144
20	120	144	264	309	355	292	273	260
30	163	198	366	432	513	409	384	365
40	209	253	477	559	658	530	499	476
50	252	308	585	693	801	649	612	585
60	294	359	685	798	940	764	721	687
70	335	409	784	905	1069	872	825	788
80	380	465	891	1035	1210	992	938	897
90	424	517	991	1168	1345	1106	1048	1002
100	463	566	1088	1296	1485	1215	1151	1098

5 Summary

In this paper, we consider the hinge beam of hinged cubic press as the research subject. First, the distribution and the value size of the internal stress of hinge beam are obtained by way of employing ANSYS to carry out the static analysis and calculation of hinge beam. Consequently, The simulation results demonstrate that the internal stress of hinge beam mainly concentrates on the upper surface of the lug roots connecting the outer cylinder with the both lugs. And the stress gradients are larger in these areas. So it is more likely to generate the fracture if the lug roots appear the defects in the process of manufacturing. Furthermore, based on the experiment methods of electrical stress measurement, we have a list of the actual tests in the lug roots of hinge beam. The results show that the maximum tensile stress 311.85MPa is located at the position GF5. Compared to the data of finite element analysis, the relative error 4.18% between the test results and the simulation ones verifies the fact that the establishment and analysis of the finite element model is reasonable and accurate. Moreover, this paper provides a referential bases for the optimization of the following structures.

Acknowledgements

The research in this paper is funded by Innovation Project of GUET Graduate Education under Grant No.YJCXS201508 and National Students' Project for Innovation and Entrepreneurship Training Program under Grant No. 201510595001 and the Based Capacity Promotion Project of Middle-aged and Young Teachers in Colleges and Universities of GuangXi in 2016 under Grant No. ky2016YB147.

References

1. M.JI. DaiQi-ke,Chinese Journal of Theoretical and Applied Mechanics,*Modern situations of electrical measuring and stress technology*,**3**(2). 97-110, (1959).
2. J.Z. Nong,Z.S.Xie, Y.Wu, Superhard Material Engineering, *The failure analysis on the hinge beam of the diamond synthesis cubic press*, **23**(6),5-9, (2011).
3. Z.R.Li,C.X.Wen,Diamond&Abrasives Engineering,*The hydraulic system failure analysis of the cubic press*,**3**(93),36-38, (1996).
4. L.J. Huang,D.B. Cen,Modern Manufacturing Engineering,*Stress-test and finite element analysis for the giant-type hydraulic vulcanizing machine*, (7),126-129, (2010).
5. C. Zhang,Jilin University,*The analysis and optimum design of China-type cubic high-pressure apparatus by FEM*, (2010).
6. Q.G. Han, X.P. Jia, H.A. Ma, Acta Physica Sinica, *Finite element simulations of thermal-stress on carbide anvil used in cubic high pressure cemented tungsten apparatus*,**58**(7),4812-4816, (2009).
7. D.C. Qin,Y. Liang, L.P. Chen, Journal of Mechanical Strength ,*Research for hinge SLEEVE'S structural optimal design based on fatigue intensity*, **28**(2),306-310, (2006).
8. C.S. Baviskar,R.M. Tayade, V. G. Patil, Infernahonal Journal of Mechanical Engineering and Robofics Research,*Determination of failure strength of curved plate weld joint using finite element analysis*,**1**(2012).
9. X. Sun,B. Wang,Y. Zhang,Journal of Guilin University of Technology,*Hinge sleeve strength calculation in cubic press*,**33**(2),322-324, (2013).

MATEC Web of Conferences 77, 01037 (2016) DOI: 10.1051/matecconf/20167701037
ICMMR 2016

Physical and Tribological Properties of Nitrided AISI 316 Stainless Steel Balls

Shicai Yang[1], Quanshun Luo[2], Hailin Sun[1], Matthew Kitchen[2] and Kevin Cooke[1]

[1]*Teer Coatings Ltd., Miba Coating Group, West Stone House, Berry Hill Industrial Estate, Droitwich, WR9 9AS, UK*
[2]*Materials and Engineering Research Institute, Sheffield Hallam University, Howard Street, Sheffield S1 1WB, UK*

Abstract. AISI 316 austenitic stainless steel balls (diameters 5.0 and 12.0 mm, typical hardness 250 $HV_{0.3}$) and flat samples (20×20×2.0 mm) were nitrided by a pulsed glow discharge Ar/N_2 plasma. Hardness of the ball surfaces was analysed using Vickers indentation. Thermal stability of the nitrided balls (diameter 12.0 mm) was studied using a furnace to heat them in air for 8 hours at temperatures up to 700.0 °C and then, after cooling to room temperature, the surface hardness of the heated balls was re-measured. Scanning electron microscopy and X-ray diffraction were used to study the microstructures, composition and phase formation of the nitrided sublayers. Unlubricated pin-on-disc wear testing was used to evaluate the wear resistance of nitrided stainless steel balls (5.0 mm diameter) and the results were compared with similar testing on hardened Cr-Steel balls (5 mm diameter with hardness of about 650 $HV_{0.3}$). All the test results indicated that the nitrided AISI 316 austenitic stainless steel balls have advantages over the hardened Cr-Steel balls in terms of retaining high hardness after heat treatment and high resistance to sliding wear at room temperature under higher counterpart stress. These properties are expected to be beneficial for wide range of bearing applications.

1 Introduction

Austenitic stainless steels are known to have outstanding toughness, biomedical compatibility, corrosion resistance and oxidation resistance [1, 2], and therefore have been widely used in industries, such as food [3], biomedical devices [4-6], petroleum and chemical engineering [7], steam turbines [2], and nuclear energy generation [8]. A recent review of the composition, microstructure and functions of stainless steels can be found in Ref. [9]. The major weakness of austenitic stainless steels is their low hardness [10], typically 150 – 250 HV, which limits the wear resistance of components such as ball bearings. The surface hardness of stainless steel can be enhanced significantly to 1000 HV by means of nitriding treatments in a gaseous or plasma medium [11, 12]. These enable the production of piston rings and gears with outstanding tribological properties. Moreover, surface nitrided stainless steel could also find applications in ball bearings to be used at elevated temperatures owing to the material's combined high surface hardness and excellent oxidation resistance. To date, however, little research has been published on the plasma nitriding of austenitic stainless steel balls. One factor may be the complexity involved in supporting multiple balls while achieving an effective plasma treatment, and to provide uniform nitriding over the entire ball surface.

In this paper, we report recent research on the characterization and tribological evaluation of plasma nitrided balls made from austenitic stainless steel AISI 316, to explore its feasibility for ball bearing applications.

2 Experimental

Austenitic stainless steel AISI 316 was employed in the research, including balls of both 5 mm and 12 mm in diameter and square flat samples of 20 × 20 × 2.0 mm in size. The balls were manufactured with a grade 10 surface finish and the flat samples were manually polished with 1200# SiC abrasive paper, all samples being ultrasonically cleaned in acetone and dried before plasma treatment.

The nitriding treatment was completed using an industrial pulsed DC glow discharge plasma process [13], at input power 2.5 kW, pulse frequency 360 kHz and potential -550 V The gaseous medium in the nitriding chamber was controlled at a pressure of 1.33 Pa and with an Ar/N_2 flow ratio of 1:3. The nitriding process was controlled in a range 450 – 500 °C. In the nitriding chamber, each ball was supported with a specially made jig to ensure effective exposure to the nitriding atmosphere for most of the spherical surface, except a small area hidden by the jig, Fig. 1. After the first 3-hour nitriding process, the chamber was vented to atmosphere and each ball was turned by an angular rotation sufficient to allow the hidden area to be nitrided in a second 3-hour nitriding process.

Figure 1. A photograph taken inside the nitriding chamber, showing balls supported on the individual jigs.

A field emission SEM, FEI Nova-200 FEG-SEM, was employed to analyse the as-nitrided surfaces and cross-sectional microstructure. For the latter, cross-sectioned samples were ground and polished to mirror finish before chemical etching in a Kailling's No. 2 solution. Energy dispersive X-ray (EDX) spectroscopic chemical analysis was also applied along with the SEM observation. A Philips X'Pert X-ray diffractometer with radiation Cu-$K_{\alpha 1}$ (λ = 0.154056 nm, at power of 40 kV and 40 mA) was employed to analyse the crystallographic properties.

The hardness properties were measured using two methods, namely, Vickers hardness of the nitrided surfaces at a load of 0.3 kg and Knoop hardness profile on polished cross-sections at an indenting load of 0.025 kg. The depth of the nitrided layers was measured on polished cross-sections using both SEM and optical microscopy.

Pin-on-disk sliding wear tests were carried using two configurations by using the nitrided ball samples and hardened Cr-steel balls with hardened and ground M42 tool steel flat samples as the counterpart. The wear tests were conducted at a constant applied normal load of 5 N and linear sliding speed of 0.2 $m \cdot s^{-1}$. Each test ran for 60 minutes, before the ball and flat samples were taken off to measure the wear volumes.

3 Results and discussion

3.1 Microstructural characterization

The as-nitrided surface is shown to be slightly rougher than the as-polished surface, in Fig. 2a, as a result of lattice expansion induced plastic deformation of the austenitic grains. Some details of the crystalline slips are shown in the inserted image at higher magnification. Fig. 2b shows an etched cross-section of the nitrided layer, in which the nitrided layer can be recognised with a clear interface to the austenite substrate, associated with the deep-etching line near the centre of the image.

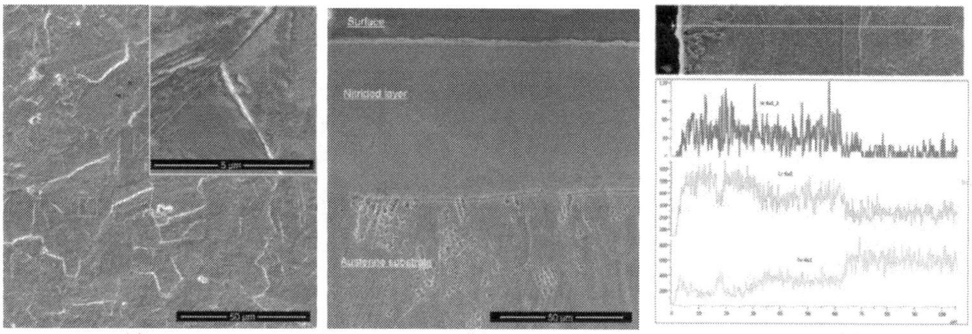

(a) (b) (c)

Figure 2. SEM and EDX analyses of the nitrided AISI316 steel ball: (a) As-nitrided surface; (b) Polished and chemically etched cross-section; and (c) EDX line scans of the N, Cr and Fe concentrations.

Table 1 Chemical composition of nitrided surface determined by SEM-EDX.

	N	Cr	Mo	Fe	Ni	Si	Mn
at%	26.77	13.77	0.87	48.68	8.03	0.84	1.05
wt%	8.43	16.11	1.87	61.16	16.60	0.53	1.30

Fig. 2c is the line-scan profiles of the characteristic X-rays N-K, Cr-K and Fe-K, to show the variation of N, Fe and Cr concentrations in the nitrided layer. The line-scans indicate a distinct profile of nitrogen in the nitrided layer, along with enriched Cr and reduced Fe concentration in comparison to the austenite substrate, which supports the Cr-trapping nitriding mechanism [9]. Meanwhile, EDX analysis also reveals the chemical compositions of the nitrided layer (acquired on the sample surface), in Table 1. Note that the N concentration is double that of the Cr concentration, indicating highly supersaturated nitrogen.

Fig. 3 shows X-ray diffraction curves of the nitrided and as-polished samples. The nitrided sample shows strong diffraction peaks of both the {111} and {200} planes as evidence of a substantially expanded austenite lattice (also called the S-phase [6, 9, 11]). In addition, several low-intensity diffraction peaks suggest precipitation of Cr- and or Fe-nitride phases.

3.2 Thickness, hardness property and thermal stability

To evaluate the thickness profile of the nitrided balls, a ball was sectioned diametrically across the region of the jig-obscured circle, and subsequently mounted, ground and polished. The nitrided layer depth was measured at equal intervals of around 1 mm with a total of 22 readings between the minimum 46.43 μm and the maximum 59.03 μm, leading to a statistical average of 54.25 ± 3.29 μm. The measurements indicate a relatively uniform thickness distribution throughout the ball surface, whereas the jig-obscured area shows a reduced layer thickness of 39.73 ± 2.25 μm. Fig. 4 shows hardness profiles across the whole nitrided depth. The nitrided layer exhibits high hardness of $HK_{0.025}$ 1100-1200. It also shows that, the double 3-hour nitriding process brought about only a marginal further increase in the hardness.

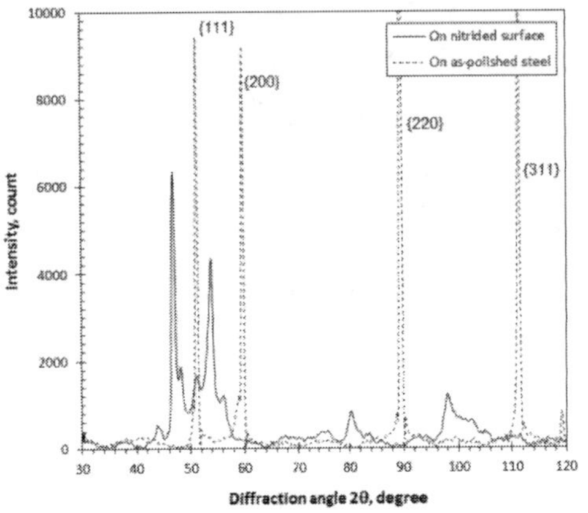

Figure 3. X-ray diffraction curves of nitrided and as-polished 316 steel samples.

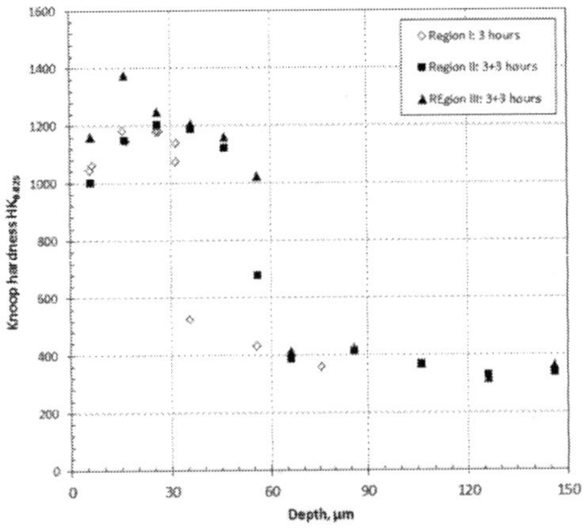

Figure 4. Microhardness depth profiles of as-nitrided ball.

Selected ball samples of both the nitrided 316 steel and hardened 100Cr steel were subjected to annealing treatments for 8 hours in air at an elevated temperature between 300 to 700 °C. After the annealing treatments, Vickers' hardness was measured on the ball surface, as shown in Fig. 5. The hardness of the 100Cr steel dropped significantly from over HV 600 to below HV 400 following the tempering transformation of its martensitic structure. In contrast, the nitrided 316 steel maintained high hardness when the annealing temperature was up to 500 °C, while the hardness was decreased when the annealing temperature was increased from 500 to 700 °C. The different hardness properties of the annealed balls are also shown in Fig. 6. In Fig. 6, the hardness profile of the 300 °C annealed nitrided ball was comparable to the as-nitrided ball, while annealing at 500 °C resulted in a hardness drop at a depth of about 20 μm. The hardness of the 700 °C annealed sample was significantly reduced. Nevertheless, the nitrided 316 steel exhibited superior thermal stability when compared to hardened 100Cr6 ball bearing steel, which implies its potential suitability for bearing components to be used at elevated temperatures.

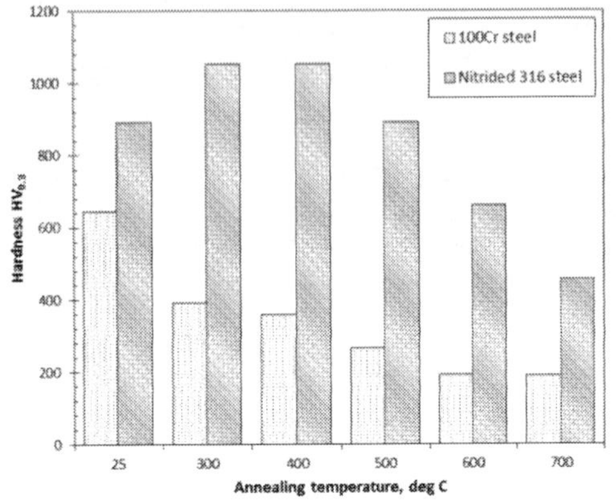

Figure 5. Effect of annealing temperature on the hardness property of nitrided 316 steel and hardened 100Cr steel.

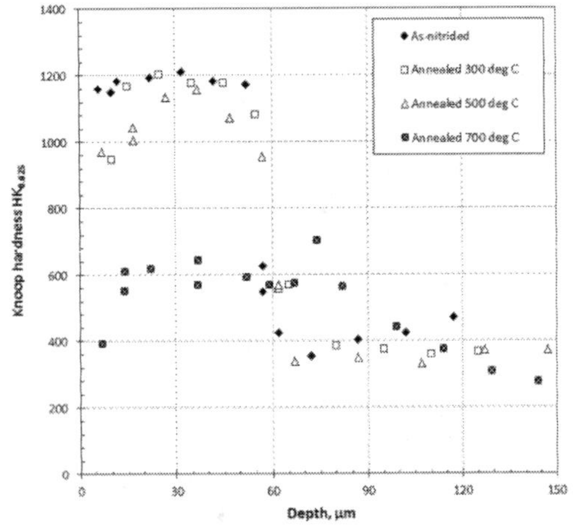

Figure 6. Microhardness depth profiles of as-nitrided and annealed balls.

3.3 Wear and friction properties

Details of the wear properties are summarised in Table 2. The 100Cr steel exhibited 12 times higher wear rate than the nitrided 316 steel when the applied load was 2 N. When the load was 5 N, the two "as received" materials' wear rates were more similar whereas the wear of the nitrided 316 was still four times lower.

DOI: 10.1051/matecconf/20167701037

Table 2 Wear depth (d) and specific wear rate (k) of the tested samples

balls	d_{ball} (μm)	k_{ball} (m^3/Nm)	k_{disc} (m^3/Nm)
At 5 N, run for 60 minutes			
N-ASS	37.2	3.1×10^{-15}	4.3×10^{-16}
Cr-Steel	78.2	1.4×10^{-14}	2.6×10^{-16}
At 5 N, run for 30 minutes			
N-ASS	23.7	2.5×10^{-15}	6.2×10^{-16}
Cr-Steel	52.6	1.2×10^{-14}	4.2×10^{-16}
At 2 N, run for 60 minutes			
N-ASS	19.2	2.0×10^{-15}	1.1×10^{-15}
Cr-Steel	66.2	2.4×10^{-14}	6.1×10^{-16}

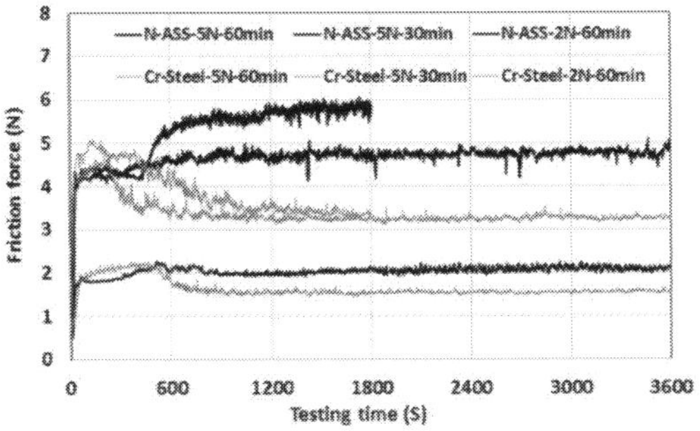

Figure 7. Friction corves of the tested samples.

Fig. 7 shows a collection of friction curves acquired in the ball-on-disk wear tests, in which "N-ASS" stands for nitrided 316 austenitic stainless steel and "Cr-Steel" for hardened 100Cr steel. The first test was carried out at an applied load 5 N for a total sliding time of 60 minutes, when the nitrided layer was worn through resulting in wear of the substrate. The sliding time was then reduced to 30 minutes to ensure that wear was limited to within the nitrided depth. Regardless of the applied loads and sliding distances, both the nitrided 316 and hardened 100Cr6 steels exhibited high friction coefficients, which could be explained by the adhesive steel-to-steel contact which occurred without the presence of lubrication.

Fig. 8a shows low-magnification images of the resulting wear scars. The wear rates of the nitrided 316 and 100Cr were dramatically different. Fig. 8b is the worn surface observed by SEM at high magnification, taken in the region of worn scar on nitrided ball after 60-minute and 5-N test. The SEM observation indicates that the rubbed nitrided ball surface exhibits typical sliding wear mechanisms of grooving deformation. Such wear is normally caused by hard abrasive particles. In this case, it is believed that the un-dissolved carbide particles in the high-alloy M42 tool steel should have played the role of hard abrasives, causing abrasive wear. When such wear mechanism dominates the sliding wear process, the wear rate decreases with the increase of hardness of the tested material, which can explain the striking difference between the wear rates of the nitrided 316 steel and the hardened 100Cr6. Fig. 9 shows the wear measurements of the counterpart M42 tool steel disks.

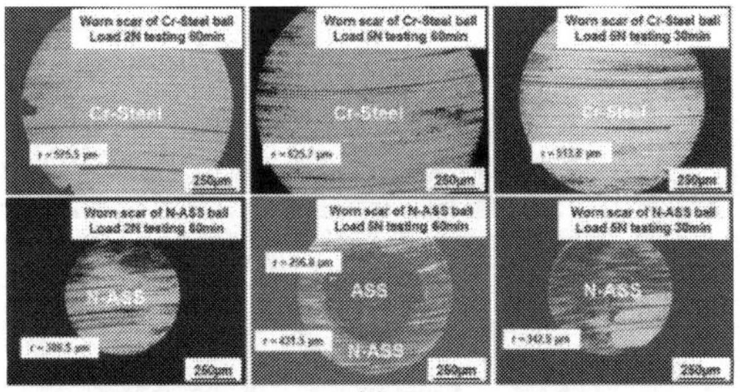

(a)

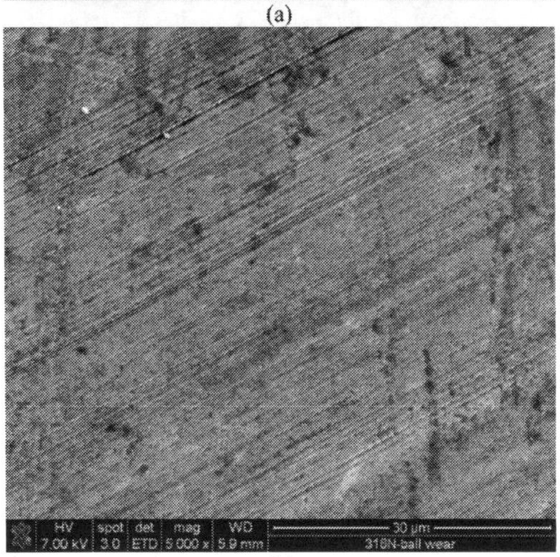

(b)

Figure 8. (a) Low magnification images of ball wear scars of nitrided 316 steel and hardened 100Cr steel; (b) A high magnification SEM image of the worn surface of the nitrided 316 ball.

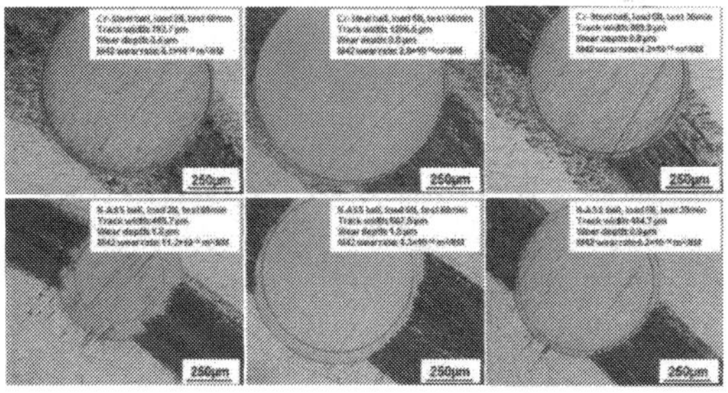

Figure 9. Low magnification images of the ball-crater measurements on wear tracks of the counterpart M42 steel disks.

The dramatically increased wear resistance of the nitrided 316 steel can be related to its increased hardness and hence increased resistance to abrasive wear and plastic deformation, and also by its enhanced thermal stability which helped to prevent softening due to the frictional heating. Further detailed investigation of the worn surfaces, e.g. by SEM examination, will be carried out later.

The main motivation of this work was to investigate the possibility that plasma nitrided austenitic stainless steel would be suitable for bearing applications. The results achieved so far do support the suggestion that, the combination of a hard and strong shell with a soft and tough core can provide significantly improved wear resistance. The sliding wear performance of the nitrided steel balls was significantly better than that of the hardened 100Cr steel balls under identical testing parameters. Meanwhile, it is also noteworthy that the nitrided balls caused more wear of the counterpart M42 steel than the hardened 100Cr steel balls. This could be attributed to the stress induced high wear on the counterpart surface. It was estimated according to load divided by ball worn scar area that the pressure at the counterpart surface for the N-ASS balls was two times higher than that for the hardened 100Cr steel balls. Obviously N-ASS balls were shown to have a higher capability to resist stress induced plastic deformation in an abrasive environment that would cause fatigue and eventually wear at sliding interface. It would be interesting to study the rolling tribology performance of the N-ASS balls but this will need to be addressed in further research.

4 Conclusions

Nitrided AISI 316 austenitic stainless steel balls were characterised as having high hardness, high capability to withstand stress and high sliding wear resistance and outperformed hardened Cr-Steel balls and hardened M42 steel discs under the identical pi-on-disc testing.

Up to a maximum hardness of $HV_{0.3}$ 1000 can be retained on the nitrided AISI 316 austenitic stainless steel balls at temperatures up to 400°C, whereas in comparison the hardness of hardened Cr-Steel balls was reduced from the $HV_{0.3}$ 650 of the original to $HV_{0.3}$ 350 after heat treated.

All the results indicated that AISI 316 austenitic stainless steel balls were significantly improved in their tribological properties after nitriding suggesting the potential applicability of these nitrided austenitic stainless steel balls in the ball bearing industry.

References

1. C. Allen, A, ball, Wear **74**, 287-3905 (1981-1982).
2. P. J. Maziasz, B. A. Pint, J. P. Shingledecker, K. L. More, N. D. Evans, E. Lara-Curzio, *Proceedings of ASME Turbo Expo* (2004) 2.
3. S. Cvetkovski, M. Matyer. Eng. **18** 283-293(2012).
4. K. Bordjih, E. Y. Jouzeau, D. Mainard, E. Payan, J. P. Delagoutte, P. Netter, Biomaterials **17** 491-500(1996).
5. T. Newson, Stainless steels for hygienic applications, *BSSA Conference-Stainless solutions for a sustainable future. Magna Science Advanture Centre*, Sheffield (2003).
6. J. Buhagiar, T. Bell, R. Sammons, H. S. DongS. J Mater Sci **22**,1268-1278 (2011).
7. R.D. Kane, M.S. Cayard, Corrosion **98,** 1-28 (1998).
8. S. Baindur, Bulletin of the Canadian Nuclear Society **29** 32-38 (2008).
9. Q. Luo, S. Yang, Int. J. Nanomed Nanosurg **1.1** 1–10 (2015).
10. E. T. Akinlabi, S, A. Akinlabi, Int J. Mechanical, Aerospace, Industrial, Mechatronic, and Manufacturing Engineering **8** 256-261 (2014).
11. Y. Sun, T. Bell, Wear **218** 34 (1998).
12. V. Toshkov, R. Russev, T. Madjarov, E. Russeva, J. Achiev. Mater. Manuf. Eng. 25, 71-74 (2007).
13. S. Yang, K. Cooke, H. Sun, X. Li, K. Lin, H. Dong, Surf. Coat. Technol. **236** 2-7 (2013).

MATEC Web of Conferences 77, 01038 (2016) DOI: 10.1051/matecconf/20167701038
ICMMR 2016

Dynamics and Energy Conversion of Aircraft Landing Gears at Touchdown

W. J. Wang

University of Sussex, United Kingdom

Abstract Dynamics of aircraft landing gears with and without pre-rotation has been analysed. The ground friction work is been formulated which raises the temperature of touch point to the rubber smoke point. The sliding time, distance, power and energy released by the friction at touch point have been presented. Simulation of target speed has been shown for different levels of pre-rotation torque.

1. Introduction

Hundred thousands of aircraft take off and make their landing everyday worldwide. Up to date, during the touchdown, all aircraft generate smoke, blacken the runway and cause significant tyre wear. No doubt, this behaviour of landing applies damage to environment and increases costs on aircraft operation. Since 1940s, some inventors have made attempt to develop pre-rotation devices and a number of them have been patented [1-2]. However, so far after 80 years past, surprisingly, none of such devices has been put into practical use. Some of the landing gear manufacturers concern about the complicity and additional weight by adding such a device. The landing touchdown smoke is still a standard scene to which most people have got accustomed. With increasing pressure on environment in this 21st century, action apparently is needed. Aircraft operators provided experience and data for investigators [3]. Rubber tires for vehicles and their friction behaviours are still attractive topics [4-9]. Some have intensively investigated rubber combustion [10]. However, fundamental understanding to the touchdown problem and the measure to reduce the landing smoke are seriously lacking. In recent years, the authors have proposed some fundamental approach to analyse and simulate the dynamic behaviours [11-12] for aircraft touchdown processes. This paper is reflecting the novel but approach to the problem by the author, aiming for eventually getting rid of the landing smoke, increasing the tyre life cycle and protecting the environment.

2. Landing gear sliding with pre-rotation

Instantly after the landing gear touches the runway, it is for the ground friction to provide a torque to accelerate the static wheel to match the landing speed within a fraction or a couple of seconds.

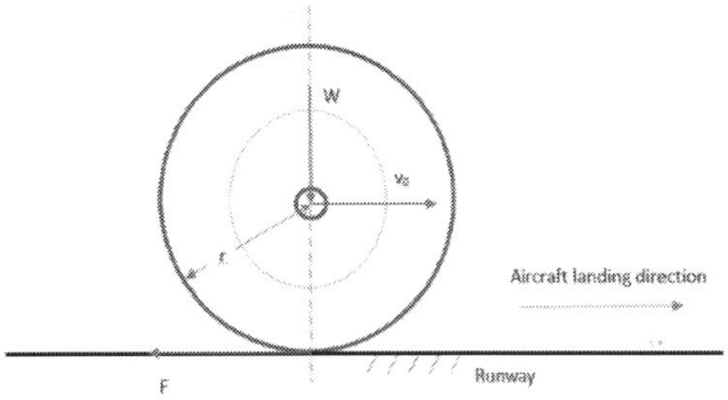

Figure 1. Landing gear wheel at touchdown

Let v be the aircraft travel speed and ω angular speed of the landing gear wheel. For a landing speed $v = v_0$, by a 100% matching, the tangential speed of wheel at touch point is also $v_0 = r\omega_0$. Assuming the friction provided by the runway ground $F = \mu W$, where μ is the friction coefficient, depending on type of tyre, contact area and runway condition; W is the vertical force on the wheel shaft due to weight of aircraft as shown in Fig. 1. Normally there are a number, from several to dozens, of wheels for an aircraft. After touchdown, using Newton's second law of motion for rotational circumstances, with a resistance torque T_r which includes those in shaft, air viscosity and ground rolling resistances,

$$F\,r - T_r = I\frac{d\omega}{dt} \tag{1}$$

where I is the moment of inertia of the wheel, r is the radius of wheel. This radius can be the effective radius, taking into account of tyre deflection. Letting ε be the pre-rotation ratio, i.e., the percentage of landing speed v_0 being pre-rotated by the device, and letting t_s be the sliding time, also assuming the angular acceleration is uniform from static or a slow speed to the matching landing speed v_0 / r,

$$\frac{d\omega}{dt} = \frac{(1-\varepsilon)v_0}{rt_s}. \quad (2)$$

It is the friction between the ground and tyre that accelerates the wheel to a landing rotational speed within t_s by limited torque $F \cdot r$. The wheel needs a period of time to accelerate to ω_0 as the ground can only provide a limited, not unlimited, friction *F*. Therefore, a sliding is inevitable. The sliding time t_s = 0 only if the ground can provide infinitive friction to generate an infinitive acceleration from static to ω_0, which is impossible. During this sliding period the wheel is assumed being accelerated uniformly. A side wind turbine is suggested as in the patents to fit on the wheel, on which an air dynamic torque is generated to accelerate the wheel to achieve a tangential speed before the touchdown. After the touchdown, no energy or just smaller energy is required to accelerate the wheel to the matching speed $v_0 = r\omega_0$. From Eqs.(1) and (2)

$$F\,r - T_r = I\frac{(1-\varepsilon)v_0}{rt_s} \quad (3)$$

Therefore, the wheel sliding time is

$$t_s = \frac{(1-\varepsilon)v_0 I}{r(Fr - T_r)} \quad (4)$$

If the pre-rotation level $\varepsilon = 0$, $t_s = \frac{v_0 I}{r(Fr-T_r)}$; if $\varepsilon = 1$, $t_s = 0$. i.e. no sliding exists nor smoke will be generated. The sliding distance, or aircraft travelled distance during sliding is

$$s_s = v_0 t_s = \frac{(1-\varepsilon)v_0{}^2 I}{r(Fr - T_r)} \quad (5)$$

Since the wheel is in the state of sliding, the tangential motion of wheel at touch point pushed by ground friction s_r is less than s_s.

$$s_r = \frac{(1+\varepsilon)v_0}{2}t_s = \frac{(1-\varepsilon^2)v_0^2 I}{2r(Fr - T_r)} \quad (6)$$

The tangential displacement caused by the forces acted on the touch point during the whole sliding period ($0 \le t \le t_s$) is

$$s = v_0 t - \frac{(1+\varepsilon)v_0}{2}t = \frac{(1-\varepsilon)v_0}{2}t \quad (7)$$

The rate of work done by ground friction, or the friction power, is

$$\Delta\dot{E} = F\frac{(1-\varepsilon)v_0}{2} \quad (8)$$

It is the power density during the sliding generated by the ground friction on a single tyre. In the case of no pre-rotation, $\varepsilon = 0$, it yields $\dot{E} = \frac{Fv_0}{2}$. Assuming $F=1.449\times10^4$ *N* and $v_0=80.8$ *m/s*, $\Delta\dot{E}=585$ *kW*. This is very powerful figure. In the case of full pre-rotation, $\varepsilon = 1$, then $\dot{E} = 0$, i.e. the ground friction does not need to do any work on tyres. Using the displacement Eqs. (5) and (6), the ground friction work at the touch point is

$$\Delta E_g = F(s_s - s_r) = \frac{(1-\varepsilon)^2 v_0^2 IF}{2r(Fr - T_r)} \quad (9)$$

According to the first law of thermodynamics, the energy is conserved. This work will be totally converted to heat energy during the sliding. Using the public domain data from Boeing 747-400 aircraft without pre-rotation, the moment of inertia of real wheel $I=46.19$ *kg-m²*, the tire radius $r=0.622$ *m*, the landing velocity $v_0=80.8$ *m/s*, ΔE_g will be 433 kJ within the sliding time $t_s=0.74$ seconds. Most of it will raise temperature of the rubber near local contact area to its smoke point, generating a large quantity of smoke. The remaining part will be dissipated against the friction T_r, which is a relatively small number. A tyre is made up of carbon polymer. Primarily, the sliding between tyre and runway surface causes rubber loses. In addition, since the conduction of rubber is poor, this energy will be converted to heat to rise local area temperature at the tyre-ground interface significantly. As a result, the high temperature is to burn the tyre circumference layer to cause the tyre rapid wear and landing smoke. The rubber heat capacity is in average 0.48KJ/KgK. This energy can cause a high temperature up to 500 ^oC on the rubber beyond its ignition point. As a polymer, the flammability depends on oxygen supply and ignition source. Although self-ignition usually does not occur due to strong air flow, it has already entered a temperature range high enough to start chemical oxidization reaction on rubber, causing a large quantity of oxidization smoke. On the other hand, the wheel has gained energy through the sliding from part of the friction work. The kinetic energy gained by the whole wheel is

$$\Delta E_w = \frac{1}{2}(1-\varepsilon^2)I\left(\frac{v_0}{r}\right)^2 \quad (10)$$

In the case of zero pre-rotation, $E_w = \frac{1}{2}I\left(\frac{v_0}{r}\right)^2$. This energy will be eventually dissipated in the subsequent aircraft braking or in air and runway frictions. The total work done by ground friction in sliding is

$$\Delta E = \frac{(1-\varepsilon)^2 v_0^2 IF}{2r(Fr - T_r)} + \frac{1}{2}(1-\varepsilon^2) I \left(\frac{v_0}{r}\right)^2 \quad (11)$$

If without pre-rotation, $\varepsilon = 0$, the total energy will be $\Delta E = \frac{v_0^2 IF}{2r(Fr - T_r)} + \frac{1}{2} I \left(\frac{v_0}{r}\right)^2$. In the case that T_r is small and can be ignored, $\Delta E = I\left(\frac{v_0}{r}\right)^2$. i.e. in this case, the work done by the ground friction is the twice of the kinetic energy gained by the wheel during the sliding.

3. Pre-rotating torque before touchdown

After the landing gear is dropped and locked into place, with a pre-rotation device fitted, it is designed to produce a constant torque *T*, when the aircraft is still in air, Newton's law of motion, again, for the wheel is applicable:

$$T - T_a(\omega) = I\frac{d\omega}{dt} = \frac{I}{r}\frac{dv}{dt} \quad (12)$$

where $T_a(\omega)$ is the resistance torque from the shaft bearing due to wheel's aerodynamic drag and weight, and the viscosity from surrounding air. The task of pre-rotation torque *T* is to accelerate the wheel from 0 to $\varepsilon\ v_0 = \varepsilon\ r\omega_0$. Ideally, when v_0 is fully reached, the wind turbine torque is just balanced by the friction $T_a(\omega_0)$. Assuming the resistance torque applied on the wheel is non-linear $T_a(\omega) = \eta\omega^2$, from Eq.(12)

$$T - \eta\left(\frac{v}{r}\right)^2 = \frac{I}{r}\frac{dv}{dt}\ . \quad (13)$$

The above can be deformed to a different appearance as

$$\frac{dv}{dt} = \frac{r}{I}\left(T - \eta\frac{v^2}{r^2}\right) \quad (14)$$

This non-linear pre-rotation system in Eq.(14) can be solved by SIMULINK modelling in MATLAB as shown in Fig.2. Using the data from Boeing 747-400 aircraft, the moment of inertia of real wheel $I=46.19\ kg\text{-}m^2$. The tire radius $r=0.622\ m$, the landing velocity is $v0=80.8\ m/s$. Pre-rotating torques $T=200\ Nm$ and $500\ Nm$, respectively, are applied. In Fig.3, for a relative low pre-rotating torque, the wheel velocity has not reached the matching speed when the resistance has reached the maximum. In the case of $T=200\ Nm$, the final velocity of wheel can reach to 63% of the landing speed. i.e. the pre-rotation ratio $\varepsilon = 0.63$. The temperature raising energy ΔE_g will be dropped by 86%.

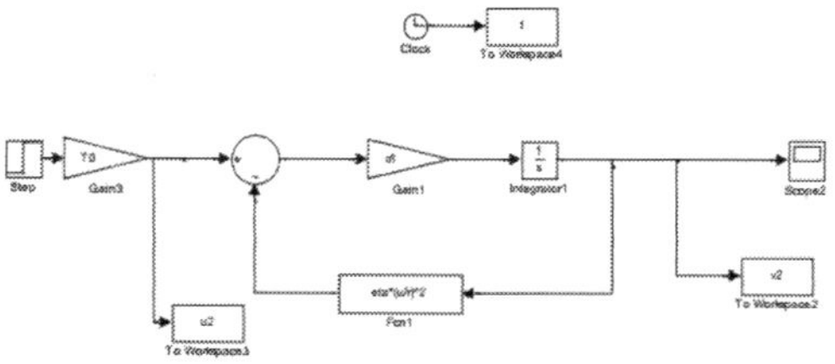

Figure 2. Simulink model for the motion during approaching with non-linear resistance torque

When applying a higher torque $T=500\ Nm$, the wheel will reach the matching speed 100% within approximately 35 seconds as shown in Fig. 4. Further increasing the torque will be over pre-rotated and is unnecessary.

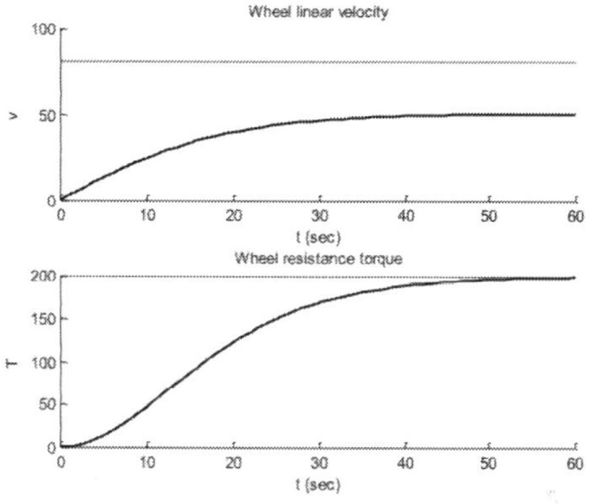

Figure 3. Wheel velocity reaches 63% of matching speed as resistance torque reaches the peak ($\eta=0.03$, $T=200\ Nm$)

Form Eqs (4-6), when the pre-rotation level is $\varepsilon = 1$, it can be seen $t_s = 0$, $s_s = 0$, and $s_r = 0$ In the full range of $0 \le \varepsilon \le 1$, i.e. including all the cases that the pre-rotation exists, does not exist more or less exits, t_s, s_s and s_r have been shown in Fig. 5. Both t_s and s_s are linear and monotonically declined to zero with the increase of ε. In addition, the friction work at the touch point, wheel kinetic energy increment and total friction work against pre-rotation ratio ε are shown in Fig. 6. It can be seen, if the pre-rotation is applied by even only 50%, the temperature raising energy caused by the friction work will be reduced by 75%.

4. Conclusion

The dynamic model for simulating landing sliding process is established to analyse the phenomenon of aircraft landing smoke and successive tyre wear. The sliding time, distance and friction work under different levels of pre-rotation are presented. Numerical examples with different pre-rotation torques are given. Ideally, a

100% pre-rotation ratio will cancel totally the landing smoke. A partial pre-rotation will have an enlarged effect on the friction heat reduction. The interpretation of the results show that the findings are reasonable and meaningful for understanding the dynamics and energy conversion during the landing, and also for helping the design of the pre-rotation devices. Up to date, the work represents a fundamental approach for understanding the landing smoke problem and an effective solution to it, which is novel and worth further investigation.

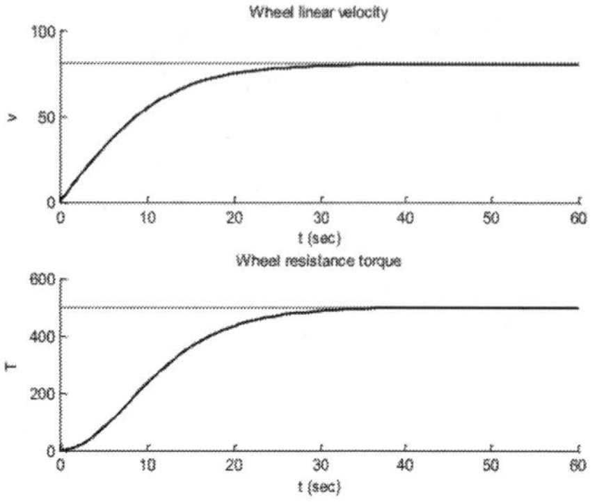

Figure 4. Wheel's tangential velocity has matched the landing speed as resistance torque reaches the maximum *(η=0.03, T=500 Nm)*

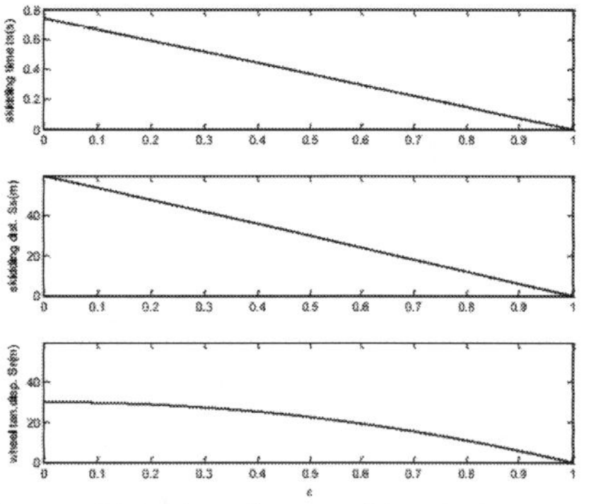

Figure 5. Sliding time, distance and the wheel tangential displacement at touch point.

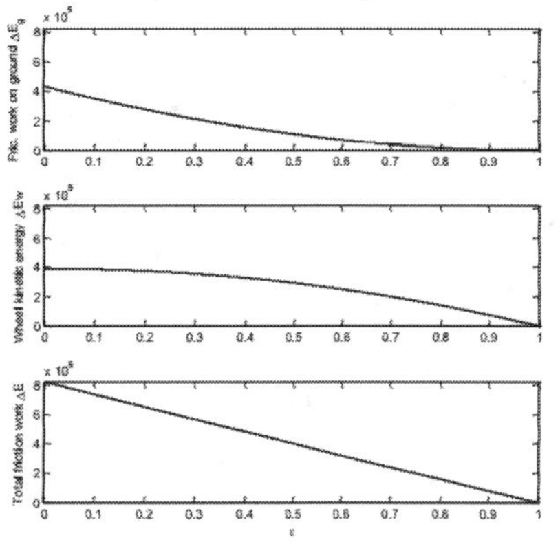

Figure 6. Friction work on the touch point, wheel kinetic energy increment and total friction work

References

1. Beazley, R. H. Aircraft wheel spinner and control. U.S. Patent, Publication No. US2414849 A. Washington, DC: U.S. Patent and Trademark Office. Retrieved from http://www.google.com/patents/US2414849 (1947).
2. Khal, S., & Khal, A. Apparatus for pre-rotating aircraft tires. U.S. Patent, Publication No. US20130112809A1, (2013).
3. Boeing Commercial Airplane Co. Approach speeds for Boeing airplanes. Retrieved from http://www.boeing.com/assets/pdf/commercial/airports/faqs/ arcandapproachspeeds.pdf, (2011).
4. Dahl, P. R. A solid friction model. The Aerospace Corporation. Technical report. Los Angeles: Los Angeles Air Force Station. Retrieved from www.dtic.mil/dtic/tr/fulltext/u2/a041920.pdf, (1968).
5. Daidzic, N. E., & Shrestha, J. Airplane landing performance on contaminated runways in adverse conditions. Journal of Aircraft, **45(6)**, 2131-2144, (2008).
6. Persson, B. Rubber friction: role of the flash temperature. Journal of Physics: Condensed Matter, **18(32)**, 7789-7823, (2006).
7. Tong, G., & Jin, X. Study on the simulation of radial tire wear Characteristics. WSEAS Transaction on Systems, **11(8)**, 419-429, (2012).
8. Hans B. Pacejka, Tyre and Vehicle Dynamics, Butterworth-Heinemann Ltd, **ISBN-10:** 0750669187.
9. Rajesh Rajamani, Vehicle Dynamics and Control, Springer.
10. Zuzana Koreňová, Mohammad Juma, et al., Experimental Study of waste Rubber Pyrolysis and Combustion, Petroleum & Cole, **47(2)**, 1-19, (2005).
11. Abdurrhman A. Alroqi and Weiji Wang, Comparison of Aircraft Tire Wear with Initial Wheel Rotational Spee, International Journal of Aviation, Aeronautics, and Aerospace, **2(1)**, (2015).
12. Abdurrhman A. Alroqi and Weiji Wang, Torque of Rotating Device Prior an Airplane Landing, International Journal of Engineering Research and General Science, **2(2)**, Feb-Mar (2014).

Adaptive Nonsingular Terminal Sliding Model Control and Its Application to Permanent Magnet Synchronous Motor Drive System

Yue Liu[1], Shuo Zhou[2]

[1]*School of Science, Shenyang Jianzhu University, Shenyang 110168, China*
[2]*Department of Mechanical and Electronic Engineering, Guidaojiaotong Polytechnic Institute,Shenyang 110023,China*

Abstract. To improve the dynamic performance of permanent magnet synchronous motor(PMSM) drive system, a adaptive nonsingular terminal sliding model control((NTSMC) strategy was proposed. The proposed control strategy presents an adaptive variable-rated exponential reaching law which the L1 norm of state variables is introduced. Exponential and constant approach speed can adaptively adjust according to the state variables' distance to the equilibrium position.The proposed scheme can shorten the reaching time and weaken system chatting. The method was applied to the PMSM speed servo system, and compared with the traditional terminal-sliding-mode regulator and PI regulator. Simulation results show that the proposed control strategy can improve dynamic, steady performance and robustness.

1 Introduction

Permanent magnetic synchronous motors have widely been used in AC driving system due to the high efficiency, high power density and good controllability. However, it is still a challenging problem to control the PMSM to get the perfect dynamic performance, because the dynamic model of PMSM is multivariable, strong coupling, nonlinear and parameters variable [1-5]. So the study on speed control of PMSM has attracted considerable attention among the scholars. Many different control methods have been proposed such as adaptive control, robust control, backstepping method, direct torque control, fuzzy logic control, etc [6-10].

In recent years, many scholars have devoted their research to applying sliding mode variable structure control (SMC) to PMSMs because SMC has good robustness to parameter perturbations, external disturbances and inaccurate mathematical description [11-15]. In [11], a variable parameter SMC was studied using reaching law approach. This method can improve dynamic and static performances, but chattering problem is not suppressed effective. A neuron-fuzzy sliding-mode controller was reported for PMSM control system in [12]. This method weakens the effects of switching control and chattering. In [13-14] a sliding mode observer (SMO) was proposed to estimate rotor position and velocity of PMSM using current and voltage of stator. But due to the presence of chattering, the output of the SMO needs to be passed through the LPF, which causes a time delay and requires extra compensation for the rotor position estimation. In [15] a terminal sliding mode controller of PMSM was designed for alleviating the chattering. However, the convergence rate and dynamic performance are not satisfied.

In this paper, an adaptive variable-rated exponential reaching law is proposed, where the L1 norm of state variables is introduced. In order to shorten the reaching time, exponential approach speed can be adjusted according to the state variables' distance to the equilibrium position. Constant approach speed can also be adaptively adjusted, which can weaken the system chattering. In order to verify the validity of proposed control strategy in PMSM, control effect will be compared with traditional terminal sliding mode method and PI regulator. Simulation and experimental results show that the proposed control strategy can improve dynamic,steady performance and robustness.

2 Performance analysis of adaptive nonsingular terminal sliding model control

Consider the following second-order nonlinear dynamic systems:

$$\begin{cases} \dot{x}_1 = x_2 \\ \dot{x}_2 = f(x) + b(x)u + g(t) \end{cases} \tag{1}$$

where $x = [x_1, x_2]^T \in R^2$ are state variables. $b(x) \neq 0$; $g(t)$ is an uncertain term, the uncertain part is assumed to be bounded as $|g(t)| \le l$, l is constant number;

u is input of system.For system (1), the conventional exponential reaching law is as follow:

$$\dot{s} = -\varepsilon \operatorname{sgn}(s) - ks, k > 0, \varepsilon > 0 \tag{2}$$

The reaching law contains two parts. One is the exponential approach $\dot{s} = -ks$. Another is constant approach $\dot{s} = -\varepsilon \operatorname{sgn}(s)$. In equation (2) , the coefficient k, ε do not have adaptive function. The convergence properties cannot achieve the best performance for different position state variables.In this paper, an adaptive variable-rated exponential reaching law is proposed as:

$$\dot{s} = -\varepsilon \frac{1}{1 + c\|x\|_1} \operatorname{sgn}(s) - (k + c\|x\|_1)s \tag{3}$$

where $\|x\|_1 = \sum_{i=1}^{n} x_i$ is the L_1 norm of state variables. $k > 0, \varepsilon > 0, c > 0, n > 0$.

The exponential and constant approach speed can adaptively adjust according to the state variables' distance to the equilibrium position. In equation (3) , the transient solution of exponential approach is $s = s(0)e^{-(k+c\|x\|_1)t}$ .The proposed exponential decay rate is faster than one's in equation (2) when $\|x\|_1$ is large. The reaching time can be shortened. The constant reaching law is $\varepsilon \frac{1}{1 + c\|x\|_1}$, which is much smaller than ε in equation (2). The reaching time can also be shortened by adjusting coefficient c when $\|x\|_1$ is small. At the same time the system chattering can be weakened. According to the nonlinear dynamic systems (1), a terminal sliding surface is proposed as:

$$s = x_1 + \frac{1}{\beta} x_2^{p/q} \tag{4}$$

The control law is designed as follow:

$$u = -b^{-1}(x)[f(x) + \beta \frac{q}{p} x_2^{2-\frac{p}{q}} + l \operatorname{sgn}(s) + \varepsilon \frac{1}{1 + c\|x\|_1} \operatorname{sgn}(s) + (k + c\|x\|_1)s] \tag{5}$$

where $p, q \in N$ are the odd number, $1 < p/q < 2$. $\beta > 0$, $l > 0$, $k > 0$, $\varepsilon > 0$, $c > 0$.

In order to prove the stability of the scheme presented in (5), the Lyapunov function candidate V is defined as

$$V = \frac{1}{2} s^2 \tag{6}$$

The time derivative of (6) is given by

$$\dot{V} = s\dot{s} = s(x_2 + \frac{1}{\beta} \frac{p}{q} x_2^{\frac{p}{q}-1} \dot{x}_2) = \frac{1}{\beta} \frac{p}{q} x_2^{\frac{p}{q}-1} \{s[g(t) - l \operatorname{sgn}(s)] - \varepsilon \frac{1}{1 + c\|x\|_1} |s| - (k + c\|x\|_1 s^2)\} \le \frac{1}{\beta} \frac{p}{q} x_2^{\frac{p}{q}-1} [-\varepsilon \frac{1}{1 + c\|x\|_1} |s| - (k + c\|x\|_1) s^2] \le 0$$

Therefore, the reaching condition $s\dot{s} < 0$ will be satisfied. The system can converge to the equilibrium in finite time.

In order to analysis performance of proposed controller, taking (7) as a typical system expressed as follows:

$$\begin{cases} \dot{x}_1 = x_2 \\ \dot{x}_2 = -25x_2 + 133u + 0.1\sin(20t) \end{cases} \tag{7}$$

The initial condition of system is $x^{\mathrm{T}}(0) = [10, 10]^{\mathrm{T}}$. The result of comparison between proposed control method and traditional terminal sliding mode regulator is shown in Fig. 1.The convergence speed of proposed method is faster significantly than the traditional terminal sliding mode regulator's.

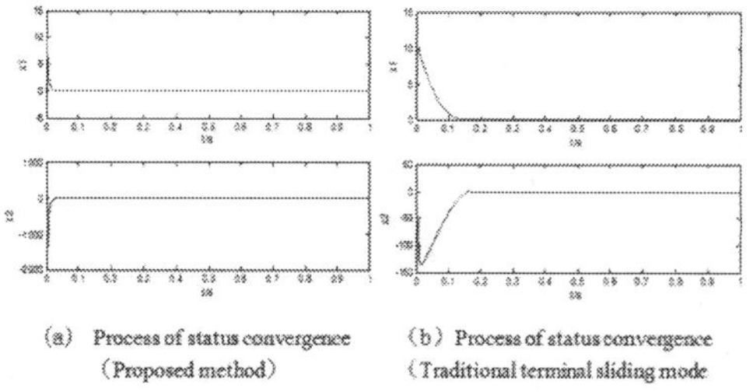

(a) Process of status convergence (Proposed method)
(b) Process of status convergence (Traditional terminal sliding mode
(c) Phase trajectory (Proposed method)
(d) Phase trajectory (Traditional terminal sliding mode method)

Figure 1. Performance comparisons of proposed method and traditional terminal sliding mode method.

3 Application of adaptive nonsingular terminal sliding model control method on PMSM

3.1 PMSM modeling

Assume that magnetic circuit is unsaturated, hysteresis and eddy current loss are ignored and the distribution of the magnetic field is sine space. In d-q coordinates the model of PMSM can be expressed as follows:

$$\begin{cases} u_d = ri_d - \omega L_q i_q + \dfrac{d\psi_d}{dt} \\ u_q = ri_q + \omega L_d i_d + \dfrac{d\psi_q}{dt} \\ T_e = \dfrac{3}{2} p \psi_f i_q \\ T_e - T_L = \dfrac{J}{p}\dfrac{d\omega}{dt} \end{cases} \tag{8}$$

Flux equations are as follows:

$$\begin{cases} \psi_d = L_d i_d + \psi_f \\ \psi_q = L_q i_q \end{cases} \tag{9}$$

where u_d, u_q represent d and q axes stator voltages respectively; i_d, i_q are d and q axes currents respectively; L_d, L_q are d and q axes stator inductance; r is stator resistance; p is number of pole pairs; ω is electrical angular velocity; J is rotational inertia; T_e is electrical magnetic torque; T_L is load torque; ψ_f permanent magnet flux.

3.2 Design of NTSMC speed controller

State variables of PMSM are as follows:

$$\begin{cases} x_1 = \omega^* - \omega \\ x_2 = \dot{x}_1 = -\dot{\omega} \end{cases} \tag{10}$$

where ω^* is reference velocity, ω is actual velocity. According to (8) and (10), state variables can be described as follows:

$$\begin{cases} \dfrac{dx_1}{dt} = -\dfrac{d\omega}{dt} = -\dfrac{p}{J}(\dfrac{3}{2} p \psi_f i_q - T_L) \\ \dfrac{dx_2}{dt} = -\dfrac{d^2\omega}{dt^2} = -\dfrac{3p^2}{2J}\psi_f \dfrac{di_q}{dt} \end{cases} \tag{11}$$

Considering the uncertain term $g_1(t)$ of PMSM, the differential equation (11) is rewritten as

$$\begin{cases} \dfrac{dx_1}{dt} = -\dfrac{d\omega}{dt} = -\dfrac{p}{J}(\dfrac{3}{2} p \psi_f i_q - T_L) \\ \dfrac{dx2}{dt} = -\dfrac{d^2\omega}{dt^2} = -\dfrac{3p^2}{2J}\psi_f \dfrac{di_q}{dt} + g_1(t) \end{cases}$$

$$\Rightarrow \begin{cases} \dot{x}_1 = x_2 \\ \dot{x}_2 = -bu + g_1(t) \end{cases} \tag{12}$$

where $b = \dfrac{3p^2}{2J}\psi_f, u = \dfrac{di_q}{dt}$, $|g_1(t)| \le l_1$.

The sliding surface is considered as

$$s_1 = x_1 + \frac{1}{\beta_1} x_2^{p_1/q_1} \tag{13}$$

The control input i_q is designed as follows:

$$i_q = \frac{2J}{3p^2\psi_f}\int_0^t \beta_1 \frac{q_1}{p_1} x_2^{2-\frac{p_1}{q_1}} + l_1 \operatorname{sgn}(s_1) +$$

$$\varepsilon_1 \frac{1}{1 + c_1\|x\|_1}\operatorname{sgn}(s_1) + (k_1 + c_1\|x\|_1)s_1]dt \tag{14}$$

The block diagram of PMSM speed-regulation system is shown in Fig. 2.

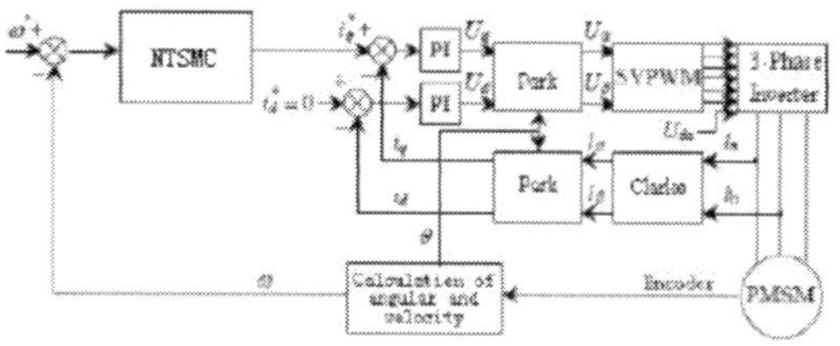

Figure 2. The block diagram of PMSM speed-regulation system.

4 Simulation results

To demonstrate the effectiveness of the proposed approach, simulations of proposed method, the PI method and the traditional terminal sliding mode method in one PMSM system were made. The related parameters of PMSM are shown in Table 1.

Table 1. Parameters of PMSM

Parameter	Value
Stator Resistance (r)	2.46 Ω
Stator Inductor(L_d , L_q)	6.35 mH
Flux (ψ_f)	0.175 Wb
Rotor Inertia (J)	1.02e-3 $kg \cdot$ m2
Poles (p)	4
Rated speed (ω^*)	3000 $r/\min$

The comparison curve of speed response among adaptive nonsingular terminal sliding model regulator, PI regulator and traditional terminal sliding mode regulator are showed in Fig. 3. From the simulation results, it can be observed that PI controller has the fluctuation and overshoot obviously. The response time of traditional terminal sliding mode controller is longer than proposed

DOI: 10.1051/matecconf/20167701039

controller. The response curve of speed, load torque and current are be shown in Fig. 4 (a)-(c) when load torque is added suddenly form 1 $N \cdot m$ to 7 $N \cdot m$ at $t = 0.05s$ .It can be observed that the proposed method has a smaller speed fluctuation and a shorter settling time compared with the PI method when the load torque changes. Moreover, the dynamic responses of speed and current using PI method are much sensitive to loading.

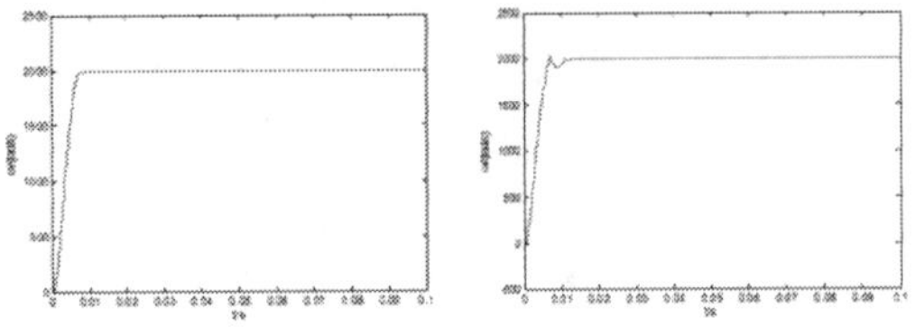

(a) Proposed method (b) PI method

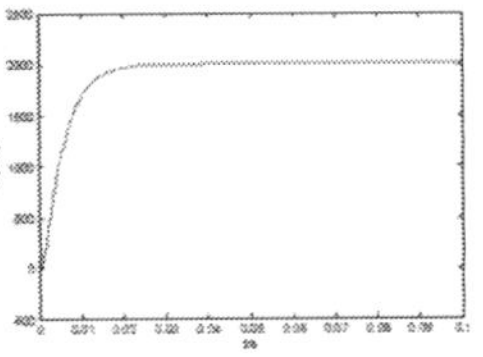

(c) Traditional terminal sliding mode method

Figure 3. Speed response when system starting

5 Conclusion

In this paper, an adaptive variable-rated exponential reaching law is presented which the L1 norm of state variables is introduced. Exponential and constant approach speed can adaptively adjust according to the state variables' distance to the equilibrium position. The proposed scheme can shorten the reaching time and weaken system chatting. The simulation results demonstrate the effectiveness of the proposed method.

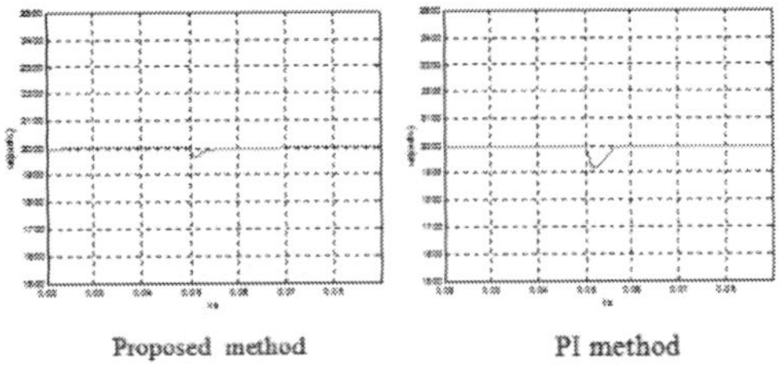

Proposed method PI method

(a) Speed response adding load suddenly

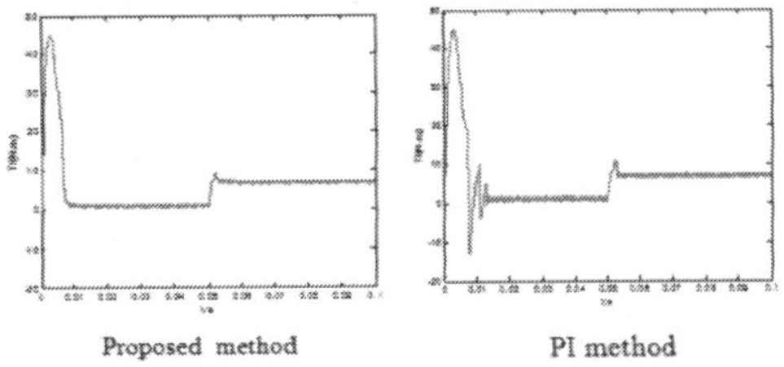

Proposed method PI method

(b)Torque response adding load suddenly

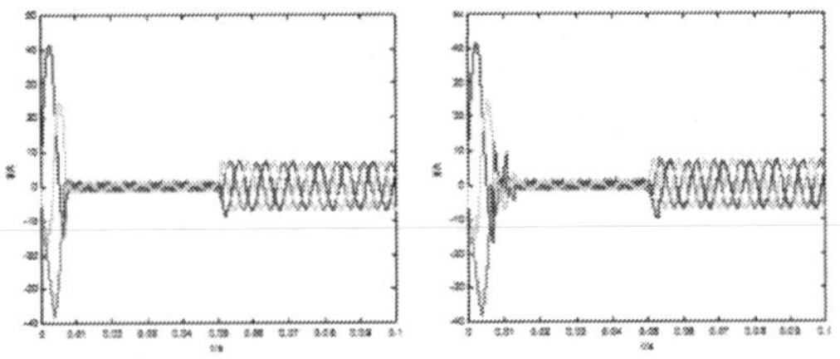

Proposed method PI method

(c)Current response adding load suddenly

Figure 4. Simulation results of system dynamic response to loading

References

1. Sun Qiang, Cheng Ming, Zhou Er, Proceedings of the CSEE, **23,6** (2003)
2. Wang Jiang,Wang Jing,Fei Xiangyang, Proceedings of the CSEE. **25**, 7, (2005)
3. Uddin M N,Abido M A,Rahman M A, IEEE Transactions on Industry Applications, **40**, **1**, (2004)
4. WANG Baoren,ZHANG Chengrui,JIA Lei, Electric Machines and Control,**11**,**3**, (2007)
5. Miller C E, Van Zyl A W, Landy C F, IEEE Africon Conference(2002)
6. QIU Jun,ZHAO Guang-zhou, Electric Machines and Control, **13**, 6(2009)
7. YANG Shu-sheng, ZHONG Yi-sheng, Proceedings of the CSEE, **29**,3(2009)
8. ZHOU Guo-rong, LI Yong-feng, Control Engineering of China,**16**,1(2009)
9. Yongchang Zhang,Jianguo Zhu, IEEE Transactions on Power Electronics,**26**,1(2011)
10. Stone Cheng,Chi-Wei Li, Control Engineering Practice, **19**,4(2011)
11. Wang Haibo,Zhou Bo,Fang Si-chen, Transactions of China Electrotechnical Society, **24**,9(2009)
12. Cetin Elmas, Oguz Ustun, Control Engineering Practice, **16**,6(2008)
13. SHANG Zhe, ZHAO Rong-xiang, DOU Re-zhen, Proceedings of the CSEE, **27**,3(2007)
14. Liu Jun, Wang Gang, Yu JinShou, Proceedings of the 8th World Congress on Intelligent Control and Automation, (2010)
15. WANG Yan-min, FENG Yong, LU Qi-liang, Electric Machines and Control, **12**, 5(2008)

MATEC Web of Conferences 77, 01040 (2016) DOI: 10.1051/matecconf/20167701040
ICMMR 2016

A Piezo-Driven Compliant Nanopositioning Stage with Large Stroke for Micro/Nano Manipulation

Xiaobo Zhu, Zhijie Wen and Pinkuan Liu

State Key Laboratory of Mechanical System and Vibration, School of Mechanical Engineering, Shanghai JiaoTong University, Shanghai 200240, People's Republic of China

Abstract. This paper presents the design, analysis, and testing of a single-axis compliant nanopositioning stage with large stroke driven by piezoelectric actuator for micro/nano manipulation. In the developed stage, a hybrid displacement amplifier integrating two kinds of flexure-based input/output decoupling modules have been developed to achieve a large stroke and precise positioning. The analytical modelling and finite element analysis are given to validate the mechanical performance and a prototype of the proposed stage is fabricated for performance tests. The experimental results show that the developed stage has the travel stroke of 214 μm with a resolution of 8 nm. It is demonstrated that the stage can be adequate for micro/nano manipulation.

1 Introduction

Recently, compliant nanopositioning stages with a large stroke are frequently required in different fields of micro/nanomanipulation.[1, 2] Meanwhile, piezoelectric (PZT) actuators with nanometer resolution, excellent dynamic characteristics, small size, and high driving force have been usually employed in the nanopositioning stages.[3, 4] However, the drawback of the piezoelectric actuator is its limited travel range which are not suitable for requiring a large workspace.

Consequently, a lever-type flexure-based displacement amplifier is widely applied to amplify the limited stroke of PZT in many compliant nanopositioning stages.[5] While is can easily induce parasitic motion errors.[6] Another type of commonly used amplifier refers to bridge-type amplifier, which has compact size, symmetric structure and sufficient amplification ratio.[4, 7] Unfortunately, the nanopositoning stages based on bridge-type amplifier have relatively poor dynamics. In addition, duo to a lateral displacement at the driving end of the stage increases with the increasing of the input displacement, a decoupling mechanism is needed for protect PZT from the lateral forces and bending torques.[8] Therefore, it is a challenge working to design a compliant nanopisitioning stage possessing both low cross-coupling and large stroke simultaneously.

In this paper, a novel compliant piezo-driven nanopositioning stage with large stroke is proposed. A hybrid amplifier, which is composed of double flexure-based amplifers, is developed to achieve a large stroke. In contrast with the commonly amplifier, the novel amplifier is characterized by compact size with large amplification ratio, high dynamics and low cross-coupling.

Meanwhile, two types of compliant decoupling modules are developed to reduce the input and output couplings.Based on the configration of the stage, the analysis models and experimental tests validate that the developed stage achieved the large stroke with nanoscale resolution, low input coupling and compact size. It is demonstrated that a considerable potentical of the stage for the applications in various micro/nano manipulations.

2 Mechanical design

The novel compliant nanopositioning stage is shown in Fig.1. It is composed of a symmetrical flexure-based input decoupling module(ABCD), a hybrid amplifier(EFGH), a flexure-based motion guide module(IJKL), and a PZT actuator. Corner-filleted flexure hinges are used in the stage .The PZT actuator adopted in the developed stage can generate a nominal displacement of up to 38μm at the max mum drive voltage of 100 V, and can provide a maximum driving force of 1000 N.

As shown in Fig.1(b), a hybrid-type amplifier, which is a combination of lever-type and bridge-type amplifier, has a compact size with a large amplification ratio and high dynamics. Meanwhile, duo to the symmetric configuration, the amplifier can eliminate the lateral offset error of the output end and gurantee a precise linear output. A couple of input decoupling modules are both mounted between two sides of the PZT. It can reduce the lateral movement perpendicular to the PZT, and protect the PZT from lateral forces and bending torques. In addition, a motion guide module is integrated with the

 DOI: 10.1051/matecconf/20167701040

moving platform can further ensure a precise positioning of the output end.

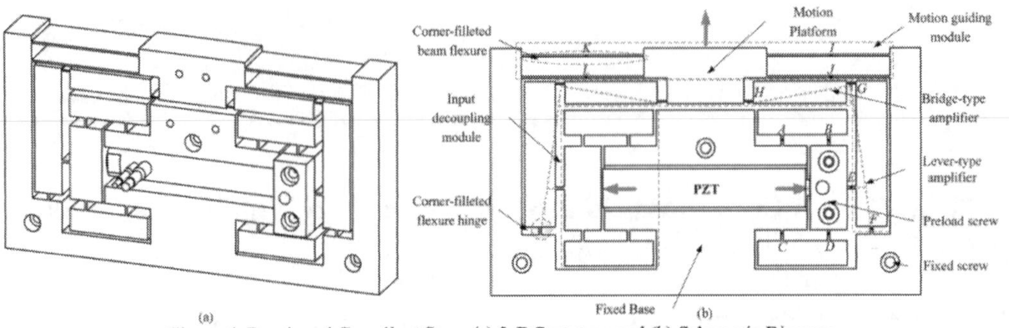

Figure 1. Developed Compliant Stage (a) 3-D Prototype and (b) Schematic Diagram

The motion of the nanopositioning stage is described as follows. When the PZT exerts on the stage, it results in the horizontal displacement of the input decoupling module, which is subsequently amplified by the lever-type and bridge-type amplifiers (EFGH) , and in turn, to prompt a precise linear displacement of the moving platform by the motion guide module (IJKL).

3 Modeling and analysis

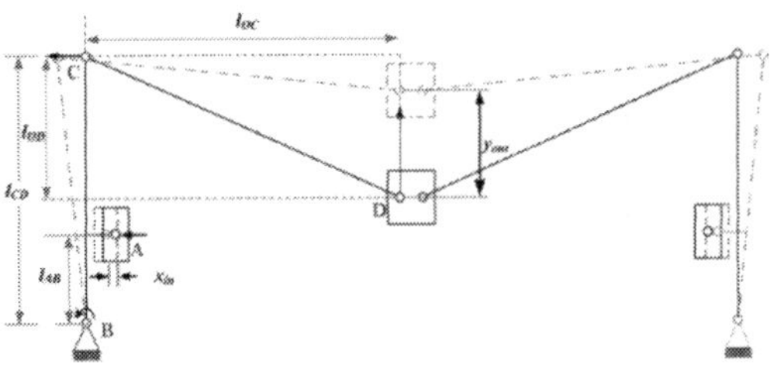

Figure 2. Kinematics of the developed compliant stage

The fundamental structure of the nanopositioning stage is a special five-bar linkage mechanism as illustrated in Fig.2. The input and output displacement are resented by x_{in} and y_{out}, respectively. The theoretical displacement amplification radio R_{amp} of the DAM using geometric relations is defined as:

$$R_{amp}=\frac{2\left(l_{OD}-\sqrt{l_{OD}^{2}-\left(\frac{l_{BC}}{l_{AB}}x_{in}+2l_{OC}\right)\frac{l_{BC}}{l_{AB}}x_{in}}\right)}{x_{in}} \tag{1}$$

According to Eq. (7), it can be seen that the stage has a constant amplification radio, which only depends on the geometric parameters of l_{AB}, l_{BC}, l_{OC} and l_{OD}, but has no relations with flexure elements.

The static model of the compliant nanopositioning stage is based on the strain energy method.[9] For Castigliano's second theorem, the total strain energy of the developed stage is

$$U=\sum_{i=1}^{10}\left(\int_0^{li}\frac{N_i^2}{2EA}dx_i\right)+\sum_{i=1}^{10}\left(\int_0^{li}\frac{M_i^2}{2EI}dx_i\right)+\sum_{i=1}^{10}\left(\int_0^{li}\frac{\alpha S_i^2}{2GA}dx_i\right) \tag{1}$$

where E is Young's modulus; A is the cross-section area of the flexure hinges; I is the inertia moment; G is the shear modulus. The normal force N_i , shearing force S_i and bending moment M_{bi} can be expressed by static equilibrium.

For strain energy method, the input displacement x_{in}, output displacement x_{out}, input stiffness of the stage can be obtained as, respectively

$$x_{in}=\frac{\partial U}{\partial F_{in}}=\frac{\delta}{\delta F_{in}}\left|\sum_{i=1}^{10}\int_0^{l_i}\left[\frac{N_i^2}{2EA}+\frac{M_{bi}^2}{2EI}+\frac{\alpha S_i^2}{2GA}\right]dx_i\right. \tag{2}$$

$$x_{out}=\frac{\partial U^*}{\partial F_{11y}^*}=\frac{\partial}{\partial F_{11y}^*}\Bigg|_{F_{11y}^*=0}\left(\sum_{i=8}^{10}\int_0^{l_i}\left[\frac{N_i^{*2}}{2EA}+\frac{\alpha S_i^{*2}}{2GA}+\frac{M_{bi}^{*2}}{2EI}\right]dx_i\right) \tag{3}$$

$$k_{in}=F_{in}/x_{in} \tag{4}$$

It is known that a piezo actuator is compressed when a force is applied. Hence, the stiffness of the PZT and the stage must be taken into account when deducing the maximum output displacement of the stage and can be expressed as

$$x_{act}=\frac{k_{pzt}}{k_{pzt}+k_{in}}x_{pzt}R_{amp} \tag{5}$$

where k_{pzt} is the stiffness of the PZT, x_{pzt} is the maximum nominal displacement of the PZT.

 DOI: 10.1051/matecconf/20167701040

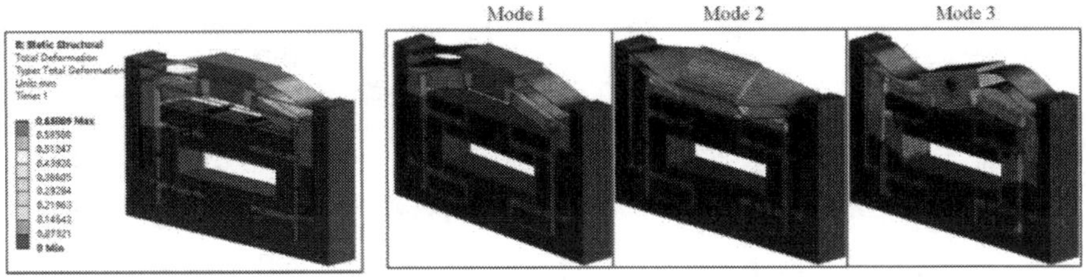

(a) (b)

Figure 3. FEA results of (a) Deformation Behaviour and (b) Mode Shapers from 1st to 3rd

Fig. 3 shows the FEA results of the developed stage by commercial software ANSYS. The stage can generate an output displacement of 208 μm with the input displacement of 18 μm as shown in Fig.3 (a) Thus, it can be deduced that the proposed stage has a displacement amplification ratio of 12.1. And it can be inferred that the input stiffness of the stage is 28.2N/μm and the output stiffness of the stage is 2.45N/μm. In addition, the relevant maximum stress of 52 MPa is much smaller than the yielding stress of the material of aluminum alloy 7075. Fig.3 (b) demonstrates that the first three natural frequencies of the proposed stage are 202 Hz, 1161 Hz and 1410 Hz, repectively.

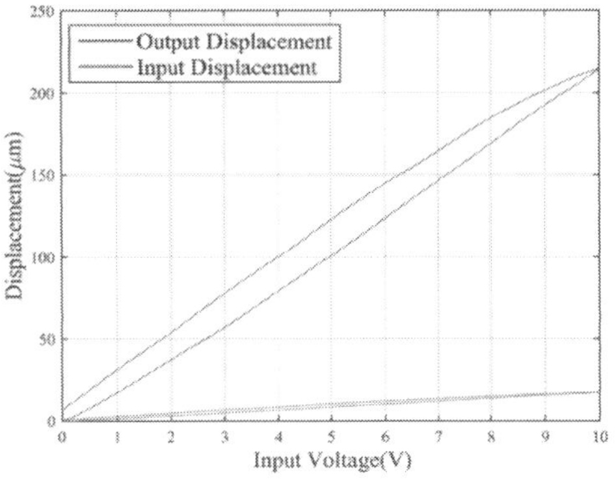

(a)

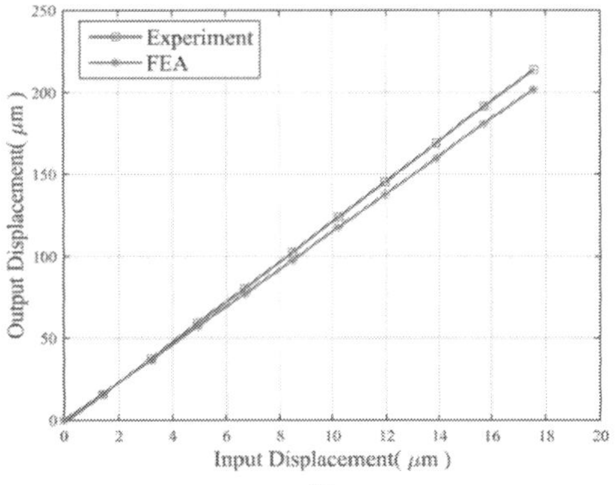

(b)

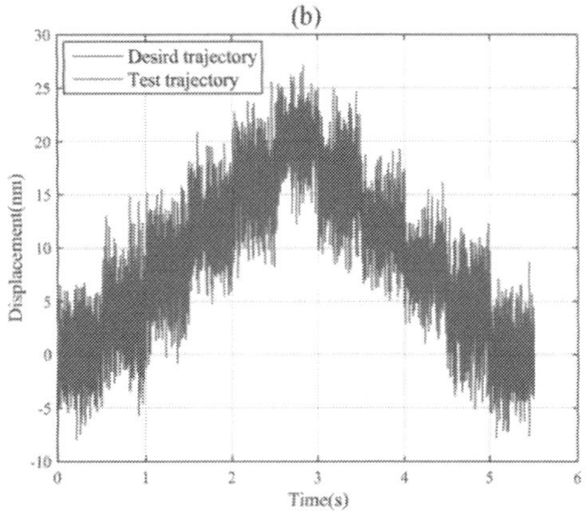

(c)

4 Experiment

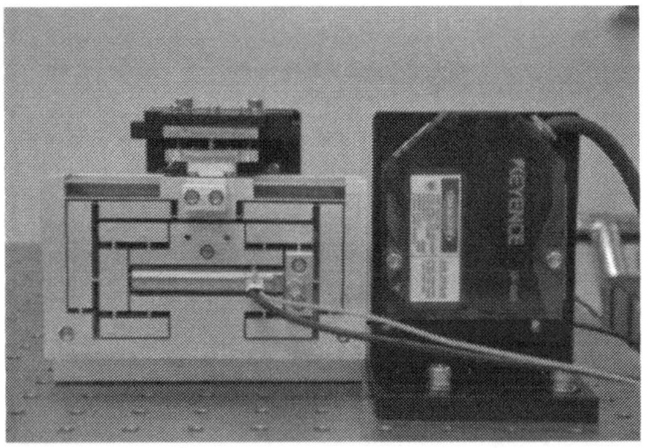

Figure 4. The Manufactured Nanopositioning Stage

Fig. 4 shows that a prototype of the proposed stage with compact size of 92×18×50 mm, which is fabricated using wire electrical discharge machining (WEDM) technique with Aluminum 7075. The positioning stage is driven by a piezo actuator (model: P-841.30, from Physik Instrument, Inc.) with a muti-channel PZT amplifier (model: E-501 from Physik Instrument, Inc.) providing driven voltage (0–100V) for the PZT. A capacitive sensor (model: 4810, from MicroSense) is integrated with the stage as a position feedback. A compactRIO 9022 from National Instrument Co. equipped with a 16-bit ADC/DAC is applied to carry out a real-time control for the proposed stage.

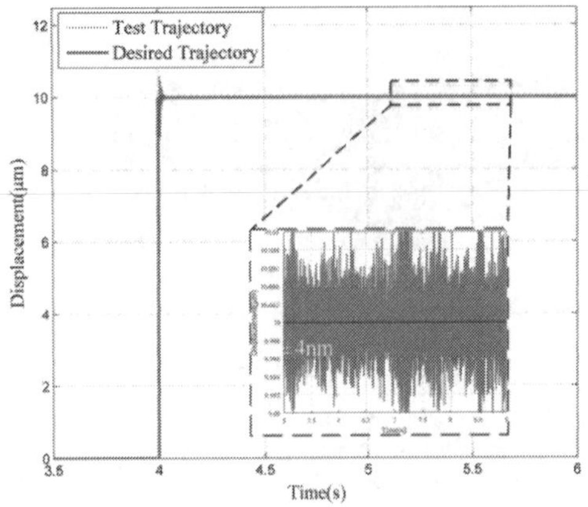

(d)

Figure 5. The Test Results of the Proposed Stage (a)Displacement for Input Voltage from 0 to 100V, (b) Amplification Ratio, (c) Resolution of 8 nm and (d) Stepwise Response of 10μm

As shown in Fig.5(a), when a 100 V voltage signal is provided to drive the piezoelectric actuator, the maximum input displacement of the stage is approximately 214 μm with respect to the input displacement of 17.7μm.The effective amplification ratio of 12.1 is shown in Fig.5(b).

Meanwhile, in close-loop positioning performance test, a minimum resolution of 8 nm is clearly attained as shown in Fig.5(c). Besides, Fig.5(d) show s a step response of 10 μm, which demonstrats that the stage has a rapid transient response (less than 30ms) and steady state error of ±4 nm.

5 Conclusion

A pizeo-driven compliant PZT-driven nanopositioning stage with large stroke is developed in this paper. To design an optimized prototype, we establish the kinematic model and perform the FEA validation. Finally, a prototype of the stage is fabricated to carry out the performance test of the stage. The test results show that the proposed stage with a minimum resolution of 8 nm has a high amplification ratio of 12.1 and a maximum displacement of 214 μm. It demonstrated that the proposed compact stage is capable of utilizing in the field of micro/nano manipulations.

References

1. C.-L. Chu, S.-H. Fan, A novel long-travel piezoelectric-driven linear nanopositioning stage, Precision Engineering, **30** (2006) 85-95.
2. P. Gao, S.-M. Swei, Z. Yuan, A new piezodriven precision micropositioning stage utilizing flexure hinges, Nanotechnology, **10** (1999) 394.
3. Y. Yong, S. Moheimani, B.J. Kenton, K. Leang, Invited review article: High-speed flexure-guided nanopositioning: Mechanical design and control issues, Review of scientific instruments, **83** (2012) 121101.
4. J.H. Kim, S.H. Kim, Y.K. Kwak, Development of a piezoelectric actuator using a three-dimensional bridge-type hinge mechanism, Review of scientific instruments, **74** (2003) 2918-2924.
5. S. Choi, S. Han, Y. Han, B. Thompson, A magnification device for precision mechanisms featuring piezoactuators and flexure hinges: Design and experimental validation, Mechanism and Machine Theory, **42** (2007) 1184-1198.
6. J.-J. Kim, Y.-M. Choi, D. Ahn, B. Hwang, D.-G. Gweon, J. Jeong, A millimeter-range flexure-based nano-positioning stage using a self-guided displacement amplification mechanism, Mechanism and Machine Theory, **50** (2012) 109-120.
7. Q. Xu, Y. Li, Analytical modeling, optimization and testing of a compound bridge-type compliant displacement amplifier, Mechanism and machine theory, **46** (2011) 183-200.
8. S. Awtar, A.H. Slocum, Constraint-based design of parallel kinematic XY flexure mechanisms, Journal of Mechanical Design, **129** (2007) 816-830.
9. N. Lobontiu, Compliant mechanisms: design of flexure hinges, CRC press, 2010.

The Design and Research of the Operation Status Detector for Marine Engine Room Power Plant Based on Noise

Hang Li, Guangdong Han and Wenhua Li

Marine Engineering College, Dalian Maritime University, Dalian, China

Abstract. Designed in this paper, based on the noise of ship engine room power plant running status of detector, is mainly used in the operation of the power plant of acoustic shell size to determine when the machine running state, this device is composed of signal disposal and alarm display adjustment part of two parts.Detector that can show the size of the voice, if exceed the set limit alarm value, the detector can sound an alarm, to remind staff equipment fails, it shall timely inspection maintenance, improve the safety of the operation of the ship.

1 Introduction

In the most important part of the detector is the part of signal conditioning, signal conditioning circuit module at present widely used in industrial control, instrumentation, etc of sensor signal detection and processing applications. It is A sensor and A/D converter device such as A follow-up of the interface. It is the sensor output very weak and low level signal is amplified by A variety of noise interference, processing, noise elimination and unified format, making it suitable for A/D converter device such as A subsequent request signal. For different object to be tested and used by many different kinds of sensors, the signal conditioning circuit is put forward different requirements. The follow-up device such as A/D converter combining with technical indicators to design A good performance of signal conditioning circuit has become the key content of system design. With noise sensor as an example, this paper discusses the design method of the corresponding signal conditioning circuit. [1]

Strong interference resistance, noise in a device measuring the noise decibels are vulnerable to the surrounding noise sound, this would require the receiver noise has voice recognition function, but it is not so, according to the design is too complicated, actually this problem with a low resolution microphone microphones can solve, economical and practical to do so.

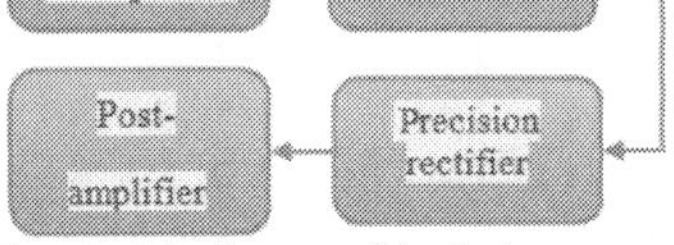

Figure 1. The principle diagram of the device

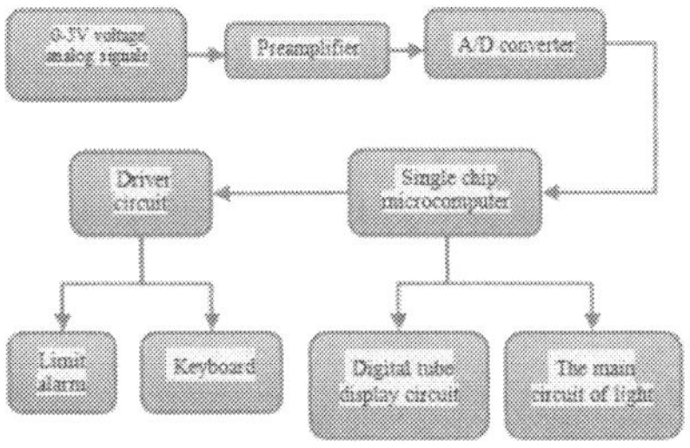

Figure 2. The principle diagram of the alarm display control part

2 The basic working principle

Introduction of the principle:

Noise in front of the microphone microphones will environment into a faint millivolt level analog voltage signals, the voltage signal through the front part stage amplifier amplification, enlarged by precision rectifier circuit, the AC voltage signal is converted into DC signal, the signal again after the amplifier section, zoom into 0 to 5V analog voltage signal, and then sending the signal to alarm display part adjustment.Can display the alarm display section size, if more than upper limit alarm value can sound an alarm, to remind staff equipment failure, should be timely check the maintenance, the range and the zero point can be adjusted through the potentiometer.[2]

3 Unit analysis

3.1 Signal conditioning part

This part by stage amplifier, precision rectifier part before and after the amplifier components. Detector receiver microphones will be in the environment noise into a faint millivolt level analog voltage signals, the voltage signal through the signal conditioning part into 0 to 5V analog voltage signal, and the analog alarm display adjustment to the section.

3.1.1 Before the amplifier application

Stage amplifier part before the electret microphone acquisition of speech signal, convert its signal amplitude is millivolt level of electrical signals, system level before the treatment, as far as possible improve signal-to-noise ratio at the input to the amplifier, guarantee the gain amplifier circuit with precise, stable. Is using the operational amplifier LM386 stage amplifier circuit diagram (see Fig. 3), 2 for inverting input pin, 3 for the same phase input pin, pin 5 for output, pin 6 and 4 respectively the power and ground, pin 1 and 8 set for voltage gain, when used in pin 7 and indirect bypass capacitors in the land. In indirect capacitance of pins 1 and 8 (pin 1 pick capacitance is level) to change the gain. [3]

Capacitor C3 is pin 7 bypass capacitors, it can have the effect of filter out noise;

Resistance of R3 and resistor R1, bleeder circuit for MIC relatively stable power supply;

Capacitor C5 is the output coupling capacitor, the capacitor is used to partition the dc voltage and coupling of audio communication signals, it constitutes the first-order high-pass filter and load;

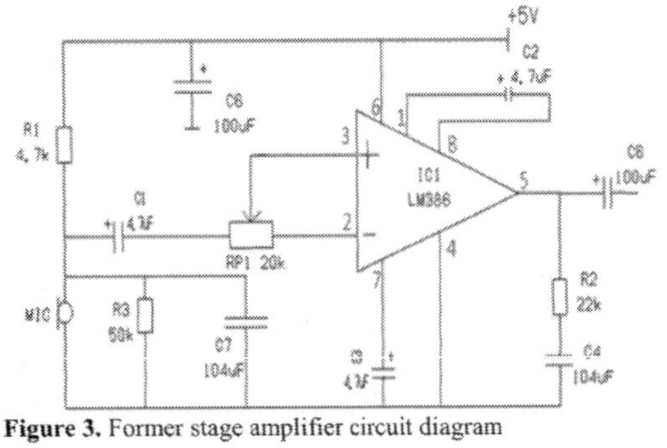

Figure 3. Former stage amplifier circuit diagram

3.1.2 Precision rectifier part

To convert AC signal into DC signal must adopt rectifier, the conventional method is to use diode rectifier directly, but due to the diode rectifier threshold voltage (0.3V or 0.7V), small signal rectifying error is big, can't even work.To weak alternating current (AC) is converted into unidirectional pulse of electricity, it is important to adopt precision rectifier, so the circuit must be composed by op-amp and diode to achieve precision rectifier.

In this section we adopt is the operational amplifier LM358 precision rectifier circuit (see Fig. 4), and its internal consists of two independent, high gain, internally frequency compensated dual operational amplifier, two channels to deal with the positive and negative half cycle, regardless of the input is a positive or negative, the output is positive voltage.

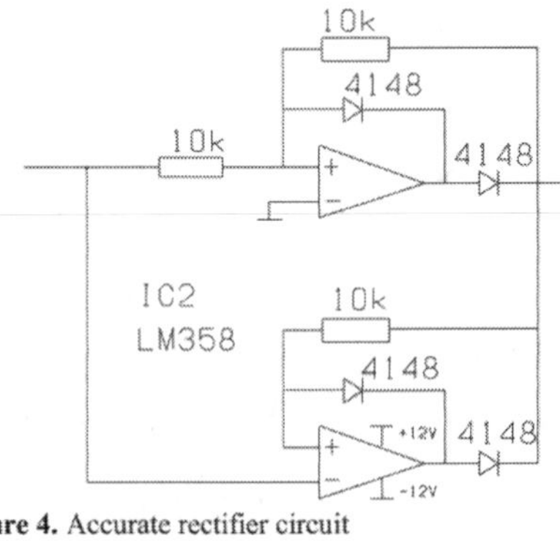

Figure 4. Accurate rectifier circuit

3.1.3 After the amplifier application

After stage amplifier adopts differential amplifier to achieve the required drive ability, its advantage lies in the differential amplifier can get higher common mode rejection ratio, strengthen the anti-interference circuit. [4]

As shown in Fig. 5 the amplifier circuitry, difference amplifier differential amplifier AD620 was adopted to realize. Internal is actually a difference amplifier AD620, the small disturbance voltage current, common mode rejection ratio is high, so deal with weak signal, which is amplified and eliminate noise, excellent performance, it can gain by adjusting the resistance between pins 1 and 8, its gain G = 1 + (49.4 k Ω/Rg) (Rg in the resistance between pins 1, 8).

AD620 of 6 pins at the back of the part is the conventional RC low-pass filter.R304 and R305 can be considered a current limiting function, can also be considered to be the separation.

OP07 precision op-amp is zero reference source, mainly serving the AD620, provide a reference for AD620 voltage, OP073 pin number at the back of the part for precision voltage regulator, to OP07 provide a stable source of unknown to zero.

Through the above three parts, we can send the electret microphone acquisition of speech signal is converted into 0 to 5V analog voltage signal, and then send it to display debug apparatus.

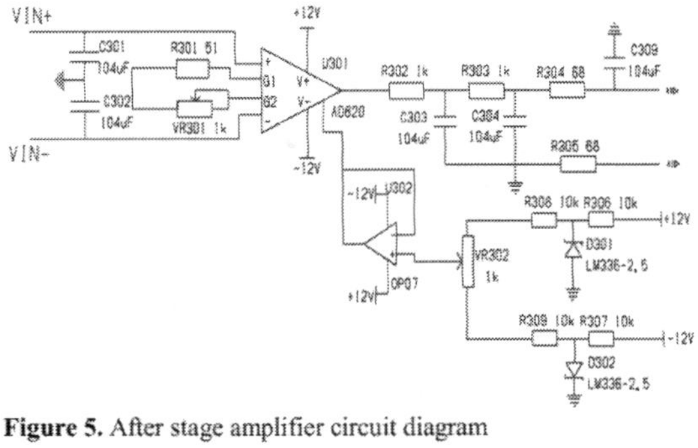

Figure 5. After stage amplifier circuit diagram

4 Alarm display control part

This part of the signal conditioning module output of 0 to 5V voltage signal is converted to a digital signal, and through the SCM displayed on the digital tube display, which shows the size of the displayed the voice of the environment, that part of the range and alarm value can be set through key artificially, depending on the equipment we can by setting the corresponding value to monitoring equipment, when sound is more than the alarm value set, will sound an alarm, to remind staff, equipment failure, should be lifted. [5]

That part of the physical diagram as shown in Fig. 6 and Fig. 7.

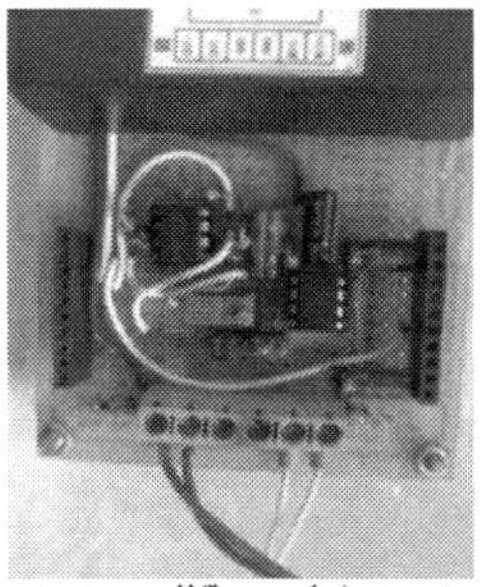

Figure 6. Former stage amplifier and Accurate rectifier physical diagram

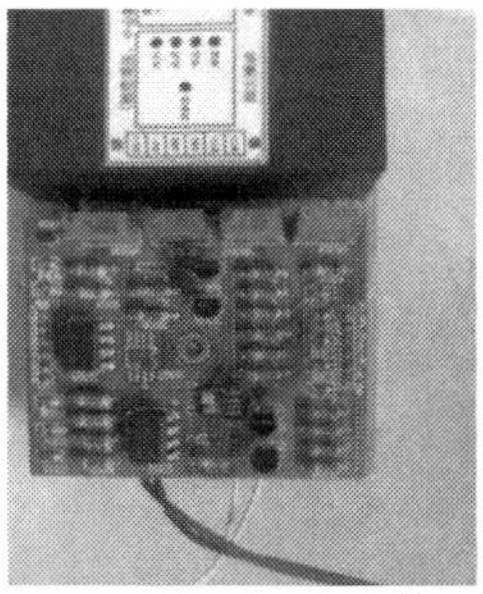

Figure 7. After stage amplifier physical diagram

5 Experimental data

In the experiments, the collected Marine auxiliaries a audio noise, the parameters for quantitative figures for 6000Hz, 16 (resolution), or long for 5s (about thirty thousand sampling points), it is the time domain waveform (see Fig. 8 time domain noise figure), obviously can not see the signal from the relevant characteristics or parameters. In view of the signal processing is shown in Fig. 8, get the waveform Fig. 9 after signal processing, including the original signal in black, a red signal after conditioning, obviously, it can be seen from the figure that the cycle of the original signal characteristics are not obvious, but after the signal has the obvious cyclical and smoothness.

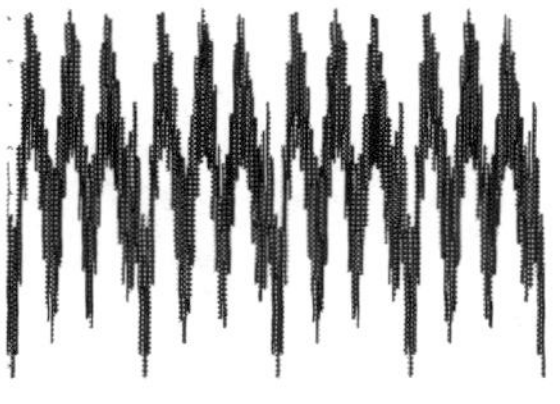

Figure 8. Time domain noise figure

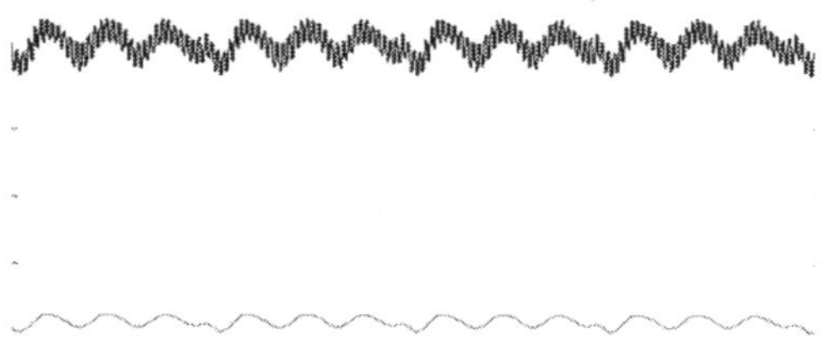

Figure 9. After the signal processing of waveform figure

In the above the noise of the cases, the noise of the detector set in the range of 0-120dB to get the following results: does not appear within the range set alarm, as shown in figure, the noise detector to 50, numerical display normally, does not appear alarm, when numerical more than set range, for example in number 150, detector alarm, remind engineer equipment that there might be a failure, should be handled in a timely manner.

Figure 10. Normal condition

Figure 11. Alarm condition

6 Conclusion

Noise analysis is presented in this paper, on the basis of ship engine room are discussed in detail the design of the utility equipment running status detector circuit, points out the structure and the parts of the circuit components selection, gives the experiment result after testing. Experiment showed that this detector has stable performance, high reliability, strong flexibility and programmable advantages, can set the range of detector and alarm of the upper and lower limits, and abnormal sound when equipment failure, and sound an alarm, to remind staff to repair in time, ensure the safe operation of the equipment, and improve the safety of the operation of the ship.

Mechanical equipment fault diagnosis technology based on the noise signal processing, relative to other diagnostic techniques have signal collection and more convenient, to adapt to a wider frequency range, more suitable for online monitoring, etc, the key lies in how to extract the characteristics of the signal, the more obvious of periodic signal, the signal conditioning unit for the signal to noise amplification, rectifier, filter and post-processing, such as: some components of the signal

 DOI: 10.1051/matecconf/20167702001

attenuation or inhibit, weak signal extraction, etc.This analysis is accomplished through regulating part processing work.In practice, if with other analysis tools, such as Fourier transform, combining can also further speed up the troubleshooting, improve the analysis accuracy. [6]

References

1. Hou Jian, Liyong-hong. Design of Signal Processing Circuit in ECG Monitoring System[J] . Industrial Control Computer, **22** (1) : 81~84, (2009).
2. Chen Jiang-hua. Analysis and Study on the Noises in DigitalVISI Circuits[J]. Journal of Shandong University of Technology, **30** (4) : 367~372, (2000).
3. Liu Jian, Qin Hui-bin. MAX260/261/262 Programmable Switched-Capacitor Filters and Its Application[J]. Journal of Hangzhou Institute of Electronic Engineering, **23** (6): 19~22, (2003).
4. Zeng Qing-yong. Weak Signal Detection[M]. Hangzhou: Zhejiang University Press, 1996. 1~83.
5. Zhang Jin-li, Jing Zhan-rong, LIANG Liang. Weak Signal Conditioning Circuit Design and Noise Estimation[J]. Electronic Measurement Technology, **30** (11): 40~42, (2007).
6. Wu Gang, Zhang Xiao-ping. Research of Resistance Thermal Noise[A]. Ordnance Industry Automation, **25** (4): 86~87, (2006).

MATEC Web of Conferences 77, 02002 (2016) DOI: 10.1051/matecconf/20167702002
ICMMR 2016

Simulation and Experiment study of ball defects of a two rotors-ball bearing-gear coupling system

Lu Xin[1], Zhang Junhong[1], Dai Huwei[1], Ma Liang[2] and Wang Jun[2]

[1] *State Key Laboratory of Engine, Tianjin University, 300072 Tianjin, China;*
[2] *Ren'ai College, Tianjin University, 301636 Tianjin, China*

Abstract. A model of a two rotors-ball bearing-gear coupling systems which considering ball defects on bearing was established. In this model the meshing force of gear coupling and ball defects on bearing was considered. A test facility was constructed to verify the simulation results. After simulation and experiment, it was found that there are sideband frequencies beside VC frequency and its multiple frequencies can be found in the faulted rotor system in the spectrum. From the acceleration signal of experiment, it was found the signal of the bearing fault performs as impacts in the high frequency range without any patterns. The characteristics of ball defects of bearing on a two rotors-ball bearing-gear coupling systems were analyzed. It can help analyze the stability of the system and the fault characteristic of ball bearing.

1 Introduction

Rolling Bearings are widely used in rotating machinery as important supporting component. The failure of bearings may lead to failure of the entire machinery [1]. Depending on research, bearing faults comprise 41% of failures in induction motor[2].Bearing defects are common faults of rotor-bearing systems including off-sized rolling elements, waviness of contacting surfaces, and those developed during operations such as cracks and pits [1].

Various numerical models have been developed to study the dynamic characteristics of rotor [3]-[9], many researches are focused with different on ball bearing faults. Changqing developed a bearing model considering the effects of distributed off-size rolling elements and waviness [5]. Kankar [1] focused on the accurate performance prediction due to the localized defects of micro level on the bearing components. McFadden and Smith [5] extended a single point defect model to a multiple points defects model in order to investigate the effects of bearing geometry, speed, load distribution, transfer function and the decay of vibration. Lynagh [8] presents a detailed model of bearing vibration, including the effect of contact spring non-linearity in balls-to-raceways' contacts. Eren [9] analyzed stator via wavelet packet decomposition to detect bearing defects. The proposed method enables the analysis of frequency bands that can accommodate the rotational speed dependence of the bearing defect frequencies.

In this paper, a two-rotor-bearing-gear coupling system model with ball defects on ball bearing was established to study the characteristics of ball defects of ball bearing. A test facility was built to verify the simulation results.

2 Model of Simulation

2.1. Ball bearing model of ball defects

Based on the study of Kankar [1], the model of inner race defect on the ball bearing is established in this section. The defects on the ball of ball bearing are shown in Figure 1.When the defects of ball contact with inner ring or out ring, the additional clearance μ_b can be written as:

$$\mu_b = \frac{D_w - \sqrt{D_w^{\ 2} - b_r^{\ 2}}}{2} \tag{1}$$

The triggering condition of the ball defects contact with inner ring and out ring can be written as:

$$\begin{cases} -\beta \le \mathrm{mod}(\dfrac{\theta_b}{2\pi}) \le \beta(inner) \\ \pi - \beta \le \mathrm{mod}(\dfrac{\theta_b}{2\pi}) \le \pi + \beta(out) \end{cases} \tag{2}$$

where $\beta = \arcsin\left(\dfrac{b_r}{D_w}\right)$

The displacement of ball can be given by:

$$\delta_{jb} = \delta_j - \mu_b = \delta \cdot \cos j\varphi - \mu - \mu_b \tag{3}$$

where μ is the radial clearance of ball bearing, δ_j is the deformation of bearing, b_r is the diameter of the ball; D_w is the pitch diameter of the bearing.

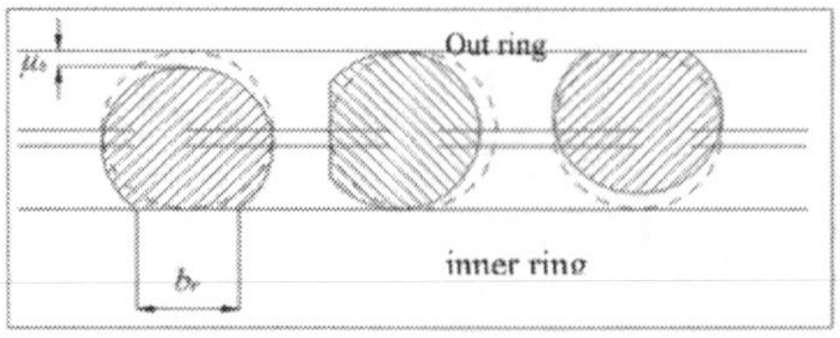

Figure 1. The ball defect of the bearing

According to the Hertzian contact theory, the local Hertzian contact force and deflection relationship for ball bearing can be written as:

$$F_b = k(\delta_{jb})^{3/2} \tag{4}$$

where k is constant for Hertzian contact elastic deformation.

2.2. Gear coupling

In the gear coupling, the meshing in the male and female couplings is asymmetric (as showed in Fig. 2), gear teeth of the two half-couplings mesh tightly in the right parts and loosely in the left parts. Consequently, the meshing distance of every tooth is changing. According to the research of Zhao [10], the meshing force in the gear coupling consist of two parts: (a) the meshing force caused by torque and (b) the meshing force caused by vibration.

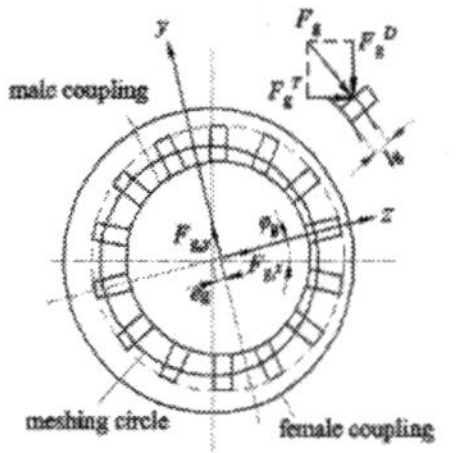

Figure. 2. Schematic diagram of the gear coupling.

2.2.1 Meshing force caused by torque.

When gear teeth are meshing tightly to transmit a large torque, deformation of a gear tooth is inevitable and the following meshing force is related to the deformation amount, meshing distance and meshing rigidity. From the study of ref [9], the torque-induced meshing force is given by

$$F_g^T = \phi_g l_g k_g \tag{5}$$

where F_g^T is the meshing force due to torque, ϕ_g is the deformation angle of gear tooth, l_g is the equivalent meshing distance of gear tooth, and k_g is the meshing rigidity of gear tooth. The ϕ_g of all gear teeth are the same when the coupling is driven by a static torque.

2.2.2 Meshing force caused by vibration.

when torque is transmitting by gear teeth, a relative movement between the two meshing teeth takes place induced by the system vibration at the same time. As a result, the displacement of meshing nodes causes a deformation of each tooth so that another meshing force is produced and expressed as

$$F_g^D = e_g k_g \sin\varphi_g \tag{6}$$

where F_g^T is the meshing force caused by vibration, φ_g is the position angle of gear tooth, and e_g is the misalignment of gear coupling.

If there is no torque to transmit, the meshing force will be zero regardless of the vibration displacement, and the meshing force will not be negative. Accordingly, the total meshing force of the ith tooth can be obtained

$$F_{g,i} = \begin{cases} \left(\phi_{g,i} l_{g,i} + e_g \sin\varphi_{g,i}\right) k_{g,i}, & \phi_{g,i} l_{g,i} + e_g \sin\varphi_{g,i} > 0 \\ 0, & \phi_{g,i} l_{g,i} + e_g \sin\varphi_{g,i} \le 0 \end{cases} \tag{7}$$

Then, the reactions of meshing force in the y-axis and z-axis directions of the rotor system are as follows:

$$\begin{cases} F_{g,y} = \sum_{i=1}^{n} \left(\phi_{g,i} l_{g,i} + e_g \sin\varphi_{g,i}\right) k_{g,i} \sin\theta \\ F_{g,z} = \sum_{i=1}^{n} \left(\phi_{g,i} l_{g,i} + e_g \sin\varphi_{g,i}\right) k_{g,i} \cos\theta \end{cases} \tag{8}$$

2.3 Model of rotor

The rotor system studied in this paper (shown in Fig. 3) conducts of two shafts, two disks, three ball bearings, a SFD, and a gear coupling. Each shaft has a disk on it, and the two shafts are connected by the gear coupling. The rotor is supported in the "0-2-1" form which means no bearing is mounted in front of the left disk, two bearings are mounted in the middle of the two disks, and one bearing mounted behind the right disk. The two bearings in the middle are both located on the left shaft. In this paper, Bearing 1 was assumed to the defects bearing. The equations of Model can be written as:

$$\begin{cases} M_{d1}\ddot{y}_{d1} + C_d\dot{y}_{d1} + K_r(y_{d1} - y_{r1}) = M_{d1}\omega^2 e_{d1}\sin\omega t - M_{d1}g \\ M_{d1}\ddot{z}_{d1} + C_d\dot{z}_{d1} + K_r(z_{d1} - z_{r1}) = M_{d1}\omega^2 e_{d1}\cos\omega t \\ M_{d2}\ddot{y}_{d2} + C_d\dot{y}_{d2} + K_r(y_{d2} - y_{r6}) + K_r(y_{d2} - y_{r7}) = M_{d2}\omega^2 e_{d2}\sin\omega t - M_{d2}g \\ M_{d2}\ddot{z}_{d2} + C_d\dot{z}_{d2} + K_r(z_{d2} - z_{r6}) + K_r(z_{d2} - z_{r7}) = M_{d2}\omega^2 e_{d2}\cos\omega t \\ M_{r1}\ddot{y}_{r1} + C_b\dot{y}_{r1} + K_r(y_{r1} - y_{r2}) + K_r(y_{r1} - y_{d1}) = F_{b,y1} - M_{r1}g \\ M_{r1}\ddot{z}_{r1} + C_b\dot{z}_{r1} + K_r(z_{r1} - z_{r2}) + K_r(z_{r1} - z_{d1}) = F_{b,z1} \\ M_{r2}\ddot{y}_{r2} + C_b\dot{y}_{r2} + K_r(y_{r2} - y_{r1}) + K_r(y_{r2} - y_{r3}) = F_{b,y2} + F_{g,y} - M_{r2}g \\ M_{r2}\ddot{z}_{r2} + C_b\dot{z}_{r2} + K_r(z_{r2} - z_{r1}) + K_r(z_{r2} - z_{r3}) = F_{b,z2} + F_{g,z} \\ M_{r3}\ddot{y}_{r3} + K_r(y_{r3} - y_{r2}) + K_r(y_{r3} - y_{r4}) = -F_{g,y} - M_{r3}g \\ M_{r3}\ddot{z}_{r3} + K_r(z_{r3} - z_{r2}) + K_r(z_{r3} - z_{r4}) = -F_{g,z} \\ M_{r4}\ddot{y}_{r4} + K_r(y_{r4} - y_{r3}) + K_r(y_{r4} - y_{r5}) = -M_{r4}g \\ M_{r4}\ddot{z}_{r4} + K_r(z_{r4} - z_{r3}) + K_r(z_{r4} - z_{r5}) = 0 \\ M_{r5}\ddot{y}_{r5} + K_r(y_{r5} - y_{r4}) + K_r(y_{r5} - y_{r6}) = -M_{r5}g \\ M_{r5}\ddot{z}_{r5} + K_r(z_{r5} - z_{r4}) + K_r(z_{r5} - z_{r6}) = 0 \\ M_{r6}\ddot{y}_{r6} + K_r(y_{r6} - y_{r5}) + K_r(y_{r6} - y_{d2}) = -M_{r6}g \\ M_{r6}\ddot{z}_{r6} + K_r(z_{r6} - z_{r5}) + K_r(z_{r6} - z_{d2}) = 0 \\ M_{r7}\ddot{y}_{r7} + K_r(y_{r7} - y_{d2}) = F_{b,y3} - M_{r7}g \\ M_{r7}\ddot{z}_{r7} + K_r(z_{r7} - z_{d2}) = F_{b,z3} \end{cases} \tag{9}$$

where M_{d1} and M_{d2} is the mass of Disk 1 and Disk 2, M_{r1}–M_{r6} are the mass of rotor points, C_d and C_b are viscous damping factor of disk and ball bearing, K_r is the stiffness of rotor, $F_{b,y}$ and $F_{b,z}$ are supported forces of ball bearing in the y-axis and z-axis directions, $F_{g,y}$ and $F_{g,z}$ are meshing forces in the y-axis and z-axis directions ,e_{d1} and e_{d2} are eccentricity of Disk1 and Disk 2.

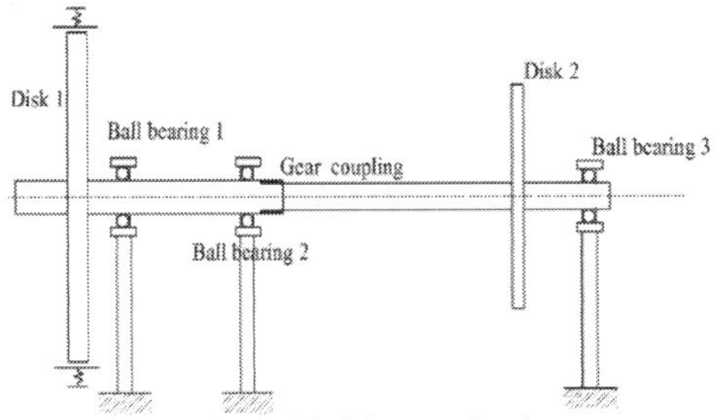

Figure 3. The model of the rotor-bearing system.

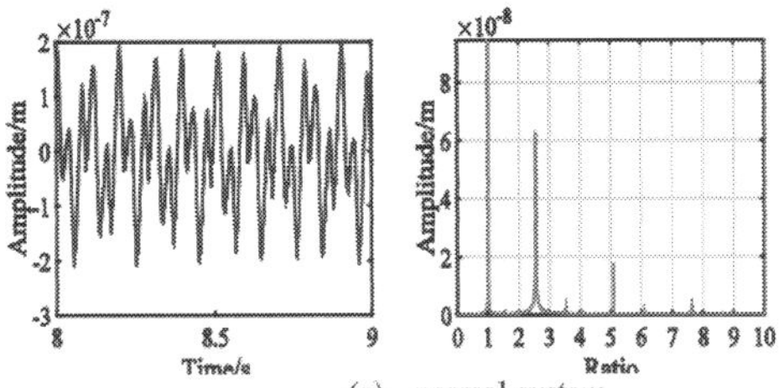

(a) normal system

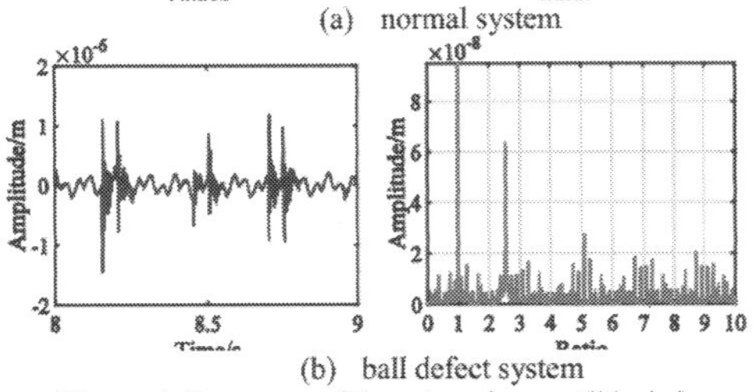

(b) ball defect system

Figure 4. Responses of the rotor when ω=600 r/min.

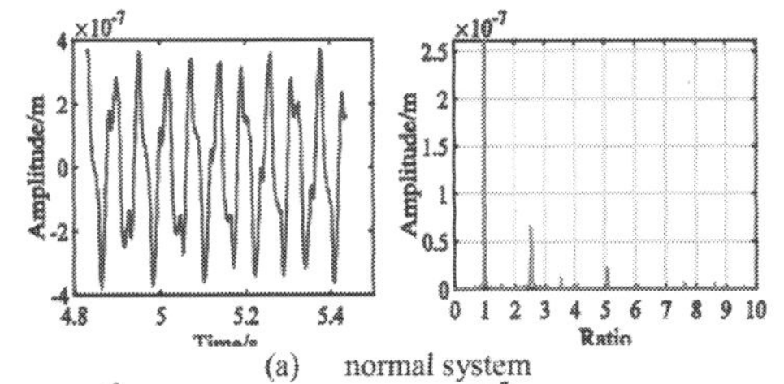

(a) normal system

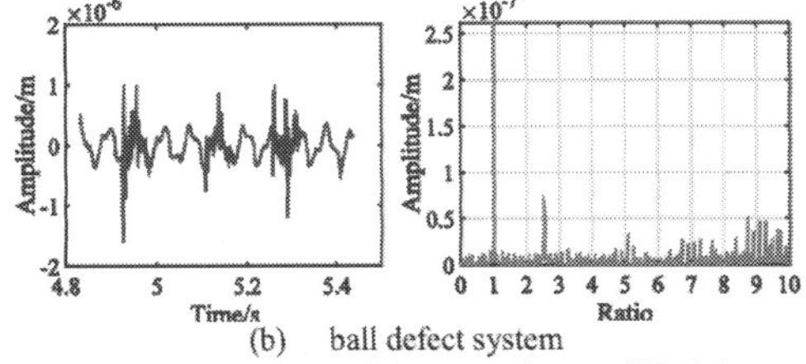

(b) ball defect system

Figure 5. Responses of the rotor when ω=600 r/min

2.4 Results of Simulation

In this study, the differential equations of motion are solved by the fourth order Runge-Kutta method. The time step of the iterative procedure is Δt=1×10-5s. The time varying data corresponding to the first 500 periods generated by the numerical integration is deliberately excluded in order to discard the transient solutions.

Fig. 4 shows the response of a normal system and ball defects on bearing system when ω=600 r/min. In Fig. 4(a), it is easy to find the VC frequency and 2VC frequency which represents the ball passage frequency, the motion of the rotor is periodic. In Figure 4(b), the time domain signal of the system has great change, the amplitude of vibration has much increase at times and then decreases sharply. In the spectrum, there are sideband frequencies beside VC frequency and its multiple frequencies can be found in the faulted rotor system.

Fig. 5 shows the response of a normal system and ball defects on bearing system when ω=1000 r/min. The characteristics of the signal in Figure 5 are similar to it in Fig. 4. With the increasing of speed, the amplitude of high frequency which is caused by ball defects is almost unchanged compared to the increase in fundamental frequency.

3 Experimental verification

3.1. Experimental rotor system

In this paper, the test facility is built to simulate a compressor rotor system. Fig. 6 shows the overall test setup and the details of the ball bearing.

As discussed in Fig. 6 system component in the test rig is the same. The AC motor is used to drive the rotor and the rotating speed is controlled by a programmable logic controller. Accelerometers and Eddy current sensors are distributed to the experiment bench to get the displacement signal and acceleration signal.

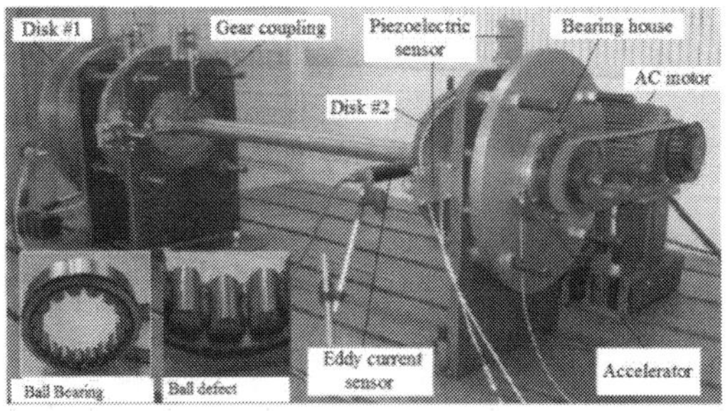

Figure 6. Experimental rotor system.

Table 1. Parameters of the NSK NU211EW bearing

Size(mm)					Number of ball
D_i	D_o	H	D_w	b_r	
55	100	21	78	12	17

The NSK NU211EW bearings are used on the rotor test. The parameters of bearing are shown in Table 1. In order to simulate the ball bearing fault, a groove is on the surface of ball by electrical discharge machining,which is shown in Fig. 5.

3.2 Results of Experiment

Fig. 7 shows the Displacement signal and Acceleration signal of rotor system with ball defects when ω=600 r/min, and Figure 8 shows the responses of the system at 1000r/min. From Figure 7(a), the motion of the rotor is periodic and the energy of frequency is focused on the low frequency which between 0 to the fundamental frequency. From Figure 8(a), the largest frequency is between 0 to the fundamental frequency and a frequency appears between fundamental frequency to 2*fundamental frequency. The characteristic of ball defects is not easy to find from figures. After analysis, it is found that, the displacement signal of experiment is not suitable to study the ball defect faults. Therefore the acceleration signal is selected to study the fault characteristic.

From Fig. 7(b), it can be found that the signal of the bearing fault performs as impacts in the high frequency range without any patterns, from Figure 8 (b), it can be found that the amplitude of signal increase obviously and the energy of frequencies moves to high frequency. The characteristics are seemed to the simulation results.

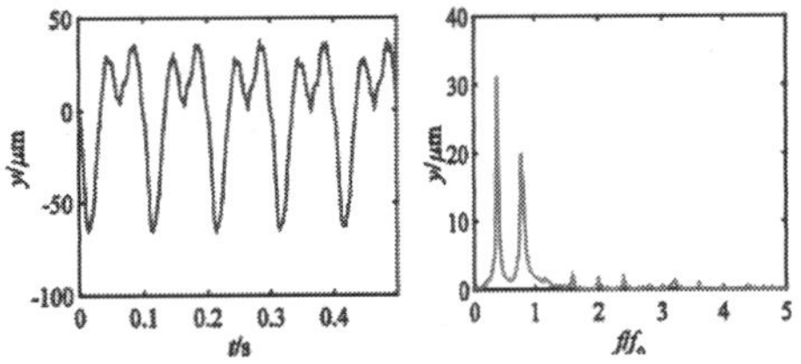

(a)Displacement signal

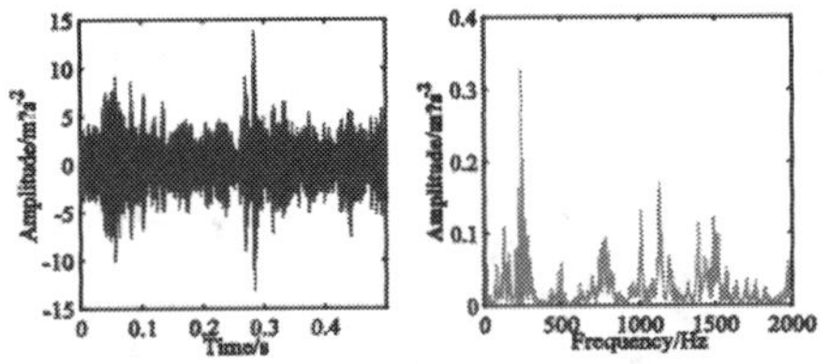

(b)Acceleration signal

Figure 7. Responses of the rotor with ball defects when ω=600 r/min

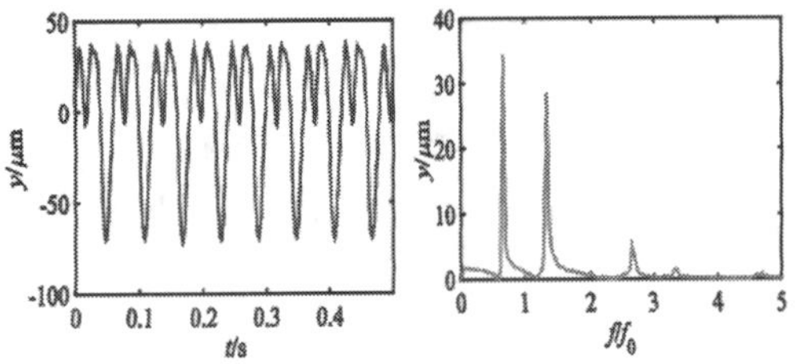

(a)Displacement signal

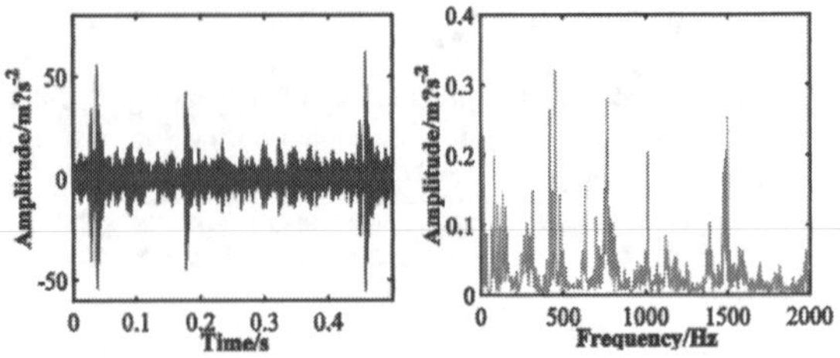

(b)Acceleration signal

Figure 8. Responses of the rotor with ball defects when ω=1000 r/min

4 Conclusion

In this paper, a two-rotor-bearing-gear coupling system model with ball defects on ball bearing to study the characteristics of ball defects of ball bearing. A test facility is built to verify the simulation results. After simulation and experimental analysis, some conclusions are obtained:

1) For the rotor system with the inner race defects on ball bearing, it is found that there is significant change on the dynamic characteristics of the rotor. According to the spectrum diagram of rotor system, there are all kinds of frequency especially high frequency appears under the ball bearing faults.
2) From the results of experiment, it is found that the characteristics of ball defects are not obvious on displacement signal, but acceleration signal reflects the characteristics of ball defects.

The results show the dynamic characteristics of ball defects on the two rotors system. It can help analyze the stability of the system and the fault characteristic of ball bearing.

Acknowledgement

We are grateful to the Specialized Research Fund for the Doctoral Program of Higher Education (20130032130005)for providing financial support for this work

References

1. P. K Kankar, Satish C Sharma, S. P Harsha, ,"Vibrationbasedperformancepredictionof ball bearings caused by localized defects". Nonlinear Dynamics. vol.3 no.69,pp: 847-875, 2012.
2. B Raison, G Rostaing, O.Butscher, and C. S Maroni , "Investigations of Algorithms for Bearing Fault Detection in Induction Drives,"IECON02, pp. 1696–1701 2002,.
3. Poornima V D, Adinarayana S, Rao B V A. Two plane balancing of a conical rotor driven by vertical belt system designed to reduce gyro effect[J]. International Journal of Mechanical Engineering and Robotics Research, 2012, 1.
4. B. Changing, Q. Xu, "Dynamic model of ball bearings with internal clearance and waviness".

Journal of Sound and Vibration. vol.294 pp:23-48, 2006,

6. B Raison, G Rostaing, O.Butscher, and C. S Maroni, "Investigations of Algorithms for Bearing Fault Detection in Induction Drives,"IECON02, pp. 1696–1701 2002,.
7. McFadden, P.D., Smith, J.D, "The vibration produced by multiple points defects of a rolling bearing" vol.98 pp:263-275, 1985
8. Lynagh N, Rahnejat H, Ebrahimi M, et al. Bearing induced vibration in precision high speed routing
5. McFadden, P.D., Smith, J.D, "The vibration produced by multiple points defects of a rolling element bearing " vol.98 pp:263-275, 1985
spindles[J]. International Journal of Machine Tools and Manufacture, 2000, 40(4): 561-577.
9. Eren L, Devaney M J. Bearing damage detection via wavelet packet decomposition of the stator current[J]. Instrumentation and Measurement, IEEE Transactions on, 2004, 53(2): 431-436.
10. G. Zhao, Z. Liu, F. Chen, Meshing force of misaligned spline coupling and the influence on rotor system, International Journal of Rotating Machinery, 2008 (2008)

MATEC Web of Conferences 77, 02003 (2016) DOI: 10.1051/matecconf/20167702003
ICMMR 2016

Co-Simulation of PCB and Enclosure to Predict Radiated Emission

Yifu Ding, Guangyu Zhang , Xu Zhang, and Zaiyuan Wu

China Automotive Technology & Research Center, Tianjin 300300, China

Abstract. With highly requirement for electromagnetic compatibility (EMC) target, effective controlling of radiated emission of equipment or subsystem is of considerable significance. An emission equivalent model of Printed Circuit Board (PCB) is set up in this paper. Prediction of resonance characteristics, far field radiation of the PCB is carried out and analysed. Then, a 3-D model of a typical equipment enclosure is set up. Resonance characteristics and radiating characteristics of the enclosure is carried out and analysed. A co-simulation method to predict radiated emission is proposed and an example of co-design of PCB and enclosure using HFSS and Siwave verified this method.

1 Introduction

Electromagnetic compatibility (EMC) problems have been found to be a tricky problem when large numbers of electronic equipment were installed on space-limited electric platform. As an important technical target, EMC has great impact on the efficiency and reliability of electric system. Simulation and prediction of radiated emission effectively in the stage of system design is of great significance to the improvement of EMC performance of system [1].

This paper aims at introducing a method to predict radiated emission of electric equipment in order to optimize the EMC characteristic of electric system. We introduce a co-simulation method to predict the radiated emission of the whole equipment. The co-simulation method is able to control over radiated emission of equipment effectively in the stage of design [2].

2 Modeling of pcb radiated emission characteristic

Radiated emission of PCB originated from several types of sources: high frequency signal generated by RC oscillator or crystal oscillator coupling on the traces on PCB and then forming an effective antenna, current and voltage-driven common-mode coupling, ground bounce and etc. Two methods are able to reduce radiated emission of PCB: One method is to control the radiated emission caused by circuit module and traces, the other method is to avoid the resonance between ground and power distribution network (PDN) [3], [4].

Modelling of PCB radiation sources could equivalent the trace or network of highly radiated efficiency to an antenna. Two types of radiation sources could be modelled to equivalent the radiated characteristic of PCB: one is common-mode radiation and the other is different-mode radiation.

2.1 Different-Mode Radiated Emission

Different-mode currents flowing through a loop result in different-mode radiated emission. The currents loop could be equalled to magnetic dipole antenna as the different-mode radiated emission model on PCB. An idealized magnetic dipole antenna constitutes of a small currents loop and its radius a meets the inequality: a<<λ . We can derive the vector and magnitude of E-field and H-field radiated from the model [5]-[7].

The vector representations of E-field and H-field are:

$$E_{\varphi} = -j\frac{\eta_0 SIk}{4\pi r}(jk+\frac{1}{r})sin\theta e^{-jkr} \quad (1)$$

$$H_{\theta} = \frac{IS}{4\pi r}(-k^2+\frac{jk}{r}+\frac{1}{r^2})\ sin\theta e^{-jkr} \quad (2)$$

$$H_r = \frac{IS}{2\pi r^2}(jk+\frac{1}{r})\ cos\theta e^{-jkr} \quad (3)$$

$$|E_{\varphi}| = -\frac{\eta_0 SI\pi}{r\lambda^2}\sqrt{1+\frac{\lambda^2}{4\pi^2 r^2}}sin\theta \quad (4)$$

$$|H_{\theta}| = \frac{IS\pi}{r\lambda^2}\sqrt{1-\frac{\lambda^2}{4\pi^2 r^2}+\frac{\lambda^4}{16\pi^4 r^4}}\ sin\theta \quad (5)$$

$$|H_r| = \frac{IS}{r\lambda^2}\sqrt{1+\frac{\lambda^2}{4\pi^2 r^2}}\ cos\theta \quad (6)$$

2.2 Common-mode radiated emission

Common-mode radiated emission originates from common-mode noise voltage. The noise voltage could be driven in two ways: one is currents-drive voltage noise and the other is voltage-drive voltage noise. Currents-drive voltage noise caused by voltage drop derives from time-varying different-mode currents flowing through the loop impedance. Voltage-drive voltage noise results from different-mode voltage in circuit network. It is practicable to equal the common-mode radiated emission to an electric dipole antenna. We can derive the vector and magnitude of E-field and H-field radiated from this model.

The vector representations of E-field and H-field are:

$$E_r = -j\frac{Idl\cos\theta}{2\pi\omega\varepsilon_0 r^2}(jk+\frac{1}{r})\sin\theta e^{-jkr} \quad (7)$$

$$E_\theta = -j\frac{Idl\cos\theta}{4\pi\omega\varepsilon_0 r^2}(-k^2+\frac{jk}{r}+\frac{1}{r^2})\,e^{-jkr} \quad (8)$$

$$H_\varphi = \frac{Idl\sin\theta}{4\pi r}(jk+\frac{1}{r})\,e^{-jkr} \quad (9)$$

$$|E_r| = -\frac{Idl\cos\theta}{2\pi r^2}\sqrt{\frac{\mu}{\varepsilon_0}}(j+\frac{1}{rk}) \quad (10)$$

$$|E_\theta| = \frac{Idl\eta_0\sin\theta}{2\lambda r}\sqrt{1-\frac{\lambda^2}{4\pi^2 r^2}+\frac{\lambda^4}{16\pi^4 r^4}} \quad (11)$$

$$|H_\varphi| = \frac{Idl\sin\theta}{2\lambda r}\sqrt{1+\frac{\lambda^2}{4\pi^2 r^2}} \quad (12)$$

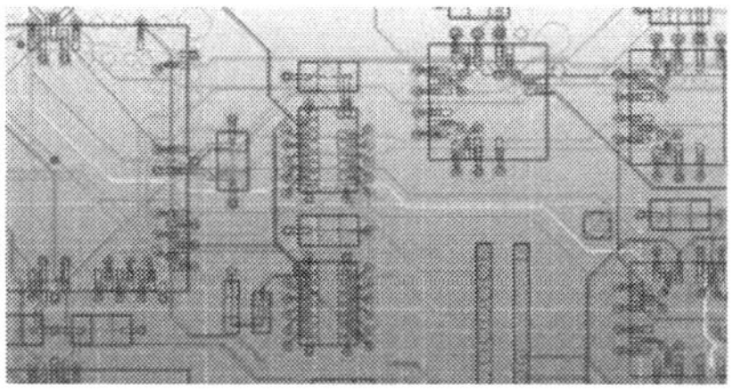

Figure 1. Key Network of a PCB.

2.3 Modelling of PCB as Radiated Emission Source

To model PCB radiated emission in a more accurate way, a thorough analysis of the key networks in PCB is necessary. The key networks should include high-speed signal network and power distribution network (PDN)[8]. A key network of an 8-layer PCB analysed in this paper is shown in Fig. 1, the highlighted wire is the high-speed signal network in this board. We put a 1V voltage source with 0-1GHz frequency swept in the network where the coordinate of positive pole is (X, Y) = (1150, 3775) and negative pole is (X, Y) = (1150, 3775). The positive is on the surface layer of the PCB and the negative is on the GND layer of the PCB. The 1m far-field radiated emission of PCB is predicted using method of FDTD and the results is shown in Fig. 2.

3 Co-Simulation of PCB and Enclosure to Control Radiated Emission

3.1 Prediction and Optimization to Resonance Character of PCB and Enclosure

This paper uses an 8-layer PCB shown in Fig. 2 and a typical enclosure as an example to verify the feasibility and effectiveness of the co-simulation method to control radiated emission. We predict the resonant characteristic of the PCB in frequency range of 0.25GHz to 2GHz and the calculation results of 5 resonance points shown in Chart 1. Resonance frequency of enclosure depends on its physical shape and the vents, slots, buttons on it. These factors would affect the resonance frequency considerably. In the stage of pre-design, we could simulate and calculate the resonant frequency under the consideration of both resonance and shielding effectiveness in order to reduce radiated emission of the equipment and provide information to the arrangement of PCB and cable inside the enclosure. A typical enclosure is modelled and the dimension and material properties are shown in Table 1. And then we predict the resonant characteristic of the enclosure in frequency range of 0.25 to 2GHz and the calculation results of 5 resonance points is shown in Chart 2.

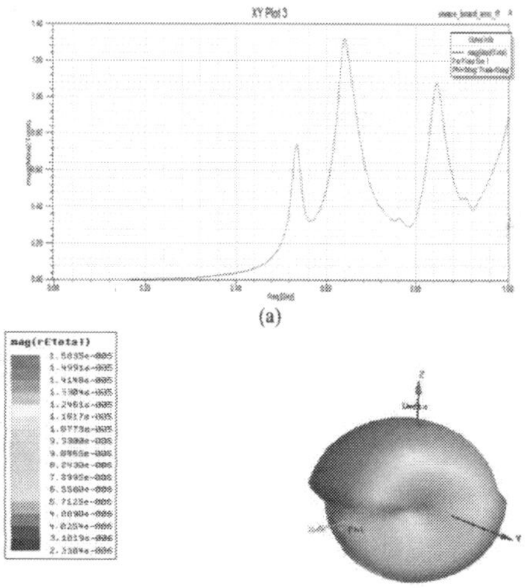

Figure 2. The 1m Radiated Emission of PCB

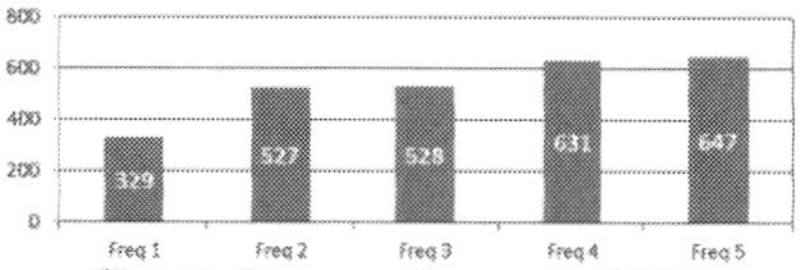

Chart 1. Resonance Frequency of PCB

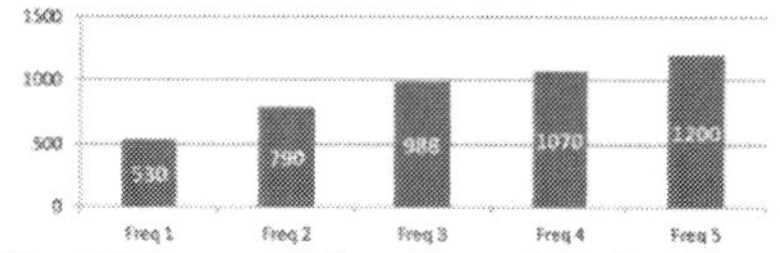

Chart 2. Resonance frequency points of enclosure.

Table 1. Dimension and material properties of enclosure

Dimension	200mm*200mm*100mm
Thickness	2mm
Material	Aluminum
Radius of aperture	10mm
Position of aperture	(198,100,60)

3.2 Radiated emission of equipment and co-design

In order to predict the radiated emission of the equipment (PCB inside enclosure), we modelled the PCB as an equivalent antenna (in chapter II) as a radiated source and then the enclosure is modelled as a 2nd time radiated source. This method avoids the inaccuracy computation of direct division to the enclosure and PCB. Then we modelled an equivalent PCB as a material of metallic surface and dielectric interlayer inside the enclosure with the dimension of 200mm length, 100mm width and 10mm thickness. The extracted equivalent radiated source is put on the position of (0, 80, 0). Simulation result of the E-field distribution is shown in Fig 3.

Table 2. Optimized Enclosure Settings

Vents Distribution	2×2
Shape of Vents	Round
Coordinates of Vents	(198,50,20) (198,50,40) (198,30,20) (198,30,40)
Radius of Vents	2.5mm

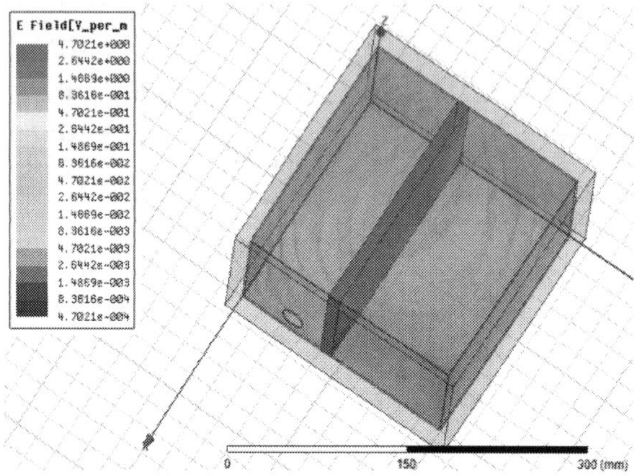

Figure 3. Distribution of E-field of the equipment.

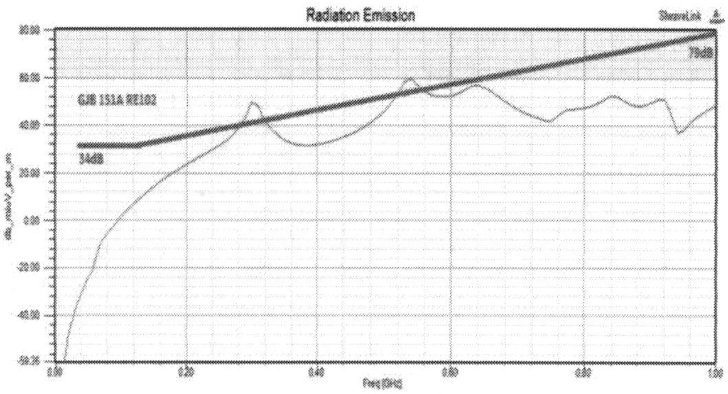

Figure 4. Simulated RE102 test results.

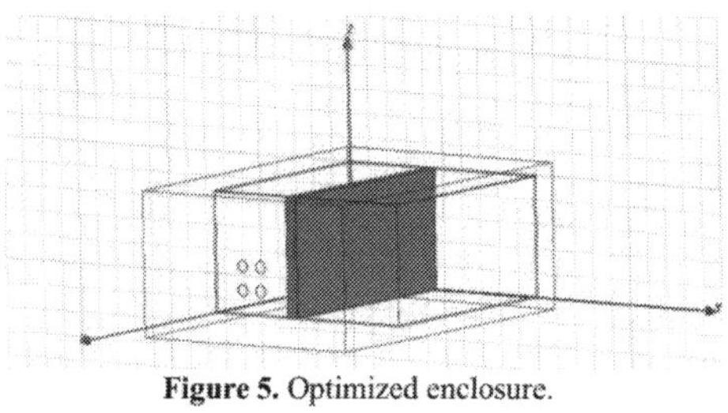

Figure 5. Optimized enclosure.

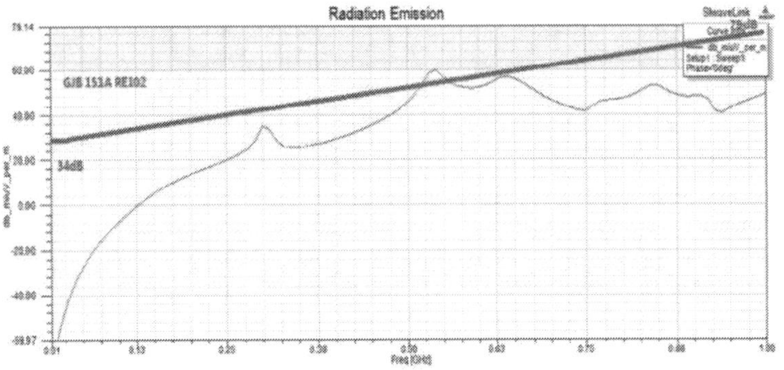

Figure 6. Simulated RE102 test results.

In order to verify the co-simulation method in design of equipment, an optimization to the radiated emission of the equipment by the reference of radiated emission limits of GJB151A is carried out. The simulation result is shown below in Fig. 4.

Obviously, there exists two frequency points of over-radiated emission under the RE102 test: a 7dB over radiated emission on the frequency of 240MHz and a 6dB over radiated emission on the frequency of 570MHz. The equipment cannot be certified in this level of radiated emission and an optimization is necessary. In order to maintain the same level of heat output performance, the whole area of the vent cannot be changed and according the theory of the effects of vent. The dimension and distribution of vents of the optimized enclosure is shown in Table 2 and the optimized enclosure is shown in Fig. 5.

The radiated emission of the optimized equipment is shown in Fig. 6. Therefore, the optimized equipment is able to be certified in the RE102 test and the radiated emission in the frequency point 340MHz is 33dBμv/m, which has a 15dB reduction from the original one.

4 Summary

In this paper, we conclude the influence of resonance character to the radiated emission of equipment and modelled the equivalent antenna to PCB radiated emission source. Using the source, as an excitation inside the enclosure, to predict radiated emission of the equipment. Thus, a co-simulation method to predict radiated emission is proposed and an example of co-design of PCB and enclosure using HFSS and Siwave verified this method. The simulation and prediction results were contrasted with radiated emission maximum limits required by GJB151A Standard and verified the correctness and efficiency of the method. This method is able to provide theoretical foundation and design method for the radiated emission control in the stage of pre-design of electric equipment and system.

References

1. Ji ZHANG, Juan SHI, Tengfei Yang. Radiated emission test and its reduction of PCB [J]. Journal of Jiamusi University, 2012, v.30; No.12005:706-709+713.
2. Yang ZHAO, Wei YAN, Bo ZHAO, Y.C LUO, Shijin LI. Evaluation and Tests of Circuit Radiated EMI [J]. DIANGONG JOURNAL, 2010, v.2510:6-13.
3. Christian Poschalko, IEEE TRANS ON EMC [J]. Vent Model for the Slot Radiation of an Enclosure Excited by Printed Circuit Board Traces with Different Loads.
4. M. Leone, "Design expressions for the trace-to-edge common-mode inductance of a printed circuit board," IEEE Trans. Electromagnetic. Compatibility. vol. 43, no. 4, pp. 667–671, Nov. 2001.
5. H.-W. Shim and T. H. Hubing, "Model for estimating the radiated emissions from a printed circuit board with attached cables due to voltage driven sources," IEEE Trans. Electromagnetic Compatibility, vol. 47, no. 4, pp. 899–907, Nov. 2005.
6. D. M. Hockanson, J. L. Drewniak, T. H. Hubing, and T. P. Van Doren,"FDTD modeling of common-mode radiation from cables," IEEE Trans on Electromagnetic Compatibility., vol. 38, no. 3, pp. 376–387, Aug. 1996.
7. D. M. Hockanson, X. Ye, J. L. Drewniak, T. H. Hubing, T. P. Van Doren, and R. E. Du Broff, "FDTD and experimental investigation of EMI from stacked-card PCB configurations," IEEE Trans. Electromagnetic Compatibility. vol. 43, no. 1, pp. 1–10, Feb. 2001.
8. J. E. Bracken, S. Poltyanko, S. Raman, and Z. J. Cendes, "Efficient full wave simulation of high-speed printed circuit boards and electronic packages," in Proc. IEEE Antennas Propagation. Soc. Int. Symp. 2004, Jun. 20–25, vol. 3, pp. 3301–3304.

MATEC Web of Conferences 77, 02004 (2016) DOI: 10.1051/matecconf/20167702004
ICMMR 2016

Analogue simulation and analysis of non-circular gear reentry error

Jianqing Dai, Shengze Wang

School of mechanical engineering, Donghua University, Shanghai 201620, People's Republic of China

Abstract. The studies on non-circular gear cutting and using errors were generally based on theory of cylindrical gear errors, and the similarities and differences between cylindrical gear and non-circular gear were combined to study the change rule of non-circular gear contacting line and obtained corresponding conclusions. In this paper, from a different perspective, by starting from the non-circular gear engagement principle, on the basis of analyzing the concrete geometric error conditions, new type tooth profile algorithm routine is reasonably modified, and needed non-circular gear tooth profile error contrast figure is obtained. Furthermore, theoretical derivation is carried out by incorporating analytical method and contacting line incremental method, validity of computer simulation is verified, and the characterization of the tangential error characteristic code is obtained, Summed up the error sources for error genetic which possess enormous significance to direct practice generation.

1 Introduction

Non-circular gear in terms of its essence can be considered as a new type of institution which integrates the advantages of both cam and gear [1, 2]. Non-circular gear mechanism not only inherited the numerous congenital advantage of the gear mechanism, but also has predictable characteristics of cam mechanism transmission rule. Therefore the non-circular gear is a kind of gear mechanism can achieve predictable variable speed ratio transmission [3].

Non-circular gear machining exist unique two errors, respectively are tool clamping error [4, 5] and reentry error of serrated knife, tool clamping error is closely related to the operator in practice, tool clamping error more belongs to the mistake error [6], and reentry error is related to tooth knife motion mechanism.

Both non-circular gear machining and cylindrical gear are subject to the basic principle of gear engagement, namely the serrated knife and gear do pure rolling, that is to say, translational distance of serrated knife should be in accordance with the curve that it passes through the pitch line. In the actual production, if the translational distance of serrated knife doesn't keep in accordance with the curve that it passes through the pitch line because of movement mechanism, therefore, tangential error generates at this moment. Due to the vast majority of tangential errors come from the eccentric movement of worm gear, essence is a harmonic gear error created by movement inharmonious between workpiece and tool, we only consider sine function of the turbine in a variety of harmonic errors.

$$\Delta y = k \sin\theta \tag{1}$$

Anticlockwise corner is positive value, reflectedin serrated knife translation motion is shown as follow.

$$L' = L + K \sin\theta \tag{2}$$

L is theoretical translation distance, L' is practical translation distance, and translational error conforms to sine function. According to relative motion principle, we considered that tooth germ doesn't move, serrated knife does reentry error movement.

2 Reentry error model

As shown in Figure 1, zero position of serrate knife just aligned the collinear with the zero position of work piece at a time of zero, in later actual production, according to relative motion principle, serrate knife generated Δy error which changed according to the sine function along axis y, resulting in the unequal translation section of serrated knife like the curve that passed through the section curve, this is reentry error model.

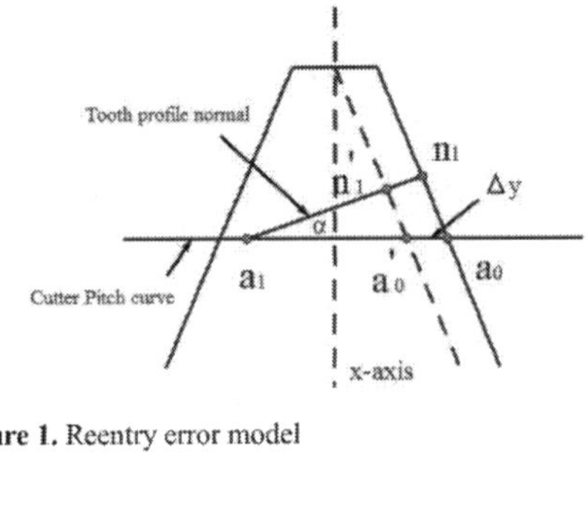

Figure 1. Reentry error model

3 Computer simulations

We used beforehand programmed non-circular gear tooth profile algorithm program. We carried out a generative simulation of the serrated knife according to the operation condition of the zero alignment, and we got the theory tooth profile diagram without the error, then we added a loop statement to x coordinate value of serrated knife small coordinate system in subroutines for simulating y coordinate figure remained immobility, to reflect reentry error conditions, then running the program again to get the tooth profile pattern when reentry error appeared. Finally, two pictures were combined in the same coordinate system to form Figure 2- Figure 5, that is, reentry error tooth profile comparison diagram (The circles are theory tooth profile, the stars are the tooth profile error).

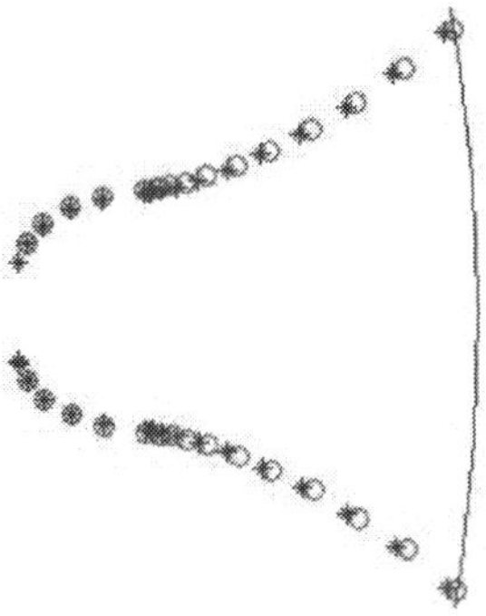

Figure 2. Comparation of initial alveoli reentry error tooth profile

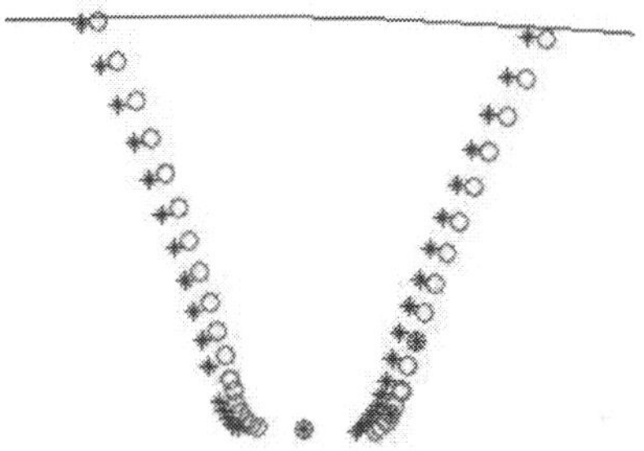

Figure 3. Comparation of reentry error tooth profile in short axis direction

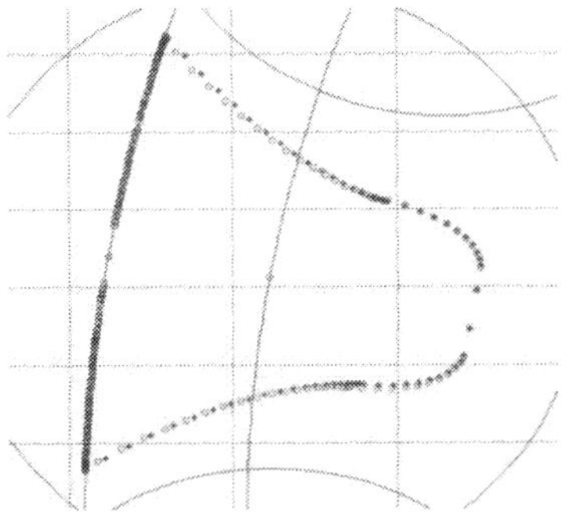

Figure 4. Comparation of reentry error tooth profilein long axis negative direction

Figure 2 and Figure 4 show the initial positions, that is, comparations of alveolar reentry error tooth profiles in long axis positive direction and short axis direction. Figure 3 and Figure 5 show the comparations of alveolar reentry error tooth profiles nearby the non-circular gear short axis positive direction and negative direction.

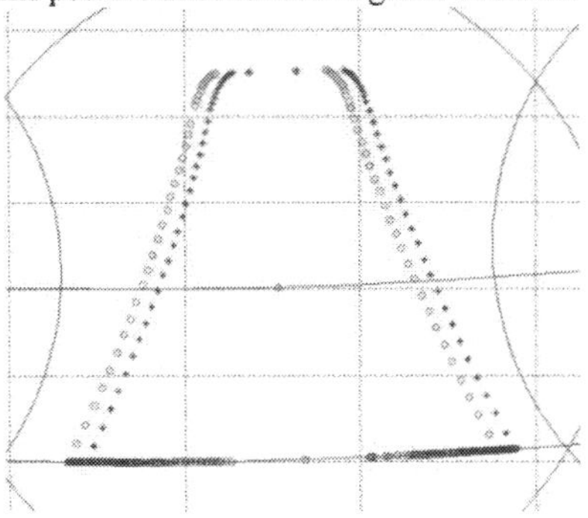

Figure 5. Comparation of reentry error tooth profile in short axis negative direction

Small red circles shows the theory tooth profile without error in from Figure 2 to Figure 5, while blue stars are tooth profile graphs that have reentry error.

4 Simulation conclusions

We can observe from the picture that the tooth profile deviation of tooth profile comparation Figure 2 and Figure 4 are not obvious, and the tooth profile excursion of the tooth profile contrast diagrams Figure 3 and Figure 5 are the most obvious. Tooth profile deviation of tooth profile contrast Figure 3 is toward to counter-clockwise direction, the tooth profile deviation of tooth profile contrast Figure 5 is toward to the clockwise direction, roughly the same size, in the opposite direction.

The formed tooth profile and the theory tooth profile after the excursion of tooth profile contrast diagrams Figure 3 and Figure 5 are different, that is, error tooth profile isn't the simple equidistant line of theory tooth profile.

The pitch curve of serrated knife tangent in the normal direction and alveoli bottom are basic coincidence, there is no obvious change, and that is, the radial error is zero.

5 Theoretical verification

In order to set the tooth profile normal distribution regulation of the non-circular beforehand for convenience, it's determined by the following [7].

$$\lambda - \mu = \text{constant} \qquad (3)$$

λ is the angle of section curve diameter and corresponding tooth profile normal, μ is the angle of section curve diameter and the tangent line positive direction.

Because of the special nature of noncircular gear pitch curve, leading to correspond exactly to the correlation does not exist between the theoretical tooth profile and tooth profile error, according to the principle of gear meshing in Figure 1 has two non circular gear to find the way. The first kind is to generate the error after a_1n_1 does not move, its new tooth shape intersect at n_1' point, points n_1 and point n_1' to form the first class correlation point; The second kind is to produce the error after a_1n_1 is also corresponding to move, the length of the same,and its new tooth shape intersect at n_1' point, points n_1 and point n_1' to form the first class correlation point, here we first study the first type of correlation point. Fom Figure 1, we can get

$$a_1n_1 = a_0a_1 * \cos\alpha \tag{4}$$

According to the theory of engagement, when the instantaneous rotation center for serrated knife is a_1 point, n_1 point of rack tooth profile corresponds with n point on conjugate tooth profile, when serrated knife offsets to the left for Δx, the point of a_1 respectively corresponds with n_1 point on rack tooth profile and n_1' point on conjugate tooth profile, on the other hand, the section curve from the instantaneous rotation center to original point is determined by the following conditions.

$$S_1 = a_0a_1 \tag{5}$$

The serrated knife excursions also comply with the above rule, but the arc length changes. In Figure 2, vector equation of the tooth profile can be expressed by the following [8, 9].

$$\overline{r_f} = \overline{r_g} + \overline{an} \tag{6}$$

Right tooth profile equationis obtained when converting into rectangular coordinates.

$$\begin{cases} x_R = r_g \cos\theta + an\cos(\theta + u - \alpha) \\ y_R = r_g \sin\theta + an\sin(\theta + u - \alpha) \end{cases} \tag{7}$$

Left tooth profile equation

$$\begin{cases} x_L = r_g \cos\theta + a'n'\cos(\theta + u + \alpha - \pi) \\ y_L = r_g \sin\theta + a'n'\sin(\theta + u + \alpha - \pi) \end{cases} \tag{8}$$

The tooth profile normal length is acquired by the following

$$an = a_0a\cos\alpha_u = S * \cos\alpha_u \tag{9}$$

$$a'n' = a_0'a'\cos\alpha_u = S * \cos\alpha_u \tag{10}$$

The analytical formula for theory tooth profile is above one offset Δy in Figure 1 is as following after theory tooth profile

$$a_0a_0' = \Delta y \tag{11}$$

Simplified as error vector quantity

$$a_1n_1 = a_0a_1\cos\alpha_u \tag{12}$$

Formula is got after being converted into rectangular coordinate

$$\begin{cases} x_n' = x_n \mp a_0a_0'\cos\alpha_u tg\alpha_u \\ y_n' = y_n \mp a_0a_0'\cos\alpha_u ctg\alpha_u \end{cases} \tag{13}$$

Due to the preceding error model, we can get the tool clamping error is an actual case of reentry error. Tool clamping error is got when Δy is a constant value, when Δy is a variate value, in principle, we need to determine the relevant Δy value that each gear cutting angle of tooth profile should correspond with, and consider the variate value Δy as the serrated knife offset of corresponding angle. In reentry error conditions, offset Δy changes follow the change of the cutting angle and comply with the sine function law $\Delta y = k\sin\theta$, here, in order to simplify the problem, and highlight the contrast effect, we set the variation cycle of Δy as $T = 2\pi$, with the movement function that changes with angle, we can conveniently find the offset Δy that every knives of reentry error correspond, again in the program, create a loop statement of serrated knife offset that changes with angle and Δy, it can simulate the tooth profile diagram of the reentry errorand summarize the error code of reentry error from it. Actually, if the offset Δy changes according to other rules, we only need to focus on the regular function, although the final result is different, but the nature of the problem is still the same.

6 Theoretical explanations

We can learn from Figure 1 that tangential error changes generally in accordance with Δy and it is generally in accordance with the law of sine function, it also explains why tangential error is small when it is the long axis, and short axis is larger and the offset changes with sine function. In the range of $0 \sim \pi$, tangential error make the tooth profile migrate anticlockwise, in the range of $\pi \sim 2\pi$, tangential error make the tooth profile migrate clockwise, and because the values of Δy Maximum value are the same, tooth profile migrations are about the same size, and in the opposite direction.

By Eq. (13), we know due to the various curvature radiuses of points on the section curve, leading to the each included angles u of each point radius vector and the tangent line are not identical, the vectorial direction of the offset is not exactly the same, so it can be proved the correctness of the simulation conclusion.

From Figure 1 that cutter pitch curve and ellipse pitch curve tangent, and no method to offset, so the simulation results will be no radial error.

7 Experiment and classification

From the above analysis, we know the reason and theoretical explanation of the tool setting error, In the actual test, we can process the non-circular gear solid model which is shown in Figure 6 through the rapid prototyping system, Using the RoyalArm RA315 measuring arm (measurement accuracy of 0.05mm) which shows in Figure 7 on the key points of non-circular contrate profile actual coordinates value and theoretical coordinate value difference judgment whether to produce tool setting error or other tangential error.

The experimental method is as follows

(1) Non-circular gear processing must have a zero, that is, the initial processing position. In the model above the left end elliptic focus as the origin of coordinates, the right level for the positive direction of axis, vertical axis direction, establish coordinate system.

(2) Because of the large number of non-circular gear tooth, large quantities of one of the measurement of each tooth space for all positions of the geometric point in the actual production is not acceptable, so we at the end of the ellipse long axis and short axis at both ends of each selected a tooth, coordinate measurement of tooth top apex and tooth root value.

(3) The actual coordinates the collection value and coordinate theory are compared to the values, the zero cogging data must be collected, the case of zero cogging in long axis is the direction and the results are shown in Table 1.

(4) According to the results of the comparison of the results in Figure 9 can be classified to learn about the causes and sources of error. Because this article is only relates to cut to the field of error, Figure 9 relates to the radial error and the geometric deviation theory proof in my other are published articles are presented in detail. This only reference related to cut to the part of the conclusion of the error.

Figure 6. Solid model of non-circular gear

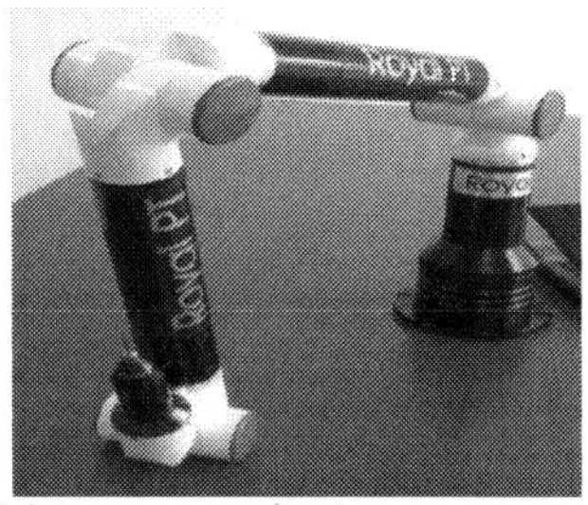

Figure 7. Joint arm measuring instrument

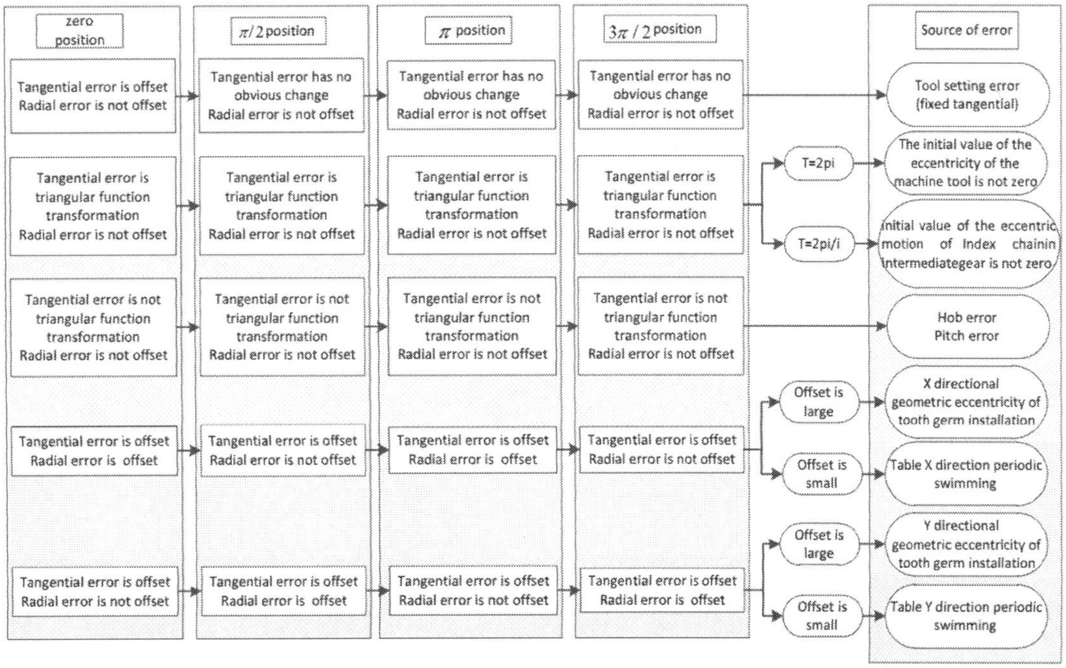

Figure 8. Tangential error factors of error genetic diagram

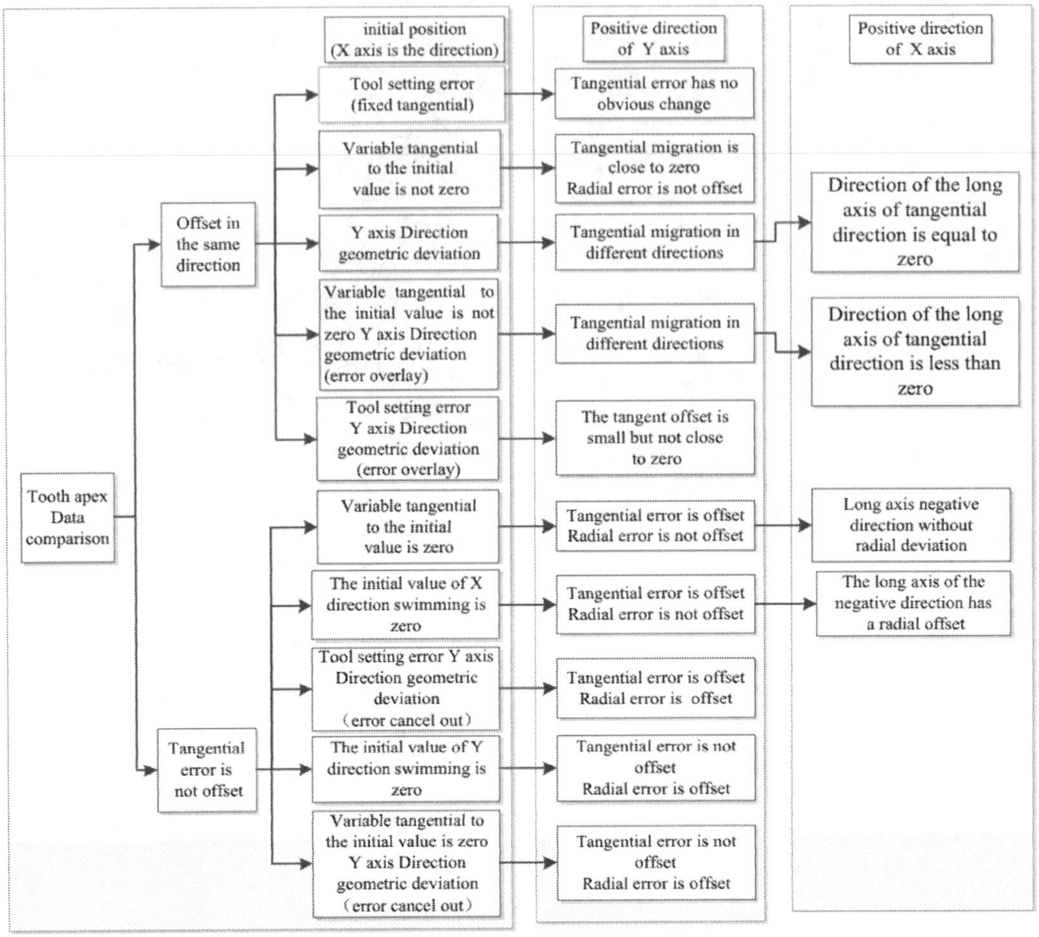

Figure 9. Characterization of characteristic code

Table 1. Tooth profile data sheet

	Long axis normal direction		Short axis normal direction		Long axis negative direction		Short axis negative direction	
X- theoretical tooth top apex	35.2954	35.2954	16.7847	14.3307	-5.4062	-5.9078	14.3250	16.7847
Y- theoretical tooth top apex	-1.4003	1.4003	13.9980	13.9331	3.0883	0.3393	-13.9980	-13.9330
X- actual tooth top apex	35.30	35.30	16.70	14.45	-5.40	-5.90	14.40	16.90
Y- actual tooth top apex	1.30	1.45	13.95	14.00	3.00	0.25	-14.00	-13.95
X- theoretical tooth of root	33.0952		15.4782		-3.5371		15.4745	
Y- theoretical tooth of root	0		11.7433		1.3606		-11.7436	
X- actual tooth of root	33.10		15.40		-3.55		15.55	
Y- actual tooth of root	0.10		11.75		1.25		-11.70	

In Ttable 1 at the coordinate data of zero tooth apex the conclusion can be obtained that the tooth profile is not offset, and the tangential direction is shifted in the same direction. From Figure 9, we know that this phenomenon may have five reasons. Tool setting error, the initial cutting error is not zero; the direction of geometrical deviations, the initial cutting error is zero and the direction of the geometric deviation of these two factors superimposed; tool setting error and the orientation geometric deviation of these two factors superimposed. The data of the short axis normal direction can be obtained in the radial direction, and there is no significant change in the tangential migration, from Figure 9 we know the model belongs to the typical causes of tool setting error.

From this example we get the test method of the origin of the tangential error and the characterization of the genetic code. In the actual production we are more concerned about how to quickly troubleshoot what are the process error factors that cause the error, so in Figure 9 based combined with common cause of cut to error of process factors are summarized Figure 8 non-circular gear tangential error factors of the error genetic diagram. So in Figure 9 based on combination of common cause of cutting error of process factors to get Figure 8 non-circular gear tangential error factors of the error genetic diagram. This will provide a powerful tool for the production of the cause of the problem quickly.

8 Conclusions

Via the system analysis and the theoretical proof of the computer simulation results for non-circular gear reentry error, we may obtain the conclusion that the reentry error is a tangential error that changes by trigonometric function, it's beneficial for the research complement of the non-circular gear error system and lay a solid foundation for the subsequent individual error of other conditions, and got the condition code of the reentry error excursion, providing theory support for timely finding out the cause of the gear errors when the factory inspection, thus improve the machining precision of the non-circular gear.

References

1. Baoni, L. I., Zhang, Y. The application and development trend of noncircular gear. Machine Tool & Hydraulics. **4**, 098 (2008)
2. Guoxing Sun, Chuanqiong Sun, Yongde Liu. The application prospect of non-circular gear on infinitely variable speed, Equipment maintenance technology, **1**, 1-4 (2009)
3. Xueqiang Gao. Transmission technology overview of non-circular gear, Mechanical drive, **3**, 5-8 (2003)
4. Huiping Yan, Chibing Hu, Limei Liu. The research of straight tooth non-circular gear hobbing machining error analysis, Mechanical drive, **4**,18-20 (2007)
5. Fangyong Tian, Chibing Hu, Yancang Jiang. An automatic tool setting method of non-circular gear hobbing, Mechanical research and application, **4**, 41-43 (2010)
6. Liqun Wu, Hongxia Bi. The determination of axial special-shaped gear NC hobbing machining adjusted cutter point coordinates, Coal mine machinery, **6**, 109-112 (2008)
7. Yongping Liu, Peng Wang, Jia Li. The elliptical gear tooth profile design via analytical method, Machinery Design and Manufacture, **2**, 214-216 (2015)
8. Fusheng Li. Non-circular gear, *China Machine Press* (1975)
9. Mingtong Wang. Involute gear accuracy, *China Machine Press* (1983)

MATEC Web of Conferences 77, 02005 (2016) DOI: 10.1051/matecconf/20167702005
ICMMR 2016

Definition of Turbulent Boundary-Layer with Entropy Concept

Zhao Rui[1], Rong Jili[1], LI Haibo[2] and Ren Fang[2]

[1] *School of Aerospace Engineering, Beijing Institute of Technology, Beijing 100191, P. R. China*
[2] *Beijing Institute of Structure and Environment Engineering, Beijing 100076, P.R. China*

Abstract. The relationship between the entropy increment and the viscosity dissipation in turbulent boundary-layer is systematically investigated. Through theoretical analysis and direct numerical simulation (DNS), an entropy function f_s is proposed to distinguish the turbulent boundary-layer from the external flow. This approach is proved to be reliable after comparing its performance in the following complex flows, namely, low-speed airfoil flows with different wall temperature, supersonic cavity-ramp flow dominated by the combination of free-shear layer, larger recirculation and shocks, and the hypersonic flow past an aeroplane configuration. Moreover, f_s is deduced from the point of energy, independent of any particular turbulent quantities. That is, this entropy concept could be utilized by other engineering applications related with turbulent boundary-layer, such as turbulence modelling transition prediction and engineering thermal protection.

1 Introduction

Entropy, in addition to energy, is a central physical quantity in thermodynamics. It serves as a measure of the irreversibility of a process and a criterion describing the thermal equilibrium of a system. It also represents the degree of disorder within a system [1]. Because of its general features, the entropy concept has been extended to many other fields, such as thermodynamic optimization, information theory and computational techniques [2]-[6].

As for the turbulent flow, McEligot et al. [7] concluded that about two-thirds or more of entropy generation occurs in the turbulent boundary-layer. Then, Zhao et al. attempted to denote the turbulent boundary-layer by the entropy concept. Through Direct Numerical Simulation (DNS) of supersonic compression ramp flows, they found that the value of entropy is much larger in the boundary-layer and keeps decreasing monotonously in the wall normal direction [8]. Based on this character, they revised the length scale of Baldwin-Lomax model, enhancing its robustness in simulating supersonic complex flows. Although these achievements indicate another latent capacity of entropy, the definition of entropy expression for turbulent boundary-layer is still not comprehensive.

This paper continues the previous work, and tries to extend this entropy concept from both the theoretical and numerical aspects. We hope this investigation here could be served as another approach for turbulence modeling, transition prediction, etc., where the calculated turbulent boundary-layer is always necessary.

2 Entropy expression

Virtually all flows of practical engineering interest are turbulent. Due to the irregular motion of turbulence, the transportations of mass, momentum and energy are enhanced, while extra energy is dissipated. In the near wall region, as for the fierce turbulent fluctuation and wall frication, a portion of mechanical energy is irreversibly transformed into internal energy, i.e. the entropy increase. It should be noted that a similar process also occurs due to the wall heat flux or the deceleration of flows where shock waves appear.

For the perfect gas model, the relation of entropy increment and energy change is revised by

$$\mathrm{d}s = \frac{\mathrm{d}q}{T} \tag{1}$$

where s is entropy, q is the heat transfer, and T is the local temperature. This form of the definition is different from the usual one [9] in two aspects: (1) q is the actual heat added on the system, no longer limited to the one caused by the reversible process only. (2) Correspondingly, the entropy increment ds is proportional to the exact differential dq.

With the conservation law of energy, also called First Law of Thermodynamic

$$\mathrm{d}q = \mathrm{d}u + p\mathrm{d}(\frac{1}{\rho}) \tag{2}$$

substituted into Eq. (1), we can finally get

$$ds = c_V \mathrm{d}(\ln \frac{p}{\rho^\gamma}) \tag{3}$$

where $u=c_VT$ is the internal energy, $c_V=R/(\gamma-1)$ is the specific heat at constant volume, R is the gas constant, $\gamma=1.4$ is the specific heat ratio, and p and ρ are the local pressure and density, respectively. As Eq. (3) shows, s is proportional to p/ρ^γ. We define entropy in this research as

$$s = \frac{p}{\rho^\gamma} \tag{4}$$

As aforementioned, the increment of s is proportional to the heat transfer q, no matter where the heat comes from (Eq. (1)). However, in order to employ the increment of s to discern the boundary-layer, the first step we should do is to separate the heat caused by viscosity dissipation from the total, especially from the wall heat flux due to both are mainly produced in the near wall region.

2.1 Process I: Entropy increment produced by viscosity dissipation

One of the commonly used temperature-velocity relationships for zero-pressure gradient boundary layers is Walz's equation [10].

$$\frac{T}{T_\infty} = \frac{T_w}{T_\infty} + \frac{T_r - T_w}{T_\infty}(\frac{u}{u_\infty}) + \frac{T_\infty - T_r}{T_\infty}(\frac{u}{u_\infty})^2 \tag{5}$$

in which the subscript ∞ denotes the free-stream values and the recovery temperature $T_r=T_\infty(1+r(\gamma-1)Ma_\infty^2/2)$ with the recovery factor $r=0.9$.

Differencing each side at the wall by multiplying the heat conductivity coefficient k, we can get the Reynolds analogy relationship

$$-k\frac{\partial T}{\partial y}\Big|_w = \frac{T_w - T_r}{(\gamma-1)\mathrm{Pr_t}\, Ma_\infty^2 T_\infty} \times u_\infty \mu \frac{\partial u}{\partial y}\Big|_w \tag{6}$$

where $k=\mu c_p/Pr_t$, $c_p=R\gamma/(\gamma-1)$ is the specific heat at constant pressure, and $Pr_t=0.9$ is the turbulent Prandtl number.

Obviously, the left term in Eq. (6) is the heat caused by the wall heat flux (Q_2) and the second term on the right is approximately the heat produced by the viscosity dissipation of kinetic energy (Q_1). Thus, in the near wall region, we may separate the sources of heat by defining a new parameter σ,

$$\sigma = \frac{T_w - T_r}{(\gamma-1)\mathrm{Pr_t}\, Ma_\infty^2 T_\infty} \approx \frac{Q_2}{Q_1} \tag{7}$$

As Eq. (7) implies, σ is determinate once the boundary condition is given. The behavior of σ is portrayed in Fig. 1. When $\sigma>0$, it means the wall heats the fluid up; $\sigma=0$ means the adiabatic wall condition; $\sigma<0$ means the heat in the fluid transfers into the wall. The influence of wall heat flux is obvious at low-speed flows. With the free-stream Mach number increased, the value of σ is tending to be constant, i.e. -0.496, which suggests the proportion of wall heat flux to the viscosity dissipation rate of kinetic energy is to be constant in hypersonic flows.

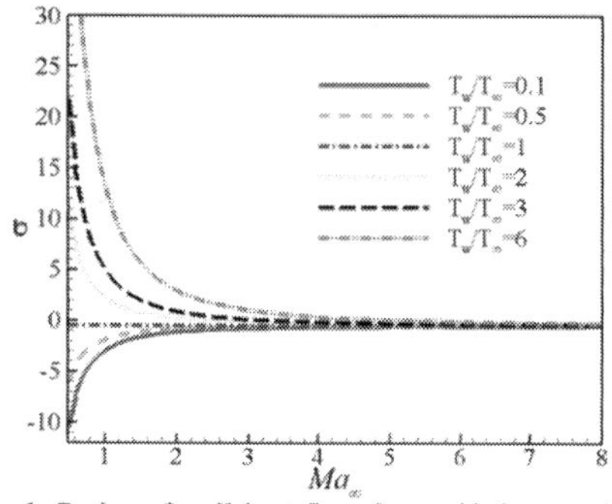

Figure 1. Ratios of wall heat flux along with increased Mach numbers

Assuming σ affects the whole boundary layer and its impact is uniform, in the near wall region, the nondimensional entropy increment caused by viscosity dissipation solely could be roughly evaluated by

$$\Delta s_{\mathrm{vis}} = \frac{1}{\sigma+1}(\frac{s}{s_\infty} - 1) \tag{8}$$

Table 1. Simulation conditions for boundary-layer flows

	Ma_∞	Re	T_w/T_∞	T_w/T_r
Case1	0.7	5.0×10^4	1.098	1.003
Case2	2.25	6.35×10^5	1.9	0.961
Case3	6	2.0×10^6	6.98	0.878

2.2 Process II: Entropy increment ratio

When the external potential flows pass the vehicle surface, the velocity is gradually decreased to zero due to the viscous frication. Hence, it is natural to foresee that the values of entropy increment near the wall are directly related with inflow speeds. In order to get rid of this dependence on Mach numbers, a new entropy concept, named entropy increment ratio ($\bar{s}$), is introduced.

$$\bar{s} = \frac{\Delta s_{\mathrm{vis}}}{Ma_\infty^2} \tag{9}$$

$\bar{s}$ represents the viscous dissipation rate of per unit mechanical energy. We extract corresponding quantities from DNS results to evaluate its character. Table 1 shows the simulation conditions of DNS for flat plates. The incoming flows are subsonic, supersonic and hypersonic, respectively. As aforementioned, the values of Δs_{vis} near the wall are quite different, varying about one order of magnitude from $Ma=0.7$ to $Ma=6$ (Fig. 2a). On the other hand, $\bar{s}$ exhibits a consistent behavior depicted in Fig. 2b. The values near the wall are all falling into the range of

0.1-0.5, unrelated with Mach numbers. This character is convenient as we always hope a model or physical parameter is general for various flow conditions. Additionally, we could also find that the range where $\bar{s}>0$ is almost equal to the extent where $u/U_\infty<1.0$, which means that $\bar{s}$ could be used to discern the boundary-layer for these simple flows.

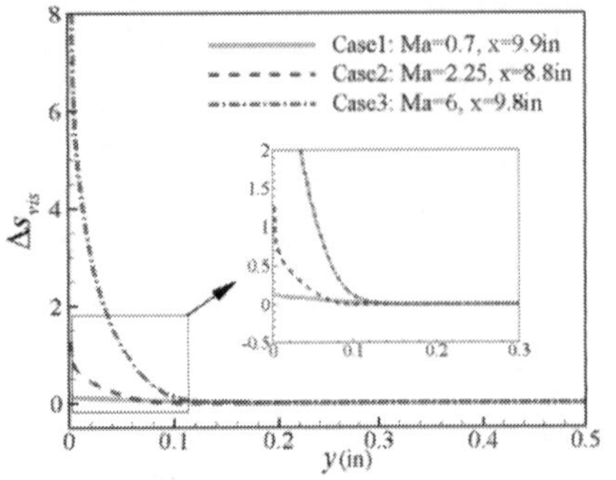

a. Entropy increment profiles normal to the wall

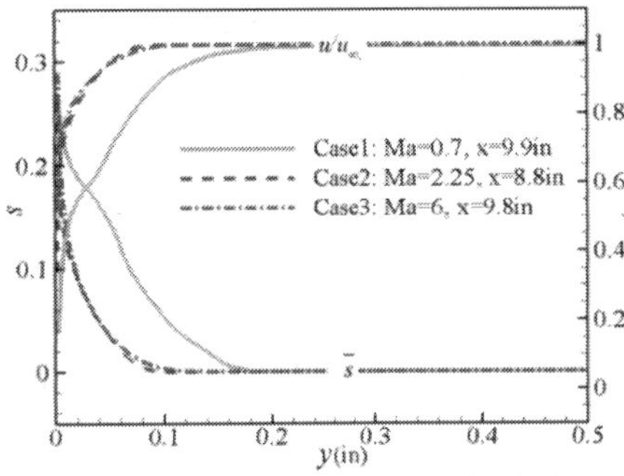

b. Comparison of entropy increment ratio and velocity profiles

Figure 2. Entropy and entropy increment ratio profiles normal to the wall of different Mach numbers (DNS results)

2.3 Process III: Entropy function f_s

After the above two steps, the proposed entropy increment ratio $\bar{s}$ could actually denote the boundary-layers for simple attached flows and even relatively complex separated flows. However, at the external flow region of a typical super/hypersonic flowfield, the values of $\bar{s}$ across shocks or free-shear layers would be in the same magnitude as that in the boundary-layer. To alleviate this inconvenience, we introduce the entropy function f_s by

$$f_s = 1.0 - \tanh(\bar{s}/l_s^2) \quad (10)$$

where l_s is the subgrid length-scale, which is designed to be less than 1.0 in the boundary-layer and increase quickly in the external flows. We construct l_s following the spirit of RANS/LES hybrid strategy [11].

$$l_s = C_g \kappa d_w / C_{DES}\Delta \quad (11)$$

in which $C_g = 3\sqrt[3]{\Delta_x\Delta_y\Delta_z}/(\Delta_x+\Delta_y+\Delta_z)$ is the grid deformation ratio, κ=0.41 and C_{DES}=0.65 are von Karman's and Smagorinsky constants respectively, and Δ is the grid spacing defined by Δ=max(Δx, Δy, Δz). C_g is less than 1.0 in the anisotropy grids near the wall and around 1.0 in the typical LES region. The combination of κd_w represents the Prandtl-van Driest RANS length scale, while $C_{DES}\Delta$ is the LES subgrid scale employed by DES-like methods.

The character of entropy function f_s is depicted in Fig. 3, by redrawing Fig. 2b according to Eq. (10). Compared with the velocity profiles, f_s could denote the boundary-layer with the range $f_s<1.0$ while it exhibit much smoother profiles than those of $\bar{s}$.

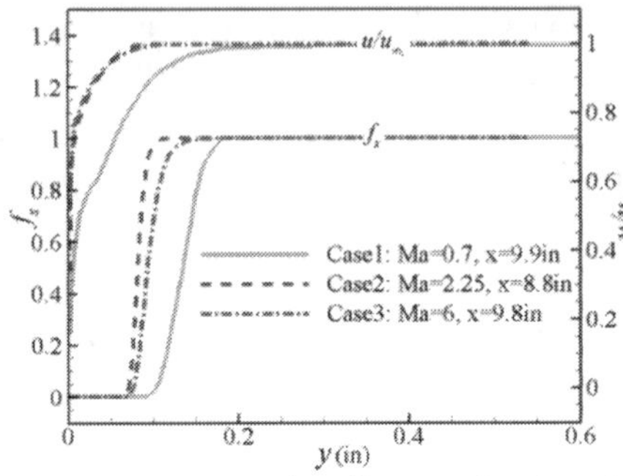

Figure 3. Entropy function f_s profiles normal to the wall (DNS results)

3 Applications

By including more test cases, this section provides detailed complements to the derivation of the entropy function f_s and evaluates its performance in complex flows. The governing equations describing the mean flowfield are the time-dependent, compressible Reynolds-averaged Navier-Stokes equations, along with one-equation Spalart-Allmaras turbulence model. The three dimensional RANS equations are solved by the finite volume method with structured grids and advanced in time by LU-SGS method. The second-order Roe scheme is used for the discretization of the inviscid terms, while the viscous terms are centrally differenced.

3.1 Low-speed A-airfoil flow (*Ma*=0.15)

As clarified in Process I, the portion of entropy increment caused by the wall heat flux is more prominent in low-speed flows (Fig. 1). This case is chosen to exhibit the influence of wall temperature and verify the effect of σ.

The details of inflow conditions and grid construction are available in [12]. The temperature of the incoming flow is 276.48 K and we prescribe three kinds of wall temperature for comparisons (Table 2). In the cold wall condition, the heat is transferred from the fluid to the wall and the total entropy increment Δs becomes negative in the near wall region (Fig. 4a). Comparatively, the hot wall heats the fluid up and Δs is much larger than that in the adiabatic wall condition (Fig. 4b, c).

After removing the effect of wall heat flux by introducing the parameter σ, the values of entropy increments Δs_{vis} fall to the same level, as can be seen in

Fig. 5. However, due to the large separation bubble in the hot wall condition, the area where $\Delta s_{vis}>0$ seems too large to denote the boundary-layer. Nevertheless, with the final Process III, the entropy function f_s could predict a proper boundary-layer region (Fig. 6).

Table 2. Three kinds of wall temperature for the A-airfoil

Wall conditions	T_w (K)	T_w/T_r	σ
Cold wall	100	0.36	-79.34
Hot wall	1000	3.6	322.54
Adiabatic wall	277.68	1	0

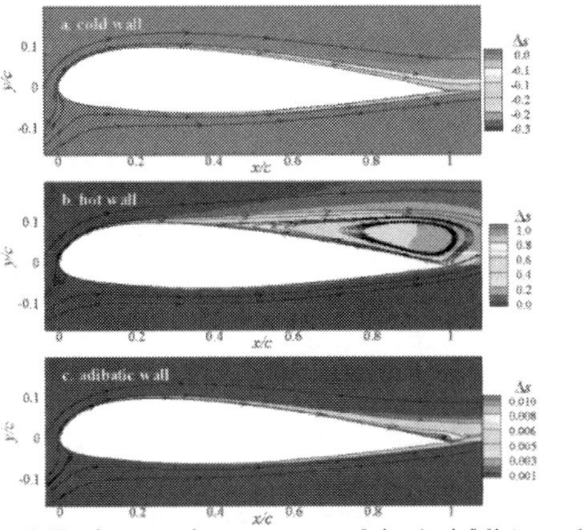

Figure 4. Total entropy increment around the A-airfoil (caused by both viscosity dissipation and wall heat flux)

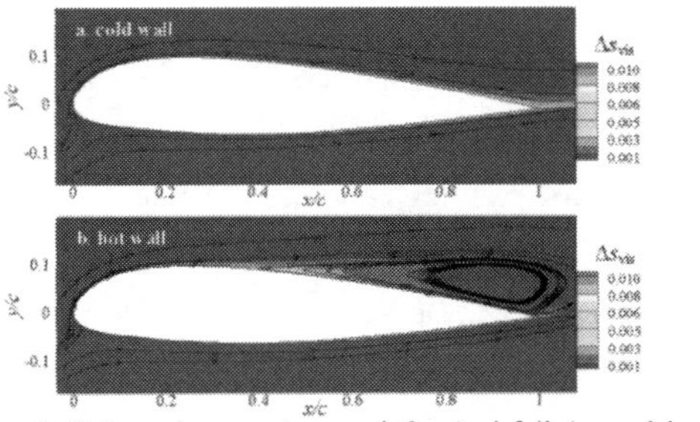

Figure 5. Entropy increment around the A-airfoil (caused by viscosity dissipation only)

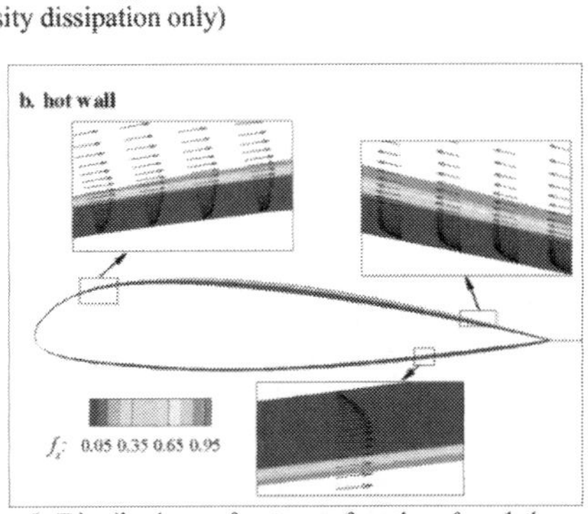

Figure 6. Distributions of entropy function f_s and the velocity vectors (only depict f_s<0.95 for convenient)

3.2 Supersonic cavity-ramp flow (*Ma*=2.92)

This case is chosen to evaluate the performance of f_s in a typical flowfield. The dominant feature is the supersonic free-shear layer, under which a large recirculation zone fills the whole cavity. On the ramped portion of the cavity, the free-shear layer reattaches to form a turbulent boundary-layer while a compression fan coalesces into a shock wave downstream of the reattachment point, depicted in Fig. 7.

The detailed inflow conditions could be found in [13]. As the wall boundary is adiabatic, we focus on the entropy increments in the external flow.

In this complex flow, entropy is not only increased in the boundary-layer, but in the free-shear layer, the recirculation and across the shocks. To alleviate these unfavorable interactions, the hybrid strategy is adopted through Process III. In general, f_s could detect an acceptable boundary-layer region, compared with the velocity profiles (Fig. 8). But in the cavity where the large recirculation fills, the thickness of the boundary-layer is overestimated by f_s. In this area, the value of $\overline{s}$ is almost uniform, and then the distribution of f_s is mainly depending on the local grid resolution according to Eq. (10).

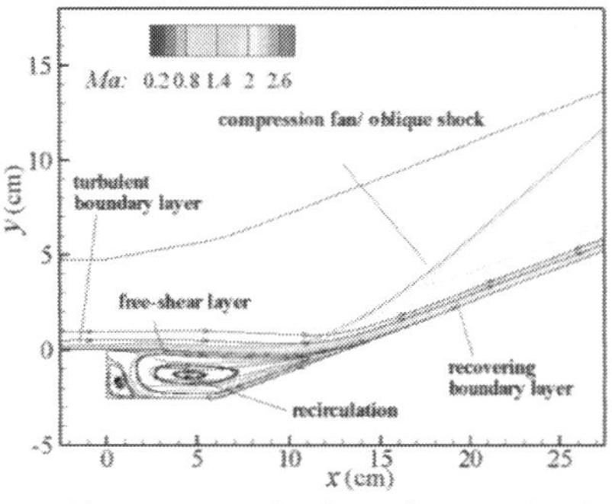

Figure 7. Flow structures for the cavity-ramp (Mach number contours)

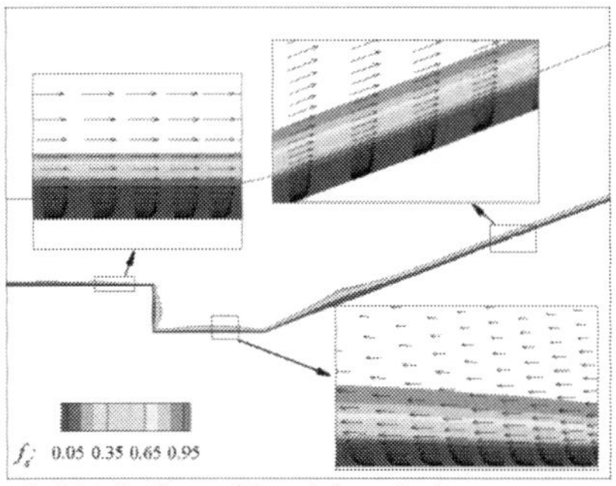

Figure 8. Distributions of fs and the velocity vectors around the cavity-ramp (only depict fs<0.95 for convenient)

3.3 Hypersonic aeroplane flow (*Ma*=6)

For engineering interests, the flow quantities at the edge of the boundary-layer, such as pressure P_e and temperature T_e, are essential for thermal shielding design.

DOI: 10.1051/matecconf/20167702005

This case is served as an attempt of f_s in denoting the boundary-layer for the engineering application. The flight conditions of this aeroplane configuration are summarized in Table 3. We clustered the grids to all solid surfaces and in the junctions of the body and the wings, ensuring the distance of the first grid line to the wall y^+ is less than 1.0.

As for this configuration, the flow structure is so complex that it is hard to find a sound principle to describe the boundary-layer quantitatively. We choose the streamwise velocity profiles as a compromise. From the comparisons of the three slices at different locations (Fig. 9), the entropy function f_s seems qualified to predict a proper boundary-layer region.

Table 3. Flight conditions of the aeroplane configuration

Ma_∞	H(km)	α(°)	T_w (K)
6	30	0	300

* H: flight height; α: angle of attack

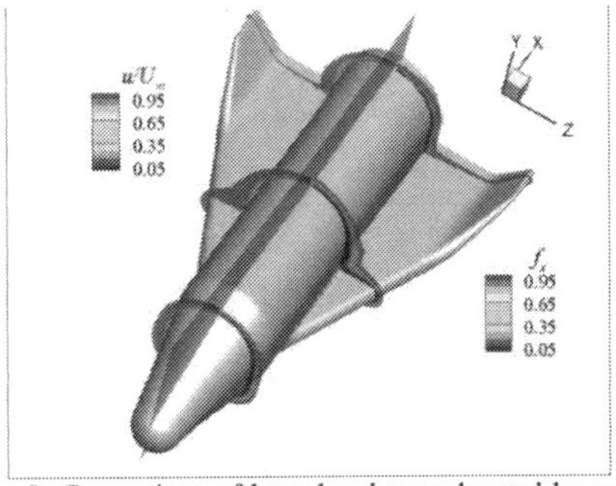

Figure 9. Comparison of boundary-layers denoted by velocity (left, only depict u/U_∞ <0.95 for convenient) and f_s (right, only depict f_s<0.95 for convenient) respectively

4 Conclusion

(1) The entropy expression for turbulent boundary-layer is derived and discussed. First, in the near wall region, we remove the effect of wall heat flux through Walz's temperature-velocity relationships. Second, the entropy increment ratio ($\bar{s}$) is introduced with the benefit of Mach number independence, proved by DNS results. Finally, the entropy increment in the external flow is avoided by including the subgrid length-scale l_s according to the hybrid strategy. Then, the entropy increment in the flowfield is only related with the viscosity dissipation in the boundary-layer.

(2) With the above procedures, the proposed entropy function f_s could reliably denote the boundary-layer region for complex flows. However, in the recirculation zone, the performance of f_s is mainly affected by the local grid resolution. This defect is left for our future work.

(3) This entropy function f_s is proposed from the energy aspect, independent on any particular turbulent quantities. Although it cannot be that the absolute best formulation was found to denote the boundary-layer, the present one for f_s appears simple and permanent enough to be recommended.

References

1. G. J. Hu, B. Y. Cao, Z. Y. Guo. Entransy and entropy revisited. Chin Sci Bull, **27**, 2974-2977 (2011).
2. G. F. Naterer, J. A. Camberos. The role of entropy and the second law in computational thermofluids. *AIAA paper*, AIAA-2001-2758, (2001).
3. H. B. Gillian, J. L. Mark. Principle of minimum entropy analysis applied to unsteady flow in hypersonic three-dimensional inlets. *AIAA paper*, AIAA-2011-2307, (2011).
4. C. E. Shannon. A mathematical theory of communication. Bell Syst Tech J, **27**, 379-423, (1948)
5. K. J. Fidkowski, P. L. Roe. Entropy-based mesh refinement, I: the entropy adjoint approach. *AIAA paper*, AIAA -2009-3790, (2009).
6. D. W. Zaide, P. L. Roe. Entropy-based mesh refinement, II: A New Approach to Mesh Movement. *AIAA paper*, AIAA-2009-3791, (2009).
7. R. Zhao, C. Yan, J. Yu. A new kind Baldwin-Lomax turbulence model under the limit of entropy. J. Beijing Univ. Aero. Astro, **38**(2):175-179, (2012).
8. D. M. McEligot, E. J. Walsh, E. Laurien, et al. Entropy generation in the viscous parts of turbulent boundary layers. J Fluids Engrg, **130**:61205, (2008).
9. A. R. Paterson. *A First Course in Fluid Dynamics* (Cambridge University Press), (1983).
10. A. Walz. *Boundary Layers of Flow and Temperature*. (MIT Press), (1969).
11. P. R. Spalart. Detached-eddy simulation, Annu. Rev. Fluid Mech., **41**, 181-202, (2009).
12. N. Durrani, N. Qin. Behaviour of detached-eddy simulations for mild airfoil trailing-edge separation. J. Aircraft, **48**(1), 193-202, (2011).
13. G. S. Settles, D. R. Williams, B. K. Baca, et al. Reattachment of a compressible turbulent free shear layer. AIAA J., **20**(1): 60-67, (1982).

MATEC Web of Conferences 77, 02006 (2016) DOI: 10.1051/matecconf/20167702006
ICMMR 2016

Capacitance Online Estimation Based on Adaptive Model Observer

Zhaohui Cen

Qatar Environment and Energy Research Institute, HBKU, Qatar Foundation, 5825, Doha, Qatar

Abstract. As a basic component in electrical and electronic devices, capacitors are very popular in electrical circuits. Conventional capacitors such as electrotype capacitors are easy to degradation, aging and fatigue due to long-time running and outer damages such as mechanical and electrical stresses. In this paper, a novel online capacitance measurement/estimation approach is proposed. Firstly, an Adaptive Model Observer (AMO) is designed based on the capacitor's circuit equations. Secondly, the AMO's stability and convergence are analysed and discussed. Finally, Capacitors with different capacitance and different initial voltages in a buck converter topology are tested and validated. Simulation results demonstrate the effectiveness and superiority of our proposed approach.

1 Introduction

Due to their penetration in electrical and electronic circuits, capacitors especially super-capacitors play a key role in information technologies and electrical engineering. There are many types of capacitors in the market, their conductors may be thin films, foils or sintered beads of metal or conductive electrolyte, etc. Also, the capacitors will be used in different application scenarios based on different capabilities such as filtering and energy storage. Because of its manufacturing drawbacks, long-time running and outer damages such as mechanical and electrical stresses, capacitors are easy to degrade and fatigue, which will cause the rated capacitance vary and probably cause failure for overall system[1], [2]. For example, many computers cannot be restarted because the capacitors in the mainboard are damaged in a form of capacitor plague. In some cases, the capacitors may cause an exploration due to the electrolyte chemistry reaction, which is a possible cause for some disasters. Therefore, it is necessary to monitoring the capacitor's status and conditions in real-time in some applications, especially power system applications such as power converters. Regarding capacitor's condition parameters, mostly used parameter is the measured capacitance value because the measured capacitance can directly indicate whether a capacitor is working in the rated capacitance, and how much the capacitance variation is because of degradation and aging.

Currently, there are few approaches to measure and monitor the capacitors' parameters and conditions. The measurement parameters mainly include capacitance, internal resistance and impedances, where capacitance is mostly adopted to measure and evaluate the status of a capacitor [3]. Regarding capacitance measurement approaches, we can classify existed methods into two classifies: offline measurement and online measurement [4]–[8]. The off-line measurement is easy to implement because most multi-meters can measure the capacitance by touching two pins of a capacitor, the drawbacks of this measurement is that, the capacitor need to be off-line and even uninstalled from the board, because the multi-meter need to use its self-generated stimulating signal to acquire and calculate the capacitance. In some improved off-line approach (may be called as quasi-online method), the off-line and online mode may be switched in a short time by the test logic control in order to compute the capacitance, but actually it is still not the best solution because the capacitor still need to be offline more or less, which is not applicable in some power applications. Aimed at the offline method drawbacks, some online capacitance estimation methods are proposed to estimate the capacitance of capacitors in wind turbo applications [9]. By injecting stimulating signal at some specified frequencies into power source or voltage of the capacitor, the capacitance can be estimated based on the response of capacitor voltage in the specified frequency. The drawback of this approach is that, the operation is very complicated, and also how to set frequency for capacitors with different capacitance and how to filtering the voltage signal is not clear for a real-world applications. Actually, the best desirable approach is that we don't need to change anything for the test capacitor and keep it online for capacitance estimation based on existed measurement signal and devices such as voltage and current sensors.

Motived from the demand in this regarding, the main contribution of this paper is that a novel adaptive model observer method is proposed to address the online capacitance estimation challenges. Compared with existed capacitance measurement approaches, the proposed approach has an obvious advantage: it can keep

 DOI: 10.1051/matecconf/20167702006

the test capacitor online and only need the voltage and current signals from the capacitor pins.

2 Problem to be addressed

In this paper, we will address the capacitance drafting problem. We use a typical case of capacitance shown in Fig. 1. The rated capacitance in the capacitor is 1000uF, but the actual capacitance is 839.2uF, which means the capacitance drafting is around 15%.

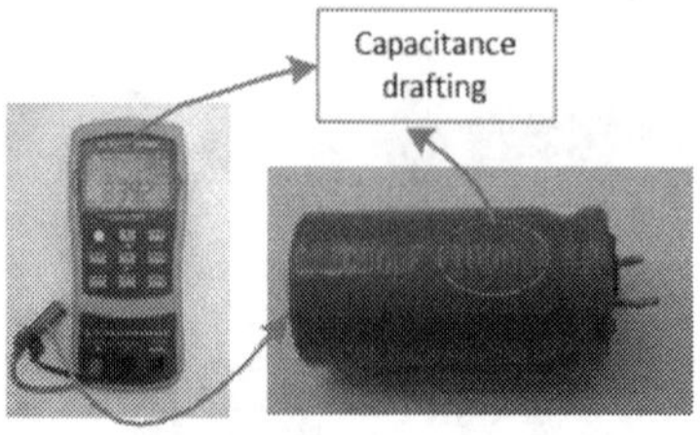

Figure 1. Capacitance drafting

So, it is needed to monitor the capacitance in real-time and online conditions. For capacitors, they can be generally denoted as its Current–voltage relation equation as following.

$$I(t) = \frac{dQ(t)}{dt} = C\frac{dV(t)}{dt} \tag{1}$$

For measurement, the capacitor voltage $V(t)$ and current $I(t)$ are available variables by senor online measurement. The problem is how to estimate the C of (1) in real-time.

3 Capacitance estimation scheme based on adaptive model observer

In order to estimate the capacitance in real-time, we designed a special adaptive observer to tracking and estimate the state and capacitance.

3.1 Adaptive model observer

Based on the current-voltage relationship in (1), we propose an adaptive model observer in the format as following.

$$C\frac{d\hat{V}(t)}{dt} - I(t) - K(V(t) - \hat{V}(t)) = 0 \tag{2}$$

where K is the observer feedback gain. If we select the proper K , the observer state will follow as $V(t) - \hat{V}(t) \to 0$.

So, based on convergence principle as shown above, a capacitance estimation scheme can be derived as following.

$$\begin{aligned} &\hat{C}\frac{d\hat{V}(t)}{dt} - I(t) = 0 \\ &\hat{C} = K(V(t) - \hat{V}(t)) \end{aligned} \tag{3}$$

By the feedback control from K , the capacitance can be estimated based on available variables such as the capacitor voltage $V(t)$ and current $I(t)$

3.2 Online estimation scheme

The online estimation scheme based on (3) can be shown in Fig. 2. Fig. 2 depicts the capacitance estimation scheme for a general capacitor, which can be used for real-time monitoring for any circuit topologies once voltage and current measurement is available. As can be seen from Fig. 1, the capacitor voltage and current need to be measured and sent into DSP controller for processing. The adaptive model observer estimates the online capacitance based on the observer feedback loop and adaptive estimation law. In this scheme, different initial

Capacitance under different initial value of the integer 1/s will converge into the estimated capacitance under the adaptive estimation law.

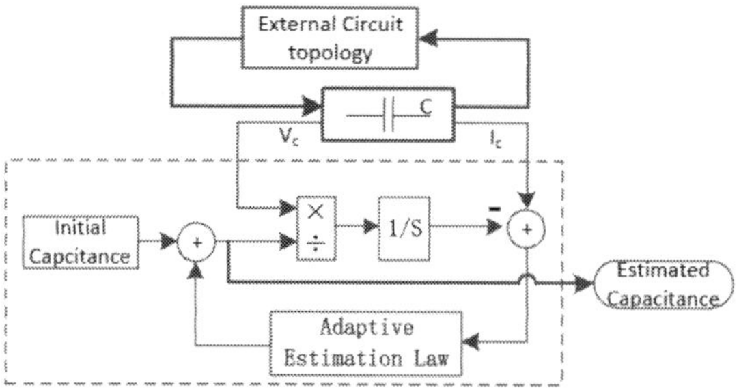

Figure 2. Online estimation scheme for general capacitors

4 Simulation results

In order to validate the effectiveness of our proposed method, we use a typical converter topology which include two capacitors (Shown in Fig. 2) to validate our proposed scheme. In this topology, two capacitors C1 (350uF) and C2 (100uF) are used for stabilize the input voltage and output voltage respectively.

The capacitance online estimation result of the capacitor C1 and C2 can be seen in Fig. 4 and Fig. 3 respectively. As can be seen from Fig. 4 (a) and (b), the capacitor voltage and current are variable with time and the transient characteristics are very obvious. As can be from Fig. 4 (c), the voltage residual can converge very quickly under the feedback gain control. As can be seen from Fig. 4(d), the estimation can track the real capacitance in a fast speed. Fig. 5 shows the online estimation result for capacitor C2. As can be seen from Fig. 5 (a) and (b), although the voltage and current is stable and different from C1, the online estimation scheme can also track the capacitance value 100uF very quickly.

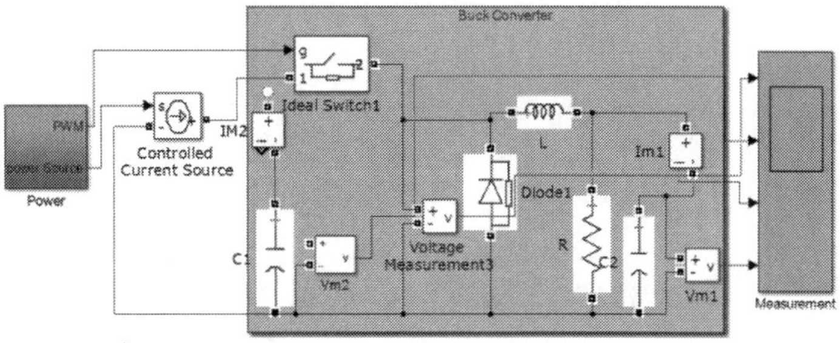

Figure 3. Test Buck converter topology

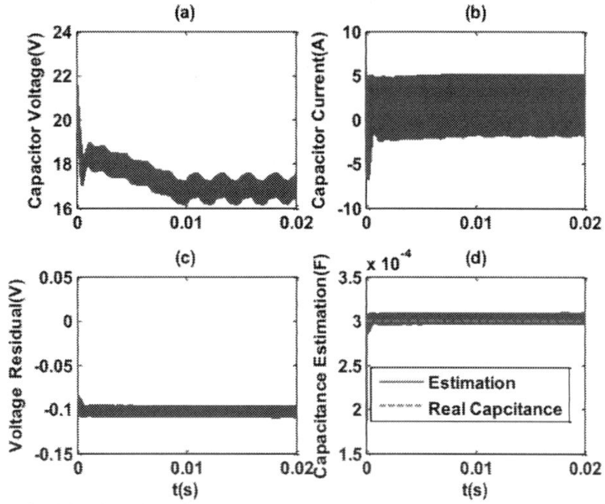

Figure 4. Online estimation result of the Capacitor C1

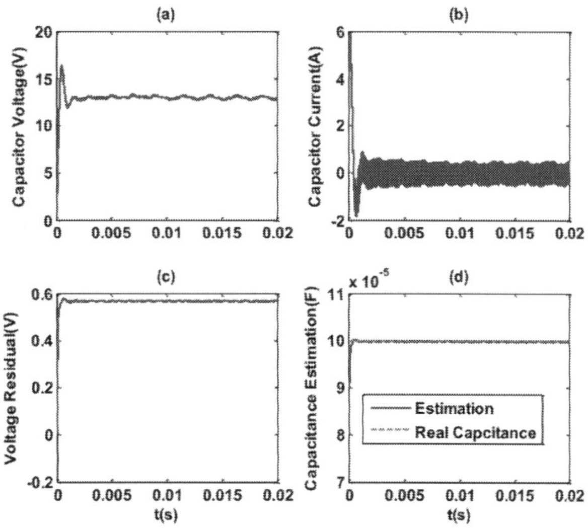

Figure 5. Online estimation results of the Capacitor C

5 Conclusions

In this paper, a novel capacitance estimation scheme is proposed to monitor capacitor s and estimate their capacitance. A buck converter consisting of two capacitors are employed to test and validate the effectiveness of the proposed scheme. Simulation results demonstrate that the capacitance online estimation scheme can track the real capacitance in a quick speed. Also, this scheme is applicable for capacitors in any topology no matter that it is online. Compared with existed approaches, our approach is more general for any capacitors and implements real-time monitoring. Our future work will focus on experiment demonstration and also consider the internal resistance's effects.

References

1. A. M. Imam, T. G. Habetler, R. G. Harley, and D. M. Divan, "Condition monitoring of electrolytic capacitor in power electronic circuits using adaptive filter modeling," in 2005 IEEE 36th Power Electronics Specialists Conference, (2005).
2. D.-C. Lee, K.-J. Lee, J.-K. Seok, and J.-W. Choi, "Online capacitance estimation of DC-link electrolytic capacitors for three-phase AC/DC/AC PWM converters using recursive least squares method," in Electric Power Applications, IEE Proceedings-, **152**, no. 6, pp. 1503–1508, (2005).
3. C. Kulkarni, G. Biswas, X. Koutsoukos, J. Celaya, and K. Goebel, "Integrated diagnostic/prognostic experimental setup for capacitor degradation and health monitoring," in AUTOTESTCON, 2010 IEEE, pp. 1–7, (2010).
4. G. Yifeng, "Analysis and design of the super capacitor monitoring system of hybrid electric vehicles," Procedia Eng., **15**, pp. 90–94, (2011).
5. P. XUE, J. GAO, Y. ZHANG, X. WANG, and G. WANG, "Discussion about the Way to Measure Capacitance and dielectric Loss of Circuit Breaker Capacitor [J]," Power Capacit., **4**, (2007).
6. B. R. Lee, "Method for monitoring the capacitance of a capacitor while adjusting its capacitance." Google Patents, 30-May-1972.
7. R. A. Kumar, M. S. Suresh, and J. Nagaraju, "Time domain technique to measure solar cell capacitance," Rev. Sci. Instrum., **74**, no. 7, pp. 3516–3519, (2003).
8. A. D. Kompelien, "Capacitance monitoring bridge circuit for an enthalpy responsive device." Google Patents, Dec. 17, (1985).
9. A. G. Abo-Khalil and D.-C. Lee, "DC-link capacitance estimation in AC/DC/AC PWM converters using voltage injection," Ind. Appl. IEEE Trans., **44**, no. 5, pp. 1631–1637, (2008).

MATEC Web of Conferences 77, 02007 (2016) DOI: 10.1051/matecconf/20167702007
ICMMR 2016

Some Applications of Generalized Char-Sets of Ordinary Differential Polynomial Sets

Farkhanda Afzal

School of Mathematics and Systems Science, Beihang University, Beijing, China

Abstract. The notion of characteristic sets, which are a special kind of triangular sets, is introduced by J. F Ritt and W.T. Wu. Wu extended Ritt's work and developed the characteristic set method not only in theory but in algorithms, efficiency and its numerous applications. Triangular sets are widely considered as a good representation for the solution of polynomial systems. After the introduction of characteristic sets by Ritt, triangular sets have become an alternative tool for representing the ideal besides the Gröbner bases. This paper is about implementation and applications of generalized characteristic sets of ordinary differential polynomial sets defined by author.

1 Introduction

The key inclination in the information era is the mechanization of mental labor with the assistance of computers. Partial systematization of mental labor allows scientists and engineers to free themselves from tedious and sometimes human unreachable tasks, to ponder on high-level innovative activities and hence to greatly enhance social productivity. The objective of mathematics mechanization is to cultivate symbolic algorithms for manipulating mathematical objects. It also helps in proving and discovering theorems in a mechanical mode. Applications and multidisciplinary studies are major advantages for attraction of scientists in mathematics mechanization.

J. F. Ritt [11] introduced the concept of a characteristic set of a finite or infinite set of differential polynomials. One of his objective was to provide a method to solve systems of differential equations. Characteristic sets of prime ideals have good properties but Ritt's process involves factorization in field extensions. Wu Wen-tsun used Ritt's work to provide an algorithm for solving systems of algebraic equations by means of triangular sets which only requires pseudo-remainder computations. Therefore Characteristic Set Method of Wu has released Ritt's decomposition from polynomial factorization, opening access to a variety of discoveries in polynomial system solving. This method and the methods for geometry reasoning and computation have various applications.

The work of Wu has been extended towards more powerful decomposition algorithms and applied to different types of polynomial systems or decompositions. In addition to be a powerful tool for Wu's general theory and method of mechanical theorem proving, the characteristic set method has proved efficient for solving a wide class of problems in geometry and algebra.

Dongming Wang [4] proposed an effective strategy to improve Ritt-Wu's algorithm of characteristic sets and a comprehensive implementation of Ritt-Wu's method in the Maple system was described by him. Wang widespread Ritt-Wu's algorithm by means of one-step pseudo-reduction with strategies for the selection of reductends and optimal reductors. In addition to Ritt-Wu's methods, there are many other proficient methods for decomposing systems of multivariate polynomials. Kalkbrener [13] familiarized the notion of regular chain and given a way for decomposing any algebraic variety into unmixed-dimensional components represented by regular chains.

Meng et al [12] follow the work of Wang to present a new algorithmic scheme for computing generalized characteristic sets competently. By familiarizing admissible reductions other than pseudo-division, this is possible to control the swell of coefficients of intermediate polynomials and to compute with smaller polynomials.

We have computed the generalized characteristic sets of Wang for the ordinary differential polynomial sets [7]. Let C be a d-char set of d-polynomial P. A necessary condition is that all the d-polynomials in the set P have d-pseudo remainder zero 0 w.r.t the set C. To make the concept of d- char generalized, this necessary condition is weakened by the replacement of d-polynomial P with an arbitrary d-polynomial set which can generate the same ideal as the d-polynomial set. In [8] we gave a comparison of d-char sets with generalized d-char sets and results of preliminary experiments show that generalized d-char sets perform better than d-char sets based on d-pseudo division in efficiency and simplicity of outcome.

2 Preliminaries

We can classify triangular decompositions algorithms for polynomial systems into several ways. One may first consider the relation between the input polynomial system and the output triangular system. In that perspective, we have two types of triangular decompositions that are basically different. Those systems for which output triangular system can encode all the points of the zero set of input polynomial system. The second are those for which triangular system represents only the generic zeros of the irreducible components of the input polynomial system. Similarly, triangular decomposition algorithms can be classified with the help of algorithmic principles on which they are relying. From this angle we have two different types of triangular decompositions. One is followed by variable elimination i.e., reduce the solving of a polynomial system in n unknowns to a system of $n-1$ unknowns. Second is the type that proceed inclemently i.e., reduce the solving of a polynomial system in m equations to that of a polynomial system in $m-1$ equations. The characteristic set method fits in to first type in each classification describe above. Mostly, works on triangular decomposition algorithms focus on incremental solving as this principle is relatively attractive, because it allows controlling the properties and growth of the intermediary computed objects.

2.1 Generalized differential characteristic sets

The char-set method of polynomial equations-solving is naturally extended to the differential case which gives rise to an algorithmic method of solving arbitrary systems of differential equations.

We use notion of the differential medial set (d-medial set) of the d-polynomial set $\mathbb{P}$. Let $\mathbb{P}$ be a non-empty d-polynomial set $\mathcal{K}\{x\}$.

Definition 2.1:A d- ascending set $\mathbb{M}$ is called a d-medial set of $\mathbb{P}$, if $\mathbb{M} \subset \langle\mathbb{P}\rangle$ and $\mathbb{M}$ has ranking not higher than the ranking of any d-basic set of $\mathbb{P}$.

The d-medial sets are the d-ascending sets with rank not higher than that of the d-basic set of $\mathbb{P}$ and in which all the d- polynomials are linear combinations of the d-polynomials in $\mathbb{P}$ with d-polynomial coefficients. Therefore any d-basic set itself is a special d-medial set of the d-polynomial set. It has been proved that in Ritt's original algorithm, the d-basic set can be replaced by the d-medial set.

For any non-empty d-polynomial set $\mathbb{P}$ in $\mathcal{K}\{x\}$, an ascending set $\mathbb{C}$ is called a generalized differential characteristic (generalized d-char) set if it satisfied the following two conditions

(1) $\mathbb{C} \subset \langle\mathbb{P}\rangle$
(2) There exists a d-polynomial set $\mathbb{Q} \subseteq \mathcal{K}\{x\}$ such that $\langle\mathbb{Q}\rangle = \langle\mathbb{P}\rangle$ and d-prem $(\mathbb{Q}, \mathbb{C}) = 0$

If we recall, for C to be a d-char set of a d-polynomial set P, a necessary condition is that all the d-polynomials in P have pseudo-remainder 0 w.r.t. C. This compulsory condition is weakened by replacing the given d-polynomial set with an arbitrary d-polynomial set which generates the same ideal as the input d-polynomial set, leading to the concept of generalized characteristic set [12]. This set may have d- polynomials of degrees smaller than the degrees of those in the input set and consequently may take less computing time for pseudo-reduction to 0. Therefore, an algorithmic scheme for computation of generalized characteristic sets is presented in [7].

This scheme is defined by several admissible d-reductions. For this purpose following admissible reductions have been introduced [8].

2.1.1 Univariate GCD d-reduction.

Univariate GCD d-reduction is defined as follows d-Rem $(\mathrm{P}, \mathrm{Q}, \mathfrak{D}_{\mathrm{UG}}) :=$

$$:= \begin{cases} [0, \mathrm{gcd}(\mathrm{P}, \mathrm{Q}, \mathrm{x_q})] & (i) \\ [\mathrm{P}, \mathrm{Q}] & (ii) \end{cases}$$

(i) If P, Q are univariate polynomials in $\mathrm{x_q}$,
(ii) otherwise ,

where $\mathrm{x_q}$ is some variable in x.

2.1.2 One-step d-pseudo-division reduction.

One-step d-pseudo-division reduction is defined as follows:

$$\text{d-Rem}(\mathrm{P}, \mathrm{Q}, \mathfrak{D}_{\mathrm{SP}}) := \begin{cases} [\mathrm{d-stprem}(\mathrm{P}, \mathrm{Q}), \mathrm{Q}] \\ [\mathrm{P}, \mathrm{Q}] \end{cases}$$

if P be d-reducible w.r.t Q; then
d-stprem $(\mathrm{P}, \mathrm{Q}) \in \langle \mathrm{P}, \mathrm{Q} \rangle$.
Therefore
$\text{d-Rem}(\mathrm{P}, \mathrm{Q}, \mathfrak{D}_{\mathrm{SP}}) \subseteq \langle \mathrm{P}, \mathrm{Q} \rangle$.
Otherwise
$\text{d-Rem}(\mathrm{P}, \mathrm{Q}, \mathfrak{D}_{\mathrm{SP}}) = [\mathrm{P}, \mathrm{Q}] \subseteq \langle \mathrm{P}, \mathrm{Q} \rangle$ is obvious.

2.1.3 One-step d-division reduction.

The division operation can also be viewed as an admissible reduction. So we define the one-step d-division reduction.

$$\text{d-Rem}(\mathrm{P}, \mathrm{Q}, \mathfrak{D}_{\mathrm{SD}}) := \mathrm{P} \begin{cases} \left[\mathrm{P} - \frac{\mathrm{M}}{\mathrm{lead(Q)}} \cdot \mathrm{Q}, \mathrm{Q}\right] & (i) \\ \left[\mathrm{P} - \frac{\mathrm{M}}{\mathrm{lead(Q')}} \cdot \mathrm{Q'}, \mathrm{Q}\right] & (ii) \\ [\mathrm{P}, \mathrm{Q}], & (iii) \end{cases}$$

(i) if there exists a monomial M of P and such that $\mathrm{lead(Q)} \mid M$,
(ii) if there exists a monomial M of P, such that $\mathrm{lead(Q')} \mid M$,
(iii) otherwise.

2.1.4 Subresultant d-PRS reduction.

We introduce the following d-reduction for d-polynomials.

$$\text{d-Rem}(\mathrm{P}, \mathrm{Q}, \mathfrak{D}_{\mathrm{PD}}) := \begin{cases} [0, \mathrm{P_r}] & (i) \\ [\mathrm{P_r}, \mathrm{P_{r-1}}] & (ii) \\ [\mathrm{P}, \mathrm{Q}] & (iii) \end{cases}$$

(iv) if lead (P) = lead(Q), ld(P) = ld(Q), and d-res (P, Q,lead (Q)) = 0,

(v) if lead (P) = lead(Q), ld(P) = ld(Q), and d-res (P, Q, lead (Q)) ≠ 0,

(vi) otherwise.

With the help of admissible d-reductions above, an algorithmic scheme NewdCharSet for computing generalized d-char sets for ordinary d-polynomial sets is introduced with two sub sequent algorithms dMedSet and dFind3R (Reader may refer to [7] for complete details).

2.2 Applications of generalized differential characteristic sets

The previous sections introduced generalized d-char sets without their applications taking into consideration. However, this approach permitted to introduce the majority of the necessary formalisms. In this section we describe the purpose of d-char sets, which is to attain information about sets or systems of differential equations d-charset methods are applied to set or systems of differential equations in order to obtain information about them. What type of information is gained does not only depend on the differential polynomial equations, but also on the used differential ranking on the indeterminates. in general, d-char set computations are used to decouple indeterminates, eradicate high orders, and segregate dependencies. However, d-char sets do not yield solutions to set or systems of differential polynomial equations. Their computations lead to simpler sets or systems which are in general easier to solve. Thereby, the computations assist in solving set and systems of differential equations but do not solve the set or systems directly. Just as Grbner bases help in solving systems of algebraic equations and do not resolve the systems directly. D-char set computations transform a set system of differential equations into another set or system. Obviously, the solutions should not change.

Example 3.1: [2]. Use the same coordinate system and differential equations for Kepler's law. Suppose we have same H_1, H_2, H_3 as

$$H1 = r2 - x2 - y2 = 0,$$
$$H2 = a2 - x'\ '\ 2 - y'\ '\ 2 = 0,$$
$$H3 = x'\ '\ y - y'\ '\ x = 0 \text{ and}$$
$$K = x''r' - r''x'.$$

Let we have d-polynomial set

$\mathbb{E} = \{H_1, H_2, H_3, K\}$ with $r \prec a \prec x \prec y$.

By computing a d-char set of a d-polynomial set $\mathbb{E}$ by the, one obtain the following d-char set:

$\{r^2x^2(r'\ '^4r^2-2r'\ '^2r^2a^2+r^2a^4+r'\ '^4x^2-r'\ '^2x^2a^2), x(ry'\ '-r'\ '\ y), x^2r^2r'\ '^2(r^2a^2+r'\ '^2y^2-y^2a^2-r^2r'\ '^2, r^2x^2(-r'\ '^2+a^2))\}$

Example 3.2: (Bertrand curves) A pair of space curves having their principal normals in common are said to be associate Bertrand curves. Let C and $\bar{C}$ be Bertrand pair of curves in one-to-one correspondence with arc lengths s, S as parameters in the ordinary metric space. Attach the trihedral (X, e_1, e_2, e_3) to C at point X and trihedral $(\bar{X}, \bar{e_1}, \bar{e_2}, \bar{e_3})$ to $\bar{C}$ at corresponding point $\bar{X}$. Denote the curvature and torsion of C and $\bar{C}$ by κ, τ and by $\bar{\kappa}, \bar{\tau}$ respectively. We have the following theorems.

Schell's Theorem. The product of τ and $\bar{\tau}$ is a constant.

Betrand's Theorem. There exist a linear relation between κ and τ with constant coefficients.

Mannheim's Theorem. The cross-ratio of X, $\bar{X}$ and the centers of κ, $\bar{\kappa}$ is a constant.

Using the orthogonality relations (see details [23]), one obtain a set of fourteen d-polynomials.

$U_1 = \bar{s}\, u_{11} + a_2\ \kappa\ -1,\quad U_2 = -a_2',$
$U_3 = \bar{s}\, u_{13} + a_2\ \tau,\quad U_4 = -u_{11}'$
$U_5 = \bar{s}\bar{\kappa} - \kappa\ u_{12} + \ \tau\ u_{13},\quad U_6 = -u_{13}'$
$U_7 = \bar{s}\bar{\kappa}\, u_{11} - \bar{s}\bar{\kappa}, u_{31} - \kappa,\quad U_8 = \bar{s}\bar{\kappa}\, u_{13} - \bar{s}\bar{\kappa}, u_{33} + \ \tau,$
$U_9 = -u_{31}',\quad U_{10} = -\bar{s} - \kappa\ u_{31} + \ \tau\ u_{33},$
$U_{11} = u_{31}',\quad U_{12} = u_{11} - u_{33},$
$U_{13} = u_{13} + u_{31},\quad U_{14} = u_{11}^2 + u_{13}^2 - 1$

With the order

$$\bar{s} \prec a_1 \prec a_2 \prec a_3 \prec u_{11} \prec u_{12} \prec u_{13} \prec u_{21} \prec u_{22} \prec u_{23} \prec u_{31} \prec u_{32} \prec u_{33} \prec \kappa \prec \tau \prec \bar{\kappa} \prec \bar{\tau}$$

Let

$$\mathbb{U} = \{U_1, U_2, \ldots, U_{14}\}$$

be a d-polynomial set. If we compute a d-char set of a d-polynomial set $\mathbb{U}$ then the output consists of the following d-char set:

$\{-a'\ ^2, s'\ '\ s'\ u^{11}(u_{11}^2 - 1), s'\ '\ u_{11}(-s'\ - u_{12} + u_{12}s'\ u_{11} + u_{11}), s'\ '\ u_{11}\, u_{13}(-u_{11} + s'\ ), u_{31}\, u_{11}\, s'\ '\ (-u_{11} + s'\ ), u_{11} - u_{33}, s'\ u_{11} + a_2\ \kappa\ -1, s'\ '\ a_2\ \tau\ u_{11}(-u_{11} + s'\ ), s'\ '\ u_{11}(a_2\ s'\ \bar{\kappa} + s'\ - u_{11}), s'\ '\ a_2\, s'\ \bar{\tau}\, u_{11}(-u_{11} + s'\ )\}.$

Example 3.3: The y-axis projection of the velocity of a particle moving within a plane is a constant c. If p is the curvature radius of the orbit of the particle. By a formula for curvature in differential geometry, we have:

$$V_1 = a - x'\ '\ = 0,$$
$$V_2 = v^2 - x'\ ^2 - c^2 = 0,$$
$$V_3 = p^2\, c^2\, x'\ '\ ^2 - (y'\ ^2 + x'\ ^2)^3 = 0.$$

Consider a d-polynomial set $\mathbb{V} = \{V_1, V_2, V_3\}$. Fix an order $c \prec x \prec v \prec a \prec p$ with c as constant parameter. Compute a d-char set of the set $\mathbb{V}$ by the algorithm NewdCharSet. We get following d-polynomial set as the output:

$\{a - x'\ '\ , -v^2 + x'\ ^2 + c_2, p^2c^2x'\ '\ ^2 - y'\ ^6 - 3\, y'\ ^4x'\ ^2 - 3\, y'\ ^2x'\ ^4 - x'^6\}.$

Example 3.4: Consider the d-polynomial set W which consists of the following four d-polynomials:

$$W_1 = c_1\, x'\ '\ ^2 + c_2\, x'\ '\ + c_3,$$
$$W_2 = y^2 - (e^2 - 1)\, x^2 - 2\, p\, e\, x - p^2,$$
$$W_3 = r - p - e\, x,$$
$$W_4 = r^2 - h.$$

where c_1, c_2 and c_3 are polynomials of p, e, h, x and x'. Consider the following order $p \prec e \prec h \prec x \prec y \prec r \prec r$ with p, e and h as constant. After computing a d-char set of a d-polynomial set $\mathbb{W}$ by the NewdCharSet, one gets the following d-char set in the output:

$\{-h + ap^2 + 2\, paex + e^2x^2a, -r + p + ex, c_1\, x'\ '\ ^2 + c_2\, x'\ '\ + c_3, -y^2 + x^2e^2 - x^2 + 2\, pex + p^2\}.$

Example 3.5: The equations describing a motion under central force can be formulated as below:

$$x'' y - y'' x = 0,$$
$$r^2 - x^2 - y^2 = 0,$$
$$h - x^2 y' + x' y = 0,$$
$$r^2 w' - h = 0.$$

where h and w stands for the angular momentum per unit mass and the inclination of the radius to the x-axis respectively.

If we take the following order $x \prec\prec y \prec r \prec h \prec w \prec v \prec a$ and consider
$\mathbb{X} = \{x'' y - y'' x, r^2 - x^2 - y^2, h - x^2 y' + x' y, r^2 w' - h\}$.

Then a d-char set of the d-polynomial set $\mathbb{X}$, is given as follows:

$$\{w' x^2 + w' y^2 - xy' + x' y, -x'' y + y'' x, h - xy' + x' y, r^2 - x^2 - y^2\}$$

The examples described above have been solved by using epsilon package in Maple 14 on a laptop having windows 7 ultimate, Intel(R)Core(TM)2 Duo CPU 2.4 GHz with 2.00 GB memory.

Besides the determination of differential solvability and differential radical ideal membership, we list a number of applications related to differential characteristic sets.

The differential characteristic set method generalized a number of well-known algorithms in symbolic computations

They can be used to compute the differential dimension of a differential algebraic set.

The capabilities of the non linear differential sets or system solver, dsolve, that is improved in maple 6 by using the diffalg package. In order to solve a single differential equation, the characteristic decomposition demonstrates the differential equations for their singular solution. If they are resolved, the complete solution set is currently returned by dsolve. The straightforward design for solving ordinary differential sets or systems is to attain advantage of the solver for single non linear differential equation. The contributed sets or system is therefore decomposed with respect to an elimination ranking. Afterward to solve the sets or system in close form by solving iteratively single differential equations in a distinctive indeterminate.

One more application of differential characteristic decomposition algorithms in the direction of solving ordinary differential equations is the suggestion is to generate new classes of ordinary differential equations, for example of first order, which may be solved jointly with a way of distinguishing whether a specified ordinary differential equation is in the class or not.

The variants of the differential characteristic decomposition algorithms are used to examine the several biological models that are arising in the literature.

The work of several mathematician showed the relevance of constructive differential algebra in control theory.

A foundation of over-determined sets or systems of partial differential equations comes up to symmetry analysis. The seminal equations of the insignificant generators of the Lie symmetry group of a differential equation form a set of linear partial differential equations. Reducing this set permits to establish if we have symmetry, to investigate the configuration of the group and apply that knowledge to classification troubles or in order to discover a reduction of the original equation. Since the system is linear, there are a number of substitute algorithms that may be applied.

3 Tables

The Table I provides information about the test examples. We have number of terms (nops) in each d-polynomial which appears in d-char set of each example, the time in seconds for computing d-char set for each example. Table II provides information about degree tuple of each d-polynomial obtained in the output of d-char set

Table 1. nops and timing in the outputs

No	nops	time
1	5, 2, 4, 2	0.062s
2	1, 2, 4, 2, 2, 2, 3, 2, 3, 2	0.187s
3	3, 2, 5	0.016s
4	4, 3, 3, 5	0.031s
5	4, 2, 3, 3	0.032s

Table 2. deg tuple of d-polynomial set in the output of NewdCharSet

No	degree tuple
1	4, 4, 4, 0],[1, 0, 1, 1],[4, 2, 2, 2],[2, 2, 2, 0]
2	[0,..., 0], [0, 0, 0, 0, 3, 0,..., 0],[0, 0, 0, 0, 2, 1, 0,..., 0] [0, 0, 0, 0, 2, 0, 1, 0,..., 0], [0, 0, 0, 0, 2, 0, 0, 0, 0, 0, 1, 0,..., 0] [0, 0, 0, 0, 1, 0,..., 0, 1, 0, 0, 0, 0],[0, 0, 1, 0, 1, 0,..., 0, 1, 0, 0, 0] [0, 0, 1, 0, 2, 0,..., 0, 1, 0, 0], [0, 0, 1, 0, 2, 0,..., 0, 1, 0], [0, 0, 1, 0, 2, 0,..., 0, 1]
3	[0, 0, 0, 1, 0],[2, 0, 2, 0, 0], [2, 0, 0, 0, 2]
4	[2, 2, 1, 2, 0, 0, 1],[1, 1, 0, 1, 0, 1, 0], [0, 0, 0, 0, 0, 0, 0], [2, 2, 0, 2, 2, 0, 0]
5	[2, 2, 0, 0, 0, 0, 0], [1, 1, 0, 0, 0, 0, 0], [1, 1, 0, 1, 0, 0, 0] [2, 2, 2, 0, 0, 0, 0]

4 Conclusion and future work

We conclude the paper with the the discussion that computation of generalized differential characteristic set leads to simpler sets or systems of differential equations for which find solution is much easier as compare to the original set or systems.

A differential characteristic set of differential ideal is a finite subset of d-polynomial sets from which many properties are often obtain by the inspection of its elements. Computation of differential characteristic sets is constructive in principle.

We have generalized the differential characteristic sets for ordinary differential case. This concept will be further improved and compared with more schemes by different strategies in the forthcoming papers. To extend the idea of generalized differential characteristic sets to partial differential case in also in our considerations. By a numerical superscript corresponding to the same superscript after the name of the author concerned. Please ensure that affiliations are as full and complete as possible and include the country.

Acknowledgment

I would like to thank my supervisor Dongming Wang for his help and support. I also want to thank anonymous referees for their suggestions.

References

[1] S.C. Chou, X.S. Gao. Automated reasoning in mechanics using Ritt-Wu's method. University of Texas at Austin, USA. (1989).

[2] S.C. Chou, X.S. Gao Automated reasoning in mechanics using Ritt-Wu's method; PartIII. Proceedings of the IFIP International Work- shop on Automated Reasoning Elsevier Science Publishers Beijing 1-12. (1992)

[3] S.C. Chou, X.S. Gao. Automated Reasoning in Differential Geom- etry and Mechanics using characteristic method IV Bertrand curves. J. of Sys. Math 6:186-192. 216–228. (1993)

[4] D. Wang. A generalized algorithm for computing characteristic sets,World Scientific Publishing Company, Singapore, 165–174. (2001)

[5] D. Wang. Elimination methods, texts and monograms in symbolic com- putations, Springer-Verlag, Wien New York. (2001)

[6] E. Hubert. Notes on triangular sets and triangulation-decomposition algorithms II: diferential systems, Lecture Notes in Computer Science, Springer-Verlag, Berlin, 2630, 40–87. (2003)

[7] F. Afzal. Generalized char sets for ordinary differential polynomial sets, Science International, **28 (3)**, (2016).

[8] F. Afzal. Comparison of Char-Sets with Generalized Char-Sets of Ordinary differential Polynomial Sets, International Conference on Applied Computing, Mathematical Sciences and Engineering (ACME), Johor Bahru, Malaysia. (2016)

[9] Gallo, G., Mishra, B, Efficient algorithms and bounds for Wu-Ritt characteristic sets, In: Mora, T., Traverso, C. (Eds.), Efective Methods in Algebraic Geometry, Progress in Mathematics, Birkhauser, Berlin, **94**, 119–142. (1991)

[10] G. Gallo, B. Mishra, Wu-Ritt. characteristic sets and their complexity. Discrete Mathematics and Theoretical Computer Science, American Mathematical Society, Providence, **6**, 1110–136. (1999)

[11] J. F. Ritt. Differential algebra, New York: AMS Press. (1950)

[12] J. Meng, L. Xiaoliang, D. Wang. A new algorithmic scheme for com- puting characteristic sets, Journal of Symbolic Computation, **50**, 431–449. (2013)

[13] Kalkbrener, M. A, generalized Euclidean algorithm for computing triangular representations of algebraic varieties, Journal of Symbolic Computation, **15**, 14–167. (1993)

[14] M. Fliess, Generized controller canonical forms for linear and nonlinear dynamocs, IEEE Trans.Automat. Control **35**, 994-1001. (1990)

[15] Li, Y, Some properties of triangular sets and improvement upon al- gorithm CharSer, In: Calmet, J., Ida, T., Wang, D. (Eds.), Artificial Intelligence and Symbolic Computation, Lecture Notes in Computer Science. Springer-Verlag, Beijing, **4120**, 82–93. (2006)

[16] S. Diop and M. Fliess, Nonlinear observability, identifitability and persistent trajesctories, in: Proc. 30th IEEE confernce on Decision and control(IEEE Press), New York, 714-719. (1991)

[17] T. Sasaki, A. Furukawa. Theory of multiple polynomial remainder sequence, Publ. RIMS, Kyoto Univ, **20**, 367–399. (1984)

[18] W. T. Wu. Mathematics-mechanization research preprints, No 1-6, MM Research Center, Academia Sinica, **986**. (1987-1991)

[19] X.S. Gao, J. Van der Hoeven, Y. Luo, and C. Yuan. Characteristic set method for differential-difference polynomial systems. Journal of Symbolic Computation, **44**, 1137–1163. (2009)

[20] X.S. Gao and Z. Huang, Characteristic set algorithms for equation solving in finite fields and applications in cryptanalysis, Journal of Symbolic Computation, **47**, 655–679. (2012)

[21] Y. Chen, X. S. Gao. Involutive characteristic sets of algebraic partial diferential equation systems, Science in China Series A: Mathematics, **46**, 469–487. (2003)

[22] Z. W. Gan, M. Zhou,Decomposition of Reflexive Differential-Difference Polynomial Systems, Applied Mechanics and Materials, Vols 380-384, 1645-1648. (2013)

The Stochastic Synthesis of the Adaptive Filter for Estimating the Controlled Systems State Based on the Condition of Maximum of the Generalized Power Function

Andrey Aleksandrovich Kostoglotov [1], Sergey Valerievich Lazarenko [2], Anton Aleksandrovich Kuznetsov [3], Vitalii Aleksandrovich Losev[2] and Igor Vladimirovich Deryabkin [1]

[1] *Rostov State Transport University, Department of Informatics, Rostov-on-Don, Russian Federation*
[2] *Don State Technical University, Department of Radio Engineering and Electronics, Rostov-on-Don, Russian Federation*
[3] *Air Force Military Training and Research Center "Air Force Academy named after Professor N.E. Zhukovsky and Yu.A. Gagarin", Department of Metrology and Metrological support, Voronezh, Russian Federation*

Abstract. The proposed procedure for the synthesis of the filter of the state estimation is based on a new mathematical model of the dynamic controlled system. It is based on the maximum condition of the function of the generalized power. The optimization boundary problem can be solved using the invariant embedding procedure. The result differs from the known ones as it has lower dimensionality, non-linear structure and provides a more accurate estimation.

1 Introduction

Constructing mathematical models of controlled systems involves the problem to keep balance between simplicity and adequate representation. Greater number of measurements and accuracy can compensate the mismatch between the observed movement and the model. But there are a number of important practical applications where it is impossible or extremely expensive. For example, in radiolocation and navigation the poor mathematical model increases estimation errors. This is due to the fact that the traditional approach to the construction of the model for the perturbation of the estimated process involves the expansion of the state space on the basis of the accepted hypothesis for the movement pattern of the controlled system [1, 2], which hypothesis in many cases is not true.

The difficulties of such processes simulation are often due to nonlinearity of the processes and complexity of their mathematical description. In the case of random factors influence the processes simulation involves the solution of the stochastic differential equations which are even more complex since they lead to the need to solve the multi-dimensional Fokker-Planck-Kolmogorov equation, and in the presence of observations — to the Stratonovich equation. Using the Gaussian approximation can reduce the simulation complexity and let to go to the equations for the extended Kalman filter. In practice further simplification is performed by use of the linearized model for the evolution of the studied process parameters. So there is a known contradiction between the requirement of maximum details of the mathematical model to improve the reliability and accuracy of simulation and the requirement for constructibility that dictated by the researchers' capability. To find a compromise within the framework of the traditional approach is a very difficult problem since the linearized models for the stochastic filters of the state estimation have virtually exhausted reserves for the accuracy increasing.

However, the description of the controlled process being in the quasideterministic approximation can be based on the prime principle of the dynamics, the Hamilton-Ostrogradski variational principle. This method has been successfully applied in the invariance theory for the perturbations. In particular, in the papers of L.I. Rozonoer and V.V. Velichenko the variational approach was used for the synthesis of non-linear perturbation-invariant systems.

The present paper studies the new mathematical model for the controlled process which is developed on the methodology of the combined-maximum principle [3, 4, 5, 6]. The approach developed by the authors assumes that the state change is the result of an unknown generalized force which is determined by the minimum of the Hamilton action [6, 7, 8]. The generalized force structure can be obtained from the condition of maximum of the generalized power function. The new model is the basis for the stochastic synthesis of the estimation equations using the invariant imbedding procedure [9, 10].

The purpose of this research is to develop a mathematical model for the control system based on the combined-maximum principle methodology and perform synthesis of the stochastic estimating filter to effectively solve the problems of filtration under conditions of regular perturbations.

2 Definition of the problem of synthesis of the mathematical model for the controlled system

In a sufficiently general form the dynamics of the continuous stochastic systems can be described by the unstrict Langevin equation

$$\dot{\mathbf{x}} = \mathbf{f}(\mathbf{x}, \mathbf{u}, t) + \mathbf{g}(\mathbf{x}, t)\boldsymbol{\zeta}(t), \tag{1}$$

where $\mathbf{x} \in R^r$ is the state vector, $\mathbf{u} \in R^k$ is vector of the controls defining the nonstationarity of the processes, $\mathbf{f}(\mathbf{x}, \mathbf{u}, t)$ is the continuous vector function of the corresponding arguments and the time t, $\mathbf{g}(\mathbf{x}, t)$ is the matrix function of the dimension $r \times l$, $\zeta(t)$ is l-dimensional random process of the white noise kind.
Equation (1) is based on the Newton's second law taking into account the stochastic forces. The structure $\mathbf{f}(\mathbf{x}, \mathbf{u}, t)$ is determined by the deterministic forces and often has a cumbersome expression. So its approximation is necessary because the stochastic simulation problem may be insoluble in real time. Under the conditions of a priori uncertainty about the nature of the acting forces the structure can be associated with the perturbation dynamic based on the variational principles [8].

Let the dynamics of the controlled object satisfies to the Hamilton-Ostrogradski principle [8]. So we can write the Lagrange equations of the second kind in the phase space of the generalized coordinates and velocities $(\mathbf{q}, \dot{\mathbf{q}}) \in R^{2n}$

$$\frac{d}{dt}\left(\frac{\partial T}{\partial \dot{q}_s}\right) - \frac{\partial T}{\partial q_s} = Q_s, \qquad \mathbf{Q} \in \overline{G}_Q,\ s = \overline{1, n}, \tag{2}$$

where $T = \frac{1}{2}\sum_{s=1}^{n} a_{ss}\dot{q}_s^2$ is the kinetic energy of the system, a_{ss} are the matrix elements of the quadratic form of the kinetic energy, n is the number of degrees of freedom, Q_s are the generalized forces, $\overline{G}_Q$ is the admitted region.

From the Hamilton-Ostrogradski principle it follows that eq. (2) is a result of stationarity of the action integral

$$\delta' S = \int_{t_0}^{t_1} (\delta T + \delta' A)dt = 0, \tag{3}$$

where $A = \int_{q_s(t_0)}^{q_s(t_1)} \sum_{s=1}^{n} Q_s(\mathbf{q}, \dot{\mathbf{q}}) dq_s$ is the work of the generalized forces; t_0, t_1 are the start and end time of observation.

The observation equation is as follows:

$$\mathbf{y}(t) = \mathbf{h}(\mathbf{q}, t) + \boldsymbol{\xi}(t), \tag{4}$$

where $\mathbf{h} \in R^n$ is known vector-function; $\xi \in R^n$ is a white noise vector with known level.

The objective functional has the form

$$J = \frac{1}{2}\int_{t_0}^{t_1} [\mathbf{y}(t) - \mathbf{h}(\hat{\mathbf{q}}, t)]^T \mathbf{R}_\xi^{-1} [\mathbf{y}(t) - \mathbf{h}(\hat{\mathbf{q}}, t)]dt, \tag{5}$$

where $\mathbf{R}_\xi \in R^{n \times n}$ is the diagonal weight matrix characterising the noise level in the observation channel, and the symbol ^ denotes estimating.

The construction of the mathematical model can be presented as the optimal control synthesis problem: it is necessary to determine the vector of generalized forces $\hat{\mathbf{Q}}$ as a function of the generalized coordinates, as well as the corresponding trajectory providing a minimum of functional (5) under the constraints (3).

3 The stochastic synthesis of the estimating filter of the controlled system state based on the maximum condition of the generalized power function

It was found in [5] that the minimum of functional (5) is determined by the maximum condition of the generalized power function

$$\Phi = \max_{\mathbf{Q} \in G_Q} \sum_{s=1}^{n}\left[\lambda Q_s(\mathbf{q}, \dot{\mathbf{q}}) - R_{\xi ss}^{-1}\frac{\partial h_s}{\partial \hat{q}_s}[y_s(t) - h_s(\hat{\mathbf{q}}, t)]\right]\dot{q}_s, \quad \lambda = \text{const}. \tag{6}$$

Thus the structure of equations for the controlled system model is as follows:

$$\ddot{\hat{q}}_s = \frac{1}{\lambda a_{ss}}\left[-\frac{|\dot{\hat{q}}_s|\dot{\hat{q}}_s}{L_s|\hat{q}_s|} + R_{\xi ss}^{-1}\frac{\partial h_s}{\partial \hat{q}_s}[y_s - h_s(\hat{\mathbf{q}}, t)]\right], s = \overline{1, n}; \tag{7}$$

where λ is the undetermined Lagrange multiplier, $L_s \geq 0$ is a constant of the switching curve [5]. We can rewrite eq. (7) in the form:

$$\begin{aligned}\ddot{\hat{q}}_s &= \frac{-|\dot{\hat{q}}_s|\dot{\hat{q}}_s}{a_{ss}L_s\lambda|\hat{q}_s|} + \frac{R_{\xi ss}^{-1}}{\lambda a_{ss}}\frac{\partial h_s}{\partial \hat{q}_s}[y_s - h_s] = \\ &= f(q_s, \dot{q}_s) + \frac{R_{\xi ss}^{-1}}{\lambda a_{ss}}\frac{\partial h_s}{\partial \hat{q}_s}[y_s - h_s].\end{aligned} \tag{8}$$

The sequential estimating problem [9] can be formulated as follows: for the random observation $y(t)$ we need to find estimation $q(t)$, $t \in [t_0, t_1]$ according the criterion of the minimum of the mean square error with t_1 increasing.

For $n=1$ we write down the model of the controlled system (8) movement as a vector differential first-order equation [1]

$$\dot{\mathbf{x}}=\mathbf{f}(\mathbf{x})+\mathbf{G}\boldsymbol{\eta}(t), \quad \mathbf{x}(t_0)=\mathbf{x}^0, \tag{9}$$

where $\mathbf{x}=\begin{bmatrix} x_1 \\ x_2 \end{bmatrix}$, $x_1=\hat{q}$, $x_2=\dot{\hat{q}}$, and

$$\mathbf{f}(\mathbf{x})=\begin{bmatrix} x_2 \\ \dfrac{-|x_2|x_2}{\lambda aL|x_1|} \end{bmatrix} \tag{10}$$

is a vector-function of the system, $\mathbf{G}=\begin{bmatrix} 0 & 0 \\ 0 & \dfrac{1}{\lambda aR_\xi}\dfrac{\partial h}{\partial \hat{q}} \end{bmatrix}$,

$\boldsymbol{\eta}(t)$ is a vector-function of the stochastic process centered with respect to x_1 with the intensity matrix $\mathbf{R}_\eta$.

Thus the problem of synthesis of the filter of state estimating is reduced to the problem to obtain the estimation $\hat{\mathbf{x}}$ from the minimum condition of the extended objective functional [9]

$$J_2=J+\frac{1}{2}\int_{t_0}^{t_1}\boldsymbol{\eta}^{\mathrm{T}}(t)\mathbf{R}_\eta^{-1}\boldsymbol{\eta}(t)\,\mathrm{d}t. \tag{11}$$

According [9] the corresponding estimation formally satisfies the two-point boundary value problem

$$\dot{\hat{\mathbf{x}}}(t)=\mathbf{f}(\hat{\mathbf{x}})-\mathbf{G}\mathbf{R}_\eta\mathbf{G}^{\mathrm{T}}\boldsymbol{\eta}(t), \tag{12}$$

$$\dot{\boldsymbol{\eta}}(t)=\frac{\partial \mathbf{h}^{\mathrm{T}}(\hat{\mathbf{x}}(t),t)}{\partial \hat{\mathbf{x}}}\mathbf{R}_\xi^{-1}[\mathbf{y}(t)-\mathbf{h}(\hat{\mathbf{x}}(t),t)]-\frac{\partial \mathbf{f}^{\mathrm{T}}(\hat{\mathbf{x}})}{\partial \hat{\mathbf{x}}}\boldsymbol{\eta}(t), \tag{13}$$

With boundary conditions

$$\hat{\mathbf{x}}(t_0)=\mathbf{x}^0+\mathbf{P}^0\boldsymbol{\eta}(t_0), \quad \boldsymbol{\eta}(t_1)=0, \tag{14}$$

where $\mathbf{P}^0$ is an analogue for the covariance matrix of the filtering errors [1, 9], $\mathbf{x}^0$ is an initial value of estimation.

Using the invariant imbedding procedure [9] we obtain the following equation:

$$\begin{aligned}
\dot{\hat{\mathbf{x}}}&=\mathbf{f}(\hat{\mathbf{x}})+\mathbf{P}\frac{\partial \mathbf{h}^{\mathrm{T}}(\hat{\mathbf{x}})}{\partial \hat{\mathbf{x}}}\mathbf{R}_\xi^{-1}[\mathbf{y}(t)-\mathbf{h}(\hat{\mathbf{x}},t)],\\
\dot{\mathbf{P}}&=\mathbf{P}\frac{\partial \mathbf{f}^{\mathrm{T}}(\hat{\mathbf{x}})}{\partial \hat{\mathbf{x}}}+\frac{\partial \mathbf{f}(\hat{\mathbf{x}})}{\partial \hat{\mathbf{x}}}\mathbf{P}+\mathbf{G}\mathbf{R}_\eta\mathbf{G}^{\mathrm{T}}-\\
&-\mathbf{P}\frac{\partial}{\partial \hat{\mathbf{x}}}\left[\frac{\partial \mathbf{h}^{\mathrm{T}}(\hat{\mathbf{x}},t)}{\partial \hat{\mathbf{x}}}\mathbf{R}_\xi^{-1}[\mathbf{y}(t)-\mathbf{h}(\hat{\mathbf{x}},t)]\right]\mathbf{P},\\
\mathbf{P}(t_0)&=\mathbf{P}^0, \quad \hat{\mathbf{x}}(t_0)=\hat{\mathbf{x}}^0.
\end{aligned} \tag{15}$$

4 The evaluation of the synthesized filter efficiency

The evaluation of the efficiency of the developed filter was carried out based on the mathematical simulation of the process of estimating of the stochastic process state with non-stationary perturbation.

The state equation (8) is as follows:

$$\dot{\mathbf{x}}=\mathbf{f}(\mathbf{x})+\mathbf{G}\boldsymbol{\eta}(t), \quad \mathbf{x}(t_0)=\mathbf{x}^0. \tag{16}$$

The observation equation (4) has the form:

$$y=\mathbf{H}\mathbf{x}+\xi, \tag{17}$$

where $\mathbf{H}=[1 \quad 0]$, ξ is white Gaussian noise of observations.

According (15) the equations of the suggested algorithm can be written in the form:

$$\begin{aligned}
\dot{\hat{\mathbf{x}}}&=\mathbf{f}(\hat{\mathbf{x}})+\mathbf{P}\mathbf{H}^{\mathrm{T}}R_\xi^{-1}[y-\mathbf{H}\hat{\mathbf{x}}],\\
\dot{\mathbf{P}}&=\mathbf{P}\frac{\partial \mathbf{f}^{\mathrm{T}}(\hat{\mathbf{x}})}{\partial \hat{\mathbf{x}}}+\frac{\partial \mathbf{f}(\hat{\mathbf{x}})}{\partial \hat{\mathbf{x}}}\mathbf{P}+\mathbf{G}\mathbf{R}_\eta^{-1}\mathbf{G}^{\mathrm{T}}-\\
&-\mathbf{P}\mathbf{H}^{\mathrm{T}}R_\xi^{-1}\mathbf{H}\mathbf{P},\\
t_0&=0, \mathbf{P}(0)=\mathbf{P}^0, \quad \hat{\mathbf{x}}(0)=\hat{\mathbf{x}}^0,
\end{aligned} \tag{18}$$

where

$$\frac{\partial \mathbf{f}(\hat{\mathbf{x}},t)}{\partial \hat{\mathbf{x}}}=\begin{bmatrix} 0 & 1 \\ \dfrac{|\hat{x}_2|\hat{x}_2 sign(\hat{x}_1)}{\lambda aL|\hat{x}_1|^2} & \dfrac{-sign(\hat{x}_2)\hat{x}_2-|\hat{x}_2|}{\lambda aL|\hat{x}_1|} \end{bmatrix}. \tag{19}$$

The mathematical simulation is performed using dimensionless units. Initial parameters are assumed as $\hat{x}(0)=x(0)=0, \dot{\hat{x}}(0)=\dot{x}(0)=0, \ddot{\hat{x}}(0)=\ddot{x}(0)=0$. The mean-square deviation of the noise of observations is 100. The simulated perturbation is

$$x(t)=1000h((t-70)(80-t)), \tag{20}$$

where $h(\cdot)$ is the Heaviside step function.

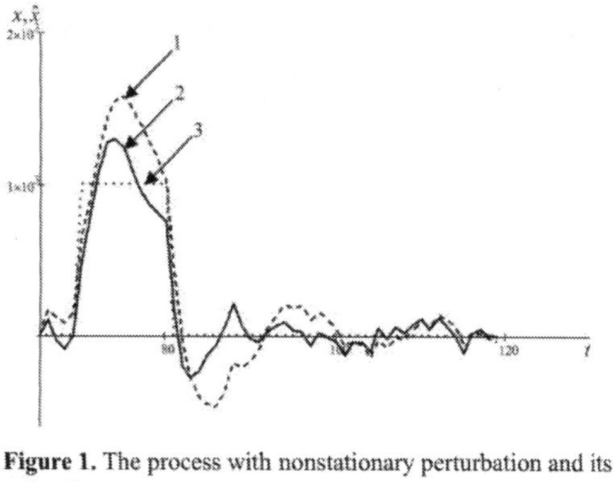

Figure 1. The process with nonstationary perturbation and its estimation.

Fig. 1 illustrates mathematic modelling results for estimating using the new and traditional algorithms. The results are: the curve 1 is the Kalman filter estimation with traditional model [1]; the curve 2 is the developed algorithm (18) estimation; the curve 3 is the process for estimation.

This computational experiment shows that the use of new estimating filter (18) for controlled systems in comparison with the Kalman filter with the traditional model [1] under conditions of regular perturbations provides increase the estimation accuracy by 10% on average. In addition, there is effect of increasing the limit of stability in the case of increasing perturbation frequency. The elementary operations count shows that application of the algorithm requires less computational effort when the operations are addition, subtraction, multiplication and division.

5 Conclusion

This model developed based on the methodology of the unified principle of maximum differs from those obtained using the forming filter by less dimensionality and by non-linear structure. Its use as a base for synthesis procedure on the basis of the invariant imbedding method leads to a new filter for estimation of the controlled systems state. It decreases the estimation errors in comparison with the classical algorithm and demands less computational resources.

Acknowledgement

The paper has been accomplished with the support of Russian Federal Property Fund grants No. 15-08-03798, 15-38-20835, 16-37-60034, 16-38-00665.

References

1. R.A. Singer and K.W. Behnke, "Real-time tracking filter evaluation and selection for tactical applications," IEEE Trans. Aerosp. Electron. Syst, vol. AES-7, 7, pp. 100 – 110 (1971).
2. X.R. Li and V. Jilkov, "A Survey of Maneuvering Target Tracking. Part I: Dynamic Models," IEEE Trans. Aerosp. Electron. Syst, 39(4), pp. 1333-1364 (2003).
3. A.A. Kostoglotov and S.V. Lazarenko, "Nonsmooth analysis in measurement processing," Measurement Techniques, **52**, pp. 117 – 124 (2009).
4. A.A. Kostoglotov, A.I. Kostoglotov, and S.V. Lazarenko, "Joint maximum principle in the problem of synthesizing an optimal control of nonlinear systems," Automatic Control and Computer Sciences, **41**, pp. 274 – 281 (2007).
5. A.A. Kostoglotov, A.I. Kostoglotov, and S.V. Lazarenko, "The Combined-Maximum Principle in Problems of Estimating the Motion Parameters of a Maneuvering Aircraft," Journal of Communications Technology and Electronics, **54**, pp. 431–438 (2009).
6. A.A. Kostoglotov, A.I. Kostoglotov, S.V. Lazarenko, and B.M. Tsennykh, "Method of an estimation of parameters of movement of the operated flying machine on the basis of an combined-maximum principle with construction of a basic trajectory," Uspehi sovremennoj radiojelektroniki. Zarubezhnaja radiojelektronika, **6**, pp. 61 – 66 (2012).
7. A. Majka, "Trajectory Management of the Unmanned Aircraft System (UAS) in Emergency Situation," Aerospace, **2**, pp. 222 – 234 (2015).
8. A.I. Lur'e, Analytical mechanics [in Russian.]. Moscow: Gos. Izd. Fiz.-Matem. Liter. (1961).
9. A.P. Sage A.P. and J.L. Melsa, System Identification. New York and London: Academic Press (1971).
10. A.A. Kostoglotov and V.N. Taran, "Suboptimal Parameter Estimation in Dynamic Systems," Avtomatica i Telemekhanica, **4**, pp. 85–94 (1997).

MATEC Web of Conferences 77, 02009 (2016)
ICMMR 2016

DOI: 10.1051/matecconf/20167702009

The simulation analysis for cartridge proportional flow valve

Peng Dong, Shengdun zhao, Yuanzhe Dong and Shuqin Fan

Xi'an Jiaotong University, No. 28, Xianning West Road, Xi'an, Shanxi, 710049, P.R. China

Abstract. Mechanical and electrical integration is an important direction for future development of the industry. There are many products in hydraulic filed. By connecting electrical and mechanical, we can obtain more precise displacement, velocity and force of controlled object. The simulation can reduct the model and evaluate the performance of the controlled system. The overall development time and costs saved. The aim of the paper is to build a model of cartridge proportional flow valve with LMS Imagine.lab -AMESim and validate the accuracy of the built model by comparing with products catalogue.

1 Introduction

The technology which is widely used in productions is benefit from combining mechanical and electrical. The hydraulic technology is wildly used in the fields of aerospace, automobile, machine tools etc. Combining electrical and mechanical and hydraulic will improve production efficiency and obtain higher control precision. Cartridge proportional flow valve is the example of the combination.

The cartridge proportional valve was used in little flow situation at first. Now, it develops into the style of using pilot proportional to control main spool and used to control high flow. With a LVDT works as feedback unit, the control accuracy of cartridge proportional flow valve is raised [1].

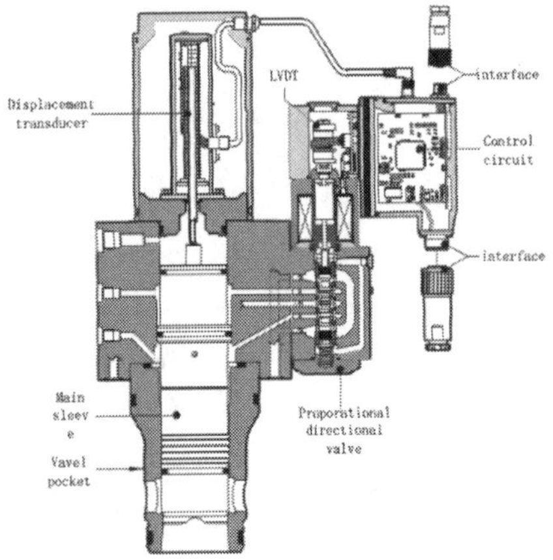

Figure 1. Cartridge proporational flow valve sketch.

The structure of cartridge is shown in figure 1. The main spool is used to control flow by using orifice, which controlled by matching main spool with valve sleeve. The displacement transducer (LVDT) is installed at the top of main spool, this unit can translate the displacement of main spool to electrical signal and feedback to control system. The proportional directional control valve is the leader of main spool, it control the displacement of main spool. Control circuit include the receiver of LVDT, and it communicate with main control system [2]. And the hydraulic system principle diagram is shown in figure 2.

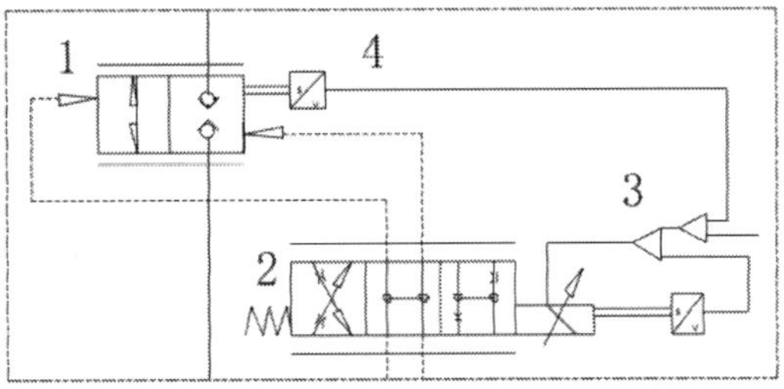

Figure 2. Hydraulic system principle diagram of cartridge proporational flow valve.

As the hydraulic system principle diagram of cartridge proportional flow valve shown, the cartridge valve simple modle as number 1 shown. The controlled flow is in its hands. And the modle of pilot proportional valve is as number 2 shown. Number 3 and 4 represent for the control circuit.

2 The mathematical modle of controlled displacement of main spool

The displacement of main spool determines path area of orifice. And the path area of the orifice influence the flow crossed by. The displacement of main spool is controlled by pilot proportional direction control valve. And the LVDT will feedback the signal of displacement to control system. The modle of pilot valve control main spool can be simplified as figure 3 shown [3, 4].

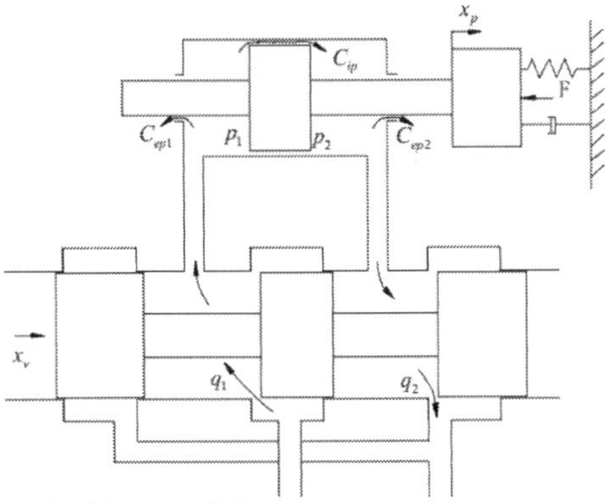

Figure 3. Valve control piston.

In this case, it is base modle of spool control cylinders.

The fundamental equations of the modle are given.

The flow equation passed by orifice is

$$q_L = K_q x_v - K_c p_L \tag{1}$$

where K_q is coefficient of flow, K_c is flow-pressure coefficient .

The flow continuity equation is

$$q_L = A_p \frac{dx_p}{dt} + C_{tp} p_L + \frac{V_t}{4\beta_e}\frac{dp_L}{dt} \tag{2}$$

where C_{tp} is total leakage coefficient of piston, V_t is total volume of fluid under compression.

The forces of the piston is

$$A_p p_L = m_t \frac{d^2 x_p}{dt^2} + B_p \frac{dx_p}{dt} + Kx_p + F_L \tag{3}$$

where K is load spring gradient

After laplace transform, the transfer function can be written as

$$X_p = \frac{\frac{K_q}{A_p} X_V - \frac{K_{ce}}{A_p^2}(1 + \frac{V_t}{4\beta_e K_{ce}} s) F_L}{s\left[\frac{m_t V_t}{4\beta_e A_p^2} s^2 + (\frac{m_t K_{ce}}{A_p^2} + \frac{B_p V_t}{4\beta_e A_p^2}) s + 1\right]} \tag{4}$$

$$X_p = \frac{\frac{K_q}{A_p} X_V - \frac{K_{ce}}{A_p^2}(1 + \frac{V_t}{4\beta_e K_{ce}} s) F_L}{s\left[\frac{s^2}{\omega_h^2} + \frac{2\zeta_h}{\omega_h} s + 1\right]} \tag{5}$$

where $\omega_h = \sqrt{\frac{4\beta_e A_p^2}{V_t m_t}}$ represent for natural frequency of hydraulic. And $\zeta_h = \frac{K_{ce}}{A_P}\sqrt{\frac{\beta_e m_t}{V_t}} + \frac{B_p}{4A_p}\sqrt{\frac{V_t}{\beta_e m_t}}$.

It represent for hydraulic damping ratio, it influence the dynamic response of the system. This mathematical model express the dynamic performance of valve control piston [5, 6].

3 The simulation modle of controlled displacement of main spool

3.1 The built of simulation modle

First step, the detailed simulation modle of cartridge proportional flow controlled valve was built by using components from Hydraulic Component Design (HCD) [7] library, Mechanical library , Hydraulic library and Signal Library [8]. Refer to the structure and the operating principle of the cartridge proportional flow controlled valve. [9, 10] The simulation modle was built as figure 4. This modle requires geometrical parameters like the mass of the spool, the diameter of the piston, the diameter of internal fixed orifices, the flow of the pilot proportional direction controlled valve [11].

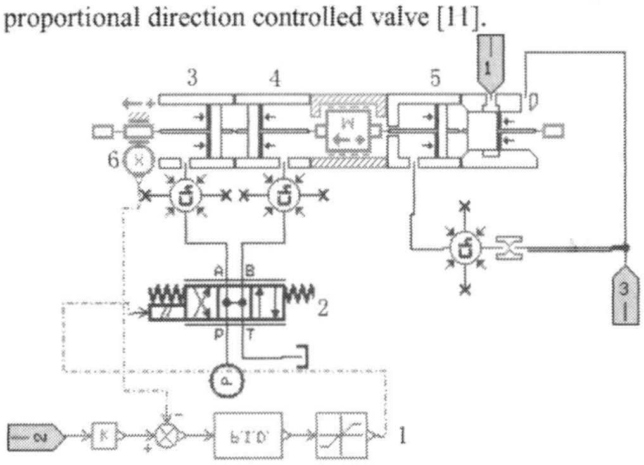

Figure 4. Simulation modle of cartridge proportional flow controlled valve.

As figure 4 shown, the simulation modle of cartridge proportional flow control valve consist of amplifier circuit (1), proportional pilot direction control valve (2). Left piston (3), right piston (4), Cartridge seating type orifice (5).

The amplifier circuit make up of one comparator which compares feedback signal with interface signal and output signal to PID controller, and one PID controller which process input signals and output to proportional solenoid. The pilot valve modle come from HYD library. It's a ideal modle which have perfect performance with control signal input. The main spool will move make orifice close when the left piston charged. And the orifice will open when the right piston charged. The flow area of orifice is depend on the diameter of main spool and the displacement of main spool [12].

 DOI: 10.1051/matecconf/20167702009

3.2 The parameters of simulation modle

The simulation modle was build and the parameters of main modle was settled as table 1 shown.

Table 1. Main parameters of modle.

Name of modle	Parameters	Unit
Flow of pilot valve	90	L/min
Natural frequency of piolt valve	70	Hz
System pressure	350	bar
Diameter of main spool	75	mm
Diameter of Left cylinder	85	mm
Diameter of left piston	76	mm
Diameter of right cylinder	85	mm
Diameter of right piston	76	mm
Mass of main spool	9.6	Kg
Parameter of P	11.7	Null

The parameters of modle are set. The simulation signal is set as figure 5 shown. It includes displacemet and the corresponding time parameter. The displacemet of the spool is zero before 1second, and a ramp signal is settled during one to six seconds. This goal signal is correspond to the working condition of the spool.

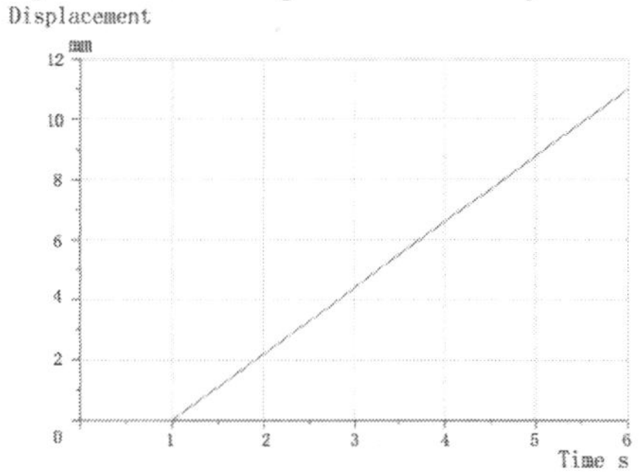

Figure 5. Goal-setting signal.

4 The ansysis of simulation results

With the input signal as figure 5 shown, the output of simulation modle is similar to input signal. The difference curves are given in figure 6 for comparing the input and the output signal.

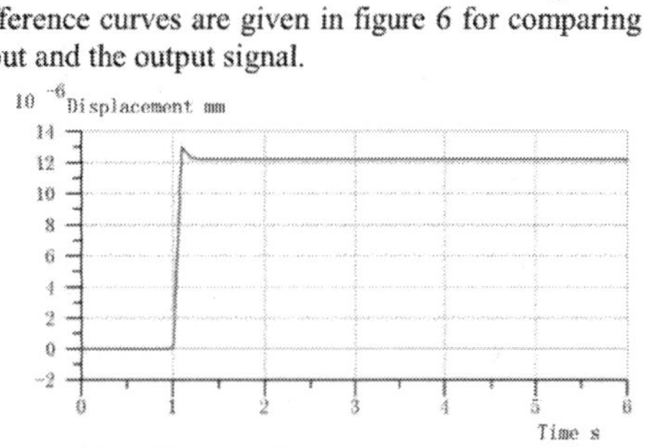

Figure 6. The difference of input and output signal.

As figure 6 shown, the difference of input and output is occurred at the first second, and the difference signal is rapidly increased. After the increasing, this signal remains stable at12 × 10^{-6}mm. This difference curves means the simulation modle have a good stability.

According to the design principle of proportional flow control valve, the flow throughout the orifice is controlled by the displacement of main spool and the flow proportional to displacement. The flow curve in the catalog is shown in figure 7[13].

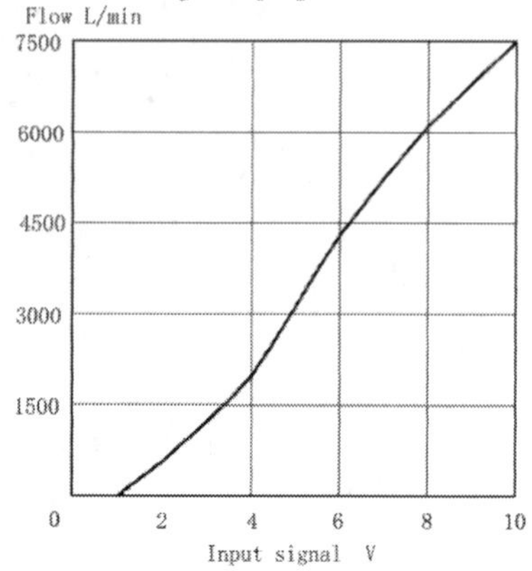

Figure 7. The flow curve given in catalog.

The curve in catalog has a dead band. After the dead band, the output flow is increased. But the trace is not straight line. The curve of the catalog reflect the law of control signal and the output flow. And the dead band is occurred at the begin of the curve for the reason of static friction between spool and valve jacket or the other reasons. With a accurate input signal, the valve will export corresponding output flow.

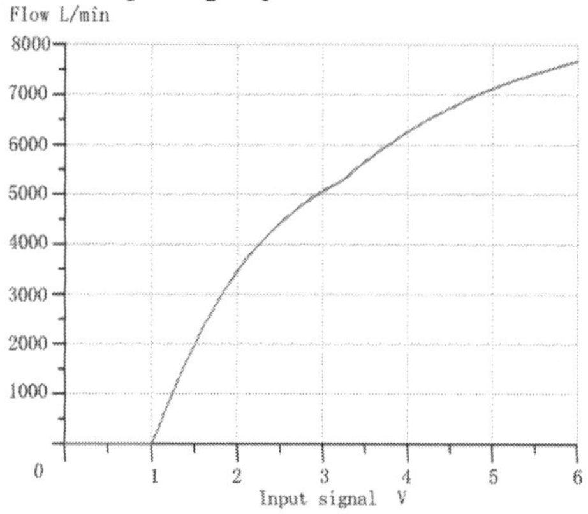

Figure 8. The output of simulation modle.

As figure 7 shown, the flow curve is similar to straight line and proportional to input signal. Because of the unstable difference pressure between orifice and other nonlinear factors, the flow curve is not straight line but similar to straight line. The output of the simulation modle is shown in figure 8.

The displacement of spool in the simulation modle is stable and the output flow curve is stable. The displacement of spool have little difference to input signal.

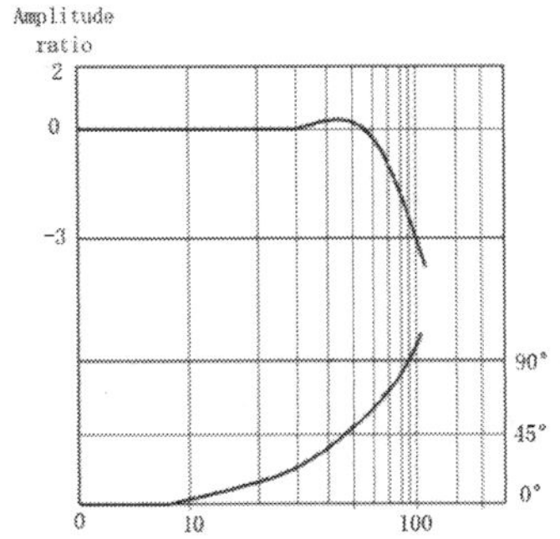

Figure 9. BODE curve given in catalog.

Compared with input signal, it is not straight line too, but it similar to input signal. Compared with flow curve given in catalog, there is some difference between them, but it does not matter. For compare the dynamic performance of the simulation modle, the BODE curve was given in figure 9 and figure 10.

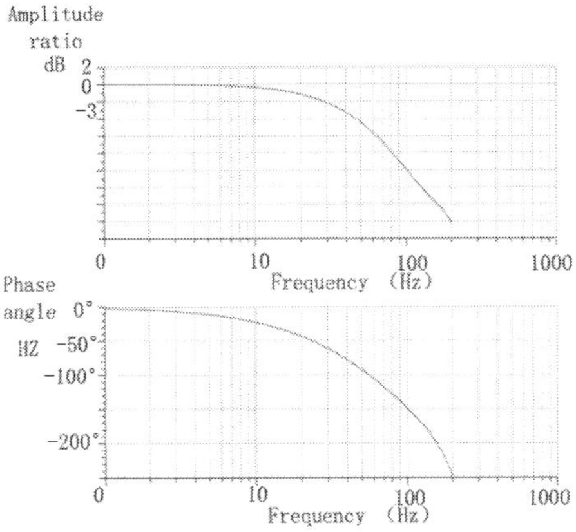

Figure 10. BODE curve output by simulation modle.

Compared the BODE curve shown in figure 10 with the BODE curve given in catalog, the difference is small and we can accept it .so, the simulation modle built in the AMESim is acceptable, and can be used to simulate the properties of the cartridge proportional flow control valve.

5 Conclusions

The control of displacement or velocity will be more precise with combination electronic technology and mechanical. Added with hydraulic technology, the electrical signal is turn into flow or press signal. There are a lot of possibilities to build a simulation modle. The simplified of cartridge proportional flow control valve turn out to be right. And the advantages of combine electrical, mechanical and hydraulic are obvious.

Acknowledgements

The authors gratefully acknowledge the contribution of the Major National Science and Technology Project of China (Grant No. 2009ZX04005-031), National Natural Science Foundation of China for key Program (Grant No. 51305333), and Natural Science Foundation of Shaanxi Province (Grant No.201407-23).

References

1. Liberfarb, Z., *Proportional pilot-operated flow control valve*. 2005, Google Patents.
2. Xu, B., et al., *Modeling and dynamic characteristics analysis on a three-stage fast-response and large-flow directional valve*. Energy conversion and management, 2014. **79**: p. 187-199.
3. Jelali, M. and A. Kroll, *Hydraulic servo-systems: modelling, identification and control*. 2012: Springer Science & Business Media.
4. Mihajlov, M., V. Nikolić, and D. Antić. *Position control of an electro-hydraulic servo system using sliding mode control with fuzzy boundary layer*. in *8th conference on Vehicle System Dynamics, Identification and Anomalies*. 2002. Citeseer.
5. Bonchis, A., et al., *Variable structure methods in hydraulic servo systems control*. Automatica, 2001. **37**(4): p. 589-595.
6. Merritt, H.E., *Hydraulic control systems*. 1967: John Wiley & Sons.
7. Dragne, F.D., et al., *ABS valve model reduction by amesim*. Proceedings of the Romanian Academy A: Mathematics Physics Technical Sciences Information Science, 2009. **10**(2): p. 189-196.
8. Xue-le, Q., et al., *Modeling and Analysis of Vehicle ESP Hydraulic Control Device Using AMESim [J]*. Machine Tool & Hydraulics, 2005. **8**: p. 115-116.
9. You-Guan, Y., G. Guo-fang, and H. Guo-Liang, *Simulation technique of AMESim and its application in hydraulic system [J]*. Hydraulics Pneumatics & Seals, 2005. **3**: p. 28-31.
10. MA, C.-l., X.-x. HUANG, and L. HAO, *Simulation and optimization studies of electro-hydraulic servo system based on AMESim [J]*. Hydraulics Pneumatics & Seals, 2006. **1**: p. 010.
11. Lynn, A., et al. *Modeling hydraulic regenerative hybrid vehicles using AMESim and Matlab/Simulink*. in *Defense and Security*. 2005. International Society for Optics and Photonics.
12. Dasgupta, K., J. Watton, and S. Pan, *Open-loop dynamic performance of a servo-valve controlled motor transmission system with pump loading using steady-state characteristics*. Mechanism & Machine Theory, 2006. **41**(3): p. 262-282.
13. Inc., A., *Atos Master Catalog*. http://www.atos.com/english/catalog/hydraulic, 2016-03-30.

MATEC Web of Conferences 77, 02010 (2016) DOI: 10.1051/matecconf/20167702010
ICMMR 2016

Modelling and Simulation of Free Floating Pig for Different Pipeline Inclination Angles

Dereje Engida Woldemichael, Fakhruldin Mohd Hashim, Mark Ovinis, Wen Ching Lee and Muhammad Hazim bin Mohd Halim

Universiti Teknologi Petronas, Mechanical Engineering Department, 32610 Bandar Seri Iskandar, Perak, Malaysia

Abstract. This paper presents a modelling and simulation of free floating pig to determine the flow parameters to avoid pig stalling in pigging operation. A free floating spherical shaped pig was design and equipped with necessary sensors to detect leak along the pipeline. The free floating pig does not have internal or external power supply to navigate through the pipeline. Instead, it is being driven by the flowing medium. In order to avoid stalling of the pig, it is essential to conduct simulation to determine the necessary flow parameters for different inclination angles. Accordingly, a pipeline section with inclination of 0°, 15°, 30°, 45°, 60°, 75°, and 90° were modelled and simulated using ANSYS FLUENT 15.0 with water and oil as working medium. For each case, the minimum velocity required to propel the free floating pig through the inclination were determined. In addition, the trajectory of the free floating pig has been visualized in the simulation.

1 Introduction

Pipelines are the most efficient and low cost fluid transportation (oil, gas, or water) means over a long distance. Pipeline operators conduct periodic inspections to find flaws and damage which might lead to leakage of the product to the environment. Early detection of leaks in pipelines is essential to reduce product loss, damage to the environment, and high clean-up costs. Therefore, it is necessary to develop a robust system into leak monitoring techniques and apply them in field. There are a variety of leak detection methods with varying degree of accuracy.

Pipeline Inspection Gauge commonly known as pig is a device that is inserted into a pipeline and propelled by the pressure of the product flow or driven by external power. Pig can be used to physically separate different fluids, clean the pipeline, collect data, or inspect the condition of the pipeline. In oil and gas industry pigs are mainly used for cleaning and inspection purposes.

Unlike conventional pigs, a free floating pig reported in this paper is a spherical shaped pig that does not occupy the whole space in the pipe and does not require its own dedicate source of energy to drive through the pipeline. It is being driven by the medium while floating without interrupting the normal operation. Thus, the movement of the pig is highly dependent on many aspects such as pressure, inclination angle, temperature and type of fluid. This study is aimed to identify flow parameters required for the pig to travel through the pipeline at different inclination angles using simulation model.

Free floating pigs can be used to inspect most pipelines including unpiggable pipelines having small radius bends, change in diameter across the pipeline, tee joints and branched lines [1]. Even in a piggable pipelines, operators use pigs as their last option due to the increased risk of "stalled" pig and disruption of the normal operation in addition to cost of locating and removing the pig. Free floating pigs can be used to address this issue since the pig diameter is smaller than the pipeline and spherical shape can easily move through small bends provided that the pipeline is clean.

A spherical shaped free floating pig was designed to conduct leak detection in pipelines. Unlike smartball [2], [3] which uses acoustic signal transmitted through the medium for leak detection, the free floating pig reported in this paper uses pressure signal and combination of other sensors to detect and locate leak.

If pig get stuck during pigging operation, it can seriously affect the normal operation and result in high intervention costs [4]. Therefore, detailed pigging simulation is very important to study the effect of different parameters before the pigs are deployed. Yu et al [5] conducted simulation study using OLGA software for deepwater flow lines before pigging operation to predict the minimum stable flow rate. Xu & Gong [6] developed a simplified pigging model for predicting the pigging operation in gas condensate horizontal pipelines and compared with simulation result from OLGA software. Tolmasquim and Nieckele [7] developed simulation model to assist in the control and design of pig operations through pipelines and predict the process variables. Esmaeilzadeh et al [8] modelled and simulated the pig in gas and liquid pipelines to predict the pig position, pig velocity, upstream optimum flow rate, and time to reach its destination. A computational fluid dynamics simulation using FLUENT was used by Zhu et al [9] to study the leaked oil flows from damaged submarine pipelines. From the above literatures, it is evident that conducting simulation study is critical before filed trial.

In practical application, the mini pig has to travel through a pipeline having a number of bends and varying inclination angles depending on the topography of the location where the pipeline is installed. The modelling and simulation of the free floating pig will help us to visualize the flow pattern, determine minimum pressure and flow rate required for the pig to freely float and driven by the medium for different inclination angles. In this study, commercial software package ANSYS FLUENT 15.0 is used. Oil and water were used as products in the pipeline.

2 Simulation model

The fluid flow model is created using the k-epsilon model ANSYS Fluent. The input parameters are the fluid type, diameter, mass and density of the free floating mini pig, and the inlet velocity. The geometry of a cut out section of the pipeline is created followed by meshing in ANSYS. The simulation results were visualized and analysed to determine the minimum velocity required to propel the free floating pig through the inclination and visualize its trajectory. The pressure and velocity profile of the fluid along the pipeline will also be visualized.

Two types of fluids namely oil and water were used as the working medium for the simulation with the properties shown in Table 1. Both oil and water are assumed to be incompressible fluids with fully developed flow pattern. Thus, "Simple" scheme was selected with "Least Square Cell Based" gradient for simulation of single phase flow. Discrete Phase Model (DPM) is chosen to simulate the movement of the free floating Pig. DPM is a solution that is designed to track the trajectory of the particle in a fluid flowing motion. DPM uses Lagrangian reference frame as the main reference where the position and velocity of individual particle are tracked independently.

Table 1. Properties of fluids considered in the simulation.

Medium	Density [kg/m^3]	Viscosity [kg/m-s]
Water	998.2	0.001003
Oil	830	0.00332

2.1 Geometric Modeling of the Inclined Pipeline

The For numerical simulation, we considered a cut out section of a pipeline where the pipeline experience change in elevation at varying angle connected with straight section at both ends. Seven inclination angles were considered namely: 0°, 15°, 30°, 45°, 60°, 75°, and 90°. The pipe diameter is set to be 100 millimetres (4 inch) and the length of the pipe is 1 meter at each section of the pipe.

Once the geometries are created for each inclination angle, the geometric models were meshed. Two key factors that need to be considered during meshing are aspect ratio and element quality. Aspect ratio is defined as the ratio between the longest dimensions to the shortest dimension of the quadrilateral element in the mesh and element quality basically determine the quality of the mesh. Table 2 shows the results on the meshing of the geometry.

Table 2. Mesh parameters

Inclination Angle [degree]	Number of nodes	Number of elements	Average element quality	Aspect ratio
0	123000	118558	0.6	3.5
15	1420288	1450626	0.8	2.7
30	159259	170384	0.6	3.6
45	226334	242574	0.6	3.5
60	433645	455227	0.7	3.4
75	486707	490363	0.7	3.3
90	2051650	1932039	0.7	2.9

After the meshing has been done, the necessary parameters and fluid properties were set up in the system. The simulation was started with initial velocity of 0.5 m/s and if the free floating pig was unable to propel through the inclination, the setup for the fluid velocity will be increased with an increment of 0.1 m/s. The steps will be repeated until the ideal initial velocity have been achieved.

2.2 Modelling the free floating pig

The free floating pig is modelled as a particle. The trajectories of the particle is computed in a Lagrangian frame using discrete phase models (DPM). The particle force balance equation given in equation 1 is integrated to get the particles trajectory.

$$\frac{du_i^p}{dt} = F_D(u_i - u_p) + \frac{g_i(\rho_p - \rho)}{\rho_p} + \frac{F_i}{\rho_p} \quad (1)$$

where

$F_D(u_i - u_p)$ is drag force per unit particle mass

$\frac{g_i(\rho_p - \rho)}{\rho_p}$ is gravity force and

$\frac{F_i}{\rho_p}$ is additional forces: pressure gradient, Saffman lift, or other user defined force

In equation (1) u_i is the fluid phase velocity, u_p is the particle velocity, μ is the molecular viscosity of the fluid, ρ is the fluid density, ρ_p is the density of the particle, and *Dp* is the particle diameter.

The forces acting on the free floating pig are gravitational, buoyance, drag, and lift forces. The gravitational and buoyance forces are constant. As depth increases, the pressure will increase. Therefore, buoyancy force is created due to the difference in pressure of the top side of the free floating pig and the bottom side;

 DOI: 10.1051/matecconf/20167702010

creating the upward thrust. In contrast, the drag and lift forces depend on the flow condition. Drag force is the force acting opposite to the direction of fluid flow due to the resistance acting on the free floating pig while the lift force is the force acting perpendicularly upward on the free floating pig relative to the motion of the fluid flow due to the pressure difference from opposite side of the free floating pig as a result of fluid flow past the pig. By increasing the velocity, the upward forces will be increased thus overcoming the downward force which will enable the free floating pig to propel through the inclination.

The simulation of the discrete phase trajectory in FLUENT assist us to determine the discrete phase inertia, hydrodynamic drag, and the force of gravity, for both steady and unsteady flows. Particle parameters have been set according to the parameters of the free floating pig such as the diameter of the pig, the density of the pig and flow rate of the transporting medium to simulate the situation as close as possible.

3 Results and Discussion

3.1 Velocity profile and trajectory of the free floating pig

The velocity profile for both fluids (oil and water) shows separation region at the bend for all inclination angles. This is because when fluid flows towards the pipe bends, there will be a radial force acting inward on the flow resulting in inertial force leading to increase velocity. Accordingly, a high velocity region is observed in the middle of the pipe and the region near the wall have a lower velocity due to friction for all inclination angles simulated. For both fluids, as the inclination angle increase, the separation angle also increase.

The trajectory of the free floating Pig at 30° and 60° inclination angles propelled by the flow of oil are shown in Fig. 1. In all cases the minimum fluid velocity to propel the free floating pig was determined and the trajectory of the free floating pig visualized. The average velocity of the free floating pig can also be determined from the simulation. For instance for the 30° inclination shown in Fig. 1 (a), the average speed of the free floating pig ranges from 1.41 m/s to 2.1 m/s, which is mainly in the green and yellow region of the velocity. Since the free floating pig is denser and heavier than the oil, the velocity of the free floating pig is slightly lower than the velocity of the transporting medium.

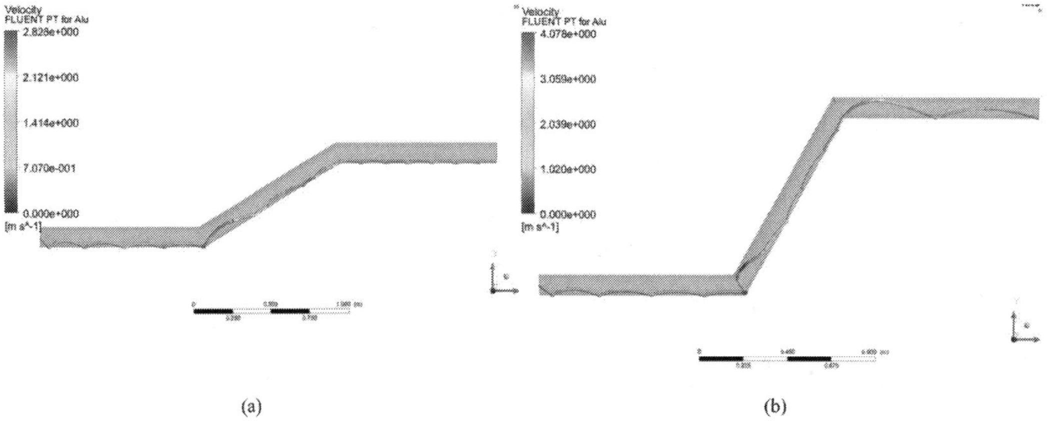

Figure 1. Free floating pig trajectory and velocity contour for (a) 30° pipe geometry (b) 60° pipe geometry

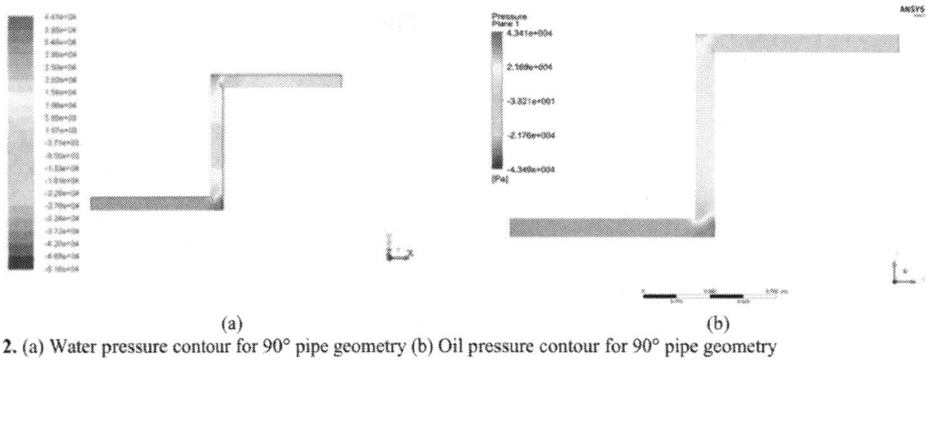

Figure 2. (a) Water pressure contour for 90° pipe geometry (b) Oil pressure contour for 90° pipe geometry

3.2 Pressure distribution

The pressure profile in the pipeline changes from the inlet to the outlet as theoretically predicted with high pressure region at the bends for both fluids. This is due to the sudden change in direction leading to inertial effect and resistance from the pipe wall at the bend. Fig. 2 shows the pressure contour for 90° pipe geometries with water and oil as working medium. Similar trends have been observed for other inclination angles. The higher the inclination angle, the higher pressure experienced at the bends.

4 Conclusions

The modeling and simulation of free floating pig for different inclination angle was presented in this paper. The simulation result assists in determining the critical flow parameters for the free floating pig to avoid stalling during operation. The simulation result assists us in visualizing the flow pattern, determine minimum pressure and flow rate required for the pig to freely float and driven by the medium for different inclination angles considered.

Acknowledgement

The authors would like to thank Universiti Teknologi PETRONAS, for the financial support in conducting this research under URIF grant.

References

1. Pikas, J. *Beyond ECDA For Evaluating'Unpiggable'Pipelines: Advances In Internal And External Tools*. in *Corrosion 2012*. 2012. NACE International.
2. Kurtz, D., *Developments in a Free-Swimming Acoustic Leak Detection System for Water Transmission Pipelines*, in *Pipelines 2006*. p. 1-8, (2006).
3. Felipe Oliveira, et al., *Smartball: A new pipeline leak detection system, and its survey of two Petrobras/Transpetro pipeline field tests*, in *Rio Pipeline 2011 Conference & Exposition*. (2011).
4. Fung, G., et al. *To pig or not to pig: the marlin experience with stuck pig*. in *Offshore Technology Conference*. 2006. Offshore Technology Conference.
5. Yu, X., et al. *Pigging Simulation Analysis of Deepwater Gas Tieback Flowlines in South China Sea*. in *The Twenty-third International Offshore and Polar Engineering Conference*. International Society of Offshore and Polar Engineers. (2013).
6. Xu, X.-X. and J. Gong, *Pigging simulation for horizontal gas-condensate pipelines with low-liquid loading*. Journal of Petroleum Science and Engineering. **48**(3–4): p. 272-280, (2005).
7. Tolmasquim, S.T. and A.O. Nieckele, *Design and control of pig operations through pipelines*. Journal of Petroleum Science and Engineering. **62**(3–4): p. 102-110, (2008).
8. Esmaeilzadeh, F., D. Mowla, and M. Asemani, *Mathematical modeling and simulation of pigging operation in gas and liquid pipelines*. Journal of Petroleum Science and Engineering. **69**(1–2): p. 100-106, (2009).
9. Zhu, H., P. Lin, and Q. Pan, *A CFD (computational fluid dynamic) simulation for oil leakage from damaged submarine pipeline*. Energy. **64**: p. 887-899, (2014).

MATEC Web of Conferences 77, 02011 (2016) DOI: 10.1051/matecconf/20167702011
ICMMR 2016

Modeling of Main Steam Parameters in Large Thermal Power Plant

Wen Xiaoqiang[1] and Jian Shuguang[2]

[1]*simulation center, Northeast Dianli University, Jilin, China*
[2] *School of Automation Engineering, Northeast Dianli University, Jilin, China*

Abstract. To solve the problem of main steam temperature abnormal of the 300 MW subcritical coal-fired boiler, the main steam temperature model was built based on running data. Firstly, the correlation analysis of the data was carried out based on the SPSS software to determine the main influence factors of the main steam temperature. Then, a nonlinear modeling method based on PLS was proposed, which is used to determine the number of the extracted components. At last, the simulation of the partial least squares regression model of the main steam temperature was realized by using MATLAB software. The partial least squares regression equation of the standard variable and the original variable was constructed by using the training samples and the accuracy and correctness of the model were tested by using the test samples.

1 Introduction

The development of information technology in power plant provides a good platform for data driven operation optimization research. Among them, based on the power plant real time & historical database, the method of modeling of the complex thermal system has gradually become one of the hot topics and it has become the basis of solving the key problems such as operation, control, performance evaluation, fault diagnosis, simulation and so on [1]. However, running data is different from the experimental data, which has many characteristics that are not conducive to modeling. For example, there are multiple correlation among variables, uneven distribution of the working conditions, and the existence of process nonlinearities, etc. All the problems are seriously hindering the development and application of data modeling methods [2]. Therefore, here the main steam temperature prediction model was built based on the Partial Least-squares Regression (PLS) algorithm.

2 Review of research methods and correlation analysis

2.1 Review of research methods

The main steam temperature of boiler has the obvious characteristics of nonlinearity, large inertia and long time-lag and in many cases, it is difficult to establish an accurate mathematical model. To solve the problem, Deng built the on-line fuzzy identification system of steam temperature based on T-S fuzzy rule model and the recursive least-square (RLS) [3]. Zhao developed an advanced Pareto multi-objective optimization with genetic algorithm based on a novel vector module adaptive function, which was applied to the PID controller parameters tuning in the boil superheated steam temperature cascade control system. The simulation results show that the improved multi-objective genetic algorithm (GA) optimization can obtain different optimal controller parameters with respect to the corresponding performance targets [4]. Zhang proposed the super-heater steam temperature model using on-line data to improve the conventional testing method and physical law methods. Based on on-line data, the identifiability of closed-loop control system of super-heater steam temperature could be met and dynamic models of super-heater steam temperature were obtained in the disturbance of super-heater spray flow. Simulations show that the model could reflect the facts of super-heater steam temperature [5]. Luan designed a new cascade control system based on the characteristics of superheated steam temperature of a boiler [6]. To solve the problem of reheat steam temperature (RST) abnormal of the 300MW power station boiler unit, Huang presented a method based on support vector regression (SVR). Based on the data sampled on spot, RST was analyzed using support vector regression method. The RST model was based on the statistical characteristics of the operating parameters and could reflect the potential relationship between RST and the operating parameters. For the units considered here, the RST was low and the temperature of reheater tube wall was high, tilt angles and desupreheater spray, etc. were taken as the tuning parameters and as the features of SVR model. The prediction results on test data with SVR-RST showed high regression coefficient with low complexity, which means SVR-RST model had excellent robustness, which was important to further optimize the operating parameters for higher efficiency and security[7].

2.2 Correlation analysis

There are many factors that can affect the temperature of the main steam, such as steam parameters, water spray temperature, flue gas temperature and flow rate, ash fouling status and so on. In order to determine the main influencing factors, here the correlation analysis was used.

There are three kinds of calculation of correlation coefficient: Pearson, Spearman and Kendall. Here we selected the Pearson correlation coefficient. The formula of Pearson correlation coefficient is as followed:

$$r = \frac{\sum_{i=1}^{n}(x_i - \bar{x})(y_i - \bar{y})}{\sqrt{\sum_{i=1}^{n}(x_i - \bar{x})^2 \sum_{i=1}^{n}(y_i - \bar{y})^2}} \tag{1}$$

where the r is regarded as the correlation coefficient. The $\bar{x}$ and $\bar{y}$ are the corresponding mean values and x_i and y_i are the ith observe value.

According to the Table 1, the factors whose correlation coefficient was greater than 0.6, were chosen to build the partial least squares regression model.

Table 1. Correlation analysis results

Influencing factors	Correlation coefficients	Influencing factors	Correlation coefficients
Main steam temperature	1.000	Feed water flow	-0.129
Pressure of main steam	0.811	Front and rear differential pressure of the main feedwater gate	-0.920
Flow of main steam	0.923	Speed of feed water pump	-0.114
Oxygen content in flue gas	-0.701	Coupler opening of feedwater pump	-0.240
Total air flow	0.834	Low pressure side opening	0.340
Inlet baffle opening of primary air fan	0.686	High pressure side opening	0.332
Inlet baffle opening of forced draught fan	0.310	Outlet temperature of low pressure side	0.436
Turbine speed	0.092	Outlet temperature of high pressure side	-0.148
Generator power	0.165	Regulating valve opening of oil fuel atomizing steam	0.643
Drum water level	0.759	Speed of powder feeder	0.137

3 Partial least-squares regression

3.1. Partial least-squares regression

The PLS method has a tremendous success in chemometrics and chemical industries for static data analysis. It integrates the principal component analysis (PCA) and canonical correlation analysis (CCA) together naturally and is convenient for the analysis of the multi-dimensional complexity system. In its general form, PLS creates components by using the existing correlations between different sets of variance while also keeping most of the variance of both sets. Now, we provide some of the formal ingredients of the method [8].

Consider a general setting of the PLS algorithm to model the relation between two data sets. Let $\boldsymbol{X} = \{x_1, x_2, \ldots, x_p\}$ denote a p-dimensional vector of variables in the first block of data, and similarly let $\boldsymbol{Y} = \{y_1, y_2, \ldots, y_q\}$ denote a vector of variables from the second set. Observing n data samples from each block of variables, PLS decomposes $\boldsymbol{X} = (x_{ij})_{n\times p}$ and $\boldsymbol{Y} = (y_{ij})_{n\times q}$ into the form:

$$\boldsymbol{X} = \boldsymbol{TP}^T + \boldsymbol{F} \tag{2}$$

$$\boldsymbol{Y} = \boldsymbol{UQ}^T + \boldsymbol{G} \tag{3}$$

where $\boldsymbol{T}$ and $\boldsymbol{U}$ are $n\times r$ matrices of the extracted r components, the $p\times r$ matrices $\boldsymbol{P}$ and $\boldsymbol{Q}$ represent matrices of projections, and the $n\times p$ matrices $\boldsymbol{F}$ and $\boldsymbol{Q}$ are the matrices of residuals. The PLS method, in which the classical form is based on the nonlinear iterative partial least square algorithm, finds projection axes w and c such that:

$$\max(\boldsymbol{Xw}, \boldsymbol{Yc}) \tag{4}$$

$$s.t.\begin{cases} \boldsymbol{w}^T\boldsymbol{w} = 1 \\ \boldsymbol{c}^T\boldsymbol{c} = 1 \end{cases}$$

The solution to this optimization problem is given by the following Eigen value problem:

$$\boldsymbol{X}^T\boldsymbol{YY}^T\boldsymbol{Xw} = \xi\boldsymbol{w} \tag{5}$$

The $\boldsymbol{X}$-latent-component vector $\boldsymbol{t}$ can be obtained as $\boldsymbol{t} = \boldsymbol{Xw}$, which is also the first Eigen vector about the equation $\boldsymbol{XX}^T\boldsymbol{YY}^T\boldsymbol{t} = \xi\boldsymbol{t}$. The $\boldsymbol{Y}$-latent -component vector $\boldsymbol{u}$ can be estimated as $\boldsymbol{u} = \boldsymbol{YY}^T\boldsymbol{t}$.

Furthermore, the above algorithm can be repeated l times with the deflation of $\boldsymbol{X}, \boldsymbol{Y}$, defined as $\boldsymbol{X}\text{-}\boldsymbol{tt}'\boldsymbol{X} \to \boldsymbol{X}$, $\boldsymbol{Y}\text{-}\boldsymbol{tt}'\boldsymbol{Y} \to \boldsymbol{Y}$ then $\boldsymbol{T}$, $\boldsymbol{U}$, the $p\times l$ matrix $\boldsymbol{W}$ and the $q\times l$ matrix $\boldsymbol{C}$ can be created, consisting of the columns from the vectors $\{\boldsymbol{t}_i\}^p_{i=1}$, $\{\boldsymbol{u}_i\}^p_{i=1}$, $\{\boldsymbol{w}_i\}^p_{i=1}$, $\{\boldsymbol{c}_i\}^p_{i=1}$, respectively.

The PLS form as regression model can be written in matrix as:

$$\hat{\boldsymbol{Y}} = \boldsymbol{XB}_{pls} \tag{6}$$

where coefficient matrix $\boldsymbol{B}_{pls}$ can be calculated as $\boldsymbol{B}_{pls} = \boldsymbol{W}(\boldsymbol{P}^T\boldsymbol{W})^{-1}\boldsymbol{C}^T$, and $\boldsymbol{P}$ is the $p\times l$ matrix consisting of loadings vectors $\{\boldsymbol{p}_i = \boldsymbol{X}^T\boldsymbol{t}_i/(\boldsymbol{t}^T{}_i\boldsymbol{t}_i)\}^p{}_{i=1}$.

 DOI: 10.1051/matecconf/20167702011

3.2 Cross validation

Accuracy of PLS regression is verified by cross-validity:

$$Q_h=1-(S_{press,h} / S_{ss,h-1}) \quad (7)$$

where $S_{press,h}$ is the sum of squares of prediction errors after h components and $S_{ss,h-1}$ is sum of squares of errors after h-1 components. As if $S_{press,h}$ is smaller than $S_{ss,h-1}$ to certain extent, adding component t_h is considered to be useful for the improvement of regression accuracy, that is, if $Q_h \geqslant 0.0975$, the quality of model could be improved by adding a component. Otherwise, the number of component is enough. The cross-validity Q_h is the criterion for determining the number of components.

4 Partial least squares regression model

According to the correlation analysis above, the main influencing factors of main steam temperature are pressure of main steam x_1, flow of main steam x_2, oxygen content in flue gas x_3, total air flow x_4, inlet baffle opening of primary air fan x_5, drum water level x_6, front and rear differential pressure of main water gate x_7 and regulating valve opening of oil fuel atomizing steam x_8.

According to the partial least squares regression algorithm, the modeling process is as follows:

• Extracting the 1st principal component t_1

Combination coefficients (weights), vector:

w_1= (0.4346, 0.4403, -0.3566, 0.3153, -0.0898, -0.4031, -0.4686，-0.0556)^T

Regression coefficients, scalar:

r_1 = 0.3348

Regression coefficient, vector:

p_1 = (0.4106, 0.3903, -0.4038, 0.3961, -0.2912, -0.4148, -0.3933, -0.0781)^T

while Q_1 = 0.6124 > 0.0975, continue to calculate.

• Extracting the 2nd principal component t_2

Combination coefficients (weights), vector:

w_2 = (0.0869, 0.2067, 0.1948, -0.3340, 0.8322, -0.0483, -0.3108, 0.0933)^T

Regression coefficients, scalar:

r_2 = 0.6749

Regression coefficient, vector:

p_2 = (0.0591, -0.3047, -0.1635, 0.00741, -0.0971, -0.0069, -0.1734, 0.9026)^T

while Q_2 = 0.8035 > 0.0975, continue to calculate.

• Extraction of the 3rd principal component t_3

w_3 = (0.1440, 0.2067, 0.1948, -0.3340, 0.8322, -0.0483, -0.3108, 0.0933)^T

Combination coefficients (weights), vector:

r_3 = 0.1146

Regression coefficients, scalar:

p_3= (0.0951, -0.1414, -0.1487, 0.0146, -0.0771, -0.0573, -0.0750, 0.9961)^T

while Q_3 = 0.0854 < 0.0975, stop calculating.

• The partial least squares regression

Thus, when the extraction of 2 components, the model's predictive ability is best, and the standardized variable regression equation is:

$$y^*=0.2737x_1^*+0.3589\,x_2^*-0.0462\,x_3^*-0.0683\,x_4^*+ 0.5169\,x_5^* + 0.1682\,x_6^*-0.4432\,x_7^*+0.0353\,x_8^* \quad (8)$$

For the Eq. 8, the original partial least squares regression equation is:

y = 492.1938 + 0.8351x_1+0.0139 x_2 - 0.1339 x_3 - 0.0588 x_4 + 0.7045 x_5 + 0.0151 x_6 - 3.8666 x_7 + 0.0426 x_8

• Evaluation of the model

The average relative error of the training sample was 0.3%. The test samples were substituted into regression equation and the comparison of the predicted results with the experimental results was shown in Fig. 1. Fig. 2 showed the relative error of test samples. The average relative error of the test samples was 0.28%. It is obvious that the error meets the engineering requirements.

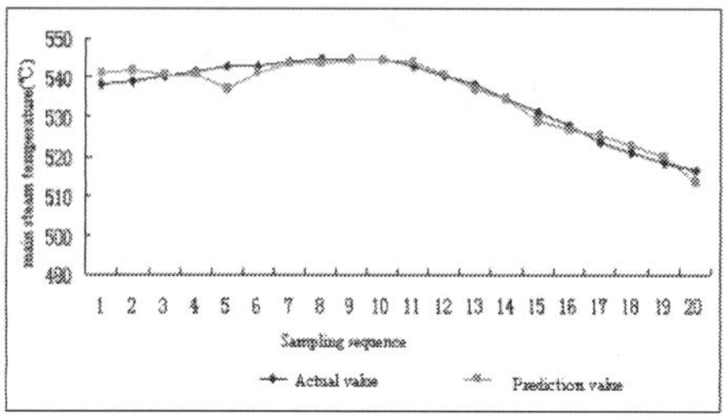

Figure 1. Prediction results of test samples

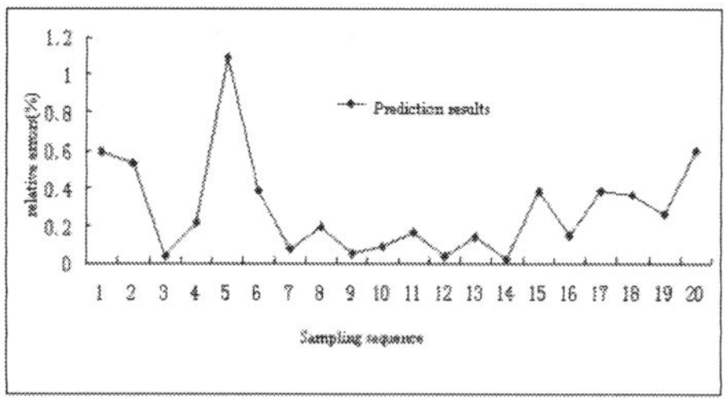

Figure 2. Relative errors of test samples

5 Conclusion

Here, the model of main steam temperature was built based on partial least-squares regression. The main influencing factors of main steam temperature were determined through correlation analysis. The prediction results showed that the model has high accuracy and can predict the main steam temperature in large thermal power plant.

Acknowledgement

This project Supported by the National Natural Science Foundation of China (51176028), the KEY Scientific, Technological Project of Jilin Province of China (20140204006SF) and Science and technology development plan of Jilin City (201464061)

References

1. Zhang X. T., Ni W. D., Li Z., Zheng S., Identifiability of building thermal system models

using on-line data. J. Tsinghua Univ. (Sci & Tech), **44**, (2004).

2. Jin T., Fu Z. G., Liu G., Yang Y. P., Application of reversed modeling method in power plant critical parameters modeling. J. Power Engineering, **29**, (2009).
3. Deng L C, Wang G J, Chen H. On-line Fuzzy Identification of the Steam Temperature Object of Boiler, Proceedings of the CSEE, **26**, (2006).
4. Zhao L., Ju G., Lv J. H., An improved genetic algorithm in multi-objective optimization and its application, Proceedings of the CSEE, **28,** (2008).
5. Zhang X T, Ni W D, Li Z. Dynamic modeling study of super heater steam temperature based on principal component analysis method and online data. Proceedings of the CSEE, **25,** (2005).
6. Luan X. C., Li S. Y., A study on multiple models predictive control for superheated steam temperature based on neural network models, Proceedings of the CSEE, **24,** (2004)
7. Huang J T., Ma L. H., Mao J. B., Qian J. X., Modeling research of the reheat steam temperature of 300MW boiler based on support vector regression. Proceedings of the CSEE, **26,** (2006).
8. Wen X Q. Analysis of carbon in coal-fired based on partial least squares regression algorithm, J. Northeast Dianli University, **32,** (2012).

DOI: 10.1051/matecconf/20167702012

A Simulation Study of Recessed Type Hydrostatic Journal Bearing

Shih-Ming Wong, Ta-Hua Lai, Min-Cheng Lin, and Shih-Chieh Lin

Department of Power Mechanical Engineering, National Tsing-Hua University, Hsinchu 300, Taiwan

Abstract. In this paper, a simulation model is established to study effects of the bearing system parameters on bearing performance. The static performance (such as load capacity, stiffness and flow) about various dimensions of multi-recess type hydrostatic journal bearings with constant-resistant restrictor compensation was investigated. The parameters studied includes length/diameter ratio, number of recess, recess length ratio, recess width ratio, initial ratio between recess pressure to supply pressure, and eccentricity.

1 Introduction

Hydrostatic bearing has been characterized with high damping, low friction and low wear. For high precision machining, hydrostatic bearing is widely adopted in recent years.

Many researchers had been devoted to the study of hydrostatic bearing systems. Dowson [1] studied the effects of the depth of recess on pressure distribution over the recess area. O'Donoghue [2] adopted the finite difference method to study the distribution of pressure over the bearing area. Majumdar [3] studied the relationship between the restrictor and load capacity. Gosh [4-6] studied the loading capacity and flow rate for multi-pad bearing. Row [7] studied the effects of pocket design on torque. Ho [8] conducted experiments to confirm simulations results. Liang [9] studied the hydrodynamics effects. Donoghue [10], Ghigliazza [11] and Elsherbiny [12] proposed ways for hydrostatic bearing system design.

In this study, a simulation model is established to study effects of the bearing system parameters on bearing performance. The static performance (such as load capacity, stiffness and flow) about various dimensions of multi-recess type hydrostatic journal bearings with constant-resistant restrictor compensation was investigated. The parameters studied includes length/diameter ratio, number of recess, recess length ratio, recess width ratio, initial ratio between recess pressure to supply pressure, and eccentricity.

In the following section, the fundamental theories for the simulation models are briefly reviewed. A series of simulations are then conducted to studied effects of parameters on system performances. Finally, conclusions are made based on the results.

2 Simulation model

The recess dimension parameters of hydrostatic journal bearings, as shown in Fig. 1, includes bearing diameter D, bearing length L, land width a, and arc length of recesses b (related to recess size).

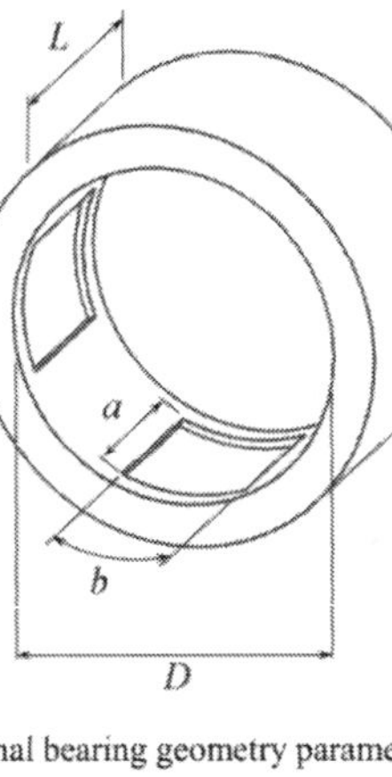

Figure 1. Journal bearing geometry parameters.

To obtain the pressure distribution of the oil film in the hydrostatic journal bearing, the generalized 2-dimensional Reynolds equation was used[1,2,3)] (neglecting the time dependent terms).

$$\frac{\partial}{\partial x}\left(\frac{\rho h^3}{12\mu}\frac{\partial p}{\partial x}\right)+\frac{\partial}{\partial y}\left(\frac{\rho h^3}{12\mu}\frac{\partial p}{\partial y}\right)=0 \qquad (1)$$

where μ is coefficient of absolute viscosity, and h is the film thickness.

The bearing surface is unwrapped in the finite difference form, represented in the x-y plane, as shown in Fig. 2. The pressure at any point (x,y) was solved iteratively using Finite Different Method (FDM), which satisfy the following boundary conditions:

$$p(\theta,\pm 1) = 0$$
$$p = p_r \text{ at the } r\text{th reces} \quad (2)$$
$$p(\theta,y) = p(\theta + 2\pi, y) \text{ for periodicity}$$

where θ is angular coordinate, p_r is recess pressure at rth recess.

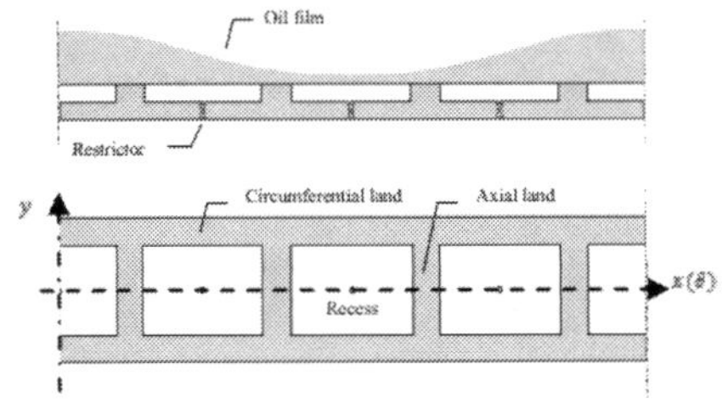

Figure 2. Hydrostatic bearing surface unwrapped.

The solution satisfying boundary conditions yields the pressure distribution of the oil film, which leads to the determination of load capacity, stiffness, and the power consumption of the bearing.

3 Simulation conditions, results and discussions

In order to study effects of the bearing system parameters on bearing performance, a series of simulations had been studied. The system parameters and levels are listed in table 1. The performance of the bearing system includes dimensionless loading capacity, $\bar{W} = w/(P_s LD)$ stiffness, $\bar{K} = kh_0/(P_s LD)$ and flow rate, $\bar{Q} = q\eta/(P_s h_0^3)$.

Table 1. System parameters and levels for simulation tests.

Parameters	Symbols	Levels
length/diameter ratio	L/D	0.5、1.0、1.5
Recess length ratio	a/L	0.25、0.50、0.75
Recess width ratio	Nb/(πD)	0.25、0.50、0.75
Number of recess	N	4、5、6、7
Eccentricity ratio	ε	0.1、0.3、...、0.9
Initial pressue ratio	β_0	0、0.1、···、1.0
Attitude angle	ϕ	0, π/N

3.1 Dimensionless loading capacity

The simulation results show that the effects of initial pressure ratio between the recess pressure and the supplied pressure on the loading is the most significant, and then the eccentricity ratio, the recess length ratio, length/diameter ratio, number of recess, recess width ratio and attitude angle.

Fig. 3 shows the interaction effects of the initial pressure ratio and the eccentricity ratio on loading capacity. It was observed that a maximum loading can be derived when the initial pressure ratio is around 0.5. And the loading capacity is generally increased as an increase in eccentricity ratio. In other words, the increase in load will increase the eccentricity

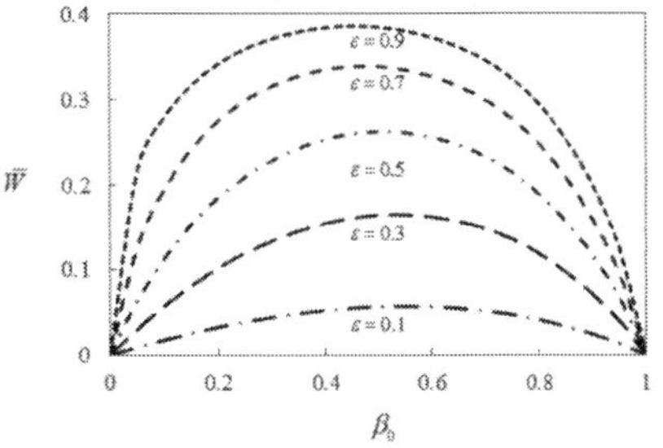

Figure 3. Interaction effects of the pressure ratio and the eccentricity ratio on loading capacity.

Fig. 4 shows the interaction effects of the recess length ratio and the length/diameter ratio on loading capacity. It was observed that the loading increased as the recess length ratio increased. And the loading capacity is generally increased as a decrease in length/diameter ratio. However, the decrease in length will decrease the load capacity.

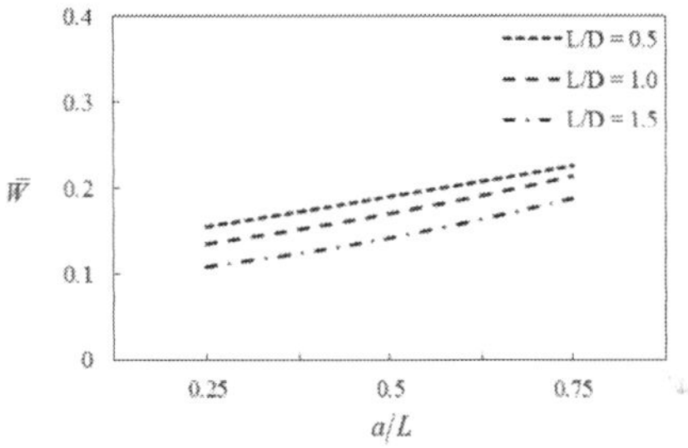

Figure 4. Interaction effects of the recess length ratio and the length/diameter ratio on loading capacity.

Fig. 5 shows the interaction effects of the recess width ratio and the number of recess on loading capacity. It was observed that the loading capacity is slightly increased as the number of recess is increased. And the effects of recess width ratio on the loading capacity is not so significant. However, a higher loading can be achieved with a ratio of 0.5.

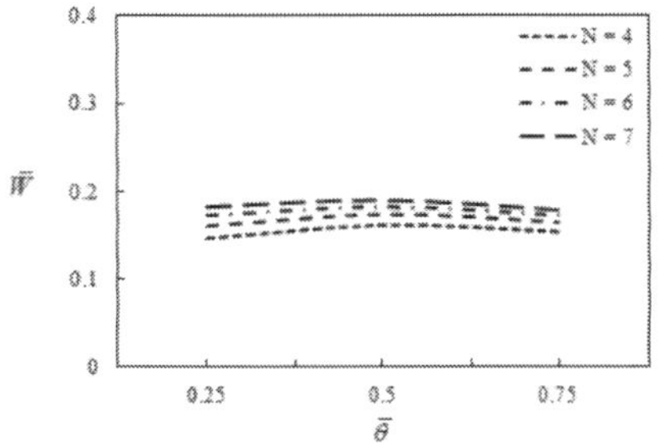

Figure 5. Interaction effects of the recess width ratio and the number of recess on loading capacity.

3.2 Dimensionless stiffness

The simulation results show that the effects of initial pressure ratio between the recess pressure and the supplied pressure on the stiffness is the most significant, and then the eccentricity ratio, the recess length ratio, length/diameter ratio, number of recess, recess width ratio and attitude angle.

Fig. 6 shows the interaction effects of the initial pressure ratio and the eccentricity ratio on stiffness. A strong interaction effect of the initial pressure ratio and the eccentricity ratio on stiffness can be observed. When the initial pressure ratio is larger than 0.5, the increase of eccentricity ratio will results in a decrease in stiffness. The dimensionless stiffness for a pressure ratio of 0.5 with no eccentricity is around 0.55 while that is reduced to 0.25 with an eccentricity of 0.8.

When the initial pressure ratio is smaller than 0.4, the increase of eccentricity ratio will results in an increase in stiffness in the beginning and a decrease in stiffness with the further increase in the eccentricity ratio. In other words, when the initial pressure is low, the increase in loading will result in an increase in stiffness in the beginning and then a decrease in stiffness if the loading is further increased.

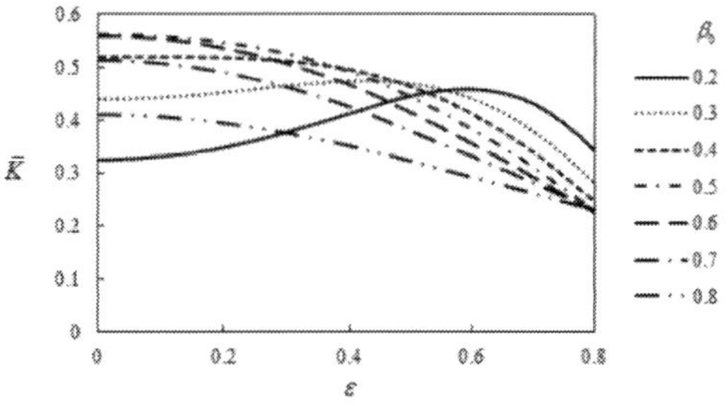

Figure 6. Interaction effects of the pressure ratio and the eccentricity ratio on stiffness.

Fig. 7 shows the interaction effects of the recess length ratio and the length/diameter ratio on stiffness. It was observed that the stiffness increased as the recess length ratio increased. And the stiffness is decreased as the length/diameter ratio increased.

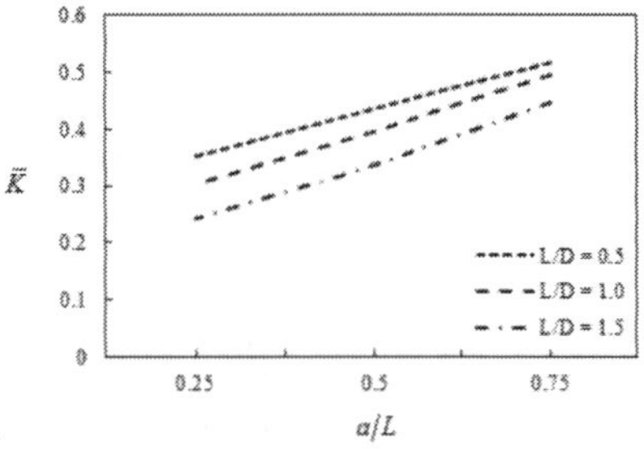

Figure 7. Interaction effects of the recess length ratio and the length/diameter ratio on stiffness.

Fig. 8 shows the interaction effects of the recess width ratio and the number of recess on stiffness. It was observed that thestiffness is slightly increased as the number of recess is increased. And the effects of recess width ratio on the stiffness is not so significant. However, a higher stiffness can be achieved with a ratio of 0.5.

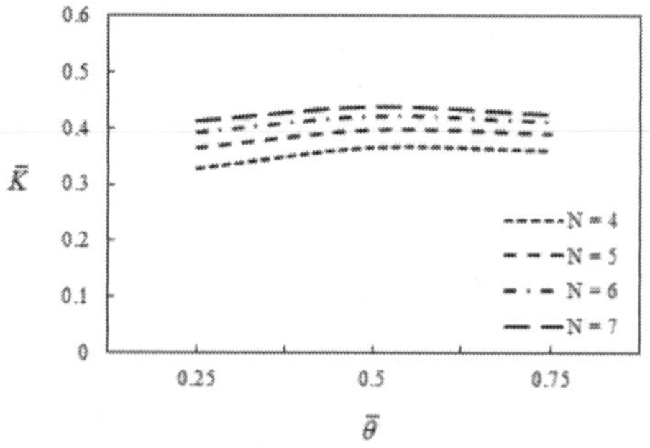

Figure 8. Interaction effects of the recess width ratio and the number of recess on Stiffness.

3.3 Dimensionless flow rate

The simulation results show that the effects of initial pressure ratio between the recess pressure and the supplied pressure on the flow rate is the most significant again, and then the recess length ratio, length/diameter ratio, recess width ratio, number of recess, the eccentricity ratio, and attitude angle.

Fig. 9 shows the interaction effects of the initial pressure ratio and the recess length ratio on flow rate. It was observed the flow rate increased as the initial pressure ratio or the recess length ratio increased. The increase in pressure ratio is accompanying with a lower flow resistance of the restrictor in simulated cases. And a larger recess length ratio also means that the land area of the bearing is thinner and a lower flower resistance for the oil to flow out.

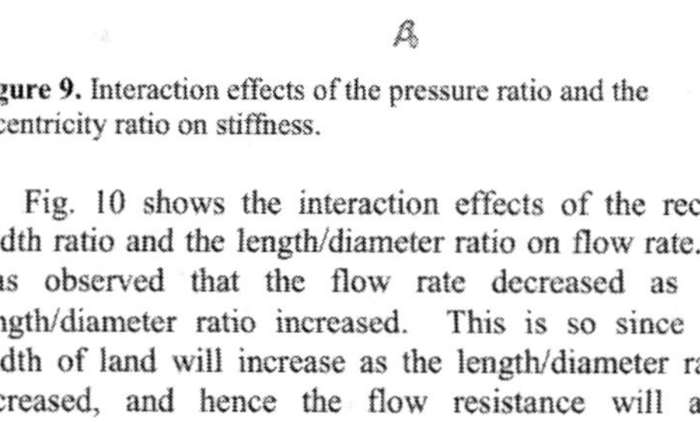

Figure 9. Interaction effects of the pressure ratio and the eccentricity ratio on stiffness.

Fig. 10 shows the interaction effects of the recess width ratio and the length/diameter ratio on flow rate. It was observed that the flow rate decreased as the length/diameter ratio increased. This is so since the width of land will increase as the length/diameter ratio increased, and hence the flow resistance will also increase as the length/diameter increased. And the flow rate is slightly increased as the increase in recess width ratio. This is so since the pocket area is increased and the land area is reduced as the recess width increased. The decrease in the land area generally results in a lower flow resistance.

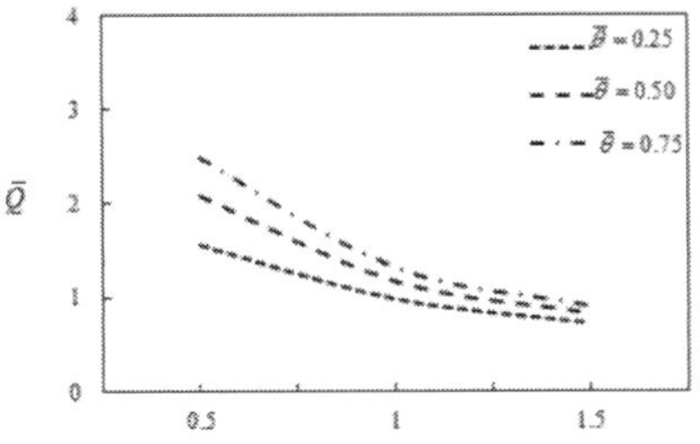

Figure 10. Interaction effects of the recess width ratio and the length/diameter ratio on stiffness.

Fig. 11 shows the interaction effects of the eccentricity ratio and the number of recess on flow rate. It was observed that the flow rate is slightly increased as the number of recess is increased. And the flow rate is also slightly increased as the eccentricity rate is increased.

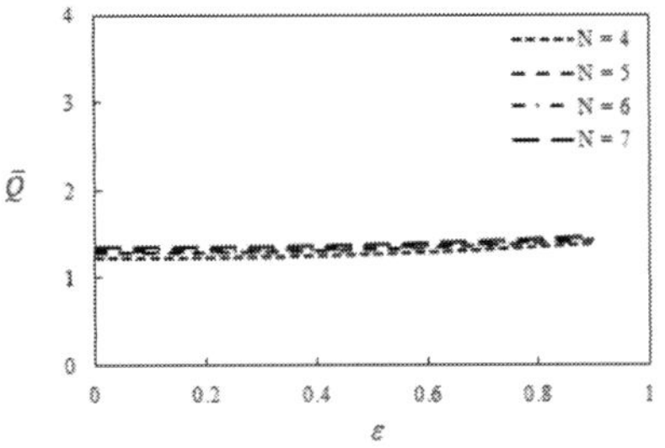

Figure 11. Interaction effects of the eccentricity ratio and the number of recess on Stiffness.

4 Conclusions

In this paper, a simulation model is established to study effects of the bearing system parameters on bearing performance. The static performance (such as load capacity, stiffness and flow) about various dimensions of multi-recess type hydrostatic journal bearings with constant-resistant restrictor compensation was investigated.

The simulation results show that the effects of initial pressure ratio between the recess pressure and the supplied pressure on all three performance index is the most significant. It was observed that a maximum loading can be derived when the initial pressure ratio is around 0.5.

A strong interaction effect of the initial pressure ratio and the eccentricity ratio on stiffness can be observed. When the initial pressure ratio is larger than 0.5, the increase of eccentricity ratio will results in a decrease in stiffness. The dimensionless stiffness for a pressure ratio of 0.5 with no eccentricity is around 0.55 while that is reduced to 0.25 with an eccentricity of 0.8.

It was observed the flow rate increased as the initial pressure ratio or the recess length ratio increased. The increase in pressure ratio is accompanying with a lower flow resistance of the restrictor in simulated cases. And a larger recess length ratio also means that the land area of the bearing is thinner and a lower flower resistance for the oil to flow out.

References

1. D. Dowson and C. Taylor, ASME, **83**, 2 (1960)
2. J. O'Donoghue and W. Rowe, Tribology, **1**, 4 (1968)
3. B. Majumdar, Wear, **14**, 6, (1969)
4. B. Ghosh, Wear, **21**, 2 (1972)
5. B. Ghosh, Wear, **23**, 3 (1973)
6. M. Ghosh and B. Majumdar, Tribology International, **13**, 2 (1980)
7. W. Rowe and D. Koshal, Wear, **43**, 1 (1977)
8. Y. HO and N. Chen, Wear, **52**, 2 (1979)
9. P. Liang, C. Lu, J. Ding and S. Chen, Tribology International, **67** (2013)
10. J. O'Donoghue, W. Rowe and C. Hooke, Wear, **14**, 5 (1969)
11. R. Ghigliazza, R. Michelini and L. Rossi, Tribology, **3**, 3 (1970)
12. M. El-Sherbiny, F. Salem and N. El-Hefnawy, Tribology International, **17**, 3 (1984)

MATEC Web of Conferences 77, 02013 (2016) DOI: 10.1051/matecconf/20167702013
ICMMR 2016

Automated Negotiation for Decentralised Scheduling

Haixiao Liu, Zhichao Shao, Xianglin Tan

Wuhan Mechanical Technology College, Wuhan, Hubei, 430075, China

Abstract. This paper is concerned with inter-company scheduling between manufacturers of similar products who wish to pool resources for their mutual benefit, but without sharing commercially sensitive situation such as order lists and production technology with one another and without the need for a third-party brokerage infrastructure. The ultimate objective of this work is to enable small-scale manufacturers, which may be competitors in certain markets, to achieve some of the economies of scale of their larger counterparts. To accomplish this, the paper introduces a simple agent-negotiation approach that is able to identify mutually beneficial resource transactions.

1 Introduction

Individual organisations may benefit from coordinating their planning and scheduling activities for various reasons. For example, inter-company scheduling can allow manufacturers or service providers to achieve objectives that they cannot accomplish on their own by pooling resources with their competitors [1], such as fulfil larger orders or offer more services out of their individual capabilities. This paper demonstrates the work by means of a manufacturing scenario but it is applicable to the service provision as well.

An increasing number of papers examine developing agent-based negotiation approaches for inter-company scheduling in manufacturing [2-11]. In them, the auction-based protocol has widely been used to guide agent negotiation behaviours [12]. A key requirement of these approaches is that they rely on the use of one agent to act as a central authority. This central authority coordinates the negotiations between agents and maintains information about all the other agents. Even if this central authority can ensure that all participating agents benefit from any suggested transactions, such an approach can be unacceptable to some or all participating agents for two reasons.

Firstly, the participating agents may not be prepared to trust another agent or a third-party with the information required to produce effective inter-company schedules. Secondly, in a competitive market, it can be contrary to an agent's long-term profit to support the profitability of its competitors. In other words, if the central authority of an inter-company scheduling system produces one solution that is beneficial to an agent, but even more beneficial to its competitors, then that solution reduces the agent's market share, which adversely affects the agent's long-term profitability. In such circumstances, negotiation through bilateral negation remains the only alternative approach to inter-company scheduling.

This paper introduces such an approach in a simple two-company, single product, single resource, one-off transaction scenario. The organisation of the paper is as follows. In Section 2 our scenario is defined that involves only two manufacturers. Next, Section 3 proposes our negotiation approach for inter-company scheduling. There the agent interactive protocol and individual strategy are illustrated. Finally, our future work towards decentralised agent-based inter-company scheduling is described in Section 4.

2 Scenario

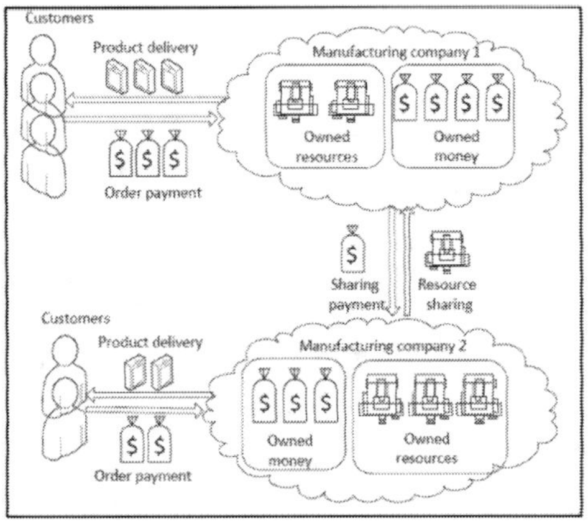

Figure 1. Our scenario

Our scenario is depicted in Figure 1. It consists of two manufacturers 1 and 2, each denoted by i aiming to maximise its profit. To simplify the problem, we assume that there is only one type of resource. A company's production capacity c_i is shown as the production resources it possesses. A company can increase the number of resources available for production by purchasing additional resources from the other company. Conversely, a company that sells resources to the other reduces the resources available to it.

This scenario takes into account only one type of product with market price m per product. To manufacture one unit of the product, r_i units of resource are required in manufacturer i. Two manufacturers are assumed to use similar amount of resources for production. Otherwise, the one with much bigger r_i may lose competitiveness in a long term.

The manufacturer receives a set of orders O_i from customers for the solely product and also obtains income by fulfilling orders. In order to fulfil an order, the manufacturer must produce as many products as stipulated in the order. No income is received for orders that are only partially fulfilled. Each order $q \in O_i$ represents the quantity of the product required to fulfil the order. For each order $q \in O_i$ that the manufacturer i fulfils, $q * r_i$ units of resources are allocated and an income $q * m$ is received.

Three additional variables are introduced to represent resource sharing between manufacturers. Boolean variable ω_q describes whether or not the order $q \in O_i$ is accepted by manufacturer i. The integer variable sq_i indicates the number of units of resource purchased by manufacturer i. Negative values of sq_i indicate that i sells $-sq_i$ units of resource. The other integer variable sp_i represents the overall price of the transaction.

If manufacturer i does not buy or sell resources, its maximum profit is computed as:

$$P_i = \max_i \sum_{q \in O_i} \omega_q * q * m \quad (1)$$

s.t.
$$\sum_{q \in O_i} \omega_q * q * r_i \le c_i \quad (2)$$

A manufacturer unable to meet all its orders is willing to acquire additional resources provided that the purchasing cost is less than the marginal profit realised from those additional resources. In other words, a resource purchase must increase total profit.

Let the manufacturer i offer to purchase sq_i units of resource with sp_i payment. Then these two integer variables in resource purchase must satisfy the constraints in Equations (3) and (4). The profit of this purchase proposal can be obtained through Equation (5).

$$sq_i \ge \sum_{q \in O_i} \omega_q * q * r_i - c_i \quad (3)$$

$$sp_i < \sum_{q \in O_i} \omega_q * q * m - P_i \quad (4)$$

$$PS(sq_i, sp_i) = \max \sum_{q \in O_i} \omega_q * q * m - sp_i \quad (5)$$

Similarly, a manufacturer will not sell resources unless the marginal profit lost through the sale is less than the total income received from transaction. Formally, i only sells $-sq_i$ units of resource at an overall price sp_i if constraints (3) and (6) are met. Its corresponding profit can be calculated via Equation (7).

$$sp_i > P_i - \sum_{q \in O_i} \omega_q * q * m \quad (6)$$

$$PS(sq_i, sp_i) = \max_i \sum_{q \in O_i} \omega_q * q * m + sp_i \quad (7)$$

3 Negotiation approach

In our scenario, manufacturers are autonomous decision makers. For this reason, each company is represented by an agent that can allocate resources for production or remain them unused. When its resources are insufficient for orders, it intends to purchase more from the other agent. Oppositely, the agent may sell out excessive resources for profit maximisation. The quantity and payment of resource sharing need to be discussed for smooth transaction.

A variety of negotiation mechanisms have been posed to guide agent interactive behaviours. The alternating-offers protocol is the most popular one where agents take turns to propose offers until agreements on given issues are achieved. The agent receiving an offer can accept the offer, make a counter-offer, or abort the negotiation.

However, this approach is crucially dependent on a central authority to decide which agent makes the initial offer. Moreover, the manufacturer may be able to achieve a better price through providing feedback than proposing counter-offers. This is because in the former situation the negotiation quantity and payment leave open.

This section introduces a modified alternating-offers protocol that governs agent negotiation behaviours from an overall view. The protocol is used to reach an agreement on resource transactions, with a view to maximise profits. We then demonstrate the strategies guiding agent individual actions in terms of their roles in the transaction.

3.1 Negotiation protocol

The agent lack of sufficient resources can identify the marginal profit from what it offers to purchase. This profit is considered as its maximal price. On the contrary, the agent with excessive resources is uncertain about the exact profit the purchaser can make. Therefore, it may miss some more profitable solutions when beginning negotiation via an offer with comparatively low selling price. This is owing to the fact that the buyer is not to pay the price higher than what the seller proposes. Thus, our approach assumes that the agent with purchase intention starts a negotiation thread. In addition, the negotiation process consists of two parallel threads when both agents require extra resources to fulfil their orders.

The negotiation protocol is illustrated in Fig. 2. First, the agent considering acquiring more resources starts a negotiation thread with a purchase *offer*. Within the context of this negotiation thread, the agent proposing the initial offer is called *buyer* and the other is *seller*. The seller then evaluates the proposal and responds in two ways. The seller can *accept* the proposal when agreeing with the terms. Alternatively, if the seller disagrees with terms of the offer, it provides *feedback* suggesting the buyer the direction of another offer.

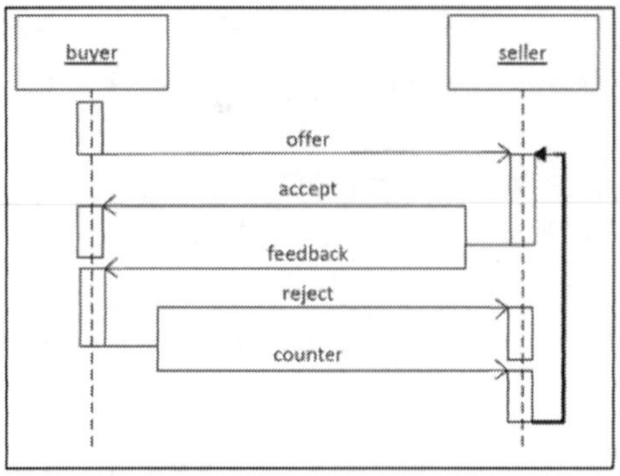

Figure 2. The negotiation protocol

Here, the seller does not reject the negotiation or make a counter-offer in face of an *offer* for two reasons. Firstly, it wishes to sell excessive resources at a reasonable price so that it would like to await an acceptable offer rather than abort the negotiation process. Secondly, the seller is unsure about the maximal price that the buyer accepts for some resources. Then it is better to leave the offer open by providing feedback rather than arbitrarily decide the payment by making counter-offer.

The seller's feedback can provide two types of advice to update the proposal. To be precise, the seller may suggest *lower_quantity* when the quantity in the received offer exceeds its production capacity. Also it can require *higher_payment* for some units of resource if the agent can meet the quantity but it disagrees with the proposed payment. The seller does not have to suggest a higher quantity of resource sharing since the *higher_payment* implies not only that it can provide the quantity of resource in the corresponding proposal but also that a higher quantity may be feasible. Therefore, this protocol only takes into account *lower_quantity* and *higher_payment* two types of feedback.

When the seller replies with *feedback*, the buyer can take two alternative actions. It can *reject* the negotiation when recognising no possibility for sharing resources even with further discussion according to all received feedback. Or the buyer can provide another offer as a *counter* based on all received feedback aiming for future agreement. Here the buyer cannot accept ambiguous *feedback* due to the fact that two agents have to reach an unambiguous agreement to share resources. Meanwhile, we do not consider the situation that the buyer provides feedback for received *feedback* for simplification.

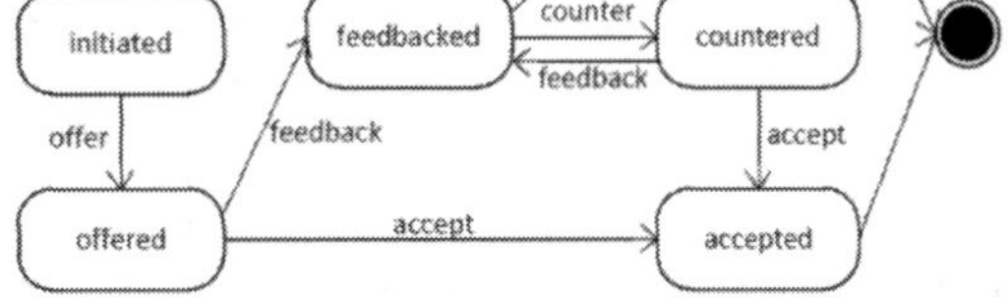

Figure 3. The negotiation states

The transition of agent states in one negotiation thread is described in Fig. 3. Two agents can hold two parallel negotiation threads. However, further discussion in one thread becomes unnecessary when the other achieves an agreement on resource sharing. Thus, the overall negotiation process terminates when in one thread the agents reach an agreement or in both threads the buyers determine to abort the negotiation.

3.2 Negotiation strategy

Prior to negotiation, each agent i computes the maximum profit P_i it can obtain without sharing resources. This imposes a lower limit on expected total profit as a result of resource negotiations. The agent starts a negotiation thread when unable to fulfil all its orders by means of possessed resources. As mentioned above, each negotiation thread consists of a buyer and a seller agent. We present negotiation strategies for each type of agent.

3.2.1 Buyer

In the negotiations, a prospective buyer i would like to purchase resources to meet as many orders as possible for maximal income from production. Moreover, for the same amount of resource that it intends to purchase, the buyer is willing to pay as little as the seller agrees for minimal expenses. Meanwhile, it may increase the payment slower when more offers are proposed in case that it pays much more than the seller's minimal price.

Let α_i be the payment proportion function of manufacturer i. Specifically for any sq_i units of resources, the prospective buyer initially offers to pay $\alpha_i(0)$ of marginal profit. When its proposal for sq_i is refused at the nth time, the buyer may increase the payment to $\alpha_i(n)$ of marginal profit unless it reaches the upper limit in Equation 4.

The buyer agent first generates the most profitable proposal of resource purchase that is then delivered as the first *offer* to the other agent. This proposal offers to purchase additional resources for fulfilling more orders with $\alpha_i(0)$ of the marginal profit as payment. The negotiation terminates with a transaction when the seller *accepts* the offer.

When the seller replies with *lower_quantity* feedback, the buyer searches for another quantity so that the number of resources requested to purchase is lower than that in the latest offer. It offers to pay $\alpha_i(0)$ of marginal profit as well. The buyer proposes a *counter* when receiving the *higher_payment* message. This negotiation thread is aborted when the buyer fails to provide a *counter* in the face of *lower_quantity* or *higher_payment.*

The *counter* is selected as the most profitable one with highest quantity from alternatives that are generated in two ways. One the one hand, the buyer searches for an offer with lower quantity than in the late proposal and $\alpha_i(0)$ of marginal profit when receiving *lower_quantity* reply. One the other hand, the agent classifies its previous offers into groups according their quantities. For each group, it finds out the number of times n that the proposals are refused. The agent then adds *counter* options with the quantity of this group and the payment of $\alpha_i(n)$ of marginal profit as long as the new payment is lower than the corresponding upper bound.

3.2.2 Seller

In the negotiation, a prospective seller i' wishes to share spare or occupied resources at the price as high as it can to maximise its profit. Also it charges not less when more resources are required by the other. This is because that if the buyer has offered sp_i for sq_i units of resource, the agent i' realises that the former can benefit not lower than sp_i for more than sq_i units. In addition, the seller can estimate the buyer profit since they two are supposed to be equipped with similar production technique. Thus, it can refuse some offers with low payment compared to estimated profit.

The seller agent has to terminate the negotiation thread when receiving a *reject* message from the buyer. Otherwise, it can either accept an *offer* or *counter* or provide *feedback* for further discussions. Since the agent is unable to sell resources more than possessed, it examines whether its capacity is lower than the sharing quantity in the proposal. If so, the agent replies with *lower_quantity* suggestion. If not, the agent computes the profit PS_i of selling resources as described in the received proposal.

The prospective seller then carries out three comparisons. Firstly, it examines profit PS_i considering resource sharing and P_i based on local assignment. Equation (8) indicates more potential benefit for i' when selling resources. If so, the agent then searches for $\mathrm{sp}(sq_i)$ that is the highest payment of i' for no more than sq_i units of resource in the negotiation history. It is set zero initially. Next, the seller takes the second comparison of sp_i and $\mathrm{sp}(sq_i)$. Equation (9) indicates that the purchase payment is regarded reasonable according to negotiation history. The last comparison occurs between sp_i and the seller's estimated profit $\mathrm{ep}(sq_i)$ for sq_i resources. The latter is decided with reference to its own production technique.

$$PS_i \geq P_i \tag{8}$$

$$sp_i \geq \mathrm{sp}(sq_i) \tag{9}$$

$$sp_i \geq \mathrm{ep}(sq_i) \tag{10}$$

The seller takes two alternative actions according to comparison results. When Equations (8), (9) and (10) are satisfied, the agent replies with an acceptance for sharing commitment. Otherwise, it deliveries *higher_payment* feedback for the quantity of resource in the *offer* or *counter*.

4 Conclusions

This paper has proposed a novel decentralised agent-based inter-company scheduling approach. The negotiation mechanism has been applied to prevent sharing of potentially commercially sensitive information. This approach suits scenarios consisting of schedule coordinating companies remain competitors. Then they are unwilling to disclose explicit information with one another or with third party agents.

Obviously the scenario employed in this paper is a simple one. Thus, future work will examine the effect of such extensions. In particular, the introduction of a temporal dimension has many implications, not only for the simulation but the negotiation mechanism as well. Another important generalisation of the work is the introduction of additional agents. Again, this affects the negotiation in a variety of ways. Especially, adding agents makes negotiations multilateral instead of bilateral, which has implications on both the negotiation protocol and an individual agent's negotiation strategies.

References

1. W. Shen, Q. Hao, H. Yoon, and D. Norrie, Applications of agent-based systems in intelligent manufacturing: An updated review, Advanced Engineering Informatics, **20(4)**, 415–431, (2006).
2. J. Reaidy, P. Massotte, and D. Diep, Comparison of negotiation protocols in dynamic agent-based manufacturing systems, International Journal of Production Economics, **99(1-2)**, 117–130, (2006).
3. M. Wooldridge, An introduction to multiagent systems, Wiley, 2009.
4. B. An, V. Lesser, and K. Sim. Strategic agents for multi-resource negotiation. Autonomous Agent and Multiagent Systems, **23(1)**:114-153, 2011.
5. D. Niyato, A. Vasilakos, and Z. Kun. Resource and revenue sharing with coalition formation of cloud providers: Game theoretic approach. In Proceedings of the 11th IEEE/ACM International Symposium on Cluster, Cloud and Grid Computing, 215-224, 2011.
6. S. Parsons, J. Rodriguez-Aguilar, and M. Klein. Auctions and bidding: A guide for computer scientists. ACM Computing Surveys, **43(2)**:10:1-10:59, 2011.
7. T. Rahwan, T. Michalak, E. Elkind, P. Faliszewski, J. Sroka, M. Wooldridge, and N. Jennings. Constrained coalition formation. In Proceedings of the 25th AAAI Conference on Artificial Intelligence, pages 719-725, 2011.
8. M. Barbati, G. Bruno, and A. Genovese. Applications of agent-based models for optimization problems: A literature review. Expert Systems with Applications, **39(5)**:6020-6028, 2012.
9. S. Chen and G. Weiss. An efficient and adaptive approach to negotiation in complex environments. In Proceedings of the 20th European Conference on Artificial Intelligence, 2012.
10. L. Duan, M. Dogru, U. Ozen, and J. Beck. A negotiation framework for linked combinatorial optimization problems. Autonomous Agent and Multiagent Systems, **25(1)**:158-182, 2012.
11. A. Elomri, A. Ghaffari, Z. Jemai, and Y. Dallery. Coalition formation and cost allocation for joint replenishment systems. Production and Operations Management, **21(6)**:1015-1027, 2012.

DOI: 10.1051/matecconf/20167702013

12. X. Zhang and V. Lesser. Meta-level coordination for solving distributed negotiation chains in semi-cooperative multi-agent systems. Group Decision and Negotiation, **22(4)**:681-713, 2013.

A materials demand model with ordering quantity of past and recency of ordering time

Hui Hsin Huang

32,Zhenli St., Tamsui Dist., New Taipei City, Taiwan

Abstract. This paper proposes a statistic model to forecast the materials demand quantity of downstream manufactures in manufacturing process. The author considers the ordering quantity of past record pattern following the normal distribution and the recency of ordering time is a renew process of statistic. The empirical data are used to estimate the parameters of proposed model and its validation. The results can be applied to predict different kind of materials demand

1 Introduction

In the manufacturing process, it is an important issue for downstream manufactures to predict the demand quantities of materials form upstream firm [1-3]. When forecasting material ordering quantity, downstream manufactures can avoid overproduction or product shortages and can also save the cost of inventory [2,4]. Thus this paper proposes a stochastic model for detecting the materials demand quantity before the upstream firm ordering.

For the purpose of this model, the information of past ordering is one of the crucial factors that may impact the probability of future ordering quantity [1, 4]. The materials consuming is often stable or following some regular pattern. Therefore, it can provide a baseline reference when predicting demand quantity in the next time. Thus, we include this variable in our model [3].

The concept of recency of ordering time stands on the view of point in RFM which is a method used for analyzing customer value [5-9]. R index is to explore how recently the customer makes his purchase behavior [5, 8, 9]. F index is to discuss how often this customer purchase [6-7]. R index is demonstrated by how much money this customer spends [6, 8, 9]. We extend the R index to the recency of ordering time which is the time length between the last ordering (purchase) time from the upstream in the observation period and the observing deadline [8]. The recency of ordering time reveals the clue that upstream manufactures may gradually run out or use up their inventory of materials if the recency of ordering time is longer[2,4,8]. This information can help the downstream manufactures to predict the demand quantity they need. Then, we consider this variable into our model.

The materials demand model is derived by the ordering quantity of past and the recency of ordering time. The paper is organized as follows. In the next section, we develop the materials demand model. We will derive the cumulative distribution function (cdf) and the probability density function (pdf) of this model with combining the ordering quantity of past and and the recency of ordering time. In Section 3, the empirical data analysis is demonstrated. It shows the results of parameter estimation and validation of real data. Finally, we offer our conclusions in section 4.

2 The Materials Demand Model

We consider the materials demand model is composed by the ordering quantity of past (denoted as Q) and the recency of ordering time (denoted as R).

$$D = R \cdot Q \tag{1}$$

2.1.1 The Ordering Quantity of Past.

We consider the past ordering quantities(denoted as Q) is a random variable(denoted as q) which follows lognormal distribution with the parameters of μ, σ^2 and its probability density function(pdf) is

$$f_Q(q|\mu,\sigma^2) = \frac{1}{q\sigma\sqrt{2\pi}} \exp\left[-\frac{(\log q - \mu)^2}{2\sigma^2}\right] \tag{2}$$

Then $\log Q$ follows normal distribution and its pdf is

$$f_{\log Q}(y|\mu,\sigma^2) = \frac{1}{\sqrt{2\pi\sigma^2}} \exp\left[-\frac{(y-\mu)^2}{2\sigma^2}\right] \tag{3}$$

2.1.2 The Recency of Ordering Time.

We consider the recency of ordering time (denoted as R) is demonstrated as renew process [9]. Then, according to the renew process in statistic [9-12], we can calculate its pdf and cumulative distribution function (cdf). First, we assume the purchase interval time is a random variable which follows exponential distribution [8]. After that, the result of pdf and cdf in the recency of ordering time (denoted as r) are computing as

$$f_R(r|\theta) = \frac{1}{\theta^2}\exp\left(-\frac{r}{\theta}\right) \tag{4}$$

and

$$F_R(r|\theta) = \frac{1-\exp\left(-\frac{r}{\theta}\right)}{\theta} \tag{5}$$

2.1.3 The Cumulative Distribution Function of Materials Demand Model.

According to equation (1), the materials demand model is composed by "the ordering quantity of past(Q)" multiplying "the recency of ordering time(R)". In order to calculate the cdf of full model, we denote random variable X as logD and is demonstrated in (6).

$$X \equiv \log D = \log R + \log Q \tag{6}$$

Then, its cdf is

$$\begin{aligned} F_X(x) &= P(\log R + \log Q < x) \\ &= \int_0^\infty P(\log R + \log Q|\log R = z, \log Q = y) f_{\log R}(z|\theta)dz \\ &= \int_0^\infty P(\log Q < x - \log R) f_{\log R}(z|\theta)dz \\ &= \int_0^\infty \int_{-\infty}^{x-r} \frac{1}{\sqrt{2\pi\sigma^2}}\exp\left[-\frac{(y-\mu)^2}{2\sigma^2}\right] dy\ f_{\log R}(z|\theta)dz \\ &= \int_0^\infty \int_{-\infty}^{\infty} \Phi\left(\frac{x-r-\mu}{\sigma}\right) dy\ f_{\log R}(z|\theta)dz \end{aligned} \tag{7}$$

In equation (7), the pdf of logR can be derived by equation (4).Then its pdf, $f_{\log R}(z|\theta)$, is

$$\begin{aligned} f_{\log R}(z|\theta) &= \frac{d}{dz}F_{\log R}(z|\theta) \\ &= \frac{d}{dz}P(\log R < z) \\ &= \frac{d}{dz}P(\log\theta^{-2} - \frac{r}{\theta} < z) \\ &= \frac{d}{dz}P(r < \theta\log\theta^{-2} - \theta z) \\ &= \frac{d}{dz}\left[\frac{1}{\theta}\int_0^{\theta\log\theta^{-2}-\theta z}\frac{1}{\theta}\exp-\frac{r}{\theta}dr\right] \\ &= \frac{d}{dz}\left\{\frac{1}{\theta}\left[1-\exp(\theta\log\theta^{-2}-\theta z)\right]\right\} \\ &= \exp(\theta\log\theta^{-2}-\theta z) \end{aligned} \tag{8}$$

2.1.4 The Probability Density Function of Materials Demand Model. Based on equation (7), we can compute the pdf of the proposed model. It shows in equation (9).

$$f_X(x) = \frac{d}{dx}F_X(x) = \frac{d}{dx}\int_0^\infty\int_{-\infty}^{\infty}\Phi\left(\frac{x-r-\mu}{\sigma}\right)\exp(\theta\log\theta^{-2}-\theta z)\,dy\,dz \tag{9}$$

3 The Empirical Data Analysis

In the empirical data analysis, we divide the empirical data into two parts. First parts are used to estimate the parameters of the proposed model. Then, according to the results of parameters estimation, the simulation data are obtained. Finally, we calculate the root-mean-square deviation (RMSD) to show the results of comparison between empirical data and simulation data.

The analysis process is demonstrated in Figure 1.

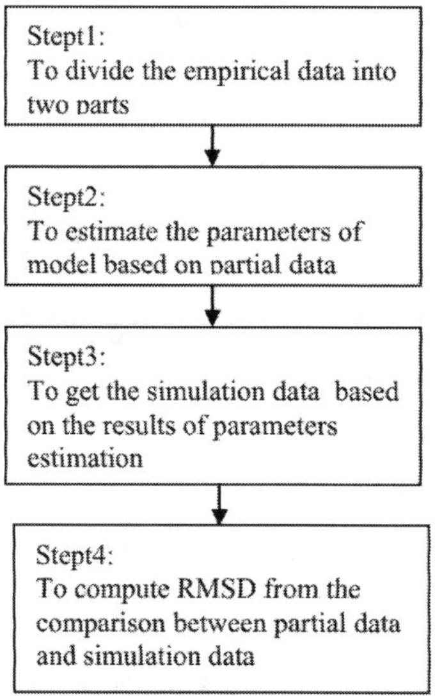

Figure 1. The steps of analysis

3.1.1 The Parameters estimation. We use MLE (maximum likelihood estimate) to estimate the parameters. Let x_j denote the materials demand quantities by upstream j, that is:

$$L(\mu,\sigma^2,\theta)=\prod_{j=1}^{n} f_X(x_j)=\frac{d}{dx}\int_0^{\infty}\int_{-\infty}^{\infty}\left[\Phi\left(\frac{x-r-\mu}{\sigma}\right)\right]^n \exp\left[n\theta(\log\theta^{-2}-z)\right]dy\,dz \tag{10}$$

We differentiate $L(\mu,\sigma^2,\theta)$ respectively regarding μ ,σ^2,θ and set them equal to zero.

That is,

$$\begin{cases} \frac{\partial\mu}{\partial}L_j(\mu,\sigma^2,\theta)=\frac{d}{dx}\int_0^{\infty}\int_{-\infty}^{\infty}\Phi\left(\frac{x-r-\mu}{\sigma}\right)\frac{n(y-\mu)}{\sigma^2}\exp\left[n\theta\,(\log\theta^{-2}-z)\right]dy\;dz=0 \\ \frac{\partial\sigma^2}{\partial}L_j(\mu,\sigma^2,\theta)=\frac{d}{dx}\int_0^{\infty}\int_{-\infty}^{\infty}\Phi\left(\frac{x-r-\mu}{\sigma}\right)\frac{4n(y-\mu)^2}{\sigma^3}\exp\left[n\theta\,(\log\theta^{-2}-z)\right]dy\,dz=0 \\ \frac{\partial\theta}{\partial}L_j(\mu,\sigma^2,\theta)=\frac{d}{dx}\int_0^{\infty}\int_{-\infty}^{\infty}\Phi\left(\frac{x-r-\mu}{\sigma}\right)\exp\left[n\theta(\log\theta^{-2}-z)\right]\left[n(\log\theta^{-2}-2-z)\right]dy\,dz=0 \end{cases} \tag{11}$$

The estimation results show in table1.

Table 1. The results of estimation.

Parameters	
μ	320
σ^2	2.356
θ	43.27

3.1.2 The Model Valiadation. We calculate the root-mean-square deviation (RMSD) to make comparison between empirical data and simulation data. The result shows in table 2.

Table 2. The results of RMSD.

RMSD*	**0.879**

* RMSD= $\frac{\sqrt{\sum_{i=1}^{m}(v_i-\hat{v}_i)^2}}{m}$

m is the simple size

v_i is the i-th empirical data

$\hat{v}_i$ is the i-th simulation data

The results of RMSD is smaller than 1. Then the proposed model has acceptable fitness with real data.

The Conclusion

This paper proposes a stochastic model to forecast the materials demand quantities. Different from previous researches, author considers the recency of ordering time as a factor to detect how long the upstream firm does not make ordering. This may impact its quantities demand but rarely be included in the demand model developing process.

In the future, researches can explore other probability distributions to try fitness between model and real data to improve the predictive power. For example, researchers can consider Weibull distribution or Erlang distributioin to instead of exponential of the recency variable.

Acknowledgements

The author would like to thank the National Science Council of Taiwan (105WFDA450005) and Aletheia University (AU-AR-104-014) for supporting this research.

References

1. Y. Xia, *Responding to supplier temporary price discounts in a supply chain through ordering and pricing decisions*, *Int. J. Prod. Res.* (2015)1-13.
2. H. H. Huang, *A Predicting Model for Demand Quantities of Downstream Manufacturer*, Applied Mechanics and Materials 624(2014) 694-697.
3. H. Kim, J. C. Lu, P. H. Kvaml, Y.C. Tsao, *Ordering quantity decisions considering uncertainty in supply-chain logistics operations*, *Int. J. Prod. Econ.* 134(2011) 16-27.
4. S. C. Chen, L. E. Cárdenas-Barrón, J. T. Teng, *Retailer's economic order quantity when the supplier offers conditionally permissible delay in payments link to order quantity*, *Int. J. Prod. Econ.*155 (2014) 284-291.
5. E. Nikumanesh, A. Albadvi, *Customer's life–time value using the RFM model in the banking industry: a*

case study, Int. J. Elec. Cust. Relat. Manag. 8(2014)234-245.

6. J. H. Kao , F. Lai, H. T. Liaw, P.H. Hsieh, *Study of Customer Value and Supplier Dependence with the RFM Model, Lecture Notes in Electrical Engineering*331(2015) 283-296.
7. K. Coussement, F. A.M. Van den Bossche, K. W. De Bocka, *Data accuracy's impact on segmentation performance: Benchmarking RFM analysis, logistic regression, and decision trees, J. Bus. Res.* 67 (2014) 2751-2758.
8. H. H. Huang, *A detection model of customer alive in information management application, Adv. Mater. Res.*684 (2013) 505-508.
9. Y. S. Cho , S. C. Moon, S. Jeong, I.B. Oh, K. H. Ryu, *Clustering Method Using Item Preference Based on RFM for Recommendation System in U-Commerce, Ubiquitous Information Technologies and Applications* 214 (2013) 353-362.
10. M.S. Ross, *Stochastic Process, Wiley, New York*, (1995).
11. M.S. Ross, *Applied Probability Models with Optimization Applications, Dover Publication, New York*, (2013).
12. M.S. Ross, *Introduction to Stochastic Dynamic Programming, Academic Press, London*, (2014)

Numerical Simulation of NO_x and Particular Matter Distribution from Urban Street in Beijing China

Xiang Liu, Shangtong Li, and Xiuling Wang

Mechanical Engineering Department, Purdue University Calumet, 46323, Hammond, IN, US

Abstract: In recent years, air quality has been a nation-wide issue in Beijing, China with frequent appearance of haze. Disappointingly, most detectors and sensors are mounted in suburb regions that are over ten kilometers away from the center of Beijing. Additionally, most researches are focusing on a general air flow in large scale instead of a specific community. It is important to be aware of the air quality at living communities in urban areas due to a large population. In this study, computational fluid dynamic (CFD) technologies were used to analyze the distribution of NO_x and particular matter (PM) from urban street in Beijing to evaluate the air quality at a certain building. As most air pollutions are caused by vehicle emissions in urban areas, containing NO_x and particular matters, traffic emissions were considered as the only source of contaminants in this study. Commercial software ANSYS Fluent® was used to simulate a number of dispersion scenarios under different boundary conditions to quantify the pollution level for the selected living environment in Beijing, China. Mass fraction, isosurfaces and streamlines of contaminant were presented to analyze the pollution distribution around the area.

1 Introduction

Urbanization transportation is accelerating in China during recent decades; however, the rapid development raises the economic level as well as pollution level in cities, which has aroused public attention [1]. In urban areas of developed cities with a large population, the air quality is much worse than small towns because of the modern tall buildings that prohibit the straight wind and create turbulence in ground level [2]. Due to the large population, a report mentioned that in China the number of cities under heavily polluted are more than other countries [3]. The major pollutant source in large city is vehicle emission [4]. There are four types pollutant generated by vehicle engines that are CO, Unburned Hydrocarbon (UHC), soot and NOx [5]. NOx is a generic term for the mono-nitrogen oxides NO and NO_2. And particular matters are microscopic solids or liquid matters suspended in the earth's atmosphere. It has been demonstrated that over 75% of NOx and 55% particular matter are from vehicles; besides, the automotive emission has a low discharge height, which is highly influencing the air quality on the ground-level. Hence, the analysis of vehicle emission has practical significance and the effect of other emission source could be neglected when urban area is being focused on [6].

Vardoulakis S has attempted to figure out the pollutant concentration in urban areas during one day and one week [7]. MUST Experiment [8] and the Nantes' 99 Experiment [9] have also presented good patterns about pollution dispersion in a large scale. Both of them have provided a close agreement between simulation results and other validated models. However, most previous works have focused on the air flow and pollutants concentration in large regions instead of the influence on a certain building at ground level. The analysis of pollutants dispersion on large scales could make good predictions on the general pollutants distribution. But it could not precisely predict the effects of pollutants on residential areas. The height of the simulation results and the location of monitored data are also significant to provide valuable information for residences.

In this research, the NO_x and particular matters distribution of crossed streets and concentration of a selected building in a community were both tracked to explore the practical influence of the pollutants on people living areas. As Beijing is a world famous city and has attracted nation attention for its haze issue, the west Third-Ring Road in urban areas was considered as the research domain because it is a typical region through two major streets in Beijing. Due to lack of measurements in urban areas, computational fluid dynamic (CFD) was utilized to present the flow pattern and pollutants distribution based on a three-dimensional model. The commercial software Fluent® was used to conduct this three-dimensional, two phases (gas and solid) steady-state simulation. Buildings were simplified by cuboids neglecting the details of outside surfaces, and the transportation emissions were treated as the only pollutant resources with uniform wind velocities. Transportation emissions were calculated based on analytical equations with the assumption that the transportation rate was at the upper limit of the nation standard. Simulation models were adjusted by comparing results with validated models through a two-dimensional simulation.

The concentration distribution of NO_x and particular matters in the whole domain during a morning was observed, and the concentration of pollutants around the selected building was analyzed to present the influence of pollutants on residential buildings. Simulation results were compared with nation standard as suggestions to the city development guidelines for the government.

2 Methodology

The main numerical model in this study is K-epsilon turbulence model. In some other similar works, even very recent studies, concerning on the CFD investigation of a fluid flow, they have employed the standard k-ε model as well and reported it could be the best turbulence model in some similar works, especially in terms of computational cost [10]. In this study, the wind flow filed between buildings shows high turbulence intensity. And the averaged Reynold term could be easily solved by standard k-epsilon model. The standard k-epsilon model is used to simulate turbulence, which contains an alternative formulation for the turbulent viscosity and conducts a modified transport equation for the dissipation rate derived from an exact equation for the transport of the mean-square vorticity fluctuation. The modeled transport equations are [11]:

$$\frac{\partial(\rho k)}{\partial t}+\frac{\partial(\rho k u_i)}{\partial x_i}=\frac{\partial}{\partial x_j}\left[\frac{\mu_t}{\sigma_k}\frac{\partial k}{\partial x_j}\right]+2\mu_t E_{ij}E_{ij}-\rho\epsilon \quad (1)$$

$$\frac{\partial(\rho\epsilon)}{\partial t}+\frac{\partial(\rho\epsilon u_i)}{\partial x_i}=\frac{\partial}{\partial x_j}\left[\frac{\mu_t}{\sigma_\epsilon}\frac{\partial\epsilon}{\partial x_j}\right]+C_{1\epsilon}\frac{\epsilon}{k}2\mu_t E_{ij}E_{ij}-C_{2\epsilon}\rho\frac{\epsilon^2}{k} \quad (2)$$

3 Initial data and assumptions

This paper is focusing on the analysis of the air quality of Western Third Ring-Road area, Beijing. Simulation domain is a square area which has a 200m side length taking the intersection of Huayuan Road as the center. In this area, the NO_x distribution and particular matter have been simulated and compared with nation standard to represent the air quality. The structures in this area are complicated, which contain different types of buildings. Residential building, which is greatly affected by pollutant because of the ventilation by opening windows, will be observed primarily.

The major pollutant NO_x source in this domain is exhausted gas from vehicles in both two roads. From the investigation done by Jiang Meng [12], the data of vehicle flow rate in simulation domain during working days especially traffic rush hours is necessary to analyze the pollution condition. From the 2006 Beijing Statistical Yearbook, it could be seen that the number of vehicle in Beijing is up to 5.43 million in 2014 [13]. In this study, the morning rush hours were considered as the major period to study. The average vehicle flow rate calculated in 2014 is showed in Table 1.

Table 1. Vehicle Flowrate in Simulation Domain in 2014 [13]

Time	Average vehicle flow rate
6:00-7:00	1746
7:00-8:00	2210
8:00-9:00	1081
9:00-10:00	1812
10:00-11:00	1823
11:00-12:00	1812

The vehicle which cannot meet upper emission limit of China IV regulation standard was neglected in this simulation. In addition, driving vehicle which has heavy-duty diesel engine in the crossroad is also forbidden. An assumption will be made that vehicles in Western Third Ring-Road could be considered as those satisfy China IV gasoline engine regulation. According to the regulations standard, the emission should be lower than the level showed in Table 2.

Table 2. China IV Gasoline Engine Regulation Standard [13]

CO(g/kWh)	HC(g/kWh)	NO_x(g/kWh)	PM(g/kWh)
1.5	0.46	2.0	0.02

Considering the transportation condition comprehensively, the engines of vehicle driving in urban area can barely reach full load but low-duty. Due to the exhaust quantities of family cars are not larger than 2.5 liters per vehicle generally, it is reasonable to assume that the power of every vehicle is 15 kW, which could be the basic data for calculating the NO_x amount on the road.

$$Q = nPq/t \ [14]$$

where:
Q: Total Pollutant Generation Rate
n: Number of Vehicle
P: Power of Vehicle Engine
q: Pollutant Generation Rate
t: Time

From the equation above, the pollutants generation rate in the crossed roads are showed in Table 3 and Table 4.

Table 3.NO_x Generation Rate in Simulation Domain

Time	NO_x in Working days(kg/h)
6:00-7:00	52.38
7:00-8:00	66.30
8:00-9:00	32.43
9:00-10:00	54.36
10:00-11:00	54.69
11:00-12:00	54.36

Table 4.PM Generation Rate in Simulation Domain

Time	PM in Working days(kg/h)
6:00-7:00	0.52
7:00-8:00	0.66
8:00-9:00	0.32
9:00-10:00	0.54
10:00-11:00	0.55
11:00-12:00	0.54

To analyze pollutants dispersion, wind velocity and wind direction are required. The wind velocity of Beijing is 1.8 m/s in January and is 1.9 m/s in July [15]. The wind direction representation method is using sixteen wind directions based on wind rose, which is the general representation method. In January the wind direction is NNW, and it is SE in July [15].

Since there are plenty of different situations to calculate, a simplification work must be done. From 7:00 am to 12:00 pm during weekdays is the time period which is analyzed in this investigation. January and July, which is the representative of winter season and summer season respectively are selected to obtain the simulation data. The contaminants generation rates and boundary conditions in the simulation are showed in Table 5.

Table 5. Simulation Parametric Study Boundary Conditions

Case	Wind Velocity (m/s)	Wind Direction	Time Period	NO_x Generation Rate (kg/h)	PM Generation Rate (kg/h)
1	1.8	NNW	6-7	52.38	0.52
2	1.8	NNW	7-8	66.30	0.66
3	1.8	NNW	8-9	32.43	0.32
4	1.8	NNW	9-10	54.36	0.54
5	1.8	NNW	10-11	54.69	0.55
6	1.8	NNW	11-12	54.36	0.54
7	1.9	SE	6-7	52.38	0.52
8	1.9	SE	7-8	66.30	0.66
9	1.9	SE	8-9	32.43	0.32
10	1.9	SE	9-10	54.36	0.54
11	1.9	SE	10-11	54.69	0.55
12	1.9	SE	11-12	54.36	0.54

This paper will focus on the pollution level of different surfaces of the residential building at both windward and leeward in different months under morning rush hours. And then the pollutant concentration will be compared with the Ambient Air Quality Standards of China that the NO_x concentration in urban area must be under 250 μg/m^3, the concentration of PM2.5 must be under 35 μg/m^3, and the concentration of PM10 must be under 50 μg/m^3 by one hour average measurement method [15]. In addition, pollutants moving velocity and concentration in the whole domain have also been observed.

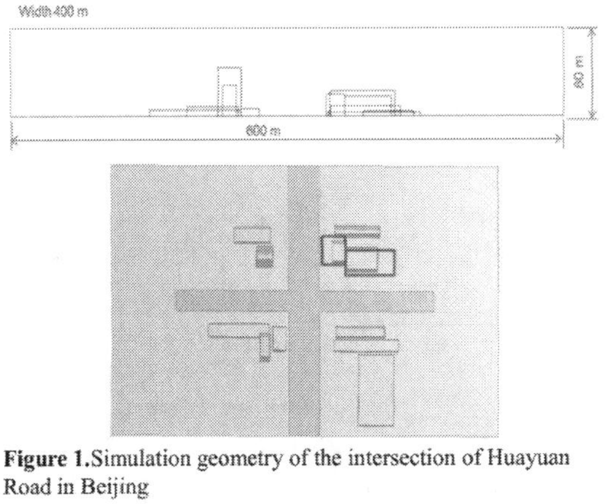

Figure 1. Simulation geometry of the intersection of Huayuan Road in Beijing

The research domain in this study is a square domain which has a 200m side length taking the intersection of Huayuan Road as the center in Beijing. The detailed simulation geometry is shown in the Fig. 1. This is a three-dimensional domain with several buildings around the sides of two main streets. The mesh for this geometry is around one million elements using both structural mesh for most parts and unstructured mesh for the domain around buildings. The red marked building in Fig. 1 is the research target because it is a living building in a big community with a large number of people affected by the pollutants every day. The simulation solver is a double-precision pressure-based solver with that the relaxation factors for pressure, velocity, and turbulence are 0.3, 0.9, and 0.7.

4 Simulation method and validation

Commercial software FLUENT was used in this study to simulate the contaminants concentration distribution using integrated equations and theories in the software. From the investigation of Efisio Solazzo [16] about the pollutant analysis of Birmingham City in England, a similar model was chosen as the validation of this paper. Base on the geometry in the reference, an optimized model was created, which is showed in Fig. 2. In this model, the inlet boundary and out let boundary were same as those for the model provided by the reference author. The measure line is at the same location as the NO_x monitors in the experiments. Besides, the emission source is the position of vehicle lanes, which are the two directions in both sides. The mesh for this geometry is nineteen thousand elements generated by automated method in ANSYS.

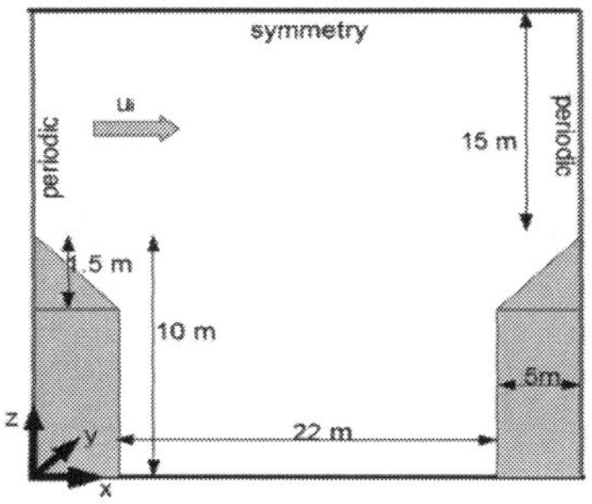

Figure 2. Optimized validation model [16]

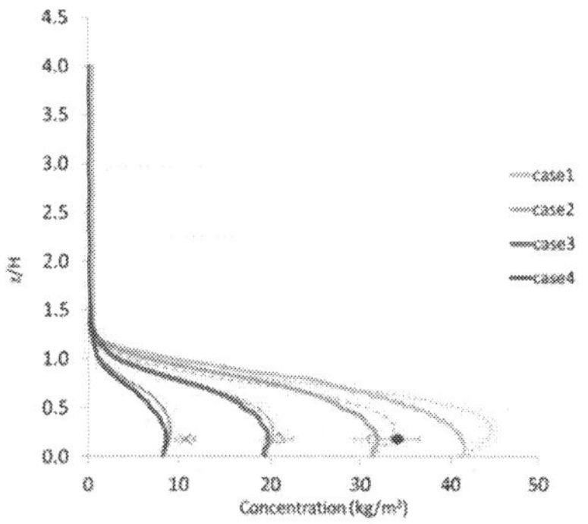

Figure 3. Comparison of simulation results between reference (dash lines) and present study (solid lines)

With the optimized model, using same boundary condition which the reference provided, the simulation of comparison has been conducted. From the result showed in Fig. 3, which has little deviation compared with the data in reference, this simulation model was proved to be credible. The different cases are for contaminant transport under different wind speed in the same wind direction [16].

5 Simulation results

The simulation results were divided into two parts as different conditions in January and July. Both two cases were studied to comprehensively analyze the annual pollution distribution in Beijing. At the same time, the different pollutants emission rates in various time period in one day are also considered to analyze the emission differences during the morning rush hours. In order to compare the results with the standard, the average value during the morning rush hours were calculated to compare to the standard values. As shown in Fig. 4 (a), the wind from north-west with the velocity of 1.8m/s in January was taken as a representative case to evaluate the wind speed changes through the height. As the height increases, the blockage effects of buildings would decrease with the wind velocity increases behind those buildings.

Fig. 4 (b) shows the average NO_x mass fraction distribution at different heights in January. Basically the NO_x distribution is consistent with the wind velocity because the wind motion would be the only force to drive the pollutants movement. As seen in the Fig. 5, the lower height area has higher NO_x mass fraction because it is closer to the emission sources.

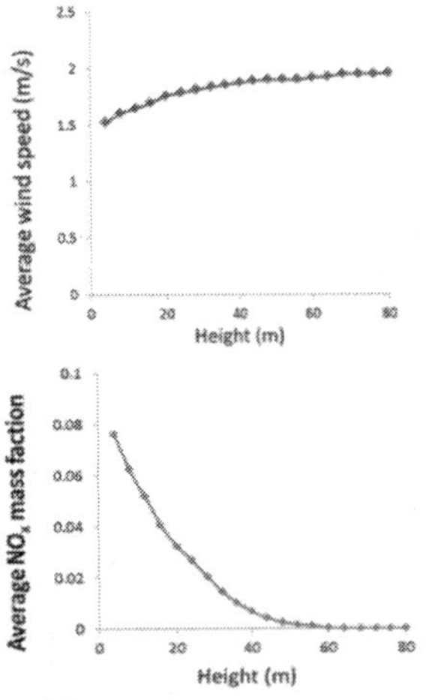

(a) (b)

Figure 4. Wind speed and NO_x mass fraction distribution in Jan.

As mentioned before, only one of those buildings which are the living building with a large number of people in the big community is the target object in this study. The building is marked by red circle in Fig. 5. It shows the iso-surface when NO_x concentration is 250 ppm. As shown in Fig. 5, the target building is outside the iso-surface, which means the NO_x concentration near the target building is lower than 250 ppm. The building is not surely affected by the NO pollutants. The table in Fig. 5 also verifies the result that the average NO_x concentrations in different directions of the building are all under the national standard.

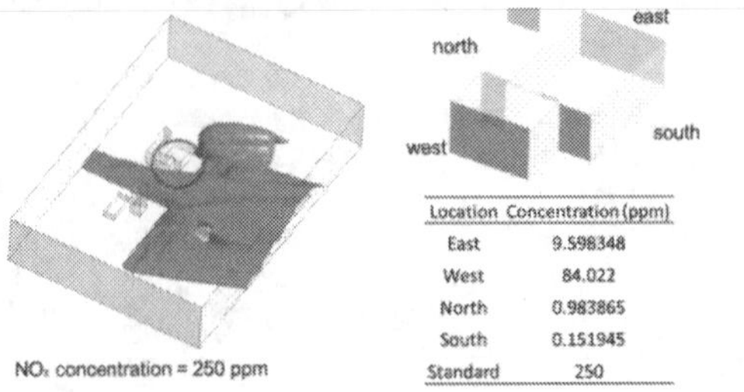

Figure 5. Iso-surface of NO_x concentration during morning rush time in January

From Fig. 6, the particle streamlines are very similar with the NO_x distribution because it is also mainly affected by the wind motion. The average particle concentrations in different directions of the target building are all under the standard that the concentration is 35 ug/m3.

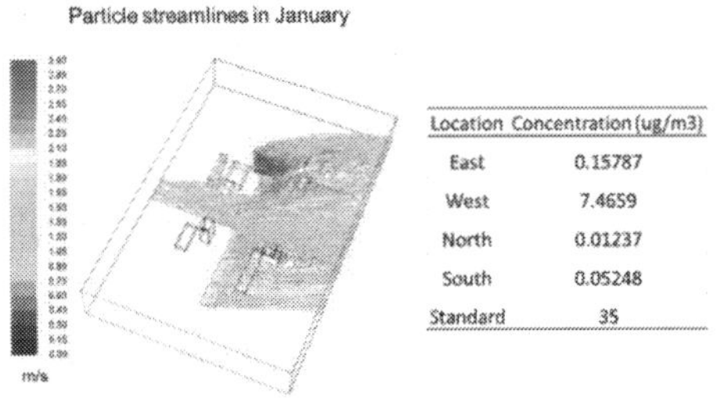

Figure 6. Average PM streamlines during morning rush hours in January

Similarly, the NO_x distribution is also affected by the wind direction in July. And the result is very similar between the case in January and July. As a result, the figures for wind speed and NO_x mass fraction distribution with height changes are not presented here. However, it has an impact on the NO_x concentration near the target building because the wind comes from the direction that the emission pollutants could directly goes into the south surface of the target building.

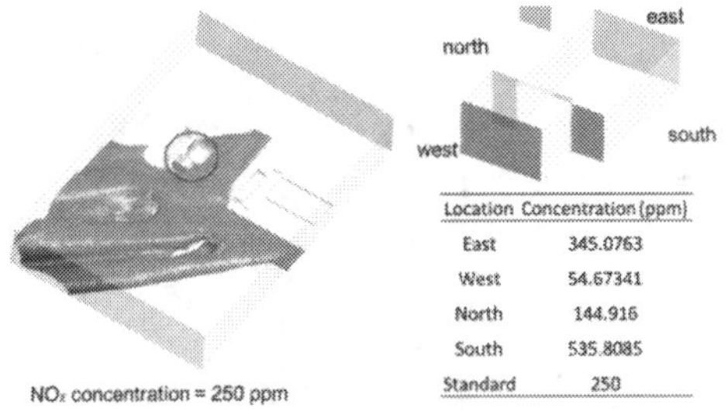

Figure 7. Iso-surface of average NO_x concentration during morning rush time in July

Fig. 7 shows the iso-surface when NO_x concentration is 250 ppm. The target building is inside the iso-surface,

 DOI: 10.1051/matecconf/20167703001

which means the NO_x concentration around the building is over the standard. This condition is different with the condition in January. And from the table in Fig. 7, it is important that the average NO_x concentration of the east and south surface could be much higher than the standard.

From Fig. 8, the particle streamlines are very similar with the NO_x distribution because it is also mainly affected by the wind motion. However, compared with that the NO_x concentration is much higher than the standard, the particle concentration is still safe to people's health in this condition.

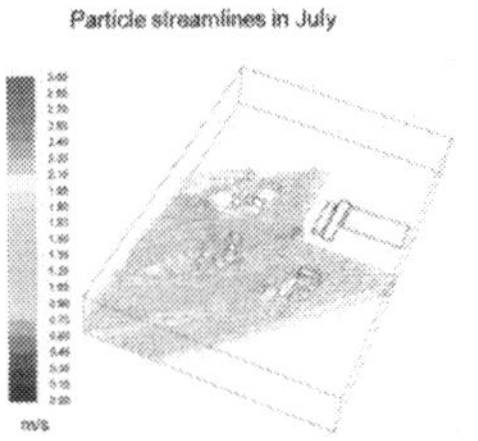

Location	Concentration (ug/m3)
East	11.25485
West	0.036286
North	5.410637
South	11.57239
Standard	35

Figure 8.PM streamlines during morning rush hours in July

6 Conclusion

From the simulation results and discussion above, the conclusion can be drawn that the NO_x and particle distribution is strongly affected by the wind direction. For the target building, the surfaces facing the streets which are the emission sources are more likely to be affected by the pollutants. This is guide for people to choose the place where to live and which room to live. Also, the result reminds us of the significance of trees because the blockage to wind has a great impact on the pollutant concentration reaching the building.

Based on the simulation results, another conclusion can be drawn is that the NO_x pollution is highly related to the cars rate and the emission from those cars. However, the particle concentration seems hardly affected by the cars rate because the particle concentration is always under the standard in this study. This could be easily understood relating to the real conditions. Actually the particle pollution is mainly caused by the emission from heavy metal factories and burning grows in specific conditions

References

1. Harrison R., Stedman J., Derwent D., AE, **42**, 603-606, (2008).
2. Woodfield N.K., Longhurst J.W.S., Beattie C.I., Chatterton T., Laxen D.P.H., EMS, **21**, 595-599, (2006).
3. World Health Organization, *Investigation of Polluted Cities in World,* (1998).
4. Ye Wu, Renjie Wang, Yu zhou, Bohong Lin, Linxin Fu, Kebin He, Jiming Hao, *On-Road Vehicle Emission Control in Beijing: Past, Present, and Future*, EST, **45**, 147-153, (2011).
5. John Haywood, *Internal Combustion Engine Fundamental*, (1998).
6. Tinghui Li, HIT, Study on Urban Traffic Pollution Monitoring and the Method of Limits Traffic Volume, (2012).
7. Vardoulakis S., Valiantis M., Milner J., ApSimon H., AE, **41**, 4622-4637, (2007).
8. Buccolieri R., Di Sabatino S., *MUST Experiment Simulations Using CFD and Integral Models*, IJEP **44**, 376-384, (2011).
9. Louka P., Vachan G., Sini J-F. Mestayer P.G., Rosant J-M., WASP, **14**, 351-364, (2002).
10. J.H. Amorim, V. Rodrigues, R. Tavares, J. Valente, C. Borrego, STE, 461–462, 541–551, (2003).
11. Versteeg, Henk Kaarle; Malalasekera, Weeratunge, *An introduction to Computational Fluid Dynamics: The Finite Volume Method.* Pearson Education, (2007).
12. Jiang Meng, Hua Fu, Xiaolin Zhao, Weiwei Zhang Meixia Song, CNU, **27**, 89-92.
13. Beijing Statistical Yearbook, *People's Government of Beijing Municipality*, (2014).
14. *2013 Beijing Weather Report, China Meteorological Bureau.*
15. 2012 Ambient Air Quality Standards of People's Republic of China, *Environment Protection Ministry of China.*
16. Efisio Solazzo, Sotiris Vardoulakis, Xiaoming Cai, AE, **45**, 5230-5239.

An investigation of dynamic mechanical behaviour of Ti6Al4V titanium alloy at room temperature

Chun Ran, Pengwan Chen, Baoqiao Guo, Ling Li, and Haibo Liu

State Key Laboratory of Explosion Science and Technology, Beijing Institute of Technology, Beijing100081, China

Abstract. To study the high strain rate shear behaviour of Ti6Al4V titanium alloy, a series of dynamic compression experiments has been performed by split Hopkinson pressure bar (SHPB) using Flat Hat-shaped specimen at room temperature. Macro true shear stress-true strain curves were obtained under different strain rate loading conditions at room temperature. The effects of strain hardening and strain rate hardening on the dynamic mechanical properties of Ti6Al4V titanium alloy were discussed. Results indicate that a) The higher the strain rate, the higher the flow stress, therefore, the material has obvious strain rate hardening effect, b) It is ductile failure for Ti6Al4V titanium alloy under quasi-static loading condition, c) For dynamical tests, the values for true shear stress increase with increasing true strain till the maximum true shear stress, on the contrary, the values for true shear stress decrease with increasing the true strain after the maximum true shear stress and d) The flow stress increases with increasing the true strain under quasi-static loading condition during the plastic deformation.

1 Introduction

Ti–6Al–4V, an α+β titanium alloy, is world widely used to manufacture the engine fan and blade in aerospace industries due to its highly attractive material parameters, such as low mass density, high strength and high temperature stability ([1]). A profusion of studies on the dynamic mechanical behaviour of titanium and titanium alloys have been carried out ([2]). Chichili *et al* ([3]) and Dong *et al* ([4], [5]) investigated dynamic mechanical behaviour of α-Ti at room temperature by using split Hopkinson pressure bar (SHPB) and found that the strain hardening is strain rate and temperature dependence. Rittel *et al* ([6]-[9]) proposed a new failure criterion based on mechanical energy (the stored energy of cold work as the driving force for microstructural rearrangement by DRX) rather than a critical strain proposed by Recht ([10]). A series of corresponding work have been carried out by Bai, Meyer, Xue, Peirs, Staker ([11]-[15]). Hat-shaped (HS) specimen was invented in 1977 by Hartmann and Meyer at the Fraunhofer Institute (IFAM, Bremen, Germany). Due to the unique geometry of HS specimen, extremely high strains at high strain-rate can be reached. However, in fact, the shear stress measurement is influenced by edge effect, and it is impossible to track the shear band, detect its temperature and measure the local strain. To reduce edge effect and measure the temperature rise during the deformation, the Flat Hat-shaped (FHS) specimen was designed by Clos ([16, [17]).

In this paper, the dynamic mechanical behaviour of Ti6Al4V titanium alloy at room temperature has been studied by FHS specimen. The purpose of this work is to bring recent experimental results on the dynamic mechanical behaviour of Ti6Al4V titanium alloy. The paper is organized as follows: First, the experimental techniques are introduced. Next, experimental results and discussion are presented, followed by concluding remarks.

2 Experimental techniques

The Ti6Al4V titanium alloy is provided by Beijing Institute of Aerial Materials. The split Hopkinson pressure bar technique is used to load the specimens dynamically. The FHS specimen is sandwiched between two bars (Fig. 1). An impact pulse is provided with a striker driven by a light gas gun. It should be noted that the setup used in this work is nickel (18Ni) bars with diameters of 14mm and the lengths of the striker, incident and transmission bar are 0.2m, 1.2m, 1.2m, respectively. The dynamic mechanical behaviour of Ti6Al4V titanium alloy at room temperature under 4000~7400s^{-1} strain rate loading conditions have been measured. Meanwhile, the quasi-static properties of Ti6Al4V at room temperature have been performed using MTS test machine (INSTRON 5985) based on GB/T 7314-2005. It should be noted that the strain rate is $10^{-3}s^{-1}$ and at least three samples were tested for each strain rate loading condition, being always found good reproducibility of results. Also it should be pointed out that the cylindrical specimens, 6mm in diameter and 10mm in length, were used in the quasi-static tests.

Fig. 2 shows the in-plane dimensions of the FHS specimen used in this study. In an axi-symmetric FHS

DOI: 10.1051/matecconf/20167703002

specimen, shear strains are concentrated in a narrow region (shown in Fig. 3). The specimen can be divided into three parts: the upper hat part, the lower edge part and the shear region where extremely high strain rate, strain and temperature generate and propagate.

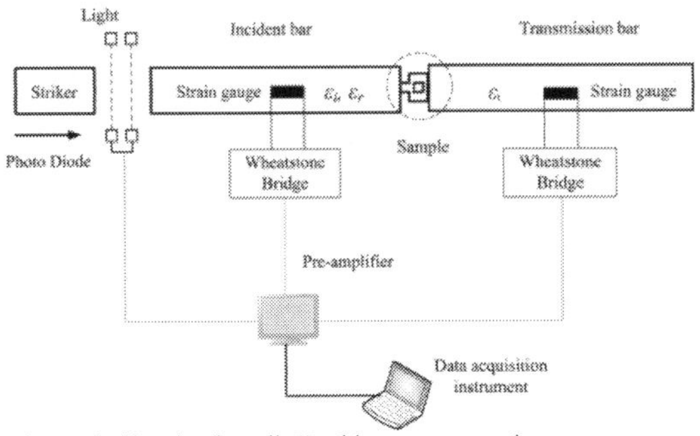

Figure 1. Sketch of a split Hopkinson pressure bar set-up

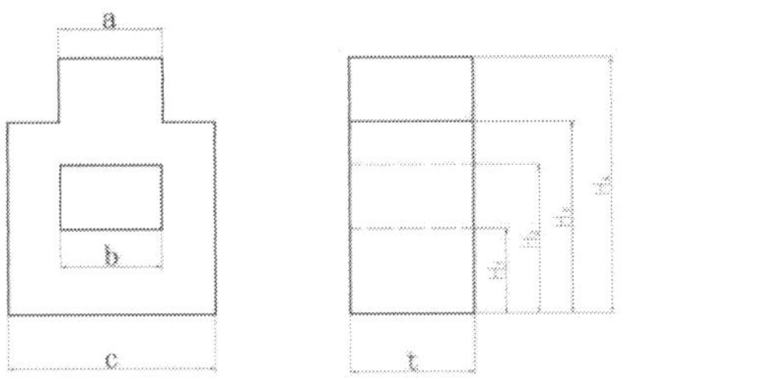

Figure 2. In-plane dimensions of FHS specimen： a=6mm, b=5.6 mm, c=13mm, H1=3mm, H2=6mm, H3=9 mm, H4=11 mm, t=6mm.

When one-dimensional stress-waves in the bars are achieved and the specimen is in a state of uniform stress, the strain rate, strain and stress histories in the specimen can be computed as:

$$\dot{\varepsilon}_s(t) = -2\frac{C_0}{h}\varepsilon_r(t)$$
$$\varepsilon_s(t) = -2\frac{C_0}{h}\int_0^t \varepsilon_r(t)dt \quad (1)$$
$$\sigma_s(t) = \frac{A_0}{A_s}E_0\varepsilon_r(t)$$

where A_0 is the cross-sectional of the bars, E_0 and C_0 are Young's modulus and elastic bar wave speed in the bar material, respectively. Here, As and h are the initial cross-sectional area and thickness of deformation section of the FHs specimen, respectively. And $\varepsilon_r(t)$ and $\varepsilon_t(t)$ represent reflected and transmitted strain histories in the bars at the specimen ends, respectively.

The stress and strain in the compression-shear section of the FHS specimens can be estimated approximately, although they are complicated. As shown in Fig. 3, the load F can be considered as two parts, compression and shear. Then the loading, shear stress, displacement and strain histories in the specimen can be computed as follows:

$$F_s(t) = \frac{E_0A_0\varepsilon_t(t)}{2} \quad (2)$$

$$\tau_s(t) = \frac{E_0A_0\varepsilon_t(t)}{2A_s} \quad (3)$$

$$A_s = \sqrt{(a-b)^2 + (H_3-H_2)^2} \times t \approx ht \quad (4)$$

$$\Delta l_s(t) = -2C_0\int_0^t \varepsilon_r(t)dt \quad (5)$$

$$\gamma_s(t) \approx \Delta l/_h = {-2C_0\int_0^t \varepsilon_r(t)dt}/_h \quad (6)$$

Here t is the thickness of the FHS specimen.

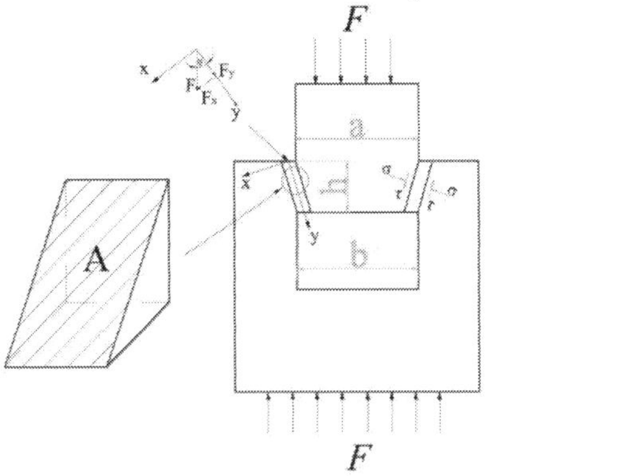

Figure 3. Schematic plot of stress state of the FHS specimen: h=H3-H2=3mm, a=6mm, b=5.6mm. A denotes the area of the shear region

3 Results and discussion

The typical stress wave tested by gauges on split Hopkinson pressure bars at room temperature is shown in Fig. 4. Then the true shear stress-true strain can be calculated based on eqs.(2)-(6).

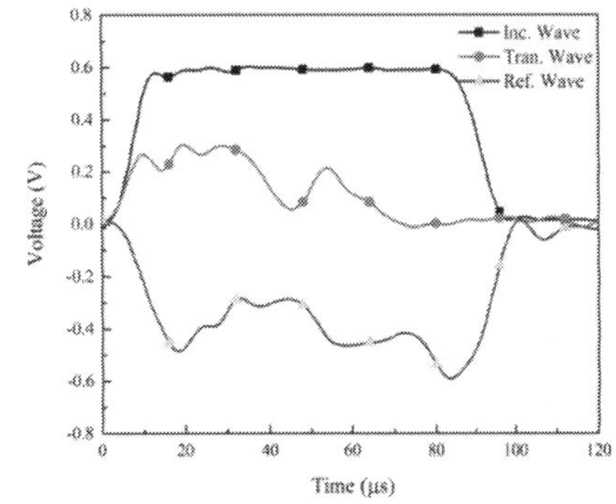

Figure 4. Stress wave tested by gauges on Hopkinson bars

3.1. Stress and strain curves

The resultant true shear stress - true strain for 6 different strain rate, ε=0.001, 4100, 4800, 5400, 6600 and 7400 s^{-1} are plotted in Fig. 5. It is seen that, the yield stress is about 350MPa for quasi-static loading (ε=0.001 s^{-1}) condition and it is about 550MPa for 7400 s^{-1} loading condition. The yield stress increases apparently with increasing of the strain rate. Therefore, the strain rate hardening effect for Ti6Al4V titanium alloy is very obviously.

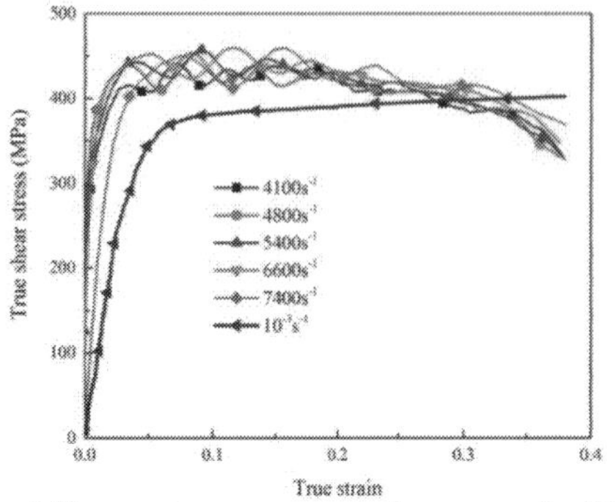

Figure 5. The true shear stress-true strain curves under different strain rate loading conditions at room temperature.

The typical true shear stress and strain rate as functions of true strain for Ti6Al4V titanium alloy at room temperature is shown in Fig.6. It is seen that, a) the maximum true shear stress is located at true strain γ=0.12. Then, the true shear stress decreases with increasing the true strain, b) the value for strain rate becomes approximately constant when the true strain $\gamma \geq 0.12$ and c) the shear failure occurs when the true strain γ=0.36.

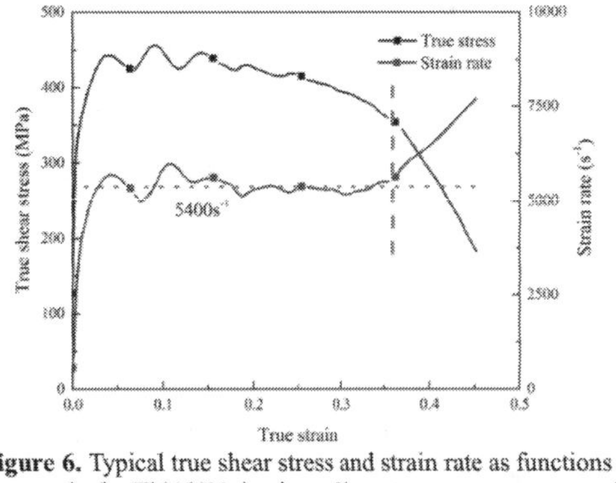

Figure 6. Typical true shear stress and strain rate as functions of true strain for Ti6Al4V titanium alloy at room temperature, the strain rate is 5300s^{-1}.

As can be seen from Figs. 5 and 6, for dynamical tests (ε=4000~7400 s^{-1}), the values for true shear stress increase with increasing true strain till the maximum true shear stress, on the contrary, the values for true shear stress decrease with increasing the true strain after the maximum true shear stress. That is, when the strain rate is fixed, such as ε=5300 s^{-1}, a) for the former part, the strain hardening effect is the much higher than the thermal soften effect so that the flow stress increases with increasing the true strain and b) for the latter part, the thermal soften effect is much higher than the strain hardening effect so that the flow stress decreases with increasing the true strain. It should be pointed out that the flow stress increases with increasing the true strain when strain rate ε=0.001s^{-1} during the whole plastic deformation.

2.2 Fractography

Fig.7 is the typical macro-morphology of Ti6Al4V titanium alloy after dynamic compression, and the typical micro-morphology of Ti6Al4V titanium alloy after quasi-static compression is shown in Fig.8. Fig.8 shows the ductile dimples areas, it should be pointed out that there are some elongated dimples along the maximum shear stress direction. Therefore, it is ductile failure for Ti6Al4V titanium alloy under quasi-static loading condition.

Figure 7. The typical macro-morphology of Ti6Al4V titanium alloy after dynamic compression. The strain rate is 5300s^{-1}.

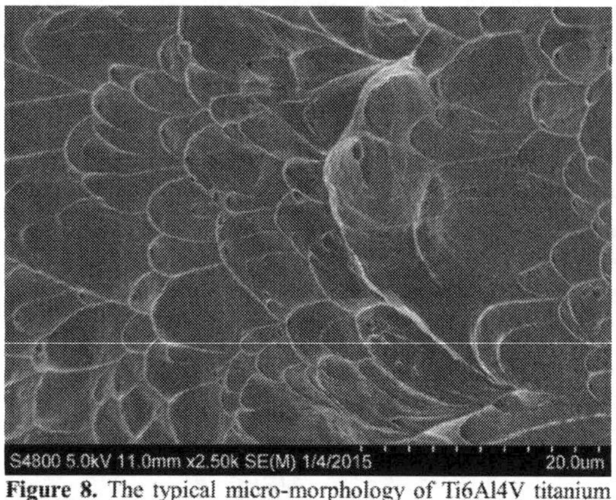

Figure 8. The typical micro-morphology of Ti6Al4V titanium alloy after quasi-static compression. The strain rate is 0.001s^{-1}.

4 Conclusions

To study the high strain rate shear behaviour of Ti6Al4V titanium alloy, a series of dynamic compression experiments has been performed by split Hopkinson pressure bar (SHPB) using Flat Hat-shaped specimen at room temperature. And the quasi-static compression tests have been performed by MTS test machine (INSTRON 5985) based on GB/T 7314-2005. The following conclusions can be drawn from the present work:

1) The higher the strain rate, the higher flow stress of Ti6Al4V titanium alloy, therefore, the material has obvious strain rate hardening effect.

2) For dynamical tests, the values for true shear stress increase with increasing true strain till the maximum true shear stress, then, on the contrary, the values for true shear stress decrease with increasing the true strain after the maximum true shear stress.
3) The flow stress increases with increasing the true strain when strain rate ε=0.001s^{-1} during the whole plastic deformation.
4) It is ductile failure for Ti6Al4V titanium alloy under quasi-static loading condition.

References

1. M.K. McQuilan, Metall. Trans. A. 7A, 22 (1944)
2. Y. B. Xu, Y.L. Bai, Y. J. Chen. Mater Sci Eng A, 299(1), 287 (2001).
3. D.R. Chichili, K.T.Ramesh, K.J. Hemker. Acta Mater., 46(3), 189 (1998)
4. X.L. Dong, Y.Q. Fu. Acta Armamentarii. 35(7), 1016 (2014).
5. Y.Q. Fu, X.L. Dong. Rare Metal Materials and Engineering, 45(1), 102 (2016).
6. D. Rittel, Z.G. Wang, M. Merzer. Physical review letters, 96(7), 075502 (2006).
7. D Rittel. Journal of Physics D: Applied Physics, 42(21), 214009 (2009).
8. D. Rittel, P. Landau, A. Venkert. Physical review letters, 101(16), 165501 (2008).
9. D. Rittel, Z.G. Wang. Mech. Mater. 40(8), 629 (2008)
10. R. Recht. Journal of Applied Mechanics, 31(2), 189 (1964)
11. L.H. Dai, Y.L. Bai. Int. J. Impact Eng., 35(8), 704 (2008).
12. F. Pursche, L.W. Meyer. Engineering transactions, 59(2), 67 (2011).
13. Q. Xue, M.A. Meyers, V.F. Nesterenko. Mater Sci Eng A, 384, 35 (2004).
14. J. Peirs, P. Verleysen, J. Degrieck, et al. Int. J. Impact Eng., 37(6), 703 (2010).
15. M.R. Staker. Acta Metallurgica, 29(4), 683 (1981)
16. R. Clos, U. Schreppel, P. Veit, J. Phys. IV France, 10, 257 (2000).
17. R. Clos, U. Schreppel, P. Veit, J. Phys. IV France, 110,111 (2003).

MATEC Web of Conferences 77, 03003 (2016) DOI: 10.1051/matecconf/20167703003
ICMMR 2016

A new efficient algorithm for multi-scale rough surfaces contact mechanics

Jun Zhang and Yiyong Huang

Aerospace Science and Engineering College, National University of Defense Technology, 41000 Changsha Hunan, China

Abstract. Traditional multi-scale rough surfaces contact algorithms face both computation accuracy and efficiency problems. In this paper, we present a new efficient algorithm, Sliding Window based Gauss-Seidel Algorithm under BEM framework to solve the problems. We slide a window of small size on the contact region to reduce linear equation dimensions. After investigating the influence coefficient matrix, we find that the deformation is only influenced by the pressures around, so our algorithm will be very accurate. By taking the Persson Model results as the baseline to compare all kinds of numerical methods, our algorithm performs better than the conventional algorithms in both accuracy and efficiency.

1 Introduction

Contact mechanics is a basic problem in mechanical engineering, whether the contact performance is good or not directly affects the normal use of seal components, brake parts and so on. Real surfaces are never ideally flat and roughness is present at different scales, from the specimen size down to the interatomic distance. Hence, when two bodies are pressed against each other, contact takes place at the asperities and the real contact area is a fraction of the nominal one. Therefore, the analysis of two surfaces contact problems must be taken into account when scale effect.

The earliest study of contact mechanics can be traced back to 1882, Hertz proposed a contact model between the elastic sphere, and gives the corresponding analytical solutions. In 1956, the multi scale effect should be considered when the elastic contact of rough surface is studied by Archard. In 1966, Greenwood and Williamson proposed the GW model, the model took rough surface description into a series of randomly distributed spherical bulge, they assumed that each asperity deformation independently of each other and used Hertz theory to solve single asperity contact problem. [1] In 1990, Majumdar and Bhushan proposed a model based on fractal contact, with fractal dimension and fractal roughness parameter to characterize a surface roughness, the MB model predicted that the total load was linear with the contact area when load is small. [2] There is an important issue in the models above that the interaction between asperities is not considered. In 2001, Persson proposed a kind of multi scale statistical model, and gave the analytical model of the full contact problem and the approximate solution of the contact problem. Persson theory regarded surface roughness as a stochastic process to analyze, used power spectral density to characterize surfaces, and concluded the probability density function of load and resolution should conform to the diffusion equation, given the diffusion equation of the boundary conditions and initial conditions can be calculated analytical solution. In a very wide range of loads, the load was linear with the contact area, but the ratio was not the same as the result of the multi asperities model and the numerical method. [3]

The comparison and verification of the existing model results contribute to the progress in understanding the limits of the current approaches and further research. Experimental study on the measurement precision of the problem, it is difficult to make accurate approximation. Therefore, numerical simulation is very important in the study of contact problem. In 2004, Hyun used the Finite Element Method to study the elastic contact problem between fractal rough surfaces. Pei extended the Hyun model to the elasto-plastic contact problem in 2005. In 2006, Persson and Yang verified the contact stress along with the variation of the observed scale in the Persson theory by numerical methods. The key lies in the Boundary Element Method for solving large-scale linear equations, Polonsky and Keer in 1999, MLMS algorithm is proposed to accelerate the solving of equations. In 2001, Borri-Brunetto proposed the ICARUS algorithm based on greedy Gauss-Seidel algorithm and the idea of the real contact area changes with the resolution. [4] The same year, Karpenko, Polonsky and Keer applied Fast Fourier Transform technology to the Boundary Element Method, and achieved good results.

With the increase of the number of boundary element grid, the dimension of the linear system of equations is increasing exponentially, which leads to the ill conditioned system of equations and can not be solved. Therefore, it is necessary to reduce the dimension of the equations. In 2008, Xia put forward a scheme based on block Gauss Seidel method for study on the contact problems, but the resolution improved to a high level, the algorithm lost its computation efficiency and calculation accuracy. [5] In this paper, a Gauss-Seidel algorithm based on sliding window is proposed to solve the problem of multi scale elastic contact problem.

This paper is structured as follows: in Section 2, we describe the mathematical model of elastic contact problem. Section 3 introduces two conventional algorithms, the Direct Gauss-Seidel algorithm and Block Gauss-Seidel algorithm. Section 4 introduces our algorithm, Sliding Window based Gauss-Seidel algorithm. Section 5 designs examples to compare the algorithms.

2 Mathematical Model

2.1 Fractal surfaces

There are lots of algorithms to generate fractal surfaces, such as Random Midpoint Displacement(MPD), W-M Function Method, Fast Fourier Transform technology etc. Here we use MPD to get random rough surfaces.

MPD is often used in terrain modelling, through the fractal dimension D and the number of iterations n to control the roughness of the generated surface [6].

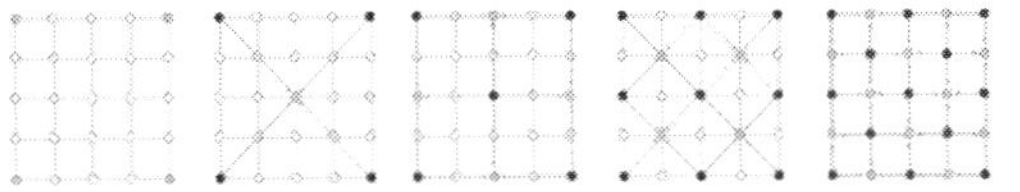

Figure 1. Yellow points mean the generating point in current iteration loop, black points represent the generated points in the iteration loop before.

By MPD we generate a three-dimensional fractal surface, as shown in Fig. 2.

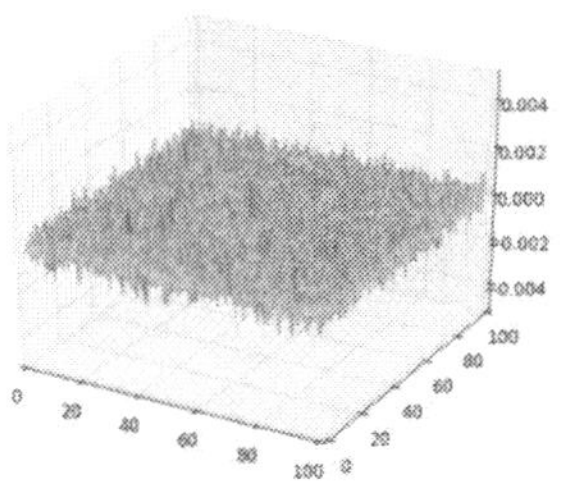

Figure 2. A fractal surface generated by MPD.

2.2 The model

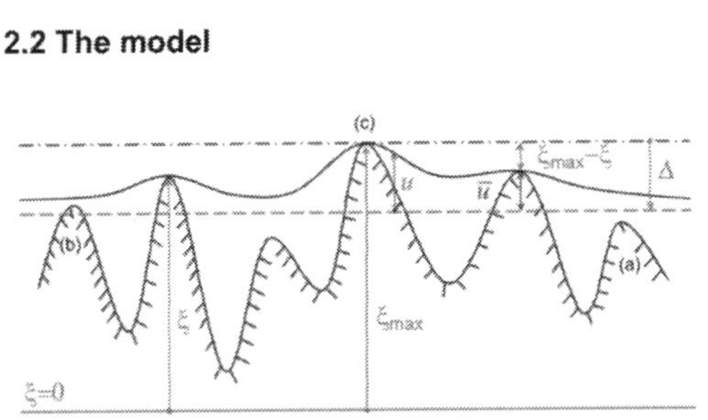

Figure3. Sketch of the contact problem between a rigid rough surface and an elastic half-plane. Its deformed configuration corresponding to the imposed far-field displacement Δ is depicted with a black solid line. The red dashed line corresponding to a rigid-body motion of the half-plane identifies the heights to be included in the initial trial contact domain.

In this paper, we assume that one of the contact surfaces is a flat elastic one, another is a rough rigid surface with fractal characteristics.

By imposing a far-field closing displacement Δ to the bodies in contact, the displacement at each point of the contact area is related to the contact pressures as follows [7]:

$$u(x) = \int_S H(x,y)p(y)dS \qquad (1)$$

where $u(x)$ is the displacement at the surface point defined by the position vector x, $H(x,y)$ is the displacement at x due to a unit pressure acting at y, and S is the apparent contact area. Assuming linear elastic isotropic materials, the influence coefficients are given by [7]

$$H(x,y) = \frac{1-v^2}{\pi E}\frac{1}{\|x-y\|} \qquad (2)$$

where E and v denote, respectively, the composite Young's modulus and Poisson's ratio of the materials of the bodies in contact, respectively.

Upon discretization of the surface as a grid where each nodal height defined by the indices (i,j) is modelled as a square punch with an elevation $\xi_{i,j}$ above a reference plane coincident with the level of the smallest surface height, the contact problem requires the simultaneous solution of the following set of equations and the satisfaction of a set of inequalities [7]:

$$u_{i,j} = \sum_{k=1}^{N_c}\sum_{l=1}^{N_c} H_{i-k,j-l}\, p_{k,l}, \quad 1 \le i \le N_c,\ 1 \le j \le N_c$$

$$u_{i,j} = \bar{u}_{i,j} = \Delta - \xi_{i,j}, \quad \forall\, (i,j) \in I$$

$$p_{i,j} > 0, \quad \forall\, (i,j) \in I$$

$$u_{i,j} > \bar{u}_{i,j}, \quad \forall\, (i,j) \in \bar{I}$$

$$p_{i,j} = 0, \quad \forall\, (i,j) \in \bar{I}$$

where I is the domain of boundary elements in contact, $\bar{I}$ is the domain of boundary elements not in contact, and Δ is the imposed far-field displacement.

3 Conventional Methods

The focus and difficulty of solving the section 2 model is to solve the equations $Hp = \bar{u}$ rapidly and accurately. This section will introduce two conventional methods.

3.1 Direct gauss-seidel algorithm

The algorithm considers that the deformation of each point is affected by all the points in the plane, and the influence decreases with the increase of the distance from the point. [8]

Algorithm description: Direct **Gauss-Seidel Algorithm**

Input: coefficient matrix H, vector $\overline{u}$, initial pressure vector p, initial contact set I_0, maximum number of iterations K_{max}, error limit ε.

Algorithm Procedure:

Step 1 $k := 0, I := I_0$

Step 2 $while \begin{pmatrix} k \le K_{max}\ and \\ \min(p) < \text{-}\varepsilon\ or \\ k = 0 \end{pmatrix} do:$

(2.1) Gauss-Seidel algorithm is used to solve the linear equations $Hp = \overline{u}$ to get the solution vector p

(2.2) $for\ (i,j) \in I\ do:$

(2.2.1) $if\ p_{i,j} < \text{-}\varepsilon\ then$

$p_{i,j} := 0\ and\ I := I \backslash \{(i,j)\}$

Step 3 $p_{opt} := p$

Step 4 $\bar{I} := \{(1, \ldots i, \ldots N_c) \times (1, \ldots i, \ldots N_c)\} \backslash I$

Output: dimensionless contact area

The main steps of the contact algorithm are summarized above, where H denotes the matrix collecting the influence coefficients, the vector $\overline{u}$ collects the imposed displacements, and the vector p collects the contact forces. The boundary elements subject to tensile forces are eliminated from the contact domain at the step (2.2.1). After convergence, the optimal solution is stored in the vector p_{opt} in the step 3. The set of boundary elements not in contact $\bar{I}$, is updated in the step 4.

3.2 Block gauss-seidel algorithm

The overall process of the Block Gauss-Seidel algorithm is consistent with 3.1, the difference is that the huge linear equations are divided into block processing, the dimension of the H matrix is reduced to $N_c \times N_c$, and then apply the Gauss-Seidel algorithm to solve. The algorithm considers that the deformation of each point is only related to all points of the row of the point, which is not related to the point of other rows. [5]

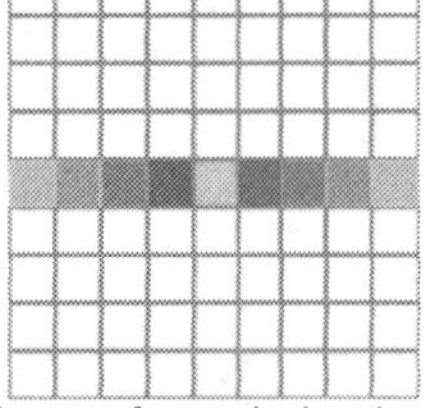

Figure 5. The influence of one point is only related to points in this row

4 Our Algorithm

Linear equations of the dimension disaster problem exist in the Direct Gauss-Seidel algorithm, when the dimension of the matrix reaches a high level, it can't be solved. The Block Gauss-Seidel algorithm using block calculation to reduce the dimension, but to each point on effects into consideration is not comprehensive, resulting in the decline of accuracy. This section will first analysis the characteristics of the coefficient matrix H, then describe Gauss-Seidel algorithm based on sliding window.

4.1 Analysis of influence coefficient matrix

Investigate the influence coefficient matrix H, $H_{i,j}^{k,l}$ denotes the influence between point (i,j) and point (k,l), which are known from the literature [7]

$$H_{i,j}^{k,l} = \alpha \left\{ (\bar{x}+a)\ln\left[\frac{(\bar{y}+a)+\sqrt{(\bar{y}+a)^2+(\bar{x}+a)^2}}{(\bar{y}\text{-}a)+\sqrt{(\bar{y}\text{-}a)^2+(\bar{x}+a)^2}}\right] + (\bar{y}+a)\ln\left[\frac{(\bar{x}+a)+\sqrt{(\bar{y}+a)^2+(\bar{x}+a)^2}}{(\bar{x}\text{-}a)+\sqrt{(\bar{y}+a)^2+(\bar{x}\text{-}a)^2}}\right] + (\bar{x}\text{-}a)\ln\left[\frac{(\bar{y}\text{-}a)+\sqrt{(\bar{y}\text{-}a)^2+(\bar{x}\text{-}a)^2}}{(\bar{y}+a)+\sqrt{(\bar{y}+a)^2+(\bar{x}\text{-}a)^2}}\right] + (\bar{y}\text{-}a)\ln\left[\frac{(\bar{x}\text{-}a)+\sqrt{(\bar{y}\text{-}a)^2+(\bar{x}\text{-}a)^2}}{(\bar{x}+a)+\sqrt{(\bar{y}\text{-}a)^2+(\bar{x}+a)^2}}\right]\right\} \quad (4)$$

where $\alpha = \frac{1-v^2}{\pi E}$,

a is a single grid length, $\bar{x} = x_i\text{-}x_k, \bar{y} = y_j\text{-}y_l$

Inspection point (6,6) influence coefficient change law, make k, l from 1 to 11, get the figure below

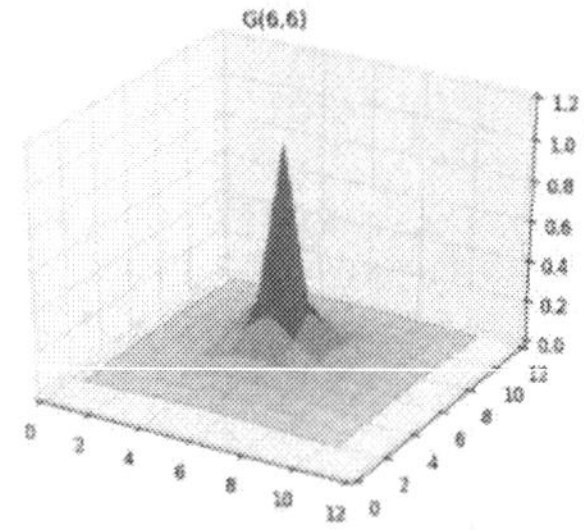

Figure 6. The distribution of influence coefficient of Point (6,6) on the plane.

From the figure can be drawn: the deformation is mainly related to the pressure near the point of several points on a certain point, the influence of further points is very small, based on the results, our algorithm is proposed.

4.2 Sliding window based gauss-Seidel algorithm

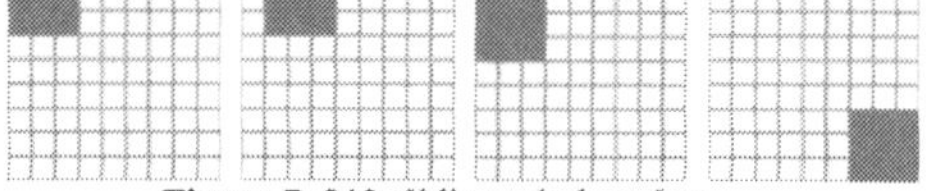

Figure 7. 3*3 sliding window demo.

Algorithm **description**: **Sliding Window Based Gauss-Seidel Algorithm**

Input: coefficient matrix H, window size w, vector $\bar{u}$, initial pressure vector p, initial contact set I_0, maximum number of iterations K_{max}, error limit ε.

Algorithm Procedure:

Step **1** $k := 0, I := I_0$

Step 2 $while \begin{pmatrix} k \le K_{max}\ and \\ \min(p) < \text{-}\varepsilon\ or \\ k = 0 \end{pmatrix} do:$

(2.1) Using Gauss-Seidel algorithm to solve the linear equations of dimension $w \times w$, the solution is the vector p within the window

(2.2) $for\ (i,j) \in I\ do:$

(2.2.1) $if\ p_{i,j} < \text{-}\varepsilon\ then$

$p_{i,j} := 0\ and\ I := I \backslash \{(i,j)\}$

Step 3 $p_{opt} := p$

Step **4** $\bar{I} := \{(1, \dots i, \dots N_c) \times (1, \dots i, \dots N_c)\} \backslash I$

Output: dimensionless contact area

Direct method in each iteration need solving $N_c \times N_c$ groups of $N_c \times N_c$ dimension linear equations, block method in each iteration requires solving N groups of $N_c \times N_c$ dimension linear equations, compared to them, the algorithm in this paper uses a sliding window of a small size sliding solution on the surface, in each iteration requires solving $(N_c\text{-}w + 1) \times (N_c\text{-}w + 1)$ groups of $w \times w$ dimension linear equations. When w is much smaller than N_c, our algorithm can approximate solution $N_c \times N_c$ groups of $w \times w$ dimension linear equation group. Not only to reduce the dimension of the purpose, but also to consider the deformation of each point by the window range of the impact of each point, which greatly improves the computational efficiency, but also the accuracy of the calculation.

Section 5 will design examples to compare the algorithms.

5 Example and Analysis

5.1 Fractal surfaces with different parameters

In this paper, the MPD algorithm is used to generate fractal surface, and the surface is generated by setting the fractal dimension D and the resolution m. The relationship between the resolution and the number of iterations of the generated surface in section 2 is $m = n\text{-}1$

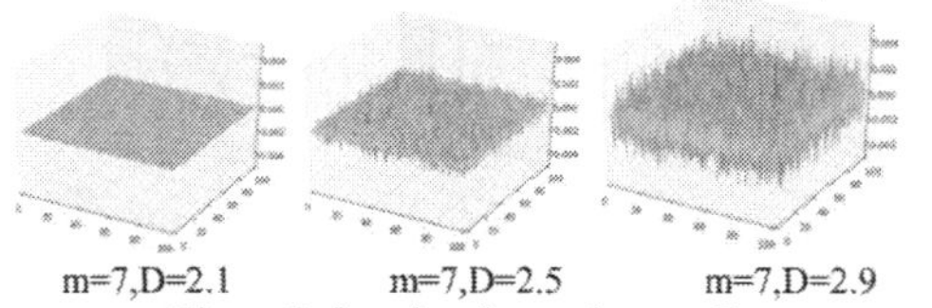

m=7,D=2.1 m=7,D=2.5 m=7,D=2.9

Figure 8. fractal surfaces where m=7

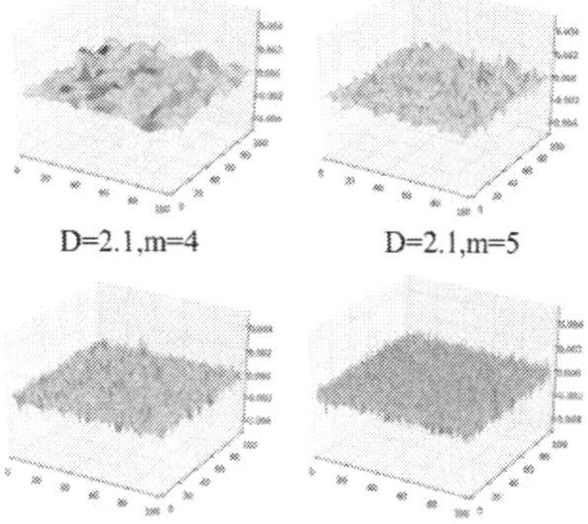

D=2.1,m=4 D=2.1,m=5

D=2.1,m=6 D=2.1,m=7

Figure 9. fractal surfaces where D=2.1

We investigate both the effect of the fractal dimension D, see examples in Figure 8, and the effect of the surface resolution by varying the generation parameter m, see Figure 9. The larger m and the larger Dwill make the surface rougher.

5.2 Comparison of algorithms

The simulation process is controlled by the imposed far-field displacement Δ, and the max Δ is the difference between ξ_{max} and the reference height see Figure 3. This paper selects a range of Δ, and is divided into 15 steps to carry out simulation. Definition of dimensionless pressure $p^* = \frac{pL}{\sigma E}$ and dimensionless contact area $A^* = A_r / A_0$, here p is the total pressure, L is the side of the contact surface, σ is r.m.s. of the distribution of surface heights, E is Young's modulus of elasticity, A_r is the real contact area, A_0 is the nominal contact area.

To investigate the effectiveness of our algorithm, we run groups of simulations on different parameters including sets of m and D, to compare Direct method, Block method and our algorithm (here we let window size=5) with the theoretical solution of Persson Model [4], see Figure 10, the results of D=2.1 m=4,5,6. Intuitively, when m=4, in a narrow range of p^*, errors of three numerical methods are very small, when p^* reach larger, Direct method performances best. But when m reaches larger to 5 and 6, our algorithm gets state-of-the-art results, outperforms the conventional algorithms.

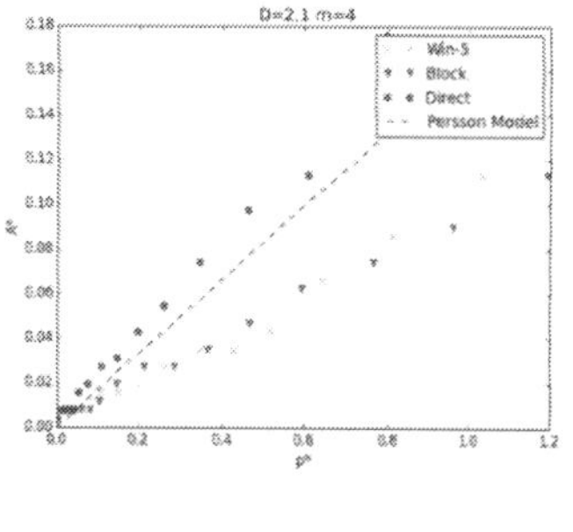

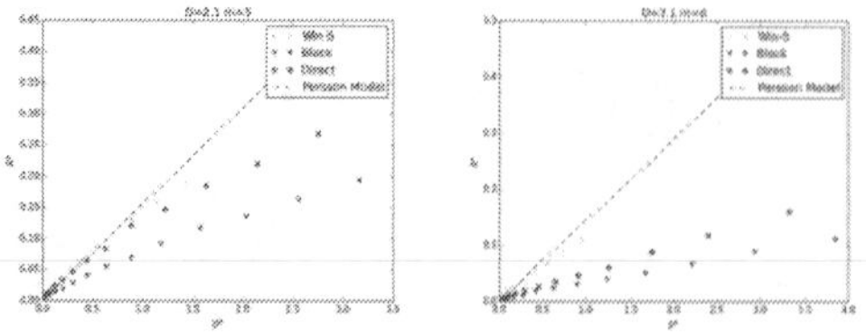

Figure 10. The comparison among algorithms where D=2.1 m=4,5,6

5.3 Effect of window size

The window size determines the accuracy and efficiency of our algorithm, if too large, it will affect the calculation speed; if too small, it will affect the calculation accuracy. It's meaningful to study the effect of window size. Here, we set the window size from 3 to 7, to compare the accuracy of different window sizes, we let the Persson Model and Direct Method as the baselines, to contrast the efficiency, we compare the log CPU time of different window sizes with the Direct Method.

5.3.1 Accuracy comparison

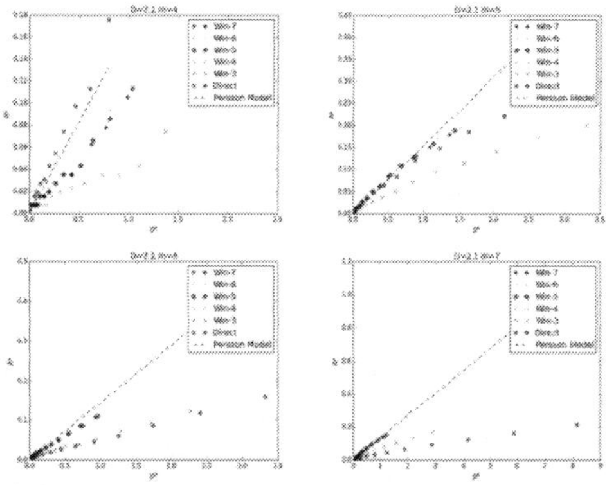

Figure11. The accuracy comparison among different window sizes, Direct Method, Persson Model where D=2.1 m=4,5,6,7

See Fig. 11, the accuracy increases with the increase of the window size as expected, but the difference is not big when window size is more than 3, especially when m is larger than 5.

5.3.2 Efficiency comparison

See Fig. 12, the computation is faster when the window size is smaller. The Direct Method is computationally faster than our algorithm when window size is larger than 3 when m equals to 4, but with the growth of m, the Direct Method computation time increases exponentially, is far more than our algorithm.

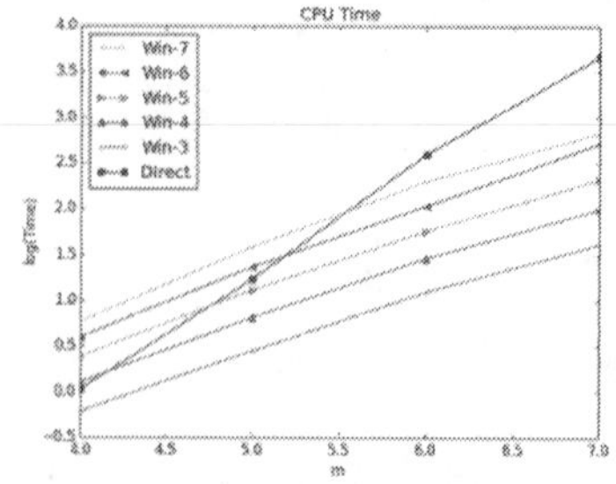

Figure 12. The efficiency comparison among different window sizes and Direct Method

5 Conclusions

We present a new efficient numerical algorithm for solving multi-scale rough surfaces contact mechanics problem under Boundary Element Method framework considering the drawbacks of both conventional methods, Direct Gauss-Seidel Method and Block Gauss-Seidel Method. In this paper we consider both the dimension reduction and accuracy, propose the Sliding Window based Gauss-Seidel algorithm. By taking the Persson Model results as the baseline to compare all kinds of numerical methods, our algorithm performs better than the conventional algorithms in both accuracy and efficiency.

References

1. J.A. Greenwood, J.B.P. Williamson, Proc. R. Soc. Lond. Ser. **A 295**(1966)
2. A. Majumdar, B. Bhushan, ASME J. Tribol. **112** (1990)
3. B.N.J. Persson, J. Chem. Phys. **115** (2001)
4. M. Borri-Brunetto, A. Carpinteri, B. Chiaia, Int. J. Fract. **95** (1999)
5. Xia Boqian, Li Ying, Journal of MeChanical Strength, **30**(5) (2008)
6. H.O. Peitgen, D. Saupe, The Science of Fractal Images, Springer-Verlag, New York (1988)
7. J.R. Barber, Elasticity, 3rd edn., Springer, Dordrecht (2010)
8. M. Paggi, Q.-C. He Wear **336-337** (2015)

MATEC Web of Conferences 77, 03004 (2016) DOI: 10.1051/matecconf/20167703004
ICMMR 2016

Hydrodynamics of Biomass Gasification in a Dual Chamber Circulating Fluidized Bed Reactor

Haifa Wahyu[1], Imam Djunaedi[2], M. Affendi[2] and M. T. Sugiyatno[2]

[1]*Research Centre for Physics, Indonesian Institute of Sciences - Lembaga Ilmu Pengetahuan Indonesia, Kompleks LIPI, Jln. Sangkuriang 21/154D, Bandung 40135 - Indonesia*
[2]*Research Centre for Electric Power and Mechatronics, Indonesian Institute of Sciences - Lembaga Ilmu Pengetahuan Indonesia, Kompleks LIPI, Jln. Sangkuriang 21/154D, Bandung 40135 - Indonesia*

Abstract. This paper presents work on hydrodynamics of several types of biomass mixture in a dual chamber circulating fluidized bed. In designing the CFBreactornecessary to know the distribution of solid particles radially and axially influenced by fluidizing gas velocity, particle size, solid circulation flux, reactor diameter and height of the reactor. These factors will affect pressure drop along the riser of the reaction chamber. Pressure drop is an important factor in the study of hydrodynamics of particle flow. The pressure drop was measured using mathematical model compared to experimental results done on a cold mode. Since it was found that both results were consistent which means that the model can be used to predict the operating parameters of CFB design.

1 Introduction

Flow hydrodynamics is an important aspect in designing circulating fluidized bed systems. Knowledge on the particle distribution along the reactor column will help determining the operational parameters of the reactors especially the pressure loss. Modeling of the flow structure in circulating fluidized bed has been studied by de Diego et al (1995) [1]. They used a mixture of sand and coal to study the particle distribution at axial position along the riser. They observed that there were dense zone and dilute zone. In the dilute region, the upward and downward particle fluxes decreased as the height of the riser increased. This was due to the net transfer of solid fluxes from the core to the annulus. Based on this observation they developed a mathematical model of a core-annulus flow structure in the dilute region. At an earlier stage, Hannes (1996) [2] applied bubble fraction in modeling the circulating fluidized bed for the combustion of coal and sand. By studying the bubble fraction and solid concentration along the riser, distribution of solid particles along the riser can be determined.Studies on hydrodynamics of a biomass gasifier semi dual fluidized bed in a cold-rig was done by Ngo et al (2013) [3]. A semi-dual fluidized bed is a novel design of dual fluidized bed where there was an internal mixing of solid particles between the riser and the gasifier. It was developed to enhance the heat and mass transfer in the system. This was shown by 17 % of back mixing obtained during the experiment. The use of gas injection and liftover in a dual fluidized bed was studied by Loffler et al (2003) [4] in an 8 MW gasifier.

The purpose is to determine the effect of secondary gas injection in the fluid dynamics of the system. It was found that the gas injection the riser has the effect on the gas flowrate.

The study used mathematical model and experiments.Kaiser et al (2003) [5] in the Part II of the studies of the hydrodynamics of a dual fluidized bed gasifier determined the solids circulation and the distribution of the solid hold up of a CFB gasification system under different conditions. The behavior of the system was then analyzed using solid concentration and variations of geometry. The concept of hydrodynamics was also applied for catalytic studies for the treatment of wastes as shown by San Jose et al (2009) [6]. The aim was to achieve stable operating conditions by differentiating the geometric factor of the conical contactor and of the contactor-particle system. Further investigation of solids circulation in a circulating fluidized bed was done by Lim et al (2012) [7] where they tried to understand how operating parameters influence solids circulation. Solids circulation affects the heat transfer that sustained the reactions in the systems. Fluidization properties of torrefied biomass were also studied by Rousset et al (2012) [8] to predict the behavior of amorphous and multi-dimensional particles of the biomass. The indicator was bed pressure drop. Continuous radial profiles of velocities and solids hold-up of sewage sludge gasification was studied in three-dimensional model of a circulating fluidized bed gasifier by Petersen et al (2005) [9]. They studied the number of feeding points for different size of gasifiers. It was found out that better mixing was obtained if the number of feeding points was larger and the column diameter was wider as well.Gungor et al (2013) [10] used particle based approach and two dimensional numerical computation to predict the hydrodynamics of biomass gasifier. The model was used to simulate the radial and axialprofiles of the bed temperature and the volumetric fraction of H_2, CO, CO_2 and CH_4 as well as tar concentration. Other researchers who also involved in using hydrodynamics concept to study fluidized bed reactor and circulating fluidized for gasification were among others Sadak et al (2002) [11], Gungor (2008) [12], Svoboda et al (2009) [13], Kersten et al (2009) [14], Kaushal et al (2008) [15], Zhang et al (2012) [16], Fushimi et al (2011) [17], Charitos et al (2010) [18], and Bi et al (2010) [19]. Most of the researchers work used coal, sand, or biomass alone as the fluidized material. In this work, we used several types of biomass in a mixture of sand in a dual chamber circulating fluidized bed. The purpose is to determine the flow behavior or the hydrodynamics of different characteristics of biomass at a certain operating condition. The experiments were conducted in cold mode.

2 Theoretical background and mathematical formulation

In designing the CFB reactor necessary to know the distribution of solid particles radially and axially influenced by fluidizing gas velocity, particle size, solid circulation flux, reactor diameter and height of the reactor. By looking at the profile of solids flowin CFB reactors, optimum operating conditions can be determined. Modeling for solids flow structure was built when the solids fluxes out of the area with a high density. To determine the flux of external circulation requires the amount of flux solids on the surface of the solid areas. High area dilute is calculated by taking the balance of the pressure loss of the entire rise rand exponential equations to determine the voidat different heights in this area. To determine the pressure loss of the riser, this work uses mathematical formulation taken from the work of Hannes (1996) [2] and Karmakar et al (2010) [20]. The mathematical formulation is applied to different type of biomass mixture. The results were compared to that of experiments done at the dual chamber circulating fluidized bed rig.

Karmakar [20] defined the riser into three zones, thatare dense where bubbling started, splash zone where the particles started to emerge from the dense bed and transport zone where the particles released from the bed and flow into the exit duct of the chamber. The net voidage in the dense zone is given by the following equation:

$$\varepsilon_{d,z} = \delta_b + (1-\delta_b).\,\varepsilon_{m,f}$$

where $\varepsilon_{d,z}$ is the net voidage, $\delta_b$is the bubbling fraction, and $\varepsilon_{m,f}$is the voidage at minimum fluidization velocity.

$$\delta_b = \left(\frac{1}{1 + \frac{1.3\left(0.15 + U_{pa} - U_{mf}\right)^{0.35}}{0.26 + 0.27\, exp\left(-3.3d_p\right)} \left(U_{pa} - U_{mf}\right)^{-0.8}} \right)$$

U_{pa} is the primary air velocity, and $U_{mf}$is the minimum fluidization velocity, d_p is the solid or the particle diameter.

Solids concentration in a hold up at minimum fluidization velocity was calculated using Archimedes number and sphericity of particles.

$$\varepsilon_{s.mf} = 1 - 0.586\emptyset^{-0.72} Ar^{-0.029} \left(\frac{\rho_g}{\rho_s}\right)^{n0.021}$$

$\emptyset$is particle sphericity, Ar number is calculated in the following manner:

$$Ar = \frac{d_s^2 \rho_g \left(\rho_s - \rho_g\right) g}{{\mu_g}^2}$$

where $\rho_g$is the gas density (kg/m^3), and μ_g is the gas viscosity (kg/ms).

Pressure loss is calculated using the following equation where $\Delta P_{dz}$is the pressure loss (Pa) in dense zone, $\rho_s$is the particle density, h_{dz} is the height of the dense zone and g is gravity.

$$\Delta P_{dz} = (1 - \varepsilon_{dz})\rho_s h_{dz}\, g$$

The behaviour of particles in the splash zone is shown by the bed voidage calculated by taking a decay factor into account. Bed voidage in splash zone equals to

$$\frac{\varepsilon_{sz} - \varepsilon}{\varepsilon_{dz} - \varepsilon} = exp[-k\,(h_{sz} - h_{dz})]$$

k is the decay factor taken from Karmakar[20]:

$$k = \frac{CU_t}{U_{pa}}$$

where C = 10 m^{-1} is a constant. The pressure loss is the integration of the voidage along the height of the splash zone.

$$\Delta P_{dz} = \int_{h_{dz}}^{h_{sz}} (1 - \varepsilon_{sz})\rho_s\, g\; dh$$

ε_{sz} is the voidage in the splash zone, h_{dz} is the height of the dense bed and h_{sz} is the height of the splash zone.

Bed voidage in transport zone is calculated by incorporating decay factor a and voidage at infinity ε_∞.

$$\frac{\varepsilon_{tz} - \varepsilon_\infty}{\varepsilon_{sz} - \varepsilon_\infty} = exp[-a\,(h_{tz} - h_{sz})]$$

a is the decay factor of solids fraction and h_{tz} is the height of any solid in transport zone. U is the fluidization velocity and U_t is the terminal velocity. D the riser diameter.

$$a(U - U_t)^2 D^{0.6} = 0.88 - 420\, d_p$$

The infinite voidage in transport section, ε_∞ equals to

$$1 - \varepsilon_\infty = \frac{K_\infty}{\rho_s(U - U_t)}$$

where K_∞

$$K_\infty = \rho_s \alpha_t (U - U_t)$$

$$\alpha_t = 1 - \left(1 + \frac{f_s(U - U_t)^2}{2\, g\, D}\right)^{\frac{1}{4.7}}$$

where α_t is the velocity head coefficient at cyclone inlet. The coefficient of friction, f_s, equals to

$$\frac{f_s \rho_s}{d_p^2}\left(\frac{\mu_g}{\rho_g}\right)^{2.5} = 5.17\left[\frac{\rho_g(U - U_t)dp}{\mu_g}\right]^{-1.5} D^2$$

for

$$\left[\frac{\rho_g(U - U_t)dp}{\mu_g}\right] \le \frac{2.38}{D}$$

or

$$\frac{f_s \rho_s}{d_p^2}\left(\frac{\mu_g}{\rho_g}\right)^{2.5} = 12.3\left[\frac{\rho_g(U - U_t)dp}{\mu_g}\right]^{-2.5} D$$

for

$$\left[\frac{\rho_g(U - U_t)dp}{\mu_g}\right] \ge \frac{2.38}{D}$$

Pressure loss is then calculated using the following equation, where ΔP_{dz} is the pressure loss in transport zone, h_{tz} is the height of the transport zone, ε_{tz} is net voidage of the transport zone. U_s is the solid velocity in riser.

$$\Delta P_{dz} = \int_{h_{sz}}^{h_{tz}} (1 - \varepsilon_{tz})\rho_s\, g\, dh + \int_{h_{sz}}^{h_{tz}} f_s \frac{U_s^2}{2}\frac{4}{D}(1 - \varepsilon_{tz})\rho_s\, g\, dh$$

$$U_s = U - U_t\sqrt{\left(1 + \frac{f_s U_s^2}{2gD_s}\right)\varepsilon_{tz}^{4.7}}$$

where

$$f_s \frac{\varepsilon_{tz}^3}{1 - \varepsilon_{tz}} = 0.0126\left[(1 - \varepsilon_{tz})\frac{U_t}{U_s}\right]^{-0.979}$$

for

$$\frac{U_t}{U_s} > 1.5$$

or

$$f_s \frac{\varepsilon_{tz}^3}{1 - \varepsilon_{tz}} = 0.0410\left[(1 - \varepsilon_{tz})\frac{U_t}{U_s}\right]^{-1.021}$$

for

$$\frac{U_t}{U_s} < 1.5$$

The amount of concentration of solids in the annulus, α_{cor}, of the riser is determined using the following equations. Where A is the area of the riser and G_s is the solids flow in the riser.

$$\bar{\varepsilon}_s A = \varepsilon_{s.cor}\alpha_{cor} A + \varepsilon_{ann}\alpha_{ann} A$$

$$G_s A = u_{s.cor}(1 - \varepsilon_{cor})A\rho_s\alpha_{cor} - u_{ann}(1 - \varepsilon_{ann})A\rho_s\alpha_{ann}$$

$$\varepsilon_{.ann} \overset{!}{=} 2$$

$$\alpha_{ann} = (1 - \alpha_{cor})$$

$$\alpha_{cor} = 1 - \frac{\varepsilon_s - \frac{G_s}{\rho_{s.U_s}}}{\varepsilon_{s.ann}\left(1 + \frac{w_{s.ann}}{u_{s.cor}}\right)}$$

Table 1. Reactor design parameters

Item	Dimension
1. Gasification Unit	
Diameter, m	0.2
Height, m	2.250
2. Combustion Unit	
Diameter, m	0.3
Height, m	1.750
3. Operational Parameters	
Fluidization velocity, m/s	0.7
Minimum fluidization velocity, m/s	0.07
The number of nozzle	4
Sand bed height, m	0.02
Material flow input, kg/hour	Max. 33.02
Sand sphericity( )	0.78
Biomass sphericity ( )	0.49
4. Bed zones	
Dense zone	20 cm height
Splash zone	60 cm height
Transport zone	2.7 m height

DOI: 10.1051/matecconf/20167703004

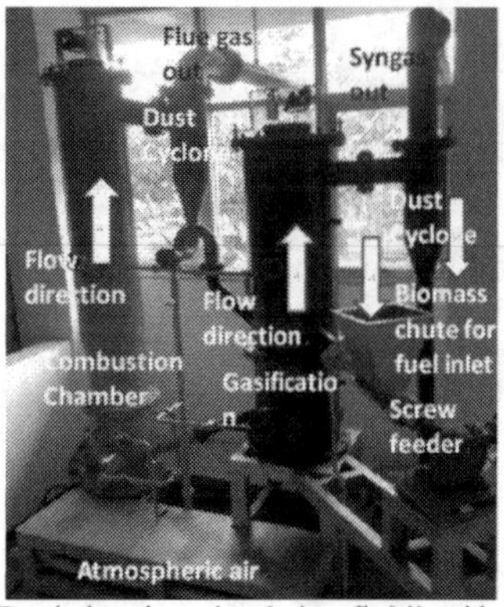

Figure 1. Dual chamber circulating fluidized bed with material flow direction inside the reactor

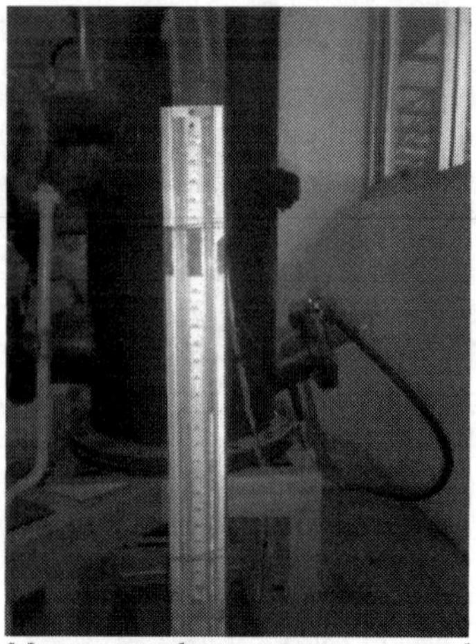

Figure 2. Manometer tube to measure pressure loss

3 Methodology

The aim of this work is to determine the hydrodynamics of a dual chamber circulating fluidized bed with the capacity of 30 kg/hour for biomass gasification. We employed mathematical equations similar to those in the papers to determine the amount of pressure loss in the system. The results were compared with theexperimental data obtained from the cold test of the equipment on many type of biomass. The equipment is shown in Fig. 1, while design parameters are shown in Table 1.

Pressure loss was measured using manometer tube differences along the riser of the gasification reactor. The following picture shows the manometer attached to the gasifier.

Materials used for this experiment were biomass and quartz sand. The following table shows the list of the materials and their properties.

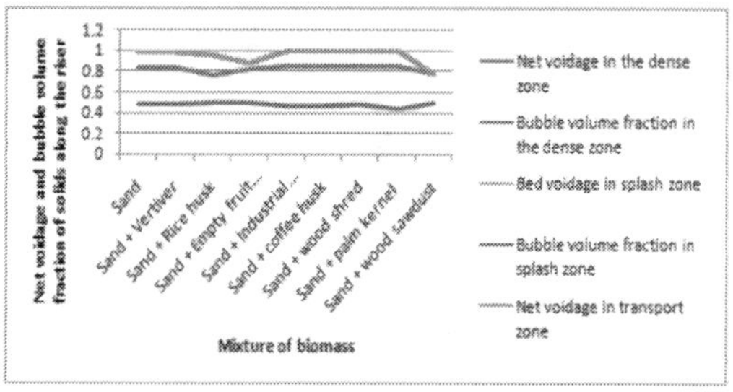

Figure 3. Net voidage of bed in the gasification column

Table 2. Materials used for the experiment on hydrodynamics and their properties

Material	Mixture Density, kg/m^3	Particle diameter, m	Amount, kg/hour
Sand	2650	0.002	3.9
Sand + Vertiver	829	0.005	1.2 (50% : 50%)
Sand + Rice husk	389	0.005	1.2(50% : 50%)
Sand + Empty fruit bunch	80	0.004	1.2 (50% : 50%)
Sand + Industrial Sludge	721	0.008	1.2 (50% : 50%)
Sand + coffee husk	561	0.01	1.2 (50% : 50%)
Sand + wood shred	240	0.015	1.2 (50% : 50%)
Sand + palm kernel	1254	0.02	1.2 (50% : 50%)
Sand + wood sawdust	240	0.0001	1.2 (50% : 50%)

4 Results and discussions

Using the parameter above, the net voidage of bed in the gasification column can be determined.

Fig. 3 shows net voidage in gasification column for where the results are similar for all types of biomass. The net voidage in the dense zone is almost similar to all types of biomass. Slight differences were observed at the splash and transport zone where the net voidage is lower for biomass with higher density. Dense zone has higher solid concentration, therefore the voidage is less compared to splash and transport zone when applying the same air velocity. Fig. 4 shows fluidization velocity that less that the terminal velocity for each biomass.

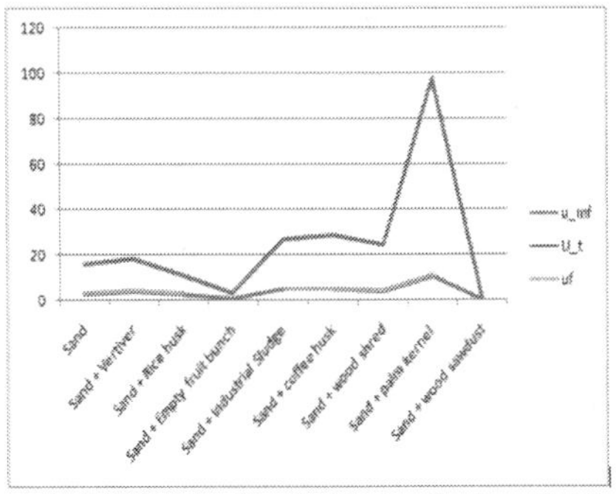

Figure 4. Minimum fluidization velocity, fluidization velocity and terminal velocity

The net voidage has the relationship with the pressure drop in each zone as shown in Fig. 4. Pressure differences is linked to the concentration of solids in each zone.

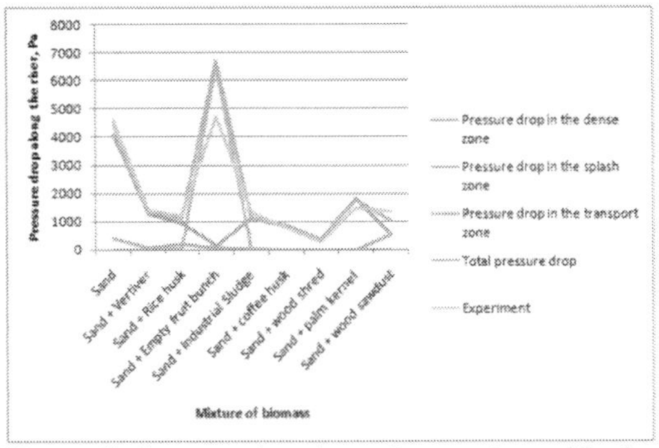

Figure 5. Pressure drop in each zone in gasification column

Pressure drop in each zone is quite consistent with the net voidage of the bed for every biomass. Pressure drop decreases with the increasing of bed height, and thus the fluidization characteristics of biomass particles.However, the results show that the pressure drop from the mathematical calculation is much lower than that of the experimental results. The experiments show that biomass fluidization requires about twice as much as pressure to generate the hydrodynamics of biomass particles in the gasification reactor.

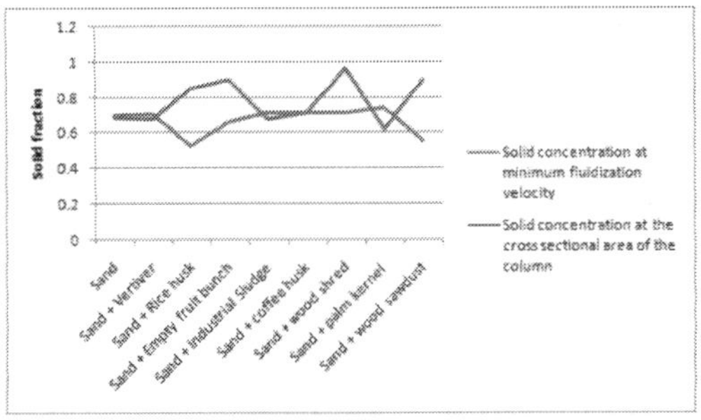

Figure 6. Concentration of solid at minimum fluidization velocity at the core of the reactor column.

Fig. 6 shows the distribution of solids along the reactor column at minimum velocity. α_{cor} represents the concentration of solids in the middle of the column at the core. At the minimum velocity, most solids are in the core of the cross sectional area of the column. The results are similar for all types of biomass with slightly difference for particle density. If the particle density is high, then the concentration of solids is lower due to the acceleration and decelaration of particles during the fluidization process.

4 Conclusions

It was found out that the mathematical model used to determine the pressure loss was consistent with the pressure loss measured during experiments. Differences in density of biomass does not affect solids fraction in the dense zone significantly, but it does affects the solids distribution at the splash and transport zone. This is due to the condition of the particle distribution at this height where more pressure is required to maintain the suspension of solids in this region. Size of biomass does not have the effect, because the size was almost similar due to the requirements of the feeding screw of the biomass. This model can be used to calculate the pressure loss and thus the operating conditions of a dual chamber circulation fluidized bed for mixture of different types of biomass.

Acknowledgements

The authors would like to thank Research Centre for Physics, Indonesian Institute of Sciences for the opportunity to conduct this research and experiments. We are very grateful with the facilities, including financial, support staff and other infrastructures so that this study can be conducted successfully. We would also like to thank other people who have helped and involved in this work since the beginning.

Nomenclature

a	decay constant, m^{-1}
A	the area of the riser
Ar	Archimedes number
C	constant, 10 m^{-1}
d_p	solid particle diameter (m)
f_s	coefficient of friction,
g	gravity (m/s)
G_s	solids flow in the riser
h_{dz}	height of the dense zone (m)
h_{sz}	height of the splash zone (m)
h_{tz}	height of the transport zone (m)
k	decay factor
K_∞	particle elutriation rate constant
U	fluidization velocity, m/s
U_{pa}	primary air velocity, m/s
U_{mf}	minimum fluidization velocity, m/s
U_s	solid velocity in riser, m/s
U_t	terminal velocity, m/s

Greek symbols

α_{cor} concentration of solids in the annulus
α_t velocity head coefficient at cyclone inlet
δ_b bubbling fraction
ΔP_{dz} pressure loss (Pa) in dense zone
ΔP_{sz} pressure loss in splash zone
ΔP_{tz} pressure loss in transport zone
$\varepsilon_{d,z}$ net voidage in dense zone
ε_{sz} net voidage in the splash zone
ε_{tz} net voidage of the transport zone
$\varepsilon_{m,f}$ the voidage at minimum fluidization velocity
ε_{∞}. voidage at infinity
$\emptyset$ particlesphericity
ρ_g gas density (kg/m^3)
ρ_s particle density (kg/m^3)
μ_g gas viscosity (kg/ms)

References

1. L. F. De Diego, P. Gayan, and J. Adanez, (1995) Modelling of the flow structure in circulating fluidized beds, Powder Technology 85, page 19-27.
2. J. P. Hannes, (1996), Mathematical Modelling of Circulating Fluidized Bed Combustion, Thesis Diplom-Ingenieur, Verfahrenstechnik (RWTH Aachen, Duitsland), Duitsland
3. S. I. Ngo,Y. I. Lim, B. H. Song, U. D. Lee, C. W. Yang, Y. T. Choi, and J. H., (2013), Hydrodynamics of cold-rig biomass gasifier using semi-dual fluidized-bed, Powder Technology 234, page 97–106.
4. G. Loffler, S. Kaiser, K. Bosch, and H. Hofbauer (2003), Hydrodynamics of a dual fluidized-bed gasifier – Part I simulation of a riser with gas injection and diffuser, Chemical Engineering Science, 58, page 4197-4293.
5. S. Kaiser, G. Loffler, K. Bosch and H. Hofbauer, , Hydrodynamics of a dual fluidized-bed gasifier – Part II: simulation of solid circulation rate, pressure loop and stability, Chemical Engineering Science, **58**, pp. 4215–4223, (2003).
6. M. J. San Jose, S. Alvarez, A. Morales, L. B. Lopez, and A. O. de Salazar, Hydrodynamics of catalyst in conical spouted beds, Catalysis Today 147S, pp. S162–S169, (2009).
7. M. T. Lim, S. Pang and J. Nijdam, Investigation of solids circulation in a cold model of a circulating fluidized bed, Powder Technology, **226**, pp. 57–67, (2012).
8. P. Rousset, T. Petithuguenin, and A. C. Azevedo, The fluidization behaviour of torrefied biomass in a cold model, Fuel, **102**, pp. 256–263, (2012).
9. I. Petersen, and J. Werther, Three-dimensional modeling of a circulating fluidized bed gasifier for sewage sludge, Chemical Engineering Science, **60**, pp. 4469 – 4484, (2005).
10. A. Gungor and U. Yildirim, Two dimensional numerical computation of a circulating fluidized bed biomass gasifier, Computers and Chemical Engineering, **48**, pp. 234–250, (2013).
11. S. S. Sadak, A. E. Ghaly and M. A. Sabbah, Two phase biomass air-steam gasification model for fluidized bed reactors: Part I—model development, Biomass and Bioenergy, **22**, pp. 439 – 462, (2002).
12. A. Gungor (), Two-dimensional biomass combustion modeling of CFB, Fuel, **87**, pp. 1453–1468, (2008).
13. K. Svoboda, S. Kalisz, F. Miccio, K. Wieczorek, M. Pohořelý, Simplified modeling of circulating flow of solids between a fluidized bed and a vertical pneumatic transport tube reactor connected by orifices, Powder Technology, **192**, pp. 65–73, (2009).
14. S. R. A. Kersten, W. Prins, B. van der Drift and W. P. M van Swaaij, Principles of a novel multistage circulating Fluidized bed reactor for biomass gasification, Chemical Engineering Science, **58**, pp. 725 – 731, (2003).
15. P. Kaushal, T. Proll, and H. Hofbauer, Model for biomass char combustion in the riser of a dual fluidized bed gasification unit: Part 1 — Model development and sensitivity analysis, Fuel Processing Technology, **89**, pp. 651–659, (2008).
16. H. Zhang, R. Xiao, D. Xiao, J. Cho, G. He, S. Shao and J. Zhang, Hydrodynamics of a novel biomass autothermal fast pyrolysis reactor: Solid circulation rate and gas bypassing, Chemical Engineering Journal 181– 182, pp. 685– 693, (2012).
17. Ch. Fushimi, G. Guan, Y. Nakamura, M. Ishizuka, A. Tsutsumi, S. Matsuda, H. Hatano, and Y. Suzuki, , Hydrodynamic characteristics of a large-scale triple-bed combined circulating fluidized bed, Powder Technology, **209**, pp. 1–8, (2011).
18. A. Charitos, C. Hawthorne, A. R. Bidwe, L. Korovesis, A. Schuster and G. Scheffknecht, Hydrodynamic analysis of a 10 kWth Calcium Looping Dual Fluidized Bed for post-combustion CO2 capture, Powder Technology, **200**, pp. 117–127, (2010).
19. X. T. Bi, and X. Li., High density and high solids flux CFB risers for steam gasification of solids fuels, Fuel Processing Technology, **91**, pp. 915–920, (2010).
20. M. K. Karmakar and A. B. Datta, Hydrodynamics of a dual fluidized bed gasifier, Advanced Powder Technology, **21**, pp. 521-528, (2010).

MATEC Web of Conferences 77, 03005 (2016) DOI: 10.1051/matecconf/20167703005
ICMMR 2016

Comparison of the Accuracy and the Efficiency among Different Molecular Dynamics Methods for Calculating the Viscosity

Yanlu Huang and Quandong Huang

Institute of Mechanical and Automotive Engineering, South China University of Technology, No. 381 Wushan Road, Tianhe District, Guangzhou 510640, China

Abstract. The viscosity is one of the important parameters which determine the characteristics of the fluid flow. In this paper, we employed three molecular dynamics methods so-called the periodic poiseuille flow method, the shear flow method and the stress autocorrelation method to calculate the viscosity of the Lennard-Jones fluid. We have found that, among the three, the periodic poiseuille flow method has the greatest convergent speed and the highest accuracy; the shear flow method with a small shear rate has the smallest convergent speed. The accuracy of the stress autocorrelation method evidently depends on the computational time. With 10^5 time-steps it is as accurate as the shear flow method, but with double time-steps it is more accurate than the shear flow method.

1 Introduction

With the developments of the micro- and nano-technologies, the fluid flow in micron or nano scale is paid more and more attentions in many scientific and industrial fields. The viscosity is one of the important parameters which determine the characteristics of the fluid flow and thus greatly affects the accuracy of the theoretical and engineering calculations. However, there is no exact theoretical expression for calculating the viscosity [1]. An experimental measurement or a numerical computation is always performed to get the viscosity value. The way of experimental measurements has the drawbacks of high cost and complex implementing procedure. As the developments of physics and the technologies concerning to computing, the numerical methods are put into realities with considering more complex models with high efficiency and high accuracy. So the way of adopting numerical computations has come to be one of the most effective methods for calculating the viscosity. The numerical models mainly include the ones based on the molecular dynamics [2], the dissipative particle dynamics [3] and the Lattice Boltzmann algorithms [4], [5].

Different numerical methods based on the molecular dynamics of calculating the viscosity can be divided into two categories so-called the equilibrium and the non-equilibrium methods. For equilibrium methods, the viscosity is obtained according to the stress-stress autocorrelation function [6] in an equilibrium state without the disturbance of external forces. In non-equilibrium methods, the external forces are exerted to the system and the non-equilibrium steady state of the fluid flow reflects the effects of the viscosity which can be obtained by calculating the shear stress. The common ways to form steady shear flows include the method with a sliding boundary and the one with periodic shear flow [6]. J. A. Baker [7] proposed the periodic poiseuille flow method in which the periodic boundary condition was adopted as a substitute of the wall boundary condition [8]. The problems of density oscillation and velocity slippage near the walls were solved out in his method, and the viscosity could be obtained by calculating the average flow velocity.

This paper discusses the three methods of the periodic poiseuille flow, the shear flow and the stress autocorrelation in more details. Each of them is applied to calculate the viscosity of the simple Lennard-Jones fluid and the three methods are evaluated through comparing their results.

2 The periodic poiseuille flow method

In molecular dynamics simulation, the poiseuille flow can be obtained by exerting a body force to the system, that is, a corresponding force to every particle. For example, two plates parallel to the xz-plane are located with the distance D and to the fluid between them is applied a body force in x- direction. The resulting flow field has a parabolic velocity pofile

$$v_x(y) = \frac{\rho g_x}{2\eta}(yD - y^2) \tag{1}$$

Integrating Equation (1) and the expression of the average velocity through the yz-plane is given as

$$\langle v_x \rangle = \frac{1}{D}\int_0^D v_x(y)dy = \frac{\rho g_x D^2}{12\eta} \quad (2)$$

where $\langle \cdots \rangle$ denotes the ensemble average. The viscosity can be calculated by Equation (2).

In order to get a zero velocity upon the boundaries, the wall model is used, which yet leads to the density oscillation and the velocity slippage near the walls. J. A. Baker [7] proposed the method of periodic poiseuille flow, in which the length of the fluid region in y-direction is increased to be 2*D*, and the walls are replaced by the periodic boundary conditions. The system is divided into two domains by the symmetric plane in y- direction and each is exerted a body force in the opposite directions. Thus a zero velocity on the boundaries and on the symmetric plane can be obtained; and two opposite parabolic velocity profiles are formed at the both sides of the symmetric plane.

3 The shear flow method

The shear flow method is also a kind of non-equilibrium method in which a steady shear rate should be reached by the common method of sliding wall. For example, in a system, two parallel plates of the xz-plane are located in a distance H and the upper wall moves in a velocity V_{shear} while the lower wall is fixed. When the system achieves the steady state a stable Couette flow is formed. The resulting flow field has a linear velocity profile.

$$v_x(y) = \frac{V_{shear}}{H} y \quad (3)$$

In the steady state, the fluid deforms because the velocities in every layer are different, which results in the production of the inner friction. According to Newton's inner friction law, in a laminar flow, the shear stress which can be given by the elements of the pressure tensor should be in direct proportion to the velocity gradient at a certain point [9]. The law can be represented as

$$p_{xy} = -\eta \frac{\partial v_x(y)}{\partial y} \quad (4)$$

After putting the Equation (3) into the Equation (4), the expression of the viscosity can be formulated as

$$\eta = -\frac{p_{xy}}{\frac{V_{shear}}{H}} \quad (5)$$

where p_{xy} denotes the xy component of the pressure tensor. The method can be used for the non-Newtonian fluids. However, it would take a long time to finish the simulation to get the viscosity when the shear rate is small.

4 The stress autocorrelation method

The stress autocorrelation method is a kind of equilibrium method. The viscosity is related to the oscillation of the off-diagonal elements of the pressure tensor. In an equilibrium simulation, the viscosity is obtained by integrating the Green-Kubo equation [6], which can be represented as

$$\eta = \frac{V}{k_B T}\int_0^\infty \langle P_{xy}(0)P_{xy}(t)\rangle dt \quad (6)$$

where $\langle \cdots \rangle$ denotes the ensemble average, V is the volume, k_B is the Boltzmann constant, T is the absolute temperature, t is the time and p_{xy} is the xy component of the pressure tensor.

The method can be used for calculating the viscosity of non-Newtonian fluid in zero shear rate.

5 Computational details

In this paper, the three methods are applied to the simple Lennard-Jones fluid-argon. The Lennard-Jones potential function used in the simulation, is shown as Eq. (7).

$$U(r_{ij}) = 4\varepsilon\left[\left\{\frac{\sigma}{r_{ij}}\right\}^{12} - \left\{\frac{\sigma}{r_{ij}}\right\}^{6}\right] \quad (7)$$

The system is non-dimensionalized by using the reduced units [2], which are widely used in the molecular dynamics simulation. For all the models, the energy parameter and the length parameter among the atoms in the fluid are all set to be 1.0, i.e., σ_{ll} and ε_{ll} are both equal to 1.0. For the shear flow method the model includes the rigid walls in which the atoms are located in the structure of face-centered cubic (FCC) with a lattice constant of 2.467. The energy parameter and the length parameter among the atoms in the walls and those in the fluid are set to be 1.0, i.e., σ_{wl} and ε_{wl} are both equal to 1.0, ensuring that the liquid wets the wall.

In all the simulations, the number density of the particles is $\rho = 0.8442$ and the temperature is $T = 0.722$. The neighbour list is updated every time step with a cut-off radius of 2.5. The time step is set to be 0.00233. The size of the simulation box is shown in Table 1.

Table 1. The size of the simulation box

Method	$L_x \times L_y \times L_z$
Periodic poiseuille flow	$12.13\times18.69\times12.13$
Shear flow	$12.09\times21.73\times12.09$
Stress autocorrelation	$12.13\times18.69\times12.13$

The periodic boundary condition is applied to all the three directions in the periodic poiseuille flow method and the stress autocorrelation method, but only to the x- and z- directions in the shear flow method. When the viscosity is calculated by the non-equilibrium methods, the viscous heating would be brought to the system by the external force. Hence, a thermostat should be chosen to

DOI: 10.1051/matecconf/20167703005

keep the temperature of the system constant. The DPD thermostat is used with the advantage of Galilean invariance which does not disturb the hydrodynamic interaction [10]. For the stress autocorrelation method, the DPD thermostat should also be applied to the system in equilibrium to make sure that the three methods use the same thermostat. In this way, the results of the three methods can be compared.

For the periodic poiseuille flow method, different external forces are applied to the system and the relevant temperature profiles are shown in Fig. 1. For example, the minimum temperature of the system with the body force 0.12 is about 0.74, which is 2% more than the input value of 0.722. It is clear that the temperature of the system rises with increasing the external forces. For the simple DPD algorithm, the temperature of the system depends strongly on the time step [11]. The DPD thermostat is not suitable for a system with the large external force because it means that a very small time step should be taken. Moreover, the computation with very small time step would cost too much time to assure the system reaching the state we want. Correspondingly, for a given time step, the body force should not be too large. In this paper, the body force 0.02 is chosen and the relevant temperature profile is shown in Fig. 2. The deviation from the expected value is very small and the viscous heating effect can be neglected.

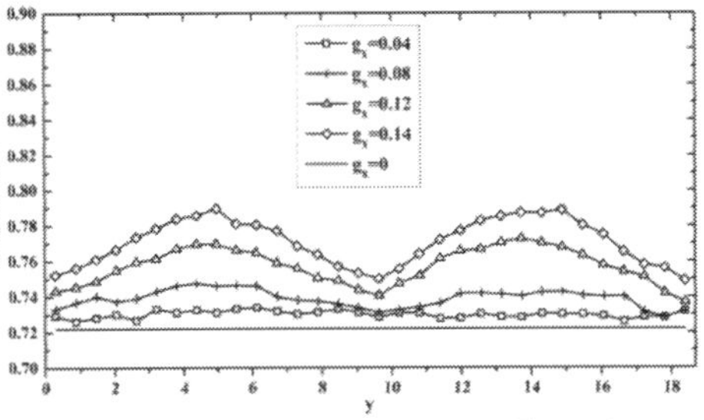

Figure 1. Comparison of the temperature profiles under different body forces

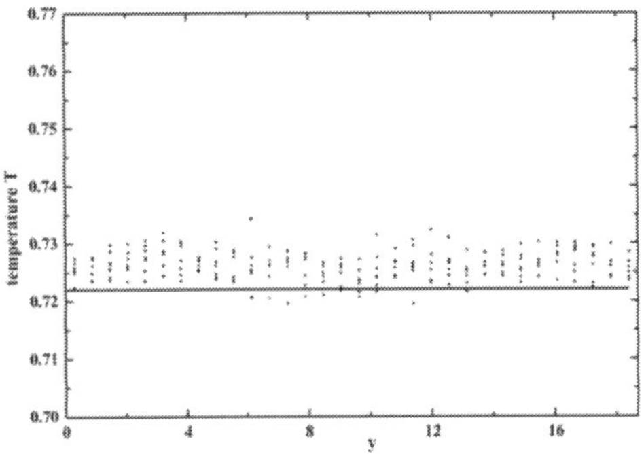

Figure 2 Temperature profile under the body force 0.02: Measurements (solid dots) and the input value (solid line)

The similar researches are undertaken when we choose the shear rate in the shear flow method. Because of the artifact of the simple DPD algorithm, the small shear rate should be chosen.

In short, the body force is set as 0.02 for the periodic poiseuille flow method, and the velocity of the upper wall is set as 0.04 for the shear flow method.

6 Viscosity of a Lennard-Jones fluid

For the periodic poiseuille flow method, there are 2313 fluid atoms which are initially located in the FCC lattice in the cubic box; and the velocities of the atoms are set randomly according to the given temperature. When the system reaches its steady state, the velocity distribution of the particles in the system will become the steady Maxwell distribution. At the beginning of the simulation, the NVE ensemble and the DPD thermostat are applied to the system. The body force is not applied to the system until it reaches the equilibrium state. The velocities of the atoms are corrected every 100 time steps to make sure that the momentum of the system is zero. The velocity profile is shown in Fig. 3. The measurements are accomplished in each part and are fitted with a quadratic function by the least square method with the correlation coefficients larger than 0.97. The velocity profile agrees well with the analytic laminar poiseuille flow pattern from Eq. (1).

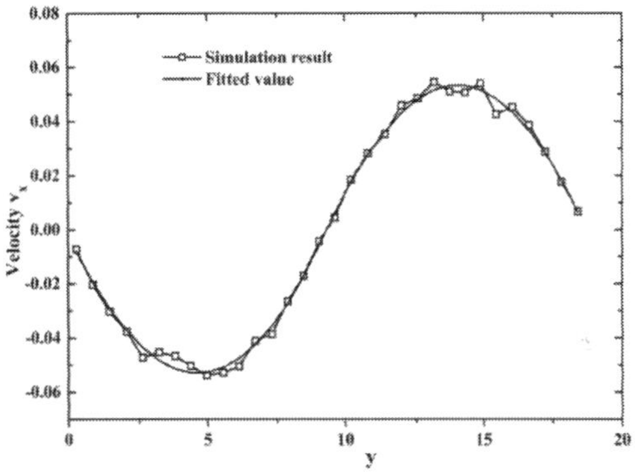

Figure 3 Velocity profile of the periodic poiseuille flow

The velocities are output every time step and the results are used to calculate the viscosity every 10^5 time steps. The process is repeated ten times. The dimensionless viscosity is 3.47 ± 0.18.

For the shear flow method, there are 1324 wall atoms and 2321 fluid atoms in the cubic box. The rigid wall has a FCC lattice. Initially, the fluid atoms are located in the FCC lattice and the velocities are set randomly according to the given temperature. At the beginning of the simulation, the NVE ensemble and the DPD thermostat are applied to the system and the upper wall moves at a constant velocity of 0.04. The viscosity is not calculated until the system reaches the steady state. The xy component of the pressure tensor is output every time step and the results are used to calculate the viscosity every 10^7 time steps. The process is repeated ten times. The method has a slow convergence and it takes about 10^7 time steps to get a reasonable result. The reason is that the shear rate we have adopted is very close to zero which amplifies the oscillation of the pressure tensor. The dimensionless viscosity is 3.99 ± 0.71.

The velocity profile in Fig. 4 shows that the steaming velocity of the particles near the fixed wall in the distance range 0 to 7 in y direction is almost zero, deviating from the theoretical values largely. A stable shear rate is formed far away from the fixed wall, which is larger than the theoretical one from Equation (3). But for all that, the method can get a viscosity comparable with that obtained by using the periodic poiseuille flow method.

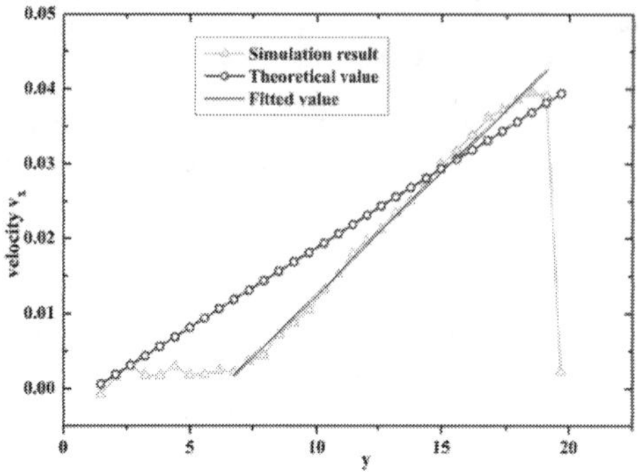

Figure 4 The velocity profile of the shear flow

For the stress autocorrelation method, there are 2313 fluid atoms which are initially located in the FCC lattice in the cubic box and the velocities of the atoms are set randomly according to the given temperature. At the beginning of the simulation, the NVE ensemble and the DPD thermostat are applied to the system. The viscosity can be calculated when the system reaches the equilibrium state for it is essentially an equilibrium method. Two simulations with different computation time are performed. One has 10^5 time steps and the other has double. Both simulations are repeated ten times. The viscosities this paper have gotten from the two simulations are 3.97 ± 0.90 and 3.72 ± 0.67, respectively.

7 Comparison of the three methods

The viscosities calculated by the three methods are in agreement, and also accord with that by J. A. Baker [7]. All the results are shown in Table 2.

Table 2. Comparison among our results and J. A. Baker's

Methods	Our results	Baker's results
Periodic poiseuille flow	3.47 ± 0.18	3.4 ± 0.2
Shear flow	3.99 ± 0.71	3.4 ± 0.8
Stress autocorrelation	3.97 ± 0.90	3.7 ± 1.2

In terms of the convergent speed, the reasonable results can be obtained with a simulation time of 10^5 time steps by the periodic poiseuille flow method and the stress autocorrelation method. However, a simulation time of about 10^7 time steps is required to get a reasonable result for the shear flow method. The shear flow method has the slowest convergent speed, while the periodic poiseuille flow method and the stress autocorrelation method have comparable ones. In terms of the accuracy, the periodic poiseuille method gives a more accurate result, which coincides well with the viscosity of 3.4 ± 0.2 given by J. A. Baker. The result obtained by the stress autocorrelation method with 10^5 time steps is as accurate as that obtained by the shear flow method, while the result with 2×10^5 time steps is more accurate than the one by the shear flow method. In short, in terms of convergent speed and accuracy, the periodic poiseuille flow is better than the methods of stress autocorrelation and shear flow.

8 Conclusion

In this paper, the three methods for calculating the viscosity with molecular dynamics simulation have been compared. The methods differ in applicability, accuracy and computational time. For a Lennard-Jones fluid, the three methods can give comparable results.

For the periodic poiseuille flow method, it can only be used to calculate the viscosity of the Newtonian fluid. However, both of its convergent speed and accuracy are the best among the three. For the shear flow method, it can be used to calculate the viscosity of the non-Newtonian fluid. When the shear rate is very small, its convergent speed is the slowest among the three methods. For the stress autocorrelation method, it can be used to calculate the non-Newtonian fluid with a zero shear rate. With a computational time of 10^5 time steps, it is as accurate as the shear flow method, but while with double time steps it is more accurate than the shear flow method.

Acknowledgement

This work was supported by the National Natural Science Foundation of China (No.51075157).

References

1. C. A. Marsh, G. Backx, M. H. Ernst, Rev. E. **56**(2), 1676, (1997).
2. D. C. Rapaport, *The art of molecular dynamics simulation* (2004).
3. P. J. Hoogerbrugge, J. Koelman, EPL. **19**(3), p. 155, (1992).
4. He X, Chen S, Zhang R, J. Comput. Phys. **152**(2), 642-663 (1999).
5. S. Succi. *The Lattice-Boltzmann Equation* (2001).
6. M. P. Allen, D. J. Tildesley. *Computer simulation of liquids* (1987).
7. J. A. Backer, C. P. Lowe, H. C. J. Hoefsloot, et al., J. Chem. Phys. **122**(15), 154503 (2005).
8. F. Varnik, K. Binder, J. Chem. Phys. **117(13)**, 6336-6349 (2002).
9. N. Kikuchi, C. M. Pooley, J. F. Ryder, et al., J. Chem. Phys. **119**(12), 6388-6395 (2003).
10. C. A. Marsh, J. A. Yeomans, EPL. **37**(8), 511 (1997).
11. D. I. Dimitrov, A. Milchev, K. Binder, Phys. Rev. Lett. **99**(5), 054501 (2007).

DOI: 10.1051/matecconf/20167703006

Research of Influence of Small Asymmetries Construction on the Dynamics Motion of Space Landing Vehicle with Considering Resonance

Vsevolod V. Koryanov and Victor P. Kazakovtsev

Bauman Moscow State Technical University, Dynamics and Flight Control of Motion of Rockets and Spacecraft Department, 105005, Moscow, Russian Federation

Abstract. This article provides an analytical methodology rapid assessment of the impact of small asymmetries on motion dynamics of space landing vehicle in a resonant motion. The methodology allows to analyze the value of the design parameters and the aerodynamic coefficients on the degree of their influence through the asymmetry of a deviation from the longitudinal axis of the vehicle velocity vector. The methodology is based on the assumption that the rapid development of resonant modes of motion. An example of the calculations for the proposed method, which showed the adequacy of the results according to the calculations for modeling the spatial movement.

1 Introduction

Aspects of the dynamics and sustainability of the movement is not controlled spacecraft, with rotation about the longitudinal axis, the presence of small asymmetries devoted sufficient number of scientific publications [1,2,3,4, etc.]. In [5] and [8], in more detail the methods for calculating the parameters of angular motion of space landing vehicles with an inflatable braking device.

Is shown in these that the presence of small asymmetries may experience resonant mode of motion, in which the longitudinal axis of the landing vehicle is significantly deviates from the velocity vector. Under certain conditions, the landing vehicle may lose stability their movement. In these papers presented the mathematical model of motion of the landing vehicle, taking into account the influence of small asymmetries.

The most complete picture of the spatial movement of the landing vehicle can be obtained by numerical integration of complex nonlinear mathematical models. However, this way of solving the problem is possible only in the final stages of design and gives only a partial (numerical) results.

In this paper we present a methodology for assessing the impact of small asymmetries in the longitudinal axis of the deflection space landing vehicle from the velocity vector in a resonant mode of motion, for use in the early design stages of the landing vehicle.

2 Modelling

Small asymmetries caused additional moments that define the dynamics of change of angular motion of the landing vehicle. Consider the equation of angular motion of the landing vehicle, taking into account the influence of small asymmetries in the projections on the axes related to the coordinate system (OXYZ) presented in [2], [6], [7]:

$$\dot{\omega}_x = \frac{d\omega_x}{dt} = \frac{1}{J_x}\Bigg[J_{xy}\cdot(\dot{\omega}_y - \omega_x\cdot\omega_z) +$$

$$+J_{xz}\cdot(\dot{\omega}_z + \omega_x\cdot\omega_y) + J_{yz}\cdot(\omega_y^2 - \omega_z^2) +$$

$$(J_y - J_z)\cdot\omega_y\cdot\omega_z + q\cdot S\cdot\ell\cdot\left(m_{x_0} + m_x^{\bar{\omega}_x}\cdot\frac{\omega_x\cdot\ell}{V} + C_y\cdot\frac{\Delta z}{\ell} - C_z\cdot\frac{\Delta y}{\ell}\right)\Bigg]$$

$$\dot{\omega}_y = \frac{d\omega_y}{dt} = \frac{1}{J_y}\Bigg[J_{xy}\cdot(\dot{\omega}_x + \omega_y\cdot\omega_z) + J_{xz}\cdot(\omega_z^2 - \omega_x^2) +$$

$$+(J_z - J_x)\cdot\omega_x\cdot\omega_z + J_{yz}\cdot(\dot{\omega}_z - \omega_x\cdot\omega_y) +$$

$$+q\cdot S\cdot\ell\cdot\left(m_{y_0} + m_y^{\beta}\beta + m_y^{\bar{\omega}_y}\cdot\frac{\omega_y\cdot\ell}{V} + C_x\cdot\frac{\Delta z}{\ell}\right)\Bigg] \tag{1}$$

$$\dot{\omega}_z = \frac{d\omega_z}{dt} = \frac{1}{J_z}\Bigg[J_{xy}\cdot(\omega_x^2 - \omega_y^2) + J_{xz}\cdot(\dot{\omega}_x - \omega_y\cdot\omega_z) +$$

$$+J_{yz}\cdot(\dot{\omega}_y + \omega_x\cdot\omega_z) + (J_x - J_y)\cdot\omega_x\cdot\omega_y +$$

$$+q\cdot S\cdot\ell\cdot\left(m_{z_0} + m_z^{\alpha}\alpha + m_z^{\bar{\omega}_z}\cdot\frac{\omega_z\cdot\ell}{V} - C_x\cdot\frac{\Delta y}{\ell}\right)\Bigg].$$

where: $\omega_x, \omega_y, \omega_z$ - the projections of the angular velocity of the landing vehicle on the axis of the related coordinate system;

$q = \frac{\rho V^2}{2}$ - dynamic pressure;

V - velocity of landing vehicle;

S - mid-section area;

ℓ - length of landing vehicle;

J_x, J_y, J_z - the principal moments of inertia;

C_x, C_y, C_z - aerodynamic coefficients of axial and transverse forces;

m_{x0} - aerodynamic moment coefficient relative to the longitudinal axis;

m_y^β, m_z^α - derivatives of aerodynamic coefficients on the angle of attack (α) and glide (β);

$m_x^{\omega_x}, m_y^{\omega_y}, m_z^{\omega_z}$ - the derivatives of aerodynamic coefficients on the projections of the angular velocity of the landing vehicle on the axis related coordinate system.

Small asymmetry: $\Delta y, \Delta z$ - the lateral displacement of the real center of mass relative to the longitudinal axis of the of the landing vehicle; J_{zy}, J_{xz}, J_{yz} - the centrifugal moments of inertia; m_{y0}, m_{z0} - the aerodynamic coefficients of transverse moments at zero values of angles of attack and glide angle.

Numerous studies of resonant modes of movement is not operated vehicles [2, 7, 11] have shown that the development of resonant mode (the maximum deviation of the longitudinal axis of the landing vehicle of the velocity vector) is held for a short time of a few seconds.

Based on this, we transform the second and third equations of (1) under the following assumptions: neglect the influence of gravity in the dynamic equations of translational motion; We assumed to be constant all the aerodynamic coefficients and their derivatives; neglected as a second small quantities, works of angular velocity in the dynamic equations of rotational motion. Thus, we can get the next formulas:

$$J_{yz} = 0; J = J_y = J_z; C_z^\beta = -C_n^\alpha; |m_y^\beta| = |m_z^\alpha|;\ m_y = -|m_z^\alpha| \cdot \beta;$$

$$m_z = -|m_z^\alpha| \cdot \alpha;$$

$$C_x = C_\tau, C_y = C_n^\alpha \cdot \alpha, C_z = -C_n^\alpha \cdot \beta \cdot$$

It further indicated - C_n^α a derivative of the aerodynamic coefficient of the normal force on the angle of attack.

Using an algorithm for transforming equations proposed in [1]. After transformation of the second and third equations in the system (1) according to the introduced estimates we obtain the second order differential equation for the angles of attack and glide angle in the following form:

$$\alpha = -\left[\frac{S}{m}\left(C_n^\alpha - C_\tau\right) + \frac{Sl^2}{J}\left|m_z^\omega\right|\right]\frac{q}{V}\alpha -$$
$$-\left[\frac{qSl}{J}\left|m_z^\alpha\right| - \frac{J - J_x}{J}\omega_x^2 + \frac{q^2S^2l^2}{mJV^2}\left(C_n^\alpha - C_\tau\right)\left|m_z^\omega\right|\right]\alpha -$$
$$-\frac{2J - J_x}{J}\omega_x\beta - \left[\frac{Sl}{J}\left|m_z^\omega\right| + \frac{J - J_x}{J}\frac{S}{m}\left(C_n^\alpha - C_\tau\right)\right]\frac{q}{V}\omega_x\beta +$$
$$+ h_\beta\frac{J - J_x}{J}\omega_x^2 + \frac{qSl}{J}\left|m_z^\alpha\right|\alpha_a - \frac{qSC_x}{J}\Delta y,$$

$$\beta = -\left[\frac{S}{m}\left(C_n^\alpha - C_\tau\right) + \frac{Sl^2}{J}\left|m_z^\omega\right|\right]\frac{q}{V}\beta - \tag{2}$$
$$-\left[\frac{qSl}{J}\left|m_z^\alpha\right| - \frac{J - J_x}{J}\omega_x^2 + \frac{q^2S^2l^2}{mJV^2}\left(C_n^\alpha - C_\tau\right)\left|m_z^\omega\right|\right]\beta -$$
$$-\frac{2J - J_x}{J}\omega_x\alpha - \left[\frac{Sl}{J}\left|m_z^\omega\right| + \frac{J - J_x}{J}\frac{S}{m}\left(C_n^\alpha - C_\tau\right)\right]\frac{q}{V}\omega_x\alpha +$$
$$+ h_\alpha\frac{J - J_x}{J}\omega_x^2 + \frac{qSl}{J}\left|m_z^\alpha\right|\beta_a - \frac{qSC_x}{J}\Delta z.$$

where additionally indicated:

α_a, β_a- wheel balancers the angles of attack and glide caused a small asymmetry of shape;

$h_\alpha = \frac{J_{xz}}{J - J_x}, h_\beta = \frac{J_{xy}}{J - J_x},$ relative centrifugal moments of inertia.

Transform the equation in the system (2), using complex variables for the angular motion parameters of the landing vehicle and asymmetries:

$$\left.\begin{array}{c} \delta = \beta + i \cdot \alpha \quad h = h_\beta + i \cdot h_\alpha \quad \omega = \omega_y + i \cdot \omega_z \\ \delta_0 = \beta_0 + i \cdot \alpha_0 \end{array}\right\} \tag{3}$$

where: δ - the spatial angle of attack;

$$\left.\begin{array}{ll} \alpha_0 = -\alpha_a - \frac{C_\tau}{|m_z^\alpha|} \cdot \frac{\Delta y}{\ell}; & \alpha_a = -\frac{m_{z_0}}{|m_z^\alpha|}; \\ \beta_0 = -\beta_a + \frac{C_\tau}{|m_z^\alpha|} \cdot \frac{\Delta z}{\ell}; & \beta_a = -\frac{m_{y_0}}{|m_z^\alpha|}; \end{array}\right| \tag{4}$$

After the transformations obtain the differential equation of angular motion of the landing vehicle, describing the angular fluctuations of landing vehicle in the complex form:

$$\ddot{\delta} + K_1 \cdot \dot{\delta} + K_2 \cdot \delta = K_3 \tag{5}$$

where:

$$K_1 = \frac{1}{V}\left[q \cdot S \cdot \left(\frac{C_n^\alpha - C_\tau}{m} + \frac{|m_z^{\bar{\omega}}|\ell^2}{J}\right)\right] + i \cdot \left[\left(2 - \frac{J_x}{J}\right) \cdot \omega_x\right]$$

$$K_2 = \left[\frac{qS\ell}{J} \cdot |m_z^\alpha| - \frac{J - J_x}{J}\omega_x^2 + \frac{q^2S^2\ell^2}{mJV^2} \cdot |m_z^{\bar{\omega}}|(C_n^\alpha - C_\tau) + \right] +$$
$$+ i \cdot \left[\frac{J - J_x}{J} \cdot \frac{qS}{mV} \cdot (C_n^\alpha - C_\tau) \cdot \omega_x + \frac{qS\ell^2}{JV} \cdot |m_z^{\bar{\omega}}| \cdot \omega_x + \dot{\omega}_x\right]$$

$$K_3 = \left[\frac{q \cdot S \cdot \ell}{J} \cdot |m_z^\alpha| \cdot \beta_0 + \frac{J - J_x}{J} \cdot \dot{\omega}_x \cdot h + \right] +$$
$$+ i \cdot \left[\frac{q \cdot S \cdot \ell}{J} \cdot |m_z^\alpha| \cdot \alpha_0 + \frac{J - J_x}{J} \cdot \omega_x^2 \cdot h\right]. \tag{6}$$

In Eq. (5) K_1 coefficient determines the damping properties of the of the landing vehicle, K_2 coefficient determines the amplitude of the deviation of the spatial angle of attack, K_3 coefficient characterizes the small asymmetries. We analyze the terms in the equation for the K_2 coefficient.

The calculations for the of the landing vehicle show that the values of the first two terms of about two orders of magnitude greater than the values of all other terms. The maximum value of the spatial angle of attack will be the minimum value of coefficient K_2, which is obtained from the condition that the first two terms.

For steady motion spatial angle of attack b is determined from (5) by the following relationship:

$$\delta_Б = \frac{K_3}{K_2}. \quad (7)$$

In this formula the coefficient K_3 depends on the value of existing small asymmetries, for given asymmetries values: $K_3 = Const$. Thus, the change of the angle is determined by the coefficient value K_2. In order to get $K_2 \to 0 \cdot \delta_Б \to \max$. Therefore, the condition $K_2 = \min$ is as follows:

$$\frac{qS\ell}{J}|m_z^\alpha| - \frac{J - J_x}{J}\omega_x^2 = 0 \quad (8)$$

The first term in the equation K_2 coefficient is the square of the frequency of natural transverse fluctuations of the landing vehicle. Formula (8) transform to a new kind:

$$\frac{qS\ell}{J - J_x}|m_z^\alpha| - \omega_x^2 = 0 \quad (9)$$

Denote in the formula (9) the first term as the resonant frequency:

$$\omega_{rez} = \sqrt{\frac{qS\ell|m_z^\alpha|}{J - J_x}} \quad (10)$$

Thus, the necessary condition for maximum spatial angle of attack is the coincidence of the angular velocity of the landing vehicle relative to the longitudinal axis and the resonance frequency:

$$\omega_x = \omega_{rez}. \quad (11)$$

Under the conditions of the balanced movement of the landing vehicle module integrated balancing attack angle b can be determined by the following formula:

$$|\delta_Б| = \sqrt{(\mathbf{Re}\,\delta_Б)^2 + (\mathbf{Im}\,\delta_Б)^2} \Rightarrow |\delta_Б| = \sqrt{\beta_Б^2 + \alpha_Б^2} \quad (12)$$

where: $\alpha_Б, \beta_Б$ - angles of attack and glide angle in a balanced movement.

After substituting the equations (2), (6) and carrying out transformations we obtain a formula for determining the maximum value of the angle of attack of the module integrated in a balanced movement in the presence of resonance modes:

$$|\delta_Б|_{max} = K_\delta \cdot \sqrt{(K_m m_{y0} + K_\Delta \overline{\Delta z} - K_J h_\alpha)^2 + (K_m m_{z0} - K_\Delta \overline{\Delta y} + K_J h_\beta)^2} \quad (13)$$

where: K_δ - coefficient of influence of small asymmetries in the value of the spatial angle of attack in a resonant mode of movement of the landing vehicle.

Coefficients of the relative asymmetry influence:

$$K_m = 1, K_\Delta = C_\tau, K_J = |m_z^\alpha|\frac{J}{J - J_x}. \quad (14)$$

Dimensionless value of the asymmetry:

$$\overline{\Delta y} = \frac{\Delta y}{\ell}, \overline{\Delta z} = \frac{\Delta z}{\ell}, h_\beta = \frac{J_{xy}}{J}, h_\alpha = \frac{J_{xz}}{J}. \quad (15)$$

Consider the example of estimating the relative impact of a variety of small asymmetries in the maximum value of the spatial angle of attack in a resonant mode for of the landing vehicle with the following characteristics:

$$|m_z^\alpha| = 0{,}343, C_\tau = 0{,}72, J = 8{,}75, J_x = 1{,}5$$

The coefficients of the relative influence of asymmetries, calculated according to the formulas (14) are as follows: $K_m = 1, K_\Delta = 0{,}72, K_J = 0{,}41$

Thus, for the landing vehicle, for equal values of asymmetries greatest impact on the maximum deviation of the spatial angle of attack will have a form of asymmetry external device [13, 14, 15]. The least affected by centrifugal moments of inertia. For equal values of the quantities of small asymmetries $m_{y0} = \overline{\Delta z} = h_\alpha$ ratio of the maximum values of the spatial angle of attack for the lateral deviation and the center of mass of the centrifugal moment of inertia for maximum spatial angle of attack due to the asymmetry of the form will be equal to the coefficients $K_\Delta = 0{,}72, K_J = 0{,}41$.

3 Results

For confirmation results of the calculations on the proposed analytical method of determining the maximum deviation will hold the spatial angle of attack of the landing vehicle by integrating the spatial system of differential equations in the landing vehicle at resonance motion conditions.

At Fig. 1, Fig. 2, Fig. 3 shows graphs of these calculations for the following quantities of small asymmetries $m_{y0} = \overline{\Delta z} = h_\alpha = 4 \cdot 10^{-3}$. In all cases, the start time of the resonance mode is six seconds, and the time to maximum spatial angle of attack equal to fourteen seconds.

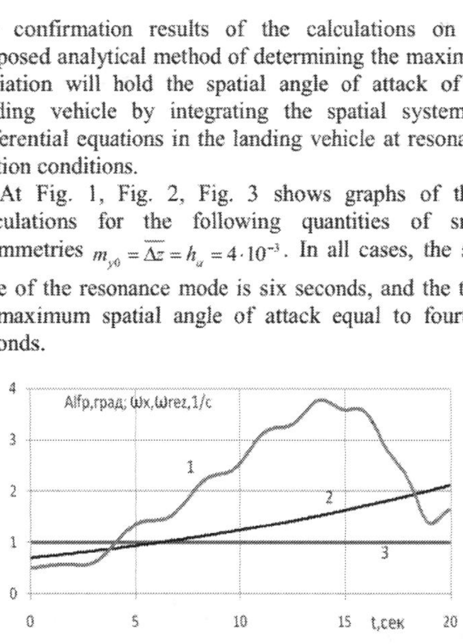

Figure 1. Graphics of spatial angle of attack (1), the resonance frequency (2) and the angular velocity of the descent the landing vehicle about the longitudinal axis (3) to effect only the asymmetry form the landing vehicle ($\alpha_{pmax1} = 3{,}78^0$).

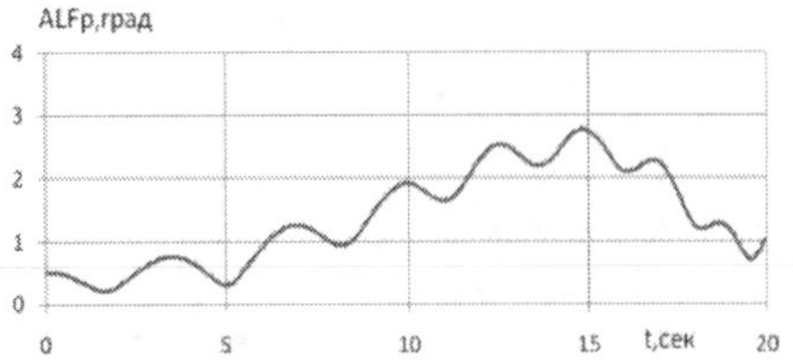

Figure 2. Graphic of spatial angle of attack to affect only the lateral displacement of the center of mass of the landing vehicle ($\alpha_{max2} = 2{,}77^0$).

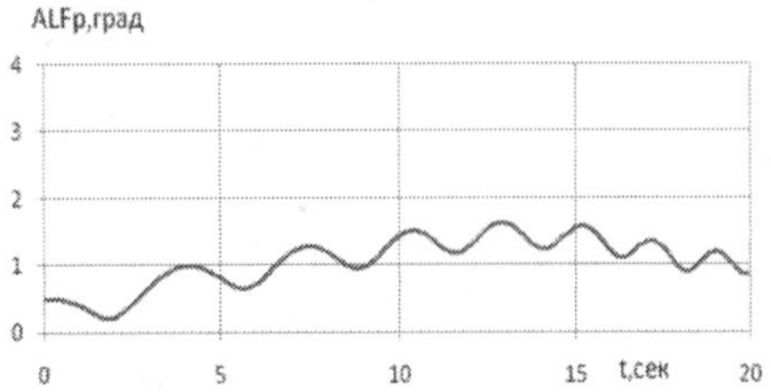

Figure 3. Graphic of spatial angle of attack to affect only the centrifugal moment of inertia ($\alpha_{pmax3} = 1{,}62^0$).

Relationship maximum values of spatial angles of attack for the lateral deviation and the center of mass of the centrifugal moment of inertia for maximum spatial angle of attack due to the asymmetry of shape, will be as follows:

$$\frac{\alpha_{I\,\max 2}}{\alpha_{I\,\max 1}} = 0{,}73, \quad \frac{\alpha_{I\,\max 3}}{\alpha_{I\,\max 1}} = 0{,}42$$

It is evident that they are almost equal coefficients $K_\Delta = 0{,}72, K_J = 0{,}41$.

Thus, confirmed the possibility of using the presented techniques for rapid assessment of the relative impact of a variety of small asymmetries in the value of spatial angle of attack of the landing vehicle in a resonant motion.

4 Conclusions

1. An analytical methodology of rapid assessment of the impact of small asymmetries in the amount of deviation of spatial angle of attack of the landing vehicle in a resonant motion. The methodology allows to analyze the value of the design parameters and the aerodynamic coefficients on the degree of their influence through the asymmetry of a deviation of the longitudinal axis of the landing vehicle of the velocity vector.

2. An example of the influence of small asymmetries in the resonance conditions for of the landing vehicle designed for the descent into the Martian atmosphere. It is shown that the solution of this problem by analytical procedure is almost identical with the results of numerical modeling of spatial movement of the landing vehicle.

References

1. Dmitriyevskiy A.A., Bogodistov S.S. Analysis of the stability of the rotational motion of the asymmetric spacecraft, returning in the atmosphere // Trudy V Scientific Conference on Astronautics. – M.: "USSR Academy of Sciences". p. 81-88, (1981).
2. Goman M.G. Analysis of the resonant modes of the spatial movement of spacecraft, which has symmetry plane in atmosphere. "Trudy TSAGI", Vol. 1789, 41.2, (1976).
3. Goman M.G. Transient resonance modes of movement uncontrolled spacecraft during flight in the atmosphere. - "Scientific notes of TSAGI", t.8, **6**.59, pp. 67-80, (1977).
4. Dmitriyevskiy A.A., Lysenko L.N. External Ballistics: A tutorial for university students. – 4th ed., Revised. And ext. - M.: "Engineering". -608 p., (2005).
5. Kazakovtsev V.P., Koryanov V.V. Metod issledovaniya dinamiki uglovogo dvizheniya kosmicheskikh spuskayemykh apparatov s naduvnym tormoznym ustroystvom // Vestnik Moskovskogo gosudarstvennogo tekhnicheskogo universiteta im. N.E. Baumana. Seriya Mashinostroyeniye. [Research technique of the attitude dynamics of a landing unmanned space vehicle with an inflatable braking device // Herald of the Bauman Moscow State Technical University. Series Mechanical Engineering], **3**(88), pp. 39-46, (2012).
6. Kazakovtsev V.P. Alahmad Ahmad Barakat. The method of assessing the impact of asymmetries uncontrolled spacecraft to the spatial angle of attack // All-Russian scientific and technical journal "Flight". **3**, pp. 44-52, (2006).
7. Kostrov A.V. Dvizheniye asimmetrichnykh ballisticheskikh apparatov [Movement asymmetric ballistic devices]. – Moscow, Mashinostroyeniye.. – 272 p., (1984).
8. Koryanov V.V., Kazakovtsev V.P. Metody rascheta parametrov dvizheniya spuskayemykh apparatov // Yestestvennyye i tekhnicheskiye nauki [Methods for calculating the motion parameters landing vehicles // Natural and Technical Sciences]. **9-10**(77). pp. 179-184, (2014).
9. Pichkhadze K.M., Vorontsov V.A., Zashchirinskiy A.M., Ponomarev V.A. Sistema spuska s orbity i avariynogo spaseniya na osnove naduvnogo tormoznogo ustroystva // Polet. Obshcherossiyskiy nauchno-tekhnicheskiy zhurnal [System descent from orbit and emergency rescue based on an inflatable braking device // Flight. All-Russian scientific and technical journal]. **8**. pp. 9-13, (2003).
10. Koryanov V.V. Issledovaniye dinamiki dvizheniya spuskayemogo apparata pri zhestkoy posadke na poverkhnost' planety // Polet. Obshcherossiyskiy nauchno-tekhnicheskiy zhurnal [Study of the dynamics of motion of the descent capsule during a hard landing on the planet's surface // Flight. All-Russian scientific and technical journal]. **1**. pp. 42-49, (2010).
11. Murphy C. H., "Symmetric Missile Dynamic Instabilities," U.S. Army Armament Research and

Development Command, Aberdeen Proving Ground, Md., Journal of Guidance and control, Vol. 4, № 5, Sept.-Oct. 1981, pp. 464-471; see also AIAA Paper 80-0320R, Jan. (1980).
12. Tolosko R. "Re-entry dynamics of a trimmed body with constant spin," – Journal of Spacecraft and Rockets, **8**.11, January 1971.
13. Koryanov V. Research of the dynamics motion of landing vehicle with inflatable braking device in the planet atmosphere // Proceedings of the International Astronautical Congress, IAC, **8**. pp. 5831-5836, (2013).
14. Aleksashkin S.N., Pichkhadze K.M., Finchenko V.S. Printsipy proyektirovaniya spuskayemykh v atmosferakh planet apparatov s naduvnymi tormoznymi ustroystvami // Vestnik FGUP NPO im. S.A. Lavochkina [Design principles in planetary atmospheres reentry vehicles with inflatable braking systems // Herald Federal State Unitary Enterprise Scientific and Production Association named after Lavochkin]. **2**. pp. 4-11, (2012).
15. Michele Iacovazzo, Valerio Carandente, Raffaele Savino, Gennaro Zuppardi. Longitudinal stability analysis of a suborbital re-entry demonstrator for a deployable capsule // Acta Astronautica. Volume 106, January–February, pp. 101–110, (2015).
16. Kazakovtsev V.P., Koryanov V.V., Danhe C. Dynamics of landing a spacecraft on a celestial body with small field of gravity // 2015 5TH International Workshop on Computer Science and Engineering: Information Processing and Control Engineering, WCSE 2015-IPCE (Information Processing and Control Engineering, Programme. 2015), pp. 191-195.

MATEC Web of Conferences 77, 03007 (2016) DOI: 10.1051/matecconf/20167703007
ICMMR 2016

Foil Bearing Stiffness Estimation with Pseudospectral Scheme

Balaji Sankar, S Jana, and Sadanand Kulkarni

National Test Facility for Rolling Element Bearings, Propulsion Division, CSIR- National Aerospace Laboratories, INDIA

Abstract. Compliant foil gas lubricated bearings are used for the support of light loads in the order of few kilograms at high speeds, in the order of 50,000 RPM. The stiffness of the foil bearings depends both on the stiffness of the compliant foil and on the lubricating gas film. The stiffness of the bearings plays a crucial role in the stable operation of the supported rotor over a range of speeds. This paper describes a numerical approach to estimate the stiffness of the bearings using pseudo spectral scheme. Methodology to obtain the stiffness of the foil bearing as a function of weight of the shaft is given and the results are presented.

1 Introduction

Oil free turbo machinery technology is one in which the rotor supports do not employ oil lubrication like the conventional rolling element bearings. Turbo compressors, turbo generators and aircraft air-cycle machines already employ these novel rotor support systems. Of these support systems, self-acting gas foil bearings are increasingly being used for many low load and high speed applications [1]. The structure of the gas foil bearings is given in Fig 1.

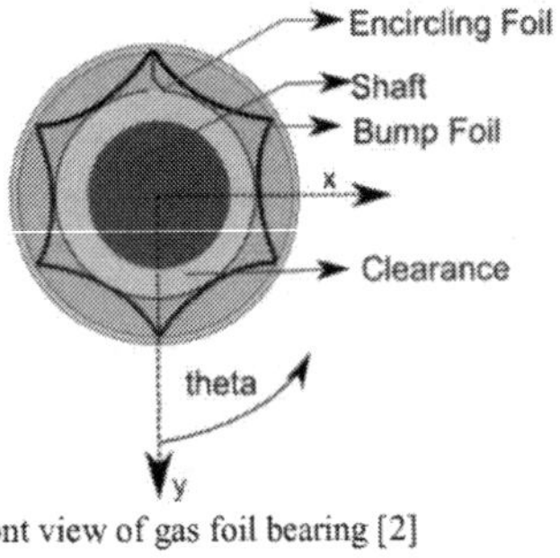

Figure 1. Front view of gas foil bearing [2]

In the gas foil bearing, the shaft is surrounded by the encircling foil. The encircling foil is supported by the bump foil. The bump foil rests on the firm foundation. There is a clearance between the rotating shaft and the encircling foil. Typically, the shaft is at an eccentric location in the bearing. As the shaft rotates, a convergent wedge is formed and the air is dragged into this wedge. As a result, there is a pressure increase in the convergent portion of the wedge. The shape of the edge depends on the location of the shaft in the bearing and on the deflection of the encircling foil due to the pressure force. When the force due to this pressure exceeds the shaft weight, the shaft levitates and rides on a film of air. More information regarding the structure and operation of the gas foil bearing is given in references [1-4].

Numerous publications have been made on the modelling of the pressure distribution in the foil bearing using multiple techniques. Bisson in Ref [5] gives an overview of conventional journal bearings and gas lubricated journal bearings. Ref [6] presents a comprehensive model covering the thermal energy transport in the bearing along with the rotor - bearing dynamics. Feng has presented an approach in which the bump foil of the bearing is modeled as a link spring structure in Ref [7]. A similar model has been used by Feng to study the behavior of the bearing in unbalanced conditions and unstable operation domains in Ref [8]. Piekos in Ref [9] has applied the pseudospectral scheme to journal bearings for micro gas turbines. This work and Ref [2, 4 and 10] are on the application of the scheme to compliant film gas foil bearing, where the deflection of the foil is also taken into account.

In this work, the gas foil bearing pressure distribution at a given operating condition is solved using the pseudo spectral scheme. By perturbing the eccentricity of the shaft and re-computing the pressure distribution, the change in force exerted by the bearing on the shaft is obtained. This change in force, when decomposed in x and y directions, gives us the kx and ky values of the bearing. The following section describes the governing equation and the solution procedure. The stiffness estimation for the chosen bearing configuration is presented next. The computed stiffness values are shown in the results section and concluded.

2 Governing equations

The simulation of the foil bearing involves iterative solution of the fluid model and structure model. The fluid model describes the pressure distribution in the fluid film surrounding the shaft. The structure model describes the deflection of the bump foil for a given pressure distribution in the encircling foil. An iterative solution is required because the computation of the pressure field depends on the film thickness, which in turn depends on the bump foil deformation. Detailed information on both the fluid and structure model is given in Ref [2]. A summary is given here for completeness of the paper.

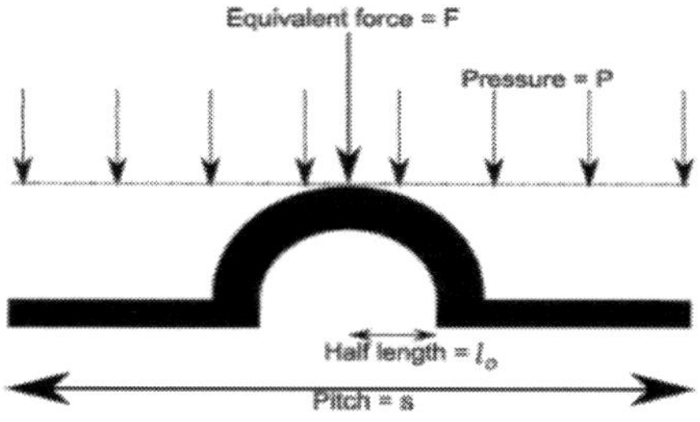

Figure 2. Pitch length of the bump foil

3 Reynolds' equation

The Reynolds' equation is given by

$$\frac{\partial}{\partial x}\left(ph^3\frac{\partial p}{\partial x}\right)+\frac{\partial}{\partial y}\left(ph^3\frac{\partial p}{\partial y}\right)=6\mu U\left\{\frac{\partial(ph)}{\partial x}\right\} \quad (1)$$

In this equation, p : pressure at a location in film, h : Film thickness at a point in the film, x : Tangential coordinate, y : Axial co-ordinate, μ : Viscosity of air and U : tangential surface speed of the shaft. Upon normalizing the quantities (same variables have been used for simplicity), the non-dimensional Reynolds' equation is given by

$$\frac{\partial}{\partial x}\left(ph^3\frac{\partial p}{\partial x}\right)+\left(\frac{1}{(L/D)^2}\right)\frac{\partial}{\partial y}\left(ph^3\frac{\partial p}{\partial y}\right)=\Lambda\frac{\partial(ph)}{\partial x} \quad (2)$$

where c is the initial clearance, R is the radius of the bearing, pa is the ambient pressure and

$$\text{Bearing Number} = \Lambda = \left(\frac{6UR\mu}{C^2P_A}\right)$$

Ambient pressure has been used for normalizing pressure, half length of the bearing for axial co-ordinate, radius of the bearing for the tangential co-ordinate and initial clearance of the bearing for film height.

4 Structure model

The structure model gives the film thickness distribution in the bearing for a given shaft location and pressure distribution. It is adapted from [11]. Multiple models describing the deflection of the bump foil structure are available. This model is chosen due to its simplicity. The non-dimensional film thickness in the foil bearing is given by

$$h = 1 + e * \cos(\theta - \phi_o) + \alpha(p - 1) \quad (3)$$

The non-dimensional compliance used in the equation is

$$\alpha = \frac{s * p_a}{H * \frac{Et^3}{12(1-\nu^2)l_o^3}}$$

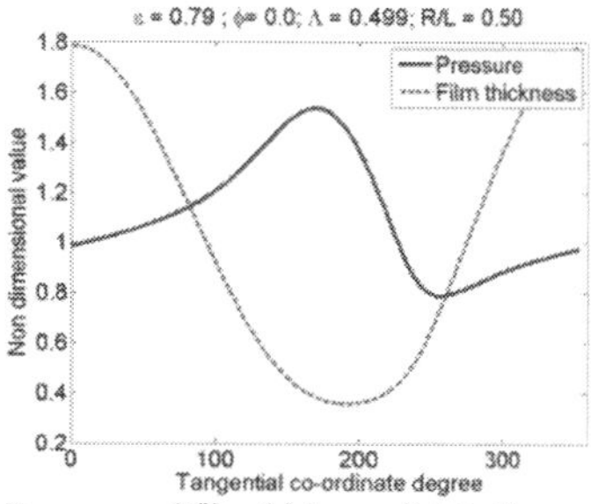

Figure 3 Pressure and film thickness distribution

where, t is the thickness of the bump foil, s is the pitch length of the bump foil, lo is the half length of the bump foil, l_o is the Poisson's ratio and E is the Young's modulus of the material. The pitch length and half-length are shown in Fig. 2.

5 Solution procedure

Detailed procedure on using the pseudo spectral scheme to solve the loosely coupled system of equations presented here is given in Ref[10]. A brief summary is given here. An initial pressure distribution is first assumed and the film thickness is calculated for this pressure distribution. The dependent variables are expressed as a series, in terms of Fourier series in tangential direction and as Chebyshev's series in the axial direction. Writing the dependents in the form of these series, enables us to estimate their derivatives using global approximations. On the contrary, using finite difference techniques to estimate the derivatives uses local approximation. Once spatial derivatives are estimated using series expansions, a time marching scheme is used to estimate the pressure distribution, while fixing the shaft location at a given eccentricity and attitude angle. The time marching is carried with recursive solving of the pressure and film thickness distribution until a steady state solution of the pressure distribution is reached. The same validated code is used for stiffness estimation in this work.

6 Stiffness estimation

Consider a bearing number of Λ = 0:4985, radius of 19.05 mm and $\frac{R}{L}=\frac{1}{2}$. For this configuration, the non-dimensional pressure and film thickness distribution are shown in Fig. 3. To visualize the pressure distribution and forces better, the plot is made in polar co-ordinates for the same configuration in Fig. 4. In this plot, the blue circle represents the location of the shaft and arrow mark shows the direction of the force and the magnitude. As the attitude increases from 0 degrees, the force increasingly becomes vertical and at a certain attitude (approximately 47 degrees), the force becomes fully vertical and supports the weight of the shaft without any horizontal force. Fig. 5 shows variation of forces with attitude angle. Only when the vertical force is equal to shaft weight and there is no horizontal force, stationary position of the shaft is theoretically possible.

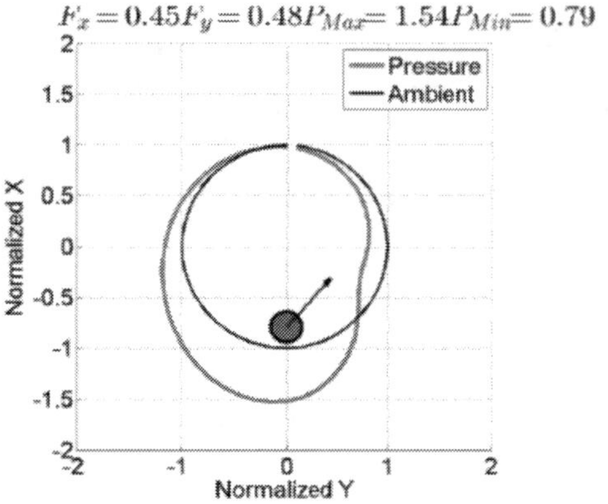

Figure 4. Polar plot of pressure distribution, ϕ= 0 degrees, ε= 0.79

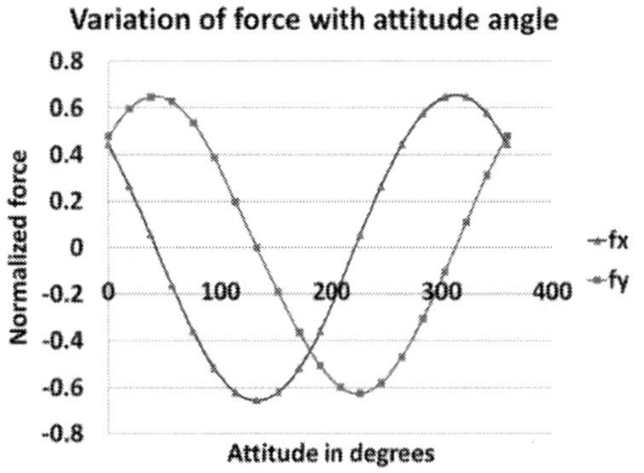

Figure 5. Variation of forces with attitude angle

Table 1. Attitude and eccentricity for different loads

S No	load (N)	Epsilon	Phi
1	9.9	0.19	72.2
2	15	0.27	69.8
3	20	0.35	67.2
4	24.9	0.43	64
5	29.9	0.51	60.6
6	35	0.59	56.6
7	39.9	0.66	52.6
8	45	0.73	48.1
9	49.9	0.79	44

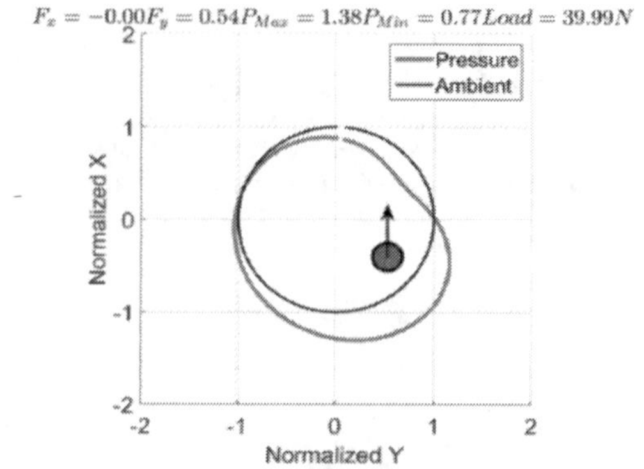

Figure 6. Balanced location for 40 N load

In this work the stiffness calculation is done for a range of load values at various eccentricities and attitude angles. In order to estimate the stiffness at a particular load, the eccentricity and attitude angle for that load are found first. The eccentricity and attitude angles for various loads are shown in table I and shown pictorially for a few load levels in Fig 6, and 7. After estimating the eccentricity and attitude for a given load, the eccentricity value is slightly increased by 1% while keeping the attitude same. This results in a change of x coordinate of the shaft, y co-ordinate of the shaft and the forces in the x and y direction. From the changes in co-ordinates of the shaft and changes in force, the 4 stiffness values are calculated as given in Table II.

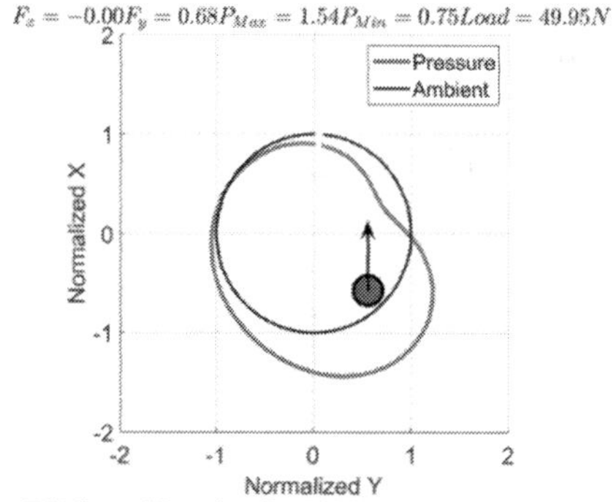

Figure 7 Balanced location for 50 Newtons

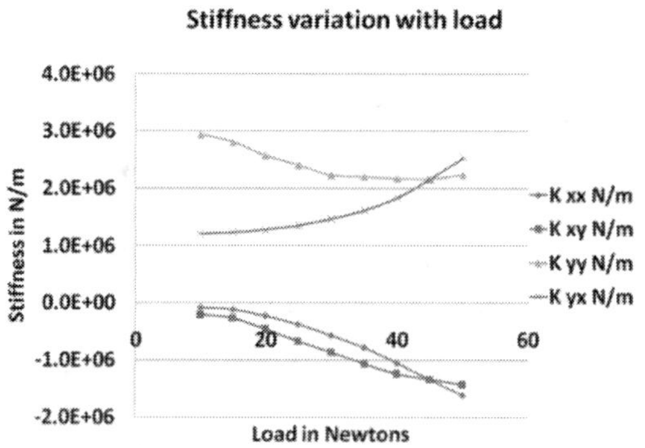

Figure 8 Stiffness variation with load

Table 2. Stiffness calculation

Epsilon	Fy	DFX	DFY	X	Y	DX	DY	K xx	K xy	K yy	K yx
-	N	Newton	Newton	micron	micron	micron	micron	N/m	N/m	N/m	N/m
0.18	10	-7.35e-3	1.03E-01	8.30E-06	3.54e-06	8.50e-8	3.50e-8	-8.65e+04	-2.10e+05	2.94e+6	1.21e+6
0.27	15	-1.47e-2	1.54e-1	1.23e-05	5.62e-06	1.25e-7	5.50e-8	-1.18e+05	-2.67e+05	2.81e+6	1.24e+6
0.36	20	-3.68e-2	2.06e-1	1.60e-05	8.05e-06	1.60e-7	8.00e-8	-2.30e+05	-4.60e+05	2.57e+6	1.29e+6
0.44	25	-7.35e-2	2.65e-1	1.93e-05	1.09e-05	1.95e-7	1.10e-7	-3.77e+05	-6.69e+05	2.41e+6	1.36e+6
0.53	30	-1.25e-1	3.24e-1	2.21e-05	1.43e-05	2.20e-7	1.45e-7	-5.68e+05	-8.62e+05	2.23e+6	1.47e+6
0.61	35	-1.91e-1	3.97e-1	2.43e-05	1.81e-05	2.45e-7	1.80e-7	-7.80e+05	-1.06e+6	2.21e+6	1.62e+6
0.68	40	-2.72e-1	4.78e-1	2.58e-05	2.22e-05	2.60e-7	2.20e-7	-1.05e+6	-1.24e+6	2.17e+6	1.84e+6
0.75	45	-3.53e-1	5.74e-1	2.66e-05	2.64e-05	2.65e-7	2.65e-7	-1.33e+6	-1.33e+6	2.16e+6	2.16e+6
0.81	50	-4.34e-1	6.84e-1	2.69e-05	3.04e-05	2.70e-7	3.05e-7	-1.61e+6	-1.42e+6	2.24e+6	2.53e+6

7 Results

The variation of the 4 stiffness components as the weight of the shaft is increase is shown in Fig. 8.The negative stiffness in the x direction is due to the increased suction pressure to the right side of the shaft, as the gap between the shaft and encircling foil decreases. Increase in load causes higher eccentricity of the shaft, as given in table I. It is also seen that absolute value of the stiffness of all 4 components increase with increase in load. This increase in stiffness with increase in load alters the natural frequencies of the system. This change in natural frequencies must be accounted for while designing the system for the given operating condition.

References

1. C. DellaCorte. Tribological limitations in gas turbine engines a workshop to identify the challenges and set future directions. Technical report,NASA, (2000).
2. BalajiSankar. Numerical simulation and experimental validation of compliant surface gas film bearings for high speed applications, Master's thesis, Academy of Scientific and Innovative Research, (2012).
3. A.Cameron. Basic Lubrication Theory.Longman, (1971).
4. Balaji Sankar and SadanandKulkarni.Non linear time dependent simulations of high speed gas lubricated bearings-an object oriented model,presented in International conference on advanced research in mechanical engineering, ICARME-2012, (2012).
5. Edmond E Bisson and William J Anderson. Advanced bearing technology, Nasa sp-38. NASA Special Publication, **38**, (1964).
6. Luis San Andreas, Thermohydrodynamic analysis of bump type gas foil bearings: A model anchored to test data. Journal of Engineering for Gas Turbines and Power, 132:042504–1 to042504–10, (2010).
7. Kai Feng. Analytical model of bump-type foil bearings using a linkspring structure and a finite-element shell model. Journal of Tribology, 132:0210706–1 to 0210706–11, (2010).
8. Kai Feng. Parametric studies on static performance and nonlinear instability of bump-type foil bearings. Journal of System Design and Dynamics, **6**, (2010).
9. Edward Stanley Piekos.Numerical Simulation of Gas-Lubricated Journal Bearings for Microfabricated Machines.PhD thesis, MIT, (2000).
10. BalajiSankar and SadanandKulkarni.Modeling of gas lubricated compliant foil bearings using pseudo spectral scheme. Tribology Online, **10(5)**, pp. 295–305, (2015).
11. Z.C. Peng and M.M. Khonsari, Hydrodynamic analysis of compliant foil bearings with compressible air flow. Transactions of the ASME, 126:542–546, (2004).

MATEC Web of Conferences 77, 04001 (2016)
ICMMR 2016
DOI: 10.1051/matecconf/20167704001

A Remote Monitoring System for Greenhouse Based on the Internet of Things

Zhenfeng Xu, Junjie Chen, Yahui Wang and Zhen Fan

School of Instrument Science and Engineering, Southeast University, China

Abstract. The Internet of Things (IOT) is considered as a great opportunity for the development in the information field nowadays, and has been applied widely in many fields. The IOT can be applied to monitor and control the microclimate factors of greenhouse remotely. In this paper, a wireless monitoring network is designed in the perception layer of the IOT. The nodes are developed based on the Mica2 hardware and the TinyOS software. The LPL (low power listening) technology is adopted to reduce the energy consumption of the relay node which is powered by a solar panel. The ACK (Acknowledgement) mechanism is used in the software to improve the quality of wireless communications. A remote monitoring terminal is developed by using Java technology. The monitoring terminal is easy to operate with good interactivity. The system has been installed in a glass greenhouse. The actual operation results show that the system is stable and reliable, which lays a good foundation for the development of remote control strategies in future.

1 Introduction

As an important component of protected agriculture, the greenhouse cultivation can achieve efficient production. In China, the average cultivated area per person is about only 40% of that in the world [1], so it is an effective way to alleviate the conflict between more people and less arable land area by promoting greenhouse cultivation techniques. In recent 30 years, the greenhouse cultivation in China has achieved great development and the area of greenhouse cultivation ranks first in the world. However, the automation of greenhouse monitoring and control system is very poor, which leads to a waste of time and human resource. The Internet of Things (IOT) is considered as a great opportunity for the development in the information field. In recent years, the IOT has been being applied widely in many fields, such as logistics, industrial environmental monitoring and smart grid, etc [2-4]. The IOT will also be helpful for improving the automation of greenhouse monitoring and control system. Consequently, we have decided to start a big project on the remote monitoring and control system for greenhouse based on the IOT. The paper is a part of this project. In this paper, we introduce the remote monitoring system, which includes a wireless monitoring network and a remote monitoring terminal.

The paper is organized as follows. In section 2, the architecture of remote monitoring system for greenhouse based on the IOT is described. In section 3, a wireless monitoring network is designed. The LPL (low power listening) technology is adopted to reduce the energy consumption of the relay node which is powered by a solar panel. The ACK (Acknowledgement) mechanism is adopted to improve the wireless communications quality. In section 4, a remote monitoring terminal is developed based on Java technology. In section 5, the system is installed in a glass greenhouse successfully. At last, the paper is concluded.

2 System architecture

The architecture of remote monitoring and control system for greenhouse based on the IOT contains three levels, i.e., perceptual layer, backbone transport layer and application layer, as shown in Fig. 1. Wireless sensor networks are usually adopted in the perceptual layer. The function of sensor nodes is to monitor environmental factors of greenhouse, such as indoor air temperature and relative humidity; while the function of controller nodes is to control the facilities, such as fans, windows. All data acquired by sensor nodes will be gathered by a base station node, which is connected with a local monitoring terminal. The function of the local terminal is to process and store the information of greenhouse environmental factors and facilities, and to upload the information to the backbone transport layer and receive control commands from it. The backbone transport layer is the middle layer, and is usually composed of 2G/3G networks, the Internet, Cloud computing service platforms, etc. Its function is to achieve the remote data transmission and sharing. The application layer is the top layer, and it includes various display devices connected with the backbone transport layer, such as PDAs, PCs, remote display terminals, etc. Growers can view the information of environmental

 DOI: 10.1051/matecconf/20167704001

factors of greenhouse and control the facilities remotely through the application layer.

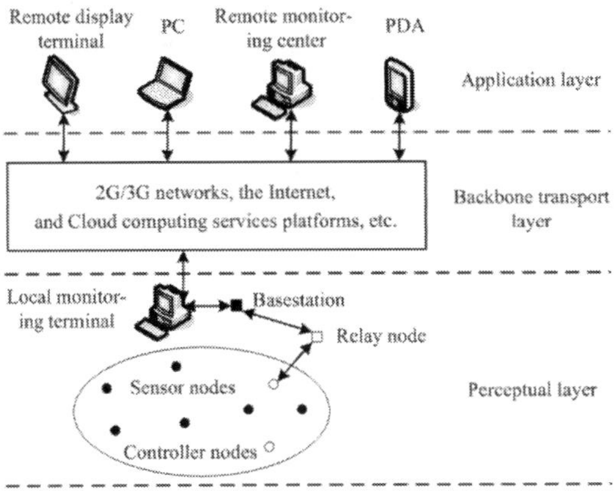

Figure 1. The architecture of remote monitoring and control system for greenhouse based on the IOT.

In the backbone transport layer of the IOT, the mature communication networks are adopted, so the researchers can pay more attention to develop the perceptual layer and the application layer. In the perceptual layer, wireless sensor networks are used more and more widely [5-7]. Compared with the wired network, the wireless network owns many advantages, such as convenient layout, easy extension of nodes, low cost of installation, etc. However, the wireless communication quality is much poorer than that of wired communication. Therefore, the researchers should consider the reliability of wireless communication properly. In section 3, the wireless monitoring network developed in our project is introduced. In section 4, we will introduce our remote monitoring terminal in the application layer, although there are some monitoring terminals reported [8, 9].

3 Wireless monitoring network

Our V-type glass greenhouse is 60m long and 30m wide. The environmental factors needed to be measured are the air temperature, relative humidity and solar illumination both inside and outside the greenhouse. The purpose of measuring outside environmental factors is to provide the basis for the future development of the control strategy. A wireless monitoring network is designed. The distribution of nodes is shown in Fig. 2. There are four sensor nodes with three placed inside and one placed outside. As we have not yet developed suitable control strategy, the controller node here is just used as a relay node in this paper. The base station node and the local terminal are placed in a monitoring room. As the distance between the greenhouse and the monitoring room is too long, about 300m, a relay node is designed. The base station is wired with the local monitoring terminal via RS-232 serial line. The local monitoring terminal is an industrial personal computer connected with the Internet.

The Mica2 hardware with CC1000 as the wireless communication device is adopted to design the wireless nodes. The working frequency of the wireless monitoring system is set to 433MHz, which has better diffraction transmission capacity than 2.4GHz. The digital sensor SHT11 is used to measure air temperature and relative humidity; while TSL2561 sensor is used to measure solar illumination intensity. The four sensor nodes and the controller node are all powered by electricity, which is convenient in greenhouse. So the power consumption of these nodes is not necessary to be considered. However, the relay node is powered by a solar panel. Therefore, it's important to reduce the energy consumption of the relay node as much as possible.

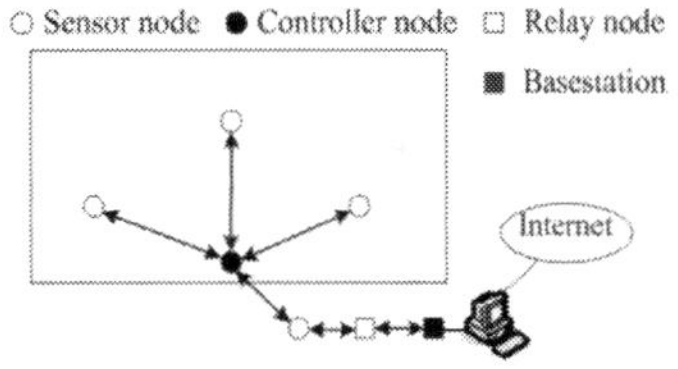

Figure 2. The wireless monitoring and control network for greenhouse.

The LPL (low power listening) technology is adopted in the software development of the relay node, in order to reduce its power consumption. The LPL technology is an asynchronous MAC protocol used for reducing the power consumption of wireless communication nodes [10]. During the most time of one period, the wireless module is turned off and only turned on for a very short time to detect RF signals. If there are RF signals, the wireless module keeps active state; otherwise returns to sleep state. The duty cycle of the active state can be reduced to below 1%, so the LPL technology is effective to reduce the power consumption of wireless nodes. The node software is developed in the TinyOS operating system, which is a special operating system for wireless sensor networks. The LPL command *setLocalWakeupInterval(T)* in the component *LowPowerListening* is called to realize the low power listening function.

In order to improve wireless communication quality, the ACK (Acknowledgement) mechanism is adopted [11]. The basic process of data transmission by using the ACK mechanism is as follows. When the transmitting node sends a data packet, if the receiving node receives the data packet, it will send back an ACK message. If the transmitting node receives the ACK message, it will stop to send the data packet repeatedly; otherwise it will continue to send the data packet until it receives the ACK message or the sending times reach the maximum set threshold. The ACK mechanism is an effective way to improve wireless communication quality. Assume that the packet loss rate of wireless transmission is 50% and the maximum threshold of sending data packet is set to ten. It's easy to know that the packet loss rate can be reduced to below 0.1%. Therefore, the usage of ACK mechanism is an effective way to improve the wireless communications quality. In the TinyOS operating system, the port *PacketAcknowledgements* in the communication component *ActiveMessageC* provides the ACK request command. When a transmitting node sends a packet, the corresponding event *sendDone* will be triggered and the transmitting node will check whether the ACK message

 DOI: 10.1051/matecconf/20167704001

is received in the *sendDone* event. If the transmitting node does not receive the ACK message, it will send the data message repeatedly.

In order to improve the anti interference ability of the sensor nodes, the software watchdog is adopted. When the microprocessor is ready to reads the data from sensors, the software watchdog is started. A long enough time is set for the watchdog to guarantee the data in sensors can be read during this time. After reading, the watchdog is closed. If the microprocessor can't read normally during this time, it resets automatically. Therefore, the sensor nodes have strong anti-interference ability.

4 Remote monitoring terminal

A remote monitoring terminal is developed by using Java technology for growers to view the environmental information and the operation states of facilities in the greenhouse on remote display terminals. The monitoring terminal software includes three parts, i.e., server, database and client, shown in Fig. 3. The functions of server are to achieve the data exchange with the base station, and store the data into the database. The functions of database are to storage and backup all information of the system. The functions of client are to display the environmental information and the operation states of facilities, as well as achieve human-computer interaction. Tomcat 6.0 is adopted as the web server and Microsoft SQL Server 2005 as the database.

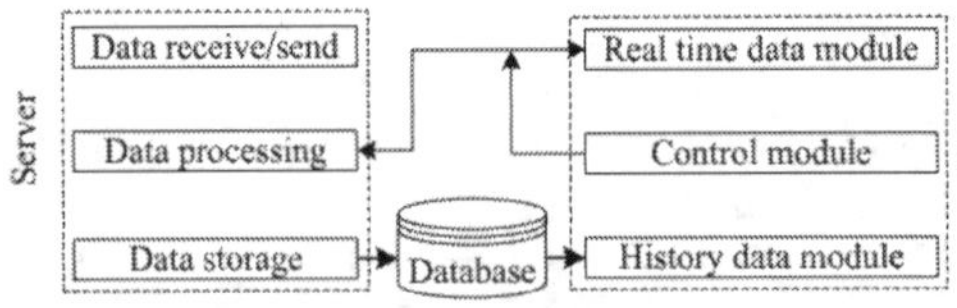

Figure 3. The overall structure of monitoring software.

The remote monitoring terminal software is installed in the local monitoring computer which is connected with the Internet. The authorized growers can use any display device connected with the Internet to view the relevant information about the greenhouse by visiting the web site of this system. The main interface of remote monitoring terminal is shown in Fig. 4. The environmental data acquired by the four sensors are all displayed in the main interface. The operation states of the facilities such as fans, humidifier and wet pad are also given at the bottom of the main interface. They are several function selection bars on the left side. In the curve fitting bar, the data of environmental factors is displayed in the form of curves. In the historical data bar, the growers can query historical data of environmental factors. In the monitoring mode bar, there are two modes that can be chosen, i.e., manual mode and automatic mode. The upper and lower limits of indoor air temperature and relative humidity can be set in this bar, as the suitable ranges of temperature and relative humidity are different for different kinds of crops. The intelligent decision bar is reserved for the development of control strategies in future.

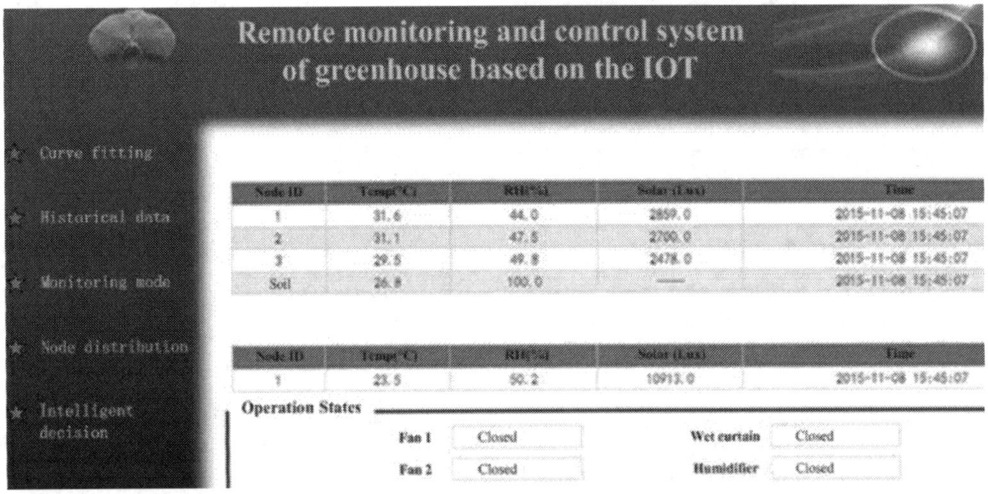

Figure 4. The main interface of the remote monitoring terminal.

5 Field experiment

The remote monitoring system has been installed in our greenhouse located in the suburbs of Nanjing, Jiangsu province, China. Some nodes installed in the field are shown in Fig. 5. The outdoor sensor node is installed near the greenhouse, so it is powered by electricity. Only the relay node is powered by a solar panel. The system works very well until now.

After login of the system web site through remote terminals, the users can view the main display interface just as shown in Fig. 4. After clicking the curve fitting bar, both indoor and outdoor environmental factors can be selected to display in curve form. Only the indoor air temperature is shown in Fig. 6, on Nov. 6, 2015, 8:00-18:00. The sampling period of the system is ten minutes. It can be seen that there is no data packet lost, which indicates that the ACK mechanism can guarantee good wireless communication quality. The continuous work of the relay node indicates the LPL technology is useful for saving power and it's suitable for wireless nodes powered by solar panels or batteries. There is a short delay of data transmission by using the LPL technology, and the delay is closely related to the dormancy period. However, the monitoring system of greenhouse is not strict with high real-time performance and the dormancy period is usually less than one minute, so the short delay does not affect the performance of the monitoring system.

(a) Internal sensor node

(b) External sensor node

(c) Relay node

(d) Controller node

Figure 5. Some nodes installed in the field.

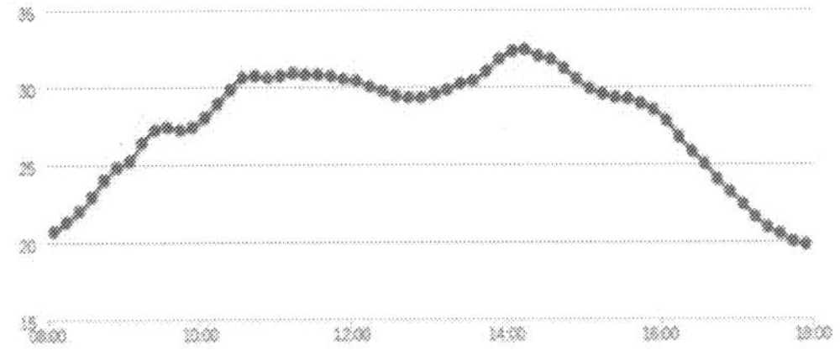

Figure 6. The greenhouse air temperature on Nov. 6, 2015, 8:00-18:00 (Abscissa: time; Ordinate: temperature/°C).

6 Discussion and the further research

The remote monitoring system of greenhouse based on the IOT is designed. The good wireless communication quality is guaranteed by using the ACK mechanism. The relay node can run continuously with very low power by using the LPL technology. The anti-interference ability of sensor nodes is improved by using the software watchdog. The remote monitoring terminal is easy to operate with good interactivity. The field experiment shows that the system is very stable and reliable, which lays a good foundation for the future research about remote control strategies.

It's well known that the heating and active cooling both consume a lot of energy, so the energy consumption accounts for a large proportion of the production cost of greenhouse. It's a challenging research to save energy as much as possible on the premise that the air temperature and relative humidity are kept in suitable ranges [12]. The energy-saving methods of control strategies are the emphases of our research in future. The structures and sizes of various greenhouses are very different, so are the facilities in greenhouse. Therefore, we will also consider the universality of new control strategies at the same time.

Acknowledgements

This work is supported by National Science-technology Support Plan Project (2014BAD08B03), Science & technology Support Plan Project of Jiangsu Province (BE2013729, BE2014312) and Science & technology special fund of North Jiangsu Province (BN2014085).

References

1. Z. Li, G. Wang, F. Qi. Current situation and thinking of development of protected agriculture in China. Chinese Agricultural Mechanization **1**: 7-10 (2012).
2. L. Atzori, A. Iera, G. Morabito. The Internet of Things: A survey. Comput. Netw. **54(15)**: 2787-2805 (2010).
3. D. Miorandi, S. Sicari, F.D. Pellegrini et al. Internet of things: Vision, applications and research challenges. Ad Hoc Netw. **10(7)**: 1497-1516 (2012).
4. Q. Sun, J. Liu, S. Li, et al. Internet of Things: Summarize on Concepts, Architecture and Key Technology Problem. Journal of Beijing University of Posts and Telecommunications **33(3)**: 1-9 (2010).
5. O. Mirabella, M. Brischetto. A hybrid wired/wireless networking infrastructure for greenhouse management. Instrumentation and Measurement, IEEE Transactions on **60(2)**: 398-407 (2011).
6. A. Han, X. Guo, Y. Sun. Event-driven scheduler in monitoring and controlling system for greenhouse based on wireless monitoring networks. Transactions of the Chinese Society for Agricultural Machinery **41(7)**: 186-190, 204 (2010).
7. R. Zhang, K. Li, B. Bai, et al. Effect of packet loss probability of transmission on greenhouse WSN measurement and control system. Journal of jiangsu university **30(4)**: 383-386 (2009).
8. M.T. Maliappis, K.P. Ferentinos, H.C. Passam, et al. Gims: A Web-based Greenhouse Intelligent Management System. World Journal of Agricultural Sciences **4(5)**: 640-647 (2008).
9. J. Lu, L. Qin, C. Shi, et al. Design and implementation of modern greenhouse remote monitoring system based on Web. Proceedings of the 32th Chinese Control Conference (Xi'an, 2013).
10. S. Hu, J. Wang. Smart low power listening for wireless sensor networks. Journal on Communications **30(9)**: 95-101 (2009).
11. G.W. Lee, N.T. Dung, E.N. Huh. Reliable transmission on wireless sensor networks with delegated acknowledgement. Proceedings of the 4th International Conference on Ubiquitous Information Management and Communication (Suwon, 2010).
12. P.J.M.V. Beveren, J. Bontsema, G.V. Straten, et al. Minimal heating and cooling in a modern rose greenhouse. Appl. Energ. **137** : 97-109 (2015).

DOI: 10.1051/matecconf/20167704002

Developing a STEP-Compliant Database System in Supporting Product Life Cycle Applications

Yusri Yusof[1], Chen Wong Keong[2]

[1]Department of Manufacturing and Industry Engineering, Faculty Mechanical & Manufacturing Engineering, Universiti Tun Hussein Onn Malaysia, Parit Raja, Johor, 86400, Malaysia
[2]Department of Mechanical Engineering, Polytechnic Kuching Sarawak, Kuching, Sarawak, 93050 Malaysia

Abstract. This paper presents details of developing a STEP-compliant database system which has the ability to support the implementation of product life cycle applications. Currently, most of CAD systems have the ability to generate geometrical and topological data in STEP file format. This developed STEP-compliant database management system has shown its ability to convert this generated STEP file format to database format, which is more suitable for the downstream of manufacturing applications, especially for process planning and machining processes based on STEP.

1 Introduction

STEP, the Standard for the Exchange of Product Model Data, is a large and powerful set of ISO (International Organisation for Standardisation) standards, all under ISO 10303. The overall objective of STEP is to provide a mechanism that describes a complete and unambiguous product definition throughout the life cycle of product. STEP provides both broadly useful data modelling methods and data models focused on specific industrial uses. The STEP standards contain several dozen separate documents [1]. Although EXPRESS is a powerful object-oriented data model descriptive language and independent of any platform. However, it is a kind of descriptive language rather than a programming language. This brings difficulty to implement the EXPRESS data model on a computer and it also undoubtedly brings a big problem with the STEP data model processing [2].

2 Translation methods

STEP database systems have been built by various research groups and commercial software companies. They can be categorized according to the type of the DBMS they used and mapping methods [3]:

1. with respect to the backend DBMS, such as relational DBMS, object-oriented DBMS, and in-house persistent data storage system
2. with respect to the mapping methods, such as mapping each STEP entity (class) into corresponding database tables-all entity attributes are stored in a same table as tuples or mapping each STEP data type into tables.
3. with respect to the programming language binding style.

3 Feature recognition

Many alternative techniques have been devised for creating feature-based models. These can be divided into two main categories based on whether features are extracted from geometry (FfG), or whether geometry is created from the feature (GfF). Currently, [4] developed a blend faces and holes recognition algorithm for detecting blend faces first, and then search for blend faces and holes in accordance with the loop data in the B-rep model. Meanwhile, [5] developed a method to search for all kinds of inner loops on a CAD model, including the ones within a face and those across multiple faces. This is because a feature locating on multiple surfaces is quite common in real CAD models, but it has not been received extensive study yet [5]. A new methodology was proposed by [6] to recognize manufacturing features of prismatic parts from STEP AP203 file. In an effort to use of database optimization technologies to feature recognition in CAD, [7] showed that a declarative feature definition can be turned into an SQL query, and database engine coupled to a CAD modeler can be used to find instances of entities satisfying the predicates which make up features. Besides that, [8] developed an automatic feature recognition system by using object oriented approach.

In this paper, STEP geometrical and topological file was used as an input for the generation of part geometrical database and which isuseful for subsequent product life cycle applications, especially used for machining feature recognition.

DOI: 10.1051/matecconf/20167704002

4 Proposed framework

Fig. 1 shows the proposed project framework with developing a relational database management system which could import geometrical and topological product data in Part-21 file from CAD. From text file-part21, EXPRESS model is mapped to relational database product model format. Direct mapping is a powerful way of translating entity and its attributes to the relational database format.

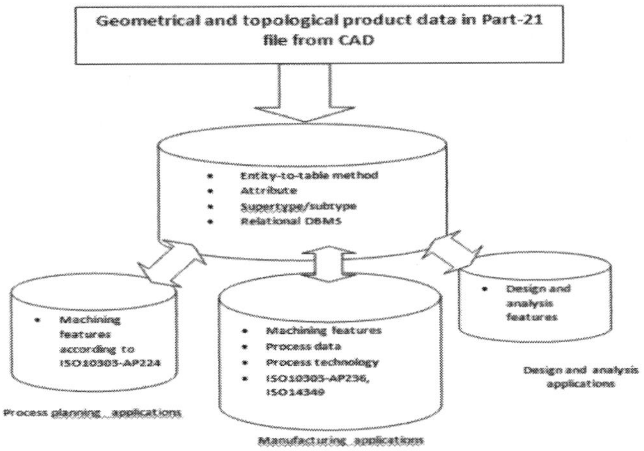

Figure 1. Proposed Project Framework

5 Implementation

Mapping EXPRESS model to relational model. Schema mapping method for object oriented to relational has several variations. Mapping alternative (R1) separates attribute to each entity type. (R1) makes largest number of mapping relations among the alternatives considered, which makes hard to manage all the relations efficiently [3]. Alternative (R2) duplicates all the supertype attributes in the subtype. (R2) separates attributes by entity type, so one relation carries an entity type and its attribute values [3]. Alternative (R3) includes every inherited entity type in one relation. The jobtype attribute is used to specify entity type. (R4) is based on attribute type rather than entity type, which is quite distinctive from other alternatives. Number of relation is reduced to the number of basic attribute types, which is relatively quite small to number of entity type [3].

This proposed relational database management system uses entity-to-table method (R1) rather than type-to-table method (R4). The two mapping methods have their own strengths and drawbacks. However, we choose the first method for our system for the following reasons: entity-to-table mapping is considered simpler to implement because an entity and its attribute can be mapped into a corresponding table; in contrast, in type-to-table mapping entity and its attributes are stored in different tables so that additional mapping rules have to be specified to assemble dispersed attributes into an entity or vice-versa.

Translation of the EXPRESS entity attributes. Translation of EXPRESS attributes are done by one-to-one mapping, as most EXPRESS attributes types including defined types are reduced to basic data types. However, translation of aggregation types requires more translation steps than one-to-one mapping.

EXPRESS Aggregation types such as arrays and lists cannot be stored in one tuple because traditional data models don't have such data types. If the class has a fixed number of elements (e.g. an array), they can be translated as a series of attributes in a tuple. If the number of elements is indefinite (e.g. a list), however, we cannot store them in a single tuple and a separate table is required. In addition, EXPRESS aggregation class can also be nested indefinitely. This proposed database system translates aggregation classes as table external to their own entity.

Querying instance data from RDBMS. Query instance data is to access data stored in the RDBMS database. SQL (Structured Query Language) is a standard DML (Direct Manipulation Language) in relational database. However, the user should be familiar with the schema mapping and the schema itself to manipulate the database. Through this access, the user will have fast access and total control over the database. The other way to access database is to use system-generated SQL. System generate SQL for a specific instance type as well as the data on the fly. Response time is slower than user-generated SQL but the user does not need the knowledge regarding schema mapping to database.

6 Case study

In this project, STEP file is created from CAD system Autodesk Inventor. At first, a design part is created on Autodesk Inventer, as shown in Fig. 3 and then save the file as *.stp,. Then, the STEP AP203 text file in the form of B-Rep can be generated in Microsoft Notepad as shown in Fig. 4.

Implementation procedures are shown in the following steps.

Step 1: Design a part using CAD (Autodesk Inventor) and save the file as STEP file (*.stp) as shown in figure 3.

Step 2: Open the STEP file with Notepad and save as text file (*.txt) as figure 4. Copy and paste the text file into the RDBMS

Step 3: Import the data into the created database system

Step 4: Mapping EXPRESS model to relational model

Read each line of character from the import STEP file

Extract each line of geometry and topological data entity and then save them into different types of entity tables with entity-to-table method (e.g. advanced-face table, edge loop table, face table, edge table, point table and so on)

Step 5: Display all geometry and topological data in different entity tables, for example, as shown in Fig. 5.

Figure 3. Part design

```
example part iso14649 part12 hole1.txt - Notepad
File  Edit  Format  View  Help
 * Generated by software with PDE/Lib inside                          *
 * PDElib Version v51a, created Thu 01/14/2010                        *
 * ITI Transcendata (http://www.transcendata.com/)                    *
 *********************************************************************/
FILE_DESCRIPTION((''),'2;1');
FILE_NAME('C:\\Users\\User\\Desktop\\4pocket.stp','2015-12-01T15:28:30',('User'),(''),'Autodes
FILE_SCHEMA(('AUTOMOTIVE_DESIGN { 1 0 10303 214 1 1 1 1 }'));
ENDSEC;
DATA;
#5=APPLICATION_CONTEXT('automotive design');
#6=APPLICATION_PROTOCOL_DEFINITION('International Standard','automotive_design',2001,#5);
#7=PRODUCT_CONTEXT('',#5,'mechanical');
#8=PRODUCT('Part1','Part1',$,(#7));
#9=PRODUCT_RELATED_PRODUCT_CATEGORY('part',$,(#8));
#10=PRODUCT_DEFINITION_FORMATION('',$,#8);
#11=PRODUCT_DEFINITION_CONTEXT('part definition',#5,'design');
#12=PRODUCT_DEFINITION('',$,#10,#11);
#18=(NAMED_UNIT(*)PLANE_ANGLE_UNIT()SI_UNIT($,.RADIAN.));
#19=DIMENSIONAL_EXPONENTS(0.0,0.0,0.0,0.0,0.0,0.0,0.0);
#20=PLANE_ANGLE_MEASURE_WITH_UNIT(PLANE_ANGLE_MEASURE(0.017453292500000),#18);
#24=(CONVERSION_BASED_UNIT('DEGREE',#20)NAMED_UNIT(#19)PLANE_ANGLE_UNIT());
#28=(NAMED_UNIT(*)SI_UNIT($,.STERADIAN.)SOLID_ANGLE_UNIT());
#32=(LENGTH_UNIT()NAMED_UNIT(*)SI_UNIT(.MILLI.,.METRE.));
#34=UNCERTAINTY_MEASURE_WITH_UNIT(LENGTH_MEASURE(0.010000000000000),#32,'DISTANCE_ACCURACY_VAL
#36=(GEOMETRIC_REPRESENTATION_CONTEXT(3)GLOBAL_UNCERTAINTY_ASSIGNED_CONTEXT((#34))GLOBAL_UNIT_
#37=AXIS2_PLACEMENT_3D('',#38,#39,#40);
#38=CARTESIAN_POINT('',(0.0,0.0,0.0));
#39=DIRECTION('',(0.0,0.0,1.0));
#40=DIRECTION('',(1.0,0.0,0.0));
#41=SHAPE_REPRESENTATION('',(#37),#36);
#42=PRODUCT_DEFINITION_SHAPE('','',#12);
#43=SHAPE_DEFINITION_REPRESENTATION(#42,#41);
#49=(NAMED_UNIT(*)PLANE_ANGLE_UNIT()SI_UNIT($,.RADIAN.));
#50=DIMENSIONAL EXPONENTS(0.0,0.0,0.0,0.0,0.0,0.0,0.0):
```

Figure 4. Part-21 text file

qry_ADVANCED_FACE

Field1	database format.Field2	Field3	Field4	Field5	Field6	Field7	Field8	Field9
#109	ADVANCED_FACE	''		#108		#72	.F.	
#140	ADVANCED_FACE	''		#139		#114	.F.	
#173	ADVANCED_FACE	''		#172		#145	.F.	
#204	ADVANCED_FACE	''		#203		#178	.F.	
#396	ADVANCED_FACE	''		#395		#370	.F.	
#237	ADVANCED_FACE	''		#236		#209	.F.	
#429	ADVANCED_FACE	''		#428		#401	.F.	
#460	ADVANCED_FACE	''		#459		#434	.F.	
#502	ADVANCED_FACE	''		#501		#465	.F.	
#524	ADVANCED_FACE	''		#523		#507	.F.	
#557	ADVANCED_FACE	''		#556		#529	.F.	
#579	ADVANCED_FACE	''		#578		#562	.F.	
#621	ADVANCED_FACE	''		#620		#584	.F.	
#268	ADVANCED_FACE	''		#267		#242	.F.	
#301	ADVANCED_FACE	''		#300		#273	.F.	
#323	ADVANCED_FACE	''		#322		#306	.F.	
#109	ADVANCED_FACE	''		#108		#72	.F.	
#365	ADVANCED_FACE	''		#364		#328	.F.	
#396	ADVANCED_FACE	''		#395		#370	.F.	
#429	ADVANCED_FACE	''		#428		#401	.F.	
#140	ADVANCED_FACE	''		#139		#114	.F.	
#460	ADVANCED_FACE	''		#459		#434	.F.	
#502	ADVANCED_FACE	''		#501		#465	.F.	

Figure 5. Advanced_Face Table (Entity-To-Table Method)

Product life cycle application: machining feature recognition. Machining feature is a high level data which is useful to product life cycle applications, especially in process planning, and useful in generating STEP-compliant CNC codes. SQL (Structured Query Language) is a standard DML (Direct Manipulation Language) in relational database. Through this access, the user will have fast access and completing the convexity test on every single edge curve entity and identify pocket feature with the result shown in Fig. 6 and Fig. 7.

UNION OF OE WITH AF D WITH CONVEXITY TESTS

ORIENTI	X1	Y1	Z1	OE1	EC	OE2	ORIENTI	X2	Y2	Z2	X3	Y3	Z3	X4	Y4	Z4	CONVE)
#1036	0	1	0	#1026	#1025	#1136	#1139	0.0	0.0	1.0	1	0	0	-1	0	0	-1
#1107	0.0	0.0	1.0	#1098	#136	#137	#140	-1	0	0	0	-1	0	0.0	1.0	0.0	-1
#1107	0.0	0.0	1.0	#1100	#200	#201	#204	0	-1	0	1	0	0	-1.0	0.0	0.0	-1
#1107	0.0	0.0	1.0	#1102	#264	#265	#268	1	0	0	0	1	0	0.0	-1.0	0.0	-1
#1123	0.0	0.0	1.0	#1114	#392	#393	#396	0	-1	0	1	0	0	-1.0	0.0	0.0	-1
#1123	0.0	0.0	1.0	#1118	#568	#569	#579	0	1	0	-1	0	0	1.0	0.0	0.0	-1
#1139	0.0	0.0	1.0	#1130	#904	#905	#908	-1	0	0	0	-1	0	0.0	1.0	0.0	-1
#1139	0.0	0.0	1.0	#1134	#1080	#1081	#1091	1	0	0	0	1	0	0.0	-1.0	0.0	-1
#1155	0.0	0.0	1.0	#1148	#712	#713	#716	1	0	0	0	1	0	0.0	-1.0	0.0	-1
#1155	0.0	0.0	1.0	#1150	#824	#825	#835	0	1	0	-1	0	0	1.0	0.0	0.0	-1
#1155	0.0	0.0	1.0	#1152	#769	#770	#780	-1	0	0	0	-1	0	0.0	1.0	0.0	-1
#1195	1.0	0.0	0.0	#1186	#1185	#1338	#1343	0	0	-1	0	1	0	0	1	0	1
#1195	1.0	0.0	0.0	#1192	#1191	#1269	#1279	0.0	1.0	0.0	0	0	1	0.0	0.0	1.0	1
#1226	0.0	-1.0	0.0	#1222	#1221	#1339	#1343	0	0	-1	1	0	0	1	0	0	1
#1226	0.0	-1.0	0.0	#1223	#1177	#1178	#1195	1.0	0.0	0.0	0	0	1	0.0	0.0	1.0	1
#1257	-1.0	0.0	0.0	#1254	#1215	#1216	#1226	0.0	-1.0	0.0	0	0	1	0.0	0.0	1.0	1
#1279	0.0	1.0	0.0	#1276	#1246	#1247	#1257	-1.0	0.0	0.0	0	0	1	0.0	0.0	1.0	1
#1331	0.0	0.0	1.0	#1285	#1169	#1170	#1195	1.0	0.0	0.0	0	1	0	0	1	0	1
#1331	0.0	0.0	1.0	#1286	#1267	#1268	#1279	0.0	1.0	0.0	-1	0	0	-1	0	0	1
#1331	0.0	0.0	1.0	#1287	#1238	#1239	#1257	-1.0	0.0	0.0	0	-1	0	0	-1	0	1
#1331	0.0	0.0	1.0	#1288	#1207	#1208	#1226	0.0	-1.0	0.0	1	0	0	1	0	0	1
#1331	0.0	0.0	1.0	#1293	#319	#320	#323	0	1	0	-1	0	0	-1.0	0.0	0.0	1
#1331	0.0	0.0	1.0	#1295	#250	#251	#268	1	0	0	0	1	0	0.0	1.0	0.0	1
#1331	0.0	0.0	1.0	#1307	#378	#379	#396	0	-1	0	1	0	0	1.0	0.0	0.0	1

Figure 6. Convexity test on each edge curve

POCKET

BASE OF POCKET.A	CountOfEDG
#1107	4
#1123	4
#1139	4
#1155	4

Figure 7. Pocket Feature Recognition

7 Conclusion

In conclusion, a relational database management system has been developed successfully used for mapping geometrical and topological data in ISO 10303 part-21 file into database format. Database system has been used for storing all the geometrical product design data is more suitable then initial file data format. Besides, database optimization technologies to feature recognition in CAD through SQL which is a standard DML (data manipulation language), has been used successfully for supporting product life cycle development applications.

Acknowledgment

The work reported in this paper has been supported by Research Fund of Universiti Tun Hussein Onn Malaysia (UTHM), and scholarship for Doctoral Program of Ministry of Education (MOE) of Malaysia.

References

1. T. Kramer and X. Xu, in: *STEP in a Nutshell* in Advanced Design and Manufacturing Based on STEP, edited by X. Xu and A. Y. C. Nee, Eds., ed: Springer London, pp. 1-22, (2009).
2. Y. ChunYan, W. Minghui, L. Nairuo, Z. Yueting, and P. Yunhe: *Translating EXPRESS language model into C language model*, submitted to SIGPLAN Not., **38**, pp. 30-39, (2003).
3. S.-J. You, D. Yang, and C. M. Eastman: *Relational DB Implementation of STEP based product* model, submitted in CIB World Building Congress, (2004).
4. J. Y. Lai, Z. W. You, Y. K. Chiu, M. H. Wang, C. H. Hsu, Y. C. Tsai, and C. Y. Huang: *On the Development of Blend Faces and Holes Recognition Algorithm for CAE Applications*. in *Key Engineering Materials*, Trans Tech Publ, (2015).
5. M. H. Wang, J. Y. Lai, Y. K. Chiu, C. H. Hsu, Y. C. Tsai, and C. Y. Huang: *On the Development of Virtual Inner Loops for Feature Recognition*. in *Key Engineering Materials*, Trans Tech Publ, (2015).
6. Venu, B., V.R. Komma, and D. Srivastava, *A Methodology to Recognize Manufacturing Features of* Prismatic *Parts from STEP AP 203 File*. Materials Today: Proceedings. **2**(4): 3170-3177, (2015).
7. Z. Niu, R. R. Martin, M. Sabin, F. C. Langbein, and H. Bucklow: *Applying Database Optimization Technologies to Feature Recognition in CAD*. Computer-Aided Design and Applications. **12**(3): 373-382, (2015).
8. Nasr, E.S.A., et al., A Feature Recognition System Using Geometric Reasoning. Procedia CIRP. **18**: 238-243, (2014).

MATEC Web of Conferences 77, 04003 (2016) DOI: 10.1051/matecconf/20167704003
ICMMR 2016

Rule-Based Storytelling Text-to-Speech (TTS) Synthesis

Izzad Ramli[1], Noraini Seman[1], Norizah Ardi[2] and Nursuriati Jamil[1]

[1]*Digital Image, Audio and Speech Technology, Faculty of Computer and Mathematical Sciences, Universiti Teknologi MARA, 40450 Shah Alam, Selangor, Malaysia*
[2]*Academy of Languages Studies, Universiti Teknologi MARA, 40450 Shah Alam, Selangor, Malaysia*

Abstract. In recent years, various real life applications such as talking books, gadgets and humanoid robots have drawn the attention to pursue research in the area of expressive speech synthesis. Speech synthesis is widely used in various applications. However, there is a growing need for an expressive speech synthesis especially for communication and robotic. In this paper, global and local rule are developed to convert neutral to storytelling style speech for the Malay language. In order to generate rules, modification of prosodic parameters such as pitch, intensity, duration, tempo and pauses are considered. Modification of prosodic parameters is examined by performing prosodic analysis on a story collected from an experienced female and male storyteller. The global and local rule is applied in sentence level and synthesized using HNM. Subjective tests are conducted to evaluate the synthesized storytelling speech quality of both rules based on naturalness, intelligibility, and similarity to the original storytelling speech. The results showed that global rule give a better result than local rule

1 Introduction

Speech synthesis is the process of converting written text to spoken audio, also known as text-to-speech (TTS). Expressive speech is a form of speech synthesis aimed to improve the naturalness of neutral speech output of standard TTS [1]. Storytelling speaking style is a type of expressive speech that is important towards the development of digital storytelling [2] and a humanoid robot [3]. Earlier work of storytelling is done by incorporating basic emotions to instil naturalness [4]. For instance, Silva et al. [5] synthesized emotions such as happy, sad, surprise and fear to create a virtual storyteller resembling human storyteller. On the other hand, developed an emotion transformer to alter the neutral prosody into desired emotive prosody as well as transform the voice quality of the synthetic speech. The emotive TTS is further used in the humanoid audio-visual avatar.

Rule-based prosody modifications have been a popular approach to incorporate emotions in storytelling speaking style. This approach was undertaken by many expressive TTS in Dutch [6], English [7], Catalan [8], Spanish [9], Indian [10], German [11] and Korean [12] languages. In [13], global rules are used to modify prosodic parameters of pitch, intensity, overall speech tempo and pause duration to convert neutral speech into storytelling. In a more recent work, local rules are employed at phrase-level instead of sentence-level to synthesize storytelling speaking style in Bengali, Hindi, and Telugu. Roekhaut et al. [14] also proposed local rules for varying prosodic parameters to synthesize storytelling style. The rules are applied to modify the prosody based on syllable position (initial, middle, end) in an utterance. Six-syllable categories are identified based on their positions and whether they are prominent/non-prominent syllables. Then, style conversion is applied to each of the syllable categories in a sentence.

In this paper, local rules and global rules of prosody modifications are further investigated to determine their performance in synthesizing storytelling in the Malay language. This paper is structured as follows. In Section II, an in-depth analysis of storytelling speaking style during neutral speech is conducted. Section III presents the process of deriving the global and local rule-sets for converting neutral speech to storytelling style speech. Evaluation of the synthesized storytelling speech using subjective tests is shown in Section IV. The conclusion and summary of the results are described in Section V.

2 Speech data acquisition and analysis

In this section, the data acquisition and prosodic parameters of neutral and storytelling speech are explained and analyzed, respectively.

2.1. Speech data collection

Two types of speech that is neutral speech and storytelling speech data are recorded from one male and one female experienced storytellers. Neutral speech is recorded from both storytellers as they read the scripts with no intonation. On the other hand, storytelling

 DOI: 10.1051/matecconf/20167704003

speaking style is recorded using storytelling intonations of the storytellers. In storytelling literature, storytelling speech may be conveyed in narrative, descriptive or dialogue mode [15]. Narrative storytelling is mainly used to inform the listener about the actions that are taking place and the characters affecting the story. Descriptive storytelling, on the other hand, describes a character or events to the listener so that they feel the character or the storyline depicted. Meanwhile, dialogue storytelling is when the storyteller typically modifies his/her voice into a more exaggerated register of expressions, where full-blown emotions may be manifested. In this work, storytelling is done in a narrative mode. Figure 1 shows the recording sessions of the female and male storytellers.

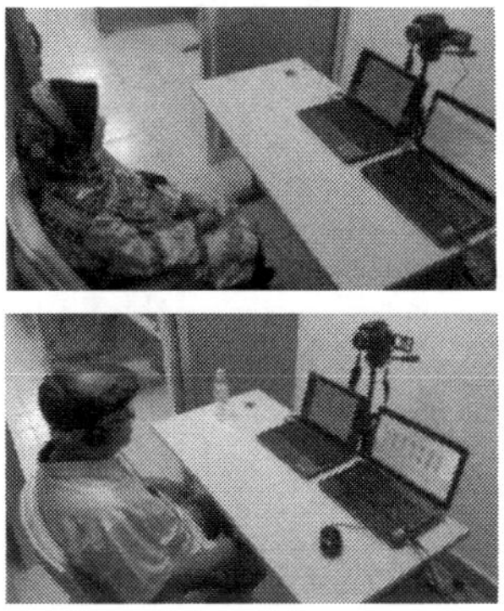

Figure 1. Recording sessions of the narrative storytelling.

In this preliminary investigation, one children's Malay folklore entitled "*Si angsa yang bertelur emas*" is selected to be narrated by the storytellers. The story is consists of 12 sentences, 113 words, and 276 syllables. The recording session is done in a quiet room, and the storyteller is seated in a comfortable chair approximately 2 feet away from the video camera. The recorded speech is stored in .wav format at a sampling rate of 41Khz with 16-bit resolution. The storytellers are given ample time to practice before recording session and they may repeat the recording until they are satisfied. All the neutral and storytelling speaking speech data are further transcribed at the sentence, word, and syllable levels using Praat [16] tool.

2.2. Prosodic parameter analysis

Analysis of prosodic parameters is important to examine which aspects of the prosody are pertinent to each other. This knowledge is used to transform neutral speech into storytelling speaking style. A total of 48 sentences, 452 words and 1,104 syllables are transcribed from the neutral and storytelling speech data of both storytellers. Five prosody parameters are further extracted from both speech data; they are pitch, intensity, duration, tempo and pauses. Table 1 tabulated the average pitch, intensity and duration of all syllables from the neutral and storytelling speech of both storytellers.

Table 1. Average prosodic parameters of both storytellers.

		Duration	Intensity	Pitch
Neutral speech	Female	0.24	61.23	171.74
	Male	0.24	76.94	144.21
Storytelling speech	Female	0.2	68.99	238.61
	Male	0.19	77.59	150.6

The analysis shows that average intensity and pitch for storytelling speech at 68.99 and 238.61 are larger than a neutral speech at 61.23 and 171.74, respectively for the female storyteller. The average intensity and pitch for the male storyteller also show a slight increase in storytelling speech. Duration, on the other hand, shows a lower average for storytelling speech at 0.2 compared to neutral speech at 0.24. Hence, this indicates that the speech tempo for storytelling speech is faster than neutral speech. The same is also true for the male storyteller.

The pause feature is analysed at phrase and sentence level. In our work, a phrase is defined as a collection of words and determined by the symbol comma (,) that exist in a sentence. Table 2 shows the average pause at phrase and sentence level. It shows that neutral speech has a longer average pause for phrase and sentence level compared to storytelling speech for both storytellers. At the sentence level, both storytellers tend to pause much longer than at phrase level before continuing to the next sentence.

Table 2. Average pause at phrase and sentence levels.

		Phrase Level	Sentence level
Neutral speech	Male	0.46	0.87
	Female	0.40	0.82
Storytelling speech	Male	0.42	0.60
	Female	0.25	0.67

Further analysis also showed that several words, typically adjectives and adverbs contain accented syllables that are stressed during pronunciations. Accented syllables tend to have increased pitch, duration, and intensity. Table 3 presents an example of an adjective word '*berat*' containing accented syllables /be/ and /rat/. It can be seen that tremendous increase of 320% for the duration, 12 % for intensity and 109% of pitch occurred during storytelling for accented syllable /be/. Similar phenomena is also shown for syllable /rat/. However, the rate of escalation is different depending on the position of the accented syllable in the word. Syllable /be/ that is located at the beginning of the word has a large increase for the duration. Meanwhile, syllable /rat/ that is positioned at the end of the word has the largest increase in pitch. Upon more inspections of other accented syllables, we discovered that the prosody parameters of storytelling speaking style varied according to the syllable positions: initial, middle or end.

Table 3. Prosody analysis for accented syllables.

Word	Syllable	Prosody	Neutral speech	Storytelling speech
'berat'	/be/	Duration	0.15	0.63 (320%)
		Intensity	77.37	86.70 (12%)
		Pitch	110.38	231.11 (109%)
	/rat/	Duration	0.32	0.47 (46%)
		Intensity	75.74	77.61 (2.5%)
		Pitch	116.18	219.96 (89%)

3 Development of prosodic rules for storytelling speaking style

This section discusses the process of converting neutral speech to storytelling style speech as shown in Figure 2.

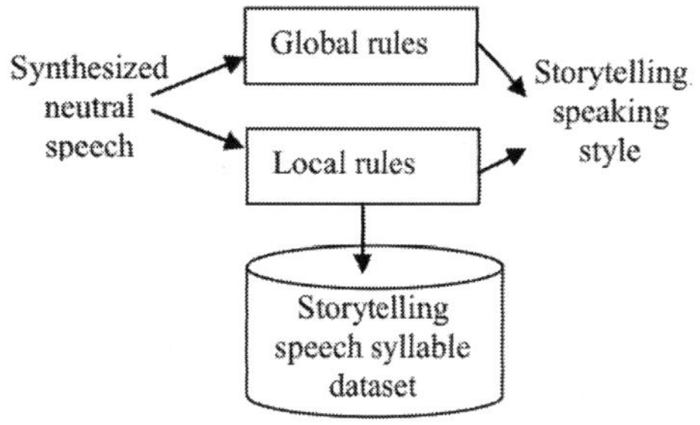

Figure 2. Rule-based storytelling speech synthesis.

The synthesized neutral speech is generated using a Malay language TTS [17]. As the aim of this paper is to compare the performances of local and global rules for synthesizing storytelling speaking styles, two set of rules are constructed to produce the storytelling speaking style speech. Global rules are set to modify prosodic parameters of accented syllables only, leaving other syllables unchanged. Global rules also alter overall speech tempo and pause duration. On the other hand, local rules modify both accented and unaccented syllables of the neutral speech to produce the storytelling speech. As mentioned earlier, prosody parameters of the accented syllables are modified based on the positions of the syllables in the word (i.e. initial accent, middle accent, final accent) [14]. The local rules also altered unaccented syllables locally based on the average prosody of the targeted speech. To construct the local and global rules of modification factors, prosody parameters of the storytelling speech syllables are referred and analyzed.

Modification factors for the global and local rules are determined by pitch contour, duration, intensity patterns, tempo and pause of both neutral and storytelling speeches. The factors are calculated by dividing the average desired (i.e. storytelling) prosodic values with the average neutral prosodic values [13]. For example, global rule modification factor of accented syllables for pitch in female storyteller is calculated as follows:

$$\frac{\text{Average pitch of accented syllables for storytelling}}{\text{Average pitch of all syllables in neutral speech}}$$

$$\frac{256.92\text{Hz}}{171.74\text{Hz}} = 1.5$$

Modification factors for duration and intensity of accented syllables for the female storyteller are calculated similarly and tabulated in Table 4. Note that pitch and intensity parameters of unaccented syllables remain unchanged (i.e. NC = no change) for unaccented syllables. Tempo (syllable per second) or speaking rate in the global rule is applied to unaccented syllable only while duration is used for accented syllables. Modification factor of tempo is calculated as such:

$$\frac{\text{Total syllables in storytelling speech}}{\text{Total duration without pause in storytelling speech}}$$

$$\frac{276}{66.3} = 4.1$$

Modification factors for pauses are average pauses computed at phrase and sentence level of the storytelling speeches. The factor at phrase level for the female storyteller is 0.25 and at the sentence level is 0.67 as depicted in Table 4.

Table 4. Global rules modification factors of the female storyteller.

Syllable type	Prosodic parameter	Modification factor
Accented	Pitch	1.5
	Duration	0.9
	Intensity	1.13
Unaccented	Pitch	NC
	Intensity	NC
	Tempo	4.1
NA	Pause (phrase level)	0.25
NA	Pause (sentence level)	0.67

NC – no change; NA – not applicable

Table 5. Local rules modification factors of female storyteller.

Syllable type	Prosodic parameter	Modification factor		
		Initial	Middle	End
Accented	Pitch	2.63	1.72	2.23
	Duration	1.16	NC	1.5
	Intensity	1.1	1.2	1.08
Unaccented	Pitch	1.3		
	Intensity	1.11		
	Tempo	4		
NA	Pause (phrase level)	0.25		
NA	Pause (sentence lev.)	0.67		

NC – no change; NA – not applicable

In Table 5, modification factors of local rules are presented. Factors for local rules are calculated similarly as global rules. However, local rules of accented syllables are segregated based on the syllable's position in the word (i.e. initial, middle, end). Another set of global and local rules are calculated for the male storyteller and are listed in Table 6 and 7, respectively.

Table 6. Global rules modification factors of male storyteller.

Syllable type	Prosodic parameter	Modification factor
Accented	Pitch	1.47
	Duration	1.14
	Intensity	1.01
Unaccented	Pitch	NC
	Intensity	NC
	Tempo	5.4
NA	Pause (phrase level)	0.42
NA	Pause (sentence level)	0.6

NC – no change; NA – not applicable

Table 7. Local rules modification factors of male storyteller.

Syllable type	Prosodic parameter	Modification factor		
		Initial	Middle	End
Accented	Pitch	2.5	1.38	1.4
	Duration	0.7	1.1	1.5
	Intensity	0.8	1.03	1.06
Unaccented	Pitch	NC		
	Intensity	0.98		
	Tempo	5.4		
NA	Pause (phrase level)	0.42		
NA	Pause (sentence lev.)	0.6		

NC – no change; NA – not applicable

3.1. Pitch

Pitch varies significantly in storytelling speech compared to neutral speech. Analysis of pitch contour in accented syllables of storytelling speeches shows that a falling or raising contour patterns. Therefore, to formulate this rise-fall pattern of pitch contour, non-linear functions are explored. In this paper, the pitch contour patterns are formulated using sine functions [18]. The formulation multiplies all pitch values within a relevant time domain $[t_1, t_2]$ for a given modification factor which is governed by the variable α . The calculation is shown in equation (1).

$$s' = s(t)*(1+(\alpha -1) \times \sin \{[(t - t_1)/(t_3 - t_2)] \times 0.5 \times \pi\} \quad (1)$$

where

s Original pitch values at time t

s' Manipulated pitch values at time t

α Pitch modification factor

The pitch modification factor for global rules represented by variable α is substituted with 1.5 (see Table 4) and 1.47 (see Table 6) for female and male, respectively. Meanwhile, pitch modification factors for local rules (see Table 5 and 7) are applied based on the positions of the syllables.

3.2 Duration

In order to efficiently impose the dynamics of storytelling speech prosody on neutral speech, duration should be modified to appropriate levels. For the global rule, two duration constant modification factor are designed. These duration modification factors are uniformly applied to accented and unaccented syllable respectively. In local rule, each position (initial, middle and final syllable) of accented syllable are scaling with different constant modification factors. Unaccented syllable is scale locally proportional to the tempo. The entire modification factor for duration parameters in both rules are showed in rows 4 in Table 4 and Table 5.

3.3 Intensity

Based on the analysis, the average intensity of a storyteller turns out to be higher as compared with neutral speech. Intensity for a speech signal may vary between speakers. Also, intensity may vary due to the recording device and distance to the speaker. In this work, the same recording device applied, and distance recording devices to the speaker are unchangeable. The rules derived from converting intensity of neutral into storytelling style speech are given below.

$$y'(t) = y(t) \times \alpha \quad (2)$$

where:

y Original intensity values at time t

y' Manipulated intensity values at time t

α Intensity modification factor

The intensity modifications factor for female and male as showed at rows 5 in Table 4 and Table 5 respectively. For local rule intensity modification factor at initial, middle and final of the syllable is given in the last three columns with a different value in both Tables.

3.4 Pause

The pause is analyzed from sentences of storytelling and neutral style speech at phrase and sentence level. In this work, the stories recorded by the storyteller have less duration of pauses as compared to neutral speech. Female storyteller average pause length of neutral at phase level is 0.4 s and storyteller 0.25 s. At sentence-level, average pause length of neutral is 0.82 s and storyteller 0.67 s. On the other hand, the male storyteller has the same pause at phase level at 0.4 s. However, average pause length of neutral is 0.87 s at sentence level which is longer than storyteller with an average pause at 0.6. The pause duration (phrase and sentence level) applied to both storytellers as showed last two columns in Table 4 and Table 5.

4 Evaluations and discussions

A subjective test is carried out to evaluate the quality of the synthesized speech. The criteria are considered (naturalness, intelligibility, similarity) compared to the original storytelling speech. In this test, subjects have to listen and evaluate the synthesized speech sentences. 5 native Malay-speaking subjects in the age group of 21-51 participated in the listening test.

For listening test, a complete text story is divided into sentences. The sentences consist of 5 to 21 words. The sentences are converted to synthesize storytelling speech using global and local rule for female and male storyteller. It synthesized using prosodic modification methods HNM. A total of 96 synthesized storytelling speech sentences are collected, which is 48 sentences of each storyteller. Here, the subjects were presented with both original storytelling speech and synthesized storytelling speech. The subjects are given time to judge the quality of the synthesized storytelling speech by comparing the original expressive speech. The evaluation has been done in a quiet place and using proper headphone to prevent external sound. Before evaluation, the subjects are explained about the criteria of naturalness, intelligibility, and similarity. They can replay the synthesized speech for better evaluation. The higher quality synthesized speech was given a mean opinion score of five and poor quality synthesized speech were given of one. The results for the prosodic rule and modification methods for female and male are given in Table 3 and Table 4 respectively.

Table 8. Evaluation result for synthesized female storyteller

Prosodic rule	Natural-ness	Intelligibi-lity	Similarity	Mean (μ)
Global	2.8	3.0	3.0	2.9
Local	2.4	2.95	2.85	2.73

Table 9. Evaluation result for synthesized male storyteller

Prosodic rule	Natural-ness	Intelligibi-lity	Similarity	Mean (μ)
Global	2.98	3.8	3.05	3.27
Local	2.8	3.55	2.96	3.1

Based on Table 3 and Table 4, global rules outperform compared to local rule for both storyteller. Mean result for global rule of female storyteller is 2.9 is better than local rule at 2.73 out of 5. While, mean results of male storyteller is 3.27 and 3.1 for global and local rule. Hence, the global rule is used for modification yields a better result than the local rule in both storytellers.

The global rules do not manipulate the entire syllable but only at accented syllable to produce much better quality of speech. It is because modification of syllable sometimes is over exaggerated and quality degenerates that happen in the local rule which modified the entire syllable to the target speech. The values of modification factor in local rule derived based on average are the cause of this matter. These modification factors are not compatible to apply in the entire syllable context. Therefore, in selecting the modification factor, the result should be considered. We can modify the prosody, however, at a certain value it meets a limitation. Better technique needs to explore to find the perfect modification factor that can match with every syllable context in converting neutral to storytelling style speech.

5 Conclusions

In this paper, the process of converting neutral speech from standard text-to-speech (TTS) system to storytelling style speech is carried out by using global and local rule. Five prosodic parameters, namely, pitch, duration, intensity, tempo and pause are analyzed for modification. The modification factors are derived by analyzing the difference between prosodic parameters of neutral and storyteller speech sentences. The subjective test results for naturalness, intelligibility, and the similar value indicated that the global rule was effective in imposing the story style on neutral speech and maintain the quality of speech. However, the used of average to determine the modification factor to suit with all syllable context is not the best idea. Further research needs to explore to improve the synthesized speech in the global and local rule. We believe that local rule can give a better result if the modification factors are applied in the correct circumstances for every syllable. The rules may be extended to develop others speaking styles such as news reading, debate, and conversation.

References

[1] D. Govind, Prasanna, and Mahadeva, "Expressive speech synthesis: a review," *International Journal of Speech Technology*, vol. 16, 237–260, Oct. (2012)

[2] B. C. Lunce, "Digital Storytelling as an Educational Tool," *Inidiana Libriaries*, 30, 77–80, (2007)

[3] R. Gelin, C. D'Alessandro, and Q. Le, "Towards a storytelling humanoid robot," *AAAI Fall Symposium Series on Dialog with Robots*, 137–138, (2010)

[4] J. Adell, A. Bonafonte, and D. Escudero, "Analysis of Prosodic Features Towards Modelling of Emotional and Pragmatic Attributes of Speech," in *Procesamiento del lenguaje natural*, 277–283, (2005)

[5] A. Silva, M. Vala, and A. Paiva, "The Storyteller: Building a Synthetic Character That Tells Stories," in *In Workshop on Representing, Annotating, and Evaluating Non-Verbal and Verbal Communication Acts to Achive Contextual Embodied Agents, at Autonomous Agents Conference*, 53–58, (2001)

[6] S. J. L. Mozziconacci, "Speech variability and emotion: production and perception," Technical University Eindhoven, 1998

[7] J. Y. Zhang, A. W. Black, and R. Sproat, "Identifying Speakers in Children ' s Stories for Speech Synthesis," 2041–2044, (2003)

[8] I. Iriondo, F. Alías, J. Melenchón, and M. Llorca,

"Modeling and synthesizing emotional speech for Catalan text-to-speech synthesis," *Affective Dialogue Systems*, 197–208, (2004)

[9] J. M. P. M Montero, J M Gutierrez-Arriola, S Palazuelos, E Enriquez, S Aguilera, "Emotional speech synthesis from speech database." (1998)

[10] P. Sarkar, A. Haque, A. K. Dutta, G. R. M, D. M. Harikrishna, P. Dhara, R. Verma, N. P. Narendra, S. K. S. B, J. Yadav, and K. S. Rao, "Designing Prosody Rule-set for Converting Neutral TTS Speech to storytelling style speech for Indian Languages : Bengali , Hindi and Telugu," 0–4, (2014)

[11] M. Schröder, "Dimensional Emotion Representation as a Basis for Speech Synthesis with Non-extreme Emotions," *Affective Dialogue Systems*, 3068, 209–220, (2004)

[12] H.-J. Lee, "Fairy Tale Storytelling System: Using Both Prosody and Text for Emotional Speech Synthesis," in *In Convergence and Hybrid Information Technology*, Springer, 317–324, (2012)

[13] M. Theune, K. Meijs, D. Heylen, and R. Ordelman, "Generating expressive speech for story telling applications," *IEEE Transactions on Audio, Speech, and Language Processing*, **14(4)**, 1099–1108, (2006)

[14] S. Roekhaut, J. Goldman, A. C. Simon, U. D. M. Umons, U. De, D. De Linguistique, U. De Genève, and I. Langage, "A Model for Varying Speaking Style in TTS systems," 4–7, (2010)

[15] R. Montaño, F. Alías, and J. Ferrer, "Prosodic analysis of storytelling discourse modes and narrative situations oriented to Text-to-Speech synthesis," *8th ISCA Workshop on Speech Synthesis*, 171–176, (2013)

[16] P. Boersma and W. David, "Praat, doing phonetics by computer," (2015)

[17] I. Ramli, N. Jamil, N. Seman, and N. Ardi, "An Improved Syllabification for a Better Malay Language Text-to-Speech Synthesis (TTS)," in *International Symposium On robotics and intelligent sensors*, 417–424, (2015)

[18] R. Verma, "Conversion of Neutral Speech to Storytelling Style Speech," (2015)

MATEC Web of Conferences 77, 04004 (2016) DOI: 10.1051/matecconf/20167704004
ICMMR 2016

Realization of wire selection platform based on three dimensional visualization technology

Peng Huang, Rui Liu, Zhenggang Fang and Hongbing Shen

China Electric Power Research Institute, China

Abstract. The three dimensional wire selection platform of the power transmission line is constructed by modelling technology and three dimensional scene render technology in combination with characteristics of the digital elevation, the digital ortho image and the earth geography information system. Through application of design and wire selection in some 330kV I and II circuit line engineering, it shows the design technical person can see scene of the three dimensional power transmission line clearly and livingly when the three dimensional visualization technology is applied in the optimized wire selection practice of the power transmission line, optimizes route of the line, saves investment of the line, reduces labour strength of person and finally improves design quality of the line.

1 Introduction

Route selection and investigation are extremely important key loops during the power transmission line design process, reasonability of the wire selection plan plays an important function on economic index, technical index and construction, operation conditions of the line. Selection of the power transmission line takes the space geographical information as basis. The traditional wire selection method takes the middle and small scale topography map as basis for rough selection, which realizes refinement and optimization of the plan through field investigation and measurement treatment on basis of rough selection. Because the current 1:10000 and 1:50000 topology maps applied in our country are mostly measured in 1980s or early, which are very obsolete and can' t reflect current ground status during design[1]. At same time the topology map isn't intuitive and size of the map is limited, the design person only find the optimum line within the scope wide 30km along the line. It is difficult for the traditional wire selection method to select the ideal route plan, which can't meet demand of power construction. How to overcome disadvantage of wire selection in the power transmission map in the topography map is a problem which is paid attention by the power designer in our country.

Following development of the modern investigation and measurement technology and appearance of the versatile business investigation and measurement image data, we are easy to obtain the demanded data. If the three dimensional visualization technology is applied in the wire selection of the power transmission line, when the investigation and measurement designer sees the ground model in the computer, it is convenient and intuitive as seeing ground from the air, which can optimize the routine plan within the large scope, it can see the relative relationship between the line and various buildings, the road and the railway etc, and obtain the effect which is difficult to obtain by the traditional wire selection method and reduce labour strength of the investigation and measurement design person[2].

2 Three dimensional visualization technology

At present, the three dimensional visualization technology is gradually applied in every field of the society and the living, such as digital city construction, military application, environment monitoring, scene spot planning, geography and mine activity, hydro geologic activity, traffic monitoring, estate development, medical aid etc. aspects. Appearance of OSG based on the OpenGL technology etc. three dimensional display platform, Quest 3D, Virtual tools, Skyline based on the DirectX 3D technology etc. system platforms greatly pushes forward development of the three dimensional visualization simulation and virtual technology.

Three dimensional wire selection of the power transmission line realizes effective mixture of the digital elevation model, the digital ortho image, the three dimension data and other GIS data information based on the three dimensional GIS platform, which re-appears real condition in the power construction environment of the power transmission line.

2.1 Geographic information system

Geographic Information System is an important tool to obtain, process, manage and analyze Geographic space data. Function mainly includes data input and edit, data management, data processing, data display and output etc.

The three dimensional GIS platform can realize integration expression of the ground, earth surface, underground, under water information, which solves seamless integration of the three dimensional CAD design achievement and the three dimensional space information platform, two dimensional and three dimensional integration etc problem. The EV-Globe platform integrates the latest geographic information and the three dimensional software technology, which has the large scope, massive, multiple source data integration management and quick three dimension real-time roaming function, it supports inquiry, analysis and calculation of the three dimensional space, it can integrate with the traditional GIS software and provides the fundamental image documents within the global scope, it can construct the three dimensional space information service system conveniently and quickly, which can complete expansion from the two dimensional GIS system to the three dimensional system quickly, it is a new generation of the large space information service platform.

2.2 Establishment of three dimensional model

2.2.1 Parameterization modelling technology

Parameterization modelling technology is a technology which controls size and information of the model through parameter data. Main characteristic is to apply size driving, i.e., size of the parameterization model is expressed by the corresponding relationship; it isn't expressed by the determined value. When one parameter value is changed, all sizes related to it will be changed automatically, and a new same type of the model will be generated.

The user will realize modification of the model through modifying the size, which will generate a part model series with same shape and different specification. Parameterization model expresses the geometrical graph as the geometrical restraint model which consists of the geometrical element and its restraint relationship through seizing the restraint relationship between the geometrical elements in the model. The key in the parameterization modelling is to establish the geometrical restraint relationship, i.e., topology restraint and size restraint. Topology restraint is quantitative description on structure of the product, it expresses relationship between the geometrical element topology and the structure, such as parallelism, symmetry and verticality etc. These relationships maintain unchanged during size driving process of the graph. Size restraint expresses position relationship between the geometrical element through the restraint expressed by the size indication, such as distance size, angle size and radius size etc, is shown in fig 1. It is the object of parameterization driving. Its features are to realize serialization design through modifying size of the graph on basis maintaining topology relationship of the original graph unchanged.

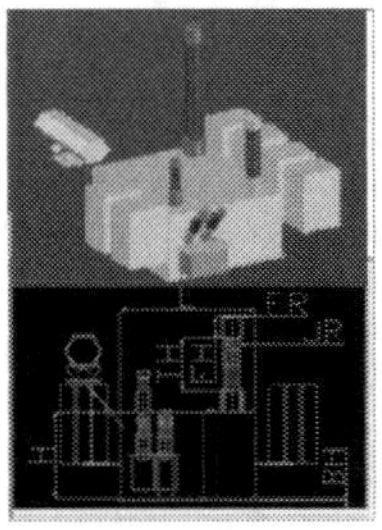

Figure 1. Parameterization modelling sample of transformer.

2.2.2 Graphic modelling technology

Graphic modelling technology is an important part during construction of the three dimensional model. Some equipment with complicated and versatile shape can realize model construction through the parameterization modelling. The graphic modelling way can make up disadvantage of the parameterization modelling, which becomes an essential model construction way. Secondly, when parameterization modelling can't realize high precision drawing of the complicate servant of some equipment and simulation requirements on the equipment are very high, it is only realized through the graphic modelling technology.

The model which is constituted by the graphic modelling technology shall have the corresponding property parameters, which can include two types of the parameter information. One is geometrical parameters, which can record space relationship of the nodes in the three dimension model, such as the appearance size, the diameter of the inner hole etc relevant parameters of the three dimensional model; another is electrical parameters, which can record the electrical property parameters of the three dimensional model, such as type of the equipment, material information etc relevant parameters, is shown in fig 2.

Figure 2. Sample of graphic modelling.

2.3 Construction of power transmission line scene

Optimization design of the graphic rendering algorithm is carried out as three parts during large scene rendering process of the power transmission line, including rendering of terrain and topology three dimensional scene model, rendering of the crossing object three dimensional model and rendering of the electrical equipment model[3-5].

2.3.1 Rendering of terrain and topology three dimensional scene model

Geographical information data for rendering of terrain and topology consists of the DEM elevation data and the DOM video data. The method to reduce consumption of a large quantity of data on the memory etc hardware device resource is to apply the blocking and grading rendering technology to solve visualization issue of the field. In order to realize switching of the geographic information data in different grade and different block quickly, the pyramid model and the four fork tree structure are applied during pre-processing and memory of the original geographic information data, is shown in fig 3.

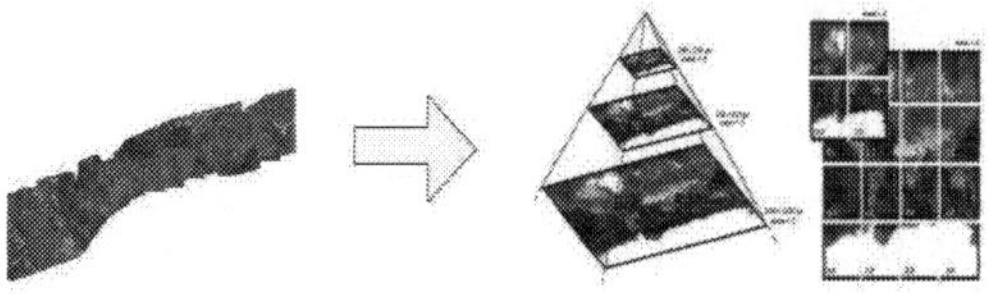

Figure 3. Processing flow of DEM and DOM data.

2.3.2 Rendering of the crossing object three dimensional model and electrical equipment model

Besides the geological information data in the three dimension simulation scene, data of the additional object on the floor are massive. It includes the crossing object (such as the house, the road, the power line etc) and the equipment in the power transmission line (such as the power transmission tower and the fitting etc). Geographical information data and these additional models constitute the complete three dimension simulation scene of the power transmission grid engineering. The rendering efficiency of the large scene scope and the detail effect of the model shall be considered during rendering of the model.

The rendering of the model is processed by the sectional loading, grading display method. Carry out sectional loading of the model in the whole three dimensional scene through position distance of the viewing angle and scope of the flat section head scope. The same model can be controlled precisely through observing the distance of the viewing angle. Only the low precision model and the outer profile frame of the model are rendered during long distance display. When observation distance is too short, load and render the finest detail characteristic of the model, the visualization display in the same scene can be realized in the large scale of the model by this way, is shown in fig 4.

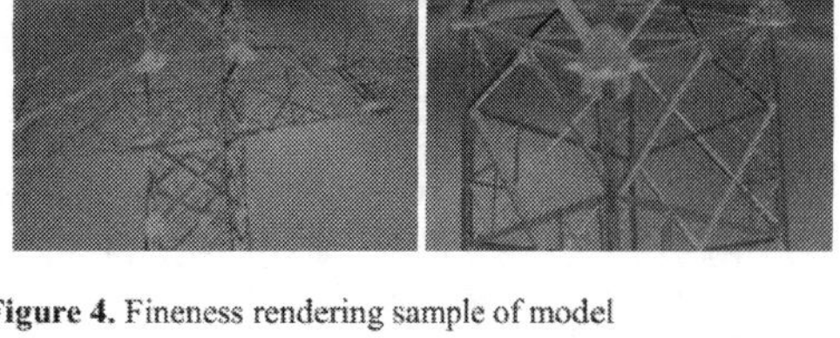

Figure 4. Fineness rendering sample of model

3 Engineering application conditions

3.1 Engineering general

For the newly built 330kV I, II circuits line engineering, the whole line consists of two single circuits which are parallel set up. The whole length of the preliminary design route is 72.5km; terrain unit along the line takes the aggraded valley plain, desert as primary, is shown in table 1.

Table 1. Table of route length and terrain distribution condition

Full name	Altitude (m)	Route length (km)	Optimized adjustment plan
Gebi plain	1270~1850	61.3	84%
Desert and hill	1570~1710	11.2	16%

3.2 Import of geographical information data

Geographical information data are GIS graphical data of the GIS map which serves to wire selection of the overhead line, it consists of the vector data and grid data from types, it is classified as 3 data types from category, which as basic geographical data, power special data and engineering data respectively.

Three dimensional route selection of the power transmission line starts from processing of the three dimensional geographic information data. Three dimensional geographical information data source in the engineering area is expanded by the data processing tools. According to known coordination system information of this engineering, longitudinal and latitude parameter of ground is inputted in date process setting, and then set the central longitudinal and input the corresponding coordination mode, selection of the banding method and the central longitudinal line. After the three-dimensional digital wire selection system is opened, the imported image can be positioned in the GIS three dimensional scene [6].

3.3 Route optimization

Take advantage of the three dimensional visualization technology based on the GIS platform in combination with the satellite image and navigation image, optimize the preliminary design route plan according to document collection and investigation at site, is shown in table2 and table 3. The following principles shall be considered during optimization.

Comprehensively consider construction and operation etc factors, try to construct and operate the line close to the road; Avoid the house effectively and optimize the route in the section with great demolition volume of the house; Reasonably select position of the various crossing, selection of the road, rive, power line etc crossing positions has great influence on the engineering cost, the built and on building power transmission line with crossing shall be reduced as possible, in particular to the power transmission line at high voltage class; Sufficiently consider terrain and forest covering along the line, apply the tree crossing and forest plan, reduce chewing of the forest and protect ecology environment.

Table 2. Route adjustment comparison table

Route section	Item	Preliminary designed route plan	Optimized adjustment plan
I section	Route length	1.85km×2	1.55km×2
	Traffic status	General	Better
II section	Route length	6.95km×2	6.65km×2
	Number of angle	3×2	1×2
III section	Route length	0.9km×2	0.8km×2
	Number of angle tower	2×2	1×2

Table 3. Application condition comparison table of power transmission tower

Comparison category	Length (km)	Base number/proportion of power transmission tower(%)	Base number/proportion of strain tower(%)	Kilometre	Consumption per km（t/km）
Preliminary design	72.5	169/85.4	29/14.6%	2.73	26.85
Optimization design	71.8	167/87.9	23/12.1%	2.65	26.77

Table 4. Statistic table of main crossing

No	Name of crossing object	Class	Quantity	Crossing way
1	Power line	750kV	2×2	Drill
2	Power line	110kV	2×4	Span
3	Power line	35kV	2×3	Span
4	Power line	10kV	2×16	
5	Communication line	Three classes	2×23	
6	Weak electricity line	Low voltage line	2×8	
7	Railway		2×2	Span
8	National road		2×2	Span
9	Land road		2×5	
10	Chop tree		64	
11	House		15	
12	River		3	
13	Geological gas pipe		2×2	

In three dimensional power transmission line sense, main crossing in the area where this engineering passes through is clearly master, which can make up disadvantage of wire selection of the power transmission line in the two dimensional terrain figures, it plays an important reference function in the tower arrangement and positioning of the line, is shown in table 4.

4 Conclusion

Carry out optimization of local route of the line in this engineering and length of the preliminary design line is 2×72.5km, the length of the optimized line is 2×71.8km, which is 2×0.7km shorter than original route, and 6 pieces of the angle two circuits are cancelled, length of the line passing Woods is reduced about 300m. In the three dimensional power transmission line scene, main crossing in the area where the line passes is clearly master, which makes up disadvantage in wire selection of

DOI: 10.1051/matecconf/20167704004

the power transmission line in the traditional two dimensional geology graph, which plays important reference function on tower arrangement and positioning of the line. The three dimensional GIS wire selection platform makes operation efficiency of the design person to improve greatly, which is beneficial to improve design quality and guarantee implementation of engineering technical index.

References

1. Junqiang Xu, Lixing Xing, Jun Pan. Remote Sensing Technology and Application, 21 (2006)
2. Yongsheng Zhang. Phys. *Remote sensing image information system* (Science Press, china, 2000)
3. Yingshi Zhao. Phys. *Analysis principle and method of remote sensing application* (Science Press, china, 2003)
4. Shulong Zhu, Baoshan Zhu, Hongwei Wang. Phys. *Remote sensing image processing and application* (Science Press, china, 2006)
5. Tianhua Li, Chonggao Liao, Wunian Yang. Journal of Geotechnical Investigation & Surveying, 12 (2006)
6. Huajun Shi, Minzhong Wan. Electric Power Survey & Design, 12 (2004)

Comparison and analysis of methods for migrating legacy systems to SOA

LIU Dong[1], ZHANG Caihuan[2]

[1]*Dept. of Information Technology, Luoyang Normal University, 471022 Luoyang Henan, China*
[2]*Dept. of Mathematics, Luoyang Normal University, 471022 Luoyang Henan, China*

Abstract. Migration of legacy systems to SOA has caught lots of attention in both academic and industry. People have proposed many methods, such as wrapping, reengineering etc. However, various characteristics of legacy systems can make the migration complicated, and different enterprises can afford different level migration cost. In this paper, the existing methods and options for evolving legacy systems are compared, and their advantages and disadvantages are identified respectively. Different adaptive situations are listed to assist in making migration decision.

1 Introduction

The generic term legacy systems has been coined to refer to existing IT assets that have been deployed in the past, and in many cases, may have come into the enterprise as a result of a merger or acquisition. No standard definition exists. There are a set of common features: large scale(often millions of lines of monolithic code); old age; obsolete languages such as assembler; lack of consistent (or any) documentation; poor management of data; degraded structure following years of modification; and finally, a support team which is totally reliant on the expertise and knowledge of expert individuals. These factors together cause the cost of maintenance support to be very high, and the applications backlog to be unacceptable [1]. However, it is an undeniable fact that legacy systems generally consist of invaluable assets with embedded business logic representing many years of coding, developments, enhancements, and modifications. In addition, legacy systems is generating unusually high profit margins in many cases and thus, are responsible for a large amount of a company's operating profit. Consequently, many business and IT benefits are realized from reusing legacy assets, evolving them to integrate with or migrate to modern platforms and architectures like Service Oriented Architecture (SOA).

SOA has become an increasingly popular paradigm at present for achieving interoperability between systems because it provides a standard based conceptual framework. At the core of an SOA is a service. A service is a coarse-grained, discoverable, and self-contained software entity that interacts with applications and other services through a loosely coupled, often asynchronous, message-based communication model [2]. Proponents suggest a long list of advantages [3]:

- Simple standards that define the available interfaces and structure of data that is conveyed across those interfaces.
- Platform and language-independent interfaces based on these standards, which allow application to invoke services operating on any device supporting the SOA regardless of the hardware platform, operating system, or implementation language.
- Clear separation of service interface from implementation, thus allowing many service updates to occur without impact on service users.
- • Message-based communication allowing distribution across a wide area.
- Loose coupling between services, minimize interdependencies and facilitating reuse.
- Mechanisms for discovery of services available and for establishing connections with services.

Clearly, one of the most attractive features of SOA to many organizations is that SOA offers the promise of enabling legacy systems to expose their functionality as services without making significant changes to the legacy systems themselves.

Over the past few years, migration of legacy systems to SOA has caught lots of attention in both academic and industry. People have proposed many methods, such as wrapping, reengineering etc. However, various characteristics of legacy systems, such as age, language, and architecture, as well as the target SOA environment, can make the migration complicated. In addition to this, different enterprises can afford different level migration cost. So it may not be immediately obvious whether to migrate or not and which method be suitable for a specific organization. In this paper, we compare the

DOI: 10.1051/matecconf/20167704005

existing methods and options for legacy systems, especially legacy systems with many applications, and identify their advantages and disadvantages respectively. Different adaptive situations are listed to assist in making migration decision.

The remainder of the paper is organized as follows. In Section 2, a few existing methods of migration to SOA are listed. Their advantages and disadvantages are compared and analysed. Accordingly, their adaptive situations are identified respectively. Section 3, illustrates the related works and highlights the main contributions, and Section 4 concludes the paper.

2 Comparison and analysis of the existing methods

2.1 requirements and background

Generally, the primary motivation for migration to SOA is based on one or more of the following requirements: reduce the high cost of ownership; accelerate time to market; monolithic architecture with little or no modularity together with redundant code. closed and obsolete technology; decreasing developers skilled in existing systems; lack of vendor support; lack of consistent documentation.

In the following subsection, we will base our comparison and analysis on the aforementioned requirements. Before presenting our findings, let's consider a situation: an enterprise has many applications. Recently, the enterprise has decided to adopt SOA and developed a SOA infrastructure. A question to be addressed is which application should be migrated to SOA, which method be suitable for it and how. The answer will be various for different application and organization. Our findings provide baseline for making decision through comparison and analysis of different methods.

2.2 Freeze

The term "freeze" means that the organization decides that no further work on the legacy application should be performed. Although freezing seems to sound absurd, it may make sense to deal with some specific application. In fact, people always justify a project at first according to the value and the state of the art of an application as well as cost-benefit analysis. Manager would prefer to have an application maintain its current state in many cases.

The freezing approach has four advantages:

- No additional cost input.
- No time consuming.
- No risk of failure.
- Continue to keep profit generated by the legacy application.

Of course, the disadvantages of this approach are also evident. All requirements mentioned above can't be met. The main pain points organization faced with cannot be solved. However, There are some cases where the freezing approach may be a good option, if:

- The size of the application is small.
- The cost of ownership is low.
- Critical business is not embedded in the application.
- The application is generating low profit margins for organization.
- The application changes infrequently.
- The application has low integration needs with other ones.

2.3 Rewrite

Rewriting means to retire the application and outsource or re-develop from ground-up. Actually, Rewriting is not one of the methods advocated by SOA proponents. But, it may make sense because of the following advantages:

- It can deliver a complete customized solution that can be built exactly to meet the organization's need.
- It is a "big-bang" approach to solve the main problem.
- New and standards-compliant technologies are introduced.
- The organization can gain the control of software source code base.
- New added values are brought by leveraging SOA.

There are a few shortcomings that make many consider the approach incompliant with the idea of SOA:

- Legacy assets cannot be reused so that investment in the past is lost.
- May be risky and long time consuming.
- May be high cost.

Although these shortcoming evidents, in the situations below, this may be an appropriate method.

- It is difficult to understand the business rules embedded in the application.
- The size of the application is small.
- The application involves obsolete or difficult to maintain technologies.
- The application needs to be customized.
- Organization needs the control of software source code base.
- Reliable outsource partner or people of development and maintenance skilled in SOA exist.

2.4 Replace

Replacement sometimes is easy to be confused with rewriting. Replacement here refers to replace an application with an Commercial off-the-shelf (COTS) package. Some common advantages and disadvantages exist between replacement and rewriting methods, for example, both belong to "big-bang" approach, both bring new added value by leveraging SOA, both introduce new and standards-compliant technologies, however ,both throw all the legacy assets away and lose the investment in the existing application. A particular advantage is less risky and time consuming. Some disadvantages differentiating from rewriting method are listed below.

- Easy to be lock in specific vendor.
- Maintenance and modification may be difficult and expensive.
- Organizations lose control of the source code base.
- Consistent functionality with the original cannot be guaranteed.
- We identify the appropriate situation for the method with the following characteristics:
- The business rules included in the application are clear, common and well understood in the organization.
- No core business is involved in the application.
- Small size application and acceptable purchase cost.
- Must be delivered in a short term.
- Infrequently changes after deployment.
- Available COTS package to meet the organization's needs.

2.5 Wrap

Wrapping method adds a new SOA interface to existing application, making it easily accessible by other software components or services. This is a black-box technology since it just surrounds application with a new interface while hiding the complexity of its internals. Wrapping gives legacy application the following benefits in a quick and a simple manner:

- It leverages the investment in existing application by taking advantage of its capabilities.
- It minimizes the risks and costs in migration.
- It provides rapid return on investment.
- It improves the interoperability by interacting with new SOA-enabled applications.
- It can be used as a gradually transition phrase to SOA.

However, wrapping method does have a few pitfalls. Firstly, wrapping method does not change the fundamental characteristics of the legacy application. Secondly, most of the technology pain points mentioned above cannot be solved through wrapping the application, such as problems in maintenance, upgrading and agility etc. Therefore, wrapping method can just be fit for the following situations:

- The size of the application is relatively small and reusable.
- A fast, cost-effective solution is needed.
- The application has a high business value and good quality code.
- Infrequently changes after deployment.
- The application has high integration needs with other ones.

2.6 Reengineer

As defined by Chikofsky and Cross [4], reengineering is the examination and alteration of a subject system to reconstitute it in a new form and the subsequent implementation of the new form. Reengineering generally includes some form of reverse engineering (to achieve a more abstract description) followed by some more form of forward engineering or restructuring.

For some being considered application in our context, Reengineering involves reverse engineering (analysing the existing application, identifying the best candidates for migration to SOA), transformation (restructuring the reusable components to services), and forward engineering (implementing services and services composition by considering the new requirements, goals, etc.) three sub-processes. So reengineering is an expensive and complex process, but it greatly increases the legacy systems control and evolution to SOA to meet future business requirements. Advantages and disadvantages of reengineering method are listed in detail below.

Advantages:

- It leverages the investment in existing application by reusing application components.
- All the requirements for migration to SOA can be fulfilled.
- It improves interoperability and agility.
- Operational costs will be reduced due to SOA adoption in the long term.
- It improves end user experience and increases customer satisfaction.

Disadvantages include: it is higher risk of failure; the costs of reengineering are for many organizations too high to afford; the reengineering process may be a long time consuming.

Accordingly, the method is suitable for the following occasions:

- The size of application is large.
- The application has high business value and carries on core business process for the organization.
- The application needs to be migrated to distributional circumstance and be integrated with many other applications may be on heterogeneous platform.
- The application needs to change frequently to meet increasingly volatile business requirements.

3 Related works

Much research has been done in the domain of legacy system analysis and migration. We divide the literature into two categories: process-based and technology-based research.

For the process-based research, a well-known "SOA-Migration Horseshoe" has been proposed [5]. This approach applies reverse engineering techniques to extract a legacy enterprise model from the legacy code, then applies enterprise modelling technologies to create a consolidated enterprise model from which services are identified using forward engineering technologies.

Finally, legacy code is transformed to services. Following the horseshoe model, a conceptual framework (SOA-MF) is proposed which embraces a holistic illustration of a migration process, and facilitates to characterize and isolate the properties of migration approaches in terms of processes it supports, artefacts included, activities carried out, and types of knowledge exploited. The Software Engineering Institute (SEI) proposes a method (SMART) for determining the feasibility of migrating legacy systems to an SOA environment; it accounts for the importance of business drivers and characteristics of the legacy system. IBM's Service Oriented Modelling and Analysis (SOMA) also embrace migration of legacy systems to SOA environments and has some portions that address legacy reuse [6]. Reijnders et al. proposes a legacy to SOA migration method by using method engineering approach [7]. Sneed presents a legacy application reengineering planning approach and suggests a cost-benefit model to support decision making [8]. Based on Sneed's work, Umar et al. presents a decision model for SOA reengineering projects that combines strategic and technical factors with cost benefit analysis for integration versus migration decision [9]. In [10], a two-phase model (SABA) is presented to assist organization to decide rationally among very different options. All the focus of the researches mentioned above has been on the whole process and strategy of migration and decision while pay no attention to the approaches themselves. The work by Almonaies et al. is the closest to our research because it also discusses the various approaches and methods and highlights their strengths and weakness [11]. However, comparison has been done between technologies in different literature with respect to a single approach rather than comparison between these approaches which is what we place emphasis on. In addition, we also disagree with Almonaies et al. classification of the approaches where rewriting, outsource developing and replacing with COTS are classified as one category, i.e. replacement. In our view, these approaches have their own characteristics and may be suitable for different applications or organizations. Furthermore, we insist that reengineering is different from redevelopment, and is one of methods of migrating legacy applications to SOA environment.

For the technology-based research, most of works focus on wrapping method. Sneed's approach presents a migration process at lower level through focusing on technical details [12, 13]. Legacy software assets can be preserved by means of wrapping technology. A black-box, wrapper-based migration process, making the interactive functionality of legacy systems accessible as web services, is described in [14]. In [15], different from black-box wrapping method, a reengineering approach is proposed which applies an improved agglomerative hierarchical clustering algorithm to restructure legacy code and to facilitate legacy code extraction for web service construction. Cetin et al. propose a mashup-based approach for migration of legacy systems to service-oriented computing environment [16]. Furthermore, some other works also focus on reshaping the existing legacy elements to the service-based elements, as in [17-20].

4 Conclusions

It is hard to say which method is the best In practice, the information systems of an organization may consist of many applications, and a mixture of methods here is used to form a strategy based on the specific needs of the given organization. The main contribution of this paper is the comparison and analysis of existing methods about their advantages and disadvantages. We also identify the most appropriate situation for each method in order to provide managers with a baseline for making decision of migrating legacy application to SOA. This is a start point of our further work. We will develop an decision model supported by intelligent decision support system for migration to SOA, and conduct case study to evaluate it.

Acknowledgement

This work is supported by the Key Research Program of Higher Education, Henan Province of China, under the grant No. 15A520024.

References

1. Bennett, K.H., M. Ramage,M. Munro, Decision Model for Legacy Systems. IEE Proceedings Software, June 1999. 146(3): p. 153-159.
2. Brown, A., S. Johnston,K. Kelly,Using Service-Oriented Architecture and Component-Based Development to Build Web Service Applications, Rational Software Corporation,R.S. Corporation2002
3. Lewis, G., E. Morris, O.B. L., S. D.,L. Wrage,SMART: The service-oriented migration and reuse technique, CMU/SEI-2005-TN-029,Software Engineering Institute,CMU,2005
4. E.J., C., C.I. J.H., Reverse engineering and design recovery: A taxonomy. IEEE Software, 1990. 7(1): p. 13-17.
5. Winter, A.,J. Ziemann. Model-Based Migration to Service-Oriented Architectures. in Proceedings of the International Workshop on SOA Maintenance Evolution (SOAM 2007), 11th European Conference on Software Maintenance and Reengineering (CSMR 2007). March 20-23, 2007. Amsterdam, the Netherlands: IEEE Computer Society.
6. Arsanjani, A., S. Ghosh, A. Allam, T. Abdollah, S. Ganapathy,K. Holley, SOMA: A Me-thod for Developing Service-Oriented Solutions. IBM Systems Journal, 2008. 47(3).
7. Reijnders, G., R. Khadka, S. Jansen,J. Hage,Developing a legacy to SOA Migration Method, UU-CS-2011-008,Department of Information and Computing Sciences, Utrecht University,2011
8. Sneed, H., Planning the reengineering of legacy systems. IEEE Software, 1995: p. 24–34.
9. Umar, A., A. Zordan, Reengineering for service oriented architectures: A strategic decision model for integration versus migration. Journal of Systems and Software, 2009. 82(3): p. 448-462.

10. H., B.K., R. M.,M. M., Decision Model for Legacy Systems. IEE Proceedings Software, June 1999. 146(3): p. 153-159.
11. Almonaies, A.A., J.R. Cordy,T.R. Dean, Legacy system evolution towards service-oriented architecture, in SOAME 2010: International Workshop on SOA Migration and Evolution. 2010: Madrid, Spain p. 53-62.
12. Sneed, H.M. Wrapping legacy COBOL programs behind an XML-interface. in WCRE '01:Proceedings of the EighthWorking Conference on Reverse Engineering (WCRE'01) 2001. Stuttgart, Germany
13. Sneed, H.M. Integrating legacy software into a service oriented architecture. in CSMR '06:Proceedings of the Conference on Software Maintenance and Reengineering. 2006. Bari,Italy
14. Canfora, G., A.R. Fasolino, G. Frattolillo,P. Tramontana, A wrapping approach for migrating legacy system interactive functionalities to service oriented architectures. J. Systems and Software, 2008. 81(4): p. 463-480.
15. Zhang, Z., H. Yang. Incubating services in legacy systems for architectural migration. in APSEC '04: Proceedings of the 11th Asia-Pacific Software Engineering Conference. 2004. Busan, Korea
16. Cetin, S., N.I. Altintas, H. Oguztuzun, A.H. Dogru, O. Tufekci,S. Suloglu, A mashup-based strategy for migration to service-oriented computing, in IEEE International Conference on Pervasive Services. 2007. p. 169–172.
17. Li, S.H., D.C. Yen,C.C. Chang, Migrating legacy information systems to web services architecture. Journal of Database Management, 2007. 18(4): p. 1-25.
18. Chen, F., S. Li, H. Yang, C.H. Wang,W. Cheng-Chung Chu, Feature analysis for service-oriented reengineering, in Software Engineering Conference. 2005.
19. Liu, Y., Q. Wang, M. Zhuang,Y. Zhu, Reengineering legacy systems with RESTful web service. Computer Software and Applications, 2008: p. 785-790.
20. Zhang, Z., H. Yang,W. Chu, Extracting reusable object-oriented legacy code segments with combined formal concept analysis and slicing techniques for service integration. Quality Software, 2006: p. 385-392.

 DOI: 10.1051/matecconf/20167704006

Mechanical Engineering of Leg Joints of Anthropomorphic Robot

Nikita Pavluk[1], Alexander Denisov[1], Andrey Kodyakov[2] and Andrey Ronzhin[1,2]

[1]*SPIIRAS 39, 14 line, St. Petersburg, 199178, Russia*
[2]*SUAI, 67, Bolshaya Morskaia str., St. Petersburg, 190000, Russia*

Abstract. The problem of design engineering of anthropomorphic robot legs is considered. An overview of the existing anthropomorphic robots and an analysis of servomechanisms and bearing parts involved in the assembly of robot legs are presented. We propose an option for constructing the legs of the robot Antares under development. A two-motor layout, used in the knee, ensures higher joint power along with independent interaction with the neighboring upper and lower leg joints when bending. To reduce the electrical load on the main battery of the robot, the upper legs are provided with a mounting pad for additional batteries powering servos. Direct control of the servos is also carried out through the sub-controllers, responsible for all 6 engines installed in the articular joints of the robot legs.

1 Introduction

Among problems, which developers of mobile general-purpose and special-purpose robots face nowadays, are issues of robot's flotation ability on the rugged terrain, autonomous movement and control of kinetic equipment [1]. A lever-hinge system of human and animals mobility, created by nature, is the most adapted to the natural earth's surface and is suitable for use during movement of the anthropomorphic robot [2].

Because of the lack of a unified methodology and software for engineering lever-hinge systems for anthropomorphic robots, developers are forced to create their own software in the design process of each individual robot [2]-[4].

The aim of this article is to analyze the existing solutions to the construction of lever-hinge mechanisms of the lower extremities (legs) of anthropomorphic robots and to develop the rational structure of legs of the robot Antares.

One of the simplest structures of the biped robot is described in [4]. It is made up of two-millimeter aluminum sheet, includes six servos operated by EyeBot controller, and weighs 1.11 kg. When walking the robot reaches a speed of 120 m/h at a maximum angle of 60 degrees between the hips. Such robot architecture with six servos is used in [5] to study the operating angles of the joints of the knee, ankle and hip.

The anthropomorphic robot in the HanSaRam series, which regularly participates in the FIRA league since 2000, is discussed in [6]. HanSaRam-VIII (HSR-VIII) robot has 28 servos, weighs 5.5 kg and can move at a speed of up to 12 cm/s.

In [7], the anthropomorphic robot Lola, having 7 degrees of freedom per leg, weighs 55 kg with the 180 cm height. The problems of the stability of the robot after stopping as well as a gradual contact of foot parts with the surface when walking are discussed. Elastic materials of the toe and heel of the robot foot ensure reduction of impact force on the robot structure during touching the surface.

For moving on complicated uneven surfaces, impassable for tracked or wheeled robots, more sophisticated non anthropomorphic structures are also being developed, with one [8], six [9], [10] and a large number [11] of lower extremities.

Based on the conducted analysis of structures of anthropomorphic robots, the robot Poppy of the French company INRIA Flowers and Darwin-OP robot of the company Trossen Robotics [12] were determined to be the closest analogues to the robot Antares under development. Let us consider these robots in more detail.

Modular robot design helps the researcher to change the movement of any robot's limb by isolating the desired limb from the rest of the body, almost without affecting the performance. The structure is specially designed for the installation of additional sensors and connection cables. In addition, such a design facilitates periodic robot maintenance service. However, the center of mass of Poppy is located in the solar plexus, which adversely affects load distribution. The robot becomes unstable and cannot move independently (only with the help of human).

Poppy has the same number of degrees of freedom as Antares in the pelvic region, distributing them in a different way, which makes it not so mobile in the knee and ankle joints. The leg joints have only 3 motors (one in each ankle, knee and hip joints). This adversely affects the overall mobility of the robot, judging by the video footage and design files of the robot available from the developers.

The robot Darwin-OP is a robot platform intended for research and development within the framework of educational process. DARwin OP has high performance and dynamic characteristics and a wide range of sensors. The robot communicates with people by using loudspeakers, microphones, cameras, tactile sensors, LEDs, hand gestures. It possesses 20 Dynamixel actuators that ensure free movement of limbs with a given accuracy and strength margin, as the gears are made of metal. The center of mass is located in the center of pelvis, which ensures the correct distribution of the load during walking and inertia, especially in the extremities. A modular robot design helps the researcher to change the movement of any limb of DARwin OP. The structure also allows the installation of additional sensors.

2 Design of leg joint of the anthropomorphic robot antares

Designing Antares included several steps associated with the development of joints of legs, arms, torso and head. First priority is given to designing the leg joint because of the following reasons: the high complexity of the layout of joint parts; necessity of this joint for robot movement in space; complexity of calculating joint unit due to the assumed highest load of parts relative to all other joints.

Total toe structure includes pelvic joint and two identical leg joints consisting of simpler attachment point to the pelvic, hip, knee, lower leg, ankle and foot units. Joints of robot pelvic and legs are developed in accordance with the proportions of the human body, adjusted for assumed growth. Their total length is 510.7 mm. The length of ankle and hip joints is 20 cm.

Since in the kinematics of the robot these joints act as a lever arm, this constructive solution is reasonable. The difference in length is achieved by connecting lower leg joint to the ankle joint which is, in turn, connected with the foot, while the upper leg joint is linked to the hip joint that is connected to the point of attachment to the hip unit.

Auxiliary batteries (an installation site is provided) will be installed in the upper legs to reduce the electrical load on the main battery of Antares as well as to control actuator powering, which will help to avoid power supply problems. To save the processing power of the main controller and the computer located in the torso, auxiliary controllers will also be installed in the upper leg joints. These controllers are responsible for the work of all six engines of robotic leg joints. The given design complicates the calculation of the kinematics of the robot motion, but provides more complex movements.

Figure 1. A two-motor knee

Through the use of twin-engine layout in the knee (Fig. 1) a separate leg joint is obtained, which interacts with neighboring upper and lower leg joints, allowing them to be independent of each other when bending (Fig. 2). In addition, the use of two-motor knee simplifies the selection of servo, as in this type of knee the load is divided between two separate motors. Another advantage of this unit is that it facilitates design engineering of the above-mentioned joints of the upper and lower legs.

Interconnected plate lines, linked by screw couplings and cross plates, make up the basis of the leg structure of Antares (Fig. 3). For parts manufacturing, 2, 4 and 10 mm thick aluminum sheets were used. 6-mm-thick aluminum rods were applied for the crossties.

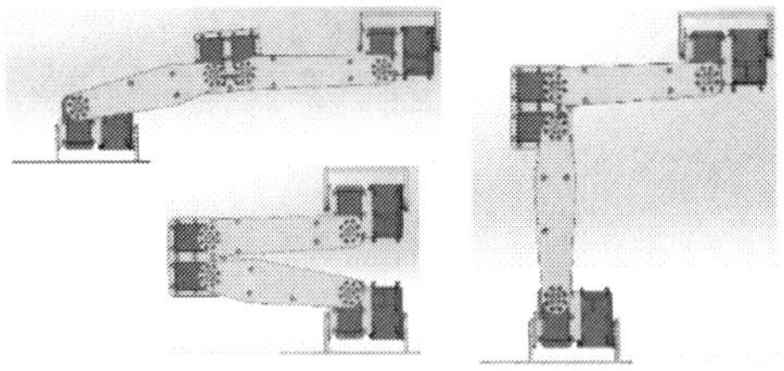

Figure 2. Functional capabilities of the two-motor knee

Parts made from 2 mm thick aluminum sheet are the basis of the joints of lower and upper legs, foot and pelvic attachment. The joints parts made from 2 mm sheet are structurally designed for power loads and pressure from top. Parts made from 4-mm-thick aluminum sheets are used as stiffening ribs intended for the torsional loads. From these sheets we made transverse upper and lower leg struts and mounting plates of the hip (4 mm thick) and foot. It is also intended to use them to locate the internal components, such as an auxiliary battery, actuators controller of the entire leg joint. Along with the transverse plates, aluminum bars are used as stiffening ribs but only as elements of structural reinforcement.

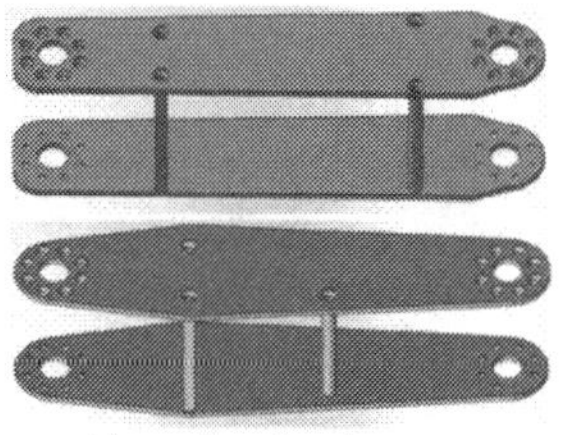

Figure 3. Upper and lower leg structures

In the tibial joint a broader transverse plate is used in order to achieve sufficiently reliable structure. This was necessary to ensure that the ankle could be used as efficiently as possible, which required the free use of the internal space of ankle joint. Screw coupling in this case is not only a stiffening rib, but also an arresting stop so that the ankle joint could not be broken and lead to the damage of other components while working. From 10 mm thick aluminum sheets one type of parts is made — a special bearing plate used for the assembly of hip and ankle joints.

For motors Dynamixel MX-64, used in the leg joints, the flanges were made that are necessary for linking separate joints parts and components to provide mobility and stability. The flanges are located in a special socket on the motor housing and the bearing and secured by a special cover which prevents the collapse of the structure during motion and because of vibration. An obligatory requirement for the bearing is a height of 3.5 mm, to comply with the centering of axial arrangement of engines in the overall structure, which is important during robot movement.

Pelvic mechanism is located in the lower part of the torso of Antares and is designed for the axial legs rotations as well as for accommodating the main battery. The construction includes two bushings, gears with gear ratio 1: 1, two motors Dynamixel MX-28, two 2- mm-thick plates, and the flange bearings 8×14 (Fig. 4). The choice of less powerful servos, compared with those used in the construction of the other legs joints, is substantiated by the fact that for axial rotations the high power of engines is not required.

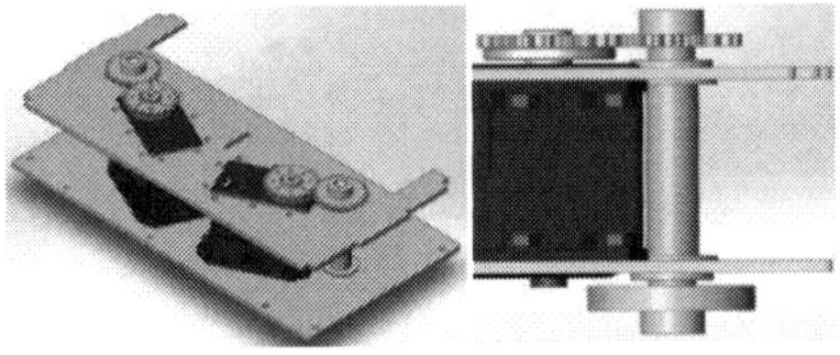

Figure 4. Pelvic mechanism

The hip and ankle joints are formed by pairwise motor connection with metal inserts; in the motor housing extra strong plastic or aluminum are used in order to withstand physical loads on the motor housing during movement (Fig. 5).

a) b)

Figure 5. Joint assemblies of: a) hip; b) ankle

Pairwise motor connection provides space saving in the construction, in order to avoid excessive massiveness. This is necessary to ensure the mobility and flexibility of the assembled joints. In addition, the construction of the hip and ankle joints is designed to the highest possible reproduction of functionality of human ankle and hip joints which are similar to a spherical joint with a limited angle of rotation. Thus, to reproduce the human joint structure, it was decided to introduce to these joints two cylindrical hinges with mutually perpendicular axes shown in Fig. 5.

This solution is applied because, at the moment, it is not possible to repeat structure of the spherical joint and make it controllable to the full for the anthropomorphic mechanism, sufficiently reliable and compact, relatively inexpensive and not requiring constant maintenance. Achieving all of this will complicate both the construction itself and the control of the robot. The implemented solution, in turn, simplifies the structure of the robot ankle without compromising ankle capabilities.

3 Experimental results and discussion

The leg construction provides large steering angles for the motors, which ensures greater flexibility and ductility of the joint compared with the mentioned analogs. The robot can easily do the splits, raise the leg straight or at an angle, bend it at the knee, while keeping the foot parallel to the floor, if necessary. There are a variety of sit-ups rather than some specific predetermined movements seen in the analogues, including touching the floor by the pelvic mechanism without the threat of damaging its joints. Pelvic mechanism provides axial rotation of the robot lower limbs, which has a positive effect on its portability and allows it to rotate on the way with a minimum turning radius, turn in the necessary direction on the spot and perform a wider range of different leg movements than those available to humans.

Pelvic mechanics allows the robot to rotate each leg at 360° around its axis; however, at this stage we limited it up to 270°, since there was no need for such a large range of rotation. The angle of flexion at the hip joint is 120°. The maximum extension angle is 45°. The side lunge of the leg is + 90°. Similar to humans, it is impossible to fully bring the leg back, as the lower limbs will touch each other. The maximum bringing back of the leg is 55°. The legs are brought back up to the moment when they almost touch each other, and this limits the range of the stroke. The side lunge of the leg is limited to an angle of 135°. It was decided to set limits up to 90°. This is due to the fact that the robot should resemble the structure of the human, and this angle allows the robot to do the splits. In this case, there is no need to have a greater range of variations of the leg position in the hip joint. The knee joint comprises two servos combined into one unit. This design solution increases the mobility and strength torque of the knee. This allows the actuator to bend the knee at an angle of 162°. The use of two actuators in the knee joint entails constant synchronization of the motors relative to each other, since one actuator does not provide the same large angle of the displacement in a joint without a significant loss in power and mobility of the structure. The result is that the location of motors corresponds to the perpendicular position relative to the frontal plane of the robot. The ankle design provides the eversion — the displacement of the foot inwards to the sagittal plane as well as the inversion — shifting of the foot outwards from the sagittal plane. Changing the generalized coordinate of the ankle joint in roll represents a rotation of the foot relative to the upper limb from the neutral position. The angle of the foot position changes: eversion is 90° and inversion is 90°. Moving toe upwards relative to the neutral position is taken as a positive angle, and downwards — as the negative angle. In this case, the pitch angle of the foot deviates from + 88° to -180°. Fig.

6 shows the results of work to change the generalized coordinate of the robot foot. Movement is carried out in full compliance with design feature that limits the angles of pitch and roll.

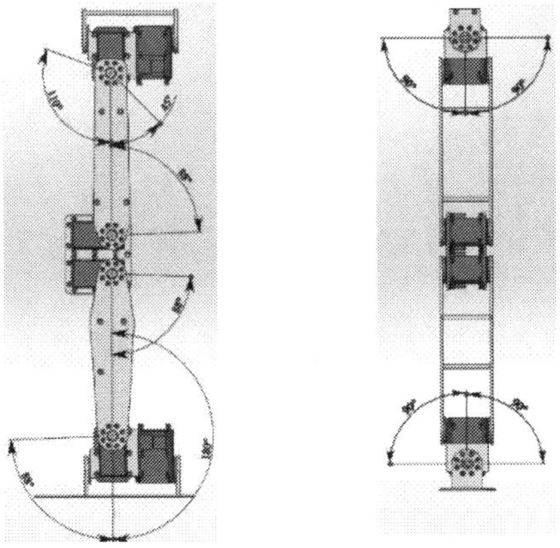

Figure 6. Turning angles of the leg

Apart from the pelvic mechanism responsible for axial rotations of legs and being a separate unit, in the legs structure we applied actuators Dynamixel MX-64. Each Dynamixel has a unique ID for connection to the common data line, supports the connection TTL, RS485 and others, can be connected to the common control bus; available LED or emergency shutdown (torque-off) functions may be set to predetermined values of temperature, current and voltage. These actuators can be adjusted to move more smoothly. Dynamixel servos can be controlled from a PC or a microcontroller, which is a great advantage in the development of prototypes. Without the battery and the controller installed in the hip joint, the estimated total weight of the metal frame, servos and flange connections is 1.07 kg, of which 756 g account for the Dynamixel MX-64 servos. Construction of the two joints of the leg and pelvic mechanism weighs 2.44 kg.

Table 1. The comparison of the angles of the human and the robot Antares

	Human	Antares
Range by the pitch of foot	from -50° to 40°	from -88° to 180°
Range by the roll of foot	from -30° to 20°	from -90° to 90°
Range of flexion-extension, for the hip	from +120° to 20°	from +110° to 45°
Range of flexion-extension, for the knee	from 0° to 110°	from 0° to 162°
Range of the side lunge of the leg in the hip	from 0° to 45°	from 0° to 90°
Range of the relative bringing back of the leg in the hip	from 0° to 30°	from 0° to 55°
Range of rotation for the hip	from -45° to 45°	from -45° to 45°

Table 1 shows a comparison of the different angles of deflection of legs positions of the human and developed prototype of an anthropomorphic robot Antares. Based on this table, we can conclude that the various actions typical of the human and the robot can be executed by the latter, as the ranges of changes of human positions angles lie within the ranges of changes of robot angles.

It should be noted that, except for the range of the foot roll, all other angles are artificially limited in order to avoid unnecessary contact of the parts lying in a single plane at the time of motion. Since the robot represents a robotic platform intended for scientific research and development within the framework of educational process, its modular nature can help the researcher to change the movement of any limb by isolating the desired limb from the rest of the body, almost without affecting performance. The structure is specially designed for the installation of additional sensors and connection cables. In addition, such a design facilitates periodic robot maintenance service.

Load weight balancing in the legs of the robot taking into account the maximum supposed weight of 8 kg occurs according to the following formula:

$$F = P/2 = 80/2 = 40\ N$$

where, F is the force applied to the lower leg assembly; P is a weight, which keeps the lower leg; the denominator — 2, since all the weight is distributed over 2 legs. Fig. 7 shows the torque supplied to the motor shaft of the ankle in the process of raising the legs. The graph shows that the ankle motor has a maximum torque when the robot is in a sitting position. The tension in the construction of the lower and upper leg with a total weight of the robot equal to 8 kilograms is 7.44 * 10 ^ 5 N/m ^ 2.

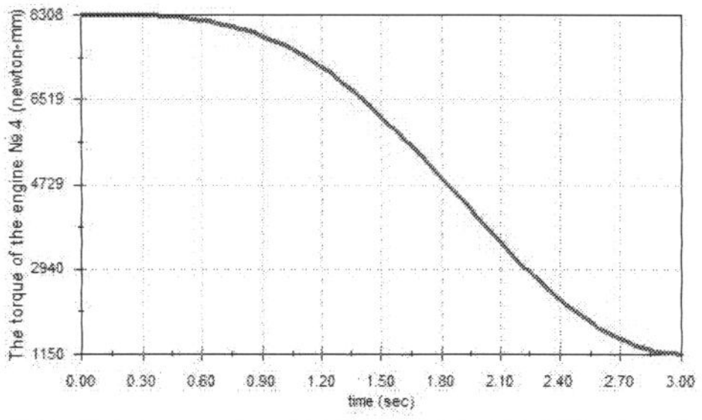

Figure 7. Torque supplied to the motor shaft

The conducted study on the maximum load revealed that the lower leg can withstand a load of 1040 Newtons. Summarizing, we can say that the lower and upper leg individually can withstand 13-fold vertical load on the structure, without succumbing to deformation and breakage. It was taken into consideration that fasteners for flange connections were fixed and immobile.

This research devoted to mechanical engineering of leg joints of anthropomorphic robot is focus on development of technical means and mobile robots, in order to modeling collaborative work of users, robots and distributed means of cyberphysical environment [13]-[17].

4 Conclusion

The conducted analysis showed the presence of anthropomorphic robot models from 30 cm to 180 cm high with a different number of degrees of freedom and kinematic schemes. It was concluded that the closest analogues to the robot Antares under development are Poppy and Darwin-OP robots. A two-motor layout, used in the knee, ensures higher joint power along with independent interaction with the neighboring upper and lower leg joints when bending. The electrical load on the main battery of Antares is reduced by the use of auxiliary batteries installed in the upper legs and powering the servos. The direct servo control is also performed by the auxiliary controllers responsible for the work of all six engines of the leg joints. Studies of the prototype design have demonstrated that individual components and parts have a more than ten-fold safety margin. The developed robot is intended for creation of assistive technologies via human-robot interaction based on multimodal interfaces in cyberphysical intelligent environment.

The study was performed through the grant of the Russian Science Foundation (project no. 16-19-00044).

References

1. V.B. Kudryashov, V.S. Lapshov, V.P. Noskov, I.V. Rubtsov. Problems of robotization for military ground technics, Izvestiya SFedU. Engineering Sciences, **3/152**: 42-57, (2014).
2. A.K. Kovalchuk, D.B. Kulakov, S.E. Semenov, V.V. Yarots, A.A. Vereikin, B.B. Kulakov, L.A. Karginov. Method for designing spatial tree-like actuators of walking robots, Engineering bulletin of the Bauman MSTU, **11**: 6-10, (2014).
3. V.V. Lapshin. About the walking motion stability, Science and Education of the Bauman MTSU, **6**: 319-335, (2014).
4. S. Warnakulasooriyaa, A. Bagheria, N. Sherburnb, M. Shanmugavel. Bipedal Walking Robot – A developmental design, Procedia engineering, **41**: 1016-1021, (2012).
5. S.C. Lima, G.H. Yeapa. The Locomotion of Bipedal Walking Robot with Six Degree of Freedom, Procedia Engineering, **41**: 8-14, (2012).
6. Yoo J.K., Lee B.J., Kim. J.H. Recent Progress and Development of the Humanoid Robot Hansaram, Robotics and Autonomous Systems, **57**: 973-981, (2009).
7. T. Buschmann, S. Lohmeier, H. Ulbrich. Humanoid Robot Lola: Design and Walking Control // Journal of Physiology, **103**: 141-148, (2009).
8. X. Yua, C. Fub, K. Chen. Modeling and Control of a Single-legged Robot, Procedia Engineering, **24**: 788-792, (2011).
9. H. Rostro-Gonzalez, P.A. Cerna-Garcia, G. Trejo-Caballero, C.H. Garcia-Capulin, M.A. Ibarra-Manzano, J.G. Avina-Cervantes, C. Torres-Huitzil. A CPG System Based on Spiking Neurons for Hexapod Robot Locomotion, Neurocomputing, **170**: 47-54, (2015).
10. P.S. Pan, C.M. Wu. Design of a Hexapod Robot with a Servo Control and a Man-Machine Interface, Robotics and Computer-Integrated manufacturing, **28**: 351-358, (2012).
11. R. Vidoni, A. Gasparetto. Efficient Force Distribution and Leg Posture for a Bio-Inspired Spider Robot, Robotics and Autonomous Systems, **59**: 142-150, (2011).
12. «ROBOTIS» catalogue. http://en.robotis.com/index/product.php?cate_code=101011
13. A.A. Karpov, A. Lale, A.L. Ronzhin. Multimodal assistive systems for a smart living environment. SPIIRAS Proceedings, **19**: 48-64, (2011).
14. A.I. Motienko, A.L. Ronzhin, N.A. Pavlyuk. Modern developments of rescue robots: possibilities and principles of their application, Science Bulletin of the NSTU, **3/60**: 147-165, (2015).
15. A.L. Ronzhin, V.Yu. Budkov. Multimodal Interaction with Intelligent Meeting Room Facilities from Inside and Outside, NEW2AN/ruSMART 2009, LNCS 5764: 77–88, (2009).
16. R.M. Yusupov, A.L. Ronzhin. From Smart Devices to Smart Space, Herald of the Russian Academy of Sciences, **80/1**: 45-51, (2010).
17. A.A. Karpov, A.L. Ronzhin. Information Enquiry Kiosk with Multimodal User Interface, Pattern Recognition and Image Analysis, **19/3** (Moscow: MAIK Nauka/Interperiodica): 546-558, (2009).

Flow Features of Three Side-by-side Circular Cylinders at Low Reynolds Number

Liu Junkao, Hao Yaqiang, Dong Dibo, Chen Weishan

State Key Laboratory of Robotics and System, Harbin Institute of Technology, Harbin, Heilongjiang Province, China

Abstract. In order to study the fluctuation of kinetic parameter of cylinder matrix in incompressible stationary flow, the flow fluid around three side-by-side circular cylinders are simulated using Immersed Boundary–Lattice Boltzmann method (IB-LBM). Drag and lift force of the three cylinders are investigated as the interval between each cylinder varied from zero to five times of the cylinder diameter. Five flow patterns are defined according to the vortices structure in the downstream of the cylinders. Power spectrum analysis of lift force is developed to explain the vortex patterns. Through the research, we find the strength and phase of the gap flow play an important role in the vortex formatting process. The vortices shedding from different cylinders neutralize and combine in the near wake, contributing a lot to the variation of forces.

1 Introduction

The interaction between solid and fluid, which has much significance in aviation, energy harvesting and equipment manufacturing, has attracted many attentions these years. Flowing past bluff body is the most common model in dealing with this kind of research, because of its wide existence, for example, power transmission lines, offshore work platform, bridge pier, etc. Wake structure interaction, vortex shedding and jet flow are the main characteristics of this kind of flow [1]. Many methods, based on the theory study, experiment and numerical simulation, have been developed to make the flow patterns of multiple cylinders in steady flow clear. Experiment study starts in an earlier time. Lift and drag coefficients, frequency of vortex shedding, pressure around the cylinders, as well as the wake model, are the main factors helping us to comprehend the nature of the flowing.

Previously, many people have done a very significant job. Surface pressure measurements, drag and lift forces measurement and the wake observation give us two explicit knowledge [2-5]. One is the flow patterns of the wake behind two side-by-side circular cylinders include three main regimes as the interval (T/D) increasing. When $T/D < 1.2$, the two cylinders act as single one forming one single street; when $1.2 < T/D < 2.0$, the flow interplay and the wake are asymmetrical; for $T/D > 2.0$, the coupled streets occurs. The other knowledge is the periodic vortex-shedding contribute to the fluctuated forces on cylinders. Vortex-shedding changes the distribution of hydraulic pressure, and the cylinders are subjected to a lift and drag force as a result. The fluctuation of lift force synchronizes to the vortex shedding.

Till now, to find out the nature of fluid and structure, many experiments are still undertaken with some new technology. The hot-wire, laser-induced fluorescence flow visualization and Particle Image Velocimetry (PIV) techniques are usually used to examine the flow filed, while fiber grating and piezo-electric technologies are employed to measure the dynamic features. And the sharp and configurations of cylinders also are settled in a more complicated way [6, 7]. Alam and Zhou identified four flow regimes of flow behind two side-by-side square cylinders using flow visualization method and the rules of street-merging were discussed [8]. The investigation presents similar but different results. The flow around four cylinders in a square configuration is observed by Lam and Chan in 2003 using laser-induced fluorescence (LIF) visualization and particle image velocimetry (PIV). The transformation of flow patterns were investigated as the angle incident varying from $\alpha=0°$ to $45°$. Like two side by side cylinders, the wake also is classified to three flow regimes according to α. Meanwhile, the configuration of cylinders leads to strong vortex-induced vibration at $\alpha=15°$ because of the jet flow between the near wake of upstream cylinder and downstream cylinder [9]. The flow filed around three side-by-side cylinders was investigated by Zhang and Zhou in 2001, arming to find out the effect of different interval on the flow patterns. A comparison between equal and unequal cylinder spacing was made on pressure, drag and lift coefficient, and vortex formation mechanisms [10].

Numerical simulation is a suppositional way to imitate the flow field significantly. Meneghini and Saltara carried out computations for flow interference between two circular cylinders in abreast and tandem arrangement for Re ranging from 100 to 200 through the finite-element method. The study presented vorticity around the cylinders and force evolution with time [11]. Unsteady three-dimensional flow simulations were implemented on different gap/diameter ratios. The growth of vertical structures and different modes of flow bifurcation along their spanwise extended core lines were discussed [12]. As to square cylinders, Burattini and Agrawal computed the two-dimensional flow field around two side-by-side square cylinders while the lateral separations between the cylinders T/D ranges from 0.5 to 6. Three regimes are measured during the varying of intervals, in-phase, in-antiphase and quasi-periodic, which has been confirmed by experiment [13]. Supradeepan classified the flow pattern more minutely. Five different flow patterns were found for the flow behind two side-by-side cylinders through periodic variation of flow, frequency spectrum analysis and phase diagrams [14].

There are few researches on three circular cylinders arranged side-by-side. This paper focus on the parameters fluctuating and the flow regime of the flow field around three circular cylinders arranged side-by-side. Immersed Boundary-Lattice Boltzmann Method is used to compute the fluid date. The relationship between fluid regime and lift force of middle cylinder are discussed.

2 Numerical simulation method – LBM

Unlike other common numerical system focused on molecular dynamics or macro-kinetics, Lattice Boltzmann method utilizes dispersed lattice and dispersed time based on microscopic models as well as macro-kinetics equations, taking macro-scope motion as the response of the collective behavior of the dispersed particles [15]. LBM takes the space and time as dispersed regular lattices, the fluid as dispersed particles. Particles' movement is restricted by lattices in particular rules. The motions of particle are consist of two progress, collision and propagation, each of which stands for the dynamics and kinematics progress. Microscopic statistics method is also used to get macro-scope parameters, like velocity and pressure.

The use of macro-scope dynamics and microscopic statistics on the dispersed microscope model provides many advantages to LBM, which involve the clear operating process, convenient dispose of complicated boundary conditions and easy access to parallel algorithms. As the development of parallel computer technics, LBM gets more attention those years.

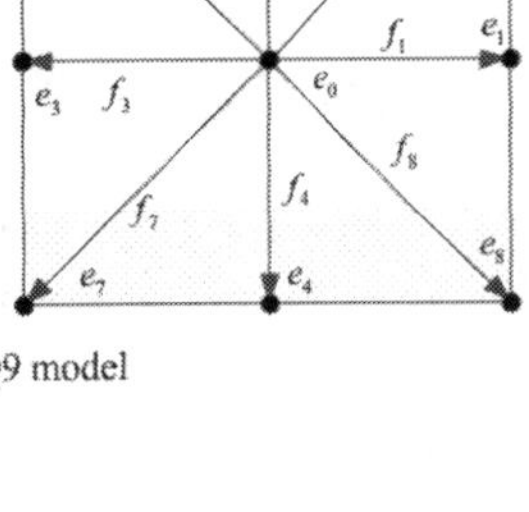

Figure 1. D2Q9 model

Different dispersed-model stands for different ways of particles collision and propagation, whereas all of them obey the law of mass and momentum conservation. In this paper, we use D2Q9 model [16], where D means the space dimensions and Q means the number of particles. The model of D2Q9 is shown in Fig. 1.

The evolution equation of LBM dispersed by two-dimensional viscous incompressible D2Q9model is as follow:

$$f_i(x+c_i\Delta t, t+\Delta t) - f_i(x,t) = \Omega_i \tag{1}$$

where:

$$\Omega_i = -\frac{f_i(x,t) - f_i^{eq}(x,t)}{\tau} \tag{2}$$

$$f_i^{eq} = \rho\omega_i[1 + \frac{c_i u}{c_s^2} + \frac{(c_i u)^2}{2c_s^2} + \frac{u^2}{2c_s^2}] \tag{3}$$

$$c_i = \begin{cases} (0,0)c & i=0; \\ (\cos\frac{i-1}{2}\pi, \sin\frac{i-1}{2}\pi)c & i=1,2,3,4; \\ (\sqrt{2}\cos(\frac{i-5}{2}+\frac{1}{4})\pi, \sqrt{2}\sin(\frac{i-5}{2}+\frac{1}{4})\pi)c & i=5,6,7,8. \end{cases} \tag{4}$$

where i is the direction of discrete velocity, i=0,1,2…8; Δt refers to the discrete time step; $f_i(x,t)$ is the particle velocity distribution functions, defined on the particle location x and time t; τ is the single-relaxation-time parameter which controls the rate of approach to equilibrium; c_i represents the particle velocity in direction; Ω_i is the collision operator, which stands for the rate of $f_i(x,t)$ after collision; f_i^{eq} is expressed as the local equilibrium distribution function; ρ and u are the macro-scope density and velocity; $c_s=c/\sqrt{3}$ is the sound speed; while ω_i is weight coefficient, ω_0=4/9, ω_i=1/9 (i=1,2,3.4), ω_i= 1/36 (i=5,6,7,8).

The macro-scope velocity and momentum are defined as follows in D2Q9 model.

$$\rho = \sum_i f_i = \sum_i f_i^{eq}, \rho u = \sum_i f_i c_i = \sum_i f_i^{eq} c_i \tag{5}$$

This paper takes those three cylinders (with equal diameters of 25) and the flow field (the size is 1400×800) as research objects. As showing in Fig. 2, take the left bottom of flow field as the origin of coordinates, the flow and cross flow direction as x and x axis. The initial and boundary conditions must be specified in advance. The three cylinders are named C1, C2 and C3 in sequence.

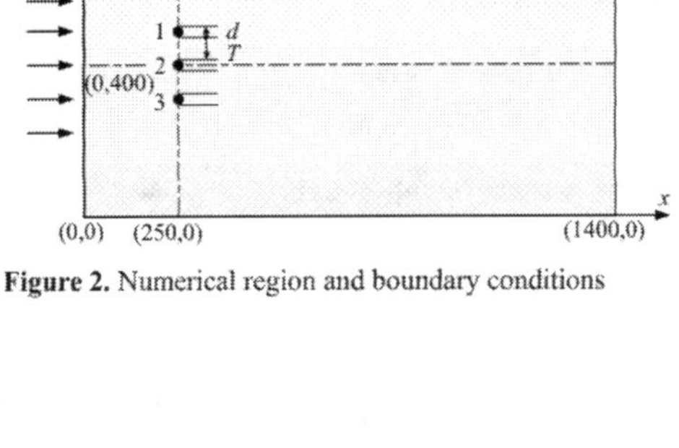

Figure 2. Numerical region and boundary conditions

DOI: 10.1051/matecconf/20167704007

where d means the cylinders diameter; u, v is the velocity on x, y directions; Re=ud/v represents the Reynolds number, v is kinematic coefficient of viscosity; T stands for interval coefficient where $T=D/d$.

When time $t = 0$, the flow field is assumed to be static and a uniform and constant free-stream flow is introduced into the domain from left boundary after that. Upper and lower boundaries are disposed with non-equilibrium extrapolation scheme to eliminate the impact on the flow regime. Non-slip boundary condition is invoked to all cylinder boundaries. Considering the significance role of the lattice structure in the numerical model, Immersed Boundary Method is applied to the cylinder boundaries. The boundary conditions are as follows:

Inlet: $u = U_\infty$, $v = 0$

Outlet: $\partial u / \partial x = \partial v / \partial x = 0$

Upper and lower boundary: $\partial u / \partial y = 0, v = 0$

Cylinder boundary: $u = v = 0$.

In this paper, the interval ranges from 0 to 5. We get the conclusion on the influence of interval to flow pattern, and interaction between fluid and solid structure, through the analysis of the static condition parameters, like the lift, drag force coefficient and vortex structure. Furthermore, in order to reveal the nature of vortex shedding of the three cylinders, FFT algorithm is developed to analysis the vortex shedding frequency.

3 Results and discussion

In this paper, the flow behavior around three circular cylinders arranged side-by-side with the same interval at Re (Reynolds number) =80 and 100 are investigated. The mean drag and the VPP lift force of C2 are shown in Fig. 3. The statistic value is compared with that of a single cylinder.

From the chart we can find both the lift and drag force reaches the maximum value when T=0. The drag coefficient decrease step by step when T > 0, reaching a valley when T=1.0. T=1.0 is a peak with a value a little higher than T=0.5. The drag coefficient decrease slightly with the value inching closer to the value for a single cylinder. As to the lift coefficient, there is a sharp decrease on lift coefficient when T=0.5 with the value evidently smaller than on single cylinder. After that, the value increases, reaching the peak twice when T=1 and T=3.5. When T > 3.5, the lift force decrease step by step, approaching the value of one single cylinder. To explain the fluctuation with T, the flow regime is investigated. Five flow patterns are defined according to the vorticity contours and the fluctuating of mean drag and VPP lift force.

Referring to the review the predecessors of advances [16], the five defined patterns includes the single solid model, interim model, in-phase model, diff-phase model and separate model. Meanwhile, contrast to Re=80, the certain pattern condition occurs at a smaller interval coefficient T at Re=100.

As shown in Fig. 4, the flow pattern behind three side-by-side cylinders performs as the single model when T=0, 0.5 at Re=80 and 100.When T=0, there is no fluid flowing through the cylinder gaps. The vortex shedding from these cylinders is rounded and periodic as shown in Fig. 4 (a). When T=0.5, the interval is so narrow that the vortex from the gaps of these three cylinders are much weaker than these from the top of C1 and the bottom of C3, which we can called the main-vortices. There is no obvious vortex shedding in the gap flow. As the Fig. 4 (b) showed, the main-vortices are spindly in the near weak but rounded and regular and in the far wake, similar to the single cylinder wake.

From the Fig. 4 (b) we may get the cause of the sharp decrease when T=0.5. As shown, the interplay between the vortices from the gaps and the main-vortices is evident in the near wake. Part of the negative vortex from the top of C2 and the positive vortex from the bottom of C1 neutralize in a limited region and the remanent parts with the same polarity trend to combine together forming single larger vortex street on the action of hydrodynamics.

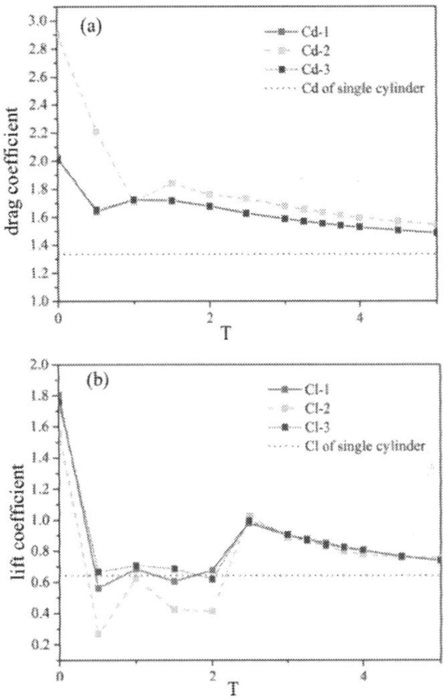

Figure 3. The fluctuating of mean drag (a) and VPP lift (b) force of C2 with T at Re=100

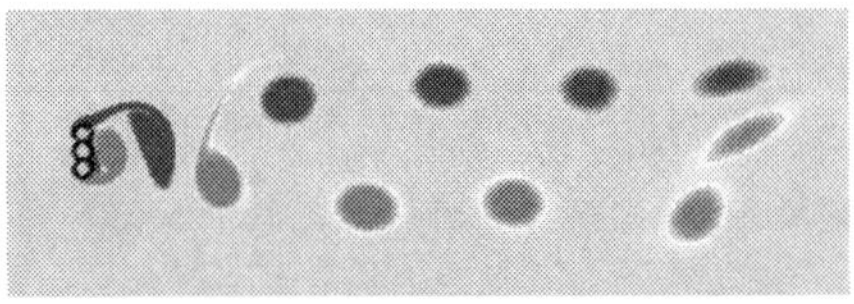

(a) Re=100 T=0

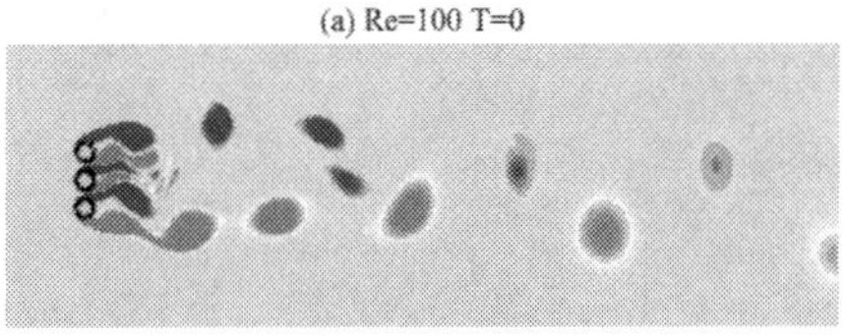

(b) Re=100 T=0.5

Figure 4. The vorticity contours for interval coefficients when T=0 and T=0.5 at Re= 100

Fig. 5 shows the pattern of interim model of three circular cylinders arranged side-by-side, which are computed when T=1.0, 1.5 at Re=80 and T=1.0 at Re=100. The spindly vortex fragmented when flowing away from the cylinders before combined, forming many irregular and aperiodic less-intense vortices. Many small vortices interact with each other, formatting the chaotic nature of this stage, which is evident in Fig. 5 (b). As the size of cylinders interval keeps growing, the vortex from the gap flow cannot be ignored and the vortex stretches longer in the near wake, resulting in much interference between vortices. For example, the positive vortex from the top of C2 trends to combine with the main vortex of same sign from the top of C1 on the action of hydrodynamics, while there is a negative wake on the way.

Furthermore, this kind of disorganized nature of the vorticity field also is observed by Sumner from the PIV experiments [16]. With the comparison of Fig. 5 (a) and Fig. 5 (b), we find the interim model occurs at smaller gap size when Reynolds number is lager.

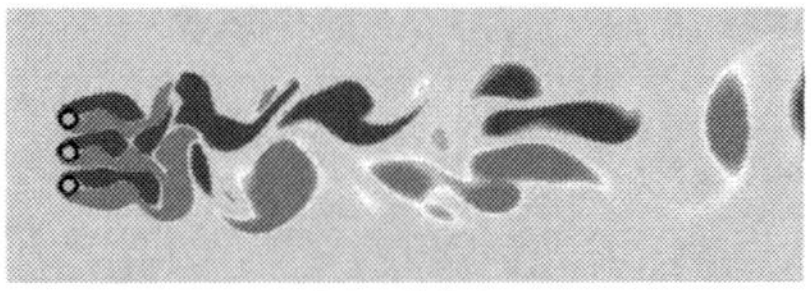

(a) Re=80 T=1.5

(b) Re=100 T=1.0

Figure 5. The vorticity contours for interval coefficients when T=1.5 at Re=80 and when T=1.0 at Re=100

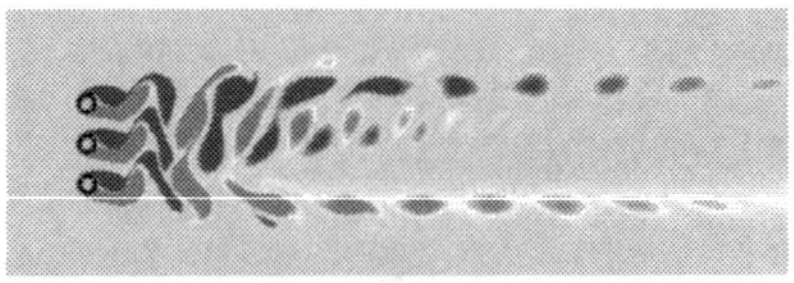

(a) Re=80 T=2.0

(b) Re=100 T=1.5

Figure 6. The vorticity contours for interval coefficients when T=1.5 at Re=80 and when T=1.0 at Re=100

Fig. 6 shows the pattern of in-phase model of three circular cylinders arranged side-by-side, when T=2.0 at Re=80 and T=1.5 and 2.0 at Re=100. In this stage, gap flow vanishes. The vortices flow from the three cylinders shedding in phase with each other in the near cylinder area, which is evident in the time history of force coefficients, as shown in Fig. 8. In the further downstream area of the cylinders, reacting of vortices occurs, resulting in a wide wake. The vortices of opposite sign neutralize and the vortices of same sign combine together. As the distance is beyond a certain value, the interference effect weaker and the flow subside. The wake divides into two parts of the opposite sign, forming two trains of vortices staggered in the water flowing direction, as shown in Fig. 7 (a).

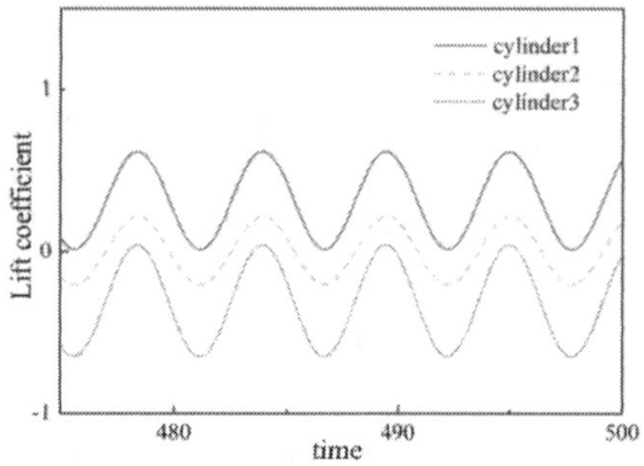

Figure 7. The stable time history of lift and drag coefficient when T=1.5 at Re=100

Fig. 8 shows the pattern of diff-phase of three circular cylinders arranged side-by-side, which is computed when T ≥ 2.5. Each of the vortex shedding regimes of the three cylinders is similar to that of the single cylinder, with little interaction between each other in the near weak. Maybe that is why the lift and drag coefficient are close to the single cylinder value.

While in phase, the asynchrony behavior is observed evidently. The vortices shedding from C1 and C3 are symmetrically synchronous with opposite phase, while the vortex of center cylinder is a little phase-leading. As shown in Fig. 8 (a) and Fig. 9, the weak of cylinder 3 is anti-phase to cylinder 1 but almost in-phase to C2. Fig. 8 (b) (D=3.5) indicates, in the far weak of the three cylinders, the vortices with the same sign from C2 and C3 combine, forming a string of irregular vortices along the flow direction. Meanwhile, the vortices from C1 are stable, showing little interaction with the vortices from C2 or C3. The vorticity contours of diff-phase matches the experiment of Williamson, revealing that the vortex shedding anti-phase is stable [4]. C1 and C3 have the same vortex shedding frequency according to the frequency spectrum of lift force, as shown in Fig. 11. The vortex shedding of C2 is a little frequenter, as shown in the Fig. 10.

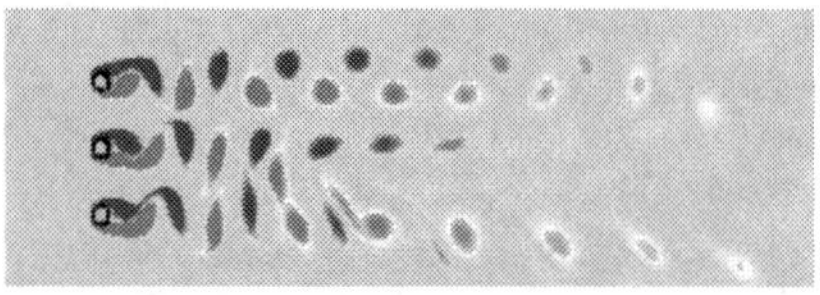

(a) Re=100 T=2.5

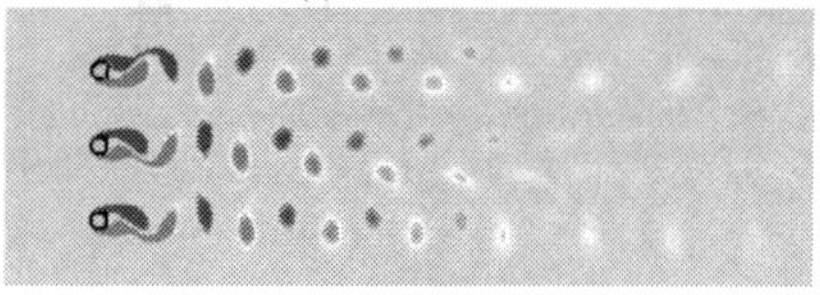

(b) Re=100 T=3.5

Figure 8. The vorticity contours for interval coefficients when T=1.5 at Re=80 and when T=1.0 at Re=100

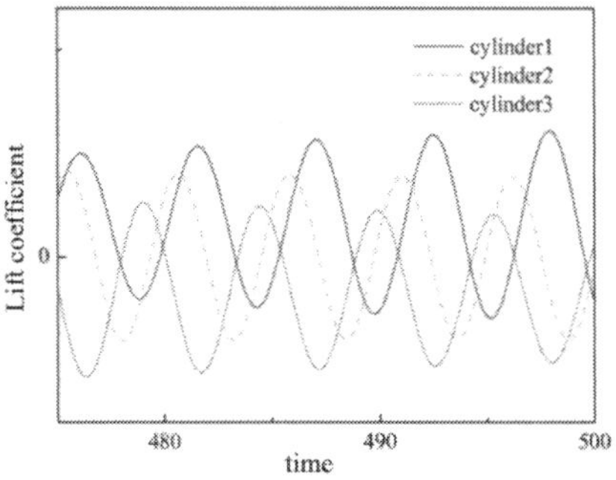

Figure 9. The stable time history of lift coefficient when T=2.5 at Re=100

Fig. 11 shows the pattern of separate model of the three circular cylinders arranged side-by-side, which are computed when T=5.0. The vorticity contour indicates the interaction among the vortices of cylinders can be ignored in the near and far wake. The flow pattern of each cylinder can be seen as the single cylinder flow regime with the same lift and drag coefficient and vortex-shedding frequency.

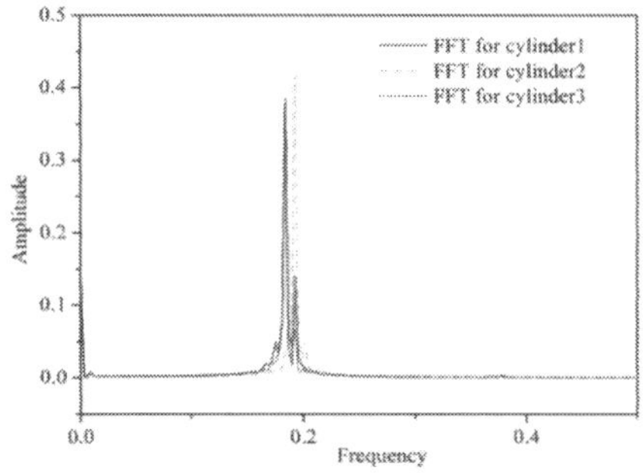

Figure 10. Power spectra analysis of lift coefficient when T=2.5 at Re=100

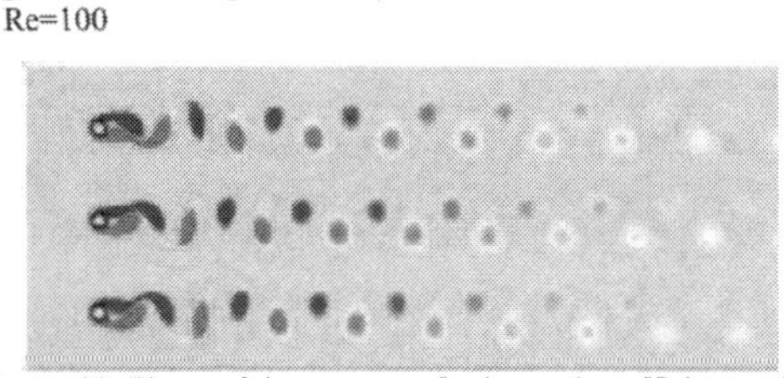

Figure 11. The vorticity contours for interval coefficients when T=5.0 at Re=100

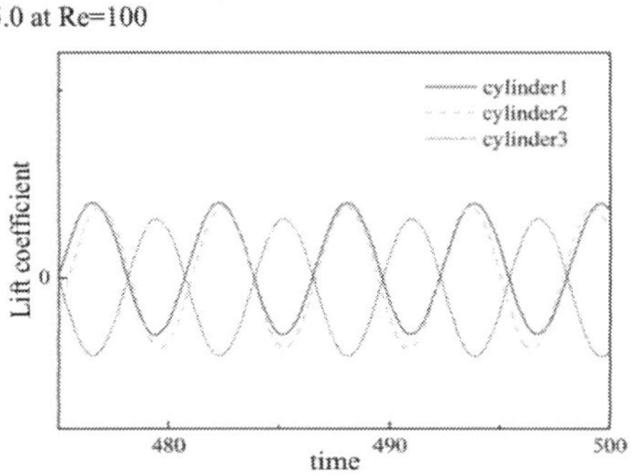

Figure 12. The stable time history of lift and drag coefficient when T=1.5 at Re=100

4 Conclusion

Flow behaviors around three circular cylinders arranged side-by-side at low Reynolds number of 80 and 100 are computed using the Immersed Boundary-Lattice Boltzmann Method in two dimensional spaces. The research focuses on the reveal of the nature between the cylinders interval and the flow parameters. The wake structure, lift and drag coefficient and the vortex-shedding frequency are investigated.

As the intervals between cylinders increase from 0 to 5.0, five flow patterns (namely the single solid model, interim model, in-phase model, diff-phase model and separate model) are defined, based on the study of vorticity contours of these three paratactic cylinders. The vortex shedding of C1 and C3 is synchronous and symmetric, while the vortex shedding of C2 is a little phase-leading, when the interval coefficient T $\geq$ 2.5. In addition, higher Reynolds number leads to a thinner sheer layer, which has the similar consequence with wider interval.

At interim model, there is a sharp decrease in the lift coefficient of C2, and the gap flow around C2 is very weak and the neutralization occurs between the positive and negative vortices, while the frequency have little change. This result may be used to reduce the vibration strength of rigid structure in the flowing fluid environment. The interaction between blunt-body and flexible-body requires further discuss.

References

1. Islam S U, Rahman H, Ying Z C, Journal of the Brazilian Society of Mechanical Sciences and Engineering, **38**, pp. 843-858, (2014)
2. Bearman P W, Wadcock A J, Journal of Fluid Mechanics, **61**, pp. 499-511, (1973)
3. Zdravkovich M M, Journal of Fluids Engineering, **99**, pp 618-633, (1977)
4. Williamson C H K, Journal of Fluid Mechanics, **159**, pp 1-18, (1985)
5. Wang Z J, Zhou Y, International journal of heat and fluid flow, **26**, pp. 362-377, (2005)
6. Zhou Y, Wang Z J, So R M C, Journal of Fluid Mechanics, **443**, pp 197-229 , (2001)
7. Lam K, Li J Y, So R M C, Journal of Fluids and Structures, **18**, pp 305-324, (2003)
8. Alam M M, Zhou Y, Physics of Fluids, **25**, 085106, (2013)
9. Lam K, Li J Y, Chan K T, Journal of Fluids and Structures, **17**, pp. 665-679. (2003)
10. Zhang H J, Zhou Y, Physics of Fluids, **13**, pp. 3675-3686, (2001)
11. Meneghini J R, Saltara F, Siqueira C L R, Journal of fluids and structures, **15**, pp 327-350, (2001)
12. Sau A, Hsu T W, Ou S H, Physics of Fluids, **19**, 084105, (2007)
13. Burattini P, Agrawal A, Computers & Fluids, **77**, pp. 134-142, (2013)
14. Supradeepan K, Roy A, Physics of Fluids, **26**, 063602, (2014)

15. Kadanoff L P, Physics today, **39**, (1986)
16. Sumner D, Wong S S T, Price S J, Journal of Fluids and Structures, **13**, pp 309-338, (1999)

MATEC Web of Conferences 77, 05001 (2016) DOI: 10.1051/matecconf/20167705001
ICMMR 2016

Design of Health Monitoring Program for Filling System based on Data Level Fusion

Cheng Long[1], Xing Xiaochen[2], Xie Weiqi[1], Cheng Rui[3] and Wang Lei[3]

[1]*Department of Space Equipment, Equipment Academy, Beijing 101416, China*
[2]*Department of Equipment Acquisition, Equipment Academy, Beijing 101416, China*
[3]*Department of Graduate Management, Equipment Academy, Beijing 101416, China*

Abstract. Aiming at filling system health monitoring, the health monitoring type partition based on the data level fusion is studied. The health monitoring type is mainly divided into two parts, fusion-threshold monitoring based on single sensor data and fusion monitoring based on multi-sensor data of the same type. On this basis, Single sensor fusion monitoring based on RTS-TA algorithm and Multi-sensor fusion monitoring based on improved weighted fusion algorithm are designed. For multi-sensor data fusion, the fusion schemes are designed respectively according to the different situation whether the initial accuracy is known or not.

1 Introduction

Test and launch ground system is mainly in charge of the spaceport ground facilities. To make a statistical of 70 typical failures during a period of test and launch ground system, which is shown as Fig. 1. During these failures, there are 48 ground facilities equipment failures (including software failures), 68.6% of all. And during the ground facilities equipment failures, the number of filling system failures is 25, accounting for more than half of all the equipment failures.

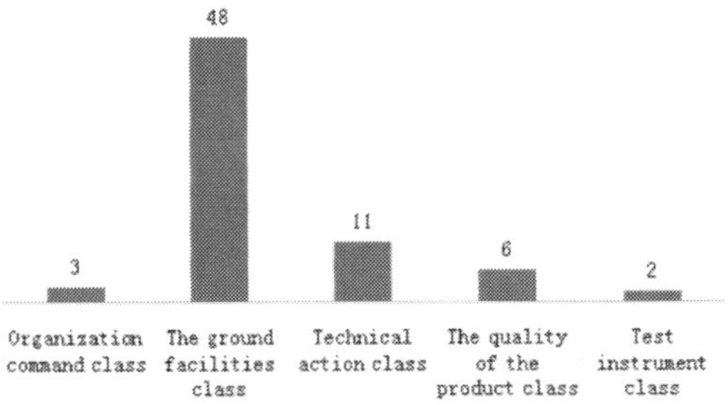

Figure 1. Statistical data of test and launch ground system typical failures

At the same time, as the propellant filling is often done before the launch within 24 hours. In this time, all the systems in the carrier rocket are in standby state. Whether the filling process can be completed accurately and safely within the given time or not, will directly affect the launch process. Once the filling failure appears, the loss is immeasurable. So the propellant filling is a key link in the process of launch site filling, and filling system is an important part of launch site ground equipments. Therefore, it is necessary to carry out the research on filling system health monitoring.

In order to complete the health monitoring of simple components such as tank, storage tank, valve and pipeline of filling system, the data information of pressure, temperature, liquid level and pipeline flow is collected. And the threshold discrimination is used for the health monitoring. The interior environment of filling system is very complex, so the single sensor test may be not useful because of the test or transmission errors. In this condition, it's difficult to guarantee the accuracy of the health monitoring for the appropriate component. To reduce the false alarm rate, the setting of threshold value is usually relatively broad compared with the normal components state. But in many cases, although the sensor component state information didn't exceed the threshold value, the component state has already been in a state of degradation in fact. In another word, person can conclude that the threshold value of monitoring method is not sensitive to the state of the component degradation.

In order to solve the problem of the low reliability of health monitoring for the threshold monitoring components of filling system, some sensors of the same type can be set up, which are used to monitor the component state at the same time. Adopting the data level fusion algorithm from information fusion technology to fusion the multi-sensors information, which can reduce the risk brought by he sensor test or transmission fault, and effectively improve the reliability and accuracy of component threshold monitoring.

In order to solve the difficult problem of threshold type component status deterioration monitoring of filling system, the forecasting technology is introduced into this paper, and single-sensor test information is used for the fusion in time dimension to finish the component predict health monitoring. Based on the state of a certain time period before the actual test information, the predict value of the time can be received. Using some rule to complete the predict value judgment, to make sure whether there is a clear prediction value mutation and whether the data can constitute an exception. For the anomaly data trends has been obtained, the state monitoring of component deterioration can be got based on some rules.

In summary, the health monitoring program for filling system based on data level fusion is designed in this paper.

2 The health monitoring type partition based on the data level fusion

The filling system health monitoring based on data level fusion technology can be divided into two types, one of them is fusion-threshold monitoring based on single sensor data and another one is fusion monitoring based on multi-sensor data of the same type.

Typically, the former has the high precision for sensor, and the threshold value is often set broader. In another conversation, when the parameters can be monitored for the same plurality of sensors, sensor accuracy requirement is general. At this time, in order to make sure that the monitoring information can truly reflect the actual health status of the component which is monitored, the multi-sensor data should be used.

2.1 Fusion-threshold monitoring based on single sensor data

Because of the physical constraints, some critical information of certain components in filling system can not meet (or not meet yet) the monitoring condition. In this conversation, the single sensor threshold monitoring type can be used only. The test data of a single sensor can be fused in time dimension, and the component health monitoring can be achieved by setting some judgment rule.

2.2 Fusion monitoring based on multi-sensor data of the same type

In order to collect the exact critical monitoring information of some components in filling system, such as the temperature, the pressure, the flow and the current information of the filling pump or the temperature, the level of propellant, multiple sensors will be set.

Usually, it can be divided into two parts. Aiming at the propellant level information, the health state can be monitored by threshold monitoring type. Aiming at the temperature and current information of the filling pump, the accuracy of the information collection can be improved based on multiple sensor data fusion.

The health monitoring type partition based on the data level fusion is shown as Fig. 2.

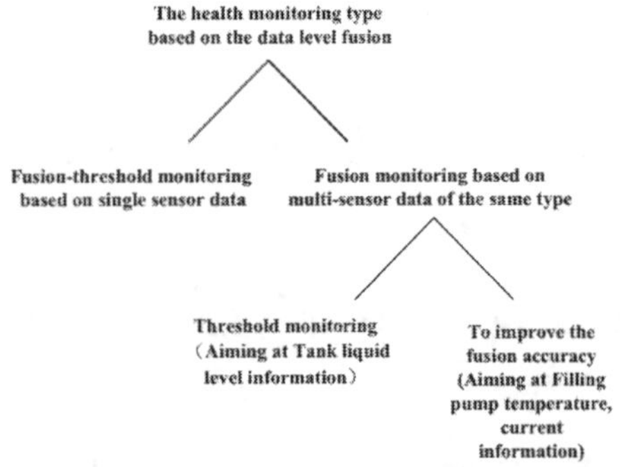

Figure 2. Statistical data of test and launch ground system typical failures

3 Single sensor fusion monitoring based on RTS-TA algorithm

At present, the health monitoring of some components in filling system still adopts manual testing and interpretation, therefore, it is necessary to add more operating positions and test personals. The increasing number of personnel and position adds the probability of the occurrence of malfunction caused by human factor, which brings harm to the implementation of the filling task. At the same time, the sensitivity of the artificial interpretation of health monitoring method for the sudden malfunction is poor, and the test efficiency is also relatively low. Therefore, by fusing the test data of the part state before the current moment and basing on prediction model, conducting the automatic prediction and interpretation for the status data in the follow-up unknown time is significant to the health monitoring for the threshold monitoring components of filling system. For the single sensor threshold monitoring, when the component state testing parameters are not yet exceeds a threshold, the prediction way to identify abnormal data, to analysis the trend of abnormal data, to complete the component state deterioration monitoring is adopted.

Time series analysis is a process based on the data information acquired of the present moment. It will analysis the inherent characteristics between data information, the relationship between data information and system disturbance, then establish a prediction model and predict the unknown data of the future unknown[1]. AR (p) model is one of the most widely used models in data prediction based on time series analysis. This model can be used to predict time series data when the data is stable, normal and zero mean. However, when using AR (p) model to predict the data, it is usually not predicted accurately and is pulled ahead for a few steps. Therefore, many scholars proposed using some algorithms and models to optimize the existing data before predicting and modelling for the data. Common optimization algorithms are such as neural network, support vector machine, genetic algorithm, particle swarm algorithm, ant colony algorithm and so on. The optimization algorithms above usually require large training sets, there are some shortcomings, such as local minimum, over learning and lack of theoretical guidance. It is difficult to adapt to real-time requirements and is not suitable for long-term data monitoring. The literature [2] proposed to apply Roll Time Series (RTS) algorithm for multi-step prediction of the condition monitoring data, using the latest predictive value continually to replace the data which first entered the prediction model, then conduct real-time correction to the prediction model parameters and improve the accuracy of the multi-step prediction.

In order to reduce the false alarm rate, the setting of threshold, combining with its own characteristics of the components to be monitored, is generally broader with respect to the normal component parameters for fusion threshold monitoring based on single sensor data. This mode of setting leads that, in some cases, although sensor test data has not yet exceeded the threshold, the trend of components' status monitoring information show that their health status has been degraded. It means that single sensor threshold monitoring isn't sensitive to the state deterioration or minor faults. When the component state parameter exceeds the threshold value, it usually indicates that an irreversible serious fault of the component is already happening.

When the component is normal, its condition monitoring parameter changes smoothly. By constructing some prediction model, a high-precision multi-step prediction of subsequent unknown parameters can be implemented. When the component state begins to degenerate, the state monitoring information will show a trend. This trend will break the autocorrelation of the condition monitoring information before and after, and get an inaccurate prediction result after the deteriorating state moment. Therefore, the method based on prediction can detect the deterioration of components state. And the occurrence of fault (or serious fault) can be avoided by take timely measures.

As the setting of threshold value is on the basis of comprehensive theoretical analysis and expert experience, its accuracy and reliability are high. In order to identify predicted outliers sensitively, the decision condition of in predictive algorithm is usually more demanding, and lead to that some normal state data is determined to be outliers. In fact, only the outliers with the presence of "close to trend" can be judged as a true exception. Based on the foregoing analysis, in order to meet the real-time, accuracy, high efficiency and other needs of priming system component health monitoring, this paper, combining the rapidity of AR (p) model with the accuracy of rolling algorithm and the analysis of the trend of data, proposes a kind of "rolling time series - trend analysis" (RTS-TA) algorithm. First, identifies the abnormal data of component state monitoring on the basis of the RTS algorithm. Then, analyses the trend of abnormal data. When abnormal data trend meet certain setting conditions, it means that an abnormal event has occurred, namely that component state appears deterioration.

4 Multi-sensor fusion monitoring based on improved weighted fusion

With the rapid development of information processing technology and computer technology, multi-sensor information fusion technology has received widespread attention in the remote sensing detection, intelligent transportation, fault diagnosis, pattern recognition and other fields and also has made many achievements[3-6]. Using multiple sensors to measure and integration the same parameter can significantly eliminate the negative consequences of sensor uncertainty and external disturbance noise and generate more accurate reliable test values than a single sensor [7-9]. Among them, the data fusion method based on weighted fusion, has high precision without a priori knowledge about the measurement data in the process of integration and fusion, and has a widely attention becaese of its optimal, unbiased and minimum mean square error (mse). How to make a good use of weighted fusion algorithm ,the key part is to determine the weights of each sensor data for the data fusion.

When the rocket filling system is performing filling task, The filling accuracy should be higher than 1%. Especially in the filling taskof low temperature filling system, the low temperature propellant of liquid hydrogen or liquid oxygen susceptible to loss. When a large number of gas accumulated in the pipeline, it is very easy to cause a measure error of flow sensor, which can reduce the accuracy of sensor monitoring data and quantitative calculation, and bring hidden trouble for the execution of filling task. In addition, the document [10] pointed out that the failure of measurement device and sensor in filling system is a kind of failure mode which is not dllowed to ignore. Therefore, to conduct the same type multi-sensor data fusion technology research also has important practical significance. The present research condition of the same type multisensor data weighting fusion algorithm is summarized and analyzed, and an improved weighted fusion algorithm considering the sensor initial accuracy is proposed. The fixed distance evidence in evidence theory is introduced into the data fusion weight value calculation, which is used for the weighted fusion of multi-sensor.

According to three different conditions, the improved weighted fusion algorithm model can be divided into the following three parts, which is shown as Fig. 3:

1) The sensor initial accuracy is known;
2) The sensor initial accuracy is unknown, but can be calculated through the actual measurement data. Using the actual measurement data to calculate sensors accuracy also need to meet the following requirements:
   (1) The state of the object under test in a long period of time domain basic will not change, and it meets (or approximate meets) the static measurement condition;
   (2) The number of measured data which are used for calculating is enough, usually more than 20;
3) The sensor initial accuracy is unknown, and also cannot be calculated through the actual measurement data. In this case, regardless of the sensor precision, only to adaptive weighted fusion based on actual measurement data.

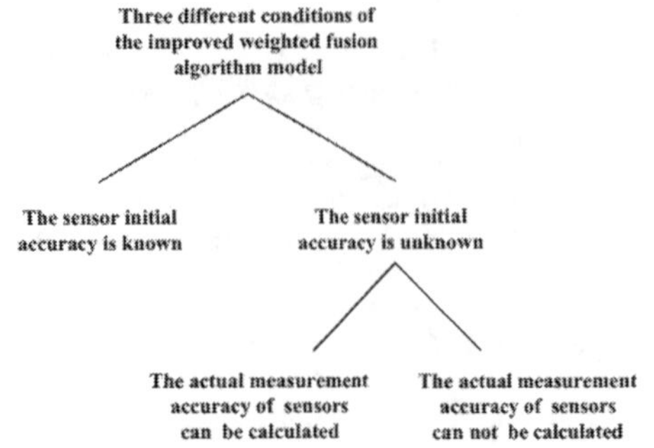

Figure 3. Three different conditions of the improved weighted fusion algorithm model

Suppose there are n homogeneous sensors to measure an object (this paper considers only homogeneous sensor). The initial accuracy of these n sensors are respectively $\sigma_1^2, \sigma_2^2, \cdots, \sigma_n^2$. The true value of the test object state is X, each sensor measurement data are respectively $x_1, x_1, \cdots, x_n$, the measurement data aevrage is $\bar{x}$, the estimate data after fusion is $\hat{x}$. The sensor can complete measurements alone, and the distribution of measurement data are independent of each other's condition.

According to the above three conditions, the improvement weighted fusion algorithm model is designed, which mainly includes the following steps, which is shown as Fig. 4:

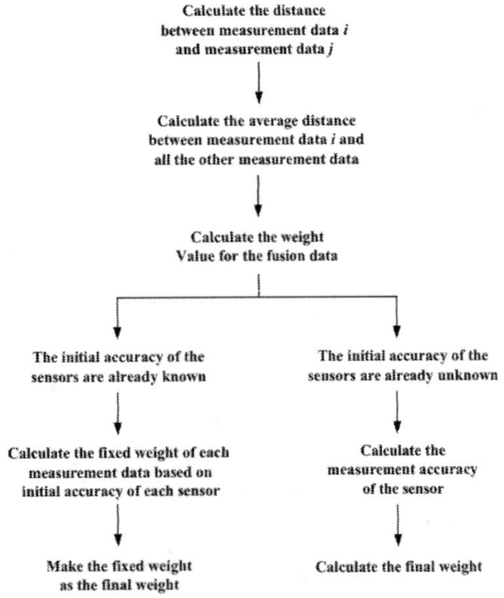

Figure 4. The improvement weighted fusion algorithm model

1) Calculating the distance between measurement data i and measurement data j based on fixed evidence distance, and person can get $d_{[c]i,j}$

$$d_{[c]i,j} = \sqrt{\frac{x_i^2 + x_j^2 - 2x_i x_j}{x_i^2 + x_j^2}} \qquad i,j = 1,2,\cdots,n \tag{1}$$

2) Calculate the average distance between measurement data i and all the other measurement data based on $d_{[c]i,j}$, and person can get $d_{[c]i}$

$$d_{[c]i} = \frac{1}{n-1}\sum_{j=1}^{n} d_{[c]i,j} \qquad i,j = 1,2,\cdots,n \tag{2}$$

3) Calculate the weight value for the fusion data based on all the measurement data at the current moment, and person can get $w_{[c]i}$

$$w_{[c]i} = \left(\frac{1}{d_{[c]i}}\right)^3 \Big/ \sum_{i=1}^{n}\left(\frac{1}{d_{[c]i}}\right)^3 \qquad i,j = 1,2,\cdots,n \tag{3}$$

4) When the initial accuracy of the sensors are already known or the measurement accuracy can be calculated, person can get the final weight value w_i based on measurement weight $w_{[c]i}$ and fixed weight. $w_{[g]i}$. If the initial accuracy of the sensor is unknown and also can't be calculated, make measurement weight to be the final weight value.

Among them, the fixed weight value calculation is divided into the following two cases:

(1) The initial accuracy of the sensors are already known

In this case, person can get the fixed weight of each measurement data based on initial accuracy of each sensor

$$w_{[g]i} = \frac{1}{\sigma_i^2} \Big/ \sum_{i=1}^{n}\frac{1}{\sigma_i^2} \qquad i = 1,2,\cdots,n \tag{4}$$

(2) The initial accuracy of the sensors are unknown but the calculation conditions are met

Colecting the historical measurement data about sensor i, and in this paper person define the number as 20. The historical measurement data are respectively $x_{i(j)} (i = 1,2,\cdots,n; j = 1,2,\cdots,20)$. The average data of the 20 historical measurement data is $\overline{x_{[h]i}}$, then the measurement accuracy of the sensor $\sigma_{[c]i}^2$ can be calculated, and then the fixed weight of each measurement data $w_{[g]i}$ can be calculated

$$\overline{x_{[h]i}} = \frac{1}{20}\sum_{j=1}^{20} x_{i(j)} \qquad i,j = 1,2,\cdots,n \tag{5}$$

$$\sigma_{[c]i}^2 = \frac{1}{20}\sum_{j=1}^{20}\left(x_{i(j)} - \overline{x_{[h]i}}\right)^2 \qquad i = 1,2,\cdots,n \tag{6}$$

$$w_{[g]i} = \frac{1}{\sigma_{[c]i}^2} \Big/ \sum_{i=1}^{n}\frac{1}{\sigma_{[c]i}^2} \qquad i = 1,2,\cdots,n \tag{7}$$

5) Then person can get the final fusion value $\hat{x}$ based on the weighted fusion of each measurement data using the final weight w_i

$$\hat{x} = \sum_{i=1}^{n} w_i x_i \qquad i = 1,2,\cdots,n \tag{8}$$

5 Conclusion

This paper has made a study of the health monitoring classification based on data level fusion, which are mainly divided into fusion-threshold monitoring based on single sensor data and fusion monitoring based on multi-sensor data of the same type. Aimming at single sensor data fusion-threshold monitoring, a kind of "rolling time series - trend analysis" (RTS-TA) algorithm based on A(R) model has been proposed. Aimming at fusion monitoring based on multi-sensor data of the same type, an improved weighted fusion algorithm considering the sensor initial accuracy is designed, which is divided into two parts: The sensor initial accuracy is known or unknown.

References

1. YANG Shu-ying, WANG Li-xian, NIU Ting-wei, et al. Multi-step prediction of rolling time series based on particle filter optimization. Journal of Systems Engineering and Electronic, **34(6)**:1097-1101, (2012)
2. LIU Hui, TIAN Hong-qi, LI Yan-fei. Short-term forecasting optimization algorithm for wind speed from wind farms based on wavelet analysis method and rolling time series method. Journal of Central South University (Science and Technology), **41(1)**:370-375, (2010).
3. Zhang Ting. Stress Identification on the Key Region of Large Span Space Structure, Harbin institute of technology, 2010.
4. Xiao Lizhi, Yang Dongping. The optimization algorithm of aviation equipment maintenance cost forecast and its applied research. Advanced Materials Research, **760**:1851-1855, (2013)
5. Haibo Huo, Yi Ji, Shiming Wang, et al. The research on AdaBoost-BPNN model of point absorber wave energy converter. 2014 IEEE International Conference on Mechatronics and Automation, **8**: 1762-1766, (2014).
6. Arkhom Songkroh, Rerkchai Fooprateepsiri, Woraphon Lilakiataskun. An intelligent risk detection from driving behavior based on BPNN and fuzzy logic combination. 13th International Conference on Computer and Information Science, 6, (2014).
7. Zhou Yao, Jiang Li-ping, LIU Yang. Data Fusion Algorithm Based on Support Degree under Interferential Environment. Fire Control & Command Control, **39(3)**:12-14,19., (2014)
8. Mahler R. "Statistics 101" for multisensor, multitarget data fusion. IEEE A&E SYSTEMS MAGAZINE, **19(1)**:53-64, (2004)
9. CHEN Shi-ming, CHEN Xiao-ling, XU Qing-gang, et al. Adaptive weighted fusion based linear distributed filtering algorithm. Control and Decision, **30 (4)**:605-610, (2015)
10. WU Xiao-tao, YANG Meng, YUAN Xiao-hui, et al. Bearing fault diagnosis using EEMD and improved morphological filtering method based on kurtosis criterion. Journal of Vibration and Shock, **34(2)**: 38-44., (2015)

MATEC Web of Conferences 77, 05002 (2016) DOI: 10.1051/matecconf/20167705002
ICMMR 2016

High precision scanning system and test technology of remote sensor

Yang Xiaole, Kang Jianbing, Wang Dongjie and Zhao Xin

Beijing institute of space Mechanics & Electricity china

Abstract. Reliability of scanning system is more important for scanning remote sensor. The sensor also has a higher request for angle measuring precision, control precision and non-linearity as the raise of resolution and S/N. A new high reliability and precision scanning system are designed. Choosing for shafting, motor and angle measuring sensor, scanning control circuit design and control scheme are introduced. High precision system needs high precision test system, a high precision non-contact test system is introduced finally. The result of test shows that angle measuring precision is smaller than 1arcsecond and non-linearity is less than 0.5%.

1 Introduction

Scanning system which drives and controls scanning mirror to swing linearly through the vertical flight direction is the key system of scanning optical remote sensor. The scanning system consists of scanning mirror, bearing, motor, angle sensor and control circuit etc. Scanning system is usually evaluated by scanning non-linearity, control precision, angle measuring precision, scanning efficiency and so on. Scanning system is a typical mechanical and electrical system, whose precision is decided by mechanical system and electronic control. With the increasing of spatial resolution and aperture, scanning mirror size and moment of inertia also increase, so control difficulty increases. This paper introduces high precision and high reliability scanning system based on the beryllium mirror, flexible pivot, inductosyn and voice coil motor in detail from two aspects [1].

2 Scanning mirror

Scanning mirror ensures strength and rigidity of the mirror body, meanwhile, the mass and moment of inertia is minimized, so it can reduce influence on support element life and disturbing torque to the satellite platform. Scanning mirror size determined by the aperture of optical system is 600mm × 400mm, so low density, high specific rigidity, high specific strength and thermal stability materials is chose to use as mirror substrate. Beryllium is recognized as the best material, and SiC is also commonly materials [1]. Beryllium is light than SiC but toxic, and still need lightweight. Scanning mirror can reduced weighing by 65%, only 6kg through lightweight. Corresponding moment of inertia is only $0.13kg.m^2$.

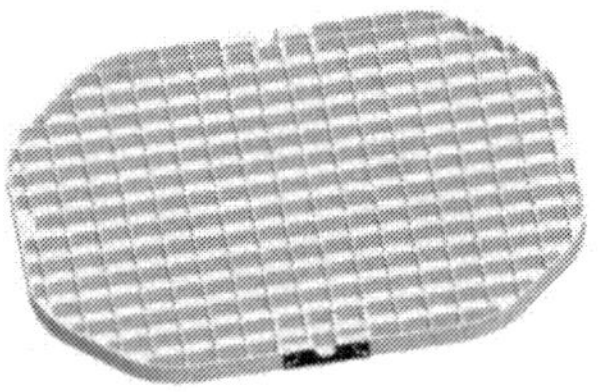

Figure 1. Configuration diagram of scanning mirror

3 Scanning shafting

The scanning shafting which is the key component provides support for scanning mirror, and the support rigidity and precision directly affect system life and control precision.

Bearing is used commonly, but oil lubricated bearings used in space application has the problem of vacuum cold welding and the harm of optical mirror pollution. It needs to use ultra low volatile oil and seal structure, and the reliability and long life is a difficult problem. Solid lubricated rolling bearing may decreases precision or has deadlocking problem due to friction when working for a long time [1].

Flexible pivot used in this paper is frictionless and elastic support, and achieving relative rotation relies on elastic material deformation around a fixed axis. No metal to metal contact, only intermolecular friction, so friction moment is lowest. Pivot structure is showed in figure 2, and it consists of three cross tablet spring (narrow tablet spring I, II and wide tablet spring) and two identical cylindrical hollow tablet spring seats. The spring and spring seat use welding technology.

Keeping flexible pivot in position needs torque motor, because of restoring moment [2].

Restoring moment coefficient of flexible pivot is 1.5025N·m/rad, maximum restoring moment is 0.132N.m accords to maximum angle. Each shafting uses two flexible pivots, so the maximum restoring moment is 0.264N·m. In the practical application, we should be as far as possible to filter out two pivot with the same restoring moment coefficient, meanwhile, motor and angle measuring sensor are respectively connected with two pivot, so as to avoid motor interfering sensor.

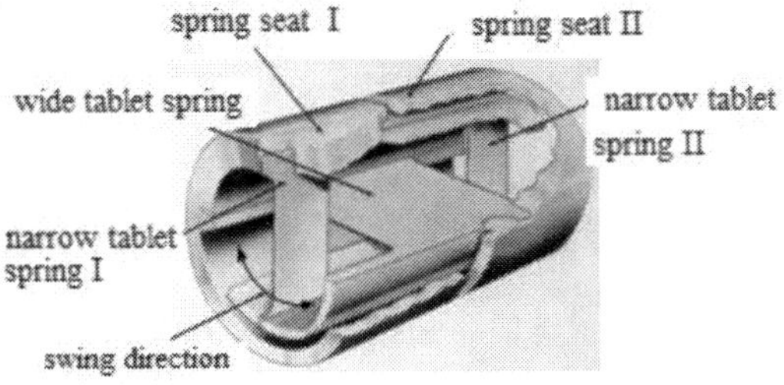

Figure 2. Structure diagram of flexible pivot

The main factor affecting scanning mechanism life is shafting life, according to using state, pivot is mainly affected by radial force during ground debugging, at the same time, radial force caused by nonaxiality should be considered. Pivot radial load ratio is calculated, according to pivot life curve in figure3, pivot (ground debugging + orbit application) is infinite life in the whole life cycle, the abscissa is load ratio, longitudinal coordinates is torsion angle.

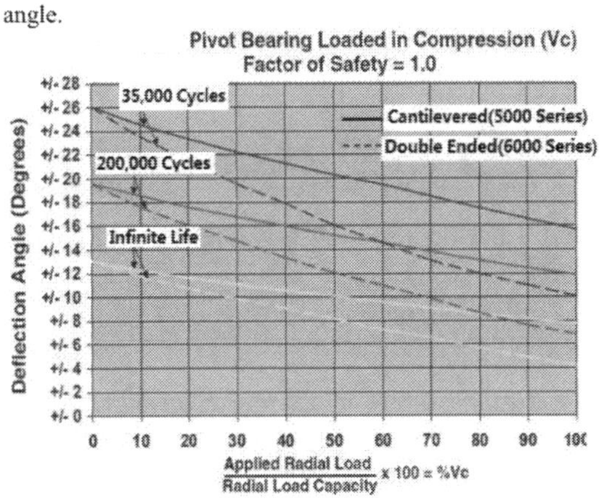

Figure 3. Pivot life curve

4 Motor

Scanning mirror is driven directly, this means it is directly connected to rotating axis. Motor is required to swing at a constant speed, so the requirements of the motor are [3]:

(1) Small moment ripple;
(2) Linearity of the torque current characteristic is better;
(3) Small moment of inertia, quick dynamic response;
(4) Big moment;
(5) Long life.

Scanning mirror is driven by voice coil motor, which is a special motor and has the characteristic of simple structure, small size, high speed, fast response and high reliability etc. Voice coil motor is considered to be almost perfect servo drive device, and exercise frequency is 500-1000Hz or even higher. Its working is based on principle of ampere force. The simplest linear voice coil motor is a tubular coil winding located in radial electromagnetic field. Bending two ends of linear voice coil motor into arc, and it will become rotary voice coil motor [4].

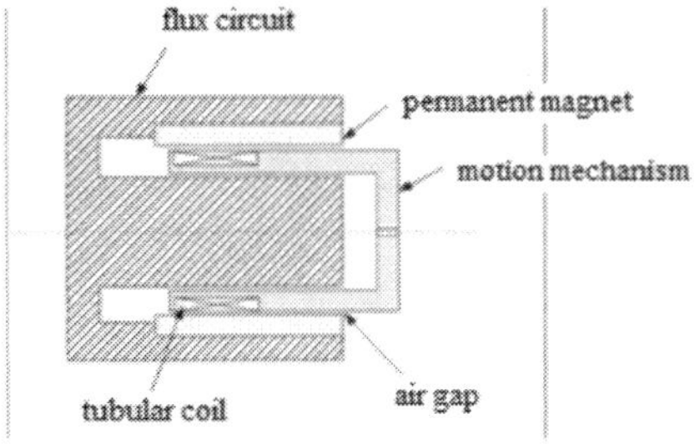

Figure 4. Structure diagram of voice coil motor

The voice coil motor focus on index of continuous torque, peak torque, current torque coefficient, angle range and weight etc.

The maximum torque requirement in this project is 0.15N·m, in the linear and nonlinear junction, continuous torque of motor is 0.64N·m, so safety factor is 4.3 and safety margin is 3.3.

The temperature of the coil should be strictly controlled, because increasing temperature will change motor output torque.

5 Sensors

The sensor used in servo control system is mainly photoelectric encoder, inductosyn and resolver etc. Photoelectric encoder has large weight and volume, and ability of anti vibration and shock is weak. Principle of inductosyn and resolver is basically same. Requirement of installation precision is relatively low and the precision and repeat performance is not affected by temperature. Inductosyn has higher precision, but has higher requirement for signal processing circuit due to weaker out signal [5]. In this project, swing angle range is 10°, absolutely angle measuring precision is better than ±1", so Absolute 360 pole inductosyn whose angle measuring precision is ±1" between 60° is used. The wire is leaded from rotor, and plate is 8mm thick aluminum with diameter of 130mm. Mechanical installation needs to ensure stator and rotor parallelism is better than 0.03, coaxial degree is better than Φ 0.03.

6 Scanning control system

The mechanical structure which consists of beryllium mirror, voice coil motor, flexible pivot and inductosyn guarantee the system has minimum moment of inertia, minimal disturbing torque and highest angle measuring precision. Scanning control system makes scanning mechanism swing stably and accurately according to the given speed. In this paper, scanning control system is divided into angle measuring circuit and control circuit. Because output signal of inductosyn is only millivolt level, so scanning angle measuring circuit is put as much as possible near inductosyn to reduce signal transmission distance. Angle information is transmitted to control circuit after converting into digital quantity.

6.1 Scanning angle measuring circuit [6]

Angle measuring circuit based on absolute inductosyn consists of excitation module, signal conditioning module, axis angle digital conversion module, data processing module and interface module and so on, and takes FPGA as the system core.

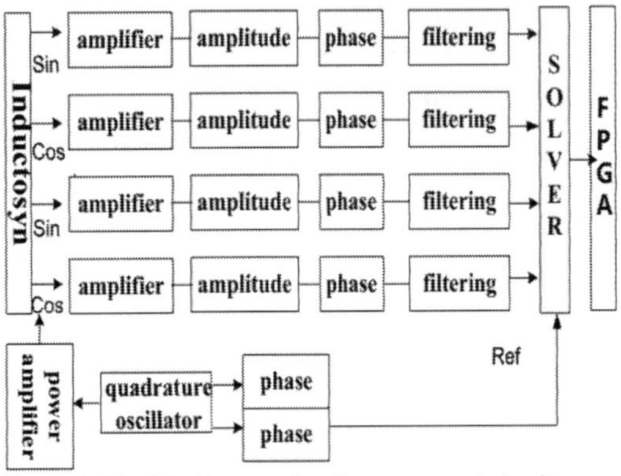

Figure 5. Principle diagram of angle measurement circuit

6.1.1 Exciting circuit

Excitation circuit selects dual channel series excitation mode, and uses quadrature oscillator as oscillation source. In order to increase driving capability, power amplifier is added behind oscillator; otherwise it will appear distortion waveform. It is best to use shielded twisted pair cable to reduce the interference of excitation signal. The excitation frequency is 10 KHz.

6.1.2 Signal conditioning circuit

Signal conditioning circuit consists of preamplifier, main amplifier, amplitude modulation, phase modulation and filtering. In order not to affect angle measuring precision, signal conditioning circuit noise must meet noise demand of axial angle solver chip AD2S80. Preamplifier circuit chooses amplifier with low noise and small temperature drift, otherwise noise will further enlarge by main amplifier. RC filter is used after preamplifier to filter out high frequency noise, and AC coupling is used to connect main amplifier to eliminate the effects of DC bias and zero drift. Filter circuit uses band-pass filter, and bandwidth is set to 8 KHz~12 KHz to further reduce noise.

6.1.3 Crude and precision fusion and error correction

The crude and precision channel of inductosyn use AD2S80 axial angle solver chip which is tracking amplitude discrimination mode. When choosing solver chip, resolution, speed and bandwidth indexes are comprehensive considered. Crude and precision fusion algorithm based on FPGA use error larger crude channel absolute angle to find output angle cycle of precision channel and then use precise angle of precision channel to find current precise angle. The error of angle measuring system is usually system error, and repeat precision is higher than absolute precision, so this paper corrects error by broken-line approximation method to improve measuring precision of the system.

6.2 Scanning control circuit

Scanning control is based on position feedback control method. First of all DSP produces position command which is satisfied system motion law, and measuring circuit sends angle information into the DSP processor. In a control cycle, DSP obtain deviation of current position and predetermined position, then calculates correction torque command to drive motor. By whole digital closed loop control, disturbing torque will be effective decreased through the feedback loop and scanning linearity and control precision are effectively guaranteed [7].

The curve of the scanning motion includes linear and reversing segments. Theoretical reversing segment motion curve should be triangular wave, but the system has first order discontinuity point.

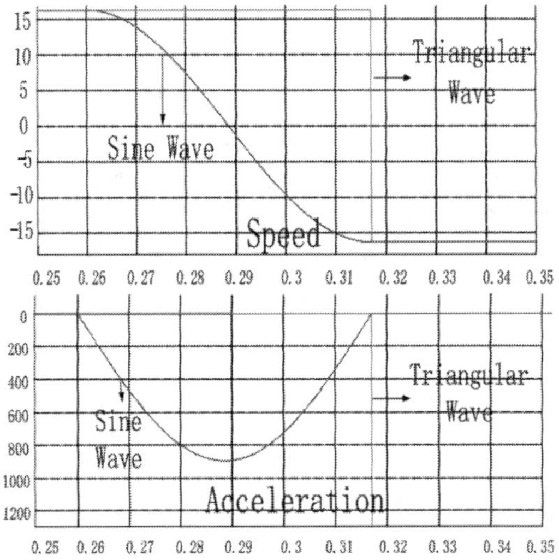

Figure 6. Comparison of speed and acceleration of sine curve and triangle curve

So sine wave is used to smoothing brake and accelerate process of reversing segment [8][9]. Velocity and acceleration of triangular wave and sine wave is showed in figure 6. The Planning sine wave as follows:

$$\theta = A \sin wt \quad (1)$$

$$V = -Aw \cos wt \quad (2)$$

$$a = -Aw^2 \sin wt \quad (3)$$

In this project, scanning frame time is 2.5s; scanning efficiency is 80%; so nonlinear time is 0.5s and sine period is 1S, w = 2π. The torque balance relationship between scanning is:

$$T(t) - T_l(t) = Ja \quad (4)$$

$T(t)$ is motor output torque, $T_l(t)$ is recovery torque of pivot (sum of two pivot), J is total moment of inertia of the motor coil and load, a is acceleration. So when $a>0$, system is acceleration; When $a<0$, system is deceleration; when $a=0$, system is uniform.

When $t = 0$, from linear segment into non-linear segment, $\theta = 0$, $V=-Aw$, V is the scanning speed, $a = 0$; so as to obtain $A= V/w$; At this time $T(t)=T_l(t)=2\times$ maximum angle of linear segment $\times$ pivot tensional rigidity, so as to obtain maximum torque requirements of motor.

When $t=T/4$, at this time scanning mirror in the maximum swing angle, $\theta = A$, $V = 0$, $a = -Aw^2$; Non-linear maximum angle and maximum system acceleration can be calculated, so as to find out the maximum disturbing torque;

When $t=T/2$, from non-linear segment into linear segment, $\theta = 0$, $V = Aw$, $a=0$; At the same time, $T(t)=T_l(t)=2\times$ maximum angle of linear segment $\times$ pivot tensional rigidity.

Motor output torque and disturbing torque curve is showed in figure 7 and figure 8:

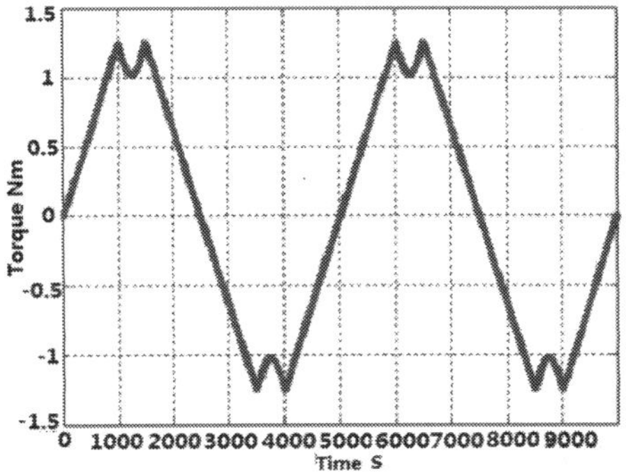

Figure 7. Motor output torque curve

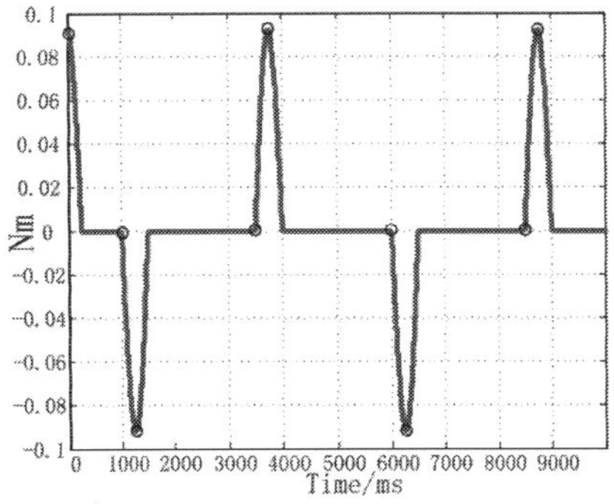

Figure 8. System disturbing torque curve

7 Scanning precision test

Common Angle testing methods are contact and non-contact type. Non-contact type does not need to be connected with scanning mechanism, so it will not generate disturbing torque and its precision will not influenced by scanning mechanism [10]. In this paper, the measuring precision of the scanning system requires 1 arcsecond, so precision of the test system needs to reach 0.5 arcsecond.

The non-contact measuring system is based on photoelectric autocollimator, and photoelectric autocollimator precision can reach 0.1 ". The disadvantage is that angle range is too small, so this paper uses two sets of photoelectric autocollimator and turntable combination scheme. Scanning mechanism is placed on the turntable, and plane mirror which is used for photoelectric autocollimator 1 is placed in the upper of rotating shaft. Multi-tooth dividing table is placed on turntable, and plane mirror which is used for photoelectric autocollimator 2 is placed in the upper of multi-tooth dividing table.

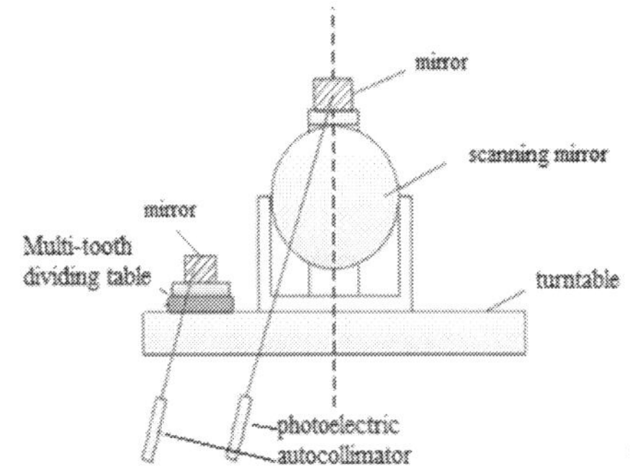

Figure 9. Principle diagram of test system

During the test, first of all two sets of photoelectric autocollimator auto collimate by mirror, then making scanning mirror rotate a certain angle, and turntable rotates the same angle in the opposite direction, so photoelectric autocollimator 1's readings backs to initial readings, this means that turntable and scanning mirror rotate the same angle. Secondly, rotating multi-tooth dividing table on the turntable, photoelectric autocollimator 2 auto collimates. The exact turn angle of turntable can be obtained by the angle of multi-tooth dividing tablee and photoelectric autocollimator 2 readings, which subtracts the angle of angle measuring circuit to obtain system angle measuring precision.

Angle measurement test results are as follows. The measuring error is less than 1 arcsecond.

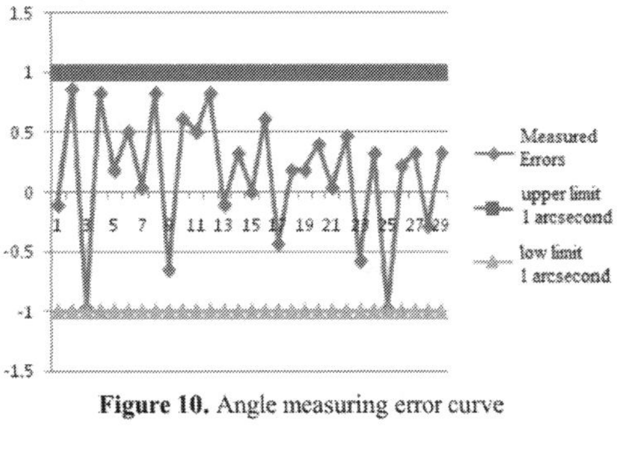

Figure 10. Angle measuring error curve

Scanning non-linearity is the difference of angular velocity of scanning mirror at any position and average velocity divides average velocity, and scanning speed is fixed time actual output angle difference getting from angle measurement circuit divides time interval. The nonlinearity of scan test is less than 0.5%. Control precision which is the maximum deviation of theory angle position and actual angle position is better than 3 arcsecond.

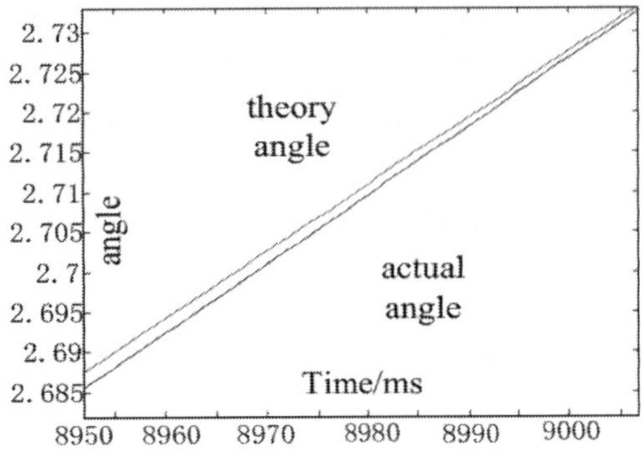

Figure 11. Scanning control precision curve

8 Conclusions

In this paper, the mechanism uses beryllium mirror, voice coil motor, flexible pivot and absolute inductosyn to guarantee that the system has minimum moment of inertia, minimal disturbing torque , highest angle measuring precision and high reliability. In the control aspect, angle measuring circuit and control circuit architecture of is used. Low noise angle measuring circuit ensures angle measuring precision and digital closed loop control algorithm guarantees that the system has sufficient servo stiffness. Scanning reversing segment uses sine wave, and system disturbing torque, motor output torque and reversing angle calculation method is given. Finally, a non-contact measuring method proves that system angle measuring precision can reach 1 arcsecond, non-linearity can be less than 0.5%, control precision is better than 3 arcsecond.

References

1. Yuan L, Yang L, "Research on low speed, big Inertia and high precision scanning mechanism of satellite camera,"The twelfth session of the National Conference on space and motor control technology, 2000.
2. Sun D, "Optimum design and experimental study of flexible support system of directed structure," Harbin Institute of Technology,2009 .
3. Qin H, Wang P, "Research on high accuracy swing scanning motor and control system,"The twelfth session of the National Conference on space and motor control technology,2000.
4. ZHANG D,FENG X. "The Technical Principle of Voice-Coil Actuator," Journal of north university of china, Papers 27(3), 224-228 (2006).
5. Huo S, "Round inductosyn turntabl eangular measuring system based on dsp and cpld," Harbin Institute of Technology,2009 .
6. QIU Z,LI W,XU B, "Absolute Inductosyn Angle Measurement System Based on DSP and AD2S80 ,"Instrument Technique and Sensor, Papers 8(8), 59-62 (2010).
7. Cai Q, "The Research and Design of Scanning Control Systembased on DSP," Shanghai Jiao Tong University,2008.
8. Luo Q,Yuan L,Yang L. "A method for Driving Scan Mirror," Spacecraft Recovery & Remote Sensing, Papers 27(1), 36-41 (2006).
9. Wu Q,Wang Q,Peng Q, "Wide bandwidth control of fast-steering mirror driven by voice coil motor," Opto-Electronic Engineering, Papers 31(8), 15-18 (2004).
10. YANG X,ZHANG T,ZHANG J, "Design of a system for testing scanner's motion control characteristics," Electronics Optics & Control, Papers 19(8), 78—81 (2012).

MATEC Web of Conferences 77, 05003 (2016) DOI: 10.1051/matecconf/20167705003
ICMMR 2016

Research of long-term properties of high-strength fiberglasses by means of the built-in FBG sensors

Peter V.Mikheev [1], Andrey A. Dalinkevich [2], Sergey A. Gusev [1], Timofey N. Igonin [2], Ludmila B. Maksaeva [2] and Tatyana A. Nenasheva [2]

[1] *Bauman Moscow State Technical University: 2, bd. 1, 2nd Baumanskaya Street, 105005 Moscow, Russia,*
[2] *Russian academy of sciences A.N. Frumkin Institute of Physical chemistry and Electrochemistry RAS (IPCE RAS) 119991, Leninsky blvd. 31. Moscow, Russia,*

Abstract. In this work the possibility of use of the built-in fiber-optical sensors (Bragg's sensors) of deformation for control of a condition of polymeric fiberglasses on epoxy polymer matrix ED-I cured at the increased temperature was investigated. Sensors were inserted in different layers of the layered plastic reinforced by biaxial fiber glass fabric. The package of layers had quasiisotropic structure. The sample made by method of vacuum impregnation was affected by saturated steam (90 °C) within 6 months. Operability of system of measurement has been as a result shown and swelling parameters in different layers of FGRP were recorded. For measurement measuring systems of two types — on a superbright light-emitting diode and on the tunable laser were used.

1. Introduction

Influence of atmospheric moisture is the most important of climatic factors by which the polymeric composite materials (PCM) at operation of products are affected in nature. As a result of diffusive penetration of water into a polymeric matrix and into the interface between the fiber and matrix in fibrous PCM the part of molecules of water is immobilized on polar groups of polymer, weakening thereby interchain interaction in a matrix and adhesive interaction on limit of the section polymer matrix - fibers. At the same time molecular mobility (effect of plasticization) is released and the relaxation processes leading to restructurings on molecular proceed and supramolecular level.[1,2]

Highly sensitive fiber Bregg grating sensors (FBG-sensors) of deformations are successfully used to an assessment of external and internal mechanical and thermal tension in various materials and elements of composite detail [3-7]. Unlike metal FBG strain gages – sensors do not rust and therefore they can be applied in the damp hot environment. Besides their accuracy is higher than at wire strain gages.

The purpose of work consisted in clarification of opportunities to use fiber-optical sensors of Bragg (FBG-sensors) for studying micromechanical moist arising in PCM at the accelerated heat-moist influence which is widely applied to modeling of climatic aging of PCM.

The problem of influence of moisture on fibreglass is important, so this material is perspective for application in networks of water supply and oil pipelines. Fiberglass pipes of less steel are subject to corrosion.

2. Experimental

For production of a sample biaxial fiber glass fabric is used (specification TU 1916-033-61664530-2013, produced by JSC Prepreg-SKM with area density of 860 g/m2 and the direction of fibers 0 °/90 °.

Multiaxial fabric has been chosen with the purpose to imitate the pipes made by winding. The quasiisotropic structure is chosen as the most applicable.

Matrix epoxy - ED-I (specification TU B3-774-2010) consist of composition from ED-20 epoxy (100 m of h) of IMTGFA hardener (isomethyltetrahydrophthalic anhydride, 20 m of h) and the accelerator unitary enterprise 606/2 (2.4,6, Tris (dimethylaminomethyl) of phenol. This type of a matrix is used in the industry and differs in high resistance to moisture. Samples of binding have been made by method of filling of images in a form. And then their mechanical properties have been defined. Before filling binding was maintained under vacuum not to allow hit of vials of air in a sample.

FGRP samples on the basis of biaxial fabric and epoxy resin ED-I produced by an infusion method. Specimen was cured at 120 °C.

Completeness of hardening polymer matrix was controlled by method of the differential scanning calorimetry, glass temperature of the sample cured binding (without reinforcing fibres) has made 132 °C. (Figure 1) .

FBG sensors have been located between layers at a different depth of a sample. When laying layers, the

 DOI: 10.1051/matecconf/20167705003

second layer has been turned by 45 degrees to create structure close to isotropic structure.

Previously the method for output of optical sensors out of plastic has been worked out. For this purpose samples of fiberglass plastic with optical fibre but without sensors were prepared. So we picked up a way to protect the optical fiber from being damaged at the edges of the plastic and at the junction of the measuring fiber with patch cords. Initially the stresses arising during cooling of the finished fiberglass plate either destroyed the sensor or bent it. This distortion leads to the loss of the optical signal below the level of sensitivity threshold of the system based on high-brightness LEDs (the method of Bauman Moscow State Technical University, BMSTU).

After picking up a set of heat shrinking tubing the problem was solved. At the same time on the plates with an optical fiber without sensors was selected the temperature for preserving patch cords connectors from damage or from resin casting.

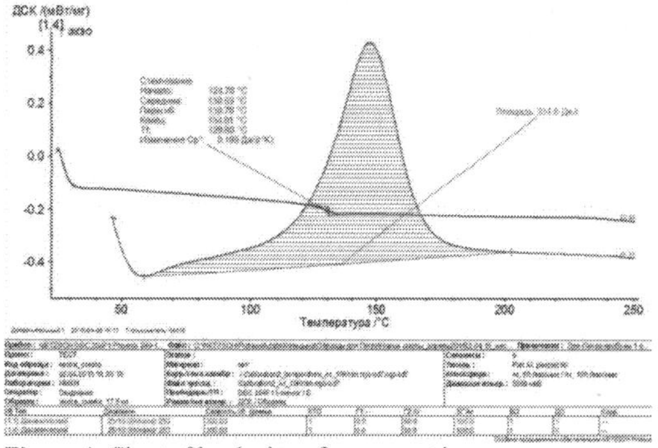

Figure 1. Chart of hardening of epoxy matrix

Sensors have been integrated into a sample only when was obtained the attenuation of a light signal accepted for measurements.

Mechanical properties of epoxy binding have been checked on standard samples "a dog bone". Results are given in Tab. 1. Binding has shown fragile linear behavior up to destruction.

Table 1 Properties of polymeric binding

№	Property	Value	Variation factor, %
1	Elasticity module, GPA	2.84	1.75
2	Strength, MPa	38.3	14.5
3	ultimate strain,%	1.4	15.7
4	Dencity, gm/cm 3	1.22	5.0

Parameters of FBG – sensors optical (produced by "Inversion Sensor") were the following:

top layer (marking No. 1341) Wavelength second layer (marking No. 13801542,28

second layer (marking No. 1380) Wavelength 1550,23

center (marking No. 1340) Wavelength 1554,06

The arrangement of sensors is given in layers on rice.2. On Fig. 3 process of laying of optical fibers with sensors between layers of industrial fiber-glass fabric before impregnation of a sample binding is shown.

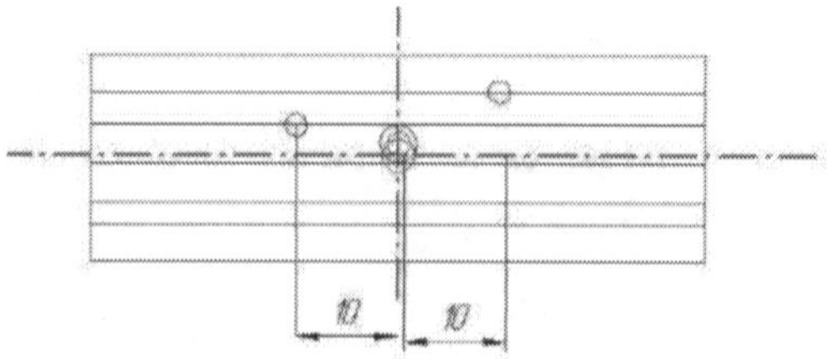

Figure 2. The scheme of an arrangement of sensors in sample layers

The sensors located in different layers have been carried on width not to create distortion of structure. For production of samples the following devices were used:

- Installation of vacuum impregnation Vacmobile INFUSION 28/2 (New Zealand);
- France Etuves XXL04.5 furnace (France).

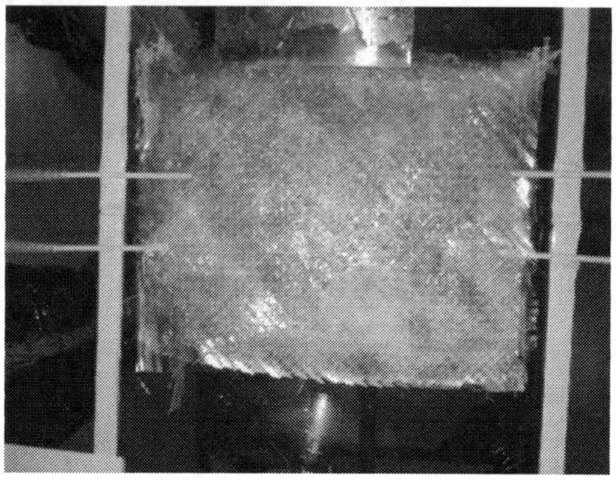

Figure 3. Process of laying of sensors between fiber glass fabric layers

For determination of temperature of vitrification binding the differential scanning calorimeter Netzch DSC 204 F1 Phoenix (Germany) was used.

For determination of mechanical properties of a sample of the witness the universal test equipment Zwick-Roel Z-100 (Germany) was used.

The endurance of samples was carried out in the certified thermo-moist camera Quantel the MHK 408 CL model (Singapore) Figure 4.

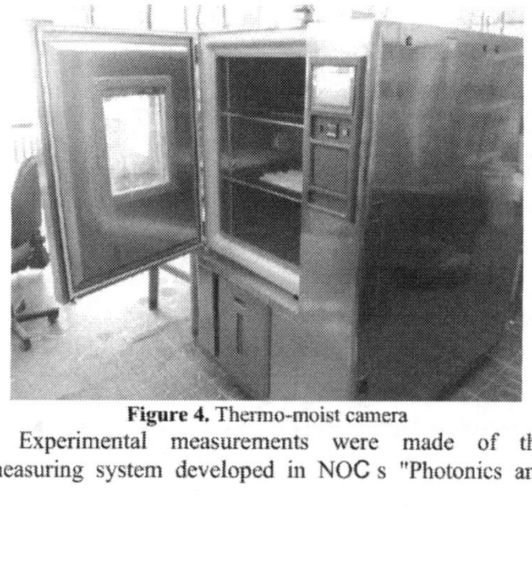

Figure 4. Thermo-moist camera

Experimental measurements were made of the measuring system developed in NOC s "Photonics and

IK-equipment" of BMSTU and the Astro A312 device of production "Inversion Sensor". These measurements received by means of both systems corrections for their use in one measuring system are compared among themselves.

The measurement system constructed in BMSTU was based on source of the white light, technically more simply, but the Astro system based on the tunable laser has big sensitivity and bigger speed.

One of optical spectra of the sensor is given in Figure 5 .

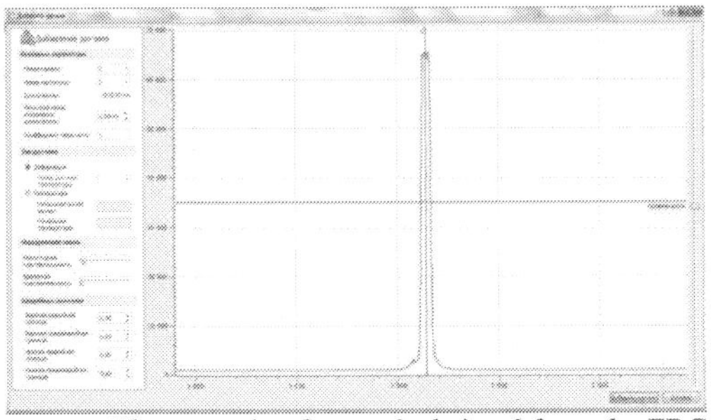

Figure 5. The example of an optical signal from the FBG-sensor received on the equipment Astro.

The scheme of sensors arrangement is provided in on Fig 2 and in Table 1. To reveal spatial unevenness of penetration of water into a sample, sensors were put on different depth (layers 1, 2 and 3). The increased temperature of hardening of polymeric binding has demanded special protection of optical fibers in places of an exit from FGRP.

Table 1. Arrangement of sensors in layers of FGRP

Material	SM-42019 (+0/90°)
Layer №	laying
1	SM-42019 (+0/90°)
FBG-sensors 1	Measuring layer
2	SM-42019 (+45/-45°)
FBG-sensors 2	Measuring layer
3	SM-42019 (+0/90°)
FBG-sensors 3	Measuring layer
4	SM-42019 (+0/90°)
5	SM-42019 (+45/-45°)
6	SM-42019 (+0/90°)

Dynamics of change of weight of a sample due to absorption of moisture is shown in Figure 7. In the same Figure the schedule of measurement of deformations is shown.

Special attention has been paid to protection of connectors against the increased temperature and humidity. Influence of hot water was so strong that all factory marking of sensors has been washed away.

The doubling sample on the same epoxy matrix ED-I and fabrics, but without sensors has been in parallel made. Samples for determination of density (hydrostatic weighing) and some physicomechanical characteristics (Tab. 2) have been cut out from FGRP plates.

Measurements of mechanical properties of a doubling sample have shown that it possesses good mechanical properties.

The doubling sample has been cutted near the sensor location. The photo on Fig.6 shows the end face of sensor. It is visible that there are no intervals between the sensor and matrix, and the structure of a composite is not broken.

Table 2. Properties of the made fibreglass

№	Property	Value
1	Elasticity module at a bend, GPA	21,7
2	Elasticity module at stretching, GPA	20,6
3	Strength at bending , MPa	397
4	Strength at shift, MPa	38,3
5	Density of plastic, g/cm3	1,9
6	Volume fraction of the reinforcing fibers, %	55
7	Density of of the used polymer matrix, g/cm3	1,22

The FGRP sample containing sensors has been subjected to heat-moist aging in a test climatic chamber at the temperature of 80 °C and relative humidity of 95%. Accuracy of maintenance of parameters of tests made ±1°C for temperature and ±1,5% for humidity.

Moisture content in the studied sample of FGRP was controlled by method of periodic weighing with an accuracy of 10-3 g.

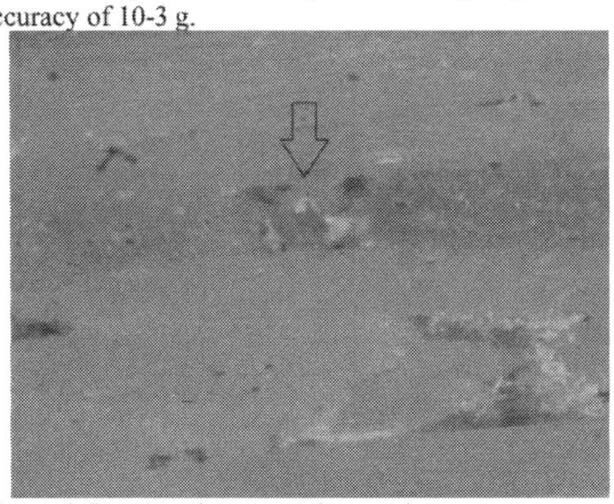

Figure 6. Sensor cut in a composite sample

3. Results and discussion

From Fig. 7 it is visible that sorption of vapors of water FGRP has abnormal character and does not submit to Fick's law. Existence of a maximum and step on kinetic curve sorption demonstrates course in a model of relaxation processes, i.e. change of internal microstructural tension and according to deformations which, in turn, are caused by diffusive and sorption penetration and interaction of water with polymer [8-11].

High sensitivity to temperature and deformation of a FBG-sensors also allow to use it as a sensitive element of the measuring system. Measuring properties of a FBG-sensors depend on effective index of refraction and on the lattice period on a formula (1). And the period of a lattice is directly connected with its deformation:

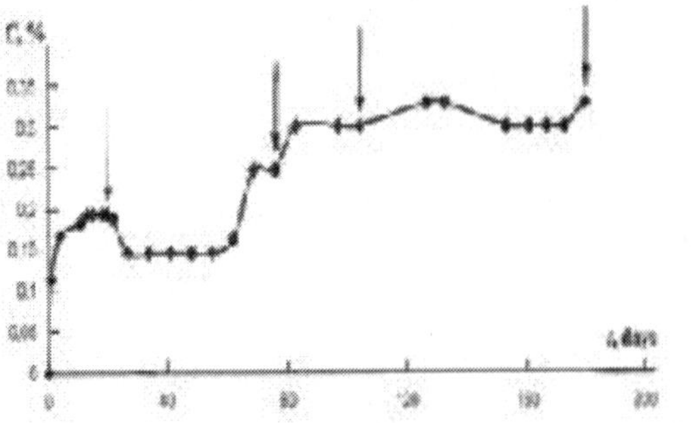

Figure 7. Water sorption by fiberglass. Arrows showed the moments of measurement of deformations

$$\Delta\lambda_B = 2 \cdot (\lambda \frac{\partial n_{ef}}{\partial l} + n_{ef} \frac{\partial \lambda}{\partial l}) \cdot \Delta l + 2 \cdot (\lambda \frac{\partial n_{ef}}{\partial T} + n_{ef} \frac{\partial \lambda}{\partial T}) \cdot \Delta T \quad (1)$$

The first part of equation gives value of shift of wavelength depending on deformation (lengthening), the second depending on temperature. Dependence of the central reflected wavelength on deformation can be also presented in the following form [5]:

$$\Delta\lambda_B = \lambda_B \cdot (1 - p_e) \cdot \varepsilon_z \quad (2)$$

where p_e — the constant of deformation of optical fiber defined from the following expression

$$p_e = \frac{n_{ef}^2}{2} \cdot [p_{12} - \nu \cdot (p_{11} - p_{22})] \quad (3)$$

where p_{11} и p_{12} — Pokkels's coefficients in a tensor of optical tension, — Poisson's coefficient. For standard fiber p_{11}=0,113, p_{12}=0,252, =0,16 and n_{ef} =1,4682 [6]

Proceeding from these values sensitivity for wavelength $_B$ =1550 nanometers make 12 nanometers/%.

Deformation of sensors after endurance in water calculated on the basis of (3) is given in Table 3.

Measurements were taken once a month and until optical measurement the sample was stored in the damp atmosphere. Thus we interfered with probable deformations when drying.

Table 3. Deformation increment from the moment of GFRP producing

Deformation in time, %, %	15.07.15	10.09.15	09.10.15	23.12.15
Upper layer	0,007	-0,017	-0,014	-0,074
Second layer	0,013	0,003	0,007	-0,001
Middle layer	0,015	0,003	0,009	-0,006

In Figure 8 changes of deformation of layers in the form of the histogram are presented.

In drawing it is well visible that external layers absorb water stronger than internal. But during such a long period (6 months), process becomes stationary and does not lead to destruction of a composite.

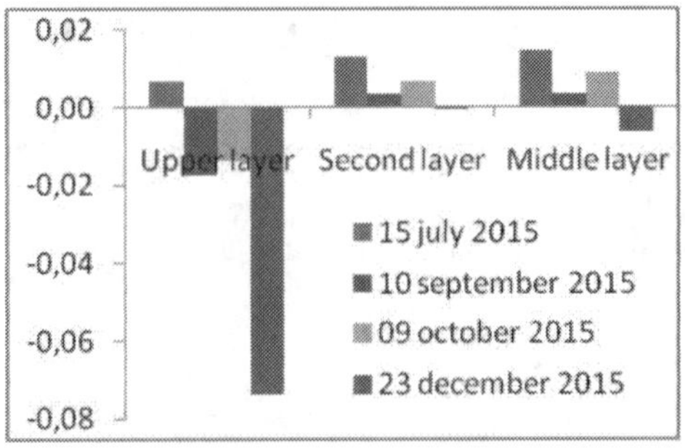

Figure 8. Deformations of layers of fibreglass at endurance in the damp atmosphere.

4. Conclusions

The possibility of fixing of thermomoist deformations in fiberglass of quasiisotropic structure is shown for a long time. Two types of measuring equipment are used: the BMSTU experimental system on the basis superbright light-emitting diode and the Astro A312 system on the basis of the tunable laser.

Deformations of swelling in different layers of fiberglass are recorded. In general epoxy resin ED-I has shown high resistance at influence of temperature and moisture.

The equipment of the built-in sensors can be used for a prediction of resource of the polymeric detail working in the conditions of long influence of water. Bragg's sensors are not subject to corrosion and therefore are suitable for such task. In our research ED-I epoxy has shown high resistance to hot water

Certain results of this work were obtained as a part of research works upon Agreement with the Ministry of Science and Education of the Russian Federation concerning the provision of grant № 14.577.21.0103 dated 16.09. 2014. The research project ID RFMEFI57714X0103

References

1. A. Cusano, A. Cutolo, J. Albert. "*Fiber Bragg Grating Sensors: Recent Advancements, Industrial Applications and Market Exploitation*", Editors, Bentham Science Publishers, 2010, PP. 35-52.
2. A.A. Dalinkevich, K.Z. Gumargalieva, S.S. Marakhovsky, A.V. Soukhanov, Modern Basalt Fibrous Materials and Basalt Fiber-Based Polymeric// JOURNAL OF NATURAL FIBERS, 2009,VL 6, № 3, P 248-271,
3. Yu.N Kulchin. The distributed fiber-optical sensors and measuring networks Vladivostok// Dalnauka, 1999, PP. 115-118.(In Russian)
4. M. Ramakrishnan, G. Rajan, Y. Semenovaand, G. Farrell, Overview of Fiber Optic Sensor Technologies forStrain/Temperature Sensing Applications in

Composite Materials, Sensors// 2016, № 16, V. 99; doi:10.3390/s16010099
5. S. Hannuscha, M. Stockmanna, J. Ihlemanna, Experimental method for residual stress analysis with fibre Bragg grating sensors, Materials Today: Proceedings 3 (2016), PP. 979 – 982
6. K.Yükseln, D. Pala, Analytical investigation of an ovelinterrogation approach of fiber Bragg grating sensors using Optical Frequency Domain Reflectometry// Optics and Lasers in Engineering 2016, № 81, PP.119–124
7. D. Abbaneo and other, Fiber Bragg Grating(FBG) sensors as flatness and mechanical stretching sensors//, Nuclear Instrumentsand Method sin Physics Research, 2016, V.824, PP. 493–495
8. P.V. Mikheev, A.V. Artemyev, V.A. Lazarev, A.B. Pnev, V.A. Nelyub, "Experience of creation of system of the built-in control of a condition of a design on the basis of fiber-optical the breggovskikh of sensors", the *Collection of reports of the All-Russian conference on tests and researches of properties of the materials TEST MAT - 2013.* Publishing house: All-Russian Research Institute of Aviation Materials (Moscow.) 2013, PP. 28-29. (In Russian)
9. V.V. Shishkin, A.E. Churin, D.S. Kharenko, I.S. Shelemba, "System of monitoring of the bearing designs of a football arena on the basis of fiber-optical sensors"// *Foton-Express*, 2013, № 6, V.110, PP. 22-23 (2013) (In Russian)
10. V.V. Shishkin, A.E. Churin, D.S. Kharenko, M.A. Zheleznova, I.S.Shelemba, "Structural health monitoring system of soccer arena based on optical sensors", Proc. SPIE 9157, *23rd International Conference on Optical Fibre Sensors*, 9157D3 (2 June 2014); doi: 10.1117/12.2071269, (2014)
11. B.S. Sarbayev, A.A. Smerdov, L.P. Tairova, V.A. Seleznyov, S.V. Sokolov, G.Ya. Buymistryuk, V.I. Izotov, A.M. Rogov, "Research of the deformed condition of designs from composite materials by means of fiber optical sensors"// *Bulletin of Bauman Moscow State Technical University*. 2011. N.SP, PP. 39-51. (In Russian)
12. E. Bar, *Constructional properties of plastic (*physical *and chemical bases of application*). The lane with English M., "Chemistry", 1967. PP. 231-235. (In Russian
13. Yu.A. Mikhaylin *Fibrous polymeric composite materials in equipment*. – SPb.: "Profession", 2013. – 752 pages. (In Russian)*Modern physical* methods *of research of polymers*. Under the editorship of Slonimsky G. L. M, "Chemistry", 1982. PP. 198-208. (In Russian)

 DOI: 10.1051/matecconf/20167706001

Study on Hydraulic Characteristic of Submerged Jet with Depression Angle Dived into Plunge Pool from the Underwater Surface

Chuncai Zhang[1, 2], Lijie Wang[1] and Xiaobing Dai[1]

[1]PowerChina ZhongNan Engineering Corporation Limited, 410014 Changsha Hunan, China
[2]School of Hydraulic Engineering, Changsha University of Science & Technology, 41004 Changsha Hunan, China

Abstract. Experiment research on submerged jet with depression angle dived into plunge pool from near-surface was carried out. The results showed that this type of energy dissipation type has good effect in improving the energy dissipation rate, reducing the surface fluctuation and reducing near-bed velocity and so on. Calculation formulas of conjugate depth and energy dissipation were obtained through theory analysis of submerged hydraulic jump in water cushion. Formula analysis showed that for high Froude number flow at a certain downstream depth, to obtain stable submerge hydraulic jump through adjusting size of jet exit, by decreasing its width is better way than by increasing height. When flow Froude number is larger sudden expansion ratio is more sensitive to energy dissipation, therefor to improving effect of energy dissipation by decreasing jet exit's width or increasing plunge pool's width can be used.

1 Introduction

In the last thirty years of reform and opening up, the country's hydropower has made great achievements attracting worldwide attention. Especially since the implementation of the western development strategy, hydropower projects of southwest reign have started construction, dam construction in these areas mostly on mountain rivers, with head high, large flow, narrow valleys and other features. Traditional underflow energy dissipation cannot meet requirements of spillway with large unit power, near-bed flow velocity is high and bottom stability is poor in plunge pool [1, 2]; ski-jump energy dissipation of flood discharge has good effect, but atomization is difficult to avoid the impact on the environment [3-6]. For this reason, Sichuan University and Mid-South Design and Research Institute proposed horizontal submerged jets energy dissipation in setting the vertical drop and sudden enlargement of stilling pool and applied in Xiangjiaba hydropower project, and good results were achieved [7-12]. In addition, Mid-South Design and Research Institute in conjunction Tuoba dam energy dissipation proposed submerged jet with depression angle dived into plunge pool from the underwater surface and conduct experimental studies [13], but this is a new type of energy dissipation type, water structure is complex and engineering practical experience is lack, so it is necessary to carry out in-depth research. In this paper, experiment was conducted to observation flow field and flow pattern of submerged jet with depression angle, hydraulics basic theories were used to establish submerged jet calculation formulas of conjugate water depth and energy dissipation, and transformation laws of plunge pool water depth and energy dissipation rate variation of Froude number were analysed in different condition of depression angle, water cushion expansion ratio and vertical drop. Findings enriched the study results of high-dam spillway energy dissipation and provided theoretical supports for engineering applications.

2 Flow pattern of submerged jet with depression angle

As shown in Fig. 1, physical model in the laboratory was established to format submerged jet with depression angle dived into plunge pool from the underwater surface, model including steel tanks, test section (jet segment and plunge pool), backwater reach and underground storage reservoir is self-circulation system. Pressure jet from jet segment inject into water cushion from the underwater surface with depression angle, and plunge pool dissipate its energy by the surrounded violent eddies. So two flow regimes of submerge jet and ternary water jump were observed through experimental in water plunge pool. As shown in Fig. 2, water cushion can be divided into several regions including submerged jet zone, impact jet zone, upstream and downstream wall attached jet zones, upstream and downstream eddy zones, left and right side eddy zones. Upstream and downstream eddies are the horizontal axis vortexes at the top and bottom edges of jet, and left and right eddies are the vertical shaft vortexes on both sides of the sudden enlargement plunge pool. Shear layer formed by vortexes can significantly increase the rate of energy dissipation units of water. After the end of

DOI: 10.1051/matecconf/20167706001

the mainstream area submerged submarine jets in water cushion pool bottom, the range of water surface fluctuation is little. Energy dissipation effect of shear between submerged jet and eddies is significant, as shown in Table 1, the maximum near-bed velocity in wall attached jet zone downstream is about half of the jet exit velocity.

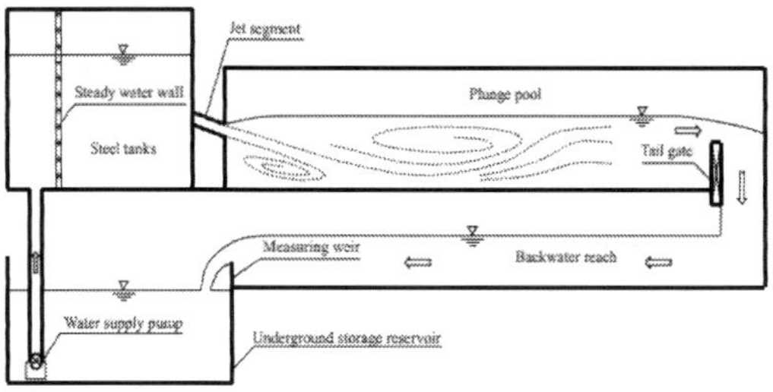

Figure 1. Model layout sketch of submerged jet with depression angle dived into plunge pool from the underwater surface.

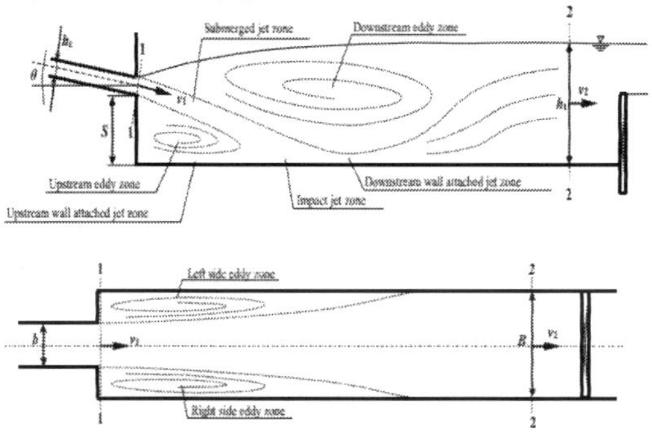

Figure 2. Flow state sketch of submerged jet.

Table 1. Average Flow Velocity of Jet Exit and Maximum Near-bed Velocity of Wall Attached Jet Zone.

Flux Q / m³·s⁻¹	Average flow velocity of jet exit v_1 / m·s⁻¹	Maximum near-bed velocity of wall attached jet zone v_{0m} / m·s⁻¹	Relative value $v_{0m}\cdot v_1^{-1}$ / %
4.33×10⁻²	3.8510	1.5909	41.31
4.54×10⁻²	4.0382	1.6673	41.29
4.75×10⁻²	4.2218	1.7421	41.27
4.96×10⁻²	4.4054	2.0593	46.75
5.15×10⁻²	4.5757	2.1383	46.73
5.34×10⁻²	4.7468	2.2168	46.70

Note: Depression angle of model is 10°, the jet exit width × height is 0.150m × 0.075m, plunge pool width is 0.30m, vertical drop height is 0.40m.

3 Theoretical formula of submerged jet

3.1 Conjugate water depth

As shown in Fig. 2, the jet exit width is b and height is h_c, depression angle is θ, vertical drop height s, plunge pool width is B, water cushion depth of submerged jet in plunge pool is h_t. Section 1-1 of the jet outlet and section 2-2 of downstream depth of submerged hydraulic jump were selected to apply momentum equation:

$$\frac{1}{2}\gamma h_c^2 b\cos\theta+\gamma BS\left(h_c\cos\theta+\frac{S}{2}\right)-\frac{1}{2}\gamma h_t^2 B=\rho\left(\frac{Q^2}{h_t B}-\frac{Q^2\cos\theta}{h_c b}\right) \quad (1)$$

where γ and ρ are the water density and density, Q is discharge flow. The formula rearranging above is:

$$\overline{H}=\frac{2Fr_1^2\left(\overline{H}\cdot\cos\theta-\frac{1}{\overline{B}}\right)}{\overline{B}\left[\overline{H}^2-\left(\overline{S}+\cos\theta\right)^2+\frac{\cos\theta}{\overline{B}}\left(\overline{B}\cdot\cos\theta-1\right)\right]} \quad (2)$$

where Fr_1 is the jet outlet Froude number; $\overline{H}$ is sequent depth radio, $\overline{H}=h_t/h_c$; $\overline{B}$ is sudden expansion ratio of water cushion, $\overline{B}=B/b$; $\overline{S}$ is the radio between the vertical drop height and the jet outlet, $\overline{S}=s/h_c$.

3.2 Energy dissipation ratio

In Fig. 2, section 1-1 and section 2-2 were selected to apply energy equation:

$$E_1=H_1+s=h_c\cos\theta+\frac{v_1^2}{2g}+s \quad (3)$$

$$E_2=H_2=h_t+\frac{v_2^2}{2g} \quad (4)$$

$$\Delta E=H_1+s-H_2 \quad (5)$$

where v_1 is the section 1-1 average velocity of jet outlet; v_2 is the section 2-2 average velocity of the submerged jet downstream; ΔE is the energy loss of submerged jet. Then energy dissipation rate of submerged jet is η:

$$\eta=\frac{\Delta E}{E_1}=\frac{H_1+S-H_2}{H_1+S}=1-\frac{\overline{H}+\frac{Fr_1^2}{2\overline{B}^2\overline{H}^2}}{\cos\theta+\overline{S}+\frac{Fr_1^2}{2}} \quad (6)$$

4 Formulas analysis

Fig. 3 shows the curves of water cushion depth dimensionless number $\overline{H}$ and the jet exit Froude number Fr_1 at different depression angle θ. From this, the same as the width of water cushion, vertical drop height, width and the height of jet exit, with increasing Froude number, water depth required to form stability submerged jump increases significantly. Fig. 4 shows the curves of energy dissipation rate η and jet exit Froude number Fr_1 at different jet depression θ. From this, under certain export jet outlet and water cushion size, energy dissipation with increasing Froude number increases, and the changes of depression angle little effect on the rate of energy dissipation; less Froude number impact energy dissipation rate significantly greater than the larger Froude number. When the Froude number is equal to 6, submerged jet energy dissipation rate is 70%; when the Froude number is equal to 10, the energy dissipation rate is over 80%. Integrated Fig. 3 and Fig. 4 shows, when other conditions are the same and the change of the depression angle less than 15°, the change of the depression angle has little effect on the water cushion depth and energy dissipation rate.

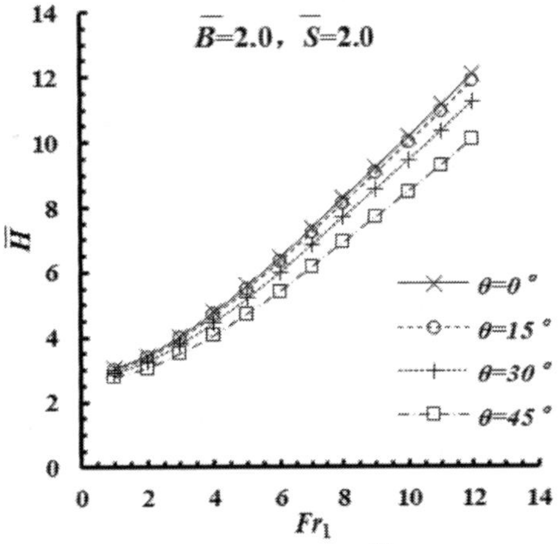

Figure 3. Relationship between Fr_1 and $\overline{H}$ in different θ.

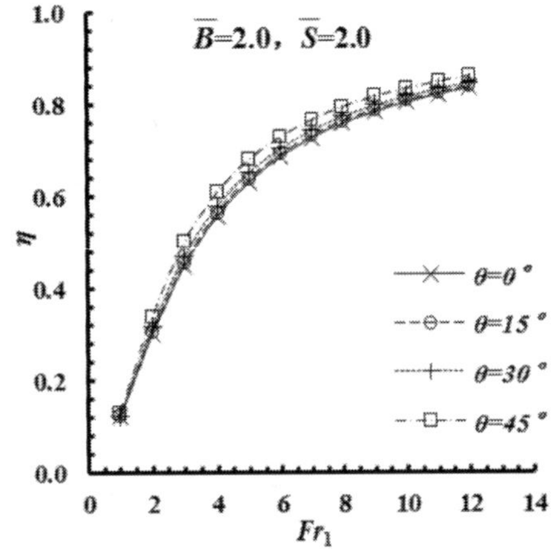

Figure 4. Relationship between Fr_1 and η in different θ.

Fig. 5 and Fig. 6 show the curves of water cushion depth dimensionless number $\overline{H}$ and the jet exit Froude number Fr_1. Seen from Figure 5, the same as the Froude number, depression angle, the height of vertical drop and the jet exit, the water cushion width is smaller or the jet exit width is larger, downstream water depth required to formed stability submerged jump increases. It is seen from Fig. 6, the same as the Froude number, depression angle, the width of water cushion and jet exit, the greater height of vertical drop or the smaller height of jet exit, downstream water depth required to formed stability submerged jump increases. Comprehensive seen in Fig. 5 and Fig. 6, when the Froude number is less than 6, jet exit width is more sensitive than vertical drop height impacted on downstream water depth. Therefore, under certain water depth conditions, if the formation of a stable submerged jump through adjusting the size of the jet exit, it is effective by increasing the height of jet exit when the Froude number is smaller and reducing the width of jet exit when the Froude number is larger.

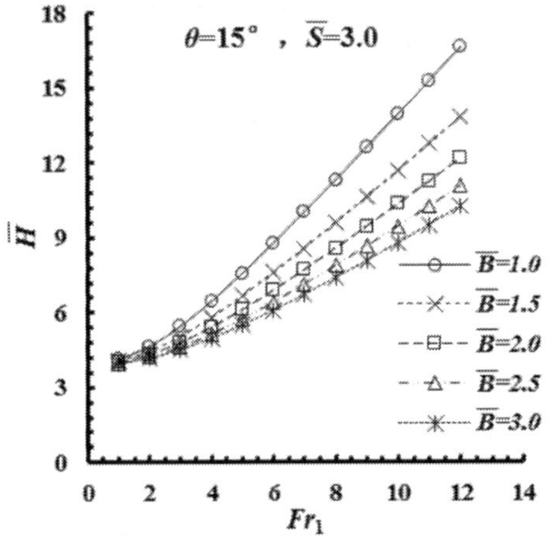

Figure 5. Relationship between Fr_1 and $\overline{H}$ in different $\overline{B}$.

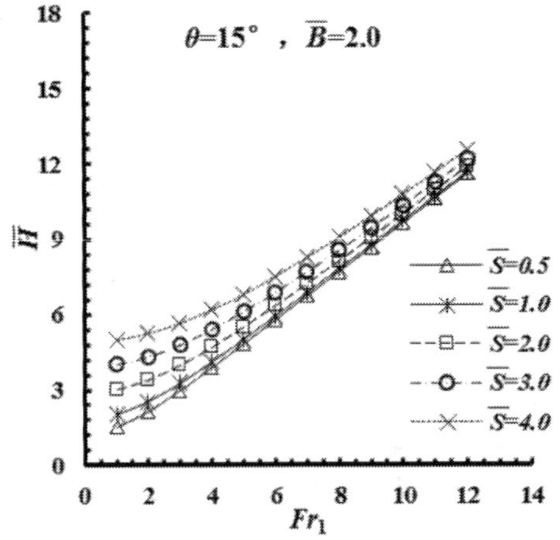

Figure 6. Relationship between Fr_1 and $\overline{H}$ in different $\overline{S}$.

Fig. 7 and Fig. 8 show the curves of energy dissipation rate η and jet exit Froude number Fr_1. Seen from Fig. 7, the same as the Froude number, depression angle, the height of vertical drop and the jet exit, the water cushion width is larger or the jet exit width is smaller, energy dissipation rate is improved; the bigger ratio of water cushion sudden expansion, the smaller impact of increasing energy dissipation rate. It is seen

DOI: 10.1051/matecconf/20167706001

from Fig. 8, the same as the Froude number, depression angle, the width of water cushion and jet exit, the greater height of vertical drop or the smaller height of jet exit, energy dissipation rate is improved. Integrated Fig. 7 and Fig. 8 show that the sudden expansion ratio is more sensitive than the radio between the height of vertical drop and the jet outlet impacted on energy dissipation especially when the Froude number of jet outlet is larger. Therefore, if the jet by adjusting the size of the outlet to obtain a higher rate of energy dissipation, it is effective by reducing the width of the jet outlet when Froude number is larger.

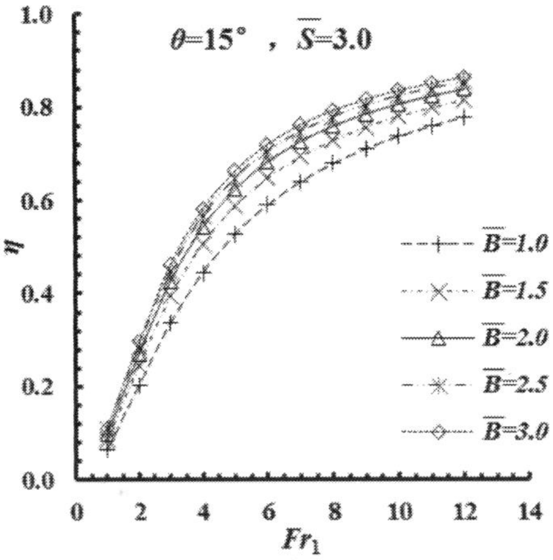

Figure 7. Relationship between Fr_1 and η in different $\overline{B}$.

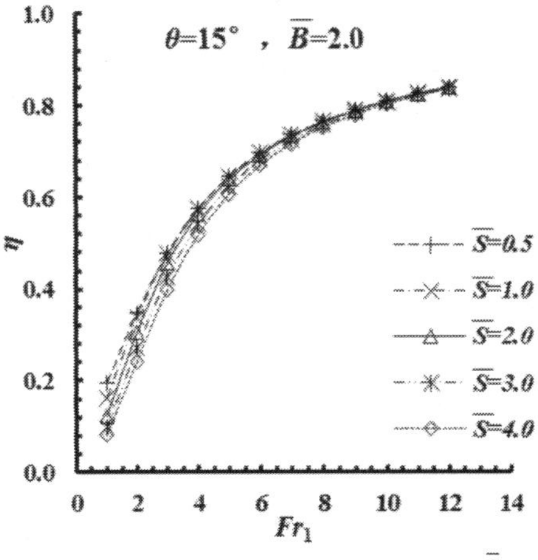

Figure 8. Relationship between Fr_1 and η in different $\overline{S}$.

5 Conclusions

Submerged jet with depression angle dived into plunge pool from the underwater surface is a new energy dissipation type, and has a complex flow structure. Hydraulic characteristics of one jet flow field were carried out in the paper.

(1) Experimental study on flow field of plunge pool results indicate that, the existing more eddies can improve the unit water energy dissipation rate, the mainstream of the dive reduces water fluctuations, and significantly reduces the near-bed velocity of plunge pool.

(2) Through theoretical analysis, calculation formulas of conjugate water depth and energy dissipation, relationships between the conjugate water depths, energy dissipation and the jet exit Froude number. Analysis showed that other things being equal to form stability submerged jump downstream water depth increases significantly with increasing Froude number, and when the depression angle less than 15° it changes little effect on downstream water depth. Under certain water depth conditions, if the formation of a stable submerged jump by adjusting the size of the jet exit, it is effective by increasing the height of jet exit when the Froude number is smaller and reducing the width of jet exit when the Froude number is larger. The sudden expansion ratio is more sensitive impacted on energy dissipation when the Froude number is larger, and therefore, it is effective by reducing the width of the jet outlet or increasing the width of water cushion to obtain a higher rate of energy dissipation.

References

1. P. Q. Liu, J. Z. Gao, G. F. Li, J. Hydraul. Eng. 1 (1999)
2. P. B. Yue, C. Zhou B. L. Jiang, J. Hydroelec. Eng. E **32**, 6(2004)
3. S. Y. Wang, X. B. Liu, R. Chen, J. of Yangtze River Sci. Res. Inst., E **32**, 2(2015)
4. S. K. Shuang, J. CHIN Inst. Water Res. and Hydropower Res., E **7**, 2(2009)
5. H. C. Dai, W. L. Xu, J. Hydroelec. Eng., E **35**, 1(2009)
6. J. P. Zhou, Z. Y. Yang, G. F. Chen, J. Hydraul. Eng., E **37**, 12(2006)
7. J. M. Zhang, Y. R. Wang , Y. Q. Yang, Adv. Water Sci., E **16**, 1(2005).
8. Q. J. Huang, S. R. Feng, Y. N. Li, CHIN J. Hydrodyn., E **23**, 6(2008)
9. Y. L. Li. *PhD thesis* (Sichuan Uni. CHIN., 2004)
10. V. Katakam, R. Seetha P. Rama, J. Hydr. Eng., ASCE, E **124**, 6(1998)
11. Y. L. Li, Y. Q. Yang, J. Hydrodyn. 3(2002)
12. J. G. Chen, M. J. Zhang, W. L. Xu, S. Li, X. L. He, J. Hydrodyn, E **25**, 4(2015)
13. Y. N. Li, X. B. Dai, P. Q. Wu, (PCZNECL, Changsha, Hu'nan, 2014)

MATEC Web of Conferences 77, 06002 (2016) DOI: 10.1051/matecconf/20167706002
ICMMR 2016

Optimization Analysis on Upper Filter System of Mechanical Water Treatment Equipment Based on Strength

Ying Gao[1], De-zhen Feng[2] and Jian-hau Sun[2]

[1]*School of Management, University of Jinan, Jinan 250022, China*
[2]*School of Mechanical Engineering, University of Jinan, Jinan 250022, China*

Abstract. Domestic water treatment is very important field. Now, mechanical water treatment technology is getting wide use in production. In the process of life water treatment, filter process is a very important step. In this paper, the strength and deformation of upper filter system which includes upper filter frames and upper filter plates were analyzed with ANSYS and useful results are got. Through the analysis on strength, the paper found the shortcomings of production and design. After analyzing and comparing the stresses and deformations of three different design schemes, the paper provided the optimized design plan of upper filter system at last.

1 Introduction

As everyone knows, the average water resource of every person is very low, especially in northern China. Further more, the most water of our country is surface water and the components of the water are very complicated. This will threaten the safety of the water for life especially for rural water for life. So it is a very important to protect and apply the surface water scientifically. On the other hand, most parts of our country are mountains and desert. The available land resource is not abundant. With the increase of the population, the development of the economy and the quickening of urbanization process, the land resources will get more and more precious. So how to save in water treatment project has become an important subject.

Now the conventional water plants of our country use steel concrete structures as the main constructions which need large land and large building project which will cost long time and much money. On the other hand, the conventional water plants are not flexible in work. So the modular mechanical water treatment plants are getting wider and wider use because they need low invest, short time limit, small land and have the advantages of flexibility, high automation and easy operation. The water treatment plant consists of several independent treatments which are composed of three parts: flocculation system, sedimentation system and filter system, as are shown in Fig. 1 and Fig. 2.

According to the characteristics of modular mechanical water treatment equipment, this paper has analyzed the strength and stiffness of the upper filter system, discussed the stress and the deformation of the filter frame and plate. At last, combining theory with practice, the paper has improved and optimized the design of filter system.

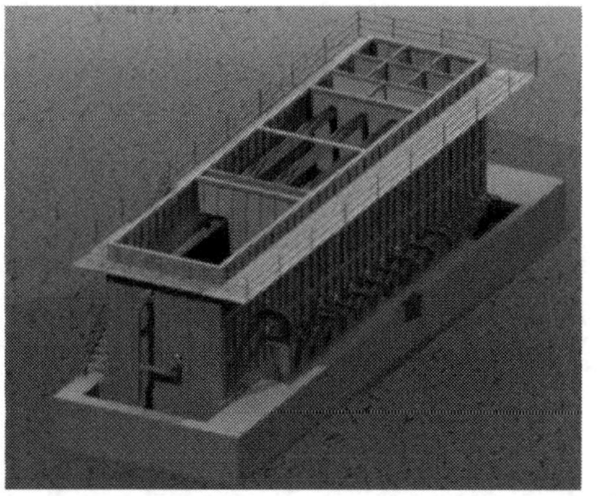

Figure 1. Mechanical water treating equipment

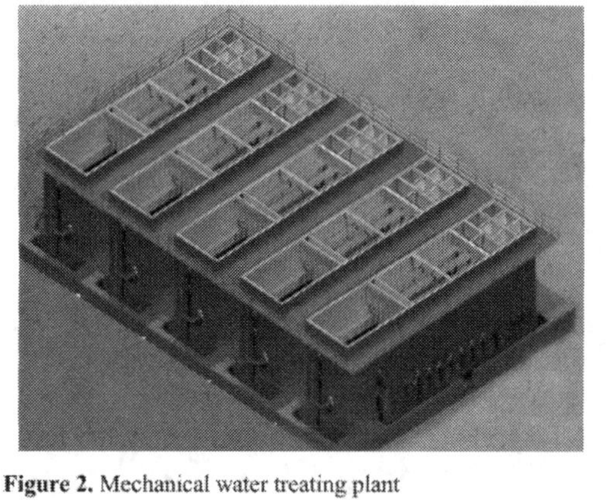

Figure 2. Mechanical water treating plant

According to the characteristics of modular mechanical water treatment equipment, this paper has analyzed the strength and stiffness of the upper filter system, discussed the stress and the deformation of the filter frame and plate. At last, combining theory with practice, the paper has improved and optimized the design of filter system.

2 Structure of filter systems

According to the requirement of practical production, the filter systems should be changed in time to adapt to the need of production process. The filter systems are composed of upper and lower filter systems. In the upper filter system, filter plate uses 30mm PVC plate on which a number of filters are mounted, the filter frame uses 3*9 layout plan, i.e. three 120*80*5mm longitudinal steel tubes and nine 120*80*5mm transverse steel tubes which all are welded to the longitudinal tubes. Under the middle longitudinal beam, there are four uniformly distributed 80*80*5mm upright columns, as shown in Fig. 1. The lower filter frame uses 3*9 layout plan, i.e. three 80*80*5mm longitudinal beam and nine 80*80*5mm transverse beams. The supports columns are twelve (3*4) uniformly distributed 80*80*5mm upright steel tubes. The paper only studies the upper filter system.

In practice, the water treatment plant uses one layer or two layers homogeneous filter materials in one filter chamber which includes frames, filter plates, filter materials etc. Filter frames support the filter plates and filter plates bear the gravity of filter materials. The upper layer filter materials are quartz sands whose density is about 2.6g/ cm3 and stacking density is 1.6 g/ cm3. The filter plates are rigid plastic PVC plates whose thickness is 30mm. The filter frames are composed of Q235B steel tubes whose size is 120*80*5mm and the supporting tubes' size is 80*80*5mm.

3 Strength analysis of original upper filter system

In the upper filter system, the filter plate uses 30mm PVC plate in which a number of filters are mounted. The filter frame uses 3*9 layout plan, i.e. three 120*80*5mm longitudinal tubes and nine 120*80*5mm transverse steel tubes. There are four uniformly distributed 80*80*5mm upright square tubes that support the longitudinal beam, as shown in Fig. 1. In transverse direction, the frame uses nine 120x80x5mm rectangular steel tubes which all are welded to the longitudinal beam. The load of quartz sand is about 0.02785 MPa.

For the rectangular sheets with usual constraints, we can use theoretical formulae to compute [1, 2]. But for the complicated combination of beams and sheets, the theoretical formulae can not be used to analyze the strength. In this case, numerical calculation must be used. In this paper, the FEM software ANSYS is used to calculate the stress and deformation of complicated structures [3-6]. Fig. 3 is the FEM model of the upper filter system, Fig. 4~6 are the analytical results of upper filter system under normal work load. Where, Fig. 4 is the stress picture of upper filter system, Fig. 5 and Fig. 6 is for stress pictures of the filter plate frame and filter plate respectively. In the analysis, the modulus of elasticity of the filter plate is 3000MPa.

When the periodic back washes are needed, the load will increase to 0.05MPa. Fig. 7~8 are the analytical results of upper filter system under back wash load.

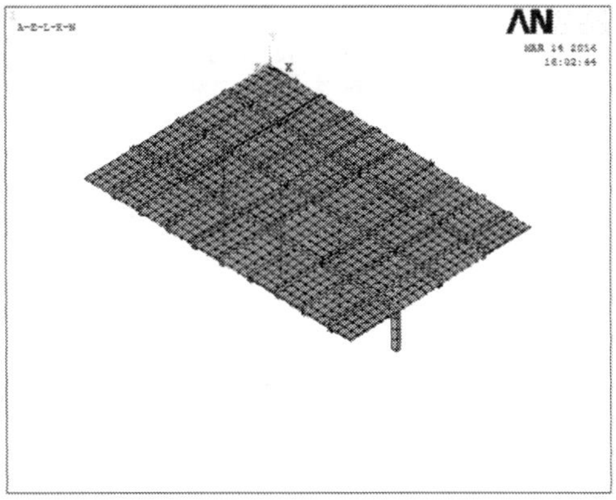

Figure 3. Model of upper filter system

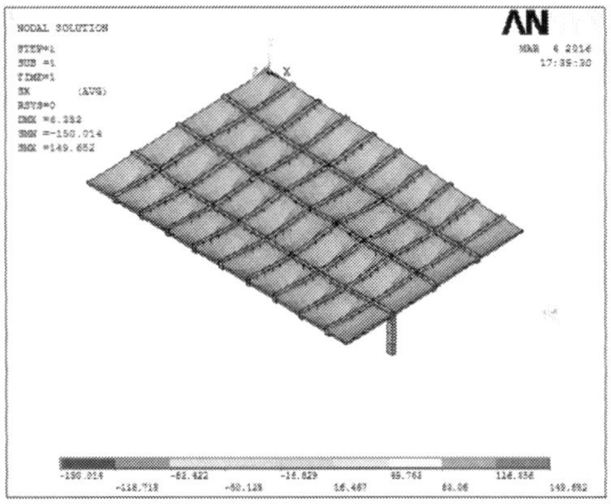

Figure 4. Stress of upper filter system

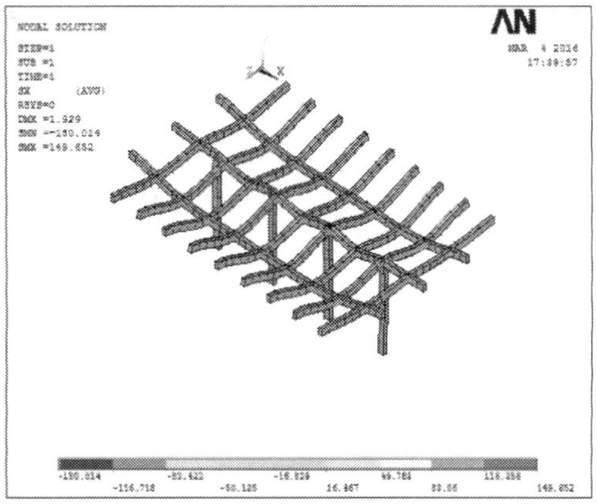

Figure 5. Stress of upper filter frame under work load

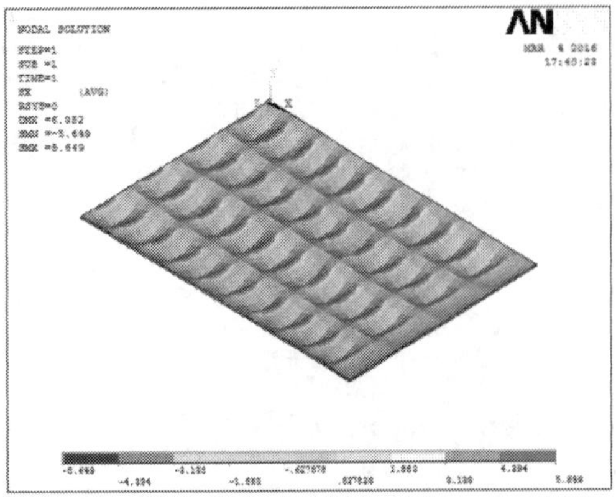

Figure 6. Stress of upper filter plate under work load

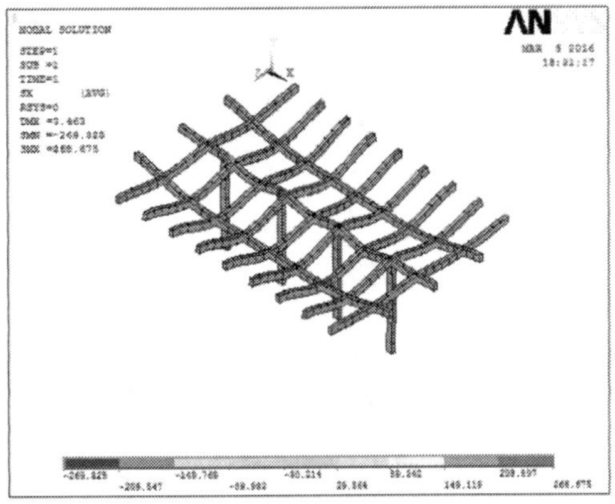

Figure 7. Stress of upper filter frame when back wash

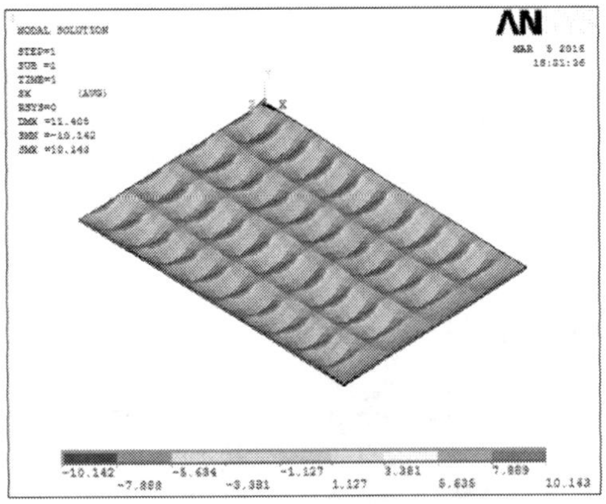

Figure 8. Stress of upper filter plate when back wash

From Fig. 5~6 we find that the largest deformation of the upper filter plate is about 6.4mm and the largest stress of the filter plate is about 5.6 MPa. The largest deformation of the upper filter frame is about 1.9mm, and the largest stress of the frame is about 150.0 MPa. At the same time, the results show that the middle longitudinal beam is most dangerous because the largest stress occurs in the middle longitudinal beam.

From Fig. 7~8 we know that the largest deformation of the upper filter plate is about 11.4mm and the largest stress of the filter plate is about 10.1 MPa. The largest deformation of the upper filter frame is about 3.5mm, and the largest stress of the frame is about 269.3 MPa. Just like the case in the normal work load, when back wash, the middle longitudinal beam is most dangerous too.

Under the back wash load, the stress of upper frames is larger than the yield limit of Q235B steel and not safe. That is to say, the upper filter frames must be reinforced. There are two reinforcing ways: changing the middle longitudinal beam with lager beam or increasing the number of the supporting tubes.

4 Strength analysis of the first reinforced upper filter system

The first reinforcing plan is changing the middle longitudinal beam from 120*80*5mm to 120*120*5mm. Fig. 9~10 are the analytical results of the first reinforced upper filter system under work load and Fig. 11~12 are the analytical results of the first reinforced upper filter system under back wash load.

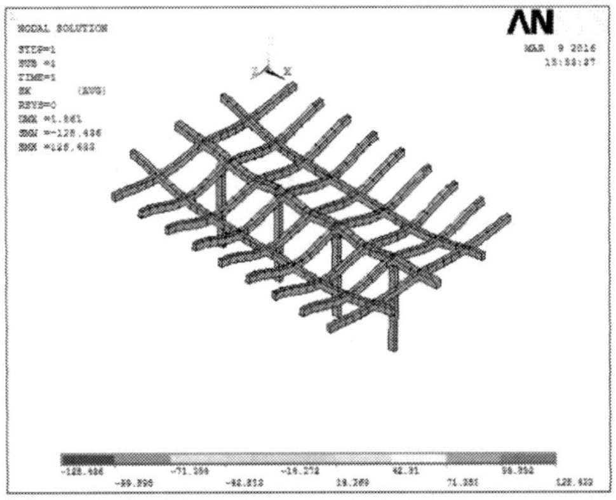

Figure 9. Stress of upper filter frame under work load

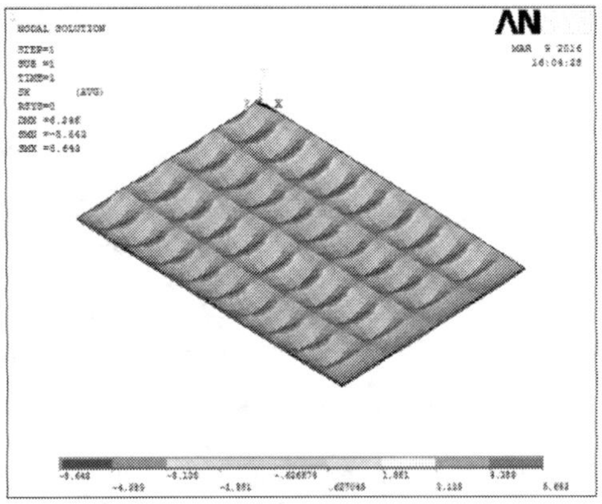

Figure 10. Stress of upper filter plate under work load

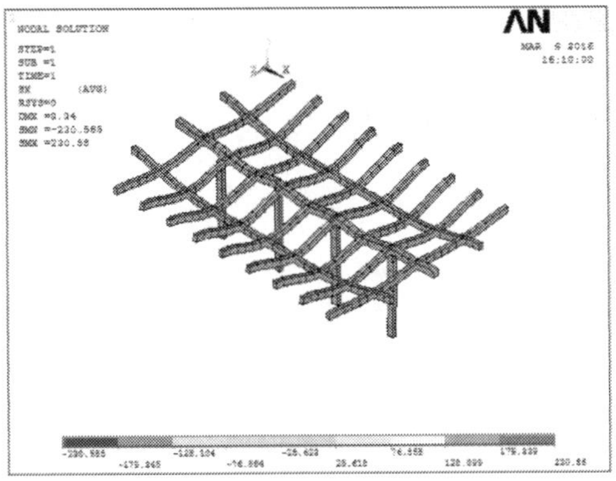

Figure 11. Stress of upper filter frame when back wash

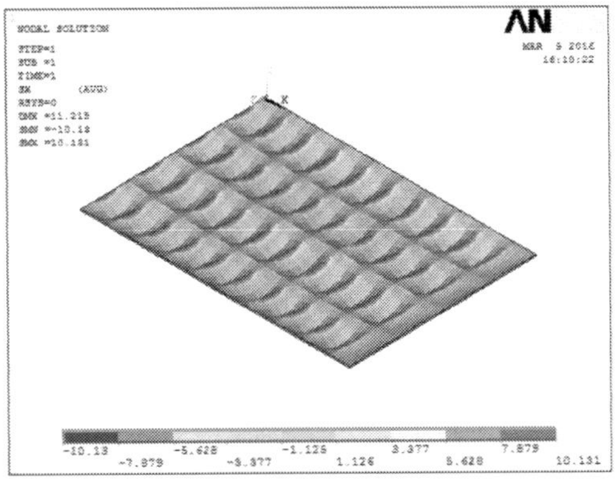

Figure 12. Stress of upper filter plate when back wash

5 Strength analysis of the second reinforced upper filter system

The second reinforcing plan is increasing another 4 supporting tubes under the flank longitudinal beams. As are shown in Fig. 13. Fig. 13~14 are the analytical results of the second reinforced upper filter system under work load and Fig. 15~16 are the analytical results of the second reinforced upper filter system under back wash load.

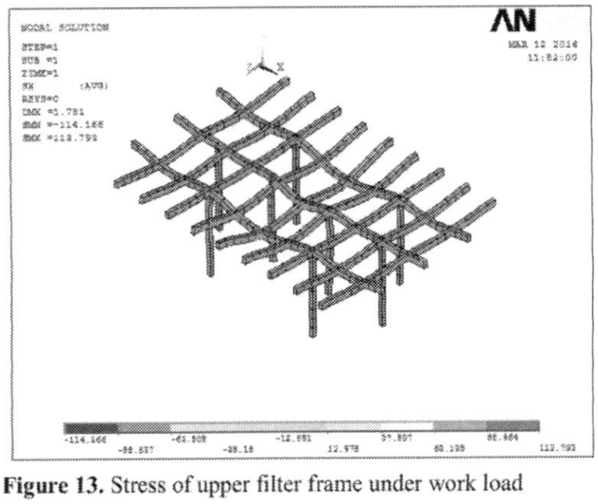

Figure 13. Stress of upper filter frame under work load

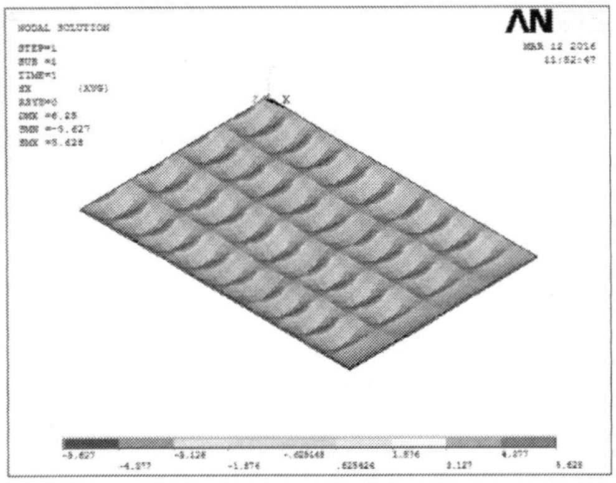

Figure 14. Stress of upper filter plate under work load

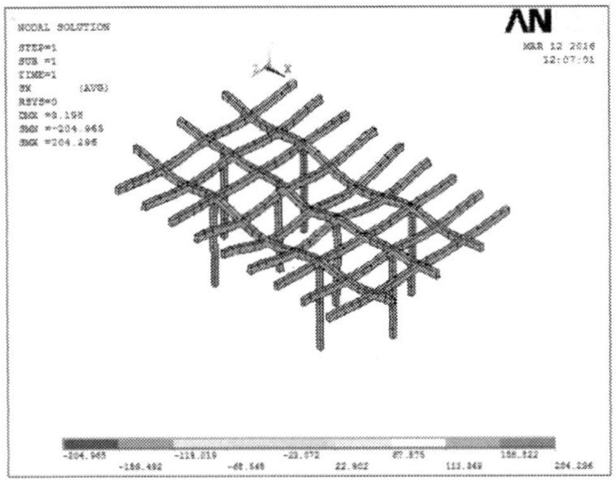

Figure 15. Stress of upper filter frame when back wash

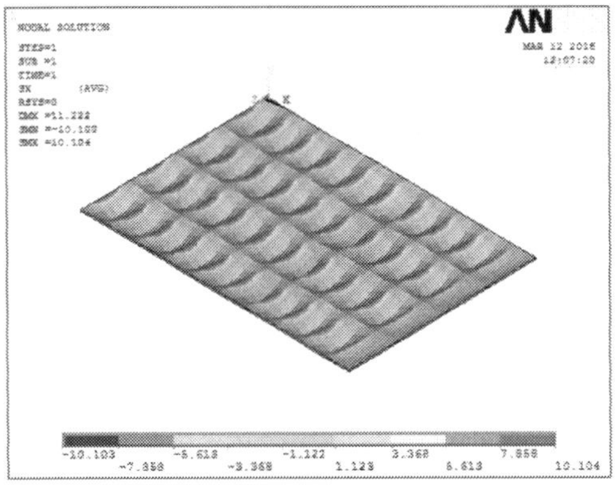

Figure 16. Stress of upper filter plate when back wash

6 Summary

From the analytical results we know that the upper filter system is in safety in normal work load. But when back wash, the original design has the largest stress 269.3MPa, the first reinforced design is 230.6 MPa, and the second reinforced design has the smallest stress 205.0 MPa. The second reinforced plan is the best design.

The deformations and stresses of the upper filter system under the normal work load and back wash load are shown in Table 1 and Table 2 respectively.

 DOI: 10.1051/matecconf/20167706002

Table 1. Deformations and stresses of the upper filter frames

Plan No.	Work load		Back wash load	
	Deformation	Stress	Deformation	Stress
1	1.9	150.0	3.5	269.3
2	1.9	128.4	3.3	230.6
3	1.8	114.2	3.2	205.0

Table 2. Deformations and stresses of the upper filter plates

Plan No.	Work load		Back wash load	
	Deformation	Stress	Deformation	Stress
1	6.4	5.6	11.4	10.1
2	6.2	5.6	11.2	10.1
3	6.3	5.6	11.2	10.1

References

1. Cheng Daxian. Mechanical Design Handbook. 5th Edition. Beijing: Chemical Industry Press, 2008
2. Xu Zhilun, Elasticity Mechanics. 4th Edition. Beijing: Higher Education Press, 2006
3. Chen Jingyi. Manual of Computer Aided Engineering Analysis of ANSYS. 1st Edition. Beijing: China Railway Press, 2001
4. T. R. Chandrupatla. Introduction to Finete Elements in Engineering. 1st Edition. Beijing: China Machine Press, 2008
5. ANSYS Inc. ANSYS Structural Analysis Guide Release 5.4 Third Edition. SAS, IP Inc., 1997
6. Zienkiewicz O C, Taylor R L. The Finite Element Method: The Basis. 5th Edition. Volume 1, Butterworth-Heinemann, Oxford, 2000

MATEC Web of Conferences 77, 06003 (2016) DOI: 10.1051/matecconf/20167706003
ICMMR 2016

Study on Wellhead Growth in Deep Well

Gao Baokui, Ren Jingwei and Gao Liang

Department of Petroleum Engineering, China University of Petroleum Beijing, Beijing, 102249, China

Abstract. Wellhead growth has occurred in many deep wells in China recently. In the paper, casing axial force and wellhead movement are calculated and compared in regard to moments that free sections of surface casing appear. The parameters considered include preliminary casing tension, free section length and change in wellbore temperature. The moment that surface casing free section comes into being has great influence on wellhead movement, so it is discussed in detail. According to comprehensive investigation on deep well and ultra-deep well accidents, wellhead growth is based on the fact that before production the cement sheath of surface casing has already failed, and during production wellhead grows up due to wellbore temperature increases. The most possible reason of surface casing cement sheath failure is bad cement job quality. Bad cement sheath cannot support casing axial loads and separates from casing. The free section of surface casing will experience large axial force cycle which is harmful to wellbore integrity. Effective measures to remedy wellhead growth is supporting the wellhead with a large base set on ground, rather than pads fixed on conductor.

1 Introduction

Wellhead growth is usually occurred in thermal wells and offshore self standing platform wells. In a normal thermal well, only production casing is connected with wellhead and in most cases the well is cemented to surface. When injecting steam or hot water, wellbore is heated, cement sheath may separate from casing and wellhead growth will take place. After injection being stopped the casing will shrink and wellhead will return to its original position, even subside[1, 2]. In an offshore oil well, surface casing has unavoidable free section from mud line to wellhead, therefore some factors, such as casing temperature and annulus pressure, can induce wellhead movement during subsequent constructions. For high level safety, study and achievements are mainly in offshore wells, but calculation methods are generally alike[3-5]. There are little open reports about wellhead growth or subsidence in onshore deep wells, and there is no mechanism study. In fact, accidents of wellhead growth or subsidence in deep wells do exist[6, 7]. In an abnormal deep well all casing strings are top cemented, so the wellhead growth mechanism is more complex[8-10]. As an important component of wellbore integrity, wellhead growth in deep well must be solved.

With regard to deep well construction procedure, change in wellhead position and casing axial force is calculated in this paper. Parameters as casing pretension, free casing length and thermal effect are considered. The moment and after effect that surface casing section becomes free are investigated in detail.

2 Models of wellhead growth

2.1 Wellbore structure of deep wells

A typical wellbore sketch of deep well in China is drawn in Fig. 1.

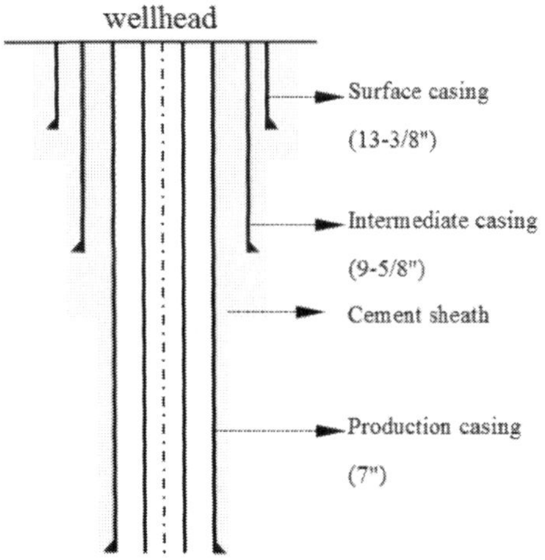

Figure 1. Typical wellbore sketch of deep well.

Generally, wellbore structure has the following characters:
- The first pipe is conductor with depth ranged from 100m to 300m, cemented to surface and will not connect with wellhead.

- Surface casing is cemented to surface, and it should directly support wellhead loads.
- Based on drilling conditions, intermediate casing and production casing may be top cemented or not top cemented (has free section).
- The liner, if there is, doesn't participate in wellhead growth.
- The main wellbore consists of surface casing, intermediate casing and production casing; these casings joint together at wellhead.

2.2 Mechanism of wellhead growth

Treat wellhead as a rigid body, then three layers of casings would grow up simultaneously at top ends. The stiffness of surface casing is much larger than others, so an obvious wellhead growth means that surface casing must have a considerable section that can move axially. Other casings can also have such intervals, and they are termed as free sections throughout this paper. Normally, free section of surface casing comes from bad cement job quality.

The principal work to be done is to find the moment that surface casing free section appears. But wellhead growth is generally observed during production, so the task is very difficult. In order to solve this problem, five moments are proposed as follows:

- Case-1, before hanging intermediate casing,
- Case-2, when hanging intermediate casing,
- Case-3, when hanging production casing,
- Case-4, when hanging tubing,
- Case-5, during production.

It is impossible to give precise conditions of cement sheath failure. So in the following investigation, primary work is performed under the assumption that surface casing free section appears outright while in different moments. Casing axial force and wellhead growth are main parameters to be calculated.

In engineering, it is hard to predict the moment when surface casing free section appears. The most optimistic expectation is to monitor it timely. In view of study and prevention, any possibility should be considered to find out the causality.

Wellhead growth reported now is a height change compared with previous position. If free section forming moment is not identified, it is impossible to trace wellhead movement before the so called growth. In this paper, the reference position of wellhead is the height of surface casing top just after it is cemented. This position doesn't distinguish whether surface casing has free section.

2.3 Calculation method

2.3.1 Initial axial force at top end of casing

Every casing will be connected to wellhead after cementing. If top of cement reaches to surface and cement job quality is good, there will be solidified pretension in casing wall. It can be regarded as tension of casing to wellhead if this casing appears free section afterward. If casing has scheduled free section, casing will have suspended load when setting at wellhead, and the load is controllable, which is the axial force at top of free casing section.

2.3.2 Formulas for calculation wellhead growth

If wellbore temperature changes, free casing section will have thermal effect, and the formula is

$$d(\delta l)=\alpha \cdot \Delta T \cdot \Delta L \tag{1}$$

where α is linear expansion coefficient of casing, ℃$^{-1}$; ΔL is micro-segment of free casing section, m; ΔT is change in temperature of casing micro-segment, ℃; $d(\delta l)$ is elongation of casing micro-segment due to thermal effect, m.

The temperature difference is different with depth, so integral is needed to obtain total heating effect.

Both temperature effect and wellhead growth of free casing section will associate with the relationship between casing axial force and deformation, which can be written as

$$dF=(EA/L)\cdot dL \tag{2}$$

where E is Young's elastic modulus, kPa; A is cross sectional area of casing wall, m^2; L is free casing length, m; dL is axial deformation of free casing section, m; dF is change in axial force of free casing section, kN.

Take wellhead as rigid body, the resultant force of wellhead at vertical direction is zero, the equation can be expressed as

$$F_1+F_2+\cdots+F_n=0 \tag{3}$$

where n is the number of forces acting on wellhead; F_i (i=1,2,···,n) is ith axial force, kN.

More formulas can be found in reference 1.

2.3.3 Temperature field

Temperature increment is the motivator of wellhead growth. So it is necessary to calculate change in temperature in every free casing section. But the calculation is very complicated, which involves wellbore structure, flow rate, fluid components, formation original temperature, flow period, and so on. In the following calculation some results are used directly without analysis in detail.

2.3.4 Simplification and hypothesis

In following calculations, some simplifications and hypotheses are used as follows:

- Wellbore is vertical,
- The influence of conductor is neglected,
- The influence of installation and remove of blowout preventer is neglected,
- The length of every free section is given, and its potential extension is neglected.

3 Case study

In this section a deep well in west China is taken as example, casing axial force and wellhead growth are calculated and compared. The length of casing free section is just a possible value in the field, namely is not necessarily the real value because real result is impossible to be identified.

3.1 Basic parameters

This deep well is top cemented for every layer of casing, and the wellbore structure is shown in Figure 1, detailed parameters are listed in Table 1.

Table 1. Detailed parameters of wellbore structure.

Casing	Hole diameter (mm)	Casing outer diameter (mm)	Casing wall thickness (mm)
13-3/8"	444.5	339.7	12.19
9-5/8"	311.2	244.5	11.05
7"	215.9	177.8	10.36

3.2 Initial axial force in casing

If cementing job quality is good, axial tension will be solidified in casing. The tension depends on casing size, density of slurry and displacement liquid, and holding pressure during cement curing. According to engineering operation, initial top axial forces (F_0) of every casing are listed in the second column in Table 2. These axial forces are different from actions between casing and wellhead. They can't be changed by subsequent loads on wellhead if cement job quality is good, unless temperature changes.

If surface casing has free section before production due to cementing quality, then top axial force of every casing will change. For example, surface casing appears free section immediately after cementing, then top axial force is zero and the free section has to self stand. The axial force of whole free section will change with subsequent loads acting on wellhead.

Based on above analysis, axial force of top casing is equal to its action with wellhead only when surface casing has free section.

3.3 Parameters of casing free section

In the following calculation, it is assumed that if surface casing has free section, intermediate casing and production casing are deemed to have free sections. According to this principle and cement job quality, empirical free section lengths (L_0) are given as that in the third column in Table 2.

Without loss of generality, if surface casing there exist free section, it must support the suspended loads imposed on subsequent casings when they are connected to wellhead. For intermediate casing and production casing, let setting loads be top axial tensions, as listed in the second column in Table 2.

Basing on casing size and length of free section, relationship of axial force and deformation, namely spring coefficient (k), is obtained and listed in the fourth column in Table 2.

The well depth is 6600m, in situ temperature is about 190℃, wellhead temperature is about 90℃ when gas flow rates is $3.0\times10^5 m^3/d$. Based on comprehensive analysis of flow rates, components, wellbore structure, flow time and free section length, free elongations (ΔL_T) from thermal effect of each casing are listed in the fifth column in Table 2.

Table 2. Casing parameters for wellhead growth calculation.

Casing	F_0 (kN)	L_0 (m)	k (m/10 kN)	ΔL_T (m)
13-3/8"	2416.3	170	0.000661	0.0306
9-5/8"	762.57	170	0.000977	0.0714
7"	1313.7	500	0.007970	0.6480

Tubing initial suspending force is 200kN, and ground equipment is simplified as 100kN vertical load.

3.4 Analysis of calculation results

According to above parameters and methods, results are obtained in accordance with above five typical moments denoting surface casing free section appearing.

For each case, six working steps are used to analyze variations of wellhead position and casing axial force. The steps are as follows:

S(1)—13-3/8" surface casing has cemented;

S(2)—9-5/8" intermediate casing has sit on wellhead;

S(3)—7" production casing has sit on wellhead;

S(4) — Tubing has hung and Christmas tree has installed;

S(5)—During production, wellbore temperature has risen but wellhead does not move (This step is hypothetical, for comparing with other steps.);

S(6)—During production, wellbore temperature has risen and wellhead growth has reached balance.

Wellhead position and casing top axial force are calculated for discussion.

3.4.1 The change of wellhead position

Fig. 2 shows the change of wellhead position of the four first cases, the horizontal ordinate is working steps and the vertical ordinate is wellhead position. The case-5 doesn't produce wellhead growth based on calculation, so it isn't drawn in Fig. 2. Conclusions are as follows:

- The case-1 is unique in style with a special wellhead move route. In other cases, wellhead positions have little difference from step 4.
- The changes of wellhead position from step 5 to step 6 are identical, and this is the so-called wellhead growth. This phenomenon indicates that wellhead growth is only associated with change of parameters before and after production, including temperature and pressure.
- Based on calculation, wellhead growth does not take place in case-5. This indicates that as long as surface casing doesn't have free section before production, wellhead growth cannot occur.

DOI: 10.1051/matecconf/20167706003

• Casing suspended loads have influence on wellhead position, but they don't affect the value of wellhead growth.

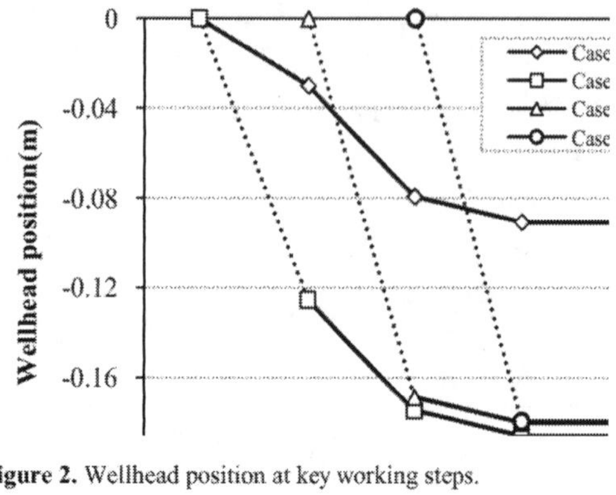

Figure 2. Wellhead position at key working steps.

3.4.2 The change of casing axial force

Fig. 3 to Fig. 5 show top axial forces of casing 13-3/8", 9-5/8" and 7" respectively, five cases are all considered. The vertical ordinate is top axial force of casing. Some unimportant information is neglected in these figures. It is necessary to distinguish between top axial force and acting force with wellhead, they are the same only when casing has free section, as solid part of curves in these figures.

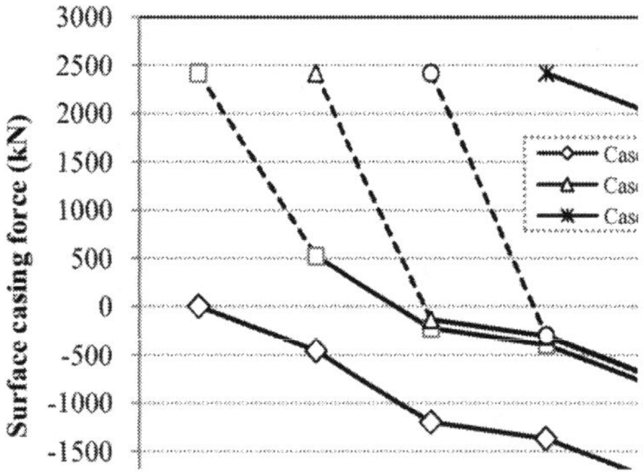

Figure 3. Axial force of 13-3/8" casing at key working steps.

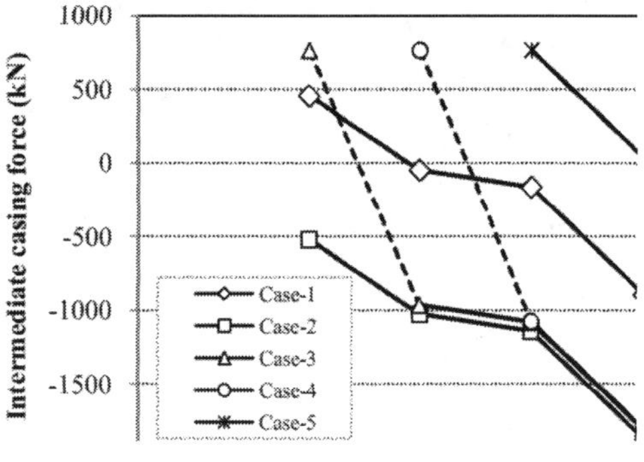

Figure 4. Axial force of 9-5/8" casing at key working steps.

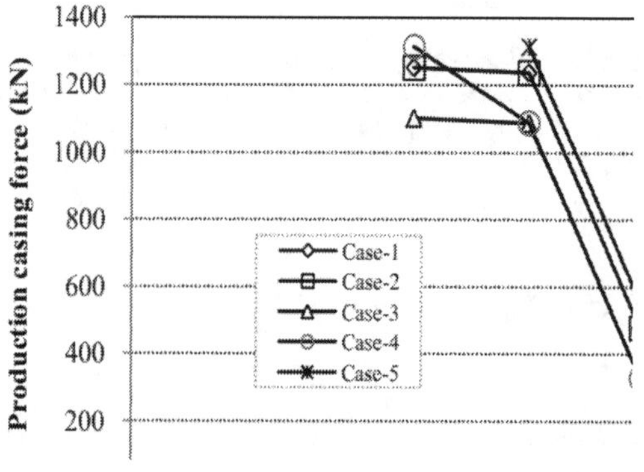

Figure 5. Axial force of 7" casing at key working steps

Combining with calculation parameters, conclusions are as follow:

• 13-3/8" casing is unique in style with a special force route in case-1. In case-2 to case-4, axial forces have little difference.

• 13-3/8" casing has the smallest axial force at step 5 and the force is press. It indicates that surface casing undertakes compressive load at wellhead, namely supports the wellhead.

• 9-5/8" casing is unique in style with a special force route in case-1. From step 2, top axial force is less than setting load due to wellhead subsidence during setting, and the decrease is transmitted to 13-3/8" casing.

• Top forces of 9-5/8" casing in case-2 to case-4 have little difference, which indicates that these three cases have little influence on wellhead growth and wellbore safety. Axial forces of four first cases are less than zero at step 5 which indicate the casing boosts wellhead upward.

• 7" casing has small size and longer free section, so its top axial force change is less and is all tension. This indicates that 7" casing does not boost wellhead.

• Axial forces of 7" casing is identical in case-1 and cases-2 due to wellhead position before setting is regarded as reference position; in case-3 and case-4 the force are close from step 4 but not identical.

• As to case-5, calculation shows that wellhead undertakes 4800kN downward loads at step 4. If this force is transmitted to surface casing by accident factors, cement sheath of surface casing may be destroyed. Wellhead gets 2400kN upward load at step 5 which mitigates about half of downward load. Therefore, wellhead doesn't move in this case.

3.4.3 Analysis of remedial measures

According to analysis above, if surface casing doesn't have free section before production, there is little possibility for wellhead to grow up during production. Therefore, measures of preventing and remedying wellhead growth are as follows:

• Top-quality cement job is the best way, especially for surface casing.

• Reinforce surface casing as soon as its cementing quality is found bad. It is recommended that an enhancing base should be made to bear wellhead load, and the base should be placed on ground rather than on conductor.

• If wellhead growth is found during production, a bracket should be made to support wellhead, and the bracket should be placed on ground too. Meanwhile, flow rates should be controlled strictly.

3 Conclusions and suggestions

• Bad cement job quality of surface casing is the root of wellhead growth in deep wells; otherwise, wellhead growth doesn't take place, except in abnormal high temperature wells.

• Wellhead growth reported is a height change when wellbore temperature rises during production. So it is impossible to draw the moment of surface casing free section appearing.

• If surface casing has free section, setting loads of subsequent casings can only change axial force of free sections while wellhead growth keeps constant.

• When surface casing has free section, axial force in each casing has great change and a chain reaction may be induced to prolong the free sections.

• Countermeasures should be taken immediately if surface casing cement job quality is not ideal. Considering great loads from subsequent casings, tubing and Christmas tree, bracket for supporting wellhead should be placed on ground rather than on conductor.

Acknowledgments

This work is supported by "National Basic Research Program of China" (973 Program, 2015CB251205).

References

1. G.B. Lepper. Production Casing Performance in a Thermal Field. The Journal of Canadian Petroleum Technology, **37**, 57-62 (1998)
2. Z. Jianbing, L. Jinyong, W. Yinsheng, M. Yanping, F. Xiuqin, T. Zhenchun. Study on wellhead ascension & inclination and measures. Oil Drilling & Production Technology, **32** Sup, 167-169 (2010)
3. A.S. Halal, R.F. Mitchell, R.R. Wagner. Multi-string Casing Design with Wellhead Movement. SPE 37443 (1997)
4. A.C. McCabe. Well Vertical Movement on Platform Wells. SPE 19241 (1989)
5. L. Xiuquan, C. Guoming, C. Yuanjiang, Z. Lei, Z. Weiguo, X. Hua. Multistring analysis of wellhead movementand uncemented casing strength in offshoreoil and gas wells. Petroleum Science, **11**, 131-138 (2014)
6. K. A. Anderson. Support systems for high wellellhead loads. The Journal of Canadian Petroleum Technology, **23**, 76-78 (1984)
7. S. Jing, Z. Yanru, L. Zhiyong, F. Haitao, J. Zeju. Casing Elongation Research of Conventional Wellsin the Processof Steam Injection. Oil Field Equipment, , **36**, 23-25 (2007)
8. G. Baokui, G. Deli. Mechanism of Radial Separation of Cement Sheath and Casing during Temperature Cycling. Petroleum Science, 2006.
9. G. Baokui, G. Deli. Study on Well Bore Damage by Temperature Change of Underground Natural-Gas Storage Caverns. Science Press USA Inc., 360-365 (2005)
10. G. Baokui, G. Deli. Shearing Damage of Casing-Cement Sheath Interface from Temperature Changing in Well Bore. Science Press USA Inc., 2443-2447 (2006)

MATEC Web of Conferences 77, 06004 (2016) DOI: 10.1051/matecconf/20167706004
ICMMR 2016

The Wireless Environment Monitoring Alarm System Based on Self-organizing Network

Zhang Huawei [1], Li Nan [2] and Jiang Zhanjun [1]

[1]*School of Electronic and information engineering, Lanzhou Jiao tong University, Lanzhou, Gansu, 730070, China*
[2]*The experiment central of the computer science and technology, Lanzhou Jiao tong University, Lanzhou, Gansu, 730070, China*

Abstract. Under complicated conditions, it is necessary for environmental monitoring to design a wireless monitoring alarm system which can replace the wired system or as a supplement. The system discussed here bases on ARM7 microprocessor named LPC1114 and transceiver module named CC2530. With ZigBee, CSM/GPRS, this system uses multiple sensors to self-organized form a data acquisition and monitoring network system with variety of sensors fusion in the region. The system has some characteristics such as quick, convenient and accurate. Combining with the GSM SMS or GPRS alarm, the system can accurately and reliably monitor temperature, humidity and other environmental factors, and realize remote monitoring in large area and the complicated environment. Thus, this system has high practical value.

1 Introduction

With the rapid development of economic and the unceasing enhancement of social modernization degree, the environment monitoring systems have played an important role in various industries. Environmental monitoring systems can be divided into two ways of wired environmental monitoring systems and wireless environment monitoring systems.

The wireless environment monitoring alarm system is is easy installed and reacts quickly. It monitors, perceives and gathers various detection objects' information in real-time which belongs to the regional environment of network, transmitting the information to the user or management platform for processing. The wireless environment monitoring alarm systems are widely used in personal families, residential quarters, public places, office buildings and warehouses, etc. The multi-node, more jump, peer-to-peer and wireless self-organized network (mobile AD hoc network) system is especially suitable for some areas where people cannot safely enter into, such as inflammable and explosive dangerous sites, as well as some places that cable network cannot reach, such as forest, grassland, remote areas, etc. [1].

2 Environmental monitoring system design

The system is composed of the main console of ARM and the ZigBee sensor network [2]. ZigBee sensor network collects data and timely feedbacks to the main console, the main console can display the current position distribution and status information from each terminal node, processing data, alarming the abnormal data. If there are no staffs there, it will inform administrator via the GSM network remotely, and then handle abnormal data in a timely manner [3]. The design block diagram about the system is shown as Fig. 1.

2.1 Main console hardware design

The main console is composed of LPC1114 microprocessor, keyboard, 12864 LCD module, AT24C02 storage module, ISD1420 voice module and buzzer alarm module [3].

NXP LPC1114 is the key of Main console, connecting ZigBee coordinator via the UART serial port for data exchange. The main console is composed of LPC1114 microprocessor, keyboard, 12864 LCD module, AT24C02 storage module, ISD1420 voice module and buzzer alarm module.

LPC1114 is produced by NXP, and it is the world's most popular 32-bit ARM Cortex-M0 microprocessor, using 32 KB Flash and 8 KB SRAM. Comparing with the existing 8/16 bits MCU, it has the advantages of high cost performance, low power consumption, simple instruction set and small code length.

ISD1420 is a single chip voice recording circuit, which is composed of the oscillator, voice storage unit, preamplifier, automatic gain control circuit, anti-jamming filter and output amplifier. It is the alarm sound module of information output in the system. While alarming happens, the recorded alarm sound will be read under the action of low level pulse.

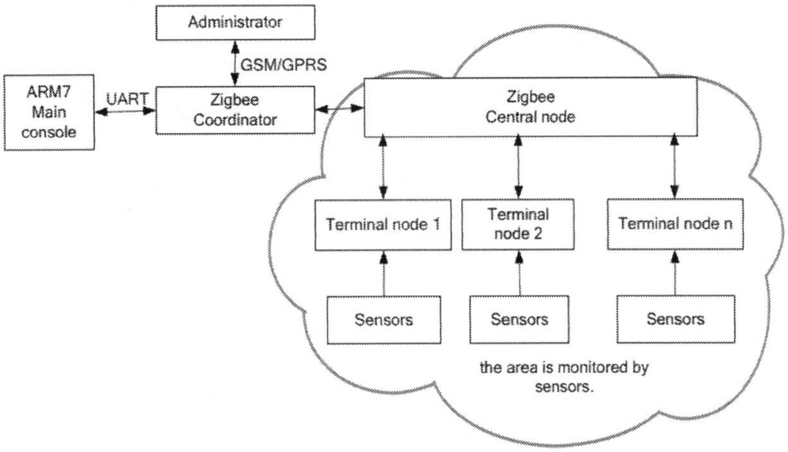

Figure 1. The structure of wireless environment monitoring alarm system.

NXP LPC1114 is the key of Main console, connecting ZigBee coordinator via the UART serial port for data exchange. The main console hardware design is shown as Fig. 2.

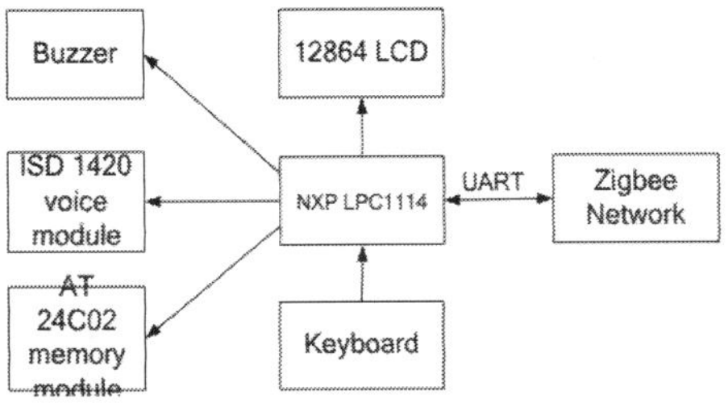

Figure 2. The design of main console based on ARM

2.2 ZigBee sensor network design

ZigBee is widely used now as a wireless network transmission technology. It uses the IEEE802.15.4 protocol standard which uses public frequency 2.4 GHz of global sharing. ZigBee wireless indoor transmission distance is 30-50m and can be raised up to 100m outdoor. As a result, according to change of the circumstance, the user will form a short distance wireless communication subnet, using wireless transceiver to transmit data quickly and easily. It does not only reduce the network cabling cost, but also improves the flexibility and extensibility of the application, and economizes human resources.

The wireless-sensor network applies a star topology, includes ZigBee coordinator, ZigBee central node, ZigBee terminal nodes, and each sensor node responsible for collecting data. No matter what kind of network topology is adopted, the coordinator node can have the only one, which is located in the upper layer of sensor nodes, acts as a gateway, as a whole to monitor and manage all the devices in the wireless-sensor networks. In monitoring area, a large amount of sensors are placed, constituting a monitoring area network by wireless self-organized way, timely collecting data by ZigBee terminal nodes to upload to the ZigBee central node. According to the agreed scope of data, central node determines whether the received data is normal and displays data through the liquid crystal or not. If there is any abnormal data, it will send alarm information to the console, during the prescribed period if the abnormal information has not been processed and is resend. After be sent three times, if central node has not received the processing message, it can send SMS (phone) by GSM/GPRS module to the predefined staff's phone number. This system should possess the following sensors driver according to the needs as follows: DHT90 temperature and humidity sensors, smoke sensors, CO sensors, the human body infrared sensors , etc [4].

2.3 ZigBee coordinator nodes' hardware design

ZigBee coordinator node is the rule maker of the whole sensor networks. The hardware design block diagram is shown as Fig. 3. The coordinator node can take the initiative to scan all the sensors of coverage areas in the wireless network, and be responsible for receiving data from sensors, transmitting to the main console via a serial port, connecting ZigBee sensor network, and completing the node joining and deleting as well as the conversion of network protocol and so on.[5]

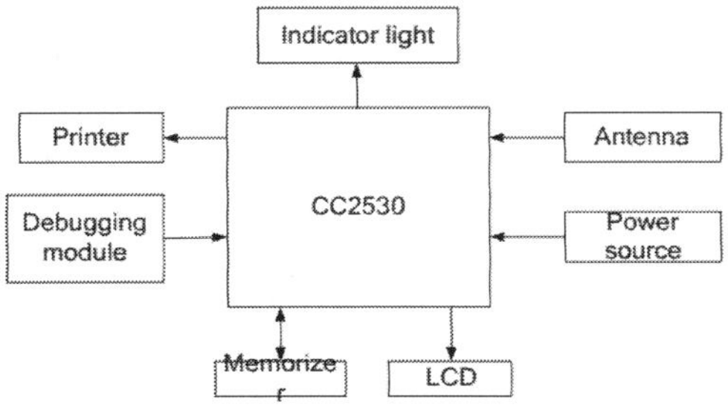

Figure 3. The hardware design of ZigBee coordinator node

ZigBee coordinator node is equipped with a rod module, which is crucial for data transmission; Power supply module is 5V power supply, which is transformed to 3.3 V by TPS73033 and MIC29302 chip, supplying the power to a CC2530 chip. Liquid crystal display module

DOI: 10.1051/matecconf/20167706004

displays each terminal sensor signals. Memory module backup necessary data in order to provide check for the administrators.

2.4 ZigBee central node and terminal node hardware design

ZigBee central node is located between ZigBee coordinator and ZigBee terminal nodes. It is necessary to monitor environmental parameters around, but also as a transit point to support communication between ZigBee central node and ZigBee terminal nodes. Moreover, ZigBee central node is connected to the GSM module and will timely send an abnormal message to the phone of the administrator to ensure that administrators have the real-time understanding of abnormal information in the case of the main console unattended [6]. ZigBee terminal devices are tiny nodes running on battery, taking own sensor be responsible for collecting environmental data around, regularly uploads data to ZigBee central node.

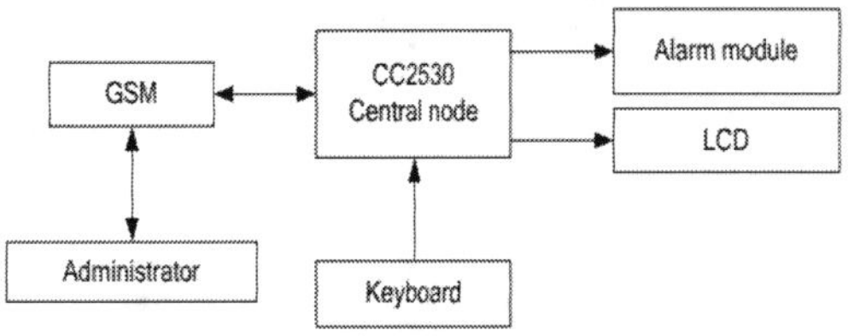

Figure 4.1 Hardware of ZigBee central node

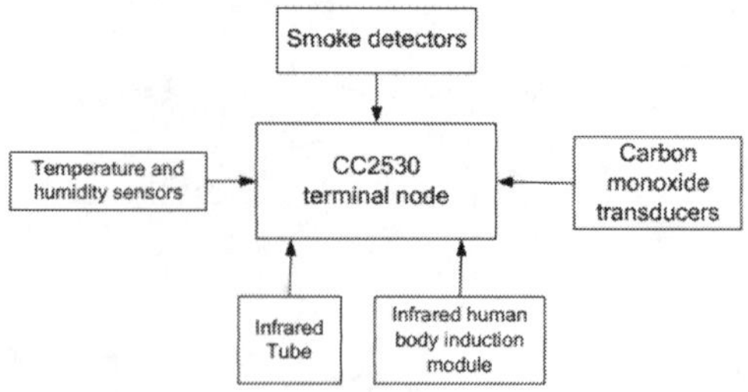

Figure 4.2 Hardware of ZigBee terminal node

3 System software design

The system software design is composed of two parts which are the main console software design and ZigBee node software design. ZigBee node software design is divided into ZigBee terminal nodes for collecting data, ZigBee central node for forwarding the data and ZigBee coordinator node for receiving the data. The followings only describe the software design of ZigBee coordinator node [7].

3.1 Main console software design

LPC 1114 microprocessor is embedded by Linux operating system as a support, adopting MDK + ULINK2 as compilers and emulators, writing the main program and interactive interface of the main console in QT / Embedded environment, using JTAG interface of the LPC1114 to easily debug program and download code through emulator ULINK2, and realizeing system functions ultimately.

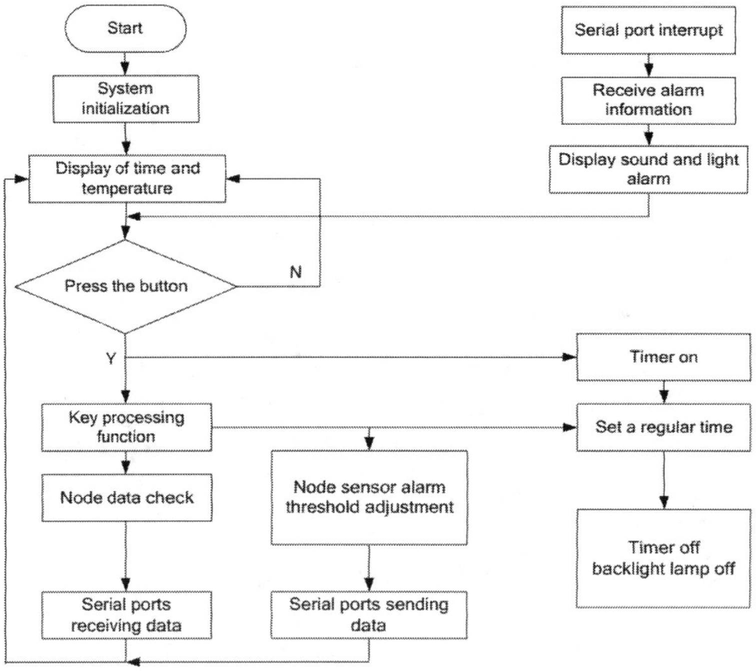

Figure 5. The software diagram of main console

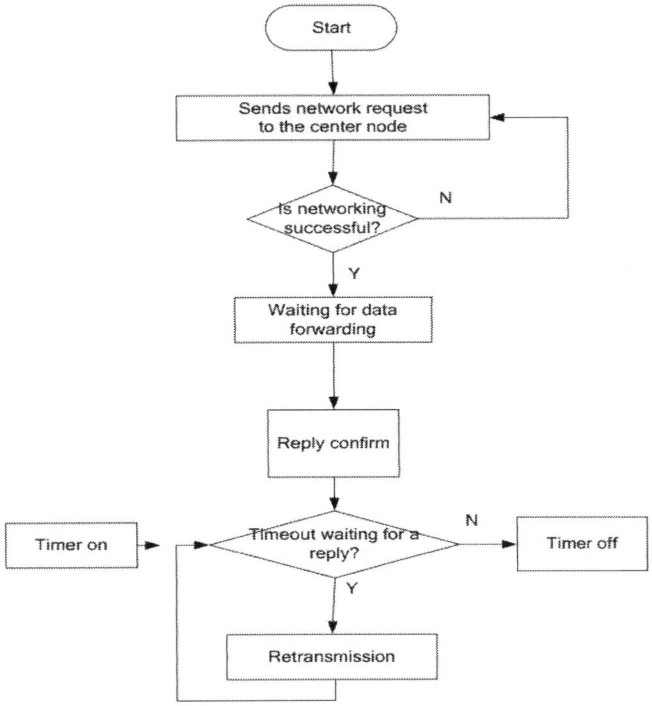

Figure 6. The software diagram of ZigBee coordinator node

It is primary that main console is started, system is initialized and time and temperature are displayed. Through the keys, time can be adjusted, the temperature alarm thresholds are set and all node data are checked. CC2530 forwards data ZigBee terminal node transmitting to the main console [8]. The block diagram of main console program design is shown as Fig. 5.

3.2 ZigBee coordinator software design

After ZigBee coordinator starts, ZigBee coordinator sends a network message to the central node. If the networking is successful, ZigBee coordinator waits for forwarding data of the main console or ZigBee central node, transmitting a confirm information, opening timer, and then waits for a reply. In the event of timeout, retransmitting will happen [9]. Its specific software flow chart is shown as Fig. 6.

4 System performance testing and results

The wireless environmental monitoring system is tested from the reliability, transmission distance and real-time performance of data. Firstly, temperature values collected and air-conditioning temperature values displayed are basically the same. The error does not exceed 0.5℃ when the node is placed conditioned room. Obviously, the deviation is in the allowable range. Humidity difference is 3%RH, some of the testing data are shown as Tab.1. On the top, it is installed the human body infrared sensor outside the door. The side-mounted is reflection infrared sensor, which makes accuracy reach more than 99% in the day and at the night. The alarm accuracy rate of sensitivity of CO sensors and smoke sensor at room is more than 99%. When the main console is unattended, after one minute, GSM short message is successfully sent (1000 tests). Secondly, the transmission distance of actual measured data is about 100 meters in case of accessibility, it reaches about 50 meters in case there is an obstacle. In addition, two nodes are placed in different places of the open outdoors, where the distance between two nodes is in the range of the wireless sensor network (the distance between two nodes is 200m). The nodes can receive data of all nodes that send packets in about 20 seconds. The difference is in the allowable range. The test results above show that wireless environment monitoring alarm system designed in this paper can complete the intended functions.

5 Conclusions

This paper presents a design and implementation scheme of wireless sensor networks which can be used in the harsh environment monitoring. It uses low-power ARM7 microprocessor LPC1114 as the main control console, consists of a variety of sensor by the ZigBee, CSM/GPRS, forming a multi sensor fusion of the data acquisition and monitoring network system. The system overcomes the bottleneck of traditional transmission rate of wireless sensor network. The ZigBee nodes in the system are powered by battery, so each monitoring point will sent data to the main control console when it receives command of the main control console or monitors the environment abnormality, or, it will not send data. So it reduces the energy consumption of the power supply. The main console can receives the ZigBee terminal nodes' data in real time and check the monitoring area sensors' data at any time. According to the practical running tests, it shows that this system has some characteristics such as high reliability, low cost, strong adaptability, easy expansion and high practical value.

Table 1. Partial information data collected by temperature and humidity sensors

temperature (℃)	16.3	24.7	30.5	humidity (%RH)	29.5	34.4	40.4
Node 1	16.5	24.6	30.6	node 1	30.2	34.6	40.1
Node 2	16.3	24.8	30.5	node 2	28.7	33.9	40.4
Node 3	16.4	24.8	30.4	node 3	30.1	34.7	40.2
Node 4	16.4	24.8	30.6	node 4	29.6	34.1	40.5

Acknowledgements

I would like to express my gratitude to all those who helped me during the writing of this thesis. My deepest gratitude goes first and foremost to Professor Jiang Zhanjun, for his instructive advice and useful suggestions on my thesis. This work is supported by the Gansu natural science foundation (145RJZA206) and the LanZhou JiaoTong University Youth Science Fund (2012005).

REFERENCES

1. Kozo Ishida. The Present Status and Development Trends of Environmental Monitoring. Environmental Monitoring in China, **1**: 4 -7(Ch) ,(2005)
2. Lina Tang,Xiancao Zheng, Xiaodan Su, etc. Experimental mobile environmental monitoring and real-time analysis as an initial application of EIoT in town villages in China. International Journal of Sustainable Development & World Ecology, **20(3)**, pp.267-271, (2013)
3. M. Martinelli, D .Moroni, G. Pieri, M. Tampucci. A Proactive GIS Infrastructure for Environmental Monitoring. Seventeenth International Water Technology Conference (IWTC17). Nov.2013.
4. Xiaofang Li, Lizhong Xu, Huibin Wang, Jie Song, Simon X. Yang. A Differential Evolution-Based Routing Algorithm for Environmental Monitoring Wireless Sensor Networks. Sensors, **10 (6)**, pp.5425, (2010)
5. Hui Wang, Sai Shen.Design and Research of Internet of Things in Environmental Monitoring Based on ZigBee. Hans Journal of Wireless Communications, **03** (05), pp.115-121, (2013)
6. J. Guevara, F. Barrero, E. Vargas, J. Becerra, S. Toral.Environmental wireless sensor network for road traffic applications. IET Intelligent Transport Systems, **6 (2)**, pp.177–186, (2012)
7. Du xinhui,Lu Xiaoqian,Bai Xueyan.Research on Intelligent Energy Saving System based on Zigbee and ARM. Journal of Taiyuan University of Technology. **43.6**, pp.762-765, (2012)
8. Liu Hong-tao, Cheng Liang-lun. Design of environmental mornitoring system with mobile sink. Computer Engineering and Applications.2010, vol.46. pp.7-9 [8] De Brito L M P L, Peralta L M R, Santos F E S, et al. Wireless sensor networks applied to museums environmental monitoring [C] //The 4th International Conference on Wireless and Mobile Communications, ICWMC 2008,2008:364-369.
9. Pingyou Hong, Ma Jin, Following Design Things and efficient environmental monitoring system based on LabVIEW sensors and micro-systems, **32 (2)**: 128-130, (2013)

MATEC Web of Conferences 77, 06005 (2016) DOI: 10.1051/matecconf/20167706005
ICMMR 2016

Time Series Model of Wind Speed for Multi Wind Turbines based on Mixed Copula

Dan Nie,Le Cao and Wei Yan

School of Economics and Management, North China Electric Power University, Changping District, Beijing102206, China

Abstract. Because wind power is intermittent, random and so on, large scale grid will directly affect the safe and stable operation of power grid. In order to make a quantitative study on the characteristics of the wind speed of wind turbine, the wind speed time series model of the multi wind turbine generator is constructed by using the mixed Copula-ARMA function in this paper, and a numerical example is also given. The research results show that the model can effectively predict the wind speed, ensure the efficient operation of the wind turbine, and provide theoretical basis for the stability of wind power grid connected operation.

1 Introduction

With the continuous development of economic globalization, energy has gradually increased to a high degree of impact on national security. In order to achieve the sustainable development of human society, to actively develop new energy sources of clean and renewable energy, to seek and explore new energy technologies has become the world's most important strategic task [1], [2]. Compared to other sources of energy, wind power is the most clean energy sources, which will not bring the acid rain, fog and haze caused by traditional energy, and radiation and other hazards caused by nuclear energy [3], [4]. Because of the uncontrollable nature of wind energy, the output power of wind power is unstable, and the volatility and intermittent are obvious. This year, the wind power installed capacity continues to increase, and the security, stability and economy power of grid system will be affected to varying degrees by the impact of wind power [5], [6]. If there is a more objective understanding of the current China's wind power industry status, a scientific identification of influence factors of the intermittency of the wind power generation, and effective measure of stochastic process of large-scale wind power generation, these can provide a very valuable reference for the development of large-scale wind power industry in our country [7].

Wind power is intermittent mainly caused by the random fluctuation of wind speed [8]. Therefore, when study intermitten tdynamic stochastic process of large-scale wind power generation, we should focus on two things. One is that the universality of wind speed time series model for multi wind turbines, the other is why there is less study ontime series model of wind speed for multi wind turbines home and abroad at present [9], [10]. In this paper, we will use time series theory and Copula theory to build the time series model of wind speed for multi wind turbines to study the spatial and temporal correlation of them.

2 Establishment of wind speed time series model for multiple wind turbines

Copula function scientifically and effectively seperated marginal distributions of random variable from the correlation structure of it, and simulate the change of wind speed of wind turbine through the ARMA. Copula function can reflect the correlation of the wind speed time series of multi wind turbines. The process is as follows:

Multi dimensional standardized wind speed time Series $\{y=(y_{1,t},y_{2,t},\ldots\ldots,x_{m,t}),t=1,2,\ldots\ldots,T\}$, Its each one dimensional time series is represented by the ARMA time series model, and the wind speed time series model based on Copula-ARMA multi wind turbines can be expressed as:

$$\begin{cases} y_{z,t}=\sum_{i=1}^{p}\alpha_{z,i}\,y_{z,t-i}+\beta_{z,t}-\sum_{j=1}^{p}\varphi_{z,j}\,y_{z,t-j} \\ \left(y_{1,t},y_{2,t},\ldots\ldots,y_{x,t}\right)\sim C_a\left(F_1\left(y_{1,t}\right),\ldots,F_x\left(y_{x,t}\right)\right) \\ v_{z,t}=\mu_{z,t}+y_{z,t}\times\sigma_{z,t} \end{cases} \tag{1}$$

where C_a represents the Copula function describing the correlation structure between wind speed sequences. For the wind speed Copula-ARMA time series model, it not only reflects the time characteristics of the wind speed series, but also combines the spatial correlation of wind speed of the wind turbine, which can fit the actual wind speed well. A deterministic ARMA time series model is composed of a sequence and a sequence. The numerical value of the sequence is generated by the first few moments, and the numerical value is generated by the Gauss white noise in the sequence. The numerical values in the sequence which are generated by the Gauss white noise are generated by the numerical values of the first few moments according to the fixed expressions. So the uncertainty of ARMA sequence is mainly generated by Gauss white noise, and in every simulation, $\theta_{i,t}$ is the only uncertainty factor. Therefore, it can reflect the correlation of the whole wind speed time series through the correlation of Gauss white noise sequence. The correlation degree of the Gauss white noise time series can be expressed by the function:

$$(\theta_{1,t},...,\theta_{x,t}) \sim C_b\left(F\left(\frac{\theta_{1,t}}{\sigma_1}\right),...,F\left(\frac{\theta_{x,t}}{\sigma_x}\right)\right) \quad (2)$$

The $(\theta_{1,t},...,\theta_{x,t})$ indicates the multidimensional Gauss white noise sequence; $(\sigma_1,\sigma_2,...,\sigma_x)$ indicates that the standard deviation of each one dimension $\{\theta_{i,t}\}$ sequence; $\Phi(\cdot)$ indicates the standard normal distribution function.

In the ARMA time series model, because the $AR(m)$ model is fixed by the first few numerical values, the perturbation is only related to the Gauss white noise $\theta_{i,t}$. According to the model expression: $\frac{\partial y_{i,t}}{\partial \theta_{i,t}} = 1 > 0$.

Due to the transformation invariance theorem of the Copula function, the Copula function is invariant when the multi-dimensional variable is changed unilaterally, so it has the following relations:

$$C_a(F_1(y_{1,t}),...,F_x(y_{x,t})) = C_b\left(\Phi\left(\frac{\theta_{1,t}}{\sigma_1}\right),...,\Phi\left(\frac{\theta_{x,t}}{\sigma_x}\right)\right) \quad (3)$$

The correlation structure between the multi dimensional wind speed sequence can be expressed by the correlation structure of the Gauss white noise sequence in the wind speed series.Thus, the formula (1) can be converted to:

$$\begin{cases} y_{z,t} = \sum_{i=1}^{p} \alpha_{z,i}\, y_{z,t-i} + \beta_{z,t} - \sum_{j=1}^{p} \varphi_{z,j}\, y_{z,t-j} \\ (\theta_{1,t},\theta_{2,t},......,\theta_{x,t}) \sim C_a\left(\Phi\left(\frac{\theta_{1,t}}{\sigma_1}\right),...,\Phi\left(\frac{\theta_{x,t}}{\sigma_x}\right)\right) \\ v_{z,t} = \mu_{z,t} + y_{z,t} \times \sigma_{z,t} \end{cases} \quad (4)$$

The wind speed time series model based on the hybrid Copula-ARMA is constructed, and the simulation process is as follows:

(1)The wind speed of the wind turbine is simulated by the ARMA time series model, and the standard wind speed sequence $\{x_{k,t}\}$ is obtained;(2)The generation of T Gauss white noise sequences which obey m-dimensional Copula function;(3)Combined with the standard wind speed sequence and the Gauss white noise sequence, the wind speed simulation sequence can be obtained according to the formula 4.

3 Example analysis

In this paper, the wind speed measured data of G provincial wind farm in China in March 2012 was analyzed. The installed capacity of the wind farm is 102MW, which is composed of 120 doubly fed wind generators with rated output of 850KW. The cut in wind speed, rated wind speed and cut out wind speed are 3m/s, 12m/s, 21m/s, and wind energy resource is good. Examples are analyzed for a total of 720 historical wind speed values in March. Sample a point every hour, and then do one hour ahead forecast of the 24 wind speed in March 31st through the model, and then compared with the actual value of the same day analyzing the curve fitting error. The wind speed curve of the area over the past period of time is shown in Fig. 1.

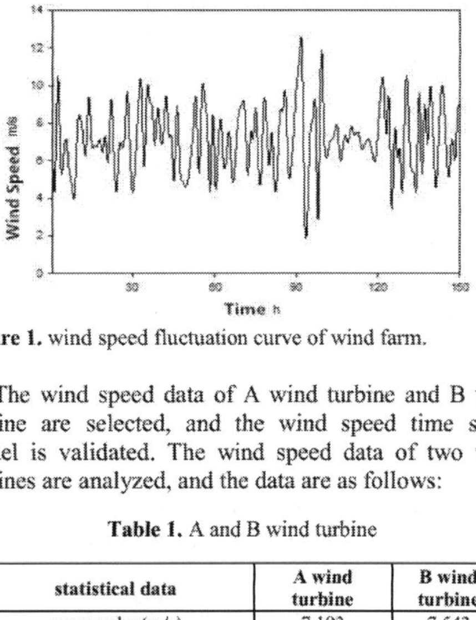

Figure 1. wind speed fluctuation curve of wind farm.

The wind speed data of A wind turbine and B wind turbine are selected, and the wind speed time series model is validated. The wind speed data of two wind turbines are analyzed, and the data are as follows:

Table 1. A and B wind turbine

statistical data	A wind turbine	B wind turbine
mean value(m/s)	7.193	7.543
Standard deviation(m/s)	1.973	2.341
Linear correlation coefficient	0.840	
Kendal	0.635	

Firstly, the ARMA model is validated by taking the A wind turbine as an example.

Because the wind speed sequence is a non stationary random sequence, its mean value is not zero. In the application of wind speed time series model, we first use the zero mean method to make the wind speed the standard sequence:

$$y_t = \frac{v_t - \mu_t}{\sigma_t} \quad (5)$$

where: v_t ——Original wind speed sequence of wind turbine generator; μ_t ——Mean value of wind speed unit; σ_t ——Variance of wind speed per year.

Then, the wind velocity ARMA time series model is made by using the AIC model,and the appropriate model order is selected at the same time. The AIC values of the model are shown in Table 2 under different orders.

Table 2. AIC value of different orders of ARMA time series model

R	M	AIC	R	M	AIC
1	1	47460.802127	3	4	44243.426530
1	2	47390.124775	3	5	44117.291021
1	3	47258.386528	4	1	47178.982975
1	4	47199.134515	4	2	44198.880355
1	5	47157.225788	4	3	47116.117534
2	1	47392.673425	4	4	44663.617039
2	2	47221.150570	4	5	47115.671914
2	3	47165.735587	5	1	47171.780603
2	4	47158.443695	5	2	47293.746394
2	5	47148.776002	5	3	47326.484630
3	1	47235.365075	5	4	47220.946265
3	2	47114.494617	5	5	47151.153943
3	3	44114.421911			

According to Table 2, when order number R=3,M=3, AIC reaches the minimum value of 44114.421911, then the wind speed simulation model can be expressed by ARMA (3, 3), that is:

$$P_t = 2.878P_{t-1} - 2.827P_{t-2} + 0.946P_{t-3} + \theta_t - \beta_1\theta_{t-1} - \beta_2\theta_{t-2} - \beta_3\theta_{t-3} \tag{6}$$

The parameters of the model are estimated by maximum likelihood, and the parameters of the model are obtained $a = \{2.878, -2.827, 0.946\}^T$; moving average parameter $\beta = \{-1.978, 1.092, -0.050\}^T$ .The first step wind speed sequence simulation model can be expressed as:

$$P_t = 2.878P_{t-1} - 2.827P_{t-2} + 0.946P_{t-3} + \theta_t + 1.978\theta_{t-1} - 1.092\theta_{t-2} + 0.050\theta_{t-3} \tag{7}$$

Among them $\theta \in NID(0, 0.4181^2)$,and then the model parameters are re- calculated according to the obtained fitting values.

Fitting the obtained values as known wind speed for the parameter estimation, and thus the regression parameters and the moving average parameters per a simulation are gained. According to the formula (6) the standard sequence is converted into the actual wind speed sequence:

$$v_t = x_t \times \sigma_t + \mu_t \tag{8}$$

The comparison of the 1h actual wind speed and the fitting curve as shown in Fig. 2. The approximate wind speed values are obtained by the wind speed time series model, and compared with the actual wind speed. The result of wind speed fitting and its error are shown in Table 3.

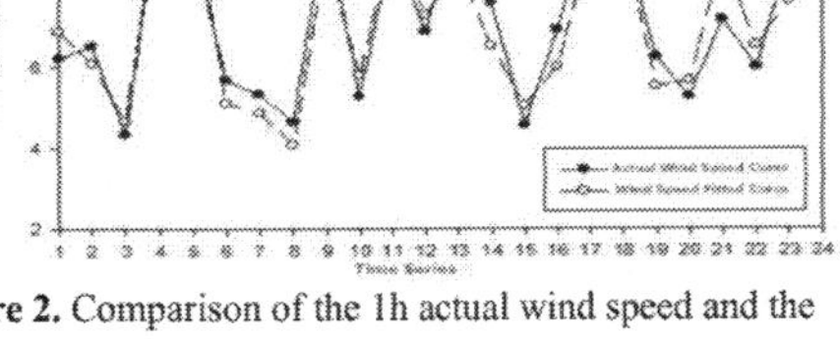

Figure 2. Comparison of the 1h actual wind speed and the fitting curve.

Table 3 shows that the maximum relative error of the model is 21.34%, the minimum relative error is 1.12%, the mean error is 9.39%, and the error standard deviation is 4.50%. Model fitting effect is good, suitable for the simulation of wind speed.

Secondly, the wind speed forecasting series of B wind turbine can be obtained by the same method:

$$P_t = 1.244P_{t-1} - 1.791P_{t-2} + 1.528P_{t-3} + \theta_t + 0.854\theta_{t-1} - 0.370\theta_{t-2} + 0.627\theta_{t-3} \tag{9}$$

where $\theta_t \in NID(0, 0.5818^2)$, So that the velocity sequence of wind speed sequence based on Copula-ARMA is obtained. The wind speed data of A and B wind turbines are statistically analyzed, as shown in Table 4.

The wind speed sequence obtained by Copula-ARMA model is compared with the numerical value of the original wind speed sequence(comparison between Table 1 and Table 4). The mean error of A wind turbine is 1.14% and the standard deviation is 4.66%; the mean error of B wind turbine is 3.95%; the standard deviation error is 7.43%; the linear correlation coefficient error is 0.14%, and the Kendall rank correlation coefficient error is 2.13%. The comparison results show that the correlation index of the original wind speed series can be kept well.

In addition, some colour figures will degrade or suffer loss of information when converted to black and white, and this should be taken into account when preparing them.

4 Conclusion

In this paper, the wind speed time series model of a multi wind turbine generator is constructed. By using the mixed Copula function to describe the spatial correlation of wind speed between different wind turbines, and then the wind speed time series model of the hybrid Copula-ARMA is constructed and the numerical example is analyzed. The model can be applied to the power system to arrange the wind turbine maintenance and its plan reasonably. The peak generation scheme can be adjusted to avoid effective wind, so that the peak load capacity of power network can be improved and the efficient operation of the unit can be ensured . Also the waste wind and the power loss of the wind turbine can be reduced, and as a result, the efficiency of wind power generation is improveand and the power consumption of the grid electricity is increase.

Table 3. results of 1H wind velocity fitting

time series	actual value	predicted value	relative error	time series	actual value	predicted value	relative error
1	6.23	6.89	10.65%	13	9.88	8.49	14.08%
2	6.52	6.08	6.74%	14	7.60	6.52	14.13%
3	4.34	4.66	7.45%	15	4.56	5.04	10.43%
4	9.95	10.06	1.12%	16	6.91	6.00	13.20%
5	9.14	10.23	11.93%	17	9.73	8.90	8.58%
6	5.67	5.11	9.97%	18	9.31	9.33	0.24%
7	5.33	4.87	8.58%	19	6.25	5.51	11.76%
8	4.65	4.08	12.27%	20	5.26	5.64	7.38%
9	9.61	8.75	8.96%	21	7.16	8.69	21.34%
10	5.27	5.80	10.02%	22	5.99	6.52	9.00%
11	8.97	9.21	2.61%	23	8.68	7.63	12.06%
12	6.87	7.29	5.98%	24	8.76	9.36	6.84%
Mean of error is 9.39%				Mean of error is 4.50%			

Table 4. A and B wind turbine

	A wind turbine	B wind turbine
Mean value（m/s）	7.111	7.245
standard deviation（m/s）	1.881	2.167
Linear correlation coefficient	0.839	
Kendal	0.622	

Acknowledge

This study is supported by National Natural Science Foundation of China (Granted No. 71471058) and the Beijing Education Committee of co-construction project.

References

1. Q. Zhao, W. Du. Development trend of the world energy industry in 2014[J]. Energy of China, **5**:39-44, (2015).
2. L. Ye, Y. N. Zhao. A review of wind power prediction based on spatial correlation[J]. Automation of Electric Power Systems, **14**:126-135, (2014).
3. Z. Song, J. Yu, Z. Zhang. Short-term wind speed forecasting with Markov-switching model[J]. Applied Energy, **130**(5):103-112, (2014).
4. S. She, L. I. Zheng, X. Cai. Research on Wind Speed Distribution Model of Wind Farm Based on Its Dynamic Space-Time Relation[J]. Power System Technology, **38**(6):1432-1438, (2014).
5. R. Zárate-Miñano, M. Anghel, F. Milano. Continuous wind speed models based on stochastic differential equations[J]. Applied Energy, **104**(2):42-49, (2013).
6. R. Carapellucci, L. Giordano. A methodology for the synthetic generation of hourly wind speed time series based on some known aggregate input data[J]. Applied Energy, **101**(1):541-550, (2013).
7. J. Tang, A. Brouste, and K. L. Tsui. Some improvements of wind speed Markov chain modeling[J]. Renewable Energy, 81:52-56, (2015).
8. Y. Q. Xu, L. L. Wang, L. Zhang. The wind speed dependence model of wind farm is built with rattan Copula[J]. Proceedings of the CSU-EPSA, **5**:62-66, (2015).
9. C. Qu, X. L. Wang, S. Y. Xie. Influence of different wind speed models and reliability indices on wind power reliability assessment[J]. Power System Technology, **37**(10):2896-2903, (2013).
10. X. Zhang, W. Q. Wang, H. Y. Wang. Wind turbine wind speed model considering wind shear and tower shadow effect[J]. Electrical Measurement & Instrumentation , 2015, 8(8):56-60.Luigi T.De Luca, Propulsion physics (EDP Sciences, Les Ulis), (2009).

Design and Implementation of PLC-Based Automatic Sun tracking System for Parabolic Trough Solar Concentrator

Jinping Wang[1,2], Jun Zhang[2] ,Yanfeng Cui[2] and Xiaolong Bi [2]

[1] *College of Energy & Environment Engineering, Southeast University EDP Sciences, Editorial Department, 211167, China*
[2] *School of Energy & Power Engineering, Nanjing Institute of Technology, 210096, China*

Abstract. A sun-tracking system for parabolic trough solar concentrators (PTCs) is a control system used to orient the concentrator toward the sun always, so that the maximum energy can be collected. The work presented here is a design and development of PLC based sun tracking control system for PTC. Sun tracking control system consists of a Programmable Logic Controller (PLC) and a single axis hydraulic drives tracking control system. Hydraulic drives and the necessary tracking angle algorithm have been designed and developed to perform the technical tasks. A PLC unit was employed to control and monitor the mechanical movement of the PTC and to collect and store data related to the tracking angle of PTC. It is found that the tracking error of the system is less than 0.6 °. Field experience shows that tracking algorithm act stable and reliable and suit for PTCs.

1 Introduction

A parabolic trough is a type of solar thermal collector that is straight in one dimension and curved as aparabola in the other two, lined with a polished metal mirror [1]-[3]. The aperture angle in parabolic trough solar collectors (PTC) is the angle between the axis of the parabola and the line connecting the focus on one end of that parabola [4]. A receiver tube with a fluid flowing inside that absorbs concentrated solar energy from the tube walls and raises its enthalpy is installed in this focal line. The collector is provided with one-axis solar tracking to ensure that the solar beam falls parallel to its axis. When a collector (its aperture normal) is not pointing directly at the Sun, some of the energy that could be collected is being lost [5]. Sun-tracking systems are designed in a way to track the sun on a single axis or in a way to track the sun on two axes. Over the years, several researchers have studied the solar tracking systems with different modes and electromechanical module to improve the efficiency of solar systems. Soteris A. Kalogirou uses three Light-Dependent Resistors (LDRs) as sensors to detects whether the collector is focused [6]. Dhanabal R, Bharathi V et al.make comparison of efficiencies of sun-tracking systems with static panel single-axis tracking system and dual-axis tracking system with fixed mount [7]. Ali Al-Mohamad presents a Sun-tracking design, whereby the movement of a photo-voltaic module was controlled to follow the Sun's radiation using a Programmable Logic-Controller (PLC) unit. All electronic circuits and the necessary software have been designed and developed to perform the technical tasks. A PLC unit was employed to control and monitor the mechanical movement of the PV module and to collect and store data related to the sun's radiation [8]. Chong K K, Tan M H use two sun-tracking methods for the heliostat, which are conventional Azimuth-Elevation and revolutionary Spinning-Elevation methods [9]. However, for parabolic trough solar collectors (PTCs), it can only use direct solar radiation, called Direct Normal Irradiance (DNI). PTCs are designed to operate with tracking about only one axis [10]. A good sun-tracking system must be reliable and able to track the sun at the right angle even in the periods of cloud days. Although the degree of accuracy required to depend on the specific characteristics of the solar concentrating system being analyzed, generally the higher the system concentration the higher the tracking accuracy will be needed [11].

In this paper, a PLC-based sun-tracking system for parabolic trough solar concentrator which could track the sun along one axes was designed and implemented. In the system, the tracking angle was calculated in real time based on solar position by the PLC controller. The PLC program was administered according to these calculated tracking angle, and by controlling the hydraulic drives of the parabolic trough solar collector with the rotary encoder signal received from the input module of the PLC, it was ensured that the PTC moves along one-axe, and maximum benefit was derived from solar energy by providing that the PTCs system are oriented at a right angle to the sun.

2 Design of the PLC-Based sun-tracking for parabolic trough solar concentrator

2.1 Overview of PLC-Based sun-tracking

The overall structure of PLC-based automatic sun-tracking system is shown in Fig. 1. The PLC based sun-tracking has the following components: PLC control system and field equipment. The difference between angle sensor to measure the PTC position and angle by PLC calculated is calculated. If this difference is greater than the threshold set in the program execution cycle, the PLC control system will produce hydraulic driven action commands enable parabolic trough solar concentrator running from east to west or west to east and track the sun. The control system by PLC controller system and field equipment constitute a feedback closed-loop control system based a rotary encoder.

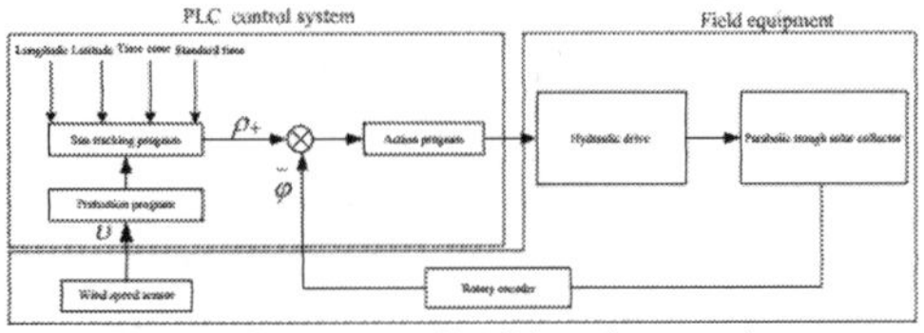

Figure 1. Overall structure of PLC-based automatic sun-tracking system

2.2 Tracking angle Algorithm

Parabolic trough solar collectors are designed to operate with tracking about only one axis. A tracking drive system rotates the collector about an axis of rotation until the sun central ray and the aperture normal area are coplanar. Fig. 2 shows how the rotation of a collector aperture about a tracking axis *r* [12].

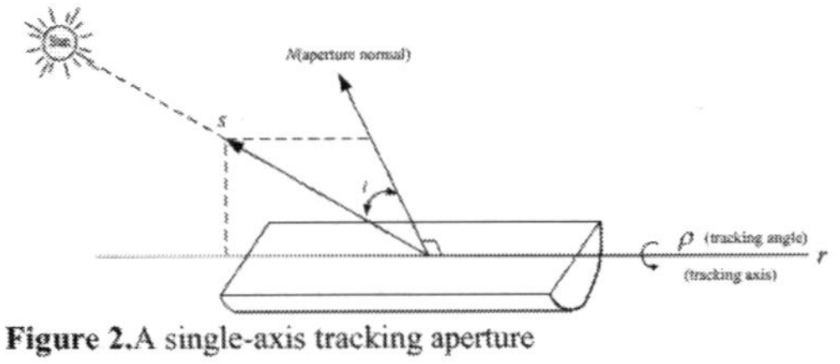

Figure 2.A single-axis tracking aperture

The tracking axis brings the sun ray unit vector *S* into the plane formed by the aperture normal and the tracking axis. *i is* the cosine of incidence angle, ρ is the tracking angle. Fig. 3 shows the *u-r-b* coordinate system, where *r* is the tracking axis, *b* is the axis that always remains parallel to the earth surface and *u* is the third orthogonal axis [13].

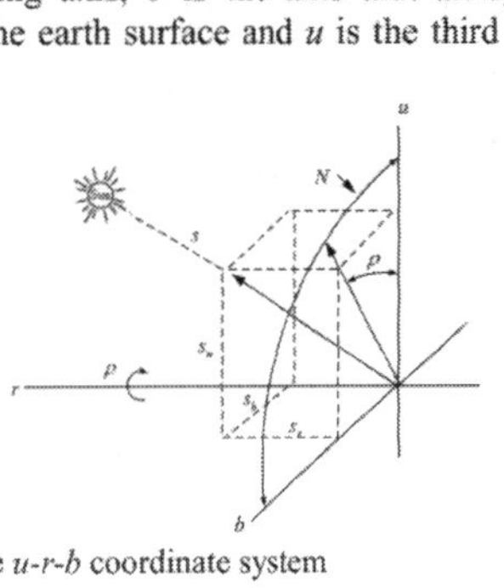

Figure 3. The *u-r-b* coordinate system

The tracking angle is written as Eq. (1) [12]:

$$\rho = \tan^{-1}(-\frac{S_u}{S_b}) \tag{1}$$

The cosine of incidence angle is written as Eq. (2) [12]:

$$i = \cos^{-1}(\sqrt{S_b^2 + S_u^2}) \tag{2}$$

where S_u is projection of the sun ray unit vector *S* on the *u* axis, S_b is projection of the sun ray unit vector *S* on the *b* axis. Because of the parabolic trough collectors is horizontal tracking axis in general, the tracking angle and cosine of incidence angle can also be written as Eq. (3) [13]:

$$\rho = \tan^{-1}\left[\frac{\sin(\gamma - \alpha_s)}{\tan\alpha}\right] \tag{3}$$

α_s is azimuth angle, α is the altitude angle.

3 Experimental and results

To test the aforementioned methodology, a prototype of an parabolic trough solar collector withlength of 50 m and aperture of 5.77m has been constructed in the Hengxi town, Jiangning District, Nanjing City, Jiangsu Province, China (located at latitude 31.45° and longitude 118.46°; see Fig. 4).

Figure 4. A prototype of parabolic trough solar collector that has been constructed at HengXi town in China

We have selected 4 typical dates to carry out experimental research,these 4 days are: March 20(vernal equinox), June 21(summer solstice), September 23(autumnal equinox), and December 22(winter solstice) in 2014. The solar altitude angle varies throughout the day as shown in Figure 5. The elevation is 0° at sunrise, reaches maximum value when the sun is directly overhead, then it begin to fall, is down to 0° at sunset. The solar altitude angle reaches the maximum value of the year in the summer solstice (June 21) at solar noon time, the maximum of solar altitude angle is 81.03°, however, the value is 34.89° in the winter solstice (December 22) at solar noon time, the difference between the two values is 34.89°. The azimuth angle is the compass direction from which the sunlight is coming . At

DOI: 10.1051/matecconf/20167706006

solar noon, the sun is always directly south in the northern hemisphere and directly north in the southern hemisphere. The azimuth angle varies throughout the day as shown in Figure 1,2. The azimuth angle reachs180°at solar noon time. The range of azimuth (sunrise to sunset) reached the maximum value of the summer solstice (June 21), is 236.08°. The value is 124.92°in the winter solstice (December 22). From the change regulation of the angle of the azimuth angle and the solar altitude angle.

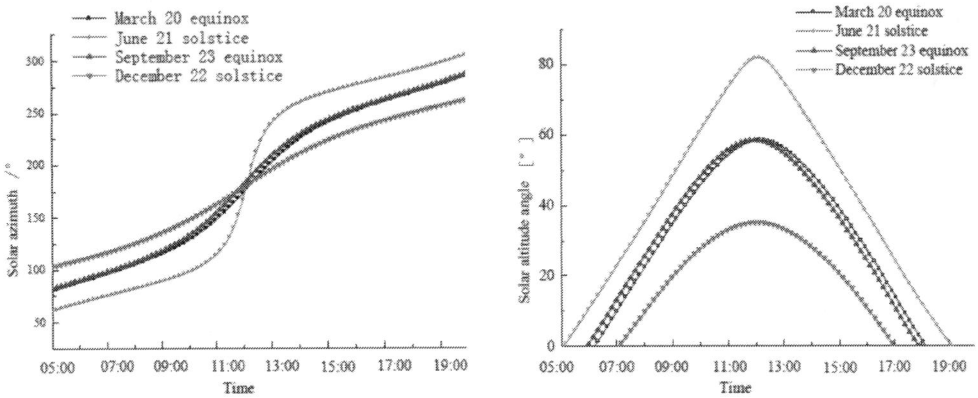

Figure 5. Variation of solar altitude angle And azimuth angle

Fig. 6 is variation of the tracking angle with 4 days in 2014.we can see that the tracking angle range of PTC is run within the range of 0 to 180°.

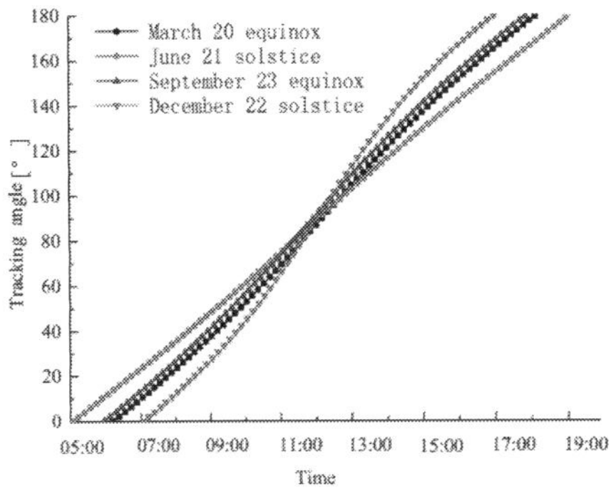

Figure 6. Variation of the tracking angle

The trajectory of the tracking angle is more complex, and the daily movement law is not the same. The range of the rotational trajectory of PTC reaches maximum in the summer solstice, reaches minimum in the winter solstice. Figure.7 is variation of change rate of tracking angle of PTC. We can see that although the tracking angle range of PTC reaches minimum in the winter solstice, but the change rate of tracking angle of PTC is the fastest, is 0.398°/min at solar noon. Although the tracking angle range of PTC reaches maximum in the summer solstice, but the change rate of tracking angle of PTC is running slower than other dates, the maximum value of the day is 0.232°/min, the difference of these two days is 0.166°/min at solar no one in the summer solstice and winter days .

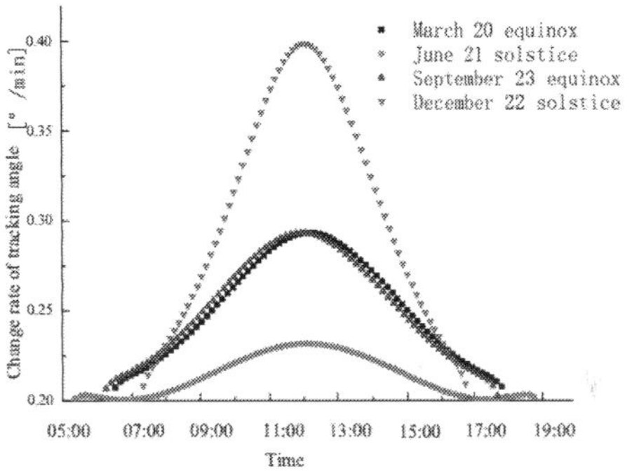

Figure 7. Variation of change rate of tracking angle of PTC

The vernal equinox and autumnal equinox had little difference. The change rate of tracking angle of PTC runs fastest at solar noon in every day .Through the above analysis, we can know the trajectory of the parabolic trough solar concentrator is very complicated, so the sun-tracking control system for parabolic trough solar collectors needs to be designed according to the nonlinear control system . A procedure for a Solar Position Algorithm (SPA) to calculate the solar zenith and azimuth angle with uncertainties equal to ±0.0003° in the period from the year -2000 to 6000 in Literature [14]. We compare SPA algorithm with the algorithm designed in this paper. Fig. 8 is the contrast curve of calculated tracking angle by the PLC system , the encoder output angle and the angle by SPA algorithm calculated in 1 hour. The maximum deviation of the angle of the SPA algorithm and the PLC system calculated is 0.12°. The maximum deviation from the angle of the encoder output angle and the PLC system calculated is 0.4°. Therefore, the tracking error of the system is less than 0.6 °.

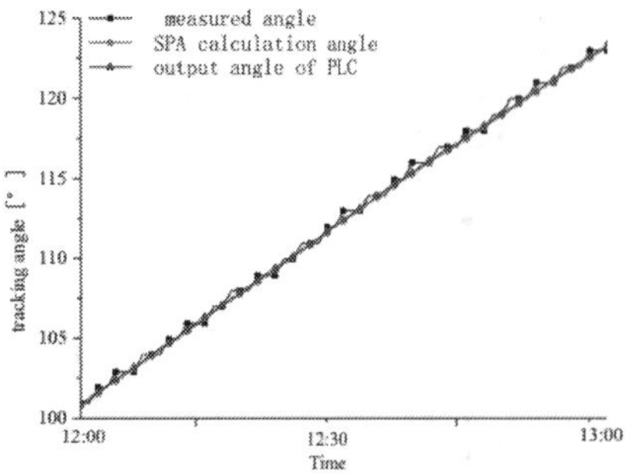

Figure 8. Contrast change curve with different angles

4 Conclusions

In this research work design of sun tracking system for parabolic trough solar concentrator has been considered by using PLC and hydraulic drive. Using this automatic sun-tracking system can make the parabolic trough solar concentrator automatic, reliable and does not depend on the weather when tracking the sun. The tracking angle algorithm presented in this paper has a high degree of accuracy,experimental results show that the tracking error of the system is less than 0.6°. The sun tracking system developed in this study provides easy installation, simple mechanism and less maintenance.

Acknowledgments

This study was simultaneously supported by Key Laboratory Open Fund Project of Jiangsu Provincial Key Laboratory of Solar Energy Science and Technology, the Natural Science Foundation of the Jiangsu Higher Education Institutions of China(Grant No.14KJD470003) and Youth Fund Key Project of Nanjing Institute of Technology(Grant No.QKJA201402).

References

1. H. Lund. Renewable energy strategies for sustainable development[J]. *Energy*, **32**(6):912-919, (2007).
2. F. Chen, N. Duic, L. M. Alves, et al. Renewislands—Renewable energy solutions for islands[J]. *Renewable & Sustainable Energy Reviews*, **11**(8):1888–1902, (2007).
3. J. P. Wang, J. Wang, X. L. Bi, and X. Wang, "Performance Simulation Comparison for Parabolic Trough Solar Collectors in China," *International Journal of Photoenergy*, vol. 2016, Article ID 9260943, 16 pages, (2016).
4. W. Jinping, W. Jun, Z. Yaoming, and B. Xiaolong, "Analysis of heat transfer characteristics for parabolic trough solar collector," Transactions of the Chinese Society of Agricultural Engineering, **31**, no. 7, pp. 185–192, (2015) (Chinese).
5. E. Zarza, L. Valenzuela, J. León, et al. Direct steam generation in parabolic troughs: Final results and conclusions of the DISS project [J]. *Energy*, **29**, pp. 635-644, (2004).
6. S. A. Kalogirou. Design and construction of a one-axis sun-tracking system[J]. Solar Energy, **57**(6), pp. 465–469, (1996).
7. R. Dhanabal, V. Bharathi, R. Ranjitha, et al. Comparison of Efficiencies of Solar Tracker systems with static panel Single-Axis Tracking System and Dual-Axis Tracking System with Fixed Mount[J]. *International Journal of Engineering & Technology*, **5**, pp. 0975-4024, (2013).
8. A. Al-Mohamad. Efficiency improvements of photovoltaic panels using a Sun-tracking system [J]. *Applied Energy*, **79**, pp. 345-354, (2004).
9. K. K. Chong, M. H. Tan. Comparison Study of Two Different Sun-Tracking Methods in Optical Efficiency of Heliostat Field [J]. *International Journal of Photoenergy*, pp. 483-490, (2012).
10. M. Ozturk. Exergy Analysis of Low and High Temperature Water Gas Shift Reactor with Parabolic Concentrating Collector [J]. *Clean Energy Systems and Experiences*. pp. 107-130, (2010).
11. K. K. Chong, and C. W. Wong. General formula for on-axis sun-tracking system and its application in improving tracking accuracy of solar collector [J]. *Solar Energy*, **83**, pp. 298–305, (2009).
12. W. B. Stine, R. W. Harrigan. Solar energy fundamentals and design: with computer applications [M]. Wiley, (1985).
13. R. V. Padilla. Simplified Methodology for Designing Parabolic Trough Solar Power Plants [D]. Dissertations & Theses - Gradworks, (2011).
14. I. Reda, A. Andreas, I. Reda, et al. Solar position algorithm for solar radiation applications [J]. *Solar Energy*, **76**, pp. 577–589, (2004).

MATEC Web of Conferences 77, 06007 (2016) DOI: 10.1051/matecconf/20167706007
ICMMR 2016

Energy Consumption and Greenhouse Gas Emission Evaluation Scenarios of Mea Fah Luang University

Onn Laingoen, Sitang Kongkratoke, Pannipha Dokmaingam

School of Health Science, Mae Fah Luang University, Chiang Rai, 57100, Thailand

Abstract. In Thailand, quantity of the educational institutes building shared one fourth of commercial building. Among the energy consumption and conservation in the building in Thailand are mostly study in typical office and resident building. Mea Fah Luang University (MFU) was selected to represent the educational institutes building where located in the northern part of Thailand. The average temperature in the northern is lower than other parts of Thailand. This study was firstly collected the data about quantity and behaviour of energy consumption in MFU based on the energy audit handbook. Although MFU is located in the northern of Thailand. The highest energy consumption is in the part of air condition. When the energy efficiency appliances and energy conservation building are implemented, the cost of energy will be saved around 15,867,960 Baht. Furthermore, the greenhouse gas emission is also reduced about 72.01 kg CO_2, equivalent/m^2/year.

1 Introduction

As the final energy consumption static in Thailand, the first two energy consumption sectors are industrial and commercial sectors, respectively. The electricity is consumed in commercial sector around 48,927.8 GWh/year during (2005-2012). This results in 29,638.8 Mton CO2 emission/year. [1]. Most of electricity is spent to operate the building in this sector. [2]. Noted that official buildings presented the highest energy consumption by surface [3]. The pervious works recommended that more than 70% of energy consumption in the building is used for air condition and lighting system. [2]-[5]. These could be said that energy conservation in the building is one of significant method for reduce greenhouse gas emission. However, the climate and the season should be investigated during developed the convection campaign [6], [7]. The temperature play the major affected to energy loading of air condition/heating system. Moreover, daytime in each season is also influence to the energy consumption in lighting system.

Universities are one of interested section that can reduce energy consumption and mitigate GHG emission [8]-[10]. Saidur [5] collected the data of the building age, operating hour and quantity of electric equipment in the office building. Then these data was used to calculate evaluate the energy saving cost and greenhouse gas mitigation under different energy saving measures. It was found that housekeeping, increasing of thermostat setpoint temperatures and setting standby mode are none investment cost for energy conservation. Replacement of incandescent with compact fluorescent and the use of insulation are the saving method with very short payback compared to lifespan. Escobedo, Briceño [11] evaluated quantity of energy consumption and greenhouse gas emission in each build of the National Autonomous University of Mexico. Then, investment cost for energy conservation was calculated to compare with energy saving bill. Finally, they suggested that replacement of incandescent with compact fluorescent T12 and T8 lamp, changing magnetic ballast to electronic ballast, installation of hybrid energy system are worth for investment. Moreover the university should support the energy saving campaign among student and occupants. Azar and Menassa [12] study the possibility of energy saving by increasing the efficiency of the commercial building operation. Then they developed the mathematical model to forecast the energy consumption in different building scale. Moreover, the human behavior if each build was also studied. They conclude that reinforce the role of human actions in energy conservation, and support efforts to integrate operation-focused solutions in energy conservation policy framework influence to reduce energy consumption around 21%.

In Thailand, quantity of the educational institutes building shared one fourth of commercial building [13]. The Green University policy is also implemented. During 2011-2015, number of Thailand's Green University, which is ranked by UI GreenMetric World University Ranking, is increased continuously [14]. Among the energy consumption and conservation in the building in Thailand are mostly study in typical office and resident building. [13] This study was developed to provide alternative solution for institute in Northern Thailand where the winter period is longer than other regions.

2 Methodology

This study was firstly survey the general information of energy consumption in Mea Fah Luang University. Then, a representative sample of buildings are selected in order to estimate energy consumption by end use. The representative of the sample was selected based on the difference building age. It should to note that MFU kicked off the Green University project in 2012. After that, the buildings were designed under the criteria and standards for energy conservation according to the National Energy Conservation Promotion Act. The energy of the selected building were audited follow the guideline of energy audits handbook [15]. Based on data collected from energy audits, an indicator of energy use by end use per occupant was estimated. Since this study was developed in 2014, the information of energy consumption is also used as base year for comparison with alternative scenarios. As mention above, the building which established before 2012 was not fully relied on energy conservation criteria. Thus, E1 building was chose as representative. On the other hand, the E3A building was also selected to represent the building that design under energy conservation criteria. All electric appliances are high energy efficiency which are labeled No.5.

2.1 CO_2 emission calculation

The emission is calculated by using the IPCC's guideline [16], Eq.1. Meanwhile the emission faction is based on the national information [17]. It should to note that the electricity that supply to MFU is generated from coal.

$$CO_2eq\ Emission = Energy\ consumption(watt)\ x\ EF \quad (1)$$

where CO_2 eq is CO_2 equivalent

EF is émission factor $\left(\frac{CO_2eq}{watt}\right)$

2.2 Energy cost saving cost calculation

To calculate the energy cost saving in the build, the following approaches as proposed by Staniaszek and Lees [18] will be used: Firstly, electric machines were categories into 4 main groups consisting (i) Electrical Lighting (ii) Air condition (iii) Office Appliances and (iv) Other Electrical Appliances. After that, unit of each electrical appliance was calculated by following formula, Eq. 2.

$$\text{Electricity Unit}\left(\frac{kWh}{day}\right) = \frac{No.power\ supplies\ x\ P\ x\ t}{1000} \quad (2)$$

where No. power supplies is number of power supplies
P is power (watt)
t is time (hour)

Then we find annual unit (kWh) by multiply with 22 days per months for electrical appliances that were not used on weekend and 30 days per months for electrical appliances were used on weekend. Furthermore, we can calculate the annual cost of electrical used by multiplying the annual unite (kWh) by the electricity rates. So, annual energy cost saving of each group of appliances was determined by Eq. 3.

$$Anu.en.saving = Anu.en\ B1 - Anu.\ en\ B2 \quad (3)$$

where Anu.en. is annual energy cost (Baht).
B is building

The future value of annual costs saving are estimated by using Eq.4-5.

$$Forcasted\ anu.en = anu.en\ saving \times fut.val\ factor (4)$$

$$fut.val.factor = (1+i)^n \quad (5)$$

where i = discount rate and n = number of periods

In addition, we conducted the future value cumulative or total cost saving of each group of appliances by sum the future value of annual costs saving on period of appliances useful life. Finally, we will get the efficiency of electrical appliances which were implemented at the selected building in term of energy costs saving.

3 Results and discussion

3.1 The university characteristics

Mea Fah Luang University is located in mountainous area. The infrastructures in the university are official buildings, dormitories, sport complex, laboratories, and facilities, etc. The academic year is divided into 2 semesters. The 1st semester is during June-September. The 2nd semester is during November-February. According to the temperature distribution in Northern of Thailand during 1999-2004 is presented in Fig. 1. The average temperature of the first and the second semester are 28.54 °C and 23.64 °C, respectively. This correspond to amount of energy consumption though the year, Fig. 2. It could be seen that energy consumption in the 1st semester is higher than the 2nd semester. However, the annual energy consumption is different since the increasing of building area and number of staffs and students. However, the energy intensity during 2011-2014 is deceased since the energy conservation policy of MFU. This also results in lower carbon emission from energy consumption as shown in Table 1.

Table 1. Energy consumption and GHG emission in MFU during 2011-2014

Year	Energy consumption (kWh/person)	GHG emission ($TonCO_2eq$/person)
2011	1181.94	0.72
2012	1136.20	0.69
2013	802.12	0.49
2014	857.63	0.52

 DOI: 10.1051/matecconf/20167706007

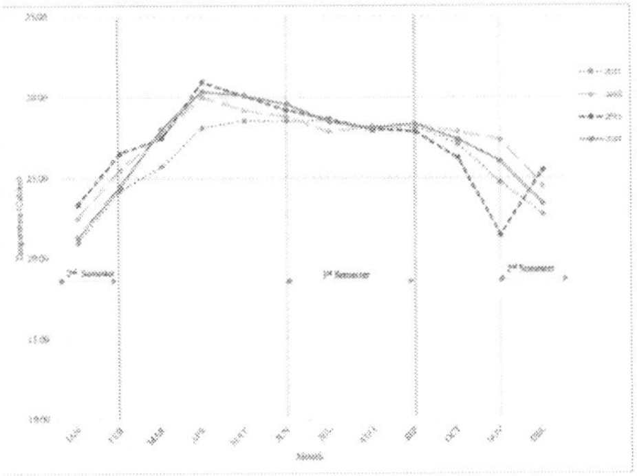

Figure 1. Mean temperature in Northern of Thailand during 2009-2014.

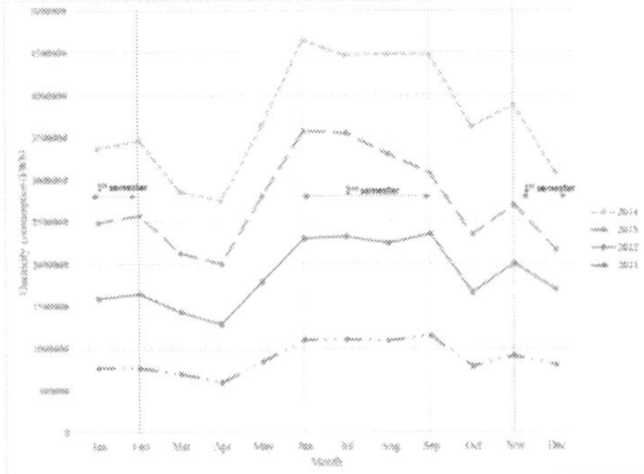

Figure 2. Energy consumption in Mea Fah Luang University during 2009-2014.

3.2 Energy consumption behavior

One parameter that affects to quantity of energy consumption is energy consumption behavior of occupants. Thus, the behavior was randomly survey in six buildings difference age. As seen in Fig. 3, the air condition is the largest energy consumption. This might be the effect of the climate as mention above. Moreover, the officers did not play attention to switch off/unplug the electric appliances even though there are not used.

3.3 Energy saving potential

3.3.1 Base case

The study was developed in 2014 where the total of energy consumption is 12,036,777.82 kWh. Total Floor area is 327,264.18 m^2. So the energy consumption per area is 36.78 kWh/m^2. The CO_2 emission is 22.4 $kgCO_{2,}$ eq/m^2.

3.3.2 Alternative scenario

Firstly, quality of energy consumption in E3A Building and E1 Building was identified. Then, the annual costs saving when compared between E3A building and E1 building was calculated as four catalogs. The annual costs saving and the future value of annual costs saving on period of appliances useful life of electrical lighting, air condition, office appliances and other electrical appliances are presented in Table 2-5, respectively. Finally, we will get the efficiency of electrical appliances which were implemented at E3A Building and E1 Building in term of energy costs saving.

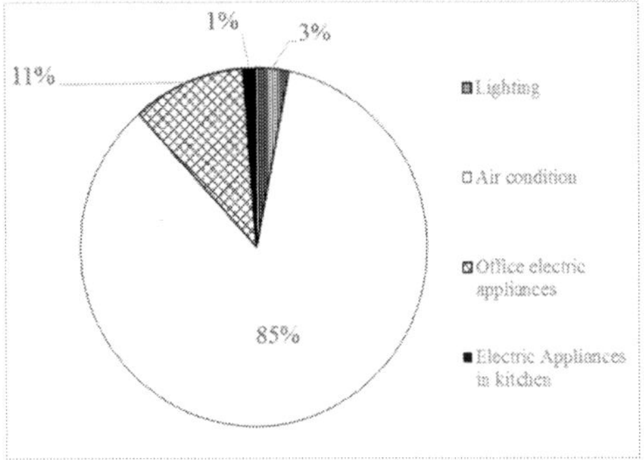

Figure 3. Energy consumption ratio of Mea Fah Luang University during academic year 2014

Table 2. Energy cost saving for electrical lighting

Periods (n)	Annual Costs saving ($x10^3$Baht)	Future Value Factor, 10% Interest Rate	Future Value of Annual Cost Saving ($x10^3$Baht)	Future Value Cumulative Costs Saving ($x10^3$Baht)
0	148.68	1.00	148.68	148.68
1		1.21	179.90	328.59
2		1.22	181.54	510.13
3		1.33	197.90	708.02
4		1.46	217.67	925.69
5		1.61	239.53	**1,165.,22**

The electrical lighting saving cost, Table 2, for this activity calculated on the basis of annual costs using in 2014. Lighting bulb was estimated to have a useful life of 7 years. In this study, the discount rate of 10 percent was used to find the future value of annual cost saving. The annual costs saving of electrical lighting compare between E3A Building E1 and Building in 2015 was 148,681 Baht. Therefore, at the completely useful life of lighting bulb, the cumulative costs saving will be equal to 1,428,680 Bath.

In the case of air condition, it was estimated 10 years useful life and 10 percent discount rate, total cumulative costs saving of air condition. As seen in Table 3, the annual cost saving in 2014 was 766,142 which is the highest energy cost saving among 4 groups of electrical appliances. The table 4 illustrates office appliances which consist of printers, fax machines, computers and photocopy machines. The annual cost saving was 439,691 Baht, at the estimating of a useful life of office appliances were 5 years with 10 percent discount rate, the cumulative costs saving accounting for 2,737,522 Baht

The other electrical appliances, Table 5, including electric kettles, microwaves, and refrigerators. The annual cost saving was -96,769 Baht, at the estimating of a useful life of other electric appliances were 5 years with 10 percent discount rate, the cumulative costs saving accounting for -602,489 Baht. It is indicated that E1

Building has more efficiency energy saving than E3A on group of other electrical appliances.

Finally, it was found that the highest energy cost saving at the end of useful life of electric appliances was air condition (12,304,247 Baht), office appliances (2,737,522 Baht) electrical lighting (1,428,680 Baht) and other electrical appliances (-602,489 Baht) respectively. The total energy cost saving when comparing between E3A and E1 was equaled to 15,867,960 Baht.

In the part of greenhouse gas emission, replacement of high energy efficiency are influenced in GHG reduction around 72.01 kg CO_2,eq./m^2/year.

Table 3. Energy cost saving for air condition

Periods (n)	Annual Costs saving (**x10^3**Baht)	Future Value Factor, 10% Interest Rate	Future Value of Annual Cost Saving (x10^{3}Baht)	Future Value Cumulative Costs Saving (x10^{3}Baht)
0	766.14	1.00	766.14	766.14
1		1.21	927.03	1,693.17
2		1.22	935.46	2,628.63
3		1.33	1,019.74	3,648.37
4		1.46	1,121.63	4,770.00
5		1.61	1,234.26	6,004.26
6		1.77	1,357.60	7,361.86
7		1.95	1,493.21	8,855.07
8		2.14	1,642.61	10,497.68
9		2.36	1,806.56	**12,304,25**

Table 4. Energy cost saving for office appliances

Periods (n)	Annual Costs saving (Baht)	Future Value Factor , 10% Interest Rate	Future Value of Annual Cost Saving (**x10^3**Baht)	Future Value Cumulative Costs Saving (**x10^3**Baht)
0	439.69	1.00	439.69	439.69
1		1.21	532.03	971.72
2		1.22	536.86	1,508.58
3		1.33	585.23	2,093.81
4		1.46	643.71	**2,737.52**

Table 5. Energy cost saving for other electrical appliances

Periods (n)	Annual Costs saving (Baht)	Future Value Factor $(1+i)^n$, 10% Interest Rate	Future Value of Annual Cost Saving (Baht)	Future Value Cumulative Costs Saving(Baht)
0	-96.77	1.00	-96.77	-96.77
1		1.21	-117.09	-213.86
2		1.22	-118.16	-332.02
3		1.33	-128.80	-460.82
4		1.46	-141.67	**-602.49**

4 Conclusion

The quantity and behavior of energy consumption in MFU were collected based on the energy audit handbook. The energy consumption during 1st and 2nd semesters is higher than summer periods. Although MFU is located in the northern of Thailand where the average temperature is lower than other parts of Thailand. The highest energy consumption is in the air condition sector. When the energy efficiency appliances and energy conservation building are implement, the cost of energy will be saved around 15,867,960 Baht. Furthermore, the GHG emission is also reduced about 72.01 kg CO_2,eq/m^2/year. This is also benifited for university in Northen of Thailand. The energy conservation of the air conditons is also important though the temperature distribtion in the Northen is lower than 31 °C.

Acknowledgements

The authors wish to show grateful thanks to Mea Fah Laung Unversity, Thailand, for financial support for this work

References

1. DEDE, *Thailand Energy Statistics 2012*, D.o.A.E.D.a. Efficiency, Editor, **72**, (2012).

2. N.Yamtraipat, et al., Energy Policy, **34**(7), 5 (2006).
3. B. Limmeechokchai and P. Chaosuangaroen, Thammasat Int. J. Sc. Tech.,**13**(2),9 (2008).
4. S. Chirarattananon and J. Taweekun, ENERG CONVERS MANAGE, , **44**(5), 19 (2003).
5. R. Saidur, Energy Policy, **37**(10), 9 (2009).
6. The Energy Conservation Center, Japan, (2010).
7. M.C. Katafygiotou and D.K. Serghides, Energy Buildings, **72**(0), 8 (2014).
8. X. Li, H. Tan, and A. Rackes, J. Clean. Prod. , **106** (11), 11 (2015).
9. W.-H. Hong et al., J. Asian. Archit. Build., **10**(2), 7 (2011).
10. L. Vásquez et al., J.Clean. Prod. (to be published)
11. A. Escobedo et al., J. Sustain. Dev. Energy, **18**(0), 6 (2014).
12. E. Azar & C.C. Menassa, Energy Policy, 2014. **67**(0), 13 (2014).
13. O.F. Kofoworola & S.H. Gheewala, Energy Buildings, **41**(10),17 (2009).
14. S.M. Library, *UI GreenMetric World Universities Ranking*. 2015 [cited 2015; Available from: http://stang.sc.mahidol.ac.th/text/ranking.htm#uigreen
15. Thumann, A. and W.J. Younger, *Handbook of Energy Audits, Seventh Edition*. 2007: Taylor & Francis
16. IPCC, *2006 IPCC Guidelines for National Greenhouse Gas Inventories* 2006, IGES, Japan The National Greenhouse Gas Inventories Programme
17. TGO, *Summary Report The Study of emission factor for an electricity system in Thailand*, (2010).
18. D. Staniaszek & E. Lees, *Determining energy savings for energy efficiency obligation schemes*. Montpelier, Vermont: ECEEE, RAP (2012).

 DOI: 10.1051/matecconf/20167706008

Efficiency of Passive Utilization of Ground "Cold" in Adaptive Geothermal Heat Pump Heating and Cooling Systems (AGHCS)

Vasilyev G.P.[1], Gornov V.F.[1] , Kolesova M.V.[1], Dmitriev A.N.[2], Silaeva V.G.[1]

[1] *JSC «INSOLAR-INVEST», Bol'shaya Filevskaya str. 32-3, Moscow, 121433 Russia*
[2] *Department of Economics and Managing the City Construction, Engineering and Economic Faculty of the Russian Plekhanov Economic Academy, Moscow, Russia*

Abstract. This article deals with estimating a potential and efficiency of utilization of passive ground "cold" for cooling buildings in climatic conditions of Moscow (Russia). The article presents results of numerical analysis to assess the efficiency of reducing peak cooling loads of the building equipped with AGHCS, through the utilization of natural cold of wells for passive cooling and cold storage in summer at night (off-peak) time with its subsequent consumption in the day time, both in passive mode, and with heat pumps. The conclusions of the article set out the basic principles of passive cooling in the design of AGHCS.

1 Introduction

Today, geothermal heat pump heating systems (GHCS) are already widely used in temperate regions of the world (North America, Europe, and China [1]). Extensive global experience of practical utilization of geothermal heat pump heating and cooling systems (GHCS) gained over the past decades identified the most urgent challenges for their mass implementation both abroad and in Russia.

The basic force constraining widespread application of GHCS are high initial capital investments in the system caused not only by high investments in construction of borehole heat exchangers (BHEs) directly ensuring low-grade geothermal heat recovery, but also by significant costs associated with increase in the electrical load of a building and by throughput capacity of electrical networks of a building. In recent years, due to global warming, even in the middle of Russia, summer load on the building cooling system is becoming increasingly important, and such load is often comparable and sometimes higher than winter heating load. As a result, in some cases, actual summer cooling load of a building determines the rated capacity of electric drive of heat pump heating and cooling systems. Moreover, costs associated with increase in the electrical load of a building and throughput capacity of electrical networks of a building are decisive when deciding on the application of GHCS.

At the same time, majority of Russian regions are characterized by low natural ground temperatures and long-lasting heating period, and therefore the long-term operation of GHCS in ground and climatic conditions of the most regions of the Russian Federation is almost always associated with freezing/defrosting the ground surrounding BHEs and storage of a great deal of cold in the ground, which should be utilized in summer for cooling a building. Research presented in this article is focused on addressing this challenge and estimating a potential and efficiency of utilization of passive ground "cold" for cooling buildings in geo-climatic conditions of Russia.

Distinction of these numerical experiments was the need to take into account the processes of freezing/defrosting of pore water in the ground surrounding BHEs when simulating processes accompanying AGHCS operation. Study of existing models of heat transfer processes in the "ground-BHE" system [2, 3] showed virtual absence of generally accepted algorithms for solving this problem and, in particular, consideration of phase transformations of pore water in the ground. In summary, numerical experiments have been carried out with the use of the software package "INSOLAR.GSHP.12" developed by the Russian company INSOLAR-INVEST. The software package "INSOLAR.GSHP.12" allows to simulate the thermal conditions of the ground mass of low-grade geothermal heat collection system in the long-term operation of AGHCS [4], taking into account phase transformations processes of pore water in the ground [5, 6].

2 Baseline building

The baseline building used during the numerical experiments was the real experimental building located at the address: Moscow, Bolshaya Filevskaya Street, estate 22, build. 2, equipped with a pilot AGHCS. Total

DOI: 10.1051/matecconf/20167706008

building area is 278 sq. m. The building has three floors. The experimental building has the shape and plan of volute. The basic technical characteristics of the baseline experimental building are shown in Table 1.

Table 1. Basic technical characteristics of the baseline experimental building

Baseline experimental building area: 278 m^2		
Building envelope name	Reduced heat-transfer resistance R, m2x°C/W	Area, m^2
Exterior walls	3,7	210
1st floor coating	4,5	122
Coating	4	170
Windows	0,8	36
Stained glass	0,6	40

Photo of the baseline experimental building is shown in Fig. 1, and its sectional view is shown in Fig. 2.

Figure 1. Photo of the baseline experimental building

Figure 2. Sectional view of the baseline experimental building

3 Numerical analysis

Numerical analysis was carried out for the baseline building equipped with AGHCS, as described in section 1. Building envelope and technical characteristics of the baseline building are shown in Table 1 and did not changed during numerical analysis. The only changes were modes of operation and, consequently, loads on engineering systems of buildings.

Numerical experiments were carried out for climatic conditions of Moscow. The calculated parameters of outside air were taken in accordance with SP 131.13330.2012 "SNIP 23-01-99* Building Climatology": Summer period - design outside dry bulb temperature is +26,3°C, enthalpy is 56.8 kJ/kg (relative humidity 56%).

As a typical climate year, 2013 year was selected as the most close to year with figures, averaged over the past 30 years.

In conducting numerical analysis, models of the baseline building AGHCS were considered in a variety of configurations using natural cold of BHEs for passive cooling and cold storage in the summer period at night (off-peak) time with its subsequent consumption in the day time both in the passive mode, and with heat pumps.

3.1 Numerical experiment 1. GHCS of the baseline building using heat pumps in summer in the reverse mode with air-cooled condenser and heat release into the atmosphere

Numerical analysis 1 is actually a basic version of AGHCS configuration, which efficiency was compared with various AGHCS configurations using natural cold of BHEs for passive cooling and cold storage.

In order to cover the building cooling loads in summer (June-August), this analysis did not account a ground. Capacity of the installed electrical equipment was determined as the sum of electric capacities of heat pumps + peak power cinch + circulating pumps, fans, etc. Domestic hot water supply was not available.

Fig. 3 shows charts of coolant temperature variation at the inlet (coldest point) of the low-grade geothermal heat collection system for 1, 3 and 5 years of AGHCS operation, and Fig. 4 shows resizing of ice lens frozen around BHE.

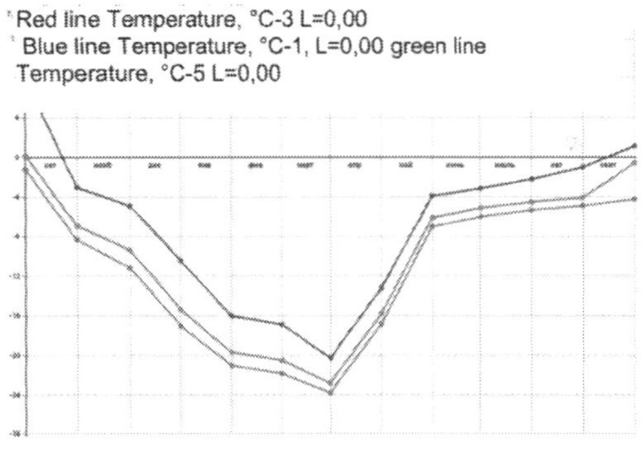

Figure 3. Variation of temperature of coolant of the ground low-grade heat collection system for 1, 3 and 5 years of GHCS

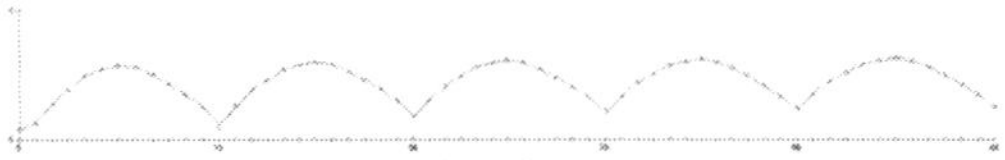

Figure 4. Resizing of ice lens frozen around BHE (X-axis - time in months, Y-axis - ice lens radius, m)

Analysis of charts and graphs shown in Figs. 1 and 2 shows that in this AGHCS configuration the coolant and thus the ground temperature decreases with each year of operation and by the end of March of the 5th year of operation, the minimum coolant temperature in the ground heat collecting system reaches -24 °C. This situation, in turn, leads to grow of ice lenses around BHE which are decreased in size in summer, but not defrosted completely and by the end of March of the 5th year of operation reach 1,-1.2 м diameter. This situation results to relatively low AGHCS coefficients of performance (COP) for the 5th year of operation, which values are shown in Fig. 5. It is obvious that the "cold" stored in ice lenses, can be effectively used in summer for passive cooling.

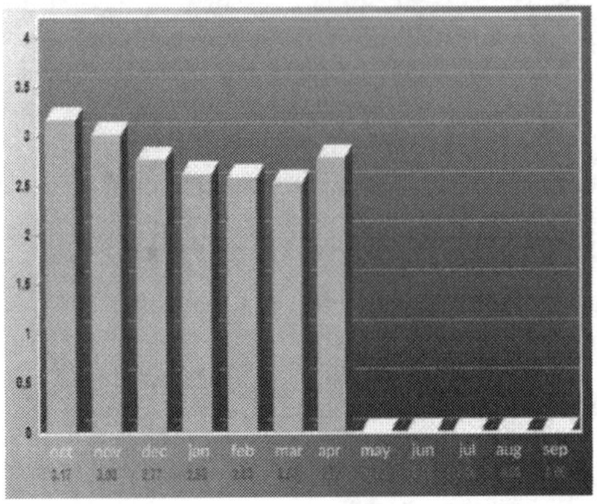

Figure 5. GHCS coefficients of performance by the 5th year of operation

Table 2 shows results of Analysis 1. During the summer period, the building cooling is ensured by heat pumps, ground cooling capacity is not used.

Table 2. Results of analysis 1

Capacity of the installed electrical equipment (HPS + peak power cinches + circulating pumps), Wyst	5.86 kW
Electric energy consumption (for the 5th year of operation) for GHCS (heating), Ws	17397.15 kW·h
Electric energy consumption (for the 5th year of operation) for GHCS (cooling), W_{hol}	1117 kW·h

3.2 Numerical experiment 2. GHCS of the baseline building with passive utilization of "cold" stored in BHEs in winter for building cooling in summer

The numerical experiment 2 studied efficiency of passive utilization of "cold" stored in BHEs in winter for building cooling in summer. In order to cover the building cooling loads in summer (June-August), this experiment accounted a ground contribution. Capacity of the installed electrical equipment was determined as the sum of electric capacities of heat pumps + peak power cinch + circulating pumps, fans, etc. Domestic hot water supply was not available.

Fig. 6 shows charts of coolant temperature variation at the inlet (coldest point) of the ground low-grade heat collection system for 1, 3 and 5 years of AGHCS operation, and Fig. 6 shows resizing of ice lense frozen around BHE.

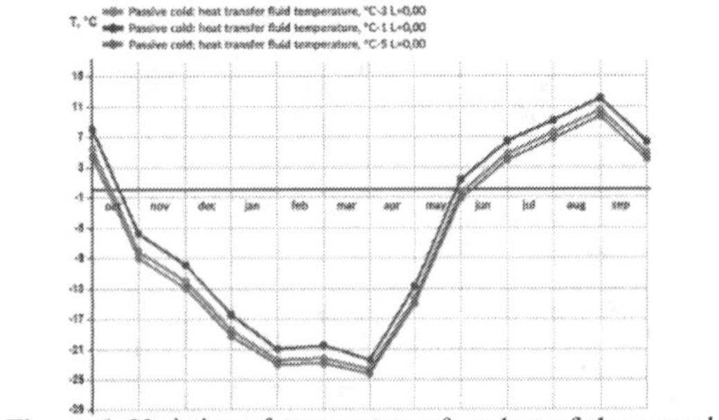

Figure 6. Variation of temperature of coolant of the ground low-grade heat collection system for 1, 3 and 5 years of AGHCS

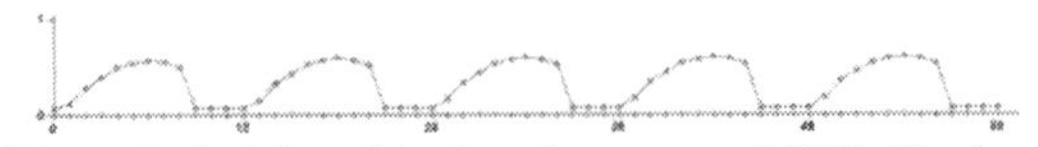

Figure 7. Resizing of ice lens frozen around BHE (X-axis - time in months, Y-axis – ice lens radius)

Analysis of charts shows that in this analysis, the ground substantially recovers its temperature potential by the beginning of each new heating season, and even reaches the zone of positive temperatures by the beginning of the 6th heating season. Consumption of "cold" from the ground and related heat energy "release" into the ground, in turn, lead to a complete defrosting of ice lenses grown around BHEs in winter. This situation results to increased (in comparison with analysis 1) AGHCS coefficients of performance (COP) for the 5th year of operation, which values are shown in Fig. 8.

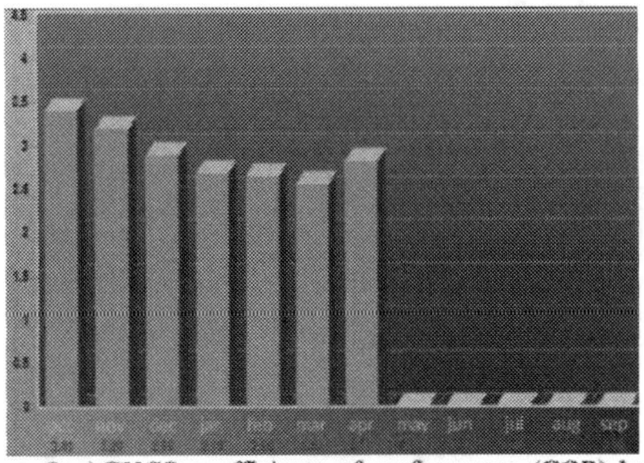

Figure 8. AGHCS coefficients of performance (COP) by the 5th year of operation

Table 3 shows the main results of Experiment 2. During the summer period, building cooling is provided from ground without heat pumps - only circulation pumps of the low-grade heat collection system and the building cooling system operate during this period.

Table 3. Main results of experiment 2

Design capacity of the installed electrical equipment (HPS + peak power cinches + circulating pumps), Wyst	5.86 kW
Electric energy consumption (for the 5th year of operation) for GHCS (heating), Ws	17009.25 kW·h
Electric energy consumption (for the 5th year of operation) for GHCS (cooling), W_{hol}	215 kW·h

3.3 Numerical experiment 3. GHCS of the baseline building using natural cold of BHEs for passive cooling and cold storage in the summer period at night (off-peak) time with its subsequent consumption in the day time both in the passive mode, and with heat pumps

The numerical experiment 3 studied efficiency of utilization of natural cold of BHEs for passive cooling and cold storage in the summer period at night (off-peak) time with its subsequent consumption in the day time both in the passive mode, and with heat pumps. Capacity of the installed electrical equipment was determined as the sum of electric capacities of heat pumps + peak power cinch + circulating pumps, fans, etc. Domestic hot water supply was not available. The experiment considered the temperature mode of coolant of the low-grade geothermal heat collection system in the warmest months of the year, from May to July. Variation of outside air temperature in May of the typical climate year is shown in Fig. 9.

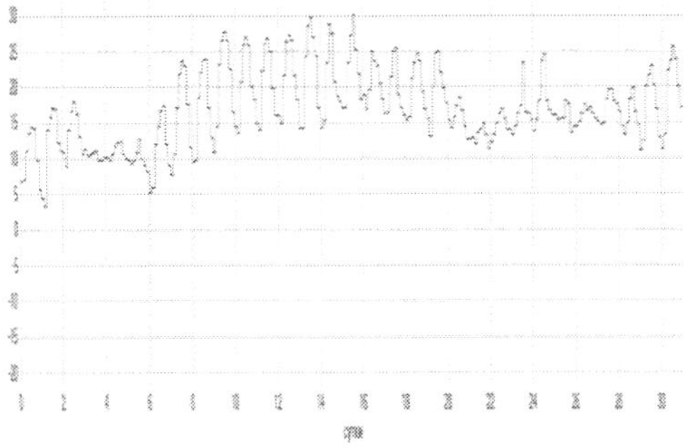

Figure 9. Outside air temperature in May

Fig. 10-Fig. 12 show variation of the coolant temperature at the inlet/outlet of the ground low-grade geothermal heat collection system in May and June when using natural ground cold for passive cooling of the building. In this case, heat pumps did not run and building cooling was fully ensured by the natural ground cold.

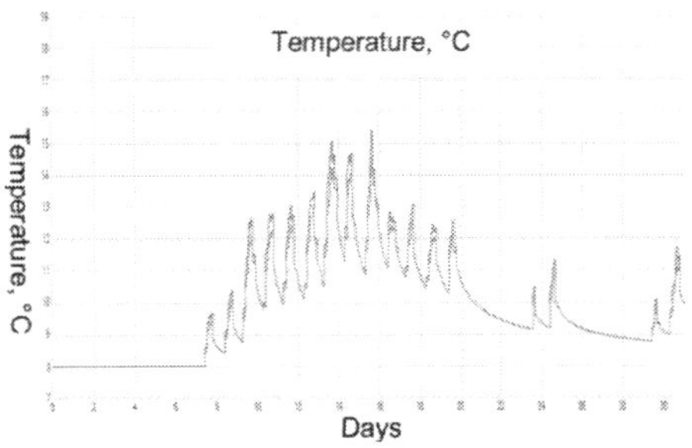

Figure 10. Coolant temperature at the inlet/outlet of the low-grade geothermal heat collection system in May when passive cooling by natural ground cold without heat pumps

Analysis of these charts shows that in the climatic conditions of the Moscow region, cooling of the baseline building can be fully achieved through the natural ground cold. As it can be seen from Figure 10, even in the warmest day, the collection system coolant temperature does not rise above 17°C that is sufficient to provide the building air-conditioning system with cold.

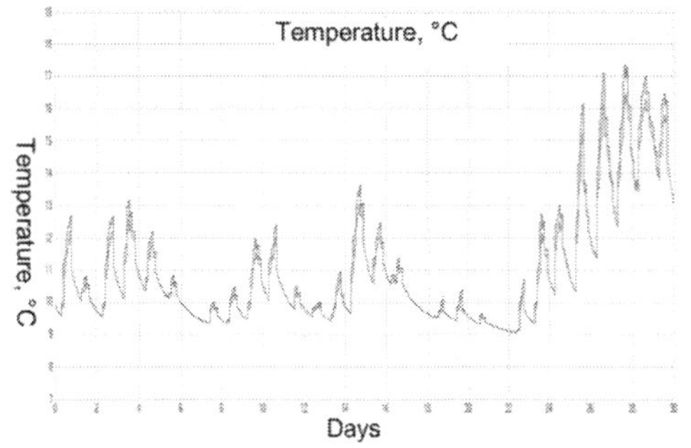

Figure 11. Coolant temperature at the inlet/outlet of the low-grade geothermal heat collection system in June when passive cooling by natural ground cold without heat pumps

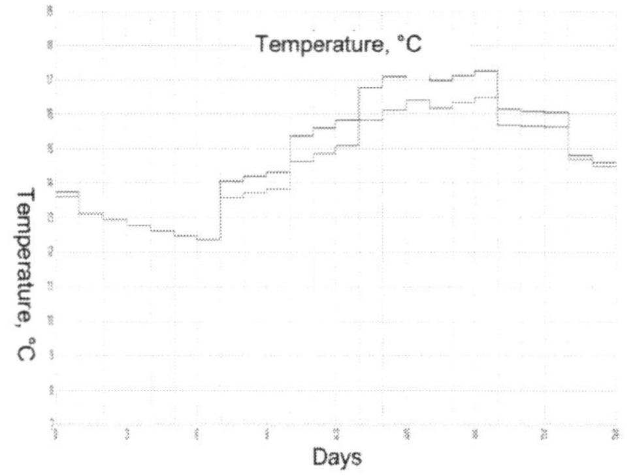

Figure 12. Coolant temperature at the inlet/outlet of the ground low-grade geothermal heat collection system in the warmest day of the year (June 28) when passive cooling by natural ground cold without heat pumps

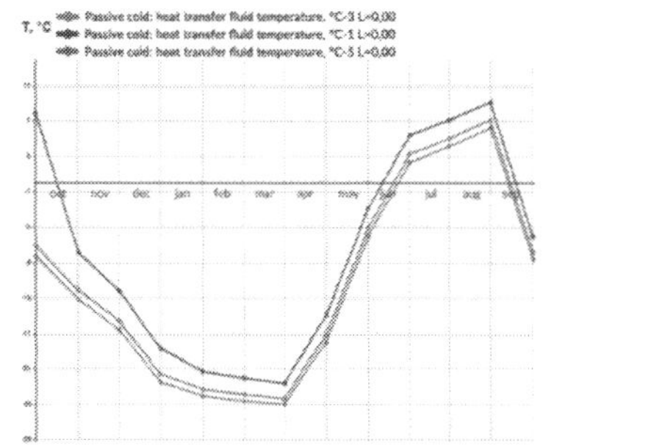

Figure 13. Coolant temperature at the inlet of the low-grade geothermal heat collection system by 1, 3 and 5 years of operation when passive cooling with the use of heat pumps

Fig. 13-Fig. 16 show charts of the heat collection system coolant temperature variations with passive use of cold stored in the ground and with the use of heat pumps.

As we can see from these charts, this modification very effectively uses "cold", generated by heat pumps in winter and stored in the ground. As it can be seen from Fig. 15 and Fig. 16, temperature of the coolant supplied to the building air conditioning system, does not exceed 12°C even in the warmest time of the year. A very important factor is the fact that, as it can be seen from charts, load release to the air conditioning system at night results to "cold charging" of the heat collection system. The cold is supplied, or "pumped", from the layers of ground surrounding BHE. This effect is clearly expressed in the charts. If necessary, this effect can be enhanced by powering heat pumps at night and fast "cold recharging" of the heat collection system.

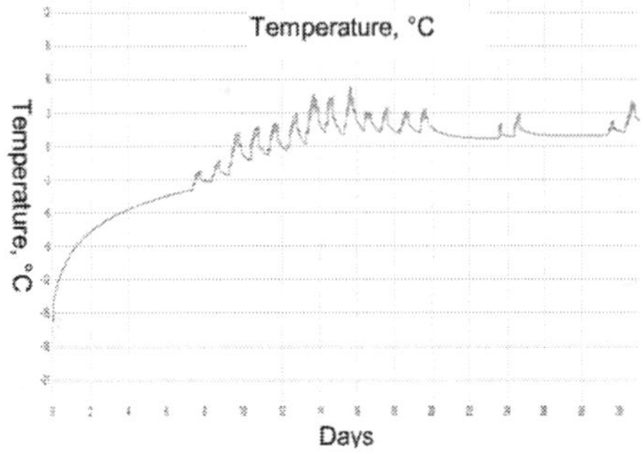

Figure 14. Coolant temperature at the inlet/outlet of the low-grade geothermal heat collection system in May when passive cooling with the use of heat pumps

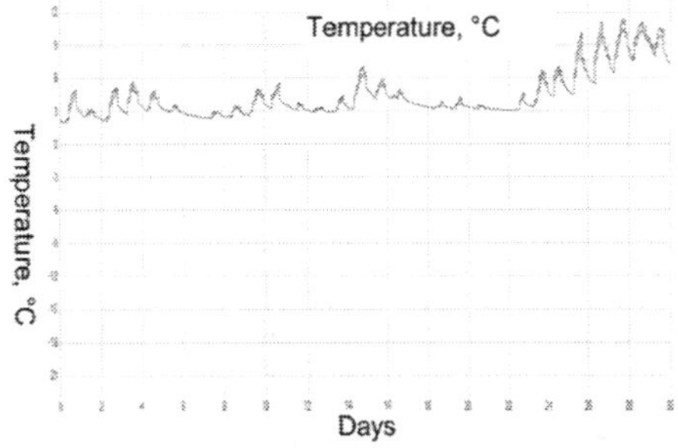

Figure 15. Coolant temperature at the inlet/outlet of the low-grade geothermal heat collection system in June when passive cooling with the use of heat pumps

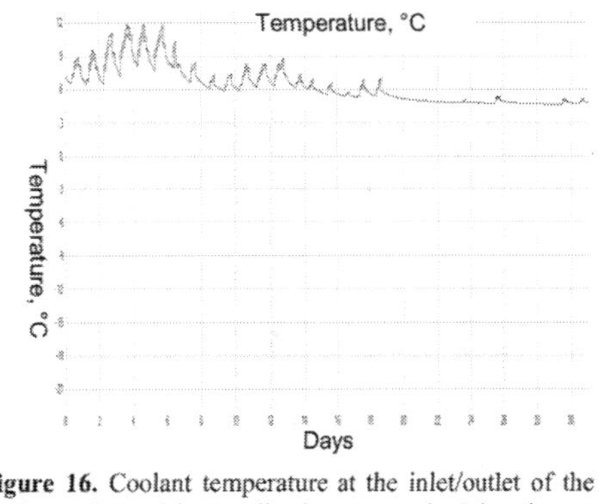

Figure 16. Coolant temperature at the inlet/outlet of the low-grade geothermal heat collection system in July when passive cooling with the use of heat pumps

4 Conclusion

Research presented in this article allowed to formulate the following principles for passive utilization of ground "cold" in the design of AGHCS.

1. When designing AGHCS, it is necessary to make the best use of passive cooling by cold stored in the ground in winter that not only allows for the building cooling without powering the heat pumps, but also significantly increases its (building) energy efficiency and energy savings by about 7%.
2. When using the cold stored in the ground of the heat collection system in summer, it is necessary to seek to its passive utilization in the air conditioning system of the building without heat pumps, while in case of lack of the ground cooling capacity heat pumps could be powered in the night time (with off-peak electricity rates) for "cold charging" of the heat collection system.
3. Use of cold stored in the ground of the heat collection system for direct cooling heat pump condensers during the summer period is impractical.
4. When designing GHCS, it is necessary to take into account the effect of "cold addition" from the ground layers surrounding BHE observed in the night time when load release to the air conditioning system. If necessary, this effect can be enhanced by powering heat pumps at night and fast "cold recharging" of the heat collection system.

Acknowledgements

The research was conducted by JSC "INSOLAR-INVEST" with the financial support of the Ministry of Education and Science of Russia. Unique identifier of the project RFMEFI57914X0026.

References

1. Lund J.W., Freeston D.H., Boyd T.L. Direct utilization of geothermal energy 2010 worldwide review. Geothermics, **40**, (2011) 159-180.
2. Yang H., Cui P., Fang Z. Vertical-borehole ground-coupled heat pumps: A review of models and systems, Applied Energy, **87**, (2010) 16-27.
3. Lamarche L., Kajl S., Beauchamp B. A review of methods to evaluate borehole thermal resistances in geothermal heat-pump systems. Geothermics, **39** (2010) 187-200.
4. G.P. Vasilyev, N.V. Peskov, M.M. Brodach, V.A. Lichman, A.N. Dmitriev, V.F. Gornov, M.V. Kolesova, "Estimation of the thermal effect of ground moisture condensation on heat transfer outside a geothermal borehole", Energy and Buildings, **82** (2014) 795-798.
5. G. P. Vasilyev, N. V. Peskov, V. A. Lichman, V. F. Gornov, M. V. Kolesova, "Simulating the thermal operating conditions in the thermal wells of ground source heat pump heat supply systems," Part I: Accounting the Porous Moisture Freezing Processes

in Soil, Thermal Engineering, **62(8)**, (2015) pp. 547-552.
6. Carslaw H.S., Jaeger J.C. Conduction of heat in solids. Oxford UK: Claremore Press, (1946).
7. Ingersoll L.R., Zobel O.J., Ingersoll A.C. Heat conduction with engineering, geological, and other applications. New York: McGraw-Hill, (1954).

Modelling of Thermal Behavior of Borehole Heat Exchangers of Geothermal Heat Pump Heating Systems

Gornov V.F. [1], Peskov N.V. [1,2], Vasilyev G.P. [1], Kolesova M.V. [1]

[1] *JSC «INSOLAR-INVEST», Bol'shaya Filevskaya str. 32- 3, Moscow, 121433 Russia*
[2] *Lomonosov Moscow State University, Leninskie Gory 1, str. 52, Moscow, 119991 Russia*

Abstract. This article reports results of comparing the accuracy of the software package "INSOLAR.GSHP.12", modeling non-steady thermal behavior of geothermal heat pump heating systems (GHCS) and of the similar model "conventional" using finite difference methods for solving spatial non-steady problems of heat conductivity. The software package is based on the method of formulating mathematical models of thermal behavior of ground low-grade heat collection systems developed by INSOLAR group of companies. Equations of mathematical model of spatial non-steady thermal behavior of ground mass of low-grade heat collection system obtained by the developed method have been solved analytically that significantly reduced computing time spent by the software complex "INSOLAR.GSHP.12" for calculations. The method allows to turn aside difficulties associated with information uncertainty of mathematical models of the ground thermal behavior and approximation of external factors affecting the ground. Use of experimentally obtained information about the ground natural thermal behavior in the software package allows to partially take into account the whole complex of factors (such as availability of groundwater, their velocity and thermal behavior, structure and arrangement of ground layers, the Earth's thermal background, precipitation, phase transformations of moisture in the pore space, and more), significantly influencing the formation of thermal behavior of the ground mass of a low-grade geothermal heat collection system. Numerical experiments presented in the article confirmed the high convergence of the results obtained through the software package "INSOLAR.GSHP.12" with solutions obtained by conventional finite-difference methods.

1 Introduction

Today, geothermal heat pump heating systems (GHCS) are already widely used in temperate regions of the world (North America, Europe, and China [1]). However, practically used mathematical models describing the ground mass thermal behavior in GHCS operation are still far from perfect. Description in models of such complex multi-capillary-porous structures, as the ground, is an extremely difficult task. In the strict sense, in addition to implementation of heat and mass transfer processes, the ground mass thermal behavior models shall take into account the chemical and mineralogical nature of the ground skeleton, mechanical structure of solids material, degree of dispersion of the medium, shape and size of particles and pores, number of phases, quantitative relationships between phases and their mutual arrangement in the medium filling the pore space, as well as many other physical and chemical parameters. Detailed account of these factors in the formulation of a mathematical model is a very serious problem, and its solution often is virtually impossible with the use of modern mathematical apparatus. Analysis of existing models of heat transfer processes in the "ground-BHE" system [2, 3] showed virtual absence of generally accepted algorithms for solving this problem and, in particular, consideration of phase transformations of pore water in the ground.

The software package "INSOLAR.GSHP.12" developed by the Russian company INSOLAR-INVEST allows to simulate the thermal behavior of the ground mass of low-grade geothermal heat collection system in the long-term operation of GHCS [4], taking into account phase transformations of pore water in the ground [5, 6].

2 Mathematical modelling method

The basis of the software package "INSOLAR.GSHP.12" is the mathematical modelling method based on two fundamental assumptions:

1. Ground mass of a low-grade geothermal heat collection system is treated as quasi-homogeneous body, to which the usual heat conduction equation applies; but with the only difference that heat and mass transfer characteristics are "effective" that allows to partially take into account mass transfer processes in the model.

2. Thermal effect of a ground heat exchanger tube register on the temperature behavior of the ground mass

of the heat collection system is approximated by introduction of the model of linear heat sinks (sources) of infinitely small diameter, placed in tube locations. This assumption seems justified, since the diameter of tubes of the ground heat exchanger is much less than depth of their laying, whereby the temperature gradient in tube walls can be neglected.

The method is as follows:

Let's consider a ground mass with arbitrarily arranged ground heat exchanger tube register, limited by coordinate planes and parallel to them planes passing through the layers of the ground, non-affected by the heat collection system. In this case, the problem of formulating a mathematical model of such a system leads to the need to address the problem of temperature behavior of the ground mass with heat sinks.

In the most general case, the heat conduction equation, which solution is the unknown temperature of the ground, can be written in the following form:

$$c\rho \frac{\partial t}{\partial \tau} = \frac{\partial}{\partial x}\left(\lambda \frac{\partial t}{\partial x}\right) + \frac{\partial}{\partial \gamma}\left(\lambda \frac{\partial t}{\partial \gamma}\right) + \frac{\partial}{\partial z}\left(\lambda \frac{\partial t}{\partial z}\right) + \sum_{i=1}^{n} q_i \xi_i, \quad (1)$$

where:

t = t (x, γ, z, τ) — ground mass unknown temperature, °C;

cρ = cρ (x, γ, z, τ) — ground volumetric heat capacity, J/ (°C· m3);

λ = λ (x, γ, z, τ, t) — ground heat conductivity, W/ (m· °C);

τ — time, sec;

$q_i = q_i$ (x, γ, z, τ, t) — power of unit of length of the i-th heat sink, W/m;

n — number of heat sinks;

ξ_i — function determining the heat sink position in the ground mass, $1/m^2$.

The boundary conditions for the equation (1) are as follows:

$$\text{while} \begin{cases} x = 0 & t = \Psi_1(\gamma, z, \tau); \\ x = \alpha & t = \Psi_2(\alpha, \gamma, z, \tau); \\ \gamma = 0 & \lambda \dfrac{\partial t}{\partial \gamma} = \alpha\rho\,(t - t_n); \\ \gamma = H & t = \Psi_3(x, H, z, \tau); \\ z = 0 & t = \Psi_4(x, \gamma, 0, \tau); \\ z = d & t = \Psi_5(\alpha, \gamma, d, \tau); \end{cases} \quad (2)$$

Initial condition (3)

While $\tau = 0 \quad t = t_0\,(x, \gamma, z, 0);$

where:

αρ = α ρ (τ) coefficient of heat transfer from the ground surface, $W/(m^2 \cdot °C)$;

$t_n = t_n$ (τ) outside air temperature accounting radiation "additive" of the incident solar radiation, °C;

Ψ_1, Ψ_2, Ψ_3, Ψ_4, Ψ_5 unknown functions describing temperature at the boundaries of the mass, °C;

t_0 initial temperature distribution in the ground, °C.

It is obvious that the accuracy of implementation of the resulting mathematical model of the heat collection system thermal behavior will greatly depend on the accuracy of formulation of the boundary conditions (2) and initial temperature distribution **t_0**, i.e. from the accuracy of approximation of functions Ψ_1, **Ψ_2, Ψ_3, Ψ_4, Ψ_5, t_n** and **t_0** that, as is known, is a serious problem.

To solve this problem, let's now consider the thermal behavior of the same ground mass, but without heat sinks (sources), i.e. under natural conditions — in the absence of heat effect of the heat collection system on the ground. The heat conductivity equation in this case would look like this:

$$c^*\rho^* \frac{\partial t^*}{\partial \tau} = \frac{\partial}{\partial x}\left(\lambda^* \frac{\partial t^*}{\partial x}\right) + \frac{\partial}{\partial \gamma}\left(\lambda^* \frac{\partial t^*}{\partial \gamma}\right) + \frac{\partial}{\partial z}\left(\lambda^* \frac{\partial t^*}{\partial z}\right), \quad (4)$$

where:

* represents a parameter belonging to the problem of natural temperature behavior of the ground mass (without heat sinks or sources), and the other symbols are the same as before.

The boundary conditions in this case can be written as follows:

$$\text{while} \begin{cases} x = 0 & t^* = \Psi_1(\gamma, z, \tau); \\ x = \alpha & t^* = \Psi_2(\alpha, \gamma, z, \tau); \\ \gamma = 0 & \lambda^* \dfrac{\partial t^*}{\partial \gamma} = \alpha\rho\,(t^* - t_n); \\ \gamma = H & t^* = \Psi_3(x, H, z, \tau); \\ z = 0 & t^* = \Psi_4(x, \gamma, 0, \tau); \\ z = d & t^* = \Psi_5(\alpha, \gamma, d, \tau); \end{cases} \quad (5)$$

Initial condition (6)

While $\tau = 0 \quad t^* = t_0\,(x, \gamma, z, 0);$

It should be noted the fact that the coefficient of heat transfer from the ground mass surface αρ in both problems can be described by the same function αρ (x, γ, z, τ), since the heat removal from the ground causes a non-significant change in the surface temperature and therefore has virtually no effect on the rate of heat exchange of the ground surface with the ambient air.

Let's now consider the difference of the resulting equation systems, that is lets subtract the system of equations (1)-(6) describing the thermal behavior of the ground mass with heat sinks (sources) from the system (1)-(4), describing the thermal behavior of the ground mass in its natural state (without heat sinks (sources)).

When subtracting the equation (1) from the equation (4) we obtain

$$c^*\rho^* \frac{\partial t^*}{\partial \tau} - c\rho \frac{\partial t}{\partial \tau} = \frac{\partial}{\partial x}\left(\lambda^* \frac{\partial t^*}{\partial x}\right) - \frac{\partial}{\partial x}\left(\lambda \frac{\partial t}{\partial x}\right) + \frac{\partial}{\partial \gamma}\left(\lambda^* \frac{\partial t^*}{\partial \gamma}\right) - \quad (7)$$

$$\frac{\partial}{} \quad \frac{\partial t}{} \quad \frac{\partial}{} \quad \frac{\partial t^*}{} \quad \frac{\partial}{} \quad \frac{\partial t}{} \quad n$$

$$-\frac{}{\partial \gamma}(\lambda\frac{}{\partial \gamma}) + \frac{}{\partial z}(\lambda^*\frac{}{\partial z}) - \frac{}{\partial z}(\lambda\frac{}{\partial z}) + \sum_{i=1} q_{\xi i}.$$

Subtracting the boundary conditions (2) from the boundary conditions (4) gives

$$\text{while}\begin{cases} x = 0 & t^* - t = 0; \\ x = \alpha & t^* - t = 0; \\ \gamma = 0 & \lambda^*\frac{\partial_{t^*}}{\partial \gamma} - \lambda^*\frac{\partial_t}{\partial \gamma} = \alpha\rho(t^* - t); \\ \gamma = H & t^* - t = 0; \\ z = 0 & t^* - t = 0; \\ z = d & t^* - t = 0; \end{cases} \quad (8)$$

Subtracting the initial conditions of the system (2) from the initial conditions (5) gives
while

$$\tau = 0 \qquad t^* - t = 0. \quad (9)$$

Let's introduce a new symbol

$$T = t^* - t \quad (10)$$

and, substituting variables in the system (7) - (9) taking into account this symbol, we obtain a new equation [1]):

$$c\rho\frac{\partial T}{\partial \tau} = \frac{\partial}{\partial x}\left(\lambda\frac{\partial T}{\partial x}\right) + \frac{\partial}{\partial \gamma}\left(\lambda\frac{\partial T}{\partial \gamma}\right) + \frac{\partial}{\partial z}\left(\lambda\frac{\partial T}{\partial z}\right) + \sum_{i=1}^{n} q_i\xi_i + \left\{\frac{\partial}{\partial x}\left[(\lambda^* - \lambda)\frac{\partial t^*}{\partial x}\right] + \frac{\partial}{\partial \Upsilon}\left[(\lambda^* - \lambda)\frac{\partial t^*}{\partial \Upsilon}\right] + \frac{\partial}{\partial z}\left[(\lambda^* - \lambda)\frac{\partial t^*}{\partial z}\right] - (c^*\rho^* - c\rho)\frac{\partial t^*}{\partial \tau}\right\} \quad (11)$$

$$c\rho\frac{\partial T}{\partial \tau} = \frac{\partial}{\partial x}\left(\lambda\frac{\partial T}{\partial x}\right) + \frac{\partial}{\partial \gamma}\left(\lambda\frac{\partial T}{\partial \gamma}\right) + \frac{\partial}{\partial z}\left(\lambda\frac{\partial T}{\partial z}\right) + \sum_{i=1}^{n} q_{\xi i} \quad (12)$$

Let's make a similar substitution of variables in the boundary conditions (8):

$$\text{while}\begin{cases} x = 0 \; T = 0; \\ x = \propto \; T = 0; \\ \gamma = 0 \; \lambda\frac{\partial y}{\partial x} = \alpha\rho T - (\lambda^* - \lambda)\frac{\partial t^*}{\partial \gamma} \\ \gamma = H \; T = 0; \\ z = 0 \; T = 0; \\ z = d \; T = 0. \end{cases} \quad (13)$$

A similar substitution of variables in the initial conditions (9) allows to obtain

while

$$\tau = 0 \qquad T = 0. \quad (14)$$

Now let's consider in more detail the resulting boundary conditions (11). As discussed earlier, the ground masses addressed in the problems are limited by conventional planes, which arrangement eliminates the heat effect of ground heat exchanger on the boundary ground layers. Consequently, the ground temperatures on the boundaries of the mass, except only the surface temperatures in the general case, will be practically identical in both problems. Accordingly, heat conductivities of ground masses considered in problems will be equal as well. As a result, without reducing the model accuracy, we can conclude that the difference of ground heat conductivities **(λ^* — λ)** for all the boundaries of the mass, including in this case (for heat collection systems) the ground surface is equal to 0, i.e. **$\lambda^* - \lambda = 0$.**

In view of this point, the boundary conditions (13) will be

$$\text{while}\begin{cases} x = 0 \; T = 0; \\ x = \alpha \; T = 0; \\ \gamma = 0 \; \lambda\frac{\partial T}{\partial \gamma} = \alpha\rho T; \\ \gamma = H \; T = 0; \\ z = 0 \; T = 0; \\ z = d \; T = 0. \end{cases} \quad (15)$$

Thus, as a result of these transformations we have obtained a new mathematical model in which the required function is unknown function T, which is function of effect of heat sinks on the natural thermal behavior of the ground and equal to temperature difference of ground masses in a natural condition and with heat sinks (sources). It should be noted that in the boundary conditions of the new model (the boundary conditions (15) and the initial condition (14)) we managed to dispense with unknown functions ψ_1, ψ_2, ψ_3, ψ_4, $\psi_5$and unknown initial ground temperature distribution, as well as the need for approximation of the annual change of the ambient air temperature and the intensity of solar radiation incident on the surface of the ground mass under consideration. However, the equation (11) has unknown function **t^***, which is the solution of the "base" problem, describing the natural thermal behavior of the ground. Naturally, the accuracy of formulating this function will have the most significant effect on the accuracy of implementation of the resulting mathematical model and its adequacy to real processes occurring in the ground.

In general, the "base" problem solution can be formulated in the model with the use of hydrometeorological data on the annual change of ground temperatures at different depths, which are monitored by virtually all weather stations.

Thus, use of the method described in the formulation of mathematical models of thermal behavior of ground low-grade heat collection systems allows to turn aside difficulties associated with approximation of external factors affecting the system. A definite advantage of this method compared with the traditional approach to modelling of thermal processes occurring in such systems is the fact that the use of experimentally obtained information about the natural ground thermal behavior in the model allows to partially take into account the whole

[1] Hereinafter, it should be understood that similar substitution of variables has been performed in functions c, ρ and λ as well.

complex of factors (such as availability of groundwater, and their velocity and thermal behavior, structure and arrangement of ground layers, the Earth's thermal background, precipitation, phase transformations of moisture in the pore space, and more), significantly influencing the formation of thermal behavior of the heat collection system, which joint account in a strict formulation of the problem is practically impossible now.

Furthermore, it should be noted that, if necessary, this method allows to use data obtained as a result of theoretical studies of the natural thermal behavior of the considered ground mass as the "base" problem solution.

If necessary, the reverse transition from the function of heat sinks effect on the natural behavior conditions of the ground mass to actual temperatures observed in the mass in the course of the heat collection system operation is carried out using substitution (10).

This method of mathematical modelling of non-steady thermal behavior of ground low-grade heat collection systems has been implemented in the software package "INSOLAR.GSHP.12", providing determination of optimal parameters of a heat collection system depending on the climatic conditions of the construction area, heat-protective characteristics of the building, performances of heat pumps, circulating pumps, heating devices, as well as their modes of operation. Use of this method of mathematical modelling in thee software package "INSOLAR.GSHP.12" allowed to turn aside difficulties associated with informative uncertainty of models and approximation of external climatic effects on ground mass by using the experimentally obtained information about the natural ground thermal behavior, which allows to partially take into account the whole complex of factors (such as availability of groundwater, their velocity and thermal behavior, structure and arrangement of ground layers, the Earth's thermal background, precipitation, phase transformations of moisture in the pore space, and more), significantly influencing the formation of thermal behavior of the heat collection system, which joint account in a strict formulation of the problem is practically impossible now.

3 Accuracy assessment

To assess the accuracy of the software package "INSOLAR.GSHP.12" we compared the results of calculations performed by traditional software packages using numerical methods for modelling of thermal behavior of processes occurring in the ground, and the calculation results obtained by the software package "INSOLAR.GSHP.12".

The comparative assessment solved the same problem in both cases: the thermal behavior of the ground mass with intermittent point linear heat sink horizontally laid at a certain depth from the earth surface was considered. At the same time, the effect of a heat sink on a heat field of the surrounding ground was assessed. The sink operated periodically: it was enabled at the time zero, operated 7 months, and then it was disabled and restarted by the 2nd-5th year.

The calculations were carried out for the ground mass in the form of a parallelepiped with dimensions of 100x100x300 m (HxWxL) with a linear heat sink in the length of 70 m laid at a depth of 30 m from the surface. The sink operated periodically: it was enabled on October 1 (time zero) and operated 7 months (till April 30), then it was disabled and restarted by the second year from 13 to 19 months and so on during 5 years.

The calculations used the following thermal characteristics of the ground: heat conductivity -2.65 W/(m * C); heat capacity -0.47 Wh/(kg * C); density 2000.00 kg/ cu.m. For all boundaries, homogeneous boundary conditions were formulated. Heat flow was assumed to be zero at all boundaries except the ground mass' surface. On the ground surface the mass lost heat to the environment at a constant temperature of the outside air - the boundary condition of the third kind. The coefficient of heat transfer from the ground surface was assumed to be 23.26 W/(sq.m). The initial temperature of the ground around the mass was equal to zero.

The heat sink was a tube heat exchanger with a radius of 0.08 m. Cooling was provided by a coolant that was pumped through the heat exchanger. The coolant heat capacity was adopted equal to 1.05 W*h/(kg*oC), and its density was equal to 1020.00 kg/cu.m.

In the traditional problem solved by numerical methods based on the symmetry condition, the half of the computational domain was considered. The number of grid node was assumed to be 33177, the number of cells (tetrahedrons) was equal to 157845. The computational grid was created using the software Gmsh (http://geuz.org/gmsh/). The heat sink power was 5 kW for the entire ground heat exchanger. [7]

Figs. 1-3 show the calculation model and results of calculation according to the conventional version of calculation using numerical methods, and Fig. 4 shows the results of calculations obtained by the software complex "INSOLAR.GSHP.12".

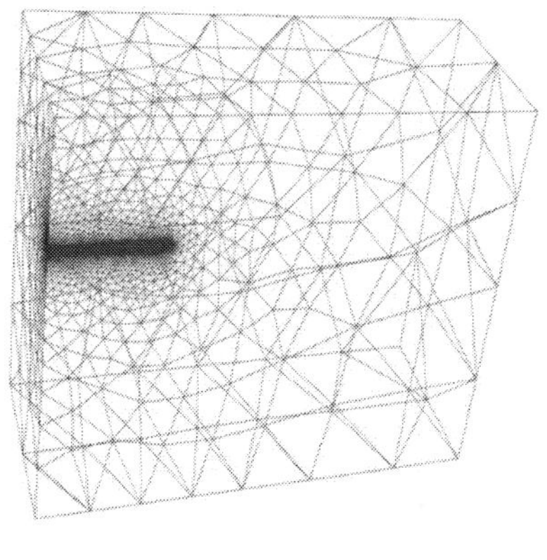

Figure 1. Computational domain (computational grid)

Comparison of charts of coolant temperature variation in the ground heat collection system shown in Figs. 3 and 4 clearly demonstrate a high convergence of the results with those obtained by conventional numerical methods,

DOI: 10.1051/matecconf/20167706009

and confirm the accuracy of the mathematic models underlying the software package "INSOLAR.GSHP.12".

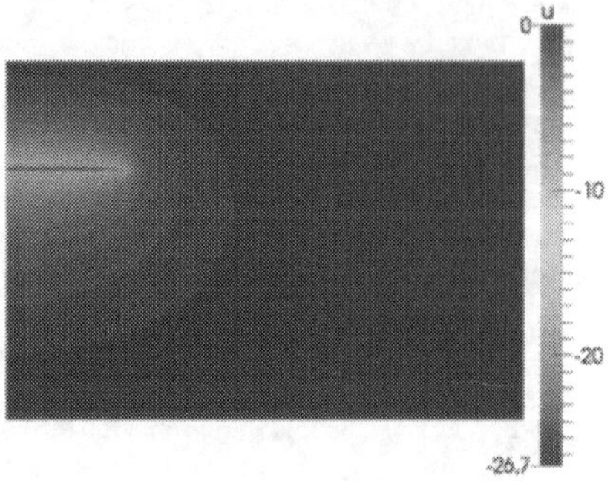

Figure 2. Temperature field in the ground mass (vertical section across the heat exchanger)

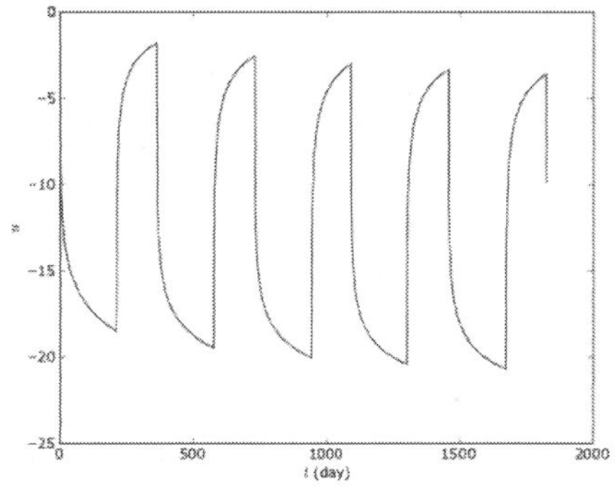

Figure 3. Ground temperature at the heat sink point

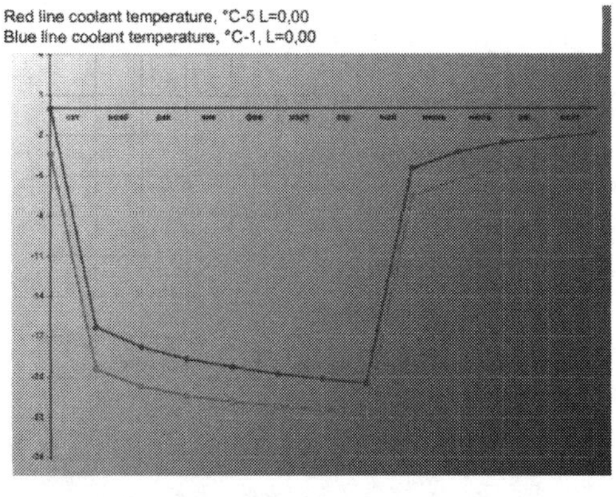

Figure 4. The coolant temperature at the heat sink point (1st (blue curve) and 5th (red curve) years of operation), calculated using the software package "INSOLAR.GSHP.12"

4 Conclusion

Results of numerical experiments presented in the article confirmed the high convergence of mathematical models underlying the software package "INSOLAR.GSHP.12" with "conventional" numerical models. Use of the mathematical modelling method described in the article allowed to describe the spatial non-steady thermal behavior of the ground mass of a low-grade geothermal heat collection system using a second-order differential equation system with homogeneous boundary conditions. The solution of this differential equation system was obtained by the analytical method of finite integral transformations. Use of analytical solution significantly reduced the "computing" time required for calculations, which in turn allowed to create the software package "INSOLAR.GSHP.12" with advanced features. The software package "INSOLAR.GSHP.12" actually solves the problem of multiparametric optimization of GHCS configuration for the specific building and construction area. At the same time the objective function of the optimization problem is a minimum annual energy costs for GHCS operation, and optimization criteria are radius of tubes of a ground heat exchanger, and its (heat exchanger) length and depth.

Acknowledgements

The research was conducted by JSC "INSOLAR-INVEST" with the financial support of the Ministry of Education and Science of Russia. Unique identifier of the project RFMEFI57915X0115.

References

1. Lund J.W., Freeston D.H., Boyd T.L. Direct utilization of geothermal energy 2010 worldwide review. Geothermics, **40**, 159-180, (2011).
2. Yang H., Cui P., Fang Z. Vertical-borehole ground-coupled heat pumps: A review of models and systems, Applied Energy, **87**, 16-27, (2010).
3. Lamarche L., Kajl S., Beauchamp B. A review of methods to evaluate borehole thermal resistances in geothermal heat-pump systems. Geothermics, **39**, 187-200, (2010).
4. G.P. Vasilyev, N.V. Peskov, M.M. Brodach, V.A. Lichman, A.N. Dmitriev,V.F. Gornov, M.V. Kolesova, "Estimation of the thermal effect of ground moisture condensation on heat transfer outside a geothermal borehole", Energy and Buildings **82**, 795-798, (2014).
5. G. P. Vasilyev, N. V. Peskov, V. A. Lichman, V. F. Gornov, M. V. Kolesova, «Simulating the thermal operating conditions in the thermal wells of ground source heat pump heat supply systems. part I: accounting the porous moisture freezing processes in soil» Thermal Engineering, (2015), **62(8)**, pp. 547-552.
6. Carslaw H.S., Jaeger J.C. Conduction of heat in solids. Oxford UK: Claremore Press, (1946).
7. Ingersoll L.R., Zobel O.J., Ingersoll A.C. Heat conduction with engineering, geological, and other applications. New York: McGraw-Hill, (1954).

Research of the Effectiveness of Using Air and Ground Low-grade Heat for Buildings Heating in Different Regions of Russia

Vasilyev G.P, Gornov V.F. Kolesova M.V.

JSC «INSOLAR-INVEST», Bol'shaya Filevskaya str 32-3, Moscow, 121433 Russia

Abstract. The article presents the results of research on zoning of the Russian Federation based on efficiency of utilization of the low-grade heat of ground and air as well as combinations thereof for heating buildings. When modeling thermal behavior of geothermal HHS in the climatic conditions of various regions of the Russian Federation we considered the effect of long-term recovery of geothermal heat on the thermal behavior of the ground, as well as the effect of the ground pore water phase transitions on the operational efficiency of geothermal heat pump heating systems. The zoning took into account temperature drop of the ground mass caused by many years of heat recovery from the ground. Ground temperatures expected for the 5th year of geothermal HHS operation were used as design ground mass temperatures.

1 Introduction

Today, heat pump heating systems (HHS) are widely used in temperate regions of the world (North America, Europe, and China [1]). Utilization of the Earth surface ground and/or air in HHS as sources of low-grade heat energy is possible almost everywhere. At the same time, the most part of the territory of Russia, in contrast to Europe and North America is characterized by lower natural ground temperatures and rather long-term heating seasons, in connection with what it is practically impossible to recover heat energy from the ground during the heating season with its (ground) positive temperatures

When operating geothermal HHS, ground mass located within the zone of heat effect of tube register of ground heat exchanger of the ground low-grade heat collection system (heat collection system) is usually subjected to repeated freezing and defrosting owing to seasonal changes in the ambient climate parameters, as well as under the influence of operating loads on the heat collection system. This, of course, is accompanied by change of the aggregation state of moisture contained in ground pores and being in general both in the liquid, and solid and gaseous phases simultaneously. Long-term operation of geothermal HHS in soil and climatic conditions in most regions of the Russian Federation is usually associated with freezing/defrosting of the ground surrounding borehole heat exchangers and these processes should always be taken into consideration when assessing the efficiency of geothermal HHS.

Configuration and type of the most "popular" HHS are usually determined by geoclimatic conditions of the country. So, HHS with water heating systems (mainly floor) [1], providing buildings heating are the most common in Central and Northern Europe, while heat pumps are more commonly used in air conditioning and air heating systems. [2] The largest number of geothermal HHS are now running in the United States, Canada, Austria, Germany, Sweden and Switzerland. The leader in terms of per capita utilization of low-grade geothermal heat is Switzerland. In Russia, unfortunately, only a few facilities equipped with geothermal HHS are maintained [1].

The Earth's surface ground in geoclimatic conditions in Russia is virtually the only widely available source of low-grade heat energy for heat pump heating systems (HHS). It is known that operation of HHS using the ground heat is usually associated with decrease in temperature potential of the recovered ground heat during the heating season. At the same time, ground low-grade heat collection systems using the "zero curtain" effect are characterized by negative temperatures of the recovered ground heat allowing to use the heat of pore moisture phase transitions in the ground. At the same time, outside air temperatures exceeding temperature of the heat recovered from the ground has been observed frequently during the heating period: thaw periods, transitional periods of the year and simply heating seasons, in which an air temperature exceeds temperature of the heat recovered from the ground.

Despite the huge world practical experience of heat pump heating systems operation, experts still debating about their efficiency in various climatic conditions. Discussions focus on the efficient utilization of various low-grade heat sources, stability of HHS operation throughout the life cycle of buildings, effect of heat recovery from the ground to its natural temperature behavior, etc. [2, 3].

For example, article [3] presents the results of studies performed in the United States and dedicated to assessment of major HHS factors, HSPF (Heating Seasonal Performance Factor) and SEER (Seasonal Energy Efficiency Ratio). HSPF and SEER are defined as the ratio of useful heat output of the heat pump during its normal seasonal use divided by the total electric energy consumed for the season. These studies analyzed efficiency of air heat pumps in the United States climatic conditions on the basis of real data about the operation of more than 9,500 heat pump systems.

Despite the conventional system of climatic zoning in accordance with ARI, the obtained data have not demonstrated homogeneity of efficiency of the air heat pump system inside the climate zones. At the same time, dependence of heat pumps efficiency on the average annual temperature and degree days of the heating season is expressed very clearly. This can be clearly seen on the map of climatic zones of the United States (Fig. 1).

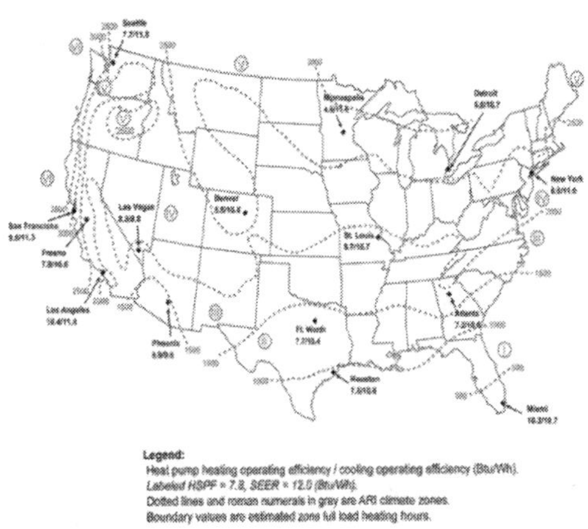

Figure 1. Dependence of heat pumps efficiencies on the average annual temperature and the heating season degree days in various climatic zones of the United States

The map in Fig. 1 shows clearly enough that heat pump systems are most effective in cooling mode than in heating mode in virtually all climatic zones of the United States.

The authors analyzed the experience of US data obtained in the article [3] and made the chart of dependence of the seasonal performance factor on the heating season degree hours.

Based on the analyzed data, the authors of [3] plotted the chart of dependence of Heating Seasonal Performance Factor on the heating season degree hours, shown in Fig 2. The chart clearly chows a tendency to HHS efficiency decrease with increase in number of heating season degree days.

2 Zoning methodology

Research of research on zoning of the Russian Federation based on efficiency of utilization of the low-grade heat of ground and air for heating buildings, presented in this paper are based on the assessment of efficiency of geothermal HHS utilization in climatic conditions of Russia [6] previously performed by its authors. In this paper, an efficiency criterion for heat pump heating systems was the climatological factor ***DHhps***,

numerically representing the number of degree hours of HHS operation in the year. This parameter is actually an analogue of well-known and widely used parameter of degree days or degree hours of the heating period, but takes into account the effect of climatological parameters on the HHS coefficient of performance (COP). Independence of the proposed climatological factor on heat protection and other characteristics of the building is an important aspect. In general, climatological factor ***DHhps*** can be determined by the following formula,

degree hours:

$$DH_{hps}=(T_k-273-h_t-t^{cp}_{н})\times z\times(T_k-T_u)/(T_k\times\eta) \quad (1)$$

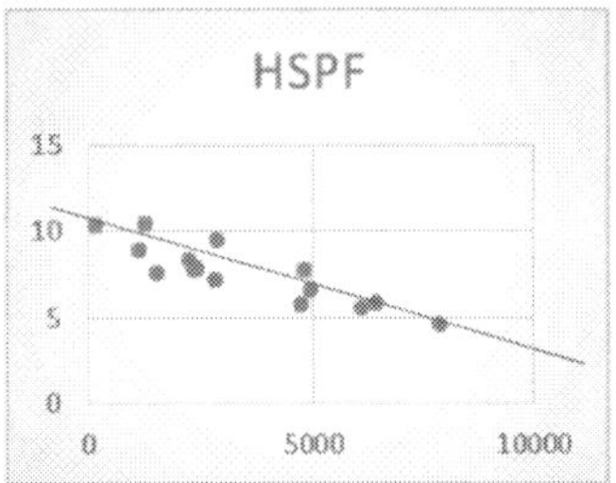

Figure 2. Dependence of Heating Seasonal Performance Factor on the heating season degree hours

where:

Tk- coolant temperature at the HHS outlet and the heating system inlet, K;

ht – temperature difference between coolant at the heating system inlet and the inside air;

$t_{н}^{cp}$- outside air temperature averaged per heating season, or the HHS operation period, is taken according to SNiP “Building Climatology” [2].

z – duration of the heating season or the HHS operation period of the year, is taken according to SNiP “Building Climatology”, hours, [2]

Ti – average temperature of the low-grade heat source for the heating season or for the HHS operation period, K. According to SNiP “Building Climatology” [2] it is equal to:

-for geothermal HHS with vertical BHEs: it is equal to the average outside air temperature increased by 1 degree

-for air systems: it is equal to the outside air temperature averaged for the heating season or the operation period;

-for combined systems: it is equal to the temperature averaged for the period of operation of each of the low-grade heat sources.

Average value η was taken equal to 0.469914 for ground, and 0.33652 for air

Values η are presented in accordance with the data of EN 14511 for the considered modes. The other used values ($t_{н}^{cp}$, z, Tu) are presented in SNiP 23-01-99 “Building Climatology”. [2]

 DOI: 10.1051/matecconf/20167706010

Zoning was performed based on two values **Tk = 308 and 323 K (35 and 50 °C).** These values correspond to the standard design modes for heat pump systems established by the European standard EN 14511. Value Tk = 308 K corresponds to floor and air heating systems.

Value h_t is assumed to be equal to **15 degrees** for Tk=308 K and equal to **30 degrees** for Tk=323 K.

According to values introduced by the formula, zoning based on climatological factor was performed. Low-grade heat source - ground. (Figs. 3-4.) Mapping was carried out for all regional centers of Russia with the use of software MapInfo Professional 10.5.

In the case of using air as the low-grade heat source, climatological factor DHhps is determined by the formula 2

$$DH_{hps}=\sum_{i=1}^{n}(T_k-273-h_t-t^{cp}_{u,i})\times z_i\times(T_k-T_{u,i})/(T_k\times\eta) \quad (2)$$

where –i -number of the period of use of the low-grade heat source, and n – number of periods of use of various sources in the year. If the source is used several periods in a year, each of the periods is assigned a number. As for the ground (as a comparison), when calculating values for air, the entire heating period by months for each city of Russia was taken conditionally.

According to values introduced, zoning based on climatological factor was performed. Low-grade heat source - air. (Figs. 5 -6.)

In the calculation of the combined use of heat sources "air+ground" for each city we used the heating period conditionally divided into ground and air, where the minimum air temperature was -5°C because at lower temperatures the use of air as a heat source is not effective and the authors suggested that at low temperatures the system will switch to ground. In the case of using combination of low-grade heat sources, climatological factor $DHhps$ is determined by the formula (2) too.

In carrying out these calculations for each administrative center of all Russian regions, we determined periods of use of each of the low-grade heat sources according to the efficiency of use: ground in the lowest values of outside air average daily temperature, air in the warmest days (up to -5°C). [5] According to these data, zoning based on the combined use of low-grade heat sources was performed. (Figs. 7-8.)

Figure 3. Zoning of the Russian Federation based on Tк=308, h_t= 15 degrees. Low-grade heat source - ground.

Figure 4. Zoning of the Russian Federation based on Tк=323, h_t= 30 degrees. Low-grade heat source - ground.

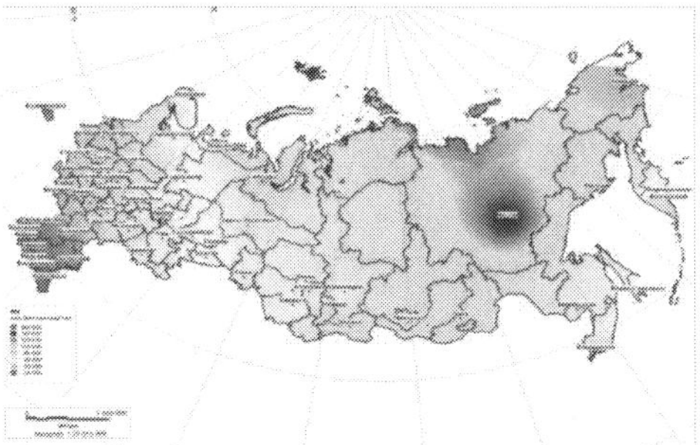

Figure 5. Zoning of the Russian Federation based on Tк=308, h_t= 15 degrees. Low-grade heat source - air

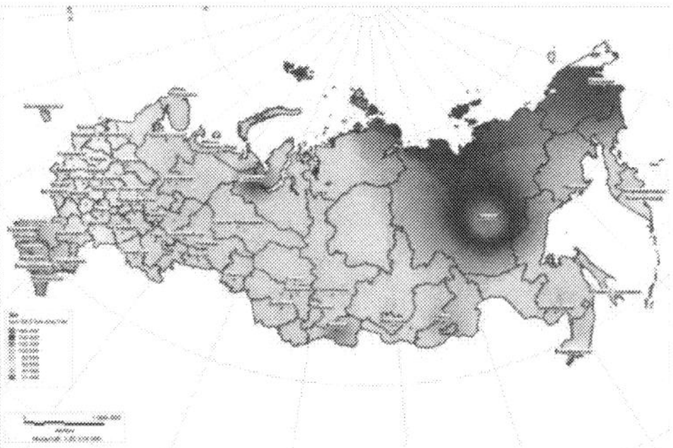

Figure 6. Zoning of the Russian Federation based on Tк=323, h_t= 30 degrees. Low-grade heat source - air

Figure 7. Zoning of the Russian Federation based on Tк=308, ht= 15 degrees. Low-grade heat source - air+ground

The calculations allowed to obtain climatological factor that allows to determine the utilization efficiency criteria for various heating systems for each region of Russia. To get the value of the actual energy consumption of a heat pump system for heating and ventilation, it is necessary to determine the specific characteristic of the building heat consumption for heating and ventilation, for example, according to SP 50.13330.2012. "Thermal protection of buildings" qотгр, W/(m2*°C) and multiply it by a climatological factor DHhps. [5, 6]

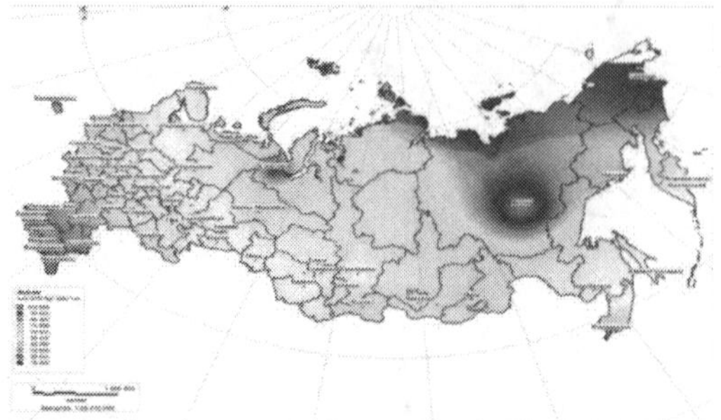

Figure 8. Zoning of the Russian Federation based on Tк=323, ht= 30 degrees. Low-grade heat source - air+ground

The results obtained were tested in the mode of air conditioning of buildings, which were carried out in field conditions at the real facility - a hotel complex in Olginka village, Tuapse. To conduct research based on the hotel power supply system, an intelligent pilot heat pump energy complex was created, including experimental heat pumps and systems of low-grade heat collection from ground and air.

3 Conclusions

Analysis of the results of numerical experiments leads to the conclusion that utilization of outside air in combination with the ground as a heat source for low-grade heat pump evaporators of the designed energy complex in the climatic conditions in Moscow and southern part of Russia is very effective. Energy saving in the application of this solution is from 13% in the climatic conditions in Moscow and 20% in southern regions of Russia.

An important feature of the results obtained is the possibility of their use in mass residential, cottage and commercial construction. Today, tens of thousands of buildings and commercial facilities (shops, stalls and so forth) are equipped with air-conditioning systems operated in winter as heat pumps for rooms heating. However, operation of air conditioners in this mode is physically possible and economically effective only until the outside temperature is above minus 15 °C. At lower temperatures, air conditioners are actually disabled and back-up heating system is required. Technical solutions developed in this article allow to solve this problem by switching to another low-grade heat source - the Earth's surface ground at low air temperatures. As a result, the developed intelligent heat pump energy complex can be effectively used all year round, both in heating and in air conditioning mode. Studies have shown that switching a heat pump evaporator from ground to air in such moments will allow 20% reduction in the required number of BHEs of a ground low-grade heat collection system, the cost of which now stands at more than 50% of the whole geothermal heat pump heating system. This provides more than 15% energy savings in comparison with HHS using only ground as a source of low-grade heat.

These data demonstrate that efficiency of ground utilization as a low-grade heat in our country is higher than efficiency of air and combined ground and air utilization, but in the south regions, these values are almost equal. [4-6]

Acknowledgements

The research was conducted by JSC "INSOLAR-INVEST" with the financial support of the Ministry of Education and Science of Russia. Unique identifier of the project RFMEFI57914X0026.

References

1. Vasilyev G.P. Geothermal heat pump heating systems and its operating efficiency in climate conditions of Russia. // AVOK – (2007) **5**, p. 58-68.
2. Climatic Atlas of the USSR. H 1.2. -M., 1960, 1962.
3. Vasilyev G.P. Gornov V.F. Peskov N.V., Kolesova M.V. Burmistrov A.A., Timofeev N.A., Leskov V.A., Yurchenko I.A. Accounting for "Zero Curtain" Effect in GSHP Simulation. Applied Mechanics and Materials, **664,** pp 243-249, (2014).
4. Vasilyev G.P. heating and cooling supplies buildings and structures using low-grade thermal energy of the surface layers of the Earth (Monograph). Publishing house "Granitca". M., "Red star", (2006) 220 P.
5. Man, Y., Yang, H., Wang, J., and Fang, Z, (2012). In situ operation performance test of ground coupled heat pump system for cooling and heating provision in temperate zone. Applied Energy, **97**, 913-920, (2012).
6. VasilyevG.P., Peskov N.V., Gornov V.F., Kolesova M.V. Yurchenko I.A. Zoning of the Territory of Russia According to the Efficiency of the Use of Low-Grade Ground Heat for Heating. MATEC Web of conferences, **30** (2015).

MATEC Web of Conferences 77, 06011 (2016) DOI: 10.1051/matecconf/20167706011
ICMMR 2016

Solar Radiation Measurement Using Raspberry Pi and Its Modelling Using Artificial Neural Networks

Shanmuga Priya Selvanathan[1], Norman Carl Freudenberg[2], and Arunabh Borkataky[3]

[1] *Department of Chemical Engineering, Manipal Institute of Technology, Manipal University, Manipal – 576104.Karnataka.India.*
[2] *Department of Physics, Carl von Ossietzky Universität Oldenburg, Oldenburg, Germany*
[3] *Department of Chemical Engineering, Manipal Institute of Technology, Manipal University, Manipal – 576104.Karnataka.India.*

Abstract. The advent of solar energy as the best alternative to traditional energy sources has led to an extensive study on the measurement and prediction of solar radiation. Devices such as pyranometer, pyrrheliometer, global UV radiometer are used for the measurement of solar radiation. The solar radiation measuring instruments available at Innovation Center, MIT Manipal were integrated with a Raspberry Pi to allow remote access to the data through the university Local Area Network. The connections of the data loggers and the Raspberry Pi were enclosed in a plastic box to prevent damage from the rainfall and humidity in Manipal. The solar radiation data was used to validate an Artificial Neural Network model which was developed using various meterological data from 2011-2015.

1 Introduction

With the increasing energy demands and depleting fossil fuels, a primary importance has been given to the development of alternate renewable energy sources. Solar energy has emerged as the best alternative not only due to its sustainability but also because it is environment friendly and an abundant energy source [1]. The applications of solar energy is endless, including agriculture, electricity generation, heating and cooling etc. A high priority has thus been assigned by many countries for solar energy harnessing and utilisation [2]. A knowledge of the solar radiation incident on the surface of the earth is of utmost importance for engineers, scientists and architects associated with the design of solar energy systems [3]. Various solar radiation measuring instruments such as pyranometer, pyrheliometer, UV radiometer, etc are employed for collection of incident solar radiation data at a specific location. Owing to high cost of installation and maintenance of these instruments, solar radiation measurement is not feasible in all locations. The data obtained from the stations installed with these solar radiation measuring instruments are used to formulate different models for the prediction of solar radiation in different places.

The Raspberry Pi is a credit-card sized computer(Fig. 1) that originated in the Laboratory of University of Cambridge in 2006 and released by Raspberry Pi Foundation in 2012 [4]. The Raspberry Pi board with a dimension of 85.60mm x 53.98mm x 17 mm, weighs only about 45g and contains a processor, graphic chip, RAM (Random Access Memory) and various connectors and interfaces for linking external devices [5]. It was developed primarily for promoting programming in schools, homes and the third world countries. The cost of a raspberry pi board is only about 35$ and has the capability of a computer costing hundreds of dollars. Even though the use of Raspberry Pi in solar radiation measurement has been limited, it has been effectively used for various tasks across different fields.

Rahim *et al.* [6] developed a solar tracker system to determine the location of the sun in the sky. A webcam was used as the main sensor which was integrated with an image processing technique and rooted into a Raspberry Pi. The processing of the image was carried out by the Raspberry Pi inplace of a computer CPU.

Deepan *et al.* [7] used a Raspberry Pi for the control and movement of an artificial intelligent hand. The system was controlled by a Raspberry Pi with an attached camera module which was able to capture and mimic the natural motion of a human hand.

Divya Priya *et al.* [8] utilised the excellent image processing and video streaming capability of Raspberry Pi for designing and implementing a low cost and an efficient underwater vehicle which has the ability to measure parameters like temperature, pressure, speed and direction.

Ganesh V Bhat *et al.* [9] described the design of an Electrocardiogram (ECG) monitoring system using Raspberry Pi. A small, portable system was designed to find the malfunctioning of the heart.

Gurjashan Singh Pannu *et al.* [10] used Raspberry Pi as a processing chip for building a monocular vision autonomous car prototype. An ultrasonic sensor with a High Definition camera was used to provide necessary data from the real world to the car. Many existing algorithms such as lane detection, obstacle detection etc. were written onto the Raspberry Pi to provide control to the car.

No prior relevant work has been carried out on the integration of Raspberry Pi with solar measuring instruments and its measurement. This was the one of the chief driving force behind this project.

Figure 1. Raspberry Pi

ANN is an AI black box model which can be effectively utilized for estimation of non-linear functions, data sorting, optimization, model predictions, etc [11]. It provides an effective way of determining associations between a number of inputs and one or more outputs which may be empirical, linear or non-linear. The ANN architecture is divided into input layer, hidden layer, output layer, connection weight and biases, activation function and summation node. This algorithm has two main stages: training (learning) and recalling (generalisation). In the training stage, network weights and biases are utilized to generate the target output by reducing the error function. A learning algorithm is used to progress through the network and it is trained by epochs. An entire cycle of all the training data existing in the network is termed as epoch. The degree of error is measured in terms of mean squared error (MSE). At the end of each epoch the MSE is determined and the modelling is continued till the error is minimised.

M. Benghanem, *et al.* [12] estimated the daily global solar radiation in Al-Madinah (Saudi Arabia) using various Back Propagation(BP) ANN models. Different combinations of air temperature, relative humidity, sunshine duration and the day of the year were used as inputs for modelling. The model with air temperature and sunshine duration had the best results with a correlation coefficient of 97.65 %.

Ozgür Çelik, *et al.* [13] developed various ANN models with Levenberg-Marquardt (LM) algorithm to estimate the solar radiation at Turkey. The models varied in terms of the number and type of input parameters, learning algorithms, number of neurons in the hidden layer, correlation between network outputs and targets. The accuracy of the developed models were tested based on statistical error analysis methods. The correlation coefficient was found to be greater than 99%.

Mohandes *et al.* [14] developed an ANN model for predicting the global solar radiation in Saudi Arabia. Latitude, longitude, altitude and sunshine duration were used as inputs for the model. Various models were obtained by varying the number of neurons in the input, hidden and output layers. Mean Absolute Percentage error varied from 6.5 to 19.1%.

Manipal (13.347°N, 74.788°E), located in south-west India has a hot and humid climate throughout the year. Even though the two data loggers of available at were created to be water proof, they failed to withstand the humid environment in Manipal. The data from the solar radiation devices could only be accessed from the computer which was connected to the data loggers. The main objective of the project was to connect the system with a Raspberry Pi so that the data could be accessed remotely and also to insulate the entire system from rain and humidity. The solar data obtained using the Raspberry Pi will be used to validate various AI model for solar radiation estimation in the future.

2 Materials and methods

2.1 Pyranometer

It is a device which measures the global radiation data on a plane surface. Solar energy incident on a unit surface area per unit time is irradiance. The Campbell Pyranometer (CMP) series of pyranometers specifically comprises of a thermopile detector mounted on a base secured by a hemispherical glass dome. The working principle of a CMP pyranometer as mentioned by Muneer *et al* [15] is as follows: The thermopile reacts to the total power, unselective to the spectral distribution of the radiation absorbed. The heat produced by the absorbed radiation moves through a thermal resistance to the heat sink. The temperature difference over the thermal resistance of the disc is converted into voltage, which is read by the computer.

2.2 Pyrheliometer

It is used for the measurement of direct or beam solar radiation with a field of view which is limited to 5°. Usually, an absorber plate is located at the base of a collimating tube having precision apertures. The apertures protect the instrument and also act as a filters which pass solar radiation between 200nm - 4000nm. The absorber receives only beam radiation and a small amount of diffuse radiation falling within the acceptance angle of the instrument.

2.3 UV radiometer

It measures the UV irradiance; either UVA(351-400 nm) or UVB(280-315) or both. UVA radiation at the earth's surface exceeds the UVB radiation by 15-20 times. Kipp and Zonen's CUV5 UV Radiometer is used in this case for measuring the global UV irradiance. UVS A-T and

DOI: 10.1051/matecconf/20167706011

UVS B-T radiometers are respectively used for measurement of UVA and UVB radiations.

2.4 Data logger

Data logger is an electronic instrument that records and reports various changes in environmental conditions over time such as temperature, humidity, solar radiation etc. The LOGBOX SD is a weatherproof data logger specifically designed to be integrated with Kipp & Zonen solar radiation instruments. Multiple instruments can be connected to it at the same time as it is an 8 channel data logger. The data collected is stored on a 512 MB SD card that is included as standard. This device has an IP 65 weatherproof enclosure and an operating temperature range from -40°C to +60°C.

Table 1 shows the materials that were used for integration of Raspberry Pi with data logger system.

Table 1. Materials required for the integrated Raspberry Pi - Data Logger system

Item	No of Units
Raspberry Pi (bare)	1
Charger to power Raspberry Pi	1
32GB class 10 SD card for OS and data storage	1
Plastic Box (waterproof container)	1
Doormat (water insulation for box)	1
Adapter to power data logger without battery	1
Pure Al Heat Sink for Raspberry Pi	1
Jumper wire male to female	1
Digital Thermometer and Humidity Sensor	1
USB to RS232 Adapter	2
Jumper Wire female to female	1
Raspberry Pi Case	1
Multi-pole power socket	1
Serial cable male female	2
85 Silica gel packs (5g)	1

All the measuring devices were connected to one of the two data loggers. These collect the data in a set interval (1 minute presently) and write it to RS232 and also to the SD cards placed inside them. Using a RS232 to USB adapter both the data loggers are connected to a Raspberry Pi that collects the data and transmits it to an internal server. The data can be accessed from any device connected to the university Local Area Network. The measured solar radiation data can be accessed from http://172.16.62.140/solar/data/. The Raspberry Pi unit can easily be accessed using a SSH client. In this case Putty was used as the client.

Fig. 2 shows the flowsheet depicting the connections made.For developing the ANN model, five input parameters were used namely minimum temperature, maximum temperature, average temperature, altitude and sunshine duration. The monthly average values of these parameters from 2011-2015 for Manipal were collected from National Renewable Energy Laboratory [16]. The monthly average solar radiation (Kwh/m2/day) for these years was used as the output to formulate the model.

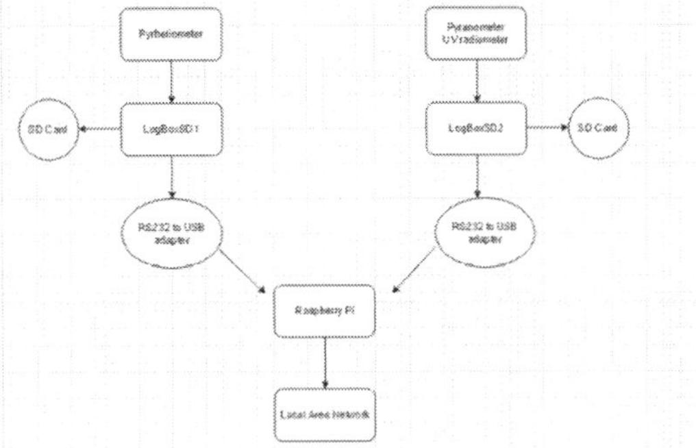

Figure 2. Flowsheet depicting the connections made

The neural network fitting tool nntool of Matlab software version 2013a was used for developing the ANN model. The model was trained using the Levenberg-Marquardt (LM) algorithm. 70% of the dataset was used for training and the rest were divided equally for testing and validation. A flowchart of the modelling process is shown in Fig. 3.

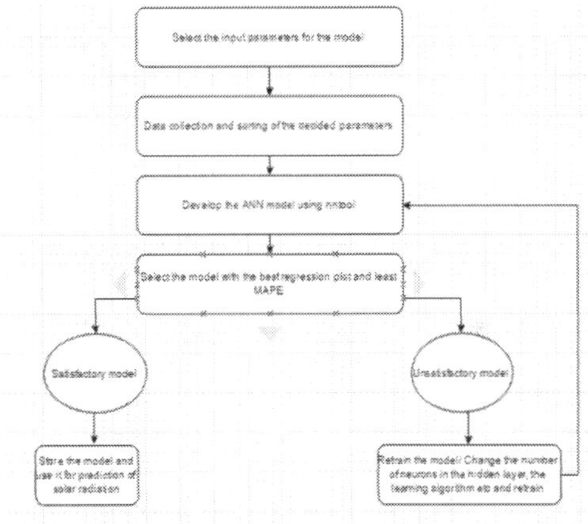

Figure 3. Flowchart of the modelling process

Figure 4. Sun tracker installed with solar radiation monitoring instruments

3 Experimental setup

The setup of the solar radiation measurement instruments located at Innovation Center, MIT Manipal is shown in

 DOI: 10.1051/matecconf/20167706011

Fig. 4. The pyranometer, pyrheliometer and UV radiometer are connected into the sun tracer assembly.

The entire system is connected to two data loggers which inturn is connected to the Raspberry Pi. Table 2 mentions about the data loggers to which instrument is connected with.

Table 2. Connections of the instruments with the data loggers

Sl. no	Instrument	Data Logger
1	Pyranometer	LogBoxSD2
2	Pyrheliometer	LogBoxSD1
3	UV Radiometer	LogBoxSD2

4 Results and discussion

The entire data logger- Raspberry Pi system was placed inside an insulated plastic box to protect the unit from extreme weather conditions as shown in Fig. 4. The box was placed few centimetres above the ground for better air circulation and to prevent direct contact with water in case of heavy rainfall. In order to sustain humidity, silica gel was placed inside the box. The level of silica gel remaining is periodically checked using the DHT22 temperature and humidity sensor. The batteries of the data loggers had started to corrode and were replaced with external grid powered voltage adapters. Instead of using a single RS232 to USB adapter two adapters are used in order to prevent interference of the two RS232 signals.The box consists of the Raspberry Pi, the two data loggers, the adapters and various components used to complete the connections.

The continuous solar radiation data obtained was documented and analysed. Fig. 5 and Fig. 6 show the solar radiation trend in Manipal for a typical sunny day. According to the obtained graphs the sunrise in Manipal is around 6:30 a.m. and the sunset is close to 7:15 p.m. From Fig. 8 the average daily global solar radiation for a typical day in March is around 620 W/m^2. The average daily UV radiation from Fig. 6 is around 25.28 W/m^2. The average values are calculated for a time range of 8 a.m. to 6 p.m.

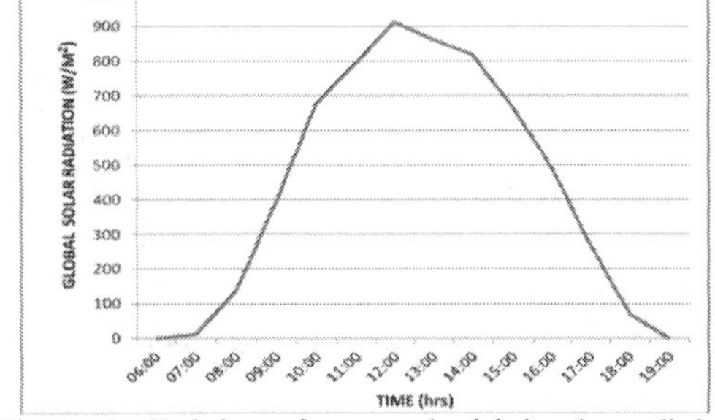

Figure 5. Variation of measured global solar radiation throughout the day

The regression plot obtained is shown in Fig. 7. It is a measure of the accuracy of the formulated model. The plot describes the relation between the measured solar radiation (target) and the estimated solar radiation (output). A regression value of 94.5% was obtained which shows a good fit for the developed model.

The performance plot is shown in Fig. 8. where the Mean Sqaure Error (MSE) decreases with the increse in the no of epochs. The Mean Absolute Percentage Error(MAPE) for the formulated ANN model was calculated to be 5.86%.

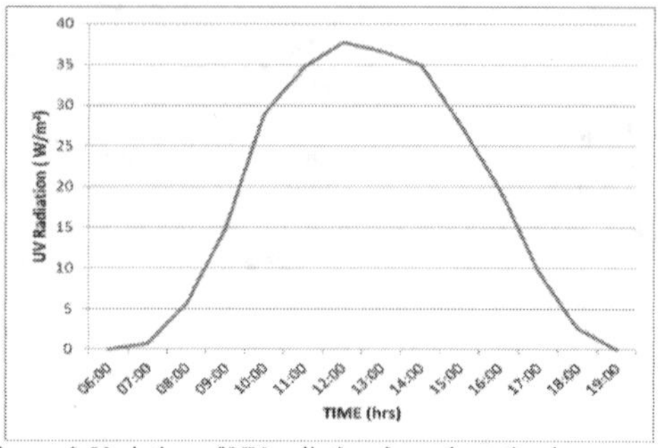

Figure 6. Variation of UV radiation throughout the day

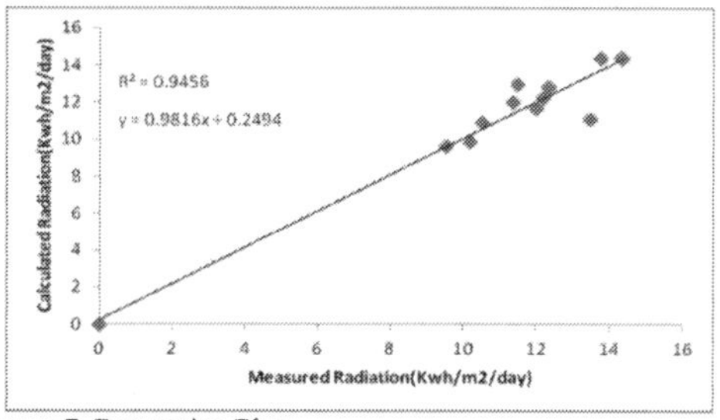

Figure 7. Regression Plot

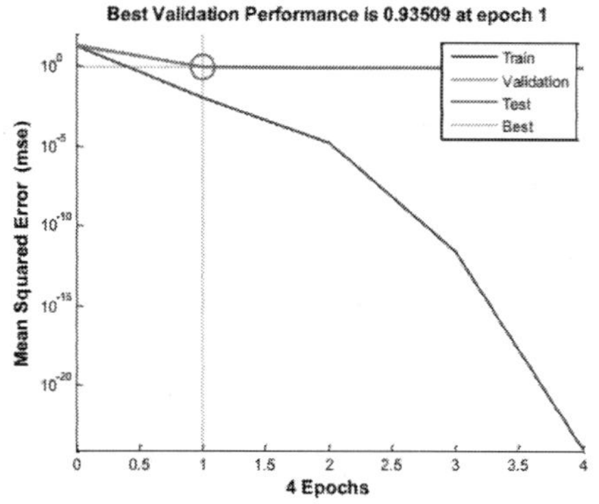

Figure 8. Performance Plot

5 Conclusions

A successful integration of the solar radiation equipments with a Raspberry Pi was made, whereby remote access to the data was enabled. The entire Raspberry Pi - Data Logger system was placed inside a plastic container which was able to withstand the hot and humid climate of Manipal.

The solar radiation data is unavailable in many meteorological stations worldwide, mainly due to high cost of calibration and maintenance of the solar radiation measurement instruments. The difficulties involved in the solar radiation measurement has led to the formulation of various models for its prediction. An AI model was developed using various meterological data and it showed fairly accurate predictions of solar radiation at Manipal with a R^2 value of 0.945 and MAPE of 5.86%.

The future work of this project is mentioned below:

- Connect the Raspberry Pi over the Wifi at MIT, Manipal so that a LAN connection is not required everytime to access the data.
- Develop different ANN models by changing the type and the number of input parameters and validate each model.
- Develop different Artificial Intelligence models such as Genetic Algorithm so that the model can be explicitly expressed in terms of its input parameters.
- Determine the effectiveness of the developed models in predicting the solar radiation of places excluding Manipal.

Acknowledgement

We would like to express our sincere gratitude to Innovation Center, Manipal Institute of Technology, Manipal for providing the required facilities for carrying out the project. We express our sincere gratitude to Manipal University for providing us the financial support for the purchase of Kipp and Zonen solar radiation instruments from Netherland and also the various components required to make the connections including the Raspberry Pi.

References

1. R.Akikur, R.Saidur, H.Ping, K.Ullah. Comparative study of stand-alone and hybrid solar energy systems suitable for off-grid rural electrification: a review. Renew. Sustainable Energy Rev., **27**(3), 738-752, (2013).
2. Kasra Mohammadia, Shahaboddin Shamshirband, Chong Wen Tong, Khubaib Amjad Alam, Dalibor Petkovic´. Potential of adaptive neuro-fuzzy system for prediction of daily global solar radiation by day of the year. Energy Conv. and Mgmt. **93**, 406-513, (2015).
3. M. Benghanema, A Mellit, S N Alamri. ANN-based modelling and estimation of daily global solar radiation data: A case study. Energy Conv. and Mgmt. **50**(7), 1644-1655, (2009).
4. J. D. Brock, R. F. Bruce, M. E. Cameron. Changing the world with a Raspberry Pi. Journal of Computing Sciences **29**(2), 151–153, (2013).
5. Vladimir Vujovic, Mirjana Maksimovic. Raspberry Pi as a Sensor Web node for home automation. Computers and Electrical Engg., **44**, 153-171, (2014).
6. R Abd Rahim, M.N.S.Zainudin, M.M.Ismail, M.A.Othman,. Image-based Solar Tracker Using Raspberry Pi. Journal of Multidisciplinary Engineering Science and Technology **1**(5), 369-373, (2014).
7. R. Deepan, Santhana Vikrama Rajavarman, K. Narasimhan. Hand Gesture Based Control of Robotic Hand using Raspberry Pi Processor. Asian Journal of Scientific Research **8**(3), 392-402, (2015).
8. Divya Priya, I.Harish. Raspberry PI Based Underwater Vehicle for Monitoring Aquatic Ecosystem. International Journal of Engineering Trends and Applications **2**(2), 65-71, (2015).
9. Ganesh.V.Bhat, Anandraddi Naduvinamani. Real Time ECG Acquisition System using Raspberry PI. International Journal of Engineering Sciences and Research Technology **3**(6), 464-468, (2014).
10. Gurjashan Singh Pannu, Mohammad Dawud Ansar, Pritha Gupta. Design and Implementation of Autonomous Car using Raspberry Pi. International Journal of Computer Applications, **113**(9), 22-29, (2015).
11. Amit Kumar Yadav, S.S. Chandel. Solar radiation prediction using Artificial Neural Network techniques: A review. Renewable and Sustainable Energy Reviews, **33**, 772-781, (2014).
12. M. Benghanema, A. Mellit, S.N. Alamri. ANN-based modelling and estimation of daily global solar radiation data: A case study. Energy Conversion and Management, **50**(7), 1644-1655, (2009).
13. Ozgür Çelik, Ahmet Teke, H. Bas Yildirim. The optimized artificial neural network model with Levenberg Marquardt algorithm for global solar radiation estimation in Eastern Mediterranean Region of Turkey. Journal of Cleaner Production journal, **1**, 1-12, (2016).
14. Mohandes M, RehmanS, Halawani To. Estimation of global solar radiation using artificial neural networks. Renewable Energy, **14**, (1-4), 179-184, (1998).
15. Muneer T, M.Asif, J.Kubie. Generation and transmission prospects for solar electricity: UK and global markets. Energy Conv. & Mgmt., **44**(1), 35 - 52, (2003).
16. http://www.nrel.gov National Renewable Energy Laboratory

MATEC Web of Conferences 77, 06012 (2016) DOI: 10.1051/matecconf/20167706012
ICMMR 2016

Atmospheric icing status and type of southwest China networks

Yawei Li[1], Xinghai Zhang[1], Peng Zhou[2], Lei Sun[2] and Longchen Liu[1]

[1]*Electric Power Research Institute of State Grid Sichuan Electric Power Company, 610072 Chengdu, China*
[2]*Graduate School at Shenzhen, Tsinghua University, 518055 Shenzhen, China.*

Abstract. Transmission line disconnection, tower collapse and insulator flashover caused by icing seriously threaten power system security. Ice type and state of transmission lines, which vary a lot with terrain and climate, in typical areas such as Daqing-ridge, Yak Mountain and Erlang Mountain in Sichuan Province in South China were investigated in this paper. It is shown that mixed-phase ice with obvious layered structure, low density, strong adhesive force and windward-side growth is the main type of ice threatening the security of transmission lines and insulators. There is more ice on the ends of insulators than other areas in severe cases, where all sheds of the insulator is bridged by ice. Besides, temperature, humidity and precipitation intensity are main factors influencing the icing process. As a result, terrain and climate play a leading role in determination of icing type and severity.

1 Introduction

In China, geographical distribution of energy presents a serious imbalance. The load centers are in concentrated distribution in the southeastern coastal strip, far away from southwest regions where clean hydropower resource is rich. Therefore, in order to meet the energy needs of economic production, EHV and UHV long distance transmission is a must for China. Under the influence of geographical environment in the south-western China, EHV and UHV transmission lines will inevitably go through high altitude areas, canyons, rivers and other regions with complex geography and climate, which leads ice-covered insulator to high flashover risk.

For example, Sichuan, a province with vast territory in south-western China, crosses the first and the second step in the three-step terrain of China, which makes the climate conditions of Sichuan power grid vary widely. In western Sichuan plateau areas, 20mm and above heavy ice areas of transmission lines widely distributes. For instance, AC 500kV Gan-Shu line go through 10mm ice area from 019 tower to 022 tower, through 20mm ice area from 022 tower to 023 tower, through 40mm ice area from 023 tower to 024 tower, through 60mm ice area from 024 tower to 033 tower, through 40mm ice area from 033 tower to 040 tower, and through 20mm ice area from 040 tower to 044 tower It shows that icing conditions of transmission lines in Sichuan region are very complex, which makes the anti-icing modification of transmission lines very difficult.

To reduce the loss of icing disaster, Sichuan Electric Power Company carried out a lot of anti-icing modification of transmission lines. By the end of 2012, the modification of 17 500kV lines and 2 220kV lines had been completed, which enhanced the anti-icing ability of Sichuan power grid. However, under the influence of regular change of climate conditions in Sichuan Province, the anti-icing modification is a long-term and arduous work in some degree. Under the influence of several times of cold-air outbreak from December of 2013 to March of 2014, ice flashover of 500kV transmission lines occurred 23 times in total, including 2 times from Gan-Shu 1st and 2nd line, 9 times from Cheng-Mu 1st and 2nd line. Since 1950s, many countries in the world, China included, have researched on icing of transmission lines and achieved many important results [1-7]. But so far, icing mechanism on transmission lines and insulator strings, flashover characteristics of ice-covered insulator strings and many other problems are still far from resolved [8-16], and there is almost no understand about the icing reasons of power grid under microtopography and micrometeorology conditions [9].

It can be seen that, under the influence of complex topography and changeable weather conditions, icing situation in south-western China power grid is complex [17-19]. Therefore, in order to understand ice forms, characteristics and formation conditions under microtopography and micrometeorology conditions in south-western China, the paper conducted the research on the icing conditions of several selected typical regions, including Daqing-ridge in Xichang City (altitude 2500m), Yak Mountain (altitude 3100m), Erlang Mountain (average altitude 3000m) and Luding County. Icing process of transmission lines and conditions in these areas is observed and recorded and icing accidents in these areas in recent are analysed and researched, in order to develop a targeted anti-ice strategy for power grid.

DOI: 10.1051/matecconf/20167706012

2 Icing morphology on transmission lines and insulators

At the initial stage of icing, ice, especially on the transmission lines, grows in bifurcation. Fig. 1(a) shows the initial stage of icing on the 500kV Cheng-Mu 1st line in Daqing-ridge.

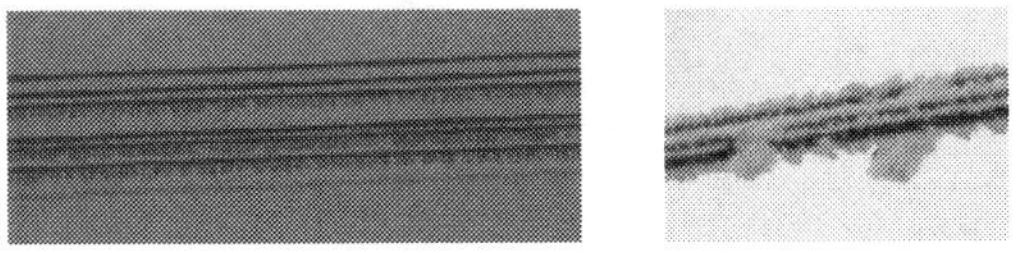

(a) initial stage of icing (b) later stage of icing

Figure 1. Structure of ice on the wire at early and late stage.

Fig. 1(b) shows the later stage of icing on the 500kV 4-bundle conductor. The icing is loose and porous with layered structure, closely sticked to conductor surface.

Ice is formed by condensation water droplets or ice crystal stacked in the object surface. Drops and ice crystal in the air are wrapped with forward movement of air (wind), so ice distribution on insulator strings has a close relationship with wind direction. The upwind side of insulator is not blocked and drops and ice crystal are more easily to cumulate, so ice on conductors and insulators is mainly distributed on their upwind surface, as shown in Fig. 2 below.

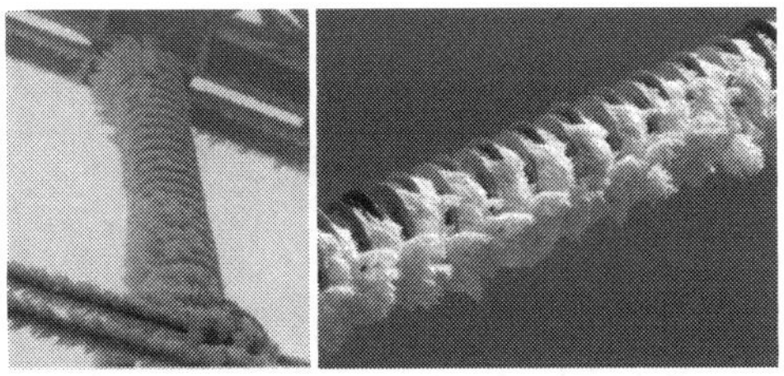

(a) "I" string (b) one hand of "V" string

Figure 2. Icing conditions of transmission lines and insulators.

The left picture of Fig. 2 is an "I" string consist of bell-shaped glass insulators on tower 131 in Shi-Ya 4th line, while the right picture is a "V" string consist of three umbrella-shaped porcelain insulators on Tower 69 in ±800kV Jin-Su line. It shows that icing is more serious in windward side than that-in leeward side

Fig. 2 also shows icicles exist in the ice, which implicates that melting and freezing due to temperature changes exist in the icing process. This is also one of the way to form-layered mixed-phase ice under natural icing conditions. This phenomenon not only exists in the icing of conductors, but also commonly exists on the surface of insulator strings, as shown in Fig. 3.

In Fig. 3, the left picture shows the windward side of a "V" string of ±800kV Jin-Su line, and the right picture shows the bottom side of an "II" string of 500kV Yue-Pu line. It can be seen that icicles exist widely along different insulator strings. The existence of icicles increases the probability of bridged insulator and flashover.

(a) V string (b) II string

Figure 3. Icicles along insulator strings.

Fig. 4 shows a flowered glass insulator string in the period of icing and melting. At early stage of not serious icing, the large sheds can prevent entire string from being bridged. However, under serious icing conditions, the large and small sheds are all covered by ice, thus the block effect of large sheds is limited.

Figure 4. Icing conditions of flower arrangement insulator strings at early and late stage.

Fig. 5 shows that icing is more on ends of insulator strings, where the electrical field is stronger.

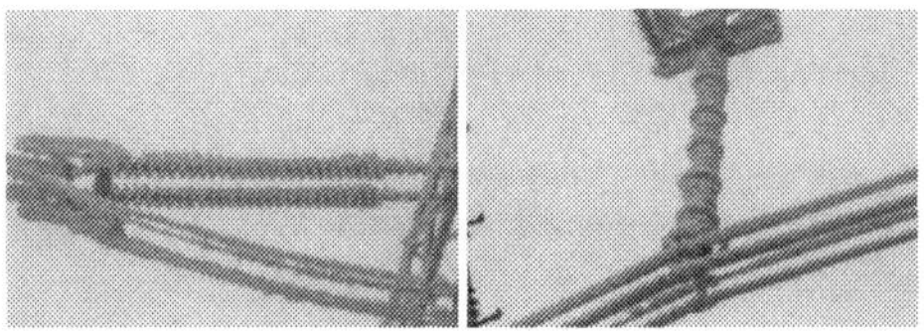

(a) tension strings (b) I string

Figure 5. Axial distribution of ice along insulator strings.

3 The type and characteristics of icing on lines and insulators

In the field research of icing circumstances on lines and insulators in Daqing-ridge, Erlang Mountain and Luding County, it is found that icing has layered structure, as shown in Fig. 6. The outer layer of ice has a loose and porous structure and white appearance, while the inner layer of ice, which adheres to conductors or insulator sheds, has a dense structure and transparent or semi-transparent appearance.

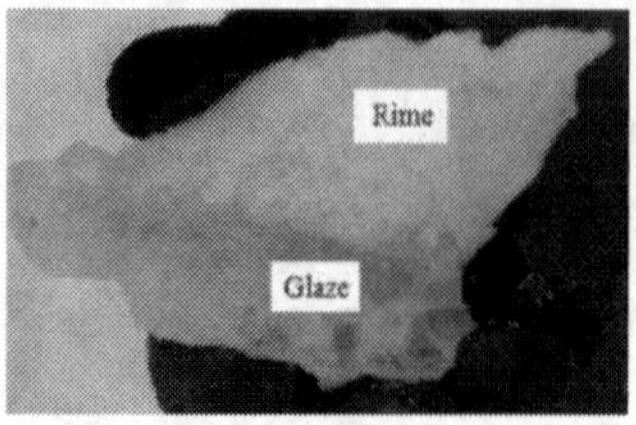

Figure 6. Structure of mixed-phase ice.

The density of ice dropped from tower was measured by drainage in field research. The results show that the density of ice ranges from 0.5 g/cm^3 to 0.6 g/cm^3.

Based on characteristics above, it can be concluded that ice on transmission lines and insulator strings mainly consists of mixed-phase ice (hard rime) [9, 16].

In the research of icing conditions in Yak Mountain, it is found that ice in this area mainly consists of rime, and its appearance is shown in Fig. 7.

Figure 7. Rime ice in the Yak Mountain Weather Stations.

The ice shown in Fig. 7 has a weak adhesion and small density, and it is very easy to drop down from conductors or insulators. The statistics of icing from 2013 to 2015 show that the thickness of ice is not more than 50mm, and no flashover accidents happened due to rime, while several accidents because of transmission lines disconnection and tower collapse happened in Daqing-ridge. At another spot, several flashover accidents of ice-covered insulators happened in Erlang Mountain.

The conductivity of ice example dropped from transmission towers in Erlang Mountain was measured. The result is shown in table 1. In the table, the melting water of glaze and rime is obtained by separating the mixed-phase ice.

Table 1. Conductivity of melting ice water of snow and ice at 20℃ (μs•cm^{-1}).

Sample spot	rime	glaze	snow
Shi-Ya 4th No.132	66.25	14.19	48.32
Shi-Ya 2nd No.115	64.22	15.13	47.54
Shi-Ya 3rd No.130	67.83	15.61	50.63

4 The analysis of ice forming conditions

4.1 Temperature and humidity

In 2014 at Shi-Ya 4th line, icing process corresponds to low temperature periods, which was observed at transmission line monitoring spots, as shown in Fig. 8. In addition, the lowest temperature in January is -6℃, and the corresponding maximum ice thickness is 6.5mm. While the lowest temperature in February is -4℃, and the corresponding maximum ice thickness is 11mm. This implicates that icing doesn't go more serious when temperature goes down.

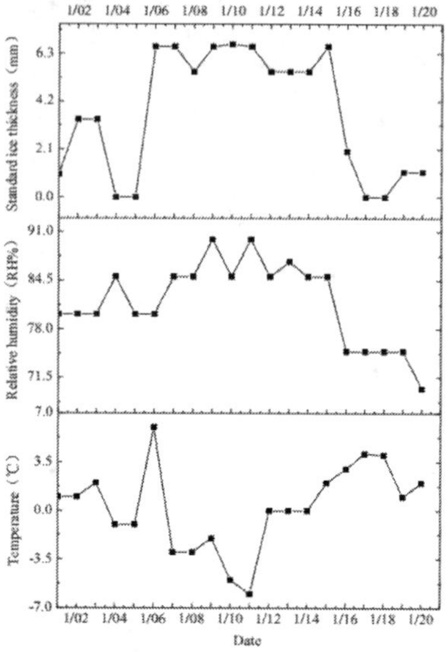

Figure 8. monitoring results of Shi-Ya 4th line.

The research of Yak Mountain 2nd line shows that, the formation and increase of ice has a close relationship with temperature change, at the same time, humidity at a relatively high level is necessary. About Jan 10th 2014 in Yak Mountain 2nd line, icing increased when temperature changed from 3℃ to 0℃, as shown in Fig. 9. In the meantime, relatively high humidity provided water droplets and ice crystal for icing. This concludes that relatively low temperature and necessary humidity are basic factors for icing. In the icing process on Feb 17th and 18th, similar phenomenon occurred that temperature fell correspondingly to icing. Despite that the lowest temperature -3℃ during the two days is lower than that of Jan 10th 0℃, however ice thickness in February 1mm is thinner than that in January 2mm.

As temperature rose, ice amount, at first rose and then fell. Maybe it's because the dry growth and wet growth is sensitive to temperature changes. At lower temperature, dry growth plays a more significant role in icing and icing water has a relatively poor mobility, while at higher temperature, wet growth become more important in icing and icing water has a stronger mobility. But at both circumstances, icing growth is inhibited. This implies that there is a temperature range where it's most favourable for icing.

Besides, the contrast of research results in January and February implies that humidity has significant influence on icing. On the whole, icing amount in a month with higher humidity is bigger than that in a month with lower humidity.

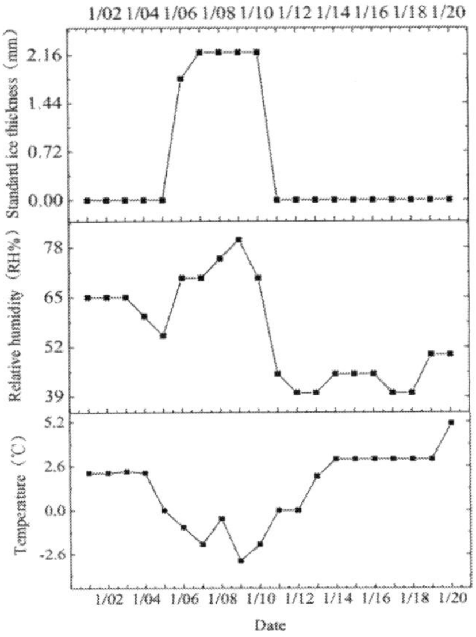

Figure 9. Monitoring results of Yak Mount 2nd line.

4.2 Precipitation and morphology

The altitude of the weather station in Yak Mountain is about 3100m, and the station is at summit where other summits around are lower. As a result, the weather station is often heavily foggy, has a little precipitation and low temperature. So icing in the area mainly consists of rime by dry growth, and wet growth seldom exists.

Daqing-ridge is a typical alpine gorge, and the gorge is open to Qionghai Lake and Xichang urban area where the terrain is more open. The air carried water vapor from Qionghai Lake is lifted and forms a quasistationary front. This results in increase precipitation in the gorge in winter and rain, snow and heavy fog appears alternately, which is beneficial to icing accumulation with dry and wet growth. As a result, icing on the transmission lines and towers is often very serious.

From Feb. 13th to Mar. 7th in 2014, 9 ice flashover occurred on 500kV Gan-Shu 1st and 2nd line. The flashover fault section is in the north side of the mountain in Erlang Mountain, where the altitude ranges from 2326m to 2889m. The terrain is at the cross of mountains and river valleys and climate change there is severe. At that spot, icing is common in winter and heavy fog is common at night. It is a typical easyily-icing microtopography and micrometeorology region.

It can be seen that icing often occurs at summits with high altitude, where precipitation is small and temperature is low. It's easier to form rime, while in the gorges and at the cross where mountains and river valleys joint, heavy mixed-phase ice is easier to form under the influence of quasi-stationary front.

5 Conclusion

Aimed at the circumstance and the diversity of icing appearance and types on power grid under complex morphology and geological condition in south-western China, the article undertook researches on icing conditions in Daqing-ridge, Yak Mountain and Erlang Mountain. The conclusions are as follows:

1)Icing on Sichuan power grid mainly consists of mixed-phase ice with inner glaze layer and outer hard rime layer.

2)As temperature rises, icing amount increases first and then decreases. At the same temperature, icing amount increases when humidity increases.

3)Rime is easier to form at summit, while mixed-phase is easier to form in gorges and the cross of mountains and river valleys.

References

1. Sun Caixin, Sima Wenxia, Shu Lichun. Atmospheric environment and electrical insulation. M. Beijing: China Electrical Press, 2002.
2. Shu Lichun, Jiang Xingliang, Tian Yuchun, et a. AC flashover performance and voltage correction of four types of iced composite insulator at altitude 4000m above. J. Proceedings of the CSEE, 2004, 24(1): 97-101.
3. Li Peng, Fan Jianbin, Su Zhiyi, et al. Insulation design in icing areas for 750 kV transmission lines [J]. Electrical Equipment, 2002, 8(3): 9-133
4. Farzaneh, M., Chisholm, W. A. Insulator icing test methods, selection criteria and mitigation alternatives [J]. IEEE Transactions on Dielectrics and Electrical Insulation, 2007, 14(6): 1462-1473.
5. Farzaneh, M. Atmospheric Icing of Power Networks. M. Berlin: Springer, 2008.
6. S. Taheri, M. Farzaneh, I. Fofana. Improved dynamic Model of DC Arc Discharge on ice-covered post insulators surfaces [J]. IEEE Transactions on Dielectrics and Electrical Insulation, 2014, 21(2): 729-739.
7. Azam Nekahi, Shahab Farokhi, Masoud Farzaneh, et al. Arc Energy and temperature during its propagation over ice-covered surfaces [J]. IEEE Transactions on Dielectrics and Electrical Insulation, 2014, 42(1): 114-119.
8. Farzaneh M., Kiemicki J. Flashover problems caused by ice build-up on insulators [J]. IEEE Electrical Insulation Magazine, 1995, 11(2): 5-17.
9. Wu Xiangdong, Xu Tianyong, Ai Zhiping. The influence of ice-covering on 500 kV transmission line in micro-terrain and micro-weather area and the countermeasure. J. High Voltage Engineering, 2002, 28(7): 55-57.
10. Ndiaye I., Farzaneh M., Fofana I., Study of the development of positive streamers. IEEE Transactions on Dielectrics and Electrical Insulation. J. 2007, 14(6): 1436-1445.
11. Farzaneh M., Chisholm A. W. Insulators for Icing and Polluted Environments. M. Hoboken, New Jersey: John Wiley & Sons, Inc., 2009.

12. Masoud Farzaneh, William A. Chisholm. 50 years in icing performance of outdoor insulators [J]. IEEE Electrical Insulation Magazine, 2014, 30(1): 14-24.
13. Zuoming Xu, Tao Xu, Wei Hu, et al. Study on dynamic flashover model of ice-covered insulator string [C]. 2013 Annual Report Conference on Electrical Insulation and Dielectric Phenomena, 2013, Shenzhen, China.
14. Kangning Wang, Tianzheng Wang, Yu Deng. The mechanism of salt migration phenomena in cooling water and its impact on ice flashover [C]. 2013 IEEE International Conference on Solid Dielectrics, 2013, Bologna, Italy.
15. S. Taheri, M. Farzaneh, I. Fofana. Equivalent Surface Conductivity of Ice Accumulated on insulator during development of AC and DC Flashover arcs [J]. IEEE Transactions on Dielectrics and Electrical Insulation, 2013, 20(5): 1789-1888.
16. Kangning Wang, Tianzheng Wang, Yu Deng, et al. Melting process of multi-Icicles and its flashover performance [C]. 2013 Annual Report Conference on Electrical Insulation and Dielectric Phenomena, 2013, Shenzhen, China.
17. László E. Kollár, Masoud Farzaneh, Pierre Van Dyke. Modeling ice shedding propagation on transmission lines with or without interphase spacers [J]. IEEE Transactions on Power Delivery, 2013, 28(1): 261-267.
18. Yu Deng, Zhidong Jia, Hao Jiang, et al. Melting process of different icing morphology and its impact on ice flashover [J]. High Voltage Engineering, 41(2): 669-679.
19. Xingliang Jiang, Bingbing Dong, Zhijin Zhang, et al. Research and development on flashover of ice-covered insulators [J]. High Voltage Engineering, 40(2): 317-335.

 DOI: 10.1051/matecconf/20167706013

Mine Ventilation System Variable Resistance Regulate Air Quantity Economy and Its Impacts on Artisanal Mine in Ivory Coast

Kouame Arthur Joseph Kouame, Fuxing Jiang, Yu Feng and Sitao Zhu

School of Civil and Environmental Engineering, University of Science and Technology Beijing, 30 Xueyuan Road, Haidian District, Beijing 100083, PR China

Abstract. The artisanal gold mining is one of the major illegal activities in Ivory Coast. Thousands of indigents and foreigners including men, women and children are involved in this dangerous activity. Contracting contagious diseases such as HIV/AIDS and multiple lung diseases are commonplace on the artisanal mining sites due to air pollution. For this reason, the differential analysis, research mine ventilation system by changing the working windage adjustment for air flow of the economy, identifies factors, obtained discriminant variable resistance air volume control economic quantitative criteria for underground actual situation of joint variable resistance method and economic variable resistance than the concept of the development of effective mine ventilation system to improve air quantity adjusting rheostat economical way. The findings of this study can help lead to better safety practices in mining in order to improve the health and safety of the miners who are involved in artisanal mining activities in Ivory Coast and other mining sites over the world.

1 Introduction

1.1 Definition

According to the definition, artisanal and small-scale mining refers widely to miners using inadequate technical materials to run mines. Due to advancement in technology nowadays, some small miners no longer use artisanal methods, and they run huge operations ([1], [2]). In order to depict small, medium, large, and informal mining activities which are mostly run by illegal miners who carried out mining with low technology or minimal machinery, it is common to use the term artisanal mining [3]. Over 100 million people around the world, mainly in developing counties, rely on this sector for income. Ivory Coast is a country in West Africa with an annual gold production that rose from 10 tons in 2014 to 23.5 tons in 2015. The artisanal mining mainly refers to mining operated by individuals, groups or communities (foreigners and indigenes) illegally due to fact that illegal miners mostly don't acquire legal status. The active involvement of many people in artisanal mining activities in Ivory Coast is mainly due to the high rate of abject poverty and the low price of agricultural produce. In order to deal with poverty and better their lives, the people engage in artisanal mining in all the parts of the country. Even when the artisanal mining sector has harmful side effects (environmental, social and financial challenges) that undermine the country's development, it has the potential to contribute to the development of the state through the provision of employment, increase of local purchasing power, stimulation of local economic growth and slowing of urban migration.

At present, for mine ventilation, air volume control studies focus mainly on the stability of the ventilation system ([4], [6], [7]) with little research on mine ventilation system and air volume control economy, which makes it difficult to judge mine ventilation system and air volume control economy. In this paper, we study mine ventilation system and air volume control to determine the impact of factors that determine air quality and to identify quantitative conditions so as to formulate reasonable measures to improve mine ventilation system and air volume control rheostat economy for efficient production and safety of coal mines and gold mines. Adopting an effective mine ventilation system is of critical benefit to Ivory Coast. Utilizing natural or mechanical air pressure, the mine ventilation system is designed to provide fresh air as well as expel foul air. And with more people getting involved in mining activies in Ivory Coast, underground mining is inevitable in the coming 10 years. There are many differences between open pit mining and underground mining. When the mining depth increases, the temperature becomes higher than the surface ground and will increase as it gets deeper. Aditionally, mining activities always generate hazardous gases and dusts—diesel emissions, blasting fumes, radiation, dusts, battery emissions, and many other contaminants—which are harmful to the workers and will cause some occupational diseases, such as silicosis, or even worse. The following is the main components of surface ground air (Table I).

Table 1. Main components of surface ground air.

Component	Mass concentration / %	Bulk concentration / %
Oxygen (O_2)	23.17	20.90
Nitrogen (N_2)	75.55	78.13
Carbon dioxide (CO_2)	0.05	0.03
Rare gases	1.23	0.94

Once the air comes into mining shafts, both its compoents and properties will change, the concentration of oxygen decreases while the concentration of carbon dioxide increases; various toxic gases and dusts mix and there is change in the air temperature, moisture and pressure. Besides O_2, N_2, CO_2 and water vapour, the main components of the underground mining air are CH_4, CO, H_2S, SO_2, NO_2, NH_3, H_2 and dusts, with CO, H_2S, SO_2 with NO_2 being the most harmful component. Thus, underground mining would not be possible without a ventilation system.

In underground mining, air is necessary not only for breathing but also to disperse chemical and physical contaminants (gases, dusts, heat, and humidity). Ventilation is the control of air movement, its amount, and direction. It is sometimes described as the lifeblood of a mine, the intake airways being arteries that carry oxygen to the working areas and the returns veins that conduct pollutants away to be expelled to the outside atmosphere. Without an effective ventilation system, no underground facility that requires personnel to enter it can opreate safely. Literature[7], [4] shows that an effective mine ventilation system possesses such assets as follows: dilution and expulsion of 80%-90% of the hazardous gas, reduction of 70%-80% of the fine particles and ousting of 60%-70% of the heat of the working faces. These merits offer protections for the mining activities and also some suggestions as to post-medical checks, training, strategy of controlling dust, dust masks wearing, and monitoring of gas in order to decrease the working incidents. Indeed, it can also improve the miners' working productivity at a lower cost.

1.2 Ventilation mining in China

China's research for ventilation system started with the establishment of the Peoples Republic of China. Since the 1950s, researches in mine ventilation theory and technology have achieved some major breakthroughs including the following:
(1) Extensive research and measurement on the roadway ventilation resistance;
(2) The establishment of a variety of work surface turbulence equations and mass transfer analysis of pollutant concentration calculation method, which provides a theoretical basis for the method of calculation of the amount of wind;
(3) application of computer calculation and analysis of complex ventilation network, which provides an effective method of analysising mine ventilation system;
(4) developed Jet Ventilation techniques and the applications to use wind flow pressure to regulate and direct airflow control technology
(5) deepening the research of unsteady flow law in the case of mine fire, establishment of a number of typical wind flow control program.

2 A mine ventilation system air volume control rheostat factors affecting economy

2.1 Qualitative analysis

In coal mines, ventilation systems for air volume often due to the mining face of change and transfer of gas emission and other natural conditions change and change the fan performance changes itself. In order to ensure the availability of air in the mine, there is need for timely supply of air volume control. Variable resistance, which includes the economic benefits of air-flow method are constrained mainly by two factors. First, to meet the demand for air volume after the economic gains which the scale of production mainly determines. The second is to change the drag caused by the additional energy consumption and cost of the project. The energy consumption is mainly used for ventilation ductwork to overcome resistance and maintain air flow stability. The law of resistance shows that only the air flow and pressure drag on.

In summary, the impact of the mine ventilation system air volume control rheostat law economic factors includes wind, drag, engineering and production scale.

2.2 Branch changes on the economic impact of wind resistance

In the mine ventilation system management process, air volume changes at the mining area producing a continuous change. When the wind through the natural points cannot meet the need, it often used to change the drag method to regulate air flow to meet mine's demand for wind. Most of our coal mining uses wings; each wing has a separate return air tunnel, the implementation of partition is airy and the simultaneous production of multiple mining areas, each with a down hole location constitutes a complex wind parallel networks. To facilitate the analysis, the mine ventilation network simplification of Fig. 1 consists of three parts:

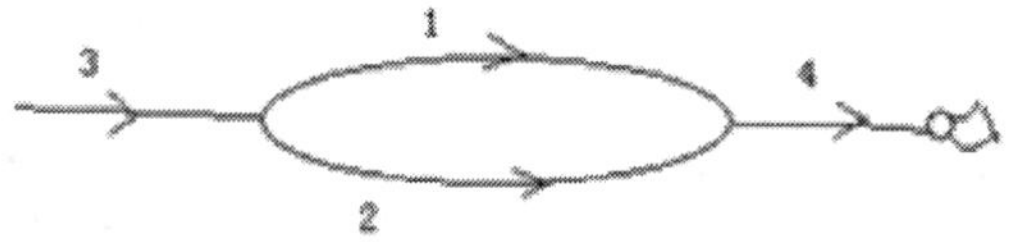

Figure 1. Simplified ventilation network graph

Mine's total inlet branch 3, equivalent drag r3, mine's total return air branch 4, equivalent drag r4, mine segment with the wind segment branch 1, 2, the equivalent wind resistance r1, r2. The need to increase the air flow branch for the branch including a large air volume r1, including the other branches of the equivalent branch r2.

Where r1, r2 respectively Mine Main Fan constitute two independent circuits.

2.2.1 The increasing resistance adjustment

To increase resistance to increased air volume regulator, that is, greater wind resistance of the branch, a large amount of air is required to increase the air volume branching approach because of mine total drag.

$$R = r_0 + \frac{1}{\left(\frac{1}{\sqrt{r_1}} + \frac{1}{\sqrt{r_2}}\right)^2}$$

where $r_0 = r_3 + r_4$; When the drag r1 Changes; r1 differential on both sides,

Available $dR = K_1 dr_1$ **(1-1)**; Where

$$K_1 = \frac{1}{r_1\sqrt{r_1}\left(\frac{1}{\sqrt{r_1}} + \frac{1}{\sqrt{r_2}}\right)^3}$$

Seen, when the drag increases, the total drag increases.

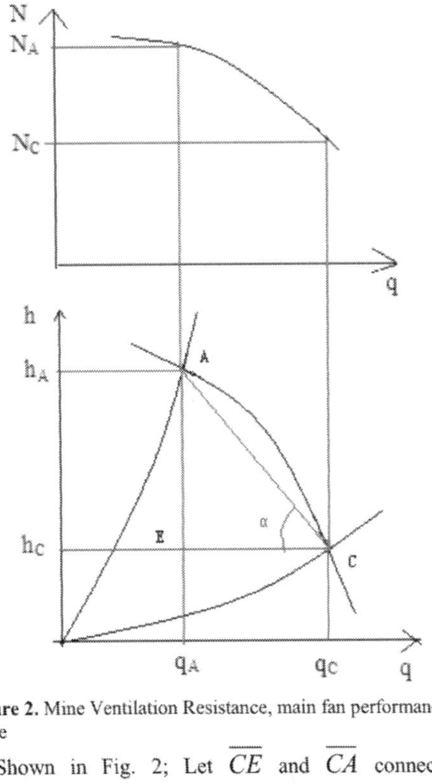

Figure 2. Mine Ventilation Resistance, main fan performance curve

Shown in Fig. 2; Let $\overline{CE}$ and $\overline{CA}$ connection angle α , the ΔAEC in the infinitesimal,

$\tan\alpha = \frac{\overline{EA}}{\overline{CE}}$; $\overline{AE} = dh_f$, $\overline{CE} = -dq_f$, So $dh_f = -\tan\alpha dq_f$, **(1-2)**; Where, q_f , h_f , respectively for the main mine fan air flow and air pressure.

Law by the wind pressure $h_f = Rq_f^2$; Differentiations $dh_f = 2Rq_f dq_f + q_f^2 dR$ (1-3)

Parallel pressure ventilation network according to law $h_1 = h_2$

Obtain $r_1(q_0 - q_2)^2 = r_2 q_2^2$; its differentiations $dq_2 = K_2 dq_0 + K_3 dr_1$ **(1-4)**

$$K_2 = \frac{1}{\sqrt{\frac{r_2}{r_1}} + 1},\ K_3 = \frac{q_0\sqrt{\frac{r_2}{r_1}}}{2r_1\left(\sqrt{\frac{r_2}{r_1}} + 1\right)^2}$$

where q_0 is the inlet and return air branch Airflow, Size equal to q_f , h_1 , h_2 respectively of the branch 1,2 pressure , q_1, q_2 respectively, the air flow rate for the branch 1,2.

Obtained by the above equations can be derived

$$dq_0 = -X_1 K_1 dr_1, (1-5)$$

$$dq_2 = (K_3 - X_1 K_1 K_2) dr_1, (1-6)$$

$$dh_f = \tan\alpha X_1 K_1 dr_1, (1-7)$$

where, $$X_1 = \frac{q_0^2}{2Rq_0 + \tan\alpha}$$

When r_1 increases, by the (1-5) shows that the total amount of wind q_0 mine be reduced,

By 1-6 combined, the actual situation shows, $K_3 - X_1 K_1 K_2 > 0$, So branch 2 air volume increases q_2, 1-7 shows that it is the work of the main fan pressure h_f increases.

Meanwhile, the 1-1 shows that, when there is increase, the main fan operating point changes.

Shown in Fig. 2, the operating point changes from point C to point A, resulting from the total air volume q_C mine decreased to q_A , the ventilation resistance increases from the point h_C to h_A, increasing the fan power set by N_C N_A.

Increasing resistance in the regulation for the centrifugal fan can increase the wind tunnel gate resistance (to reduce its opening) [5]. This method is relatively simple and implementation cost of the project is small and suitable for the production of smaller mines. However, from the above analysis, this approach will bring additional energy consumption, affecting the total air volume ventilation system stability. So adjust the time not too long.

2.2.2 The reduction in resistance adjustment

Reduction in resistance adjustment that reduces total intake, return air branch drag, increasing the total amount of wind to increase the air volume requires a larger air volume branch in two ways.

When the drag r_0 changes, the

$$R = r_0 + \frac{1}{\left(\frac{1}{\sqrt{r_1}} + \frac{1}{\sqrt{r_2}}\right)^2}$$

r_0 Differential on both sides, we get $dR = dr_0$ **(1-8)**
Seen, when the drag A1 is reduced, the total drag decreases.

And then the $r_1(q_0 - q_2)^2 = r_2 q_2{}^2$

Differentiations $dq_2 = \frac{dq_0}{1+\sqrt{\frac{r_2}{r_1}}}$; (1-9)

From $(1-2)$ $(1-3)$ $(1-8)$ $(1-9)$ were derived restaurants $dq_0 = -X_1 dr_0$ (1-10)

$$dq_2 = \frac{-X_1 dr_0}{1+\sqrt{\frac{r_2}{r_1}}} \quad (1\text{-}11)$$

$$dN = \tan\alpha X_1 dr_0 \quad (1\text{-}12)$$

When r_0 is increased by 1-10, it shows that the total amount of wind q_0 mine increased by 1-11 and shows that two branches of the increased air flow q_2 also by (1-12) shows that the main fan working pressure decreases.

Meanwhile, 1-8 shows that when r_0 decreases, R decreases, making the main fan operating point to change. Shown in Fig. 2, the operating point changes from point A to point C, resulting in the total air volume from the mine increased to q_A q_C; the ventilation resistance is reduced to the h_A h_C; fan power is reduced to the N_A N_C.

The reduction in resistance regulation in general is in the mine (or system) of the main inlet and return air duct taken to increase in parallel roadway, reduce air duct, wide brush roadway, replacement of the friction coefficient smaller shoring, and other measures to reduce local resistance ([5], [6]). From the above analysis, this approach allows the main fan running costs of the economy, but the cost of the project is large, suitable only for large-scale production of the mine.

3 Mine ventilation system for air volume adjustment rheostat economic analysis

3.1. Mine ventilation system air volume control rheostat definition of economics

When the ventilation network sees a drag strip r_i change in wind direction, assuming drag increment is dr_i, the resulting change in the total amount of drag is dR, then its total economic benefits the mine ventilation system M impact ψ_i can be expressed as $dr_i \propto \psi_i \ (i = 1,2,3,...,n)$, ψ_i is the mine ventilation system for air volume adjustment rheostat revenue degree, the size of the reaction of the drag is changed economic growth, namely economy. Set up

Where in dM D is the change in the network has the drag dr_i, when the mine ventilation system changes in the value of total economic benefits. ψ_i is larger, indicating that the variable resistance regulates air flow economy better.

3.2 Economic introduction of variable resistance than λ_e

By the previous analysis, adjusting for changing the working speed air flow method, increasing resistance adjustment and reduction in resistance regulation has advantages and disadvantages. To resolve this contradiction in the actual production, two methods can be applied: Reduce intake, return air branch drag, increased need to wind fewer branches drag, which combined increased need to wind a large branch of the air flow in order to meet production requirements [8]. The method of this regulation is defined as the amount of wind resistance combined variable air volume control method, increasing resistance and resistance reduction ratio known as the Joint variable resistance ratio.

Joint variable resistance method model shown in Figure 1-1, r_3 and r_4 required a reduction in the total resistance of the mine into the return air branch $(r_0 = r_3 + r_4)$, r_2. Thus there is the need to increase the amount of wind branch, r_1 including the need to increase resistance to other branches, including branches equivalent branch. Wherein r_1, r_2 and r_0, respectively, and the main mine fan constitute two independent loops.

$$\psi_i = \frac{dM}{|dR|} \quad (2\text{-}1)$$

Mine Main Fan constitutes two independent loops, and meets $\Delta r_0 = -\lambda \Delta r_1$. Wherein Δr_0, Δr_1 of the branch into the return air drag and wind resistance change amount Branch 1.

When λ met $\psi_i = \frac{dM}{|dR|} \to \max$, economics highest point of λ is defined as the economic variable resistance ratio, denoted by λ_e.

Insufficient air flow ventilation system can easily lead to accumulation of gas, causing suffocation and explosion, ventilation and safety effects and the indirect effects of mine efficiency. The mine ventilation network should be to ensure the mine ventilation system airflow stability [4], under the premise to maximize economy.

3.3 Mine ventilation system joint economic analysis rheostats

Joint variable resistance model for economic analysis, design total intake, return air drag r_0 and r_1, respectively, a branch change Δr_0, Δr_1 and $\Delta r_0 = -\lambda \Delta r_1$, the total drag variation ΔR, the increase in total revenue for the ΔM. Settlement revenue per unit time is as shown in Table 2-1.

Table 2-1 Rheostat adjusted income statement

Branch	1	2	3.4
Air volume to meet revenue ten thousand Yuan A 10,000 yuan \ m^3	0	$A\Delta q_2$	0
Increase power consumption B 10,000 yuan \ kw	$B\Delta N_1$	0	$B\Delta N_0$
Power consumption to maintain stable air flow C10,000 yuan \ m^3 ($\Delta q_0 \geq 0$ h $C = 0$	0	0	$C\Delta q_0$
Engineering cost D 10,000 yuan	0	0	Dr_0
ΔM	$A\Delta q_2 + B\Delta N + C\Delta q_0 + Dr_0$		

where, ΔN_1 and ΔN_0 are the branches 1 and 3 (or 4) drag to change the main cause of the mine ventilation system fan power variation, $\Delta N = \Delta N_0 + \Delta N_1$

From the previous analysis, the mine ventilation system to change the working method windage adjustable air volume gains and r_0, r_1 branch related to changes in wind resistance, assuming that only the operating point is changed, the characteristic curve did not change. To this end, the mold showed in united rheostat.

Type variable resistance caused due to economic analysis of the changes.

Let branches 1 and into the back, the wind drag each branch change Δr_1, Δr_0; Parallel networks mine total differential of total drag $dR = dr_0 + K_1 dr_1$ (2-2)

According to the law $h_1 = h_2$ parallel pressure ventilation network was $r_1(q_0 - q_2)^2 = r_2 q_2{}^2$,

Therefore $dq_2 = K_2 dq_0 + K_3 dr_1$, (2-3)

All kinds can be derived from the above $dN = h_f dq_f + q_f dh_f$

By the (1-4) know; can be obtained from Table 2-1;

$$\mathrm{dM} = \mathrm{Adq_2} + \mathrm{BdN} + \mathrm{Cdq_0} + \mathrm{Dr_0} \quad (2\text{-}4)$$

Set up $dr_0 = -\lambda dr_1$,

$$\lim \frac{dM}{|dR|} = \begin{cases} \frac{(AK_2X_1 - CX_3 - D)\lambda + A(K_3 - K_1K_2X_1) - B(1-K_1)K_4 - CX_1K_5}{\lambda - K_1}, & 0 \le \lambda < K_1 \\ \frac{(AK_2X_1 - CX_3 - D)\lambda + A(K_3 - K_1K_2X_1) - B(1-K_1)K_4 - CX_1K_5}{K_1 - \lambda}, & K_1 < \lambda \end{cases} \quad (2\text{-}5)$$

Among $K_4 = \frac{\tan\alpha}{2Rq_0 + \tan\alpha}$

The above analysis shows that when $\lim \frac{dM}{|dR|} \to \max$ is full, from the economic point of view, the system shows the most reasonable proportion of variable resistance adjustment. From (2-5), when the characteristic curve is not changed, $\tan\alpha$ are constant, economy ψ_i and λ, only the scale of production, and therefore according to the different production scale, drawing the $\psi_i - \lambda$ curve.

The analysis shows that when the mine production is smaller, it allows $\frac{A}{D} < \frac{K_1}{K_3}$

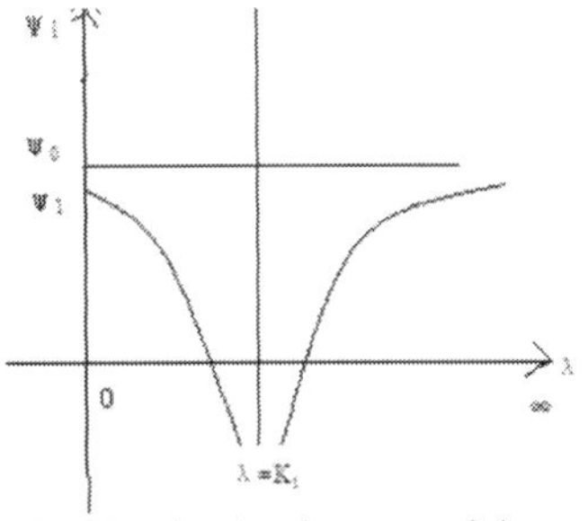

Figure 3. Small-scale mine air volume control rheostat economy curve

Smaller yield, $\psi_j - \lambda$ curve as shown in Fig. 3. When $\lambda = 0$ is, that only the method of increasing resistance adjustment, has a relatively high economic,

$$\psi_i = \frac{A(K_3 - K_1K_2X_1) - B(1-K_1)K_4 - CX_1K_1}{K_1}$$

There is no visible problem engineering expenses, mainly due to the additional energy consumption and output and total air volume changes, so adjust the time not too long, otherwise too much additional energy consumption, impacts the economy. When $\lambda \to \infty$, that resistance reduction method using only regulate $\psi_i = AK_2X_1 - D$, there is no additional visible decline in wind energy issues, due to smaller production, so the cost of the project is a major constraint on economic factors. Although the reduction in resistance of the main fan running costs the economy, but construction costs are too high and the economy with respect to increasing resistance method improves, but the magnitude is not large, so the production of smaller mines by reduction in resistance adjustment method is not practical.

When $0 < \lambda < \infty$, that combined variable resistance adjustment λ increases, ψ_i decreases and then increases, but is always less than the economic adjustment, reduction in resistance is low, when combined variable resistance ratio $\lambda = K_1$, the mine changes in the value of total drag $\Delta R \to 0$, drag curve indicated by the mine, no additional energy consumption, air flow stability [1], the existing scale of production, the reduction in resistance larger proportion of project costs, up from Fig. 1-4, shows that at this time the most economical poor. Visible production scale is small, the increasing resistance of the best economic regulation. At such production conditions, the reduction in resistance regulation or co-resistance reduction method is not advantageous.

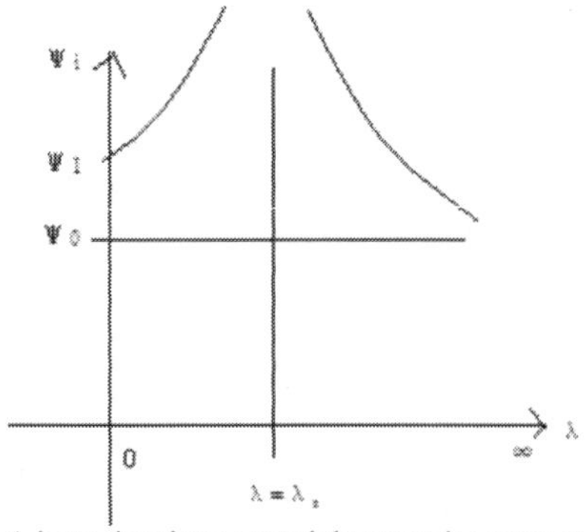

Figure 4. large air volume control rheostat mine economy curve

When the large-scale production meets the $\frac{A}{D} > \frac{K_1}{K_3}$, the production of large $\psi_i - \lambda$ curve is shown in Fig. 4. When the $\lambda = 0$ or $\lambda \to \infty$, that is, only increasing resistance adjustment or just adjust the resistance reduction, economy is relatively low. When $0 < \lambda < \infty$, i.e. a joint variable resistance, ψ_i with λ first increases and then decreases when the $\lambda = K_1$, $\psi_i \to \max$, economy best, so $\lambda_e = K_1$, this time drag mine changes in the value of total $\Delta R \to 0$, from the mine drag curve can also be seen, no additional energy consumption, and stable air flow, relative to the scale of production, the proportion of project costs are relatively small, so efficiency is relatively high, consistent with the economic analysis.

So in this production conditions, the joint variable resistance method most superiority, λ closer λ_e, mine ventilation system air volume control economy is higher.

The above analysis and reduction in resistance with increasing resistance to the regulation of air flow characteristics match. $\lambda = \lambda_e$ time of significance is the change in the value of the mine total drag $\Delta R \to 0$; mine drag curve does not change; there is no additional energy consumption, ventilation system air flow is stable and rational engineering, economy best.

4 The main conclusion

Rheostat for adjusting the mine ventilation system air flow when the mine ventilation system should be considered wind stability [1], the scale of production, energy consumption, and engineering, to choose the best method for regulating the mine ventilation system to ensure safe and economical operation.

Joint variable resistance method is suitable for large-scale production of the mine, then you can adjust the resistance reduction and comprehensive adjustment method has the advantage of increasing resistance, variable resistance proportional ratio $\lambda = \lambda_e$ variable resistance air volume control has the highest economy.

Increasing resistance mine ventilation system air volume control should be avoided in the mine well into the main, install conditioning return air duct, or affect the air flow stability [1], which indirectly affects economy. Reduction in resistance adjustment should be avoided in the construction of the tunnel with the wind resistance reduction because of the cost of the project, a large, complex operation, and can be reduced slightly, with little success, the economy is poor.

Optimization of mine ventilation system design, as far as possible, pre-designed mine ventilation system network architecture, resistance distribution and air flow with the actual situation and put into operation in line to reduce the daily mine ventilation system regulation times it had to keep the main fan working conditions point for stability, reduce the additional energy consumption and maintenance cost of the project.

5 Conclusion

The objective of any ventilation system is twofold. First, the primary ventilation must allow air through the main airways to the immediate working area out by the working faces, thus making fresh air available for face ventilation, and then return the contaminated air through return (exhaust) airways to the surface. Second, the face ventilation system must be designed to effectively utilize the available air in the immediate working area to sweep the working face, to capture and remove dust, and to dilute and carry away gas, if any, emitted during mining activities. The Ivory Coast mines need an effective ventilation system to create suitable micro-environment of the working face. Constructing such system provides solution to the heat problems and has profound significance on safety in the mine, thus making the mine activities a good venture.

References

1. Veiga, M. M., Angeloci, G., Hitch, M., & Velasquez-Lopez, P. C.. Processing centres in artisanal gold mining. *Journal of Cleaner Production*, **64**:535-544, (2014).
2. Veiga, M. M., Angeloci, G., Ñiquen, W., & Seccatore, J. Reducing mercury pollution by training Peruvian artisanal gold miners. *Journal of Cleaner Production*, **94**, 268-277, (2015).
3. Veiga, M. M. *Introducing new technologies for abatement of global mercury pollution in Latin America* . CETEM/CNP, p. 94, (1997).
4. Wang Chong-Xian, LI Xu Guo, Tan waves. Mine ventilation system air volume factors influencing the stability [J]. Coal, **33**(8):931-935, (2008).
5. Zhang Guo. Ventilation Safety Science [M]. Xuzhou: China University of Mining Press, (2007).
6. Xurui Long. Ventilation Network Theory [M]. Beijing: Coal Industry Press, (1993).
7. Yhingmin.Mine aerodynamics and mine ventilation system[M].Beijing:Metallurgical Industry Press, (1994).
8. Xie Zhenhua, Song Cun-yi. Engineering fluid mechanics [M]. Beijing: Metallurgical Industry Press, (2007).

Evaluation of the Risk Factors on Coal Dust Explosion in Warehouse

Latana Chanthaphasouk[1], Kreangkrai Maneeintr[1], Pinyo Meechumna[1], Suteera Luengwattanapong[1], Chantamanee Poonjarernsilp[2]

[1]*Carbon Capture, Storage and Utilization Research Laboratory Department of Mining and Petroleum Engineering, Faculty of Engineering, Chulalongkorn University, 10330 Bangkok, Thailand*
[2]*Department of Chemical Engineering, Faculty of Engineering, Rajamangala University of Technology Krungthep, Bangkok 10120, Thailand*

Abstract. Coal has been one of the major energy sources in the world. Many industries use coal as a main fuel. A coal dust explosion is one of the main hazards of coal utilization because of its massive damage. Coal dust explosion hazards involve the combustible fine dusts or other small particles that present a fire or deflagration hazard when suspended at a sufficient concentration in air or some other oxidizing medium. When such materials are contained in an enclosure, they present an explosion hazard. To eliminate the possibility of dust explosions by ensuring that the dust concentration does not exceed the minimum explosibility concentration (MEC) or the amount of dust per unit volume of air below which the dust cloud cannot propagate flame. Therefore, the objective of this work is to measure the MEC for coal dust explosion with the various conditions of coal storage such as the particle size, moisture of coal, degree of dispersion and delayed time of ignition source to prevent the coal dust explosion which aims to study and design the explosion safety measures for coal dust handing installations. The results show that smaller size of particle, low moisture in coal and high coal dust dispersion can increase the chance or risk of dust explosion. Also, the shorter time of dust dispersion exposes to ignition source can enhance the possibility of explosion in coal storage.

1 Introduction

Coal has been one of the major energy sources in the world. Many industries such as power plants, cement industries, pulp and paper industries use coal as a main fuel. However, there are many concerns about the environmental and safety issues. A coal dust explosion is one of the main hazards of coal utilization because of its massive damage. These hazards are recognized for the coal mines and coal processing plants. There are a number of scientific work aimed to reduce the coal dust explosion hazards in coal mines during the last century [1]. Moreover, dust explosion can occur in many industries including agriculture, food product, chemical, textiles [1]. It can happen within thermal dryer, cyclones, bag house, pulverizing fuel system, grinding mill and belt conveyor equipment [1], [2]. A dust explosion can cause of death and injuries and make the company waste of investment with combustible dust incident [3]. Dust explosion hazards involve the combustible fine dusts or other small particles that present a fire or deflagration hazard when suspended at a sufficient concentration in air or some other oxidizing medium. When such materials are contained in an enclosure, they present an explosion hazard [1], [3].

The explosions are caused by the combustion of materials. Particle properties as a fuel in the presence of adequate oxygen. Also, combustible dust in industry often pile up as dust layer or suspended as dust could otherwise the layer of dust was easy to induce fire and dust explosion often was convinced by dust could [4], [5]. The fire is ignited at a high enough energy. Moreover, in confined spaces or enclosed materials to the combustion, particles can be suspended or dust in the air for long enough [3]. There are a variety of size of particles that are burned up and dispersed. The explosion theory is based on the explosion of dust, which is called the theory of dust explosion pentagon consisting of fuel (dust), dispersion of dust, confinement, ignition source, and oxygen or air as shown in Fig. 1 [5], [6].

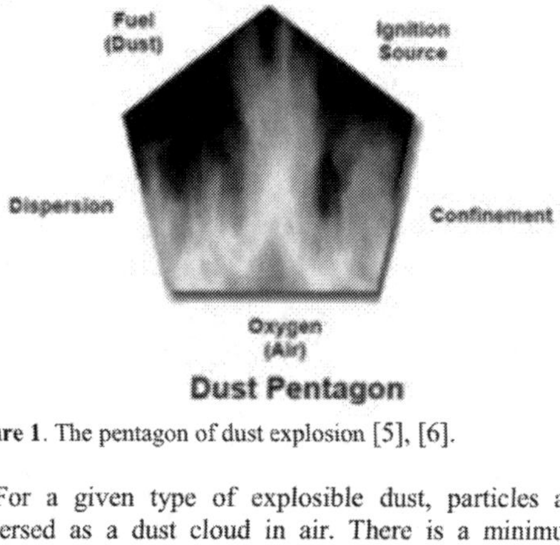

Figure 1. The pentagon of dust explosion [5], [6].

For a given type of explosible dust, particles are dispersed as a dust cloud in air. There is a minimum quantity of dust per unit volume of air below which the dust cloud cannot propagate flame. From the theory, one could eliminate the possibility of dust explosions by ensuring that the dust concentration does not exceed this

 DOI: 10.1051/matecconf/20167706014

minimum limit which is called the minimum explosibility concentration (MEC) which can be calculated by mass of particles divided by volume of the area for explosion. These amount of particles causing the explosion are usually determined with laboratory scale apparatus [7]-[9]. Tanthapanichakoon et al [9] calibrate the MEC equipment with lycopodium and compare the results with literature [10] and the results present that MEC of lycopodium is ranging from 47-50 g/m3 comparing with literature [10] at 45±5 g/m3. Furthermore, other materials such as wheat starch, polymer, toner and so on are tested for MEC. Moreover Man and Harris [3] has studied the addition of rock dust such as limestone to inert the coal dust explosion.

Therefore, the objective of this work is to measure the MEC for coal dust explosion with the various conditions of coal storage such as the particle size, moisture of coal, degree of dispersion and delayed time of ignition source to prevent the coal dust explosion. Also, this research investigates the explosion characteristics of coal dust which is undertaken as a part of the research at Chulalongkorn University, Thailand, which aims to study and design the explosion safety measures for coal dust handing process in the coal storage.

2 Experiment

2.1 Material

Coal dust for this study is obtained from the coal storage of Siam Cement Group Company. The size distribution of this dust is ranging from less than 45 micron to 53 micron.

2.2 Equipment

The experiment of dust explosion are conducted in 0.012297 cm^3 volume of stainless steel tester chamber connected with air tank to supply air for gas dispersion with the pressure up to 50 psi or 3.5 kg/cm^2. A 4 litter air compressor used for this study are compressor model OL1204 with the maximum of compressed air at 8 bar from Scientific Company Limited to feed compressed air to the tank. The high precision scale to measure the amount of coal dust comes from the OHAUS Company model PA214C with readability of 0.0001 digits and the operating conditions from 10̊c to 40 °C at 10% to 80% relative humidity. Sieve analysis model STS-S 414 from Soil Testing Siam Co. Ltd., is used to characterize and separate coal dust with different size from less than 45 micron to 250 micron. The moisture content is measured by using a moisture meter model MC7825PS with digital display indicating the moisture condition of the material. Specification of display is 4 digits, 10mm LED with measurement ranging 0% to 80%, and ±0.5% accuracy. The operating conditions are 0-50 °C of the temperature and humidity below 90% RH. Also, oven used to control the moisture of coal dust or sample derives from Memmert with maximum temperature at 200 °C. This oven is operated at more than 100 °C for this study.

2.3 Experimental procedure

Experimental procedure to obtain the MEC is described as follow. First humid coal dust sample is dried in the oven to control the humidity to the desired moisture. The known quantity of coal dust is measured and placed at the bottom of the dispersion nozzle. Coal dust is dispersed by compressed air and ignited by the high-voltage electric sparking [9]. Coal dust can be exploded or burned. The tests are performed 3 times with the same conditions to ensure that the results are correct.

The operating conditions for this study are presented in Table 1.

Table 1. Experimental operating conditions for this study.

Parameter	Value
Particle size (micron)	Less than 45, 45, and 53
Coal humidity (%)	0, 3, 6, 9 and 12
Pressure of air (kg/cm^2)	0.5, 1, and 1.8
Time (s)	0.1, 0.3, and 0.5

3 Results and discussion

3.1 Verification of equipment and procedure

The apparatus and procedure are verified and compared by using lycopodium powder with the size of 32 µm, 35% to 45% humidity [9]. The MEC of lycopodium is obtained as shown in Figure 2 which is getting along well with the previous work with 4.9% deviation [6], [9]. Therefore, this equipment and procedure are good enough to further study. The experimental data for verify of equipment with lycopodium powder as show blow:

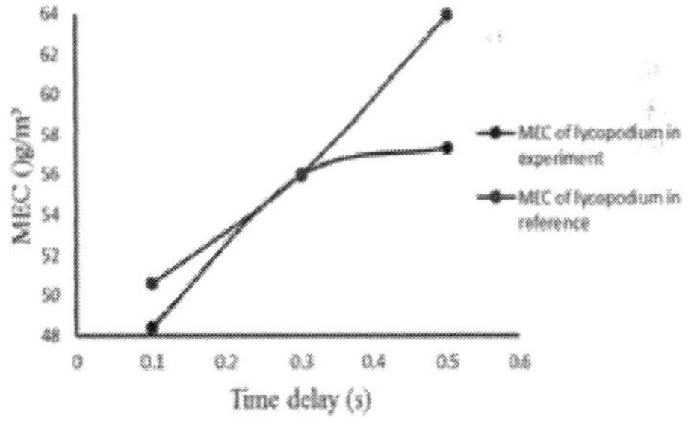

Figure 2. Verification of calibration MEC with lycopodium

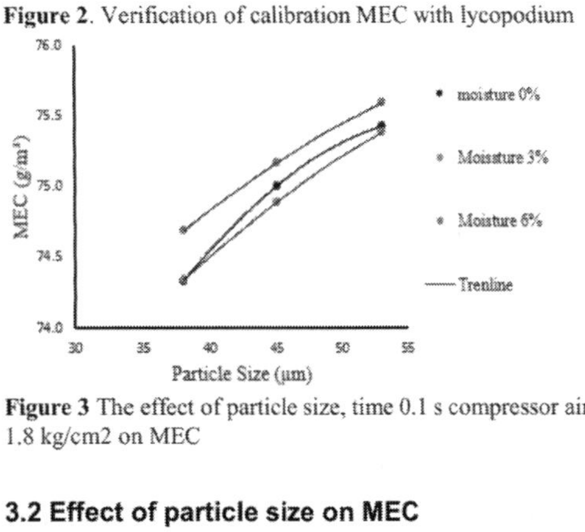

Figure 3 The effect of particle size, time 0.1 s compressor air 1.8 kg/cm2 on MEC

3.2 Effect of particle size on MEC

DOI: 10.1051/matecconf/20167706014

The experimental results as shown in Fig. 3, indicate that particle size has effect on the amount of coal dust explosion or MEC in that MEC increases with particle size of coal dust. The smaller the particles, the higher chance to have coal dust explosion. Therefore, coal dust can be agglomerated to bigger size to reduce or prevent the explosion. From the experiment result the moisture content 0 % and 6 % and the percent of deviation is 0.48 %.

3.3 Effect of coal humidity on MEC

The effect of moisture content in coal is presented in Fig. 4. From the figure, it is obvious that coal humidity has less effect on MEC measurement. However, in term of ignition, humidity in coal is the key factor for coal explosion because in many cases, coal is burned but not exploded. Therefore, with higher moisture content in coal, the risk for coal dust explosion is reduced.

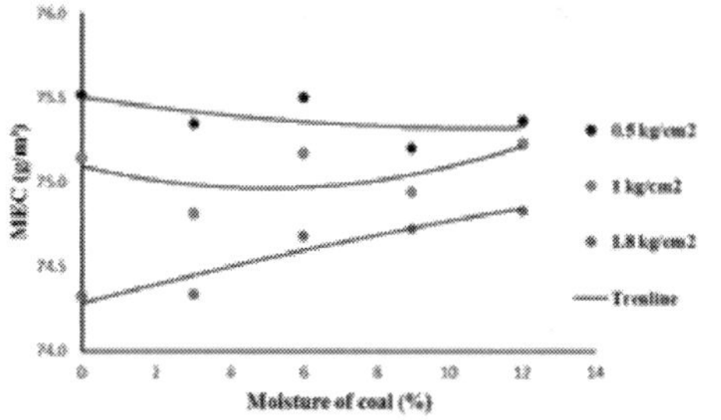

Figure 4. The effect of moisture in coal and air pressure, particle size less than 45 micron, 0.1 s on MEC.

3.4 Effect of types of air pressure on MEC

Air pressure reflects the degree of dispersion of coal in the confined space. If pressure of air injection is higher, the coal dust will be dispersed more. The results as shown in Fig. 4 indicate that at higher pressure, MEC will be lower. That means higher risk for coal dust explosion.

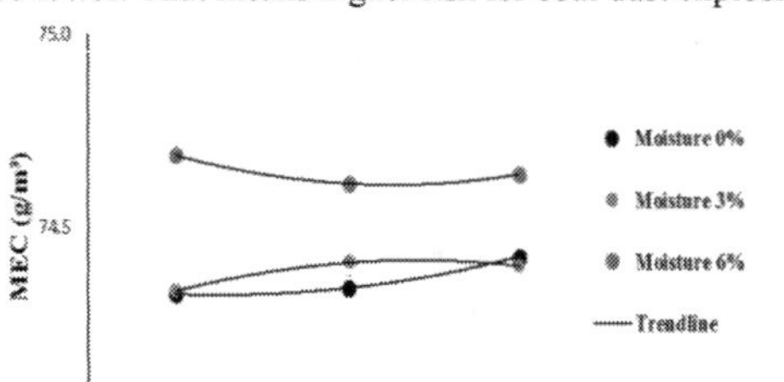

Figure 5. The effect of delayed time, particle size than 45 micron 1.8 kg/cm^2 for ignition on MEC

3.5 Effect of delayed time on MEC

The delayed time for sparking or ignition is the parameter to investigate the explosion. Less delayed time can increase the chance for coal to ignite thus increasing the risk for coal dust explosion. For this study, the effect of time as presented in Fig. 5. has less effect to measure MEC. However, more delayed time will make less dispersion of coal because coal dust will drop down and the concentration of coal in the air becomes low. Then the ignition sources will start ignition with low concentration of dust. Therefore, there is less chance of dust to get ignited.

4 Summary

Coal dust explosion can cause lots of damage. To avoid this, the amount or concentration of coal dust per unit volume or MEC is measured with the various conditions of coal storage such as the particle size, moisture of coal, degree of dispersion and delayed time of ignition source to prevent the coal dust explosion which aims to study and design the explosion safety measures for coal dust handing installations. The results present that moisture content in coal has more effect on dust explosion. The smaller size of particle, low moisture in coal and high coal dust dispersion can increase the possibility or risk of dust explosion. Also, the shorter time of dust dispersion exposed to ignition source can enhance relatively low the possibility of coal dust explosion.

Acknowledgement

The authors would like to acknowledge the ASEAN University Network/Southeast Asia Engineering Education Development Network (AUN/SEED-Net) for financial support of this research. Secondly, the author would like to thank Graduate School Thesis Grant and Chulalongkorn University to accommodate for research periods. Also, the authors would like to thank Siam Cement Group (SCG) for coal dust sample. Lastly many thanks to the research students from Rajamangala University of Technology Krungthep for helping the experiment.

References

1. R.K. Eckhoff: Dust Explosion in the Process Industries, 3rd edition, Gulf Professional Publishing (2003).
2. S.R. George: The Explosibility of Coal Dust, Washington, Government printing office, (1911).
3. C.K. Man and M.L. Harris: submitted to Journal of Loss Prevention in the Process Industries, **27**, (2014).
4. Y. Chunmiao, L. Chang, L. Gang: Coal dust explosion prevention and protection based on inherent safety, (2011).
5. F. Norman, J. Berghmans, and F. Verplaetsen: The minimum ignition energy of coal dust in an oxygen enriched atmosphere, (2013).
6. W. Tanthapanichakoon, V. Chongvisal, T. Charinpanitkul and V. Taepaisitphongse: Dust Explosive Limit Tester, Department of Chemical Engineering, Graduate School, Chulalongkorn University, (1996).
7. Information on http://www.nfpa.org

DOI: 10.1051/matecconf/20167706014

8. M. Mittal: submitted to International Journal of Advanced Engineering Technology, (2003).
9. ASTM designation E1515 – 07. Standard test method for Minimum Explosible Concentration of Combustible Dusts, (2007).
10. APPIE. Standard/APS-001. Dust explosion Relief Venting for bag Filters, The Association of powder Process Industry and Engineering, Kyoto, Japan, (1983).

MATEC Web of Conferences 77, 06015 (2016) DOI: 10.1051/matecconf/20167706015
ICMMR 2016

Research on Interactive Control of Electrolytic Aluminum Load and Wind Power Output

Ge Simin [1], Yu Kun [1, 2]

[1]*Hohai University, College of Energy and Electrical Engineering, 210098 Nanjing, China*
[2]*Jiangsu Engineering Research Center for Distribution & Utilization and Energy Efficiency, 210098 Nanjing, China*

Abstract. With the large-scale wind power generation connected to Gansu power grid, electrolytic aluminium load has reached a certain scale at the same time, the vast majority of electrolytic aluminium load directly connects to power grid using 330kV transmission line. According to the physical characteristics and historical data, the continuous adjustment characteristic of electrolytic aluminium load is analysed. Based on this characteristic, a mathematical model for the electrolytic aluminium load is established. Aiming at reducing power network loss and wind power consumption, an optimization model based on the load regulation characteristics of electrolytic aluminium is constructed, which is optimized by particle swarm optimization algorithm. Based on the case data of Gansu power grid, the optimal method based on the load regulation characteristics of electrolytic aluminium is analysed and its feasibility is verified.

1 Instruction

Gansu Province has lots of wind energy resources [1] which are mainly concentrated in Jiayuguan and Jiuquan area, and are away from the province's electricity load center. Further more, lots of wind power can not be consumed since the wind power consumptive ability in Jiayuguan and Jiuquan areas is very limited [2]. The wind power could be transmitted to the main power grid of Gansu province through double circuit 750kV transmission line in Jiuquan Hexi. However, remote delivery of large-scale active power increases the loss of power system and also reduces the voltage quality of Hexi power grid and the main power grid of Gansu province, which increase the network loss value more. In addition, due to the reverse distribution characteristic of the new energy and conventional generations [3], the power loss is huge when conventional generations are adjustable. The randomness and volatility of the new energy sources [4] may lead to a large voltage fluctuation of the transmission line, which further increase the power loss [5]. To solve this problem, paper [6] reduces the randomness of new energy output by using the coupling of different new energy sources, but it does not solve the problem of large-scale new energy consumption. Aiming at consuming the new energy, paper [7] achieves dynamic economic dispatch by optimizing the power of conventional thermal power generating units, which makes all the wind power output be consumed by power grid. However, sometimes only by coordinating with conventional power supply, the new energy output cannot be completely consumed. Paper [8] uses spot price to stimulate demand side resources to consume the wind generation output, but the analysis of demand side resources is not detailed enough.

Electrolytic aluminium industry [9-10] in Gansu Province are one of the most important industries, which has large capacity and regulation characteristic, usually use 330/750kV high voltage to supply the power directly. Based on the regulation characteristics of electrolytic aluminium load, the load cooperates with the new energy to consume as much new energy as possible and reduce the power loss. In paper [11], the characteristics of high energy load are analysed and the corresponding mathematical model is established, but there is no specific method or strategy towards the power loss reduction problem based on load features of electrolytic aluminium.

In this paper, based on the characteristic analysis of the electrolytic aluminium load, the mathematical model of electrolytic aluminium load is constructed. The interactive optimization model based on the load regulation characteristics of electrolytic aluminium is established aiming at reducing power network loss and wind power consumption. Particle swarm optimization algorithm is used to optimize the model, and the power loss reduction method based on the load regulation characteristics of electrolytic aluminium is analysed. Its feasibility is verified by a numerical example.

2 Load characteristics and mathematical model of electrolytic aluminium

 DOI: 10.1051/matecconf/20167706015

2.1 Continuous regulation characteristic of electrolytic aluminium load

Rhyolite molten salt electrolysis [12] has been used in modern aluminium electrolysis industry, with powerful direct current into, electrochemical reaction happens at a high temperature on the two poles of the electrolytic tank to produce aluminium metal. In this production process, the electrolytic tank needs to be continuously supplied with direct current, and the size of electrolytic aluminium load current can be regulated by controlling the conduction angle of thermistor. The entire production process requires a constant current. If the direct current fluctuates largely, it will significantly affect the stability of the operation state (especially thermal equilibrium) of the electrolytic tank, and will interfere normal control process of aluminium electrolysis, thus the technical and economic indicators of electrolytic aluminium production will be influenced. Therefore, when the electrolytic aluminium load is regulated, the load production reduction needs to be in the scope of the electrolytic tank, and the load current of the electrolytic aluminium can be adjusted by controlling the conduction angle of the controllable silicon.

According to the above analysis, under normal circumstances, the fluctuation of electrolytic aluminium load capacity is small, the load curve is relatively smooth, as shown in Fig. 1. Since the electrolytic aluminium load often has a number of feeders, each feeder is connected to a rectifier unit, and each unit can be adjusted within a certain range. Through the analysis of historical data, the adjustment range of electrolytic aluminium load is -10%~10%.

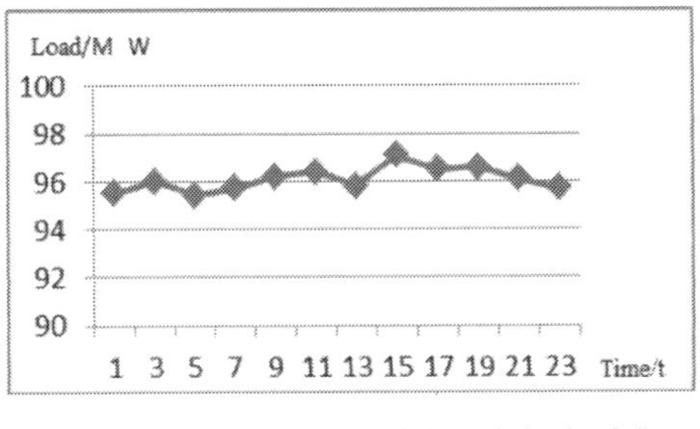

Figure 1. Typical daily load curve of electrolytic aluminium.

2.2 Mathematical model of continuous adjustable for electrolytic aluminium load

In the adjustable state of electrolytic aluminium load, when the load began to participate in the regulation, the load will change from load rating according to the ramp rate, and stop by the set value. When the set value is lower than the current value, the ramp rate is called the load reduction ramp rate. When the set value is lower than the current value, the ramp rate is called the load increase ramp rate. The ramp rate of aluminium load differs among different units.

In order to maintain the normal production of electrolytic aluminium load, the electrolytic aluminium production amount should fluctuate within the allowable range, and the set value of load should be in the allowable range as well. Use a single regulation as an example, if the load is adjusted from rating to a set value and then return to the rating, the typical electrolytic aluminium load regulation curve is shown in Fig. 2.

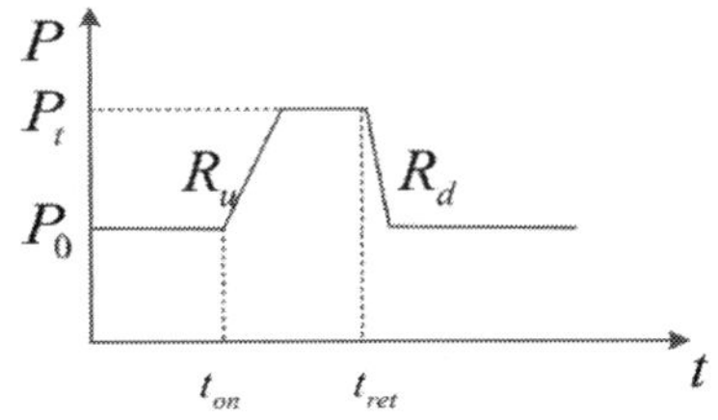

Figure 2. Typical load regulation curve of electrolytic aluminium.

The mathematical model of the load regulation characteristics of electrolytic aluminium can be established by the above regulation curve:

$$P = \begin{cases} P_0, t \le t_{on} \\ P_0 + R_u(t - t_{on}), t_{on} < t \le t_{on} + \dfrac{P_t - P_0}{R_u} \\ P_t, t_{on} + \dfrac{P_t - P_0}{R_u} < t \le t_{ret} \\ p_t - R_d(t - t_{ret}), t_{ret} < t \le t_{ret} + \dfrac{P_t - P_0}{R_d} \\ P_0, t > t_{ret} + \dfrac{P_t - P_0}{R_d} \end{cases} \quad (1)$$

$$P_{min} \le P \le P_{max} \quad (2)$$

$$c_i = \sum_{i=1}^{T} \frac{p_{i,t}}{\eta_i} \quad (3)$$

$$c_{min} \le c_i \le c_{max} \quad (4)$$

where P_0 represents the capacity rating of the load or the value of the load capacity before adjustment, t_{on} represents trigger time of load regulation, R_u represents the load increase ramp rate, t_{ret} represents the return time or the next time adjustment, R_d represents the load reduction ramp rate, P_{min} represents the minimum value of the load regulation range, P_{max} represents the maximum range of load regulation, c_i represents load production amount, $p_{i,t}$ represents the active power consumption of the load at the time of t, η_i represents the unit product power consumption, c_{min} represents the minimum value of the load production amount range, c_{max} represents the maximum value of the load production amount range.

When the load needs to be adjusted down to the set value, the mathematical model can be adjusted on the basis of the formula (1) as follows.

$$P=\begin{cases} P_0, t \le t_{on} \\ P_0 - R_d(t-t_{on}), t_{on} < t \le t_{on} + \dfrac{P_0 - P_t}{R_d} \\ P_t, t_{on} + \dfrac{P_0 - P_t}{R_d} < t \le t_{ret} \\ p_t + R_u(t - t_{ret}), t_{ret} < t \le t_{ret} + \dfrac{P_0 - P_t}{R_u} \\ P_0, t > t_{ret} + \dfrac{P_t - P_0}{R_d} \end{cases} \quad (5)$$

3 Optimization of network loss and wind power consumption based on the regulation characteristics of electrolytic aluminium load

From analysis above, electrolytic aluminium load has adjustable characteristics, but constraints of load need to be considered during the regulation. When the electrolytic aluminium load is involved in the process of power loss reduction, it should be cooperated with conventional generation to achieve active power balance.

3.1 A network loss optimization model considering the load regulation characteristics of electrolytic aluminium

3.1.1 objective function

Usually, daily load curves are similar, so optimization result of one chosen day is representative. The optimization sets the total loss on the power grid to be minimum as the goal, including line loss and transformer loss.

$$\min f \sum_{t=1}^{T} \left(\sum_{i=1}^{N} \frac{P_i^2 + Q_i^2}{U_i^2} R_{Li} + \sum_{j=1}^{M} \frac{P_j^2 + Q_j^2}{U_j^2} R_{Tj} \right) \quad (6)$$

where P_i , Q_i , U_i each represents the active power, reactive power and voltage at the first end of the branch i , and R_{Li} is line impedance. P_j , Q_j , U_j each represents the active power, reactive power and voltage at the first end of the branch j , and R_{Tj} is transformer impedance. N is the total number of lines, M is the total number of transformers, T is the total length of time.

3.1.2 constraint condition

Firstly, the active power balance of power system is considered:

$$\sum P_G = \sum P_L + \sum P_{LOSS} \quad (7)$$

where $\sum P_G$, $\sum P_L$, $\sum P_{LOSS}$ each represents the total power output of the system, the total load and total transmission loss. The total generating capacity includes the wind generation output and conventional power output.

Secondly, the electrolytic aluminium load constraints are considered, the adjustment process must be in electrolytic aluminium adjustable range, including the limit value of load:

$$P_{L\min} \le P_L \le P_{L\max} \quad (8)$$

where $P_{L\min}$ represents the minimum value of the load regulation range, $P_{L\max}$ represents the maximum value of the load regulation range.

Load ramp rate constraint:

$$\Delta P_{\max\text{down}} \le P_{adj}(t) - P_{adj}(t-1) \le \Delta P_{\max\text{up}} \quad (9)$$

where $\Delta P_{\max\text{down}}$ represents the maximum decreasing load variation, $\Delta P_{\max\text{up}}$ represents the maximum increasing load variation, $P_{adj}(t)$ represents load active power of time t , $P_{adj}(t-1)$ represents load active power of time $(t-1)$.

Since the normal production should not be affected during the regulation, it is ensured by the order production constraints:

$$c_{\min} \le c_i \le c_{\max} \quad (10)$$

where $c_{\min}$ represents the minimum load production amount, $c_{\max}$ represents the maximum load production amount.

Output constraint of conventional power supply:

$$P_{G\min} \le P_G \le P_{G\max} \quad (11)$$

where $P_{G\min}$ represents the minimum value of the conventional power regulation range, $P_{G\max}$ represents the maximum value of the conventional power regulation range.

3.2 An optimization algorithm considering the regulation characteristics of electrolytic aluminium load

According to the analysis above, the network loss optimization problem that considers the regulation characteristic of electrolytic aluminium load is a typical constrained nonlinear programming problem, using particle swarm algorithm [13] to solve and the specific steps of the algorithm are as follows..

(1) Firstly, encode the active power of electrolytic aluminium load, wind power generation output and conventional generation output, code value of electrolytic aluminium load is its active power in the power grid and code value of conventional generation and wind generation are their output. The range of code value is ensured by the given load and generation.

(2) The number of electrolytic aluminium load, wind power generation and conventional generation decides the dimension m of particle swarm. Random

function is used to encode each particle. If the number of particles is n, then there will be an n*m matrix variables, which represents the population in particle swarm algorithm. At the same time, random function is used to initialize the speed of each particle. Set historic individual optimal position zbest as current location, and set the best of group as current gbest.

(3) The population that is composed of electrolytic aluminium load active power, wind generation output and conventional generation output is the independent variable, while network loss value is the dependent variable. Set the network loss value as the fitness function, through the power flow calculation, the value of fitness function of each particle can be achieved.

(4) If the current value of fitness function of the particle is better than its historical optimal value, then the historical optimum will be replaced by the current position.

(5) If the historical optimal of the particle is better than its global optimal, then the global optimal will be replaced by its historical optimal.

(6) Update the speed and position of each particle according to the following formula.

$$v_i(k+1) = wv_i(k) + c_1r_1(p_i - x_i(k)) + c_2r_2(p_g - x_i(k)) \tag{12}$$

$$x_i(k+1) = x_i(k) + v_i(k+1) \tag{13}$$

where p_i represents individual extremum, p_g represents the extremum of the group, w , c_1 , c_2 are constants, $v_i(k)$, $v_i(k+1)$ each represents the current speed of the particle i in step k and step $k+1$ during the iteration, $x_i(k)$, $x_i(k+1)$ each represents the current position of the particle i in step k and step $k+1$ during the iteration.

(7) Iteration number plus 1. if the condition of end is not reached, then go to the third step, otherwise output the result and end.

4 Example analysis

Based on the data of Gansu power grid, an example is constructed as shown in Fig. 3 to analysis the method of network loss reduction and wind power consumption involving electrolytic aluminium load. The validity of the proposed method is verified. There are 2 wind generations, wind generation unit 1 has the active power of 60MW, and is connected to the net through node 1 grid, wind generation unit 2 has the active power of 70MW, and is connected to the net through node 9. Node 6 and node 11 are both conventional generations, and node 7 is connected with electrolytic aluminium load, whose active power is 310MW.

As the result of simulation, network loss reduces from 277MW to 268MW, the online capacity of wind generation unit1 increases from 60MW to 67MW and the online capacity of wind generation unit2 increases from 70MW to 75MW when conventional generation participates during the process.

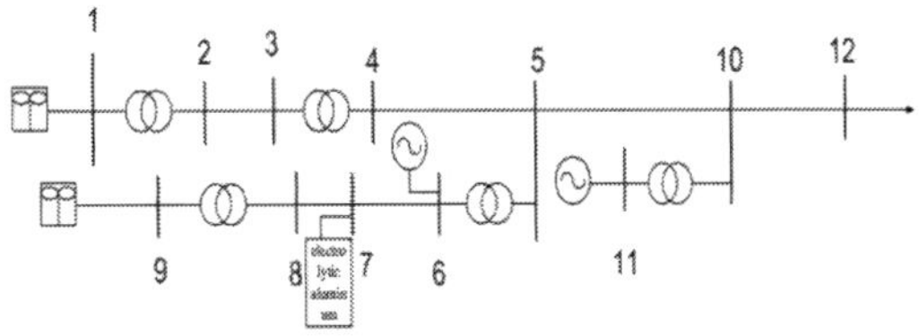

Figure 3. Simulation example wiring diagram.

On the basis of conventional generation participating in regulation, the network loss can be reduced to 256MW when considering the continuous adjustable characteristic of electrolytic aluminium load. The percentage of the reduction of network loss is 7.6%, which is more than twice as high as the regulation using conventional generation. And the online active power of wind generation raises more as well, unit1 increases 20% and unit2 increases 17%.

In other words, the network loss can be reduced and the consumption of wind generation can be increased in some degree by regulating conventional generation. And if the continuous regulation characteristic of electrolytic aluminium load is considered, the change of both network loss and wind power consumption would be more significantly.

5 Conclusions

In this paper, according the actual background of Gansu power grid, based on the analysis of the electrolytic aluminium load, the mathematical model of electrolytic aluminium load is set up. Aiming at reducing network loss, the optimization model of power network loss reduction based on the load regulation characteristics of electrolytic aluminium is established. Due to the characteristics of the model, particle swarm optimization algorithm is used to optimize the model. The example is constructed based on the data of Gansu power grid, and the power loss reduction method based on the load regulation characteristics of electrolytic aluminium is analysed. The result shows that regulate electrolytic aluminium load and conventional generations at the same time can significantly reduce the network loss.

Acknowledgements

This research is supported by the National Development and Reform Commission (the research, development and industrialization for integrated management of smart distribution grid & utilization and enhancing energy efficiency technology based on internet of things).

References

1. Xiao Chuangying, Wang Ningbo, Zhi Jing, et al, Automation of Electric Power Systems, **64** (2010)
2. Liu Chang, Wu Hao, Gao Changzheng, et al, Power System Protection and Control, **61** (2014)

DOI: 10.1051/matecconf/20167706015

3. Cui Hong, Xia Chengjun, Luo Zongjie, et al, Electric Safety, **54** (2009)
4. Zhen Xin, Yang Dezhou, Wang Liping, et al, Power System Protection and Control, **39** (2011)
5. Liu Dewei, Huang Yuehui, Wang Weisheng, et al, Automation of Electric Power Systems, **77** (2011)
6. Luo Qing, Chao Qin, Wang Yibo, et al, Electric Power Automation Equipment, **42** (2014)
7. Zhou Wei, Peng Yu, Sun Hui, et al, Proceedings of CSEE, **13** (2009)
8. Liu Xiaocong, Wang Beibei, Li Yang, et al, Power System Technology, **2955** (2014)
9. Qiu Zhuxian, Engineering Science, **41** (2003)
10. Lu Yufei, Yuannan Metallurgy, **58** (2004)
11. Chen Runze, Sun Hongbin, Jin Hongyang, et al, Automation of Electric Power Systems, **168** (2015)
12. Feng Naixiang, Peng Jianping, Wang Yaowu, et al, Journal of Material and Metallurgy, **1** (2010)
13. Chen Jianhua, Li Xianyun, Deng Donghua, et al, Power System Technology, **77** (2007)

MATEC Web of Conferences 77, 06016 (2016) DOI: 10.1051/matecconf/20167706016
ICMMR 2016

Evolving Resilient Back-Propagation Algorithm for Energy Efficiency Problem

Yang Fei , Gao Pengdong and Lu Yongquan

High Performance Computing Center, Communication University of China, Beijing, China

Abstract. Energy efficiency is one of our most economical sources of new energy. When it comes to efficient building design, the computation of the heating load (HL) and cooling load (CL) is required to determine the specifications of the heating and cooling equipment. The objective of this paper is to model heating load and cooling load buildings using neural networks in order to predict HL load and CL load. Rprop with genetic algorithm was proposed to increase the global convergence capability of Rprop by modifying a corresponding weight. Comparison results show that Rprop with GA can successfully improve the global convergence capability of Rprop and achieve lower MSE than other perceptron training algorithms, such as Back-Propagation or original Rprop. In addition, the trained network has better generalization ability and stabilization performance.

1 Introduction

Artificial intelligence techniques have emerged as powerful tools that could be used to replace time-consuming procedures. In the last decade, research into ANNs has shown explosive growth, the interest showed to them is mainly due to their ability resides in appealing properties they exhibit: adaptability, learning capability, and ability to generalize [1]. ANNs were developed initially to model biological functions, they learn from experience and rapidly solve hard computational problems. With the spread of computers, later research was directed at exploring the possibilities of using and improving them for performing specific tasks. Nowadays, ANNs are receiving a lot of attention from the international research community with a large number of studies concerning training, structure design, and real world applications, ranging from classification to robot control or vision [2]-[4]. The neural network training task is a capital process in supervised learning, in which a pattern set made up of pairs of inputs plus expected outputs, is known beforehand, and used to compute the set of weights that makes the ANN to learn it.

Back-propagation (BP) [5] is the most popular ANN training algorithm. It is a gradient descent method that aims at minimizing the total root mean squared error (RMSE) between actual and desired outputs of a network by employing gradient information. BP imposes limitations on a type of the ANN activation function. It requires the activation function to be differentiable. Although BP has been successfully applied to various problems, it has a number of drawbacks due to its gradient descent nature. Among shortcomings of BP are slow learning process, tendency to get trapped in local minimum easily, and difficulties in training large ANNs, i.e., ANNs with large number of layers and neurons [6]. Resilient back-propagation (Rprop) is considered the best algorithm, measured in terms of convergence speed, accuracy and robustness with respect to training parameters [7]. Unlike BP algorithm, Rprop do not need to calculate derivatives of the error function, it use only the sign of the derivative to determine the direction of the weight update and thus, can work with non-differentiable or even non-continuous functions. Comparing to the traditional BP algorithm, the Rprop algorithm offers faster convergence and is usually more capable of escaping from local minima [8].

However, the efficiency and efficacy of both BP and Rprop algorithm depend significantly on initial weights which are initialized to small random values in the same interval. The Genetic algorithm (GA), one of the evolutionary algorithms, is a heuristic stochastic search algorithm [9]. Due to the good global searching ability and can learn the near-optimum solution without the gradient information of error functions, More attention has been attracted on combination of evolutionary algorithms and artificial neural networks. In recent years, GA has been applied to complex problems and a lot of valuable conclusions have been obtained [10]-[12].

In this paper, we combine GA and Rprop, with more "global" information like the magnitude of the network batch error, in order to improve the learning speed. That is, to optimize the connection weights of the neural networks with GA, so as to overcome the disadvantage of getting stuck in local minimum easily. First we describe theory basis of Neural Network. Then GA and Rprop algorithm is introduced, respectively. Furthermore, the hybrid algorithm GARP is proposed to perform analysis for energy efficiency problem. Finally, our results are discussed and conclusions are drawn.

2 Artificial neural networks

Artificial Neural Networks (ANNs) was first described by Warren McCulloch and Walter Pitts [13], is a mathematical model that consists of a set of interconnected processing elements, called neurons or nodes. ANNs were inspired by biological nervous systems, which propagate information through many nerve cells connected to each other with nerve fibres. Analogously to natural nervous systems, neurons in ANNs are connected to each other into a processing network, where each neuron receives, modifies and transfers signals. Fig. 1 shows a typical model of an artificial neuron. The links between neurons are called connections. Each connection has a corresponding value, called a connection weight, which is used to modify an incoming signal.

The outgoing signal of each neuron is calculated by means of the activation function (also referred to as the transfer function) expressed by the following equation

$$y_i = f_i(\sum_{j=1}^{n} x_j w_{ij} - \theta_i) \qquad (1)$$

where y_i is the output of a neuron i, x_j is the jth input signal to a neuron i, wij is the connection weight between neurons i and j, θ_i is an internal threshold value of a neuron i, i.e., value, which weighted sum of incoming signals needs to achieve to activate a neuron. Mathematically, the threshold value shows the position of the highest increment of the monotony increasing activation function.

The function is usually chosen such that it adds nonlinearity in the work of a network, but does not change the result significantly. Most popular are S-shape (sigmoid) functions, such as logistic, hyperbolic tangent, Heaviside, etc. However, some linear functions, e.g. piecewise linear function, are also applicable to ANNs with non-gradient learning methods.

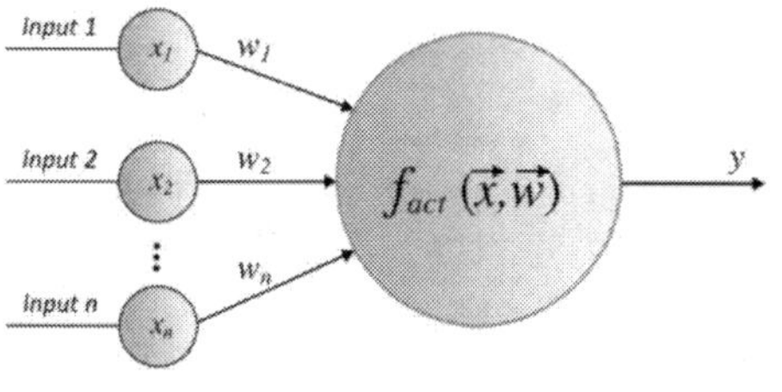

Figure 1. A model of an artificial neuron.

As a first step, the architecture of the network has to be decided. In this work we will concentrate one simple model: the feedforward networks. The feedforward model comprises networks, in which the connections are strictly feedforward, i.e., no neuron receives input from a neuron to which the former sends its output, even indirectly. To be precise, we will consider the so-called multilayer perceptron (MLP), in which units are structured into ordered layers, and connections are allowed only between adjacent layers in an input-to-output sense. Neurons in an ANN are located in layers, which are classified into input layer, hidden layer and output layer. In each neuron, we assign an activation function which will be triggered by weighted input signal. Neurons in one layer are connected, fully or partially, to the Neurons in the next layer. Feedback connections to previous layers are also possible. Fig. 2 shows a typical three layers MLP in which neurons are structured into ordered layers, and weighted connections are allowed only between adjacent layers units or neurons. The network contains two inputs, three hidden and one output neurons.

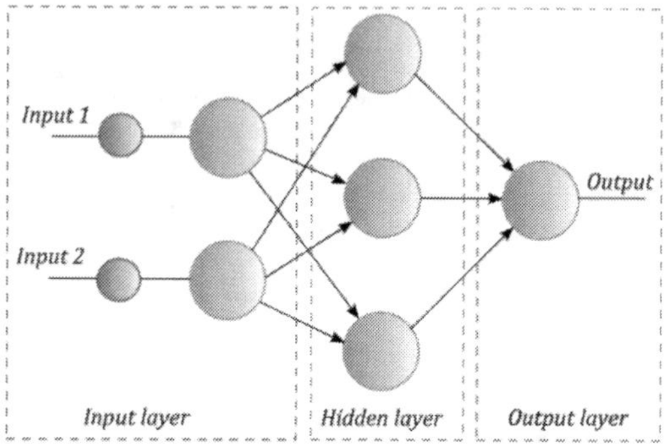

Figure 2. ANN with two input neurons, one hidden layer with three neurons in it, and one output neuron.

For any MLP, after defining the number of layers and the number of units per layer, the last step in the design is to adjust the weights of the network, so that it produces the desired output when the corresponding input is presented. This process is known as training the ANNs or learning the network weights. The network training is iterative processes which adjust weights and threshold, to reduce the error of network at a minimum or stop at a preset training step. As initially stated, we will focus on the learning situation known as supervised training, in which a set of input/desired-output patterns is available. Thus, the ANN has to be trained to produce the desired output according to these examples. The input and output of the network are both real vectors in our case. Then give the input of samples and obtain the output results. The essence of a learning algorithm is the learning rule, which determines how connection weights are changed.

3 The hybrid algorithm

3.1 Genetic algorithm

Genetic algorithms (GAs), developed by John Holland, represent the most popular type of evolutionary algorithms, a search technique used in computing to find exact or approximate solutions to optimization and search problems. GA was inspired by the adaptation and learning capabilities of nature species. Compared to other methods, GA is capable of exploring discontinued spaces of solutions with a minimum background knowledge and domain-based information. GA is directed random search techniques used to look for parameters that provide a good solution to a problem. Essentially they are nothing more than educated guessing. The "education" comes from knowing the suitability of previous candidate solutions and the "guessing" comes from combining the fitter attempts in order to evolve an improved solution.

GAs apply all genetic operators follow the classical order: first, they will select individuals for reproduction, then produce new individuals by applying crossover and finally, modify offspring by means of the mutation operator. After these steps, offspring individuals are included in the new population.

3.2 The resilient propagation algorithm

The Resilient Propagation algorithm (Rprop) is an improved adaptation of the batch backprop algorithm, and for numerous problems it converges very quickly. It uses only sign of the back-propagated gradient to change the bias/weights of the network, instead of the magnitude of the gradient itself. Therefore, the adaptation of the weight-step is not "blurred" by gradient behavior. Instead, each weight has an individual evolving update-value and the size of the weight change is determined by a separate update value. The purpose of the Rprop training algorithm is to eliminate these harmful effects of the magnitudes of the partial derivatives. Only the sign of the derivative can determine the direction of the weight update; the magnitude of the derivative has no effect on the weight update. The size of the weight change is determined by a separate update value. Rprop updates weight as follows.

$$\Delta_{ij}(t)=\begin{cases}\eta^{+}\cdot\Delta_{ij}(t-1), & if\ \frac{\partial E}{\partial w_{ij}}(t-1).\frac{\partial E}{\partial w_{ij}}(t)>0\\ \eta^{-}\cdot\Delta_{ij}(t-1), & if\ \frac{\partial E}{\partial w_{ij}}(t-1).\frac{\partial E}{\partial w_{ij}}(t)<0\\ \Delta_{ij}(t-1), & otherwise\end{cases} \quad (2)$$

Where $0<\eta^{-}<1<\eta^{+}$, and η^{+} is empirically set to 1.2 and to 0.5, respectively.

Rprop updates weight can be simply described as follows: Whenever the partial derivative of the error function E with respect to the corresponding weight wij changes its sign, it indicates that the value of last update was too big and the algorithm has jumped over a local minimum. The update-value is decreased by a factor. If the derivate retains its sign, the update-value is slightly increased by the factor in order to accelerate convergence in shallow regions.

Once the update-value for each weight is adapted, the weight-update itself follows a very simple rule: if the derivative is positive (increasing error), the weight is decreased by its update-value, if the derivate is negative, the update-value is added to the weight.

$$\Delta w_{ij}(t)=\begin{cases}-\Delta_{ij}(t), & if\ \frac{\partial E}{\partial w_{ij}}(t)>0\\ +\Delta_{ij}(t), & if\ \frac{\partial E}{\partial w_{ij}}(t)<0\\ 0, & otherwise\end{cases} \quad (3)$$

There is one exception to the rule above. If the partial derivative changes sign,

$$\Delta w_{ij}(t)=-\Delta w_{ij}(t-1), \text{if } \frac{\partial E}{\partial w_{ij}}(t-1)\cdot\frac{\partial E}{\partial w_{ij}}(t)<0 \quad (4)$$

In order to avoid a double punishment of the update value, there should be no adaption of the update value in the succeeding step. In general, this can be done by setting $\frac{\partial E}{\partial w_{ij}}(t-1)=0$ in the $\Delta_{ij}(t)$ update-rule above.

3.3 GARP mechanism

In this work, GA is used to find a good initial set of connection weights for ANN so that GA-aided Rprop (GARP) may find the global optimum even faster than Rprop only case. The evolution process starts with the generation of an initial population of individuals, where each individual represents a possible set of connection weights for a given ANN topology. After initialization, each individual is evaluated according to its worth regarding to the solving problem. Traditional GAs use binary stings to represent individuals and apply classical crossover and mutation operators. However, modern GAs tends to utilize real-valued representations, since it reduces the length of the genetic string, which is especially important for solving complex problems with many parameters. In this study, each individual (a set of weights for a given ANN) is encoded by concatenation of all connection weights of a network, whereby the connection weights to the same hidden/output node are put together.

The role of the fitness function in EANNs plays the error function, i.e., MSE or RMSE; the fitness function is employed to evaluate the quality of the solution at each epoch. In this work, fitness function is calculated by the average of the error. In the prediction problem, the error is based on the mean squared error (MSE) or root mean squared error (RMSE) in the following formula.

$$MSE=\frac{1}{N}\sum_{i=1}^{N}(y_i-\hat{y}_i)^2 \quad (5)$$

$$RMSE=\sqrt{\frac{1}{N}\sum_{i=1}^{N}(y_i-\hat{y}_i)^2} \quad (6)$$

$$Fitness=\frac{1}{MSE} \quad (7)$$

Where N is the total number of instances, y_i $\hat{y}_i$ are the actual value and the model output respectively.

DOI: 10.1051/matecconf/20167706016

Following that, typical steps of the classical EA are performed (see Fig. 3): first, individuals are selected for reproduction based on their fitness; then new individuals are formed by means of genetic operators; and finally, new population is created. The evolution process continues until some termination criteria are reached. A typical cycle of the evolution of connection weights is as follows:

- Generate a population of size k, where each individual represents a set of n connection weights;
- Evaluate fitness of each individual in the population according to the error between actual and desired outputs over the training data;
- Select parents for reproduction based on their fitness;
- Create offspring by applying genetic operators to the parents;
- Form the next generation of k offspring individuals.

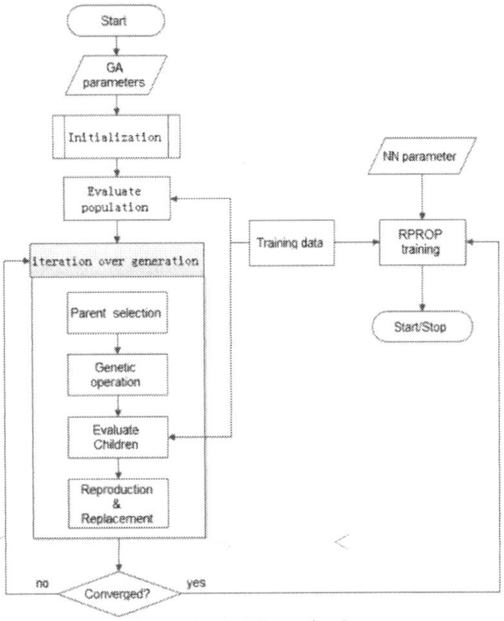

Figure 3. GARP mechanism.

4 Results and discussion

The energy efficiency data set taken from UCI database was used for comparison [14]. The dataset comprises 768 samples and 9 features, aiming to predict two real valued responses. The dataset contains eight attributes and two responses list in Table 1. The aim is to use the eight features to predict each of the two responses: heating load (Y_1) and cooling load (Y_2) [15]. The building parameters used are: relative compactness (X_1), surface area (X_2), wall area (X_3), roof area (X_4), overall height (X_5), orientation (X_6), glazing area (X_7), and glazing area distribution (X_8). We will try to predict heating load and cooling load.

Table 1. The parameters of enegry efficiency data set.

Mathematical	Representation Input or output variable
X_1	Relative compactness
X_2	Surface area
X_3	Wall area
X_4	Roof area
X_5	Overall height
X_6	Orientation
X_7	Glazing area
X_8	Glazing area distribution
Y_1	Heating load
Y_2	Cooling load

To evaluate an ANN, training used 80% of the total patterns and the rest were used for forecasting out of samples using the trained network. We used an 8-7-2 MLP architecture of input-hidden-output units yielding a population vector of 79 members (weights). The weight value range is set to [-10, 10]. The initial GA population is set to 50 and the weights and bias is generated randomly for each individual. The GA was executed for 10 generations on a population of 50 individuals. The activating function of the neurons is the sigmoid function.

In order to evaluate performance of GARP, two other algorithm BP and Rprop are involved. We trace the MSE of the training and testing pattern set after performing 50 independent runs. Table 2 shows the training error and testing error that obtained with GARP algorithm and compares them to the BP and Rprop algorithm. Comparing results are plotted shown in Fig. 4.

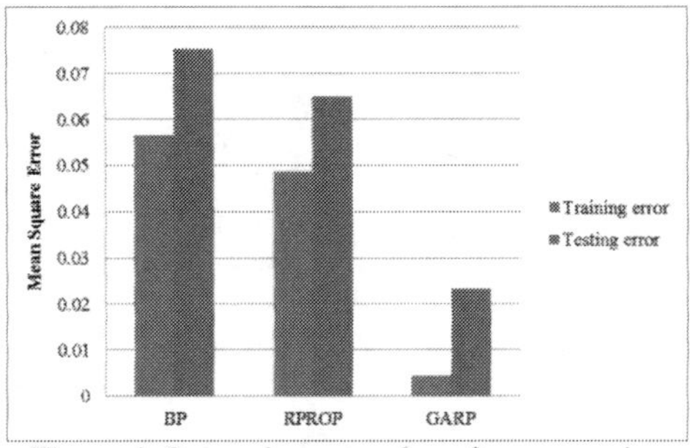

Figure 4. Comparing network performance using different algorithms.

As it has been shown, we can observe that GARP is the faster algorithm, what confirms intuition on the velocity of local search compared to BP and Rprop. The MSE of GARP algorithm is much smaller than that of BP and Rprop. The convergence analysis shows that it carries the characteristic and the advantages of them and without any side effect. In the simulation result, the global convergence capability is further improved compare with BP and the original Rprop, and the training error and testing error is 0.004349 and 0.023233 respectively. It also fixed the low convergence rate problem in Rprop with GA and some of them are even higher than Rprop with BP. Overall, it outperformed BP, Rprop in terms of convergence rate and global convergence capability.

Fig. 5 and Fig. 6 plot the heating load and cooling load from ANN and testing data, respectively. It is shown that good agreement is achieved between GARP prediction and target. The trained GARP model can be able to perform prediction task well. Specifically, starting the evolution process with more generations, Rprop with GA strategy produced more compact networks with smaller testing errors.

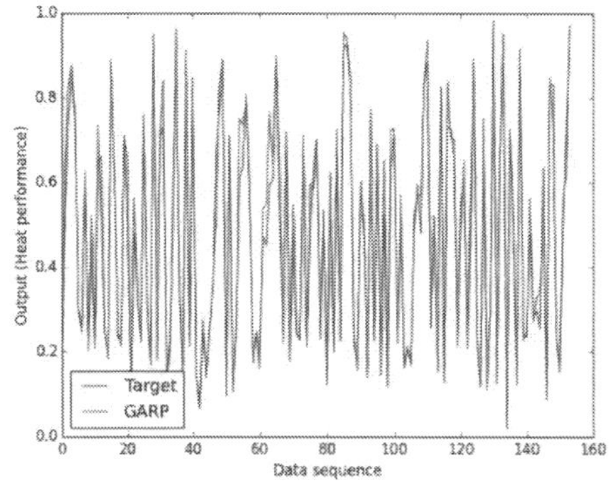

Figure 5. Output heating load from GARP and testing dataset.

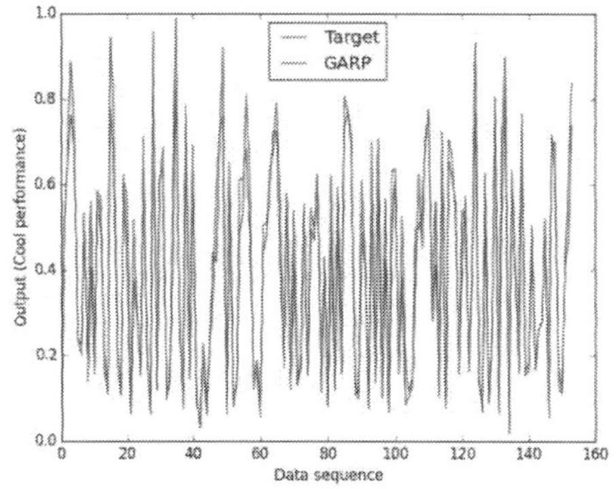

Figure 6. Output cooling load from GARP and testing dataset.

5 Conclusions

It is widely accepted that the Rprop algorithm is one of the best performing first order learning algorithms for multilayer neural networks. In this paper we introduced an efficient improvement of the Rprop algorithm that is based on genetic algorithm. Furthermore, in energy efficiency problem, the differences training error and testing error between Rprop with GA, BP and the original Rprop are statistics. Our results show that GARP algorithm gets lower MSE than the problem-specific algorithms (BP and Rprop). This will improve the convergence rate of the network and reduce the training failure, and the neural network's generalization ability is better than the algorithms that only use BP or Rprop. As a result, evolving neural networks can lead to enhanced performance and make GARP look as a promising algorithm for neural network training. As a future work we plan to add new algorithms to the analysis, and to apply them to more instances. In addition, Rprop, a batch training algorithm, can be parallelized easily to improve the computing efficiency. Therefore, future work will also focus on accelerating the training speed through parallel technique.

References

1. J.H. Holland, Adaptation in Natural and Artificial Systems: An Introductory Analysis with Applications to Biology, Control, and Artificial Intelligence., U Michigan Press (1975).
2. B. Temeyer, W.G. Jr, … Symposium, Iowa (2003).
3. H. Hasanzadehshooiili, A. Lakirouhani, A. Šapalas, Archives of Civil and Mechanical Engineering **12**, 477 (2012).
4. J. Fernández, A. Pernía, F.J. Martínez-de-Pisón, R. Lostado, Engineering Structures **32**, 3018 (2010).
5. Y. Hirose, K. Yamashita, S. Hijiya, Neural Networks **4**, 61 (1991).
6. B. Widrow, M.A. Lehr, Proceedings of the IEEE **78**, 1415 (1990).
7. M. Riedmiller, H. Braun, RPROP - A Fast Adaptive Learning Algorithm, Proc. Of ISCIS VII), Universitat (1992).
8. M. Riedmiller, H. Braun, in Neural Networks, 1993., IEEE International Conference On, IEEE **586** (1993).
9. L. Davis, others, Handbook of Genetic Algorithms, Van Nostrand Reinhold New York (1991).
10. X. Yao, Proceedings of the IEEE **87**, 1423 (1999).
11. G.-H. Kim, J.-E. Yoon, S.-H. An, H.-H. Cho, K.-I. Kang, Building and Environment **39**, 1333 (2004).
12. F. Ahmad, N.A.M. Isa, Z. Hussain, R. Boudville, M.K. Osman, in CICSyN **78** (2010).
13. J.-S.R. Jang, C.-T. Sun, Neuro-fuzzy and Soft Computing: a Computational Approach to Learning and Machine Intelligence, Prentice-Hall, Inc. (1996).
14. C. Blake, C.J. Merz, Department of Information and Computer Science **55** (1998).
15. A. Tsanas, A. Xifara, Energy and Buildings **49**, 560 (2012).

A Simplified Motion Isolated Load Control for Multi-Channel Syncro-Loading System

Zhang Yanyan, Yang Zhidong

Harbin Institute of Technology, Institute of Electro-hydraulic Servo Simulation Test System, 92 West Dazhi Street , Harbin, China

Abstract. For the multi-channel syncro-loading (MCSL) test, the specimen may be driven by a particular motion of another position channel, representing an in-service duty cycle. The loading accuracy is largely affected by the specimen motion. To reduce the cross coupling from the motion, the cross coupling characteristics between multi-channels is analyzed, and a simplified motion isolation load control (MILC) algorithm is provided, in which the velocity and acceleration of the specimen, measured by the accelerometer, are added into the forward path of the loading control channel in order to drive the actuator to match the motion of the specimen. The test results indicate that the MILC algorithm can significantly improve the loading accuracy compared with the conventional loading control method.

1 Introduction

The multi-channel syncro-loading system driven by electro-hydraulic actuators is an important facility for the structure test, and it has been extensively applied in the aerospace, train, vehicle, and material fields [1-4]. During a test, it is necessary to simulate the multi-directional load on the specimen, and the loading accuracy is very important to evaluate the strength, lifetime and performance of the specimen. However the conventional control method, which controls each actuator independently[1, 3-5], can not usually reach the desired loading accuracy, because there is the strong cross coupling among the actuators. Especially, when controlling the position and force simultaneously, the force of one channel can be affected by the position of another channel seriously.

In the paper, to eliminate this loading coupling from the position channel, and improve the loading accuracy, the cross coupling characteristics between the position channel and the force channel is analyzed, and a motion isolated load control algorithm is proposed, which is then validated in the fatigue test of the train's bogie.

2 The effect of motion on loading accuracy

The multi-channel syncro-loading system usually includes more than one actuator, which may be controlled by position or force. There are some cross coupling between the force controlled actuator and the position controlled actuator.

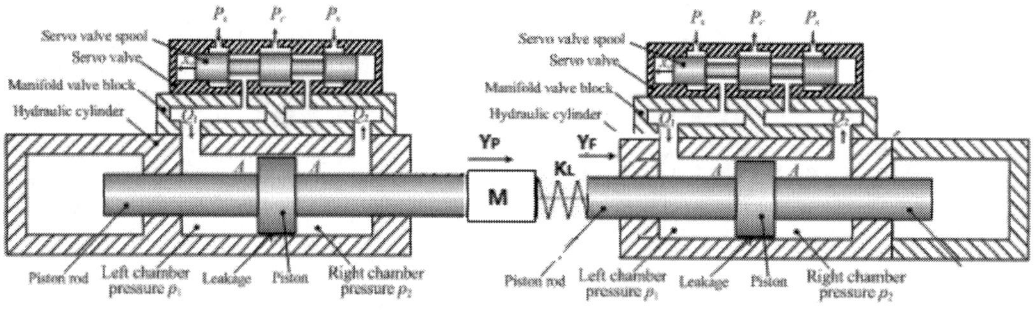

Figure 1. Actuators controlled by positon and force in MCSLs.

The actuator is mainly composed of the high-frequency electro-hydraulic servo valve and the double rod asymmetric hydraulic cylinder, the model of which is shown as Fig. 1. For the force controlled actuator, the mathematical mode can be derived.

The load flow supplied by the valve is the function of the spool displacement and the load pressure drop:

$$Q_{LF} = K_{qF} \cdot X_{vF} - K_{cF} \cdot P_{LF} \tag{1}$$

where, Q_{LF} is the load flow, P_{LF} is the load pressure drop, K_{qF} is the linearized flow coefficient, K_{cF} is the linearized flow-pressure coefficient and X_{vF} is the spool displacement of the servo valve.

Taking leakage and compressibility into consideration, the load flow of the hydraulic cylinder is:

$$Q_{LF} = A_F s Y_F + \left(C_{tcF} + \frac{V_{tF}}{4\beta e} s \right) P_{LF} \tag{2}$$

where, A_F is the piston area, Y_F is the piston displacement, C_{tcF} is the total leakage coefficient of the cylinder, which is comprised of the internal leakage and external leakage, V_{tF} is the total volume of the oil in two cylinder chambers, βe is the bulk modulus of the fluid.

The output force of hydraulic cylinder is:

$$A_F P_{LF} = m_F s^2 Y_F + B_{cF} s Y_F + F \tag{3}$$

$$F = K_L \left(Y_F - Y_P \right) \tag{4}$$

where, m_F is the equivalent weight of the piston, B_{cF} is the viscous damping coefficient of the piston, and F is output force of the hydraulic cylinder. The piston displacement Y_p of the actuator controlled by position can affect the output force, as shown in Eq.4, and it can be seen as the disturbance variable.

Combining Eq.1- Eq.4,we can get the Transfer Function (TF) from Y_p or X_{vF} to F, as illustrated in Fig .2.

It can be further simplified as Fig. 3.

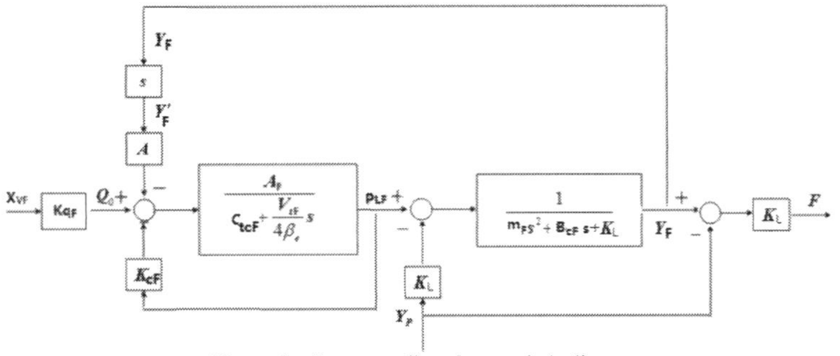

Figure 2. Cross coupling characteristic diagram

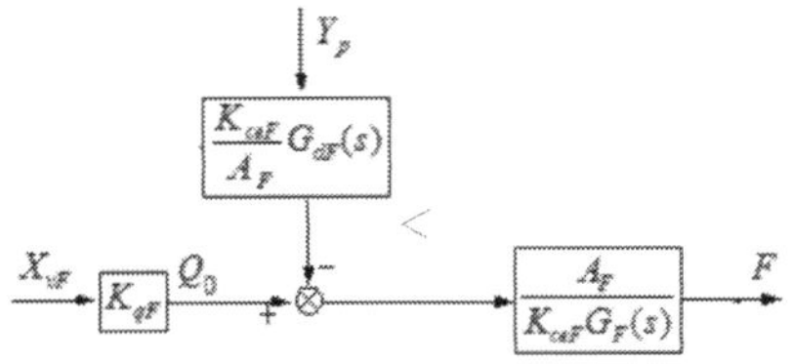

Figure 3. Simplified cross coupling characteristic diagram

Supposing X_{vF} =0, the output F derived from Y_p is the coupled force F_d , which can be seen as the loading error, and undesired. It can be shown as:

$$F_d = Y_p \cdot G_{dF} / G_F \tag{5}$$

where

$$G_{dF}(s) = \frac{A_F^2 + K_{ceF} B_{cF}}{K_{ceF}} \left[\frac{m_F V_{tF}}{4\beta_e \left(A_F^2 + K_{ceF} B_{cF} \right)} s^2 + \left(\frac{B_{cF} V_{tF}}{4\beta_e \left(A_F^2 + K_{ceF} B_{cF} \right)} + \frac{m_F K_{ceF}}{A_F^2 + K_{ceF} B_{cF}} \right) s + 1 \right] s$$

$$G_F(s) = \frac{m_F V_{tF}}{4\beta_e K_F K_{ceF}} s^3 + \left(\frac{B_{cF} V_{tF}}{4\beta_e K_F K_{ceF}} + \frac{m_F}{K_F} \right) s^2 + \left(\frac{V_{tF}}{4\beta_e K_{ceF}} + \frac{B_{cF} K_{ceF} + A_F^2}{K_F K_{ceF}} \right) s + 1$$

According to Eq.5, it can be seen that the coupled force F_d is determined by the velocity, acceleration and jerk of the position system. And the jerk is just a temporary result, the key elements are only velocity and acceleration.

3 Simplified MILC algorithm

For the velocity and acceleration of the specimen are the key elements for the coupled force F_d, in order to reduce the influence of the coupled force on the dynamic loading precision of the loading system, the simplified MILC (Motion Isolated Load Control) is proposed, in which the velocity and acceleration of the position controlled channel is feed forward on the force controlled channel to compensate influence of the coupled force, as shown in Fig. 4. If the high frequency noise is introduced by the acceleration feed forward, a low pass filter before the high pass filter. The compensation rate is limited by the maximum velocity of the actuator.

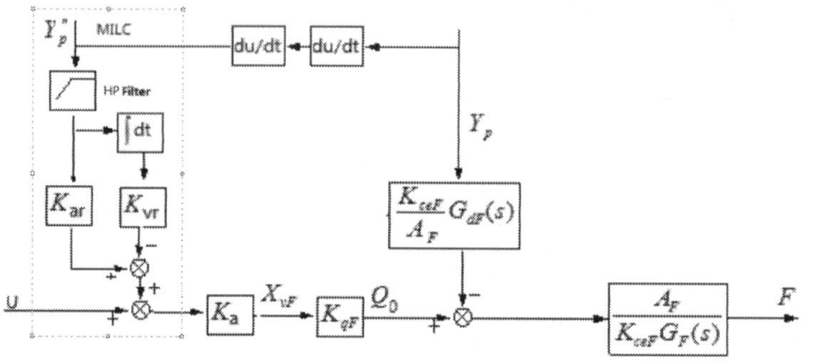

Figure 4. MILC using Acceleration and Velocity

In the simplified MILC, two hardware additions are required compared to the conventional loading control method:

(1) An accelerometer, measuring the acceleration along the axis of movement of the MILC-controlled actuator, must be attached on the specimen.

(2) A compliant element, for example a rubber bush, should be fitted between the MILC-controlled actuator and the specimen. This provides some mechanical isolation between the specimen and the actuator.

The motion of the specimen, in terms of the velocity and acceleration, is obtained though the accelerometer. Then the velocity and acceleration signals are added into the forward path of the loading channel in order to drive the actuator match the motion of the specimen.

Besides any remaining mismatch between the actuator and specimen motion causes a loading error. The size of this error is directly related to the stiffness of the compliant element. The lower stiffness is better, but reducing the stiffness will have an effect on the performance envelope of the loading system. For example, the actuator is more likely to run out of the stroke for high load due to the extra movement of the compliant member.

3.1 Tuning velocity feed forward

Give a small or zero loading command to the actuator and excite the specimen with a high amplitude, low frequency motion (e.g. 2Hz), an unwanted loading variation will be seen. This is mainly related to the velocity of the specimen, which is the dominated factor at the low frequency.

Increase the velocity gain until the amplitude of the loading variation reaches the minimum. Theoretically, the optimal value for the velocity gain is the reciprocal of the maximum velocity of the actuator.

The 2nd order high pass filter in the method normally needs no adjustment. It affects the estimation of the velocity from the acceleration measurement as follows:

The cut-off frequency should be well below (e.g. by a factor of 10) the minimum frequency of interest to minimize phase distortion for the generated velocity. And a lower damping will cause that the phase distortion diminishes more rapidly as the frequency increasing, especially for the frequency greater than the cut-off frequency, but there is more signal amplification at the cut-off frequency. And you may find there is a small loading variation at the frequency similar to the cut-off frequency of the high pass filter. This can be removed by increasing the PID integral gain, or by increasing the cut-off frequency and/or damping.

3.2 Tuning acceleration feed forward

Keep a small or zero loading command for the actuator and excite the specimen with a high frequency motion (10Hz or greater), an unwanted loading variation will be seen; if the velocity gain has been adjusted correctly, this variation will mainly be related to the acceleration of the specimen.

Start the acceleration gain with 0.001 times the velocity gain. Increase the acceleration gain until the loading variation reaches the minimum.

4 Experimental studies

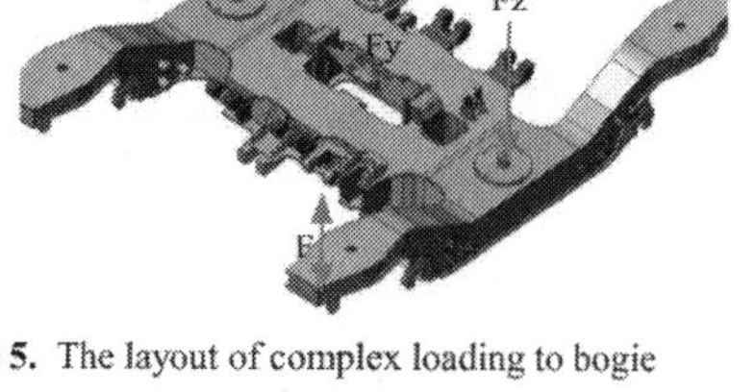

Figure 5. The layout of complex loading to bogie

As the proof of the MILC, the fatigue test of the train's bogie is carried out according to the criteria in 《UIC615-4-2003_20OR_20》[6]. The loading channels controlled by the force, which is definited in Fig.5, include two vertical forces *Fz* and one longitudinal force *Fy*. Meantime, there is another position channel to simulate the road irregularity.

Some experimental results are demonstrated in Fig.6, the feedback values of two vertical force loading channels and one position channel are recorded. During the test, the velocity and acceleration feed forward doesn't work at first, which is equivalent to the conventional loading control method, then the velocity and acceleration feed forward gain is adjusted, and the MCSL method works. It can be seen that, the jitter of the loading force has occurred at the peak of the half sinusoidal wave of the position channel for the conventional method. Adjusting velocity and acceleration feed forward gains, the jitter disappears gradually until the satisfied loading accuracy is obtained.

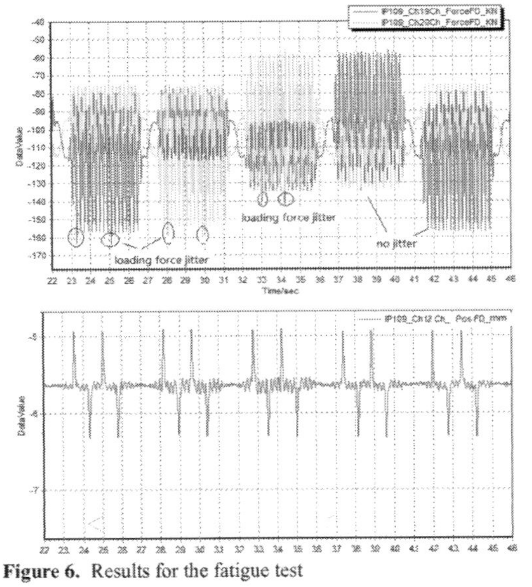

Figure 6. Results for the fatigue test

5 Conclusions

To eliminate the loading coupling from the motion of the position channel, and improve the loading accuracy, the cross coupling characteristics between multi-channels in the syncro-loading test is analyzed, and a simplified motion isolated load control algorithm is provided, in which the velocity and acceleration of the specimen are added into the forward path of the loading control channel in order to drive the actuator to match the motion of the specimen. Results of the train's bogie fatigue test show that the algorithm can significantly improve the loading accuracy compared with the conventional loading control method.

6 Acknowledgment

The author wishes to acknowledge contribution to some tests in this paper from Dr. Liu Weiya of the National Engineering Lab for High Speed Railway in Qingdao.

References

1. Dong Bo. Research on Application of Multi-channel Syncro-loading System in Strength Test of Power Bogie Frame. Railway Quality Control, **10**, 13-15 (2007).
2. Yan Junmao. Vehicle Engineering(Third Edition). Beijing. China Railway Publishing House. 213-254 (2007)
3. Li Jupeng, Wei Wenjun. Research on Electro-hydraulic Servo Loading System. Gansu Science and Technology, **21(3)** , 97-98 (2005)
4. Sui Jiefei. Design and Implementation of Electro-hydraulic Servo Loading System of Distributed Rudder. Civil Aircraft Design and Research, **38(2)**, 26-31 (2011)
5. Lu Lin,Wang Jun Zheng,Chen Cai Ke,Wang Li Tao. Application of Electro-hydraulic Force Control System in Structure Intensity Experimental Facility. Chinese Hydraulics & Pneumatics, **8**, 57-59 (2004)
6. UIC615-4-2003_20OR_20. Motive power units-Bogies and running gear-Bogie frame structure strength tests.UIC. (2003)

MATEC Web of Conferences 77, 07002 (2016) DOI: 10.1051/matecconf/20167707002
ICMMR 2016

Bursting detection by Variable-Interval Time-Averaging (VITA) on spanwise oscillated turbulent channel flow

Atilla **Altıntaş**

Karabuk University, Engineering Faculty, 78050 Karabuk, Turkey

Abstract. The main aim of the study is to detect the bursting phenomenon in spanwise oscillated Lorentz force applied turbulent channel flow. Regarding the biggest reason of the turbulence production is known as the bursting phenomenon, detecting the decrease on it could be a very effective tool as an estimation of the effectiveness of the applied control strategy to the flow. The data gained by DNS (Direct numerical simulations) at low Reynolds number ($Re_\tau = 180$ based on the wall-shear velocity and channel half width). The Variable-Interval Time-Averaging (VITA) technique has been applied to three different streamwise locations for both applied force and no-force cases. The Lorentz force applied only to the lower wall. In first step the very well known drag reduction is observed, we compared the drag reduction results for the applied force and no-force cases both for the lower and upper wall of the channel. The bursting detection results are given statistically, which clearly shows that bursting phenomenon frequency tend to decrease as a result of the applied Lorentz force. It is also has shown that associated turbulence production has similar trend.

1 Introduction

The flow control studies are attracting a great attention since Prandtl successfully controlled a boundary layer [1], many numerical, experimental and theoritical studies are performed to control the drag in turbulent boundary layers [2–7]. One of the effective methods of the drag control is known as Lorentz force excitation. In this method, the magnets and electrodes are placed on the wall and by choosing an electrically conducting fluid the Lorentz force is generated in the desired direction.

The Lorentz force applications are performed in all three directions (streamwise, spanwise and wall-normal). It seems better results are obtained for spanwise direction [6].

It is known that Lorentz force used by [2] for flow control for the first time. They applied streamwise Lorentz force to a laminar boundary layer in order to delay the transition to turbulence of the flow over submerged vehicles. A more stable Blasius boundary layer profile under the influence of the Lorentz force is obtained by [8]. Later some experimental studies performed which explored the possibility of achieving viscous drag reduction in turbulent boundary layers [9]. An another important experimental study was later performed by [4], they placed the magnets and electrodes to direct the electromagnetic (EM) body force axially downstream and axially upstream. In these experiments a turbulence suppression up to 30% was achieved for the former case with the interaction parameter ($St \sim O(1)$ is used, which represents the so-called "interaction parameter" is the ratio of the Lorentz force to the inertial force. However, with the body force axially upstream these researchers found an augmentation of the turbulence.

O'Sullivan and Biringen showed that feedback control schemes are capable to reduce skin friction in turbulent boundary layers [10].

A DNS study of a turbulent channel flow performed at different low Reynolds numbers ($Re_\tau = 100, 200, 400$), in which an idealized Lorentz force is applied for both open-loop and closed-loop configurations. In an idealized open-loop case they achieved up to 40% drag reduction. They provided a relation between St number and a given Reynolds number for the best drag reduction [6].

This paper presents DNS simulations of channel flow (Reynolds number of 180) in which spanwise oscillating Lorentz force is applied along the spanwise direction. The aim of the paper is to figure out the bursting frequency under the influence of the spanwise oscillated Lorentz force in turbulent channel flow.

The paper is organized as follows. First, the methodology and equations are presented followed by a brief description of the numerical method. In the next section the results are presented and discussed and then some conclusing remarks are given.

2 Methodology

2.1 Governing equations

The resulting force can be estimated as a body force and directly added as a body force term to the Navier-Stokes equations. Thus, the governing equations take the following nondimensional form

$$\frac{\partial \mathbf{u}}{\partial t} + \mathbf{u} \cdot \nabla \mathbf{u} = -\nabla \mathbf{p} + \frac{1}{Re_\tau} \nabla^2 \mathbf{u} + St\,(\mathbf{J} \times \mathbf{B}), \quad (1)$$

$$\nabla \cdot \mathbf{u} = 0, \quad (2)$$

where Re_τ is the Reynolds number based on u_τ (wall shear velocity) and δ (half channel height). $St = J_0 B_0 \delta / [\rho u_\tau^2]$, is the Stuart number which represents the relative strength of the Lorentz force with respect to the inertia force, where J_0 and B_0 are the current density and the magnetic flux density values at the wall, respectively.

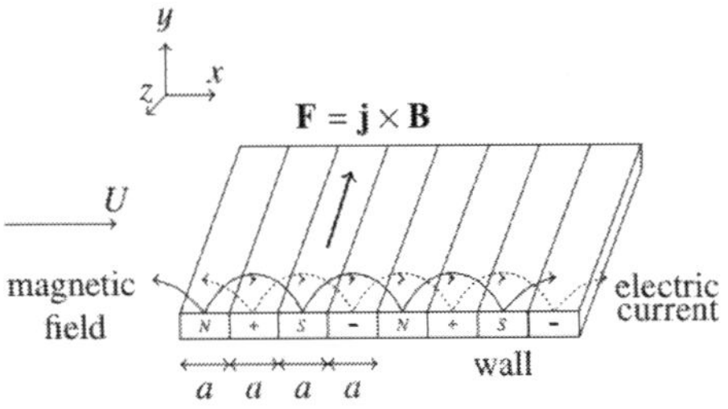

Figure 1. Illustration of magnets and electrodes arrangement for generating a Lorentz force along the spanwise direction [6].

2.2 Force

The spanwise force oscillates in time and decays exponentially in the wall-normal direction,

$$f_z^+ = St \exp\left(-\frac{\pi y^+}{a^+}\right) \sin\left(\frac{2\pi t^+}{T^+}\right). \quad (3)$$

According to [6] the rule defining the effective St value for different Reynolds numbers given as

$$\left(\frac{St\,T^+}{Re_\tau\,\pi}\right)_{opt} = 20. \quad (4)$$

We performed simulations for $Re_\tau = 180$. For $T^+ = 100$, equation 4 gives $St = 36\pi$. The parameter T^+ is the period of oscillation, and a^+ (the magnet and electrode widths, which here are assumed to be the same) sets the distance the force penetrates into the flow; it is here taken as $a^+/\pi = \Delta^+ = 10$. Lorentz force excitation is applied along the spanwise direction for forcing the flow to oscillate at a certain frequency.

The required Lorentz force can be created by placing electrodes and magnets side by side, in the streamwise direction parallel to one another, as shown in figure 1. This configuration generates Lorentz force in the spanwise direction which decays exponentially in the wall-normal direction. We apply the force in the lower half of the channel, but because of the exponential function it is active only for $y^+ = 60$.

2.3 Direct numerical simulations

An incompressible, finite volume code is used [11]. The numerical procedure is based on an implicit, fractional step technique with a multigrid pressure Poisson solver and a non-staggered grid arrangement. For the momentum equations, central differencing is used in space and the Crank-Nicolson scheme is used in the time domain.

A constant volumetric driving force is used in the streamwise momentum equation by which the frictional Reynolds number, $Re_\tau = 180$, is prescribed. Periodic boundary conditions are used in the streamwise and spanwise directions while the usual no-slip boundary conditions are enforced at the walls. The domain size is $2\pi\delta \times \pi\delta \times 2\delta$ in the streamwise, spanwise and wall-normal directions, respectively with grid size $98 \times 98 \times 98$. The superscript '+' sign indicates a nondimensional quantity scaled by the wall variables: e.g. $y^+ = yu_\tau/\nu$, where ν is the kinematic viscosity, u_τ is the wall-shear velocity. In this work u, v, w represents the streamwise, wall-normal and spanwise velocities respectively. Before applying any control, all simulations are allowed to reach a fully developed turbulent flow state.

3 Results and discussion

In figure 2 the time histories of drag forces in the vicinity of the wall, which is dominated by wall shear stress, τ_w, is demonstrated. In figure 2(a) there is not any control, in which the lower and upper wall exhibits similar drag forces. However in figure 2(b) the spanwise oscillated Lorentz force is applied to the lower wall, which exhibits a significant drag reduction compare to upper, no-force applied wall.

Figure 3(b) and figure 3(a) represents the contour plots for the spanwise velocity at the location of half the streamwise length of the channel ($x = \pi$). Comparison of the both cases suggests that in the applied force case the vortex structures near the lower wall are suppressed in applied force case compared to no-force case.

3.1 Bursting detection comparison

It is now very well known that the biggest reason of the turbulence production is the bursting phenomenon which accounts approximately 70% of the turbulence production [12–14]. Here we applied the Variable-Interval Time-Averaging, (VITA) technique to detect the bursting frequency in both the force applied and no-force cases. For a fluctuating quantity $Q(x_i, t)$ the variable-interval time average is defined by [15]

$$\widehat{Q}(x_i, t, T) = \frac{1}{T} \int_{t-\frac{1}{2}T}^{t+\frac{1}{2}T} Q(x_i, \tau)\, d\tau, \quad (5)$$

here T is the averaging time. As T becomes large, approaches to conventional averaging results. To represent a measure of the turbulent energy during time T a localized variance is defined as;

$$\widehat{var}(x_i, t, T) = \widehat{u'^2}(x_i, t, T) - [\widehat{u'}(x_i, t, T)]^2, \quad (6)$$

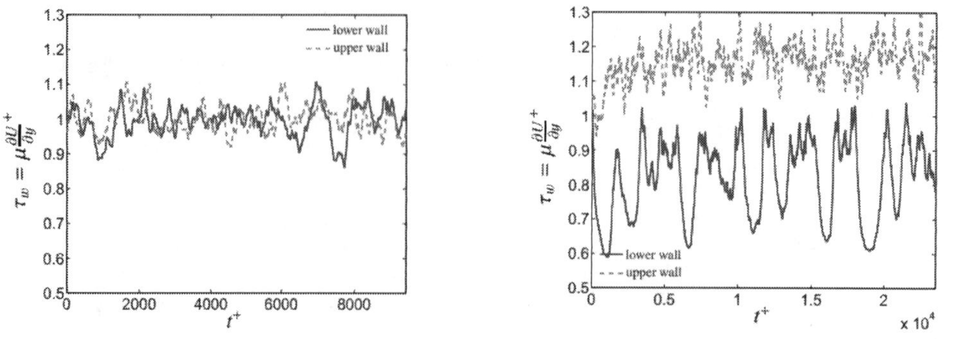

(a) No-force applied case.

(b) Force applied case.

Figure 2. Normalized drag history, Re_τ=180.

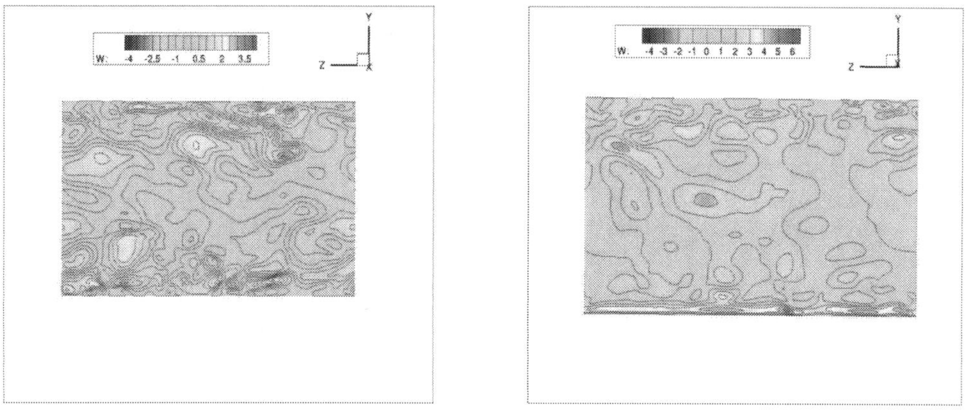

(a) No force applied case.

(b) Force applied case

Figure 3. Spanwise velocity field, contour plots from $x = \pi$. $Re_\tau = 180$.

The detection function $D(t)$ is defined as;

$$D(t) = \begin{cases} 1, & \text{for } \widehat{var} > k u'^2_{rms} \text{ and } \frac{\partial u'}{\partial t} > 0 \\ 0, & \text{otherwise} \end{cases} \quad (7)$$

where k is the threshold level, which here taken as $k = 1$ for detecting strongest structures responsible for the turbulence production [16]. The result of the bursting detections and related turbulence production are figured out in figure 4. The turbulence production values are compared with DNS data [17]. The data has been gathered at four different streamwise locations and at 10 different wall-normal locations for each streamwise locations. The streamwise locations are $x^+ = 275, 565$ and 1036. The wall-normal locations are $y^+ = 5, 10, 15, 20, 30, 40, 50, 62$.

It is observed that a big amount of the the bursting phenomenon occures at the vicinity of the wall ($y^+ < 20$) (figures 4(a),4(b),4(c)). As expected, the turbulence production results also parallel with the bursting frequency results which indicates that the most of the turbulence production occures at the vicinity of the wall (figure 4(d)).

The bursting frequency results clearly indicates that in all three different locations the bursting phenomenon occures in a much less frequency for the applied force cases compare to no-force cases (figures 4(a),4(b),4(c)). Beside the turbulence production values are lower at applied force wall for the applied force case compare to no-force case which is coherent with bursting detection results.

4 Conclusions

A DNS study of a fully developed turbulent channel flow is performed where a spanwise oscillating Lorentz force has applied near the lower wall (figure 1). Very well known drag reduction is obtained first (figure 2), then bursting phenomenon as well as related turbulence productions are investigated. The spanwise velocity contours are depicted

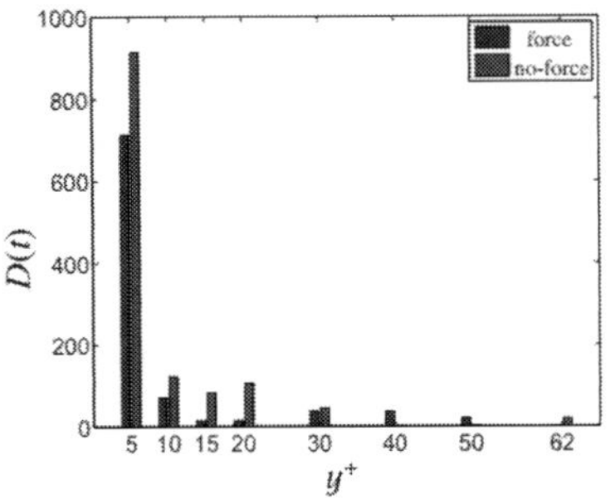

(a) Bursting detection at $x^+ = 275$.

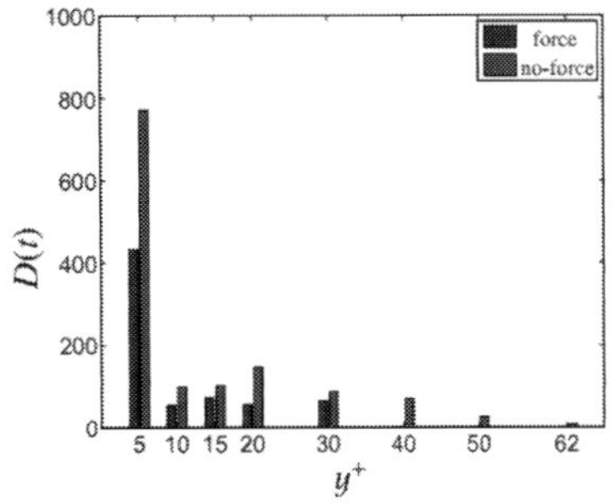

(b) Bursting detection at $x^+ = 565$.

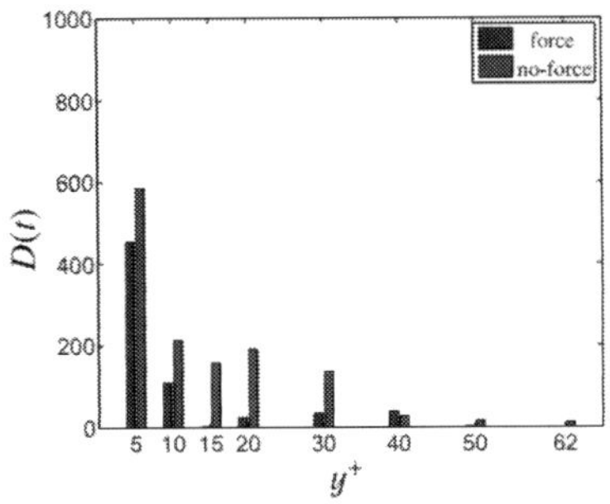

(c) Bursting detection at $x^+ = 1036$.

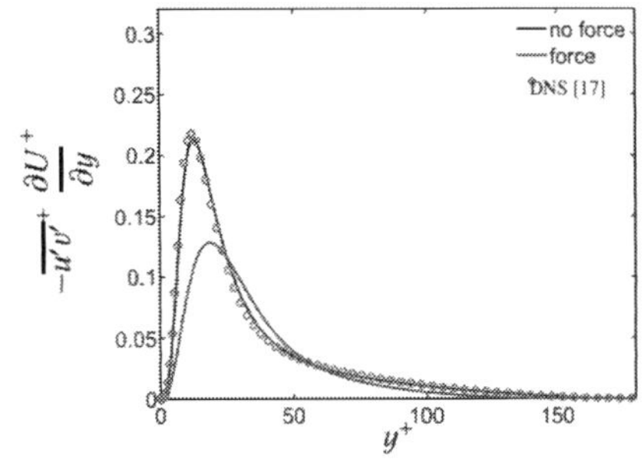

(d) Turbulence production rate, lower wall, $Re_\tau = 180$.

Figure 4. Burst detection at three different streamwise locations.

which clearly indicates that vorticies are suppressed in applied force case (figure 3). The VITA technique has been used to detect the bursting frequency which investigated at three different streamwise locations for ten different wall-normal locations.

We found that in the applied force cases bursting phenomenon occures in a less frequency compare to no-force cases at all three locations (figures 4(a),4(b),4(c)). We also support these results with related turbulence production values, which indicates that in applied force case turbulence production has lower values compare to no-force case (figure 4(d)).

References

[1] L. Prandtl, NACA TM 452 (1904)
[2] A. K. Gailitis, O. A. Lielausis, Appl. Magnetohydrodyn. Trudy Inst. **12**, 143-146 (1961)
[3] H. Choi, P. Moin, J. Kim, J. Fluid Mech. **261**, 75-110 (1994)
[4] C. Hennoch, J. Stace, Phys. Fluids **7**, 1371-1383 (1995)
[5] C. H. Crawford, G. E. Karniadakis, Phys. Fluids **9**, 788-806 (1997)
[6] T. W. Berger, J. Kim, C. Lee, Phys. Fluids **12**, 631-649 (2000)
[7] T. Albrecht, H. Metzkes, R. Grundmann, G. Mutschke, G. Gerbeth, Magnetohydrodynamics **44**, 205-222 (2008)
[8] A. B. Tsinober, A. G. Shtern, English translation: Magnetohydrodynamics **3**, 103-105 (1967)
[9] D. M. Nosenchuck, G. L. Brown, *in Near-Wall Turbulent Flows* Elsevier Science edited by R. M. C. So, C. G. Speziale, and B. E. Launder, Publishers B.V., New York, 689–698 (1993)
[10] P. L. O'Sullivan, S. Biringen, Phys. Fluids **10**, 1169-1181 (1998)
[11] L. Davidson, S.-H Peng, Int. J. Numer. Fluids **43**, 1003-1018 (2003)
[12] H. T. Kim, S. J. Kline, W. C. Reynolds, J. Fluid Mech. **50**, 133-160 (1972)
[13] J. Kim, NASA TM 85944 (1984);
[14] J. Kim, Phys. Fluids **28**, 52-58 (1985);
[15] R. F. Blackwelder, R. E. Kaplan, J. Fluid Mech. **76**, 89-112 (1976)
[16] M. Ichimiya, I. Ikuo Nakamura, S. Yamashita, Exp. Thermal and Fluid Science, **17** 37-48 (1998)
[17] R. D. Moser, J. Kim, N. N. Mansour, Phys. Fluids, **11** 943-945 (1999)

MATEC Web of Conferences 77, 07003 (2016)
ICMMR 2016
DOI: 10.1051/matecconf/20167707003

Stochastic Subspace Method for Experimental Modal Analysis

Dazhi Liu[1] Shuhong Xiang[1] Shichao Fan[1] Ye Li[1] Zheng Liu[1] and Jiangang Sun[2]

[1] Beijing Institute of Spacecraft Environment Engineering:Beijing,100094,China
[2] Ship Model Experiment Institute,Wuxi,214000,China

Abstract. The formula of stochastic subspace identification method is deduced in details and the program is written out. The two methods are verified by a vibration test on a 5-floor rigid frame model. In this test the gauss white noise generated from a shaker table to simulate the ambient vibration on the model, and the response signals are measured. Next, the response data of experiment are processed by auto-cross spectrum density method and stochastic subspace identification method respectively, the two methods are verified by comparing with the theory result. and bearing out the superiority of stochastic subspace identification method compared to auto-cross spectrum density method.

1 Introduction

Parameter identification is the emphasis of the structural healthy test, and stochastic subspace is an effective structural modal parameter identification method, its advantages are as follows:

Parameterization: From data fits, obviously, system identification arithmetic requires stated specific parameters. Only parameter required the rank number of the system in the subspace identification arithmetic as the application of state space model. But classical arithmetic needs a lot of work to confirm alleged canonical model, as the model of minor parameter number.

Convergence: Subspace identification arithmetic converges fast. QR decomposition and SVD decomposition used in the method are not required iterative operations, so its convergence is faster than classical arithmetic and usually the convergence question doesn't exist in this method. This mathematics comes from value linear algebra, so value stability is guaranteed exactly. So convergence and value stability need not be considered.

Model curtailment: As operating system analysis, as soon as possible, low rank number is always wished to be obtained. In the subspace identification, Curtailment model can obtain from I/O data,it doesn't need to calculate the model with high rank.

2 System stochastic state space model [1]

The equation of systemic vibration can be expressed as:

$$M\ddot{U}(t)+C\dot{U}(t)+KU(t)=F(t)=Bu(t) \quad (1)$$

where, M,C,K are mass matrix, damping matrix and stiffness matrix respectively; $U(t)$ is structure displacement vector, $F(t)$ is force vector; B is place vector; $u(t)$ is function of time.

Assuming the input is stochastic white noise (gauss distribution). The state equation can be written in time discretization form as:

$$\begin{cases} x_{k+1}=Ax_k+w_k \\ y_k=Cx_k+v_k \end{cases} \quad (2)$$

$$E\left[\begin{pmatrix} w_p \\ v_q \end{pmatrix}\begin{pmatrix} w_p^T & v_q^T \end{pmatrix}\right]=\begin{pmatrix} Q & S \\ S^T & R \end{pmatrix}\delta_{pq} \quad (3)$$

where, $y_k \in R^{l\times 1}$ is output vector of k sampling step (Δt) at the 1st measuring point, $x_k \in R^{n\times 1}$ is system state vector; n is system rank, $A \in R^{n\times n}$ is state matrix, $C \in R^{l\times n}$ is output matrix; E is the mathematics expectation, δ_{pq} is Kronecker function, $w_k \in R^{n\times 1}$ is process noise; $v_k \in R^{l\times 1}$ is measuring noise; w_k, v_k are white noise, and $E(w_k)=0$, $E(v_k)=0$.

3 Stochastic state subspace parameter identification theory [2-4]

In ambient vibration tests, sometimes more measuring points are required, but the number of sensors is limited, the procedure is divided into several steps. In order to unify each measuring step, the reference measuring points are important and must be selected previously. The locations of sensors on the reference points must be fixed and data must be recorded continuous. Defining the matrix Hankel block as:

$$H^{ref} \equiv \frac{1}{\sqrt{j}}\left(\begin{array}{cccc} y_0^{ref} & y_1^{ref} & \cdots & y_{j-1}^{ref} \\ y_1^{ref} & y_2^{ref} & \cdots & y_j^{ref} \\ \vdots & \vdots & \cdots & \vdots \\ y_{i-1}^{ref} & y_{i-1}^{ref} & \cdots & y_{i+j-2}^{ref} \\ \hline y_i & y_{i+1} & \cdots & y_{i+j-1} \\ y_{i+1} & y_{i+2} & \cdots & y_{i+j} \\ \vdots & \vdots & \cdots & \vdots \\ y_{2i-1} & y_{2i} & \cdots & y_{2i+j-2} \end{array}\right) \equiv \left(\frac{Y_{0/i-1}^{ref}}{Y_{0/2i-1}}\right) \equiv \left(\frac{Y_p^{ref}}{Y_f}\right) \in R^{(r+l)i\times j} \tag{4}$$

Using QR decomposition to Hankel matrix:

$$H = \begin{bmatrix} Y_{past} \\ Y_{future} \end{bmatrix} = RQ^T = \begin{bmatrix} R_{11} & O \\ R_{21} & R_{22} \end{bmatrix}\begin{bmatrix} Q_1^T \\ Q_2^T \end{bmatrix} \tag{5}$$

According to theory of the space projection, the orthogonal projection is expressed by:

$$O_i = {}^{Y_{future}}\!/\!_{Y_{past}} = Y_{future}Y_{past}^T\left(Y_{past}Y_{past}^T\right)^{-1}Y_{past} = R_{21}Q \tag{6}$$

where, ${}^{Y_{future}}\!/\!_{Y_{past}}$ is orthogonal projection of Y_{future}'s row space on Y_{past}'s row space; $\left(Y_{past}Y_{past}^T\right)^{I}$ is Moore—Pseudo inverse matrix of Y_{past}.

The velocity of the programmer's operation is increased by using QR decomposition.

$$O_i = USV^T = \begin{bmatrix} U_1 & U_2 \end{bmatrix}\begin{bmatrix} S_1 & 0 \\ 0 & S_2 \end{bmatrix}\begin{bmatrix} V_1^T \\ V_2^T \end{bmatrix} \approx U_1S_1V_1 \tag{7}$$

where, both $U \in R^{il\times il}$, $V \in R^{j\times j}$ are orthogonal matrixes, and satisfy $U^TU = UU^T = I_{il}$, $V^TV = VV^T = I_j$, $S \in R^{il\times j}$ is a diagonal matrix, and row full order in the condition of common noise disturbance, but there are several obvious 'broken points' for the element sizes on the diagonal, the system rank can be obtained from the number of the diagonal elements before the 'broken point'.

SVD is widely applied to solving the problem of main component analysis, for a set of signal from some system structure, $U_1S_1V_1^T$ contains main information of the signal, and $U_2S_2V_2^T$ contains secondary information of the signal, and which is considered as the noise of the signal. If the system isn't polluted by noise, S_2 is equal to zero. But in fact, the noise disturbance is unavoidable, so S_2 is not equal to zero; But S_2 is smaller than S_1, so $U_2S_2V_2$ can be ignored and the aim result of the noise is achieved

According to the subspace system identification theory[5,6], projection matrix O_i is also equal to the product between generalized observable matrix C and non-steady Kalman filer estimation $\hat{X}_i$ of the state sequence X_i:

$$O_i = \begin{bmatrix} C \\ CA \\ \vdots \\ CA^{i-1} \end{bmatrix}\begin{bmatrix} \hat{x}_i & \hat{x}_{i+1} & \cdots & \hat{x}_{i+j-1} \end{bmatrix} = \Gamma_i\hat{X}_i \tag{8}$$

The main purpose of Kalman filter is to obtain future state vector by optimization forecast based on fore-passed state vectors (output data before the time k), system matrix and noise covariance Γ_i is obtained by SVD using Hankel matrix.

$$\Gamma_i = U_1 S_1^{1/2}, \hat{X}_i = S_1^{1/2} V_1^T \tag{9}$$

Generalized observable matrix is calculated by SVD. To obtain systemic generalized observable matrix, SVD is done to projection matrix. The mention is row full order commonly in the condition of noise disturbance, but there are several obvious 'broken points' on the element sizes on the cross, the systemic rank number can be obtained from the cross elements' number before the 'broken point' . SVD decomposition is widely applied to solving the problem of main component analysis. a set of signal from some system structure, contains main information of the signal, and secondary information of the signal is considered as the noise of the signal.
Matrix A and C are obtained by generalized observable matrix: the following equation can be obtained by the equation (7):

$$C = \Gamma_i(1:l,:) \tag{10}$$

$$A = [\Gamma_i(1:il-l,:)]^* \Gamma_i(l+1:il,:) \tag{11}$$

where, $\Gamma_i(1:l,:)$——first row block of Γ_i;

$[\Gamma_i(1:il \quad l,:)]^*$——fake inverse matrix of Moore—Penrose of foregoing (i-1) row block of Γ_i;

$\Gamma_i(l+1:il,:)$——later (i-1) row block of Γ_i.

Systemic dynamic characteristics are divided by the eigenvalue and eigenvector of system matrix *A*. The purpose of the paper is only to indentify three mode parameters of the system: nature frequence, damping ratio and mode shape, which are only related to system matrix *A* and observable matrix *C*. So it is enough to calculate A and C, and the stochastic subspace is ignored here.

4 Structure mode analysis[7 ~ 10] and procedure

Eigenvalue decomposed by System matrix A is as follows:

$$\Lambda = \psi \Lambda \psi^{-1} \tag{12}$$

where, $A = ding(\lambda_m) \in R^{n\times n} (m = 1, \ 2, \ \cdots, \ n)$; λ_m is eigenvalue of discretion system, $\Psi \in R^{n\times n}$ is systemic eigenvector matrix.

The relation between the eigenvalue and the nature frequency ω, damping ratio ζ is as follows:

$$\lambda_i^c, (\lambda_i^c)^* = -\zeta_i \omega_i \pm j\omega_i \sqrt{1-\zeta_i^2} \tag{13}$$

where, $(\lambda^c_i)^* = \lambda^c_i$ is the conjugate form.

The nature frequency of the system is:

$$\omega_i = |\lambda_i^c|, f_i = \omega_i / 2\pi \tag{14}$$

where, $(\lambda^c_i)^R$ is real part of λ^c_i.

As the mode shape of the measuring point is observed from eigenvector of the system. The mode shape of nature equation can be written as:

$$\Phi = C\psi \tag{15}$$

The MATLAB procedure chart of stochastic subspace method is as follows (Fig. 1):
For the determination of model order, and for the decomposition of spectrum mateices. In theory, the number of non-zero singular value is rank of spectrum matrix, the order of the model. but in fact,spectrum matrix is obtained by actual test data. More or less it contain noise in actual test, and it result that singular value in the matrix is not really zero. Singular value can produce sudden drop, but fell sharply before the singular value can be thought of as order of the model.

5 Data processing with stochastic subspace method compared with the results of FEM

The natural frequency from finite element analysis [11-13] is as follows (Table 1):

Table 1. Natural frequency from finite element analysis

Order of the element	1	2	3	4	5	6
Natural frequency（HZ）	10.825	30.091	37.889	54.234	67.953	70.632

The mode shape from finite element analysis is as follows:

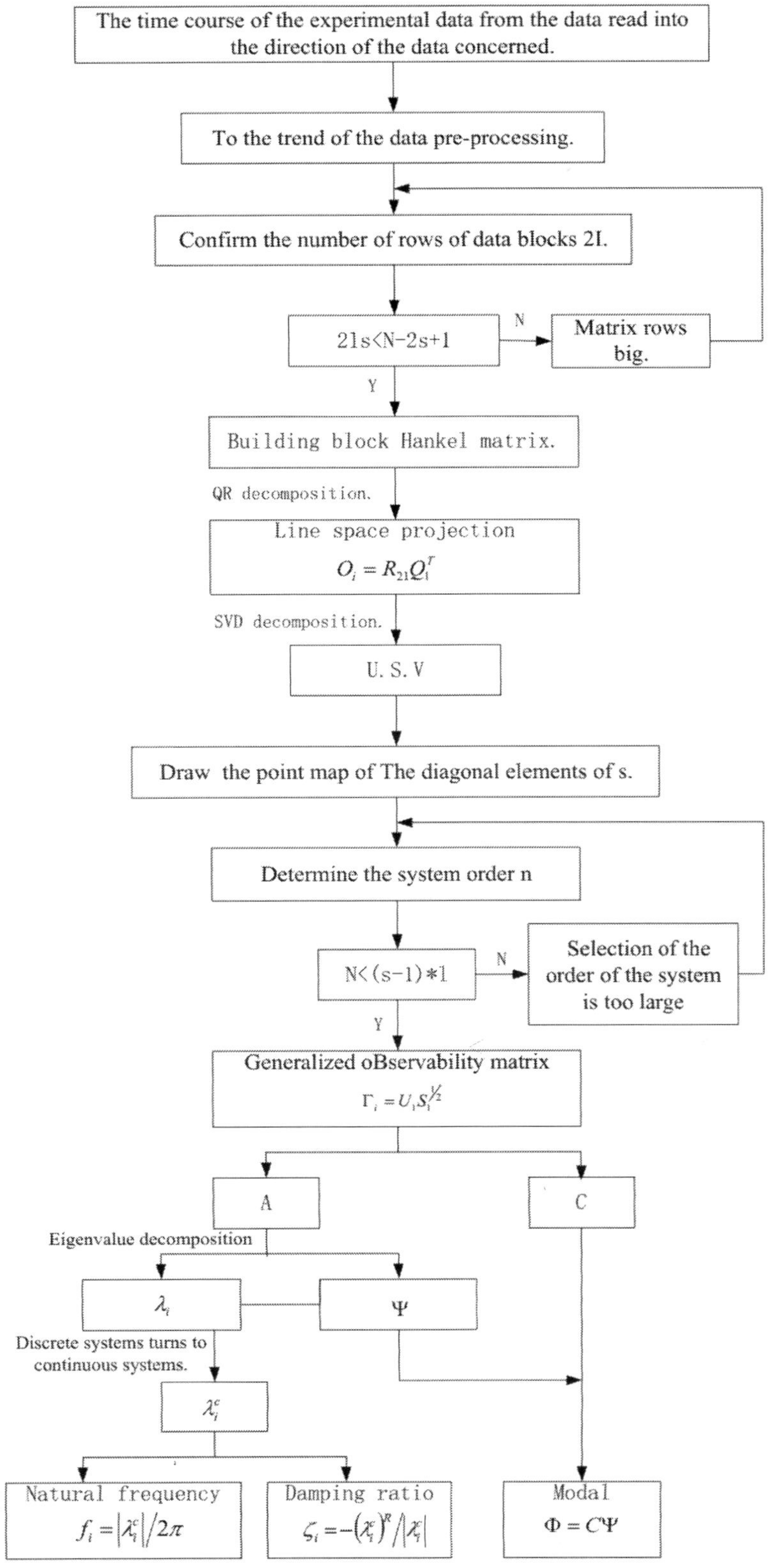

Figure 1. MATLAB procedure flow chart

DOI: 10.1051/matecconf/20167707003

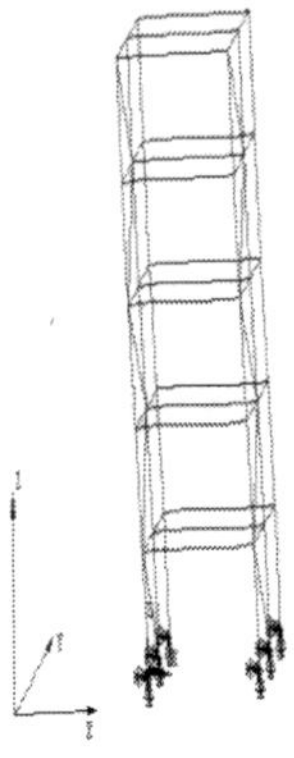

Figure 2. The first modal shape

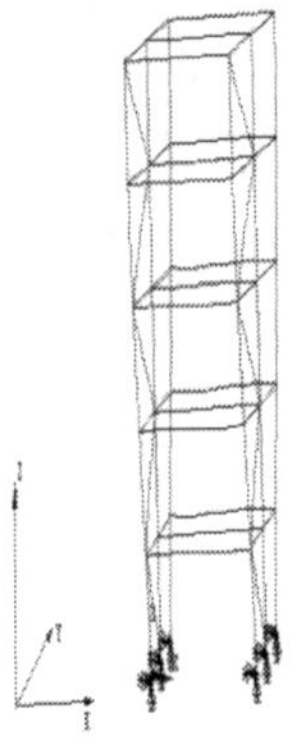

Figure 3. The second modal shape

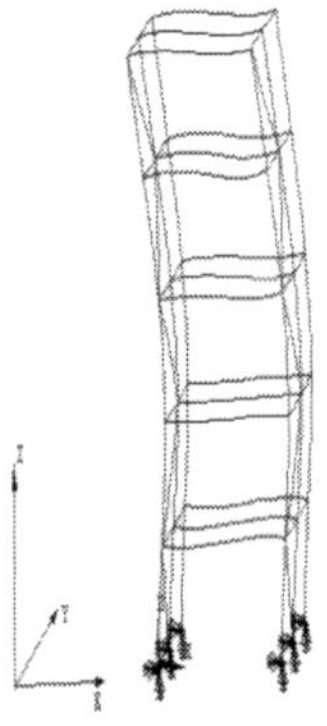

Figure 4. The third modal shape

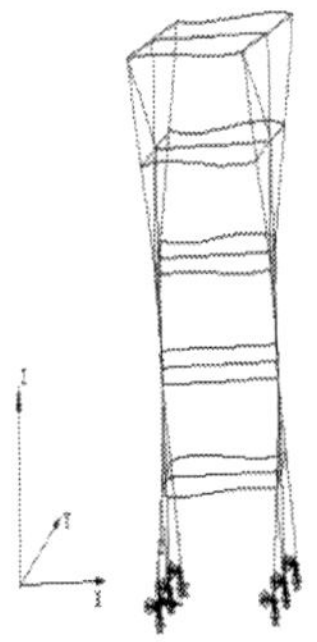

Figure 5. The first modal shape

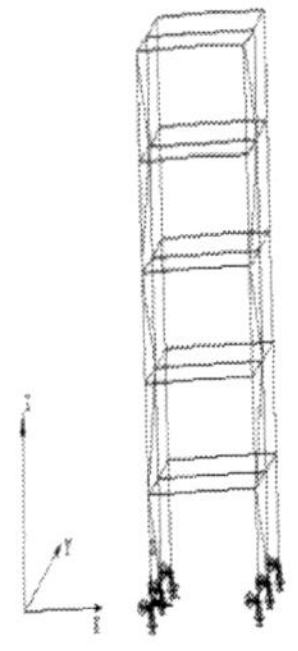

Figure 6. The second modal shape

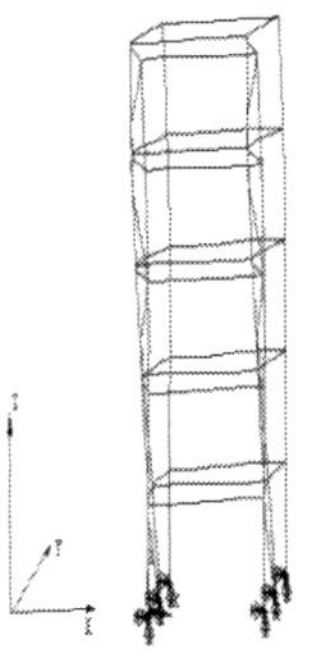

Figure 7. The third modal shape

The number of the row block of Hankel Matrix is 200 when the data is processed by the stochastic subspace method. As the calculation capacity is limited, only the data of the first 5000 discrete point is used in x and y directions. The 4 nature frequencies in x direction and the 2 nature frequencies below 100 Hz in y direction, listed in table 1. The first 6 mode shapes by stochastic subspace method list as follows:

1,3 modal shapes are without bracing the x direction (the direction of small united) of a bend, two bends(2nd bends), bending vibration mode 3 plus bracing for the y-direction (the direction of large bending rigidity) of a bending vibration mode, 2、6 vibration mode for the direction of a small twist twisting stiffness, torsional type 2, five vibration mode for the large distortions in the direction of a torsional stiffness, 4 vibration mode, respectively, for small united of the bending direction of 2 plus a small distortion torsional stiffness of 1 direction , 2 torsional united model for the small plus small distortions in the direction of 3 united of the direction of a torsional type.

It is very close that the results are obtained by the finite element analysis and the stochastic subspace method, and it proves that the stochastic subspace method is reliable.

The test data of Hankel Matrix is processed by the stochastic subspace method, and the modal parameter is identified. Then the results are compared with the finite element analysis, and it shows that the stochastic subspace method is available.

Table 2. Result of modal analysis

Order	Finite element method	cross-spectral method			Random subspace（SSI）		
	Natural frequency（Hz）	Natural frequency（Hz）	Frequency error（%）	Damping ratio（%）	Natural frequency（Hz）	Frequency error（%）	Damping ratio（%）
1	10.825	10.90	0.69	0.61	10.9612	1.26	0.6527
2	30.091	29.85	0.80	0.27	29.8193	0.90	0.4354
3	37.889	37.95	0.16	0.34	37.6066	0.75	1.0480
4	54.234	—	—	—	54.6168	0.71	0.4936
5	67.953	65.55	3.54	1.18	65.7421	3.25	1.0483
6	70.632	70.45	0.26	0.19	70.4692	0.23	0.3650

6 Conclusions

The indentified nature frequences by self mutual chart method and stochastic subspace method are close to the theory resultand identification precision satisfies the request. It shows that the two methods are effective.

The first six modals can be identified by stochastic subspace method, but cannot by self mutal. It shows that the test is successful and stochastic subspace method is more excellent than self mutual chart method. Stochastic subspace method can be used in dense mode identification broadly.

The results obtained by the two identification methods are very close, and it proves that the two methods are reliable. Although self mutual chart method is a traditional graphic method and has some shortcomings. It can identifies accurately the first several nature frequencies by data's reasonable adding window and fit. So the identified result by self mutal chart method has great reference value. And the self mutual chart method can be a supplement to the stochastic subspace method for its convenient.

References

1. Zhifang Fu, Vibration modal analysis and parameter identification. Mechanical industry publishing company Beijing (1990)
2. Hui Wang, The application of modal analysis in damage detection of pipeline.Thesis of the master scholarship in DaLian university of science and engineer (2009)
3. Michael Döhler, Xuan-Binh Lam, Laurent Mevel, "Uncertainty quantification for modal parameters from stochastic subspace identification on multi-setup measurements," Mechanical Systems and Signal Processing, April (2013)
4. Loh, Chin-Hsiung; Liu, Yi-Cheng; "Stochastic subspace identification for output-only modal analysis: accuracy and sensitivity on modal parameter estimation", Sensors and Smart Structures Technologies for Civil, Mechanical, and Aerospace Systems, Proc. of SPIE **7981**(2011)
5. B. Peeter, G. De. Roeck. Reference-based Stochastic Subspace Identification for Output-only Modal Analysis. Mech. Syst. Sign. Proc, **13** 6:855~878 (1999)
6. Van Overschee P, De Moor B. Subspace Identification for Linear Systems.*Theory-Implementation-Applications. Kluwer Academic publishers, Dordrecht,The Netherlands,* 36~54(1996)
7. B. Peeters. System Identification and Damage Detection in Civil Engineering. PhD thesis. Department of Civil Engineering, Katholieke Universiyt Leuven, Belgium, **12** 1~10 (2000)
8. Tao Guoquan, Wei Yuchen, Lü mingyun, Wu Zhe, Modal tests and properties analysis on truss structure of large scale carbon fiber Journal of Beijing University of Aeronautics and Astronautics **37** 3 (2011)
9. Zhang Guo-wen, TANG Bao-ping, MENG Li-bo, Improved stochastic subspace identification algorithm based on eigendecomposition.Journal of Vibration and Shock **31** 7 (2012)
10. Sung-Han Sim B.F.Spencer Jr. ZHANG Min XIE Huicai Sung-Han Sim B.F.Spencer Jr. Distribution modal identification based on stochastic subspace method，Journal of Shenzhen University (SCIENCE & ENGINEERING) **26 4** (2009)
11. Baoping Tang,Guowen,Libo Meng,Model energy-based distinguishing spurious modes obtained from stochastic subspace identificationJournal of HuaZhong University of Science and Technology.NATURE SCIENCE **40** 3 (2012)
12. Zhiyuan Yao, Fengquan Wang, Yan Liu, Stochastic subspace identification method based on continuous model for modal parameters of engineering structures Journal of Southeast University (NATURAL SCIENCE EDITIO. **34** 3 (2004)
13. Michael Döhler, Xuan-Binh Lam, Laurent Mevel, "Uncertainty quantification for modal parameters from stochastic subspace identification on multi-setup measurements," Mechanical Systems and Signal Processing, April (2013)

MATEC Web of Conferences 77, 07004 (2016) DOI: 10.1051/matecconf/20167707004
ICMMR 2016

Design of synchronous optimal control strategy based on double motor drive precision air suspension platform

Ling Xiang [1], Zhang Yu[1], Guo Xiao-kun[2] and Li Xi-feng [1]

[1]GuangDong Institute of Automation , Guangdong Provincial Key Laboratory of Modern Control Technology, Guangdong Modern control technology with optical and electrical Public Laboratory ,Guangzhou 510070, Guangdong, China
[2] Shenzhen Huizhicheng Technology Co. Ltd, Shenzhen 518057,Guangdong,China

Abstract. Aiming at the double motor driving structure of precision air suspension platform, the control strategy of the system is studied under the condition of various mechanical vibration and electrical signal noise. Based on controllability and observability of systems, the optimal controller is designed to realize the fast rise and fast stability of the system. Simulation results verify the effectiveness and feasibility of the control strategy, and the system performance is optimal.

1 Introduction

At present, the guide mode of traditional rotary servo motor combined with ball screw driven in precision motion platform has been gradually replaced by non friction supporting technology and non friction direct drive technology. The flotation technology oriented and bearing support form precision air motion platform is gradually in the field of high speed high accuracy servo motion emerge [1], [2], which directly driven by linear motor for the way to eliminate the system friction and dead zone nonlinear disturbance and structure with no mechanical contact, can achieve sub micron or even nanometer positioning accuracy in the long travel range.

Precision air suspension platform has the advantages of high precision, no friction, low pollution and other advantages, which are widely used in the field of lithography technology, ultra precision machining, detection of biological technology, nano surface topography measurement, and to continue to develop in the direction of high speed, high acceleration, high accuracy [3], [4]. But due to the lack of damping mechanism, direct drive vulnerable to interference and rigid flexible coupling is weak link characteristics will undoubtedly increase the difficulty of controller design. Erkorkmaz *et al.* [5] use the input shaping and state feedback controller to suppress the residual vibration of the air bearing platform and realize the precision control of the air bearing platform. Kim *et al.* [6] Combine with the approximate time optimal control and the delay control, which apply to the precision air suspension platform to suppress the residual vibration of the platform and reduce the stability time of the Jung-Jae. [7] and [8] respectively in the application of fuzzy algorithm and neural network algorithm to suppress the residual vibration and vibration control system by input shaping technique.

Although these control algorithms have certain compensation to the system, but the system performance also affected. At the same time, air suspension motion platform for external disturbance, (such as linear motor force ripple, drive electrical noise and measurement noise, cable of cable force), change the environment and the change of system parameters will directly affect the platform movement precision. Therefore, this paper for precision air suspension platform of double motor drive structure, in the presence of external disturbances, the system of modeling and optimal control strategy are designed, so as to achieve the system's rapid rise and rapid and stable, the performance of the system to achieve optimal.

2 System Modeling

The single axis motion of the double motor drive in a precision air suspension motion platform is described by a dual input dual output linear system,

$$\begin{cases} x(k+1)=A_c x(k)+B_c u_c(k)+B_{c,w} w(k) \\ y_c(k)=C_c x(k)+v(k) \end{cases} \tag{1}$$

The system is affected by the internal and external environment of the system. Form in, $k \in Z^+$ is the number of time steps, the system state variable is the displacement and speed of the two ends of the push rod: $x=[x_1,x_2,v_1,v_2]^T$. System control vector is input voltage signal: $u=[u_1,u_2]^T$, The system output vector is the displacement of the two ends of the push rod: $y=[y_1,y_2]^T$.

$w=[w_1,\cdots,w_n]^T\in R^n$, $v=[v_1,\cdots,v_p]^T\in R^p$ are external disturbance and random noise vector. Assume:

(1) $w(k)$ and $v(k)$ are the zero mean Gauss white noise vector for each other, and they are independent of each other:

$E\{w(k)\}=0$, $Cov\{w(k),w(\tau)\}=Q_k\delta(k-\tau)$; $E\{v(k)\}=0$, $Cov\{v(k),v(\tau)\}=R_k\delta(k-\tau)$; $Cov\{w(k),v(\tau)\}=0$.

Form in, $E(\bullet)$ is mathematical expectation; Q_k is non negative definite symmetric matrix for corresponding dimension, which is the covariance matrix of $w(k)$; R_k is the positive definite symmetric matrix of the corresponding dimension,which is the variance matrix of $v(k)$; $\delta(t-\tau)$ is the Kronecker δ function, that is,

$$\delta(t-\tau)=\begin{cases}1, & t=\tau\\ 0, & t\neq\tau\end{cases}$$

(2) The initial value $x(0)$ of the state vector $x(k)$ is a random variable. The statistical properties of $x(0)$ are known, that is,

$$E\{x(0)\}=x_o,\ E\left\{[x(0)-x_o][x(0)-x_o]^T\right\}=P(0)$$

(3) $x(0)$ and $w(k)$, $v(k)$ are independent of each other, that is,

$$Cov\{w(k),x(0)\}=0,\ Cov\{v(k),x(0)\}=0$$

3 Synchronous optimal control problem description

According to the establishment of the dynamic model (1), the optimal control problem for the precision air suspension platform with disturbance and noise can be expressed as:

$$u_c(k)=-L(k)\cdot x(k)=-L(k)\cdot \hat{x}(k)$$

Because the system state variable $x(k)$ is the displacement and speed of the push rod, x_1 and x_2 can be directly measured by the precision grating ruler, and the speed term v_1 and v_2 can not be directly measured. Therefore, it is needed to design the algorithm to estimate the value of the algorithm. The optimal control problem of the system is changed into:

$$u_c(k)=-L(k)\cdot\begin{bmatrix}x_1(k)\\ x_2(k)\\ \hat{v}_1(k)\\ \hat{v}_2(k)\end{bmatrix} \qquad (2)$$

That is to solve the stochastic optimal control rate makes the random two performance indicators to achieve the minimum:

$$J=E\left[x^T(N)Q_1x(N)+\sum_{k=0}^{N-1}\left(x^T(k)Q_2x(k)+u_c^{\ T}(k)Ru_c(k)\right)\right]$$

In the formula, Q_1, Q_2 are the state vector weighting matrix, which are the symmetric nonnegative definite matrix. R is the input vector weighted matrix, which is the symmetric positive definite matrix.

According to the formula of the optimal control rate (2), the control strategy can be divided into two parts: the optimal estimation of the state vector and the optimal gain of the system, as shown in Fig. 1.

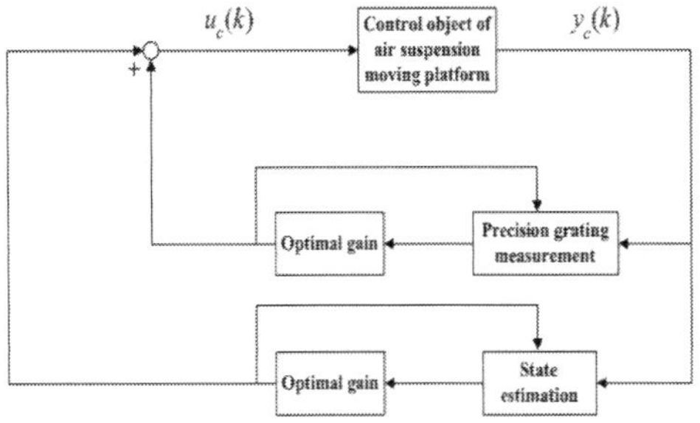

Figure 1. Synchronous optimal controller based on double motor drive precision air suspension platform

4 Synchronous optimal controller design

4.1. System controllability and observability

According to the dual motor driven precision air suspension platform system model, if the system matrix A_c is non singular matrix, and at time $k\in[0,t_f]$ range:

$$rank\left[A_c(t_f-1)A_c(t_f-2)\cdots A_c(1)B_c(0), (A_c(t_f-1)\cdots A_c(2))B_c(1),\cdots,B_c(t_f-1)\right]=n$$

and

$$rank\begin{bmatrix}C_c(0)\\ C_c(1)\cdot A_c(0)\\ \cdots\cdots\\ C_c(t_f-1)\cdot\left(A_c(t_f-1)\cdots A_c(0)\right)\end{bmatrix}=n$$, the control system is completely controllability and observability.

Therefore, the optimal control strategy of the system is designed under the premise that the control system can be controlled and can be observed.

4.2. Optimal state estimation $\hat{x}(k)$

According to the observation sequence $y_c(0)$, $y_c(1)$,$\cdots$, $y_c(k)$, the motion velocity terms in the state vector

$x(k)$ need to be estimated in order to achieve synchronization optimal control, which should be made as close as possible to the actual value, to reduce the synchronization error. Subjected to external interference of double motor driving precision gas suspension system model (1), through the discrete time Kalman filtering method of optimal state estimation $\hat{x}(k)$ is,

$$\hat{x}(k)=\hat{x}(k|k-1)+K(k)[y_c(k)-C_c(k)\cdot\hat{x}(k|k-1)] \quad (3)$$

In the formula, $K(k)$ is the optimal filtering gain, $\hat{x}(k|k-1)$ is one step optimal linear prediction estimation of state vector $x(k)$, which is obtained based on the system state equation.

$$\hat{x}(k|k-1)=A_c(k-1)\hat{x}(k-1)+B_c(k-1)u_c(k-1) \quad (4)$$

As a result, the filter estimation error of the state vector $x(k)$ is:

$$\tilde{x}(k)=x(k)-\hat{x}(k)$$
$$=x(k)-\hat{x}(k|k-1)-K(k)[y_c(k)-C_c(k)\hat{x}(k|k-1)]$$
$$=\tilde{x}(k|k-1)-K(k)C_c(k)\tilde{x}(k|k-1)-K(k)v(k)$$

According to the orthogonality theorem, the estimation error $\tilde{x}(k)$ should be orthogonal to the measured value $y_c(k)$:

$$E\{\tilde{x}(k)y_c^T(k)\}=E\{[\tilde{x}(k|k-1)-K(k)C_c(k)\tilde{x}(k|k-1)$$
$$-K(k)v(k)][C_c(k)x(k)+v(k)]^T\}$$
$$=E\{\tilde{x}(k|k-1)x^T(k)C_c^T(k)-K(k)C_c(k)$$
$$\cdot\tilde{x}(k|k-1)x^T(k)C_c^T(k)-K(k)R_k\}=0$$

The optimal filter gain matrix:

$$K(k)=P(k|k-1)C_c^T(k)[C_c(k)P(k|k-1)C_c^T(k)+R_k]^{-1} \quad (5)$$

Form in,

$$P(k|k-1)=E\{\tilde{x}(k|k-1)\tilde{x}^T(k|k-1)+\tilde{x}(k|k-1)\hat{x}^T(k|k-1)\}$$

Which is the optimal estimation error variance matrix of state vector $x(k)$. Because the optimal estimation error is:

$$\tilde{x}(k|k-1)=x(k)-\hat{x}(k|k-1)=A_c(k-1)\tilde{x}(k-1)+B_{c,w}(k-1)w(k-1)$$

Then,

$$P(k|k-1)=E\{\tilde{x}(k|k-1)\tilde{x}^T(k|k-1)\}=E\{[A_c(k-1)\tilde{x}(k|k-1)$$
$$+B_{c,w}(k-1)w(k-1)][A_c(k-1)\tilde{x}(k|k-1)+B_{c,w}(k-1)w(k-1)]^T\}$$

That is,

$$P(k|k-1)=A_c(k-1)P(k-1|k-1)A_c^T(k-1)+B_{c,w}(k-1)Q_kB_{c,w}^T(k-1) \quad (6)$$

Form in, $P(k-1|k-1)$ is the estimate the error variance matrix for the optimal filter, that is,

$$P(k-1|k-1)=E\{\tilde{x}(k-1)\tilde{x}^T(k-1)\}$$
$$=P(k-1|k-2)-K(k-1)C_c(k-1)P(k-1|k-2)-P(k-1|k-2)$$
$$\cdot C_c^T(k-1)K^T(k-1)+K(k-1)C_c(k-1)P(k-1|k-2)C_c^T(k-1)K^T(k-1)$$
$$+K(k-1)R_kK^T(k-1)$$

Then,

$$P(k|k)=P(k|k-1)-K(k)Cc(k)P(k|k-1)-P(k|k-1)C_c^T(k)K^T(k)$$
$$-K(k)C_c(k)P(k|k-1)C_c^T(k)K^T(k)+K(k)R_kK^T(k)$$

That is,

$$P(k|k)=P(k|k-1)-P(k|k-1)C_c^T(k)[C_c(k)P(k|k-1)C_c^T(k)+R_k]^{-1}C_c^T(k)P(k|k-1) \quad (7)$$

According to the characteristics of the Kalman filter due to its in the calculation process does not need to store any data, but to continue to "predict revise" push calculated optimal state estimates recursively, and computation and observation of the Kalman gain matrix $K(k)$ is independent of the value of the. Therefore, the filter gain value can be calculated in advance to reduce the actual amount of calculation. Integrated (6), (7) and (5) to obtain the optimal estimation error variance matrix recursion relations are as follows:

$$P(k+1|k)=A_c(k)P(k|k-1)A_c^T(k)+B_{c,w}(k)Q_kB_{c,w}^T(k)-A_c(k)P(k|k-1)$$
$$\cdot C_c^T(k)[C_c(k)P(k|k-1)C_c^T(k)+R_k]^{-1}C_c^T(k)P(k|k-1)A_c^T(k) \quad (8)$$

Formula (8) is called forward Riccati equation. Under the premise of ensuring the system can control and can view, step by step forward to obtain the steady solution of the forward Riccati equation according to the given initial value $P(0|0)$, and then the filter gain matrix $K(k)$ can be obtained off-line.

4.3. Optimal gain $L(k)$

After the optimal filter estimation of the state vector is obtained, the solution of the optimal gain matrix in the optimal control rate is considered. This problem can be solved by solving the linear quadratic optimal regulator for two times. In order to ensure the disturbed by the noise and the disturbance of the air suspension system can meet the performance optimal conditions, even if the random quadratic performance index reach the minimum, to achieve the rapid rise and rapid and stable, the optimal gain matrix $L(k)$ need satisfies the equation:

$$L(k)=[B_c^T(k)S(k+1)B_c(k)+R]^{-1}B_c^T(k)S(k+1)A_c(k) \quad (9)$$

In the formula, $S(\cdot)$ satisfies the dynamic backward Riccati equation (10):

$$\begin{cases} S(N)=Q_1 \\ S(k)=A_c^T(k)S(k+1)A_c(k)+Q_2 \\ -A_c^T(k)S(k+1)B_c(k)[B_c^T(k)S(k+1)B_c(k)+R]^{-1}B_c^T(k)S(k+1)A_c(k) \end{cases} \quad (10)$$

In the formula, Q_1 and Q_2 are the state vector weighted matrix in the performance index, and R is the input vector weighted matrix in the performance index. N represents the termination time. Under the premise of ensuring the system can control and can observe, step by step forward iteration to get the stable solution of the backward Riccati equation based on the value of the given end time.

5 Simulation analysis

Based on the above mentioned above, a model of the air suspension system is established, considering a double input and double output discrete time system with a shape like (1), the initial state of each parameter and system is as follows:

$$A_c = \begin{bmatrix} 0 & 0 & 1 & 0 \\ 0 & 0 & 0 & 1 \\ 0.1 & -0.3 & 0 & -0.25 \\ 0 & -0.43 & 0.2 & -0.35 \end{bmatrix}, B_c = \begin{bmatrix} 0 & 0 \\ 0 & 0 \\ 1 & 1 \\ 201 & 218 \end{bmatrix}$$

$$C_c = \begin{bmatrix} 1 & 0 & 0 & 0 \\ 0 & 1 & 0 & 0 \end{bmatrix}, B_{c,w} = \begin{bmatrix} 1 & 0 & 0 & 0 \\ 0 & 1 & 0 & 0 \\ 0 & 0 & 1 & 0 \\ 0 & 0 & 0 & 1 \end{bmatrix}$$

System disturbance: $w(k) \sim N(0, I_{4*4})$;

Random measurement noise: $v(k) \sim N(0, I_{2*2})$;

Initial state vector: $x(0) = [100,100,0,0]^T$;

State estimate initial value:

$\hat{x}(0) = [97.12 \quad 49.3 \quad 0.05 \quad 0.05]^T$;

Filtering estimation error variance initial value:

$p(0) = 4I_{4*4}$;

State vector weighted matrix: $Q_1 = Q_2 = 25I_{4*4}$;

Input vector weighted matrix: $R = I_{2*2}$.

According to the system parameters and initial conditions given above, after 7 steps of iteration, the stability of the forward Riccati equation (8) is obtained:

$$P(8) = \begin{bmatrix} 2.1023 & 0.1382 & -0.0098 & 0.2076 \\ 0.1382 & 2.2282 & -0.2661 & -0.3538 \\ -0.0098 & -0.2661 & 1.1243 & 0.1674 \\ 0.2076 & -0.3538 & 0.1674 & 1.2830 \end{bmatrix}$$

Based on the solution of the optimal estimation error variance matrix $P(\cdot)$, the kalman filter gain matrix $K(k)$ is calculated:

$$K = \begin{bmatrix} 0.7603 & -0.2253 \\ -0.2397 & 0.7747 \\ 0.0311 & -0.0934 \\ 0.1230 & -0.1530 \end{bmatrix}$$

At the same time, after 2 steps of iteration, the stable solution of the backward Riccati equation (10) is:

$$S(3) = \begin{bmatrix} 25.0098 & -0.0294 & -0.0000 & 0.0245 \\ -0.0294 & 25.0882 & 0.0000 & 0.0735 \\ -0.0000 & 0.0000 & 50.0098 & -0.0294 \\ -0.0245 & 0.0735 & -0.0294 & 50.1494 \end{bmatrix}$$

The periodic solution $S(\cdot)$ into (9), the solutions of optimal gain for linear quadratic regulator:

$$L = \begin{bmatrix} 1.2572 & -3.7463 & -0.0117 & -3.1225 \\ -1.1592 & 3.4522 & 0.0117 & 2.8774 \end{bmatrix}$$

Based on the filter gain and optimal gain of linear quadratic regulator, the optimal state estimation $\hat{x}(k)$ can be further calculated according to the formula (3) and (4), which is shown in Figure 1,and the output of the system corresponding and control sequences such as shown in Fig. 2.

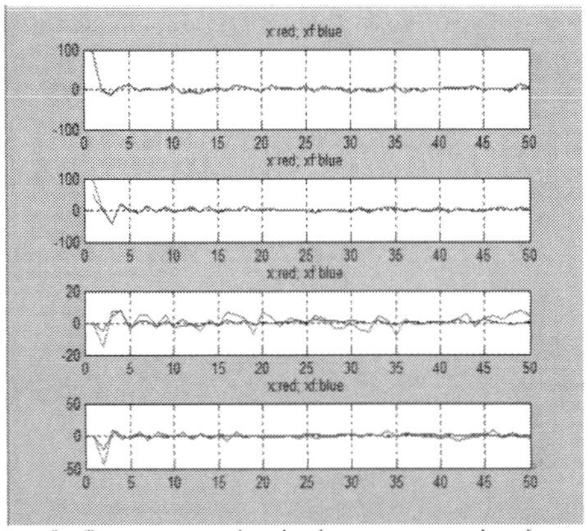

Figure 2. State vector and optimal state vector estimation

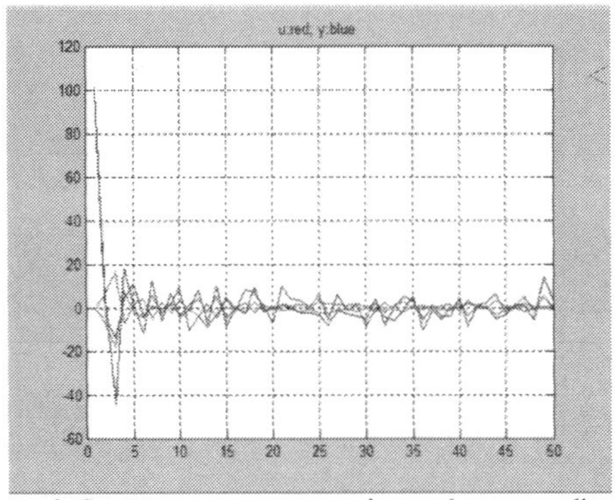

Figure 3. System output response and control sequence diagram.

It can be seen from the change trend of the curve in the chart, for a multi input and multi output air suspension system, the external disturbance has a certain effect on the control performance of the system. However, the state estimates eventually in the vicinity of the zero tends to be stable, which are estimated by the Kalman filtering algorithm. And the error between the estimated value and the actual state value is controlled within a very small range. The simulation results show that the compensation effect of Calman filtering algorithm guarantees the stability of the system to a certain extent, and it also ensures the precision accuracy of the control action.

6 Conclusion

In this paper, a dual motor drive structure for precision air suspension platform is established, and the mathematical model of the system is established under the condition of various mechanical vibration and electrical noise. Under the premise of ensuring the control system can be controlled and be observed, the design of the optimal controller is designed to realize the fast rising and fast stabilization of the system. The effectiveness and feasibility of the control strategy are verified by simulation experiment. Simulation experimental results show that the design of synchronous optimal control algorithm for a variety of system disturbance and random noise with certain compensation, makes the system has been able to maintain relatively stable output, and has high control performance.

References

1. WANG ying. high-acceleration/high-precision motion control of linear servo system for IC packing [D].Shanghai: Shanghai Jiao Tong University(2006)
2. PU donglin. Rapid vibration suppression and high precision positioning control of high acceleration air bearing motion system [D].Shanghai: Shanghai Jiao Tong University(2011)
3. LI yuntang. Study on high-acceleration/high-precision aerostatic positioning stage for IC packing [D]. Shanghai: Shanghai Jiao Tong University(2007)
4. DONG zeguang. Modeling, analysis and control of a gas-lubricated precision positioning stage [D]. Shanghai: Shanghai Jiao Tong University(2014)
5. B.K. Yoo, W.C. Ham, Adaptive control of robot manipulator using fuzzy compensator, IEEE Transactions on Fuzzy Systems, 8(2) , 186-199(2000)
6. H. A. Talebi, R.V. Patel and K. Khorasani, Control of flexible-link manipulators using neural networks, Springer Press(2001)
7. Erkorkmaz K, Gorniak J, Gordon D, Precision machine tool XY stage utilizing a planar air bearing arrangement[J]. CIRP Annals-Manufacturing Technology, 59(1):425–428(2010)
8. Viktorov V, Belforte G, Raparelli T, Modeling and identification of gas journal bearings: Externally pressurized gas bearing results[J]. Journal of tribology, 127(3):548–556(2005)

Research on the Random Shock Vibration Test Based on the Filter-X LMS Adaptive Inverse Control Algorithm

Wei Wang, Zhangwei Chen, Qiangli Luan

The State Key Lab of Fluid Power & Mechatronic Systems , Zhejiang University, Hangzhou 310013, China

Abstract. The related theory and algorithm of adaptive inverse control were presented through the research which pointed out the adaptive inverse control strategy could effectively eliminate the noise influence on the system control. Proposed using a frequency domain filter-X LMS adaptive inverse control algorithm, and the control algorithm was applied to the two-exciter hydraulic vibration test system of random shock vibration control process and summarized the process of the adaptive inverse control strategies in the realization of the random shock vibration test. The self-closed-loop and field test show that using the frequency-domain filter-X LMS adaptive inverse control algorithm can realize high precision control of random shock vibration test.

1. Introduction

Shock test [1]-[3] is mainly used in the simulation of the transient impact signal of falling, shelling and artillery firing as a common vibration test method. Since in the process of the real shock (such as the missile launch), the shock signal on the specimen is not standard combined with stochastic and shock characteristics at the same time, so the random shock vibration test is put forward, which can simulate the environment shock load on specimen.

Effective control of multi-axis hydraulic vibration test system relies on the accurate estimation of system frequency-response function (FRF) matrix, since the existence of nonlinear factors in the system, lead to the error between the actual system FRF matrix and the control matrix (FRF), to eliminate this error, there must be real-time update of the system FRF matrix to reduce the error between the control matrix and the actual matrix. This paper proposed using a frequency-domain filter-X LMS adaptive inverse control algorithm, and applied to the random shock vibration control process of two-exciter hydraulic vibration test system and summarize the process of the realization of adaptive inverse control strategies. The self-closed-loop and field test show that using the frequency-domain filter-X LMS adaptive inverse control algorithm can realize high precision control of random shock vibration test.

2 Adaptive inverse control strategy

Adaptive inverse control technology [4], [5] based on the effective control of inverse transfer function of controlled object, to realize the rapid and accurate tracking of the instructions from output to input. The reference model $M(f)$ is the requirements of control system, means the delay link, which can ensure that the stability of the control of delay system and non-minimum phase system, and ensure the controller has enough control precision to realize the precise and small error response output of control system. The transfer function $H(f)$, the inverse transfer function of controlled objects (impedance function) $Z(f)_k\ \{k=0\cdots\cdots N\}$ as the transfer function of the controller, mutual relations are represented as:

$$Z(f)=\frac{M(f)}{H(f)} \tag{1}$$

For adaptive inverse control system, external interference signal will result in the divergence of solution of system impedance function which means the formula (1) is not satisfied. To eliminate the influence of noise interference on the system and consider the effective control of the system using adaptive LMS algorithm, the paper puts forward a frequency domain of filter-X LMS adaptive inverse control algorithm, and apply it to the random shock vibration control process of two-exciter hydraulic vibration test system, the control process mainly includes the object modeling, inverse modeling and adaptive inverse control.

Object modeling H_v method was adopted to realize the initial transfer function recognition, to improve the precision of the system transfer function identification, the process of multi-times identification and average processing are conducted, until the precision meet the requirements. Frequency filter-X LMS adaptive inverse control algorithm does not affect the Wiener solution of controller, even the object identification model error exists.

Inverse modeling process using the copied object model $\hat{H}(f)$ to realize the identification of the system impedance function $Z(f)$ with adaptive LMS algorithm.

Adaptive inverse control process will regard $Z(f)$ as the controller transfer function which is applied to the actual physical model $H(f)$ to achieve effective control.

3 Random shock vibration control

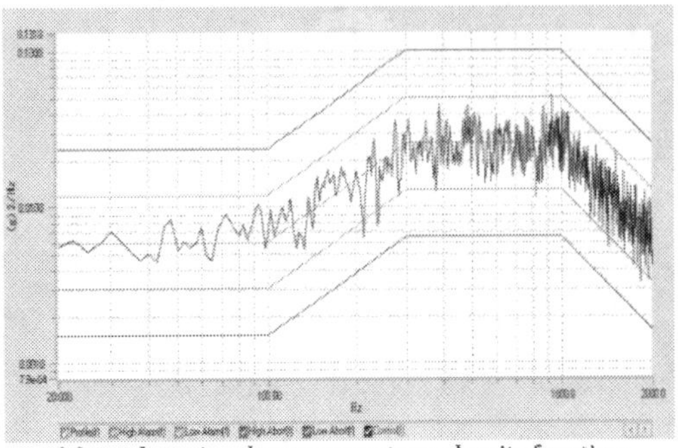

(a) random signal power spectrum density function

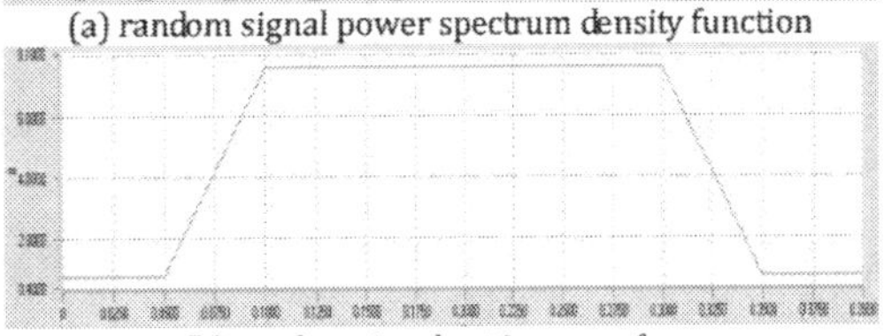

(b) random signal cutting waveform

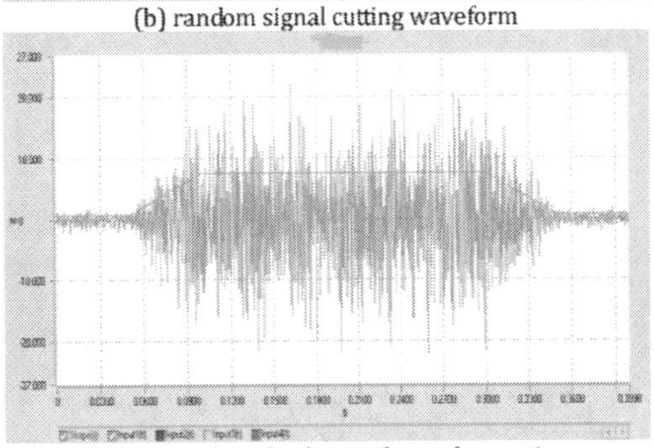

(c) the random shock waveform after cutting

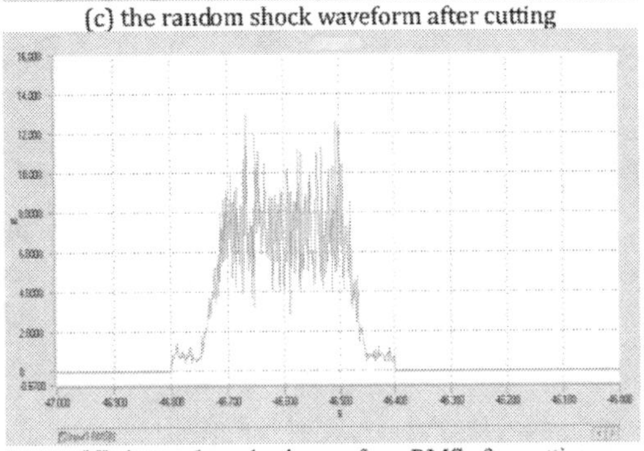

(d) the random shock waveform RMS after cutting

Figure 1. random shock vibration teat waveform.

In random shock vibration test [6], set the random signal power spectrum density function firstly (1 (a)), the reference spectrum generate driving signal power spectrum reiteration according to the test process, after cutting by window (1(b) ,then input to the vibration test system, the response (control) signal(1(c)), control signal effective value (1(d)).

4 Test system

In this paper, the control core of random shock hydraulic vibration test system is the Premax VT - 6008 servo vibration controller[7] which is based on a frequency domain filter-X LMS adaptive inverse control algorithm provided by Hangzhou ECON Technology co., LTD, it adopt integrated PID control and double closed-loop control, the control frequency range can be up to 0.1 Hz - 400 Hz and applicable to the vertical or horizontal single degree of freedom test and tri-axial six degrees of freedom test; the test subject is the two-exciter hydraulic vibration test system provided by the State Key Lab of Fluid Power & Mechatronic Systems of Zhejiang university.

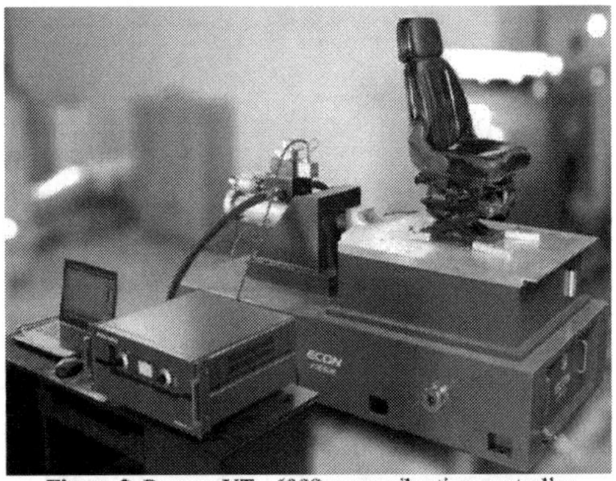

Figure 2. Premax VT - 6008 servo vibration controller

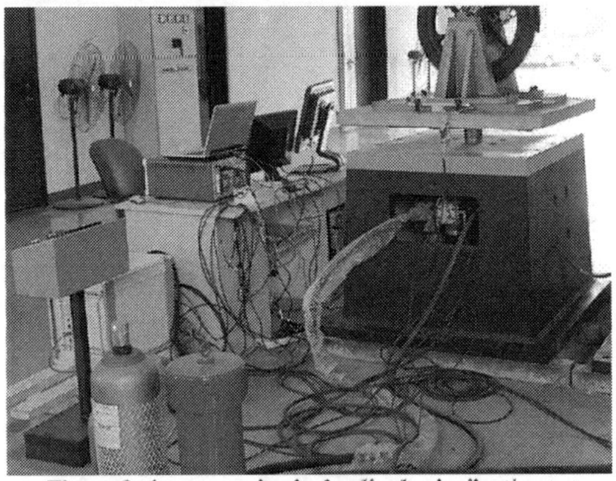

Figure 3. the two-exciter hydraulic shock vibration test system

5 Random vibration test

5.1 The self-closed-loop test

DOI: 10.1051/matecconf/20167707005

The closed-loop test chose 2 output (incentives) channels, 2 input (control) channels. The bandwidth of random power spectral is 5~100Hz, the control signal effective value is 0.15g, the correlation coefficient between signals is 0.2. Fig. 4(a) ~ (d) for show the closed-loop test system results: Figure 4(a), 5(b) show the 2 input (control) the power spectral density curve of spectrum (control) after the signal energy recovery; Fig. 4 (c), 4 (d) show the 2 input (control) in the time domain waveform curve of the signal. According to the results of the closed-loop test, the control spectrum curve can repeat the reference spectrum waveform in high precision, at the same time, the input waveform can realize the expected random shock waveform according to the shape spectrum (Fig. 1).

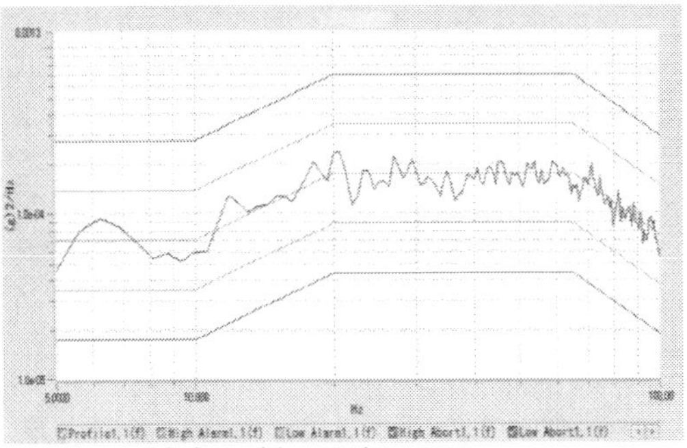

(a) channel 1 control spectrum

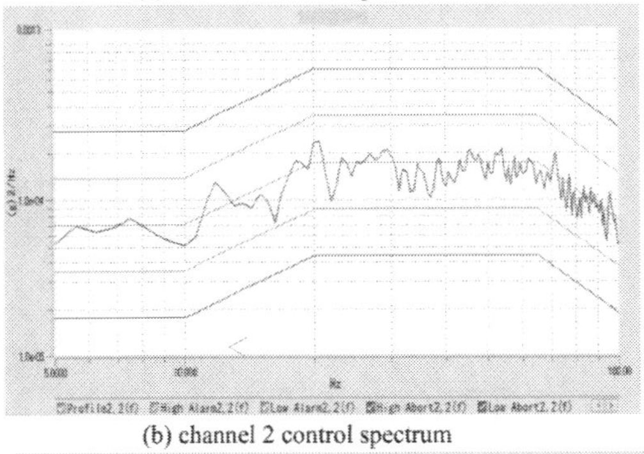

(b) channel 2 control spectrum

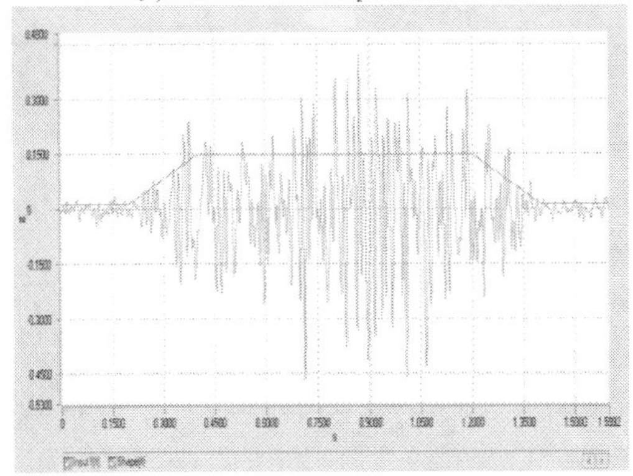

(c) channel 1 control signal

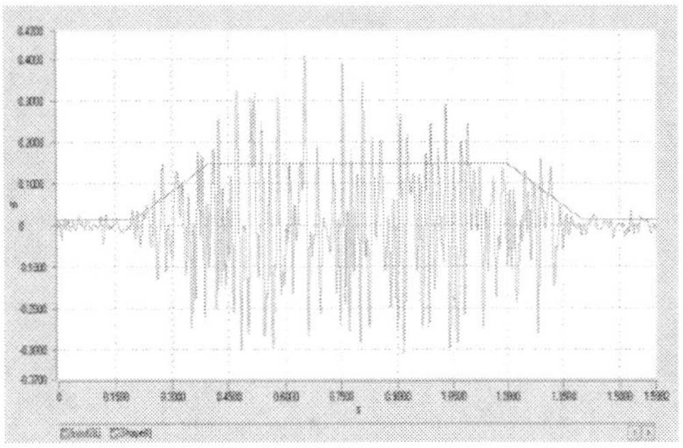

(d) channel 2 control signal

Figure 4. the result of random shock closed-loop test

The closed-loop test shows that the random shock test based on the frequency domain filter-X LMS adaptive inverse control algorithm can achieve high precision in the self-close-loop control, the spectrum control precision satisfies the requirement of experiment.

5.2 The field test

The random shock vibration test on the two-exciter hydraulic shock vibration test system for is shown in Fig. 5 in 2 points incentive, 2 point control experiment, the field test reference spectrum is the same to the closed-loop test. After 4 effective iterative calculation, the random shock vibration test of the two-exciter hydraulic shock vibration test system show in Fig. 6 (a) ~ (d). Fig. 6 (a) and (b) are the control signal (after energy recovery) power spectral density curve, Fig. 6 (c) and (d) are the control signal time domain waveform curve.

Figure 5 the two-exciter hydraulic shock vibration test system

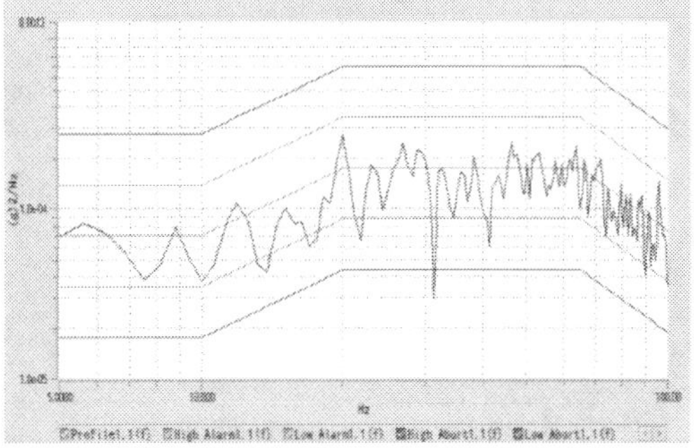

(a) channel 1 control spectrum

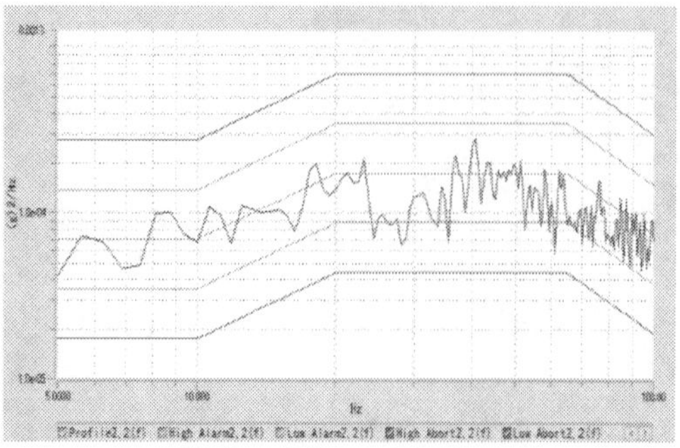

(b) channel 2 control spectrum

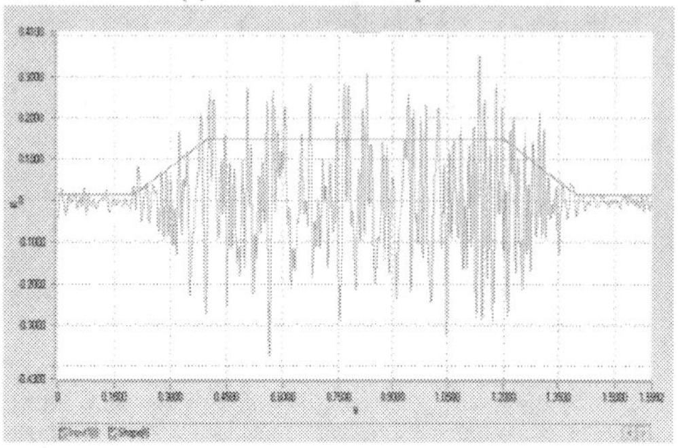

(c) channel 1 control signals

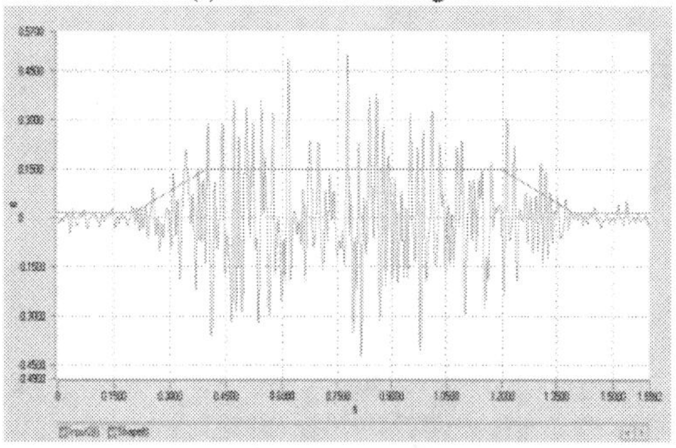

(d) channel 2 control signals

Figure 6. results of the random shock field tests

From the random shock vibration test on the two-exciter hydraulic shock vibration test system, within the scope of the test frequency width (5Hz ~ 100Hz), the scope of the error of the power spectral density curve of the control signal is control within ±1.5dB (as shown in Fig. 6 (a) and (b)), as the test progressed and the increase of iterations, the field test can achieve good control effect, the random shock test can control time domain signal according to the cutting waveform. Further, the test can realize the high precision control by using the frequency domain filter-X LMS adaptive inverse control strategy, thus proving the effectiveness of the proposed control strategy.

6 Conclusion

This paper adopts the two-exciter hydraulic shock vibration test system as the test object and Premax VT - 6008 servo controller as the control core to conduct the vibration test research based on filter-X frequency domain LMS adaptive inverse control strategy. The research respectively conduct the self-closed-loop test and field test of relevant control strategy to carry on the experiment, the experimental results showed that: in the process of random shock vibration test, by using the filter-X of frequency domain LMS adaptive inverse control strategy implementation on the system impedance matrix of the adaptive correction, effectively restrain noise interference in the system to test the influence of the system control and improve control accuracy; Further, the test can realize the high precision control by using the X-filter of frequency domain LMS adaptive inverse control strategy, thus proving the effectiveness of the proposed control strategy.

References

1. Ammanagi S, Poornima V, Sera A, et al. Development of a digitally-controlled three-axis earthquake shake table[J]. Current Science. 190-203(2006)
2. Harman C. Multi-axis vibration reduces test time[J]. EE Evaluation engineering. 44-47(2006)
3. Yugang Zhao. Shock response analysis method and its Application[D]. Zhejiang University. (2004)
4. Schoukens J, Rolain Y, Pintelon R. Leakage reduction in frequency-response function measurements[J]. IEEE Transactions on Instrumentation and Measurement. 2286-2291(2006)
5. Boroujeny F B, Chan K S. Analysis of the frequency-domain block LMS algorithm[J]. IEEE Transactions on Signal Processing. 48(2000)
6. Huijun Yu, Zhangwei Chen. Research on adaptive control used for time domain replication of road simulation[J]. Journal of Vibration Engineering. 498-501(2007)
7. Qiangli Luan. Research of key technologies of the control system in hydraulic vibration tests[D]. Zhejiang University. (2015)

Effects of Cutting Tool Parameters on Vibration

Mehmet Alper Ince and İlhan Asiltürk

University of Selcuk, Faculty of Engineering, Konya, Turkey.

Abstract. This paper presents of the influence on vibration of Co28Cr6Mo medical alloy machined on a CNC lathe based on cutting parameters (rotational speed, feed rate, depth of cut and tool tip radius). The influences of cutting parameters have been presented in graphical form for understanding. To achieve the minimum vibration, the optimum values obtained for rpm, feed rate, depth of cut and tool tip radius were respectively, 318 rpm, 0.25 mm/rev, 0.9 mm and 0.8 mm. Maximum vibration has been revealed the values obtained for rpm, feed rate, depth of cut and tool tip radius were respectively, 636 rpm, 0.1 mm/rev, 0,5 mm and 0.8 mm.

1 Introduction

Vibration is a repetitive, periodic, or oscillatory response of a mechanical system. The rate of the vibration cycles is termed "frequency." Repetitive motions that are somewhat clean and regular, and that occur at relatively low frequencies, are commonly called oscillations, while any repetitive motion, even at high frequencies, with low amplitudes, and having irregular and random behavior falls into the general class of vibration [1].

In a machining operation, vibration is problem. Vibration affects the machining performance and in particular, the surface finish and tool life. In all the cutting operations like turning, drilling and milling, vibrations occur in the machining due to a dynamic motion between the cutting tool and the work piece.

The vibration which occurs in most machines, vehicles, structures, buildings and dynamic systems is undesirable, not only because of the resulting unpleasant motions and the dynamic stresses but also because of the noise produced. Noise is generally considered to be unwanted sound, and since sound is produced by some source of motion or vibration causing pressure changes which propagate through the air or other transmitting medium, vibration control is of fundamental importance to sound attenuation. Vibration analysis of machines and structures is therefore often a necessary prerequisite for controlling not only vibration but also noise [2], [3].

Vibration control system for turning operation uses a closed-loop feedback circuit which measures the relative vibration between the cutting tool and the work piece [4]. There have been many investigations on vibration prediction and controlling based on periodic measurements of various machining conditions using accelerometer and active vibration controller.

Two generic techniques used for solving these vibration control problems are modifying stiffness or the fundamental natural frequency of the specified components/subsystems, and their damping [5].

There have been many investigations on vibration prediction and control in turning. An integrated vibration avoidance and contouring error compensation algorithm is used for multi-axis CNC machine tools [6]. They developed an integrated method which shapes the trajectory commands in such a way that they do not excite the structural modes.

Based on active and passive damping treatments, different models are developed to avoid the machine tool structure vibrations. A computational–theoretical model was developed to represent a structural system with Coulomb damping having two degrees of freedom [7]. Approximately 85 percent of the passive damping treatments in actual applications are based on viscoelastic materials [8]. They worked on the cutting tool vibrations and control of cutting tool vibration using a damping pad made up of neoprene.

Devin and Osadchii [9] proposed a new tool design with an increased vibration-damping ability, which includes special elements made of damping materials to reduce vibration amplitude and surface roughness.

The cutting-force based vibration analysis can be done to ascertain the effect of the tool entering angle on tool vibration and tool life in a titanium alloy milling operation [10]. The vibration signals of carbide tool are monitored with the help of accelerometer and these signals then send to data acquisition system and a portable computer. The main purpose of this vibration analysis is to find the features revealing tool wear. Wavelet Packet decomposition is used to find out feature frequency band which reveals tool wear. The high degree of correlation between selected band energy, energy entropy, time-domain features and tool wear are observed. Hence, it shows the effectiveness of the proposed model that facilitates machine performance and reliability estimation in terms of tool life and tool wear.

The excessive wear on cutting tools leads to distortions in dimension of manufactured components. At the same time it increases scrapped levels thereby incurring additional costs. Therefore, it is crucial to detect and monitor the wear on a cutting tool in most metal cutting processes and several research efforts have done to develop on-line tool condition monitoring systems. In online metal cutting tool condition monitoring, the cutting force (Static and Dynamic) analysis and vibration analysis is done for tool wear monitoring [11]. The main purpose of this study was to develop a Tool Condition Monitoring System based on analytical modelling of online sensor signals. The results shows that the cutting forces in Z-direction and the vibration signatures were most sensitive to tool wear.

Puertas [12] state vibration can be measured in terms of peak acceleration, r.m.s value of velocity, peak to peak displacement. Ghani [13] presented a study of tool life, surface finish and vibration, while turning nodular cast iron using ceramic tool. They found that surface finish was to be almost constant with the progression of the flank wear under different cutting conditions Bonifacio [14] presented a study on correlating tool wear, tool life, surface roughness and tool vibration in finish turning with coated carbide tools. They concluded that the feed didn't influence the vibrational signal and had a little effect on surface roughness. Thomas [15] studied the effect of tool vibration on surface roughness during lathe dry turning process on mild carbon steel samples at different levels of speed, feed, depth of cut, tool nose radius, tool length and work piece length. Safeen [16] studied the effect of cutting tool vibration on surface roughness of work piece in dry turning operation .The surface roughness of the work piece is proportional to cutting tool acceleration. This effect interact with other independent variable such feed rate, depth of cut, speed. Surface roughness of work piece increases parallel to the tool vibration with increasing tool over hang.

Ahmed Syed Adnan and Sathyan Subbiah [17] have observed reduction in cutting forces and feed forces when transverse vibrations are applied. Chip thickness is also reduced and surface finish is improved upon application of vibration. This study investigates vibrations that are applied along the cutting edge and perpendicular to the cutting velocity.

Chen et al. [18] have present on reliability estimation for cutting tools based on logistic regression model using vibration signals. The three steps of new reliability estimation approach for cutting tools are as follows. First, on-line vibration signals of cutting tools are measured during the manufacturing process. Second, wavelet packet (WP) transform is employed to decompose the original signals and correlation analysis is employed to find out the feature frequency bands which indicate tool wear. Third, correlation analysis is also used to select the salient feature parameters which are composed of feature band energy, energy entropy and time-domain features. Finally, reliability estimation is carried out based on logistic regression model. The approach has been validated on a NC lathe. Under different failure threshold, the reliability and failure time of the cutting tools are all estimated accurately.

Wang et al. [19] have present on a theoretical and experimental investigation of the tool-tip vibration and its influence upon surface generation in single-point diamond turning (SPDT). In the present study, two characteristic peaks (twin peaks) are identified and found to be corresponding to the tool-tip vibrations by power spectrum density (PSD) analyses. The vibrations possess the features of small amplitude but high frequency. A physical model is proposed to capture the dominant factor based on the characteristic and it reveals that the twin peaks are attributed by the impact between the tool tip and workpiece and the process damping effect. Hence, a geometric model of surface roughness is proposed to take account of tool-tip vibration and it is verified through a series of experiments.

Marcus et al. [20] have studied the vibration reduction using passive absorption system with Coulomb damping. This research aims at investigating the possibility of using the Coulomb damping, mainly to lessen they vibrations of structures submitted to human loadings. The great advantage in using this type of damping is that we can easily obtain high levels of damping with values well controlled and adjusted to the need of the project. A computational–theoretical model was developed to represent a structural system with Coulomb damping, containing two degrees of freedom. In order to calibrate this model some experimental tests were carried out with a cantilever beam. A parametric study was performed after the theoretical– computational model had been adjusted. The results indicate that the system is only applicable to some situations, and care should be taken in the design.

Prediction of optimum cutting parameters and operating conditions for the annealed Co28Cr6Mo ASTM F 1537 steel that has low machinability, high mechanical properties, is resistant against corrosion and temperature and generally used as prosthesis in medical field is considered as a significant problem that needs to be tackled by the manufacturing experts in the related field. In this study involves the influence of rotational speed, feed rate, depth of cut and tool tip radius on vibration was determined during machining of Co28Cr6Mo medical alloy. A computer numerical controlled (CNC) machine is used for machining Co28Cr6Mo medical alloy in the present study.

2 Machining conditions and vibration measurements

In the experimental study an annealed Co28Cr6Mo ASTM F 1537 steel having hardness of 40 HRC was used. The specimen with dimensions of Ø50x500 mm was prepared. Turning process was carried out on a TC25-L type Sogotec CNC lathe and vibration values were measured on a Kistler 8692C50 model accelerometer and Kistler 5134 B Charge Amplifier device. The tests were conducted under dry machining conditions and in every test a new cutting bit was used to machine longitudinally along the work-piece. The tool holder used was MTJNR-

L 2525 M16, cutting bits were TNMG 160404, 160408, 160412 MT form produced by Taegutec company and cladded with TiCN by the PVD method and at the quality of TT 8020. With the recommendations from the manufacturer cutting parameters given on Table 1 and the experimental set up is shown in Fig. 1.

In this study, a total of 81 physical experiments were conducted. The first parameter column of the table shows the rpm (n), the second column gives feed rate (f), the third column is for depth of cut (a) and the last column lists the tool tip radius (r)

Table 1. Cutting parameters and their levels

Symbol	Parameter	Unit	Level 1	Level 2	Level 3
n	Rotational speed	rpm	318	477	636
f	Feed rate	mm/rev	0.1	0.15	0.25
a	Depth of cut	mm	0.5	0.7	0.9
r	tool tip radius	mm	0.4	0.8	1.2

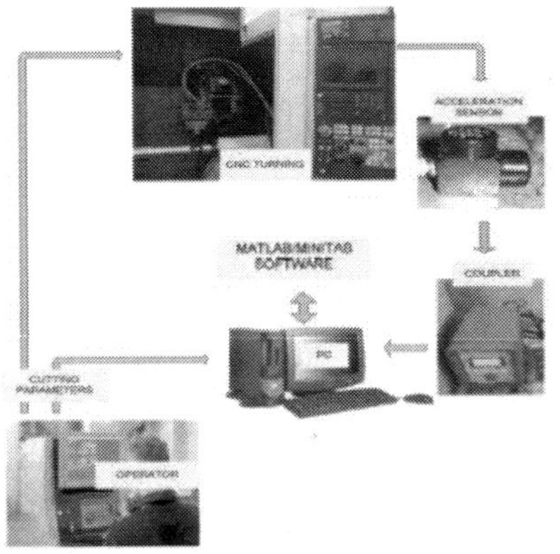

Figure 1. Experimental set-up

3 Conclusion

Vibration usually decreases with increase in tool tip radius. The minimum value of vibration obtained is 9.38 mV/g at n= 318 rpm, f= 0,25 mm/rev, a= 0,9 mm and r= 0.8 mm. Maximum value of vibration is 93.15 mV/g at n= 636 rpm, f= 0,1 mm/rev, a= 0,5 mm and r= 0.8 mm

The relationship between cutting speed and vibration is inversely proportional. Generally, increasing the cutting speed decreases the vibration. The relationship between feed rate and vibration is proportional. Generally, increasing the feed rate increases the vibration. The relationship between depth of cut and vibration is proportional. Generally, increasing the depth of cut increases the vibration. The relationship between tool tip radius and vibration is inversely proportional. Generally, increasing the tool tip radius decreases the vibration. For example, in Fig. 2, graphs of vibration are shown in the tool tip radius (0,4-0,8-1,2 mm), n = 477 rev/ min, f = 0,15 mm/rev, and a = 0,5-0,7-0,9 mm.

a= 0,5 mm

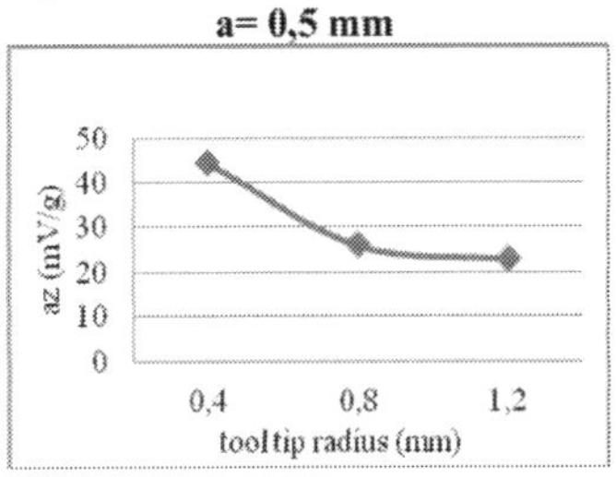

a= 0,7 mm

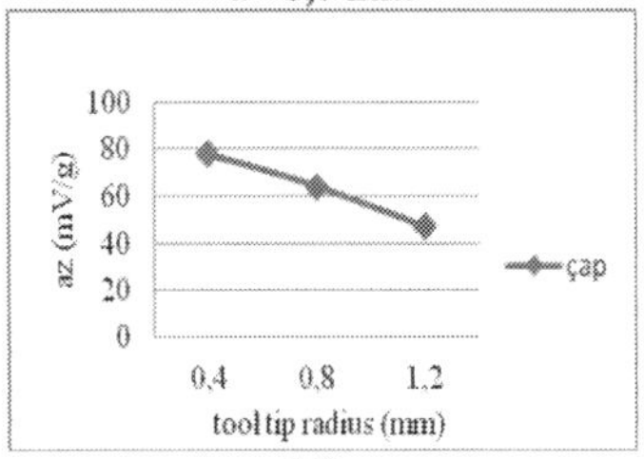

a= 0,9 mm

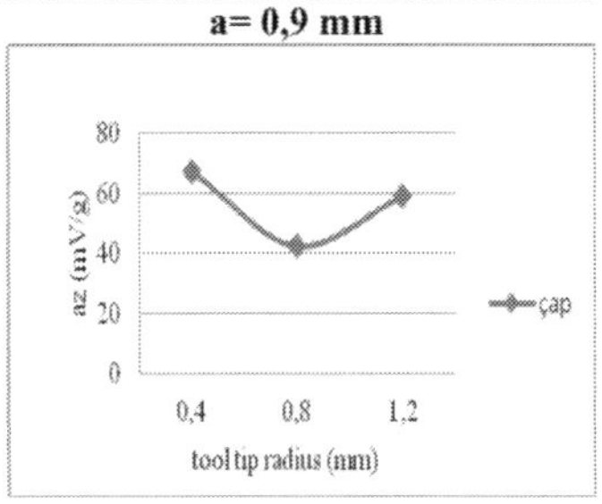

Figure 2. Graphs of Vibration for n=477 rpm, f= 0,15 mm/rev

n = 477 rev/min, f = 0.15 mm/rev, and a = 0,5 mm, tool tip radius of the tool was decreased, value of vibration was suddenly decreased.

n = 477 rev/min, f = 0.15 mm/rev, and a = 0,7 mm, tool tip radius of the tool was decreased, value of vibration was decreased.

n = 477 rev/min, f = 0.25 mm/rev, and a = 0,9 mm, tool tip radius of the tool was increased 0.4 to 0.8 mm, value of vibration was suddenly decreased; tool tip radius of the tool was increased 0.8 to 1.2 mm, there was increased in value of vibration.

Acknowledgements

This study is supported by Scientific Research Projects Coordinators (BAP) of Selçuk University. Their support is greatly appreciated.

References

1. W. de Silva Clarence, *Vibration Fundamentals and Practice*, (2000).

2. C.F. Beards, *Engineering Vibration analysis with* Application *to control Systems,* (1995).
3. H. Yunshan, G.W. Wei and Y. Xiang. DSC-Bitz method for the free vibration analysis of midline plates. *Int.J.Numer.Meth.engng,* **62:** 262-288, (2005).
4. A. Taskesen, Computer aided nonlinear analysis of machine tool vibrations and developed computer software, *Mathematical and computation Applications,* **3:** 377-385, (2005).
5. I. Eugene, *Stiffness and damping in mechanical design,* 494, (2007).
6. Y. Altintas, M.R. Khoshdarregi, Contour error control of CNC machine tools with vibration avoidance; *CIRP Annals - Manufacturing Technology,* **61,** pp. 335–338, (2012).
7. M. A. Louroza, N. Roitmanb, C. Maglutab, Vibration reduction using passive absorption system with Coulomb damping; *Mechanical Systems and Signal Processing;* **19,** pp. 537–549, (2005).
8. S. S. Abuthakeer, P.V. Mohanram, G. Mohan Kumar, Prediction and Contour of Cutting to Vibration in CNC Lathe with Anova and Ann; *International Journal of Lean Thinking,* **2,** no. 1, pp. 2526-2537, (June 2011).
9. L. N. Devin and A. A. Osaghchii, Improving Performance of CBN Cutting Tools by Increasing their Damping Properties, *Journal of Superhard Materials,* **34,** no. 5, pp. 326-335, (2012).
10. A. I. Sette, A. E. Diniz, R. Pederiva, Vibration analysis of cutting force in titanium alloy milling; *International Journal of Machine Tools & Manufacture*, **50,** pp. 65-74, (2010).
11. D.E. Dimla Sr., P.M. Lister, On-line metal cutting tool condition monitoring force and vibration analyses; International Journal of Machine Tools & Manufacture, **40,** pp.739–768, (2000).
12. I.A. Puertas, C.J. Luis Pérez, "Surface roughness prediction by factorial design of experiments in turning processes"; *Journal of Materials Processing Technology*, **143-144,** 390-396, (2003).
13. A.K Ghani , I.A. Choudhury, Husni, "Study of tool life, surface roughness and vibration in machining nodular cast iron with ceramic tool"; *Journal of Materials Processing Technology*, **127**, pp. 17–22, (2002).
14. M.E.R. Bonifacio, Diniz A.E. " Correlating tool wear, tool life, surface roughness and vibration in finish turning with coated carbide tools"; *Wear*, **173**, pp. 137–144, (1994).
15. M. Thomas, Y. Beauchamp, A.Y. Youssef, J. Masounave, " Effect of tool vibration during lathe dry turning process" , *computers Ind. Eng.*, **31,** no. 3-4, pp. 637-644,1996.
16. Y. Safeen, Kassab, K.Y. Khoshnaw, "Effect of cutting tool vibration on surface roughness of work piece in dry turning operation", *Eng. & Technology,* **25,** no. 7. pp 123-128, (2007).
17. A.S. Adnan, S. Subbiah; "Experimental investigation of transverse vibration-assisted orthogonal cutting of AL-2024"; *International Journal of Machine Tools & Manufacture;* **50,** pp. 294–302, (2010).
18. B. Chen, X. Chen, B. Li, Z. He, H. Cao, GaigaiCai;"Reliability estimation for cutting tools based on logistic regression model using vibration signals"; *Mechanical Systems and Signal Processing,* **25,** pp 2526-2537, (2011).
19. H. Wang, S. To, C. Y. Chan, C. F. Cheung, W. B. Lee; "A theoretical and experimental investigation of the tool-tip vibration and its influence upon surface generation in single-point diamond turning"; *International Journal of Machine Tools & Manufacture*, **50,** pp. 241-252, (2010).
20. A. Marcus, Lourozaa, N. Roitmanb, C. Maglutab; "Vibration reduction using passive absorption system with Coulomb damping"; *Mechanical Systems and Signal Processing*, **19,** pp. 537–549, 2005.

Design of a Low Phase Noise VCO for Rubidium Atomic Frequency Standard

Qingyun Ju[1], Xinwei Li[1], Lei Ji[1], Liang Tang[2] and Donghai Qiao[2]

[1]*School of Electronic and Information Engineering, Soochow University, 215000 Suzhou, China*
[2]*Institute of Acoustics, Chinese Academy of Sciences, 100000 Beijing, China*

Abstract. Compared to the size and the power consumption of the traditional atomic clocks, the ones based on coherent population trapping (CPT) can provide improvement by two orders of magnitude in both aspects above, making them needed urgently in many fields. Among different CPT atomic clocks, the one made with the rubidium atom are used widely, and their operating performance largely depends on its internal voltage-controlled oscillator (VCO) which is used to provide a proper microwave signal. Based on this, a small size and low phase noise 3.035GHz VCO is designed with a modified Clapp circuit topology using low-cost surface-mount components, including a coaxial resonant (COAX) with high quality factor. The designed VCO is simulated and optimized with the combined use of the negative resistance analysis method and the transmission analysis with a virtual-ground. In order to obtain the desired results, different values of key capacitor elements are tried according to their concrete influences on the VCO during the process of the circuit tuning. The test results show that the phase noises of the VCO are -60.49dBc/Hz@300Hz, -73.08dBc/Hz@1KHz and -97.48dBc/Hz@10KHz, the output power is -1.13dBm and the voltage-controlled tuning sensitivity is about 12MHz/V.

1 Introduction

With the progress of the society and the development of the science and technology, there is a growing demand for timing tools with small size, high precision and low power consumption in many fields, particularly in the military industry and the precision measurement field. Atomic clocks based on CPT can meet all the specs well [1] and they can be realized with different atoms among which the rubidium atom is most widely used. According to the operating mechanism of the CPT atomic clocks, their performance largely depends on the internal VCO [2]. The frequency of the VCO is designed to be equal to or half of the atomic ground state hyperfine splitting frequency to modulate the laser effectively [2]. An appropriate frequency tuning ability is also needed to compensate for the frequency drift caused by component tolerances, temperature changes and other factors [3]. In this paper, a low phase noise 3.035GHz VCO based on a COAX with high quality factor is proposed, which has the potential to be used in ^{85}Rb CPT atomic clocks.

In order to design the VCO, the overall circuit can be designed firstly with the negative resistance analysis method which can make the oscillator start up fast, and with the use of the virtual-ground technology, the circuit can be evolved into a form that a gain circuit in series with a resonance circuit. The oscillation frequency and the gain margin of the circuit can be determined according to the Nyquist criterion [4]. Parameters of the circuit are optimized by observing the magnitude and phase responses of the open-loop gain to obtain better performance of the designed VCO.

2 The structure and design method

2.1 The structure of the VCO

The VCO circuit is designed with a Clapper structure including a resonance circuit, a gain circuit, a voltage-controlled bias circuit and an output isolation circuit. The operating frequency is determined by the resonance circuit which is composed of a COAX with high quality factor and a varactor with good linearity. The gain circuit is not only used to provide a suitable gain for the oscillator circuit, but also to maintain the stable oscillation. And value of the gain should be large enough to achieve oscillation but be low enough to prevent the transistor from strong compression and to reduce oscillations at higher harmonics. The voltage-controlled bias circuit is used to control the reverse bias voltage of the varactor to realize the adjustment of the oscillation frequency. The output isolation circuit is used to isolate the VCO and the test instruments to reduce the bad influences of the impedance mismatch and the load pull. The attenuation value can be designed based on the needed output power of the VCO whose principle frame diagram is shown in Fig. 1.

 DOI: 10.1051/matecconf/20167707007

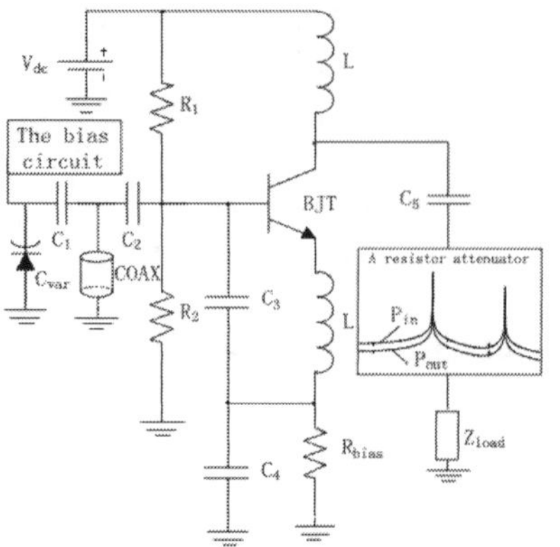

Figure 1. The principle frame diagram of VCO

2.2 Theories about the used design methods

At the beginning, the VCO is designed with the negative resistance analysis method which needs to satisfy (1),

$$Z_T = R_T + jX_T = 0 \tag{1}$$

where Z_T is the total impedance of the circuit, R_T and X_T are the real part and the imaginary part of Z_T, respectively. And they can be expressed in (2) and (3),

$$R_T = R_S + R_L = 0 \tag{2}$$

$$X_T = X_S + X_L = 0 \tag{3}$$

where R_S、X_S、R_L、X_L are the real and imaginary parts of the source impedance and load impedance, respectively. For a passive load, the real part of the impedance is always positive, so the active device should present an appropriate negative resistance when the real part of the total impedance is equal to 0. The condition in (2) mainly determines whether the oscillator can start up. When the oscillation occurs, the total resistance needs to be less than 0, and then there is a continuous growing current corresponding to the frequency in the circuit. As the current increases, the absolute value of the total resistance decreases gradually until the current of the circuit has been a stable value, meaning that the total resistance of the circuit has been equal to 0. On the other hand, the condition in (3) mainly determines the oscillation frequency of the circuit.

However, the conditions stated above are not sufficient to ensure the circuit can oscillate steadily. The stable oscillation also requires that the frequency disturbances and current disturbances caused by different factors can be gradually reduced to none. The frequency disturbances can be adjusted by the control voltage of the VCO, while the current or amplitude disturbances need to present an exponential decay. In this case, the total impedance of the circuit should meet the Kurokawa condition [5] in (4).

$$\frac{\partial R_T}{\partial I}\frac{\partial X_T}{\partial \omega} - \frac{\partial X_T}{\partial I}\frac{\partial R_T}{\partial \omega} > 0 \tag{4}$$

For a passive load, the real and imaginary parts of the total impedance of the VCO usually meet the conditions that $\partial R_L / \partial I = \partial X_L / \partial I = \partial R_L / \partial \omega = 0$, meaning that (4) can be expressed as (5).

$$\frac{\partial R_S}{\partial I}\frac{\partial}{\partial \omega}(X_S + X_L) - \frac{\partial X_S}{\partial I}\frac{\partial R_S}{\partial \omega} > 0 \tag{5}$$

Usually, $\partial R_S / \partial I > 0$ can be satisfied in the start-up process, so (5) can be well met if $\partial(X_S + X_L)/\partial \omega >> 0$, indicating that the total reactance which determines the oscillation frequency has a steep phase slope near the resonant frequency. This implies that a high quality factor circuit will result in maximum oscillator stability. In this paper, a COAX with high unloaded quality factor of about 450 is chosen to provide strong guarantee for good performance of the designed VCO.

However, on the basis of the negative resistance analysis method, the total impendence of the circuit is equal to 0 when the oscillation is stable, meaning that the quality factor is infinite. This situation is obviously impossible so that the phase noise evaluated with the Leeson formula relating to quality factor of the circuit is meaningless. The Leeson formula is shown in (6) [3-4],

$$L(\Delta f) = 10\lg\left\{\frac{FkT}{2P_{in}}\left[1 + \left(\frac{f_0}{2\Delta f Q_L}\right)^2\left(1 + \frac{f_c}{\Delta f}\right)\right]\right\} \tag{6}$$

where Δf is the frequency offset from the carrier center frequency f_0, $L(\Delta f)$ is the ratio of the sideband power in a 1Hz bandwidth at the offset of Δf to the total output power generated by the oscillator in dBc/Hz, F is the large signal noise figure of the active device in dB, k is the Boltzmann's constant, T is the equivalent noise temperature in K, P_{in} is the signal power at oscillator input in dBm, Q_L is the loaded quality factor and f_c is the flicker corner frequency of the active device.

In addition, the gain of the circuit cannot be observed directly with the negative resistance analysis method. The circuit in Fig. 1 can be evolved into a two port circuit in which a gain circuit and a resonance circuit are connected in a series form using the virtual-ground technology [6]. In this case, the amplitude and phase of the open-loop gain can be observed by the S-parameters, the resonant frequency of the circuit can be obtained by the zero phase crossings and the loaded quality factor of the circuit can be roughly estimated according to the group delay.

3 Circuit simulation

3.1 The equivalent model of the COAX

Normally, the oscillation frequency of the VCO designed with a COAX is lower than the self-resonant frequency of the COAX which can be equivalent to an inductor with high quality factor on this occasion. According to the impedance characteristics of the COAX, the impedance changes fastest near the self-resonant frequency so that

DOI: 10.1051/matecconf/20167707007

the frequency tuning range of the circuit will be restricted. For example, when the resonant frequency of the circuit is increased by reducing the capacitance of the varactor in the vicinity of the self-resonant frequency, the equivalent inductor of the COAX will be increased in the form of tangent function, then the resonant frequency of the circuit is decreased apparently, which is opposite to the adjusting direction of the varactor, and vice versa [3]. Therefore, the self-resonant frequency of the COAX needs to be properly higher than the operating frequency. In this design, the customized COAX has a self-resonant frequency of 3.104GHz and an unloaded quality factor of 458. And a large enough frequency tuning range and a high loaded quality factor can be expected at the desired frequency of 3.035GHz.

The resonance circuit composed by the COAX and a coupling capacitor is shown in Fig. 2. The TLPSC is the model of the used COAX in ADS software and its parameters can be obtained via the manufacturer. The L1 is the pin inductor of the COAX and its value is 0.6nH according to its data sheet. The C1 is the coupling capacitor and it constitutes a resonance circuit together with the COAX. The Term1 represents the input impedance of the transistor under a given bias voltage. In the simulation process, the parameters of the COAX need to be modulated slightly in the range of tolerance in order to realize resonance at the frequency of 3.035GHz.

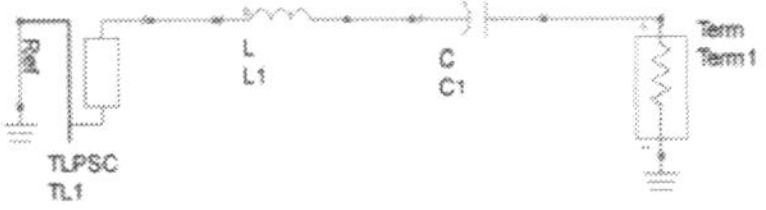

Figure 2. The equivalent model of the COAX.

Fig. 3 shows the S-parameter simulation results of the circuit. As can be seen, the circuit achieves resonance at 3.035GHz. And then the VCO can be designed based on the resonance circuit in Fig. 2.

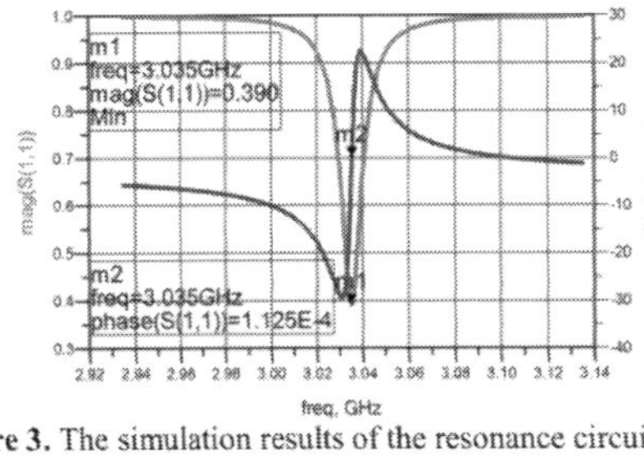

Figure 3. The simulation results of the resonance circuit.

3.2 Design with the negative resistance analysis method

The resonance circuit with high quality factor in Fig. 2 is connected to the base electrode of the transistor which has a high gain, an appropriate cut-off frequency and a low noise figure near the frequency of 3.035GHz. The chosen super-abrupt junction varactor possesses a low equivalent series resistor, contributing to obtaining a low phase noise. When the VCO is simulated, the varactor is replaced by an equivalent series RLC circuit and lightly coupled into the circuit with a proper capacitor. And then a modified Clapp circuit can be built along with the COAX. A resistor divider circuit is placed at the base electrode of the transistor to provide a proper bias voltage. In order to stabilize the static operation point of the transistor, its emitter electrode is connected with a resistor which can be used as the negative feedback component. The oscillation signal is output using a coupled capacitor after which a resistor attenuator is placed to provide the VCO some isolation from the load. The principle diagram of the VCO is shown in Fig. 4.

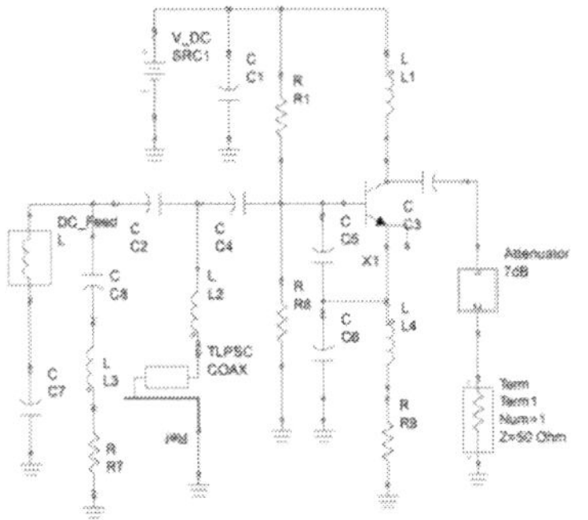

Figure 4. The principle diagram of the VCO.

Based on the negative resistance analysis method, the values of C5, C6 and R9 in Fig. 4 are adjusted and optimized so that the circuit can present a large enough reflection coefficient when seen from the base electrode of the transistor and a proper negative resistance. The simulation result of the reflection coefficient is shown in Fig. 5. It can be seen that the maximum value of the curve is 8.304dB achieved at 3.035GHz, indicating that a big enough negative resistance can be expected and the designed circuit can start up near 3.035GHz.

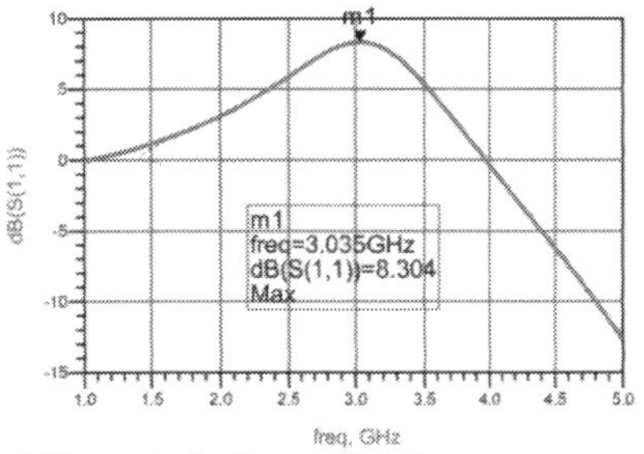

Figure 5. The port reflection coefficient.

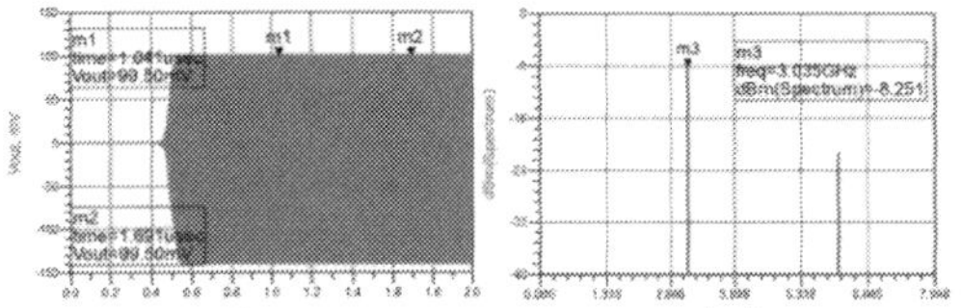

Figure 6. The transient simulation results in the time and frequency domain.

The transient simulation results in the time and frequency domain are shown in Fig. 6. It can be seen that the circuit can achieve a stable oscillation in a short time.

DOI: 10.1051/matecconf/20167707007

The fundamental frequency of the output oscillation signal is 3.035GHz and the power is -1.251dBm before attenuated by a 7dB attenuator at the end of the circuit.

3.3 Optimization with the virtual-ground method

With the use of the negative resistance analysis method, the circuit has reached a steady oscillation. However, the negative resistance analysis method lacks an effective observation of the operating mechanism of the circuit. In order to solve the problem, the transmission analysis with a virtual-ground is used. And then the gain, the quality factor and other parameters of the circuit can be evaluated. Besides, based on the simulation results, the parameters of the designed VCO can be optimized to achieve a low phase noise VCO.

According to the virtual-ground analysis method, a new reference ground point should be firstly selected for analyzing the circuit conveniently without changing the signal transmission characteristics of the original VCO circuit in Fig. 4. Generally, the emitter electrode of the transistor is selected as the virtual-ground point [6] to separate the gain module and the tuning module explicitly. This is because the emitter electrode is in both the control circuit and the controlled circuit. Then the base electrode of the transistor can be chosen as the breakpoint in the circuit to analyze the characteristics of the oscillation [7]. The schematic diagram with a virtual-ground point at the emitter electrode is shown in Fig. 7.

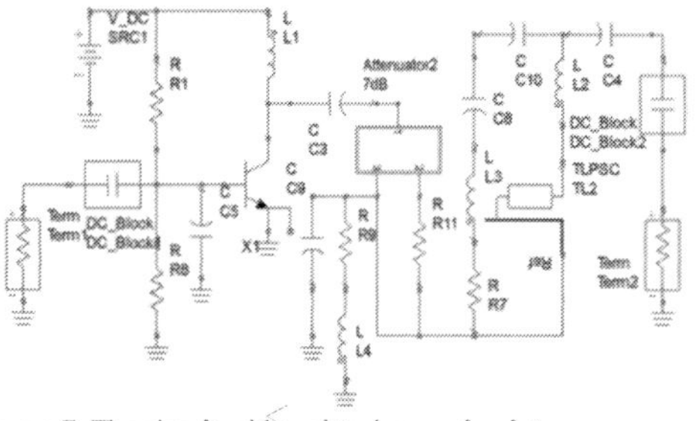

Figure 7. The circuit with a virtual-ground point

In order to achieve the maximum power transmission, the impedances of the two ports should be set as a pair of conjugate values according to the input impedance seen into the gain module. The circuit is simulated and the open-loop gain of the circuit can be calculated based on the formula in (7) [1].

$$Gain \approx \frac{S_{21}}{1-S_{11}S_{22}} + S_{12}\left[\frac{(S_{21}-2)S_{21}-(1-S_{11}S_{22})}{(1-S_{11}S_{22})^2}\right] \quad (7)$$

The simulation results are shown in Fig. 8. The curve of the open-loop gain is plotted in the polar coordinates as shown in Fig. 8(a). One net clockwise encirclement of the point (1, 0) is achieved as the frequency is increased, indicating that the Nyquist criterion can be met and the designed oscillator can start up [1]. Besides, curves of the magnitude and phase of the gain in (7) are plotted in the rectangular coordinates as shown in Fig. 8(b). Two zero phase crossings can be observed around the frequency of 3.035GHz, but only the first zero phase crossing occurs with negative slope and positive gain about 8.6dB while the second occurs with negative gain, meaning that the closed-loop network will only oscillate at the first zero phase crossing which is 3.037GHz. Though the simulated zero phase crossing is slightly higher than the desired operating frequency of 3.035GHz, mainly caused by the difference between the virtual-ground analysis and the transient analysis, the analysis of the circuit functions and the parameter optimization are not affected. It can also be seen that, the zero phase crossing on the left present a steep slope, indicating that the circuit possess a high load quality factor. Based on the equation (8), the load quality factor at the crossing point is about 139.

$$Q = \frac{\omega}{2}\frac{\partial\varphi}{\partial\omega} \quad (8)$$

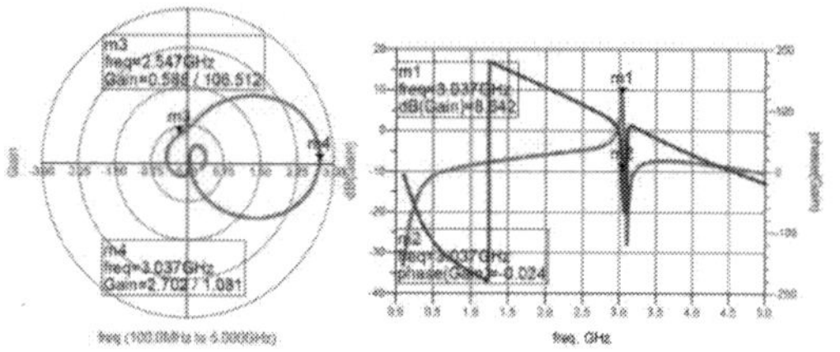

Figure 8. The curves of the gain in different coordinates: (a) the polar coordinates, (b) the rectangular coordinates.

3.4 Harmonic balance simulation results

The results of the harmonic balance simulation are shown in Fig. 9. It shows that the phase noises are -71.87dBc/Hz and -85.123dBc/Hz at the frequency offset of 300Hz and 1KHz, respectively, meeting the requirements of the ^{85}Rb atomic frequency standard well.

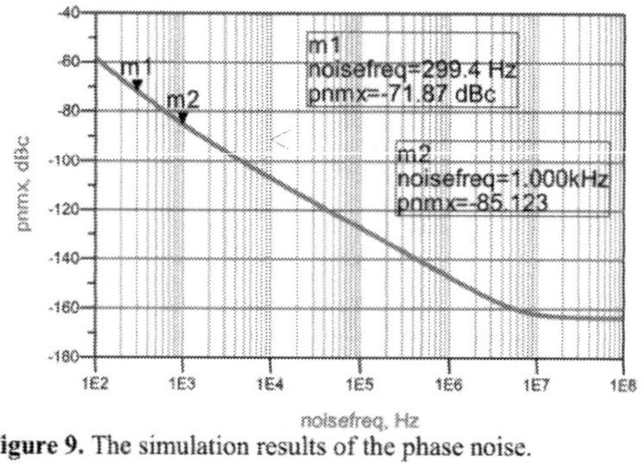

Figure 9. The simulation results of the phase noise.

4 Test results

In order to realize a VCO with a small size, 0402 SMT technology components are used and the circuit is laied out in a tightly enclosed topology. These choises resulted in a 10.5mm by 8.8mm by 0.6mm final design which also includes the COAX and a grouned metal shield. The supply voltage of the VCO is 2.8V dictated by that of the entire CSAC system which is 3.3V and by the voltage regulator used to distribute it. The fabricated VCO board

 DOI: 10.1051/matecconf/20167707007

is shown in Fig. 10 with a size (the effective size) smaller than a dime.

Figure 10. The fabricated VCO PCB.

When the VCO is tested with the same values of the passive components in Fig. 4, the test results are not in accordance with the simulation results well. Though the nonlinear simulation did not perfectly predict the VCO behaviours, it provided a good starting point to pursue optimizations. According to the simulation, performance of the designed VCO is mainly affected by capacitors of C4, C5 and C6 in Fig .4 when using the same COAX and varactor. Besides, the loaded factor quality of the VCO circuit can be augmented by increasing C6 or reducing C4, resulting in a lower phase noise. Doing so however, the output power will reduce and the VCO may not start up under all operating conditions. The test results show that the VCO can operate with better performance with an increased C6 and a reduced C4, this may benefit from the high transition frequency of the transistor used in the circuit. By some tuning on the values of the key capacitors described above, desired results were achieved. The test results show that the fundamental frequency is 3.035GHz when the varactor is reversely biased at 0.4V. The phase noises of the VCO are -60.49dBc/Hz@300Hz, -73.08dBc/Hz@1KHz and -97.48dBc/Hz@10KHz shown in Fig. 11. Moreover, the phase noise at the frequency offset of 10KHz is much better than -90dBc/Hz@10KHz claimed in the doctoral thesis [8].

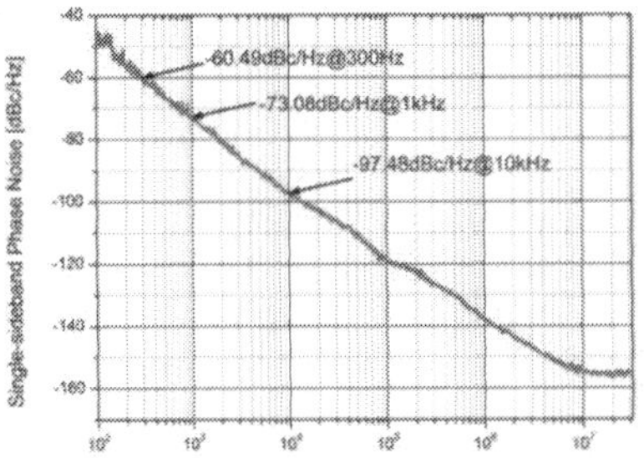

Figure 11. The test results of the phase noise.

The power of the fundamental signal is -1.13dBm before attenuated by a 7dB attenuator at the end of the circuit, meeting the requirement of the rubidium atomic clock based on the CPT phenomenon [8]. The tuning range of the reverse bias voltage of the varactor is 0~2.5V. The voltage-controlled tuning curve is shown in Fig. 12, the tuning sensitivity is about 12MHz/V when the tuning voltage changes between 0~2.0V.

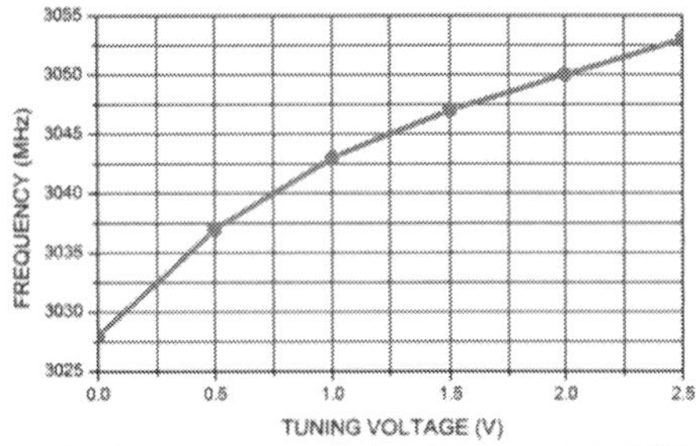

Figure 12. The voltage-controlled tuning curve of the VCO.

5 Conclusions

A small size and low phase noise 3.035GHz VCO based on a COAX with high quality factor is proposed. The circuit is designed, simulated and optimized by the combination of the negative-resistance analysis method and the virtual-ground technology. The test results show that the phase noises are -60.49dBc/Hz and -97.48dBc/Hz at the frequency offset of 300Hz and 10KHz respectively, the power of the fundamental signal is -1.13dBm and the voltage-controlled tuning sensitivity is about 12MHz/V, indicating that the designed VCO in this paper can meet the requirements of the ^{85}Rb CPT atomic clocks well.

References

1. L. Hollberg, J. Kitching, L. Liew, et al, A chip-scale atomic clock based on ^{87}Rb with improved frequency stability, Optics Express, **13**(4):1249-1253 (2005).
2. S. Knappe, V. Shah, P. D. D. Schwindt, et al, A microfabricated atomic clock, Applied Physics Letters, **85**(9):1460-1462 (2004).
3. S. Romisch, R. Lutwak, Low-power, 4.6-GHz, Stable Oscillator for CSAC, *International Frequency Control Symposium and Exposition*, 448-451 (2006).
4. M. Randall, T. Hock, General oscillator characterization using linear open-loop S-parameters, *Microwave Theory & Techniques IEEE Transactions on*, **49**(6):1094-1100 (2001).
5. D. M. Pozar, *Microwave Engineering*, John Wiley and Sons, 613-627 (2012).
6. S. Alechno, Analysis method characterizes microwave oscillations, Oscillator Analysis, Part 1, Microwaves & RF, 83-86 (1997).
7. A. Brannon, M. Jankovic, J. Breitbarth, et al, A Local Oscillator for Chip-Scale Atomic Clocks at NIST, *International Frequency Control Symposium and Exposition*, 443-447 (2006).
8. K. Deng, Experimental Study on the Miniaturized Coherent Population Trapping Atomic Clock, Ph.D. Thesis of Peking University (2011).

MATEC Web of Conferences 77, 07008 (2016) DOI: 10.1051/matecconf/20167707008
ICMMR 2016

Design of Unconstrained DMC to Improve the Distillate Product Purity of the Distillation Column

S. Bhat Vinayambika[1], I. Thirunavukkarasu[1], and S. Shanmuga Priya[2]

[1]*Dept. of Instrumentation and Control Engineering, Manipal Institute of Technology, Manipal University, Karnataka-India*
[2]*Dept. of Chemical Engineering, Manipal Institute of Technology, Manipal University, Karnataka-India*

Abstract. This paper demonstrates the use of unconstrained Dynamic Matrix Control (DMC) to control the process transfer function with time delay. The selection of tuning parameter is challenging task in predictive control algorithm. The DMC scheme is designed and it is used here to control the process transfer function, First Order Plant with Dead Time (FOPDT) model. Here, one manipulated variable (reflux) and one controlled variable (distillate) is considered for the implementation. The algorithm significantly controls the reflux to improve the distillate product purity. The simulation is done using MATLAB m-file. Both servo and regulatory responses were obtained. The simulation result validates the effectiveness of the proposed algorithm.

1 Introduction

The predictive control approach is a popular technique for the control of slow dynamical systems, such as those encountered in chemical process control in the petrochemical, pulp and paper industries, gas pipeline control, waste heat boiler, and active vibration control of railway vehicle [1]. It was developed by Shell Oil Company in 1980's. One of the basic predictive algorithm is Dynamic Matrix Control (DMC) is presented here for Single Input Single Output (SISO) system [2]. DMC has the capability to handle constraints on the manipulated and controlled variables in a systematic manner. The basic elements of DMC are: (i) a model of the plant to predict its future, (ii) a cost function which reflects control objectives, (iii) constraints on inputs and states/outputs, (iv) an optimization algorithm and, (v) the receding horizon principle [2]. The tuning of DMC for FOPDT has been simulated here. The tuning of unconstrained DMC is a challenging task since the number of adjustable parameters that effect closed loop performance. The tuning parameter (P, M, and N) needs to be selected appropriately to obtain the better performance.

S. Rajasekaran and T. Kannadasan [3] proposed simplified predictive control for controlling a temperature of a fluid stream using the shell and tube heat exchanger. The first order plus time delay transfer function model of heat exchanger have been considered for the study. R. D. Kokate and L. M. Waghmare [4] have presented Internal Model Control (IMC) based DMC control technique to control the heat exchanger process. The aim here is to eliminate the offset brought from load disturbances. Saniye Ay and Sileyman Karacan [5] had proposed the decoupling constrained model predictive control of a multi-component packed distillation column. R. D. Kokate and L. M. Waghmare proposed DMC algorithm for Single Input Single Output (SISO) system. The response of the system depends on the selection of the tuning parameters. There is various tuning technique to identify the parameters P, N and M of the DMC algorithm in literature [6]. A simple First Order Plus Dead Time (FOPDT) model is considered for study and experimentation. S. Srinivasulu Raju *et al.* [7] designed the dynamic matrix control algorithm for driven pendulum system. The second order transfer function model of the driven pendulum has been considered here to implement the control algorithm. Jose Manuel Lopez-Guede [8] had studied the effect of the prediction horizon in the performance of a Dynamic Matrix Control (DMC). It is found that low to moderate value of prediction horizon results in a better performance of the DMC controller. Lin Cong *et al.* [9] developed multivariable DMC for high purity internal thermally coupled distillation column. They have considred feed as benzene/toluene system for the study. Manimaran. M *et al.* [10] proposed DMC for a pilot scale distillation column.

The paper briefly introduces the Dynamic Matrix Control algorithm. Also the desired system response transfer matrix is proposed and it is simulated using Matlab m-file. The servo and regulatory response is studied here. The DMC can handle the load disturbance rejection effectively. The FOPDT transfer function is considered here for illustration. Results are presented so as to demonstrate the superiority of the proposed method. Finally, conclusions are given in the last section.

2 Materials and methods

2.1 Conventional dynamic matrix control

The DMC is the most basic and popular MPC algorithm used in the process control industries. The block diagram representation is as shown in Fig. 1. The DMC adopts step response approach. It works on the principle of receding horizon. The time horizon was moved one step ahead during every control cycle and the optimization problem is solved during every control cycle. The main feature of this predictive control algorithm is its ability to deal with constraints on system variables. Since constraints are usually neglected in many of control algorithms [3].

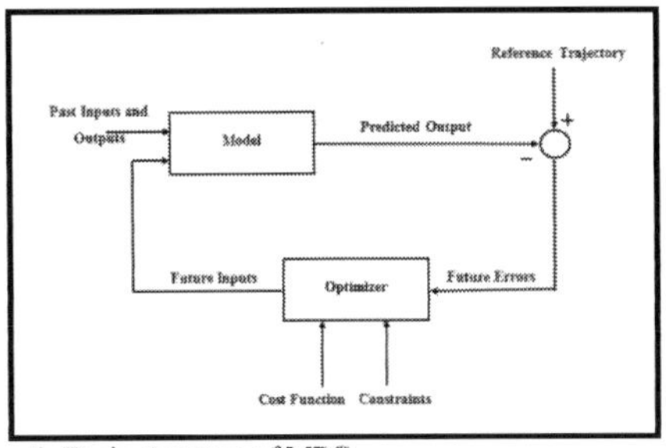

Figure 1. Basic structure of MPC

The process model consists of the information about the controlled process. According to the manipulated variables it predicts its future values. There after minimization of cost function ensures that the error is reduced. In the last step different optimization techniques are applied and the output gives the input sequence for the next prediction horizon.

The DMC algorithm basically has the following steps:

1. Step response model for the given plant.
2. Over finite predictive horizon find the quadratic performance objective.
3. Based on predicted value try to follow the set-point as closely as possible.
4. Optimal input need to be calculated based on least squares problem.

DMC control is based on a step response model that calculates a desired value of the manipulated value that remains unchanged during the next time step [4]. The new value of the manipulated variable the value that gives the smallest sum of squares error between the setpoint and the predicted value of the controlled variable. The number of time steps the DMC uses for its prediction is called the Prediction Horizon.

A large part of DMC's appeal is drawn from an intuitive use of a finite step response model of the process, a quadratic performance objective over a finite prediction horizon, and optimal manipulated input moves computed as the solution to a least squares problem.

2.2 Cost function

The cost function is a measure of performance of the process model. It is also called objective function. It is the performance criterion for controller [6]. The controller will make plant to follow a particular reference trajectory with satisfying performance criterion. This is obtained by minimizing the cost function. All objective function is having subfunctions which are required to be penalized. There are different types of cost functions like standard least square or quadratic objective function and absolute value objective function. The most commonly used objective functions is quadratic objective function. It is a sum of squares of the difference between the setpoints and the model predicted outputs and the control moves. A control move is the change in control action from one step to another step. For a P step prediction horizon and M step control horizon, the quadratic objective function is given below.

$$J = \sum_{i=1}^{P} \left(y_{r_{k+i}} - y_{P_{k+i}} \right)^2 + \lambda \left(\sum_{i=0}^{i=M-1} \left(\Delta u_{k+i} \right)^2 \right) \quad (1)$$

where

λ : Control weighing function

Δu : Change in the manipulated variable from one sample time to another sample time

y_r : Reference trajectory

y_P Predicted output

M : Model length

P : Prediction horizon

k : Sampling instants

2.3 Derivation of DMC algorithm

The main motive of Dynamic Matrix Control is to find the input signal that best satisfies to a given performance criterion, it predicts how the system will behave if the signal is applied [7, 11]. This algorithm uses the samples of step response of plant to capture plant's nature and solves a control problem for an optimal control action to track the given input [8].

Based on the step response model:

$$y_k = \sum_{i=1}^{\infty} s_i \Delta u_{k-i} \quad (2)$$

Since the step response saturates after N steps:

i.e $s_N = s_{N+1} = s_{N+2} = \ldots\ldots$

Thus, substituting (1) into (2) we have,

$$y_k = \sum_{i=1}^{N-1} s_i \Delta u_{k-i} + s_N \sum_{i=N}^{i=\infty} \Delta u_{k-i}$$

$$\text{i.e } y_k = \sum_{i=1}^{N-1} s_i \Delta u_{k-i} + s_N \left(u_{k-N} \right) \quad (3)$$

Thus, predicted output at the first time step in the future can be found from:

$$y_{k+1} = \sum_{i=1}^{N-1} s_i \Delta u_{k-i+1} + s_N \left(u_{k-N+1} \right)$$

$$y_{k+1} = s_1 \Delta u_k + \sum_{i=2}^{N-1} s_i \Delta u_{k-i+1} + s_N u_{k-N+1} \qquad (4)$$

Hence, for j[th] step in future:

$$y_{k+j} = \sum_{i=1}^{j} s_i \Delta u_{k-i+j} + \sum_{i=j+1}^{N-1} s_i \Delta u_{k-i+j} + s_N u_{k-N+j} \qquad (5)$$

The above equation can be separated into effects of past and future control moves:

$$y_{k+j} = s_1 \Delta u_{k+j-1} + s_2 \Delta u_{K+j-2} + \ldots\ldots + s_j \Delta u_k$$

Effects of current and future moves

$$+ s_N u_{k-N+j} + s_{j+1} \Delta u_{k-1} + s_{j+2} \Delta u_{k-2}$$
$$+ \ldots\ldots\ldots\ldots\ldots\ldots + s_{N-1} \Delta u_{k-N+j+1}$$

Effects of past moves

There is no control moves beyond the control horizon M steps, so:

$$\Delta u_{k+M} = \Delta u_{k+M+1} = \ldots\ldots = \Delta u_{k+P-1} = 0$$

In matrix-vector form, for a prediction horizon P and a control horizon of M steps yields:

$$\begin{bmatrix} y_{k+1} \\ y_{k+2} \\ \vdots \\ y_{k+j} \\ \vdots \\ y_{k+P} \end{bmatrix}_{P\times 1} = \begin{bmatrix} s_1 & 0 & 0 & \cdots & 0 & 0 \\ s_2 & s_1 & 0 & \cdots & 0 & 0 \\ \vdots & \vdots & & & & \\ s_j & s_{j-1} & s_{j-2} & \cdots & \cdots & s_{j-M+1} \\ \vdots & \vdots & \vdots & & & \vdots \\ s_P & s_{P-1} & s_{P-2} & \cdots & \cdots & s_{P-M+1} \end{bmatrix}_{P\times M} \begin{bmatrix} \Delta u_k \\ \Delta u_{k+1} \\ \vdots \\ \Delta u_{k+M-2} \\ \Delta u_{k+M-1} \end{bmatrix}_{M\times 1}$$

$$+ \begin{bmatrix} s_2 & s_3 & s_4 & \cdots & s_{N-2} & s_{N-1} \\ s_3 & s_4 & s_5 & \cdots & s_{N-1} & 0 \\ \vdots & \vdots & & & 0 & 0 \\ s_{j+1} & s_{j+2} & \cdots & s_{N-1} & 0 & 0 \\ \vdots & \vdots & \vdots & & & \vdots \\ s_{P+1} & s_{P+2} & \cdots & 0 & \cdots & 0 \end{bmatrix}_{P\times(N-2)} \begin{bmatrix} \Delta u_{k-1} \\ \Delta u_{k-2} \\ \vdots \\ \Delta u_{k-N+3} \\ \Delta u_{k-N+2} \end{bmatrix}_{(N-2)\times 1} + s_N \begin{bmatrix} u_{k-N+1} \\ u_{k-N+2} \\ \vdots \\ u_{k-N+P} \end{bmatrix}_{P\times 1}$$

$$\text{i.e } y_P = S_f \Delta u + S_{past} \Delta u_{past} + s_N u_P \qquad (6)$$

Thus, the difference between the reference trajectory y_r and the future prediction y_P is:

$$y_r - y_P = y_r - \left[S_{past} \Delta u_P + s_N u_P\right] - S_f \Delta u \qquad (7)$$

The above equation can be written as:

$$e^f = e - S_f \Delta u \qquad (8)$$

where the future predicted error is composed of 'free response' (e) and 'forced response' ($-S_f \Delta u$).

The least square objective function is [9]:

$$J = \sum_{i=1}^{P} \left(e^f_{k+i}\right)^2 + \lambda \sum_{i=0}^{M-1} \left(\Delta u_{k+i}\right)^2$$

$$\sum_{i=1}^{P} \left(e^f_{k+i}\right)^2 = \begin{bmatrix} e^f_{k+i} & e^f_{k+2} & \ldots & \ldots & e^f_{k+P} \end{bmatrix} \times \begin{bmatrix} e^f_{k+1} \\ e^f_{k+2} \\ \ldots \\ \ldots \\ e^f_{k+P} \end{bmatrix}$$

$$= \left(e^f\right)^T e^f$$

$$\lambda \sum_{i=0}^{M} \left(\Delta u_{k+i}\right)^2 = \begin{bmatrix} \Delta u_k & \Delta u_{k+1} & \ldots & \ldots & \Delta u_{k+M-1} \end{bmatrix} \times \begin{bmatrix} \lambda & 0 & 0 & 0 & 0 \\ 0 & \lambda & 0 & \ldots & \ldots \\ .. & .. & .. & .. & .. \\ .. & .. & .. & .. & .. \\ 0 & .. & .. & .. & \lambda \end{bmatrix} \times \begin{bmatrix} \Delta u_k \\ \Delta u_{k+1} \\ \ldots \\ \ldots \\ \Delta u_{k+M-1} \end{bmatrix}$$

$$= \Delta u^T \lambda \Delta u$$

Thus, objective function is written in the form:

$$J = \left(e^f\right)^T e^f + (\Delta u)^T \lambda (\Delta u)$$

The objective function is subjected to the modelling equation equality constraint:

$$e^f = e - S_f \Delta u$$

On expanding the objective function,

$$J = \left(e - S_f \Delta u\right)^T \left(e - S_f \Delta u\right) + (\Delta u)^T \lambda \Delta u$$

$$J = e^T e - S_f^T \Delta u^T e - e^T S_f \Delta u + \Delta u^T S_f^T S_f \Delta u + \Delta u^T \lambda \Delta u$$

$$J = e^T e - 2 S_f^T \Delta u^T e + \Delta u^T \left[S_f^T S_f + \lambda\right] \Delta u \qquad (9)$$

Thus, DMC control law solution for the minimization of the above given optimization function is given by solving $\frac{\partial J}{\partial \Delta u} = 0$ and hence find Δu.

$$\frac{\partial J}{\partial \Delta u} = 0 - 2 S_f^T e + 2\left(S_f^T S_f + \lambda\right) \Delta u = 0$$

$$\Delta u = \left(S_f^T S_f + \lambda\right)^{-1} S_f^T e \qquad (10)$$

Thus, $\Delta u = Ke$

where $K = \left(S_f^T S_f + \lambda\right)^{-1} S_f^T$ and $e = y_r - y_P$

Only one control move is actually implemented to calculate the Δu, thus only first row of K matrix is considered [10].

$$\text{i.e } \Delta u_k = K_1 E \qquad (11)$$

The flow chart of Dynamic Matrix Control algorithm is given in Fig. 2 as shown below.

2.4 DMC tuning

The DMC design technique has some design parameters (P, M and N), which can be tuned to get the desired response of the controller. These parameters are called the tuning parameters of the controller. The following guidelines are basically used to tune a DMC.

- The model length (N) should be approximately the open loop time of the process. When using step response model, it is important to make certain that the model length (N) is long enough to capture the steady state change [12].

- P determines how far into the future the control objective reaches and thus includes the main dynamic characteristics of the target. If P is increased results in more gradual control action but increases computational time.
- The control horizon (M) determines the number of the control actions calculated into the future. Small value of M makes the controller insensitive of noise. The less M is useful for controlling the stability of the system while larger M results in excessive control action and increases the flexibility, but it may lead to instability.
- Prediction and control horizons differ in length. Usually, the prediction horizon is selected to be much longer than the control horizon i.e., (N≤P≤M).
- Control weighting function (λ) is depends on prediction horizon and control horizon. It is set to zero if P>>M. As the control horizon is increased, the control moves tends to become more aggressive so a larger weight is needed to penalize the control moves.

3 Results and discussion

The transfer function model of Wood and Berry distillation column is considered here for the analysis [12].

$$\begin{bmatrix} x_D \\ x_B \end{bmatrix} = \begin{bmatrix} \dfrac{12.8e^{-s}}{16.7s+1} & \dfrac{-18.9e^{-3s}}{21s+1} \\ \dfrac{6.6e^{-7s}}{10.9s+1} & \dfrac{-19.4e^{-3s}}{14.4s+1} \end{bmatrix} \begin{bmatrix} L \\ V \end{bmatrix}$$

$$\text{i.e } G(s) = \begin{bmatrix} \dfrac{12.8e^{-s}}{16.7s+1} & \dfrac{-18.9e^{-3s}}{21s+1} \\ \dfrac{6.6e^{-7s}}{10.9s+1} & \dfrac{-19.4e^{-3s}}{14.4s+1} \end{bmatrix}$$

The plant is having two controlled variables y_1, y_2 and two manipulated variables u_1, u_2. The two controlled variables are composition of the distillate product (x_D) and bottom product (x_B). The manipulated variables are reflux (L) and reboiler flow rate (V). The FOPDT model is considered here. The input, reflux flow rate L and output, distillate product x_D is considered here to simulate the DMC. Thus, the transfer function model is given by:

$$\text{i.e } G(s) = \frac{12.8e^{-s}}{16.7s+1}$$

The above transfer function with sampling time 0.5 and after zero order discretization is given by:

$$G(z) = \frac{0.3776z^{-3}}{1-0.9705z^{-1}}$$

The following are the DMC tuning parameters of the above plant transfer function.

Simulation time = 200sec

Prediction horizon (P) = 10

Control horizon (M) = 3

Weight of control effort (λ) = 1

Model length (N) = 120

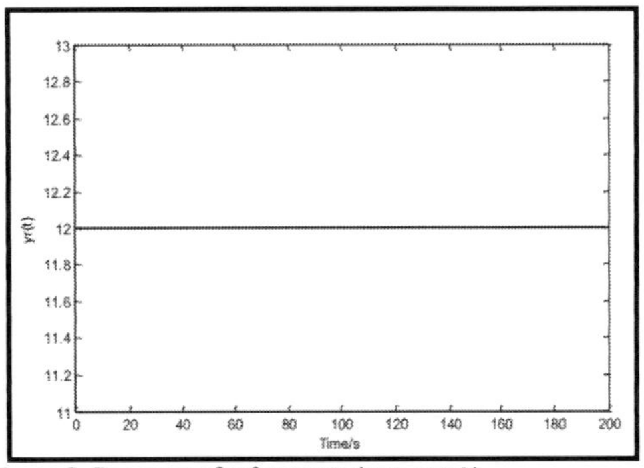

Figure 3. Response of reference trajectory $y_r(t)$

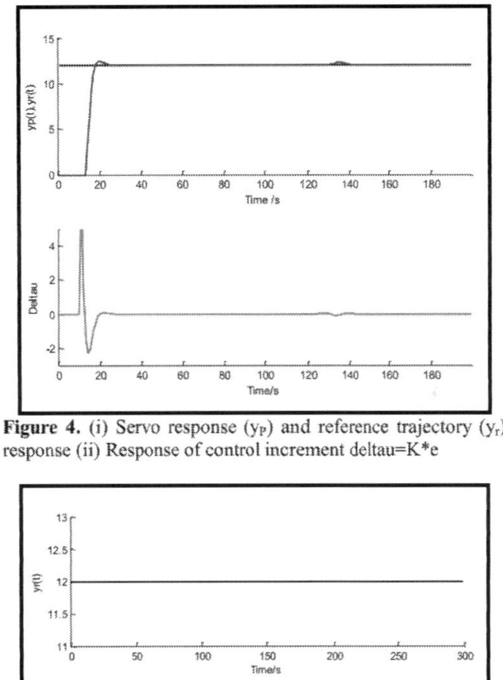

Figure 4. (i) Servo response (y_P) and reference trajectory (y_r) response (ii) Response of control increment deltau=K*e

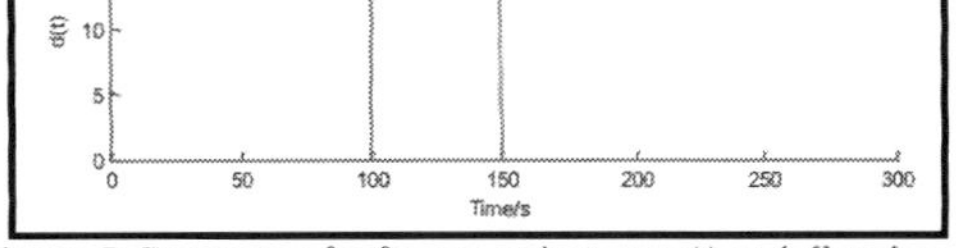

Figure 5. Response of reference trajectory $y_r(t)$ and disturbance d(t)

Thus, developed an unconstrained DMC algorithm considering one manipulated variable (Reflux flow rate) and controlled variable (Distillate product) of Wood and Berry distillation column. The response reaches the steady state, even in the presence of disturbance.

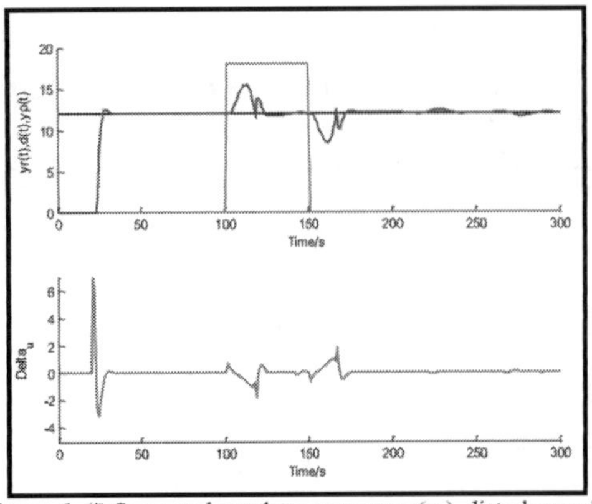

Figure 6. (i) Servo and regulatory response (y_P), disturbance d(t) and reference trajectory (y_r) (ii) Response of control increment deltau=K*e

4 Conclusions

The unconstrained DMC implemented for FOPDT model considering one manipulated variable (Reflux flow rate) and a controlled variable (distillate product) of the distillation column. It is a kind of optimization control algorithm, based on prediction model of the step response rolling conduction and marking feedback correction. It is suitable for linear time-invariant system which is asymptotically stable. The tuning parameters (P, M and N) place a major role in obtaining better performance. Both the servo and regulatory response is obtained. The simulation result shows good tracking capacity and good disturbance rejection ability. DMC has been found to be satisfactory to control the distillate product purity subject to the specified manipulated variable.

A possible direction for future work is the extension of the method to validate in real time. This topic is currently under investigation and will be reported later. Further, the result of DMC needs to be compared with the DMC-PID in simulation environment

5 Conflict of interests

The authors declare that there is no conflict of interests regarding the publication of this paper.

Acknowledgement

Vinayambika S Bhat to acknowledge the Mangalore Institute of Technology and Engineering, Moodabidri, for sponsoring her Ph.D. programme.

References

1. Manfred Morari, Jay H. Lee, Carlos E, Garcia, "Model Predictive Control," (2002).
2. William Y. Svreck, Donald P. Mahoney and Brent R. Young, A real time approach to process control, 3 ed., John Wiley and Sons Ltd., (2014).
3. S. Rajasekaran and T. Kannadasan, "A simplified predictive control for a shell and tube heat exchanger," *Journal of Engineering Science and Technology*, **2**, no. 12, (2010).
4. R. D. Kokate and L. M. Waghmare, "An IMC-DMC control for heat exchanger process," *International Journal of Computer Applications*, **2**, no. 26, pp. 96-100, (2010).
5. Saniye Ay and Suleyman Karacan, "Decoupling constrained model predictivecontrol of multi-component packed distillation column," *World Applied Sciences Journal*, **13**, no. 3, pp. 517-530, (2011).
6. R. D. Kokate and L. M. Waghmare, " Review of tuning methods of DMC and performance evaluation with PID algorithms on a FOPDT model," *International Journal of Control and Automation*, **4**, no. 2, pp. 95-110, 2011.
7. S. Srinivasulu Raju, T. S. Darshan and B. Nagendra, "Design of quadratic dynamic matrix control for driven penduum system," *International Journal of Electronicsand Communication Engineering*, **5**, no. 3, pp. 363-370, (2012).
8. Jose Manuel Lopez Guede, Borja Fernandez Gauna, Manuel Grana, Fernado Oterino, "On the influence of the prediction horizon in dynamic matrix control," *International Journal of Control Science and Engineering*, **3**, no. 1, pp. 22-30, 2013.
9. Lin Cong, Xinggao Liu and Chengyu Wang, "Dynamic matrix control of high-purity internal thermally coupled distillation column," *Canadian Journal of Chemical Engineering*, **92**, pp. 696-702, (2014).
10. Manimaran. M, Malaisamy. S, Mohamed rafiq. M, Petchithai. V, Chitra .V. S., Kalanithi. K and Abirami. H, "Parameter identification anddynamic matrix control Design for a nonlinear pilot distillation column," *International Journal of ChemTech Research*, **7**, no. 1, pp. 382-388, (2015).
11. David Di Ruscio, "Model predictive control with integral action: A simple MPC algorithm," *Modeling, Identification and Control*, **34**, no. 3, pp. 119-129, (2013).
12. B. Wanye Bequette, Process Control: Modeling, Design and Simulation, 4 ed., Prentice Hall, (2002).

Terminal Velocity Maximization of Air-to-Air Missiles in Agile Turn Phase

Min Ki Kim[1], Min-Jea Tahk[1], Bong-Gyun Park[2] and Yun-Young Kim[2]

[1]*KAIST, Department of Aerospace Engineering, 291 Daehak-ro Yuseong-gu Daejeon, Korea*
[2]*LIG Nex1, PGM R&D Laboratory, 207 Mabuk-ro Giheung-gu Yongin-si, Korea*

Abstract. Terminal velocity maximization of air-to-air agile missiles during agile turn phase is studied. It is important to make sure that agile missiles have enough velocity after agile turn stage in order to impact the target. The agile turn is operated during the boost phase. Thus, the agile turn problem is formulated as terminal velocity maximization with given final time. Optimal solutions are obtained by two different methods, pseudospectral method and parameterized costate optimization method. The parameterized costate optimization method is a part of shooting method of indirect method which is proposed to verify the solutions obtained by pseudospectral method. At first, the basic missile model with constant axial acceleration is treated in the two dimensional yaw plane. After looking into the basic missile model, missile model with aerodynamic forces and thrust is analyzed.

1 Introduction

In the last few decades, advances in aircraft maneuvering technology continued to produce new challenges for anti-air weapon systems. One of the challenges is to make air-to-air missiles impact the target at any direction and at any aspect. This demand resulted in the development of 'agile air-to-air missiles'. The agile missiles are classified into each mission phase which includes: launch, agile turn at high angle of attack, midcourse, and endgame [1]. The agile turn problems are focused on the first two stages which are the key issues of missiles' performance because these stages affect the missiles' launch envelope. These stages are accomplished during the boost phase.

The agile air-to-air missiles have lock-on after launch ability and even can engage targets in the rear hemisphere relative to the launch aircraft. Therefore, it is important to make sure that agile missiles have enough velocity after agile turn stage in order to impact the target. Solving an optimal control problem by setting cost function as minimum agile turn time may make velocity decrease significantly. It is undesirable since it cannot use relative velocity that is earned from the launch aircraft and may results into mission failure. Thus, this paper focuses on maximum terminal velocity in the agile turn stage.

The agile turn problem has been an area of research interest for the past 20 years. However, due to heavy amount of computational workloads of obtaining the optimal solution, it has not been focused on getting the optimal trajectory. Rather, it has been examined to get over the uncertainty and nonlinearity because of high angle of attack regime during the agile turn phase [2], [3].

There have been many efforts to reduce computational workloads of obtaining the optimal solution by simplifying the missile model. Lee [4] has dealt with a nonlinear adaptive autopilot design for agile missile systems that are reduced to a two dimensional plane problem. In [4], the paper proposed a longitudinal autopilot for angle-of-attack tracking based on backstepping control methodology in conjunction with the time-delay adaptation scheme. Ryu [5] introduced the design of the angle of attack controller using the pole placement approach for the high angle of attack missiles.

This paper investigates the optimal control problem by proposing 'parameterized costate optimization method'. It is one of the shooting methods of indirect method which transforms the problem from the original optimal control problem to finding initial values of costates. By using this method to a simple missile model, solutions obtained by applying pseudospectral method [6] can be verified. Also numerical optimization is conducted on missile model with aerodynamic forces and thrust.

2 Parameterized costate optimization method

There are several numerical ways to solve optimal control problems. One way is using GPOPS-II software which employed an hp-adaptive version of the Radau pseudospectral method [7]. Before applying this method, it needs to be verified to apply to various complex dynamic systems. Thus, parameterized costate optimization method is proposed in this section. It is a part of shooting method of indirect method.

A final time fixed optimal control problem can be formulated as (1) and (2) with state $x(t) \in R^n$ and control input $u(t) \in R^m$ in time interval $[t_0, t_f]$.

DOI: 10.1051/matecconf/20167707009

$$\dot{x} = f(x,u,t) \tag{1}$$

$$J = \phi(x_f, t_f) + \int_{t_0}^{t_f} g(x,u,t)dt \tag{2}$$

where x_f denotes final states. Approaching the optimal control problem with indirect method, the Hamiltonian function is defined as:

$$H = g + \lambda^T f \tag{3}$$

Then, the equations of motion can be written in Hamiltonian's form (4).

$$\begin{aligned} &\frac{\partial H}{\partial x} = -\dot{\lambda} \\ &\frac{\partial H}{\partial u} = 0 \\ &f - \dot{x} = 0 \end{aligned} \tag{4}$$

Solving (4) with boundary conditions gives the optimal solution. Since this method is applied to agile air-to-air missiles, assume that the initial conditions are given to every state. However, not all states have final conditions.

$$\{\{x_K\},\{x_F\}\} \in x \tag{5}$$

$x_K \in R^{n_K}$ and $x_F \in R^{n-n_K}$ denote states that final conditions are known and states that final conditions are free respectively. When final state is free, final costate condition, (6), needs to be satisfied from the optimal control theory.

$$\lambda_F(t_f) = \frac{\partial \phi}{\partial x_f} \tag{6}$$

where $\lambda_F \in R^{n-n_K}$ denotes costates related to states x_F. Note that if final state x_{Kf} is known, related costate's final value becomes free and when final state is free, related costate's final value has to meet (6). Moreover, all initial states are known whereas every initial costate is undetermined. Therefore, when initial costates are treated as unknown variables, solving optimal control problem becomes finding appropriate initial costates' value while final conditions are met. To illustrate mathematically, let J_2 be defined as:

$$J_2 = f(\lambda_0) = \sum_{i=1}^{n_K}(x_{K_i}(t_f) - x_{K_i f})^2 + \sum_{j=1}^{n-n_K}(\lambda_{F_j}(t_f) - \frac{\partial \phi}{\partial x_{F_j f}})^2 \tag{7}$$

The whole idea of parameterized costate optimization method is to make initial costates as parameters which are variables and find those values that satisfy $J_2 = 0$. In (7), the term $x_{K_i}(t_f)$ and $\lambda_{F_j}(t_f)$ come from the integration results of (4) with initial costate guess values. Therefore, note that J_2 is a function of initial costate values. It is important to make J_2 exactly zero so its solution is equivalent to the optimal solution.

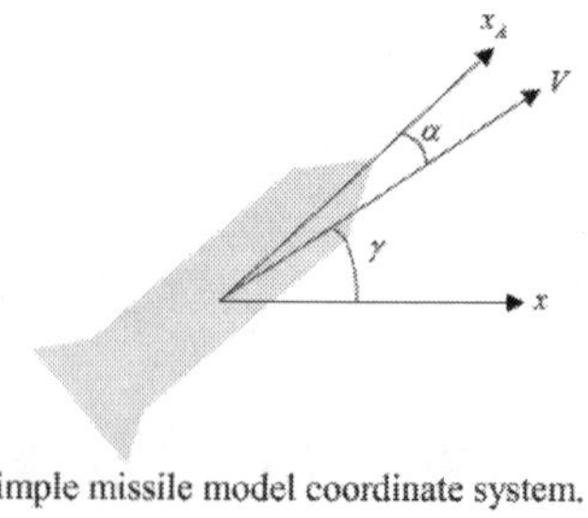

Figure 1. Simple missile model coordinate system.

2.1 Simple missile model without control bound

A simple missile model is employed in two dimensional yaw plane optimal control problem in order to verify pseudospectral method with parameterized costate optimization method.

The missile dynamics and optimal control problem are formulated as follows:

$$\dot{V} = a\cos\alpha \tag{8}$$

$$\dot{\gamma} = \frac{a\sin\alpha}{V} \tag{9}$$

$$J = \phi(x(t_f), t_f) + \int_{t_0}^{t_f} g(x(t), u(t), t)dt = -V_f \tag{10}$$

V, a, α, and γ denote velocity, acceleration, angle of attack, and flight path angle respectively. The control command is α. Since the agile turn problem is defined as maximizing terminal velocity, cost function is set to $-V_f$. Also initial and final conditions are:

$$V(0) = V_0 \tag{11}$$

$$\gamma(0) = \gamma_0, \gamma(t_f) = \gamma_{t_f} \tag{12}$$

Applying (4) and (6) to this problem derives

$$\begin{aligned} &\dot{V} = a\cos\alpha && V(0) = V_0, V(t_f) = free \\ &\dot{\gamma} = a\sin\alpha / V && \gamma(0) = \gamma_0, \gamma(t_f) = \gamma_{t_f} \\ &\dot{\lambda}_V = \lambda_\gamma a \sin\alpha / V^2 && \lambda_V(0) = parameter, \lambda_V(t_f) = -1 \\ &\dot{\lambda}_\gamma = 0 && \lambda_\gamma(0) = parameter, \lambda_\gamma(t_f) = free \end{aligned} \tag{13}$$

Such that

$$\tan\alpha = \frac{\lambda_\gamma}{\lambda_V V} \tag{14}$$

Finding λ_{V0} and $\lambda_{\gamma 0}$ that makes J_2 zero is equivalent to solving optimal control problem. J_2 is defined as:

$$J_2 = f(\lambda_{V0}, \lambda_{\gamma 0}) = (\gamma(t_f) - \gamma_{t_f})^2 + (\lambda_V(t_f) + 1)^2 \tag{15}$$

DOI: 10.1051/matecconf/20167707009

Following simulation results is conducted to compare the numerical solutions between using pseudospectral method and using parameterized costate optimization method.

Table 1. Simulation parameters

Parameters		Value	Unit
a	Constant axial acceleration	350	m/s^2
t_f	Boost time	3	s
V_0	Initial velocity	250	m/s
γ_0	Initial flight path angle	0	deg
γ_f	Final flight path angle	30	deg

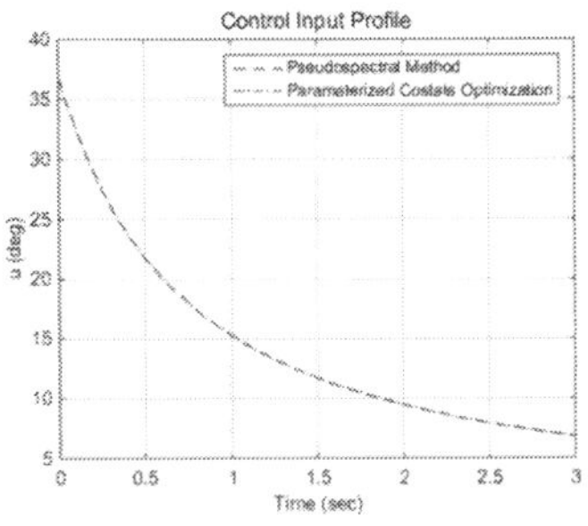

Figure 2. Control input α profile

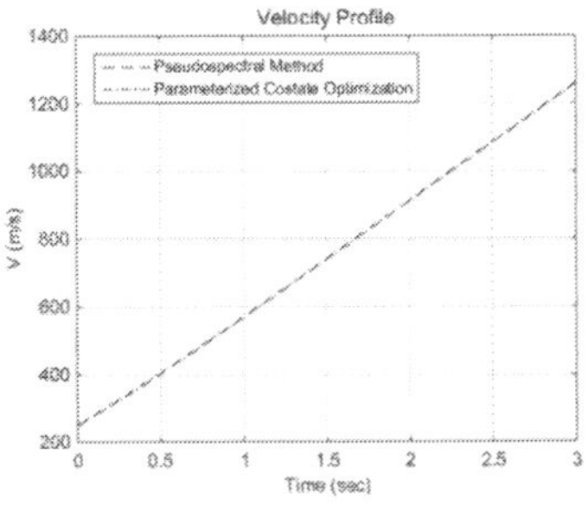

Figure 3. Velocity profile

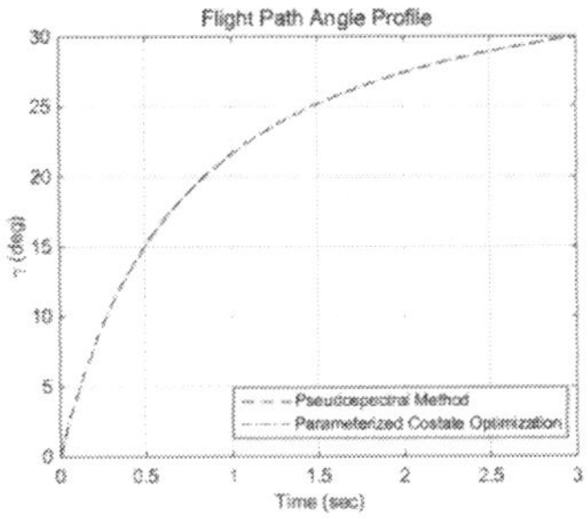

Figure 4. Flight path angle profile

The graphs in Fig. 2 and Fig. 3 show that both methods give the same results.

2.2 Simple missile model with control bound

Previous control input result shows that control input goes over 25 degrees which is too high for an actuator to follow that angle of attack command. Thus, control command needs to be bounded to a certain value which the actuator can withstand.

$$|\alpha| \leq 25\deg \tag{16}$$

Same optimal control problem as in part 2.1 is treated in this control bounded problem.

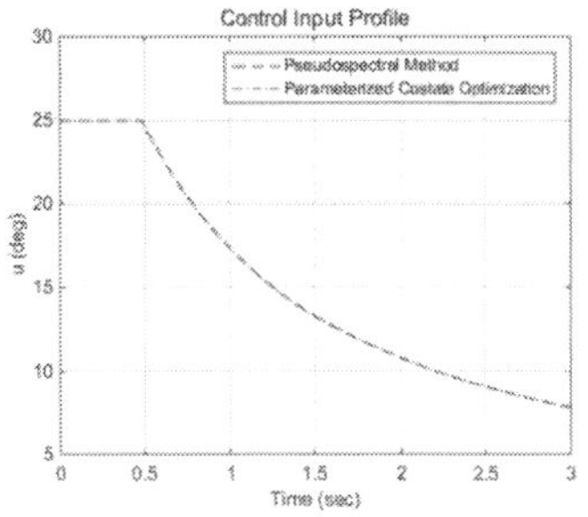

Figure 5. Control input α profile

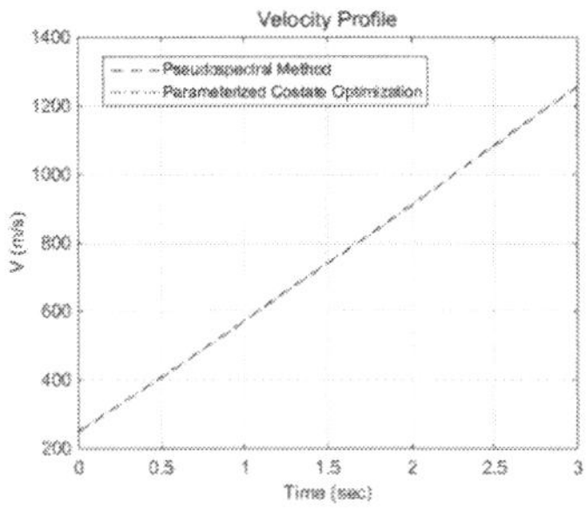

Figure 6. Velocity profile

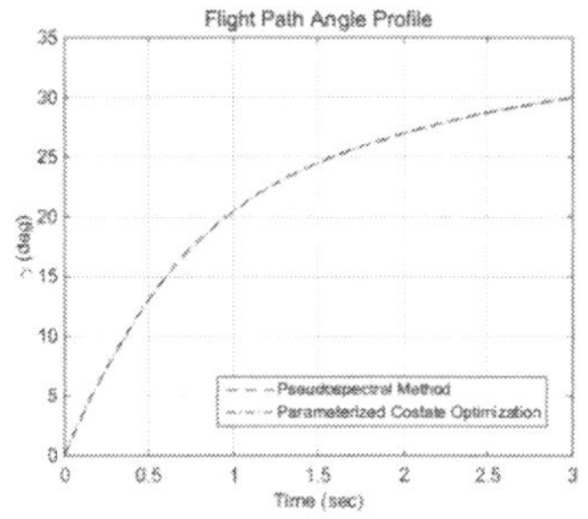

Figure 7. Flight path angle profile

When control is bounded, Pontryagin's minimum principle needs to be introduced. Fig. 4 shows bang-singular control. It starts with upper-bound bang control at first 0.5 seconds. This upper-bound bang control is determined from Pontryagin's minimum principle. However, after 0.5 seconds, it is following the singular arc which is determined from (14).

The parameterized costate optimization method verified pseudospectral method well on both unbounded control problem and bounded control problem.

3 Terminal velocity maximization

From the previous section, pseudospectral method is verified. Thus, pseudospectral method will be applied to terminal velocity maximization problem using the missile model with aerodynamic forces and thrust. This missile model considers mass variation $m(t)$; thrust $T(t)$; aerodynamic forces $L(\alpha,V)$ and $D(\alpha,V)$; and time lag τ . Equations of motion are written as:

$$\dot{V} = \frac{T(t)\cos\alpha - D(\alpha,V)}{m(t)} \tag{17}$$

$$\dot{\gamma} = \frac{T(t)\sin\alpha + L(\alpha,V)}{m(t)V} \tag{18}$$

$$\dot{X} = V\cos\gamma \tag{19}$$

$$\dot{Y} = V\sin\gamma \tag{20}$$

$$\dot{\alpha} = \frac{u-\alpha}{\tau} \tag{21}$$

$$|u(t)| \le 25\,\text{deg} \tag{22}$$

Also, aerodynamic force models are given as Fig. 7. The lift and drag coefficients versus angle of attack are plotted at Mach number 0.7 to 1.0 with 0.1 increments.

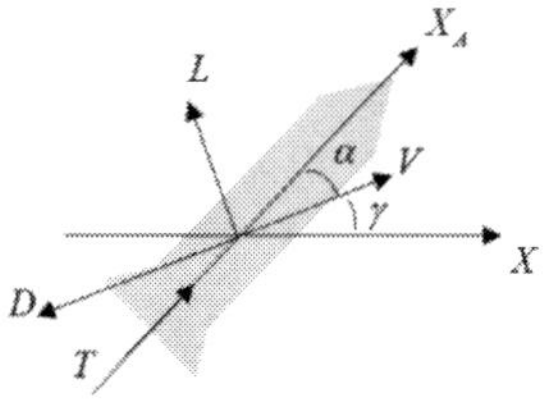

Figure 8. Missile Model Free Body Diagram

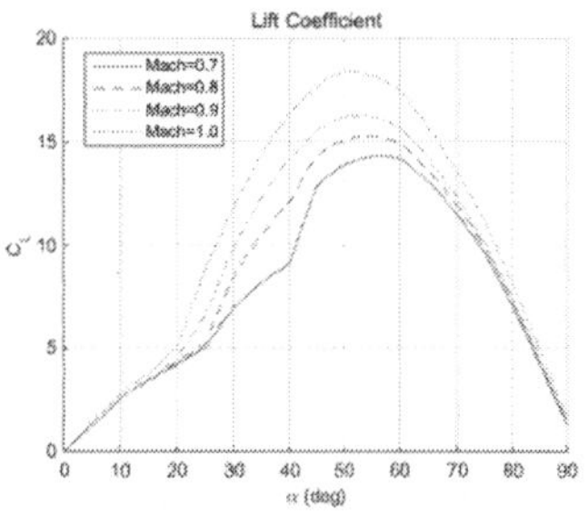

Figure 9. Lift Coefficients versus Angle of Attack

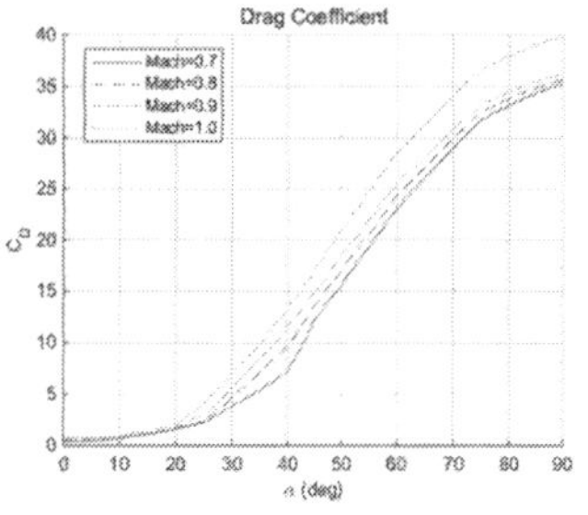

Figure 10. Drag coefficients versus angle of attack

Table 2. Missile Parameters

Parameters		Value	Unit
t_f	Boost time	2.69	s
m_0	Initial mass	92	kg
c	Mass burning rate	13	kg/s
T	Average thrust	35000	N
S_{ref}	Reference wing area	0.0182	m^2
τ	Time constant	0.3	s

Table 3. Simulation Parameters

Parameters		Value	Unit
ρ	Air density	1.0065	kg/m^3
α_0	Initial angle of attack	0	deg
V_0	Initial velocity	250	m/s
γ_0	Initial flight path angle	0	deg
γ_f	Final flight path angle	50, 75, 100	deg

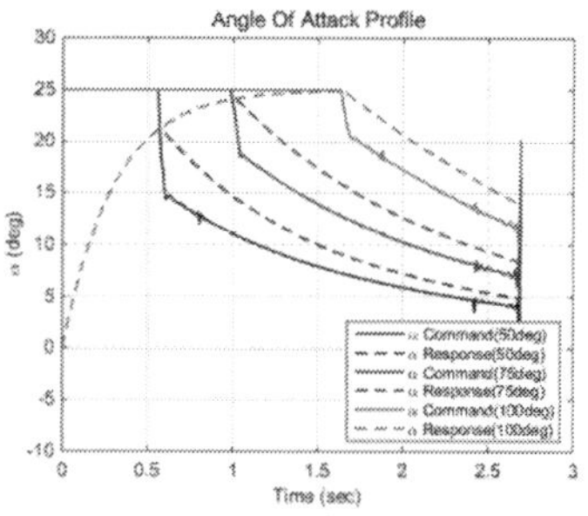

Figure 11. Angle of attack command and response profile

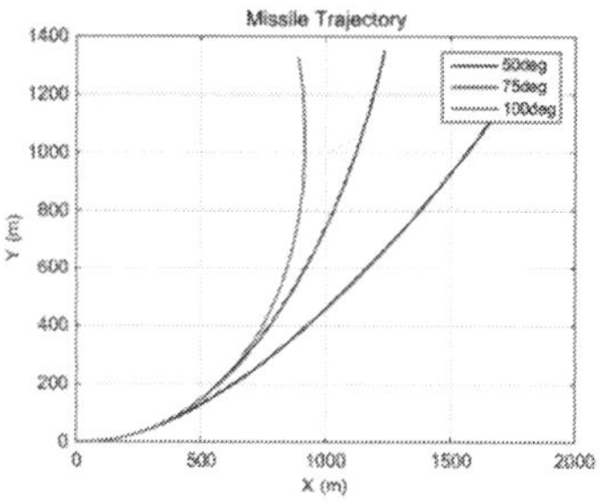

Figure 12. Missile trajectory

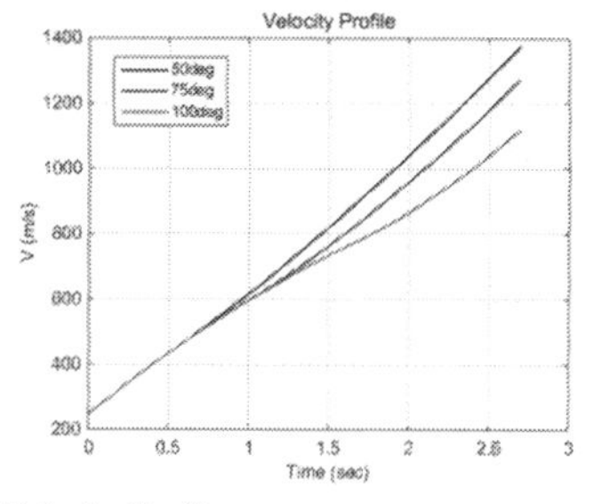

Figure 13. Velocity Profile

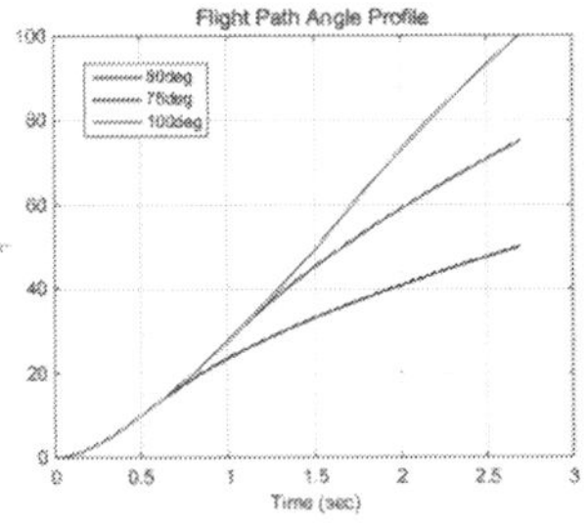

Figure 14. Flight Path Angle Profile

The missile parameters are from SRAAM model in [8]. Also MATLAB GPOPS-II software is applied to get the optimal solution of pseudospectral method. As the final flight path angle demands for bigger angles, velocity increase rate slows at the mid-time. This is because the agile air-to-air missiles use thrust more to the turning maneuver rather than increasing velocity.

The results of angle of attack command shown in Fig. 8 show that optimal control command follows bang-singular control. Since control is bounded at 25 degrees, control command remains at its maximum value for few seconds and then follows its singular arc. The bigger final flight path angle makes the bang control time longer. The bang-singular control could make velocity higher than bang-bang control since bang-bang control could lose missile's velocity when the second bang occurs.

4 Conclusion

Based on the analysis of the optimal control theory, parameterized costate optimization method is proposed as a numerical method of the optimal control problem. By applying this method, solutions obtained by pseudospectral method are verified. Those numerical methods were applied to air-to-air missiles' agile turn problem. The agile turn problem can be formulated as maximizing terminal velocity problem. Therefore, with simple missile model and missile model with aerodynamic forces and thrust, maximizing terminal velocity problem is studied. The optimal control was given as bang-singular control, which is a combination of bang-bang control and control that meets optimality necessary condition. For the future works, bang-singular control should be analyzed. This might include obtaining analytical solutions and finding bang off time.

Acknowledgment

The research is supported by LIG Nex1 (Y16-002).

References

1. K. A. Wise, and D. J. Broy, "Agile missile dynamics and control." Journal of Guidance, Control, and Dynamics, **21**, pp. 441-449 (1998).
2. M. B. McFarland, and A. J. Calise, "Adaptive nonlinear control of agile antiair missiles using neural networks." Control Systems Technology, IEEE Transactions on, **8(5)**, pp. 749-756 (2000).
3. D. M. McDowell, G. W. Irwin, and G. McConnell, "Online neural control applied to a bank-to-turn missile autopilot." In *Proceedings of the AIAA Guidance, Navigation, and Control Conference*, pp. 1286-1294 (1995).
4. C. H. Lee, T. H. Kim, and M. J. Tahk, "Agile missile autopilot design using nonlinear backstepping control with time-delay adaptation." Transactions of the Japan Society for Aeronautical and Space Sciences, **57(1)**, pp. 9-20 (2014).
5. S. M. Ryu, D. Y. Won, C. H. Lee, and M. J. Tahk, "Missile autopilot design for agile turn control during boost-phase." International Journal Aeronautical and Space Sciences, **12(4)**, pp. 365-370 (2011).
6. D. Garg, M. Patterson, W. W. Hager, A. V. Rao, D. A. Benson, and G. T. Huntington, "A unified framework for the numerical solution of optimal control problems using pseudospectral methods." Automatica, **46(11)**, pp. 1843-1851 (2010).
7. M. A. Patterson, and A. V. Rao, "*A General-Purpose MATLAB Toolbox for Solving Optimal Control Problems Using the Radau Pseudospectral Method.*" Manual for GPOPS-II, Gainesville, United States, pp. 2-3 (2013).
8. P. H. Zipfel, "*Modeling and Simulation of Aerospace Vehicle Dynamics 2nd edition*" AIAA, Inc (2010).

MATEC Web of Conferences 77, 07010 (2016) DOI: 10.1051/matecconf/20167707010
ICMMR 2016

Nonlinear Vibration Analysis of Helical Face-Gear Transmission System with Multi-Factor

Liu Bo , Zhao Jiangang, Qian Jianhua and Wang Xiaoyu

China Nuclear Power Technology Research Institute, 518031 Shenzhen, China.

Abstract. A systematic analysis of the dynamics of a helical face-gear system with 8 degree of freedom is performed in this study under complex excitation. The nonlinear dynamic system is solved by the Runge-Kutta method. The bifurcation and dynamic load characteristics of the system is identified from a series of diagrams. The effect of multi-factor on bifurcation diagrams is also analysed. The results real that with the increase of dimensionless frequency, the system undergoes the process of periodic, chaotic and periodic motions. The amplitude of the dynamic load gradually increases in a certain range and begins to decrease at a certain value. A higher mesh damping coefficient or a higher input torque or a lower gear backlash can reduce the vibration or eliminate chaotic responses.

1 Introduction

Face gear transmission has the characteristics of light weight, simple structure, high contact ratio and low meshing noise. The transmission has found application in helicopter transmissions [1] and other important aviation transmission device widely. Some domestic and foreign experts [2-6] have done the related research work in the principle of gear engagement, tooth surface contact strength, tooth root bending strength, tooth profile machining and profile modification. Due to the gear transmission system includes parameters of gear backlashes [7], time-varying mesh stiffness [8], mesh damping [9], transmission errors [10], tooth surface friction [11], the system is a strongly nonlinear system and the vibration properties of the gear system will directly affect the stability and reliability of the system [12]. In the nonlinear dynamics of gear system, the author [13] did some research relevantly and analyzed the effect of speed ratios, backlashes and load on the dynamic response of gear system. Specially, scholars [14-15] have done some research on the nonlinear vibration properties of the face-gear system, but most of them not fully consider the influence of multi-factor on the gear system. Meanwhile, studies on the nonlinear vibration behavior of the helical face-gear system is very rare at present.

In this research, a bending-torsion-axial coupled nonlinear dynamic model of helical face-gear with 8 degrees of freedom is established, including the factors of time-varying mesh stiffness, mesh damping, support stiffness, support damping, transmission errors, backlashes, tooth surface friction and external excitation. Fourth-fifth order Runge-Kutta method with variable time step is introduced to solve the dimensionless dynamic differential equations. The bifurcation and dynamic load characteristic of the system is identified from the bifurcation diagrams, the time histories, phase diagrams, Poincaré maps and the diagram of dynamic load coefficient using the dimensionless frequency as control parameters. The effect of multi-parameter on bifurcation diagrams is also analyzed.

2 System model

According to the principle of gear engagement, the model of orthogonal helical face-gear can be built shown in Fig. 1. Rectangular coordinate system is established at the two gear axis intersection point. x-axis coincides with the axis of the face-gear, the y-axis coincides with the axis of the helical pinion, tangential line of the gear is parallel to z-axis.

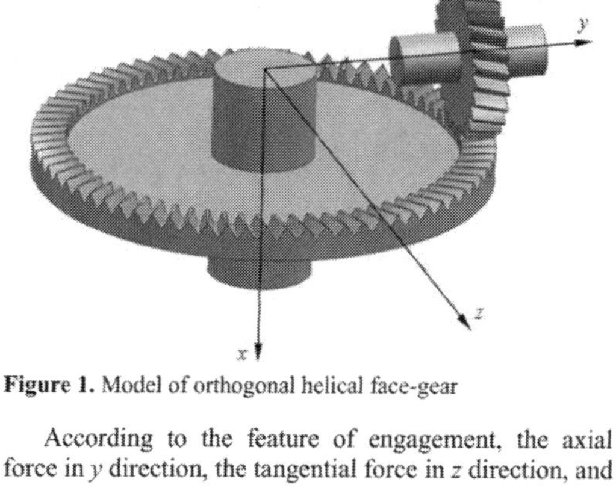

Figure 1. Model of orthogonal helical face-gear

According to the feature of engagement, the axial force in y direction, the tangential force in z direction, and radial force in x direction are applied to the helical pinion. Radial force in y direction, tangential force in z direction and axial force in x direction are applied to the face-gear. The elastic supported dynamic model of the helical face-gear transmission system is built under those forces, as is shown in Fig. 2. In the model, $k_{Ij}(I=X, Y, Z, j=\text{p, g})$ is the bearing stiffness; c_{Ij} $(I=X, Y, Z, j=\text{p, g})$ is the bearing damping; $b_{Ij}(I=X, Y, Z, j=\text{p, g})$ is the bearing backlash; K_h is the mesh stiffness; c_h is the mesh damping; f is the nonlinear friction coefficient; b_h is the gear backlash; $e(t)$ is the transmission errors and varies as a function of time. Both gears are simulated with concentrated mass and inertia, gear shafts are regarded as rigid body, and the bearing stiffness and damping for both ends of the shafts along the x, y, z-direction are treated equivalently.

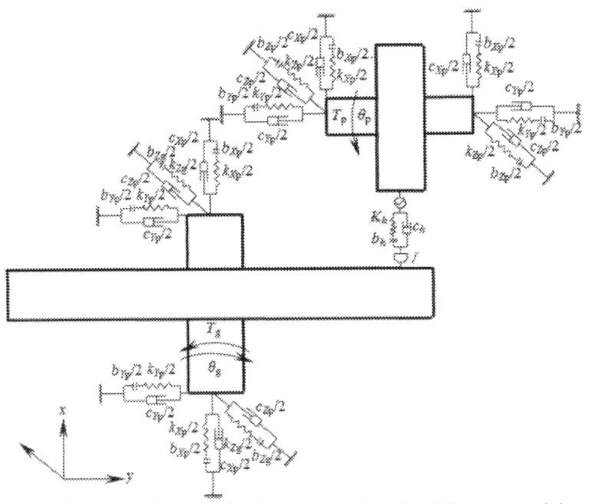

Figure 2. Dynamic model of orthogonal helical face-gear drive system

The translational vibration freedom (X_p, Y_p, Z_p, X_g, Y_g, Z_g) along x, y, z-direction and degrees of freedom in rotation around the body axial direction (θ_p, θ_g) are considered respectively. Then, there are 8 DOF in the gear system, expressed as

$$\{X_p, Y_p, Z_p, \theta_p, X_g, Y_g, Z_g, \theta_g\}^T \tag{1}$$

3 Nonlinear dynamic formulation

Assuming the following parameters

$$\begin{cases} a_1 = \sin\alpha_n \sin\beta \\ a_2 = \sin\alpha_n \cos\beta \\ a_3 = \cos\alpha_n \sin\beta \\ a_4 = \cos\alpha_n \cos\beta \\ a_5 = \sin\alpha_n \\ a_6 = \cos\alpha_n \end{cases} \tag{2}$$

in which α_n is the pressure angle. β is the helical angle.

Due to the vibration displacement and the transmission errors, the gear pair produces the relative displacement along the normal direction of the meshing point, which can be written as

$$\begin{aligned} \lambda_n &= a_4(r_p\theta_p - r_g\theta_g) + a_4(Z_p - Z_g) + a_3(Y_p - Y_g) \\ &+ a_5(X_p - X_g) - e(t) \end{aligned} \tag{3}$$

where r_p is the pitch radius of the pinion, r_g is the nominal radius of the face-gear, $r_g = mz_g/2$. θ_p and θ_g are the angle of the pinion and the face-gear. $e(t)$ is the transmission errors induced by the gear machining and installation errors and can be written as

$$e(t) = e_0 + e_r \cos(\omega_h t + \phi_r) \tag{4}$$

in which e_0 and e_r are the constant and the variable of transmission errors, ω_h is the mesh frequency, ϕ_r is the initial phase angle.

The dynamic mesh force along the normal direction can be performed as

$$F_n = K_h(t) f(\lambda_n) + c_h \dot{\lambda}_n \tag{5}$$

in which c_h is the mesh damping. λ_n is the relative displacement along the normal direction of the meshing point. $f(\lambda_n)$ is the backlash function and varies as a function of time. $K_h(t)$ is the time-varying mesh stiffness which is a cycle of periodic function for meshing period and can be expressed as

$$K_h(t) = K_m + K_r \cos(\omega_h t + \phi_r) \tag{6}$$

in which K_m is the mean mesh stiffness, K_r is the variable amplitude of mesh stiffness.

The coordinate axis component of the dynamic mesh force can be performed as

$$\begin{cases} F_x = a_5 F_n + a_6 \eta f F_n \\ F_y = a_3 F_n - a_1 \eta f F_n \\ F_z = a_4 F_n - a_2 \eta f F_n \end{cases} \tag{7}$$

in which f is the friction coefficient. η is the direction coefficient of the friction force which is determined by the nominal sliding velocity and can be expressed as

$$\eta = sign(l_p - r_{b1} \tan\alpha_n) \tag{8}$$

in which r_{b1} is the base cylinder radius of the pinion. l_p is the friction arm of tooth on any point of the actual line of contact which can be expressed as

$$l_p = \sqrt{r_{a1}^2 - r_{b1}^2} - \varepsilon P_{b1} + \omega_1 r_{b1} t \tag{9}$$

in which ε is the contact ratio of the gear pair, ω_1 is the rotational speed of the pinion, r_{a1} is the addendum cylinder radius of the pinion, P_{b1} is the base pitch of the pinion.

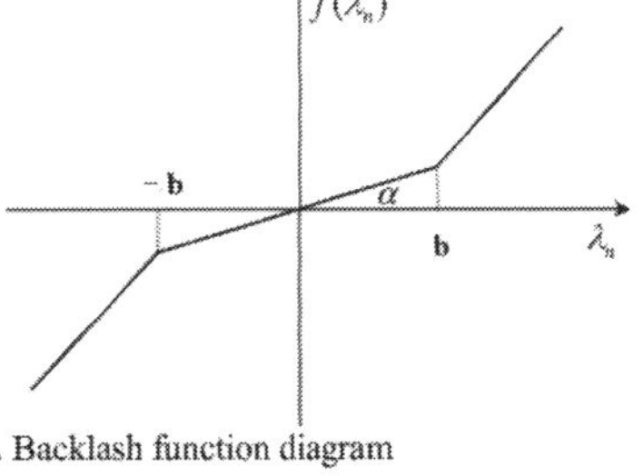

Figure 3. Backlash function diagram

The tooth backlash is one of the important factors that causes the nonlinear vibration of the gear system and is a strongly nonlinear term in the gear system dynamic equations. The effect of backlash is considered as a displacement type nonlinear function shown in Fig. 3 and described as follows

$$f(\lambda_n)=\begin{cases}\lambda_n-b_h(1-\alpha) & \lambda_n>b_h\\ b_h\alpha & |\lambda_n|\le b_h\\ \lambda_n+b_h(1-\alpha) & \lambda_n<-b_h\end{cases} \tag{10}$$

in which b_h is the half of the total gear backlash, α is the stiffness ratio between the two contact stages.

The bearing supporting shaft has a certain gap and also uses the same backlash nonlinear function as the gears to express. Bearing supporting force along x, y, z-coordinate direction can be expressed as follows

$$F_{Ij}=k_{Ij}f(I_j) \tag{11}$$

in which I_j and k_{Ij} (I=X, Y, Z, j=p, g) are the bearing backlash and stiffness along the x, y, z-direction for different gears.

According to the Newton's Second Law, the equations of the helical face-gear transmission system dynamic model under the assumption of nonlinear suspension can be written as

$$\begin{cases}m_p\ddot{X}_p+c_{Xp}\dot{X}_p+k_{Xp}f(X_p)=-F_x\\ m_p\ddot{Y}_p+c_{Yp}\dot{Y}_p+k_{Yp}f(Y_p)=-F_y\\ m_p\ddot{Z}_p+c_{Zp}\dot{Z}_p+k_{Zp}f(Z_p)=-F_z\\ J_p\ddot{\theta}_p=T_p-F_zr_p\\ m_g\ddot{X}_g+c_{Xg}\dot{X}_g+k_{Xg}f(X_g)=F_x\\ m_g\ddot{Y}_g+c_{Yg}\dot{Y}_g+k_{Yg}f(Y_g)=F_y\\ m_g\ddot{Z}_g+c_{Zg}\dot{Z}_g+k_{Zg}f(Z_g)=F_z\\ J_g\ddot{\theta}_g=-T_g+F_zr_g\end{cases} \tag{12}$$

in which T_p and T_g are the input torque and output torque which can be expressed as

$$\begin{cases}T_p=T_0+T_{e1}\sin\omega_1 t\\ T_g=-T_0+T_{e2}\sin\omega_2 t\end{cases} \tag{13}$$

in which T_0 is the mean torque, ω_1 and ω_2 are the rotational speed of the pinion and the face-gear. T_{e1} and T_{e2} are the variable amplitude of the input and the output excitation torque.

Taking the relative displacement along the normal direction of the meshing point λ_n as a new degree of freedom and merging the torsional vibration equations of Eq.(12), we can get

$$\begin{aligned}&m_e\ddot{\lambda}_n-a_5m_e\ddot{X}_p+a_5m_e\ddot{X}_g-a_3m_e\ddot{Y}_p+a_3m_e\ddot{Y}_g\\ &-a_4m_e\ddot{Z}_p+a_4m_e\ddot{Z}_g+a_4^2c_h\dot{\lambda}_n+a_4^2K_h(t)f(\lambda_n)\\ &=F_p+F_g+F_{pf}+F_{gf}-m_e\ddot{e}(t)\end{aligned} \tag{14}$$

In which m_e, F_p, F_{pf}, F_g and F_{gf} can be performed as

$$\begin{cases}m_e=J_pJ_g/(r_p^2J_g+r_g^2J_p)\\ F_p=a_4m_er_pT_p/J_p\\ F_{pf}=a_2a_4m_er_pl_p\eta fF_n/J_p\\ F_g=a_4m_er_gT_g/J_g\\ F_{gf}=a_2a_4m_er_g^2\eta fF_n/J_g\end{cases} \tag{15}$$

in which J_p and J_g are the moment of inertia of the pinion and the face-gear.

Introducing the dimensionless parameters, Eqs. (12) and (14) can be expressed as

$$\begin{cases}\ddot{x}_p+2\xi_{xp}\dot{x}_p+(a_5+a_6\eta f)\xi_{hp}\dot{\lambda}+k_{xp}f(X_p)\\ +(a_5+a_6\eta f)k_{hp}f(\lambda)=0\\ \ddot{y}_p+2\xi_{yp}\dot{y}_p+(a_3-a_1\eta f)\xi_{hp}\dot{\lambda}+k_{yp}f(Y_p)\\ +(a_3-a_1\eta f)k_{hp}f(\lambda)=0\\ \ddot{z}_p+2\xi_{zp}\dot{z}_p+(a_4-a_2\eta f)\xi_{hp}\dot{\lambda}+k_{zp}f(Z_p)\\ +(a_4-a_2\eta f)k_{hp}f(\lambda)=0\\ \ddot{x}_g+2\xi_{xg}\dot{x}_g-(a_5+a_6\eta f)\xi_{hg}\dot{\lambda}+k_{xg}f(X_g)\\ -(a_5+a_6\eta f)k_{hg}f(\lambda)=0\\ \ddot{y}_g+2\xi_{yg}\dot{y}_g-(a_3-a_1\eta f)\xi_{hg}\dot{\lambda}+k_{yg}f(Y_g)\\ -(a_3-a_1\eta f)k_{hg}f(\lambda)=0\\ \ddot{z}_g+2\xi_{zg}\dot{z}_g-(a_4-a_2\eta f)\xi_{hg}\dot{\lambda}+k_{zg}f(Z_g)\\ -(a_4-a_2\eta f)k_{hg}f(\lambda)=0\\ \ddot{\lambda}-a_5\ddot{x}_p+a_5\ddot{x}_g-a_3\ddot{y}_p+a_3\ddot{y}_g-a_4\ddot{z}_p+a_4\ddot{z}_g\\ +a_4^2\xi_h\dot{\lambda}+a_4^2k_hf(\lambda)=f_p+f_g+f_{pf}+f_{gf}-f_e\end{cases} \tag{16}$$

in which the dimensionless parameters can written as

$x_j=X_j/b_h$, $y_j=Y_j/b_h$, $z_j=Z_j/b_h$, $\lambda=\lambda_n/b_h$, $\omega_n=\sqrt{K_m/m_e}$, $\omega_{ij}=\sqrt{k_{Ij}/m_j}$, $\xi_{ij}=c_{Ij}/(2m_j\omega_n)$, $\xi_{hj}=c_h/(2m_j\omega_n)$, $\omega=\omega_h/\omega_n$, $\xi_h=c_h/(2m_e\omega_n)$, $\tau=\omega_nt$, $f_j=F_j/(m_eb_h\omega_n^2)$,

$f_{jf}=F_{jf}/(m_eb_h\omega_n^2)$, $k_h=1+\frac{K_r}{m_e\omega_n^2}\cos(\omega\tau+\phi_r)$,

$f_e=\frac{e_r\omega^2}{b_h}\cos(\omega\tau+\phi_r)$, $k_{ij}=\omega_{ij}^2/\omega_n^2$,

$k_{hj}=K_h(\tau)/(m_j\omega_n^2)$, $f(\lambda)=\begin{cases}\lambda+\alpha-1 & \lambda>1\\ \alpha & |\lambda|\le 1\\ \lambda+\alpha+1 & \lambda<-1\end{cases}$

$$f(I_j)=\begin{cases}I_j/b_h-(1-\alpha)b_{ij}/b_h & I_j>b_{ij}/b_h\\ \alpha & |I_j|\le b_{ij}/b_h\\ I_j/b_h+(1-\alpha)b_{ij}/b_h & I_j<-b_{ij}/b_h\end{cases}$$

$I=X,Y,Z$, $i=x,y,z$, $j=$ p,g .

DOI: 10.1051/matecconf/20167707010

4 Numerical simulation and discussion

The nonlinear dynamic equations presented in Eqs. (16) for the helical face-gear system were solved by the Fourth-fifth order Runge-Kutta method with variable time step. The time step in the iterative solution procedure was assigned a value of $T_m/300$ ($T_m=2\pi/\omega$, T_m is the dimensionless mesh cycle) and the iterative time was set to $300T_m$. The first hundreds of cycles were specifically excluded to ensure the analyzed data related to steady-state conditions. The chaotic motion of the system was identified by the bifurcation diagrams, time history, phase diagrams and Poincaré maps. The dynamic load characteristic of the system was analyzed by the diagram of dynamic load coefficient which was variation with time.

4.1 Bifurcation diagram

The bifurcation diagram for the dimensionless displacement of helical face-gear system using the dimensionless frequency ω as control parameter at gear backlash ($b_h=1$), meshing damping coefficient ($c_h=0.14$), input torque ($T_0=850$ N•m) is shown in Fig. 4. It can be obtained that the system displays single periodic motion at the low values of the dimensionless frequency, i.e. $\omega<0.6$. Especially, there exist the jump phenomenon in the diagram at $\omega=0.5$. As ω is increased from 0.6 to 0.9, the single-periodic is replaced by nT-periodic and the system state changes from 2T-periodic motion ($\omega=0.6$-0.8) to 4T-periodic motion ($\omega=0.8$-0.9). For values of the dimensionless frequency in the range $\omega=0.9$ to 0.96, the system performs chaotic motion. As ω is further increased from $\omega=0.96$ to 1.05, the chaotic motion transits to nT-periodic motion. However, at higher values of the dimensionless frequency ($\omega=1.05$-1.75), the gear system undergoes chaotic motion once again. Finally, for all $\omega>1.75$, the chaotic motion is replaced by 2T-periodic and single periodic motion.

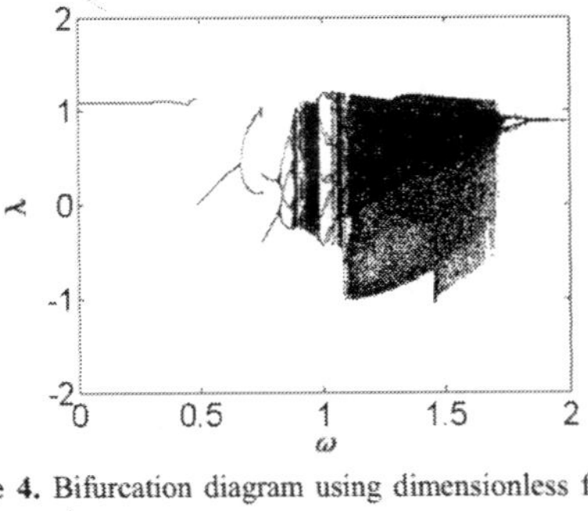

Figure 4. Bifurcation diagram using dimensionless frequency, ω, as control parameter.

4.2 Prediction of the dynamic responses

To observe the response and chaotic behavior of the helical face-gear system at some values of the dimensionless frequency, the time history, phase diagrams and Poincaré maps of the gear system as shown in Figs. 5-12 are draw at $\omega=0.3$, 0.7, 0.85, 0.95, 1.0, 1.4, 1.8 and 1.9, respectively. Significantly, when the phase diagrams form a closed curve, the Poincaré maps show n discrete points and the time history curves are highly regular, the system is in periodic motion, i.e. $\omega=0.3$, 0.7, 0.85, 1.8 and 1.9. On the contrary, the phase diagrams is highly disordered, the Poincaré maps form geometrically fractal structures and the time history curves become irregular, the system exhibits a chaotic behavior, i.e. $\omega=0.95$ and 1.4. However, at the values of the dimensionless frequency $\omega=0.3$ and 1.9, the displacement amplitude is significantly smaller than others, which are in regular single periodic motion.

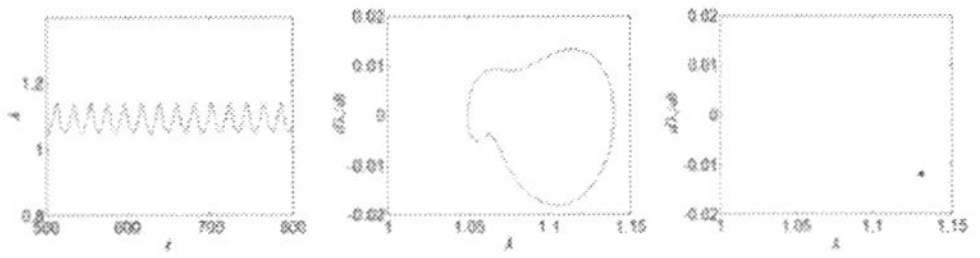

Figure 5. Simulation results with dimensionless frequency $\omega=0.3$.

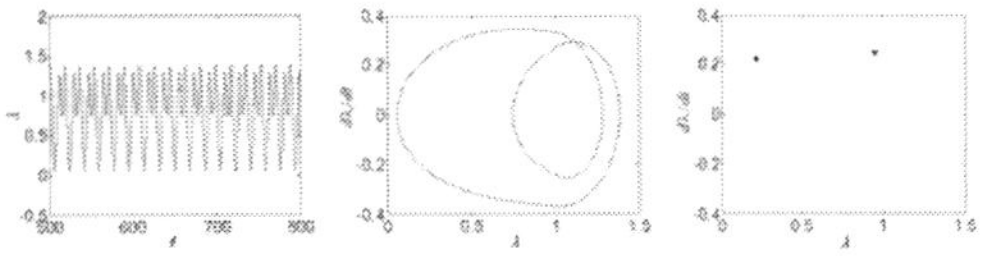

Figure 6. Simulation results with dimensionless frequency $\omega=0.7$.

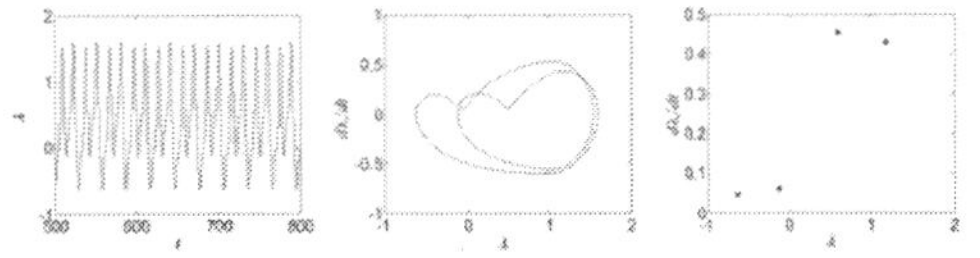

Figure 7. Simulation results with dimensionless frequency $\omega=0.85$.

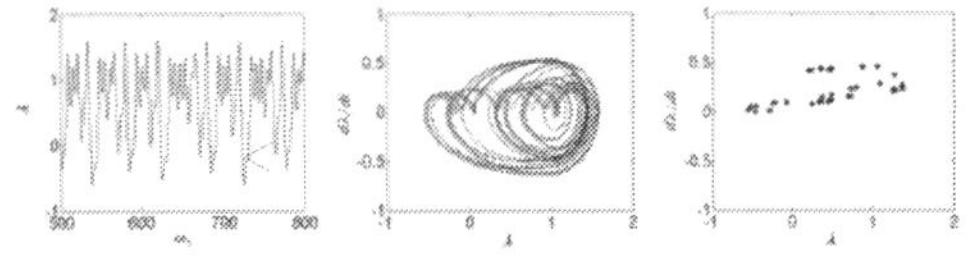

Figure 8. Simulation results with dimensionless frequency $\omega=0.95$.

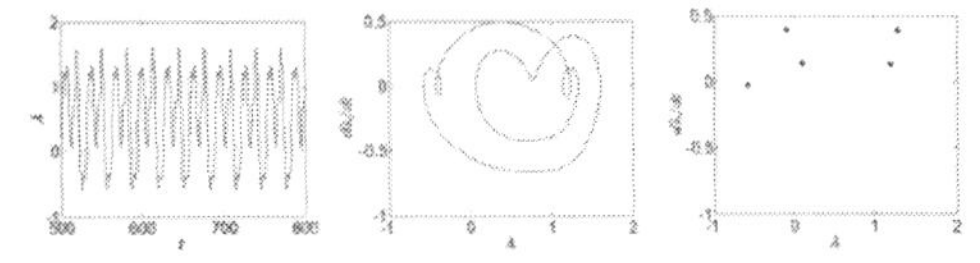

Figure 9. Simulation results with dimensionless frequency $\omega=1.0$.

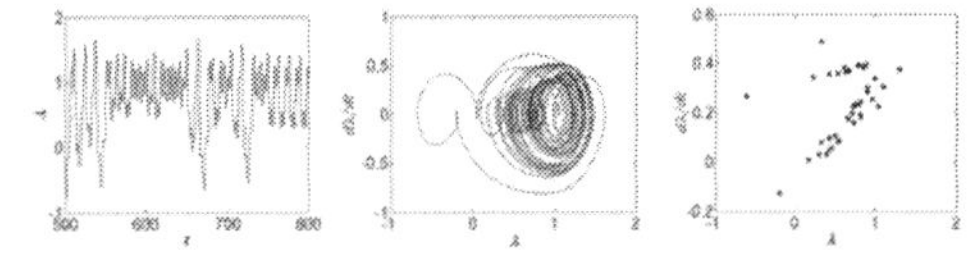

Figure 10. Simulation results with dimensionless frequency $\omega=1.4$.

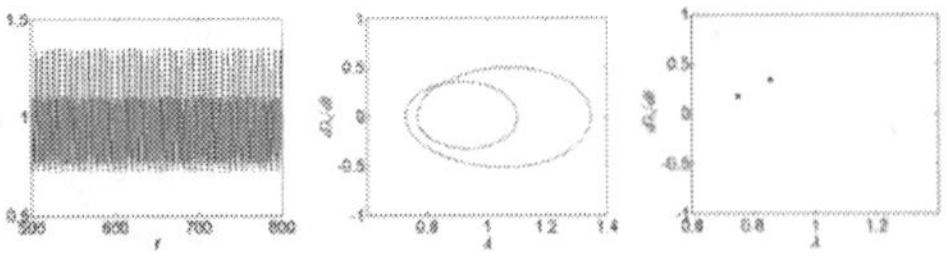

Figure 11. Simulation results with dimensionless frequency ω=1.8.

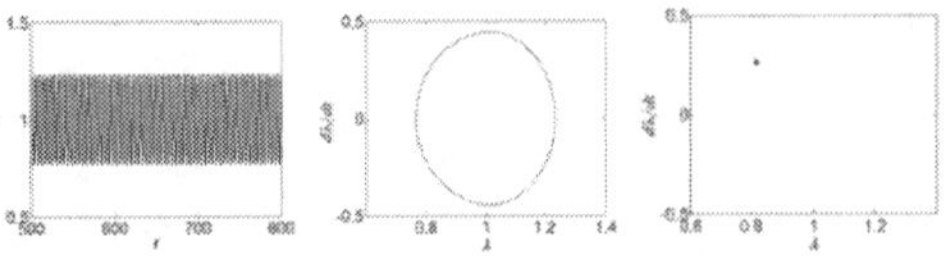

Figure 12. Simulation results with dimensionless frequency ω=1.9.

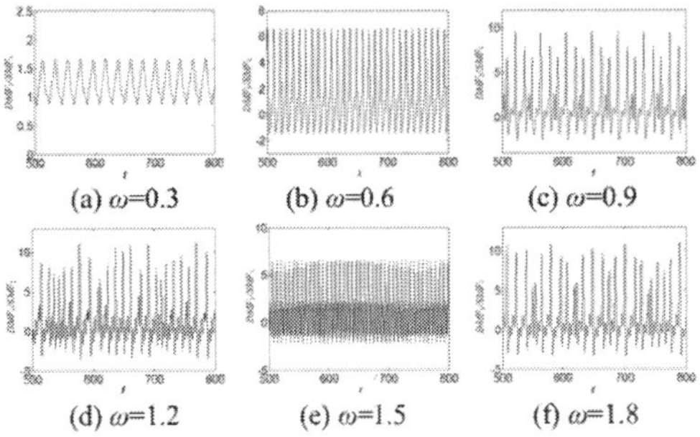

(a) ω=0.3 (b) ω=0.6 (c) ω=0.9

(d) ω=1.2 (e) ω=1.5 (f) ω=1.8

Figure 13. Dynamic load coefficient of the system with different dimensionless frequency, ω.

Furthermore, to observe the dynamic load characteristic of the system, Fig. 13 illustrates the corresponding diagram of dynamic load coefficient (DMF_1/SMF_1) at different values of the dimensionless frequency. In the range of ω=0.3-1.5, the DMF_1/SMF_1 becomes larger, as the ω is increased. However, at the high value of the ω=1.8, the DMF_1/SMF_1 is significantly small. Corresponding to the time history diagrams, the trend of time history diagrams is consistent with the trend of the dynamic load coefficient. In other words, choosing the suitable speed ratio can decrease the dynamic response and load of the gear system.

4.3 Effect of multi-parameter on bifurcation diagrams

Figs. 14-16 illustrate the corresponding bifurcation diagram of the system at different dimensionless meshing damping coefficients, gear backlashes and input torques. With the increase of the dimensionless meshing damping coefficient c_h and input torque T_0, the Chaos areas of the bifurcation diagrams become smaller gradually, which means that the system trends to be more stable. However, with the increase of the dimensionless gear backlash b_h, the chaos areas and the displacement amplitude become larger obviously. The motion of the system trends to be more complicated. So, the dynamic response of the system can be improved with a higher mesh damping coefficient or a higher input torque or a lower gear backlash. All in all, the bifurcation diagram of the gear system can be controlled by selecting right parameters.

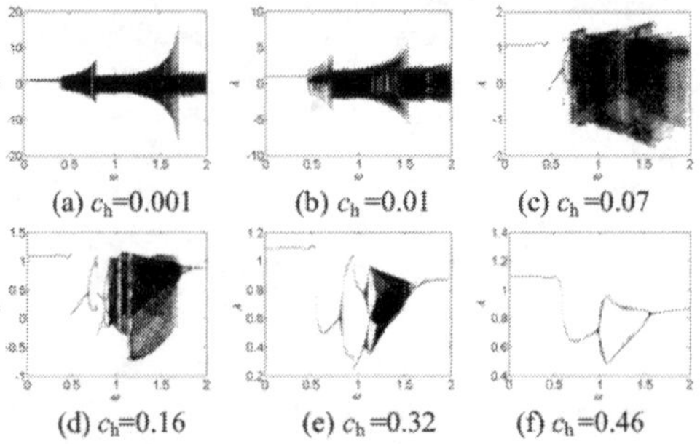

(a) c_h=0.001 (b) c_h=0.01 (c) c_h=0.07

(d) c_h=0.16 (e) c_h=0.32 (f) c_h=0.46

Figure 14. Bifurcation diagram using dimensionless frequency, ω, as bifurcation parameter at different dimensionless meshing damping coefficients, c_h.

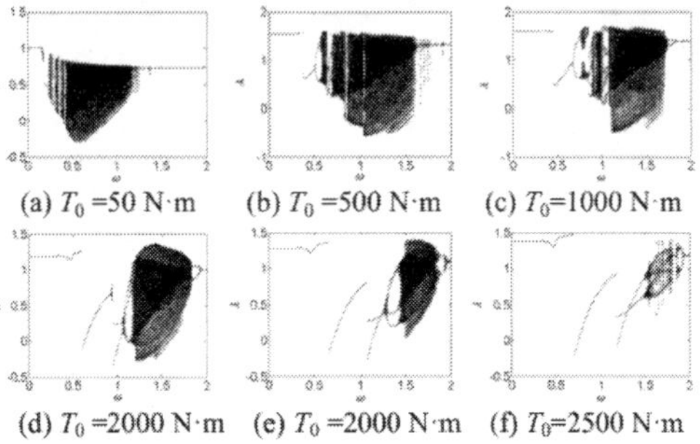

(a) T_0 =50 N·m (b) T_0 =500 N·m (c) T_0=1000 N·m

(d) T_0 =2000 N·m (e) T_0 =2000 N·m (f) T_0=2500 N·m

Figure 15. Bifurcation diagram using dimensionless frequency, ω, as bifurcation parameter at different input torques, T_0.

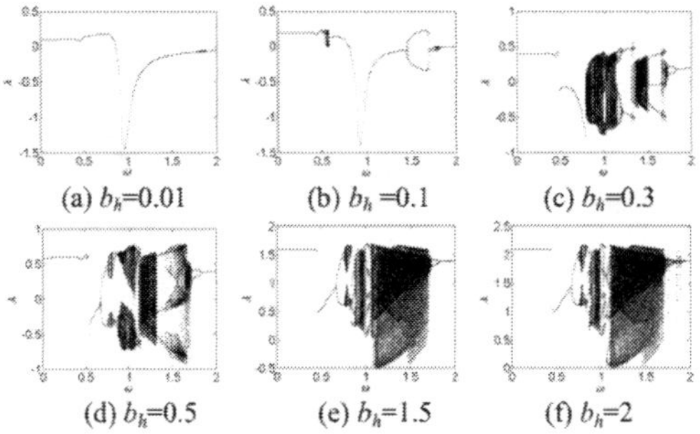

(a) b_h=0.01 (b) b_h =0.1 (c) b_h=0.3

(d) b_h=0.5 (e) b_h=1.5 (f) b_h=2

Figure 16. Bifurcation diagram using dimensionless frequency, ω, as bifurcation parameter at different dimensionless gear backlash, b_h.

5 Conclusions

In this study, the dynamic characteristics of a helical face-gear system with 8 degree of freedom have been investigated under considerations of the time-varying mesh stiffness, mesh damping, support stiffness, support damping, transmission errors, backlashes, tooth surface friction and external excitation. The chaotic motion and dynamic load characteristics of the system have been discussed by the bifurcation diagrams, time history, phase diagrams Poincaré maps and the diagram of dynamic load coefficient. The study has been analyzed the bifurcation diagrams of the system using dimensionless frequency as bifurcation parameter at different dimensionless mesh damping coefficients, gear backlashes and input torques, respectively.

(i) With the increase of dimensionless frequency, the system undergoes the process of periodic, chaotic and periodic motions. Jumping phenomenon and double Chaos areas can be found in the bifurcation diagram.

(ii) The amplitude of the dynamic response and load is small, when the system is in the regular periodic motion. So, choosing the suitable speed ratio can decrease the vibration or avoid chaotic motion.

(iii) With the increase of the dimensionless mesh damping coefficient or input torque, the system trends to be more stable. With the increase of the dimensionless gear backlash, the system becomes more complicated. So, the dynamic response of the system can be improved with a higher mesh damping coefficient or a higher input torque or a lower gear backlash.

The last, the systematic analysis of the chaotic and dynamic load characteristics of the helical face-gear nonlinear dynamic system under multi-factor conditions can offer a method to decrease the dynamic response and load of the gear system.

References

1. Litvin F L, Wang J C, Bossler R B, J. Mech. Des. N. Y **116**, 672-676 (1992)
2. Litvin F L, Zhang Y, Wang J C, J. Mech. Des. N. Y **114**, 642-647 (1992)
3. Litvin F L, Gonzalez-Perez I, Fuentes A, Comput. Method. Appl. M **194**, 3870-3901(2005)
4. Tsay M F, Fong Z H. P, I. MECH. ENG. C-J. MEC **221**, 715-725(2007)
5. Zhao N, Guo H, Fang Z D, C. J. A. P **23**, 2142-2146(2008)
6. Cui Y M, Fang Z D, Su J Z, C. J. A **23**, 1346-1351(2013)
7. Chen S Y, Tang J Y, Luo C, MECH. MACH. THEORY **46**, 466-478(2011)
8. He S, Gunda R, Singh R, J, SOUND. VIB **301**, 929-949(2007)
9. Chang-Jian C W, Chang S M, NONLINERA ANAL-REAL **12**, 979-989(2011)
10. Sun T, Hu H Y, MECH. MACH. THEORY **38**, 1371-1390(2003)
11. Vaishya M, Singh R. J, SOUND. VIB **348**, 671-694(2001)
12. Lewichki D G, Heath G F, Filler R R.A. F. P, AHS-Inter **2**, 1051-1060(2007)
13. Lin T J, Wang D H, Ran X T, J. V. S. **32**, 1-7(2013)
14. Yang Z, Wang S M, Fan Y S, J. C. U **34**, 26-35(2011)
15. Chen S Y, Tang J Y, Chen W T, MEC **49**, 1023-1037(2014)

MATEC Web of Conferences 77, 07011 (2016) DOI: 10.1051/matecconf/20167707011
ICMMR 2016

Multi-Domain Modeling Based on Modelica

LIU Jun , WANG Guochen and LUO Yanyan

Qingan group co., LTD., 710077 Xian, China

Abstract. With the application of simulation technology in large-scale and multi-field problems, multi-domain unified modeling become an effective way to solve these problems. This paper introduces several basic methods and advantages of the multidisciplinary model, and focuses on the simulation based on Modelica language. The Modelica/Mworks is a newly developed simulation software with features of an object-oriented and non-casual language for modeling of the large, multi-domain system , which makes the model easier to grasp , develop and maintain.It This article shows the single degree of freedom mechanical vibration system based on Modelica language special connection mechanism in Mworks. This method that multi-domain modeling has simple and feasible, high reusability. it closer to the physical system, and many other advantages.

1 Introduction

With the development of science and technology ,product design is becoming more and more complicated. Sophisticated product design not only involves multiple areas but also updates frequently. Therefore, in the modern industry, the simulation modeling technology plays an increasingly important role.

Generally, complex products is the product of cross subject, which will be involved in multiple areas, such as electronics, control, software, machinery, etc. Most traditional modeling software supports only a single field of modeling and simulation. In recent years, there are some solutions for multidisciplinary modeling, such as multiple domain modeling based on interface and modeling based on high level architecture[1], but these are more or less the problems. Through multidisciplinary modeling for complex product design, it not only be modeled in all areas, also need a lot of coordination simulation in the overall system level, it has brought a lot of difficulties for simulation modeling of complex product.

Multidisciplinary modeling (multi-domain modeling) is a model of different disciplines assembled into a larger simulation model. These areas generally relates to mechanical, control, electronic, hydraulic, pneumatic and other. It developed from a single to multi-domain modeling ,discrete modeling in continuous or discrete domain to hybrid modeling [2], eventually developed into the stage of object-oriented modeling.

2 The solutions of multiple domain modeling

2.1 Interface-based modeling method

It is the modeling method based on interface that when project is involved in different fields, to modeling it in the professional software of corresponding areas, then use the software provided interfaces between interaction to joint simulation. In this way, the modeling mainly focus on the real-time performance and stability of the data transmission in the interface. This presents the following questions: First, It must be such an interface between software. Second, the interface development is often a private company. There is no uniform standard, and do not open to the public. It's a heterosexual strong and difficult expansion upgrade.

2.2 HLA-based modeling method

The method that HLA (High Level of Architecture) multi-domain modeling is the Defense Modeling and Simulation Office proposed in 1995. This method that made use of HLA agreement is mapping output variables of model A to input variables of model B . This modeling method also has some problems: First, it does not solve the problem of multi-domain modeling fundamentally. It's just improved and enhanced to a method of interface-based. Component model of different areas is still simulating in corresponding simulation software. Second, different simulation software need to write specific code in order to achieve the response configuration. Thus, the problem becomes complicated.

2.3 The modeling method based on the unified language

The method that based on the Unified Modeling Language [3] solve the problems caused by the other two methods. It does not need to design additional interfaces for the different simulate software. It solves the problem caused due to the interdisciplinary modeling. In multi-domain modeling based on unified language, all models by the same language, so that the form of consistency can be achieved. Coupling between different areas become better.

The most representative multi-domain unified modeling language called Modelica [4]. Modeling by Modelica thought focus on knowledge which can be accumulated, reusable and reconfigurable. Excellent properties of Modelica language makes it have an advantage in other multidisciplinary modeling methods. It will inevitably become the mainstream of multidisciplinary physical system modeling method.

3 Multi-Domain modeling based on the modelica

3.1 Modelica language modeling thought

Modelica is a new kind of Modelica language. It contains a variety of advanced modeling theory. The basic principle of Modelica can describe multi-domain domain model is: it is based on energy conservation, use of mathematical equations to describe the model. It obey the kirchhoff's law. In general kirchhoff's law, the physical system is regarded as energy system composed of components with a port. The energy transfer between components is described by flow variables and the potential variables. Modelica language has some characteristics, such as Object oriented modeling, non-causal modeling and narrative modeling.

3.1.2 Object-oriented modeling

The real world is composed by various objects with their own internal state and movement. Modelica language uses the class as the basic unit. Class includs variables, equation and members of the class. Variables represent attributes of the class. Equations represents behavior of the class. Class through variety of ways to build the system model, such as inheritance and component connections.

3.1.2 Non-causal modeling

Characteristics of non-causal modeling idea is to use mathematical modeling equations to describe the behavior of physical objects and attributes. It has the advantage that when defining equation does not need to solution of the equation are given. This can greatly reduce the difficulty of modeling. Further, the modeling time is also reduced.

3.1.3 Narrative Modeling

Narrative modeling focus on description of physical objects. It focus on the following two parts: to describe the attributes and behavior of the object itself; to Describe the physical laws object by mathematical equations . Describes the behavior of the object is to describe the interactive interface between the objects. Modeling using this method, the topological structure and physical structure of the model are consistent.

In the Modelica, with the aid of connectors, components implemented data exchange with each other, as shown in Fig. 1. In the process of simulation, the component can use the connector to achieve communication with the other components. Take for example a RLC circuit, which embodies the characteristics of interface in Modelica. In modelica language, it is implemented using a connector, as shown in Fig. 2.

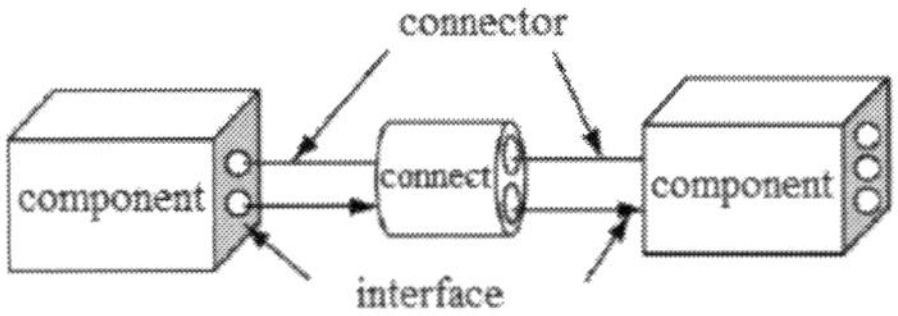

Figure 1. Relationship between component and connector

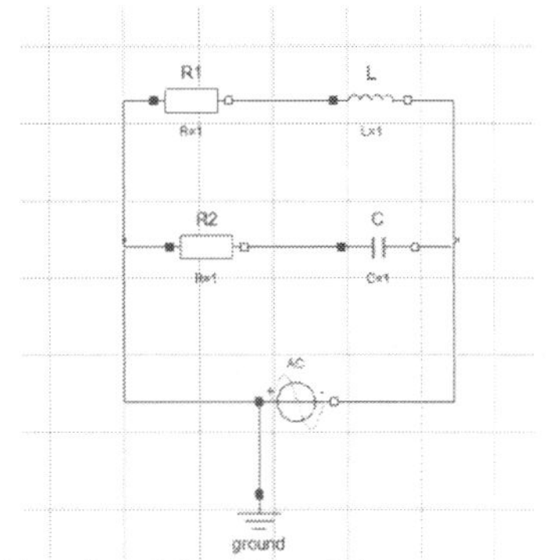

Figure 2. (a) RLC model interface window

```
model rlc
  Modelica.Electrical.Analog.Basic.Ground ground ;
  Modelica.Electrical.Analog.Basic.Resistor R1(R=1) ;
  Modelica.Electrical.Analog.Basic.Resistor R2 (R=10) ;
  Modelica.Electrical.Analog.Basic.Inductor L(L=0.1) ;
  Modelica.Electrical.Analog.Basic.Capacitor C(C=1.e-002) ;
  Modelica.Electrical.Analog.Sources.SineVoltage AC(V=220, freqHz=50) ;
protected
  Modelica.Electrical.Analog.Interfaces.PositivePin p ;
  Modelica.Electrical.Analog.Interfaces.NegativePin n ;
equation
  connect(R1.n, L.p);
  connect(L.n, C.n);
  connect(R1.p, AC.p);
  connect(R2.n, C.p);
  connect(R2.p, p);
  connect(AC.n, n);
  connect(ground.p, AC.p);
end rlc;
```

Figure 2. (b) RLC model interface window

3.2 The characteristics of Modelica language modeling

Modelica is developed by Modelica association that is a nonprofit organization in Sweden. It is an object-oriented language suitable for large-scale modeling of complex physical systems and it is free to use. Using Modelica to build a model based on their physical structure, so that different areas of the model can be consistent .It can be achieved unified multi-domain modeling . Mathematical description of Modelica models is the differential, algebraic and discrete equations. It can automatically solve equations, and no additional processing. It is good at Hardware-in-the-loop simulation and embedded control systems. Mainly based on the following consideration:

(1) Using Modelica to build a model based on their physical structure, especially suitable for Hardware-in-the-loop simulation and embedded control systems.

(2) In the Modelica language, mathematical model is differential equations, algebraic equations, and discrete equations. So, it can be used directly associated physical formulas. Without demanding logic relationship , no need for a formula for further conversion.

(3) when it is solved, the directivity equation does not need to be concerned . So compared with other simulation software, It reduces the difficulty of modeling greatly.

(4) It has object-oriented features. Relative to the assignment, the equation does not require the direction of data flow , so there is no complex causality. It reduces the difficulty of modeling, and has better support component reuse and evolutionary model.

(5) It uses mathematical formulas (equations) to describe the physical laws and phenomena of subsystems in different areas . It is based on the topology of the physical model, and using the language internal component coupling mechanism, to achieve multi-domain modeling.

(6) Modelica based modeling commonly used in two ways. One is the use of graphical modeling tool, by dragging and combining the components in the standard library. Finally, add a connection to complete the establishment of the model. Can be set in detail the corresponding parameters in the model components, the simulation model is made. Another way is to write Modelica language code using Modelica language to generate the corresponding model.

(7) Multi-domain modeling based on Modelica language, when the components modeling is completed, mainly depends on the connector for data exchange.

The following spring with a mass, for example, describe the linear damping moving under it a exciting force, as shown in Fig. 3. The tools used are Mworks, language is the Modelica language. Solving the amplitude-frequency characteristic curve and phase frequency characteristic curves [5]. Then the The following simulation results, as shown in Fig. 4, Fig. 5 and Fig. 6.

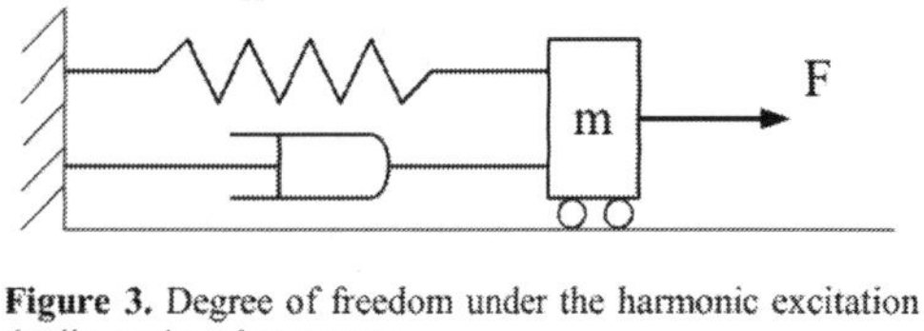

Figure 3. Degree of freedom under the harmonic excitation of the linear damping system

by Newton's law is:

$$m\ddot{x} + c\dot{x} + kx = F_0 \cos\omega t \tag{1}$$

Simplification:

$$\tan\phi = \frac{2\zeta p\omega}{p^2 - \omega^2} \tag{2}$$

$$X = \frac{X_0}{\sqrt{\left(p^2 - \omega^2\right)^2 + \left(2\zeta p\omega\right)^2}} p^2 \tag{3}$$

In the end:

$$\beta = \frac{1}{\sqrt{(1-\gamma^2)^2 + (2\zeta\gamma^2)^2}} \tag{4}$$

$$\tan\phi = \frac{2\zeta p\omega}{p^2 - \omega^2} \tag{5}$$

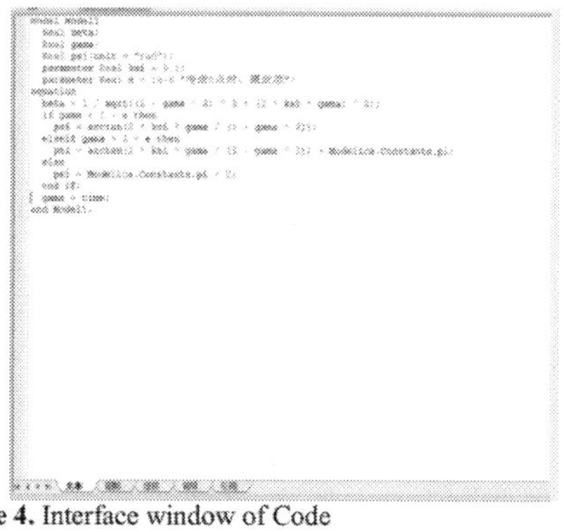

Figure 4. Interface window of Code

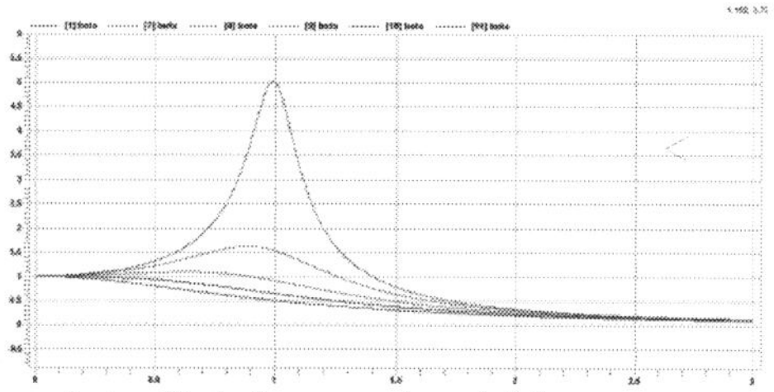

Figure 5. Amplitude-frequency characteristic curve
(It is amplitude-frequency response curves under different damping coefficient in the system)

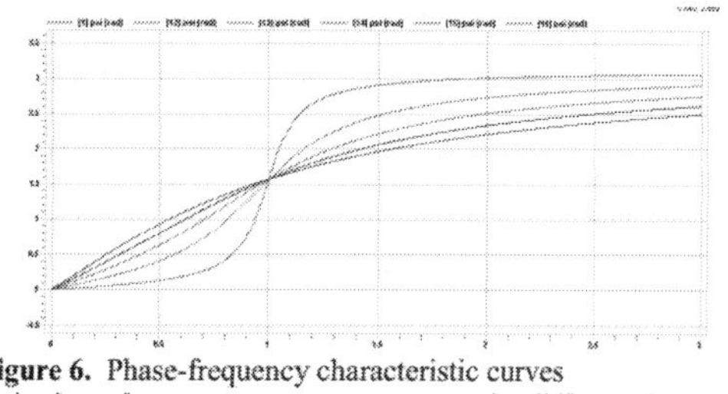

Figure 6. Phase-frequency characteristic curves
(It is phase-frequency response curves under different damping coefficient in the system)

4 Conclusion

Modelica language is now widely accepted multi-domain modeling language. C language and it have a lot of similar places. And it is easy to use. Using this method can reduce the modeling difficulty. It using the mathematics modeling, and don't have to be limited to the area and tools.Based on the foundation libraries,more and more specicalized libraies will be built in the future.It will accelerate the development of multidiscipinary modeing tu more special fields and applications.

References

1. Chen X B, Xiong G L,Guo B, et al. Research on Co-simulation Running Based on HLA[J]. Journal of System Simulation, (2003).
2. FRITZSON P, ENGELSON V. Modelica-A unified object-oriented language for system modeling and simulation [C] / / ERIC H Proc 12th European Conf on Object-oriented Programming, Brussels: Springer-Verlag, (1998).
3. Zhao J J, Ding J W, Zhou F L, etal. Modelica and Its Mechanism of Multidomain Unified Modeling and Simulation [J]. Journal of System Simulation, (2006).
4. PeterFritzson. PrinciplesofObject-OrientedModeling and Simulation with Modelica 2.1 [M]. Wiley-IEEE Press, (2004).
5. FANG TONG,XUE PU. Vibration theory and application. Northwestern Polytechnical University, (2004).

MATEC Web of Conferences 77, 08001 (2016) DOI: 10.1051/matecconf/20167708001
ICMMR 2016

Effect of Phase Angle on Fatigue Failure of 30CrMnSiA Steel under Tension-torsion Fatigue Loading

Tianqi Liu[1], Xinhong Shi[1] and Jianyu Zhang[2]

[1] *School of Aeronautic Science and Engineering, Beijing University of Aeronautics and Astronautics, Beijing 100191,China.*
[2] *School of Aeronautics and Astronautics, Chongqing University, Chongqing 400044, China.*

Abstract. Tension-torsion high-cycle fatigue experiments with solid cylindrical bar specimens of 30CrMnSiA steel were carried out under different phase angles. And the experiment data, characteristics of stresses with plane orientation and fracture appearance under different loading conditions were analyzed and investigated. Experiment results show that fatigue life increases gradually with the increase of phase angle. The fatigue life is shortest for 0° phase angle, while the fatigue life is longest for 90° out-of-phase angle. It could be found from analysis of stress parameters that as the phase angle increases, the shear stress amplitude decreases and the maximum normal stress increases gradually, which are on the maximum shear stress amplitude plane close to the axial of specimen. Macro-analysis of the fracture appearances show that when the phase angle increases, the angel between crack initiation plane and specimen axial decreases, and cracks are inclined to initial and propagate on or near the maximum shear stress amplitude plane with larger maximum normal stress under proportional loading as well as non-proportional loading.

1 Introduction

30CrMnSiA steel is a kind of quenched and tempered steel with the advantages of high strength and fatigue resistance that widely used in aircraft, such as the landing gear, engine mount and important bolt structure and components. The structure and components in practical engineering is often under multiaxial fatigue load. A large number of studies have shown that the phase angle is an important parameter in the study of multiaxial fatigue.

For the influence of phase difference on fatigue failure, McDiarmid [1,2] analysed the stress characteristics of different load cases and assumed that the existence of a phase angle will have a positive effect on the fatigue failure. Papadopoulos [3,4] assumed that the existence of a phase angle has little influence on hard metals, but for ductile metals, the influence of the phase angle on fatigue failure cannot be ignored. Susmel [5,6] thought that the existence of phase has a negative effect on the fatigue failure of ductile metals. Tovo[7] studied the fatigue failure of EN-GJS-400-15 cast iron and found that the multiaxial fatigue strength increases when phase angle exists. Anes [8] found that the existence of a phase angle has a negative effect on multiaxial fatigue failure of 42CrMo4 steel and AZ31B-F Mg alloy. Shi [9] found that the existence of a phase angle has a positive effect on the fatigue life of 2A12-T4 aluminum alloy under tension-torsion fatigue loading.

To investigate the effects of phase difference on the tension-torsion, high-cycle fatigue failure of 30CrMnSiA steel, both experiments and theoretical analysis are conducted in this paper.

2 Experiments

2.1 Material and specimens

The material used in the current experimental investigation is 30CrMnSiA structure steel. The chemical compositions of the 30CrMnSiA structure steel are: C(0.31%), Mn(0.85%), Si(0.99%), S(0.009%), P(0.01%), Ni(0.05%), Cr(0.87%), W(0.01%), Mo(0.02%), V(0.01%), Cu(0.18%), Ti(0.003%), Fe(the rest). The tension elastic modulus is 207GPa, tension yield strength is 1196MPa, and tension strength is 1334MPa. The torsion modulus is 77.2GPa, torsion yield strength is 825MPa, and torsion strength is 1040MPa.

The solid cylindrical bar specimen is adopted, which machined from the round bar of Φ=30mm. The minimum diameter of 30CrMnSiA steel specimen is D=12 mm. All of the specimens are polished with several abrasive papers up to grade 2000. The dimension tolerance is 0.02mm and surface finish of the specimen is less than 0.2μm to reduce the scatter of the experiment result. The knuckle radius of the specimen is R=220 mm. Shape and dimensions of the specimen is shown in Fig. 1.

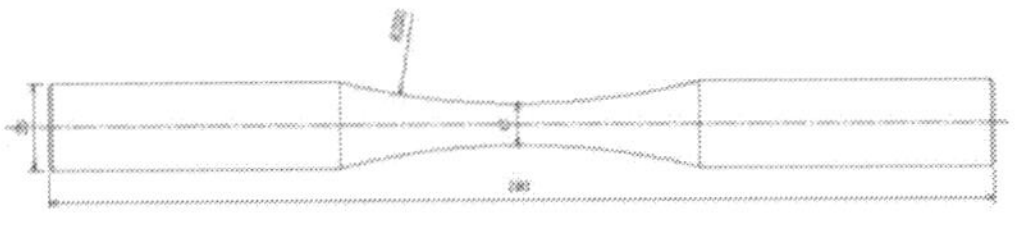

Figure 1. Shape and dimensions of the specimen.

2.2 Experiment procedure

The fatigue experiments of this study are performed on a PLS-200/1500 servo-hydraulic tension-torsion load frame. The test system, which has a capacity of 1500Nm in torque and 200kN in axial load, is equipped with electronic and computer controls, as well as data acquisition. The axial and shear loads are controlled concurrently, while the cyclic loading is fully reversed. All of the experiments are conducted at room temperature in ambient air. The loading frequency is f = 3Hz in the experiments, and in each load case, three or four specimens have been tested. The waveform of the load is described as follows:

$$\sigma_x = \sigma_{x,a} \sin\omega t \quad (1)$$

$$\tau_{xy} = \tau_{xy,a} \sin(\omega t - \delta) \quad (2)$$

where σ_x is the axial stress, $\sigma_{x,a}$ is the axial stress amplitude; τ_{xy} is the shear stress, $\tau_{xy,a}$ is torsion stress amplitude; and δ is the phase angle between tension stress and torsion stress. Both the axial stress amplitude and shear stress amplitude are 357.5MPa in the test, and four different phase angles (δ=0°, 30°, 60°, 90°) are chosen for the test. The loading paths are shown in Fig. 2.

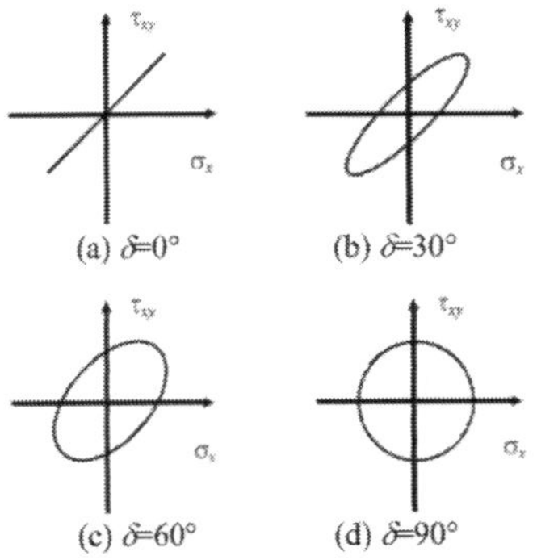

Figure 2. Loading paths for the tension-torsion fatigue tests with different phase angles.

3 Results and discussion

3.1 Fatigue life

Table 1 shows the median fatigue life of 30CrMnSiA steel under different phase angles, and the tension-torsion fatigue test results are shown in Figure 3. In the figure, tension-torsion fatigue lives of specimens are described by the open label, and the solid label means the median life in each load case. It can be seen from the figure that the median fatigue life increases with the increase of phase angle. Therefore, the existence of a phase angle has a positive effect on the fatigue life of 30CrMnSiA steel, which is same to 2A12-T4 aluminum alloy [9]. The difference between the two materials is that the fatigue life of 2A12-T4 aluminum alloy increased about three times when the phase angle increased from 0° to 90°, but for 30CrMnSiA steel, fatigue life increased only about 1.3 times when the phase angle increased from 0° to 90°. This may because that the plastic flow performance of 2A12-T4 aluminum alloy is better than 30CrMnSiA steel [10].

Tabel 1. Fatigue life under different phase angles.

δ/°	**0**	**30**	**60**	**90**
fatigue life /×10[5]	1.02	1.09	1.26	1.30

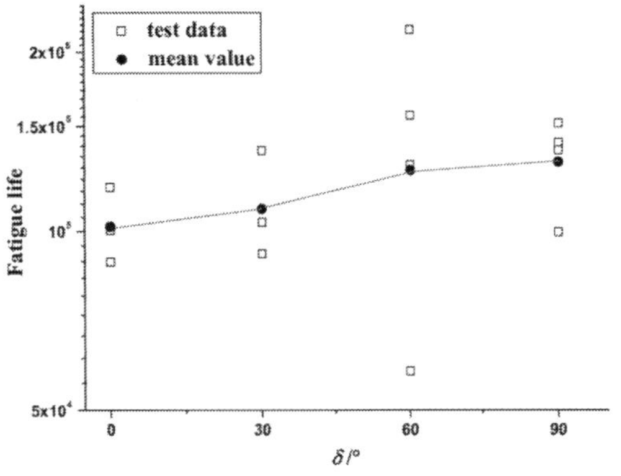

Figure 3. Relation of fatigue life and δ

3.2 Stress analysis

Fig. 4(a) shows the loading forms on the specimen. The X axis is defined as the longitudinal direction of the specimen and Y axis is perpendicular to X axis. The angle between an arbitrary plane and the X axis is defined as φ.

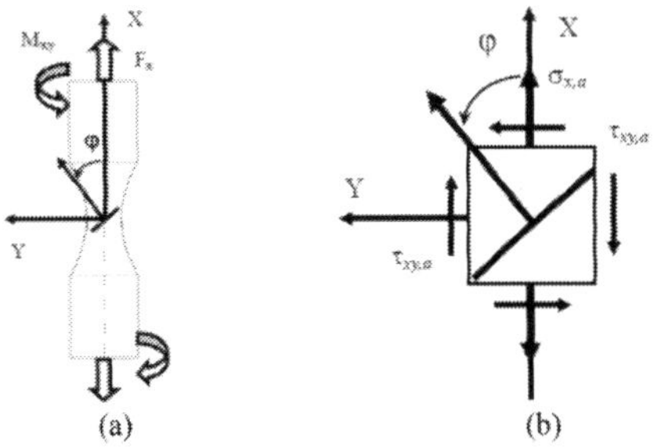

Figure 4. Stress analysis

It is widely believed that crack initiation is mainly controlled by the amplitude of alternating shear stress. Due to the derivation of Papadopoulos [4], the maximum shear stress amplitude is on the planes that perpendicular

DOI: 10.1051/matecconf/20167708001

to the XY plane. For the specimens under tension-torsion fatigue load, the critical point is on the surface of the specimen. Fig. 4(b) shows the stress element on the surface of the specimen. The changes of shear stress amplitude with different phase angles on different planes are shown in Fig. 5. It can be seen from the figure that two maximum shear stress amplitude planes exist for each load case. The maximum normal stress on different planes is shown in Fig. 6.

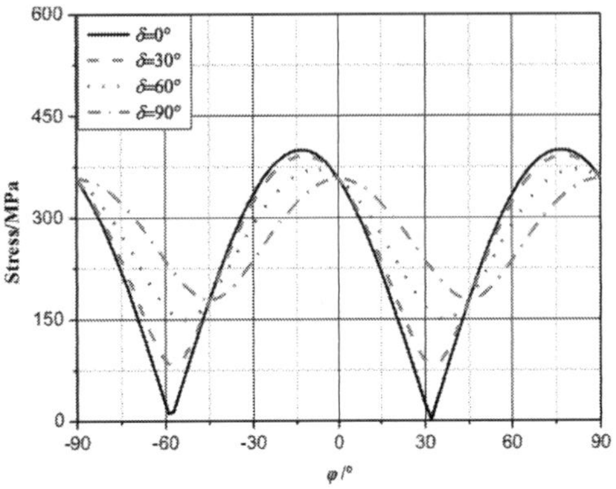

Figure 5. Variations of shear stress amplitude on different planes

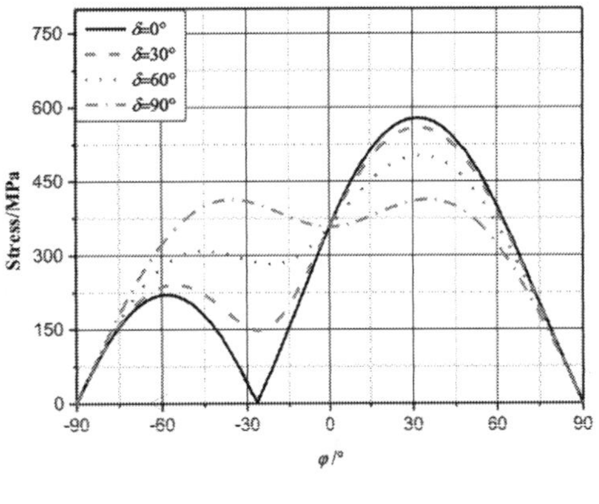

Figure 6. Variations of maximum normal stress on different planes

Fig. 7 shows the change of the maximum shear stress amplitude and the two maximum normal stresses on the maximum shear stress amplitude planes under different phase angles. The values of the stress parameters are given in Table 2 when the phase angle δ=0°, 30°, 60°, 90° respectively.

The two maximum normal stresses on the two maximum shear stress amplitude planes are defined as $\sigma_{n1,max}$ and $\sigma_{n2,max}$, respectively. It can be seen that the maximum shear stress amplitude ($\tau_{n,max}$) decreases from 399.7MPa to 357.5MPa when the phase angle increases from 0° to 90°. The maximum normal stress $\sigma_{n1,max}$ increases with the increase of phase angle while $\sigma_{n2,max}$ decreases. Therefore, as the fatigue life increases with the increase of phase angle, the maximum shear stress amplitude may be the main cause of fatigue failure for 30CrMnSiA steel under tension-torsion fatigue loading.

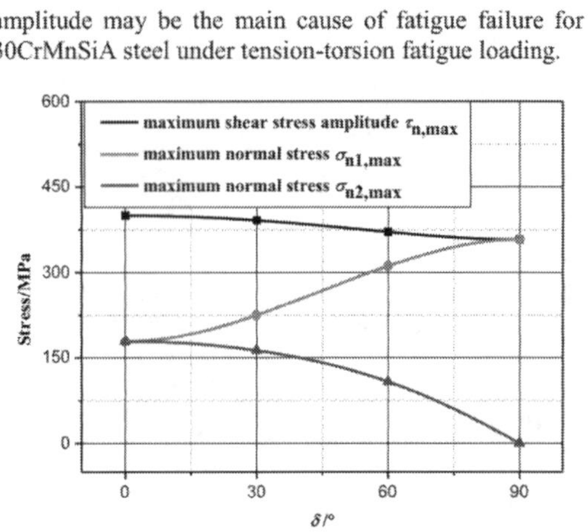

Figure 7. Variations of maximum shear stress amplitude and maximum normal stresses under different phase angles

Table 2. Directions of the maximum shear stress amplitude and the crack initiation plane.

δ /°	0	30	60	90
$\tau_{n,max}$/MPa	399.7	391.27	370.78	357.5
$\sigma_{n1,max}$/MPa	178.75	225.28	311.24	357.5
$\sigma_{n2,max}$/MPa	178.75	162.75	107.64	0

3.3 Fracture appearance

The fracture appearances of the 30CrMnSiA steel specimens under tension-torsion fatigue loading at different phase angles are shown in Fig. 8.

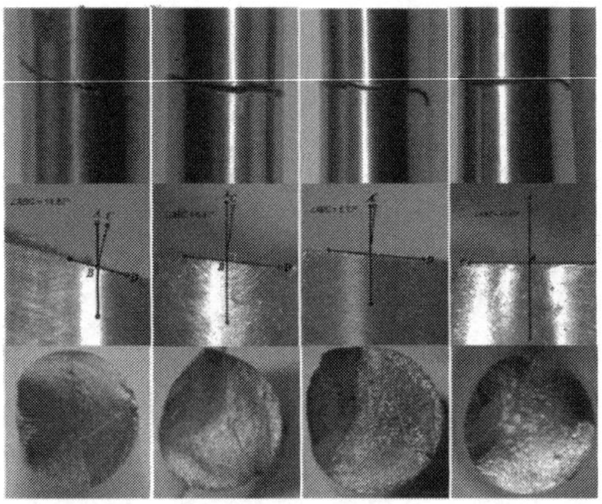

(a) δ=0° (b) δ=30° (c) δ=60° (d) δ=90°

Figure 8. Early crack propagation direction and fracture appearance under different phase angles

It can be seen from the figure that crack initiated on the surface of the specimens as indicated by a red arrow. The extension and final regions can be clearly observed. As the phase angle increases, the direction of the crack on the surface of the specimen gradually tends to become perpendicular to the axial of the specimen. The crack initiation angle is similar to the direction of the maximum shear stress amplitude plane and gradually flattens as the phase angle increases. The appearances of the fractures are smooth and the areas of the final rupture regions are nearly identical at different phase angles observed.

Table 3 shows the theoretical directions of the maximum shear stress amplitude plane (φ) and measured angle of the crack initiation plane (φ^*) under tension-torsion load with different phase angles.

Table 3. Directions of the maximum shear stress amplitude and the crack initiation plane.

δ /°	0	30	60	90
φ /°	76.7 -13.3	77.7 -12.3	81.6 -8.4	90 0
φ^* /°	-14.67	-6.81	-5.17	-0.34

When δ=0°, which is shown in Fig. 8(a), the crack initiated on the surface of the specimen, and the initial crack is in the direction of maximum shear stress amplitude (-13.3°) with a higher maximum normal stress, as shown in the previous analysis. The propagation of the surface crack is approximately along the same direction until failure occurs. The fracture appearance is smooth, and shear characteristics are clearly shown. When δ=30° and δ=60°, which are both shown in Figs. 8(b) and 8(c), respectively, the angles between the initial cracks and the axial of the specimens decrease, and the fracture appearances become flatten. Similarly, the initial crack is in the direction of maximum shear stress amplitude plane with a higher maximum normal stress. When δ=90°, which is shown in Fig. 8(d), the maximum shear stress is shown to be perpendicular to the axial direction of the specimen in this case, the crack initiated in the same direction as the maximum shear stress. The fracture appearance is also nearly perpendicular to the axial direction of the specimen.

Based on the discussion above, test results show that the fatigue life increases with the increase of phase angle. The maximum shear stress amplitude decreases with the increase of phase angle. The crack initiation and early propagation direction is close to the maximum shear stress amplitude plane with a higher maximum normal stress. Therefore, both the maximum shear stress amplitude and the maximum normal stress on the maximum shear stress amplitude plane effect the fatigue failure of 30CrMnSiA steel under tension-torsion fatigue loading at different phase angles, and the maximum shear stress amplitude seems to be the dominant factor.

4 Conclusions

The tension-torsion fatigue tests for 30CrMnSiA steel show that:

(1) with the increase of phase angle, fatigue life of the 30CrMnSiA steel increases, which means the fatigue life is shortest under proportional loading (δ=0°) and longest when the phase angle δ=90° ;

(2) the maximum shear stress is known to be the dominant factor that causes fatigue failure in 30CrMnSiA steel under tension-torsion fatigue loading at different phase angles;

(3) with the increase of phase angle, the angle between the initial crack plane and the longitudinal of the specimen decreases, and the crack initiation and early propagation is usually along the maximum shear stress amplitude plane with a greater maximum normal stress.

Acknowledgement

The National Natural Science Foundation of China is acknowledged for supporting the project (11172021).

References

1. D. L. McDiarmid, Fatig Fract Eng Mater Struct **9.6**, 457-475 (1987).
2. D. L. McDiarmid, Fatig Fract Eng Mater Struct **14.4**, 429-453 (1990).
3. I. V. Papadopoulos, Fatig Fract Eng Mater Struct **18.1**, 79-91 (1995).
4. I. V. Papadopoulos, P. Davoli, C. Gorla, M. Filippini, A. Bernasconi, Int J Fatigue **19.3**, 219-235 (1997).
5. L. Susmel, N. Petrone, *European Structural Integrity Society* **31.03**, 83-104 (2003).
6. L. Susmel, R. Tovo, P. Lazzarin, Int J Fatigue **27**, 928-943 (2005).
7. R. Tovo, P. Lazzarin, F. Berto, M. Cova, E. Maggiolini, Theor Appl Fract Mec **73**, 60-67 (2014).
8. V. Anes, L. Reis, B. Li, M. Freitas, Int J Fatigue **58**, 102-113 (2014).
9. X. H. Shi, J. Y. Zhang, R. Bao, Journal of Aeronautical Materials **36.8**, 965-968 (2010).
10. C. C. Zhang, *Fatigue life prediction of structures in HCF region under complex stress field* (2010).

MATEC Web of Conferences 77, 08002 (2016) DOI: 10.1051/matecconf/20167708002
ICMMR 2016

Absorption characteristics of multi-layered material anechoic coating on frustum-of-a-cone cavities

Liu Guo-qiang[1,2] , Lou Jing-jun[1,2], Li Hai-tao[1, 2]

[1] College of Power Engineering, Naval Univ. Of Engineering, Wuhan 430033, China
[2] National Key Laboratory on Ship Vibration &Noise; Wuhan 430033, China

Abstract. Improve the low-frequency acoustic performance of sound absorption coating of submarine is a hot issue. Based on the theory of sound wave normally impinging on the absorption layer of 2D,the 2D shaft mode of frustum-of-a-cone cavities is established by finite element COMSOL, which is test to analyze the absorption characteristics combined structural parameters. The calculation results indicate that under the condition of the same perforation rate ,the absorption performance of multi-layered material anechoic coating is better than that of single-layered material anechoic coating, especially to the performance of low-frequency; the absorption performance of frustum-of-a-cone cavities is better than that of cylindrical and ellipsoidal structures. With the perforation rate is increasing,the absorption performance of anechoic coating in low-frequency improve..Choosing the appropriate cavity structure is important. And the future acoustic optimization of the multi-layered frustum-of-a-cone cavities anechoic coating is guided by the study.

1 Introduction

Underwater anechoic coating is an important part of the stealth of underwater vehicle. With the development of active sonar technology, the frequency of detecting acoustic is getting lower and lower, so improve the low-frequency acoustic performance of sound absorption coating of submarine is a hot issue, and continuous optimization of the cavity structure could make the anechoic coating in low-frequency perform more excellent [1].

Mainly from the two aspects of the structure and material optimization to increase the absorption performance of anechoic coating in low-frequency. In terms of structure, Shang Chao [2] held that the absorption property of frustum-of-a-cone cavities which make air as backing by using finite element method, according to optimizing the design, the anechoic coating performance of the whole frequency band, especially low-frequency, is further improved; Tao Meng [3] analyzed the absorption property of anechoic coating of the combined cavity type by using waveguide finite element method, the absorption performance of anechoic coating in low-frequency greatly improve with the different embedding ratio of combined cavity type; Tang Weilin [4] put forward that using the boundary conditions of the front and back boundaries the expressions of resonance frequencies and reflection coefficients of coating are given. In terms of materials, Lü Linmei [5] gave the optimization method of how to improve the absorption effect in low-frequency, according to embedded viscoelastic absorption of different shapes of the scattering layer.

The above literature is the optimization of the structure and material of single-layered material anechoic coating. This paper focuses on the analysis of absorption performance of multilayer material of frustum-of-a-cone cavities structure. Results show that the absorption performance of multi-layered material anechoic coating is better than that of single-layered material of frustum-of-a-cone cavities structure and single cavity structure of multi-layer materials. Especially, the absorption effect in low-frequency is improved and obtain the rule about the influence of other structural parameters on the covering layer.

2. Theoretical model

Assuming that the fluid is an ideal fluid,it has the characteristics of non viscous, uniform, incompressible flow and zero flow rate. In addition, the density and pressure are assumed to be constant too.

Fluid finite element equation:

$$\left[M^{p}\right]\{\ddot{p}\}+\left[K^{p}\right]\{p\}+\rho_{0}[R]\{\ddot{\delta}\}=\{\Phi\} \qquad (1)$$

where: ρ_0——density of fluid medium

p——acoustic pressure

$[K^p]$—— global stiffness matrix;

$[M^p]$—— mass matrix;

$[R]$—— global coupling matrix;

$\{\Phi\}$——the influence of outer surface acoustic field in an unbounded fluid domin exerted by sound field within a finite domin.

The finite element vibration equation of the general solid structure can be expressed by the following formula [6]:

$$\left[M^s\right]\left\{\ddot{\delta}_{VI}\right\}+\left[K^s\right]\left\{\left\{\vec{\delta}_{VI}\right\}\right\}=\left\{F^m\right\}+\left\{F^p\right\} \tag{2}$$

where $[M^s]$—— globa mass matrix;

$[K^s]$—— global stiffness matrix。

$$\begin{cases}\left[M^s\right]=\iiint\limits_{V_{II}^e}\left[B_\delta\right]^T\left[D\right]\left[B_\delta\right]dV\\ \left[K^s\right]=\rho_s\iiint\limits_{V_{II}^e}\left[N_\delta\right]^T\left[N_\delta\right]dV\end{cases} \tag{3}$$

where ρ_s——density of structural materials;

$[N_\delta]$—— *displacement* interpolation function matrix;

$[D]$—— *constitutive* relation matrix, $[\sigma]=[D][\varepsilon]$。

combine with the boundary conditions,get:

$$\left[M^s\right]\left\{\ddot{\delta}\right\}+\left[K^s\right]\left\{\delta\right\}-\left\{R\right\}^T\left\{p\right\}=\left\{F^m\right\} \tag{4}$$

According to the finite element equation of fluid and solid, the fluid and solid coupling finite element equations of the whole system are rewritten as follows:

$$\begin{bmatrix}-[R]^T & [K^s]-\omega^2[M^s]\\ [K^p]-[C_\Phi]-\omega^2[M^p] & -\rho_0\omega^2[R]\end{bmatrix}\begin{Bmatrix}\{p\}\\ \{\delta\}\end{Bmatrix}=\begin{Bmatrix}\{F^m\}\\ \{C_0\}\end{Bmatrix} \tag{5}$$

$$\begin{bmatrix}\begin{bmatrix}X_{11}^{(1)} & X_{12}^{(1)} & X_{13}^{(1)} & X_{14}^{(1)}\\ X_{21}^{(1)} & X_{22}^{(1)} & X_{23}^{(1)} & X_{24}^{(1)}\\ X_{31}^{(1)} & X_{32}^{(1)} & X_{33}^{(1)} & X_{34}^{(1)}\\ X_{41}^{(1)} & X_{42}^{(1)} & X_{43}^{(1)} & X_{44}^{(1)}\end{bmatrix} & \begin{bmatrix}X_{11}^{(2)} & X_{12}^{(2)} & X_{13}^{(2)} & X_{14}^{(2)}\\ X_{21}^{(2)} & X_{22}^{(2)} & X_{23}^{(2)} & X_{24}^{(2)}\\ X_{31}^{(2)} & X_{32}^{(2)} & X_{33}^{(2)} & X_{34}^{(2)}\\ X_{41}^{(2)} & X_{42}^{(2)} & X_{43}^{(2)} & X_{44}^{(2)}\end{bmatrix}\\ \begin{bmatrix}X_{11}^{(3)} & X_{12}^{(3)} & X_{13}^{(3)} & X_{14}^{(3)}\\ X_{21}^{(3)} & X_{22}^{(3)} & X_{23}^{(3)} & X_{24}^{(3)}\\ X_{31}^{(3)} & X_{32}^{(3)} & X_{33}^{(3)} & X_{34}^{(3)}\\ X_{41}^{(3)} & X_{42}^{(3)} & X_{43}^{(3)} & X_{44}^{(3)}\end{bmatrix} & \begin{bmatrix}X_{11}^{(4)} & X_{12}^{(4)} & X_{13}^{(4)} & X_{14}^{(4)}\\ X_{21}^{(4)} & X_{22}^{(4)} & X_{23}^{(4)} & X_{24}^{(4)}\\ X_{31}^{(4)} & X_{32}^{(4)} & X_{33}^{(4)} & X_{34}^{(4)}\\ X_{41}^{(4)} & X_{42}^{(4)} & X_{43}^{(4)} & X_{44}^{(4)}\end{bmatrix}\end{bmatrix}\begin{Bmatrix}\begin{Bmatrix}\{p_{S_1}\}\\ \{p_{S_2}\}\\ \{p_{C_1}\}\\ \{p_{In}\}\end{Bmatrix}\\ \begin{Bmatrix}\{\delta_{S_1}\}\\ \{\delta_{S_2}\}\\ \{\delta_{C_1}\}\\ \{\delta_{In}\}\end{Bmatrix}\end{Bmatrix}=2\begin{Bmatrix}\{0\}\\ \begin{Bmatrix}\{Z_1\}\\ \{Z_2\}\\ \{Z_3\}\\ \{Z_4\}\end{Bmatrix}\end{Bmatrix} \tag{6}$$

The sound pressure P of fluid nodes and the displacement of structural nodes are in accordance with the periodic boundary conditions as follows:

$$\begin{cases}\left\{\chi_{S_3}\right\}=\left\{\chi_{S_1}\right\}e^{j\psi_x}\\ \left\{\chi_{S_4}\right\}=\left\{\chi_{S_2}\right\}e^{j(\psi_{dx}+\psi_y)}\\ \left\{\chi_{C_2}\right\}=\left\{\chi_{C_1}\right\}e^{j\psi_x}\\ \left\{\chi_{C_3}\right\}=\left\{\chi_{C_1}\right\}e^{j(\psi_{dx}+\psi_y)}\\ \left\{\chi_{C_4}\right\}=\left\{\chi_{C_1}\right\}e^{j(\psi_x+\psi_{dx}+\psi_y)}\end{cases} \tag{7}$$

In the equation of fluid structure, according to the corresponding nodes of the element, each matrix is divided into blocks. And then combined with the boundary conditions, the coupling equations can be changed into the following form:

X represents delta or P

$\vec{\delta}$——displacement vector of structure particle

By solving (7), the displacement and sound pressure of each node in the unit can be obtained. The sound pressure is expressed in sum of series respectively.

$$\begin{cases}p_r(x_i,y_i)=\sum\limits_{n=-M_x}^{M_x}\sum\limits_{n=-M_y}^{M_y}R_{nm}^p e^{-j(k_{nx}x_i+k_{my}y_i-k_{nm}z_-)}=\left\{E_r^{(i)}\right\}_{1\times N}\left\{R_{nm}^p\right\}_{N\times 1}\\ p_t(x_i,y_i)=\sum\limits_{m=-M_x}^{M_x}\sum\limits_{n=-M_y}^{M_y}T_{nm}^p e^{-j(k_{nx}x_i+k_{my}y_i+k_{nm}z_+)}=\left\{E_t^{(i)}\right\}_{1\times N}\left\{T_{nm}^p\right\}_{N\times 1}\end{cases} \tag{8}$$

$$(i=1,2,3,\cdots\cdots,N)$$

where

$$\left\{E_r^{(i)}\right\}=\left\{e^{-j(k_{nx}x_i+k_{my}y_i-k_{nm}z_-)}\right\},$$

$$\left\{E_t^{(i)}\right\}=\left\{e^{-j(k_{nx}x_i+k_{my}y_i+k_{nm}z_+)}\right\}$$

$$(n=-M_x:1:M_x;m=-M_y:1:M_y)$$

From the point of view of the calculation of acoustic energy, The reflection or transmission coefficient can be expressed as follow:

$$\begin{cases}R=\sqrt{\sum\limits_n\sum\limits_m R_{nm}^2}\Big/N\\ T=\sqrt{\sum\limits_n\sum\limits_m T_{nm}^2}\Big/N\end{cases} \tag{9}$$

So expression of sound absorption coefficient is defined as follow:

$$\alpha=1-\left|R\right|^2-\left|T\right|^2 \tag{10}$$

3. Computational model

Finite element simulation COMSOL model is shown in Fig. 1. Consider that the arrangement of the absorption performance of frustum-of-a-cone cavities has a periodic, select a unit of anechoic coating as the research object in order to facilitate the calculation. The anechoic coating is composed of a surface layer, a perforated layer, a base layer and a steel backing layer. Fig. 2 is a cross-sectional view of a frustum-of-a-cone cavity structure.

According to the symmetry, the model of the two dimensional axisymmetric model is used in this paper. As shown in Figure 1, on the boundary conditions, using COMSOL to transform the periodic boundary of the 3D model into the boundary of the two dimensional axisymmetric model base on the sound solid coupling module. Due to contact with the vehicle shell, the lower boundary displacement is zero, and the radial displacement of the model is zero because of symmetry, so boundary 6, 7 displacement is zero, the model can only axial deformation; oundary 2, 3, 4 for free deformation for simulate the change of cavity with sound pressure. Then, the plane wave is normal incident at the boundary 4, boundary 8, 5 set to the specified displacement, boundary 1 is set to a fixed constraint. Triangular mesh that be free splited are adopted when simulation calculation, the maximum size of the grid is 2.0mm. The sound absorption coefficient of the model is calculated by the sound solid coupling module. Material parameters are shown in Table 1.

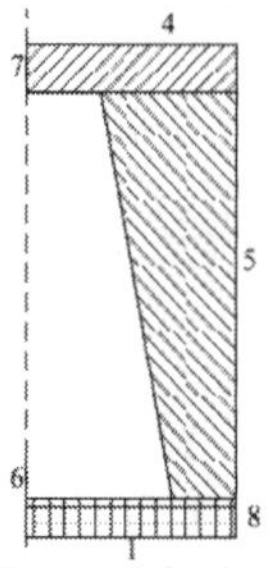

Figure 1. two dimensional axisymmetric model

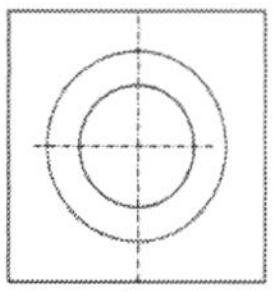

Figure 2. Frustum-of-a-cone cavity structure

Table 1 Structure and material parameters of anechoic coating

Material	R / mr	H / mr	$\rho/(kg\cdot n$		E / Pa	
Surface layer	21	5	1100	0.23	200	0.492
Perforated layer (frustum-of-a-cone cavity)	12	45	1300	0.45	300	0.493
Base layer	21	5	1400	0.6	500	0.490
Water layer	21	100				

4. Verification of model validation and analysis of results

4.1 Verification of COMSOL model

In order to verify th e validity of the finite element model, the structure parameters of the viscoelastic model are as follows: outside the cylindrical radius a is20mm, the inner radius of B is 12mm, and the thickness of L is 50mm, the thickness of the surface layer and the base layer are 3mm. The material parameters of the single layer are as follows: the elastic modulus is 400MPa, the loss factor is 0.2, the density is 1300kg/m3, and the Poisson's ratio is 0.490. Comparison of analytical formulas for the calculation of the two dimensional theory under the condition of steel backing of Ref.[7] and results of FEM is shown in Fig. 5. In the calculation of the cavity cover cone, frustum-of-a-cone cavity with cylindrical cavity structure is very similar, so in the finite element analysis, the free boundary is added in the cavity, let sound solid coupling boundary and non- reflective fluid sound absorbing boundary are not affected,so does with the the accuracy of the COMSOL model [10]. Through the analysis of the graph, it can be seen that the variation trend and the numerical value of the two curves are consistent within the scope of engineering error, which shows the feasibility of the COMSOL model of the frustum-of-a-cone cavity.

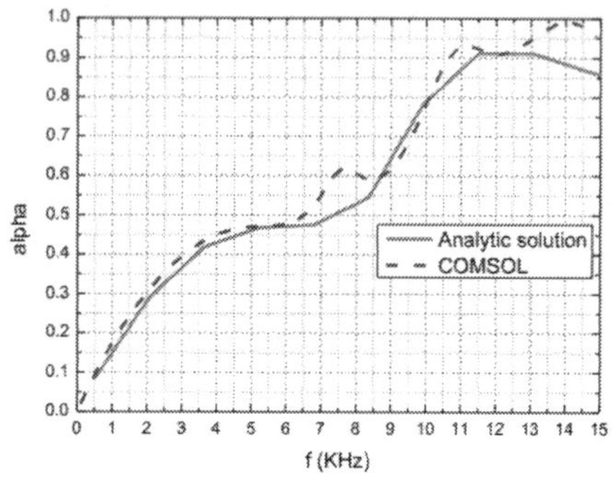

Figure 4. Comparison between two dimensional axisymmetric model and two dimensional theoretical model

4.2 Effect of materials on anechoic coating

Fig. 6 shows the effect of different layers of material and the cavity on the sound absorption coefficient or not under the same perforation ratio. The multi-layered materials are composed of these three different materials :the surface layer, the perforated layer, and the base layer ; a single layer of material is a covering of a material. From the graph, the multi-layered materials coatingis better than the single layered material coating. In low-frequency performance, the sound absorption coefficient of multi-layered material is 0.65, and the sound absorption coefficient of single layered material is 0.59 at the first sound wave crest. With the combination of different materials, the covering layer performs better at low-frequency. With the addition of the cavity, the cover layer is greatly improved at low-frequency sound absorption. That because the perforated layerp is added to the cavity, which make the plane longitudinal wave be converted to shear wave in the process of sound absorption and shear lossfactor of absorption layer is much larger than that of longitudinal wave,w hich can increase disspating the acoustic energy. Besides, the recent published articles [8]-[10] show that multi-layer materials are more efficient than the single-layer ones in terms of stress distribution. Therefore, using multi-layered material instead of single-layer materials could be a good choice for enhancing both absorption and structural performance.

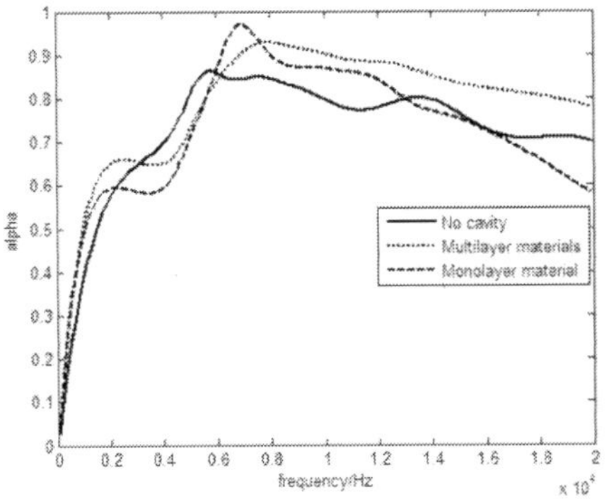

Figure 6. Comparison of sound absorption curves between different layers of material

4.3 Effect of different cavity structure on anechoic coating

The influence of cavity structure on sound absorption coefficient under the same perforation rate is discussed in this article. Fig. 7shows that omparison of sound absorption curves between cylindrical cavity, frustum-of-a-cone cavity, hybrid cavity and ellipsoidal cavity. Obviously, the low- frequency sound absorption performance of the hybrid cavity is better than that of the other cavity structures,and frustum-of-a-cone cavity is second. frustum-of-a-cone cavit in the first absorption trough much lower than the cylindrical cavity ,ellipsoid cavity, which sound absorption coefficient is 0.49 and corresponding frequency is 3950Hz. In the middle and high frequency, sound absorption performance of the four kinds of cavity structures are generally the same, and sound absorption coefficient is stable at about 0.8. In engineering application, two kinds of situations should be taken into account to choose.

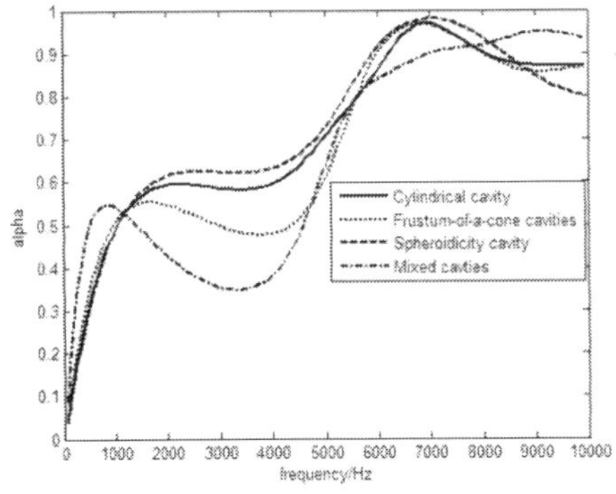

Figure 7. Comparison of sound absorption curves between different cavity structures

4.4 Influence of perforation rate on anechoic coating

This example discusses the influence of the change of the perforation rate on anechoic coating. In Fig. 8, The perforation rates of case 1,case 2,case 3 are 28 % , 46% , 60% .As known from Fig. 8, With the perforation rate is increasing, the absorption performance of anechoic coating in low-frequency improve, the first sound wave crest to the low frequency shift. In the middle and high frequency, the sound absorption effect of the covering layer of the three kinds of perforation rate structure is generally similar. With perforation rate increases ,The lifting effect for the medium- frequency and high- frequency sound absorption is very small.

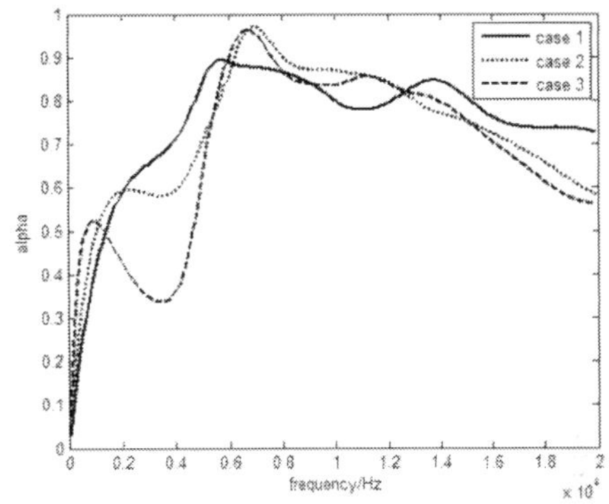

Figure 8. Comparison of sound absorption curves between different perforation rates.

5 Conclusion

In view of relatively poor situation of the absorption performance of anechoic coating in low-frequency currently, This paper, based on the theory of sound wave normally impinging on the absorption layer of 2D, build the 2D shaft mode of frustum-of-a-cone cavities which is established by finite element COMSOL,and analyze the absorption characteristics combined structural parameters, conclusions are as follows:

(1) Under the condition of the same perforation rate, the absorption performance of multi-layered material anechoic coating is better than that of single-layered material anechoic coating, especially to the performance of low-frequency; the absorption performance of frustum-of-a-cone cavities play an important role in improving the absorption performance in low-frequency of anechoic coating, which is better than that of single cavity structure.

(2) The perforation rate is essential for improving the absorption performance in low-frequency of anechoic coating. With the perforation rate is increasing,the absorption performance of anechoic coating in low-frequency improve. The relationship between the absorption performance in interm ediate-frequency/ high-frequency and the perforation rate is not very clos

(3) Choosing the appropriate cavity structure is important.When compared to combined cavity structure, frustum-of-a-cone cavitie structure provide little assistance to the improvement of the absorption performance of anechoic coating in low-frequency; compared with the cylindrical and ellipsoidal form, the opposite is the case.

References

1. MENG Xiao-yu,XIAO Guo-lin,CHEN Hong.Review of the present situation and development of acoustic stealth technology for submarines abroad.J.*Ship Science and Technology* **33**,11(2011).
2. SHANG Chao,ZHANG Jia-zhong,WEI Ying-jie.Absorption characterstics of anechoic coating embedding frustum-of-a-cone cavities.J.*Journal of Ship Mechanics* **14**,12 (2010).
3. TAO Meng,ZHUO Lin-kai.Acoustic performance of sound absorption coating containing composite cavities.J.*Journal of Shanghai Jiaotong University* **47**,3 (2013).
4. TANG Wei-lin,HE Shi-ping,FAN Jun.Two-dimensional model for acoustic absorption of viscoelastic coating containing cylindrical holes .J.*Acta Acustica* **30**,4 (2005).
5. Lü Lin-Mei,Wen Ji-Hong,Zhao Hong-Gang,et al.Dynamical mechanical property of viscoelastic materials and its effect on acoustic absorption of coating.J.*Acta Phys*.Sin **63**,15(2014).
6. HU Peng.Research of Acoustic Coating's Properities under Hydrostatic Pressure by Modular-FEM Method[D].*Shanghai:Shanghai Jiaotong University*,2008.
7. HE Shi-ping,TANG Wei-lin,HE Lin,et al.Analysis of acoustic characteristics of anechoic coating containing varying sectional cylindrical cavity .J.*Journal of Ship Mechanics* **10**,1 (2006).
8. M.M. Mirsayar, B. Takabi.Fracture of underwater notched structures.J. *Engineering Solid Mechanics* **4** (2016) .
9. B. Takabi.Thermomechanical transient analysis of a thick-hollow FGM cylinder.J. *Engineering Solid Mechanics* **4** (2016) .
10. [10] SHANG Chao,WEI Ying-jie,ZHANG Jia-zhong,CAO Wei.Absorption characteristics of anechoic coating embedding elliptic-cyclinder cavities.J.*Journal of Harbin Institute of Technology* **44**,1 (2012).

MATEC Web of Conferences 77, 08003 (2016) DOI: 10.1051/matecconf/20167708003
ICMMR 2016

Performance analysis of the plastic square plate based on the fiber reinforced PA66

Jian Wang , Junhong Zhang, Weidong Li, Zhexuan Xu

State Key Laboratory of Engines, Tianjin University, Tianjin 300072, People's Republic of China

Abstract. As the plastic component has the advantage of the lightweight, they are used widely in the industry and many researches are conducted simultaneously; In this study, combined the elastic constitutive model and the fiber orientation, the simulations of the plastic square plate with two material definitions are carried out. They are the orthotropic material and the anisotropic material respectively. Through the simulation comparisons, the anisotropic behaviors on the thermal field, the pressure field and the vibration characteristics of the fiber reinforced PA66 are all analyzed. The results show that the different material definitions have important influence on the structure analysis. It offers the guidance for the design and analysis of the plastic component.

1 Introduction

In recent years, the plastic components are applied in the mechanical industry widely, especially the automobile industry. The plastic components include the intake manifold, the oil cooler cover, the gear cover, the oil pan and so on. The engineering plastic has the advantage of the low density, high damping ratio, strong corrosion resistance, and good manufacture ability [1-3]. Therefore, the plastic component has the benefits of the lightweight, high strength, low noise and high productivity. As the density of the plastic is small and the price is equal to the metal, the cost price of the plastic component is lower compared with the metal component. With the plastic component develops, the structure analysis of the plastic component become get attentions. And the correlation studies on the plastic component become more and more.

However, the plastic materials contain many types, such as PP, PVC, ABS, PA and so on. These plastic materials have the low elasticity modulus and the poor structure strength. To improve the structure characteristic of the plastic materials, the engineering plastics are developed. Currently, the engineering plastics are reinforced with the carbon fiber, the glass fiber and other fibers. In addition, they are divided into two types: the short fiber reinforced plastic and the long fiber reinforced plastic. Currently, the plastic components are treated as the isotropic material for most researchers. However, the fiber has important influence on the material property. Therefore, the fiber orientation is almost impossible to ignore. For the common plastic component and the short fiber reinforced plastic, they can be treated as the isotropic material. However, for the long fiber reinforced plastic components, they should be treated as the an isotropic material.

The orientation and the number of the fibers in the plastic has important influence on the structure characteristic of the fiber reinforced plastic components. The geometry of the component and the injection process parameters such as the injection position, the number and timing of the injection pressure, the injection gates, the injection time, the mold temperature and the melt temperature can all influence the fiber distribution of the plastic component[4-7]. The elasticity modulus along the fiber orientation is large and the one normal to the fiber orientation is small. The difference makes that the long fiber reinforced components have obviously anisotropic mechanical property. Therefore, two simulations are conducted based on the isotropic material definition and the anisotropic material definition to study the difference of the thermal field, the pressure field and the vibration characteristics.

In this paper, the mold flow analysis is carried out first. The glass fiber orientation of the fiber reinforced PA66 plastic board is obtained. According to the material property parameters of the PA66 (GF 30%), the flexibility matrix of the anisotropic material is constructed. Combined the fiber orientation and the flexibility matrix, the anisotropic material definition is carried out. Next, the finite element model is constructed and the deformations and the stresses under pressure field and thermal field are calculated. Then the vibration characteristics of the two square plates with different material definitions are compared. Through the comparisons of the deformations, the stresses and the vibration characteristics, the difference and the cause are elaborated. It offers the guidance for the design and analysis of the plastic component. Elastic constitutive model of the orthogonal anisotropic material

2 Elastic constitutive model of the anisotropic material

An orthotropic material has three mutually perpendicular plates of symmetry at every point, with the intersections of these plates known as the principal axes of anisotropy [8]. Its material properties are, in general, different along different principal axis. The stress component and the strain component are not coupled. The stress-strain relationship can be defined based on the generalized Hook's law [9].

$$\{\varepsilon\}_m = [C]_m \{\sigma\}_m \tag{1}$$

$$\{\sigma\}_m = [C]_m^{-1} \{\varepsilon\}_m = [D]_m \{\varepsilon\}_m \tag{2}$$

$$\{\varepsilon\}_m = [\varepsilon_{11}\ \varepsilon_{22}\ \varepsilon_{33}\ \gamma_{12}\ \gamma_{23}\ \gamma_{31}]^T \tag{3}$$

$$\{\sigma\}_m = [\sigma_{11}\ \sigma_{22}\ \sigma_{33}\ \tau_{12}\ \tau_{23}\ \tau_{31}]^T \tag{4}$$

$$[D]_m = [C]_m^{-1} \tag{5}$$

Where ε is the strain and σ is the stress; ε_{ii} is the tensile strain and γ_{ij} the shear strain; σ_{ii} is the tensile stress and τ_{ij} is the shear stress; $[D]_m$ is the stiffness matrix and $[C]_m$ is the flexibility matrix.

$$[C]_m = \begin{bmatrix} \frac{1}{E_1} & -\frac{\mu_{21}}{E_2} & -\frac{\mu_{31}}{E_3} & 0 & 0 & 0 \\ -\frac{\mu_{12}}{E_1} & \frac{1}{E_2} & -\frac{\mu_{32}}{E_3} & 0 & 0 & 0 \\ -\frac{\mu_{13}}{E_1} & -\frac{\mu_{23}}{E_2} & \frac{1}{E_3} & 0 & 0 & 0 \\ 0 & 0 & 0 & \frac{1}{G_{23}} & 0 & 0 \\ 0 & 0 & 0 & 0 & \frac{1}{G_{13}} & 0 \\ 0 & 0 & 0 & 0 & 0 & \frac{1}{G_{12}} \end{bmatrix} \tag{6}$$

Where E_i is the tensile elastic modulus and G_{ij} is the shear elastic modulus; μ_{ij} is the poisson ratio. As the elastic matrix of orthotropic material is symmetrical, $E_1\mu_{21} = E_2\mu_{12}$, $E_2\mu_{32} = E_3\mu_{23}$ and $E_3\mu_{13} = E_1\mu_{31}$. Therefore, only 9 constants in equation (6) are independent. The constants of $[D]_m$ can be obtained through finding the inverse matrix of $[C]_m$. The stiffness matrix constants can be represented as equation (7) with the constants in equation (6).

$$\left.\begin{aligned} d_{11} &= \frac{1-\mu_{23}\mu_{32}}{E_2E_3\Delta} \\ d_{22} &= \frac{1-\mu_{13}\mu_{31}}{E_1E_3\Delta} \\ d_{33} &= \frac{1-\mu_{12}\mu_{21}}{E_1E_2\Delta} \\ d_{12} = d_{21} &= \frac{\mu_{21}+\mu_{31}\mu_{23}}{E_2E_3\Delta} = \frac{\mu_{12}+\mu_{13}\mu_{32}}{E_1E_3\Delta} \\ d_{23} = d_{32} &= \frac{\mu_{32}+\mu_{12}\mu_{31}}{E_1E_3\Delta} = \frac{\mu_{23}+\mu_{21}\mu_{13}}{E_1E_2\Delta} \\ d_{31} = d_{13} &= \frac{\mu_{13}+\mu_{12}\mu_{23}}{E_1E_2\Delta} = \frac{\mu_{31}+\mu_{21}\mu_{32}}{E_2E_3\Delta} \\ d_{44} &= G_{12} \\ d_{55} &= G_{23} \\ d_{66} &= G_{31} \end{aligned}\right\} \tag{7}$$

Where $\Delta = \frac{1-\mu_{12}\mu_{21}-\mu_{23}\mu_{32}-\mu_{31}\mu_{13}-2\mu_{12}\mu_{23}\mu_{31}}{E_1E_2E_3}$.

A transversely isotropic material is one with physical properties which are symmetric about an axis that is normal to a plate of isotropy [10]. This transverse plate has infinite plates of symmetry. Thus, the material properties are the same in all directions within this plate. For example, the unidirectional composite material is transversely isotropic. In a unidirectional composite, the plate normal to the fiber direction can be considered as the isotropic plan. Here the fibers would be aligned with the x1 axis, which is normal to the plate of isotropy. So the x2 axis and the x3 axis are on the plate of isotropy. Equation (8) can be obtained. Therefore the number of independent constants in the elasticity matrix is reduced to 5.

$$\left.\begin{aligned} &E_2 = E_3 \\ &\mu_{12} = \mu_{13}, \mu_{21} = \mu_{31}, \mu_{32} = \mu_{23} \\ &G_{13} = G_{12} \end{aligned}\right\} \tag{8}$$

The flexibility matrix of the transversely isotropic material can be represented as equation (9).

$$[C]_m = \begin{bmatrix} \frac{1}{E_1} & -\frac{\mu_{12}}{E_1} & -\frac{\mu_{12}}{E_1} & 0 & 0 & 0 \\ -\frac{\mu_{12}}{E_1} & \frac{1}{E_2} & -\frac{\mu_{23}}{E_2} & 0 & 0 & 0 \\ -\frac{\mu_{12}}{E_1} & -\frac{\mu_{23}}{E_2} & \frac{1}{E_2} & 0 & 0 & 0 \\ 0 & 0 & 0 & \frac{1}{G_{23}} & 0 & 0 \\ 0 & 0 & 0 & 0 & \frac{1}{G_{12}} & 0 \\ 0 & 0 & 0 & 0 & 0 & \frac{1}{G_{12}} \end{bmatrix} \tag{9}$$

Where $G_{23} = \frac{E_{22}}{2(1+\mu_{23})}$. The constants of stiffness matrix $[D]_m$ can be obtained through finding the inverse matrix of the flexibility matrix $[C]_m$.

DOI: 10.1051/matecconf/20167708003

3 Performance comparison of the orthogonal anisotropic material and the isotropic material

3.1 Model

To compare the mechanical properties of the orthogonal anisotropic material and the isotropic material, a square plate model of the unidirectional composite material is constructed (500×500×5 mm), shown in Figure 1. The material used is a type of PA66 with 30% fiber glass supplied by Lanxess Chemical Co. Ltd. The properties of the PA66 (GF 30%) are listed in Table 1.The location of the injection mold gate is in the middle, which makes the fiber orientation uniform.

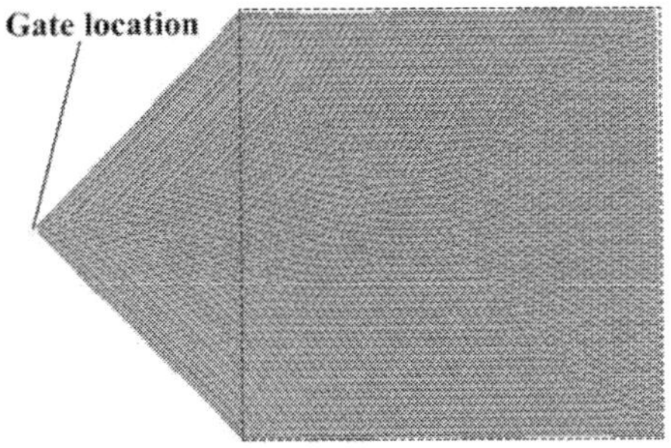

Figure 1. Model of the transversely isotropic material

Figure 2. Boundary conditions of the square plate model

Table 1. Properties of PA66 (GF 30%).

Properties	Values
Material structure	Crystalline
Glass Fiber	30%
Solid density (g/cm^3)	1.36
Melt density (g/cm^3)	1.16
E_1 (MPa)	10300
E_2 E_3 (MPa)	6800
μ_{12} μ_{23}	0.4
G_{12} (MPa)	1910
α_1 (1/K)	3E-5
α_2 α_3 (1/K)	9E-5
Ejection temperature (℃)	185
Recommended mold temperature (℃)	90
Recommended melt temperature (℃)	300

3.2 Definition of different materials

To compare the deformations and stresses of the two materials under the same pressure, a finite element model is constructed here, shown as Figure 2. The six degrees of freedoms (DOFs) of the plate border are constrained. The square in the center of the plate is the zone of pressure. The magnitude of pressure is 0.1 MPa.

Through defining the different material properties for the finite element model, the deformations and stresses of the two plates under the same pressure are computed. The results of deformation and stress are shown in Figure 3 and Figure 4 respectively. Seen from Figure 3, the deformations of the two models are similar. It is because the four sides of the plate are all constrained. However the normal elastic modulus of the transversely isotropic material is 6800 MPa, which is smaller than the elastic modulus of the isotropic material. Therefore under the same pressure, the deformation magnitude of the transversely isotropic material plate is larger. Comparing the stresses of the two plates in Figure 4, it can be seen that the stress distribution of the isotropic material plate is central symmetrical and that of the other plate has the obvious anisotropic characteristic. As the principal elastic modulus is larger than the normal modulus, the stress along the principal direction(X axis) is larger than that along the vertical direction.

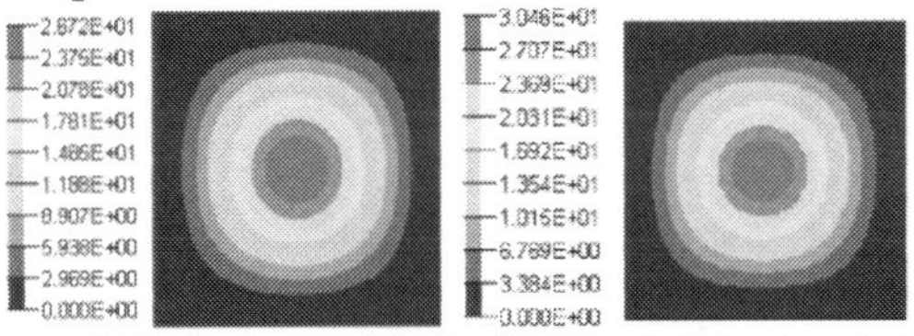

(a) Isotropic material (b) Transversely isotropic material

Figure 3. Deformation under pressure of 0.1 MPa

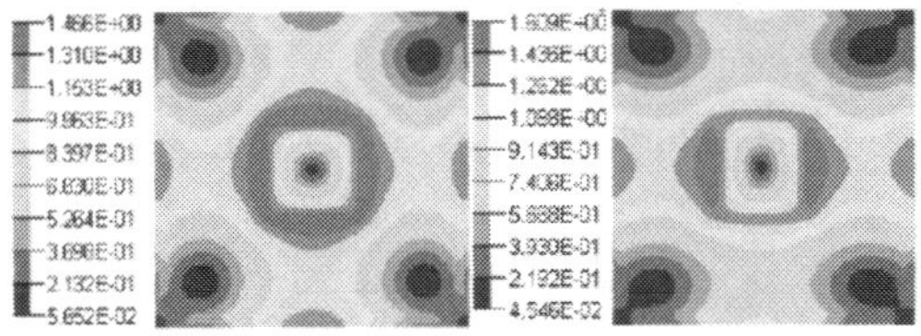

(a) Isotropic material (b) Transversely isotropic material

Figure 4. Stress under pressure of 0.1 MPa

To analyze effects of temperatures on the two materials, the boundary conditions of the square plate model are changed. The constraints are the same with Figure 2. The zone of pressure is changed to zone of T_1 and $T_1 = 323\text{K}$. The temperature of other area is set to T_2 and $T_2 = 293\text{K}$. The results of deformation and thermal stress are shown in Figure 5 and Figure 6 respectively. Seen from Figure 5 and Figure 6, the thermal deformation and stress of the transversely isotropic material plate both have the obvious anisotropic characteristic compared with the isotropic material plate. As the thermal expansion coefficient along the Y axis is larger than both that along X axis and the thermal expansion coefficient of the isotropic material, the deformation along the Y axis is larger. As the center zone has the little temperature difference, the deformation of the zone is smaller. The Figure 6 shows that maximal thermal stress of the transversely isotropic material plate is smaller than that of the isotropic material plate in the center zone. It is because the elastic modulus along the Y axis is smaller. As the thermal deformation along the Y axis is larger, the stress is larger than that of the isotropic material plate along Y axis except the center zone. In the same way that the thermal stress along X axis is smaller.

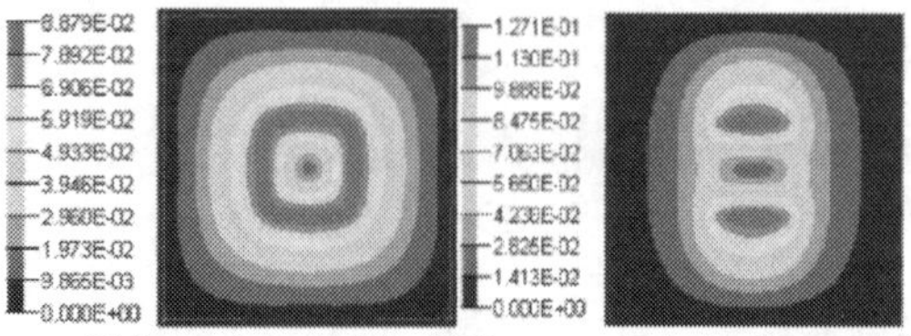

(a) Isotropic material (b) Transversely isotropic material

Figure 5. Deformation under the temperature difference of 30K

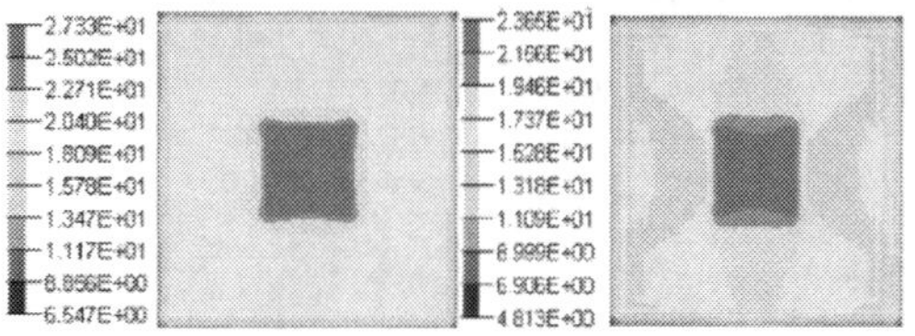

(a) Isotropic material (b) Transversely isotropic material

Figure 6. Thermal stress under the temperature difference of 30K

To compare vibration characteristics of the two square plates, the free modes are analyzed. The modal frequencies of two square plates are listed in Table 2. And the first three modal shapes are shown in Figure 7. Table 2 shows the modal frequencies of the anisotropic plate are lower than those of the isotropic plate. As the densities of the two plates are the same, it means the global stiffness of the anisotropic plate is lower than that of the isotropic plate. Figure 7 shows the first three modal shapes of the isotropic plate are all central symmetrical and the anisotropic plate is different. The first modal shape of the anisotropic plate is the first torsion and the symmetry axis is along the diagonal line (45°). As the comprehensive effects of both the principal elastic modulus and the vertical modulus, the modal shape is similar with that of the isotropic plate. However, the principal elastic modulus is larger than the vertical modulus. Therefore the second modal shape is the bending vibration along the vertical direction(Y axis) and the third modal shape is the bending vibration along the principal direction(X axis).

Table 2. Comparisons of the plate modal frequencies

Order	Isotropic	Anisotropic
1	31.29Hz	24.90Hz
2	45.92Hz	44.26Hz
3	60.59Hz	60.51Hz
4	81.93Hz	68.53Hz
5	81.93Hz	75.79Hz

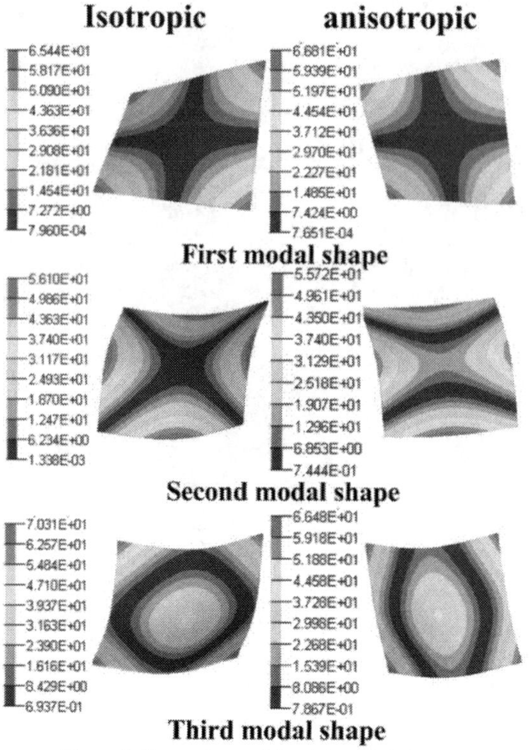

Figure 7. Comparisons of the plate modal shapes

4. Conclusions

In this study, a comprehensive performance research based on the orthotropic material was proposed. The effect of the orthotropic glass fiber reinforced PA66 characteristic on the structure performance was analyzed through the simulation of the plates with two different materials. Main conclusions can be drawn as follows:

(1)Under the same pressure, the deformation and the stress values of the transversely isotropic material plate is larger. The stress along the principal direction(X axis) is larger than that along the vertical direction.

(2)Under the same thermal boundary condition, the maximal thermal stress of the transversely isotropic material plate is smaller than that of the isotropic material plate, and the thermal deformation is larger along the vertical direction.

(3)The modal frequencies of the anisotropic plate are lower than those of the isotropic plate. The modal shapes of the transversely isotropic material plate show obvious anisotropic characteristic.

References

1. Launay A, Maitournam M H, Marco Y, et al. Cyclic behaviour of short glass fibre reinforced polyamide: Experimental study and constitutive equations[J].

International Journal of Plasticity, 2011, 27(8): 1267-1293.
2. Ku H, Wang H, Pattarachaiyakoop N, et al. A review on the tensile properties of natural fiber reinforced polymer composites[J]. Composites Part B: Engineering, 2011, 42(4): 856-873.
3. AlMaadeed M A, Kahraman R, Khanam P N, et al. Date palm wood flour/glass fibre reinforced hybrid composites of recycled polypropylene: Mechanical and thermal properties[J]. Materials & Design, 2012, 42: 289-294.
4. Gupta M, Wang K K. Fiber orientation and mechanical properties of short - fiber - reinforced injection - molded composites: Simulated and experimental results[J]. Polymer Composites, 1993, 14(5): 367-382.
5. Chung S T, Kwon T H. Numerical simulation of fiber orientation in injection molding of short - fiber - reinforced thermoplastics[J]. Polymer Engineering & Science, 1995, 35(7): 604-618.
6. Han K H, Im Y T. Numerical simulation of three-dimensional fiber orientation in short-fiber-reinforced injection-molded parts[J]. Journal of materials processing technology, 2002, 124(3): 366-371.
7. Zhang J, Wang J, Lin J, et al. Multiobjective optimization of injection molding process parameters based on Opt LHD, EBFNN, and MOPSO[J]. The International Journal of Advanced Manufacturing Technology, 2015: 1-16.
8. Deshpande V S, Fleck N A, Ashby M F. Effective properties of the octet-truss lattice material[J]. Journal of the Mechanics and Physics of Solids, 2001, 49(8): 1747-1769.
9. Muskhelishvili N I. Some basic problems of the mathematical theory of elasticity[M]. Springer Science & Business Media, 2013.
10. Aboudi J. Mechanics of composite materials: a unified micromechanical approach[M]. Elsevier, 2013.

MATEC Web of Conferences 77, 08004 (2016) DOI: 10.1051/matecconf/20167708004
ICMMR 2016

Reliability analysis on high-pressure fuel pipe of diesel engine

Weidong Li , Junhong Zhang, Jian Wang, Zhexuan Xu

State Key Laboratory of Engines, Tianjin University, Tianjin 300072, People's Republic of China

Abstract. High-pressure fuel pipe is an important component of the fuel injection system of diesel engine, it suffer complex loading from engine. The fatigue life of fuel pipe has great impact on reliability of engine. In this study, the fluid-solid coupling model which considers the action of fuel is established. Analysis on the effect of fuel's pressure oscillation on the mode is carried out. Through the simulation comparisons, the mode frequency and mode shape of each model are all analysed. Base on the fluid-solid coupling model, frequency response analysis is done to study the reliability of high-pressure fuel pipe. The results show that the pressure oscillation of fuel has important influence on the mode frequency. The maximum Stress Tensor and Vibration Velocity appeared at the position of first order mode frequency. It offers the guidance for the design and analysis of High-pressure fuel pipe.

1 Introduction

High-pressure fuel pipe connects the high pressure fuel injection pump and nozzle, and transfer the fuel to cylinder for the operation of the diesel engine at a certain pressure and velocity through the fuel pressure wave. So it is an important component of the fuel injection system of diesel engine. In recent years, the emission regulations became increasingly strict, and the requirements of engine power performance became higher. In order to solve these problems, it is important to improve the efficiency and the emission performance of diesel engine. Experts had done a lot of research in improving the engine performance. The high-pressure common rail fuel injection system with a high injection pressure was proposed as a new technology applies on the majority of diesel engine. Increased spray pressure has plenty of influence on injection performance, such as acquiring finer spray droplets, enhancing turbulent mixing rate to attain better fuel-air mixture [1], shortening ignition delay time [2,3], reducing soot emission [4,5] and improving specific power [6].

Nowadays, the spray pressure increased even as high as 280MPa, and most of the diesel engine was more than 180MPa of spray pressure. The increasing of injection pressure has a huge impact on the high pressure fuel pipe. The high-pressure fuel pipe works in a terrible condition and was under two different acting force. One is the pressure oscillations triggered by the injection event, the high pulse pressure acting on the wall inside of fuel pipe directly. Another is the vibrate of high pressure fuel injection pump and cylinder head of diesel engine, the vibration signal due to deformation of high pressure fuel pipe. Both of the acting force would due to the fatigue break of high pressure fuel pipe when the vibration of the diesel engine is getting stronger and the fuel injection pressure become higher. The fatigue failure phenomenon of high-pressure fuel pipe was still in a high frequency. In order to reduce the fatigue failure of high-pressure fuel pipe, the experts doing heave research to improve the performance of fuel pipe [7].

Bae J H [8] has doing the preform design of fuel pipe to preventing the head of high-pressure fuel pipe from folding in the heading process. In his study, the angle, curvature and outer diameter of the die and the length of the trapped part were selected as main parameters in the design of the optimal preform shape that minimizes the radius of folding. Song [9] was carry out a tensile test and a fatigue test to optimize the autofrettage process for improving fatigue life of fuel pipe. And the fatigue analysis was considered not only the compressive residual stresses of the inner surface but also the tensile residual stresses of the outer surfaces of the pipe. Bai SZ [10] analysis the reason of high-pressure fuel pipe's damage by vibration test. Most of the experts analysis the vibration performance of high-pressure fuel pipe when it was empty. But the analysis of high-pressure fuel pipe with fluid-solid coupling was still a little appeared. The fuel flow in high speed had made great impact force on fuel pipe wall. It also had influence on the vibration characteristics of high-pressure fuel pipe.

In this paper, the mode analysis of high-pressure fuel pipe is carried out first. The model of high-pressure fuel pipe will establish by the material property parameters and the shape of pipe that acquire according to the experimental measurement. Based on the fluid-solid coupling method, the mode analysis of high-pressure fuel pipe is held to compare the result of mode analysis of fuel pipe only. The influence on the mode frequency and mode of vibration of high-pressure fuel will prove to explain to the importance of fuel flow in high speed. Then the Reliability analysis will carry out to simulation the stress distribution of high-pressure fuel pipe when the diesel engine in normal operation. Through the research, a theoretical basis was provided for fatigue life prediction and fatigue resistance design.

2 Method of fluid-solid coupling

Usually the structural dynamic equations of a linear structural system without fluid-solid coupling will be [11]:

$$[M_e]\{\ddot{U}_e\}+[C_e]\{\dot{U}_e\}+[K_e]\{U_e\}=\{F_e\} \quad (1)$$

Where $[M_e]$ is the mass matrix; $[C_e]$ is the damping matrix; $[K_e]$ is the stiffness matrix; $\{\ddot{U}_e\}$ is the acceleration; $\{\dot{U}_e\}$ is the Velocity; $\{U_e\}$ is the displacement and $\{F_e\}$ is external excitation.

In the analysis of fluid-solid coupling, it needs to consider both the structural dynamics equation, fluid equation and fluid continuity equation. The fluid-solid coupling equations will be[12]:

$$\begin{bmatrix}[M_e] & [0]\\ [M^{fs}] & [M_e^p]\end{bmatrix}\begin{Bmatrix}\{\ddot{U}_e\}\\ \{\ddot{P}_e\}\end{Bmatrix}+\begin{bmatrix}[C_e] & [0]\\ [0] & [C_p]\end{bmatrix}\begin{Bmatrix}\{\dot{U}_e\}\\ \{\dot{P}_e\}\end{Bmatrix}+ \begin{bmatrix}[K_e] & [K^{fs}]\\ [0] & [K_e^p]\end{bmatrix}\begin{Bmatrix}\{U_e\}\\ \{P_e\}\end{Bmatrix}=\begin{Bmatrix}\{F_e\}\\ \{0\}\end{Bmatrix} \quad (2)$$

Where $[M^{fs}]$ is the element coupling mass matrix, $[M_e^p]$ is the fluid element mass matrix, $\{\ddot{P}_e\}$ is the acceleration of fluid, $\{\dot{P}_e\}$ is the Velocity of fluid, $\{P_e\}$ is the displacement of fluid, $[C_p]$ is the damping matrix of fluid, $[K_e^p]$ is the stiffness matrix of fluid and the $[K^{fs}]$ is the element coupling stiffness matrix. In this paper, the pressure oscillation of fuel will consider either.

3 Reliability analysis on high-pressure fuel pipe

3.1 Model

To compare the mode frequency and mode shape of the empty high-pressure fuel pipe and the fuel pipe fill with fuel, the model of each condition of high-pressure fuel pipe is establish by using hexahedral mesh, shown in Figure 1. The empty high-pressure fuel pipe is made up by 229050 nodes. And the model of fuel pipe fill with fuel is consisted of 740662 nodes. The structure of high pressure oil pipe is very simple, therefore the number of cells is enough to satisfy the accuracy of the calculation. The material of fuel pipe used is a type of seamless steel pipe which is HQ3-8×3.6-Q/LG J01. The properties of the pipe and fuel are listed in Table 1.

Table 1. Material properties table

Properties	Values	
	Pipe	fuel
Elastic Modulus(MPa)	2.1E05	-
poisson's ratio	0.3	-
Density (t/mm^3)	7.9E-09	8.4E-10
velocity of sound (mm/s)	-	1.4E06

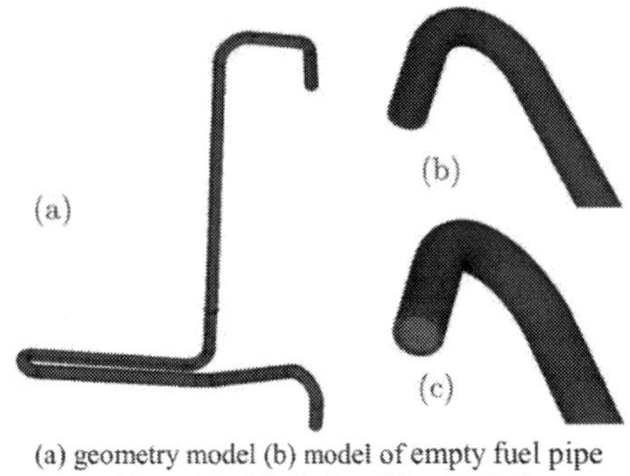

(a) geometry model (b) model of empty fuel pipe
(c) model of fuel pipe fill with fuel
Figure 1. Different models of high-pressure fuel pipe

3.2 Mode analysis of different model

In order to study the influence on mode frequency and mode shape when fuel pipe fill with fuel, a finite element model is constructed. The properties of fuel are considered and the pressure oscillation is loaded to the pipe wall. The six degrees of freedoms (DOFs) of the pipe end are constrained. A mode analysis is carried out base on the finite element model. The results of mode frequency of different model are shown in Table 2. Seen from Table 2, the mode frequency of two models has some differences. The mode frequency of fluid-solid coupling model is lower than fuel pipe model in the same order mode. It is because the mass of fuel can reduce the inherent frequency of model. The pressure oscillation which loaded to the pipe wall will produce pre-stress that can influence the mode frequency either. But the different between mode frequencies of two models is small in high order of mode. The density and rigidity of pipe's material has greater influence of mode frequency in high order of mode. Because the pressure oscillation has great influence on modal frequency of fuel pipe, it must be considered when doing the reliability analysis of high-pressure fuel pipe.

Table 2. Compare the mode frequency of different model

Order	mode frequency	
	Fuel pipe only	Fuel pipe with fuel
1	83	73.6
2	160.6	155.5
3	203.3	197.5
4	264.5	262.6
5	314.4	311.5
6	402.9	401

The compare results of mode shape of two models are shown in the figure followed. The first five mode shapes are shown in order of Figure2 to Figure 6. Seen from Figure 2 to Figure 6, the mode shape of the two models are similar. It is because the fuel distribution along the pipe, it means the shape of model does not change. The weight of high-pressure pipe has less effect in the mode shape. Compare the first five mode shape, the displacement of fluid-solid coupling model is larger own to its weight.

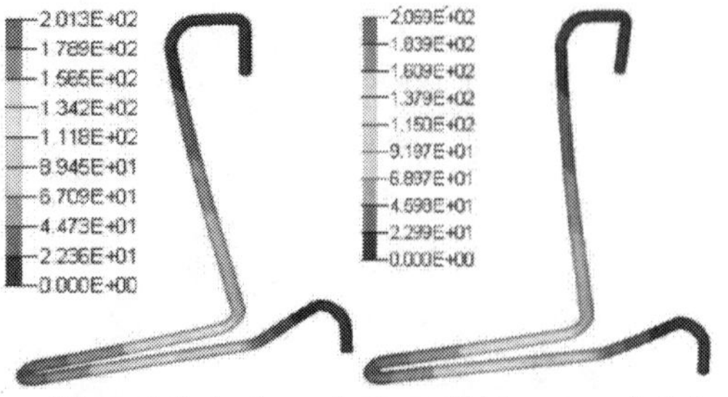

Figure 2. first order mode shape of high-pressure fuel pipe

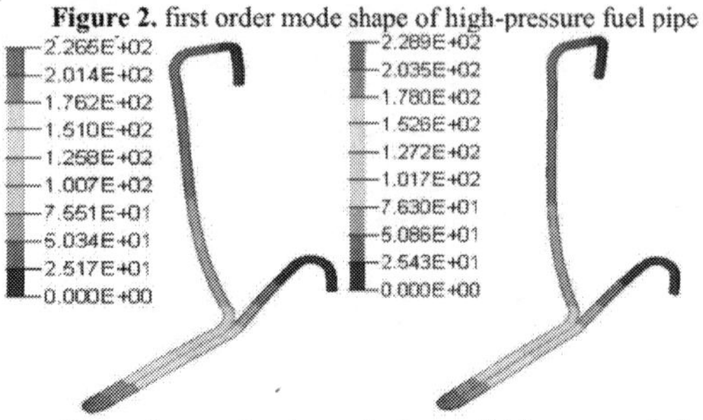

Figure 3. second order mode shape of high-pressure fuel pipe

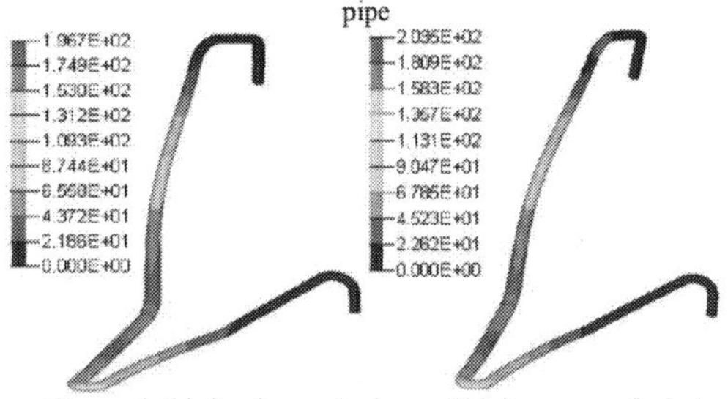

Figure 4. third order mode shape of high-pressure fuel pipe

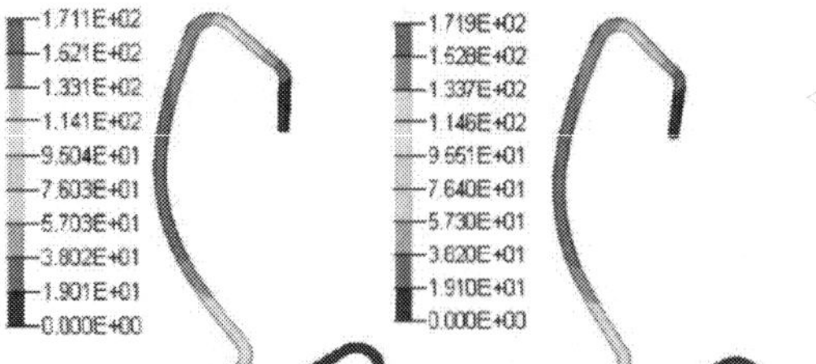

Figure 5. 4th order mode shape of high-pressure fuel pipe

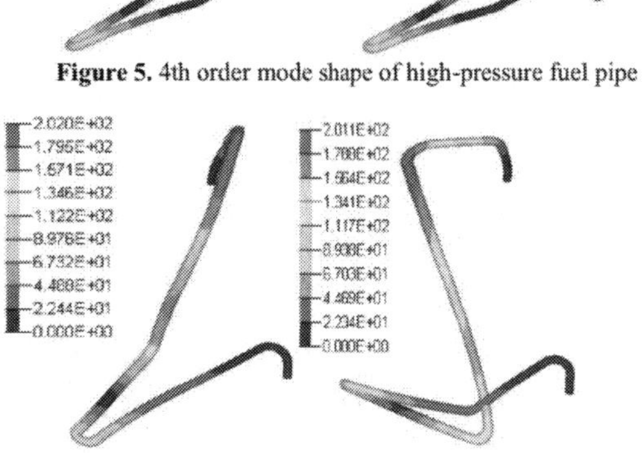

Figure 6. 5th order mode shape of high-pressure fuel pipe

3.3 Reliability analysis on high-pressure fuel pipe

To study the reliability of high-pressure fuel pipe, the frequency response analysis of fuel pipe is carried out. The excitation of high pressure fuel injection pump was measure from engine bench test that the engine operation in 2200rpm. Then the excitation is loaded to the end of fuel pipe on corresponding position. The result of stress distribution is shown in Figure 7. From the Figure 7, it can be seen that the maximum stress is distributed around the inflexion point. Because the fuel pipe has better stiffness and it is hard to be out of shape in straight part. So the maximum deformation is occur on inflexion point, and the maximum stress position is the same with maximum deformation.

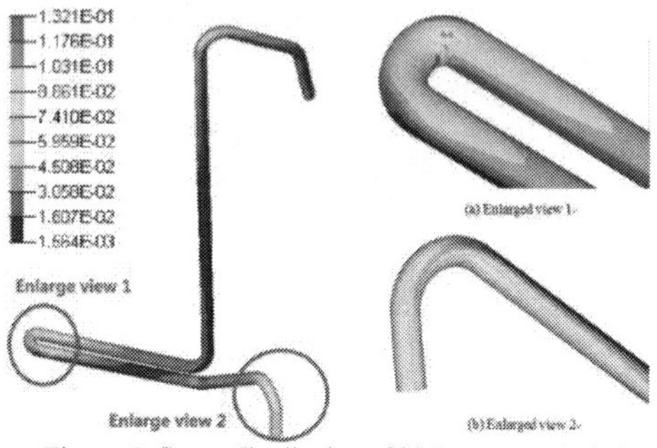

Figure 7. Stress distribution of high-pressure fuel pipe

To study the reliability and vibration characteristic of high-pressure fuel pipe, the Stress Tensor and Vibration velocity of the maximum stress position is analysed. The result of Stress Tensor and Vibration velocity are shown in Figure 8 and Figure 9. As seen in the Figure 7, there is some peak shown in the frequency of 83Hz, 203Hz and 401Hz. The maximum Stress Tensor occur in83Hz, it is closed to first order mode frequency. The resonance is happened in this place. Seen from the Figure 9, the maximum vibration velocity also happened in the same frequency. When the engine is operation in 2200rpm, the forced vibration frequency will be 73.3Hz. The probability of fatigue failure of high-pressure fuel pipe will increase rapidly when the resonance occurred. It's important to optimize the structure of fuel pipe and fixed position. In this way, the mode frequency of fuel pipe will keep away from engine's forced vibration frequency. The reliability and fatigue life of high-pressure fuel pipe can be enhancing greatly.

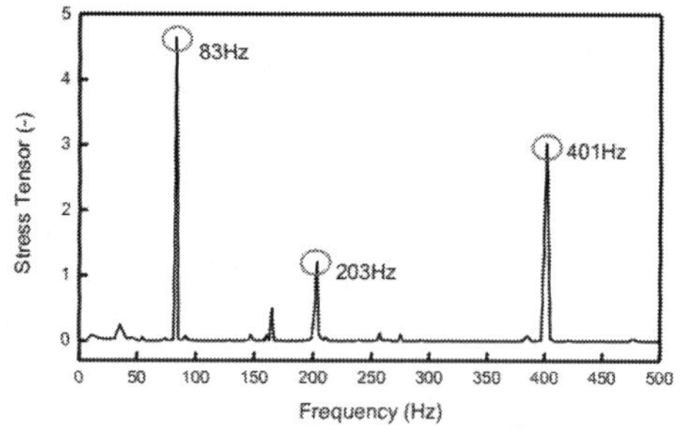

Figure 8. Stress Tensor of the point of maximum stress

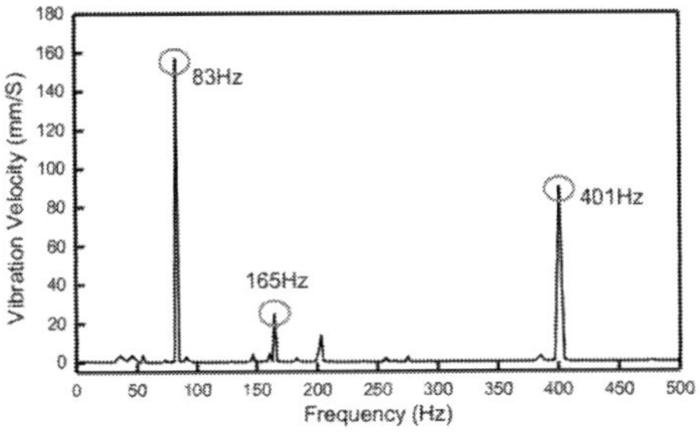

Figure 9. Vibration Velocity of the point of maximum stress

4. Conclusions

In this study, the mode analysis based on the finite element model of high-pressure fuel pipe was proposed. The effect of the pressure oscillation of fuel on the mode was analyzed through the simulation of two models. Then the reliability analysis of high-pressure fuel pipe base on fluid-solid coupling model was carried out. Main conclusions can be drawn as follows:

(1) The mode frequencies of the model which consider the fuel action are lower than another model. The mode shapes of two models are similar.

(2) The maximum stress is distributed around the inflexion point. The minimum stress is occur on the middle of each straight part.

(3) The maximum Stress Tensor and Vibration Velocity appeared at the position of first order mode frequency.

References

1. K.Nishida, W.Zhang, T.Manabe, Effects of micro-hole and ultra-high injection pressure on mixture properties of D.I.Diesel spray, SAE Technical Paper 2007-01-1890,2007,pp.1353–1361.
2. F.Tao, P.Bergstrand, Effect of ultra-high injection pressureon diesel ignition and flame under high-boost conditions, SAE Technical Paper 2008-01-1603, 2008, pp.1–22.
3. O.A.Kuti, W.Zhang, K.Nishida, etal., Effect of injection pressure on ignition, flame development and soot formation processes of biodiesel fuel spray, SAE Technical Paper2010-32-0053,2010.
4. Can Ö, Celikten I, Usta N. Effects of ethanol additio n on performance and emissions of a turbocharged in direct injection diesel engine running at different inje ction pressures[J]. Energy conversion and Manageme nt, 2004, 45(15): 2429-2440.
5. Wang X, Huang Z, Kuti O A, et al. Experimental and analytical study on biodiesel and diesel spray characteristics under ultra-high injection pressure[J]. International journal of heat and fluid flow, 2010, 31(4): 659-666.
6. Puhan S, Jegan R, Balasubbramanian K, et al. Effect of injection pressure on performance, emission and combustion characteristics of high linolenic linseed oil methyl ester in a DI diesel engine[J]. Renewable Energy, 2009, 34(5): 1227-1233.
7. Hasan S G A, Kumar G S. Finite Element Analysis a nd Fatigue Analysis of Spur Gear under Random Loa ding[J]. International Journal of Engineering Science s & Research Technology, 1(4): 513-523.
8. Song S A, Bae J H, et al. Study of Improvement in Fatigue Life of Fuel Injection Pipe of Common Rail System[J]. Transactions of the Korean Society of Mechanical Engineers A, 2013, 37(8): 991-998.
9. Bai Shu-zhan, Jia Ying-jun, et al. Analysis of Damage of Engine Fuel Injection Pipe Based on Vibration Test [J]. I. C. E &Powerplant , 2007, 99(3): 49-52.
10. Han K H, Im Y T. Numerical simulation of three-dimensional fiber orientation in short-fiber-reinforced injection-molded parts[J]. Journal of materials processing technology, 2002, 124(3): 366-371.
11. Itoh T. Damped vibration mode superposition method for dynamic response analysis[J]. Earthquake Engineering & Structural Dynamics, 1973, 2(1): 47-57.
12. Zhang Jun-hong, Guo Qian, Wang Jian, et al. Low noise optimization design of plastic oil cooler cover based on the liquid-solid coupling method and multi-objective topological optimization [J]. Journal of Vibration and Shock, 2016, 35(7): 186-191.

Design of Experimental Suspended Footbridge with Deck Made of UHPC

Marek Blank [1], Petr Tej [1], J. Kolísko[1] and L. Vráblík [2]

[1]*Klokner Institute, Czech Technical University in Prague*
[2]*Faculty of Civil Engineering, Czech Technical University in Prague*

Abstract. This paper deals with the static and dynamic design of experimental footbridge for pedestrians and cyclists in the municipality Lužec nad Vltavou in Czech Republic, Europe. This work aims to familiarize the reader with calculations carried out and the results obtained, describing the static and dynamic properties of proposed footbridge. The construction of footbridge is designed as a suspended structure with prestressed bridge deck consisting of prefabricated UHPC panels and reversed "V" shaped steel pylon with height of approximately 40 meters. The deck is anchored using 24 steel hangers in one row in a steel pylon - 17 ropes in the main span and 7 cables on the other side. Range of the main span is 99.18 meters and the secondary span is 31.9 m. Deck width is 4.5 meters with 3.0 meters passing space. The bridge is designed for the possibility of passage of vehicles weighting up to 3.5 tons. Deck panels are made of UHPC with reinforcement. At the edge of the bridge on the side of the shorter span the bridge deck is firmly connected with abutment and on the other deck it is stored using a pair of sliding bearings. The utilization of the excellent properties of UHPC allows to design a very thin and lightweight construction of the deck, which could not be achieved with the use of normal concrete.

1 Introduction

The footbridge transfers the bike path and the pedestrian trail over the river Vltava, connecting the village Lužec nad Vltavou with surrounding municipalities on the other side of the river. This footbridge will permanently replace current ferry.
The structural design according to various requirements and basic static and dynamic analysis of the structure were performed and all kinds of load regarding the nature of the structure were considered.

2 Materials

Ultra High Performance Concrete - UHPC is a new cementitious composite material, which is mainly used for reducing or optimizing a weight and dimensions of structures. Its outstanding features, mainly high compressive strength (>150 MPa), high tensile strength (>10MPa) and high durability, are generally known [1][2][3]. UHPC is a homogenous material whose mixture consists of very fine aggregates, cement, fine fillers (silica fume, fly ash), superplasticizers and randomly distributed fibres (steel, glass, polymer). A very low water-cement ratio and a fine grained mixture provide extremely low porosity [3].

3 Structural Design

The axis of the bridge leads over the river bed in approximately perpendicular direction and at the end of the bridge, on the shorter span side, passes in earth ramp located partly in a left curve of radius 30.0 meters for the connection to the existing road. Access ramps on both sides are sufficiently dimensioned to flood protection measures, but they are also made in the minimum slopes and equipped grassy vegetation in accordance with the landscape character of its surroundings.

The bridge deck height is kept in top curved surface of radius 777.5 meters. This ensures enough camber of the footbridge over the rivers canaliculate profile measuring 35.0 meters with height of 7.0 meters, which was necessary due to the requirement of waterway administrator. The rise of an arch is chosen so that the ends of the supporting structure would still have the maximum allowed slope for wheelchair use of the building.

The supporting structure consists of panels stapled to the contact joint by prestressed rods and cables. Pretension leads in outer beams with the width of 0.75 meters and a thickness of approximately 0.3 meters. These beams are connected by desk with at least 60 mm thickness, which forms a walk-bearing surface. The surface of this area is fitted with sprayed anti-slip insulation finish in natural color of the original UHPC material. Each panel is reinforced by a pair of ribs located in about ¼ the length of the panel from each end. The utilization of the excellent properties of UHPC allows to design a very thin and lightweight construction of the

deck, which could not be achieved with the use of normal concrete [4-9].
Each panel contains an embedded steel weldment at a distance of 1.25 meters from the end of the panel, in which will be anchored the hanging rope. The total length of the panels is 5.65 m. End panels of the deck will be accompanied by a monolithic part for the realization of the restraint (connection of the bridge deck and abutments) and for the installation of bridge bearings and the expansion joint on the opposite side of the bridge. The deck therefore dilates its entire length to the one abutment. Dewatering of the bridge is provided by a free water drain from the surface. The deck will be illuminated by spotlights mounted in the construction of the railing, which has the height of 1.3 meters.

4 Static Design

Basic static scheme of the footbridge is formed by a beam suspended by rope elements (able to transmit only tension force) into the steel pylon, while the last hanger on the side of the shorter span is anchored to the abutment outside the supporting structure. On this side is the desk cantilevered to the abutment, which positively influences the dynamic properties of the system. On the opposite side is the structure deposited on the sliding joints (bearings) that allow the dilation of the structure. Computing model was created in Scia Engineer 2014, Civil Midas and Midas FEA [10]. For the making of the model have been used 2D shell finite elements. The analysis was based on the basic principle of the hangers acting as elements with a sagging which axial stiffness is affected by the magnitude of a tensile force. The calculation is, given the nature of the construction, materially and physically nonlinear.

The structure is loaded by permanent and variable load effects. Permanent loads include the self-weight of the construction and all additionally built parts. Variable loads represent loads caused by traffic on the bridge - pedestrians, vehicles up to 3.5 tons, and climatic loads - wind, snow and temperature. Pedestrian movement on the structure is simulated by moving surface load with size of 4.0 kN / m^2 and vehicle load is represented by moving two axle vehicles of total weight of 3.5 tons. Both loads are considered in the model so that their effect would be maximal due to the monitored variable - thus unloading effects are neglected.

Within the primary static calculation was mainly controlled the extreme normal stress at crucial parts of the structure to maintain the pressure reserve and tension in the walls of the pylon. Furthermore, the stability of the controlled compression of the steel pylon and two fundamental methods – Euler's concept of stability and the introduction of production inaccuracies into the shape the pylon. The result of the calculation was the need to fill the foot of the pylon with concrete to reduce the otherwise high number of reinforcements and prevent buckling of the walls. Further was also observed the deformation of the structure, taking into consideration that all shifts and deflections caused by the self-weight of the structure will be removed by the camber and simultaneously by prestressing of the hangers (refer to Fig. 1).

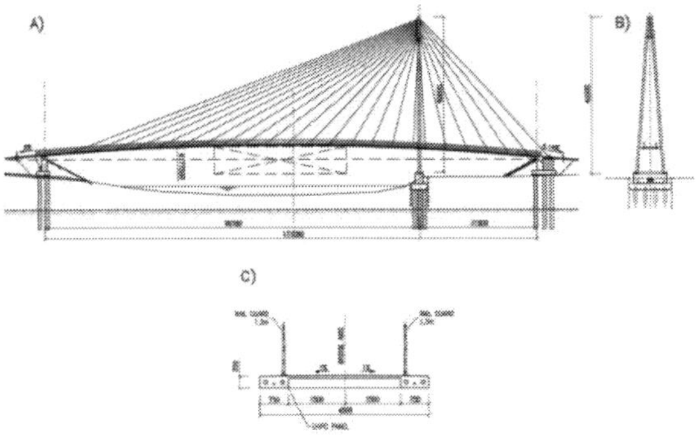

Figure 1. – A) Typical Cross-Section of the Deck, B) Longitudinal Section, C) Cross-Section.

At the very hangers occurs reducing of their axial stiffness due to their elongation. Introduced normal force is partially depleted by changing the geometry of the element until the moment when the cable is activated. From this point it can be assumed that the element behaves linearly and the full stiffness of the rope applies. This behavior is known as tension stiffening. Static design effort was to ensure such axial forces in the cables (by permanent loads, variable loads, but also the actual final tightening of the ropes) that during their minimal stress still remained sufficient tensile reserve in ropes, which would guarantee their linear behavior and thus the linear behavior of the structure. The model takes into account the influence of the construction process and the gradual tensioning hangers.

Tensioning of the cables, and the normal force in general applied in them has a beneficial effect of reducing tensile stresses arising in the deck of the footbridge from the extreme positions of the live load. Minimum required tensile strength, which ensures its linear behavior, was set for each rope by using the calculation based on its geometric characteristics. The effort is within the construction of the object to strain the ropes so that it acts almost linearly by minimizing their sag. From the static model, it is possible to determine that the ropes at the edge of the bridge on the side of his longer range will need to be prestressed, because of their angle does not allow the transfer of sufficient tensile normal force of self-weight of the structure to ensure their linear behavior (refer to Fig. 2, 3).

5 Dynamic Design

Basic static scheme of the footbridge is formed by a beam suspended by the rope elements (able to transmit only tension force) into the steel pylon, while the last curtain on the side of the shorter field is anchored to the abutment outside the supporting structure. On this side is the desk cantilevered to the abutment, which positively influences the dynamic properties of the system. On the opposite side is the structure deposited on the sliding joints (bearings) that allow the dilation of the structure. Computing model was created in Civil Midas and Midas FEA. For the making of the model have been used 2D shell finite elements. The calculation is, given the nature of the construction, materially and physically nonlinear.

Natural Frequencies and Mode Shapes of Proposed Structure are Listed below:

n.1 f=0,632 – deck bend / n.2 f=0,945 – deck torsion / n.3 f=1,098 – deck and pylon bend / n.4 f=1,508 – deck and pylon bend / n.5 f=1,864 – pylon bend / n.6 f=1,949 – deck and pylon bend / n.7 f=1,955 – deck torsion and longitudinal pylon bend / n.8 f=2,389 – deck torsion and longitudinal pylon bend/ n.9 f=2,758 – deck and pylon bend / n.10 f=2,981 – deck torsion and longitudinal pylon bend (Fig. 4).

Figure 2. Vertical Deformation of the Deck

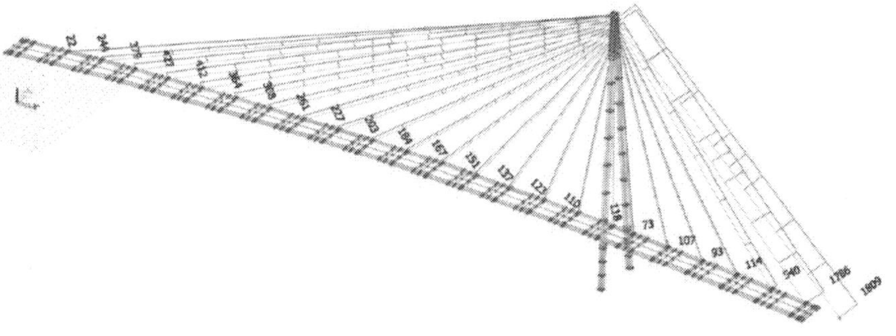

Figure 3. Normal forces in ropes from self-weight of the structure

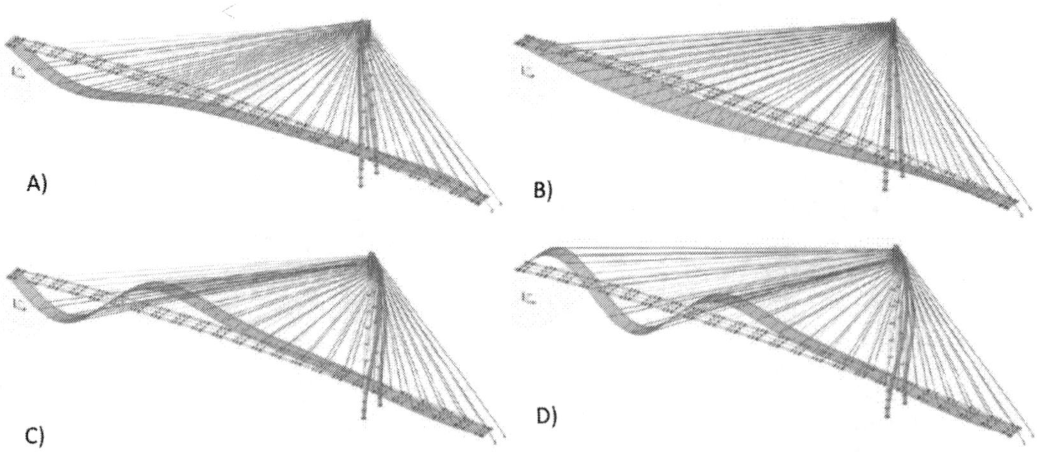

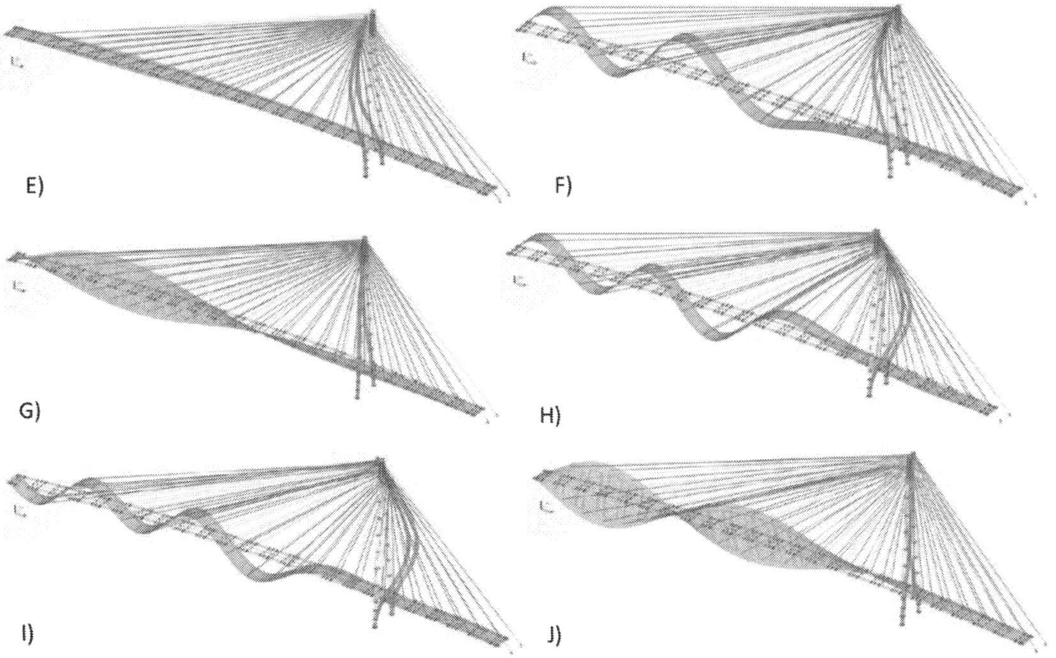

Figure 4. A) First mode shape, B) Second mode shape, C) Third mode shape, D) Fourth mode shape, E) Fifth mode shape, F) Sixth mode shape, G) Seventh mode shape, H) Eight mode shape, I) Ninth mode shape, J) Tenth mode shape

6 Conclusions

The basic static and dynamic analysis of the structure were performed and all kinds of load regarding the nature of the structure were considered. Due to the significant nonlinear behavior of components of the structure it was necessary to deal with the construction as a geometrically and physically nonlinear task.

Given the nature of the structure, it was made its basic modal analysis for the determination of natural frequencies and associated mode shapes. From the results it is possible to conclude the behavior of the structure and its response to load by pedestrians. With regard to the fact that it is a very lightweight construction, it is possible to predict from the modal analysis results, that after the dynamic load test (to determine the logarithmic decrement of attenuation) it will be necessary to fit a passive vibration damper on the construction, which will change the mass distribution on the structure.

The construction met all the criteria and it is possible to conclude, that the design is feasible and meets the requirements for this type of construction.

Acknowledgement

This research is supported by grant No. GAČR 15-2670S.

References

1. B. Graybeal, "Ultra-High Performance Concrete," FHWA Report HRT-11-038, March 2011
2. AFGC/SETRA; "Bétons fibrés à ultra-hautes performances. Recommandations." Documents scientifiques et techniques. Association Française de Génie Civil, Setra, Bagneux Codex, 2013.
3. JSCE-USC; "Recommendations for Design and Construction of Ultra-High Strength Fiber-Reinforced Concrete Structures – Draft."
4. SETRA/AFGC: Ultra High Performance Fibre-Reinforced Concretes: Interim Recommendations, 01/2002.
5. AFGC/SETRA: Ultra High Performance Fibre-Reinforced Concretes: Recommendations (06/2013). Paris CEDEX.
6. JSCE, Recommendations for Design and Construction of High Performance Fiber Reinforced Cement Composites with Multiple Fine Cracks, Rokugo K., (ed.) Concrete Eng. Series, 82, 2008.
7. Processings from HIPERMAT 3rd International symposium on UHPC and Nanotechnology for High Performance Material, Kassel, March 7-9,2012.
8. Kolísko, J., Rydval, M., Huňka, P. UHPC – Assessing the Distribution of the Steel Fibre and Homogeneity of the Matrix. fib Symposium Tel Aviv, Tel Aviv, Israel. 2013. ISBN 978-965-92039-0-1.
9. Paulík P.: The effect of curing conditions (in situ vs. laboratory) on compressive strength development of high strength concrete. Procedia Engineering, Vol. **65**, 2013, Pages 113-119.
10. FIB Model Code for Concrete Structures 2010, ©2013 fib, Ernst & Sohn.

MATEC Web of Conferences 77, 08006 (2016) DOI: 10.1051/matecconf/20167708006
ICMMR 2016

Effects of Dental Implant-abutment Interfaces on the Reliability of Implant Systems

Xiao Zhang[1], Lei Liu[1,2], Yaling Wang[1], Xianshuai Chen[1,3]

[1] *Shenzhen Institutes of Advanced Technology, Shenzhen, China, 518055*
[2]College of Mechanical Engineering, Shanghai University of Engineering Science, Shanghai, China, 201620
[3] Foshan Angels Biotechnology Co., Ltd., Foshan, China, 528000

Abstract. In this paper, by analyzing the effects of two different kinds of implant-abutment connection interfaces under the same working condition on the mechanical and fatigue performances of the implant system as well as on the surrounding bones, we intend to study such effects on the reliability of the implants and provide a theoretical basis for the design and clinical application of dental implant systems. For the purpose, we adopt a 3-D modeling method to establish the model, and use FEA (finite element analysis) to carry out static mechanic and fatigue analysis on the implant system and its surrounding bones; then we make the two implant systems, and carry out fatigue tests on a dynamic fatigue testing machine to verify the FEA results. After comparing the results from the two different systems, we find that the stress distribution and fatigue safety factor of the system which has deeper axial matching of the taper connection are better than those of the other system, that is to say, between the two major elements of a implant system, the axial length of the connecting taper and the size of the hexagon, the former has greater effects than the latter. When the axial matching is deeper, the stress distribution of the implant system will be better, the fatigue safety factor will be higher, and the implant system will be more reliable.

1 Introduction

Many factors may influence the success of dental implantation; however, the reliability of the implant system itself is one of the most important [1-4]. At present, a dental implant system is generally assembled of the implant, the abutment and the central screw, forming a two-part structure. The reliability of an implanted system lies mainly in two factors: the force bearing capacity and the fatigue resistance during continuous bites. Therefore, a good design of implant system is bound to make the system get a better stress distribution and a greater fatigue safety factor under forces. The implant-abutment connection interface, i.e. the connecting and contact means between the implant and the abutment, is therefore the key factor of the reliability of the implant system. Today most of the relatively good products on the market adopt the implant-abutment connection interface that uses a taper surface and a hexagon connection matching [5, 6]. However, there is no clear conclusion regarding how to optimize the interface design and what kind of factors having the greater effect on the reliability of implant systems.

On the basis of the research on current designs, in this paper, we carry out analysis on the effects of two different kinds of implant-abutment connection interfaces on the stability of the systems, so as to optimize the design and make choices between the two systems. First we use the CAD software SolidWorks to establish two implant system models and the corresponding alveolar bone models based on two different kinds of implant-abutment connection interfaces, then we apply ANSYS Workbench to conduct FEA analysis on the systems and alveolar bones for static force analysis and cycle fatigue analysis, and finally we make sample pieces and carry out dynamic fatigue test to study the stress distribution and fatigue resistant characteristics of the systems, and verify the influencing factors of the implant-abutment connection interface on the reliability of implant systems.

2 Design of the Dental Implant-Abutment Connection Interface and the Implant System

In the design of dental implants, we mainly take into account the interior hexagon angle positioning, the platform transfer and the Morse tapered connection. The connection interface between the implant and the abutment is just the connection between the hexagon and the taper. Figure 1(a) shows the connection interfaces of #1 implant system and #2 implant system as well as their sizes. The taper angles of both systems are 24°, the axial match heights are respectively 1.25mm and 1.65mm, the distances between opposite sides of the hexagon are respectively 2.64mm and 2.5mm, and the length and diameter of both implants as well as other dimensions of the abutments are the same. That is to say, the two

DOI: 10.1051/matecconf/20167708006

implant systems mainly differ in the axial height of the taper and the size of the hexagon.

The exterior thread of the implant is designed to be trapezoidal, with the pitch (P) of 0.6mm, the diameter of 4.3mmm and the length of 11.5mm, while the diameter of the abutment neck is 5mm; the central screw is 8mm long, with standard thread M2. The size of this design is the most commonly used on the market [6, 7]. The crown bears the main load. The simulated crown consists of a cylinder and a hemisphere: the cylinder is 10mm in diameter and 5mm in length, while the hemisphere is 10mm in diameter [8]. The modeling of the shape of the alveolar bone tissue is based on the CT scanning data of human lower jaws, including two parts: the cortical bone and the cancellous bone. The crown is wrapped with the cortical bone, which is 2mm thick [6]. By the SolidWorks software, we now establish the component models and analytical models above, as shown in Figure 1(b)(c).

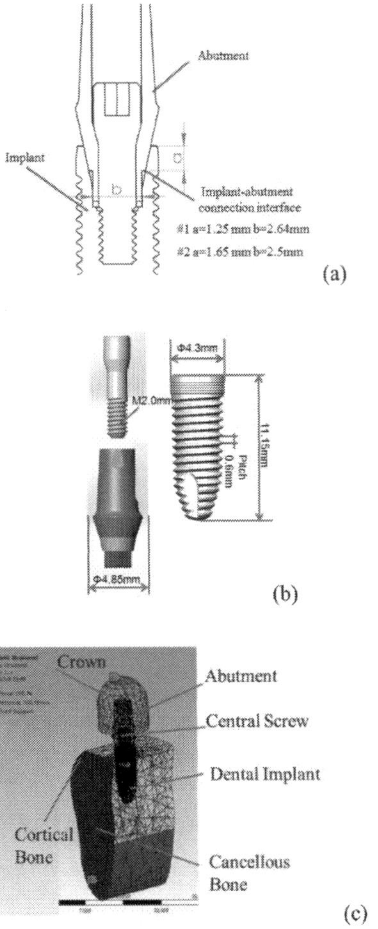

Figure 1. (a) Implant-abutment connection interface; (b) (c) Diagrams of implant system analysis models

3 Finite element analysis (FEA)

3.1 Materials and Methods.

Through the CAD interface of the ANSYS software, we import the file into the FEA software ANSYS Workbench for static and fatigue analysis. In this research we assume that all the tissue and materials in the models are continuous, homogeneous and isotropic linear elastic materials. The materials for the implant systems are titanium alloy Ti6Al4V for its best biocompatibility, and the crowns adopt zirconia [6, 9]. Specific material properties are shown in Table 1.

Table 1. Material Properties

Material	The young's modulus (E, GPa)	Poisson's ratio (V)
Cortical bone	14	0.30[6]
Cancellous bone	3	0.30[6]
Ti6A14V	110	0.35[6,9]
Zirconia	70	0.28[6,9]

3.2 Static Analysis and Fatigue Analysis on the Implant Systems

This FEA simulates the working conditions under which the implant system and the alveolar bone are in full osseointegration. The lower of the alveolar bone is a fixed constraint, the contact between the implant and the abutment, and the contact between the alveolar bone and the implant are both assumed to be fixed [5, 6]. The force direction is an angle of 30 degrees from the axial direction of the implant, and in the test we use a force of 200N, with a pre-tightening torque of 30N.CM on the central screw [5-7]. What the working condition for fatigue analysis is that under the conditions defined by the above constraint and contact, in the force direction, load the forces from 20N to 200N in sine waves, with the amplitude of 90N and frequency of 15Hz. An acceptable sample should not be damaged or yield under $5X10^6$ times of cyclic loading [8], as shown in Figure 2.

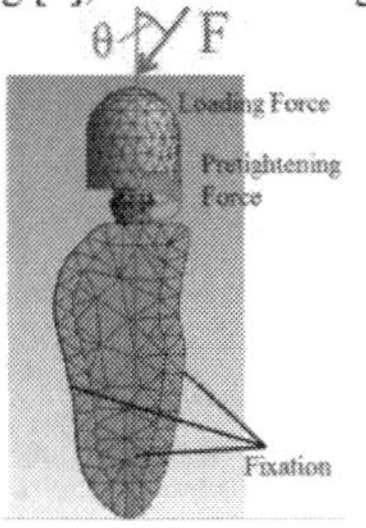

Figure 2. The FEA model and the working condition of an implant system and the alveolar bone

3.3 Results

An implant system, which has better stress distribution and greater fatigue safety factor under loading, will have better stability. As the implant-abutment connection interface is the relatively weak part of a system, so we focus on the stress distribution and the fatigue safety factor of the interface, and through the comparison analysis on the two interfaces, we study their effects on the stability of the whole implant system as well as on the stresses of surrounding bones. The two kinds of interfaces and their equivalent stress contours are shown in Figure 3, while the results of analysis on fatigue safety factors are shown in Figure 4.

As can be seen from Figure 3, the maximum stress of #1 implant system is 289Mpa, occurring at the edge of the taper of the implant-abutment connection interface. The minimum stress at the implant-abutment connection interface is 1.8MPA, where which at the connection between the abutment and the crown; the maximum stress of the alveolar bone is 48.3 MPA, where which at the edge of its upper end connecting to the implant. The maximum stress of #2 implant system also occurs at the edge of the taper of the implant-abutment connection interface. The maximum stress at the implant-abutment connection interface is 267MPA, and the minimum stress is 1.44Mpa, where which at the connection between the abutment and the crown; the maximum stress of the alveolar bone is 48MPA, where which at the edge of its upper end connecting to the implant.

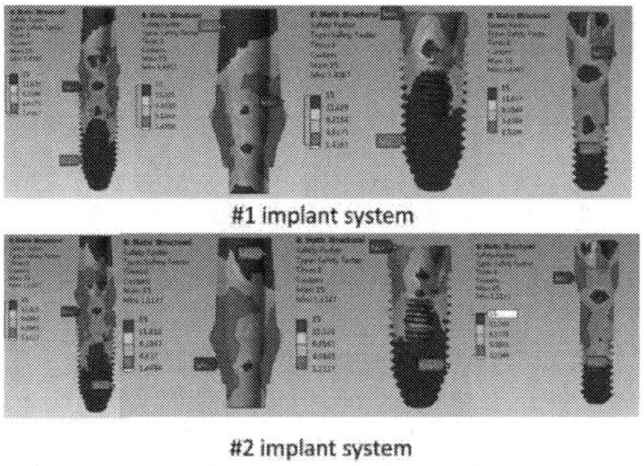

Figure 3. Two kinds of implant-abutment connection interfaces and their equivalent stress contours

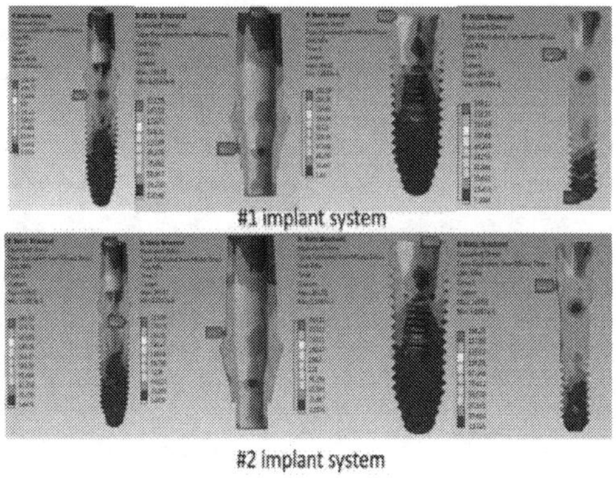

Figure 4. Fatigue safety factors

As can be seen from Figure 4, the minimum safety factor of #1 implant system is 1.1, where which at the edge of the taper of the implant-abutment connection interface, and the maximum safety factor is 15; the minimum safety factor of #2 implant system is 1.4, where which at a position the same as that of the implant-abutment connection interface of #1 implant system, and the maximum safety factor of the implant-abutment connection interface is 15.

4 Fatigue test

From the FEA analysis above, we get the diagram of effective stress distribution and the fatigue safety factor. Now we use the Willemin-Macodel 408MT 7-axis turning-milling machine center to make the two kinds of implant systems, as shown in Figure 5. Then we adopt the fatigue test method [8] which is commonly used internationally for dental implant system to carry out the fatigue test. Figure 6 shows the fatigue test program and the test site. The test machine is BOSE's dynamic fatigue test machine, 3510-AT, and the force sensor precision is ±0.5%. The rigid fixture is used to position and clamp the implant. The head bearing device is a stainless steel hemispheric cap, with a size the same as the crown in the FEA analysis. There is no constraint in the horizontal direction at the contact position between the loaded parts and the hemispheric cap, with the loaded force, frequency and cycle period the same as in the FEA working condition.

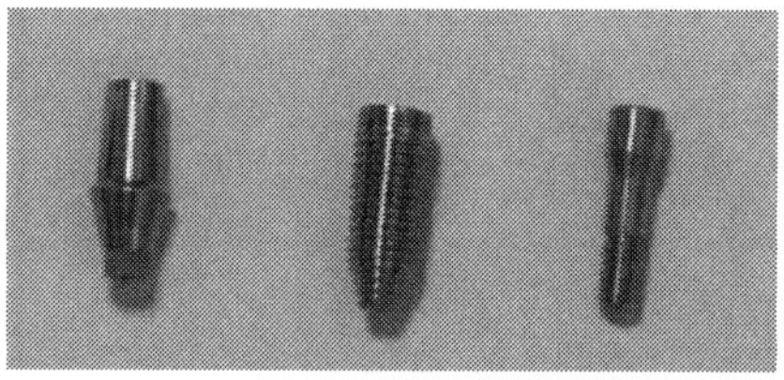

Figure 5. Components of an implant system

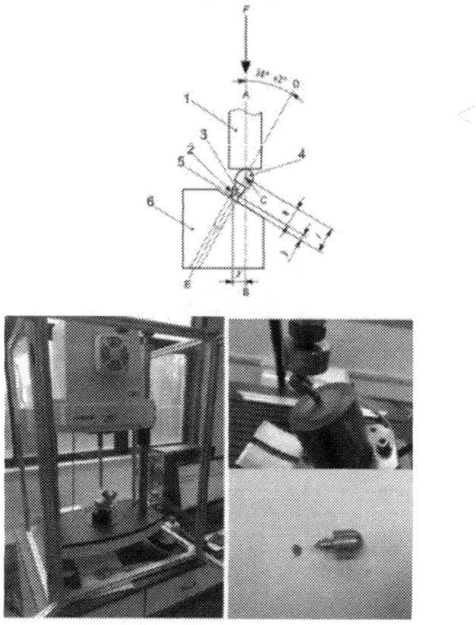

1. Loading device; 2. Nominal bone level; 3.Connecting part; 4.Hemispherical loading member; 5.Dental implant; 6. Specimen holder; 7. L=11mm; y = l×sin30 = 5.5mm.

Figure 6. Fatigue test program and the test site

The test results are shown in Table 2 below. #1 implant system is damaged, as shown in Figure 6, while #2 results in no damage. As can be seen from the comparison of FEA analysis results, #1 implant has a lower safety factor, although not yet reaching a degree of damage. The damage of the implant-abutment connection interface in #1 implant system during the test may relate to the taper matching precision, yet the overall results are consistent with the FEA analysis results, that is to say, the damaged part is basically the same as the position of the maximum stress at the implant-abutment connection interface in the FEA analysis. This not only further demonstrates the feasibility of the FEA method, but also proves that the #2 implant-abutment connection is better.

Table 2. Test experiment results

	Load (N)	Cycle(times)
#1 implant system	200	3,934,006
#2 implant system	200	5,000,000

5 Discussion

For further study of the effects of implant-abutment connection interfaces on the reliability of implant systems, we get the statistics of all the maximum equivalent stresses for the components under the above working conditions, as shown in Table 3.

Table 3. The statistics of the maximum stresses for implant systems and alveolar bone

Number	Force (N)	Force angle (degree)	Implant (Mpa)	Abutment (Mpa)	Central screw (Mpa)	Cortical bone (Mpa)	Cancellous bone (Mpa)
#1 implant system	200	30	289	221	154	48.3	9.8
#2implant system	200	30	269	215	149	48	9.8

From the above table and figure we can learn, in the implant-abutment connection interfaces, the stress of the taper bears is larger, and the stress of the hexagon bears is comparatively smaller, which means it is the taper that bears much of the loads from the outside. Meanwhile, when comparing the features, the force distribution and the fatigue characteristics between the two implant-abutment connection interfaces, we find that the axial height of the taper matching of #2 interface is 0.4mm greater than #1 and its hexagon size is 0.14mm smaller than #1, but the implant system bears smaller stress. This indicates that the anti-pressure ability of the taper completely counteracts and exceeds the effects caused by the size reduction of the hexagon, that is to say, the taper plays a more decisive role on the reliability of the implants, and furthermore, it has more obvious effect on reducing the long-term stress of the surrounding alveolar bone.

As for the above two implant-abutment connection interfaces, #2 has a better design, its implant system has better stress distribution, fatigue resistance, reliability, and better biomechanical reaction with the surrounding bones.

6 Conclusion

To sum up, by comparing the two implant systems of different implant-abutment connection interfaces, we can draw the following conclusions:

Under the same simulated occlusal loading conditions, between the two major elements of implant-abutment connection interfaces, the axial matching depth of the taper connection has greater effects on the reliability of the implants system than the size of the hexagon. To be specific, when the axial matching is deeper, the stress distribution of the implant system will be better, the fatigue safety factor higher, and the implant system more reliable;

The results of FEA fatigue analysis are consistent with that of simulation experiments, and can therefore provide a reference for the optimized design of implant systems.

Acknowledgments

This work is supported by the Science and Technology Program of Shenzhen, China (No. JCYJ20140617143643469), and Foshan Innovative and Entrepreneurial Research Team Program, China (No. 2014IT100102).

References

1. M. I. El-Anwar, S. A. Yousief, T.A. Soliman, et al: A finite element study on stress distribution of two different attachment designs under implant supported overdenture. The Saudi Dental Journal .Vol. 27 (2015), p.201–207
2. M. Degidi, G. Daprile, and A. Piattelli, Determination ofprimary stability: a comparison of the surgeon's perception andobjective measurements, The International Journal of Oral &Maxillofacial Implants, vol. 25 (2010),p.558–561
3. S.-J. Ahn, R. Leesungbok, S.-W. Lee, Y.-K. Heo, and K. L.Kang, Differences in implant stability associated with variousmethods of preparation of the implant bed: an in vitro study,Journal of Prosthetic Dentistry, vol. 107 (2012),p.366–372
4. N. Ferrigno, M. Laureti, and S. Fanali, Dental implants placementin conjunction with osteotome sinus floor elevation: a12-year life-table analysis from a prospective study on 588 ITI implants,

 DOI: 10.1051/matecconf/20167708006

Clinical Oral Implants Research, vol. 17 (2006), p.194–205

5. G. Mehdi, A. Belarbi, B. Mansouri, et al: Numerical study of effect of elastomeric stress absorbers on stress reduction in bone-dental implant interface. J Appl Oral Sci. Vol. 23 (2015), p.87-93
6. X. Zhang,Z. Xie, et al: Effects of custom-made and thread dental implant system on the stress distribution in alveolar bone: a 3-dimensional finite element analysis. Applied Mechanics and Materials. Vols. 475-476 (2014), p.1487-1493
7. Straumann Product Catalog, Information on: http://www.straumann.com/
8. International Organization for Standardization: ISO 14801-2007 Dentistry — Implants — Dynamic fatigue test for endosseous dental implants, IHS under license with ISO (2007).
9. L-Y.Jia: Investigation of Fatigue Properties and Fatigue Design Diagram of Titanium alloy Ti-6A1-4V. East China University of Science and Technology (2014).

Indigenous Design for Automatic Testing of Tensile Strength Using Graphical User Interface

Rafay Ali , Faraz Junejo, Rafey Imtiaz and Usama Sultan Shamsi

Shaheed Zulfikar Ali Bhutto Institute of Science and Technology, Mechatronics Engineering Department, Karachi, Pakistan

Abstract. Tensile Testing is a fundamental material test to measure the tenacity and tensile strength. Tensile strength means ability to take tensile stress. This Universal Testing Machine is designed using Dual Cylinder Technique in order to comply with the maximun load (tensile force) with the reduction of minimum physical effort and minimized losses.It is to provide material testing opportunity to the students of different institutions, locally and globally, at lowest price; so that they can have a comprehensive understanding of the testing procedures and examining material properties on practical grounds with a minimum expenditure. The testing mechanism is automated along with the inclusion of all necessary parameters and the movement of the members. Specimens of different engineering materials were used to carry out the test and their mechanical strength was tested and compared with the actual values.

1 Introduction

New methodologies to measure the physical properties of thin films are currently required. Particularly, reported mechanical properties of materials at these dimensions are currently controversial in the scientific literature. Thus, it is necessary to propose techniques for determining mechanical properties of thin films, such as elastic modulus, Poison's ratio and strength. Properties of materials at micro and nanoscale are of considerable interest because of the unique properties associated with small volumes.

Tensile testing is an effective way to investigate the mechanical properties of materials and it is a well-established technique for bulk sample characterization. However, tensile testing is not easily implemented for micro and nano-structured materials due to the small dimensions of the specimen. At present, there exist the testing machines for materials, are expensive and its flexibility to make modifications is limited.

An important feature of this device is the simplicity to exchange components according to the user requirements: the low cost, low machine compliance and the high resolution obtained. This machine can also be adapted for compression testing with appropriate samples and grips.

These universal testing machines are specifically designed for demanding testing situations and are suitable for applications from all fields, whether for quality control or for research projects. All Universal Testing Machines can be modified as required to suit individual requirements.

Commercial testing machines with the required capabilities are available, however they are marginally costly. The tester will need to be assembled for a fraction of the cost of an off the shelf tester while retaining the same capabilities as commercially available systems [1].

Figure 1. Real design solid work concept

2 Materials & method

Fig. 1 presents a schematic of the Mechanical support base which is assembled with the cross member shown. The cross-member slides through the cross-head by means of the guide rods. The structural components of the testing system are used to subject material specimens to tensile or compressive forces from which we can derive very useful material-specific properties in the observations known as a stress strain plot. While the motion control and data acquisition is done through the computer. A servo motor provides the necessary torque to turn the master cylinder of the hydraulic actuator [2]. The overall electrical system consists of a servo motor and driver circuit, an optical encoder, an amplifier, and a controller card. The function of the encoder is to measure the rotation of the servomotor. The controller and a PC-bus plug-in interface card connect the computer to the mechanical testing system. It is planned to connect

Analog inputs such as load cell, displacement, and strain gauge to this card using the auxiliary channels for data acquisition [3].

The machine is capable of applying a tensile force of 5KN (limiting force) and is operated using hydraulic oil. The test process involves placing the test specimen in the testing machine and slowly extending it until it fractures. During this process, the elongation of the gauge section is recorded against the applied force. The data is manipulated so that it is not specific to the geometry of the test sample. The elongation measurement is used to calculate the engineering strain, ϵ, using the following equation (Eq. 1):

$$\epsilon = \delta L/L\_0 = (L-L\_0)/L\_0 \quad (1)$$

where δL is the change in gauge length, L_0 is the initial gauge length, and L is the final length.

The force measurement is used to calculate the engineering stress, σ, using the following equation (Eq. 2):

$$\sigma = F/A \quad (2)$$

where F is the tensile force and A is the nominal cross-sectional area of the specimen. The machine does these calculations as the force increases, so that the data points can be graphed into a stress–strain curve.

The shoulders of the test specimen can be manufactured in various ways to mate to various grips in the testing machine. Each system has advantages and disadvantages; for example, shoulders designed for serrated grips are easy and cheap to manufacture, but the alignment of the specimen is dependent on the skill of the technician. On the other hand, a pinned grip assures good alignment. Threaded shoulders and grips also assure good alignment, but the technician must know to thread each shoulder into the grip at least one diameter's length, otherwise the threads can strip before the specimen fractures [4].

In large castings and forgings it is common to add extra material, which is designed to be removed from the casting so that test specimens can be made from it. These specimens may not be exact representation of the whole workpiece because the grain structure may be different throughout. In smaller workpieces or when critical parts of the casting must be tested, a workpiece may be sacrificed to make the test specimens. For workpieces that are machined from bar stock, the test specimen can be made from the same piece as the bar stock [4].

Alignment of the test specimen in the testing machine is critical, because if the specimen is misaligned, either at an angle or offset to one side, the machine will exert a bending force on the specimen. This is especially bad for brittle materials, because it will dramatically skew the results [5]. This situation can be minimized by using spherical seats or U-joints between the grips and the test machine. If the initial portion of the stress–strain curve is curved and not linear, it indicates the specimen is misaligned in the testing machine. The strain measurements are most commonly measured with an extensometer, but strain gauges are also frequently used on small test specimen or when Poisson's ratio is being measured.

3 Implementation

Working Principle. The working principle of our project includes the mechanism of rotating the screw jack mechanically using the handle provided with a key so that it remains fixed within. When the screwjack rotates, the pressure is exerted though a master cylinder, attached directly with the screw jack, and the force is transmitted through the cylinder to the crosshead.

The force member (crosshead) moves in the upward direction exerting the force on the load-cell and ultimately exerting the tensile force on the specimen attached. The force is continuously exerted till the specimen breaks and fracture strength is achieved. This stepwise increment of the force and its effects on the specimen are recorded as a data set of values and computation is being simultaneously done on every data set using our created Graphics User Interface (GUI).

The set of observations are recorded in tabular form and a stress-strain curve is plotted with the help of these values using a Graphics User Interface (GUI). The machine must have the proper capabilities for the test specimen being tested. There are four main parameters: force capacity, speed, precision and accuracy. Force capacity refers to the fact that the machine must be able to generate enough force to fracture the specimen. The machine must be able to apply the force quickly or slowly enough to properly mimic the actual application. Finally, the machine must be able to accurately and precisely measure the gauge length and forces applied; for instance, a large machine that is designed to measure long elongations may not work with a brittle material that experiences short elongations prior to fracturing[6].

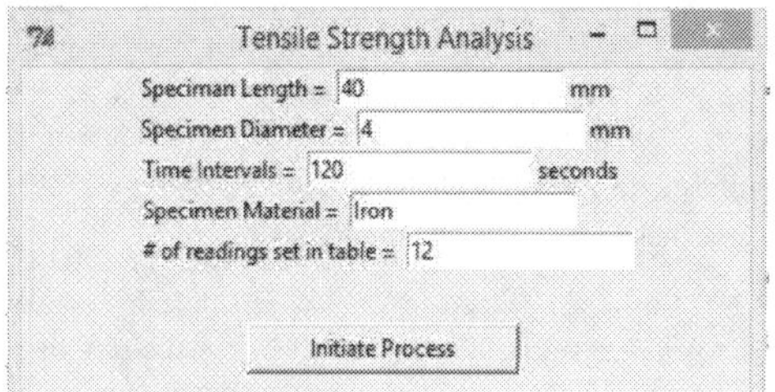

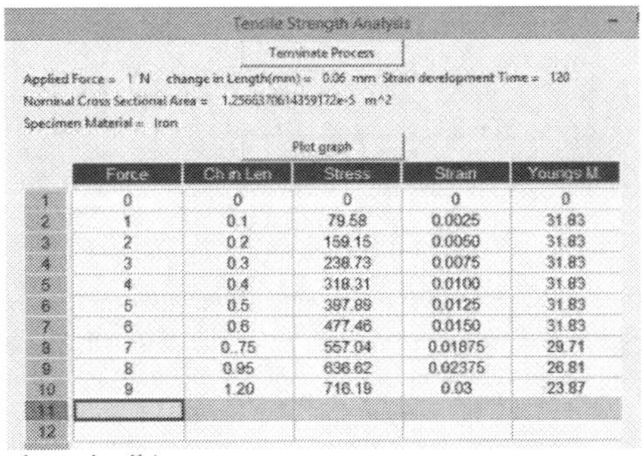

	Force	Ch in Len	Stress	Strain	Youngs M
1	0	0	0	0	0
2	1	0.1	79.58	0.0025	31.83
3	2	0.2	159.15	0.0050	31.83
4	3	0.3	238.73	0.0075	31.83
5	4	0.4	318.31	0.0100	31.83
6	5	0.5	397.89	0.0125	31.83
7	6	0.6	477.46	0.0150	31.83
8	7	0..75	557.04	0.01875	29.71
9	8	0.95	636.62	0.02375	26.81
10	9	1.20	716.19	0.03	23.87
11					
12					

Figure 2. Graphics user interface (Specimen details)

4 Analysis

Before initiating the tensile test, user-defines important experiment parameters in relevant fields.

This prompts a new window tabulating the readings based on first screen. As we see initially, the fields are empty but they will populate after each time interval passes. These are processed readings gained from the Loadcell and LVDT sensors. The Young's modulus is then calculated from the stress and strain values. The process is followed by the set of observations taken with constant increment in the force applied on the attached specimen as seen in Fig. 2.

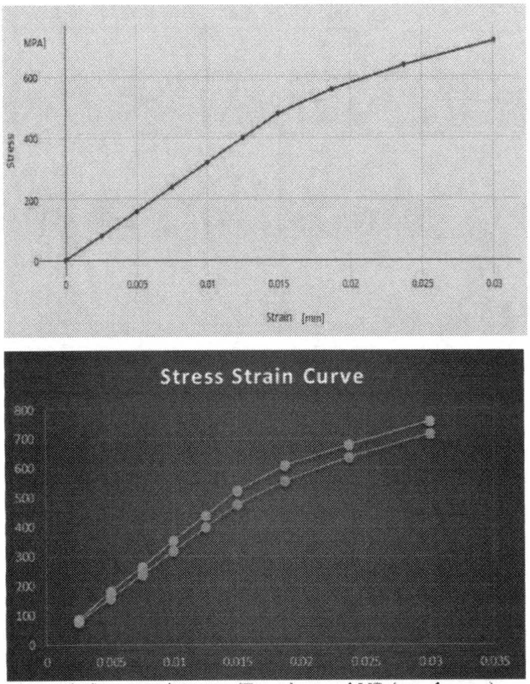

Figure 3. Stress strain curve (Experimented VS Actual curve)

Through these set of observations the stress strain curve is plotted. The graph in Fig. 3 below shows the comparison between the experimented values (blue) and the tested values (orange) as plotted.

The percentage error between the values can be calculated by using the following formula

%error=|(experimentalvalue-theoretical value)/(theoretical value)|
%error= |(79.85-85)/(85)|
%error= 6.05%

5 Results

The performance of this Universal Testing Machine indicates that it is appropriate to obtain reliable mechanical properties of compliant materials in thin and soft materials. The testing phase was completed for elastic deformation of the steel specimen by the end of the project timeline. Of these, Iron, Steel and Aluminum specimen were acquired. For a grade450 steel specimen, we can read off yield strength value as 450 Mpa. If we observe the graph generated by the machine we see that nonlinearity occurs at approximately 480 Mpa. Furthermore, the obtained observation set of values is compared to the actual available values and the error is calculated. The error is found to be 6%.However, the environment temperature, purity of material specimen and other factors could lead to slight deviation and minimizing these factors to controllable extent could increase the machine's performance and optimum results. The standard set of values has been referred from Shigley's, 8th edition [7].

6 Conclusion

This Automated UTM is designed to provide a helping hand to the students and personnel associated with the Mechatronics Engineering field and to provide them the better understanding of the UTM machine of international standard. Secondly, provide Pakistan this machine, manufacturing it locally at a very reasonable cost with exceptional functionality and automated mechanism. The stated objective was to design and build a low-cost, compact prototype for conducting axial load testing, should also act as a platform for future expansion, incorporating a modular design and flexible fixture mounting points to allow for the addition of new capabilities as needed. Furthermore, this provides us with an opportunity to be the entrepreneurs with its development to the industrial scale. This will provide a means of employment opportunities involved in the development when taking it to the mass manufacturing level.

The system was envisioned to provide additional low-end support to the students of our University. Thus, the aim, set to overcome the difficulties faced by the students in understanding the actual concept of behavior (elastic and plastic) exhibited by different engineering materials (ductile and brittle) under the action of applied load along with the integration of different electronic equipment: sensors (loadcell and LVDT), so that they can have a visual representation of the different regions of the material under load. Yet, as the project evolved and the promising prototype that we saw coming together, this project has the potential to provide undergraduate students hands-on experience in setting up and conducting mechanical testing.

References

1. Theusen, G. J. and Fabrycky, W. J. Engineering Economy. 6th edition. Englewood Cliffs, N. J.: Prentice-Hall. 1984.
2. Juvinall, R. C., and Marshek, K. M., 1991. Fundamentals of machine component design, John Wiley & Sons, Inc., NY, pp. 349.
3. Information on http://library.utm.utoronto.ca/
4. Standard test method for tensile properties of plastics, ASTM International
5. Wang, L., Lau, J., Thomas, E. L., and Boyce, M. C., 2011.Co-continuous composite materials for stiffness, strength, and energy dissipation, Advanced Materials **23**(13), April, pp. 1524-1529.
6. Billings. S. A. Identification of Nonlinear Systems: A Survey. Proc. Instn Electr. Engrs, Part D, **127**(6): 272-284, (1980).
7. Budynas, R. G., and Nisbett, J., K., 2008. Shigley's Mechanical Engineering Design, 8th edition. McGraw-Hill, NY, pp. 1006.

Si/SiC heterojunction optically controlled transistor with charge compensation layer

Pu Hongbin, Wang Xi and Chen Zhiming

Department of Electronic Engineering, Xi'an University of Technology, Xi'an 710048, China

Abstract. A novel n-SiC/p-Si/n-Si optically controlled transistor with charge compensation layer has been studied in the paper. The performance of the device is simulated using Silvaco Atlas tools, which indicates excellent performances of the device in both blocking state and conducting state. The device also has a good switching characteristic with 0.54μs as rising time and 0.66μs as falling time. With the charge compensation layer, the breakdown voltage and the spectral response intensity of the device are improved by 90V and 33A/W respectively. Compared with optically controlled transistor without charge compensation layer, the n-SiC/p-Si/n-Si optically controlled transistor with charge compensation layer has a better performance.

1 Introduction

With electronic products are wide application, electromagnetic interference (EMI) effects become a critical issue for safety and reliability in power electronic systems. Photoelectric isolation method is an effective way that used to resolve electromagnetic interference problem[1]. Although Si-based light controlled transistor has been developed for many years, its physical properties have obvious limitations in high temperature, high frequency and high power applications. With large wafer-size silicon carbide (SiC) wafer localization and mature of high quality 6H-SiC and 4H-SiC epitaxial technology, SiC power devices set off an upsurge study.

However, SiC is not optically active at the near IR wavelength range, where light sources are readily available for optical communication[2]. In order to realize light-activation of SiC power switching devices, a hybrid approach that combines a silicon photo-receiver module with a SiC power darlington transistor have been proposed[3]. In this hybrid Si-SiC device, the darlington transistor is activated by photocurrent output from an individual Si photodiode. Therefore, the EMI problem may still exist because the direct input of the darlington is actually an electronic signal. To prevent the EMI problem absolutely, we proposed light-activated SiC transistors in earlier work[4-5]. In which by using narrow bandgap material SiCGe as base layer, the device can realize light-activated power switches for common light source.

In this paper, a novel optically controlled SiC power transistor structure based on the 4H-SiC/Si heterojunction and charge compensation layer is proposed to improve spectral response range and efficiency of the device. By employing the device simulator Silvaco Atlas, we have simulated the performances of this device based on SiC/Si heterojunction and charge compensation layer.

2 Device structure

The structure schematic of n-SiC/p-Si/n-Si heterojunction optically controlled transistor with and without compensation layer are shown in figure 1. The emitter electrode and collector electrode are locating at left and right side respectively. The optical window between emitter and collector is passivated by Si_3N_4 to reduce reflectivity of light. The heavily doped 4H-SiC is used as emitter region material. The emitter region of device is 3.0μm in thickness and 2.0μm in width. The n-SiC/p-Si heterojunction in the device working as emitter junction can improve the injection efficiency of the emitter and reduce the energy of the triggering light. To wider the light response range, silicon is used as material of base region and collector region. The base region is 3.0μm in thickness and 2.0μm in width. The base current is supported by photocurrent of collector junction. The width of collector region is 17μm. In order to improve the light absorption efficiency, the thickness of the collector region must be longer than the depth of light absorption. As shown in figure 1(b), the thickness and doping concentration of compensation layer under the silicon collector region should be designed basing on charge-balance theory. Under the effect of compensation layer, there is a two-dimensional electric field in silicon collector region. The two-dimensional electric field can help improve the blocking voltage and spectral responsivity of the transistor. To avoid charge imbalance affected by the substrate-assisted depletion, the semi-insulating SiC is used as substrate.

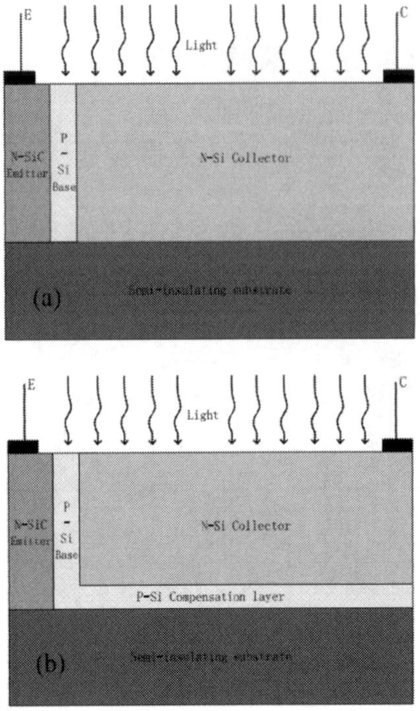

Figure 1. Structure schematic of the Si/SiC heterojunction optically controlled transistor (a) without charge compensation layer (b) with charge compensation layer.

3 Simulation and discussion

In order to achieve the realistic results, several important material parameters used in Silvaco Atlas were adjusted to obtain the closest agreement with published material data of 4H-SiC[6-8]. The major parameters used in the paper are shown in table 1.

Table 1. Major parameters used in the paper.

Parameter	value	unit
Eg300	3.26	eV
Affinity	3.8	eV
Eab	0.2	eV
Edb	0.1	eV
Augn	5×10^{-32}	cm^6/s
Augp	2×10^{-32}	cm^6/s
Permittivity	9.66	-
Gvb	4	eV
Gcb	2	eV

When the device is under blocking condition, there is a forward bias voltage fall over the collector junction. As the doping concentration of collector region is lighter than the doping concentration of base region, the depletion layer is mainly broadening in collector region. The three-dimensional electric field distributions of the optical transistors under the blocking condition both with and without compensation layer are shown in figure 2. As the comparison between the electric field of two devices in figure 2(a) and figure 2(b), the electric field distributes more uniformly in figure 2(b). Due to the existence of charge compensation layer, the depletion region in figure 2(b) is broadening in two directions. The charge compensation layer helps extend the collector junction in horizontal direction, which is helpful to form a more electric field in vertical direction. As the device with the charge compensation layer is designed according with the charge-balance theory (qN_D=-qN_A), the 2D Poisson equation can be write as

$$\nabla^2\phi = \frac{\partial^2\phi}{\partial x^2} + \frac{\partial^2\phi}{\partial y^2} = -\frac{\rho}{\varepsilon} = -\frac{qN(x,y)}{\varepsilon} = -\frac{q(N_D + N_A)}{\varepsilon} = 0 \quad (1)$$

where ρ is the charge density. The electric field in drift region can be write as

$$E = \nabla\phi = \text{constant} \quad (2)$$

which means the drift region is completely exhausted and the distribution of the electric field should be more uniform.

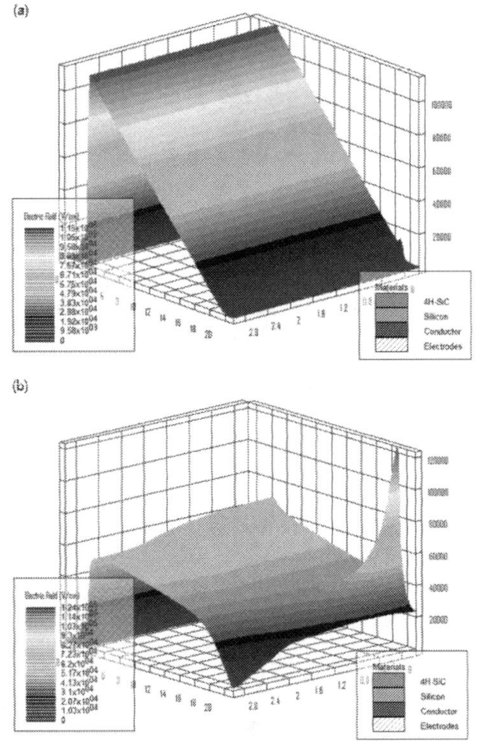

Figure 2. Electric field distribution of Si/SiC heterojunction optically controlled transistor (a) without charge compensation layer (b) with charge compensation layer.

As two-direction distributed electric field of optical transistor with charge compensation layer is more

 DOI: 10.1051/matecconf/20167708008

uniform than the one without charge compensation layer, the transistor with charge compensation should have higher breakdown voltage and the breakdown voltage should be more stable when the doping concentration of collector region fluctuating. Simulation results of blocking characteristics for two transistors with different collector region doping concentration are shown in figure 3. As the existence of charge compensation layer, the uniformly distributed electric field makes the breakdown voltage of transistor with charge compensation layer be higher than transistor without charge compensation layer. And the breakdown voltage of transistor with charge compensation layer is more stable than transistor without charge compensation layer when the doping concentration of collector region is changing from $5\text{x}10^{14}\text{cm}^{-3}$ to $1\text{x}10^{16}\text{cm}^{-3}$.

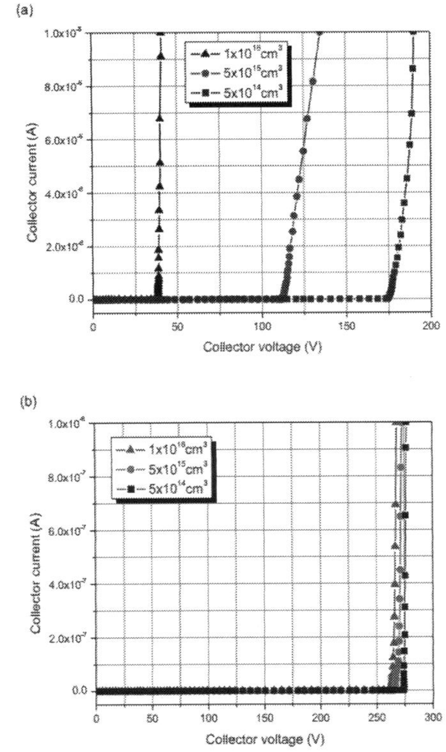

Figure 3. Blocking characteristics of Si/SiC heterojunction optically controlled transistor for different collector region doping concentration (a) without charge compensation layer (b) with charge compensation layer.

From the simulation results, the spectral response of the optically controlled transistor under 5V bias voltage with and without charge compensation layer are shown in figure 4. The device with charge compensation layer has higher spectral response than the device without charge compensation layer. With charge compensation layer, the peak value of optically controlled transistor's spectral response is improved to 130A/W at the wavelength of 550nm. As silicon is used as light absorption material, the response range contains almost all range of the visible spectrum, which is from 400nm to 1100nm.

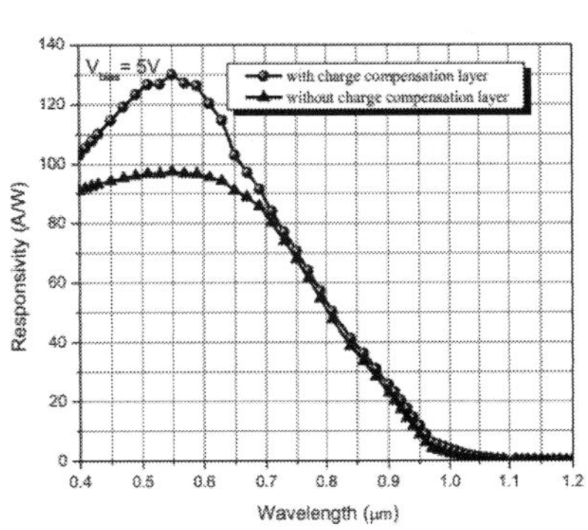

Figure 4. Calculated spectral response of Si/SiC heterojunction optically controlled transistors both with and without charge compensation layer.

Spectral responsivity depends on several inherent factors including material refractive index, absorption coefficient, device structure, doping and temperature of operation. Other external factors such as bias voltage and power of incident light also influence the spectral responsivity. Diffusion process is the main transporting mechanism of photon-generated carrier in neutral region. But the drift process in depletion region is the main process for photocurrent. The structure schematic of light absorption layer in the device with charge compensation layer is shown in figure 5. The photon-generated carrier is a function of the incident distance. The number of the photon-generated carrier in vertical direction and horizontal direction can be calculated respectively as

$$G_1 = \int_0^b (1-R)\frac{P}{h\nu}\alpha e^{-\alpha x}dx = (1-R)\frac{P}{h\nu}(1-e^{-\alpha b}) \quad (3)$$

$$G_2 = \int_e^b (1-R)\frac{P}{h\nu}\alpha e^{-\alpha x}dx = (1-R)\frac{P}{h\nu}(1-e^{-\alpha(b-e)}) \quad (4)$$

where P is the incident power, h is the Planck constant, v is the incident frequency, α is the absorption coefficient of material and R is the reflectivity. The responsivity of the device with 2D electric field in drift region can be write as

$$R = \frac{\beta I}{P_{in}} = \frac{\beta q}{P_{in}}\int_\lambda \left[c\int_0^b \alpha e^{-\alpha x}dx + (d-c)\int_e^b \alpha e^{-\alpha x}dx \right] d\lambda \quad (5)$$

where β is the gain of the transistor. The responsivity of the device with one-dimensional electric field in drift region is

$$R = \frac{\beta I}{P_{in}} = \frac{\beta q}{P_{in}}\int_\lambda \left[c\int_0^b \alpha e^{-\alpha x}dx \right] d\lambda \quad (6)$$

The responsivity of the device with a 2D electric field in drift region is higher than the device with a one-dimensional electric field in drift region obviously. The reason is that the two-dimensional electric field formed by charge compensation layer in the transistor can help create one more light absorption layer in horizontal direction. The light absorption layer in horizontal direction improves the spectral response of the optically controlled transistor.

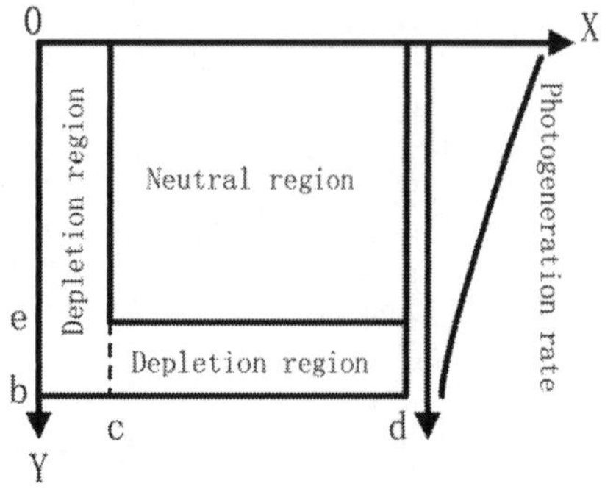

Figure 5. Structure schematic of light absorption layer.

To demonstrate the turn-on characteristics of device with charge compensation layer, the I-V characteristic curve of the device is simulated. The wavelength of incident light is 550nm and the light power density is changed from 0W/cm^2 to 1W/cm^2. The optical characteristics of 4H-SiC/Si heterojunction optically controlled transistor with charge compensation layer for different incident light power density are shown in figure 6. When the power density of the incident light is 0W/cm^2, the device is in blocking state. The device could not be triggered on, unless the power density of the incident light was larger than critical value. The photo current in the device triggered by the light is different, when the power density increased. The light with higher power density can trigger higher photo current, which can support higher base current. As the gain of the transistor is a constant value, increasing the photo current is the only way to obtain higher collector current.

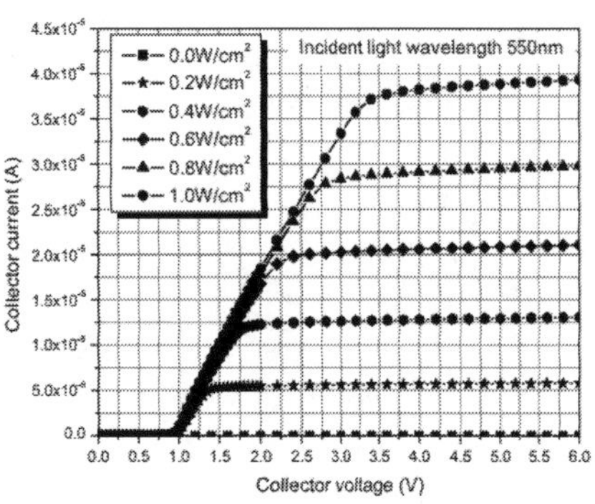

Figure 6. Optical characteristics of Si/SiC heterojunction optically controlled transistor with charge compensation layer for different incident light power density.

Figure 7 shows the transient characteristic of the SiC/Si heterojunction optically controlled transistor with charge compensation layer. As shown in figure 6, the on-state current density is about 160A/cm^2 and the off-state current density is below 0.01A/cm^2. The rising time is about 0.54μs and the falling time is about 0.66μs, which indicates a good transient characteristic of the device under the triggering by a 550nm light.

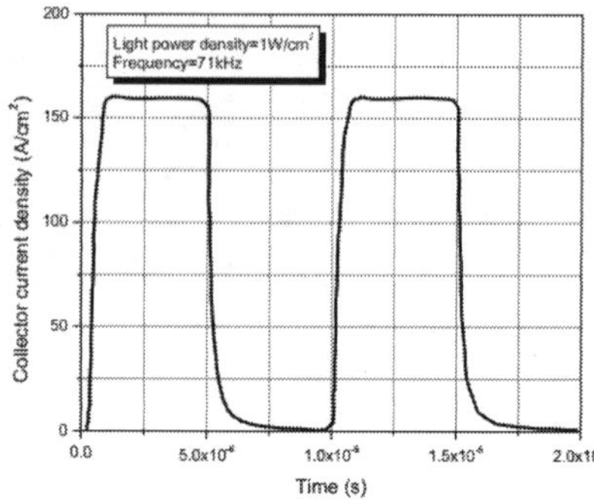

Figure 7. Transient characteristic of the Si/SiC heterojunction optically controlled transistor with charge compensation layer.

4 Conclusion

The optically controlled transistor with charge compensation layer based on 4H-SiC/Si/Si hetero-structure has wide enough spectral response range and big enough spectral response intensity to visible light. The charge compensation layer helps improve the breakdown voltage and spectral responsivity of the optical transistor obviously. The peak value of the spectral response is 130A/W at the wavelength of 550nm. The excellent optical and electrical performance of Si/SiC heterojunction optically controlled transistor makes it attractive for SiC-based optically controlled power devices' development.

Acknowledgement

The paper is financially supported by the National Natural Science Foundation of China (Grant No. 51177134) and the Natural Science Basic Research Plan in Shaanxi province of China (No. 2015JM6286).

References

1. J. R. Todd, IEEE National Aerospace and Electronics Conference, 456-463, (1992)
2. L. Wentao, C. Zhiming, Power Electronics, 45(4), 89-91, (2011)
3. P. Bhadri, D. Sukumaran, K. Ye, IEEE Midwest Symposium on Circuits and Systems, 1, 401-404, (2001)
4. C. Zhiming, P. Hongbin, F. R. Beyette, Chinese Physics Letters, 20, 430-432, (2003)
5. C. Lin, P. Hongbin, C. Zhiming, Z. Shaojiang, Materials Science Forum, 663, 494-497, (2011)

6. M. Usman, M. Nawaz, Solid-state Electronics, 92, 5-11, (2014)
7. Y. Liangchun, S. Kuang, IEEE Transactions on Electron Devices, 55(8), (2008)
8. D.M. Nguyen, C. Raynaud, N. Dheilly, M. Lazar, D. Tournier, P. Brosselard, D. Planson, Diamond and Related Materials, 20, 395–397, 2011.

MATEC Web of Conferences 77, 08009 (2016) DOI: 10.1051/matecconf/20167708009
ICMMR 2016

Vertex Combination Method for Heat Transfer Analysis of Structures with Uncertain Parameters

Zhengquan Liu , Yiwang Bao

State key laboratory of green building materials, China Building Materials Academy, 100024 Beijing, China

Abstract. Uncertainty is wide-spreads in practical projects which can be solved by interval method. Regarding the temperature field prediction of structures with uncertain parameters, a vertex combination method is presented on the basis of the sampling according to the vertex combinations of interval parameters and monotonic relationship of objective function which can effectively reduce the number of boundary combined samples. A numerical example of heat transfer problems of a panel without thermal resource is used to verify the effectiveness of the proposed method.

1 Introduction

Probability analysis method and non-probability analysis method are the main methods to deal with uncertainty problem. Among them, probability analysis method has the most effectively to express the effect of uncertain parameters of system uncertainty degree quantitatively when it has sufficient sample probability distribution information [1, 2]. However, the exact sample probability distribution information sometimes is difficult to obtain and interval probability model was proposed by domestic and foreign scholars based on interval analysis theory [3, 4], in which does not require exact probability distribution information of uncertain parameters. It only needs to know the uncertain variables in the upper and lower bounds and is easy to utilize and has been widely applied in practice [5-9]. Regarding the temperature field prediction of structures with uncertain parameters, a vertex combination method is presented on the basis of the sampling theory. The system response range can be obtained by the vertex combination method which can effectively reduce the number of boundary combined samples. Finally, a numerical example of heat transfer problems of a panel without thermal resource is used to verify the effectiveness of the proposed method.

2 Vertex combination method

For the steady state heat transfer problem, by using the finite element method the discrete equation of node temperature can be expressed as

$$\mathbf{KT} = \mathbf{F} \tag{1}$$

where $\mathbf{T}$ is node temperature vector; $\mathbf{K}$ is the thermal stiffness matrix; $\mathbf{F}$ is the equivalent thermal load vector.

In practical engineering, it is difficult and expensive to access the parameter probability density function because it needs adequate trial numbers to construct the uncertainty model. However, it gets easier to obtain the roughly range of uncertain parameters compared with the probability density function by using less uncertain information. In this case, we assume that the material properties, external loads and boundary conditions of the heat transfer model can be expressed as

$$\begin{aligned}\boldsymbol{\alpha}^I &= [\underline{\boldsymbol{\alpha}}, \overline{\boldsymbol{\alpha}}] \\ &= (\alpha_i^I)_n \\ &= ([\underline{\alpha}_i, \overline{\alpha}_i])_n \\ &= (\alpha_i^c + \Delta\alpha_i\delta)_n \\ &= \boldsymbol{\alpha}^c + \Delta\boldsymbol{\alpha}\delta \qquad i = 1, 2, \cdots, n\end{aligned} \tag{2}$$

where $\underline{\alpha}_i, \overline{\alpha}_i$ is the lower and upper bounds;
$\alpha_i^c = (\overline{\alpha}_i + \underline{\alpha}_i)/2$ and $\Delta\alpha_i = (\overline{\alpha}_i - \underline{\alpha}_i)/2$ are called the interval midpoint and interval radius respectively; interval parameter $\delta = [-1, 1]$ can be expressed simply by δ ; n is the number of uncertain parameters.

The interval parameters of this group will result in the uncertainty of the thermal stiffness matrix $\mathbf{K}$ and the thermal load vector $\mathbf{F}$ in Eq. (1). The steady state heat transfer equation with interval parameters can be rewritten as

$$\mathbf{K}(\boldsymbol{\alpha}^I)\mathbf{T}(\boldsymbol{\alpha}^I) = \mathbf{F}(\boldsymbol{\alpha}^I) \tag{3}$$

Vertex combination method is an interval analysis method to determine the upper and lower bounds of the system response by the boundary combination of interval variables according to the equation

$$\mathbf{K}^I\mathbf{T}^I = \mathbf{F}^I \tag{4}$$

where $\mathbf{K}^I = (k_{ij}^I)_{n\times n}$, $\mathbf{F}^I = (f_i^I)_n$.

Based the interval theory, the solution sets of interval linear equations Eq. (4) is convex and the extreme values are located in interval variables boundary (vertex point). Therefore, the lower bound $\underline{\mathbf{T}}$ and upper bound $\overline{\mathbf{T}}$ of the interval temperature response $\mathbf{T}^I$ can be obtained by the following expressions

$$\begin{aligned}\underline{\mathbf{T}} = (\underline{T}_j) &= \min_{\substack{1\le m\le 2^{n\times n}\\ 1\le q\le 2^n}} (T_j^{mq}) \\ &= \min_{\substack{1\le m\le 2^{n\times n}\\ 1\le q\le 2^n}} \left(\mathbf{K}_m^{-1}\mathbf{F}_q\right) \\ \overline{\mathbf{T}} = (\overline{T}_j) &= \max_{\substack{1\le m\le 2^{n\times n}\\ 1\le q\le 2^n}} (T_j^{mq}) \\ &= \max_{\substack{1\le m\le 2^{n\times n}\\ 1\le q\le 2^n}} \left(\mathbf{K}_m^{-1}\mathbf{F}_q\right)\end{aligned} \tag{5}$$

where $\mathbf{K}_m = \left(\hat{k}_{ij}^m\right)$ is the vertex thermal stiffness matrix under the condition of $\hat{k}_{ij}^m = \underline{k}_{ij}$ or $\hat{k}_{ij}^m = \overline{k}_{ij}$ $i,j=1,2,...,n;\ m=1,2,...,2^{n\times n}$; $\mathbf{F}_q = \left(\hat{f}_i^q\right)$ is the vertex equivalent load vector under the condition of $\hat{f}_i^q = \underline{f}_i$ or $\hat{f}_i^q = \overline{f}_i$ $i=1,2,...,n;\ q=1,2,...,2^n$.

The traditional vertex combination method is to select the minimum and maximum values of the response of the system by the combination of all the elements in the interval matrix and the load vector. As can be seen from Eq. (3), interval matrix and vector of each element changes depending on the parameter vector. That is the interval matrix and vector elements are related to each other and usually not possible at the same time to get the upper or lower bounds. Therefore, the results obtained by using the vertex combination method based on the interval linear equations Eq. (4) are conservative. On this basis, we propose a vertex combination method based on interval parameters.

A n-dimensional hypercube is a uncertain space by combining with n independent interval parameter $\alpha_i^I = [\underline{\alpha}_i, \overline{\alpha}_i]$. When the temperature response on the interval input parameters satisfy a certain relation, the extreme temperature response has been reached in the vertices of the hypercube. Therefore, the lower bound $\underline{\mathbf{T}}$ and upper bound $\overline{\mathbf{T}}$ of the interval temperature response $\mathbf{T}^I$ can be obtained by the following expressions

$$\begin{aligned}\underline{\mathbf{T}} &= (\underline{T}_j) \\ &= \min_{1\le m\le 2^n} (T_j^m) \\ &= \min_{1\le m\le 2^n} \left(\mathbf{K}_m^{-1}\mathbf{F}_m\right) \\ \overline{\mathbf{T}} &= (\overline{T}_j) \\ &= \max_{1\le m\le 2^n} (T_j^m) \\ &= \max_{1\le m\le 2^n} \left(\mathbf{K}_m^{-1}\mathbf{F}_m\right)\end{aligned} \tag{6}$$

where $\mathbf{K}_m = \mathbf{K}(\alpha_1^m,...,\alpha_n^m)$ is the vertex thermal stiffness matrix under the condition of $\alpha_i^m = \underline{\alpha}_i$ or $\alpha_i^m = \overline{\alpha}_i$ $i=1,2,...,n$; $\mathbf{F}_m = \mathbf{F}(\alpha_1^m,...,\alpha_n^m)$ is the vertex equivalent load vector under the condition of $\alpha_i^m = \underline{\alpha}_i$ or $\alpha_i^m = \overline{\alpha}_i$ $i=1,2,...,n$.

There is 2^n calculation times for n independent interval parameter vertex combinations. Compared with combinations of vertices of the interval linear equation shown in Eq. (5), it has less calculation and more accurate. However, when the number of uncertain parameters n is relatively large, the amount of calculation of the combination of the vertex can not be ignored.

Actually, by using the central difference method, we can obtain the first-order derivative values of a nodal temperature response T_j on a range of input parameters α_i^I, and then judge function monotone from the temperature response interval parameters.

$$\frac{\partial T_j(\alpha_i^c)}{\partial \alpha_i^I} \approx \frac{T_j(\alpha_i^c + \delta^*) - T_j(\alpha_i^c - \delta^*)}{2\delta^*} \tag{7}$$

where δ^* represents a small change in the midpoint of the parameter α_i^c.

We assume the node temperature response T_j is monotonically increasing with respect to the variable interval $\alpha_1^I, \alpha_2^I, ..., \alpha_k^I$ and monotonically decreasing with respect to the variable interval $\alpha_{k+1}^I, \alpha_{k+2}^I, ..., \alpha_n^I$. Then in solving the the lower and upper bounds ($\underline{T}_j$ and $\overline{T}_j$) of temperature response, we just consider two interval parameters combinations as following

$$\begin{aligned}&\mathbf{K}(\underline{\alpha}_1,...,\underline{\alpha}_k,\overline{\alpha}_{k+1},...,\overline{\alpha}_n)\mathbf{T} \\ &= \mathbf{F}(\underline{\alpha}_1,...,\underline{\alpha}_k,\overline{\alpha}_{k+1},...,\overline{\alpha}_n) \\ &\mathbf{K}(\overline{\alpha}_1,...,\overline{\alpha}_k,\underline{\alpha}_{k+1},...,\underline{\alpha}_n)\mathbf{T} \\ &= \mathbf{F}(\overline{\alpha}_1,...,\overline{\alpha}_k,\underline{\alpha}_{k+1},...,\underline{\alpha}_n)\end{aligned} \tag{8}$$

Compared with the direct combination of the interval parameters shown in Eq. (6), it can greatly reduce the workload by the prior judgment of the monotonic relationship between the temperature response and the parameters.

DOI: 10.1051/matecconf/20167708009

3 Simulating Process of VCM

The process of solving the upper and lower bounds of the interval temperature response with the method of vertex combination is shown in Fig. 1.It need to point out that the effectiveness of the vertex combination method is established on the basis of the monotonic relationship between the temperature response and the uncertain parameters. For many complex problems in practical engineering, the monotone of function is not strictly satisfied in result that the calculation error of vertex combination method will be larger, and even error results. In addition, when the interval uncertain parameters are more, a large number of combinations of parameter boundaries will lead to the decrease of computational efficiency. From the point of view of computational accuracy and computational cost, the vertex combination method presented in this section is more suitable for uncertain parameters of uncertain parameters.

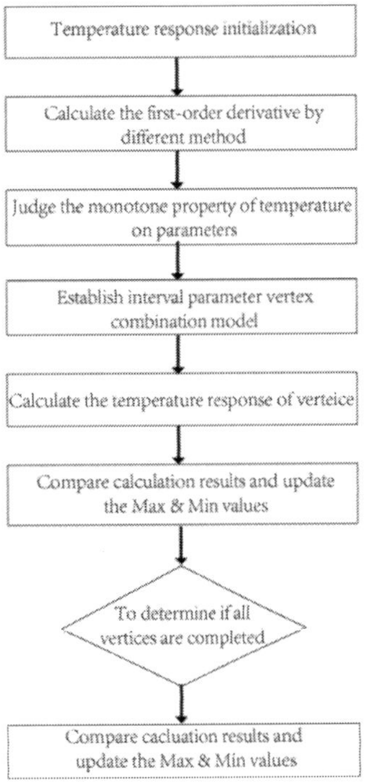

Figure 1. Numerical simulating flow chart of vertex combination method

4 Numerical example

As shown in Fig. 2, the thickness and inner heat source of the flat plate were recorded as 2δ and Q respectively. k is its thermal conductivity, the convection heat transfer between left and right bounds and fluid with temperature T_e. Surface heat transfer coefficient was expressed by h. We will discuss how to determine the temperature response of the flat structure at any position.

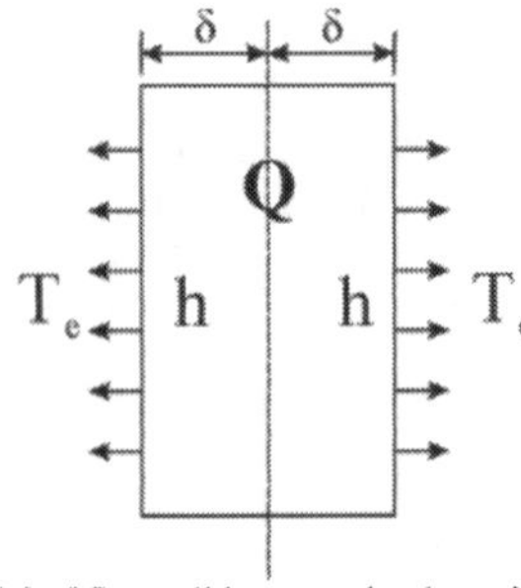

Figure 2. Model of flat wall heat conduction with heat source

The thermal conduction can be expressed by

$$\begin{aligned} & k\frac{\partial^2 T}{\partial x^2}+Q=0 \\ & x=0 \quad \frac{\partial T}{\partial x}=0 \\ & x=\delta \quad -k\frac{\partial T}{\partial x}=h(T-T_e) \end{aligned} \tag{9}$$

We can obtain the solution of the temperature response by using the integral method as following

$$T(x)=\frac{Q}{2k}(\delta^2-x^2)+\frac{Q\delta}{h}+T_e \tag{10}$$

Assuming the input parameters of the heat transfer problem is in the range of certain changes, which k ∈ [20,30] W/(m • K), Te ∈ [10,15] °C, h ∈ [2000,2500] W/(m^2·K). In addition to the material properties, loads and boundary conditions, this example also consider the uncertainty interval structure dimensions of δ ∈ [5,7] cm. The upper and lower bounds of the temperature response at each position are calculated by using the Monte Carlo simulation and the vertex combination sampling methods of the vertex combination in this section, as shown in Table 1.

Table 1. Upper and lower bounds of temperature response along thickness

Location x (cm)	Bounds	Analytical solution (°C)	Monte Carlo Method (10^5 samples)	vertex combination method Method (32 samples)
			Value (°C)	Value (°C)
0	lower	84	84	84
	upper	267	267	267
1	lower	82	82	82
	upper	263	263	263
2	lower	76	76	76
	upper	251	251	251
3	lower	66	66	66
	upper	231	231	231
4	lower	52	52	52
	upper	203	203	203
5	lower	34	34	34
	upper	167	167	167

DOI: 10.1051/matecconf/20167708009

5 Summary

In view of the steady state heat transfer problem, a numerical discrete model with uncertain parameters is established, and the interval uncertainty analysis technique is proposed based on the sampling principle. Vertex combination method is easy to implement, and the calculation results are accurate. But a large number of vertices will lead to the decrease of computational efficiency, which limits the generalization of the problem to the high dimension complex problem.

References

1. Ripley B D, *Stochastic Simulation*, John Wiley & Sons, New York, (2009)
2. Wang C, Qiu Z, He Y, INT J HEAT MASS TRAN. **91** (2015)
3. Neumaier A. *Interval Methods for Systems of Equations*, Cambridge University Press, Cambridge, (1990)
4. Ben-Haim Y, Elishakoff I. *Convex Models of Uncertainty in Applied Mechanics*, Elsevier Science Publishers, Amsterdam, (1990)
5. Qiu Z P, Elishakoff I, COMPUT METHOD APPL M. **152** (1998)
6. Rao S S, Berke L, AIAA. **35** (1997)
7. Xia B, Yu D, Liu J, J FLUID STRUCT, **38** (2013)
8. Wang C, Qiu Z, INT J HEAT MASS TRAN. **91**(2015)
9. Wang C, Qiu Z, INT J HEAT MASS TRAN. **91**(2015)

MATEC Web of Conferences 77, 09001 (2016) DOI: 10.1051/matecconf/20167709001
ICMMR 2016

A Neuromotor Device for Reducing Phantom Limb Pain in Individuals with Spinal Cord Injury

Lei Cui[1] , Gonzalo Sepulveda Astudillo[1] and Garry T. Allison[2]

[1]*Department of Mechanical Engineering, Curtin University, Australia*
[2]*School of Physiotherapy and Exercise Science, Curtin University, Australia*

Abstract. Phantom Limb Pain is a disorder that can be experienced by individuals after amputation or spinal cord injury. In spinal cord injury the paralysis or paresis is often bilateral, thus limiting the application of apparent movement as a therapeutic model for phantom limb pain. This project aimed to develop a robotic rehabilitation device that replicated apparent movement to apply the same therapeutic principles with individuals with lower limb phantom pain that have bilateral paralysis of paresis. The proposed device achieved lower limb planar motion of the knee by a six-bar linkage of a single degree of freedom (DOF). It is driven by a linear actuator while the ankle motion is achieved by a gear motor, reaching an effective 70° range of motion for both joints. The system features closed loop control using feedback from surface electromyography sensors, limit switches and position sensors with an Arduino microcontroller as the control unit. This device will be used to further our understanding of the disorder and create opportunities for robot aided treatment for individuals with phantom limb pain as a result of spinal cord injury.

1 Introduction

Spinal cord injury (SCI) is damage to the spinal cord, which could lead to loss of mobility and sensory function. A high proportion of the people with SCI experience neuropathic pain from an area of their body that is paralysed due to central nervous system damage caused by the SCI, rating it as one of the most unmanageable problems [1].

Neuropathic pain occurs in the insensate region below a complete SCI is usually predominated by central nervous system mechanisms in a way similar to phantom limb pain (PLP) following amputations [2]. The person perceives pain from parts of their body that are paralysed in the absence of actual tissue injury, which is not well managed with medications. It has been hypothesized that the illusion of limb movement in individuals with SCI (Moseley et al) as they think of moving their limb may reduce the symptoms of neuropathic pain, as the mirror technique alleviated the symptoms of PLP in amputees [3].It is likely that any successful therapeutic device relies on the coordination of the imaginary movement (specific motor cortex activation) and the concordant visual feedback of the movement. Such coordination is critical as it has been shown that imaginary movement without visual feedback of the limb moving increases temporary dysthesia or replicates localised pain memory for PLP [4, 5].

Since SCI often induces paralysis or paresis of both lower limbs, the neuromotor device needs to be able to move the knee and ankle of the patient in accordance to their thoughts of moving the limb. This effectively creates the illusion that the patient is moving their own leg. This concordant afferent and efferent cortical activation forms the basis of the proven mirror-box rehabilitation methods for individuals with amputations [3].

Lower-limb exoskeletons have been used for various purposes. Ferris et al. [6-9] built a pneumatically powered knee-ankle-foot orthosis, which was used to test this device's mechanical performance during human walking. Chu et al. [10] proposed a hydraulically driven lower-limb exoskeleton driving the hip, knee, and ankle joints to compensate for the strength and endurance of a human under a payload. Lugris et al. [11] presented an active stance-control knee-ankle-foot orthosis. Cui et al [12] proposed a one active degree-of-freedom closed-loop compliant exoskeleton to enable humans to carry loads during long distance steady locomotion. Dollar and Herr [13] reviewed the state-of-the-art of lower limb exoskeletons in terms of design, actuation, sensory, and control.

However, the aforementioned devices are not able to fulfil this specific task as they either neglect to rotate the ankle joint or do not have the ability to incorporate muscle feedback into the locomotion process. In this paper, we presented a robotic rehabilitation device capable of applying therapeutic principles similar to mirror motion with individuals that have spinal cord

DOI: 10.1051/matecconf/20167709001

injury and experience phantom limb pain by simulating limb movement.

This paper is organized as follows. Section 2 introduces the mechanical design of the neuromotor device. Section 3 presents kinematic and stress analyses. Section 4 discusses the system design and implementation of the device. Section 5 concludes the paper.

2 Overview of mechanical design

2.1 Design considerations

The design considerations defined by the scope of the project are listed in Table 1.

Table 1. Design considerations.

Customer Requirements	Technical Parameters
Able to accommodate patients of different heights	Ranging from 1.6 *m* to 1.9 *m*
Able to withstand the weight of the leg	Force and stress safety under the nominal load of 15 *kg*
Able to stop at any instant	Emergency stop implementation
Enhanced safety	Limit switches and software safety implementation
Ease of setup and use	Attached to a rehabilitation chair

2.2 Mechanical design

We proposed to use a hybrid parallel-serial linkage mechanism to actuate the knee and ankle joints. The knee joint is actuated by a six-bar linkage with a linear actuator drives the knee joint and a servo-motor driven drives the ankle joint, as in Fig. 1.

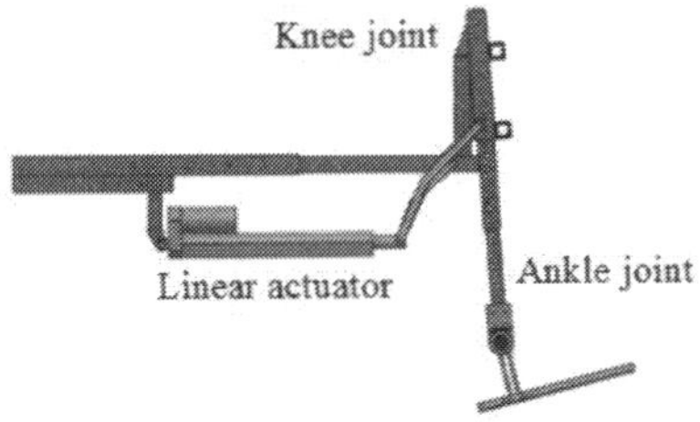

(a) Home position

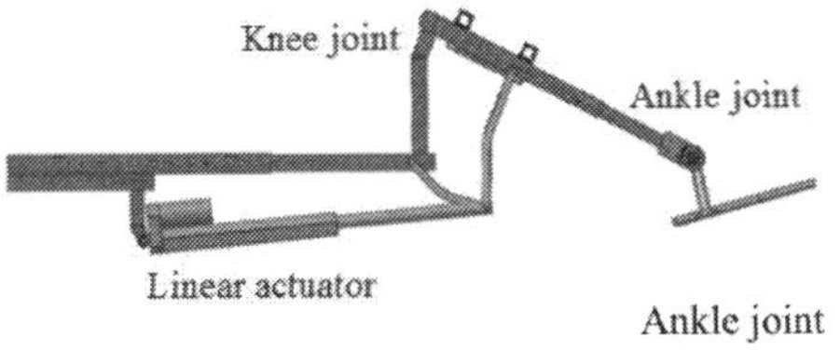

(b) Extended position

Figure 1. A hybrid parallel-serial knee-ankle orthosis in home and extended positions

The design features standard sized square hollow section tubing to construct the majority of components supporting the limb. The linear actuator provides an effective range of motion of 70 degrees at the knee joint.

3 Kinematic and stress analyses

3.1 Kinematic Analysis of the knee mechanism

The six-bar linkage consists of a ternary link and five binary links connected by five revolute joints and one prismatic joint, which is a variation of Stephenson type six-bar linkage [14], as in Fig. 2.

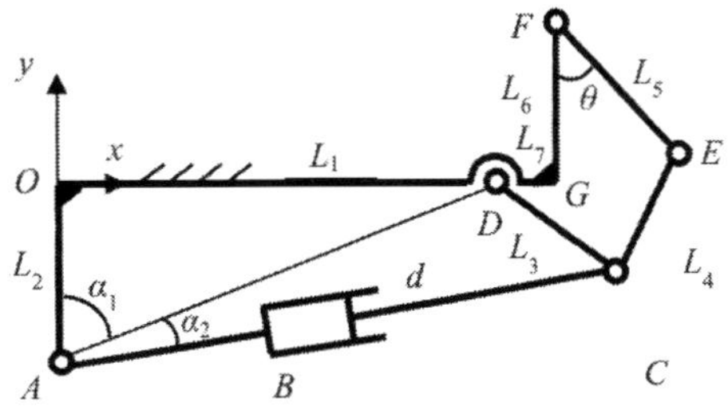

Figure 2. Schematic drawing of the six-bar knee mechanism

The structural parameters are: $OD = L_1$, $OA = L_2$, $CD = L_3$, $CE = L_4$, $EF = L_5$, $FG = L_6$, $GD = L_7$. The input is d, which is the length of AC, and the output is knee angle θ, which is the angle between GF and FE, as in Fig. 2.

The angle α_2 between DA and AC can be obtained as

$$\alpha_2 = \mathrm{acos}\left(\frac{d^2 + L_1^2 + L_2^2 - L_3^2}{2d\sqrt{L_1^2 + L_2^2}}\right) \tag{1}$$

The angle $\alpha_1 = \mathrm{atan}(L_1/L_2)$ is a structural parameter. Hence the coordinates of the joint C are

$$\begin{aligned} C_x &= d\sin(\alpha_1 + \alpha_2) \\ C_x &= -L_2 + d\cos(\alpha_1 + \alpha_2) \end{aligned} \tag{2}$$

The link lengths L_4 and L_5 constraints transform the problem of finding the knee angle θ into circle-circle intersection, where the first circle is centred at the joint C with a radius of L_4 and the second circle is centred at the joint F with a radius of L_5, as in Fig. 3.

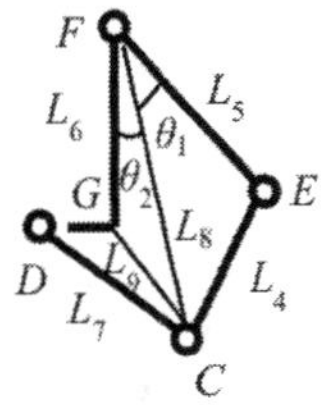

Figure 3. The knee angle θ

It follows that the knee joint θ is

$$\theta = \theta_1 + \theta_2 = \mathrm{acos}\left(\frac{L_8^2 + L_5^2 - L_4^2}{2L_5L_8}\right) + \mathrm{acos}\left(\frac{L_8^2 + L_6^2 - L_9^2}{2L_6L_8}\right) \tag{3}$$

where

$$L_8 = \|CF\| = \sqrt{(C_x - L_1 - L_7)^2 + (C_y - L_6)^2}$$
$$L_9 = \|CG\| = \sqrt{(C_x - L_1 - L_7)^2 + C_y^{\,2}}$$

3.2 Stress analysis

We conducted stress analysis in SolidWorks on key parts of the device subject to 15 *kg* load, considering the average weight of the calf and foot of a 150 *kg* adult is around 150 kg, as in Fig. 4. The weakest component has a safety factor of 1.85.

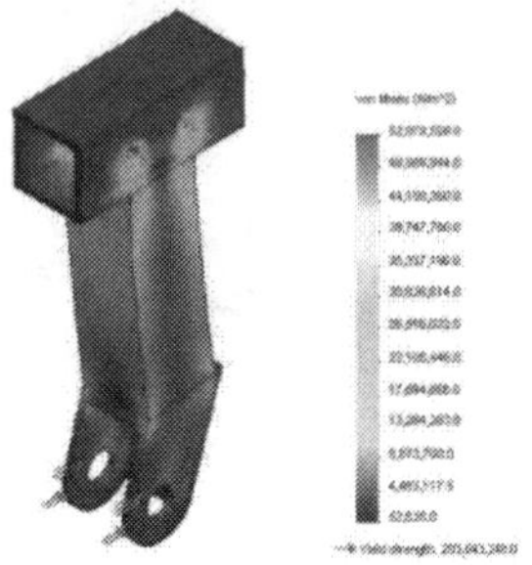

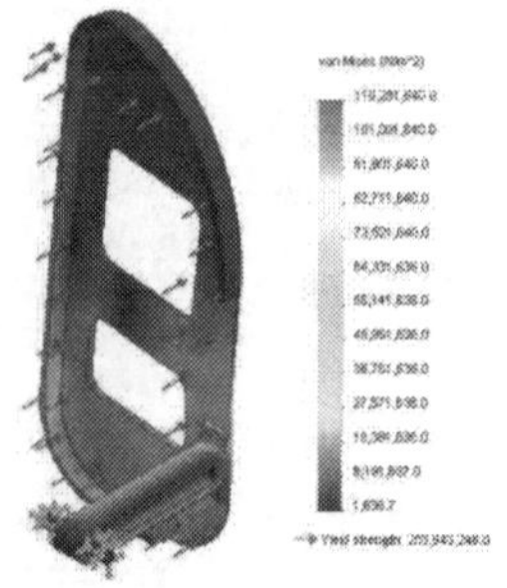

Figure 4. Stress analysis of the base link and the footpad

4 System design and implementation

Safe operation of the system was achieved through the implementation of safety elements such as position sensors, limit switches, emergency hand switch and physical motion stoppers. Also, a graphical user interface was designed to facilitate the operation of the system and to provide supervisory control to the therapist. An overview of the system architecture block diagram is shown in Fig. 5.

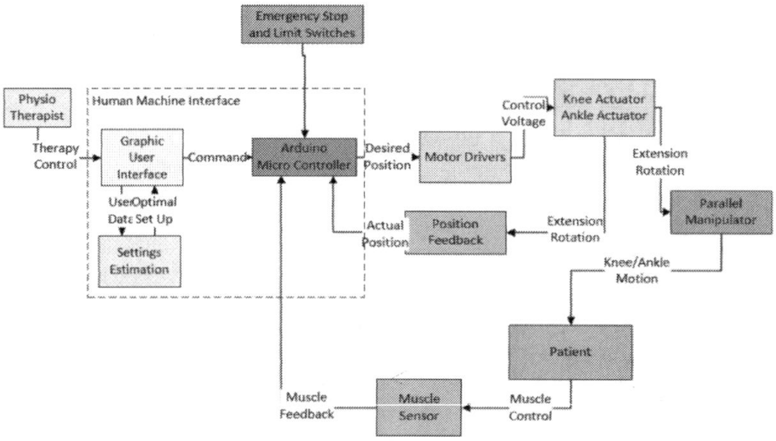

Figure 5. System architecture block diagram

A prototype was built to validate the proposed design, as in Fig. 6.

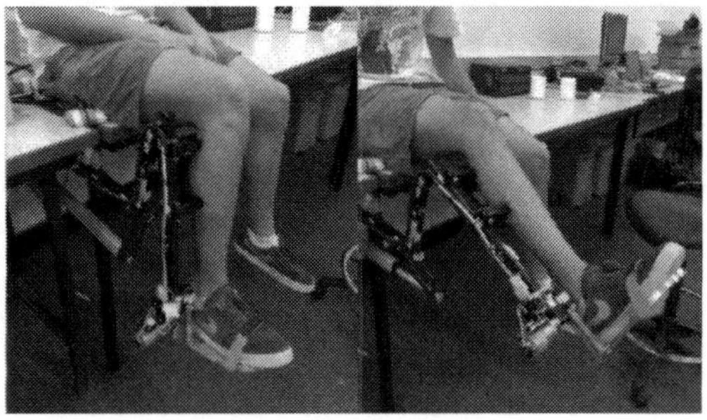

Figure 6. A prototype of the proposed design

We proposed to use diagonal activations where the controller is the EMG signal of the biceps on the upper limb. The individual was instructed to imagine a rope between the hand and ankle and that they act together resulting in the biceps EMG signal initiating the movement, as in Fig. 7.

We tested the device on one healthy subject. Surface EMG electrodes to trigger the muscle feedback function of the system were placed on biceps and the subject bent his forearm to initiate one cycle of rehabilitation training.

The results showed that the motion and speed performance was satisfactory. The lower limb was elevated at a comfortable pace, which was natural and unforced. The joint angles and sEMG sensor signals are shown in Fig. 8.

DOI: 10.1051/matecconf/20167709001

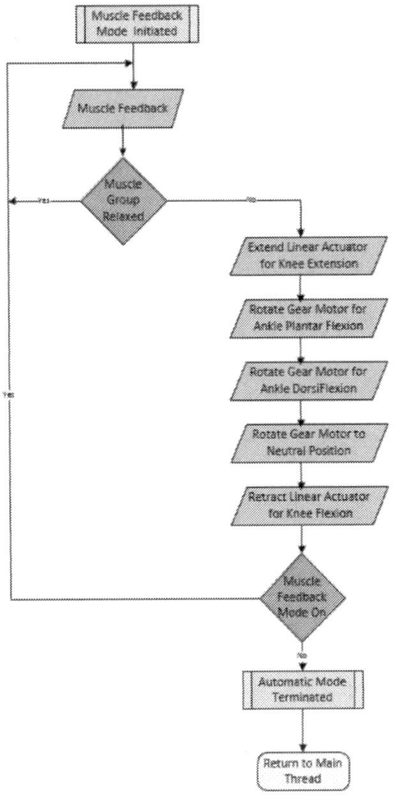

Figure 7. Control flowchart

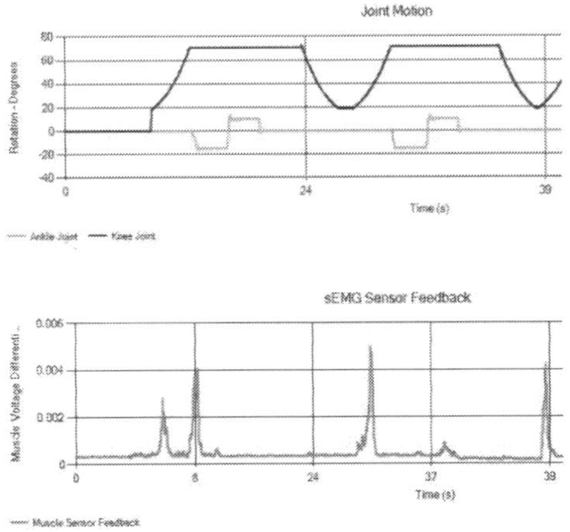

Figure 8. Knee and ankle joint displacements and the corresponding sEMG signals on the biceps

As can be seen on the joint motion plot, the actuation sequence begins with the ankle actuator, upon reaching the desired position; ankle actuation begins and is followed by knee flexion. Muscle feedback above 2 mV threshold trigger actuation sequence, as seen in both cycles, during this time rehabilitation device only records muscle feedback and signals higher than 2 mV are neglected until cycle is terminated.

5 Conclusions

We presented a lower limb rehabilitation system, which introduced planar limb motion to the paralysed limb of patients through a single degree of freedom six-bar linkage for the knee joint and a geared motor for the ankle joint in conjunction with associated control and safety elements. Further development to increase patient comfort in terms of limb support and thorough safety testing would complete the system for patient trials.

References

1. E. G. Widerström-Noga, E. Felipe-Cuervo, J. G. Broton, R. C. Duncan, and R. P. Yezierski, "Perceived difficulty in dealing with consequences of spinal cord injury," *Archives of physical medicine and rehabilitation,* vol. 80, pp. 580-586, (1999)
2. H. Flor, T. Elbert, S. Knecht, C. Wienbruch, C. Pantev, N. Birbaumer, *et al.*, "Phantom-limb pain as a perceptual correlate of cortical reorganization following arm amputation," *Nature,* **375**, pp. 482-484, (1995)
3. B. L. Chan, R. Witt, A. P. Charrow, A. Magee, R. Howard, P. F. Pasquina, *et al.*, "Mirror therapy for phantom limb pain," *New England Journal of Medicine,* **357**, pp. 2206-2207, (2007)
4. G. L. Moseley, "Using visual illusion to reduce at-level neuropathic pain in paraplegia," *Pain,* **130**, pp. 294-298, (2007)
5. G. L. Moseley, N. Zalucki, F. Birklein, J. Marinus, J. J. van Hilten, and H. Luomajoki, "Thinking about movement hurts: the effect of motor imagery on pain and swelling in people with chronic arm pain," *Arthritis Care & Research,* **59**, pp. 623-631, (2008)
6. G. S. Sawicki and D. P. Ferris, "Powered ankle exoskeletons reveal the metabolic cost of plantar flexor mechanical work during walking with longer steps at constant step frequency," *Journal of Experimental Biology,* **212**, (2009)
7. K. E. Gordon and D. P. Ferris, "Learning to walk with a robotic ankle exoskeleton," *Journal of Biomechanics,* **40**, pp. 2636-2644, (2007)
8. P. C. Kao, C. L. Lewis, and D. P. Ferris, "Invariant ankle moment patterns when walking with and without a robotic ankle exoskeleton," *Journal of Biomechanics,* **43**, pp. 203-209, (2010)
9. C. R. Kinnaird and D. P. Ferris, "Medial Gastrocnemius Myoelectric Control of a Robotic Ankle Exoskeleton," *IEEE Transactions on Neural Systems and Rehabilitation Engineering,* **17**, pp. 31-37, (2009)
10. A. Chu, H. Kazerooni, and A. Zoss, "On the Biomimetic Design of the Berkeley Lower Extremity Exoskeleton (BLEEX)," in *Robotics and Automation, 2005. ICRA 2005. Proceedings of the 2005 IEEE International Conference on*, pp. 4345-4352, (2005)
11. J. M. Font-Llagunes, R. Pàmies-Vilà, J. Alonso, and U. Lugrís, "Simulation and design of an active

orthosis for an incomplete spinal cord injured subject," *Procedia IUTAM,* vol. 2, pp. 68-81, (2011)
12. Z. Shen, M. White, G. T. Allsion, and L. Cui, "Single Degree-of-Freedom Leg Exoskeleton Assisting Human Locomotion," presented at the IEEE/RSJ International Conference on Intelligent Robots and Systems, Deajeon, South Korea, (2016)
13. A. M. Dollar and H. Herr, "Lower Extremity Exoskeletons and Active Orthoses: Challenges and State-of-the-Art," *IEEE Transactions on Robotics,* **24**, pp. 144-158, (2008)
14. K.-L. Ting and X. Dou, "Classification and branch identification of Stephenson six-bar chains," *Mechanism and Machine Theory,* **31**, pp. 283-295, (1996)

DOI: 10.1051/matecconf/20167709002

Research on the Visual Processing System of the Punch Press

Xuan Sun[1], Xianpeng Sun[1], Chenglong Tang[2], Cao Sun[1] and Changsheng Ai[1]

[1]*School of Mechanical Engineering, University of Jinan, Jinan, China*
[2]*Shandong Shuaike Mechanical Manufacturing Co. Ltd, Weifang 261200, Shandong, China*

Abstract. Most of raw materials of small hardware processing for plate scraps, and it's realized through the manual operation of ordinary punch, which way has the low production efficiency and the high labor intensity. In order to improve the automation level of production, developing and designing of a visual processing system for punch press manipulator which based on the MFC tools of Visual Studio software platform. Through the image acquisition and image processing, get the information about the board to be processed, such as shape, length, the center of gravity position and pose, and providing relevant parameters for positioning gripping and placing into the punch table positioning of the feeding manipulator and automatic programming of punching machine, so as to realize the automatic operation about press feeding and processing.

1 Introduction

With China's economic development into the new normal and put forward the strategy of "made in China 2025". As the basis of advanced manufacturing technology and equipment automation, intelligence and information technology, robots play a decisive role in the field of modern production[1-2], one of the important components of the machine vision technology. In the stamping industry of our country, although there have been a large number of CNC(computer numerical control) punch press or modified semi automatic punch for stamping parts manufacturing, But like gaskets, hinges and other small metal parts which is widely used in the production still get through the mode of production such as shown in Fig. 1. If we want to make full use of CNC machine to achieve automatic processing, there are still some problems to be solved:

(1)At present, the operation object of punch press manipulator is material waste, as shown in Fig. 2, When the manipulator adsorbent and delivery the board to clamp, easy to cause the clamp clamping effect is not ideal, even holding failure, and then affect the stamping quality.

(2)The existing CNC punch press operation of CAM(the computer-aided manufacturing), the processing plate shape is standard, size is fixed. When the stamping plate is irregular, the work for measuring the plate size will be complex, which is not conducive to the continuity of production.

In view of the above problems, we introduced the machine vision system for industrial manipulator, and give someone install the eyes[3]. Through the image processing program design by MFC(the Microsoft foundation class) with the platform of Visual Studio software, which according to analysis and processing the pixel distribution and brightness information, extraction or calculate sheet shape, length, position of center of gravity and position information, then provide the corresponding data support for operation to the feeding manipulator.

Figure 1. Manual stamping operation

Figure 2. Plate processing material

2 The design of image acquisition system

2.1 Light source selection

The source which can be used for image acquisition contains: fluorescent lamp, halogen lamp, gas discharge lamp, light emitting diode (LED) and laser diodes (LD).

During design the lighting of the machine vision, we need to according to the performance characteristics of object (metal plate) which work with the machine vision, from the three aspects into account:

(1)According to the requirements of the performance of the light source itself, need to consider the contrast, brightness, robustness and so on;

(2)According to the requirements of the shape and structure of the light source, need to consider the uniformity of a light source, the light source structure;

(3)According to the light source and the required installation location should be considered by means of the irradiation angle, mounting distance and space layout [4-5].

According to the actual conditions, determine the way of lighting with surface light source by back-light illumination, and the space design as shown in Fig. 3, the example image is shown in Fig. 4.

Figure 3. The space design of camera and light source

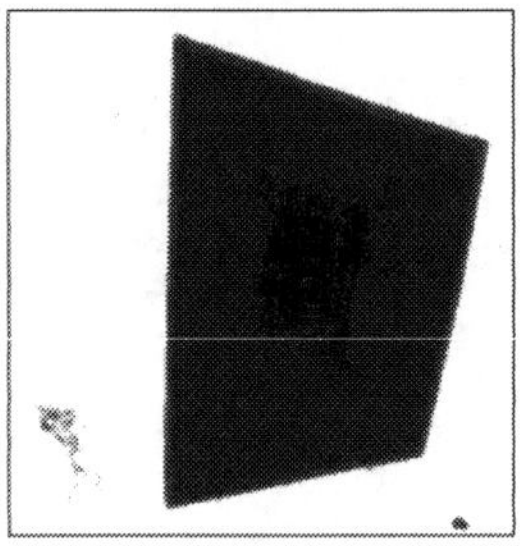

Figure 4. Experimental picture

The figure appears flocculent shadow or flecks are all the pollutants on the glass, in spite of this, the image collecting by the back-light illumination have distinct difference between the foreground and background, So the scheme have better anti-interference for the noise, pollution, etc.in the horizon.

2.2 Camera and image acquisition card

Industrial camera is a core component of machine vision, which will directly affect the the image resolution and quality of the collection. The optical lens of CCD camera like as the eye lens of the human, its role is to target aggregation in the CCD array plane, so the performance parameters of the chosen lens, will directly determine the camera can "see" image [6].

According to the features of actual conditions, existing experimental equipment and object detection (irregular sheet), Determined by the German Basler acA2500-14gm industrial CCD camera, and which pixel Depth of is12 and the maximum frame rate is 14. The selected lens is Japan Comutar lens, model is M2514-MP2, which focal length is 25mm, and the aperture is F1.4-F16C. In the meantime, the Image acquisition is the selection of Cognex MVS-8504 PCI card.

3. Image enhancement and feature extraction

After completing of shooting the board which is putted in the view of the camera, the need for image processing and analysis, including image enhancement and feature extraction. This is a key precondition of obtaining information of the target of the image.

3.1 Image enhancement

Due to the object of machine vision is Sheet material leftovers, according to the processing requirements and actual conditions, the image will contain the interference information, which because of the interference of noise, environment of the oil and impurities. Therefore, to remove the noise and weaken the interference information maximum degree by image enhancement, at the same time, ensure the contour of target image be becoming more prominent, in order to carry out post processing work [7].

Median filter is a nonlinear filtering method. It can effectively remove pulse noise interference, and ensure the details clarity of the original image, but someone has some features just like point, line, and spire will loss some relevant characteristics during processing of the image.

The Fig. 4 is the original collected images, Although the median filter can effectively remove the noise and most small spots in the image, and enhance the image edge effect [8], as shown in Fig. 5, could not remove the "burr" on the edge of the image. In order to make the edge clear, it is necessary to sharpen the edge contour by Laplasse.

Laplasse is a kind of method for edge sharpening to enhance the image by using the Laplasse operator, The basic idea is to set up as the Laplasse operator, And after two order partial derivation:

$$\nabla^2 f = \frac{\partial^2 f}{\partial x^2} + \frac{\partial^2 f}{\partial y^2} \qquad (1)$$
$$= f(i+1,j) + f(i-1,j) + f(i,j+1) + f(i,j-1) - 4f(i,j)$$

In image sharpening, we usually use a template such as $\begin{bmatrix} -1 & -1 & -1 \\ -1 & 9 & -1 \\ -1 & -1 & -1 \end{bmatrix}$,

The method is get the gradient of the neighborhood in eight directions of the center pixel, and gain the relationship between the center pixel gray and other pixel gray in the neighborhood by summing gradient, and then using the results of the gradient algorithm to adjust the pixel gray image to achieve edge sharpening. The effect of image edge sharpening, as shown in Fig. 6, the red line box in the figure is pointed out that the characteristics of the change point of reference.

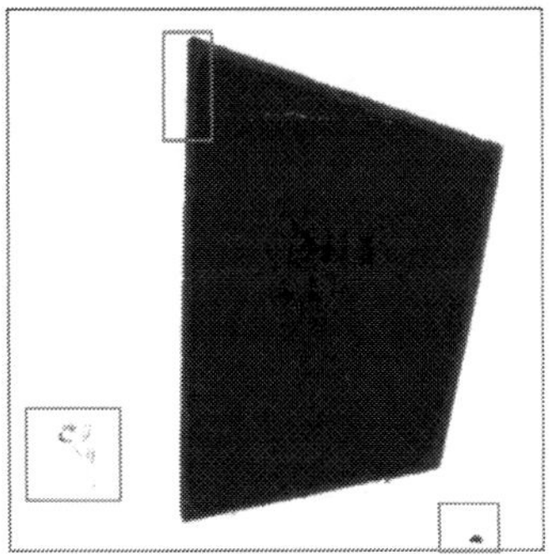

Figure 5. After median filtering

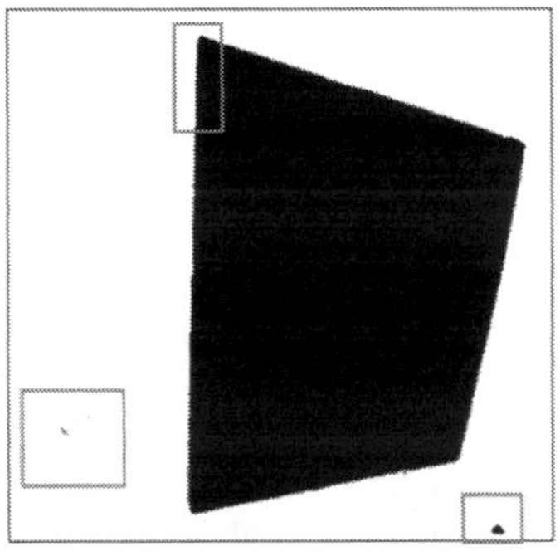

Figure 6. After laplasse sharpen

3.2 Image segmentation

Image segmentation is to separate the image of the target area and the process of extraction, the extracted features can be color, gray and so on, is the key step before image analysis[9].

Through the information of the histogram image determine image gray distribution information, and selecting one or more threshold T, the gray level of the image will be divided into several parts, and thus make the same gray value range of image distinguish with other parts, or make the non target area direct to back-grounding. So this method is especially suitable for the images with different targets and backgrounds.

In order to ensure the image segmentation effect, In order to ensure that the segmentation threshold is not distribution due to the difference of the gray level, the iterative method is used to determine the optimal segmentation threshold, which can get better image segmentation results.

3.3 Edge and Contour extraction

The edge is a set of connected pixels that make the pixel gray scale produce a step change or a change of the roof. The Edge extraction is the process of extracting the intersection line between the fore-ground and the back-ground in the image by some operators. Contour extraction is a process to obtain the process of sheet metal contour information by extracting the edge of the detected edge. By the previous work of threshold segmentation, the realization of the image of the two value of processing, so that the image only composed of the pixels which gray values only are 0 or 255 (black and white). In this paper, the shadow difference method is used to extract the contour, and the realization method is: gets the pixel value of a point A in the image $Z_A = 255$, as shown in Fig. 7, make a judgment for all of the 8 points in the neighborhood of the gray value, if every point Z_i != 255, then the pixel value Z_i is set to 0 in the 3 × 3 region, so that the area of whitening; while the pixel value of the 8 points of the neighborhood is not all 255, the point will be set as the boundary point, and the pixel value of the pixel is preserved. In the end, the Fig. 8 will be converted to Fig. 9, and the red line box in the figure points out the main change characteristic reference point.

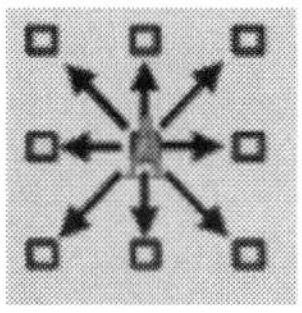

Figure 7. A feature pixel

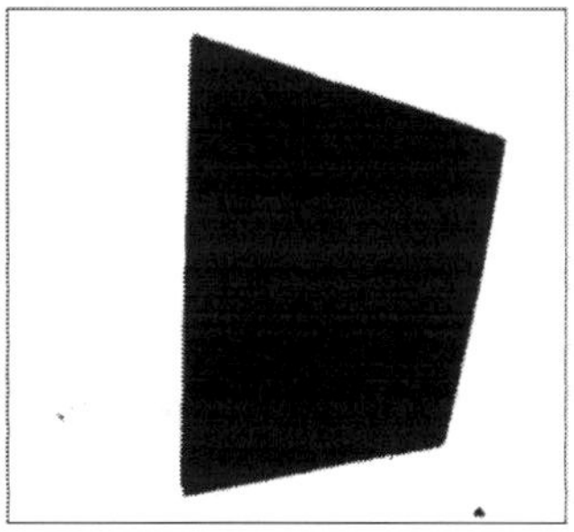

Figure 8. Threshold segmentation

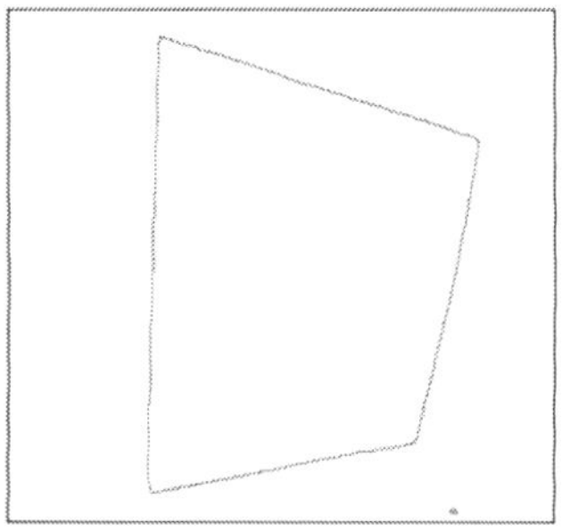

Figure 9. Contour extraction

 DOI: 10.1051/matecconf/20167709002

3.4 Image feature extraction

Edge and contour extraction is the original description of the segmentation results, which contains the parts that do not need, which needs to determine one or more features from the area or profile [10]. The characteristic of this paper is the best fitting line of the edge of the plate. Currently widely used line detection method mainly has the least square method and Hough transform method. Because the least square method is not easy to have multiple line segments to form a closed contour line extraction, So using the Hough transform method as the selection.

According to the general principle of line detection of Hough transform, there are many lines in the extracted edge contour, that corresponds to a plurality of (r,θ) points in space, each line is composed of a large number of consecutive pixels, and reflected in the space that corresponds to the number of sine curve through the point (r_i, θ_i), therefore, through the largest number of sine curve group on the composition (x, y) space in the best straight line. When a curve in (r, θ) space is also a common point, the point will be a intersection of two straight lines in (x, y) space. The specific implementation process of Hough transform to detect straight line is as follows:

(1)Initialize a array in the transform domain (r, θ) space and the number in the r direction is the number of pixels in the diagonal direction of the image, The number in the θ direction is defined as 90 (angle from 0~180, interval is 2 degrees).

(2)Search the map of all black spots, each black spot in the transform domain of the corresponding points on the plus 1;

(3)Find the maximum points in the transform domain and record, in the meantime, cleared all of points near the maximum point;

(4)According to the requirements and finding out multiple maximum points in the image in turn;

(5)Draw the straight line drawn from all the points of the maximum value, and there is the intersection of the two straight lines.

Which added a zero function in the c step, and the realization of the function: after extracting the best approximation linear will be clear of the image contour and make straight contour clear, reduce post size measurement of interference. Fig. 10 is the detection results with traditional Hough transform linear, Fig. 11 is the improved linear extraction effect, as shown below:

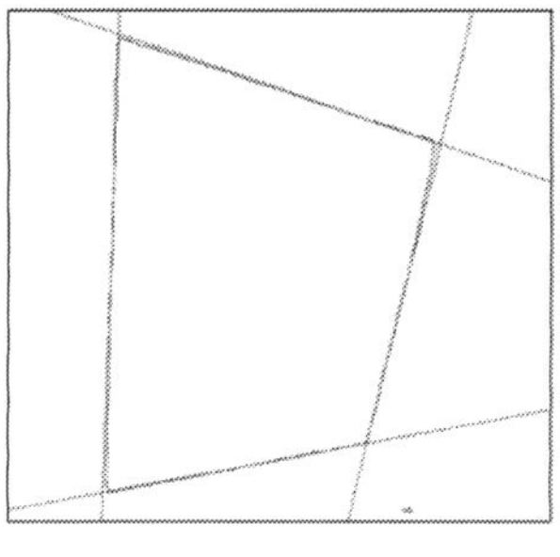

Figure 10. The effect before Improvement

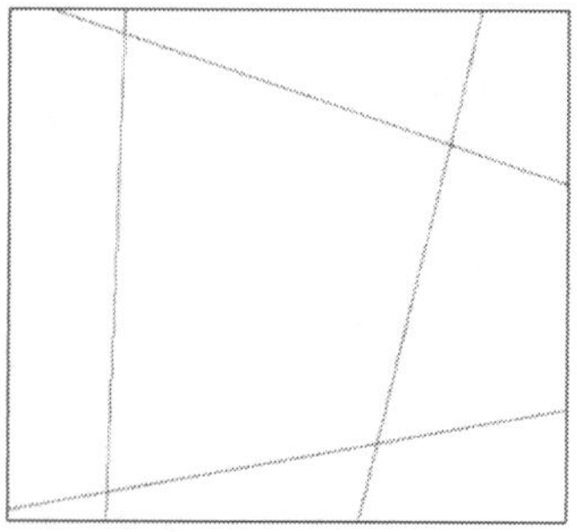

Figure 11. The effect after Improvement

4 Dimensional measurement and Display

Dimensional measurement is one of the most common techniques used in machine vision technology, including the length, angle, aperture, diameter, area and other geometric parameters of the object. In this paper, the extraction of target parameters including plate area, length and angle.

Pixel is the basic unit of image composition, but also as a representation of the size of the image unit. The image resolution is a unit of the number of points on the length and width of the image, and the size is equal to the product of the number of pixels in the length and width direction. So when the camera's resolution is determined, the area of the image is also determined. When the space position of the camera and the tested plate are fixed, there existed the ratio K between the measured plate length L and the number of pixels N_{pixel} corresponding to the edge in the image, that is $L=K \cdot N_{pixel}$. The K value can be obtained by calibrating the calibration board, such as the calibration board to take the length of a fixed point is l, the standard fixed point in the image corresponding to the number of pixels is P, then the ratio coefficient K=l/P.

In order to obtain pixels number in each side of the image, on the basis of reference to the relevant methods of current corner detection, using Moravec corner detection algorithm related ideas to extract the linear intersection point of each edge. The main ideas are as follows:

(1)Firstly, the gray value of each pixel in the image is retrieved one by one. And set a local window which size is 3x3. If the detected pixel is white, the other 8 points of values within the window are judged, and recorded as the number of white spots;

(2)Judge the number of white spots, if count != 5, count cleared, and if count == 5, put the coordinates value of the point into the corresponding array;

(3)At the same time, clear those adjacent pixels which have meet above conditions. The final extracted corners are marked with a, b, c, d, and the coordinates of each corner point are displayed.

By calculating the distance between the adjacent nodes, the length of each edge is calculated. In order to be advantageous to output the dimension parameter accurately, and the intersection point of the graph is dynamically marked. Provision that the point a is the minimum distance to the point of origin, and find out the maximum, minimum and intermediate values of each point of the coordinate system with axis I through the bubble sort method, and provides corresponding to the minimum point b, d corresponding to the maximum point, the intermediate value corresponds to point c.

After the above algorithm we can obtain the coordinates values of polygon corner, then get the number of pixels which contained in the line between two corners. After obtaining the coordinates of each corner, the angle value specified in the image can be obtained through the included angle of the two straight line slope, and the area of the measured object can be obtained.

Through calculating the distance between the two points, select the maximum edge length. By the above definition of point a, b, c, d can be known, which are marked in a clockwise direction. Because the shape and the angle of the four edges are random, the maximum side may appear on any edge of a, b, c, d. At the same time, according to the material requirements, after the plate is grabbed by the robot hand, should put the maximum edge on the clamp, and make sure that the whole plate is on the punch table. Therefore, it is necessary to adjust the angle of the detected angle to output the effective adjustment angle. In order to deal with all kinds of situations, decide that measure the angle between the maximum edge and the I axis by the method of piece wise analysis, that is analysis each line segment, for example the line segment ab is the longest side, and calculate the angle between the long side and the I axis. The algorithm flow of the design is shown in Fig. 12.

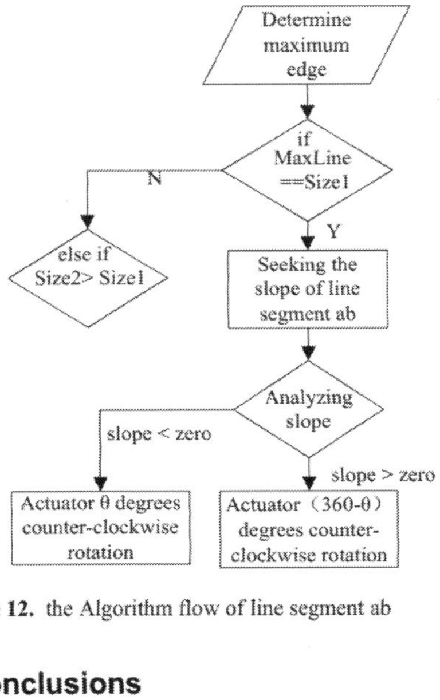

Figure 12. the Algorithm flow of line segment ab

5 Conclusions

Through the image processing and analysis for the real-time collected images, then obtain the information of plates, include contour shape, length, position and posture, which are waiting for processing, and provide real-time and accurate data support for the motion control of the automatic feeding of the robot and automatic generation of CAM machining program for punching machine.

Acknowledgment

This work is supported by Agricultural Bureau independent innovation projects in Shandong Province (the project number is 2014GJJS0401) and Agricultural machinery and equipment research and development innovation project in Shandong Province(the project number is LU NONGJI JI ZI[2015]9). In the same time, thanks for the Shandong Shuaike Machinery Manufacturing Mo., LTD. provides the experimental conditions.

References

1. S. X. Tian, Study on the material and design of blanking robot visual servo system, Donghua University, (2011).
2. J. B. Su, Application of machine vision in automatic positioning of machine parts and components, Machine Tool & Hydraulics, (2013).
3. J. X. Li, Research on the control of the mechanical and electrical integration production system based on the bus slot of CNC punch, Nanjing university of aeronautics and astronautics, (2010).
4. X. C. Mei, The development and research of the blanking robot on the stamping equipment, South China University of Technology, (2012).
5. S. T. Liu and F. L. Yin, the new development of Image segmentation method based on graph cuts, Automation journal, (2012).
6. L. Ding, Y. P. Zhang and X. Y. Zhang, Overview of image segmentation methods and performance evaluation, Software, (2010).
7. B. Zhou and B. M. Qiao, Research on the calibration method of the combination of boundary extraction and geometric invariance, Pure and applied mathematics, (2011).
8. Y. X. Sun, Boundary extraction of 2D gray image based on robust boundary tracking algorithm, Journal of ZhongKai University of agriculture and Engineering, (2009).
9. M. Pan, Y. X. Zhu, J. Li, L. Li and Y. T. Liu, Measurement of the shape and dimension of vacuum circuit breaker using visual imaging tech-nology, Manufacturing Automation, (2014).
10. G. L. Yan, J. Fu and Y. N. Yang, Part size measurement technology based on machine vision, Innovation and application of science and technology, (2013).

3D Web-based HMI with WebGL Rendering Performance

Atitayaporn Muennoi, Daranee Hormdee

Embedded System R&D Group, Computer Engineering, Faculty of Engineering, 123 Khon Kaen University, Thailand

Abstract. An HMI, or Human-Machine Interface, is a software allowing users to communicate with a machine or automation system. It usually serves as a display section in SCADA (Supervisory Control and Data Acquisition) system for device monitoring and control. In this papper, a 3D Web-based HMI with WebGL (Web-based Graphics Library) rendering performance is presented. The main purpose of this work is to attempt to reduce the limitations of traditional 3D web HMI using the advantage of WebGL. To evaluate the performance, frame rate and frame time metrics were used. The results showed 3D Web-based HMI can maintain the frame rate 60FPS for #cube=0.5K/0.8K, 30FPS for #cube=1.1K/1.6K when it was run on Internet Explorer and Chrome respectively. Moreover, the study found that 3D Web-based HMI using WebGL contains similar frame time in each frame even though the numbers of cubes are up to 5K. This indicated stuttering incurred less in the proposed 3D Web-based HMI compared to the chosen commercial HMI product.

1 Introduction

A Supervisory Control and Data Acquisition (SCADA) system is commonly used in the factory or building automation fields. A Human-Machine Interface (HMI) plays the essential role in a SCADA system. It not only presents data from sensors to a human operator, but also accepts input from the operator to send commands to control devices. Traditionally, the vendors of SCADA system provide proprietary HMI software as a part of their software product. However, along with the growth of web technology, the modern rendering technology especially, 3D graphics and the need of real-time communication can be utilized in order to exhibit production line in action [1], a 3D web-based SCADA HMI should gain popularity.

This paper studied on the performance of 3D Web-based HMI with WebGL Rendering in order to serve large 3D HMIs. Furthermore, rendering performance of 3D Web-based HMI is compared to a standard web HMI product to find out how 3D Web-based HMI can achieve a rendering performance when compared with a commercial product.

The rest of the paper is organized as follows: related work in Section 2, experimental descriptions of the 3D Web-based HMI are explained in Section 3, the performance evaluation of the 3D Web-based HMI is provided in Section 4, conclusions and ideas for further work are given in Section 5.

2 Literature Review

This section consists of previous works, evaluation metrics, 3D Web-based HMI rendering technology and Genesis64 HMI product detail, the benchmark for this work.

2.1 Previous works

Various works have been done on 3D HMI include HMI for java platform [1] and 3D HMI product which can deploy to the web [2].

An example of HMI products, Genesis64 is an HMI/SCADA automation software suite from ICRONICS Company. It presents a GraphWorX64 component that can publish HMI platforms to the web via WPF or Silverlight [3-6]. However, on Microsoft Windows, GraphWorX64 integrated with WPF for full function 3D HMI platform, DirectX and .NET Framework need to be installed then the HMI needs to be run on Internet Explorer [5, 7].

Traditional 3D web HMI usually found limitations such as specific platform and software installation required. To overcome those problems, a new approach is proposed. With 3D Web-based HMI with WebGL and WebSocket study [8], user can run the HMI on any platforms and no need to install any software. Moreover, WebSocket is used for rapid data communication between server and client instead of HTTP. The research results showed that 3D Web-based HMI could achieve a good resolution when data had a rapid changing [8].

2.2 Metrics

The indicator of successful web applications is how fast it can turn the data into pixel on screen and it still can satisfy user experience. Web HMI is also considered as one of web applications. To investigate the rendering performance of 2D/3D applications, a common metric is a frame rate or Frame per Seconds (FPS) [9-13]. Normally, web applications should produce more than 30 FPS, so human visual system cannot detect a frame skipping, which creates a good user experience [10, 11]. However, general applications are usually targeted at 60 FPS in order to be consistent with refresh rate 60Hz of most display devices [11, 12]. In other words, applications should take 16.67ms to prepare for each frame or also called frame-budget or frame time [11, 12], which frame time is the inverse of FPS. Although FPS is a common metric, but there is a defect because FPS is not linear against with the time that use to render a frame [12]. To clarify, even though applications contain the high FPS, it does not guarantee that applications can display smoothly if there is highly frame time variance. Thus, this study also analyzes the frame time in order to validate a frame time delay which causes a stuttering.

In this study, FRAPS [10, 11, 13] is used to capture the amount of FPS and frame times while web HMIs are running.

2.3 3D Rendering of 3D Web-based HMI

3D Web-based HMI uses WebGL to render a 3D scene. WebGL is a cross-platform web graphics library based on OpenGL ES 2.0 [14], which program in JavaScript. WebGL execute directly on a computer's Graphics Processing Unit (GPU) and all drawing contents are contained in the <canvas> element on HTML5 browser. To reduce complexity of using WebGL which is a low-level API, Three.js [15], a JavaScript 3D Library, is used. In Three.js, WebGLRenderer() is enabled to render a scene of 3D Web-based HMI by using a capability of WebGL. Furthermore, 3D Web-based HMI uses ability of RequestAnimationFrame() to run animations match with a display refresh rate for smooth and continuous display. [12, 16]

2.4 Genesis64, GraphWorX64 and Desktop Platform

To study rendering performance of 3D Web-based HMI, a standard 3D web HMI product was examined simultaneously in order to investigate how 3D Web-based HMI had rendering performance when it was compared with the other. Then, Genesis64, GraphWorX64 component from ICRONIC Company [7] was selected.

Genesis64 is HMI/SCADA software solution which consists of various components inside. One of those is GraphWorX64 that has a Web Publishing Wizard, which can convert a GraphWorX64 HMI files into a HTML file format and then publish the HTML file to a web server [7]. The publishing can be done in many platforms depending on a display type. This experiment focused on Desktop platform (.gdfx file) because this platform provides a full function 3D view. To use the desktop platform, client machines need to install upper .NET Framework 4.5. After this platform is published on a server, user can access the HMI via Internet Explorer [7].

3 Experiment Descriptions

The experiment was designed into 4 parts to compare our HMI against GraphWorX64;

First is Display Performance Evaluation. This was done by measuring the FPS while increasing the number of 3D cubes from 1 to 5K, where the objects could hardly move, hence, no need to explore further. Two web browsers, Internet Explorer V.11.0.26 and Chrome V.49.0.2623.110, were used. Unfortunately, GraphWorX64 can only run on IE here.

Second is Frame Time Latency Analysis. Since exploring only FPS cannot guarantee the display smoothness. Although the average FPS maybe adequate, rendering time for each frame can still swing a lot. And that could certainly deliver a terrible result with a chance of stuttering on the display. In order to analyze in depth about the rendering performance, the period each frame used for rendering for 60s was captured for rendering on only IE here. Six experiments have been conducted here for a single cube vs. 1K, 2K, 3K, 4K and 5K cubes.

Third is Execution and Rendering Time Analysis on our HMI. The purpose of this section is to identify time spent on each process. Time stamps for both execution time and rendering time were captured simply on IE.

Finally, CPU, memory and GPU usages were monitored via Windows Performance Monitor [17] and GPU-Z [18], respectively.

The entire experiment has been done on a computer with Intel® Core™ i5-3450 CPU 3.10GHz up to 3.50GHz, 8GB RAM and Graphic adapter NVIDIA Geforce GTX 550Ti with Graphics Clock 900MHz and Standard Memory Config 1024MB and Windows7 operating system.

In order to investigate the display behavior of our 3D Web-based HMI and GraphWorX64 HMI, it is necessary that both HMIs have the same initialization properties such as animation update rate, canvas size, testing object geometry, and so on. The environment of the testing in this study is defined as follow: Rendering target in section 3.1, testing regulation in section 3.2 and rendering update rate in section 3.3. All environments are for our 3D Web-based HMI and GraphWorX64 Desktop platform.

3.1 Rendering Target

In this experiment, both HMIs were specified to have the same target which could render animations continuously with rendering speed at 60FPS, which was determined in order to match with 60Hz display refresh rate of the testing display device. When the target was defined and almost-identical properties were initialized, both HMIs were ready to be investigated the rendering performance by observing the changing frame rate when both HMIs had more loads and the point where how much load effect on the frame rate to miss out of aimed target.

3.2 Testing Regulation

For the fair comparison, both HMIs must use the same testing regulations. 3D cubes were used here as the primary objects. All cubes were rendered on 1,920 x 1,080 pixel display resolution. To fit for view, five hundred cubes were stacked per level. In order to capture the changing frame rate, via one of the dynamic properties, those objects must spin around. Furthermore, this rotation movement also can illustrate on how 3D graphics have more benefit over plain 2D graphics in HMI.

3.3 Rendering Update Rate

As explained in Section 2, 3D Web-based HMI used RequestAnimationFrame() to drive animations on a screen. In the implement, it was used to increase a rotation degree of objects and it was used to update the scene and camera rendering for objects changed displaying on a screen.

Per GraphWorX64, the signal simulation was used to create an object movement and the parameters which derived from this signal used to specify an object degree. This implement uses the ramp up signal to animate the cube turns around its axis consistently. In addition, to display animations according with user requirements, GraphWorX64 needed to be setup the input property which is Update Rate. The Update Rate was used to determine the period during data will be displayed in millisecond. To study the best performance of GraphWorX64, the Update Rate 1ms was defined, which is the fastest Update Rate that GraphWorX64 can acceptable in other words GraphWorX64 can display animations continuously as fast as the HMI support.

4 Results and Discussion

Four experiments have been performed. The details are as follow.

4.1 Display Performance Evaluation

Figure 1 illustrated the results on display performance of the proposed 3D Web-based HMI on both IE and Chrome comparing with that of GraphWorX64 when increasing the number cubes. At the beginning, our HMIs can deliver the maximum Frame Rate (60FPS) for sometimes and even longer than those of GraphWorX64. And later all frame rates for all cases have dropped dramatically. Overall, our HMI on both web browsers can perform well and try to keep up with the commercial one throughout the time though not as good as GraphWorX64. Whereas at 30FPS, where human usually can detect skipping, our FPS (~1K cubes) on IE was approximately twice (only 1.3 times on Chrome) less than GraphWorX64's (~2K cubes), hence the choice of web browser is one of the factors effecting on the performance.

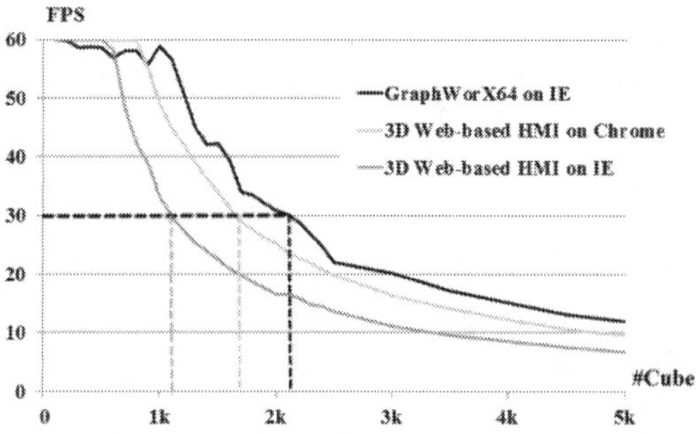

Figure 1. Frame per Second of GraphWorX64 and 3D Web-based HMI.

4.2 Frame Time Latency Analysis

Figures 2-7 shows frame time latency of 3D Web-based HMI and GraphWorX64. The display will have frequency as refresh rate of monitor (60 Hz) which means that display will be updated every 16.67ms although HMI can render frame faster than that. As shown in Figures 2, a single cube of our HMI can finish processing and rendering within less than 16.67ms. Hence, the result was displayed every 16.67ms whereas GraphWorX64 spent 16.67ms (refresh for displaying once) to 66.68ms (refresh for the same displaying four times). Spent many lengths of time rendering which means that there were different frame time as short or long frame time, so the display was inconsistent especially when the frame time was in the speed that human eyes can perceive (above 33.33ms). However, if the frame time on display is consistent (spending the same time in each frame), the display will run smoothly.

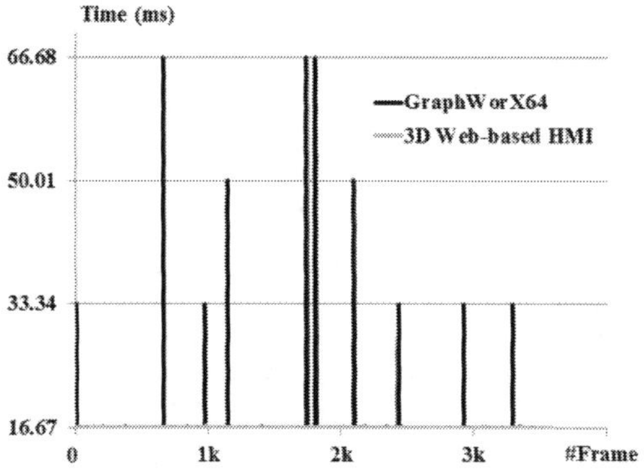

Figure 2. Frame Time Latency on 1 Cube Rendering

For 2K, 3K, 4K and 5K cubes (Figure 4-7), it can be clearly seen that even GraphWorX64 spent time producing display per 1 frame faster than our HMI (lower graph), length of time in displaying each frame was broader. That is, with the same number of objects, 3D Web-based HMI can display more smoothly except the case of 1K cubes (Figure 3) that cannot be directly compared to GraphWorX64 since any rendering that are lower than 16.67ms will only display at 16.67 ms; so, the experiment cannot detect result of GraphWorX64. On the other hand, our HMI was known that there are 2 lengths of frame time which are 16.67-33.34 and 33.34-50.01ms, and since some frames overlapped at 33.34ms, it could not finish within 2 times of refresh, and it was detected at 50.01ms (average frame time of 1K cubes is ~29ms and the time that our HMI spend for rendering per 1 cube will be explained in section 4.3).

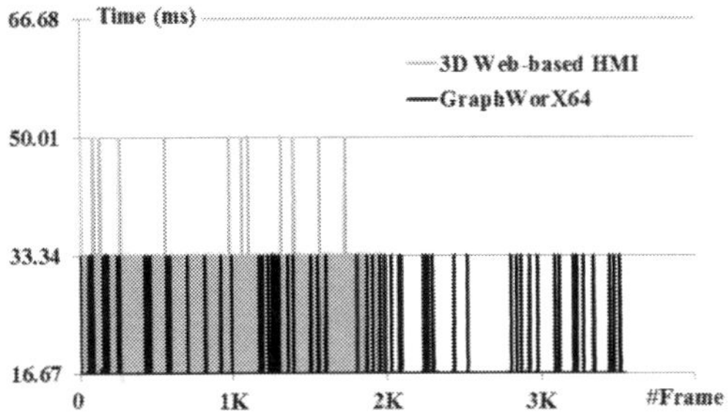

Figure 3. Frame Time Latency on 1K Cubes Rendering

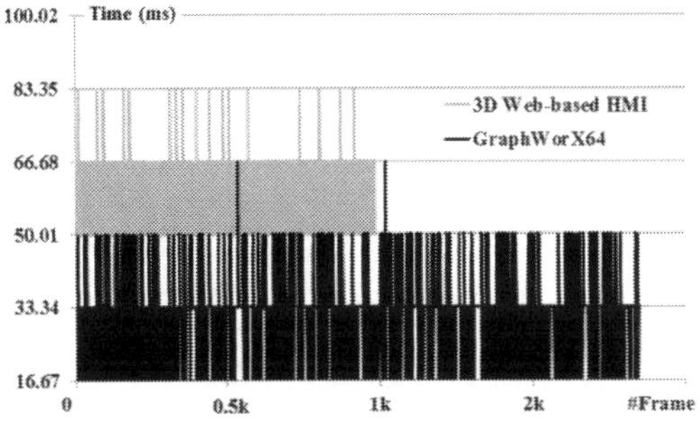

Figure 4. Frame Time Latency on 2K Cubes Rendering

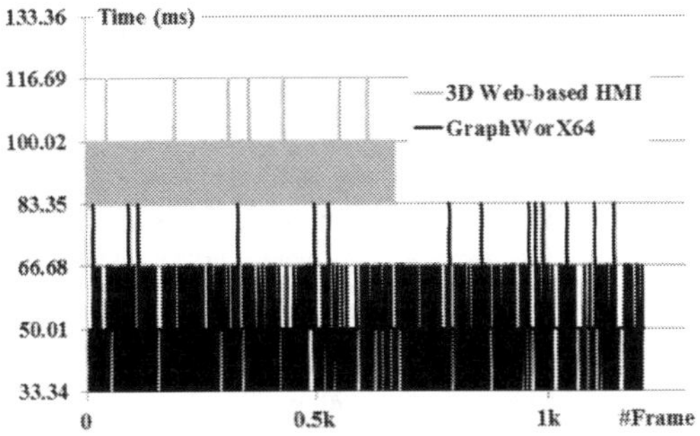

Figure 5. Frame Time Latency with 3K Cubes Rendering

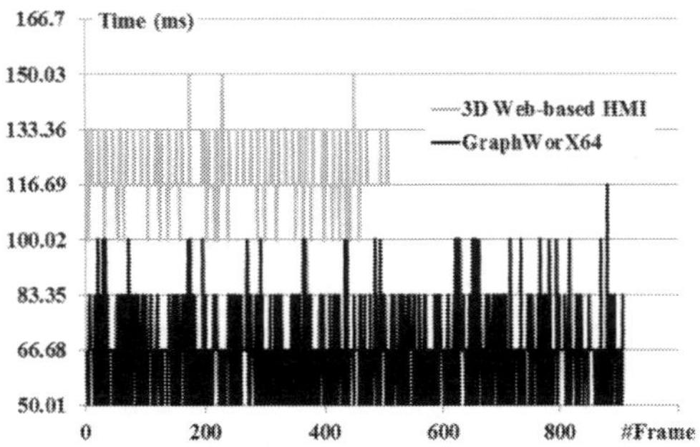

Figure 6. Frame Time Latency on 4K Cubes Rendering

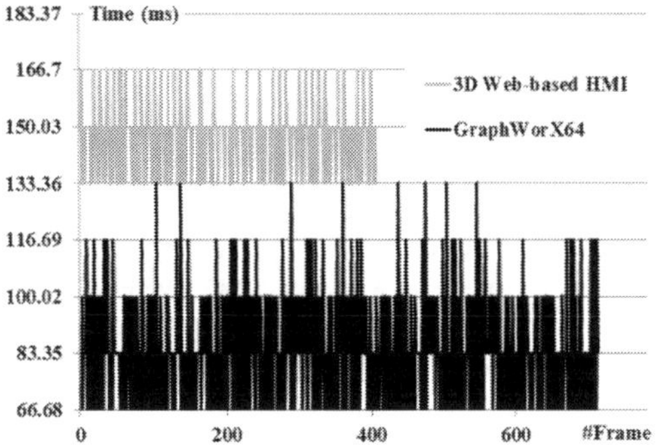

Figure 7. Frame Time Latency on 5K Cubes Rendering

4.3 Execution and Rendering Time Analysis

Examining the performance of our 3D Web-based HMI was found to be more instructive. One way to approach this is to ask, how much time does the process requires for creating the objects (Execution Time) vs rendering those (Rendering Time). Table 1 shows the results, however, execution and rendering time only for 1K-5K were large enough to be measured. Per all cases, over 96% of the time spent for displaying here went to rendering time, leaving only a small faction towards execution time. Execution Time can be reduced with a better algorithm used for resource management. While Rendering Time, caused by applying the original Three.js without any modification here, should be more focused on cause it was the majority. A well research on the code optimization, the Rendering Time can be significantly improved.

Table 1. Execution Time vs Rendering Time.

#Cube	T_{Exec} (ms)	T_{Render} (ms)	%Exec	%Render
1K	0.897	28.498	3.052	96.948
2K	1.815	57.249	3.074	96.926
3K	2.721	86.842	3.038	96.962
4K	3.655	113.928	3.109	96.891
5K	4.780	148.987	3.109	96.891

From the data, Rendering Time for a single object can then be calculated giving the results of 0.026-0.033ms. These results can endorse Figure that when the total rendering time for all objects less than 16.67ms is (refresh rate) our HMI can certainly obtain 60FPS. That complied with Figure 1 that with the number of cube under ~0.5K, our HMI can offer the maximum 60FPS.

4.4 Resource Usages

Lastly, the usages of CPU, memory and GPU have been explored. As the numbers of objects have been increased, both CPU and memory usages remained stable at 40% for CPU usage and ~240MB and ~180MB for memory for GraphWorX64 and our HMI respectively. While GPU usage seemed to be wobbly for both GraphWorX64 and our HMI, where ours used roughly 5% higher in average.

5 Conclusions and Future Work

The proposed HMI incorporated WebGL technology, a 3D graphics library commonly used in games and entertainment applications. This 3D Web-based HMI can be accessed simply via any HTML5 web browser on any platform without any specific/extra software requirement.

This paper has also presented the rendering performance evaluation of the proposed 3D Web-based HMI. The HMI contains 60 FPS when there are maximum 0.5K and 0.8K cubes (max performance) it contains 30 FPS when there are 1.1K and 1.6K cubes (minimum performance) when it was run on Internet Explorer and Chrome respectively. Frame time in each case study contain similar size, which means continuous and stable performance. For checking execution vs. rendering time, it was found that it spent most of the time in rendering as ~96%. And, it spent 0.026-0.033ms when rendering a single cube. Experiment contains CPU usage 40% for both; memory usage ~240MB and ~180MB for GraphWorX64 and our HMI respectively; GPU usage for our HMI was approximately 5% more than GraphWorX64.

As for future works, research on algorithm during execution time that can reduce the number of objects or polygons while maintaining to display the same model. The less components sent for rendering, the less loads in this process. This should yield to better HMI performance.

Acknowledgement

This work would have not been accomplished without the free product trial of Genesis64 HMI/SCADA software solution provided by ICONICS Inc.

References

1. O. Severa, M. Cech, and P. Balda, "New tools for 3D HMI development in Java," in *Carpathian Control Conference (ICCC), 2011 12th International*, 2011, pp. 342–346.
2. R. Agrusa, V. G. Mazza, and R. Penso, "Advanced 3D visualization for manufacturing and facility controls," in *2nd Conference on Human System Interactions, 2009. HSI '09*, 2009, pp. 456–462.
3. Iconics.com, *GraphWorX64 - Vector-based Graphics Representing the Real World*, 2016. [Online]. Available: http://www.iconics.com/Home/Products/HMI-SCADA/GENESIS64/GraphWorX64.aspx#.VqXYvfmLTIU
4. Iconics.com, *Mobile HMI - Instant KPIs and Alerts for Mobile Devices*, 2016. [Online]. Available: http://www.iconics.com/Home/Products/Web-Solutions/MobileHMI.aspx#.VqXZufmLTIV
5. ICRONICS, Inc, "GENESIS64 Remote Client Options." Sep-2013.
6. Iconics.com, *WebHMI - Web-Based Real-Time Automation Software*, 2016. [Online]. Available: http://www.iconics.com/Home/Products/Web-Solutions/WebHMI.aspx#.VqXcEvmLTIV
7. ICONICS, Inc, "GENESIS64 Standard Training Manual Version 10.61.".
8. A. Muennoi and D. Hormdee, "3D Web-based HMI with WebGL and WebSocket", in *International Conference on Embedded Systems and Intelligent Technology (ICESIT)*, Korea, 2014, pp. 59-64.
9. R. Hoetzlein, "Graphics Performance in Rich Internet Applications," *IEEE Computer Graphics and Applications*, vol. 32, no. 5, pp. 98–104, Sep. 2012.
10. F. Hamady, A. Chehab, I. Elhajj, and A. Kayssi, "User experience-based mechanism for preserving energy in graphics-intensive applications," in *New Circuits and Systems Conference (NEWCAS), 2013 IEEE 11th International*, 2013, pp. 1–4.
11. E. Rathgeb, K. Echtle, and B. Müller-Clostermann, *Measurement, Modelling, and Evaluation of Computing Systems and Dependability in Fault Tolerance: 15th International GI/ITG Conference, MMB & DFT 2010, Essen, Germany, March 15-17, 2010, Proceedings*. Springer Science & Business Media, 2010.
12. A. Anyuru, *Professional WebGL Programming: Developing 3D Graphics for the Web*. John Wiley & Sons, 2012.
13. Fraps.com, *FRAPS game capture video recorder fps viewer*, 2016. [Online]. Available: http://www.fraps.com
14. Khronos.org, *WebGL - OpenGL ES 2.0 for the Web*, 2016. [Online]. Available: https://www.khronos.org/webgl/
15. Threejs.org, *three.js - Javascript 3D library*, 2016. [Online]. Available: http://threejs.org/
16. Mozilla Developer Network, *Window.requestAnimationFrame()*, 2016. [Online]. Available: http://developer.mozilla.org/en-US/docs/Web/API/window/requestAnimationFrame
17. Windows Performance Monitor, *Technet.microsoft.com*, 2016. [Online]. Available: https://technet.microsoft.com/en-us/library/cc749249.aspx
18. TechPowerUp, *TechPowerUp*, 2016. [Online]. Available: https://www.techpowerup.com/gpuz

MATEC Web of Conferences 77, 09004 (2016) DOI: 10.1051/matecconf/20167709004
ICMMR 2016

Intelligent Transportation Control based on Proactive Complex Event Processing

Yongheng Wang, Shaofeng Geng and Qian Li

College of Computer Science and Electronic Engineering, Hunan University, Changsha, China

Abstract. Complex Event Processing (CEP) has become the key part of Internet of Things (IoT). Proactive CEP can predict future system states and execute some actions to avoid unwanted states which brings new hope to intelligent transportation control. In this paper, we propose a proactive CEP architecture and method for intelligent transportation control. Based on basic CEP technology and predictive analytic technology, a networked distributed Markov decision processes model with predicting states is proposed as sequential decision model. A Q-learning method is proposed for this model. The experimental evaluations show that this method works well when used to control congestion in in intelligent transportation systems.

1 Introduction

Internet of Things (IoT) technology provides dynamic global network infrastructure where objects in real world can be connected into the internet based on various kinds of sensors. Recently, with the rapid development of information and communication technologies, building an IoT application is no longer difficult. The main problem that people concern now is how to process the huge events produced by IoT applications.

Complex Event Processing (CEP) [1] is the technology to accept as input a stream of primitive events that happen in the external environment, interpret and combine them to identify higher level composite events. The high level events can be sent to the controller of the system which can execute some actions to determine the overall system's behavior. CEP has been used in many areas, such as sensor networks for environmental monitoring [2], continuous analyzing of stocks to detect trends [3], etc.

The traditional kind of CEP is called reactive event processing which means agents are triggered by events and react to change states of the system. Another kind of method is called proactive event processing, which means system has the ability to mitigate or eliminate undesired future events, or to identify and take advantage of future opportunities, by applying prediction and automated decision making technologies [4]. IoT with proactive CEP provides a new way to improve Intelligent Transportation Systems (ITS) or transportation IoT. For example, we can predict some future congestion states based on the current state and historical data, and then take some actions to avoid some future congestion states.

To support proactive CEP, besides the core CEP technology, we also need Predictive Analytics (PA) and sequential decision making technology. PA is the technology that can be used to predict future events or system states through the analysis of historical events or states. As an optimal sequential decision making process for stochastic dynamic systems, Markov Decision Processes (MDP) becomes a reasonable choice for sequential decision making in proactive event processing. However, currently there are still some challenges to support proactive CEP. In the PA part, we need to develop models and algorithms that have high accuracy and good performance for IoT applications. Sequential decision making with MDP for proactive CEP is a more challenging technology since we need to handle not only huge joint state and action space, but also predicting state which is not included in standard MDP.

In this paper, we propose a proactive CEP architecture and method (Pro-CEP) for intelligent transportation control. The proactive CEP architecture contains probabilistic event processing network, predictive analytic component and decision making component. An extended Networked Distributed Markov Decision Processes (ND-MDP) model with predicting states and corresponding Q-learning method are also proposed. The Pro-CEP method is evaluated in a simulated transportation IoT and the results show it works well when processing proactive congestion control in transportation IoT.

2 Related Works

CEP recognizes complex events from a set/sequence of raw events by continuously monitoring the data flow.

This project is sponsored by the National Natural Science Foundation of China, under grant 61371116.

Luckham and Etzion introduced the basic concepts and architecture of CEP [1,5]. Event Processing Agent (EPA) is a component that takes a set/sequence of primitive events as input and output a set of complex events according to some logic. Event Processing Network (EPN) is a network composed of EPAs, event producers, and event consumers linked by channels.

Some research work about proactive event based systems has been proposed, such as proactive management of transport processes and proactive application event notification in sensor network. Engel et al. proposed a proactive event processing framework based on CEP, PA and MDP [4]. In their framework, two new types of agents are added: predictive agents which can derive future uncertain events based on prediction models, and proactive agents which can choose the best proactive action that should be taken. Compared with our work, the work of Engel et al. lacks of implementation detail and is not optimized for large scale IoT application. We used the similar framework but we investigated novel predictive analytic and decision making method for transportation IoT.

MDP has been studied for many years and recently several variants of MDP emerged. If we use MDP in proactive CEP, one of the challenges is the NP-complete problem caused by large state space and action space. Basically there are two research directions on this problem. In the first direction the problem is simplified by using the information from the application domain. For example, Jia et al. used state aggregation methods [6]. It is difficult to find a common solution in this direction since these methods are related to specific application domain. In the second direction, approximate methods are used for large scale MDP. For example, Cao et al. used dynamic programming based on events [7]. The key idea in these methods is to approximate the value function during the optimization process.

In many IoT applications, the agent does not have complete model of the environment and does not know the exact reward function. Therefore reinforcement learning becomes a reasonable choice. Zhang et al. proposed a model-free, scalable learning approach that synthesizes multi-agent reinforcement learning and distributed constraint optimization (DCOP) [8]. Their method distributes the learning of the joint policy and employs DCOP techniques to coordinate distributed learning to ensure the global learning performance. In order to control traffic congestion, Fares et al. proposed a system that introduces a new microscopic framework at the network level based on collaborative MDP modeling and an associated cooperative Q-learning algorithm [9]. Compared to our work, the current MDP models and learning methods try to find optimal policy based on current state or history state but not predicted future state.

3 System Architecture

The goal of our system is to control traffic congestion intelligently using proactive CEP in transportation IoT. There are various sensors in the system to get the position of vehicles, such as induction loops, RFID, GPS, etc. There are also other sensors to get environment information, such as temperature and humidity sensor, wind sensor, etc. The traffic control agents try to reduce potential congestion through controlling the traffic light.

The system architecture is shown in figure 1. In order to process uncertain event, we extended the EPN framework into Probabilistic Event Processing Network (PEPN) which is composed of many connected Probabilistic Event Processing Agents (PEPA). Raw events are generated by sensors or other information system (such as weather broadcast system). The imprecise raw events are processed by PEPN to get probabilistic complex events, which represent some high level semantic such as the position or running path of a vehicle. Based on the run time prediction of congestion state by predictive analytic component, the decision maker selects optimal actions using MDP model and assigns corresponding proactive agents (PRA) to execute the actions.

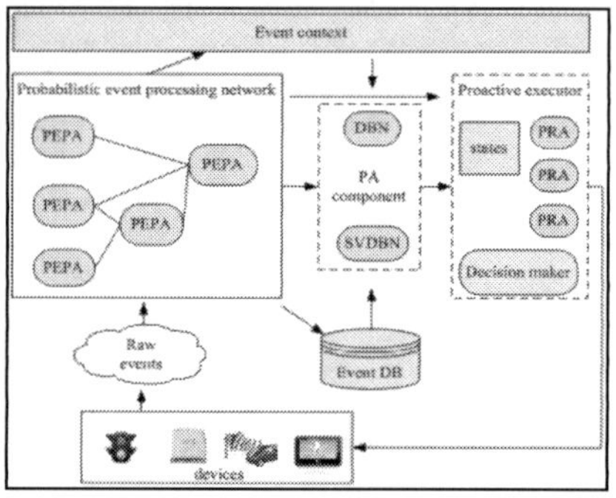

Figure 1. System architecture.

Definition 1 (primitive event): A primitive event means an atomic occurrence of interest in time which is represented as <A, T, P_r> where A is the set of attributes, T is the timestamp that the event occurs and P_r is the probability value of the event.

Definition 2 (complex event): Complex event is a combination of primitive events or complex events by some rule which is represented as <E, R, T_s, P_r> where E represents the elements that compose the complex event, R represents the rule of the combination, T_s represents the time span of the complex event and P_r is the probability value.

Event type is a specification for a set of events that have the same semantic intent and the same structure. Every event is an instance of an event type. An event type can represent either primitive events or complex events. The main complex event patterns include ALL, ANY, COUNT, SEQ, etc. The detailed meaning of the patterns is described in [5]. For example, in the area of transportation system, the COUNT pattern can be used to count the number of vehicles in a specified area during specified time span. The SEQ pattern can be used to represent the running path of a vehicle.

We have developed a high-performance Distributed Probabilistic Complex Event Processing method (DPCEP). DPCEP extends SASE [10] which is based on

DOI: 10.1051/matecconf/20167709004

NFA. Parallel algorithm is designed to improve the performance for both single and distributed streams. A query plan based method is used to process hierarchical complex event from distributed event streams. Query plan optimization is proposed based on query optimization technology of probabilistic databases.

In our work we only consider event matching EPA which is the most important one. PEPA is the extension of EPA that supports probabilistic event processing and we also call it probabilistic CEP engine. Multiple PEPAs are connected by event channel to create an EPN. An event channel is a processing element that receives events from one or more source processing elements, makes routing decisions, and sends the input events unchanged to one or more target processing elements in accordance with these routing decisions. The detailed information about DPCEP is described in [11]. The predictive analytic component is implemented using Adaptive Bayesian Network (ABN) based on the work of Pascale et al. [12].

4 Sequential Decision Making

Definition 3 (ND-MDP$^+$): A Networked Distributed Markov Decision Processes with future states (ND-MDP$^+$) is defines as $\langle I, S, \bar{S}, \vec{A}, P, R\rangle$, where I is a finite set of agents. S is a finite set of states, with distinguished initial state s_0. $\bar{S}$ is a finite set of future states which can be predicted from historical states. $\vec{A}=\times_{x\in I} A_i$ is a set of joint actions, where A_i is the set of actions for agent i. P : $S\times\vec{A}\rightarrow S'$ is the state transition function, defining the distributions of states that result from starting in a given state and a joint action by agents. R: $S\times\vec{A}\rightarrow\Re$ is the reward function for the set of agents for each set of joint actions and each state.

Being different from standard ND-MDP, ND-MDP$^+$ contains a future state set $\bar{S}$ which can be predicted using the predictive analytic method. The proactive event processing system starts from predicting future state and select a joint action to be executed by agents according to the joint policy. The execution of the joint action maximizes the total expected reward for both the future state and the current state.

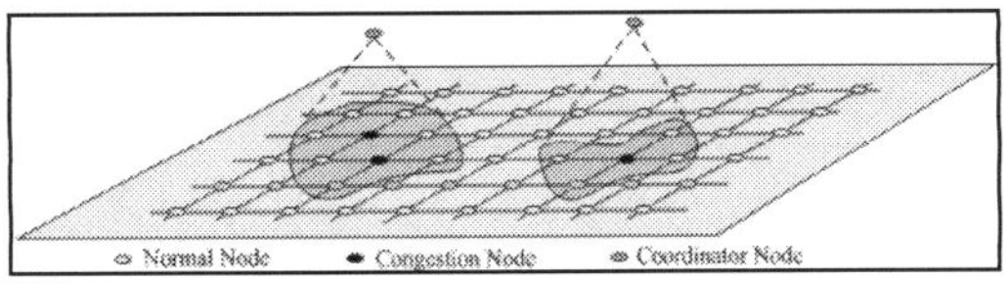

Figure 2. Transportation network partition.

In order to resolve the congestion control problem with ND-MDP$^+$, we use a dynamic network partition method to partition the traffic network as shown in Figure 2. We partition every predicted congestion node and it neighbor nodes into a group. If two groups are overlapped, we simply combine them into one group. Based on the assumption that congestion nodes are sparse, we can get a few separated groups. A coordinator node is selected for every group. We use G to represent the partitioned hyper graph and every $g\in G$ connects all the nodes in a group.

We use the reinforcement learning method Q-learning to learn the joint policy. The Q-function $Q(s,\bar{s},a)$ represents the reward of executing joint action a under state s and predicting state $\bar{s}$. The optimal joint policy π^* can be derived from $Q(s,\bar{s},a)$ by

$$\pi^*(s) = \arg\max_{a\in A} Q^*(s,\bar{s},a) \quad (1)$$

The extended Q-learning can be described by the following equation:

$$\begin{aligned} Q(s^t,\bar{s}^t,a^t) &= (1-\alpha)Q(s^t,\bar{s}^t,a^t) \\ &+ \alpha[(1-\beta)r_s^t + \beta r_{\bar{s}}^t \\ &+ \gamma \max_a Q(s^{t+1},\bar{s}^{t+1},a)] \end{aligned} \quad (2)$$

where $\alpha\in(0,1)$ is the learning rate, $\gamma\in[0,1]$ is the discount factor, r^t is the reward executing action a on state s, $Q(s^t,\bar{s}^t,a^t)$ represents the Q-value of executing action a for joint state $(s,\bar{s})$ at time t. The joint reward r^t contains two parts where r_s^t means the reward from the current state S to its next state and and $r_{\bar{s}}^t$ means the reward from predicting state $\bar{S}$ to its next predicting state. $\beta\in(0,1)$ is a weight factor.

The coordination graph (CG) framework [13] for collaborative multi-agent system assumes the action of an agent i only depends on a subset of the other agents $\Gamma(i)$. This assumption is consistent with our transportation scenario since the traffic flow of an intersection is affected by the nearby intersections. In this paper, $\Gamma(i)$ means the agents that in the same group with i. Based on this assumption, we use a decomposable Q-function $\hat{Q}(s,a)$ to approximate the global Q-function $Q(s,a)$. This $\hat{Q}(s,a)$ is defined as a sum of local Q-functions of the neighbor agents:

$$\hat{Q}(s,\bar{s},a) = \sum_{l\in G} Q(s_l,\bar{s}_l,a_l) \quad (3)$$

Where $Q(s_l,\bar{s}_l,a_l)$ is the Q-value for agent on node l by executing action a_l on joint state $(s_l,\bar{s}_l)$ and behaving globally optimally from then on in respect to maximizing $\hat{Q}(s,\bar{s},a)$. This approximation is reasonable because the global reward (congestion reduction) is the sum of local reward and $\hat{Q}(s,\bar{s},a)$ is closely related to $Q(s_l,\bar{s}_l,a_l)$. With the $\hat{Q}(s,\bar{s},a)$ approximation, the Q-learning update rule in equation (2) can be changed to

$$\begin{aligned} \sum_{l\in G} Q_l(s_l^t,\bar{s}_l^t,a_l^t) &= (1-\alpha)\sum_{l\in G} Q_l(s_l^t,\bar{s}_l^t,a_l^t) + \\ &\alpha\{\sum_{l\in G}[(1-\beta)r_{sl}^t + \beta r_{\bar{s}l}^t] + \gamma\max_a \hat{Q}(s^{t+1},\bar{s}^{t+1},a)\} \end{aligned} \quad (4)$$

In order to decompose $\max_a \hat{Q}(s^{t+1},\bar{s}^{t+1},a)$, we define $a^* = \arg\max_a \hat{Q}(s,\bar{s},a)$. Then we get

$$\begin{aligned} \max_a \hat{Q}(s^{t+1},\bar{s}^{t+1},a) &= \hat{Q}(s^{t+1},\bar{s}^{t+1},a^*) \\ &= \sum_{l\in G} Q_l(s_l^{t+1},\bar{s}_l^{t+1},a_l^*) \end{aligned} \quad (5)$$

Then the Q-learning update rule for each local group is

$$Q_l(s_l^t, \bar{s}_l^t, a_l^t) = (1-\alpha)Q_l(s_l^t, \bar{s}_l^t, a_l^t) + \alpha\{[(1-\beta)r_{sl}^t + \beta r_{\bar{s}l}^t] + \gamma Q_l(s_l^{t+1}, \bar{s}_l^{t+1}, a_l^*)\} \quad (6)$$

In order to find the joint action that maximizes the joint Q-value for both current state and predicting state, we use the max-sum algorithm proposed by Stranders et al. [14]

5 Experimental Evaluation

5.1 Simulation System

In order to evaluate our method, we created a transportation IoT simulation system which is shown in Figure 3. This system is based on the road traffic simulation package SUMO [15]. The simulation system supports getting event data from the traffic network and executing actions. SUMO supports "induction loops" which can detect vehicles that pass corresponding areas. Getting the location of every vehicle at any time is also supported. The event manager is connected to SUMO through the TraCI (Traffic Control Interface) interface of SUMO to get the induction loop variables and vehicle location variables. The values of these variables are used to simulate RFID and GPS readers.

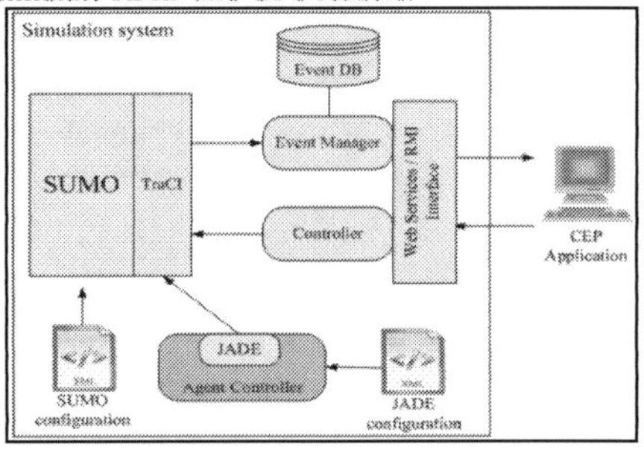

Figure 3. Simulation system architecture.

In the simulation system we set a road network of 15×15 intersections and set 80 thousand vehicles. Every road has 2 lanes and every intersection has traffic lights. In order to simulate real traffic system, a set of rules are defined. Every vehicle has a home location and an office location. A vehicle v_i runs between home and office with probability p_i. The vehicles also go to other places such as supermarket, hospital, etc., with corresponding probabilities. In order to support more intelligent control of vehicles, we use the JADE (Java Agent DEvelopment Framework) framework which can support multi-agent system. Based on this framework, the agent controller can control the moving of vehicles in SUMO with intelligent algorithm which considers the environment.

Based on this simulation system we evaluated the precision and performance of our method. We used four servers with Xeon E3 processor and 16GB memory. The operating system is Ubuntu 14.04. Three of the servers are used to run Pro-CEP and the last server is used to run the simulation system.

In order to evaluate Pro-CEP method, we partitioned the congestion value (calculated based on the waiting queues of intersections) into 10 levels where "0" means no congestion and "9" means the highest congestion. Engel et al. proposed a framework and basic model without implementation detail. We have not found other work which has the same function with ours. Therefore our method is only compared with the default simulation system and the result is shown in Figure 4. In this experiment, only directed neighbors are partitioned into the group of predicted congestion nodes. From the figure we can see the average congestion level reduced obviously when Pro-CEP method is used. The reason is that Pro-CEP method can predict the congestion state and take some actions to avoid it proactively. The figure also shows that the reduction of congestion level becomes more obvious when the congestion level becomes high. We think the reason is that when the congestion is not serious, the proactive actions can't affect the states too much.

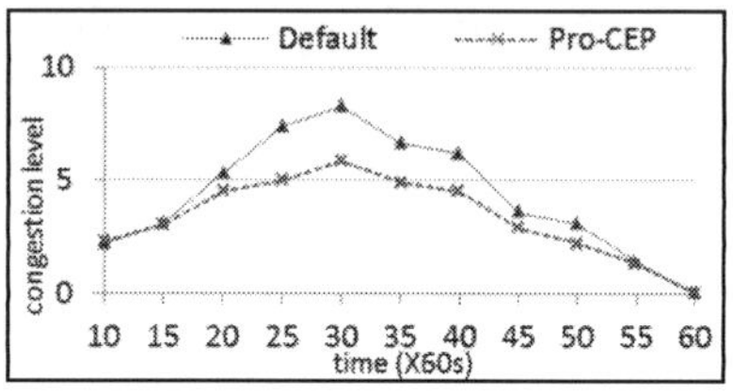

Figure 4. average congestion level over time

In the next experiment we evaluated the effect of Pro-CEP with different vehicle numbers and group partition policies. We use average reduction level to represent the effect of Pro-CEP which means the average reduced level when Pro-CEP is used comparing to default observed values. We use 3 group partition policy L_1, L_2 and L_3 where the number i in L_i means how many layers of neighbors are partitioned into the group of a predicted congestion node. The result is shown in Figure 5. As we can see, when vehicle number increases from 40 thousand, the average reduction level decreased. We think the reason is that when vehicle number is large, the density of congestion nodes becomes large. Some neighbor congestion nodes are partitioned into the same group which makes our method difficult to find optimal joint actions. We can also find that when the number i in L_i is increased, we get better result but the improvement is slight. From this result we believe in our experiment the reduction of congestion for a node is mainly related to its direct neighbor nodes.

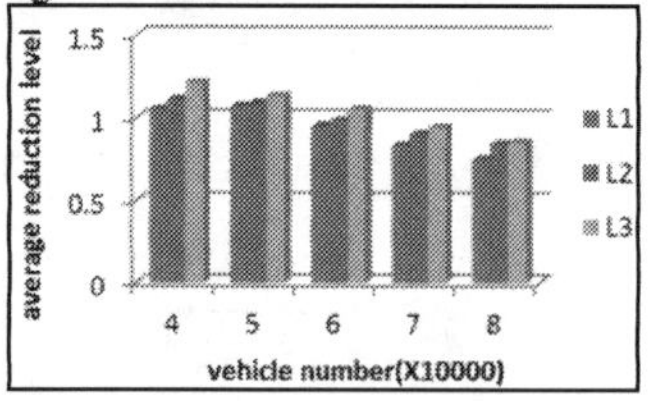

Figure 5. effect with different vehicle numbers and group partition policies

We also evaluated the performance of Pro-CEP. The average decision time with different vehicle number and group partition policy is shown in Figure 6. We can find the average decision time increases when the vehicle number becomes large. The reason is that when vehicle number is large, more neighbor congestion nodes are partitioned into same group which makes the decision process more complex. We can also find when the number i in L_i is increased, the average decision time is increased obviously. The reason is that more neighbors need to be considered and the group becomes large. From Figure 5 and Figure 6 we conclude that we should not use large i for L_i.

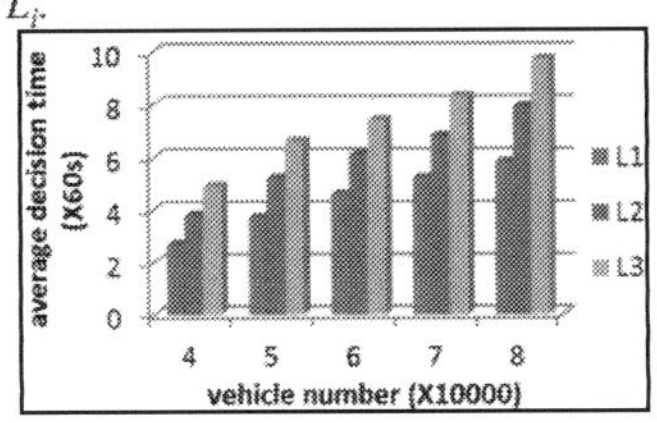

Figure 6. performance with different vehicle numbers and group partition policies

From all the experiments we can see Pro-CEP works well when processing proactive congestion control in transportation IoT. The ND-MDP$^+$ model has obvious effect and acceptable performance for congestion control in large transportation IoT.

6 Conclusion

In this paper we proposed a proactive complex event processing method for intelligent congestion control in transportation IoT. A networked distributed Markov decision processes model with predicting states is used as sequential decision model. Q-learning method is proposed to find optimal joint policy. The experimental evaluations show that this method works well for proactive congestion control in transportation IoT.

The performance and scalability of Pro-CEP still need to be improved. In the max-sum algorithm for ND-MDP$^+$, new optimization method needs to be find if we want to use this method for large scale applications.

References

1. D.C. Luckham. Event Processing for Business: Organizing the Real-Time Enterprise. Wiley Press, Dec. 2011.
2. K. Broda, K. Clark, R Miller, et al. SAGE: a logical agent-based environment monitoring and control system. Ambient intelligence, Lecture Notes in Computer Science, vol 5859. Springer, Berlin, Heidelberg, pp 112–117.
3. A.J. Demers, J. Gehrke, M. Hong, et al. Towards expressive publish/subscribe systems. Lecture Notes in Computer Science (including subseries Lecture Notes in Artificial Intelligence and Lecture Notes in Bioinformatics), v 3896 LNCS, 2006, pp. 627–644.
4. Y. Engel, O. Etzion. Towards proactive event-driven computing. In Proc. of Fifth ACM International Conference on Distributed Event-Based Systems, DEBS 2011, New York, pp.125-136.
5. O. Etzion, P. Niblett. Event Processing in Action. Manning Publications, 2010.
6. Q. S. Jia. On State Aggregation to Approximate Complex Value Functions in Large-Scale Markov Decision Processes. IEEE Transactions on Automatic Control, 56(2), pp. 333- 344.
7. Q. S. Jia and Q. C. Zhao. Strategy optimization for controlled Markov process with descriptive complexity constraint. Science China Series F: Inform. Sci., 52(11), pp. 1993–2005.
8. C. Zhang, V. Lesser. Coordinated Multi-Agent Reinforcement Learning in Networked Distributed POMDPs. In Proc. of the 25th AAAI Conference on Artificial Intelligence and the 23rd Innovative Applications of Artificial Intelligence Conference, August 7, 2011, San Francisco, CA, United states, pp. 764-770.
9. A. Fares, W. Gomaa. Multi-Agent Reinforcement Learning Control for Ramp Metering. Advances in Intelligent Systems and Computing, Vol. 1089, pp. 167-173.
10. E. Wu, Y. Diao and S. Rizvi. High-performance complex event processing over streams. In Proc. of 2006 ACM SIGMOD international conference on Management of data, June 27-29, 2006, Chicago, IL, USA.
11. Y.H Wang, K. Cao and X.M Zhang. Complex Event Processing over Distributed Probabilistic Event Streams. COMPUTERS & MATHEMATICS WITH APPLICATIONS, 66 (2013), pp1808–1821.
12. A. Pascale, M. Nicoli. Adaptive Bayesian network for traffic flow prediction. In Proc. of Statistical Signal Processing Workshop (SSP), 2011 IEEE, pp.177-180.
13. M. Papageorgiou, H. Hadj-Salem, J. M. Blosseville. Alinea: A local feedback control law for on-ramp metering. Transportation Research Record (1320) (1991).
14. R. Stranders, A. Farinelli, A. Rogers, et al. Decentralised coordination of mobile sensors using the max-sum algorithm. In IJCAI, issue 2009, pp.299–304.
15. M.Behrisch, L. Bieker, J. Erdmann, et al. Sumo - simulation of urban mobility: An overview. Proceedings of the third international conference on advances in system simulation, Barcelona, Spain, October, pp. 63–68.

 DOI: 10.1051/matecconf/20167709005

Vision and Force Feedback Control for Microassembly Process

Chen Zhi yao, Ye Xin , Zhang Zhi jing, Cai Shao peng and Li Yang

School of Mechanical Engineering, Beijing Institute of Technology, Beijing, P.R. China

Abstract. Microassembly process is one of the difficulties in the field of manufacturing. According to the structural characteristics and assembly process difficultly of 3D meso-scale micro-devices and the requirement of multi-axis control in coaxial alignment assembly, this paper developed a vision and force feedback control method oriented to Microassembly System with Coaxial Alignment. This system combines multi-axis motion control with machine vision and force sensors, which can complete nondestructive assembly of parts and increase the precision of microassembly process into micron level.

1 Introduction

In the current microelectromechanical systems (MEMS) research field, high precision MEMS composed of meso-scale micro devices is the mainstream of civil and national defence micro manufacturing systems [1].

In the core equipment with urgent needs of the science and technology project and high precision MEMS manufacturing, guide instruments in aerospace and aviation field, intelligent fuze and NIF ignition targets, all belong to the meso-scale micro device demanding assembly gaps. The assembly level of those devices highly restricts the development of the high-end manufacturing of new generation into high precision and high stability. The assembly gaps of those instruments and components must be strictly restricted in micron level, and the alignment detection precision must be restricted in submicron level.

According to the features of those components and systems, an in-house microassembly system with coaxial alignment function (MSCA) was proposed [2]. This microassembly system uses a CCD camera to detect the position and pose of parts, and then assembles in the way of man-machine cooperation. Different target parts were assembled into the base part in the assembly process.

Microassembly process with coaxial alignment function is divided into two stages: the first stage, before target part contacting with base part, the visual feedback can detect the position and variation deviation between the two parts; the second stages, withdrawing the vision system and begin to assembly, during the target part contacting with base past, the force feedback can detect the position and variation deviation between the two parts further. In order to achieve non-destructive microassembly, this paper proposed vision and force feedback control method for coaxial alignment.

This paper is organized as follows. Section 2 reviews the motion platforms in the MSCA, vision system, and control system. The vision feedback and the force feedback are discussed in section 3. Section 4 describes the experiment process and result of this work. Section 5 concludes the work and gives the future direction of our work.

2 Background

2.1 MSCA

MSCA is used for the micro scale robotic assembly operation through the way of man-machine cooperation. As shown in Fig. 1.

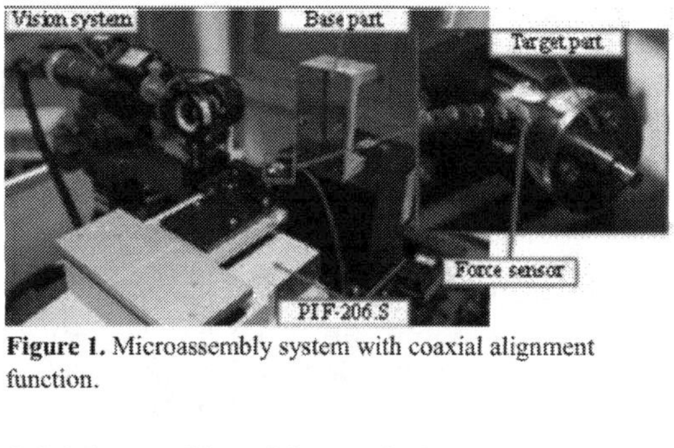

Figure 1. Microassembly system with coaxial alignment function.

2.1.1 Composition of the manipulator

This manipulator uses a combination of macro and micro movement to achieve the combination of a long working stroke and short precision feed to adjust the position and orientation of the part. Macro moving platforms, which supports two vacuum absorption micro-grippers and 6-DOF force sensors (ATI NANO17, resolution: 0.3mN,

0.004N·mm), uses THK precision module (positioning precision:0.02mm, repeated positioning precision) and grating feedback in the axial feed, closed-loop configuration open loop control mode in order to record the precise macro feed amount. Micro moving platform uses a hexapod 6-DOF precision alignment worktable, PI F-206.S, which supports a mechanical fixture. The former module can provide 100mm large motion range with a resolution of 0.02mm while the latter worktable can only translate ±5.7mm motion range along the x-, y-, and z-axes with the translational resolution of 0.1μm and can also rotate 5.5° along γ-, β-, and α-axes with the rotational resolution of 2μrad.

2.1.2 Vision system

A vision system is installed onto the MSCA to provide visual feedback for the assembly through the way of man-machine cooperation. The vision system consists of a cube prism, monochrome Basler scA1600-14gm digital camera (resolution: 1626×1236, optical size: 1/1.8'', pixel size: 4.4μm), auto zoom and focus lens remodeling by Navitar Zoom 12X. The working distance is 86, optical magnification is 0.58X~7X. Two white point light sources, OPT-RI5090 and OPT-RI5045, is used as the light source.

2.2 Control system

MSCA has to control more than 20 axis movements, and how to guarantee the good stability and coordination between different axes has become one of the difficulties in current High Precision Micro Assembly Control System. There are five kinds of typical control system architecture: ① base on CNC system; ② base on PLC; ③ base on general controller; ④ base on computer and special controller; ⑤ base on field bus [3]. Among them, the forth architecture is more excellent than others in multi-axis motion control, hardware compatibility, scalability and development platform. According to the features of MSCA, this paper developed an "industrial computer+ special controller+ UMAC" open control system based on the forth architecture.

2.2.1 Hardware selection

Advantech 610L is used as industrial control machine which has a large number of various interfaces, not only meets the current requirements, but also extends other functions through the rest interfaces according to follow-up research and development. We select Yaskawa SGD7S servo drives as the macro moving parts (high-speed, high frequency and high precision positioning control) to control. The F-206 Hex Align controller for micro moving parts (PI F-206.S) performs sophisticated six-space coordinate transformations and path planning internally, automatically. Multi-axis motion control card selects UMAC, which has good openness and scalability, can control up to 32 axes simultaneous movement.

2.2.2 Software design

In the "industrial computer+ special controller+ UMAC" architecture, industrial computer as the host, whose task is to the communicate with special controllers and data acquisition cards, get the system status in the assembly process, allocate control tasks and coordinate lower controllers simultaneously to complete the assembly task through the multi-tasking operating system. Special controllers belong to execution layer, receive host commands, and perform real-time control function according to specified parameters.
Depending on the different functions to be realized, control system application software includes modules shown in Fig. 2.

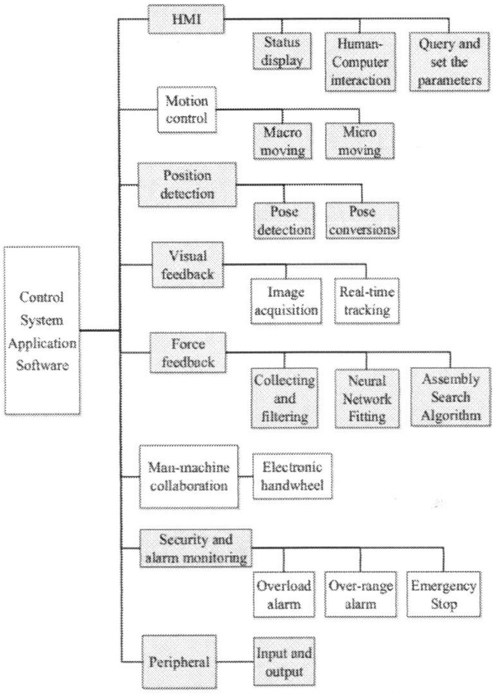

Figure 2. The architecture of control system application software.

3 Vision and force feedback

3.1 Vision feedback

Principle of coaxial optical alignment detection is shown in Fig. 3. A special prism was integrated to capture the images of the base part and target parts, which has two special surfaces: Surface 1 was a fully reflecting surface. Surface 2 was coated with a semi-anti-semi permeable. The CCD camera could capture the image of the target part and the base part simultaneously through the prism and obtain the base part and the target part's position corresponding relationship in the same image plane in different pose. As can be seen from the Fig. 3, optical path difference exists in the process of completing the image capture work of the target part and the base part. If we want to get a clear image of base part and target part simultaneously, wo must do the initial calibration before alignment to eliminate the optical path difference and obtain a clear image.

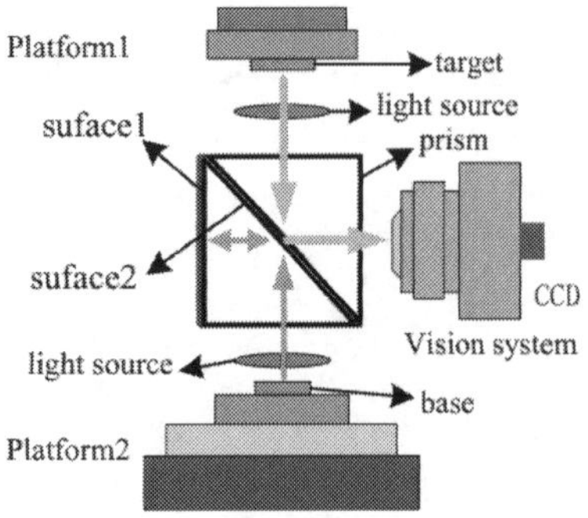

Figure 3. Principle of coaxial optical alignment detection.

We can see the optical path difference between the target part and the base part of the cube prism from Fig. 3, the light of target part reflected directly to the camera image plane through the surface2 of the prism, the light of the base part reflected to the camera image plane through the surface1 in the left of the prism after penetrating the surface 2. In that way, the light energy losses of the target part and the base part are different in cube prism. Therefore, the target part and the base part must be taken in the way of individually lighting, and the system uses time-sharing imaging control strategy. Coaxial alignment system is shown in Fig. 4.

Figure 4. Coaxial alignment system.

Lens focusing and zooming controller use Navitar's Motor driver 1-62506, This driver application allows the user to configure and control a two motor (zoom and focus) system complete with limit sensors. The controller is connected via serial communication. Light source control selects the digital voltage controller OPT-DP1024-4, its performance parameters are shown in Table 1.

Table 1. Lights source controller performance parameters.

	Parameter
Model	OPT-DP1024-4
Power supply	AC100~240(V)
Voltage output	14.0V~24.0(V)
Brightness level	256
Number of Channels	4
Single Output	0.65(A)

3.2 Force feedback

The main function of force feedback is threshold alarm during the parts contacting. As shown in figure 5, in the process, adjusting the assembly micro devices space pose, calculating the stress of the micro devices of various pose, finding the maximum stress, comparing with the permissible stresses of materials, obtaining the range of stresses that realize nondestructive assembly and calculating the corresponding range of assembly force. According the calculation results, microassembly with coaxial alignment system can realize micro devices nondestructive assembly on the premise of the assembly force in the safety range. In the assembly process, strictly monitoring the force of the specific direction whether exceeds the safety threshold or not.

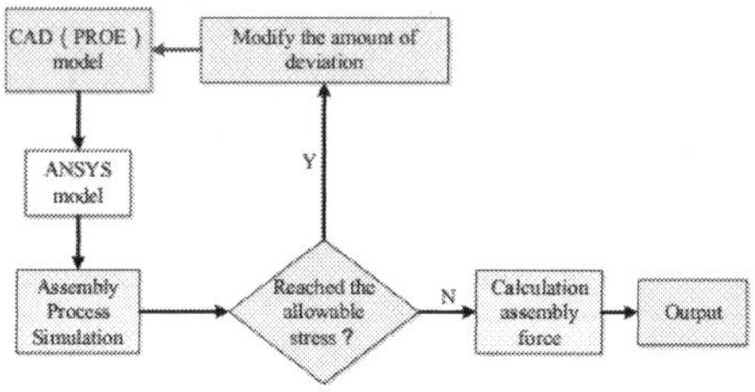

Figure 5. Step of virtual detection.

4 Experimental results and discussion

Experiment based on MSCA, the author conducted coaxial alignment assembly experiment. In order to verify the assembly accuracy of MSCA, we specially designed and processed 3mm diameter shaft-hole array assembly parts of high-precision assembly, as shown in Fig. 6. We selected the 3rd shaft and the 3rd hole as a precision verification target, which single coordinate clearance is less than 2μm. The 3rd hole' s diameter is 2.9994mm, cylindricity is 0.0021mm; the 3rd shaft' s diameter is 2.9971mm, cylindricity is 0.0012mm.

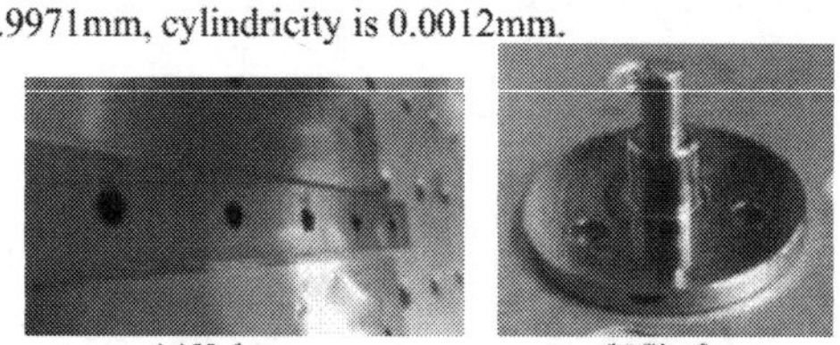

(a)Hole array (b)Shaft

Figure 6. 3mm diameter shaft-hole array assembly parts.

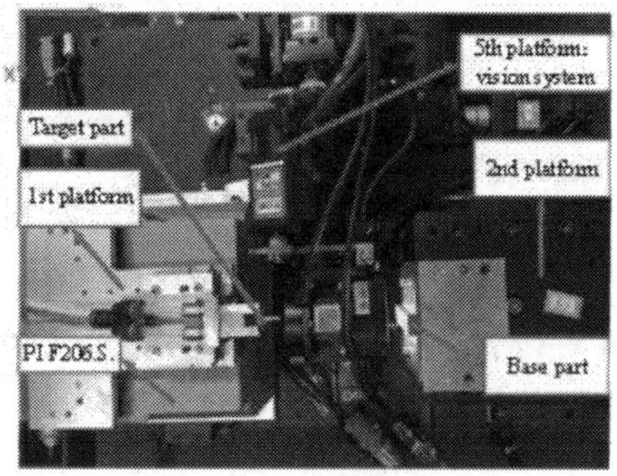

Figure 7. Assembly experimental devices.

DOI: 10.1051/matecconf/20167709005

Coaxial alignment assembly experimental devices are shown in Fig. 7. Using vacuum adsorption method to grip the parts, the target part is fixed to the 1st platform; the base part is fixed to the 2nd platform. Vision system is in the 5th platform, which perpendiculars to the 1st and 2nd platforms' console axis in a horizontal plane. PI F-206.S is placed on the 1st platform to adjust the target part's 6-DOF pose in micro-scale.

Coaxial alignment assembly experimental flow is shown in Fig. 8.

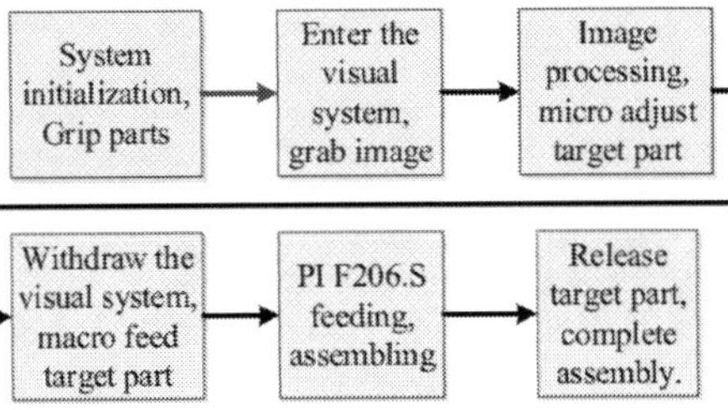

Figure 8. Coaxial alignment assembly experimental process

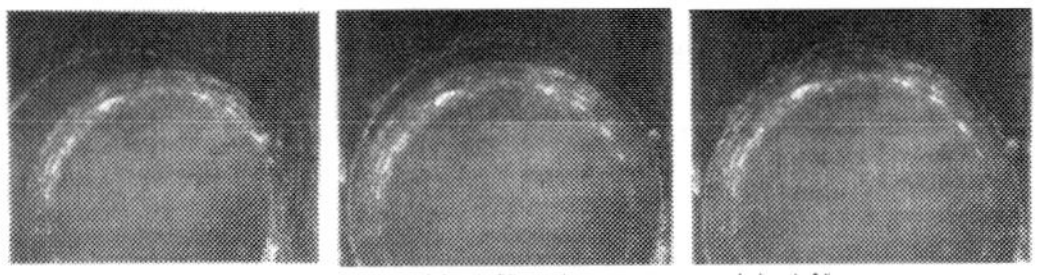

(a)Before alignment (b) Aligning (c) Alignment completed

Figure 9. Images change process in coaxial alignment experiment.

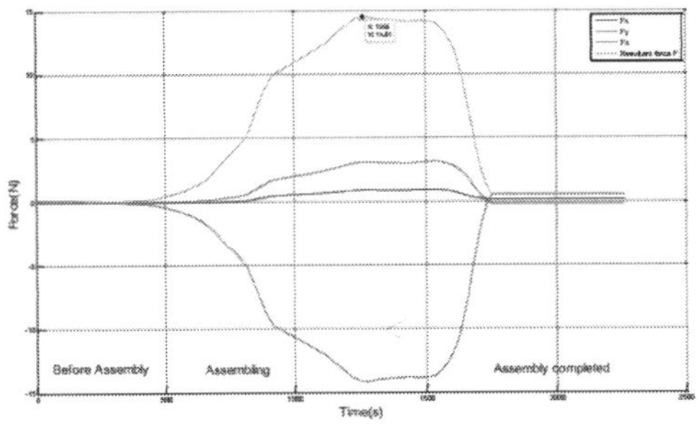

Figure 10. The changing process of the amount of the forces.

During the experiment, CCD camera captured the images of the shaft and the hole, as shown in Fig. 9. The image of the shaft and hole which is before aligned are shown in Fig. 9. (a), apparently there is deviation between the parts. Fig. 9. (b) shows the image of the shaft and hole in the process that PI F206.S is adjusting the target part's pose according to visual feedback. Fig. 9. (c) is the state of the shaft hole after coaxially aligning, which means that the target part and the base part have completed the coaxial alignment. During assembly, the changing process of the amount of the forces is shown in Fig. 10, it can be seen from the Fig. 10 that the maximum resultant force is 14.51N, which is far less than the allowable assembly force. Experiments show that, MSCA can complete nondestructive assembly of parts in the micron level.

5 Conclusion

Oriented to MSCA, the control system is capable of controlling multiple axes of motion simultaneously which was developed using "industrial computer+ special controller+ UMAC" open control system architecture. Combining with different detection equipment such as machine vision and force sensor, the system can complete nondestructive assembly of parts and increase the precision of microassembly process into micron level, through the way of man-machine cooperation. In the future work, several researches need to be done on visual servoing and finding out the relationship between the micro-force and micro-displacement. We have to do some improvements to increase the automation of the MSCA system.

Acknowledgements

This work was financially supported by the National Natural Science Foundation of China (51575052) and (51405025). Any opinions, findings, and conclusions or recommendations expressed in this material are those of the authors and do not necessarily reflect the views of the National Natural Science Foundation of China.

References

1. Ehmann.Kornel, DeVor.Richard, Kapoor.Shiv, Ni. Jun: *Workshop: Micro/Meso-Mechanical Manufacturing*, Evanston, llinois, USA, Northwestern University, (2000).
2. Xin Ye, Jun Gao, Zhi Jing Zhang, Applied Mechanics and Materials, **487**, 678-681(2014)
3. Spear.Mike, Process Engineering, **75**, 38–39 (1994).

MATEC Web of Conferences 77, 09006 (2016) DOI: 10.1051/matecconf/20167709006
ICMMR 2016

Intelligent Smartphone based system for detecting speed bumps and reducing car speed

Yousef-Awwad Daraghmi and Motaz Daadoo

Computer Engineering Department, College of Engineering and Technology, Palestine Technical University-Kadoorie, Palestine,

Abstract. Although speed bumps are used to force drivers reduce car speed for avoiding accidents, these bumps may cause car crash or accident when drivers do not notice them. Studies have proposed different methods to detect bumps and alert drivers. However, these methods have limitations and require modifications to enable accurate detection. Also these methods did not propose speed reduction approaches. Therefore, in this research, we propose a method that utilizes smartphone microelectronic mechanical technology for speed bump detection. The system uses the gravity sensor to detect the vertical vibration of cars passes over bumps and the GPS to determine the position of the bump. To give accurate detection results, data are collected from crowd, stored and processed on the cloud. The system also contains a speed reduction unit which is attached to the brake pedal and reduces the speed if a bump is detected. A small scale experiment showed that the system detected the position and the height of bumps with a very small error. The system also reduced the speed of cars at the moment they hit the bumps to a point that does not cause any harm to cars or passengers.

1 Introduction

Speed bumps or speed humps have become popular recently for forcing drivers reduce car speed to avoid accidents. In many countries particularly developing countries speed bumps are placed on roads randomly, and a traffic sign that shows there is a bump ahead may not exist [1], [2]. Consequently, the driving pattern is disturbed and drivers hit the bumps at high speed causing accidents and damage to cars. This problem can be solved by firstly detecting speed bumps and warning drivers in advance so that they can reduce speed before hitting the bump. Secondly, if the driver did not notice the warn message or did not speed down, the car speed should be reduced automatically using an intelligent speeding down system.

Previous studies have provided fundamental methods for detecting speed bumps [3]–[6]. These methods include RADAR sensors or LIDAR sensors which are used for the adaptive cruise control system [5]. These sensors increase the cost of the car, and drivers who already own cars are not willing to pay for additional devices on board. Also, ultrasonic sensors can be used for detecting speed bumps but these sensors are not very reliable as their range is very limited and they may not provide sufficient time for the driver to brake. Further, video sensors can detect speed bumps based on the color as speed bumps are usually painted in white and yellow [7], [8]. Video sensors require image processing techniques to analyze the color and determine the distance and height of the bump. However, video sensors are limited because bumps usually are not painted and video is not effective at night. Another method is the accelerometer sensors which continuously provide the x, y, and z coordinates of any point on the road [3], [4]. The accelerometer sensors can overcome the limitations of previous methods as they exist in most smartphones owned by drivers which reduces the cost. Also, these sensors can detect bumps more accurately than the aforementioned methods. This method collects big data that need efficient storage and processing before giving any good result. Existing accelerometer based methods are also limited to warning or notifying drivers about the bumps and are not accompanied with mechanical speed control solutions.

In this research, we propose an intelligent system for speed bump detection and speed reduction. The system utilizes the smartphone gravity sensor for detecting bumps and the accelerometer sensor for identifying the position of the bump. We build an application for cruise control which can be installed on android based smartphones. Every car driver equipped with a smartphone is a source of bumps related data (positions) and these data are uploaded to cloud. A Data mining algorithm called K-mean clustering analyzes the data which are collected from the crowd and determine the positions and the height of the bumps on roads according to geographical locations. Once the driver connects to internet, the smartphone uploads raw data and downloads information about the bumps. The system also contains a speed reduction unit which receives a blue tooth signal from the smartphone if there is a bump ahead and reduces

DOI: 10.1051/matecconf/20167709006

speed. This unit consists of a microcontroller controlling a DC motor which is connected to the brake pedal. If the driver does not reduce car speed, the motor will pull the pedal down gradually. The current car speed, distance to the bump and height of the bump are considered for identifying the rotation of motor and the amount of brake, i.e., the speed when the car hits the bump. This research is also important for establishing a unified database for speeds bumps in the country.

2 The proposed system

2.1 System architecture

The Architecture of the proposed system consists of three main units that are: The sensing unit, the cloud unit and the speed reduction unit. The architecture is shown in figure 1.

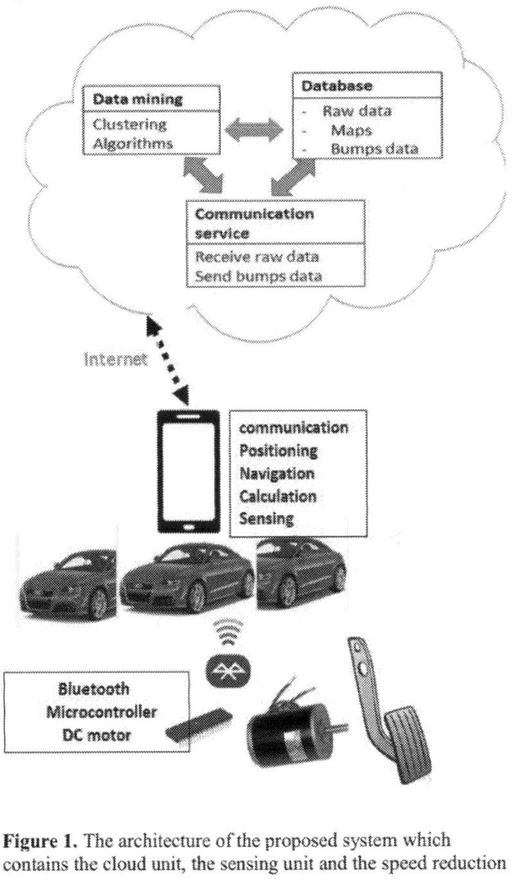

Figure 1. The architecture of the proposed system which contains the cloud unit, the sensing unit and the speed reduction unit.

2.1.1 The smartphone unit

This unit is basically the smartphone internal software and hardware components. It is responsible for detecting road bumps using the smartphone gravity sensor. We adopted the approach proposed in [4] to detect bumps. Once a bump is detected, its position is recorded in a local database and the driver is notified. This unit also contains the navigation part to guide drivers along the roads. Further, Wi-Fi is used to upload raw data (bump position, height, and timestamp) from the local database to the cloud server, and download accurate data (bump position and height) from the cloud to the local database. Also, Bluetooth is used to connect the smartphone with the speed reduction unit. The local database is used to reduce the burden of calculations, otherwise, the smartphone needs to access the cloud server for every calculation.

We developed an application called "Matabat" that utilizes the above components. The application consists of a sensing module that continuously monitors road conditions via the gravity sensor, a positioning module to identify car location by the GPS, a communication module for the Wi-Fi and Bluetooth, a navigation module for assisting drivers during trips, and reduction module which takes the car position and speed from the positioning module and obtains next bump position from the local database then calculates the value of speeding down defined as reduction value (RV). This application is installed on drivers' smartphones which form crowd sourcing. Due to unavailability of 3G on some roads and its high cost, we allowed the upload and download to be through the Wi-Fi as a batch log file before or after trips. The application uses the Bluetooth to send a message which contains the RV to the speed reduction unit when there is a bump a head.

2.1.2 The cloud unit

This unit contains three layers that are: the communication API layer which is in charge of receiving the raw data from the smartphones and sending back accurate data; the database layer which stores the raw data, road maps (google map) and bumps data; and the data mining layer which process the raw data and produce the accurate bumps data. The last layer is referred to as bumps mining which will be explained later in this section.

2.1.3 The speed reduction unit

This unit is composed of a microcontroller which receives the message containing the RV from the smartphone unit via the Bluetooth and controls the rotation of the DC motor. The DC motor is connected with the brake pedal. The DC motor rotates and pull down the brake pedal according to the RV. Indeed, this unit can be replaced by the internal mechanical braking system and car computer system which is fully controlled by the manufacturer. This unit was designed to evaluate the performance of the detection system. Fortunately, drivers did not mind to put this unit in their cars as it fits under the pedal, and it does not affect the normal usage of manual braking.

2.2 Bumps detection

As in [4], we use the gravity sensor to detect speed bumps. Firstly, the direction of gravity is identified and then the Vertical Components (VCs) are extracted from the sensor readings. Note that the sampling rate of the gravity sensor is 20Hz. The objective here is to determine the anomalies among VCs. To exclude onboard car vibrations, underdamped oscillation system is used to model the vibrations. The vibrations are represented as a long term standard deviation. When the vehicle passes over a bump, the VCs change dramatically and the standard deviation changes for a short time. We call the ratio between the long term standard deviation and short term standard deviation the Bump Index (BI),

$$BI = \frac{\sigma}{\sigma t},$$

where σ is the long term standard deviation and σt is the short term standard deviation. The BI value expresses the height of the bump. The sensing module runs the BI calculations every second which means that the short term changes are considered every second.

The amount of data obtained from the gravity sensor is very large and consumes much computational power and memory from the smartphone. To overcome this problem, we use the running average method and discard old data. To simplify the process, we keep the running average working on the last five minutes. The local database stores the BI value, the GPS coordinates at which the BI occurs and the timestamp. When the trip finishes and the smartphone connects to the internet, it uploads to the cloud the log file which contains the BIs during the entire trip.

2.3 Bumps mining

The position obtained from the GPS and stored in the database is not totally accurate so an data mining algorithm is used to reduce the error in the position. The algorithm is also necessary to ensure that the anomaly represented by the BI is a real bump. For this purpose, we use the K-means algorithm that is used for clustering. In our context, every car j submits a set of bumps positons $\mathbf{P}_j = (P_1, P_2, ..., P_n)$ where each P_i is the GPS coordinates (x_i, y_i). The coordinates are then classified to routes using Google map. If m cars pass through a road, the road will have m sets of bumps. The k-means clustering algorithm runs firstly on these points to find the mean value of the distance between points. The mean value should achieve the minimum mean square error. We consider the points within 10 meters are in one cluster that is because the GPS error is within 10 meters. The outcome of the algorithm is $b_1, b_2, ..., b_c$ where c is the number of bumps in one road. These mined data are stored in bump data table in the database.

We run the algorithm secondly on the height of the bumps and find the mean value of height of point in one cluster. The result is the height of the bump found in the first round. The height is stored in the table. Each table contains the height, x and y coordinates of each bump in each road. When a car downloads the route information, it will have all bumps data in each road.

2.4 Speed reduction

The car continuously calculates its position P_j and speed v using the GPS. The speed reduction module checks if there is a bump in the coming 100m every second. If a bump exists, the module calculates the distance d to the bump. According to the third equation of motions, the speeding down a can be found using the distance to the bump, the current (initial) speed v, and the final speed w as

$$a = -\frac{w^2-v^2}{2d}.$$

The final speed is categorized based on the height of the bump. For simplicity we assume the safe speed to pass over the bump is 30Km/hour. The reduction vale (RV) equals to the speeding value multiplied by a positive factor which depends on calculation error and mechanical calibrations for the DC motor. The DC motor should pull down the brake pedal faster when RV is bigger.

3 Experiment and results

Our experiment included three cars which drive in a route of 10Km length and contains 47 bumps in TulKarm city, Palestine. Each car was equipped with a smartphone mounted to the windshield. The cars went along the route 10 times. At the first time, the databases were empty so the cars did not download any data. At the end of the first route the cars uploaded batch log files to the cloud. Every time a file is uploaded, a new position of the bump is added to the database. The best case is when the new position equals the old position. Also, at every time, the server runs the k-means algorithm to identify the exact location of the bump. We compared the real bump position with the output of the algorithm and we found that the error is very small. We also compared the real height with the detected height and the difference is relatively small. Some values have high error due to the few number of cars since the data were not enough to produce good results. Table 1 shows a sample of the error for some bumps. The average position error for all bumps is 0.47m and average height error is .39m.

Table 1. The error in the detected position and height of some speed bumps.

Bumps ID	Position error (meter)	Height error (meter)
13	0.7	0.5
24	0.4	0.3
36	0.5	0.4
44	0.7	0.2
52	0.4	0.3
56	0.6	0.2

The second experiment was to test if the car speed is reduced by the DC motor and what is the speed when the car goes over the bump. We set the targeted speed 30Km/H for all bumps. However, the resulted speed has a small error. Table 2 shows a sample of speeds for some bumps. The initial speed is when the bump was detected from the bump and the bump speed is when the car passes over the bump.

Table 2. The initial speed when the bump is detected and the final speed when the car hits the bump.

Bumps ID	Initial speed Km/H	Bump speed Km/H
13	70	25
24	68	36
36	60	33
44	75	27
52	75	32
56	78	29

The difference between the bump speed and the targeted speed is very small and does not affect the car. However, the effect of the height on the bump was not studies in this experiment.

4 Conclusion

This research proposes a system for reducing car speed on bumps to avoid car crash or car accidents. The system utilizes microelectronic mechanical technology on smartphone such as the GPS and the gravity sensor for detecting the bumps and utilizes DC motor attached to brake pedal for reducing car speed. We conclude that using these kind of sensors requires large number of data to obtain good results. So we employed the concept of crowd source for the collection of data from as many cars as possible. We also conclude that such methods require cloud services to analyze such big data and produce the desired results.

We used a simple way to speed down the cars by pulling down the brake pedal using DC motor. Our future work will include controlling car speed using the internal computer. We believe that better results can be obtained in this way. Our future work will also include modifying the K-mean cluster algorithm so that it can give better results on big data. We will integrate this projects with other Intelligent Transportation Systems projects performed by our team [9]–[13]. Our main goal is to establish an intelligent transportation system in Palestine.

References

1. M. Jain and A. P. Singh, "Speed-BreakerEarlyWarningSystem," in *USENIX Federated Conferences Week*, 2012.
2. J. P. Hessling and P. ZHU, "Analysis of Vehicle Rotation during Passage over Speed Control Road Humps," in *International conference on Intelligent Computation Technology and Automation (ICICTA)*, 2008, pp. 304–308.
3. Y.-T. Chuang, C.-W. Yi, T.-H. Wu, and C.-Y. Chan, "Model the vibrations of rack-mounted on-vehicle devices by underdamped oscillations," in *17th International IEEE Conference on Intelligent Transportation Systems (ITSC)*, 2014, pp. 982–987.
4. C. W. Yi, Y. T. Chuang, and C. S. Nian, "Toward Crowdsourcing-Based Road Pavement Monitoring by Mobile Sensing Technologies," *IEEE Trans. Intell. Transp. Syst.*, vol. 16, no. 4, 2015.
5. C. Fernandez, M. Gavilan, D. F. Llorca, I. Parra, R. Quintero, A. G. Lorente, L. Vlacic, and M. A. Sotelo, "Free space and speed humps detection using lidar and vision for urban autonomous navigation," *IEEE Intell. Veh. Symp. Proc.*, pp. 698–703, 2012.
6. J. Martín, R. Bahena, C. A. Cruz, A. Z. López, A. Ferreyra-ramirez, C. Lindavista, and D. G. A. Madero, "Speed booms detection for a ground vehicle with computer vision," in *12th WSEAS international conference on Mathematical and computational methods in science and engineering*, 2010, pp. 258–264.
7. C. G. Keller, M. Enzweiler, M. Rohrbach, D. F. Llorca, C. Schnörr, and D. M. Gavrila, "The benefits of dense stereo for pedestrian detection," *IEEE Trans. Intell. Transp. Syst.*, vol. 12, no. 4, pp. 1096–1106, 2011.
8. A. Wedel, H. Badino, C. Rabe, H. Loose, U. Franke, and D. Cremers, "B-spline modeling of road surfaces with an application to free-space estimation," *IEEE Trans. Intell. Transp. Syst.*, vol. 10, no. 4, pp. 572–583, 2009.
9. Y.-A. Daraghmi, C.-W. Y. Yi, and T.-C. Chiang, "Space-time Multivariate Negative Binomial Regression for Urban Short-Term Traffic Volume Prediction," in *12th International Conference on ITS Telecommunications*, 2012.
10. Y.-A. Daraghmi, C.-W. Y. Yi, and T.-C. Chiang, "Mining Overdispersed and Autocorrelated Vehicular Traffic Volume," in *International Conference on Computer Science and Information Technology*.
11. Y. Daraghmi and A. M. Hasasneh, "Accurate Real-Time Traffic Sign Recognition Based on the Connected Component Labeling and the Color Histogram Algorithms," in *International Conference on Signal Processing*, 2015.
12. Y.-A. Daraghmi, C.-W. Yi, and T.-C. Chiang, "Negative Binomial Additive Models for Short-Term Traffic Flow Forecasting in Urban Areas," *IEEE Trans. Intell. Transp. Syst.*, vol. 15, no. 2, pp. 784–793, 2014.
13. Y.-A. Daraghmi and M. Daadoo, "Improved Dynamic Route Guidance based on Holt-Winters-Taylor Method for Traffic Flow Prediction," *Int. J. Traffic Transp. Eng.*, vol. 4, no. 3, 2015.

The pilot study of object following for automated guided vehicle

Jieh-Shian Young and Cheng-Yan Lin

Institute of Vehicle Engineering, National Changhua University of Education, #1, Jin-De Rd, Changhua 500, Taiwan

Abstract. This paper aims at the pilot study of the object following for automated guided vehicle (AGV).The proposed approach utilizes the laser range finder (LRF) to detect the distances of all obstacles surrounded in a specified angle range. The AGV can autonomously follow the assigned object according to the rational reflection intensity and range of the object sensed from LRF. The cruise speed control for a moving object depending on the distance between AGV and this object is considered since it is one of the crucial parameters for object following. This paper also discusses the influences of the sensor delay which is an indispensable issue in both control law and control logic. This pilot study employs a down scaled AGV to verify the feasibilities of the steering, accelerating, braking, and object following algorithms developed for AGVs.

1 Introduction

In 2009, the US National Highway Safety Administration began studying collision warning system specification. In 2011, the European Commission adopted a resolution and began to promote pre-brake system. In 2013, all commercial vehicles within Europe must be installed the pre-crash brake system. In 2015, all new cars not just commercial vehicles within Europe must be installed forward collision warning (FCW).

Many studies show that when vehicles have forward collision warning or pre-crash brake assist or autonomous pre-crash brake systems, can increase driving safety and reduce the severity of the injury [1, 2].

The automobile industry needs not only safe structures but also active safety. Moreover, traffic in Taiwan is highly congested. It may cause the car accidents as drivers do not pay attentions due to the traffic jams, tired driving, etc. In case vehicles with automatic brakes and automatic following system can raise the traffic efficiency and reduce the driving hazards. The realization of active safety systems can be achieved through the information collected by sensors that identify front environment and obstacles. Furthermore, an appropriate control mechanism applied to vehicle safety is indispensable.

Many studies have utilized a variety of techniques based on difference sensors to achieve the object following functions, such as radio-frequency identification (RFID)[3], Kinect sensor (computer vision)[4], laser range finder (LRF)[5, 6], etc. Some of them utilize multiple sensors[7]. This study proposes a technique which adopts only LRF sensor by its range information and reflection intensity. This paper shows the results of the pilot study for its feasibility of object following function in a vehicle verified by an automated guided vehicle (AGV).

2 Preliminaries for experiment setup

2.1 The AGV

The AGV adopted in this study is one of the products from Play Robots Inc., or 4W4D Combo AGV as shown in Figure 1. The AGV has size of 42x32x17cm. The wheels of the AGV are equipped by four 12V DC motors. Its maximal moving speed is approximate to 1.1m/s. The maximal payload of the AGV is 5 kg. This AGV is with the maximal climbing angle of 40° . The data communication medium for the AGV is the Bluetooth by which the control commands can be received and some requests can be also transmitted. Table 1 lists the brief specifications of the AGV. The laptop personal computer (PC) can play as a main control role for the AGV since the maximal payload weight is much less than 5 kg. In addition, PC can receive the distance information from LRF through Ethernet by proper software programming.

Figure 1. The 4W4D Combo AGV[8].

Table 1. The specifications of the 4W4D Combo AGV[8].

Model	4W4D-combo
Supply voltage	6VDC/12VDC battery
Dimension	42x32x17 cm
Power source	DC 12V motor*4
Max. moving speed	1.1m/sec
Max. climbing angle	40°
Max.load	5kg
Data medium	Bluetooth

The AGV utilizes the Arduino UNO micro control board and can receive commands from controller via Bluetooth module. The motors revolution rate of the AGV can be control by discretized speed command values ranged from 0 to 127. Figure 2 sketch the relation between speed command values and the speeds of the AGV. Y=X-2.3 is the linear approximate to the test data for different speed commands where X is the speed command and Y is the speed of the AGV.

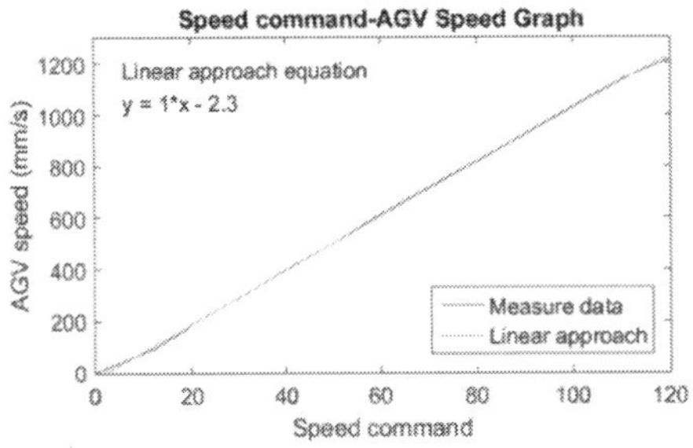

Figure 2. Relation between speed command and AGV speed.

2.2 The laser range finder (LRF)

There are many methods to measure the distance in applications such as ultrasonic sensors, computer vision with image process, LRF, etc. Table 2 lists the comparisons for these sensors. The computer vision approach has lower cost. However, their accuracies and computation time are not as good as the LRF. In addition, they are very sensitive to the environment such as image background. For the sake of the fast and accurate requirements, LRF is adopted in this paper.

Table 2. Comparisons for different sensors.

	Computer Vision	Ultrasonic Sensor	Laser Range Finder
Error Rate	Low	High	Mid
Computing	Fast	Slow	Mid
Accuracy	Mid	Low	High

The type of the LRF in this study is Hokuyo UST-20LX. The distance detected for this kind of LRF is ranged from 0.06 m to 60m with the accuracy of ± 0.04 m. Its scanning angle is from -135° to 135°, or the span angle is 270° as shown in Figure 3. The scanning speed of the LRF is 25ms/rev and one step, or the angular resolution, is 0.25°. It is intuitive that the maximal step number is 1080. The data communication field bus for the LRF is the Ethernet with the specified data communication protocol. In applications, the range of the scanning angle for this LRF can be assigned within $\pm 135^{\circ}$ by command received via Ethernet. Figure 4 shows the LRF of Hokuyo UST-20LX. Table 3 lists its specifications.

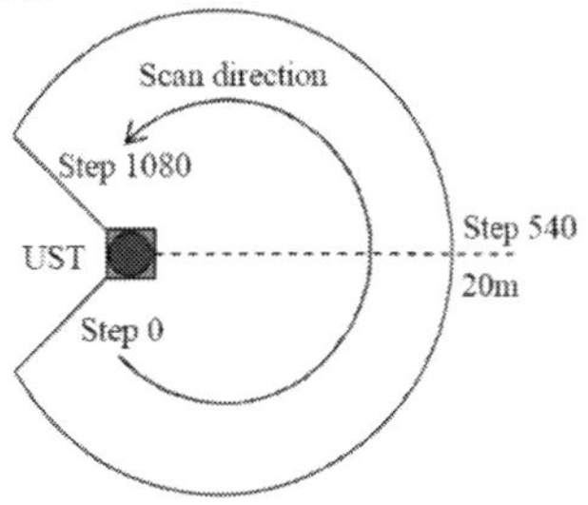

Figure 3. The scanning angle of Hokuyo UST-20LX[9].

Figure 4. The LRF of Hokuyo UST-20LX[10].

Table 3. The specification of Hokuyo UST-20LX[9].

Model	UST-20LX
Supply voltage	12VDC/24VDC
Supply current	<150mA
Detection range	0.06 m to 60m
Accuracy	±0.04m
Scan angle	270°
Scan speed	25ms(at 2400rpm)
Angular resolution	0.25°
Data Medium	Ethernet

The Hokuyo UST-20LXLRF data communication medium is Ethernet with Sensor Communication Interface Protocol (SCIP)[11]. Its sensor structure sketched in Figure 5 composes of host part, communication interface part, and target (sensor) part. The controller has to establish a TCP/IP communication mechanism in order to receive the information of the LRF including the distances, reflection intensities, and the corresponding angles. For example, the factory default valuesfor the network address settings are as follows. IP

DOI: 10.1051/matecconf/20167709007

address: 192.168.0.10; Subnet Mask: 255.255.255.0; Default Gateway: 192.168.0.1; Port Number: 10940. The typical responses of Hokuyo UST-20LX LRF are illustrated in Figure 6 as the PC transmits the "GE" command which is so-called the distance and intensity acquisition command for Hokuyo UST-20LX. The data in Figure 6 are actually the printable ASCII codes. Each distance datum comprises 3 characters and each character for distance information contains 6 bits. For instance, "1Dh" is decoded as 5432 mm. Figures 7 and 8 illustrate the decoding algorithm and the verification software program, respectively.

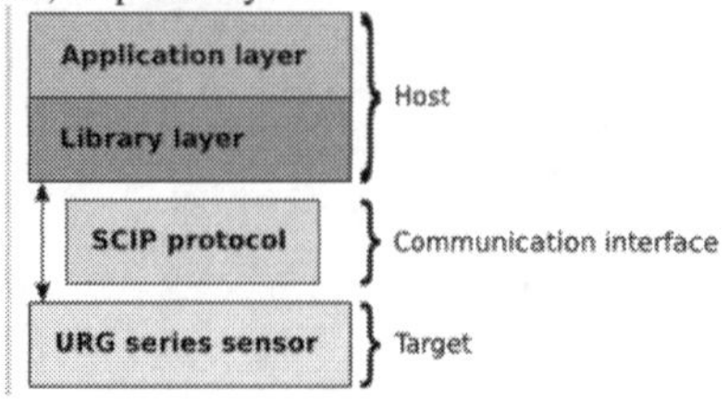

Figure 5. The sensor structure of Hokuyo UST-20LX[11].

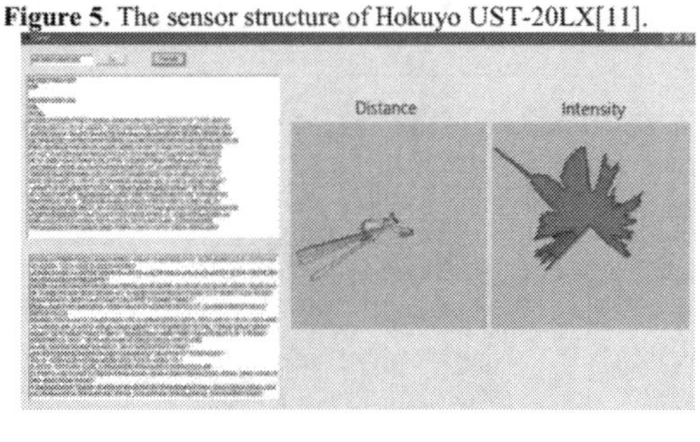

Figure 6. The typical responses of Hokuyo UST-20LX.

'1'(31h) 'D'(44h) 'h'(68h)

↓ *Subtract 30h*

1h 14h 38h

↓ *Merge*

000001 010100 111000

↓ *Decimal Value*

$2^3 + 2^4 + 2^5 + 2^8 + 2^{10} + 2^{12} = 5432$mm

Figure 7. The distance decoding algorithm[12].

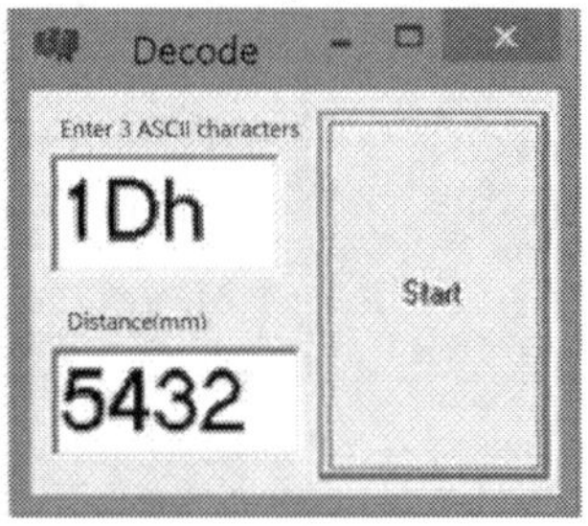

Figure 8. The distance decoding algorithm and the verification software program.

2.3 AGV control platform

In this study, the laptop PC served as a steering controller is fixed on the platform as shown in Figure 9. The LRF is installed on the forefront of the AGV. This PC can transmit the control commands to the LRF and can receive the measuring data from LRF viathe Ethernet port of LRF. The laptop PC analyzes the received data from LRF, determines the potential object, and forwards an applicable command to AGV via bluetooth after following a control rule including control law and control logic.

Figure 9. AGV control platform.

LRF provides the distances and intansities reflected from obstacles located within the assigned scanning angles span. The control platform has two main problems. One is that how to recognize the assigned object with the LRF response data. The instansity information is practical if the strongest reflection instansity is detected from the assigned object. The PC, or the steering controller, can then be easy to recognize the potential object by the different reflection intansity patterns or strengths of the object. The AGV can follow the object with the appropriate control strategy based on its corresponding distance and angle. The other problem is that what control strategy can make the object following of AGV more accurate and more efficient. This pilot study aims at the previous studied not only on the communication techniques between PC and LRF and between PC and AGV but the techniques of the LRF information analysis, the potential object determination, applicable control commands generation.

3 Control specifications

A properly distance between the object and the AGV has to be kept for the sake of the object following purpose. The proper distance definitely depends on the speed of the object. Figure 10 sketches the proposed relation between distance to target and AGV speed. Under consideration of avoiding wheel slip, moderate acceleration for the AGV is necessary. In this paper, the final speed will be achieved by separating 10 portionswithin 0.1 second from initial speed. For instance, the initial speed and final speed of AGV are 0.5 m/s and 1 m/s, respectively. The speed commands are sequentially transmitted for every 0.01 s as $V_c = 0.4 + \frac{(1.0-0.4)n}{10}$ where V_c is the speed command and n is ranged from 1 to 10.

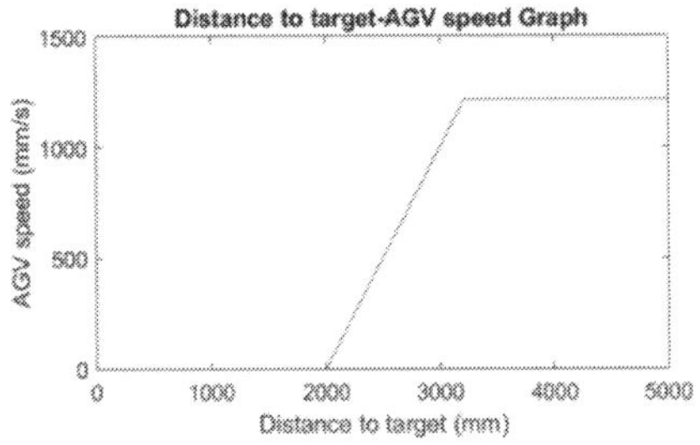

Figure 10. Distance-speed specification for AGV.

The turning mechanism of the AGV is also important for the object following purpose. This paper proposes a similar triangles rule to facilitate the turning of the AGV. Figure 11 schemes the relations of the proposed rule.This rule can be formulated as an equation with turning radius and the speed of the AGV as follows.

$$R = \frac{B}{2}\left(\frac{V_o + V_i}{V_o - V}\right),$$

where V_o is the velocity of outward wheels, V_i is the velocity of inward wheels, V is the velocity of AGV, R is the turn radius of AGV, and B is width of AGV.

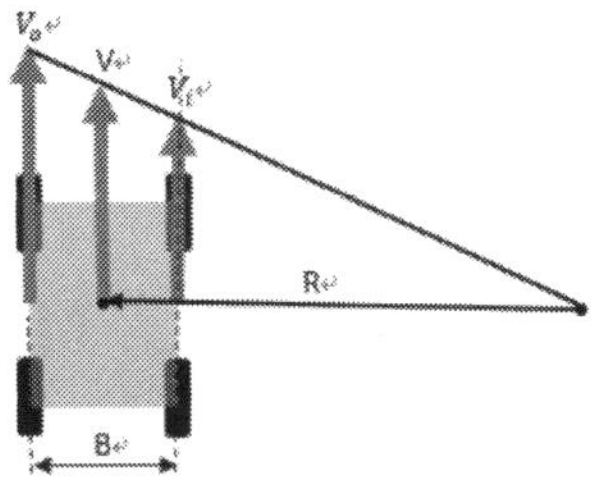

Figure 11. Similar triangle rule for AGV turning.

An accurate path to the target for the AGV is determined by the desired steering radius of the AGV according tothe distance and angle data received from the LRF. Figure 12 sketches the relation between the AGV and the target object where L is the distance between the AGV and the target object, R is the desired turning radius of AGV, and $\frac{\pi}{2} - \theta$ is the heading angle from the AGV to target object. In Figure 12, it yields

$$\frac{R}{\sin\theta} = \frac{L}{\sin(180 - 2\theta)}.$$

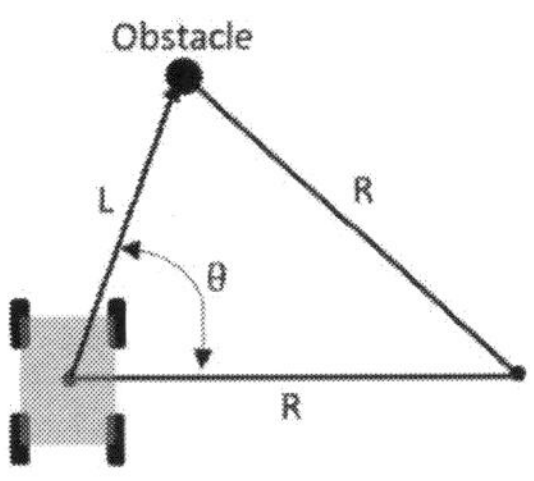

Figure 12. Relation between the AGV and the target object.

4 Verifications of pilot study

The steering controller can utilize the laser intensity and distance information from LRF to determine the potential target object. However, the intensity varies from the distance information. It is indispensable to attenuate the distance effect and that will facilitate the determining the target object. Figure 13 demonstrates the target object determining program.

Figure 13. The target object determining program.

The steering control strategy for a straight moving object is one of the most significant verifications for the object following of AGV. The target object will accelerate and decelerate as it moves straight in the verification scenarios. The AGV following the specified target as the target moves straight has been verified but the responses are not quick enough. This paper suggests that the scanning angle of the LRF can be smaller to save the time of decoding since the bottleneck of the computation is the LRF response information decoding.

5 Conclusion

This pilot study conveys almost all possible preliminary work for the target object following of AVG. This paper proposes the control specifications such as the relation of the distance and the speed, the similar triangle rule for AGV turning, and the relation between the distance and the turning radius. The future work will be carried out for all possible verifications of the target object followingincluding the target object moves in S-type route in different radii. That will verify the capability of the steering controller for different turning radius.

References

1. M. Furutani, presented at the Proceedings of Convergence International Congress & Exposition On Transportation Electronics, US, 2004 (unpublished).
2. K. D. Kusano and H. C. Gabler, IEEE Transactions on Intelligent Transportation Systems **13** (4), 1546-1555 (2012).
3. M. Kim, N. Y. Chong, H.-S. Ahn and W. Yu, presented at the Proceedings of The 3rd annual IEEE Conference on Automation Science and Engineering (IEEE CASE 2007), Scottsdale, AZ, USA, 2007 (unpublished).

4. D. Calisi, L. Iocchi and R. Leone., presented at the VISAPP (Workshop on on Robot Vision), Barcelona, Spain, 2007 (unpublished).
5. E.-J. Jung, J. H. Lee, B.-J. Yi and S. i. Yuta, IEEE Transactions on Mechatronics **19** (6), 1963 - 1976 (2014).
6. W. Chung, H. Kim, Y. Yoo, C.-B. Moon and J. Park, IEEE Transactions on Industrial Electronics **59** (8), 3156-3166 (2012).
7. N. Bellotto and H. Hu, IEEE Transactions on Systems, Man, and Cybernetics, Part B: Cybernetics **39** (1), 167-181 (2009).
8. (2016). *Play Robot*. Available: http://www.playrobot.com/robotics/666--email.html
9. Kamitani, Utsugi, Kamon and Yamamoto. (2014, Scanning laser range finder smart-URG mini UST-20LX(UUST004) specification.
10. (2009). *Hokuyo*. Available: http://www.hokuyo-aut.jp/02sensor/07scanner/ust_10lx_20lx.html
11. H. Fukuda, J. Fujimoto, D. Huang, K. Kimoto, M. Kristou, M. Nagata, K. Matsuo, M. Musashi, T. Momo, S. Watanabe and F. Wada. (2016). *URG Network*. Available: https://sourceforge.net/p/urgnetwork/wiki/Home/
12. Aoki and Hosoda. (2014, Communication protocol specification for UST series 10LX/20LX.

MATEC Web of Conferences 77, 09008 (2016) DOI: 10.1051/matecconf/20167709008
ICMMR 2016

Development of photoelectric balanced car based on the linear CCD sensor

Feng Wang[1] , Yanlei Zhao[1],Jigao Zhao[1], Haofeng Zhao[1], Hao Zhang[1], Pengfei Dong[1], Xiaoli Zheng[2]

[1]College of Physics and Electronic Information Engineering, Wenzhou University, Wenzhou, China
[2] College of Teacher Education, Wenzhou University, Wenzhou, China

Abstract. The smart car is designed based on Freescale's MC9S12XS128 and a linear CCD camera. The linear CCD collects the road information and sends it to MCU through the operational amplifier. The PID control algorithm, the proportional–integral–derivative control algorithm, is adopted synthetically to control the smart car. First, the smart car's inclination and angular velocity are detect through the accelerometers and gyro sensors, then the PD control algorithm, the proportional–derivative control algorithm, is employed to make the smart car have the ability of two-wheeled self-balancing. Second, the speed of wheel obtained by the encoder is fed back to the MCU by way of pulse signal, then the PI control algorithm, the proportional–integral control algorithm, is employed to make the speed of smart car reach the set point in the shortest possible time and stabilize at the set point. Finally, the PD control algorithm is used to regulate the smart car's turning angle to make the smart car respond quickly while the smart car is passing the curve path. The smart car can realize the self-balancing control of two wheels and track automatically the black and while lines to march.

1 Introduction

Self balancing car is an unstable and non linear system. Conventional control techniques have been widely used in the self balancing car. However, these techniques require the complete mathematical model. The paper presents the photoelectric balanced car based on the linear CCD sensor without building the mathematical model of the system.

2 The design of hardware circuit and proof

2.1 Choice of tracking sensor module

From the Table 1, CCD has advantages in the amount of information, light adaptability, useful perspective, complexity of circuit design, difficulty of debugging and cost performance. Option 3 is selected, namely the CCD is selected.

2.2 Choice of main controller

Option 1: A STC89C51 MCU is adopted as the main controller. The advantages are that its cost performance is high and its low demand to the programmer. The disadvantages are that its function is weak in Multitasking or high speed project

Option 2: A high-performance MCU such as MC9S12XS128 is adopted as the main controller.

The MC9S12XS128 MCU is an ultralow power 16-bit MCU with more ordinary I/O port, multichannel PWM module, multichannel A/D module, TIM module and serial communication module Comparing with STC89C51 MCU, MC9S12XS128 has obvious advantages in the operability and functions. Option2 is selected, namely the MC9S12XS128 is adopted as the main controller.

2.3 Choice of power supply

The design of power supply mainly involved in this is a regulated power supply. The leading indicator of regulated power supply is the stability of the output.

Power supply is a priority in the design of power supply can meet the requirements of voltage and current, the second is the efficiency. In this design in addition to the motor driver is larger, the current provided directly by the battery, other modules require all current is small, the general voltage chip can meet the requirements. The power of this design needs to include 3.3V, 5V, and 7.2V power distribution planning schemes as shown in Figure 1.

Table 1. The advantages and disadvantages of each scheme.

	Sensor Type	Amount of information	Light adaptability to the environment	Adjustable useful perspective	The complexity of circuit design	The difficulty Of debugging	relative cost
Option 1	infrared	less	weak	close	easy	easy	low
Option 2	laser	less	Strong	far	difficult	difficult	high
Option 3	CCD	more	Strong	far	easy	easy	low

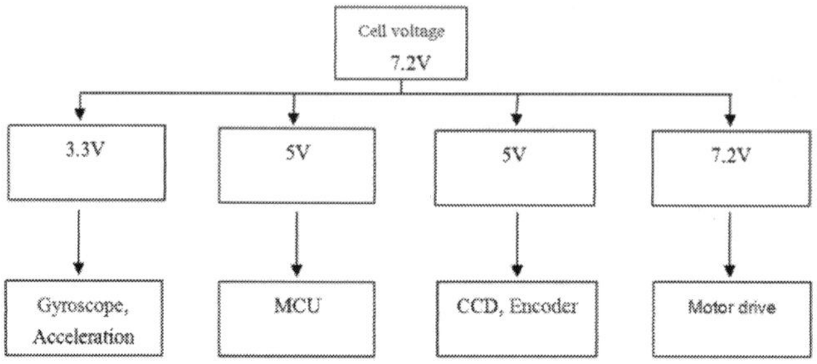

Figure 1. Power distribution planning schemes.

2.3.1 Choice of 3.3 V-voltage regulator systems

Option 1: The low dropout voltage regulator of LM1117 is selected to stable voltage.LM1117 is linear regulator chip, had the characteristics of low pressure differential, small ripple, whose peripheral as long as after several capacitance can meet the constant voltage output. Circuit is shown in Figure 2.

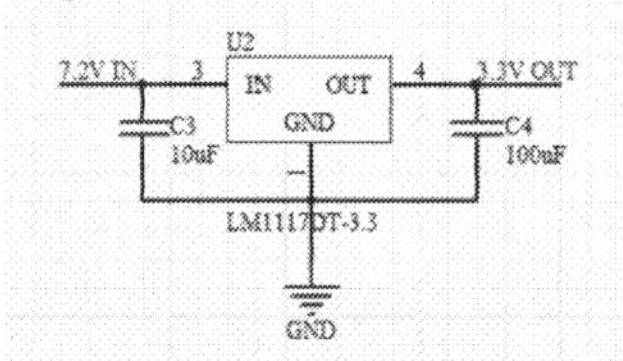

Figure 2. LM117-3.3V regulating circuit.

Option 2: Choose a switching chip of LM2576. LM2576 type is switching power supply voltage chip, and has high efficiency. But the output ripple of LM2576 is larger than LM1117, and LM2576's demand of PCB writing is higher. What's more, circuit is complex, as shown in Figure 3.

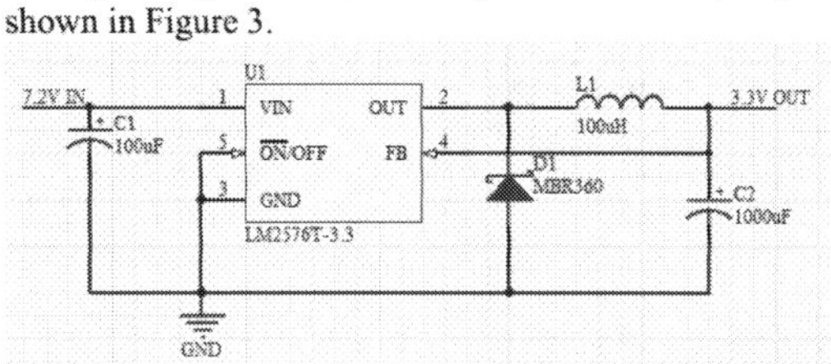

Figure 3. LM2576-3.3V regulating circuit.

Because the power supply is to power a gyroscope and accelerometer, the stability of two sensor outputs has great effect on the upright controlling of the car. In order to ensure the stable output of sensor, so we choose the scheme, which is designed by LM1117-3.3V-voltage regulator chip.

2.3.2 Choice of 5 V-voltage regulator systems

Option 1: Choose LM7805-linear regulator chip.LM7805 just need a few peripheral devices to constitute a stable voltage stabilizing belt. Its' device has protection circuit of over-current, overheat and the compensating pipe in the design, but the difference between the input and output is big what LM7805 needs. The input needs voltage which is greater than 7.5V to have a more stable output. Circuit is shown in Figure 4.

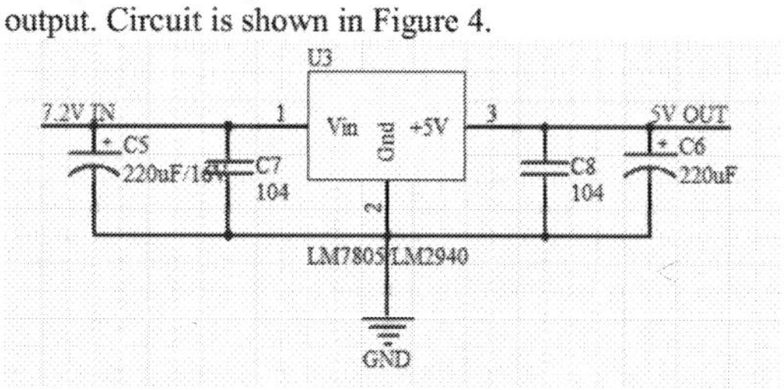

Figure 4. LM7805/LM2940 regulating circuit.

Option 2: Choose LM2940-linear regulator chip. LM2940 is the power supply of low dropout linear regulator. Its' characteristic is that the difference between input and output is low, and the input voltage is from 7V to 26V. Inside the chip has static current to reduce the circuit, at the same time with current limit, limiting of heat, meter and insert, and other functions. LM2940's distribution of pin with upright packaging is as the same as LM7805's distribution of pin, and the peripheral circuit can also be the same. Circuit is shown in Figure 4.

Option 3: Choose TPS7350-linear regulator chip. TPS7350 is the low dropout linear regulator chip with micro power consumption, it has characteristics of high precision of output voltage, low output noise and low pressure differential, and the input voltage can below 5.33V, but it costs higher than the first two schemes. The circuit diagram is shown in Figure 5.

DOI: 10.1051/matecconf/20167709008

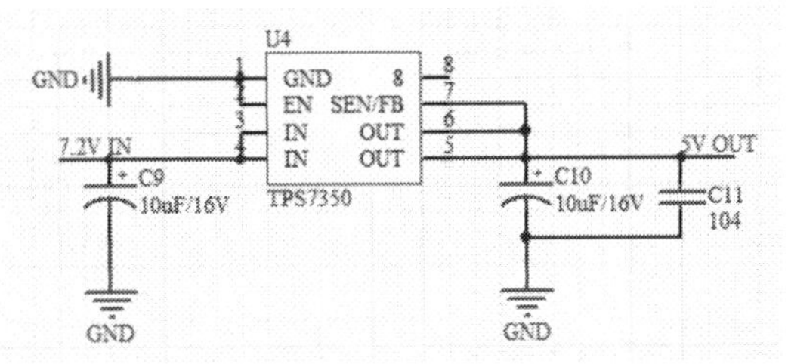

Figure 5. TPS7350 regulating circuit.

Considering the battery voltage is 7.2V, LM7805 is difficult to meet the requirements, it has very good reflecting in the process of testing. When the battery voltage falls to around 7.4 V, the battery internal resistance will be larger, voltage's output is unstable, especially in the intelligent vehicle, the effect is not ideal. TPS7350 has the very good effect, stable output, but considering the factor of cost, last 5 v regulated power system are adopted LM2940 voltage regulator, achieving the desired effect.

2.3.3 Choice of 7.2V-power supply system

The voltage of 7.2V is the battery voltage, in principle, the bigger voltage the motor supplies, can the speed of output rotational be faster, the stronger the motivation will be, therefore, battery provides directly power to drive motor.

2.4 Choice of motor driver

Option 1: The integrated chip such as MC33886 is selected. The advantages are that its technology is mature. It has a good performance on the motor control. It has perfect over-current, under-voltage, over-temperature protection, and other functions. But we must parallel slices MC33886 in order to decrease the larger internal resistance.

Option 2: The integrated chip such as BTS7960 is selected to set up the full bridge converter. The advantages are that its internal resistance is far less than MC33886, so it can produce less heat when Motor current is larger.

Option 3: The full bridge driver is set up through the MOS tube. The advantages are that the price is cheap, the structure is simple and it is easy to control. However, the output current of the transistor are smaller, so the drive capability is limited. Because it is discrete component, its stability can't be guaranteed and its volume is larger. As a result, option 3 is selected, namely the MOS tube is selected.

2.5 Choice of the acceleration and the gyroscope module

Option 1: Direct read is that sensor outputs analog signal directly and then apply MCU's analog-to-digital conversion module to read.

Option 2: Indirect reading requires indirect access to get each direction of the acceleration information through sequential control and the MCU processing.

As a result, option 1 is selected, namely the direct read is selected. Because direct analog output chip has the advantages is that information can be read easily and code can be controlled well. At the same time, programmers will meet great difficulties in code design.

2.6 Choice of velocity measurement module

Option 1: The photoelectric encoder is selected to measure the speed. It has low cost and convenient installation. Because of the manufacturing process, its precision is not high and it is unable to obtain high precision speed signals, and the signals are greatly influenced by the environment.

Option 2: The optical-electricity encoder is selected to measure the speed. Although it has high cost, its precision is higher than the photoelectric encoder. Therefore, option 2 is selected, namely the optical-electricity encoder is selected to measure speed.

3 Software design

The software design of system is shown in Figure 6.

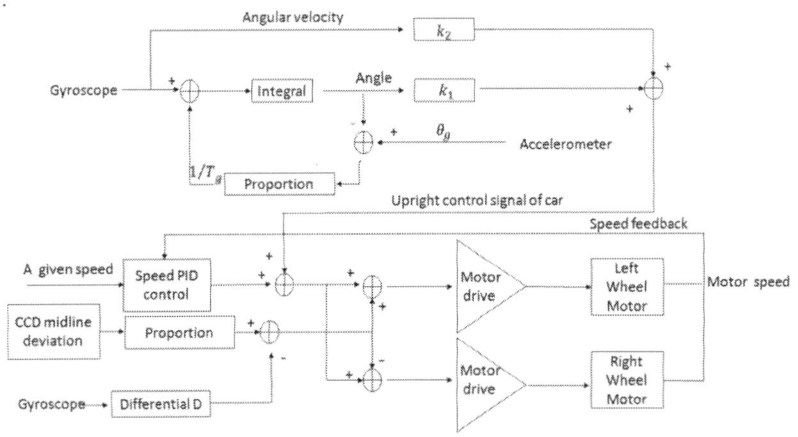

Figure 6. Block Diagram of the software system using PD, PI and PID control algorithm.

4 Experimental data and analysis

To verify the results of self-balancing control, the upright test, the speed test, the direction test and the comprehensive test are carried out.

4.1 Upright test

From the Table 2, the befitting P parameter is in the vicinity of 7.0. Therefore, P parameter is set as 7.3. There is a slight oscillation phenomenon, when parameter P=7.3. The oscillation phenomenon means that the car swings back and forth. The slight oscillation phenomenon can be inhibited when the D parameter is adjusted.

4.2. Speed Test

Captions should be typed in 9-point Times. They should be centred above the tables and flush left beneath the figures.

4.3 Direction Test

Please refer to Table 6.

Table 2. Upright control P parameter adjustment.

P parameter（D parameter =0）	0.0	3.0	6.0	7.0	8.0	11.0
Car model test phenomenon	Fall	Donot stand uprightly	Stand uprightly, sway back and forth	Stand uprightly, rocking displacement is small	Stand uprightly, but there is an oscillation phenomenon	Stand uprightly and there is a violent oscillation phenomenon

Table 3. Upright control D parameter adjustment.

D parameter（P parameter=7.3）	0.0	0.1	0.2	0.3	0.4
Car model test phenomenon	Stand uprightly and there is a vibration phenomenon	Stand uprightly, car vibration has a decreasing trend	Stand uprightly, car vibration is small	Stand uprightly and there is a trembling phenomenon	Stand uprightly and a trembling frequency speeds up

Finally, D parameter is set as 0.2.

Table 4. Speed control I parameter adjustment

I parameter（P parameter =0）	0	0.1	0.2	0.3	0.4
Car model test phenomenon	Stand uprightly, vibration is small	Stand uprightly, a car sway back and forth slowly when the external force is applied to the car	Stand uprightly, a car sway back and forth when the external force is applied to the car	Stand uprightly, a car sway back and forth quickly when the external force is applied to the car	Stand uprightly and there is a trembling phenomenon

Finally, I parameter is set as 0.2.

Table 5. Speed control P parameter adjustment

P parameter（I parameter =0.2）	0.0	2.0	4.0	6.0	8.0
Car model test phenomenon	Stand uprightly, a car sway back and forth when the external force is applied to the car, the car can't be stop soon	Stand uprightly, a car sway back and forth when the external force is applied to the car, the car can be stop	Stand uprightly, a car sway back and forth when the external force is applied to the car, the car can be stop soon	Stand uprightly, a car sway back and forth when the external force is applied to the car, angular variation of the car body is big	Donnot stand uprightly

Finally, P parameter is set as 4.1.

Table 6. Direction control P parameter adjustment

P parameter（D parameter =0）	0.00	0.25	0.50	0.75	1.00
Car model test phenomenon	Stand uprightly, car primary marching direction can't be	Stand uprightly, but direction control ability is	Stand uprightly, direction control ability is strong, there is no	Standing uprightly is unstable, there is	Standing uprightly is unstable, there is

	recovered after car marching direction is changed	weak	overshoot phenomenon	an obvious overshoot	an serious overshoot

Finally, P parameter is set as 0.6.

Table 7. Direction control D parameter adjustment

D parameter (P parameter =0.6)	0.000	0.010	0.020	0.030	0.040
Car model test phenomenon	Stand uprightly, direction control ability is strong, there is an overshoot phenomenon	Stand uprightly, direction control ability is strong, car overshoot has a decreasing trend	Stand uprightly, direction control ability is strong, there isn't an overshoot phenomenon	Stand uprightly, there isn't an overshoot phenomenon, but there is a trembling phenomenon	Stand uprightly, there isn't an overshoot phenomenon, but there is a violent trembling phenomenon

Finally, D parameter is set as 0.025.

4.4 Comprehensive Test

Table 8. Comprehensive test parameter setting

	P parameter	I parameter	D parameter
Upright control	7.5		0.21
Speed control	4.2	0.2	
Direction control	0.45		0.028

5 Conclusions

Through the above test, the photoelectric balanced car based on the linear CCD sensor has the following functions. It can reach two-wheeled self-balancing, march at a given speed, and turn on a bend smoothly. Marching at a given speed takes advantage of the negative feedback mechanism. The left and right motor speed is changed when the linear CCD sensor detects the deviation from the track in order to turn on a bend smoothly. So, the photoelectric balanced car can realize the self-balancing control of two wheels and track automatically the black and while lines to march.

6 Future works

It is difficult to provide a farther forward for the linear CCD sensor because the linear CCD sensor is extremely sensitive to external environment, but the farther forward is important to a Freescale smart car.

As for the laser sensor, the biggest characteristic is to provide a further perspective, and it has strong ability to adapt to the environment.

Acknowledgments

This work is supported by the opening laboratory project of Wenzhou University--- The design of upright smart car, Wenzhou, China, with the Project No. 15SK31A. This work is financially supported by the innovation and entrepreneurship training program of Wenzhou University--- The design of an upright balance intelligent vehicle based on multi algorithm fusion, Wenzhou, China, with the Project No. DC2014048.

References

1. P. Bai Jian and X. Gao Mei, "Based on the Research of Self-Balancing Vehicle Posture Sensor System and Design," in *Intelligent Systems Design and Engineering Applications, 2013 Fourth International Conference on*, 2013, pp. 193-196.
2. M. Bodson, J. N. Chiasson, R. T. Novotnak, and R. B. Rekowski, "High-performance nonlinear feedback control of a permanent magnet stepper motor," *Control Systems Technology, IEEE Transactions on*, vol. 1, pp. 5-14, 1993.
3. D. Gao, P. Han, D. Zhang, and Y. Lu, "Study of sliding mode control in self-balancing two-wheeled inverted car," in *2012 International Conference on Measurement, Instrumentation and Automation, ICMIA 2012, September 15, 2012 - September 16, 2012*, Guangzhou, China, 2013, pp. 2000-2003.
4. I. Kale, J. P. Mackenzie, and T. I. Laakso, "Motor car acoustic response modelling and order reduction via balanced model truncation," *Electronics Letters*, vol. 32, pp. 965-966, 1996.
5. K. Liu, M. Bai, and Y. Ni, "Two-wheel self-balanced car based on Kalman filtering and PID algorithm," in *Industrial Engineering and Engineering Management (IE&EM), 2011 IEEE 18Th International Conference on*, 2011, pp. 281-285.
6. J. Sun and H.-Z. Liu, "Two balanced car LQR algorithm achieved based on LabVIEW," in *2011 International Academic Conference on Numbers, Intelligence, Manufacturing Technology and Machinery Automation, MAMT 2011, December 24, 2011 - December 25, 2011*, Wuhan, China, 2012, pp. 592-597.
7. L. C. Wang, Z. Y. Liang, and X. Y. Ding, "Design and implementation of the two wheel self-balancing car based on single chip microcomputer of freescale," in *2013 International Conference on Energy Research and Power Engineering, ERPE*

DOI: 10.1051/matecconf/20167709008

2013, May 24, 2013 - May 25, 2013, Zhengzhou, Henan, China, 2013, pp. 1028-1032.

8. K. Xia, H.-J. Wang, Z.-G. Zhang, and J.-q. Chen, "Design of a self-balanced system based on AccuStar electronic inclinometer," in *Electrical Machines and Systems, 2009. ICEMS 2009. International Conference on*, 2009, pp. 1-4.
9. L. Zhijun, L. Yang, Y. Chenguang, and D. Nan, "Motion control of an autonomous vehicle based on wheeled inverted pendulum using neural-adaptive implicit control," in *Intelligent Robots and Systems (IROS), 2010 IEEE/RSJ International Conference on*, 2010, pp. 133-138.

Trajectory Tracking Control of Underactuated USV with Model Perturbation and External Interference

Jinbiao Liu , Jun Luo [1,2], Jianxiang Cui, Yan Peng [3]

Department of Mechatronic Engineering Shanghai University, Shanghai, China

Abstract. This paper presents an improved trajectory tracking controller based backstepping control algorithm to address the trajectory tracking problem of an underactuated Unmanned Surface Vessel (USV). A Three Degree of Freedom (DOF) underactuated maneuvering motion model for water-jet-propelled USV is established, and controller can track both straight line trajectory and curve trajectory with relatively high accuracy. The stability of USV system is proved by BIBO（bounded-input-bounded-output stable） characteristics，then simulation experiments are carried out to demonstrate the robustness and precise control performance of the controller.

1 Introduction

These Unmanned Surface Vessel (USV), also known as the water robot, which is an unmanned surface vehicle with autonomous planning and navigation capability. Generally, the main difficulties in the tracking control of USV are: system uncertainty, respectively, uncertainty within the system and external uncertainty of the system; underactuated characteristic of the system, the mainstream USV only fits with a water jet propeller, there is no horizontal drive, belong to a typical under-actuated system. how to make the system of USV in high-speed movement to maintain the desired tracking accuracy while maintaining maneuverability has become a hot research topic.

In recent years, the track control of USV has attracted wide attention in the academic field. At home and abroad, there is a deep research on the track control of the USV, and there are many kinds of research methods, such as PID and improved PID, Lyapunov direct method, Backstepping design method,etc. Godhavn et al. [1] designed a global exponential stabilization controller which can achieve good tracking performance in the external environment by using the Backstepping design method and feedback linearization, but the heading angle under the controller is not controlled. Pettersen et al. [2] designed a trajectory tracker which can make underactuated ship's heading tracking error and position tracking error achieve exponential stability on the assumption that the expected heading angle acceleration is greater than zero conditions. K. Do et al. [3] designed a trajectory tracking controller for under-actuated ship base on the Lyapunov direct method, the backstepping method and the cascade system theory, the controller can finish straight and curved global asymptotic tracking control.

In view of the USV have problems that model parameters uncertainty effect the USV path tracking, A.P. Aguiar et al. [4] combined the logic switch control with Backstepping technology, and designed the adaptive supervisory controller, this method made the path tracking error converges to any small neighborhood of the origin, the simulation results proved the effectiveness of the controller. Peng et al. [5] used the Calman filter to estimate uncertain parameters and designed the path tracking controller. In order to solve the adverse effects of ship model uncertainty and external disturbance ,Do et al. [6] proposed a robust adaptive controller base on Lyapunov direct method and Backstepping after carrying out non-linear coordinate transformation,and estimated the uncertainty of model parameters by using the Lipschitz parameter projection technique.

Though researchers have made lots of contributions and proposed many pioneering methods for trajectory tracking control of USV in the literature mentioned above, in the face of complex sea conditions, the tracking control accuracy is still unable to reach the ideal level. This paper aims to provide and prove a nonlinear backstepping trajectory tracking method which can track an arbitrary reference trajectory for underactuated USV.

2 Problem formulation

[1] Fund program: Science and Technology Commission of Shanghai Municipality, China(Grant No. 14DZ1110900)
[2] Fund program: Project of Shanghai outstanding academic leader,China(Grant No. 15XD1501800)
[3] Fund program: Supported by Natural Science Foundation of Shanghai Municipality, China(Grant No. 13ZR1454300)

2.1 USV modeling

Ships are complex motion with 6-DOF in the sea, including model parameters perturbation for the hull and the complexity of external environment interference, it's difficult to establish accurate model of 6-DOF for USV. Therefore, the research is mainly considering the movement of the surge, sway and yaw directions on the horizontal plane, ignoring the effects of heave, pitch and roll motion, only about 3-DOF motion model. Inertial reference coordinate system $O_E - X_E Y_E Z_E$ with origin defined on Earth and body-fixed reference $O_b - X_b Y_b Z_b$ are established as show in Fig. 1. the mathematical model of an underactuated USV moving in the horizontal plane can be described as follows[7]:

$$\begin{cases} \dot{\eta} = J(\eta)\upsilon \\ M\dot{\upsilon} = -C(\upsilon) - D(\upsilon)\upsilon + \tau \end{cases} \tag{1}$$

with

$\eta = [x, y, \psi]^T, \upsilon = [u, v, r]^T$, $\tau = [\tau_1, \tau_2, \tau_3]^T$,
$M = diag\{m_{11}\ m_{22}\ m_{33}\}, D(\upsilon) = diag\{d_{11}\ d_{22}\ d_{33}\}$

$$J(\psi) = \begin{pmatrix} \cos(\psi) & -\sin(\psi) & 0 \\ \sin(\psi) & \cos(\psi) & 0 \\ 0 & 0 & 1 \end{pmatrix}$$
$$C(\upsilon) = \begin{pmatrix} 0 & 0 & -m_{22}v \\ 0 & 0 & -m_{11}\mu \\ m_{22}v & m_{11}\mu & 0 \end{pmatrix} \tag{2}$$

where (x, y) are the Cartesian coordinated of USV's center of mass in $O_E - X_E Y_E Z_E$, ψ denotes yaw angle in $O_b - X_b Y_b Z_b$. μ, v, r denote surge, sway and yaw velocities expressed in $O_b - X_b Y_b Z_b$, $[\tau_1, \tau_2, \tau_3]^T$ are control inputs of USV, $\tau_2 = 0$, m_{11}, m_{22}, m_{33} are USV's inertia coefficients including added mass effects, d_{11}, d_{22}, d_{33} are hydrodynamic damping coefficients.

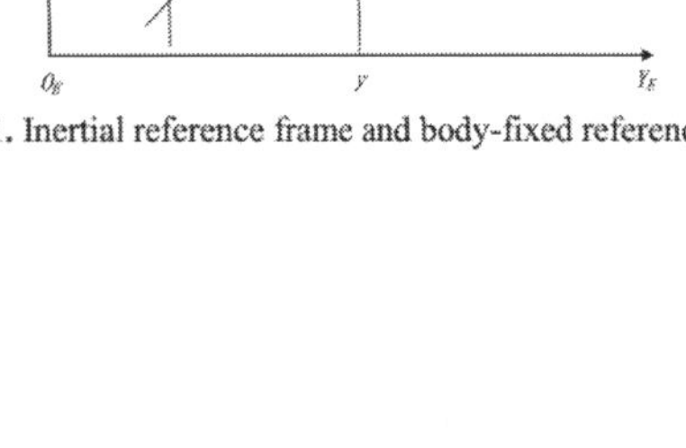

Figure 1. Inertial reference frame and body-fixed reference.

By the above formula can get the kinematics equation of the USV:

$$\begin{cases} \dot{x} = u\cos(\psi) - v\sin(\psi) \\ \dot{y} = u\sin(\psi) - v\sin(\psi) \\ \dot{\psi} = r \end{cases} \tag{3}$$

Without considering the environmental disturbance, the dynamic equation of the USV can be expressed as:

$$\begin{cases} m_{11}\dot{u} - m_{22}ur + d_{11}u = \tau_1 \\ m_{22}\dot{v} + m_{11}ur + d_{22}v = 0 \\ m_{33}\dot{r} + (m_{22} - m_{11})uv + d_{33}r = \tau_3 \end{cases} \tag{4}$$

where τ_1 is longitudinal driving force for the control input, τ_3 is yaw moment generated by the combined action of propeller and rudder.

2.2 Problem statement

In this paper ,the control problem of trajectory tracking for USV can be formulated as follows: Considering the situation that model parameter perturbation and external disturbance factors which exist in underactuated Unmanned Surface Vessel (USV), robust control is the key problem of the USV, and the control objective for trajectory tracking of USV is to ensure the tracking errors converge to an arbitrarily small neighborhood of zero as t $\to \infty$.

3. Controller design

3.1 Sliding mode control law design

Suppose an expected track of USV:

$$\begin{cases} \dot{x}_d = u_d\cos(\psi_d - v_d) \quad \sin(\psi_d) \\ \dot{y}_d = u_d\sin(\psi_d - v_d) \quad \sin(\psi_d) \\ \dot{\psi}_d = r_d \\ m_{11}\dot{u}_d - m_{22}v_d r_d + d_{11}u_d = \tau_{1d} + f_1 \\ m_{22}\dot{v}_d + m_{11}u_d r_d + d_{22}v_d = 0 \\ m_{33}\dot{r}_d + m_{22}u_d v_d - m_{11}u_d v_d + d_{33}r_d = \tau_{3d} + T_3 \end{cases} \tag{5}$$

Define the position tracking error for USV:

$$\begin{cases} u_e = u - u_d \\ v_e = v - v_d \\ r_e = r - r_d \end{cases} \tag{6}$$

Differentiate the function, the error equation:

$$\begin{cases} \dot{u}_e = \frac{m_{22}}{m_{11}}(v_e r_e + v_e r_d + v_d r_e) - \frac{d_{11}}{m_{11}} u_e \\ \quad + \frac{1}{m_{11}}(\tau_1 - \tau_{1d}) \\ \dot{v}_e = -\frac{m_{11}}{m_{22}}(u_e r_d + u_e r_e + u_d r_e) - \frac{d_{22}}{m_{22}} v_e \\ \dot{r}_e = \frac{m_{11} - m_{22}}{m_{33}}(v_e u_e + u_e v_d + v_e u_d) - \frac{d_{33}}{m_{33}} r_e \\ \quad + \frac{1}{m_{33}}(\tau_3 - \tau_{3d}) \end{cases} \quad (7)$$

Design first-order exponential stable sliding mode switching surface:

$$S_1 = u_e + \lambda_1 \int_0^t u_e(\tau) d\tau, \ \lambda_1 > 0 \quad (8)$$

Differentiate the switching surface S_1:

$$\begin{aligned} \dot{S}_1 &= \dot{u}_e + \lambda_1 u_e \\ &= \lambda_1 u_e - \dot{u}_d + \frac{\tau_{1d} + f + m_{22} v_d r_d - d_{11} u_d}{m_{11}} \end{aligned} \quad (9)$$

Make $\dot{S}_1 = 0$, the equivalent control law is obtained:

$$u_{eq} = \hat{d}_{11} u_d - \hat{m}_{11}(\lambda_1 u_e - \dot{u}_d) - \hat{m}_{22} v_d r_d \quad (10)$$

The "^" indicates the estimated value of uncertain model parameters, the same below.

In order to eliminate the phenomenon of "high frequency flutter" of the system, this paper adopts saturation function $\text{sat}(S_1/\phi_1)$ to replace the function of the desired sliding mode $\text{sgn}(S_1)$.

$$\text{sat}(S_1/\partial) = \begin{cases} S_1/\partial, & |S_1| \le \partial \\ \text{sgn}(S_1), & |S_1| > \partial \end{cases} \quad (11)$$

Define switching control law:

$$u_{sw} = -\zeta_1 \cdot \text{sat}(S_1/\partial) \quad (12)$$

Therefore, the longitudinal driving force control law of water jet propulsion USV with parameter uncertainty and external disturbance is designed:

$$\begin{aligned} f &= \hat{d}_{11} u_d - \hat{m}_{11}(\lambda_1 u_e - \dot{u}_d) \\ &\quad - \hat{m}_{22} v_d r_d - \zeta_1 \cdot \text{sat}(S_1/\partial) \end{aligned} \quad (13)$$

In order to determine the parameters of the state function of the USV, define the boundaries of hydrodynamic damping coefficient and inertia parameter which includes additional mass as :

$$|m_{ii} - \hat{m}_{ii}| \le M_{ii} \quad |d_{ii} - \hat{d}_{ii}| \le D_{ii}, \ i = 1,2,3. \quad (14)$$

Select the following Lyapunov function:

$$V_1 = \frac{1}{2} m_{11} S_1^2 \quad (15)$$

Differentiate the function and calculate $\dot{S}_1$、f in the formula:

$$\begin{aligned} \dot{V}_1 &= m_{11} S_1 \dot{S}_1 \\ &= S_1 \begin{Bmatrix} (\lambda_1 u_e - \dot{u}_d)(m_{11} - \hat{m}_{11}) + v_d r_d (m_{22} - \hat{m}_{22}) \\ -u_d(d_{11} - \hat{d}_{11}) - \zeta_1 \cdot \text{sat}(S_1/\partial) \end{Bmatrix} \end{aligned} \quad (16)$$

According to the condition of sliding, the formula(16) must be satisfied[8]:

$$\dot{V}_1 = m_{11} S_1 \dot{S}_1 \le -\eta_1 \hat{m}_{11} |S_1|, \ \eta_1 > 0 \quad (17)$$

Combine formula (16) and (17), state parameter ζ_1 is solved:

$$\begin{aligned} \zeta_1 &= M_{11}|\lambda_1 u_e| + M_{11}|\dot{u}_d| + M_{22}|v_d r_d| \\ &\quad + D_{11}|u| + m_{11}\eta_1 \end{aligned} \quad (18)$$

Design second-order exponential stable sliding mode switching surface:

$$S_2 = \dot{v}_e + 2\lambda_2 v_e + \lambda_2^2 \int_0^t v_e(\tau) d\tau, \ \lambda_2 > 0 \quad (19)$$

Differentiate the switching surface S_2:

$$\dot{S}_2 = \ddot{v}_e + 2\lambda_2 \dot{v}_e + \lambda_2^2 v_e \quad (20)$$

Differentiate the second formula of (8):

$$\begin{aligned} \ddot{v}_e &= -\frac{m_{11}}{m_{22}} \begin{pmatrix} \dot{u}_e r_d + u_e \dot{r}_d + \dot{u}_e r_e \\ + u_e \dot{r}_e + u_d \dot{r}_e + \dot{u}_d r_e \end{pmatrix} - \frac{d_{22}}{m_{22}} \dot{v}_e \\ &= -\frac{m_{11}}{m_{22}}(u_e + u_d)\dot{r}_e \\ &\quad - \frac{m_{11}}{m_{22}} \begin{pmatrix} \dot{u}_e r_d + u_e \dot{r}_d \\ + \dot{u}_e r_e + \dot{u}_d r_e \end{pmatrix} - \frac{d_{22}}{m_{22}} \dot{v}_e \end{aligned} \quad (21)$$

Make:

$$Q = -\frac{m_{11}}{m_{22}}(\dot{u}_e r_d + u_e \dot{r}_d + \dot{u}_e r_e + \dot{u}_d r_e) - \frac{d_{22}}{m_{22}} \dot{v}_e \quad (22)$$

So

$$\dot{S}_2 = Q - \frac{m_{11}}{m_{22}}(u_e + u_d)\dot{r}_e + 2\lambda_2\dot{v}_e + \lambda_2^2 v_e$$

$$= Q - \frac{m_{11}}{m_{22}}(u_e + u_d)\left\{\begin{array}{l} \frac{m_{11}-m_{22}}{m_{33}}\begin{pmatrix} v_e u_e + \\ u_e v_d + v_e u_d \end{pmatrix} \\ -\frac{d_{33}}{m_{33}} r_e + \frac{1}{m_{33}}(\tau_3 - \tau_{3d}) \end{array}\right\} \quad (23)$$

$$+ 2\lambda_2\dot{v}_e + \lambda_2^2 v_e$$

Make:

$$W = Qm_{22}m_{33} - m_{11}(u_e + u_d)\left\{\begin{array}{l} (m_{11}-m_{22})\begin{pmatrix} v_e u_e \\ +u_e v_d + v_e u_d \end{pmatrix} \\ -d_{33} r_e + (\tau_3 - \tau_{3d}) \end{array}\right\} \quad (24)$$

$$+ m_{22}m_{33}(2\lambda_2\dot{v}_e + \lambda_2^2 v_e)$$

$$N = m_{11}(u_e + u_d) \quad (25)$$

$\dot{S}_2 = 0$, the equivalent control law is obtained:

$$\hat{T}_{eq} = \frac{\hat{W}}{\hat{N}} \quad (26)$$

$\hat{W}$ and $\hat{N}$ are estimated value. So, the yaw torque tracking control law for USV is designed:

$$T = \frac{\hat{W} - \zeta_2 \cdot \mathrm{sat}(S_2 / \partial_2)}{\hat{N}} \quad (27)$$

When $\hat{N} = 0$, the control law produces a singular value. we can assume $\hat{N} > 0$.
Similar to the literature [9,10], assume the parameter h is limited, $|h - \hat{h}| \le H$, select the following Lyapunov function:

$$V_2 = \frac{1}{2} m_{22}m_{33}S_2^2 \quad (28)$$

Differentiate the Lyapunov function:

$$\dot{V}_2 = m_{22}m_{33}S_2\dot{S}_2$$

$$= S_2\left\{(N \cdot \frac{\hat{W} - \zeta_2 \cdot \mathrm{sat}(S_2 / \partial_2)}{\hat{N}}) - W\right\} \quad (29)$$

$$= S_2\left\{\begin{array}{l} (N\hat{N}^{-1} - 1)\hat{W} + \hat{W} \\ -W - N\hat{N}^{-1}\zeta_2 \cdot \mathrm{sat}(S_2 / \partial_2) \end{array}\right\}$$

Define geometric mean of parameters N:

$$\begin{cases} N_{max} = (\hat{m}_{22} + M_{22})u_d - (\hat{m}_{11} - M_{11})u > 0 \\ N_{min} = (\hat{m}_{22} - M_{22})u_d - (\hat{m}_{11} + M_{11})u > 0 \\ \hat{N} = \sqrt{N_{max}N_{min}} \\ \beta = N_{max} / N_{min} \\ 1/\beta \le N\hat{N}^{-1} \le \beta \end{cases} \quad (30)$$

If design control parameters ∂_2 as follows:

$$\partial_2 = \beta(H + \hat{m}_{22}\hat{m}_{33}\eta_2) + (\beta - 1)|\hat{W}| \quad (31)$$

So

$$\dot{V}_2 = S_2\left\{\begin{array}{l} (N\hat{N}^{-1} - 1)\hat{W} + \hat{W} \\ -W - N\hat{N}^{-1}(\beta(H + \hat{m}_{22}\hat{m}_{33}\eta_2) \\ +(\beta - 1)|\hat{W}|) \cdot \mathrm{sat}(S_2 / \partial_2) \end{array}\right\} \quad (32)$$

From the above, $N\hat{N}^{-1}\beta \ge 1$, so:

$$\begin{cases} -N\hat{N}^{-1}\beta H \cdot \mathrm{sat}(S_2\partial/_2 \cdot S_2) + \hat{W} - (W \le S_2) \\ -N\hat{N}^{-1}\beta\hat{m}_{22}\hat{m}_{33}\eta_2 \cdot \mathrm{sat}(S_2\partial/_2 \cdot S_2) \le 0 \\ -N\hat{N}^{-1}(\beta - 1|\hat{W}|) \cdot \mathrm{sat}(S_2\partial/_2 \cdot S_2) \\ +(N\hat{N}^{-1} - 1)\hat{W}S_2 \le 0 \end{cases} \quad (33)$$

So

$$\dot{V}_2 = m_{22}m_{33}S_2\dot{S}_2 \le -\hat{m}_{22}\hat{m}_{33}\eta_2|S_2| \quad (34)$$

That satisfies the sliding condition.

3.2. Stability analysis

The above design control laws satisfy the sliding mode condition, and the selected two switching surface are asymptotically stable. Therefore, the trajectory of USV approach to the switching surface with exponential function. In order to prove the stability of USV systems, often by proving the yaw motion is bounded input bounded output stability (BIBO)[11]. Define the following Lyapunov function:

$$V_3 = \frac{1}{2} m_{33}r^2 \quad (35)$$

Derivative to the time:

$$\dot{V}_3 = r(T + \tau_3 - d_{33}r + m_{11}uv - m_{22}uv) \quad (36)$$

When the function V_3 is a decreasing function, r must be a decreasing function. Accordingly, we can know that when $d_{33}|r| > |T + \tau_3 - m_d uv|$, $\dot{V}_3 < 0$, and because the variables T、τ_3、u、v are bounded, so r must be a bounded value.

In summary, the USV's yaw motion satisfies the BIBO characteristics.

4 Simulation experiment

In order to verity and illustrate the performance of the control schemes proposed for trajectory tracking control of underactuated USV, computer simulations were carried out on an USV developed by Shanghai University, simulation environment is a tool platform based on Matlab/Simulink.

The simulation experiment considers the contrast test under the condition that the USV exists model parameters perturbation and environmental disturbance. In this paper, the linear trajectory and circular trajectory are simulated and tested in the case of external disturbance and non-disturbance, and the validity and feasibility of the controller are verified by experiments.

Fig. 2 and Fig. 3 are USV 's simulation results for linear trajectory tracking(Red represents the desired trajectory of the USV, black for the actual trajectory, green ship type represents the direction of the ship's bow, the same bellow), the desired trajectory of the USV is a straight line with a slope of 0.02, its equation is:

$$x_d = 0.5 + 0.02t,\quad y_d = 0.5 + 0.02t \qquad (37)$$

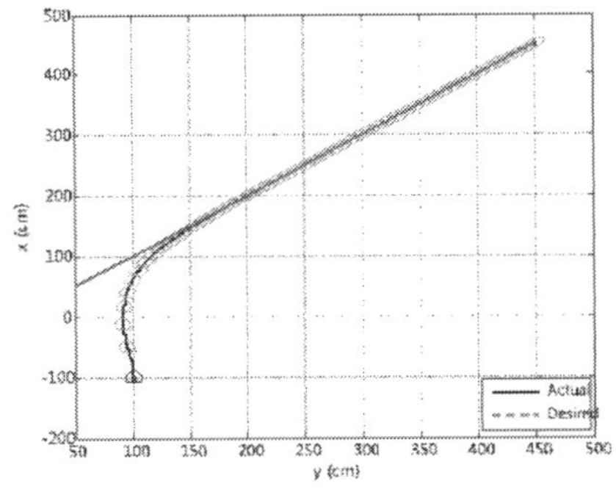

Figure 2. Desired linear trajectory and actual trajectory without interference.

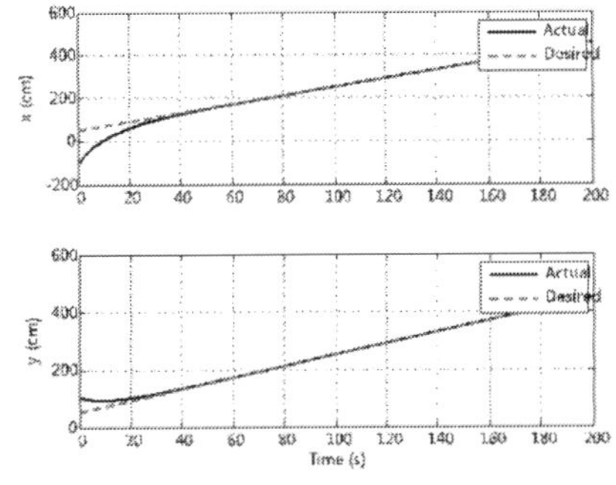

Figure 3. Desired position and actual position without interference.

Fig. 4 and Fig. 5 are USV 's simulation results for circular trajectory tracking, the desired trajectory of the USV is a circle with a radius of 2, its equation is:

$$x_d = 2\cos(\pi/t),\quad y_d = 2\sin(\pi/t). \qquad (38)$$

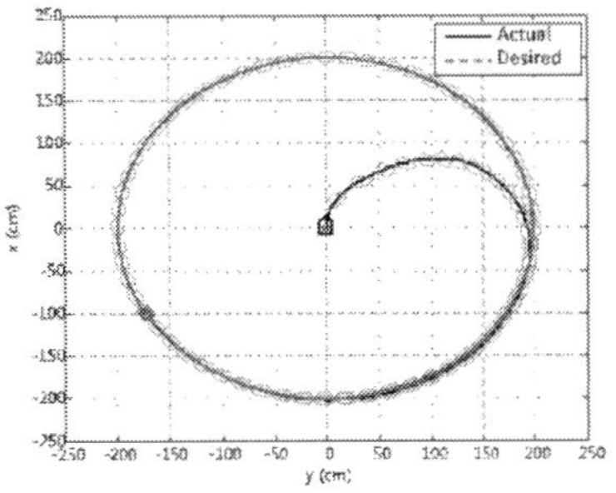

Figure 4. Desired circular trajectory and actual trajectory without interference.

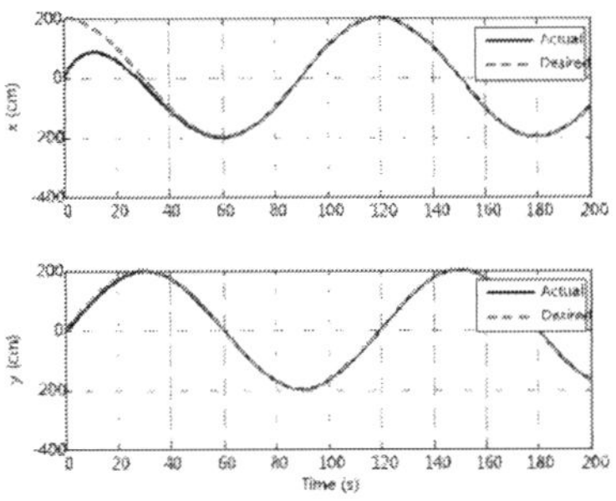

Figure 5. Desired position and actual position without interference.

From the results of the simulation test in Fig. 4 and Fig. 5, in the absence of external disturbance, USV not only can quickly track the desired trajectory in position, the direction of the ship's bow also can point to the tangential direction of movement correctly. To further verify the advantages of the controller designed in this paper, comparing the trajectory tracking of the circular motion under the condition of interference and non-interference. The overall effect of the trajectory tracking simulation under the conditions of environmental interference is shown in Fig. 6, and desired position and actual position with interference is shown in Fig. 7.

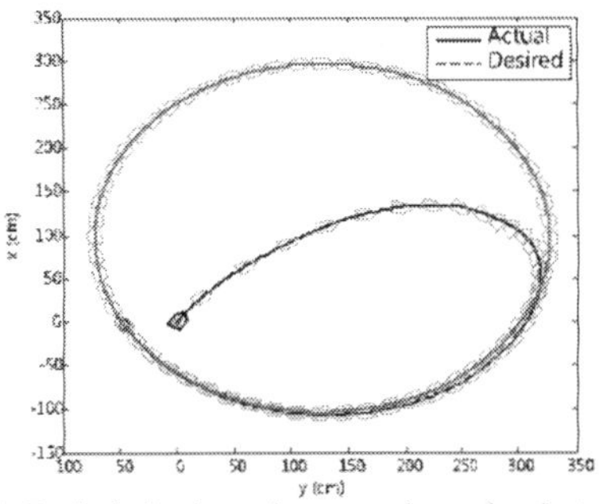

Figure 6. Desired circular trajectory and actual trajectory with interference.

DOI: 10.1051/matecconf/20167709009

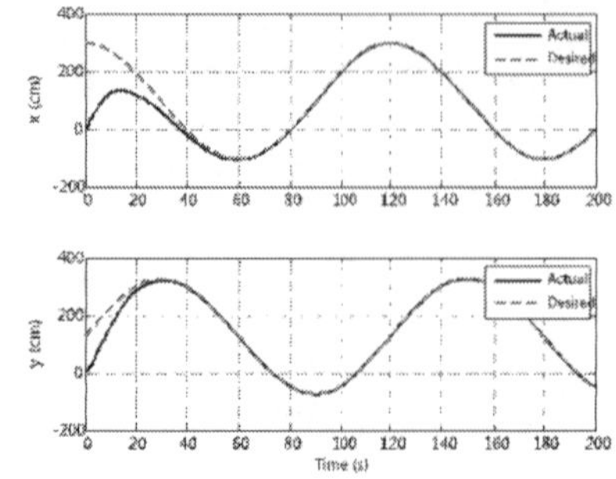

Figure 7. Desired position and actual position with interference .

Fig. 8 and Fig. 9 show response curves of surge velocity and sway velocity of USV under different conditions. it is suggested that the controller has a smooth tracking control for the desired trajectory.

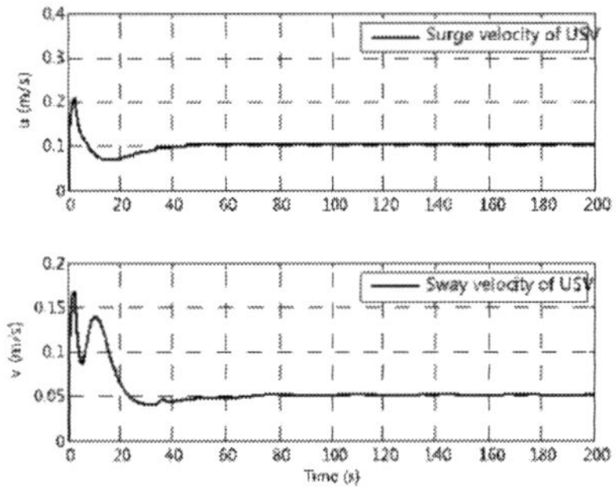

Figure 8. Surge and sway velocity with non-interference .

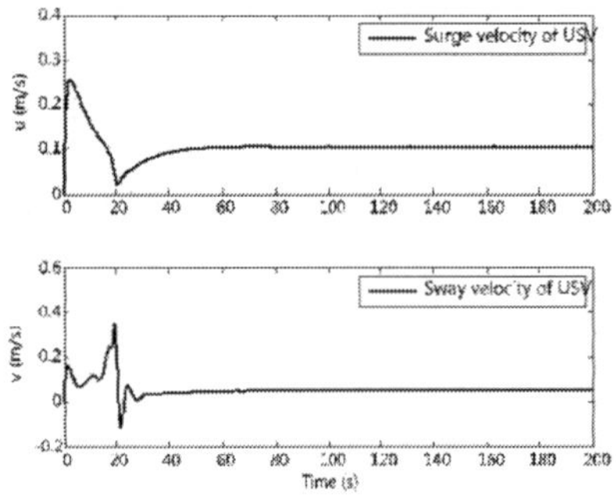

Figure 9. Surge and sway velocity with interference .

Simulation results in Figs. 2 to 9 show that:

1. The control algorithm proposed could achieve trajectory tracking control of USV without the role of side thrusters.

2. The controller guarantees that the USV can track a circle trajectory and a straight line trajectory and have the actual trajectory rapidly convergence to desired trajectory.

3. Global robustness and precise control performance of the controller designed is shown in the simulation experiment results.

5 Conclusions

An improved trajectory tracking controller is proposed in this paper for trajectory tracking control of underactuated USV, based on backstepping control algorithm, combined with sliding mode control method. This controller can track both straight line trajectory and curve trajectory with relatively high accuracy. The stability of USV system is proved by BIBO characteristics. and the control algorithm is verified and illustrated in simulation experiments.

References

1. Godhavn, J.M. Nonlinear tracking of underactuated surface vessels[C]. In proceedings of the 35th IEEE Conference on Decision and Control kobe,Japan. Dec. 1996, Vol. 1: 975-980.
2. Pettersen, K.Y., Nijmeijer H. Tracking Control of an Underactuated Surface Vessel[C]. Proceedings of the IEEE Conference on Decision and Control, Tampa, Florida, 1998: 4561-4566.
3. Do, K.D., Jiang, Z.P., Pan, J. Underactuated ship global tracking under relaxed conditions[J]. IEEE Transactions on Automatic Control, 2002, 47(9): 1529-1536.
4. Aguiar, A.P., Hespanha, J.P. Logic-based switching control for trajectory-tracking and path-following of underactuated autonomous vehicles with parametric modeling uncertainty[C]. Proceedings of the 2004 American Control Conference, Boston, USA, 2004: 3004-3010P.
5. Yan P, Jianda H, Qi S. Tracking Control of Underactuated Surface Ships: Using Unscented Kalman Filter to Estimate the Uncertain Parameters[C]. Proceedings of the 2007 IEEE International Conference on Mechatronics and Automation Harbin, China. 2007:1884-1889.
6. Do, K.D., Pan, J. State and output-feedback robust path-following controllers for underactuated ships using Serret-Frenet frame[J]. Ocean Engineering. 2004, 31: 587-613P.
7. WAN Lei, CUI Shipeng, ZHANG Guocheng, et al . Path following control of underactuated autonomous underwater vehicles [J] . Electric Machines and Control, 2013, 17(2) : 103－111.
8. J.J.E Slotine. Application of nonlinear control [M]. Beijing: press of mechanical industry, 2009.
9. Ashrafiuon H, Muske K. Sliding mode tracking control of surface vessels[C]. Seattle,Washington, USA, 2008: 556-561.
10. Ashrafiuon H, Muske K, Mcninch L, et al. Sliding mode tracking control of surface vessels[J]. IEEE Transactions on Industrial Electronics. 2008, 55(11): 4004-4012.
11. Mcninch L, Ashrafiuon H, Muske K. Optimal specification of sliding mode control parameters for unmanned surface vessel systems[C]. Proceedings of the 2009 American Control Conference, Hyatt Regency Riverfront, USA, 2009: 2350-2355.

DOI: 10.1051/matecconf/20167710001

Wear of Flame-Sprayed Ni-Cr-B-Si Powder Coating on Journal for Seal Contact

Sheng-Yen Hu, Jia-Jyun Liao and Yuan Kang

Chung Yuan Christian University, Department of Mechanical Engineering, 32023 Chungli, Taiwan, R.O.C

Abstract. Flame-sprayed techniques is used in this paper to coat Ni-Cr-B-Si powder on low-carbon steel or bearing steel materials of the journal surface. The wear tester is used to explore material properties of the binding capability, surface hardness, wear and friction within each layer depth. The normal force is applied in addition to the cladding layer by not only using bearing ball but also oil seal pieces, to explore rubber material of oil seal contact journal. In experiments to explore the material and processing conditions affect the microstructure and hardness of the cladding layer, and at the same hardness, surface roughness to affect the performance of the mill run.The results showed that spraying Ni-Cr-B-Si alloy powder in mild steel sheet to melt and run, cladding layer and the substrate has a uniform distribution of fine abrasive particles and binding effect, causing the substrate surface hardness (HRC) about promotion 10 times. While, if sprayed Ni-Cr-B-Si alloy powder to steel panels bearing surface because the surface coated compact structure, can reduce the surface roughness and the coefficient of friction, and more improve the wear resistance of the cladding layer.

1 Introduction

Flame spray coating method is a powder or wire heated by oxyacetylene flame with a high velocity gas jet to blow semi-molten powder coating material onto the substrate. In comparison with other thermal spraying, flame spraying is simplest, minimum limitation, and good performance in accumulation, especially for self-melt alloy powders cladded on a wide range of substrate materials. Moreover, the method has the advantage of pile efficiency, low cost, layer integrity, and large thickness to compare with physical vapour deposition (PVD)

Seal ring is contacting shaft journal by spring pressure with a relative velocity, although very different hardness between both materials, however combined surface pitting and agglutination injury can be induced by hard grit abrasion, long-term effect of high temperature and friction heat, and contact stress fatigue due to shaft rotation. Surface coating with Ni-Cr alloy powder on shaft journal can promote performances of hardness, strength, durability in medium-temperature, resistance to wear and corrosion, and low friction, et cetera. Adding B and Si can lower melt point of Ni and Cr because of eutectic in order to promote bonding and eliminate voids but hardness may be reduced slightly. Hence this paper studies the tribological performance and durability of shaft journal coating Ni-Cr-B-Si alloy powder by flame spray.

Wang et al. [1] studied adding cerium oxide and Cr on the high-speed gas spraying Ni-Al intermetallic layer microstructure and micro indentation and erosion of nature. Ni-Al alloy powder has strong binding, high stack of thick, base substrate for use, adding cerium oxide and chromium can help reduce the Ni-Al intermetallic in the spraying process due to too hard and brittle and cause spall. In addition to adding cerium oxide and chromium also help to improve the hardness of the coating, erosion resistance and elasticity modulus, while reducing micro-cracks and porosity of the coating.

Simm and Freti [2] studied multiphase material, the enhanced influence relative size and content of the abrasive wear behavior. Abrasive wear test results show that, in order to strengthen the WC particles Ni-Cr-B-Si thermal spray coating alloy has good scratch wear performance, when the WC particles of the content exceeds 30 %, of the WC-Ni-Cr-B-Si improve abrasion coating layer scratch wear performance is not obvious. So the abrasive particles whether it will strengthen the surface of material causing deep scratches, strengthening phase particle size is an important determinant.

Fohl et al. [3] using thermal spraying coat ceramic hard phase particles WC, TiC, Cr3C2 and Ni or Co as a bonding phase, on a steel surface. They found the hard phase helps to increase wear resistance ability, but for larger loads the wear environment, due to the brittleness and adhesion of the hard phase with insufficient strength, abrasion resistance will be lowered, but the use of an electron beam, the coating layer can be fused again and improve this problem.

The use of Taguchi orthogonal array L18 (21 × 37) planning experiments, Lin et al. [4] have selected the target atom ratio (Ti/Al), the substrate bias voltage, nitrogen pressure, the arc current, TiAl deposition time,

TiAlN deposition time, surface condition, sputter cleaning of eight factors, to explore the use of cathodic arc plasma deposition process for coating characteristics and wear behaviour of bulk TiAlN coating on the tool steel. Qiao et al. [5] utilized arc spraying to coat Fe-Cr-Ni and explore its wear characteristics of coating layer, they found that two kinds of the abrasive mechanism and material transfer mechanisms coexist.

Hiraga et al. [6] studied plasma spraying of Ni-Ti composite coating of nest (cavitation) corrosion resistant to apply this spraying technique to high temperature dielectric deposition of metal. They found that a shape memory effect of the Ni-Ti alloy layer by arc-Ti wire spraying, plasma spraying and plasma Ni-TiCo-deposition (imitation alloy) prepared, Ni-Ti shape memory effect depending on the coating and the nature of most of the spray control process.

Lee [7] used plasma spraying to coat Al2O3, Al2O3 + 3% TiO2, Al2O3 + 13% TiO2, Al2O3 + 40% TiO2 four different ceramic powders on the 6061 aluminum alloy, and then re-fuse by laser to reduce the disadvantages of the porous coating. Cracks exist because of the large differences in thermal conductivity and thermal expansion coefficient between the ceramic coating and the substrate. The generation of cracks can be reduced by using stainless steel as the substrate. Also, raising the preheating temperature can reduce porosity.

Ko and Robertson [8] in order to understand and seek to replace hexavalent chromium surface coating of the carcinogenic process, as aircraft landing gear, HVOF WC-Co coatings of abrasion tests conducted at room temp. and 21℃-40℃, found its good wear resistance.

Bartuli et al [9] in order to force the flow equation simulation WC-Co Nano HVOF spray coating process in the best parameters, explore the combustion flame and particle flow of hot gas chemical and dynamical properties of WC-Co coated nanostructures affected layer by abrasion tests found nanostructure coating having a low coefficient of friction, and having high hardness and high fracture toughness.

Based on the above studies, using the spray method can indeed improve the surface hardness of the substrate, but in the process due to incomplete powder bulk produce large voids in the coating layer or sandwiched impurities, and reduce the binding strength of the coating layer. Flame-sprayed techniques is used in this paper to coat Ni-Cr-B-Si powder on low-carbon steel or bearing steel materials of the journal surface. As used herein, manner to grind the thickness of cladding layer sequentially removed, and then use the wear tester to explore material properties of the binding capability, surface hardness, wear and friction within each layer depth. The normal force is applied in addition to the cladding layer by not only using bearing ball but also oil seal pieces, to explore rubber material of oil seal contact journal, Mount Run characteristic between the two. , explores the influence of spray material and processing conditions for tribologic characteristics of cladding layer. In experiments to explore the material and processing conditions affect the microstructure and hardness of the cladding layer, and at the same hardness, surface roughness to affect the performance of the mill run.

2 Specimen production

The work principle is shown by Figure 1(a), a powder alloy is heated by oxyacetylene flame and blown to be accelerated by gas jet to a high velocity. Flame spray equipment, spray gun, and its various nozzles in bottles for protection can be seen in Figures 1(b) and 1(c). In use, the required oxygen pressure is 4.0bar, 0.7bar acetylene and compressed air 0-6bar. This equipment can coat low-temperature metal powder, self-fluxing powder, ceramic powder, and aluminum, zinc, plastic and other low melting point materials. Spray a round shaft specimen should be held horizontally erected by a controllable speed chuck as shown by the right front of the device to cruise shaft rotating with constant speed and a flame spray gun fixed by fixture as shown on the left rear, lower left corner of the pipeline respectively is connected to a power outside of the picture, the compressed air source and oxygen-acetylene gas source, the mixed powder is packed in the top of the spray gun opening inverted tank.

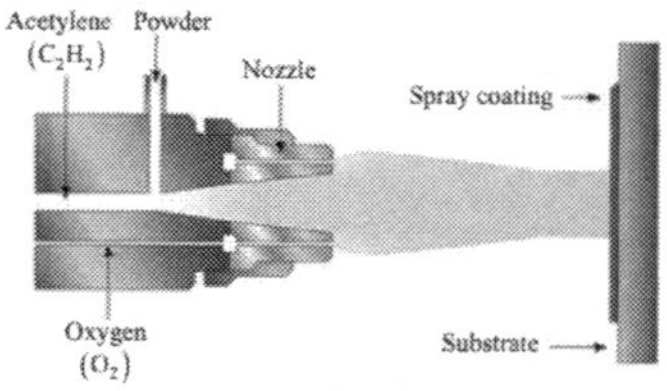

(a) principle

(b) equipment

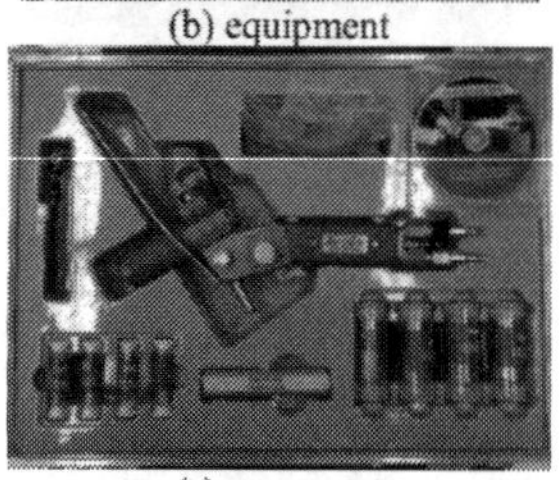

(c) spray gun

Figure 1. Illustrations of flame spray

Ni-Cr-B-Si alloy powder is a self-fluxing powder, mixed elemental powders need to use mixer devices and ultrasonic vibration machine, stir and put in a beaker, the beaker is placed in the oven by 60°C air-dried to 2 hours, and finally these finished powder into standard containers. Ni 72.4, Cr 15%, B 3.25%, Si 4.5% with C and Fe balance of powder contents are used in this paper. Both substrates are used, which are the AISI52100 bearing steel solid round ring with width 19 mm and grinded to 68 mm diameter, and low carbon steel sheet with length 73mm, width 33mm and grinded to 7 mm thickness. Required specimen fixed using a grinder and surface pre-

DOI: 10.1051/matecconf/20167710001

treatment, the purpose of pre-treatment to remove the surface layer fatigue, oil, iron black, rust and other parts, while increasing the surface roughness of the test piece, the spray coating adhered more not falling.

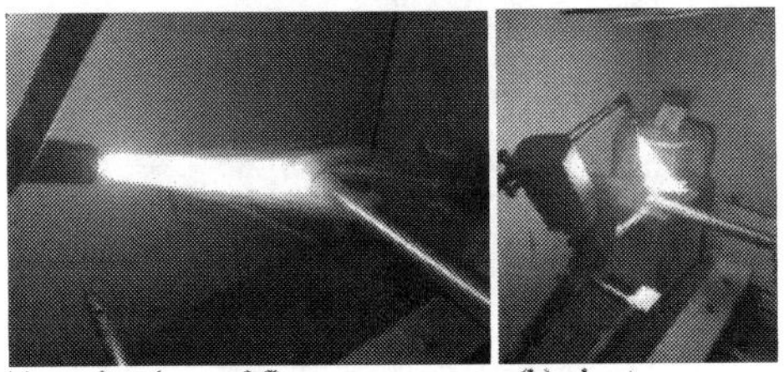

(a) pre-heating and flame-spray (b)reheat

Figure 2 flame-spray process

(a) Ni-Cr coating (b) Ni-Cr-B-Si coating

Figure 3 Comparison in finished specimens

Flame spray gun is used to preheat piece in spray coating process as shown in Fig. 2(a), spray guns is maintained a certain distance from the specimen, preheated to the operating temperature (preheating temperature depends on the spray coating material powder), after turning spray under cover switch gun powder spray an even coating to a desired thickness to start to examine the quality and size of the sprayed layer. The reheat process is required for material of sprayed self-fluxing powder after the completion of the above steps, the use of the spray gun to melt the test piece spray layer is heated again. Reheat as seen in Fig 2(b) raises coating material temperature to the melting point to achieve hardening effect by the enhancement of binding force. In this paper, cylindrical specimen diameter 70mm, spraying speed between 70-180rpm, reheat speed between 30-60rpm, both speeds adopted can be no significant influence for finished quality. After reheat, specimens are required to be cooling in room temperature. If the cooling is too fast, make shrinkage stress in coating layer can be too large to make broken off or shrinkage deformation. At last the specimen surface is finished by grinding. This article uses the flame spray to coating material Ni-Cr powder and Ni-Cr-B-Si powder on ring surface of mild steel substrate and the bearing steel AISI52100. Figs. 3(a) and 3(b) show both finished specimens which have been coated and grinding finished by Ni-Cr and Ni-Cr-B-Si.

3 Tribological performance test

The use of multi-function wear test machine of the ball-on-ring type, a three rotary degree of freedom ball is pressed on the rotating specimen ring with SAE 15W/40 oil (viscous index VI=134, kinematic viscosity is 106.2 cSt at 40°C and 14.06 cSt at 100°C) of wet test to evaluate tribological performance of the flame spray coating specimen. Also, the ball is replace by a seal rubber to simulate the spindle contacted by mechanical seal which is evaluated by is abrasion ability. Applying load 10N, 50N, 100N on ball and rotating ring with speed 300rpm, 500rpm, 700rpm in cross wear test, 100 minutes to observe the various hierarchical wear scar spray coating conditions of the trend of increase wear time.

In wear test, the ram of specimen clamp is deformed because of the friction which is measured by strain gauge to determine force which is divided by the applied load and the coefficient of friction can be obtained. While K type thermocouple including temperature recorder is used to measure temperature rise of oil in the wear test. The record of temperature rise can reflect heat generating and dissipation due to friction and abandon the influence of environmental temperature. The first use basis weight (10kgf) of Rocwell hardness testing applies on specimen and load is added sequentially and kept an appropriate time, and then the load is returned to basis. The use of indentation depth due to load 150kgf applied by diamond cone and steel ball for Rockwell hardness can be determined by $HRC = 100 - 0.5h$ and $HRB = 130 - 0.5H$, respectively. In general, HRC is suitable for iron and steel materials, HRB is suitable for measuring relatively soft material of low carbon steel, copper, aluminum etc..

4 Analysis results of the relationship between abrasion spray coating and ball

4.1 Influence of coating thickness, material, and method for surface hardness

The test piece spray evenly coated Ni-Al alloy powder which rose a thickness of 2 mm, the specimen to be done to complete spray coating each layer hardness measurement, this experiment is 0.3mm floor, ground to the desired thickness using a grinder that is using the Rockwell hardness machine hardness measurement, measurement continues to complete the above steps until the spray coating is exhausted, hardness and thickness affect the relationship shown in Figure 4(a). The results from that, due to the measurement of the hardness of Rockwell hardness, it is the ultimate measure of the depth of the indentation. If the thickness is greater than the amount of the indentation depth measured values are similar, the results of this experiment thus decided to give the anti-wear material reserve amount of 1mm thickness of the wear test as a benchmark. Since the Ni-Al alloy as a base to use just simply increase the adhesion of spray coating and the test piece. Its hardness is too soft (HRB78) so we doesn't use it. Hardness of Ni-Cr alloy although already higher than the mild steel sheet, but still less than the surface hardness of the steel to improve the bearing requirements. The results can be known from Fig. 4(b).

Ni-Cr alloy by spray coating contained after many pores so that adhesion and hardness limited rise; boron alloy powder within the Ni-Cr, elemental silicon powder and increase financial Run procedures so that the specimen surface hardness increased dramatically. The alloy microstructure of Ni-Cr and Ni-Cr-B-Si sprayed coating are shown in Figs. 5(a) and 5(b), respectively. This photo reveals that Ni-Cr alloy in the distribution of large pores and spray cover loosely organized, uneven, this is because they do not entirely caused by the accumulation of the metal spray coating process between the particles. Reversely, Ni-Cr-B-Si alloy melt due to reheat through the process gas overflow, so that the surface of dense, thereby improving the quality test strip.

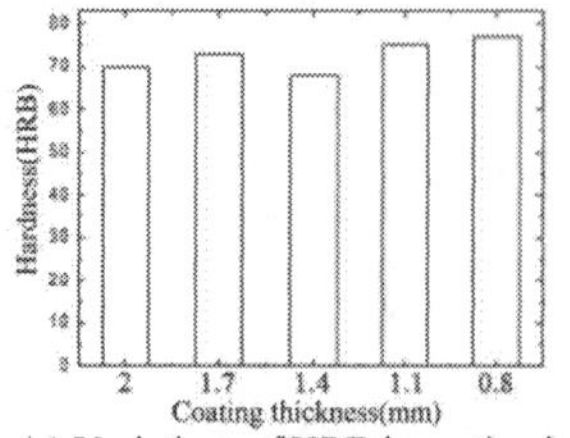

(a) Variations of HRB in coating layer

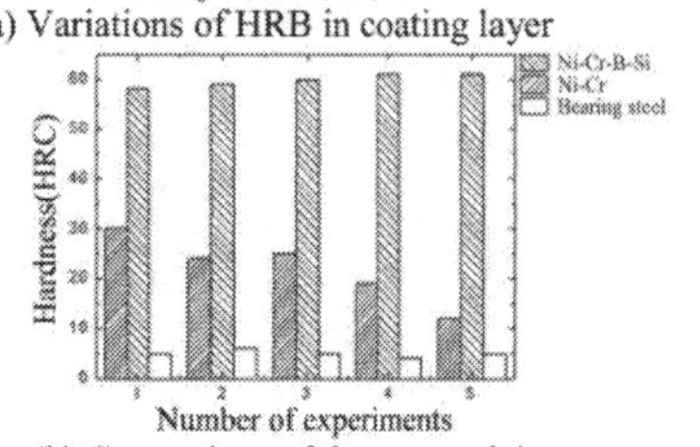

(b) Comparison of three materials

Figure 4 Hardness measurement results

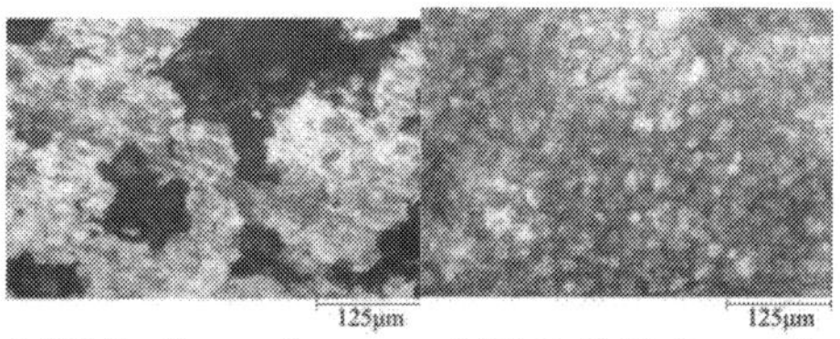

(a)Ni-Cr alloy coating (b)Ni-Cr-B-Si alloy coating

Figure 5 Microstructure photos of flame-spray coating

4.2 Wear between the ball and spray coating surface

For testing time 230 minutes with rotating speed 500 rpm and applied load, adhesive wear in contact is transferred from a softer material friction surface to another surface of the rubbing harder, causing loss of material, the main features of the traces of friction and wear of the surface of the elongated strip and tear marks on hard surfaces can stick more than the material as shown in Fig. 6. The Fig. 6(a) for coating Ni-Cr alloy black stripes on the surface of adhesive wear are shown, Figs. 6(b) and 6(c) to the original bearing steel panels and surface spray coating Ni-Cr-B-Si alloy specimen from the figure apparent steel panels bearing wear scar darker, which represents more serious wear and tear, its hardness and alloy gap Ni-Cr-B-Si small, so that the impact of the coefficient of friction material wear is important.

After the wear test was observed using an optical microscope can be seen in Fig. 7, wherein the Ni-Cr alloy due to soft texture and uneven surface, so the hard texture of the surface of the ball will be less damage and wear scar also relatively obvious, but the specimen is generated serious scratches. Ni-Cr-B-Si alloy hardness is similar to bearing steel, but its grind marks and scratches are more shallow than bearing steel. Because Ni-Cr-B-Si alloy steel bearing surface texture of good surface roughness and friction between the ball relatively peaceful, so is better than the original mill-run feature of bearing steel.

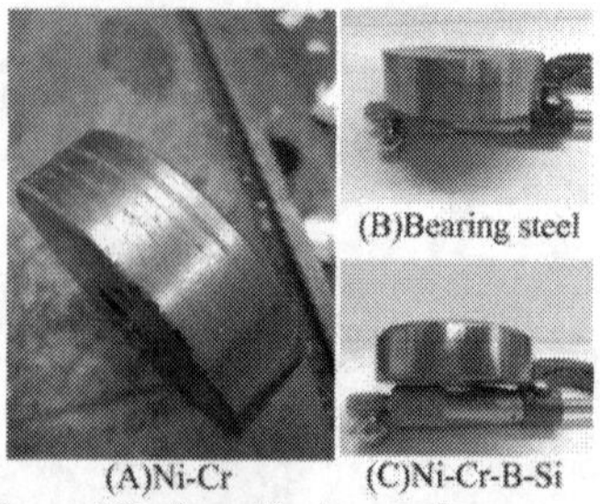

(A)Ni-Cr (C)Ni-Cr-B-Si

Figure 6 Surface adhesion on after wear test

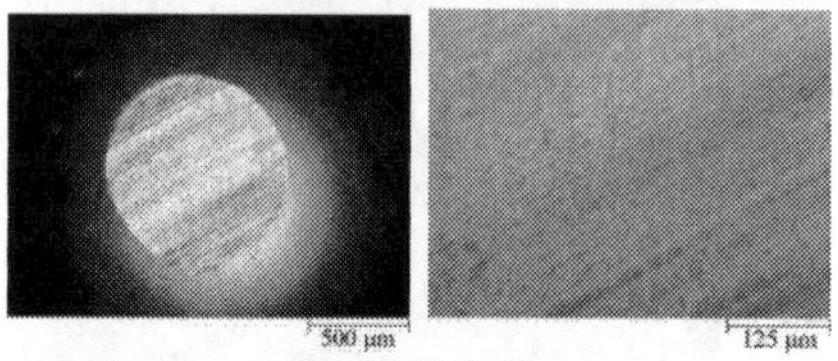

(a) Ni-Cr-B-Si

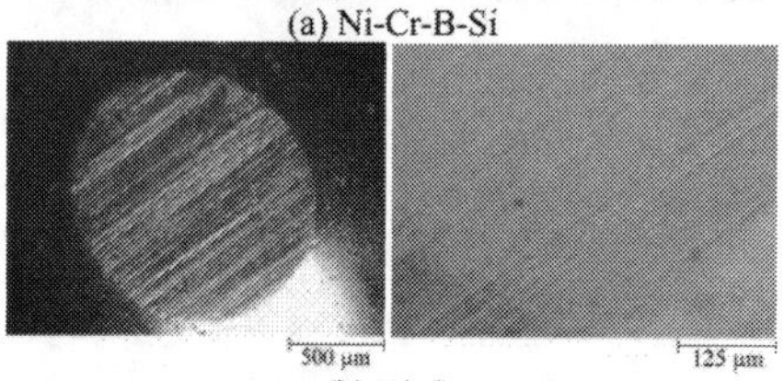

(b) Ni-Cr

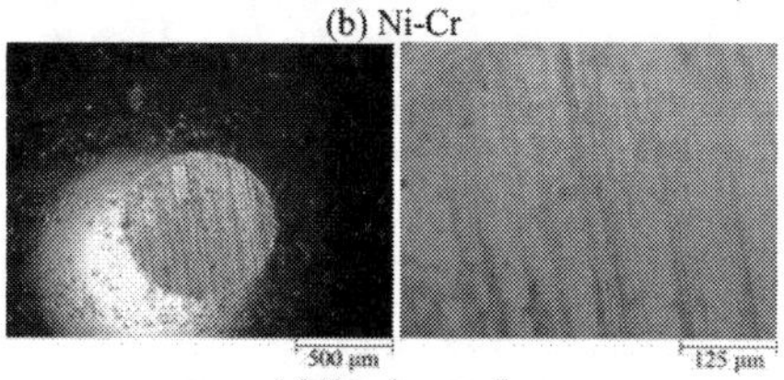

(c) Bearing steel

Figure 7 Adhesive observation after wear test (left ×200 magnification and right ×50 magnification)

4.3 Friction and surface roughness analyses for flame spray coating

Fig. 8 is a relationship between the coefficient of friction of the test piece and the ball friction reflects the energy loss caused by friction size, while reducing energy consumption in tribology research and to increase the hardness and wear resistance of the sprayed coating is just as important. Friction coefficient measurement process due to the test pieces roundness, so the test strip will produce a slight bounce up and down while rotating the final result data plotted against jitter phenomenon (friction data acquisition: 1 sec / time); Ni-Cr alloy is sometimes the reason for low friction coefficient and low surface hardness. The surface layer is scratched caused by the contact area of the grinding ball becomes smaller, resulting in a value near the strain gauge measured becomes small. For Ni-Cr-B-Si alloy compared with the bearing steel, the friction coefficient of Ni-Cr-B-Si alloy is low so the surface of Ni-Cr-B-Si alloy easier to get smooth (bright) surface using grinding machine, thereby increasing mill-run nature of the material, spray coating material to reduce surface wear, can be learned from this experiment test, the greater the friction that is greater the abrasion wear test piece, relative scratches will be more obviously (deep).

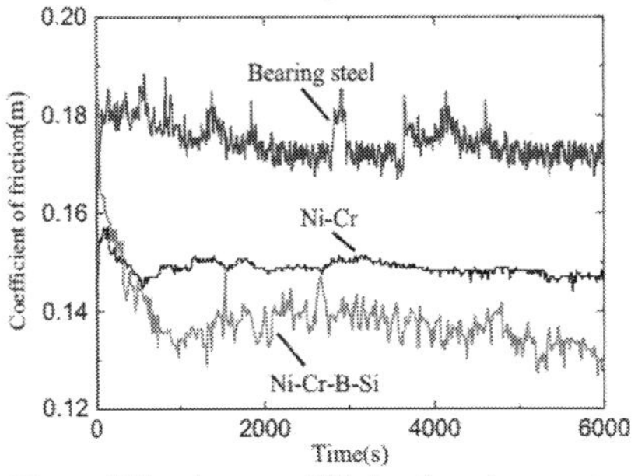

Figure 8 The changes of friction force in wear test

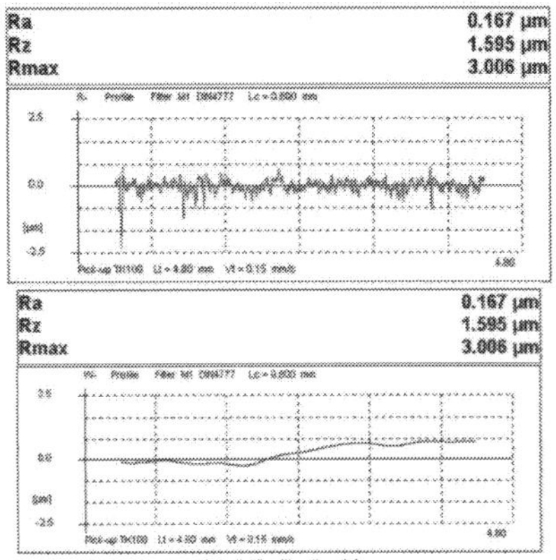

(a) Ni-Cr-B-Si

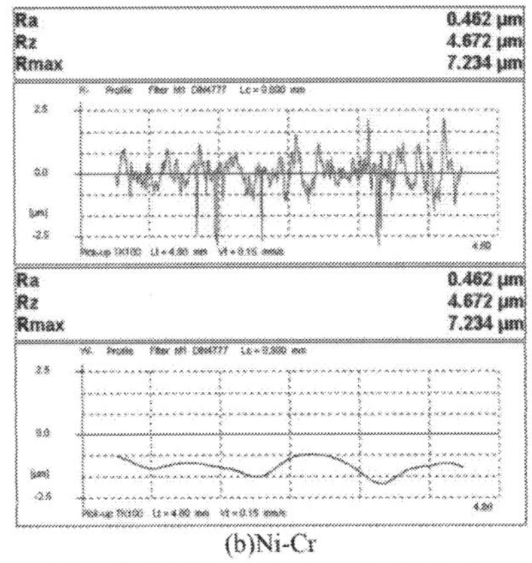

(b)Ni-Cr

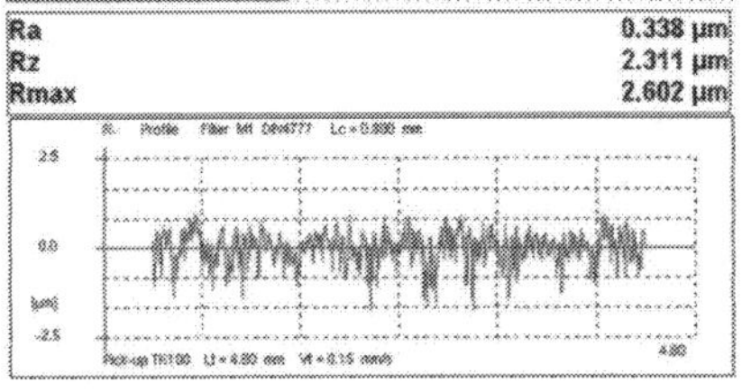

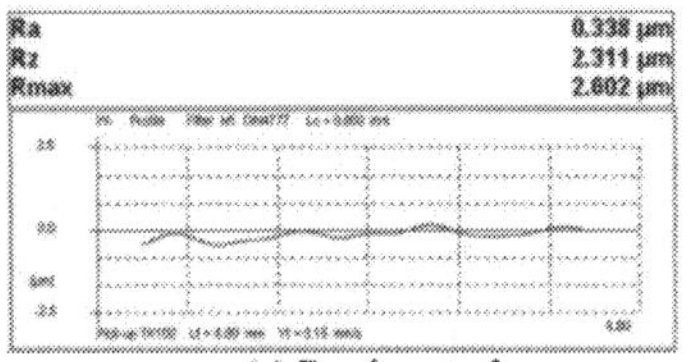

(c) Bearing steel

Figure 9 Surface roughness of flame-spray coating (upper are R contours and lower are W contours)

Fig. 9 is a result after roughness measuring machine, in which a test piece are two graphs show, respectively, the roughness parameters (R and collectively outline) and corrugation parameters (W and collectively outline). Ni-Cr alloy surface most rough, because the impurities entrained with the pores in the spray coating process, resulting in hardness and binding capacity are greatly reduced due to the abrasion test after surface pit scars, scratches, Ni-Cr-B-Si W R value map and graph alloys are bearing steel than good to come, it also proves that the gap between the small hardness, the denser the material surface roughness, the better, the better the wear resistance, the figure has continued to W Figure raise or lower the platform to the test strip or slightly tilted due to contour beyond all who desire to maintain a clear illustration of the same magnification and the commitment now, it's too rough segments would exceed the display range.

5 Wear of flame spray coating

A seal as shown in Fig. 10 is constructed by rubber body, metal ring, and Garter spring which presses the seal lip to forced contact journal. This paper uses common NBR(Acrylonitrile butadiene rubber) to apply load on cutting into bulk to be fixed on tester fixture and sandwiched abrasion test machine table, so that the test piece in contact with the seal, this test method for testing purposes only seal material and spray coating and wear characteristics of the relationship between the steel bearing comparison. Fig. 11 is a seal abrasion wear condition after the experiment and to wear the seal bearing steel has been noticeable scratches and cracks (black part), the seal of the mill and Ni-Cr-B-Si alloy it is only in case of the emergence of the rubber will produce heat after swelling phenomenon.

Figure 12 (left) and bearing steel and Ni-Cr-B-Si alloy seal comparison chart for friction wear trends, the data jitter test piece is part roundness caused, Ni-Cr-B-Si alloy Run of the mill more primitive characteristics of good bearing steel, bearing steel specimens because the surface roughness is poor, so the friction generated is larger, the seal is damaged more serious. This part of the test results represent, among spray coating materials and seal materials are not mutually exclusive phenomenon will occur, and extent of the damage and the test piece surface friction seals have a proportional relationship. Figure 12 (right) and bearing steel and Ni-Cr-B-Si alloy and seal against the grinding lubricant temperature trend comparison chart, the figure shows, spray coating test piece Ni-Cr-B-Si alloy of the wear when the oil temperature is high and the test temperature a little faster due to the coating layer containing the constituent elements of B (boron), boron ceramic material thermal conductivity (27.4) than the bearing steel test piece (46.6) is poor, so in experiments Ni-Cr-B-Si alloy specimen can not be introduced into the lubricating oil due to the heat generated by the friction of the test sheet, resulting in a lower Ni-Cr-B-Si alloy specimen temperature but higher oil; opposite bearing steel panels due to better thermal conductivity, heat easily import test lubricant film, resulting in high temperature bearing steel panels, the lower the oil temperature results.

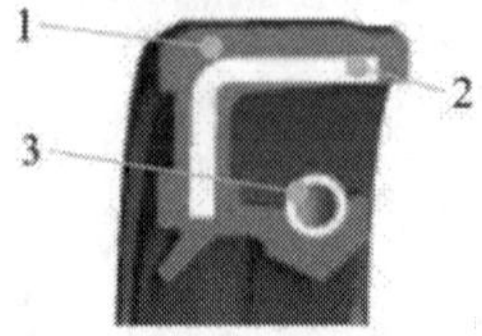

Figure 10 Oil seal structure(TC type)

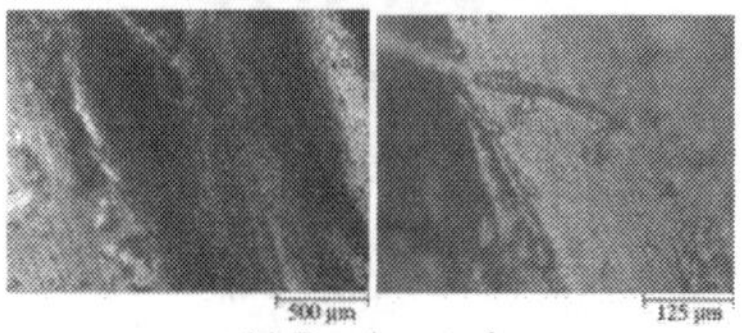

(a) Bearing steel

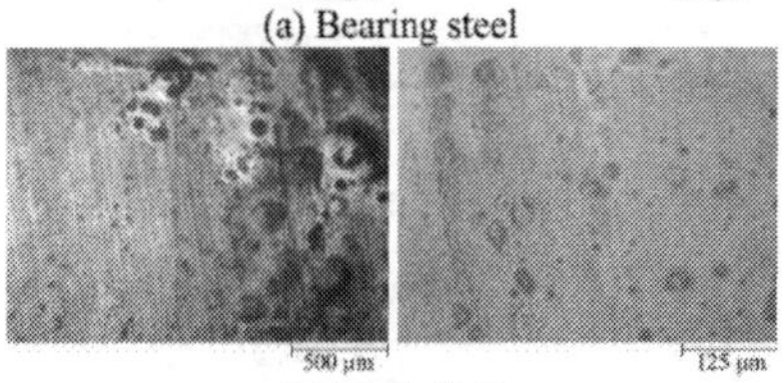

(b) Ni-Cr-B-Si

Figure 11 The extent of the test strip to seal wear loss of test conditions: seal material NBR, oil products 80W / 90, the rotational speed 300rpm, load 250N, wear time 230 minutes (left ×50 magnification and rignt ×200 magnification)

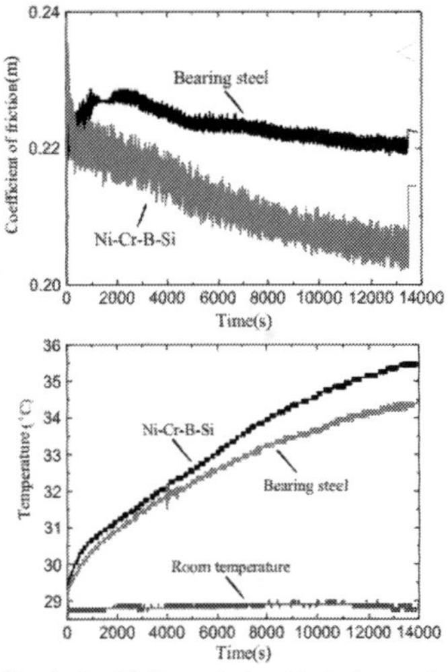

Figure 12 Top is the friction relationship between the specimen and oil seals, and bottom is the relationship oil temperature changed. Abrasion test conditions: Seal material NBR, oil products 80W / 90, the rotational speed 300rpm, load 250N, wear time 230 minutes.

6 Conclusion

The flame spray coating can improve tribological performances for both spindle journal and seal, this study has drawn the following points: (1) The use of Ni-Cr-B-Si alloy powder spray coated on mild steel and reheated coating to melt, the bonding strength thus can be enhanced, to increase substrate hardness, to raise wear resistance and to reduce friction coefficient because spray coating hard particles Cr and protection particles Ni are distributed uniformly. (2) For the same hardness, the lower surface roughness is the better ability of resistance to wear and surface scratches is. (3) Powder incomplete due to the accumulation of the coating can be influenced by the spray distance and the gas flow rates, which induces large and small pores, thereby affecting the spray coating bond strength and wear ability, the more the number of stomata spray coating mechanical properties more difference (4) spray coating Ni-Cr-B-Si alloy in the specimen, its thermal conductivity is poor, due to friction caused by rising oil temperature has little effect on the test piece (but cannot because of the heat and the oil temperature rises), is less likely to make the inter-friction surface temperature continues to rise and damage the oil seal surface structure.

Acknowledgments

The authors deeply indebted to the Minister of Science and Technology, R.O.C., under the contracts NSC102-2221-E-033-021 for their financial supports.

References

1. Y. Wang, W. Chen, L. Wang, Micro-indentation and Erosion Properties of Thermal Sprayed NiAl Intermetallic-Based Alloy Coatings, Wear **254**, 350-355 (2003).
2. W. Simm, S. Freti, Abrasive wear of multiphase materials, Wear **129**, 105-121 (1989).
3. J. Fohl, T. Weissenberg, J. Wiedemeyer, General aspects for tribological application of hard particle coatings, Wear **130**, 257-288 (1989).
4. Kwang-Lung Lin, Ming-Yeong Hwang, Cheng-Dau Wu, The deposition and wear properties of cathodic arc plasma deposition TiAlN deposits, Materials Chemistry and Physics **46**, 7-83 (1996).
5. Y. L. Qiao, B. S. Xu, S. N. Ma, X. B. Liang, J. C. Chi, Tribological Properties and Mechanism of Fe-Cr-Ni Coating, Jixie Gongcheng Xuebao/Chinese Journal of Mechanical Engineering **38**, (2002).
6. H. Hiraga, T. Inoue, H. Shimura, A. Matsunawa, Cavitation Erosion Mechanism of NiTi Coatings Made by Laser Plasma Hybrid Spraying, Wear **231**, 272-278 (1999).
7. Ying-Da Li, A Study of Al2O3/TiO2 Coatings by Plasma-Sprayed for High Dielectric Application, Master's thesis, NCTU (2001).
8. P. L. Ko, M. F. Robertson, Wear Characteristics of Electrolytic Hard Chrome and Thermal Sprayed WC-10Co-4Cr Coatings Sliding against Al-Ni-Bronze in Air at 21°C and at -40°C , Wear **252**, 880-893 (2002).
9. L. Valentinelli, T. Valente, F. Casadei, L. Fedrizzi, Mechanical and Tribocorrosion Properties of HVOF Sprayed WC-Co Coatings, Corrosion Engineering Science and Technology **39**, 301-307 (2004).

DOI: 10.1051/matecconf/20167710002

A Variable Stiffness Analysis Model for Large Complex Thin-Walled Guide Rail

Xiaolong Wang , Aimin Wang and Yanpeng Lv

Beijing Institute of Technology, Department of Mechanical Engineering, 5 South Zhongguancun Street, Haidian District, Beijing, P.R.China

Abstract. Large complex thin-walled guide rail has complicated structure and no uniform low rigidity. The traditional cutting simulations are time consuming due to huge computation especially in large workpiece. To solve these problems, a more efficient variable stiffness analysis model has been propose, which can obtain quantitative stiffness value of the machining surface. Applying simulate cutting force in sampling points using finite element analysis software ABAQUS, the single direction variable stiffness rule can be obtained. The variable stiffness matrix has been propose by analyzing multi-directions coupling variable stiffness rule. Combining with the three direction cutting force value, the reasonability of existing processing parameters can be verified and the optimized cutting parameters can be designed.

1 Introduction

Large complex thin-walled guide rail has large longitudinal dimension and small cross section which is a typical thin-walled workpiece with low rigidity and high precision requirement. Different processing parameters must be carefully considered in different machining surfaces. To solve these problems, a lot of researches are focused on the cutting simulations [1]-[4]. The optimized cutting parameters can be obtained from simulations. However, this method is not suit for large workpiece due to huge computation. As for large complex thin-walled structure and weak local stiffness, the influence on the deformation of low rigidity caused by cutting force was studied by Sveten et al. [5]. The optimization method of the blade clamping by finite element stiffness analysis simulation was also established [6], [7]. Nevertheless, these studies are all qualitative analysis of the stiffness of the thin-walled workpiece. The quantitative analysis method of the stiffness is rarely mentioned.

The aim of this study is to model the variable stiffness of large complex thin-walled guide rail. Applying one direction's cutting force on sampling points separately and analyzing the deformation on three directions, the value of stiffness can be obtained, and the variable stiffness curves can be drawn. Then, the coupling of three directions' cutting force are studied and the linear superposition relationship is proposed. The calculation method of variable stiffness matrix is given. Combined with the cutting force, the reasonability of existing processing parameters are verified, and the optimized cutting parameters are designed.

2 Guide rail structure and the clamping form

The guide rail structure is shown in Fig. 1. The material is 2A12 aluminum alloy with pre-stretching. The cross section is shown in Fig. 2. The size in longitudinal direction is 4m and the cross section size is 127*290. The guide rail is a typical thin-walled workpiece with complex cross section and low rigidity. The large end face's thickness is only 6mm. To reduce the machining deformation, the traditional processing usually adopt slow feed, small cutting depth, which can reduce the force in the process of machining. Although this method can reduce deformation, the selection of cutting parameters depends on experience, usually very conservative. So the machining efficiency is very low.

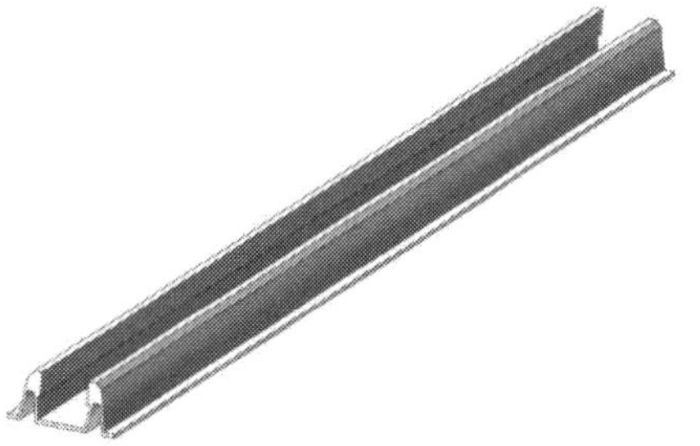

Figure 1. Guide rail structure.

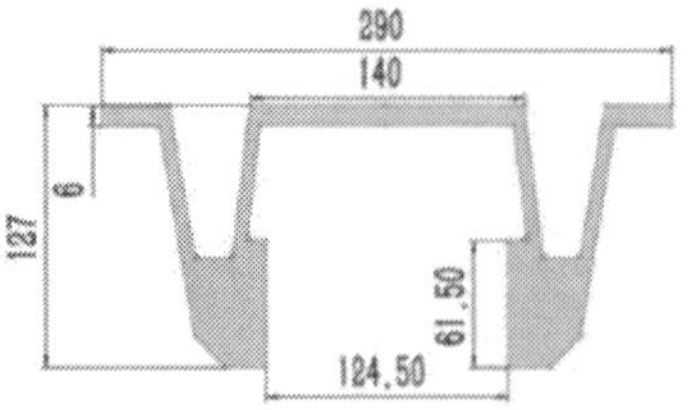

Figure 2. Guide rail cross section.

The machined surface stiffness distribution is different according to the guide rail structure. For example, the inner machined surface's stiffness is higher than the big end surface. So we can optimize its processing efficiency by analyzing its stiffness. The stiffness distribution in inner machined surface is also different among different regions. Further analysis shows even in one exact region, the three directions stiffness are also different. So it's very necessary to obtain the rule of variable stiffness and the quantitative stiffness value to optimize the cutting parameters.

Different clamping form will contribute different stiffness distribution. Fig. 3 shows the clamping form of processing inner machined surface. The big end surface is put on the processing platform, and two wings are clamped by five platens along the longitudinal direction.

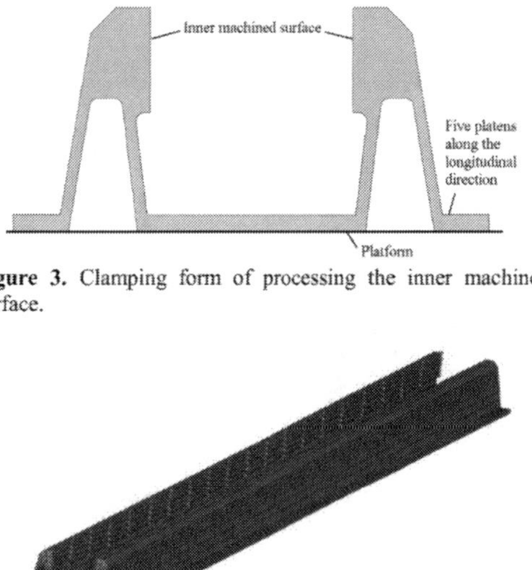

Figure 3. Clamping form of processing the inner machined surface.

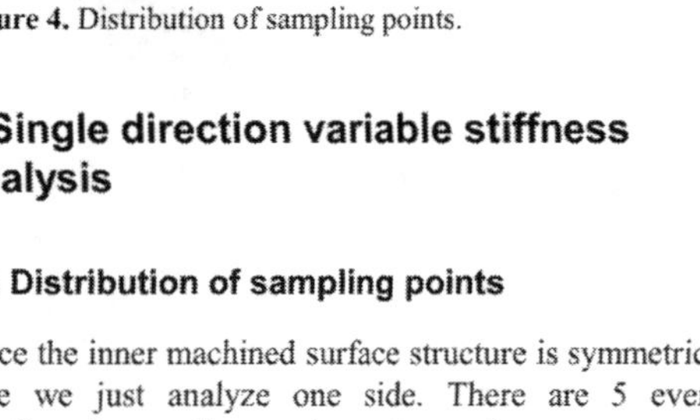

Figure 4. Distribution of sampling points.

3 Single direction variable stiffness analysis

3.1 Distribution of sampling points

Since the inner machined surface structure is symmetrical, here we just analyze one side. There are 5 evenly distributed sampling points along the cross section direction. The top sampling point's mark x equals 1. Also there are 21 evenly distributed sampling points along longitudinal direction. The left sampling point's mark y equals 1. There are a total of 105 samples, and each one has its unique mark x.y, which are shown in Fig. 4.

3.2 Variable stiffness definition

The stiffness value is variable in the inner machined surface. The three directions' stiffness of one sampling point is also different. Here we define the y direction stiffness value of a sampling point for single x direction cutting force R_{xy} is calculated as in Eq. 1.

$$R_{xy} = F_x / U_y \quad (kN/mm) \tag{1}$$

where F_x is the cutting force in x direction.

So we can describe the single x direction stiffness of a sampling point by the R_{xx}, R_{yx}, R_{zx} three values.

3.3 Single direction variable stiffness curve

Three direction simulated cutting force are applied respectively in a sampling point, and the three direction deformation of a signal direction cutting force can be obtained. According to the definition of variable stiffness, the three stiffness values are calculated. We call that the variable stiffness value curve in one direction is the single direction variable stiffness curve. It can be drawn along the longitudinal direction or cross section direction. For example, we apply the y direction simulated cutting force, with 1kN, in the sampling line 'a' which is consist by all sampling points x=a (points a.*), and the deformation in y direction can be obtained from ABAQUS. Then, the pints stiffness R_{yy} are calculated and the five variable stiffness curves along longitudinal direction are drawn as Fig. 5.

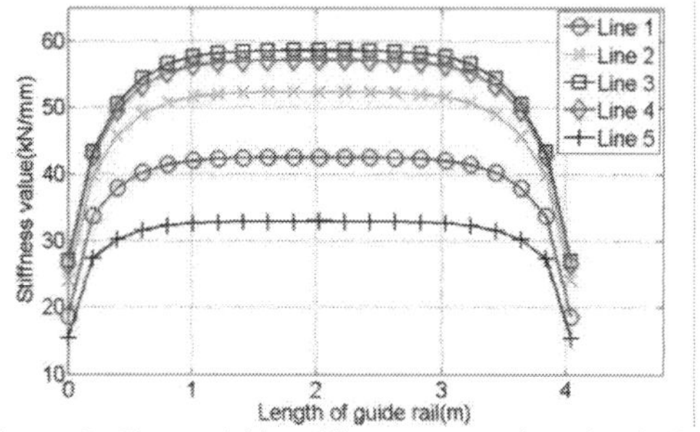

Figure 5. Five variable stiffness curves along longitudinal direction.

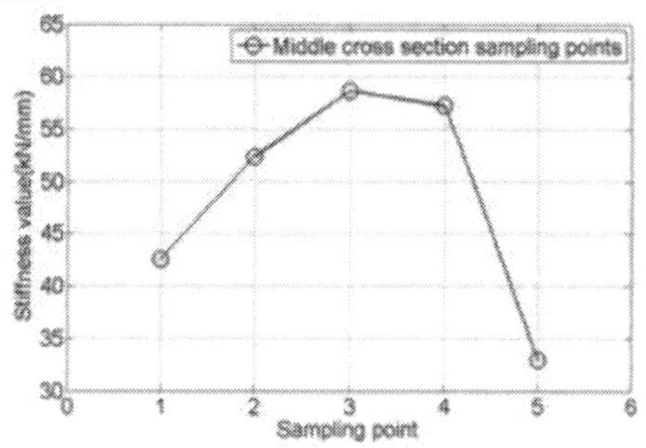

Figure 6. Variable stiffness curve along cross section.

The variable stiffness curve of middle cross section sampling points is plotted in Fig. 6. This curve is made up of the sampling points y=11 (points *.11).

The simulation results show that the stiffness was significantly lower than the middle stiffness at the ends of the guide rail in the longitudinal direction. The lowest stiffness value was just about 46% of the highest; In the cross section direction, the stiffness value increased form sampling point 1 to 3, then decreased until sampling point 5. The lowest stiffness value was just about 56% of the highest. Using this method, the same results can be obtained according to x and y direction stiffness value.

Thus, the stiffness value on the inner machined surface range is very large, which means higher cutting speed can be used and will not cause large deformation. So the cutting efficiency can be improved.

4 Multi-directions coupling variable stiffness analysis

In the process of actual cutting, the cutting force is resultant force of three perpendicular force. The single direction variable stiffness analysis shows that each direction cutting force can cause three directions deformation. So it's necessary to analyze the coupling relationship of variable stiffness. Thus, the variable stiffness model can be used as the guidance in the actual cutting processing.

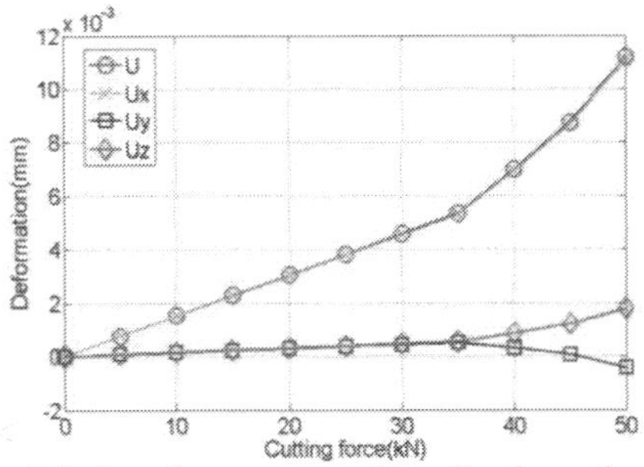

Figure 7. Deformation curves caused by x direction cutting force.

4.1 Variable stiffness linear superposition relationship

The x and z multi-directions coupling relationship is studied firstly. Here we take the sampling point 1.1 for example. First obtain the deformation caused by x direction cutting force using ABAQUS. The cutting force is range from 0 to 50 kN. The deformation curves caused by x direction cutting force are shown in Fig. 7. Where Ux, Uy, Uz are the three directions' deformation, U is the equivalent deformation. Here we use $f_x(F)$ to represent the deformation in this case. Using the same method we can obtain the deformation caused by z direction force $f_z(F)$ which is shown in Fig. 8, and the deformation caused by x and z direction force $f_{xz}(F)$ which is shown in Fig. 9. The linear coupling deformation difference value $\Delta f(F)$ curve is shown in Fig. 10, $\Delta f(F)$ is calculated by Eq. 2.

$$\Delta f(F) = f_x(F) + f_z(F) - f_{xz}(\mathrm{F}) \tag{2}$$

The curves from Fig. 7 to Fig. 10 all consist of two part, the linear part and nonlinear part. The nonlinear part is caused by the plastic deformation. Because in this part, the simulating cutting force is too big that the workpiece is yielded. The linear part indicates that the stiffness value is a constant before yield.

The curves in Fig. 7 and Fig. 8 indicate that the cutting force in different direction will cause different deformation in the same direction due to the different stiffness value.

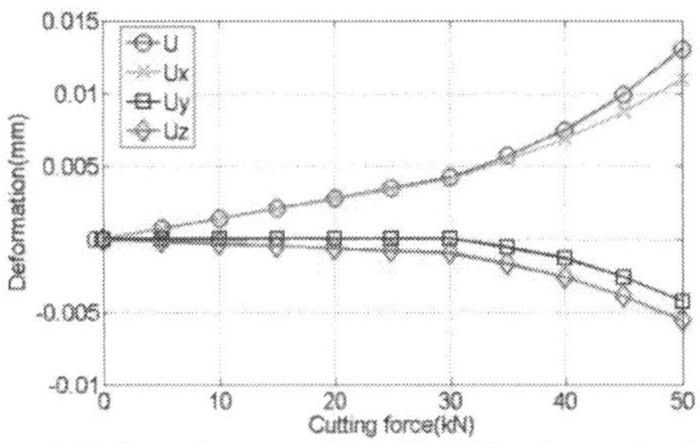

Figure 8. Deformation curves caused by x direction cutting force.

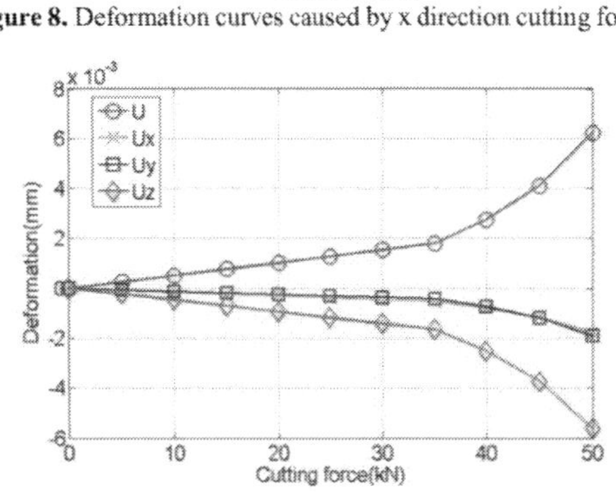

Figure 9. Deformation curves caused by x and z direction cutting force.

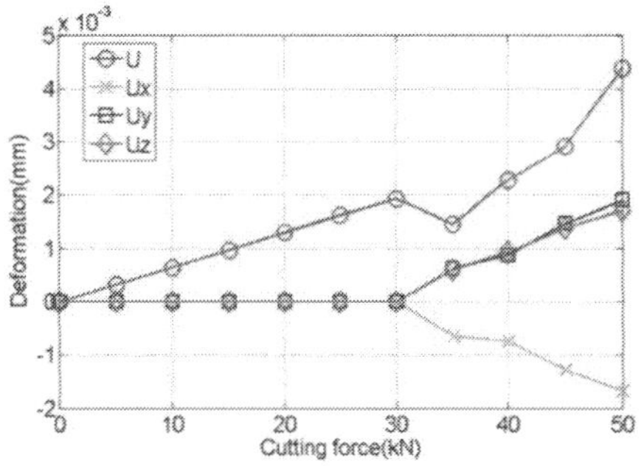

Figure 10. Linear coupling deformation difference value curves.

Fig. 10 shows two kind of coupling relationship. As for the deformation in one direction, the deformation difference value equals to 0 which means the deformation observe linear superposition relationship in elastic stage. However, the equivalent deformation doesn't observe this rule.

Further study shows that the extensional coupling relationship in three directions is also satisfied linear superposition relationship. Thus this relationship offers a simple method to studied the complex cutting processing by decomposing it into three direction and analyzing respectively. Then we can combine them by linear superposition to guide the actual cutting processing.

It should be noted that, the plastic deformation is only happened in the cutting region in actual cutting processing. Most of the workpiece is still in a low stress elastic state. The influence of low stiffness is reflected in the elastic region. So it's reasonable that we just consider the linear part of the curve.

4.2 Variable stiffness matrix analysis model

Assuming that the actual cutting force F is decomposed into three directions as in Eq. 3.

$$F=\begin{bmatrix} F_x & F_y & F_z \end{bmatrix}^T \tag{3}$$

Since each direction's cutting force has different influence in the deformation of three directions, there are 9 independent variables in a sampling point. We describe that R is the variable stiffness matrix in Eq. 4. Where the 9 variables represent the stiffness value. The variable stiffness matrix can be obtained from the static analysis simulation by ABAQUS.

$$R=\begin{bmatrix} R_{xx} & R_{yx} & R_{zx} \\ R_{xy} & R_{yy} & R_{zy} \\ R_{xz} & R_{yz} & R_{zz} \end{bmatrix} \tag{4}$$

Assuming that the deformation in three directions is u, as in Eq. 5.

$$u=\begin{bmatrix} u_x & u_y & u_z \end{bmatrix}^T \tag{5}$$

From the linear superposition relationship, the expression between cutting force, variable stiffness matrix and deformation is shown in Eq. 6.

$$\begin{bmatrix} u_x \\ u_y \\ u_z \end{bmatrix}=\begin{bmatrix} 1/R_{xx} & 1/R_{yx} & 1/R_{zx} \\ 1/R_{xy} & 1/R_{yy} & 1/R_{zy} \\ 1/R_{xz} & 1/R_{yz} & 1/R_{zz} \end{bmatrix}\begin{bmatrix} F_x \\ F_y \\ F_z \end{bmatrix} \tag{6}$$

The cutting force in three directions can be obtained by simulation or experiment. The cutting parameters are reasonable if the calculated deformation is less than cutting precision. On the other hand, if the cutting precision is given, the allowable maximum cutting force can be obtained.

5 Conclusion

Taking the large complex thin-walled guide rail as analysis object, expound the modeling process of the variable stiffness model which can be used to guide the actual cutting processing.

As for a certain clamping structure, the variable stiffness curves are the keys to evaluate the longitudinal and cross section stiffness. According to the coupling cutting force of different direction, the linear superposition relationship is used to couple the stiffness by variable stiffness matrix. The variable stiffness matrix can be obtained from the static analysis simulation by ABAQUS. The time consumption is acceptable even to analyze the large complex thin-walled guide rail.

Compared the simulating deformation which is calculated by cutting force and variable stiffness matrix with cutting precision, the reasonability can be verified, the allowable maximum cutting force can also be obtained. The cutting experiments using the variable stiffness model will be implemented in future work.

References

1. Yang Y, Wang YL, Li CH. Study on machining distortion of titanium alloy aircraft monolithic component by finite element method and experiment. ADV SCI LETT. **4**(8):3206-3210 (2011).
2. Shi GQ, Deng XM, Shet C . A finite element study of the effect of friction in orthogonal metal cutting. Finite Elem Anal Des, **38**:863–883 (2002).
3. Yong Yang , Weiwei Zhu, Study on cutting temperature during milling of titanium alloy based on FEM and experiment. Int. J. Adv. Manuf. Technol.**73**(9):1511-1521 (2014).
4. Ratchev S, Liu S, Becker A A. Error compensation strategy in milling flexible thin—wall parts[J] . J. Mater. Process. Technol.**162**(5):673-681 (2005).
5. R Svetan ,W Huang, W Liu. Modeling and simul ation en-vironment for machining of low-rigidity comp onents [J]. J. Mater. Process. Technol. **153**:67 – 73 (2004).
6. H. Wang, L. Huang, C. Yao, et al.Integrated Analysis Method of Thin-Walled Turbine Blade Precise Machining[J]. Int. J. Precis. Eng. Manuf. **16**(5):1011-1019 (2015).
7. Wang, Y., Chen, X., Gindy, N., et al. Elastic Deformation of a Fixture and Turbine Blades System based on Finite Element Analysis, Int. J. Adv. Manuf. Technol.**36**(4):296-304 (2008).

Effect of Process Parameter in Laser Cutting of PMMA Sheet and ANFIS Modelling for Online Control

Anamul Hossain, Y. Nukman and A. M. Sifullah

Department of Mechanical Engineering, Faculty of Engineering, University of Malaya, Kuala Lumpur 50603, Malaysia

Abstract. Laser beam machining (LBM) is a promising and high accuracy machining technology in advanced manufacturing process. In LBM, crucial machining qualities of the end product include heat affected zone, surface roughness, kerf width, thermal stress, taper angle etc. It is essential for industrial applications especially in laser cutting of thermoplastics to acquire output product with minimum kerf width. The kerf width is dependent on laser input parameters such as laser power, cutting speed, standoff distance, assist gas pressure etc. However it is difficult to get a functional relationship due to the high uncertainty among these parameters. Hence, total 81 sets of full factorial experiment were conducted, representing four input parameters with three different levels. The experiments were performed by a continuous wave (CW) CO_2 laser with the mode structure of TEM_{01} named Zech laser machine that can provide maximum laser power up to 500 W. The polymethylmethacrylate (PMMA) sheet with thickness of 3.0 mm was used for this experiment. Laser power, cutting speed, standoff distance and assist gas pressure were used as input parameters for the output named kerf width. Standoff distance, laser power, cutting speed and assist gas pressure have the dominant effect on kerf width, respectively, although assist gas has some significant effect to remove the harmful gas. ANFIS model has been developed for online control purposes. This research is considered important and helpful for manufacturing engineers in adjusting and decision making of the process parameters in laser manufacturing industry of PMMA thermoplastics with desired minimum kerf width as well as intricate shape design purposes.

1 Introduction

Laser beam machining is one kind of thermal process that can provide good quality product by its outstanding advantages such as non-contact, no tool wear, high accuracy, intricate shape etc. [1], [2]. In laser cutting processes, kerf width should be formed that is allowed to design an intricate shape. Research interest topic is successful cut with minimum material removal or kerf width. Li *et al.* [3] studied on laser cutting of polymer/ceramic materials and concluded that laser is a suitable fabricating processes for ceramics, glass and polymers. Caiazzo *et al.* [4] applied CO_2 Laser cutting on different polymeric plastics. He evaluated the values of kerf widths on top and bottom side, melted area, melted volume per unit time and surface roughness values on cut edges. Choudhury *et al.* [5] developed mathematical model of heat affected zone and surface roughness models by response surface methodology (RSM) of three polymeric materials namely polypropylene (PP), polycarbonate (PC) and polymethylmethacrylate (PMMA). Dubey *et al.* [6] also developed models of material removal rate and kerf width, considering the input parameters as cutting speed, gas pressure, pulse width and pulse frequency. The uses of PMMA are myriad and its applications in different fields include not only the substitute of glass but also in some essential medical equipment namely bone cement, intraocular lens, dentures, prostheses, biosensors and biomechanical device [7]. Laser machining is a complex process moreover CO_2 laser interaction with PMMA begins vaporization while it is subjected to melt shearing [2], [8]. ANFIS model has a capability to manage this kind of complex situation. However, as per our review it has not found very much model of kerf width for PMMA sheet materials including laser power, cutting speed, standoff distance, assist gas pressure together. Hence an attempt has been taken to see the influence of the laser cutting process of PMMA sheet and ANFIS model for online control by including these four input parameters.

2 Methodology

The purpose of this research is to cut PMMA sheets into small rectangle shapes with the aim of determining the minimum kerf width and to investigate the effect of the input parameters on output quality of PMMA sheet. Hence, among all the design of experiment (DOE) methods, a full factorial design was chosen to evaluate the whole area of the quality although it consumes significant amount of experimental resources and time. In the present study, four factors with three levels were employed, representing a total 81 (3^4) sets of experiments. Digital caliper was used to measure side line lengths H_1 and H_2 as shown in Fig. 1. To calculate the kerf width, H_1 and H_2 are measured three times at different location and was calculated using Eq. 1.

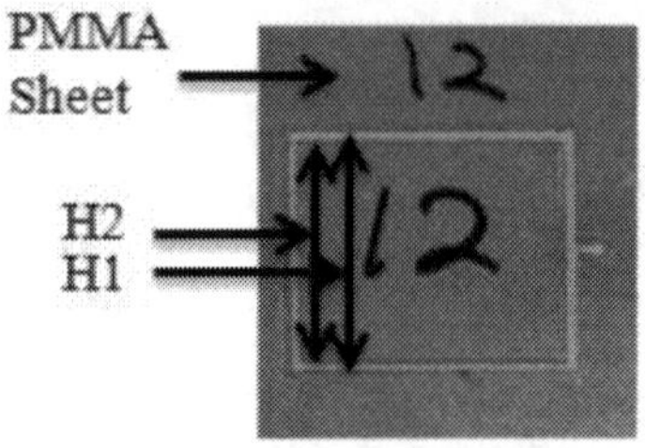

Figure 1. A schematic diagram for Kerf width measurements

$$Upper\ kerf\ width = \frac{(H_1 - H_2)}{2} \tag{1}$$

Table 1 shows the lists of levels for assist gas pressure, laser power, cutting speed and standoff distance used in the experiment.

Table 1: Input variables and their levels.

Input Parameters	Level 1	Level 2	Level 3	Units
Laser Power	100	300	500	Watt
Cutting Speed	0.2	0.7	1.2	m/min
Standoff Distance	1	5	10	mm
Assist Gas Pressure	0.5	2.5	4.5	Bars

3 Effects of laser input parameters on kerf width from experimental study

Fig. 1 (a) shows the combined effects of laser power and cutting speed on kerf width at mid-level assist gas pressure and standoff distance. Based on the Fig. 1 (a) the kerf width increases with the increase of laser power. The reported increase also depends on cutting speed. At low-level cutting speed, kerf width is bigger than that of high-level cutting speed.

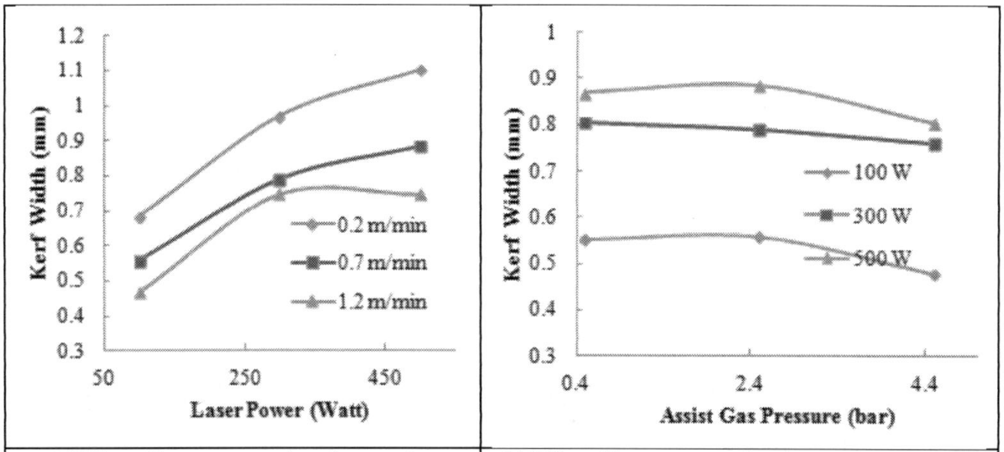

(a)Effect of laser power and cutting speed at constant Mid SOD & GP.

(b) Effect of laser power and assist gas pressure at mid standoff distance & cutting speed.

(c) Effect of standoff distance and assist gas pressure at mid laser power and cutting speed.

(d) Effect of cutting speed and standoff distance at mid laser power and assist gas pressure.

Figure 2: Effect of process parameters on kerf width in laser cutting.

 DOI: 10.1051/matecconf/20167710003

This is because at low-level cutting speed the workpiece had longer exposure to laser radiation which is reliable for vaporization. Moreover, it can be observed that the increase of kerf width is more noticeable as the power increases from 150 W to 300 W. The increase in kerf width is less significant as the power increases from 300 W to 450 W. As shown in Fig. 2(b), the effect of assist gas pressure on kerf width is not so much significant as laser power. However, it is important to blow away any harmful gases and also vaporize material from the workpiece. In addition, the measured kerf width decreases with the increase of assist gas pressure due to the cooling effect of the assist gas. As shown in Fig. 2(c), the kerf width also increases with the increase of standoff distance and assist gas pressure. This is due to the fact that with the increase in standoff distance the laser beam diameter also increases. Hence the kerf width increases. As shown in Fig. 2(d), the kerf width also increases with the increase of standoff distance.

This is due to the fact beam diameter increases as well. Meanwhile, an increase of cutting speed leads to a smaller kerf width because the localized spot on the workpiece had less exposure to laser radiation. From the analysis of the Fig. 2 it can be seen that standoff distance, laser power, cutting speed and assist gas pressure have the dominant effect on kerf width, respectively.

4 Effect of laser input parameters on kerf width from ANFIS model

Fig. 3(a) shows the surface plot of kerf width values at different laser powers and cutting speeds. It is evident that the kerf width value decreases with the increase in cutting speed. Kerf width increases with the increase of laser power as well. From Fig. 3(b) it is evident that the kerf width value tends to decrease with the increase in cutting speed and it increases with the increase in standoff distance as well.

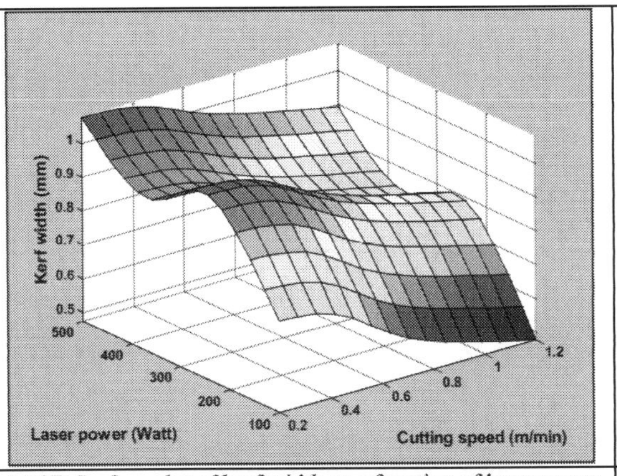

(a) 3D Surface plot of kerf width as a function of laser power and cutting speed.

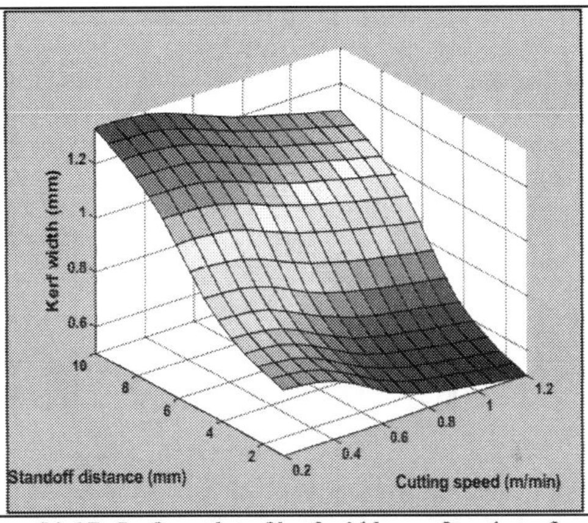

(b) 3D Surface plot of kerf width as a function of standoff distance and cutting speed.

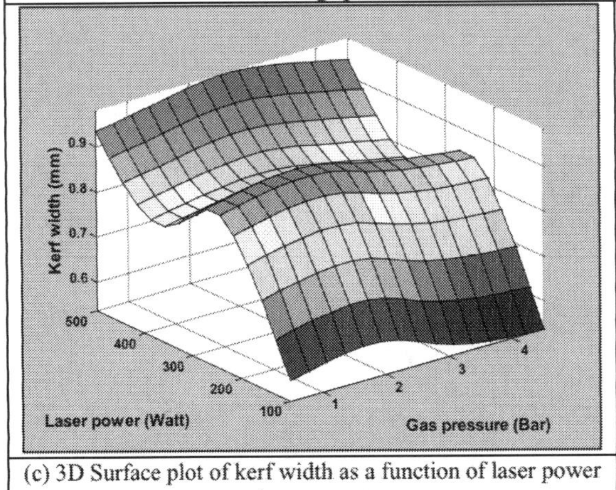

(c) 3D Surface plot of kerf width as a function of laser power and gas pressure.

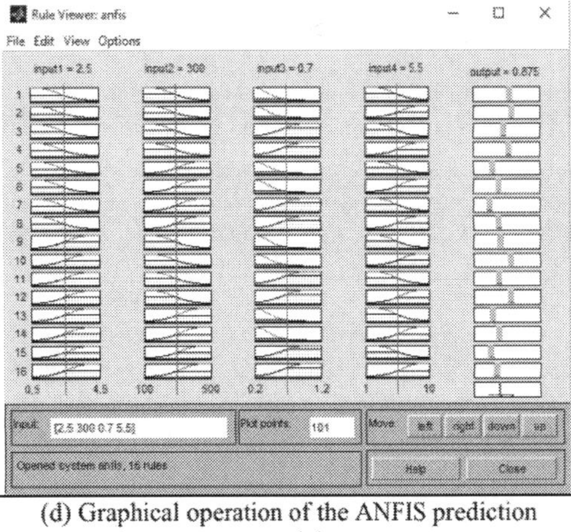

(d) Graphical operation of the ANFIS prediction model

Figure 3. Effect on kerf width from ANFIS model.

It is evident from Fig. 3(c) that the kerf width increases with the increase of laser power of up to 300 W and exhibits smaller kerf width at 400 W and it gradually decreases when it reaches 400 W and again increases up to 500 W this is due to the turbulent effect of the assist gas. Fig. 3(d) shows the graphical rule viewer for the prediction model. For prediction capability evaluation relative error and goodness of fit were used. The statistical analysis of the ANFIS result for kerf width were relative error (4.23%) and goodness of fit 0.9876 (R2= 0.9754) in training, whereas relative error (3.54%) and goodness of fit 0.9853 (R2= 0.9708) were obtained in testing.

5 Conclusions

The purpose of this research is to cut polymethylmethacrylate (PMMA) sheets into small rectangle shapes with the aim of determining the minimum kerf width. In the manufacturing process, it is important to cut the workpiece with minimum or moderate kerf width because it is related to material loss from the parent material. All the experiments were performed by varying the input parameters with different levels. Experimental data was used to analyse the results. In the study, adaptive neuro fuzzy inference system in laser machining approach of PMMA material has been proposed also to get the minimum KW with low computational cost. According to statistical evaluation criterions of predicted performance of the developed ANFIS model has been found to be a good agreement. From the study, the following conclusions can be drawn:

1. When PMMA sheets of 3 mm thick was cut using a CW CO_2 laser, the resulting kerf width becomes wider with the increase of standoff distance and power, but the kerf width decreases as the cutting speed increases.
2. The minimum kerf width was found to be 0.3167 mm at gas pressure 0.5 bar, laser power 100 W, standoff distance 1 mm and cutting speed 1.2 m/min
3. The statistical analysis of the ANFIS result for kerf width were relative error (4.23%) and goodness of fit 0.9876 (R^2= 0.9754) in training, whereas relative error (3.54%) and goodness of fit 0.9853 (R^2= 0.9708) were obtained in testing.
4. Standoff distance, laser power, cutting speed and assist gas pressure have the dominant effect on kerf width, respectively, although assist gas has some significant effect to remove harmful gas.

References

1. N. Rajaram, J. Sheikh-Ahmad, S. H. Cheraghi, CO_2 laser cut quality of 4130 steel, International Journal of Machine Tools and Manufacture, **43**, pp. 351-358, (2003).
2. J. Powell, CO_2 laser cutting, Cambridge Univ Press, (1993).
3. K. Li, D. W. Zeng, K. C. Yung, H. L. W. Chan, C. L. Choy, Study on ceramic/polymer composite fabricated by laser cutting, Materials Chemistry and Physics, **75**, pp. 147-150, (2002).
4. F. Caiazzo, F. Curcio, G. Daurelio, F. M. C. Minutolo, Laser cutting of different polymeric plastics (PE, PP and PC) by a CO_2 laser beam, Journal of Materials Processing Technology, **159**, pp. 279-285, (2005).
5. I. A. Choudhury, S. Shirley, Laser cutting of polymeric materials: An experimental investigation, Optics & Laser Technology, **42**, pp. 503-508, (2010).
6. A. Kumar Dubey, V. Yadava, Multi-objective optimisation of laser beam cutting process, Optics & Laser Technology, **40**, pp. 562-570, (2008).
7. R. Q. Frazer, R. T. Byron, P. B. Osborne, K. P. West, PMMA: An essential material in medicine and dentistry, Journal of Long-Term Effects of Medical Implants, **15**, pp. 629-639, (2005).
8. K. F. Tamrin, Y. Nukman, N. A. Sheikh, M. Z. Harizam, Determination of optimum parameters using grey relational analysis for multi-performance characteristics in CO_2 laser joining of dissimilar materials, Optics and Lasers in Engineering, **57**, pp. 40-47, (2014).

 DOI: 10.1051/matecconf/20167710004

Effect of Duty Ratio at Different Pulse Frequency during Hole Drilling in Ceramics Using Electrochemical Discharge Machining

Pankaj Kumar Gupta[1], Jai Prakash Bhamu[1], Chandra Shekhar Rajoria[1], Nitin Kumar Lautre[2], Vishnu Agarwal[3]

[1] *Department of Mechanical Engineering, Engineering College, Bikaner (Raj.) - India.*
[2] *Department of Mechanical Engineering, Visvesvaraya National Institute of Technology, Nagpur (MS.) - India*
[3] *Department of Mechanical Engineering, GITS, Udaipur (Raj.) - India*

Abstract. The Electro Chemical Discharge Machining (ECDM) is a hybridization process of electro chemical and electric discharge machining, which is used for machining of hard, brittle and nonconductive materials. The present paper investigates the effect of duty ratio at different pulse frequency on the four process responses, i.e. material removal rate, depth of penetration, surface damage and heat affected zone of the hole drilled by ECDM. The requirements of a quality drilled hole are to have less surface damage and heat affected zone along with larger depth of penetration and material removal rate. The selected alumina ceramic (Al2O3) work piece material is widely used in hi-tech electrical apparatus. The results were characterized by optical microscope and analysed by using pulse wave forms of voltage with time. The results show that the effectiveness of pulse frequency varies according to the desired response characteristics. The research findings provide the most effective duty ratio for an improved quality of drilled hole in ceramics.

1 Introduction

Electro Chemical Discharge Machining (ECDM) was invented as 'electrical discharge drilling' in 1968 for producing micro holes in nonconductive materials such as glass, ceramic, etc. [1]. ECDM is a combination of the two non-conventional machining methods, namely; electro chemical machining (ECM) and electrical discharge machining (EDM) [2]. The common non-conventional methods used to fabricate the holes in nonconductive work pieces are abrasive jet machining and ultrasonic machining [3, 4]. However, the application of these processes are restricted due to poor material removal rate, high tool wear, poor surface finish and geometrical inaccuracy [4]. Therefore, ECDM is an alternative approach to machine the nonconductive work materials. The alumina ceramic is most widely used in high temperature electrical insulators and in other electrical applications due to higher strength, stiffness, high wear and corrosion resistance, high thermal stability and excellent insulation as compared to its counter materials. The subsequent section details some of the investigations conducted on variety of ceramics by various researchers.

It was investigated that material removal in ceramics occurred by etching of grains through high temperature electrolyte and melting [5]. Gas Filled Electro discharge Electro Chemical Machining (GFEECM) was developed to machine the ceramics [6]. The ECDM was used for drilling holes on non-conducting zirconium oxide ceramic components [7]. It was reported that effect of tool tip with flat front and taper side wall provides the controlled and effective machining [8].

The literature revealed the need for accurate and precise machining of ceramics. In ECDM, the material is melted by thermal energy and removed by subsequent chemical etching [9, 10]. This thermal energy is transferred in the workpiece and creates the heat affected zone (HAZ). The HAZ results in cracks on the workpiece surface and distort the hole dimensions. The surface damage (SD) is unwanted material removal and losses the accuracy of drilled hole. Hence, reduction of the HAZ and SD with high material removal rate (MRR) and depth of penetration (DOP) are desirable to produce more accurate holes in ceramics. The HAZ was reduced in a pulse voltage supply, because the hole drilling takes place during the pulse-on-time of the cycle and the material cools during the pulse-off-time [11]. The present article investigates the results of increasing duty ratio at different pulse frequency on the MRR, DOP, SD and HAZ at desirable points of interest. From the results, a favourable machining condition with minimum SD and HAZ along with maximum MRR and DOP was reported.

2 Experimental Procedure

An experimental setup was developed in the laboratory [12]. The applied voltage was 65 V using DC pulse power supply. Rectangular pulse waves with a specific pulse width were supplied. The number of pulse cycles per unit time is pulse frequency. The ratio of 'pulse on time' to total pulse cycle time (sum of pulse on time and pulse off time) is termed as duty ratio [11]. The voltage wave forms of pulses were recorded by Digital Storage Oscilloscope (DSO). The pulse frequencies are classified in four sets (in multiple of frequency from 1000 Hz to 1 Hz) according to pulse width as given in Table 1.

Table 1. Sets of Duty ratio According to Pulse Width and Pulse Frequency

Set	Pulse width		Pulse Frequency [Hz]	Duty ratio [%]
	Pulse on time [ms]	Pulse off time [ms]		
I (X1000 Hz)	0.1	0.1-0.04-0.01	5000-7000-9000	50-70-90
II (X100 Hz)	1	1-0.42-0.11	500-700-900	50-70-90
III (X10 Hz)	10	10-4.28-1.11	50-70-90	50-70-90
IV (X 1Hz)	100	100-42.85-11.11	5-7-9	50-70-90

The NaOH electrolyte with 20% wt/v concentration was used in the experiment at room temperature. A cylindrical stainless steel tool with diameter of 700 µm was used. The auxiliary electrode made of graphite was kept at a distance of 40 mm from tool. The workpiece was a 9 mm thick alumina ceramic (Al_2O_3) material.

The schematic of a drilled blind hole produced by ECDM is shown in Fig. 1. After drilling of the ceramic workpiece, the diameters of the top side of hole were measured as HAZ and SD using an optical microscope at 20X magnification. The HAZ was calculated as the difference in diameters of HAZ and SD, while the SD was the obtained by subtracting the tool diameter from diameter of SD. The DOP was measured by a dial gauge (Mitutoyo). Three specimen were drilled at constant parametric setting. Each experiment was performed for 5 minutes and average results of responses are presented herein.

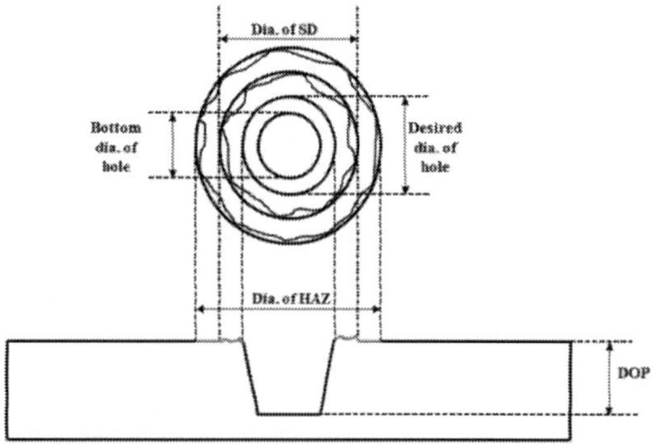

Figure 1. Schematic of the Drilled Hole

3 Result and Discussion

Effect of duty ratio was experimentally investigated on ceramic work material at four sets of pulse frequency. The experimental results revealed that the magnitude of all response characteristics i. e. MRR, DOP, SD and HAZ increased with the increase in duty ratio as shown in Figure 2 (a), (b), (c), and (d).

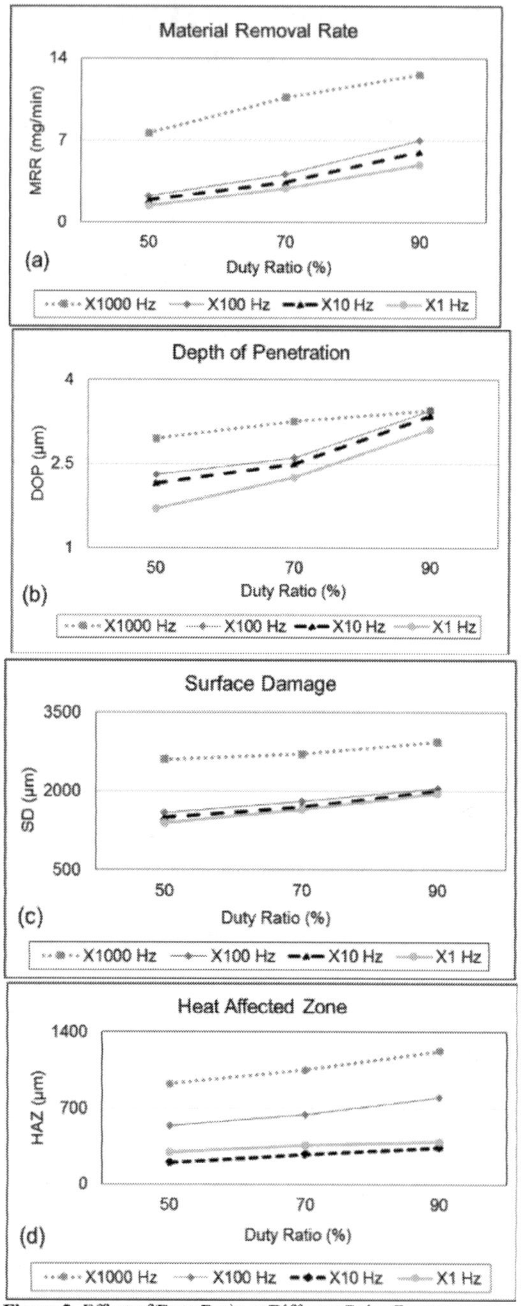

Figure 2. Effect of Duty Ratio at Different Pulse Frequency on Responses

The reason attributed is that the increase in duty ratio, increases the proportion of voltage applied time (pulse on time). Therefore, duration of thermal energy supply increased with increase in duty ratio. The maximum MRR, DOP, SD and HAZ were observed at 90% duty ratio in set I as shown in Fig. 2 (a to d)). It is evidenced to the high switching frequency as observed from pulse wave form in Fig. 3 (a) and side discharge [12, 13]. The minimum SD was at 50 % duty ratio, which is almost equal in set II, III and IV (Figure 2-c). It occurs because the effect of pulse off time and pulse on time is equal at this point. It means both pulse width plays equal role as shown in Fig. 3 (b). But HAZ was higher in set-IV as compared to set III at 50% duty ratio. It seems that longer pulse off time allows to flow the heat in the workpiece during pulse on time sufficiently in set-IV, while in set-III, the pulse off starts to play its role before the pulse on tries to propagate in the workpiece. The hole drilled with minimum HAZ with microscopic image and pulse wave form are shown in Fig. 3 (b).

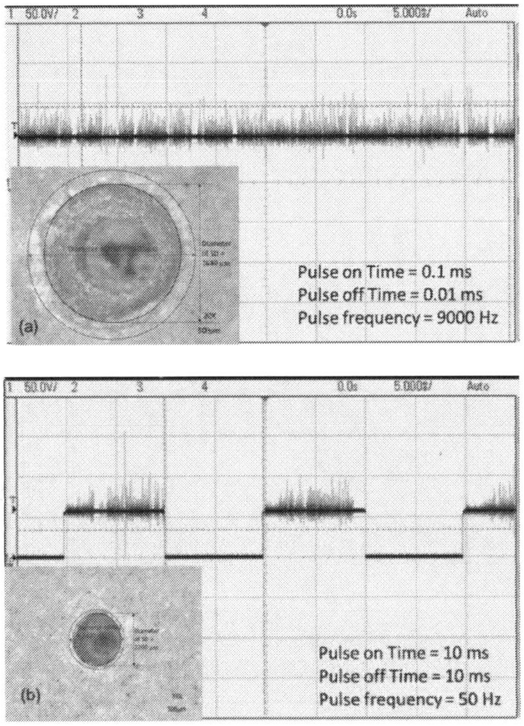

Figure 3. Pulse Wave Forms and Drilled Hole (a) at Maximum MRR Condition (b) at Minimum HAZ Condition

4 Conclusion

The holes were drilled at various pulse frequencies. The pulse width decides the frequency in a pulse power supply. All the response characteristics were found increased with increase in duty ratio. However, in order to get accurate and more precise holes with minimum SD and HAZ, compromise is required with MRR and DOP. Therefore, it is concluded that frequencies in multiples of X10 Hz are recommended for minimum HAZ and SD, while the frequency in multiples of X1000 Hz is preferable for maximum MRR and DOP for drilling of holes in alumina ceramic work material. Results of frequencies in multiples of X100 Hz, X10 Hz and X1Hz are very close for the MRR, DOP and SD.

References

1. R. Wuthrich, and V. Fascio: Machining of nonconducting materials using electrochemical discharge phenomenon – an overview, Int. J. Mach. Tool. Manuf. Vol. 45 (2005), p. 1095-1108.
2. V. K. Jain and S. Adhikary: On the mechanism of material removal in electrochemical spark machining of quartz under different polarity conditions, J. Mater. Process. Technol. Vol. 200 (2008), p. 460–470.
3. M. C. Panda and V. Yadava: Intelligent modelling and multi-objective optimization of die sinking electro-chemical spark machining process, Materials and Manufacturing Processes. Vol. 27 (2012), p. 10–25.
4. Z-P. Zheng, H-C. Su, F-Y. Huang and B-H. Yan: The tool geometrical shape and pulse-off time of pulse voltage effects in a Pyrex glass electrochemical discharge micro drilling process, J. Micromech. Microeng. Vol. 17 (2007), p. 265–272.
5. H. Tokura, I. Kondoh, M. Yoshikswa: Ceramic material processing by electrical discharge in electrolyte. Journal of Materials Science, Vol. 24 (1989), p. 991-998.
6. L. Yonghong, J. Zhixin, L. Jinch: Study on hole machining of non-conducting ceramics by gas-filled electro discharge and electrochemical compound machining, J. Mater. Process. Technol. Vol. 69 (1997), p. 198-202.
7. B. Doloi, B. Bhattacharyya and S. K. Sorkhel: Electrochemical discharge machining of non-conducting ceramics, Defence Science Journal. Vol. 49 (4) (1999), p. 331-338.
8. B. Bhattacharyya, B.N. Doloi and S. K. Sorkhel: Experimental investigations into electro chemical discharge machining (ECDM) of non-conductive ceramic materials. J. Mater. Process. Technol. Vol. 95 (1999), p. 145-154.
9. N. Gautam and V. K. Jain, Experimental investigations into ECSD process using various tool kinematics, Int. J. Mach. Tools Manuf. Vol. 38 (1998), p. 15–27.
10. A. Kulkarni, R. Sharan and G. K. Lal: An experimental study of discharge mechanism in electrochemical discharge machining, Int. J. Mach. Tools Manuf. Vol. 42 (2002), p. 1121–1127.
11. D. J. Kim, Y. Ahn, S. H. Lee, Y. K. Kim: Voltage pulse frequency and duty ratio effects in an electrochemical discharge micro drilling process of Pyrex glass, Int. J. Mach. Tools Manuf. Vol. 46 (2006), p. 1064–1067.
12. P. K. Gupta, A. Dvivedi and P. Kumar: Effect of pulse duration on quality characteristics of blind hole drilled in glass by ECDM, Materials and

DOI: 10.1051/matecconf/20167710004

Manufacturing Processes. (2015), doi: 10.1080/10426914.2015.1103857.
13. P. K. Gupta, A. Dvivedi and P. Kumar: Impact of pulse width on hole overcut in an electrochemical discharge machining of blind holes, International Conference on Processing and Fabrication of Advanced Materials XXIII,. IIT Roorkee, Vol. 1 (2014) p. 493-498

Workpiece Machining Accuracy Prediction Based on Milling Simulation

Lv Yan-peng, Wang Ai-min, Wang Xiao-Long

School of Mechanical Engineering, Beijing Institute of Technology, Beijing 100081, China

Abstract. To ensure the machining accuracy of workpiece, it is necessary to predict the workpiece deformation in machining process through establishing a high precision workpiece deformation forecast model. To solve these problems, a more efficient variable stiffness analysis model is proposed, which can obtain quantitative stiffness value of the machining surface. Applying simulated cutting force in sampling points using finite element analysis software ABAQUS, the single direction variable stiffness rule can be obtained. First of all, finite element simulation model of face milling is established with the Johnson-Cook material model and failure model of 7050 aluminum alloy. Prediction model is established based on SVM and input data is provided by the finite element software ABAQUS simulation. Results show that the model prediction relative error is less than 5%. It is concluded that the effects of milling parameters on workpiece machining deformation and practical guide for production.

1 Introduction

7050 aluminum alloy is mainly used in the aerospace field because it has some advantages such as light weight, high strength, high temperature resistance, corrosion resistance. But this component has poor rigidity and is prone to deformation in the processing. It is necessary going through correcting and aging treatment after processing to ensure the machining accuracy of workpiece. Thus how to reduce the workpiece machining deformation is a realistic problem which needs to be settled.

If the workpiece deformation can be predicted accurately before the processing according to the machining parameters to, the machining precision of workpiece can be improved. To achieve this goal, it is necessary to establish a high precision and strong generalization ability machining deformation prediction model which can provide feasible guidance to improve the efficiency of the practical production.

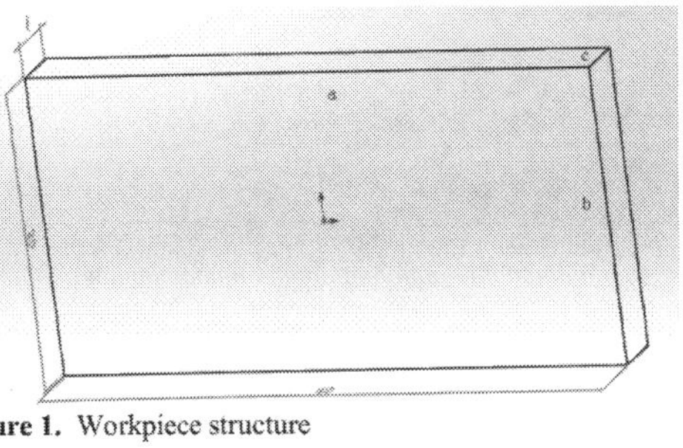

Figure 1. Workpiece structure

2 Workpiece structure and the clamping form

The workpiece structure is a cuboid as shown in Fig. 1. Its material is 7050 aluminum alloy. In Fig. 1, a is equal to 40 mm, b is equal to 20 mm, c is equal to 4 mm.

Workpiece is manufactured while the surface of edge b, c is fixed.

3 The simulation theory

Previous processing deformation prediction model was set up mainly by regression analysis method [1], the neural network [2], etc. The former method in modeling accuracy and generalization ability can't meet the requirements of machining deformation prediction, the latter method need establish a model, which cost a long training time into the local extreme value point and weak generalization ability.

3.1 SVM

SVM [3], [4] (Support Vector Machine) was first proposed by Cortes and Vapnik in 1995, which is good at solving the small sample, nonlinear and high dimensional pattern recognition. It overcomes the defect of the neural network in structure, easy to fall into local extremum problems, and combining the advantages of neural network and grey model, and can be applied to the function fitting other Machine learning problems.

The main ideas of the SVM can be summarized as two points: (1) it is to solve the problem of linear separable analysis through the use of nonlinear mapping algorithm, which will undivided linear sample low-dimensional input space into a high-dimensional feature space linear separable, so as to make the high dimensional feature space by using linear algorithm for linear analysis on the nonlinear characteristics of samples is possible. (2) it is based on structural risk minimization theory to construct the optimal hyperplane segmentation in the feature space and a certain upper bound in the sample space for the expected risk to a probability.

SVM method ingeniously solved the Curse of Dmensionality. Due to the kernel function expansion theorem, there is no need to know the explicit expression of nonlinear mapping. Because it is establishing linear machine learning in the high dimensional feature space, so compared with the linear model, not only almost does not increase the computational complexity, and to some extent to avoid the "Curse of Dmensionality". All this can be attributed to expand and calculation theory of kernel function.

This paper uses Polynomial Kernel as follow:

$$K(x,y)=\left[(x\cdot y)+1\right]^d \tag{1}$$

3.2 Johnson-cook constitutive model

Considering that the cutting parameters are the main factors of the machining deformation, I choose Johnson-cook constitutive model with the support of the theory of thermal coupling force. This model fully consides the friction between cutter and workpiece and elastic-plastic deformation to produce heat of cutting tool and the workpiece , the heat transfer between the workpiece and environmental factors. It is important that the effects of each variable be properly assessed, rather than assuming all distinguishing characteristics are due to strain rate alone.

Johnson-cook model [5]-[7] is a materials expression, which can reflect the strain rate effect and temperature softening effects of ideal elastic-plastic model. It is valid for the temperature range from room temperature to the melting point temperature range, particularly suitable for simulation under high strain rate of metal material. Its specific expression [3] is as follow:

$$\sigma=\left[A+B\varepsilon^n\right]\cdot\left[1+C\ln\left(\frac{\dot{\varepsilon}}{\dot{\varepsilon}_0}\right)\right]\cdot\left[1-\left(\frac{t-t_r}{t_m-t_r}\right)^m\right] \tag{2}$$

where A, B, C, n and m is outstanding parameter, which can be get by experiment, They respectively are yield stress, Strain hardening constant, strain rate hardening constant strengthening coefficient, stress hardening coefficient and thermal softening coefficient; ε、$\dot{\varepsilon}$、$\dot{\varepsilon}_0$ respectively are equivalent plastic strain, equivalent plastic strain rate, quasi static strain rate; t、t_r、t_m respectively are material deformation temperature, environment temperature and melting point.

3.3 Cutting force model

Through the consult the manual of metal cutting, we can know the experience formula of the milling force formula as follows :

$$F_C=\frac{C_F a_p^{x_F} a_f^{y_F} a_w^{u_F}}{d_0^{q_F} n^{w_F}} k_{F_C} \tag{3}$$

where a_p is Back cutting depth, a_f is feed engagement, a_w is nominal width of cut , d_0 is milling cutter diameter, n is milling speed. The rest of the parameters can be found through the reference books. Combined with the actual data processing technology, we can calculate the size of cutting force.

In this paper, processing tool is face milling cutter of cemented carbide.

In expression (1), n = 2000 r/min, d_0 = 60 mm, a_f = 0.6 mm/r, a_p = 2 mm, a_w = 53 mm, C_F = 7750, x_F = 1.0, y_F = 0.75, u_F = 1.1, w_F = 0.2, q_F = 1.3, k_{FC} = 0.55, we get expression (4) after taking these parameters in expression (3), as follow:

$$F_C=\frac{7750\times 2^1\times 0.216^{0.75}\times 53^{1.1}}{60^{1.3}\times 2000^{0.2}}\times 0.55=227.2N \tag{4}$$

4 Milling Simulation

The simulation is based on Johnson–Cook constitutive model as expression (1). In this paper, A = 400 MPa, B = 424 MPa, n = 1.426, C = 0.001, tr = 293 K, tm = 863 K.

Milling process simulation flow chart is shown in Fig. 2.

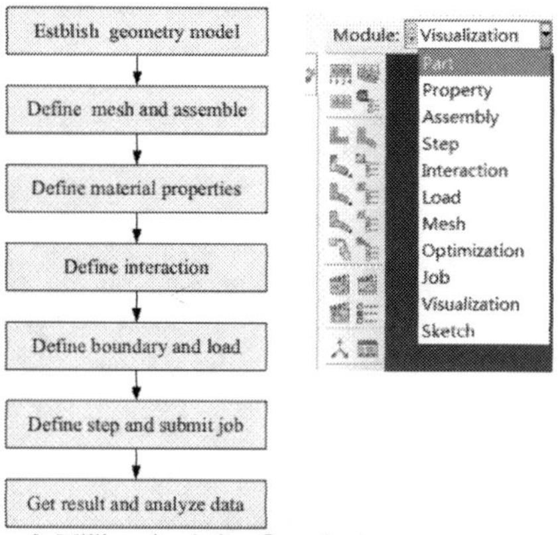

Figure 2. Milling simulation flow chart

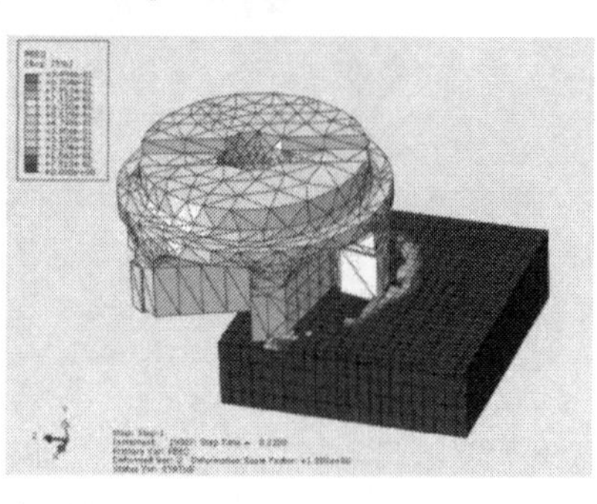

Figure 3. Milling simulation

Milling process simulation is shown in Fig. 3. The forces of three sides can be obtained the through milling simulation, as shown in the figure below.

Through the simulation of metal milling, we can know curve of the milling force as Fig. 4.

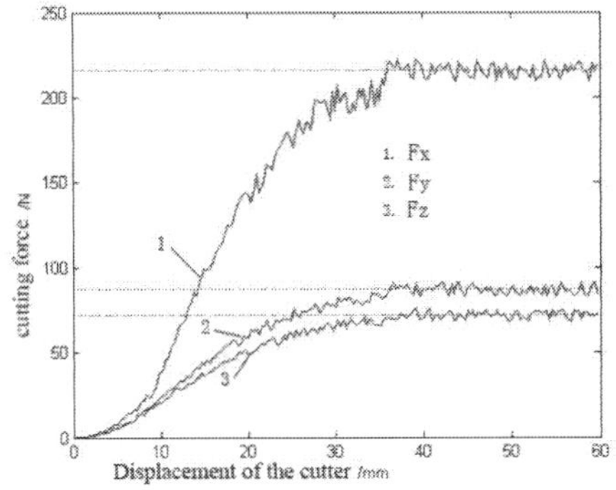

Figure 4. Cutting force curve of milling simulation

The milling force F can be get according to the results of the simulation as follow:

$$F=\sqrt{F_x^2+F_y^2+F_z^2}=\sqrt{216^2+(-72)^2+(-87)^2}=243.7\ N$$

The milling force error ε of the simulation calculation and the empirical formula is as follow:

$$\varepsilon=\frac{|F_C-F|}{F_C}\times 100\%=7.27\%$$

The simulation results is close to empirical formula calculation results. Thus we can think the results of the simulation model is accurate.

5 The experimental process

This paper in milling process, we try three main process parameters for different values through comparing the rail deformation and stress in order to get optimized process parameters.

5.1 The situation of the different feeding

Milling cutter with cutting deep 0.1 mm, the speed of 1000 r/min, the cutter feeding respectively is 1.2 mm/r, 1.8 mm and 3.0 mm/r. Contrast the guide maximum stress and deformation results are as follows:

Table 1. The stress and deformation of workpiece in different feeding

cutting deep mm	speed r/min	cutter feeding mm/r	Max stress MPa	Max defor-mation μm	average deformat-ion μm
0.1	1000	1.2	105.1	62.5	34.9
0.1	1000	1.8	97.4	55.1	31.2
0.1	1000	2.4	92.0	50.8	28.1

5.2 The situation of the different speed

Milling cutter with single cutting deep 0.1 mm, feeding is 1.2 mm/r, the mill speed of 1000 r/min, respectively 1500 r/min and 5000 r/min.Contrast the guide maximum stress and deformation results are as follows:

Table 2. The stress and deformation of workpiece in different speed

cutting deep mm	speed r/min	cutter feeding mm/r	Max stress MPa	Max defor-mation μm	average deform-ation μm
0.1	1000	1.2	105.1	62.5	34.9
0.1	1500	1.2	101.1	59.4	30.7
0.1	2000	1.2	81.4	47.2	25.3

5.3 The situation of the different cutting deep

Milling speed of 2000 r/min, feeding is 1.2 mm/r, the milling cutter with single cutting depth were taken as 0.05 mm, 0.1 mm and 0.05 mm. Contrast the guide maximum stress and deformation results are as follows :

Table 3. The stress and deformation of workpiece in different cutting deep

cutting deep mm	speed r/min	cutter feeding mm/r	Max stress MPa	Max defor-mation μm	average deform-ation μm
0.05	1000	1.2	86.4	45.6	26.8
0.1	1000	1.2	105.1	62.5	34.9
0.15	1000	1.2	108.2	72.4	38.1

6 Conclusion

1) The stress and deformation are positively related to cutting deep and cutter feeding, are negatively related to speed. Therefore, the selection of cutting depth and feed speed need to be careful in case the workpiece deformation is too large.

2) Cutting deep is the biggest factor influenced the stress and deformation. Therefore, we should pay attention to reduce the cutting depth, and appropriately increase the feed rate and cutting speed, so as to guarantee the machining efficiency and reduce the workpiece deformation.

3) We can predict the workpiece deformation according to this paper. If we add more simulation data, we will get more accurate prediction.

4) We should do some cutting test to verify the correctness of the simulation results the next step.

References

1. D. R. Leseur. Experimental investigations of material models for Ti-6Al-4V titanium and 2024-T3 aluminum. Tech. Rep. DOT/FAA/AR-00/25. US department of Transportation. Federal Aviation Administration. September, (2000).
2. G. Kay. Failure modeling of titanium 6Al-4V and aluminum 2024-T3 with the Johnson-Cook material model. Tech. Rep. DOT/FAA/AR-03/57. US

department of Transportation, Federal Aviation Administration, September, (2003).
3. Caydas U, Ekici S. Support vector machines models for surface roughness prediction in CNC turning of AISI 304austenitic stainless steel [J]. Journal of Intelligent Manufacturing, **23**(3):639-650, (2012).
4. Salgado D, Alonso F, Cambero I. In -process surface roughness prediction system using cutting vibrations in turning [J] .International Journal of Advanced Manufacturing Technology, **43**(1):43-51, (2009).
5. Yong Yang, Weiwei Zhu, Study on cutting temperature during milling of titanium alloy based on FEM and experiment. The International Journal of Advanced Manufacturing Technology, **73**(9):1511-1521, (2014).
6. Ratchev S, Liu S, Becker A A. Error compensation strategy in milling flexible thin—wall parts[J]. Journal of Materials Processing Technology, **162**(5):673-681, 2005.
7. Wang, Y., Chen, X., Gindy, N., *et al.* Elastic Deformation of a Fixture and Turbine Blades System based on Finite Element Analysis, The International Journal of Advanced Manufacturing Technology, **36**(4):296-304, (2008).

MATEC Web of Conferences 77, 10006 (2016) DOI: 10.1051/matecconf/20167710006
ICMMR 2016

Surface Improvement of Shafts by Turn-Assisted Deep Cold Rolling Process

Raghavendra Prabhu[1], S. S. Sharma[1], Jagannath K[1], Krishna Kumar[1], S. M. Kulkarni[2]

[1]*Faculty, Dept. of Mechanical & Manufacturing Engineering, Manipal Institute of Technology, Manipal University, India*
[2]*Faculty, Dept. of Mechanical Engineering, National Institute of Technology Karnataka, Surathkal, India*

Abstract. It is well recognized that mechanical surface enhancement methods can significantly improve the characteristics of highly-stressed metallic components. Deep cold rolling is one of such technique which is particularly attractive since it is possible to generate, near the surface, deep compressive residual stresses and work hardened layers while retaining a relatively smooth surface finish. In this paper, the effect of turn-assisted deep cold rolling on AISI 4140 steel is examined, with emphasis on the residual stress state. Based on the X-ray diffraction measurements, it is found that turn-assisted deep cold rolling can be quite effective in retarding the initiation and initial propagation of fatigue cracks in AISI 4140 steel.

1 Introduction

The surface layer is determined by manufacturing processes, and mainly, by finishing processes. Machining is a modest alternative process for fabricating a wide range of mechanical components, such as gears, cams, shafts, axles and others. The process of machining steel is complex and the surface generated is influenced by several variables: steel properties (elastic and plastic deformations), tool material and geometry, vibration of cutting tool, cutting speed, feed, depth of cut, lubricant, etc. Besides, previous works have shown that in the machined surface of metals a damage region is produced that is different from the bulk of the material [1]. During machining, the surface layer is subjected to elastic-plastic deformation and heating, which result in structural changes, strain hardening and residual stresses. Fillets, grooves, notches, and holes are always present in most of the engineering components, and fatigue related cracks will originate from these sites where the local stress is highest. All stress concentrators in the component can never be eliminated; however the negative effects of stress concentrators can be reduced by different means, one of which is the compressive residual stress. Compressive residual stresses in the surface region of materials with high hardness increase the fatigue life and the fatigue limit during cyclic loading compared with the materials that are free of residual stresses [2]-[4]. Residual stresses are introduced in metallic components intentionally with shot peening/prestraining, or unintentionally with metal-forming processes such as forging, rolling, milling, and welding. These processes introduce residual stresses as a result of inhomogeneous plastic deformation due to mechanically or thermally induced loads [5,] [6]. Mechanical surface treatments, such as shot peening, laser shock peening and deep rolling efficiently enhance the fatigue performance of various metallic components. This enhancement results from induced macroscopic compressive residual stresses and work hardening at the surface and in near-surface regions inhibiting or retarding crack initiation and propagation resulting in fatigue performance enhancement. This paper focusses on carrying out deep cold rolling process in a cost-effective way using turn-assisted deep cold rolling (TADCR) instead of special machines and tools. The TADCR involves conventional lathe with a follower rest and rolling attachments, to improve the fatigue life and surface properties of AISI 4140 steel.

The objective of this work is to investigate the effect of parameters in TADCR process on residual compressive stress. The effect of four parameters, namely, rolling force, ball diameter, initial roughness of the workpiece and number of passes are considered for investigation. The residual stresses are measured experimentally and are calculated using finite-element analysis. The increase in the measured residual stress was correlated with the results obtained from finite-element analysis.

2 Material and process

The workpiece material used in this study is AISI 4140 steel which is especially recommended for the manufacture of transmission shaft, gear shaft, crank shaft and also for a wide variety of automotive type applications [7]. The work pieces are received as bright cylindrical bars of 12mm diameter. The chemical composition of AISI 4140 steel in mass% is as follows: 0.40C, 0.27Si, 0.66Mn, 0.055P, 0.046S, 1.20Cr, 0.25Mo, 0.16Ni. The ultimate tensile strength and yield strength of the starting specimen at room temperature is 946MPa and 848 MPa respectively. The average initial hardness of the material measured by MATSUZAWA Micro-Vickers hardness tester and is found to be about 225HV.The specimens are prepared as per the ASTM-E466 requirements to conduct fatigue tests. Figure 1 shows the dimensions of the specimen that is used for conducting the experiments. Specimens are turned to given diameter on a conventional lathe to render a surface roughness common in turning process.

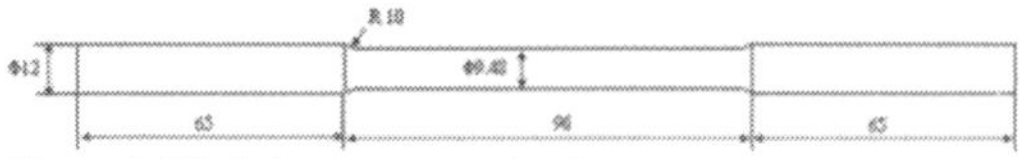

Figure 1. Workpiece geometry (mm)

The proposed TADCR set-up consists of a lathe (PSG type A 141) and an in-house designed tool and other accessories as shown in Fig. 2. A Kistler dynamometer mounted on the lathe tool post is used to measure the forces during the process. The specimen is held in a three jaw chuck at one end and supported by tail stock at the other end. The rolling force is adjusted through depth of penetration.

Figure 2. Experimental set-up of TADCR process

Table 1. Factors and levels for CCD

Factor	Unit	Level 1	Level 2	Level 3
Ball Diameter	mm	6	8	10
Rolling Force	N	250	500	750
Initial Roughness	μm	4.84	6.15	7.46
No. of Passes		1	2	3

The forces are recorded using the DynoWare software. An experimental plan with Central Composite Design (CCD) is used to investigate the influence of four parameters, rolling force, ball diameter, initial roughness of the workpiece and number of passes on residual compressive stress. Each parameter is considered in three levels and three replicates are used for each design point in the CCD. The parameters, their levels and magnitudes are shown in Table 1. Residual compressive stress measurements are made by using a Rigaku X-ray diffractometer. Samples are cut out along the feed direction and microstructure of the stressed region is acquired by an optical microscope.

3 Results and discussion

3.1 Finite element analysis of TADCR process

In the last decades, a majority of the research was centered on the experimental determination of relationships for the surface finish and state of residual stress with the burnishing and deep cold rolling process parameters for specific applications. Afshin Manouchehrifar and Kianoush Alasvand [8] conducted a comprehensive 3D finite element dynamic analysis with considering spring back effect to simulate the deep rolling process. The results revealed that increase in overlap of the rolling tracks largely increases the magnitude of the residual stress filed created in target plate. Moreover, increase in the friction coefficient between roller and target plate results in decrease in magnitude of the residual stress filed. It is found that, increase in the force intensity results in increase magnitude of the maximum residual stress. The objective of this study is to establish an FEM model for TADCR process to study the effects of process parameters on residual stresses and to validate the simulation results with results obtained from TADCR experiments. In this study, 2D FEM models of TADCR process were developed. Turning and consequent deep cold rolling experiments were conducted in order to understand the process setup, to evaluate the surface improvement and to understand the influence of process parameters. The basic model allowed the study of the influence of process parameters, such as ball diameter, depth of penetration, initial roughness of work piece and number of tool passes, on residual stresses.

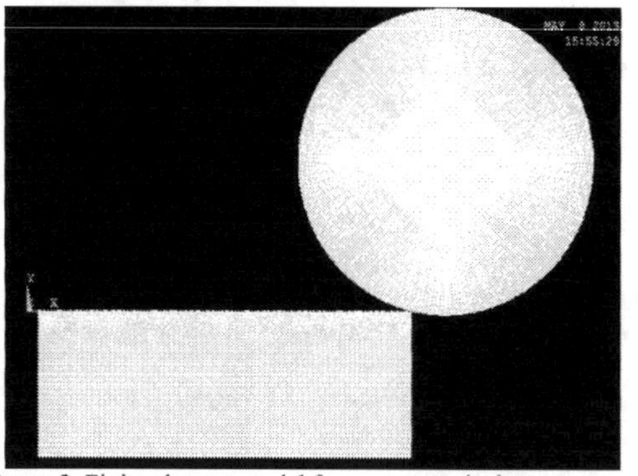
Figure 3. Finite element model for contact analysis

The FEA package ANSYS was used to generate a 2-Dimensional contact model. The dimensions of the model are taken exactly same as the workpiece and tool dimensions. This is shown in Figure 3. The dimensions of the model are taken exactly equal to the workpiece and tool dimensions.

DOI: 10.1051/matecconf/20167710006

Both the workpiece and ball were meshed with four-noded 2D solid162 elements with 6 degrees of freedom at each node. The ball is modelled as a circle and it is given the nodal translation degree of freedom in "x" direction. The workpiece boundary conditions restricted all the degrees of freedom of the three sides of the workpiece that were not in contact with the ball. Fig. 4 shows a basic diagram of the deep cold rolling 2D model with all the applied boundary conditions. The ball remains non-deformable when loads are applied, so in the finite element model a rigid material model is chosen and for the work piece kinematic plastic material model is chosen. The optimum value of deep cold rolling force determined experimentally is used in the Finite Element Analysis. The results of this analysis include the induced residual compressive stress. The applied deep cold rolling force is related to the depth of penetration, which in turn is related to the surface finish obtained and the residual stresses induced in the component. Thus depth of penetration plays an important role in the TADCR process. Surface roughness generated after the turning operation was used to model the surface roughness pattern which was further used to simulate the process.

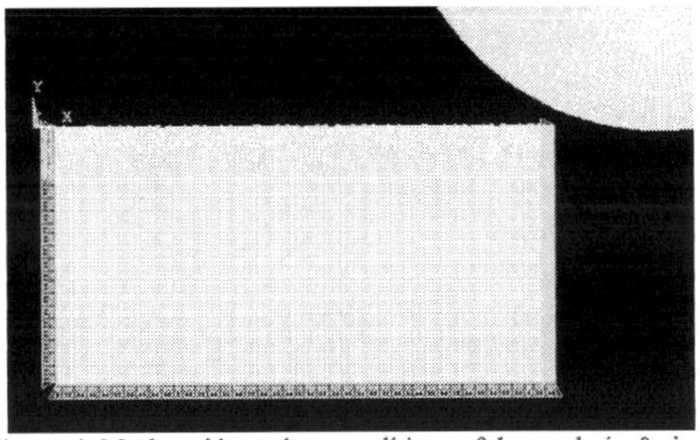

Figure 4. Mesh and boundary conditions of the workpiec0e in 2D simulation

The main effect of deep rolling is restricted to a thin layer below the surface (0.2-0.4 mm). A FE model that describes the effects on a small geometric volume on a thin layer below the ball was developed according to Fig. 5. The mesh topology was zoomed with a higher density close to the surface. The model consists of 28315 2D solid 162 elements. Mesh was made finer near the surface of the work piece.

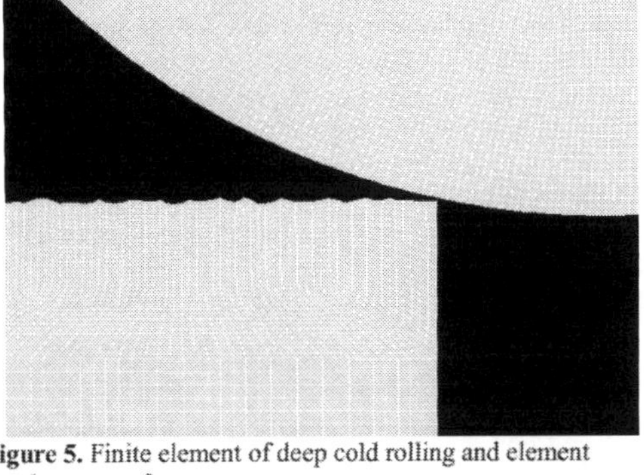

Figure 5. Finite element of deep cold rolling and element topology at surface

3.2 Finite element model validation

To obtain the stress distribution, a central area of the simulation is selected. The Finite Element results from elements just below the surface are analyzed and post processed. Figure 6 shows the effect of deep cold rolling on the simulated residual stress distribution. In fact, as the deep cold rolling force increased, higher residual stresses are obtained, and at a greater depth. Therefore, results show that when the objective of this surface treatment is the introduction of high residual stresses, the best option is to roll the surface with the maximum force; however this can lead to an increase of roughness and geometrical error.

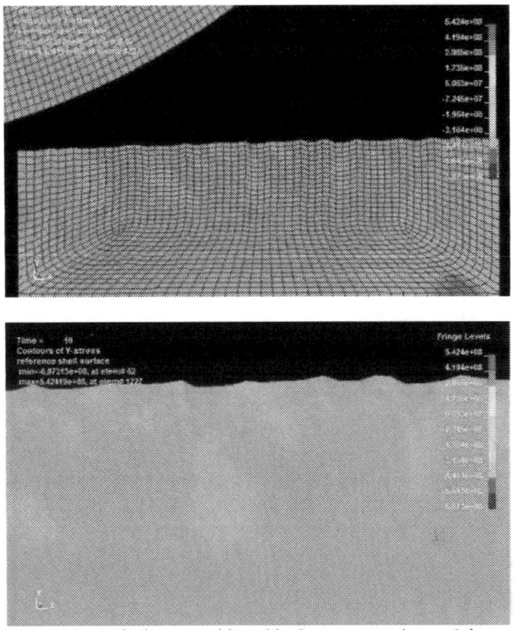

Figure 6. Analysis zone with residual stresses on the model

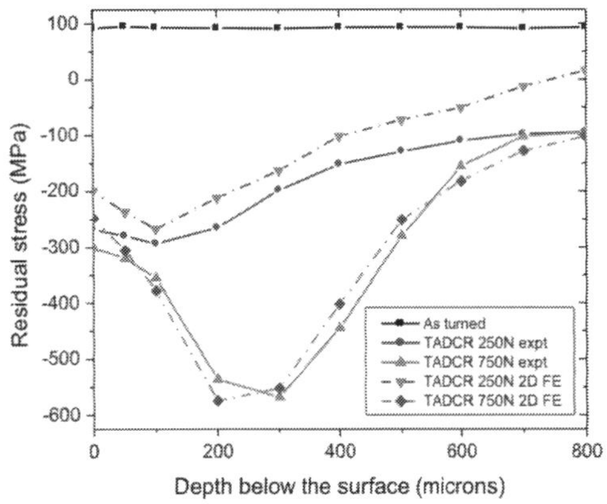

Figure 7. Residual stress evolution from simulations & experiments

In order to verify the accuracy of the finite element simulation of turn-assisted deep cold rolling, the residual stress profile was compared with the experimentally obtained data and is shown in Fig. 7. As the figure indicates, the values obtained from FEA are quite closer to the experimentally determined values because the modelling of FEA includes surface-to-surface contact in 2D between the tool and workpiece, which represents the real phenomenon occurring in actual turn-assisted deep cold rolling process.

4 Conclusions

In this study, the effects of four process parameters ball diameter, rolling force, initial roughness of the workpiece and number of tool passes on the residual compressive stress of AISI 4140 steel are investigated. Rolling force has a major influence on residual stresses. The amount of plastic deformation in the workpiece increases with force and hence the magnitude of compressive residual stresses increase with increasing rolling force. Predictions of residual stress distributions beneath the turn-assisted deep cold rolled surface, with the 2D FEM model are in good agreement with the experimental measurements obtained using X-Ray diffractions. The maximum residual stress obtained by the 2D FE simulation differed by approximately 15 % from the experimental data.

References

1. P. G. Benardos and G. C. Vosniakos: Predicting Surface Roughness in Machining: A review, International Journal of Machine Tools & Manufacture, **43**, No. 8, p. 833-844, (2003).
2. L. Bertini and V. Fontanari: Fatigue Behavior of Induced Hardened Notched Components, International Journal of Fatigue, **21**, p. 611-617 (1999).
3. H. Wohlfahrt : Shot Peening and Fatigue of Materials, Proceedings of International Conference on Shot Peening 1, A. Niku-Lari, Ed., Pergamon Press Oxford, pp. 257-262 ,(1982).
4. J. K. Gregory and L. Wagner: Selective Surface Aging to Improve Fatigue Behavior in a High-Strength Beta Titanium Alloy, Proceedings of 5th International Conference on Fatigue and Fatigue Threshold, EMAS Publishing, England, pp. 177-182 (1993).
5. G. H. Majzoobi, K. Azadikhah and J. Nemati: The effects of deep rolling and shot peening on fretting fatigue resistance of aluminium 7075-T6. Materials Science and Engineering A 516, p. 235–247, (2009).
6. D. W. Hammond and S. A. Meguig : Crack Propagation in the Presence of Shot-peening Residual Stresses, Engineering Fracture Mechanics, **37**, pp 387-390 (1990).
7. E. Oberg and R.E. Green: Machinery's Handbook, 25th edition, Industrial Press Inc., New York, (1996).
8. Afshin Manouchehrifar and Kianoush Alasvand: Finite element simulation of deep rolling and evaluate the influence of parameters on residual stress, Recent Researches in Applied Mechanics, p. 121-127, (2009).

Maintenace Engineering in Process Plants

Karthik Silaipillayarputhur

Assistant Professor, Department of Mechanical Engineering, College of Engineering, King Faisal University, Saudi Arabia

Abstract. The continuous and uninterrupted operation of process plants which is most essential for their profitable performance depends on reliability and availability of all the critical process equipment. Maintenance department provides chief support for the safe, reliable and successful functioning of a process plant. Process critical equipment must be effectively monitored and maintained so that they can support the continuous production in the plant. Some of the key elements required for the successful functioning of the maintenance team are briefly outlined in this manuscript. Though it could be exhausting to conscientiously apply all such principles regularly, it shall be in the benefit of the organization to consider and apply such concepts as much as possible.

1. Introduction

Process industry make products from raw materials and they run continuously for several months before the plant is shut down for a scheduled overhaul. All process industries generally have high technology and they run as lean as possible. Process industry have several rotating equipment such as motors, pumps, fans/blowers, compressors, gearboxes and process equipment such as heat exchangers, cooling towers, steam traps, control valves, relief valves, pressure vessels, etc. Failure of such equipment could result in total stoppage of the plant resulting in down time, un-happy customers and several thousands of dollars of monetary loss.

Thus, it is very essential to have a strong maintenance team to ensure the safe and reliable operation of the plant at all times. The main objectives of a maintenance department would be to provide dependable and safe operation of equipment at all times, cost reduction, minimize downtime and to have extensive planning and execution of shut down jobs. The maintenance department must therefore come up maintenance practices and strategy to meet the objectives.

Diverse textbooks, papers, online resources concentrating on maintenance practices are available in the literature. Mobley [1] developed a maintenance engineering hand book wherein the proven best practices for maintenance, repair, overhaul, inventory management, root-cause analysis and performance management are discussed. Niebel [2] furnishes the material necessary to provide cost effective maintenance, details the use of bar-coding in maintenance and discusses utilities management. Moubray [3] gives practical explanation on reliability centered maintenance and how it can be applied in an industry. Smith [4] examines the principles of root cause analysis and on how they could be applied to everyday situations within the process plant. Townsend [5] considers safe maintenance practices in a process plant. Silaipillayarputhur [6] discusses the available predictive maintenance techniques that could be applied to common chemical/process plant equipment. Therein the author discusses the various predictive maintenance techniques that are commonly applied to rotary and process critical equipment.

2. Maintenance techniques

Maintenance strategy have evolved over the last 60-70 years. Consider Fig. 1 describing the maintenance techniques in the last several years. To begin with, the maintenance strategy was to service the equipment after break down. This means that the equipment ran continuously until it encountered failure. This is an un-acceptable maintenance practice as when the equipment completely shuts down; it probably requires several expensive components to be replaced. Moreover, the break down could happen at an un-expected time causing the entire plant to shut down.

Thereafter, preventive maintenance techniques slowly came into practice, and they are relevant in certain situations even today. In preventive maintenance, the equipment is thoroughly re-conditioned in certain fixed intervals. This practice is very similar to changing engine oil in the car as per stipulations in the car manual. Though this practice ensures the reliability of the equipment, this practice leads to performing maintenance activities when they may not be really required. Therefore, preventive maintenance on all process critical equipment leads to wastage of valuable time and money.

Though several techniques such as reliability centered maintenance (RCM), total productive maintenance (TPM), total quality management (TQM), preventive

DOI: 10.1051/matecconf/20167710007

maintenance (PM), predictive maintenance (PDM) are all available and widely discussed, it is almost impractical to apply one such technique for the maintenance of a process plant. The newer generation maintenance strategy would be to incorporate several aspects that would ensure safe and reliable operation of the plant with the least maintenance costs. In no particular order, some of the main elements may be summarized as follows:

(a) incorporation of computerized maintenance management system (MMS)
(b) effective spares planning and control
(c) inclusion of well-defined preventive maintenance jobs in the MMS
(d) integration of well developed predictive maintenance system linked with MMS
(e) review existing maintenance practices and establish new practices/techniques available in the market
(f) optimization of available resources
(g) vendor development and import substitution without compromising on quality
(h) employment of reliability studies and process safety management for the reliable and safe operation of equipment
(i) cross training and refresher training for technicians

All of the above listed items are discussed in the following section.

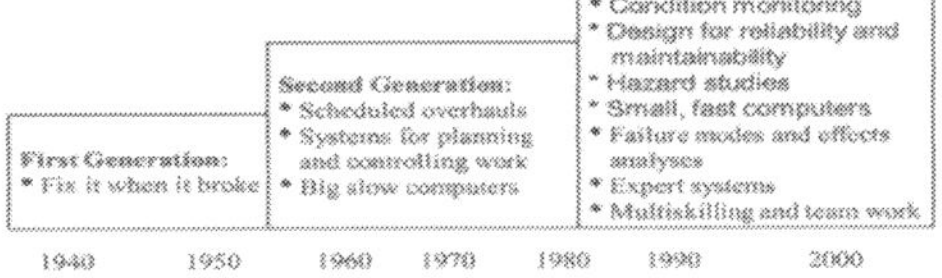

Figure 1. Changing maintenance techniques [6]

3. Maintenance engineering

As listed in the previous section, it is essential to cover various items to successfully manage a maintenance team. Only certain critical aspects are discussed in this paper and there could be several more depending on the nature of the business and the process itself. First and foremost it is very essential to have a computerized maintenance management system to handle regular maintenance and shut down activities effectively. MMS helps the engineers, supervisors, managers, technicians, operators, etc., to up-load the maintenance requests/activities in the MMS. MMS ensures that the maintenance activities are documented and not forgotten at a later time. Likewise, shut down activities must be listed in a well-defined sequence along with the required labor in the MMS. Since maintenance requests and activities are all listed in the software, the supervisor can schedule the activities based on priority, availability of resources, available working hours, and the equipment condition. A well-established MMS helps the organization in getting the work accomplished in a timely manner and in the most cost efficient way.

Spare parts planning and control helps the maintenance team to have the right spare and the required quantity on hand when it is needed. Having the correct spare part on hand helps the maintenance crew to execute the maintenance tasks without issues. The MMS can be linked with the spare part inventory so that the maintenance personnel can readily verify whether the required parts are available in the inventory to execute a given maintenance activity.

While setting up the spare parts inventory system, part details must be furnished precisely along with the preferred vendors. In addition, for every spare part, the minimum required inventory must be ascertained after careful evaluation. The order point for a given spare part must be chosen based on consumption history, lead-time, and the minimum required quantity at the plant.

A spare parts inventory system must be set up such that the parts are ordered automatically without going through the approval system. This helps in getting the parts ordered on time without any delays. It must be noted that spare parts held in the inventory costs several thousands of dollars and therefore the required quantity must be established after careful evaluation. Like food products, certain spare parts expire after a specific time. All these must be taken into account while setting up the spare parts inventory system. In the event of an unexpected total shutdown of the plant, several spare parts may be required for the shutdown activities. Therefore, it is sometimes beneficial to have an inventory system for the regular operations and a separate inventory system for shut down activities.

Spare parts must be received only after careful evaluation by the required personnel. Material test reports (MTR) and certificate of compliance (COC) must be checked before the parts are accepted in the inventory. An erroneous part can certainly create a tremendous issue, as the spare part shall not be useful when it is required the most.

Essential preventive maintenance activities in the plant should be identified after careful evaluation and must be added in the MMS system. The MMS system must alert the maintenance team when a certain item is due for routine inspection. This will help in addressing the preventive maintenance tasks in an organized manner. Regular cleaning of filters for HVAC units, cleaning of filtration units for cooling systems & pumps, checking for leaks, oil level in gear boxes, pumps, compressors, checking for hot spots in an electrical drives, etc., can all be classified under preventive maintenance tasks. The frequency of preventive maintenance for a certain activity is chosen based on performance and past experiences. The preventive maintenance activities in the MMS system must be reviewed on a regular basis to keep the list as up to date as possible. It must be noted that preventive maintenance tasks reduces the possibility of having an unscheduled maintenance, minimizes catastrophic machinery failures, and above all minimizes the possibility of having an expensive downtime.

A well developed predictive maintenance system is also required for reliable and safe operation of equipment. Due to the costs involved in performing a thorough maintenance on the equipment, and due to time constraints it is only possible to perform a maintenance activity only when it is required. Several proven predictive maintenance techniques are available for rotary and process equipment. Any rotary equipment having a defect shall exhibit vibrations. Thus, vibration analysis is the most commonly employed technique to judge the operating condition and to detect defects in the rotary equipment. In parallel with the vibration analysis, lubrication analysis can be employed to analyze the operating condition of the equipment. Likewise, ultrasound, electrical imbalance analysis and infrared thermography can also be utilized to detect the defects in the equipment. These have been described in [7]. While in operation, all process critical equipment must be analyzed using the above techniques on a regular basis. When an equipment begins to exhibit an adverse change in the performance, then that equipment must be examined more regularly until it is gets closer to the alert level. Based on the machine's condition and down time availability the equipment can be taken down for service and a thorough maintenance must be performed on the equipment. Predictive maintenance must be linked with the MMS so that operating issues are captured and recorded for any further maintenance activities.

As a part of predictive maintenance, process critical equipment in the plant may be linked to the Digital Control System (DCS). This is certainly very useful as the control room operators, engineers, and mangers can continuously monitor the performance of the equipment. In addition, since the data is available in the system, the operating trend could be charted for further evaluation.

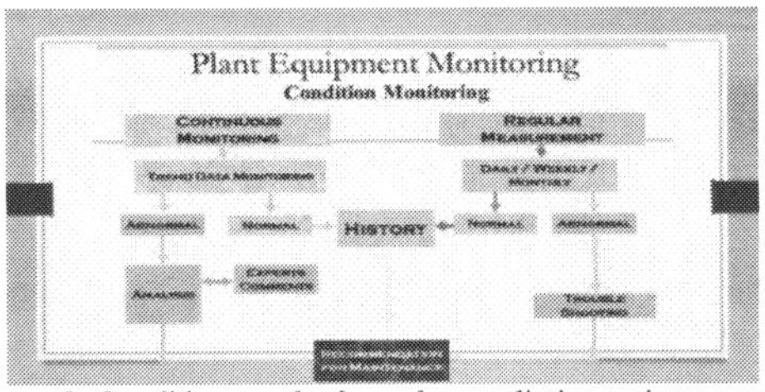

Figure 2. Condition monitoring using predictive maintenance practices

While performing a maintenance activity, it is very important to execute a thorough maintenance. The maintenance department must conduct the required inspections and activities on the equipment while it is taken down for service. Else, after the equipment start-up, it is very possible that a different issue could arise on the same equipment and pose an embarrassing situation for the maintenance team.

With the help of vendors, the maintenance engineers must review the existing maintenance practices and examine whether any new techniques are available in the market. The new technique may help the maintenance team to perform a certain maintenance activity quickly and more effectively. For example, until about a decade ago, lubrication for the fan/blower bearings was checked/dispensed using grease guns every fortnight. The new trend is to use an automatic lube dispenser wherein the technician does not have to lubricate the equipment for several months. Thus, it is certainly worthwhile to regularly check for new practices available and implement those wherever possible.

Keeping the future in mind, it is very beneficial to develop vendors in the near vicinity of the facility for maintenance tasks and for the procurement of parts consumed at the facility. Any process plant would require special cleaning facilities, furnace facilities, assembling facilities, rebuilding facilities, etc. In the long run, for faster service, it will be beneficial for the process industry to have such facilities neighboring the plant. Likewise, without compromising on quality, it may be worthwhile to check whether the long lead time items could be manufactured by local vendors. This will help in having quality parts on hard without being delayed for several months.

It is essential to integrate the principles of process safety management (PSM) and reliability engineering within the maintenance strategy. This helps in maintaining mechanical integrity of the equipment and assuring quality in maintenance activities. Reliability tools help to detect the factors contributing to the poor performance of the equipment, the interaction between the factors, expected time of failure and the effects of failure. Principles of process safety management help in identifying the root causes for an incident and issues before the startup of an equipment or a process. It also confirms contractor safety and ensures adequate training for the employees and contractors. These principles dictate that proper procedures are developed for maintenance & operations and that process hazards are identified by the application of reliability tools. PSM principles require that the procedures are regularly verified for accuracy and that adequate trainings and refresher trainings are provided to technicians. In-addition, these principles direct the maintenance team to conduct audits regularly and plan for emergencies.

4. Summary

The author, using years of maintenance experience in process plants, has summarized some of the important aspects for the effective administration of a maintenance program. Each aspect described in this paper has enough value to be discussed as a separate paper and for brevity; the discussions have been kept as short as possible. It should be noted that for a successful management of a maintenance team in a process plant, it is almost unrealistic to adopt a single maintenance technique. Various other factors such as coordination between the maintenance and other departments, application of reliability principles during design phase and regular operations, conducting energy audits, establishing workshop with required facilities, etc., are all of importance. All these are beyond the scope of this paper.

Equations should be centred and should be numbered with the number on the right-hand side.

References

1. Mobley, K., Maintenance Engineering Handbook, 8th Ed., McGraw-Hill Education, (2014).
2. Niebel, B.W., Maintenance Management, 2nd Ed., CRC Press, (1994).

3. Moubray, J., Raliability Centerd Maintenance, 2nd Ed., Industrial Press, ISBN-13: 978-0831131463, (1997).
4. Smith, W., Root Cause Analysis for Process Plants, LSA Publishing, (2012).
5. Townsend, A., Maintenance of Process Plant, 2nd Ed., Institution of Chemical Engineers, ISBN-13: 978-0852952924, 1992.
6. Silaipillayarputhur, K., Effective Maintenance Practices in a Process Plant, Journal of Applied Global Research, **6**, no. 17, p. 90-103, 2013.

Research on the platform of combined multi-functional bioreactor loading performance

Tan Qinsong, Zhang chunqiu, Gao Lilan, Men Yutao

Tianjin Key Laboratory of the Design and Intelligent Control of the Advanced Mechatronical System, School of Mechanical Engineering, Tianjin University of Technology, Tianjin 300384, China

Abstract. Mechanical environment plays an important role in the growth of cells, tissues and organs, in the study of mechanobiology and tissue engineering, and loading bioreactor acts as the basic platform. We developed a combined multi-functional bioreactor based on linear motor, and studied on the loading performance of different modules and loading methods. In the rolling / sliding loading module, in the effective distance for 50mm case, the bioreactor frequency is 0.01Hz to 0.3Hz. Under tensile and compressive loading module, the minimum amplitude of the bioreactor can reached 0.01mm while the frequency can reach 10Hz. Under the impact test module, in the effective distance for 50mm case, acceleration a=54.16m/s2canbe realized. In the high acceleration cyclic loading module, acceleration 8Gcan be realized. The combined multi-functional bioreactor provides conditions for the study related mechanobiology and tissue engineering.

1 Introduction

At present, there is a big challenge to repair the defect of human body tissue. Tissue engineering is an ideal method for repairing the tissue defect. Tissue engineering need to construct artificial tissue in vitro, and bioreactor plays a crucial role in tissue engineering, They can be constructed on the cell scaffold in a controllable mechanical environment, even large-scale industrial production [1].

Bioreactor is one of the key technologies in tissue engineering. the ability of accurately simulated tissue, organ and cell mechanics environment is the focus of attention, and each tissue type (such as skin, bone, blood vessel, and cartilage) may require designing the different bioreactor [2]. Therefore the design of special tissue bioreactor should be based on comprehensive understanding of the biological and engineering problems, and typical engineering problems such as reliability, reproducibility, scalability, and security should be considered [3,4], a lot of bioreactor systems have been developed and are expected to be high [4,5]. Cartilage tissue engineering bioreactor mainly types includes mechanical stirring bioreactor, perfusion - hydraulic type bioreactor, rotating wall bioreactor, air lift bioreactor, micro load type bioreactor and other bioreactor. Although there are various bioreactors in colleges, universities and research institutes, but linear motor driven bioreactor is relatively rare at home and abroad. In addition, compared to ball screw linear motor has the advantages of simple structure, high precision, high speed and high acceleration etc.

This paper conducted a deep study of the combined multi-functional bioreactor based on linear motor. Provided references for mechanical biology and tissue engineering experiment, and determined the suitable frequency, amplitude and maximum acceleration of the bioreactor. According to the main modules of the loading platform, study on the loading performance from the following aspects.

2 Instruments

Combined multi-functional bioreactor not only can realized rolling / sliding and combine rolling and sliding loading, but also carry on high frequency / low frequency and double frequency loading, see Fig. 1.

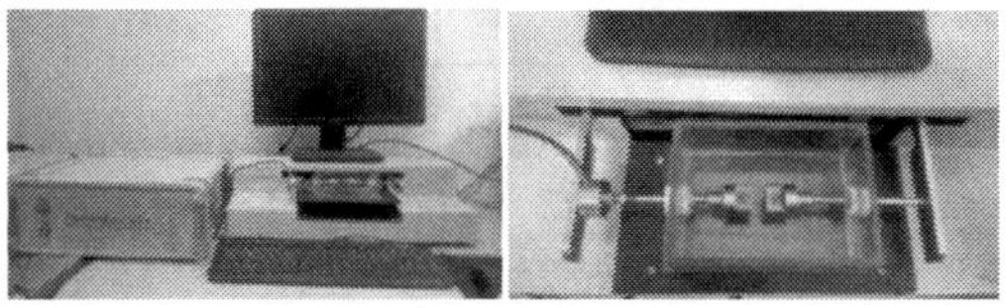

Figure 1. Combined multi-functional bioreactor.

The combined multi-functional bioreactor is divided into two parts: the hardware system and the software system. The hardware system includes the linear motors, drivers, sensors, PMAC motion control card, clamp, base and hood. The software system includes man-machine interface program and motion program in PMAC motion control card (Fig. 2).

The parameters of main components: Effective distance of the linear motor 50mm, continuous force is 38N, peak force is 111N. Resolution of grating ruler is 5 microns.

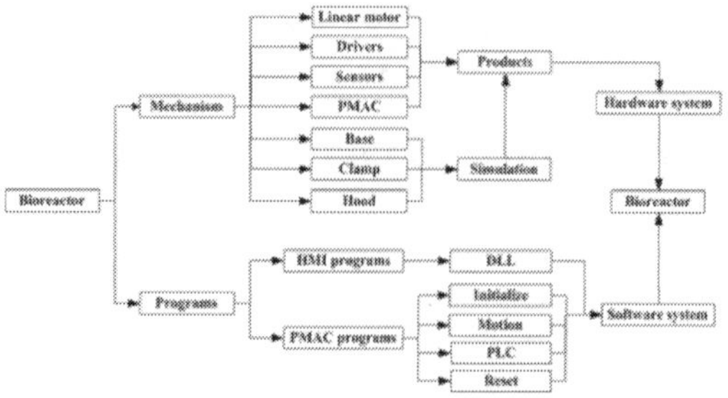

Figure 2. System diagram.

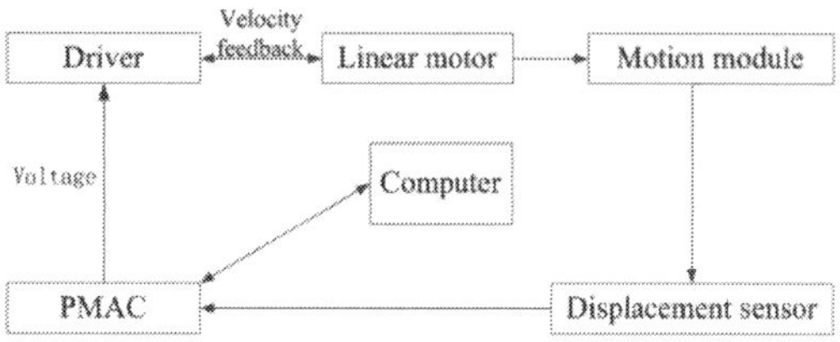

Figure 3. Control principle diagram.

The combined multi-functional bioreactor control process: Firstly, the computer send commands to the PMAC motion control card. Secondly, the result of compared input of computer and feedback of position sensor as the input of drivers. Finally, compare the input of drivers with feedback of speed sensor and the results of compared drive motor to move so that achieve the intended function in the experiment.

3 Methods

3.1 Study on the tension and compression loading module

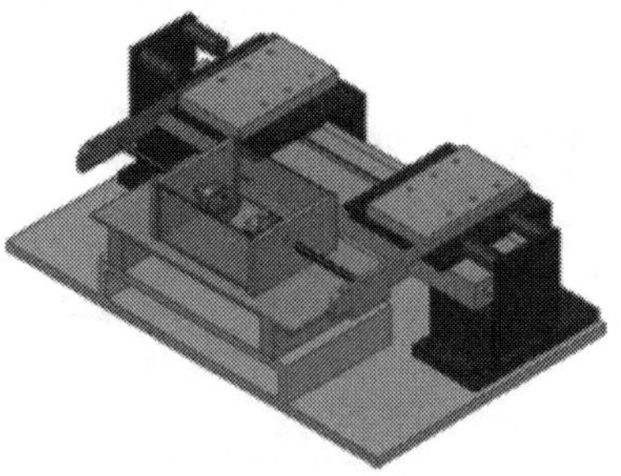

Figure 4. The tension and compression loading module

Theory analysis: the linear motor motion model is divided into triangular, trapezoidal and S type curve. S type curve motion model of linear motor be influenced by variable acceleration. It is bad for loading experiment, and the linear motor life would also be affected. Trapezoid motion curve because of the uniform motion stage, has a protective effect for the motor, but the frequency is not too high. In addition, for short stroke, linear motor is recommended to use triangular velocity mode (no uniform). The trapezoidal velocity model will be more efficient for long stroke. So we use triangular of linear motor mode for theoretical and experimental analysis (Fig. 5).

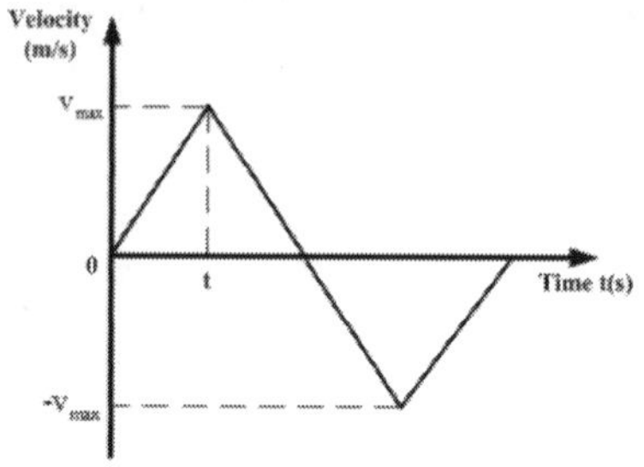

Figure 5. The triangle mode speed-time.

$f=\frac{1}{T}=\frac{1}{4}\sqrt{\frac{F}{2Sm}}=\frac{1}{4}\sqrt{\frac{2F}{mL}}$, F is the force of the linear motor. L is the distance of the linear motor. m=0.925kg is the quality of mover. So we can know the frequency f=10.134Hz, the max frequency f=17.32Hz.

The experimental results are as follows:

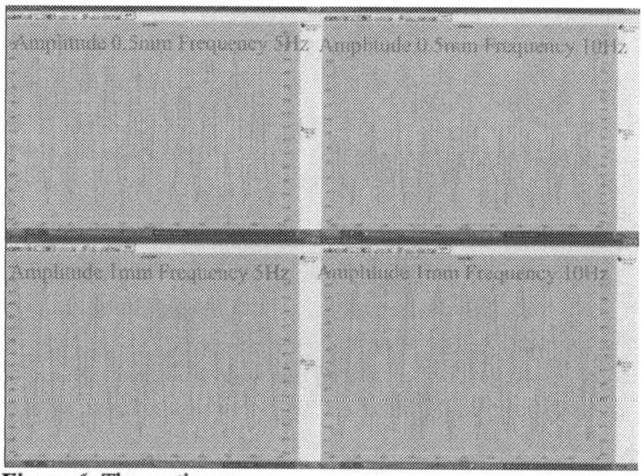

Figure 6. The motion curves.
Note: the green curve stand for displacement-time. the red curve is speed-time.

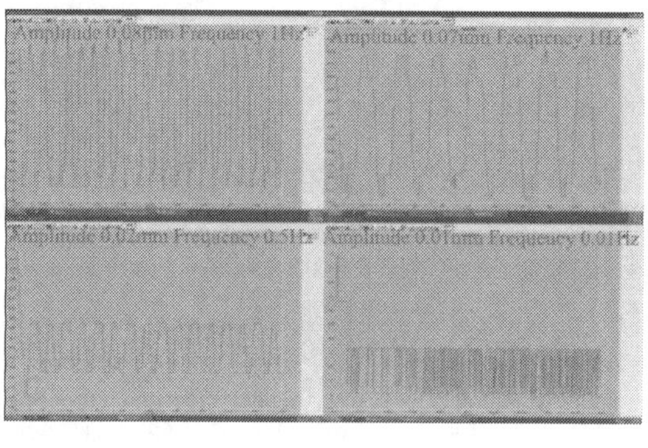

Figure 7. The motion curves
Note: the red is displacement-time curve.

Analysis:experimental results are basically consistent t o theoretical analysis results. The experiments (Fig. 6) show that the frequency of the bioreactor can reach $10Hz$. With the increase of the frequency or the decrease of the displacement, the motion characteristics of the system will be unstable. The frequency of the system is inversely proportional to the square root of the displacement. A large number of experimental data statistics show that, the system has better movement characteristics when the amplitude are $80\mu m$, $20\mu m$, $10\mu m$ and the frequency are $1Hz$, $0.5Hz$, $0.01Hz$ (Fig. 7).

3.2 Research on the rolling and sliding pressure module

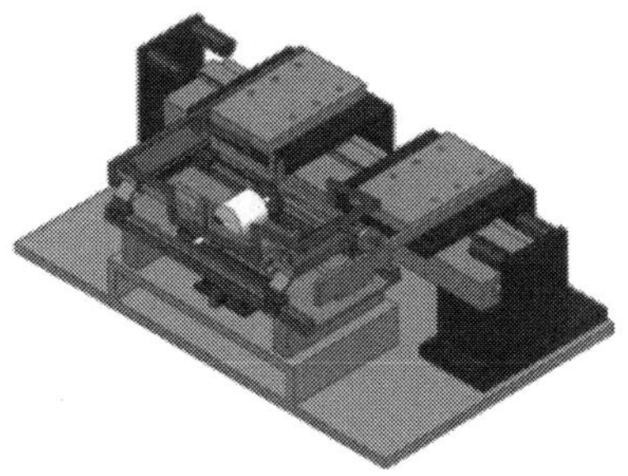

Figure 8. The Rolling and sliding pressure module.

In rolling and sliding loading test, the roller has larger displacement than tension and compression loading, in addition the effective stroke of the linear motor is $50mm$, so the given amplitude is $50mm$ and $30mm$, the experimental results are as below:

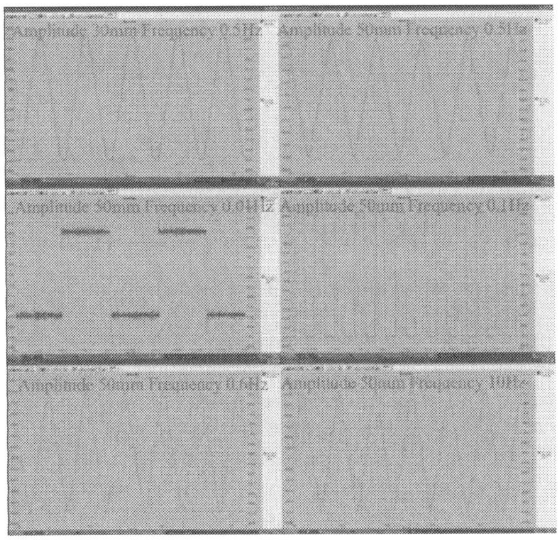

Figure 9. The motion curves.
Note: the green curve stand for displacement-time. the red curve is speed-time.

Analysis: In the same frequency, the larger the amplitude, the better the performance of the bioreactor. In the effective stroke $50mm$, the system has the accurate amplitude and frequency when the given frequency is $0.01Hz$ to $0.3Hz$. The frequency of $0.3Hz$ to $0.6Hz$, the system has accurate frequency, but the amplitude has great error. When the frequency is larger than $0.6Hz$, the actual amplitude and the actual frequency have large errors (Fig. 9).

3.3 Research on the high acceleration of cyclic loading experiment

Theory analysis: the selection of linear motor includes calculating the maximum force and continuous force. The maximum force is determined by the mass of the moving load and the maximum acceleration. Force = total mass * acceleration + friction + external stress, because of the small friction of linear motor, the experiment is carried out under no load, so:

$$F=ma=0.925a=111N \quad a=120\text{m/s}^2=12\text{G}$$

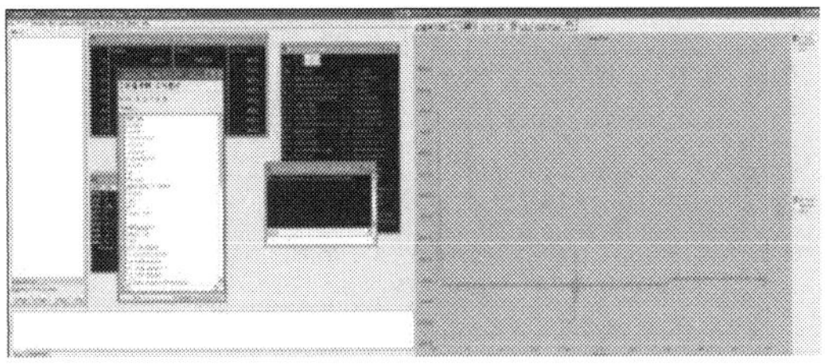

Figure 10. The programming interface and acceleration-time curves
Note: the green curve stand for theoretical acceleration-time. the red curve is actual acceleration-time.

The experimental results show that the theoretical acceleration of PMAC program design is $10G$ while the actual acceleration of the motor can reach about $8G$, and the actual motion curve and the theoretical curve change rules are in agreement.

3.4 Study on the impact test

Giving amplitude and frequency of linear motor, and simulating impact test of linear motor. Test results are as follow:

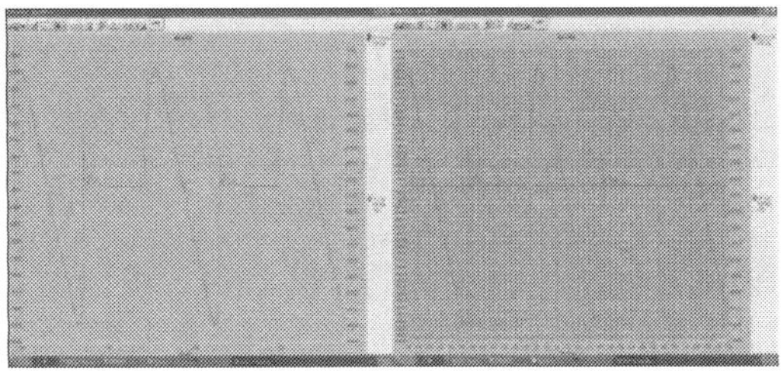

Figure 11. The motion curves.
Note: the green curve stand for speed-time. the red curve is displacement-time.

Experimental results: through analysis curve, we can know impact acceleration and impact force of the experimental platform are as follows:

$a = 54.16m/s^2$ $F = ma = 0.925 \times 54.16 = 50.1N$.

4 Conclusion

1) In tension / compression loading module, experiments results show that: With increasing frequency, the motion characteristics of the system is more unstable. The minimum amplitude of the combined multi-function bioreactor loading platform can reach $0.01mm$, the frequency range is $0.01Hz$ to $10Hz$. Theoretical analysis shows that the frequency of the system is inversely proportional to the square root of the displacement.

2) Under the rolling and sliding loading module, experimental results show that: The loading amplitude of the experimental platform can reach $50mm$. The frequency range from $0.01Hz$ to $0.3Hz$ can guarantee the accuracy of the actual amplitude and the actual frequency.

3) Under cyclic loading module, maximum acceleration of the experimental platform can reach 8G. Actual motion curves and the theory of motion curve is roughly consistent, indicating that programs and control of the experimental platform are right. If you want to achieve greater acceleration, you can choose a larger force or longer distance of linear motor.

4) Under the impact test module, impact acceleration and impact force of the experimental platform are as follows:

$a = 54.16m/s^2$ $F = ma = 0.925 \times 54.16 = 50.1N$.

Because the linear motor has many advantages such as high acceleration, high precision, simple structure and so on. So the linear motor can be used for the development of high frequency, great acceleration and impact of bioreactor, will also be more and more applied to the development of other kinds of bioreactor.

Acknowledgements

The project was partly supported by the National Natural Science Key Foundation of China (Nos.11432016) and the National Natural Science Foundation of China (Nos. 11372221,11572222,11402171).

References

1. T. Takebe, S. Kobayashi, H. Kan, H. Suzuki, Y. Yabuki, M. Mizuno, T. Adegawa, T. Yoshioka, J. Tanaka, J. Maegawa, H. Taniguchi, Human elastic cartilage engineering from cartilage progenitor cells using rotating wall vessel bioreactor, Transplant. Proc. **44** (2012) 1158–1161.
2. Ratcliffe, A. and Niklason, L. E.: Bioreactors and bioprocessing for tissue engineering. Ann. N.Y. Acad. Sci., **961**, 210–215 (2002).
3. Martin, I., Wendt, D., and Heberer, M.: The role of bioreactors in tissue engineering. Trends Biotechnol., **22**, 80–86(2004).
4. Naughton, G. K.: From lab bench to market: critical issues in tissue engineering. Ann. N.Y. Acad. Sci., **961**, 372–385(2002).
5. Griffith, L. G. and Naughton, G.: Tissue engineering — current challenges and expanding opportunities. Science, **295**,1009–1014 (2002).

 DOI: 10.1051/matecconf/20167711002

Design and building of modern two-photon laser scanning microscope for using in neuroscience

Maxim Doronin[1] and Alexander Popov[1,2]

[1]*Institute of Neuroscience, Lobachevsky State University of Nizhny Novgorod, 603950, Nizhny Novgorod, Russia*
[2]*Department of computer control systems and telecommunications, Volga State University of Water Transport, 603950, Nizhny Novgorod, Russia*

Abstract. One of the main visualization method in neuroscience is two-photon microscopy. However, scientists need to upgrade their microscopy system so regular because they are interested to get more specific data. Also this is due to the fact that this technique allows to obtain 3D images of tissues due to laser focus change, that is possible due to substantially greater penetration depth on the main wavelength into biological tissues. Self-developed microscopy system allow to modify the construction of microscope in not-complicated manner depending on specialized experimental models and scientific tasks. This work is small attempt to help researchers who are interested to build their own two-photon laser scanning microscope for using in neuroscience. The work was performed with financial support of the government represented by the Ministry of Education and Science of the Russian Federation, the unique identifier of the project is RFMEFI57814X0079, the agreement on granting a subsidy №14.578.21.0079 dated 28.11.2014.

1 Introduction

Application of modern techniques offer high resolution properties for living systems research is significant key for neurobiology progress. there are certain peculiarities arising from the specifics of experiments using this technique. In one case, for example, the analysis of the morphological cell data of hippocampal slices of the brain of rodents, other experiments *in vivo* depending on the design using the full-sphere of the virtual reality or, for example, in the form of a cylindrical hover platform. Working with selected methods require both of high-technology equipment availability and knowledge and skills to applying non-trivial mathematical algorithms for control two-photon microscopy system [1] and for well-formed data analysis [2], [3]. Two-photon microscopy offers several advantages compared with other forms of microscopy. Creation of two-photon system taking into account its specific use in neuroscience is probably the best solution. This article may be regarded as a quick reference to laboratory staff who are wishing to develop their own microscopy system for self-service and modernization of the system and in order to save the lab budget.

2 Optical part

Suggested type of microscope is module-based (Figure 1), where main module of it is body represented as metal base to which other modules are mounted. It could be stereotaxic part to fix the animals (chamber in case of in vitro experiments), scanners, shifting platform with fixed objectives nosepiece, optical elements using for work of differential interference contrast (DIC) [4] and multi-photon mode. Technically, the operation of DIC is provided with installation of a visible light source and an IR filter, an iris-shutter, a diffuser, a mirror to reflect the beam in the vertical direction and lens. After passing through the lower block and being reflected the light beam reaches the polarizer, and then passes onto a condenser with Nomarski prism. Then a light beam achieve the sample, which is fixed on the chamber. Passing through the sample, the beam achieve the optical components on a shifting platform (fixed to the frame of the microscope), and passes the second Nomarski prism and the analyzer. Here the interference of previously separated rays is performed. After passing the mentioned optical path the light beam falls onto the so-called illuminator of reflected light, that does not contain any filters or mirrors in the DIC mode operation, and then the beam passes onto IR camera that displays the image on the monitor. Galvo [5] and resonant [6] scanners are presented on Figure 1. The first one may be used for rectangular area (line-by-line) scanning, line scanning or point scanning and for uncaging some drug (e.g. glutamate, GABA), the second one – for fast frame rectangular area scanning. Maximal speed for galvo scanner is up to 2 fps (for 512x512 pixels frame), and for resonant scanner is up to 30 fps (for 512x512 pixels frame). IR Ti:Sa femtosecond pulsed lasers (tuning

DOI: 10.1051/matecconf/20167711002

wavelength range is 690–1080 nm) are using for fluorophore activation. Non-descanned detectors (NDDs) are using for detecting of fluorescence. Each NDD based on pair of bandpass filter to extract dye absorption wavelength and photomultiplier to detect fluorescent signal. Photodetectors are installed in close proximity to the aperture of lenses, besides, the fluorescence beam doesn't pass the way back through the scanning module, but directly gets onto the detector. The use of several dyes simultaneously is often the necessary requirement in multiphoton microscopy. Therefore the design of the microscope detector allows to make fast change of emission filters and dichroic mirror. The standard cube for filters mounted on the slider like "swalow-tail" that is easily inserted and taken out from the detector is used for this purpose and significantly facilitates the work of the researcher. The multiphoton detector must also be equipped with means of electronic protection and a shutter preventing the damage of the photo multiplier and increase in its performance term. Modern microscopy systems allow to mount up to 4–6 detectors depending on researcher goals.

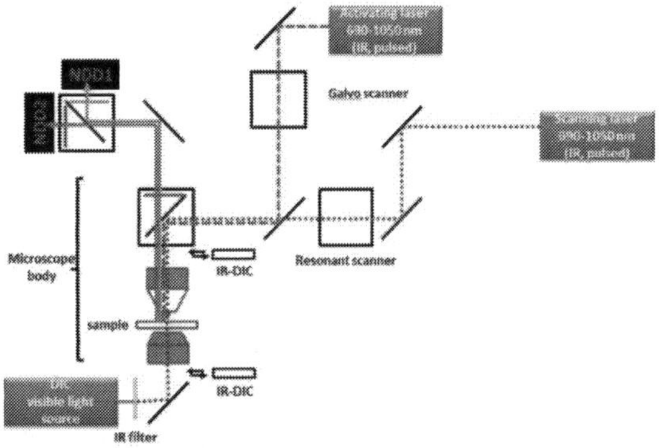

Figure 1. Optical part of two-photon microscopy system including galvo and resonant scanners.

To improve the quality of the resulting images it is necessary to use not only good program support but also to employ the advantages of the technical (physical) characteristics of the microscope, on the one hand, helping to move the lens to the sample as quickly and often as possible, and on the other hand, providing the most accurate movement when scanning a sample. The design must take into account all these features. In the first case (rapid movement to the sample) it is advisable to use a mechanical movement using a larger number of gear, and in the second case - the minimal number of the gear and electronic movement.

3 Design

Researchers should develop and design high-technology platform for realisation module-based microscope (the three-dimensional model of future microscope could be developed using software, e.g. Compass). Microscope consists of high-technology alloy allowed both to follow requirements of structure rigidity and minimal mass of microscope. All the components and elements of microscope should be developed in such a way as to have possibilities for receiving all the performance goals on the one hand and provide to evolution and addition techniques which could be used in that unique microscopy system on the other. For example if researchers are planning to have an in vivo experiments than working space (Figure 2) and, thus, the body of microscope should be higher than systems without in vivo approaches. Moving platform located above DIC elements and working space is using for mounting of objective nosepiece, filter cube(s) and NDDs. The electric drive control system for practicing a given angle or motion uses feedback sensors providing data concerning the angle or position of the output shaft of the operating engine. If the actuator motor employs a synchronous stepping motor, it is possible to dispense with the feedback sensor (DR) and to simplify the engine control system (CS).

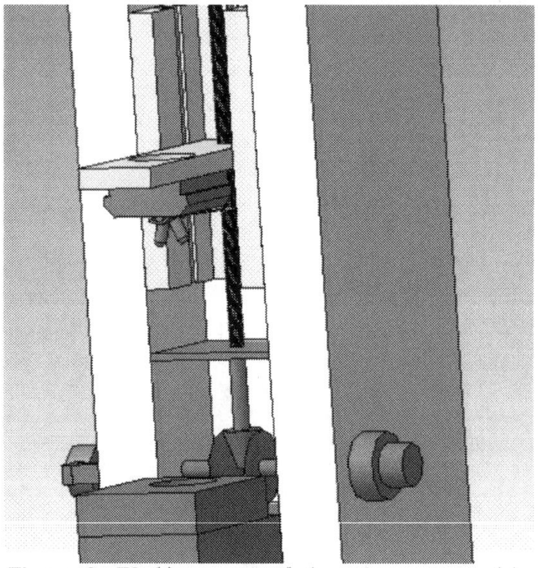

Figure 2. Working space of the microscope model developed using Kompas software.

Moving platform should be constructed as handy element in the context of ergonomics and functional. Moreover, developer should turn attention to minimal approach increment (stepper motor with at list 500 steps per rev should be used) of moving platform on the one hand, and to high-speed displacement and no slop (free motion) on the other hand. Other important aspect is accurate positioning of chamber for brain tissue (or alive animals) as well as micromanipulators for using electrophysiological and other (e.g., bolus loading) methods. U-shaped moving table can be adapted for specified type of microscope. It is possible to to set up this table independently from the microscope. For maintenance purposes it is possible to remove this table without disassembling the microscope. Motorized axes could be controlled by independent controller (control box and joystick or keyboard) or PC using especial software which could be integrated in mail control software for microscopy system.

4 Control

To improve the quality of the resulting images it is necessary to use not only advantages of the technical (physical) characteristics of the microscope, but also good program support. Control of system is realized either exactly by using digital-to-analog (DAC) and analog-to-digital (ADC) converter(s) (e.g. National Instruments boards and others) or by using control element of microscope components (e.g. USB, COM connection). However development of cross functional and convenient software to control the system is very important object for researchers. For achieving these goals developers could use commercially available software (ScanImage, Micro-Manager, YouScope or others) or create their own software using Matlab, LabView, Python or others. In case of software development researcher could create few sub-programs for each of elements controlled by software (e.g. manipulators, focus, lasers, scanners, acousto-optic modulator (AOM), detectors and others). After this step sub-programs could be combine to one main program which will control all of them and synchronize their integrated work. Having practical experience researchers could develop design of their software more friendly for end user than commercially available software because of their sector-specific goals. Software also should implement algorithms to scan region(s) of interest. These algorithms will depend on scanner type. Galvo scanner could be used for rectangular area (line-by-line, few options are presented in Figure 3) scanning, line scanning or point scanning and for uncaging some caged drug (e.g. glutamate, GABA) in sequence of points. At the same time resonant scanner could be used for fast frame rectangular area scanning in relation to feature of resonant scanner construction.

Figure 3. Few options of laser beam movement in case of rectangular area (line-by-line) scanning using by galvo scanner.

Working cooperatively, these scanners could fast scan rectangular area (resonant scanner) and activate caged drugs (galvo scanner) using different wavelength from different IR Ti:Sa femtosecond pulsed laser sources. Important aspect during any type of scanning is synchronizing between movement of laser beam by scanner and input signal from detector(s). Input signal received in one moment of time is detected fluorescent signal from the point where was directed laser beam by scanner at the same moment of time. Thus, software could construct fluorescent signal frames and simply analyse them (overlapping to each other, scale bar, construct 3D images and others). Moreover, researchers could realize more complicated data analysis allow represent data correctly. Additionally all the experimental data (include parameters of hardware such as laser intensity and wavelength, photomultipliers sensitivity and others) could be saved in especial file which could be opened both in that software to repeat experimental parameters and specially developed software to represent data and analyze it (e.g. using Matlab and others).

5 Conclusions

As a result of development of the own two-photon system taking into account all special characteristics allows to lower the expenses and to increase the quality of methods helping to improve the quality of the obtained data, enables to conduct researches on live samples with the minimum damage of fabric, makes possible to carry out researches of the data of morphological slices of mice or rats hippocampus, as well as researches on live samples, that is carrying out experiments in vivo. Due to the optimal location of photomultipliers the fluorescence from the activated objects is caught optimally, that allows qualitatively to visualize the object, and does not result in an overheat and burning down of the sample. Comparative characteristics of the self-building microscope and commercially available microscopes (Nikon AIR MP, Olympus FV1000 MPE, Carl Zeiss LSM 510 NLO) are presented in the Table 1.

Table 1. Comparative characteristics of the self-building microscope and commercially available microscopes.

Parameter	Self-building microscope	Nikon (AIR MP)	Olympus (FV 1000 MPE)	Carl Zeiss (LSM 510 NLO)
Scanning velocity	Seconds (512x512 pixels)	Seconds (512x512 pixels)	Seconds (256x256 pixels)	Seconds (256x256 pixels)
Flexibility of construction (availability *in vivo* experiments)	Yes	No	No	No
Resolution (depends on objective), μm	0.4–0.8	0.4–0.8	0.4–0.8	0.4–0.8
Light source	Ti:Sa IR femtosecond pulsed laser	Ti:Sa IR femtosecond pulsed laser	Ti:Sa IR femtosecond pulsed laser	Ti:Sa IR femtosecond pulsed laser
Detector	Photomultiplier	Photomultiplier	Photomultiplier	Photomultiplier

Detecting wavelength renge, nm	300–920	400–750	400–750	400–750

Acknowledgments

The work was performed at the Lobachevsky State University of Nizhny Novgorod with financial support of the government represented by the Ministry of Education and Science of the Russian Federation, the unique identifier of the project is RFMEFI57814X0079, the agreement on granting a subsidy №14.578.21.0079 dated by 28.11.2014.

References

1. K. Svoboda and R. Yasuda, *Neuron*, vol. 50, no. 6, pp. 823–39 (2006)
2. A. Danielyan, V. Katkovnik, and K. Egiazarian, *IEEE Trans. Image Process.*, vol. 21, no. 4, pp. 1715–1728 (2012)
3. A. Danielyan, M. Vehviläinen, A. Foi, V. Katkovnik, and K. Egiazarian, *2009 Int. Work. Local Non-Local Approx. Image Process. LNLA 2009*, no. 118312, pp. 125–129 (2009)
4. G. Nomarski, *Rev Hematol*, vol. 12, no. 4, pp. 439–42 (1957)
5. R. P. Aylward, *Sens. Rev.*, vol. 23, no. 3, pp. 216–222 (2003)
6. C. J. R. Sheppard and R. Kompfner, *Appl. Opt.*, vol. 17, no. 18, pp. 2879–2882 (1978)

DOI: 10.1051/matecconf/20167711003

Activity of Chlorhexidine Gluconate Loaded at Varying Polyelectrolyte Multilayers against Aggregatibacter Actinomycetemcomitans

Sirorat wacharanad[1], Witchayut Sasimomthon[2], Punyawut Wongyai[2], Apipon Vudhivanich[2], Kasidit Tippawan[2]

[1]*Department of preventive, Faculty of Dentistry, Naresuan University, Thailand*
[2]*Faculty of Dentistry, Naresuan University, Thailand*

Abtract. Chlorhexidine gluconate (CHX) has a bactericidal effect to Aggregatibacter actinomycetemcomitans (Aa) which causes periodontal disease. Therefore the controlled loading and release of CHX from multilayer thin film is beneficial for the local treatment of periodontitis. This study used ten bilayers of PDADMAC/PSSMA and PDADMAC/AL and loaded CHX into each film. Both un-loaded and loaded thin films were analysed using UV-Vis spectroscopy and AFM. The loaded film was thicker and rougher than the un-loaded film for both PDADMAC/PSSMA and PDADMAC/AL film. The loaded films were released in H_2O_2 for 24 h, and UV-Vis spectroscopy was measured and compared with the standard calibration curve of CHX. Results indicated that PDADMAC/PSSMA released 0.00052 M and PDADMAC/AL released 0.00071 M respectively. Both the loaded films were tested for toxicity to Aa. Neither film killed all colonies of Aa, however they reduced colony growth by 99% compared to the blank condition.

1 Introduction

Periodontal disease can affect elderly people with the loss of teeth. Clinical findings show that a periodontal pocket depth of more than 3 mm results in horizontal or vertical bone loss, clinical attachment loss and gingival recession [1]-[3]. Following treatment by scaling and root planing of the deep pockets, the dentist usually irrigates the area with a local antibiotic such as chlorhexidine (CHX) solution and inserts a biodegradable periochip [4]. This local antibiotic has a bactericidal effect on bacteria remaining in the periodontal tissue and decreases inflammation and recurrence rate [4], [5]. CHX is an antibiotic agent in the bisbiguanide group. It is composed of two chains which are linked together with hexamethylene. The chemical name is 1,6-di [4-chlorophenyl diguanido] hexane [6] (Fig. 1).

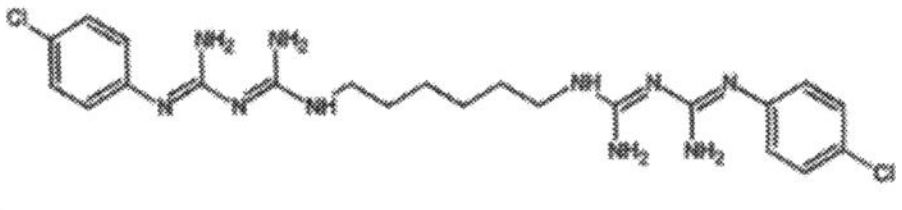

Figure 1. The chemical structure of chlorhexidine gluconate (CHX) [7].

CHX is effective against both gram-positive and negative bacteria, but shows little effect on fungi, viruses and tuberculosis (TB). It acts by destroying the cytoplasm of the bacteria which then leaks and the cell dies. Moreover high concentrations of CHX can bind with adenosine triphosphate and nucleic acid resulting in cell death after 8 –12 h. In the medical field, CHX is used as a mouth rinse for patients after surgery, rinsing during root canal treatment and a cleaning agent for medical equipment. An effective bactericidal concentration of CHX is between 0.12% and 0.20 % w/v. [8], [9].

In this study CHX was used as a model drug for loading and release into two types of polyelectrolyte multilayer (PEM) thin film. PEM thin films were developed by Decher and Hong in 1991 for modifying the surface of material [10]. PEMs were generated using the attractive force between positive and negative polyelectrolytes to form layer-by-layer thin film. The technique has been applied in many fields such as medical, engineering and industry as it is easy to use and inexpensive [11], [12]. Moreover the technique can be modified on a variety of materials including glass slides, silicon, gold, aluminium, wood and fibre [10], [13]. Positively and negatively charged polyelectrolytes are used to perform PEM. Examples of polyelectrolytes are polymers, proteins, nanoparticles and vitamins [14], [15].

This paper studied the ability of PDADMAC/PSSMA and PDADMAC/AL films to control the loading and release of CHX and their toxicity to anaerobic bacteria such as Aggregatibacter actinomycetemcomitans (Aa).

2 Experiment

2.1 Chemicals and methods

Poly(diallyldimethylammoniumchloride) (PDADMAC, medium molecular weight, 20% by wt in water, typical Mw=200,000 – 350,000), poly(sodium 4-

styrenesulphonic acid-co-maleic acid) (PSSMA, Typical Mw ~ 20,000), Alginic acid (Al, Typical Mw = 10,000 – 600,000) and poly(sodium 4-styrenesulphonate) (PSS, typical Mw = 70,000) were purchased from Sigma-Aldrich, (Thailand) Co., Ltd. Sodium chloride and sodium hydroxide were purchased from Labscan Asia Co., Ltd., Thailand. Chlorhexidine gluconate (CHX) was purchased from the Dental Hospital, Naresuan University. All chemicals and solvents were used as received without any further purification. Doubly distilled water was used in all experiments.

2.2 Multilayer thin films preparation

Ten millimolar PDADMAC and PSSMA were prepared at pH 9 and then added to 0.1M NaCl. A quartz slide was cleaned with piranha solution (H_2SO_4:H_2O_2 (70:30) and then coated with four layers of primer (PDADMAC/PSS). The multilayer thin film between PDADMAC and PSSMA was prepared by dipping the quartz slide in PDADMAC for 2 min, followed by three rinses in distilled water with pH adjusted at 9 by NaOH. The slide was then dipped in PSSMA for 2 min, followed by three rinses with distilled water. This step achieved the formation of one bilayer. Dipping was continued in PDADMAC and PSSMA until ten bilayers of PDADMAC/PSSMA had formed. The preparation of PDADMAC and Alginate thin film followed the same method [16].

2.3 Standard calibration curve of CHX

The concentration of CHX was varied at 0.00011 M 0.00056 M 0.00111 M 0.0017 M and 0.0022 M and measured by UV-Vis spectroscopy. The relation between concentration and absorbance value at 255 nm was plotted.

2.4 Loading CHX

To load of CHX in the PDADMAC/PSSMA and PDADMAC/AL films, the coated quartz slides were dipped into CHX solution (0.2% w/v, 50 ml). After dipping each film into CHX for 1, 2, 5, 10, 20, 30, 60 min and 24 h their wavelength was measured by UV-Vis spectroscopy. Finally, PDADMAC/PSSMA and PDADMAC/AL films both before and after loading were measured by atomic-force microscopy (AFM) to analyse their thickness and surface roughness.

2.5 Releasing CHX

Loaded PDADMAC/PSSMA and PDADMAC/AL films were dipped into H_2O_2 solution to release all CHX from the quartz slide. Then this solution was measured with UV-Vis spectroscopy to analyse the absorbance value at the maximum peak of CHX wavelength.

2.6 Antibacterial property against Aa

The antibacterial activity against Aa was tested using the standard method. The loaded films were exposed to 20 µl of Aa in brain heart infusion broth (2 ml). After 24 h incubation at 37 °C, 5% CO_2 condition, the bacteria/broth mixture was diluted five times. Then 50 µl of diluted bacteria was placed onto brain heart infusion agar using the spread plate method. After 24 h incubation the bacteria were counted. The result was corrected by the dilution factor to give the number of colony forming units (CFU) per millilitre. The percentage of bacterial reduction was then calculated and compared to the unloaded film.

3 Results and discussion

3.1 Multilayer thin film preparation

Twenty layers of PDADMAC/PSSMA and PDADMAC/AL were formed by the polyelectrolyte multilayer technique. The PEM films were then characterised via UV-Vis spectrophotometry and AFM. The wavelength of each PEM is shown in Fig. 2.

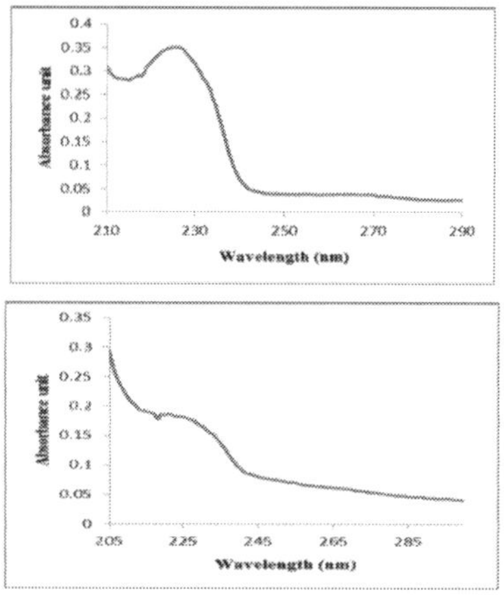

Figure 2. The UV–Vis absorbance spectra of PDADMAC /PSSMA film (upper) and PDADMAC/Alginate film (lower).

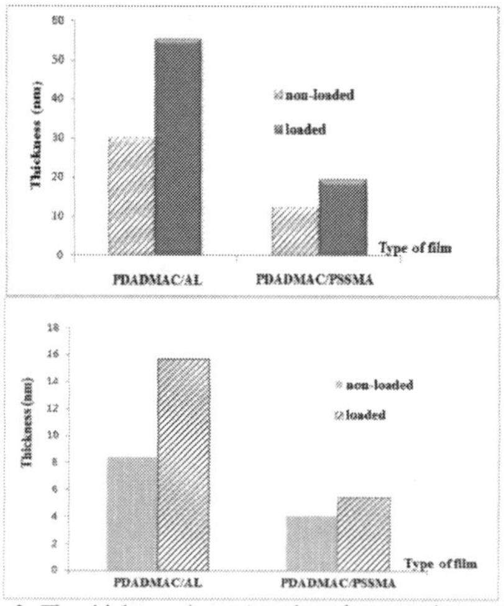

Figure 3. The thickness (upper) and surface roughness (lower) of loaded and non - loaded PDADMAC/PSSMA and PDADMAC/AL films.

The maximum absorbance value for 20 layers of PDADMAC/PSSMA and PDADMAC/AL was 226 and 210 nm as the wavelength of PSSMA and AL respectively. The PSSMA and AL were both present in the PEM thin film. [17], [18]. AFM was used to analyse the thickness and surface roughness of the multilayer thin film. Unloaded PDADMAC/AL film was thicker and rougher than PDADMAC/PSSMA film. The thickness of the PDADMAC/AL film was 30.39 ± 0.28 nm, whereas the thickness of the PDADMAC/PSSMA film was 12.54 ± 0.48 nm. The roughness values were similar to their thickness and the surface roughness of PDADMAC/AL was higher than PDADMAC/PSSMA (Fig. 3)

AFM indicated that the PDADMAC/AL thin film was thicker than the PDADMAC/PSSMA thin film as the charge density of Alginate was less than the PSSMA. Thus, the polyelectrolyte chain of Alginate had a lower negative charge than the chain of PSSMA. For this reason the chain of Alginate showed more elasticity and looping than the PSSMA chain. The PSSMA chain had a higher negative charge, so the repulsive force was higher, resulting in a straighter chain. The Alginate chain was adsorbed on the surface of the quartz slide more closely because of the lower repulsive force between the chains. Moreover, the looped Alginate chain made the multilayer thin film rougher and thicker. On the other hand, the PSSMA chain contained a higher negative charge which made this film smoother and thinner [19]. The characteristics of both chains and their adsorption on the surface are shown in Fig. 4 and the patterns of both multilayer thin films are presented as Fig. 5.

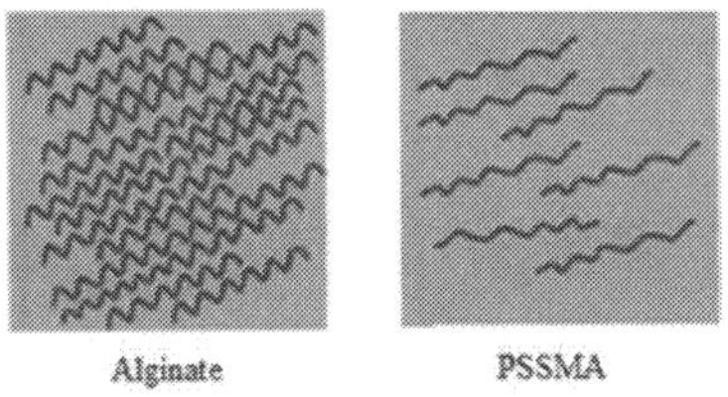

Figure 4. The characteristics of both alginate and PSSMA chains and their adsorption on the surface

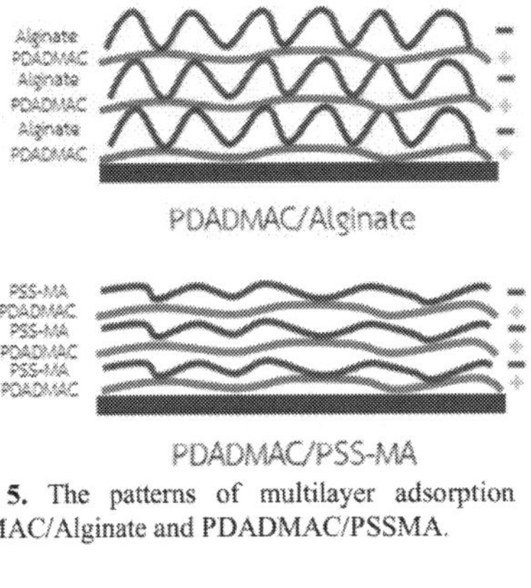

Figure 5. The patterns of multilayer adsorption of both PDADMAC/Alginate and PDADMAC/PSSMA.

3.2 Loaded CHX into PEM

The relationship between the concentration of CHX and its absorbance value was measured by UV-Vis spectroscopy and plotted as a calibration curve (Fig. 6 upper). The absorbance value increased with increasing concentration of CHX. This linear relationship was demonstrated as Eq. (1).

$$y = 491.2x - 0.0327 \quad (1)$$

After loading CHX into both PEMs by dipping in 0.2% CHX solution for 24 h, the films were measured for absorbance as a function of time. Results showed that CHX adsorbed into both PEM thin films quickly within 20 min. The absorbance then increased up to 1 h and remained constant through 24 h. The kinetic of loading CHX into thin film as a function of time is shown in Fig. 6 (lower). The absorbance of the loaded PDADMAC/AL film after 24 h was higher than the loaded PDADMAC/PSSMA film.

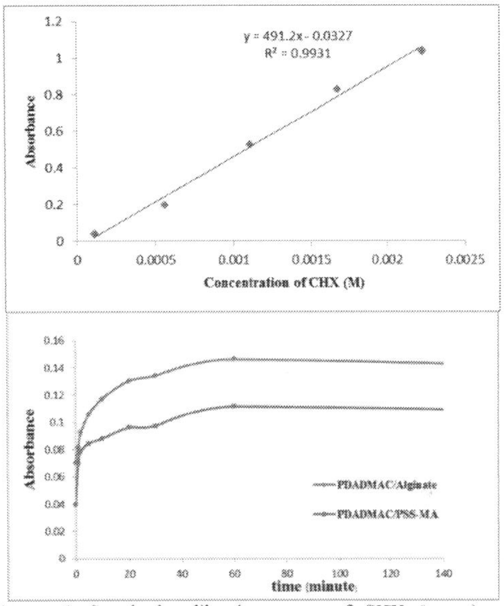

Figure 6. Standard calibration curve of CHX (upper) and kinetic curve showing the maximum absorbance of CHX when loaded into both thin films as a function of time (lower)

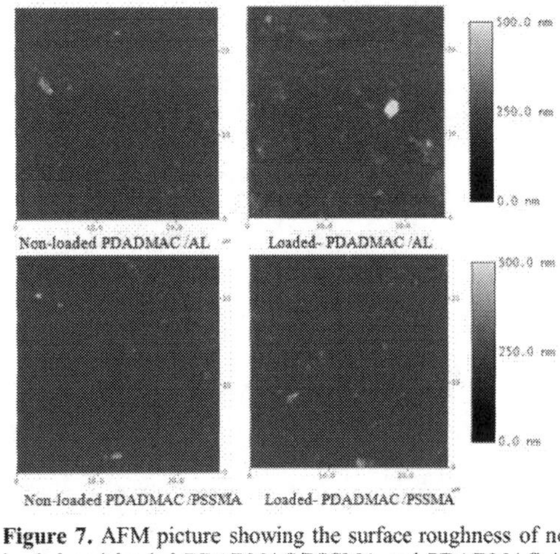

Figure 7. AFM picture showing the surface roughness of non-loaded and loaded PDADMAC/PSSMA and PDADMAC /AL films.

After loading CHX into the films, the thickness and surface roughness of both were characterized by AFM. The thickness of the loaded PDADMAC/AL film was 55.64 ± 1.25 nm, whereas the loaded PDADMAC /PSSMA was 19.73 ± 2.18 nm. The roughness of the loaded PDADMAC/AL was three fold higher than the PDADMAC/PSSMA film. The thickness and surface roughness between loaded and non-loaded PEM is shown in Fig. 3. The AFM picture comparing loaded and non-loaded film is shown in Fig. 7. After loading CHX into both films, their surface roughness increased.

The mechanism used for loading CHX into the multilayer thin film was electrostatic force. The top of the multilayer thin film was Alginate and PSSMA which had a negative charge. When loading CHX into these films, the attractive force between the negative and positive charge encouraged loading CHX into thin film multilayers. The adsorption rate of CHX was rapid during the initial 20 min, and then constant up to 24 h. This result agreed with Wei Yan et al. who studied the loading of drugs into nanoporous polymer thin films. They found that the initial absorbance value increased suddenly and then became constant [21]. Considering all the absorbance values, the thickness and roughness of both films were higher for PDADMAC/Alginate than for PDADMAC /PSSMA. The difference in the polyelectrolyte properties of the chain produced a PDADMAC/AL film with higher porosity and thickness and a greater surface area for loading CHX. PDADMAC/PSSMA had a straight chain and the multilayer thin film was smooth, thin and low in porosity, with less surface area for loading the drug.

3.3 Release of CHX in H_2O_2 after 24 h

CHX was released from both PEM thin films by immersion in H_2O_2. The released solution was measured by UV-Vis spectroscopy to calculate the remaining concentration of CHX in the PEM thin films. The absorbance of H_2O_2 after releasing CHX from PDADMAC/AL was 0.314 ± 0.013 and for PDADMAC/PSSMA was 0.225 ± 0.087. The absorbance results were calculated as concentrations using Equation 1, where y was the absorbance unit and x was the concentration of CHX. Finally the concentration of CHX loaded into PDADAMAC/AL was 0.00071 M (0.064%w/v) and in PDADMAC/PSSMA 0.00052 M (0.047% w/v). However, the effective concentration of CHX required to kill bacteria was 0.0034 M (0.12% w/v).

3.4 Antibacterial effect

After loading CHX into PDADMAC/PSSMA and PDADMAC/AL both multilayer thin films were tested for toxicity to Aa. CHX release from the multilayer thin films decreased the amount of Aa. In the blank condition, colonies of Aa increased to 7.7×10^9 CFU/ml. Loaded PDADAMAC/PSSMA decreased the growth of Aa to 1.64×10^6 CFU/ml with a percentage reduction of 99.98%. The loaded PDADMAC/AL reduced the growth of Aa to 1.02×10^7 CFU/ml, calculated at a percentage reduction of 99.87% (Fig. 8).

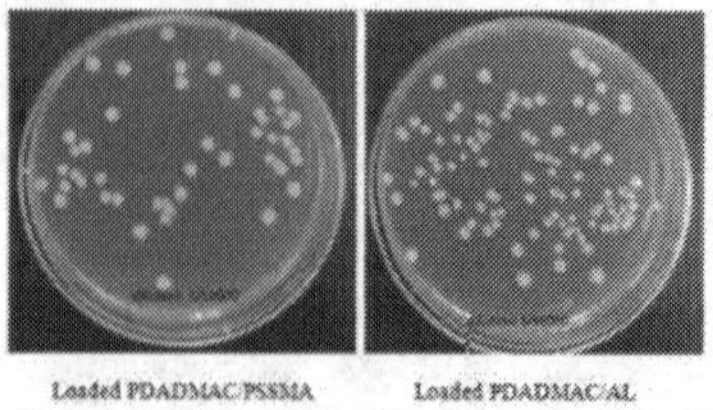

Figure 8. Colonies of Aa after exposure to both multilayer thin films after 24 h.

The result of the antibacterial testing determined that both loaded multilayer thin films decreased the growth of the Aa colony compared to the blank condition. CHX was released from both films to decrease the Aa colony. PSSMA is a polyelectrolytes composed of 50% of carboxylate groups(pK_a= 8.8) and sulfonic groups (pK_a nearly 1) [19]. Alginate is a polyelectrolytes contained carboxylic groups (pK_a= 3.2) [23], [24]. For both loaded multilayer thin film, when pH changed from 9 to 7.4 accelerate the rate of degradable of film due to the lower ionization of the carboxylate groups [19], [25], [26]. The attractive force between layer was lower resulting film was decompose and release CHX into broth. So when both loaded PEM thin films were at pH 7.4 in brain heart infusion broth, their charge and attractive force between the layers altered. This change made the film more susceptible to decomposition, and release of CHX.

However loaded film could not inhibit or bactericidal effect to Aa. They had just been decreased the growth of colony of Aa, due to their concentration of CHX which could loaded into both thin film was less than the effective concentration(0.12% w/v). Therefore they could not inhibit the Aa. So we can develop these multilayer thin films in their physical property such as the number of layer, increase dipping time or increase of concentration of CHX which used to load into film. This could be modified them more capacity to loaded concentration of CHX.

4 Conclusion

Smart PEM films can be manufactured to control loading and release of CHX. More CHX could be loaded into PDADMAC/AL than into PDADMAC/PSSMA thin films. Both loaded multilayer thin films decreased the growth of the Aa colony nearly 100 % compared to the blank condition. However, the concentration of CHX that could be loaded in both films was lower than 0.12% w/v. Thin films require further development to increase the release of drugs to inhibit bacteria. These smart thin films will be beneficial for the improved treatment of periodontal disease.

Acknowledgements

This work was supported and funded by Faculty of Dentistry, Naresuan University, Thailand and Mrs.

DOI: 10.1051/matecconf/20167711003

Suttimas Yuakyong, Faculty of Dentistry, Naresuan University.

References

1. G.C. Armitage, M.P. Cullinan, Periodontol 2000, **53**, (2010).
2. J. Highfield, Aust. Dent. J, **54**, (2009).
3. A.M. Tuomainena, M. Jauhiainenc, P.T. Kovanend, J. Metsoc, S. Pajua, P.J. Pussinena, Microb. Pathogenesis, **44**, (2008).
4. K.S. Abdellaouia, N.V. Castionib, R. Gurnya, Eur. J Pharm. Bioparm, **50**, (2000).
5. N. Jain, G.K. Jain, S. Javed, Z. Iqbal, S. Talegaonkar, F.J. Ahmad, R.K. Khar, Drug Discov. Today, (2008).
6. T.H. Lee, C.C. Hu, S.S. Lee, M.Y. Chou, Y.C. Chang, Int. Endod. J, **43**, (2010).
7. M. Moritz, M.G. Moritz, Appl. Surf. Sci., 356 (2015)
8. K.J. Anusavice, N.Z. Zhang, C. Shen, J Dent. Res., **85**, (2006).
9. B.M. Eley, Brit. Dent. J, **186**, (1999).
10. A.P. Ramos, F.G. Doro, E. Tfouni, R.R. Gonçalves, M.E.D. Zaniquelli, Thin Solid Film, **516**, (2008).
11. O. Etienne, C. Picart1, C. Taddei, P. Keller, E. Hubsch, P. Schaaf, J.C. Voege, Y. Haikel, J.A. Ogier1, C. Egles, J Dent. Res., **85**, (2006).
12. J.C. Harnet, E.L. Guen, V. Ball, H. Tenenbaum, J. Ogier, Y. Haikel, C. Vodouhe, J Mater Sci: Mater Med, **20**, (2009).
13. S. Boddohi, C.E. Killingsworth, M.J. Kipper, Biomacromolecules, **9**, (2008).
14. L. Shen, N. Hu, Biomacromolecules, **6**, (2005).
15. B.F. Abu-Sharkh, Polymer, **47**, (2006).
16. P. Kittitheeranun, N. Sanchavanakit, S.T. Dubas, W. Sajomsang, Langmuir, **26**, (2010).
17. E. Tjipto, J.F. Quinn, F. Caruso, Langmuir, **21**, (2005).
18. F. Shen, A.A. Li, R.M. Cornelius, P. Cirone, R.F. Childs, J.L. Brash, P.L. Chang, Photocrosslinked microcapsules, (2005).
19. E. Tjipto, J.F. Quinn, F. Caruso, Journal of Polymer Science, Part A, Polymer Chemistry, **45**, (2007).
20. K.Y. Lee, D.J. Mooney, Prog. Polym. Sci., **37**, (2012).
21. W. Yan, V.K.S. Hsiao, Y.B. Zheng, Y.M. Shariff, T. Gao, T.J. Huang, Thin Solid Films, **517**, (2009).
22. C. Wang, W. Ye, Y. Zheng, X. Liu, Z. Tong, Int. J Pharm, **338**, (2007).
23. J. Shi, N.M. Alves, J.F. Mano, Macromol. Biosci., **6**, (2006).
24. H.M. Mansour, M. Sohn, A. Al-Ghananeem, P.P. DeLuca, Int. J. Mol. Sci., **11**, (2010).
25. S. Wacharanad, S.T. Dubas, Adv. Mater. Res., **701**, (2013).
26. P. Sriamornsak, S. Sungthongjeen, AAPS PharmSciTech, **8**, (2007).

Bone Fracture Modeling and Analysis by Three Point Bending: A Review

Mohammad Shahril Salim, Ahmad Faizal Salleh and Ruslizam Daud

School of Mechatronic Engineering, Universiti Malaysia Perlis (UniMAP), 02600 Arau, Perlis, Malaysia

Abstract. Bones are the main structural components of a skeleton in our body. They run as a special role in the body providing its shape maintenance, protection of internal organs and transmission of forces. Their structural integrity is essential for the quality of life. Unfortunately, bones can only sustains loads until it reach a certain limit. Understanding fracture behaviour of bone is necessary for prevention and diagnosis of trauma. This paper aims to review and update readers on current published research explicitly related to bone fracture analysis conducted by three point bending test. Past studies have shown that most analysis have been done on fracture mechanism and how to find fracture toughness of the bone. Besides using an experimental approached, some of the research used Finite Element (FE) as a tool to investigate the fracture mechanism.

1 Introduction

When participating in any collision sports, there will be a risk of injury as the dynamic impact of the collision may overload the musculoskeletal system beyond its limit. The general risk factors for injuries have been addressed to be skill, Body Mass Index (BMI), playing experience, equipment, previous injury, playing position, and level of play or competition. But, the most common related injuries are fractures. Fractures are very common scenario in our life. Basically it will occur when the physical force exerted on the bone is much stronger than the bone itself. As we getting older, our bones become more brittle and we are more likely to suffer fractures from falls that would not occur when we were young [1]. Bone fracture toughness and bone resistance to crack growth is a key factor in resistance of bone to monotonic and fatigue failure. There are two main important occur in fracture toughness which are intrinsic and extrinsic mechanism. Intrinsic mechanism is about the plasticity deformation at small scale lengths while extrinsic refer to crack shielding where microstructure inhibit the crack growth.

The mechanical properties of bone are determined by their composition and structure over length – scales, and influenced by the loads it supports through adaptation and remodelling. As we all knew, bone is a living tissue. Thus it might be always response to mechanical stimuli. Because of the response, the structure of bone is result from the continous stress induced modelling. At the macroscopic scale, bone is composed of cortical bone and trabecular bone. Encouraged by the idea that an accurate evaluation of bone quality could potentially be a predictor of risk of bone fracture, recent studies on the fracture properties of bone have focused on understanding the origin of bone toughness and its resistance to crack propagation in relation to bone's multi-scale structure [2], [3]. Despite the current understanding of the origins and mechanisms of toughening in human bone, it is still not fully understood how specific disease states can affect these mechanisms and how certain therapies can improve bone toughness to reduce the fracture risk. Mostly, an evaluation will be done by using animal models of human bone to predict the outcome of therapies in humans. An understanding of how fracture mechanisms change for different bone diseases and with different therapies is critical to the design of effective treatments on any diseases [4]. The fracture toughness of human cortical bone under mixed mode loading can be calculated using the Linear Elastic Fracture Mechanics (LEFM) assessment of critical stress intensity factor K_{IC} in Mode I or K_{IIC} in Mode II. However, instead of using the LEFM, the plastic behavior part which required the adoption of nonlinear fracture mechanics can be used by determining the value of J integral based on Strain Energy Released Rate (SERR).

Current practice to determine bone tissue modulus and fracture toughness is from three-point bending test. Despite the high clinical relevance of bone fractures, there is a lack of validated Finite Element (FE) models describing the complete fracture behavior. FE model can be used to represent an analysis under similar conditions as in the experiment. FE analysis is considered nowadays as a standard tool for the assessment of the mechanical properties of bone. It is acknowledged that, the finite element models used only provide a very simplified representation of a bone segment, however their simplicity lends clarity to the observation of different responses. Computational FE models present a method of studying complex mechanical interactions without

DOI: 10.1051/matecconf/20167711004

introducing artificial foreign bodies and can provide a more continuous description of tissue behaviour. Although it provide a very simplified representation of any segment, however it simplicity lends clarity to observation of different impact responses when more or less representative organic tissue models are employed.

2 Literature Review

Bone is a great topic of interest for researchers. Understanding the structure-related properties of bone and how they are affected by aging, disease and therapies are very interesting area. Most of previous studies have used animal's bone compared to human bone. Little is known about the mechanisms of mouse bone fracture and its similarities and differences from fracture in human bone. Carrieo et al. investigated the fracture toughness during crack initiation and crack propagation for mouse bone. In the study, they monitored the extension of stable crack growth in mouse femora by simultaneously imaging the crack path during toughness measurement to discern the sources of toughness. Enviromental scanning electron microscope (ESEM) is used to measure the crack-initiation toughness and crack-resistance curve (R-curve). This has been done to identify the salient damage and toughening mechanisms [4]. Bones were tested in three-point bending with a span of 6 mm (span/femur radius ratio of ~4) using Gatan Microtest 2 kN three-point bending stage as in Fig. 1.

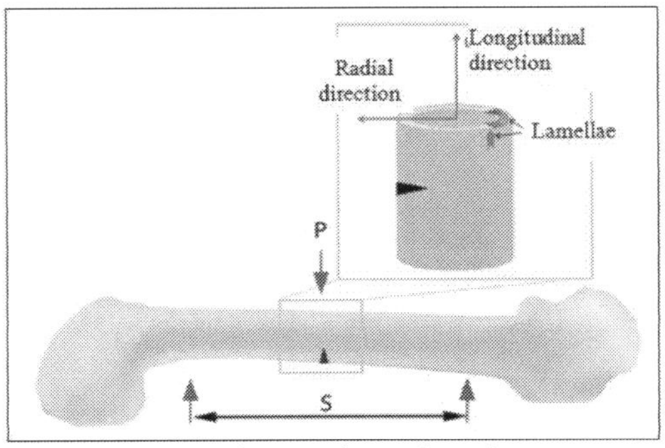

Figure 1. Fracture toughness testing performed on mouse femurs with a through-thickness sharp notch on the posterior side loaded in three-point bending in ESEM [4].

Fracture toughness values were then been evaluated by using the LEFM for a through-thickness crack in a circular thick-walled cylinder:

$$K = F_b \frac{PSR_o}{\pi(R_o^4 - R_i^4)} \sqrt{\pi R_m \theta_{inst}} \quad (1)$$

where F_b is a geometric constant for thick-walled cylinders, P is the instantaneous load, S is the span width, R_m, R_i, and R_o are the mean, inner and outer radius of bone and θ_{inst} is the instantaneous instability half-crack angle.

The majority of the works found in the literature regarding the fracture characterization of cortical bones are dedicated to mode I (opening) loading. S. Li et al. in their study considered a bovine femoral cortical bone tissue. The focused was on understanding of the principal and causes of bone fracture by considering the effect of both crack propagation direction and cortex position. The setup of the test is shown in Fig. 2.

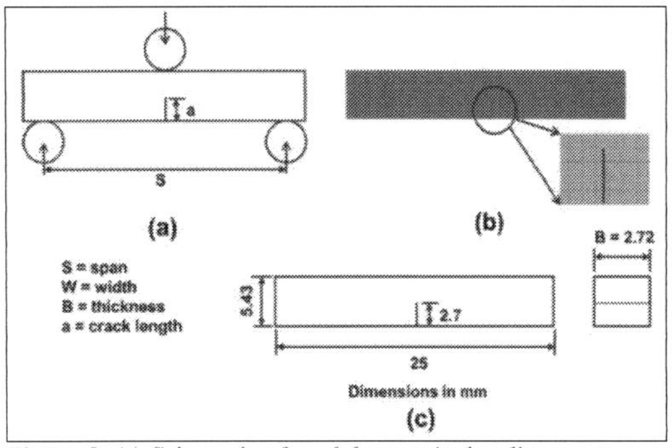

Figure 2. (a) Schematic of used three-point bending setup, distance between fixed grips is S = 4W; (b) mesh used for cortical bone specimen; (c) geometry and dimensions of cortical bone specimens used [5].

In the study, the behaviour of all specimens was predominantly non-elastic [5]. Therefore, an elastic-plastic fracture mechanics (EPFM) parameter, J-integral was calculated based on British Standard BS 7448-1 [6]:

$$J = \left[\frac{FS}{BW^{1.5}} \times f\left(\frac{a_0}{W}\right)\right]^2 \frac{(1-v^2)}{E} + \frac{2U_p}{B(W-a_0)} \quad (2)$$

where S is the bending span, F is the applied force, $f(a_o/W)$ is a function of (a_o/W), v is Poisson's ratio, E is elastic modulus, U_p is the plastic part of area under plot of force versus specimen displacement along the load-line, B is the specimen's thickness, W is the effective width of the test specimen and a_0 the average original crack length.

Fracture characterisation under mode II loading has been incomparably have less studied than mode I. This is probably because the fracture tests for this pure loading mode are very hard to perform in cortical bone. But, N. Dourado et al. have performed an end notch flexure test in the context of pure mode II fracture characterization of bovine cortical bone. To overcome the difficulties intrinsic to crack length monitoring during its propagation an identical crack method was employed as data reduction scheme. Fracture behaviour of each specimen was discovered by inverse method, considering a trapezoidal shape for the softening law. The consistency of obtained results leads to the conclusion that the trapezoidal law is sufficient to simulate fracture behaviour of bone under mode II loading and also very simple to be executed [7].

Besides using mode I and II, several researches have also been done on mixed-mode I+II loading. From a physiological point of view, fracture in bone is rarely induced by pure loading mode conditions. Therefore, crack initiation and propagation is more likely to occur under mixed-mode loading. Only a few works have been published so far regarding the mixed-mode loading. Silva et al. analysed a mixed-mode fracture characterization on human cortical bone from two tibiae of young male donors. Only the medial region (mid-shaft) of each tibia was used in the study. The main objective of their work is to verify the validity of using the Single Leg Bending (SLB) (Fig. 3) under mixed-mode I+II loading. The SLB is a simple three-point bending test on a pre-cracked specimen with a cut arm to induced mixed-mode I+II loading with nearly fixed ratio [8].

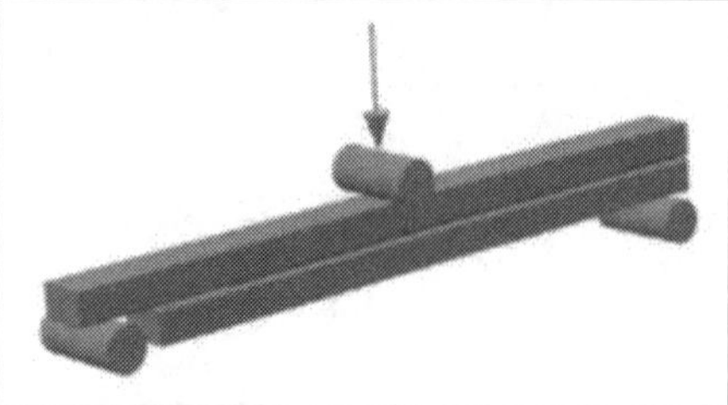

Figure 3. Single leg bending (SLB) [8]

It was verified that experimental values of fracture energy under mixed-mode I+II are very uniformly consistent with each other since a small coefficient of variation was obtained. The determination of the cohesive laws under mixed-mode can be viewed as a fundamental task in studies involving aspects that can affect bone fracture resistance.

T.L. Norman et al. have conducted a study using the compact tension method to determine the fracture toughness indices, critical energy release rate (G_c) and critical stress intensity factor (K_c) for human Haversioan bone and bovine bone under mode I loading. K_c and G_c are derived from theory based on LEFM. Results from this study show that the absolute values of K_c and G_c for human bone are lower than the absolute values of K_c and G_c for bovine bone [9]. The test was conducted over a crack range of 0.5< a/W< 1.0 while previous studies as in Table I were conducted for a/W< 0.7.

Table 1. Summary of previous study of fracture toughness of cortical bone using compact tension method

Investi-gator(s)	Bone Type	Orien-tation	K_c (MN m$^{-3/2}$)	G_c (Nm^{-1})
Wright and Hayes (1977)	Bovine femur	Longitu-dinal	3.04 – 3.95	810 – 1524
Behiri and Bonfield (1989)	Bovine tibia	Longitu-dinal	3.2	-
Norman et al. (1992)	Bovine tibia	Longitu-dinal	6.67	1191

Summary of other research are been summarized in Table II.

Table 2 Summary of previous research.

Ref.	Description	Conclusion/ Recommendation
[4]	Mouse femora were notched on the posterior surface at their mid-diaphysis and tested in three-point bending using an in situ loading stage within an ESEM	3D morphological characterization of the microcrack damage does not provide measurements of the bone toughness and its evolution in relation to the crack path and extension.
[5]	Experimental study and numerical simulations of fracture processes in a bovine femoral cortical bone tissue using single-edge-notch-bending specimen in a three-point bending setup.	Bone demonstrated a non-uniform elastic-plastic fracture process for different cortices. Fracture toughening mechanisms varied for different fracture propagation directions.
[7]	Finite element study on fracture energy of bone under pure mode II loading. Trapezoidal law is used on simulate fracture behaviour.	Proposed miniaturized end notched flexure test is appropriate for bone fracture characterization under mode II loading.
[9]	Plain strain assumptions of linear elastic fracture mechanics (LEFM) to bone were investigated. Fracture toughness tests were used to assess the crack inhibiting effect of human bone microstructure on fracture resistance.	The toughness of human and bovine bone is still questionable.
[10]	Assess the accuracy of beam theory, which were tested on mice. Structure of femora was measured by micro-computed tomography. Tissue modulus was calculated from beam theory using micro-FE results	Re-evaluation of the tissue properties obtained from three-point bending especially in mouse genetics
[11]	Development of FE model based on continuum damage mechanics (CDM) approach to simulate the damage initiation and propagation of a single trabecula till complete fracture.	Consideration of time-dependent mechanical behaviour of bone (loading rate). Further investigation is required to develop anisotropic damage models for bone at tissue level.
[12]	Experimental study of the mechanical behaviour and fracture toughness of bovine bone under tensile and bending test.	Water content affects the ability of deforming and dissipating energy under mechanical loading, leading to a more ductile material.
[13]	A reference point indentation technique (BioDent) was developed to determine bone's resistance to fracture.	Further study of the indentation distance increase (IDI) values and other BioDent indentation parameters in ex vivo models.
[14]	Investigation of the osteonal-, micro- and tissue-level mechanical behaviour of human cortical bone tissue samples from young ang elderly donors using atomic force microscope (AFM)	Further research to establish the homogenisation of nanoelastic features that is a function of ageing and whether it is a change occurring in osteoporotic bone.

3 Conclusion

Limitation associated with current clinical fracture risk assessment tools highlight the need for increased understanding of the fracture mechanisms of the bone. Additional knowledge on the quality of the bone is essential for predicting and reducing the fracture risk associated with age and diseases. This review indicates study on bone fracture mostly been conducted in human bone, mouse bone and bovine bone. By interpreting mouse and bovine bone in terms of human bone toughness can enhance the progress of pre-clinical testing for translational medicine aimed to reduce bone fracture risks. Most researchers used three-point bending test as it was easy to conduct and create initial crack for fracture toughness calculation. Hence, more research should be done by using finite element tools as it is the safe and non-invasive technique.

References

1. R. Barlett, Sports Biomechanics: Reducing Injury and Improving Performance, Routledge, London (1999).
2. K.J. Koester, J.W. Ager, R.O. Ritchie. The True Toughness of Human Cortical Bone Measured with Realistically Short Cracks. Nat. Mater. **7** (2008), 672 – 677.
3. H. Peterlik, P. Roschger, K. Klaushofer and P. Fratzl. From Brittle to Ductile Fracture of Bone. Nat. Mater. **5**, pp. 52 – 55, (2006).
4. A. Carriero, E. A. Zimmermann, S. J. Shefelbine and R. O. Ritchie. A Methodology for The Investigation of Toughness and Crack Propagation in Mouse Bone. Journal of the Mechanical Behavior of Biomedical Materials **39**, pp. 38 – 47, (2014).
5. S. Li, A.A Wahab, V.V Silberschmidt. Analysis of Fracture Process In Cortical Bone Tissue. Engineering Fracture Mechanics **110**, pp. 448–458, (2013).
6. British Standards Institue. Fracture Mechanics Toughness Tests. Method for Determination of K_{IC}, Critical CTOD and Critical J Values of Metallic Materials. BS 7448-1, (1999).
7. N. Dourado, F.A.M. Pereira, M.F.S.F. Moura, J.J.L. Morais and M.I.R Dias. Bone Fracture Characterization Using the End Notched Flexure Test. Materials Science and Engineering C33, pp. 405–410, (2013).
8. F.G.A. Silva, M.F.S.F de Moura, N. Dourado, J. Xavier, F.A.M. Pereira, J.J.L Morais and M.I.R. Dias. Mixed-Mode I+II Fracture Characterization of Human Cortical Bone Using The Single Leg Bending Test. Journal of The Mechanical Behavior of Biomedical Materials **54**, pp. 72–81, (2016).
9. T. L. Norman, D. Vashishth and D. B. Burr. Fracture Toughness of Human Bone Under Tension. J. Biomechanics, **28**, pp. 309–32(1995)0.
10. G. H. V. Lenthe, R. Voide, S. K. Boyd and R. Muller. Tissue Modulus Calculated From Beam Theory is Biased by Bone Size and Geometry: Implications for The Use of Three–Point Bending Tests to Determine Bone Tissue Modulus. Bone, **43**, pp. 717–723, (2008).
11. H. Ridha and P. J. Thurner. Finite Element Prediction with Experimental Validation of Damage Distribution in Single Trabaculea During Three-Point Bending Tests. Journal of The Mechanical Behavior of Biomedical Materials **27**, pp. 94–106, (2013).
12. F. Libonati and L. Vergani. Bone Toughness and Crack Propagation: An Experimental Study. Procedia Engineering **74**, pp. 464–467, (2014).
13. A. Carriero, J.L. Bruse, K.J. Oldknow, J.L. Millan, C. Farquharson and S.J. Shefelbine. Reference Point Indentation Is Not Indicative of Whole Mouse Bone Measures of Stress Intensity Fracture Toughness. Bone **69**, pp. 174–179, (2014).
14. O.L. Katsamenis, T. Jenkins and P.J. Thurner. Toughness and Damage Susceptibility In Human Cortical Bone Is Proportional to Mechanial Inhomogeneity At The Osteonal-Level. Bone, **76**, pp. 158–168, (2015).

Sustainable Biofuels and Other Related Bio-Products from Palm Cultivations

Farid Nasir Ani

Department of Thermodynamics and Fluid Mechanics, Faculty of Mechanical Engineering, Universiti Teknologi Malaysia, UTM 81310, Skudai, Johor Bahru, Malaysia

Abstract. Various kinds of palm trees are grown in the tropical regions of the world that provide edible oils for food consumption. They have in common in providing edible oils from the nuts and some from the mesocarp such as oil palm. More than 400 mills in Malaysia have provided more than sufficient energy from the biomass in processing the fruits to crude palm oil. The biomass are the mesocarp fibres, shells or endocarp, empty fruit bunches and palm oil mill effluents. One option of treatment of this biomass is to convert into bio-oil, bio-char, bio-adhesives and methyl ester. Recent research using microwave processing of biomass and biodiesel are given in this paper.

1 Introduction

Cultivated palm plantation produces biomass from their processing mills plays an important role in the daily activities in producing the food commodities. Generally, biomass also provides flexibility in shifting usage from food to energy and to material applications whenever needed, and has been well been demonstrated at large scale production. This paper describes the recent work on four possible routes to provide energy and other related products from biomass. The products are bio-oil, bio-char, methyl ester (B100) and biolubricants. These products can be further converted into a range of finer products with several other conversion processes.

2 Oil palm biomass availability and characteristic

The palm oil mills and its plantations produces the enormous amount of biomass that includes empty fruit bunches (EFB), shells (OPS), fibers (OPF), effluents (POME), trunks and fronds. In year-end 2012, the palm oil mills generated approximately 95.21 million tonnes of oil palm waste of which comprises of EFB, 21.90, OPF 12.38, OPS 5.71, POME 55.22 mtonnes, [1]. The composition of OPS are 31.0 cellulose, 20.0 hemicellulose and 49.0 lignin, OPF are 40.0 cellulose, 39.0 hemicellulose and 21.0 lignin and EFB are 40.0 cellulose, 36.0 hemicellulose and 24.0 lignin, [2].

Ag Agricultural wastes are usually of high moisture content and low in bulk density and consequently have relatively low calorific values. The chemical and physical characteristics of oil palm biomass are presented in Table 1a, 1b and 1c. The range for the oxygen content of biomass is 38 - 45% on moisture and ash free basis. The result of high oxygen content leads to relatively low calorific value of 14 – 20 MJ/kg.

Table 1a. Typical analysis of oil palm biomass.

Material	Elemental composition % wt (dry ash free)			
	C	H	N	O
Shells	55.35	6.27	0.37	38.01
Fibers	52.89	6.43	1.08	39.6
Bunches	47.89	6.05	0.65	45.41

Table 1b. Typical analysis of oil palm biomass.

Material	Proximate analysis % wt (air dry)			
	C	H	N	O
Shells	55.35	6.27	0.37	38.01
Fibers	52.89	6.43	1.08	39.6
Bunches	47.89	6.05	0.65	45.41

Table 1c. Typical analysis of oil palm biomass.

Material	Gross C.V. MJ kg^{-1}	Ave. bulk density kg m^{-3}
Shells	19.56	440 (size<18mm)
Fibers	19.15	-
Bunches	17.83	-

2.1 Bio-oil initiative

The interest on fast pyrolysis at UTM using fluidized bed reactor was developed with capacity of 0.6-1.0 kg/h was initiated in early 1990s, [3]. Later, a 6 kg/h continuous fast pyrolysis rig was designed, fabricated and operated,

 DOI: 10.1051/matecconf/20167711005

[4]. In 2005, Genting Sanyen Bhd, using BTG technology, installed a commercial rotating cone pyrolysis plant which runs on EFB with capacity of 2000 kg/h were operated in Malaysia, [5]. Salema and Ani [6] used microwave assisted pyrolysis of OPS and OPF with different bio-char ratio. The use of bio-char with OPS and OPF was found to initiate the pyrolysis conditions. It concluded that the low microwave technique gave significant energy saving route for bio-oil production. Abubakar introduced an overhead stirrer in the MW pyrolysis technique to reduce bio-oil deposition problem within MW reactor, [7]. The inert gas is supplied from the top side of the reactor and subsequent quenching of the vapors at the bottom side with vertical condensers. The phenols content was found much higher with 85 % under GC-MS area. Ani, Salema and Hassan characterized and compared the pyrolysis bio-oil from OPS and EFB using various pyrolysis techniques. The phenolic yield in bio-oils was found to be dependent on the process temperature. The OPS bio-oil produced higher fractions of phenol and its compounds, as compared to EFB bio-oil [8].

2.1.1 Bio-oil properties

In general, bio-oil is dark brown in colour and usually slightly heavier than water with high viscosity at room temperature. The energy content of the bio-oil obtained ranges from 16.25 to 22.5 MJ/kg. This value has lower heating value than fuel oil due to the presence of some highly oxygenated components in the bio-oil. Bio-oil has high acidity, pH of 2.42 as a result of acetic and formic acids contents. It has low pour point and high flash point due to the existence of low boiling volatile compounds with a considerably high amount. Bio-oil from oil palm shells consists of significantly high fraction of phenol and phenolic compounds, with concentrations up to 50.54 area % and 22.21 area % respectively. Bed temperature was found to be the key parameter affecting the phenol and phenolic compounds. The bio-oil is complex mixtures of various chemical compounds. Almost 50 kinds of major chemical compounds were identified from the OPS bio-oil samples from which only important components. The phenol percentage in the OPS bio-oil for second study was higher (50 wt%) compared to earlier work which was about 22 wt % at 490 °C.

The phenolic concentrations in bio-oil vary from various pyrolysis technics. The content of phenol from MW pyrolysis system was found to be the highest of about 62 area % in case of EFB and 84 area % in case of OPS. Basically, the process condition associated with the pyrolysis technologies can affect the formation of chemical compounds in the bio-oil, particularly, the bed temperature. The EFB bio-oils obtained from fluidized bed also contained a high amount of phenolic compounds, [9]. It was interesting to find that high amount of phenol can be obtained through MW pyrolysis without any prior treatment of EFB biomass.

2.1.1 Bio-oil adhesives

Bio-oil contains phenol and phenolic compounds which could be used in fine chemicals, pharmaceuticals, food favourings and biopolymers. Recent research on the utilisation of OPB bio-oil as resins and adhesives were conducted by Md Kawser and Ani [10], Wong and Ani [11], Wong and Ani [12], Lai, Ani and Hassan [13]. The bio-oil from the earlier work was extracted using solvent extraction technique to gain the useful phenol and phenolic compounds. The extracted oil-palm-shell-based phenol was used in the manufacture of phenol formaldehyde wood adhesives. It was found that, the wood bonding performances between petroleum-based adhesives and the oil-palm-shell-based adhesives, were comparable and acceptable. Wong and Ani [12]. Abubakar, Salema and Ani [7] produces phenol rich (85% area) bio-oil from OPS using microwave assisted pyrolysis and suggested that it could be used directly for resin and adhesives application. Recently, Choi et. al. suggested that bio-oil can typically substitute for fossil phenol but, the main concern in the production of phenolic resins, appears to be the substitution level of bio-oil for fossil phenol [14].

In general, bio-oil is dark brown in colour and usually slightly heavier than water with high viscosity at room temperature. The energy content of the bio-oil obtained ranges from 16.25 to 22.5 MJ/kg. This value has lower heating value than fuel oil due to the presence of some highly oxygenated components in the bio-oil. Bio-oil has high acidity, pH of 2.42 as a result

3 Oil palm biochar activities

The reflection, absorber and dielectric properties of oil palm bio-char and their composites are given in Ani et. al. [15], Yusof et. al. [16] and Salema et. al. [17]. The comparison of BET surface area between microwave and conventional heating was studied by Abas and Ani [18]. It shows that biochar produced via microwave heating technique has higher quality as compared to conventional heating pyrolysis in which significant cracks and fissure due to convective heating profiles and differences in temperature of outer and inner surfaces which leads to more fragile biochar.

The observation of BET surface area of biochar produced from microwave heating system was investigated by Guo and Lua, [19], Salema and Ani [20] and Foo and Hameed, [21]. The highest surface area (255.7 m^2.f(v hsg `udq`fd onqd rhyd ne 11-21 @ was detected from biochar from EFB which observed by Foo and Hameed, [21]. It should be noted that, the surface area of biochar produced from OPF via microwave heating which was 205.21 m^2/g, much lower than conventional heating [21] and it was observed that the lowest surface area of 194.3 m^2/g was detected in biochar from OPS by Guo and Lua, [19].

3.1. Activated carbon

The quality of biochar is enhanced using a physical activation after carbonation process. The applications of activated carbon are numerous, from air cleaning to water treatment. Both processes of carbonation and activation could be done using microwave assisted process with temperature ranging up to 900 °C. The temperature in the reactor was closely controlled by a temperature controller [22]. Studies show that chemical activation give higher BET surface areas using less MW power. Also, despite using lower MW power, other parameters also contributed signicantly to the development of the porosity in the activated carbon, such as activated time, temperature, impregnation ratio and agent.

3.2 Carbon molecular sieves

Oil palm shells have been processed to produce microporous carbon adsorbents for air separation [23]. The pore structures of CMS obtained show excellent separation capability of the O_2/N_2 selectivity was as high as 48. This kinetic selectivity of air was greatly influenced by the carbonization temperatures, although the micropore surface area and volume were lower and causing less O_2 uptake capacity. In another study, CO_2 removal from natural gas, biogas and landfill gas which enhance its calorific value, the adsorbents prepared had shown their capability in adsorbing CO_2 and CH_4 but were more selective towards CO_2. It was discovered that the char sample had shown the best molecular sieving effect for the separation of CO_2 and CH_4 where the selectivity value was the highest in the case for the separation of CO_2 from CH_4 gas mixture, [22].

3.3 Methyl ester production

Palm oil has successfully been processed into palm oil methyl ester. It had been used as B05 commercially in unmodified engines of buses, truck, taxies and cars in Malaysia. The utilization of palm oil as fuel has realized the competition between human needs in food and non-food applications. It utilization as a renewable source of fuel is a cost factor due to the planting and production cost of the crude oil. In addition, high cost of heating using cogeneration of steam and electrical bring the production cost of methyl ester higher. Thus, an alternative method, microwave irradiation was applied for the production of methyl ester. A continuous transesterification of plant oils was used in a modified microwave oven. The physical properties of waste oil methyl esters met the biodiesel standards.

An average of 2.128 kW hr of electrical energy could be generated from one kg of B100, while the production of one kg B100 needs 0.468 kWhr electricity. Thus, there is a potential for production of 1.659 kWhr extra electrical energy per kg biodiesel which it proves the sustainability of the system. A sustainable processing system was obtained using this system for diesel engines power generation [24]. The analysis was also supported by Choedkiatsakul et al [25], which utilized even lesser energy consumption of 0.1167 kWh/L of B100.

3.4 Research in bio-lubricants

The increase of worldwide concerns about health, the environment and limited petroleum resources has promoted the use of biodegradable and non toxic products for lubricants and hydraulic oils. Animal fats and vegetable oils are considered as renewable substitutes for the mineral-based oil as lubricants. Vegetable oils reveal good lubricating abilities because they produce low coefficients of frictions but has higher wears rate. However, some factors such as the oxidation of vegetable oil must be taken into consideration, Hassan et.al. [26]. In his recent study, he found that the blend of RBD palm olein with mineral engine oil showed good reliability with regard to viscosity parameter. The blend offers better performance because it has a smaller WSD and higher FTP compared to the other oil samples. It had also been proven that the blend reduces material loss of around 48.72 and 41.70% compared to the neat RBD palm olein and neat mineral engine oil, respectively. Hence, it is proved that this blend has the potential as an antiwear lubricant.

4 Conclusions

Palm cultivation holds an important asset for its food sources and other applications from its oil and biomass. However, the choice to use biomass for power, fuel, and biomaterials depends on a variety of factors, such as availability, public policy, cost of biomass, capital cost of processing equipments, infrastructure facilities and markets for alternative energy and materials. The development of advanced conversion technologies such as microwave assisted processing is likely to make biomass both energy and materials products competitive with petroleum-based products. The integrated processing plant of the palm fruits using excess energy for modular processing could be used in the future for the palm mills operation.

Acknowledgement

The author wishes to acknowledge the MOSTI and the MOHE, Malaysia for the research grants obtained and RMC, UTM for the management support during the course of these studies.

References

1. GGS (2013) http://www.ggs.my/index.php/palm-biomass (accessed 23rd Oct. 2013).
2. F.N. Ani, Sustainability and Recycling through Thermal Conversion of Bioresources, *Professorial Inaugural Lecture Series, Penerbit UTM Press.* (2012)
3. M.N. Islam, R. Zailani, and F.N. Ani, Pyrolytic oil from fluidized bed pyrolysis of oil palm shell and its characterisation, *Renewable Energy*, **17**, pp. 73-84 (1999)

4. X.Y. Lim and F.N. Ani, Continuous Fast Pyrolysis of Oil Palm Shell for Bio-Oil Production, *JSPS-VCC Group Seminar 2006, Natural Resources and Energy Environment, 11 –12 December 2006, Johor Bahru, Malaysia*, pp 98-106 (2006)
5. R.H. Venderbosch, Gansekoele, E., J.F. Florijn, D. Assink, and H.Y. Ng, Pyrolysis of palm oil residues in Malaysia, *PyNe Newsletter*, Issue 19, pp. 2-3 (2006)
6. AA. Salema and F.N. Ani, Microwave induced pyrolysis of oil palm biomass, *Bioresource Technology*, **102** (3), 3388-3395 (2011)
7. Z. Abubakar, AA Salema, F.N. Ani, A New Technique to Pyrolyse Biomass in a microwave system: Effect of Stirrer Speed, *Bioresource Technology*, **128** (0), 578-585 (2013)
8. F.N. Ani, A.A. Salema, I. Hassan, Bio-oils characteristic from oil palm biomass from different fast pyrolysis techniques. *Applied Mechanics and Materials*, 554, 266-270 (2014)
9. S.W. Kim, B.S. Koo, J.W. Ryu, J.S. Lee, C.J. Kim, D.H. Lee, G.R. Kim, and S. Choi, Bio-oil from the pyrolysis of palm and Jatropha wastes in a fluidized bed, *Fuel Processing Technology*. **108**, 118-124, 2013-Special Issue of APCRE 11. (2012)
10. J. Md Kawser and F.N. Ani, Oil Palm Shell as a Source of Phenol, *Journal of Oil Palm Research*, **Vol 12**, No 1, June, 2000, pp 86-94 (2000)
11. C.C. Wong and F.N. Ani, Properties of Oil Palm Shell Based Phenol Wood Adhesives Compared with Petroleum Based Phenol Wood Adhesives, *Journal of Oil Palm Research*, PORIM, **Vol 16**, No 1, June 2004, pp 24-35 (2004)
12. C.C. Wong and F.N. Ani, Feasibility Study on the Potential Production of Oil Palm Shell Bio-Phenol, *Oil Palm Industry Economic Journal*, **Vol 7**, No 2, 28-33 (2007)
13. J.C. Lai., F.N. Ani and A. Hassan, Water absorption of lignocellulosic phenolic composites, *Polymers and Polymer Composites*. 16, 6, p. 379-388 (2008)
14. C. Gyung-Goo, O. Seung-Jin, L. Soon-Jang, K. Joo-Sik, Production of bio-based phenolic resin and activated carbon from bio-oil and biochar derived from fast pyrolysis of palm kernel shells, *Bioresource Technology*, **178**, 99–107 (2015)
15. F.N. Ani, W.K Wan Ali, AA. Yusof, Effect of oil palm shell carb.on–polyester resin composite on microwave absorber properties, *Jurnal Mekanikal 19*, 1-10 (2005)
16. A.A. Yusof, W.K. Wan Ali, T.A. Rahman, F.N. Ani, *Microwave and reflection properties of palm shell carbon–polyester conductive composite absorber* (2005)
17. A.A. Salema, Y.K. Yeow, K. Ishaque, F.N. Ani, M.T. Afzal, A. Hassan, Dielectric properties and microwave heating of oil palm biomass and biochar, *Industrial Crops and Products* **50**, 366-374 (2013)
18. F.Z. Abas and F.N. Ani, Comparing characteristics of oil palm biochar using conventional and microwave heating, *Jurnal Teknologi (Sciences and Engineering)*. **68**, 3, p. 33-37 (2014)
19. J.Guo, and A. C. Lua.), Preparation Of Activated Carbons from Oil-Palm-Stone Chars By Microwave-Induced Carbon Dioxide Activation. *Carbon* **38**(14): 1985-1993 (2000)
20. A.A. Salema and F.N. Ani, Microwave-Assisted Pyrolysis of Oil Palm Shell Biomass Using an Overhead Stirrer. *Journal of Analytical and Applied Pyrolysis* **96**:162- 172 (2012)
21. K. Y. Foo, and B. H. Hameed, Microwave-Assisted Preparation of Oil Palm Fiber Activated Carbon for Methylene Blue Adsorption. *Chemical Engineering Journal* 166(2): 792-795 (2011)
22. J.S. Tan, Preparation of Microporous Carbon Adsorbents from Oil Palm Shell by Physical and Chemical Treatments for Gas Separations, *MEng Thesis*, Faculty of Mechanical Engineering, Universiti Teknologi Malaysia (2001)
23. J.S. Tan and F. N.Ani, Carbon molecular sieves produced from oil palm shell for air separation, *Separation and Purification Technology*. **35**, 1, p. 47-54 (2004)
24. F. Motasemi and F. N. Ani, A review on microwave-assisted production of biodiesel, *Renewable and Sustainable Energy Reviews*. **16**, 7, p. 4719-4733 (2012)
25. I. Choedkiatsakul, K. Ngaosuwan, S. Assabumrungrat, S. Mantegna, G. Cravotto, Biodiesel production in a novel continuous flow microwave reactor, *Renewable Energy*, 83, 25-29 (2015)
26. M. Hassan, F.N. Ani, S Syahrullail, Tribological features of refined, deodorized and bleached palm olein with mineral oil blend, *Tribology Transactions*, 1-8 (2015)

A Study of the Influence of Gas Channel Parameters on HT-PEM Fuel Cell Performance Using FEM Analysis

Viorel Ionescu[1,2] and Nicolae Buzbuchi[3]

[1]*Department of Physics and Electronics, Ovidius University, Constanta, 900527, Romania*
[2]*Doctoral School of Mechanical Engineering and Mechatronics, Maritime University, Constanta, Romania, 900663*
[3]*Department of Engineering Sciences, Maritime University, Constanta, Romania, 900663*

Abstract. Proton Exchange Membrane Fuel Cells (PEMFC) are highly efficient power generators, achieving up to 50-60% conversion efficiency, even in sizes of a few kilowatts. Comsol Multiphysics, a commercial solver based on the Finite Element Method (FEM) was used for developing a three dimensional model of a high temperature PEMFC that can deal with both anode and cathode flow field for examining the micro flow channel with electrochemical reaction. Cathode gas flow velocity influence on the cell performance was investigated at first. Polarization curves for three different channel widths (0.8,1.6 and 2.4 mm) and three different channel depths (1,2 and 3 mm) were computed at a cathode inlet flow velocity of 0.06 m/s. Oxygen molar concentration at cathode catalyst layer-GDL channel interface and local current density variation along the cell length were also studied for specific gas channel geometries.

1 Introduction

Low temperature proton exchange membrane fuel cells(LT-PEMFCs) are in the present used as energy converters, especially for automobile applications, being used in a clean hybrid power supply system and providing the rated power to the load [1], [2]. The high capital cost of platinum catalyst layer and of pure hydrogen fuel, reduced lifetime of 5000 working hours are limiting the wide utilization of LT-PEMFC. It is also difficult to maintain proper water and thermal management, fuel cell performance being strongly affected [3,4]. High ohmic losses in the cell were observed when membrane become dehydrated as water removal rate exceeds water generation rate [5].

There are research advances of high temperature PEM fuel cells (HT-PEMFCs) with polybenzimidazole (PBI) - based membrane for usage in combined heat and power generation systems (FC μCHP) due to various reasons [6]: enhancing of the electrochemical kinetics for the cell reactions, the simple cooling system based on the increased temperature gradient between the fuel cell stack and the coolant, increasing of the CO tolerance which enables the utilization of lower quality reformed hydrogen.

Various studies have devoted to developing mathematical bi-dimensional models for the transport of reactants and product water in a PEMFC. Anode and cathode GDL two phase flow in the fuel cell were described by Berning and Djilali with the help of multi-phase and multi-component model, using unsaturated flow theory (UFT) [7].

The 3D HT-PEMFC numerical model used in this paper is derived from 3D single-phase isothermal model developed by E.U. Ubong *et al.* [8] to predict the performance of a HT-PEMFC with a PBI membrane.

The objective of this work is to investigate the effect of cathode gas flow velocity and gas channel geometry (width and height) on the cell performance with the help of a 3D model using commercial FEM package Comsol Multiphysics (version 4.2).

2 Model Set-up

The 3D computational domain includes a section of the PBI membrane and both cathode and anode gas flow channels, GDLs, and catalyst layers (see Fig. 1).

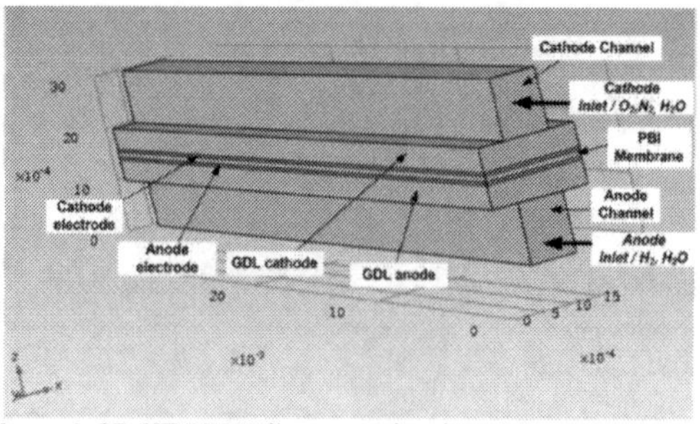

Figure 1. 3D HT-PEMFC computational model geometry.

The operating conditions and parameters of the base model are given in Table 1 and Table 2.

Table 1. Operating parameters

Parameter	Value
Operating conditions	
Reference pressure (Pa)	101325
Cell voltage (V)	0.9
Temperature (°C)	150
Oxygen reference concentration (Mol/m^3)	40.88
Hydrogen reference concentration (Mol/m^3)	40.88
Inlet H_2 mass fraction (anode)	0.743
Inlet H_2O mass fraction (cathode)	0.18
Inlet O_2 mass fraction (cathode)	0.228
Gas pair diffusivities and basic fluid parameters	
H_2-H_2O Binary diffusion coefficient (10^{-4} m^2/s)	1.603
N_2-H_2O Binary diffusion coefficient (10^{-5} m^2/s)	4.48
O_2-N_2 binary diffusion coefficient (10^{-5} m^2/s)	4.18
O_2-H_2O binary diffusion coefficient (10^{-5} m^2/s)	4.91
Anode inlet flow velocity (m/s)	0.17
Cathode inlet flow velocity (m/s)	0.42
Anode viscosity (10^{-5} Pa·s)	1.12
Cathode viscosity (10^{-5} Pa·s)	2.68

Table 2. Design parameters

Parameter	Value
Gas channel dimensions	
Cell length (cm)	2
Channel width (mm)	0.8
Channel depth (mm)	1
Width of the shoulder (mm)	0.9
Catalyst layer information [9]	
Thickness (μm)	50
Permeability (m^2)	2.36×10^{-12}
Porosity	0.2
Open volume fraction for gas diffusion	0.4
Membrane PBI/H_3PO_4 *properties [10]*	
Thickness (μm)	98
Conductivity (S/m)	1.74
GDL properties [11]	
Conductivity (S/m)	222
Bulk porosity	0.4
Permeability coefficient ($\times 10^{-11}$ m^2)	1.8
Thickness (μm)	380
Conductivity (S/m)	222

At steady state, single-phase, isothermal model of PEM fuel cell consists of five principles of conservation: mass, momentum, species, energy and charge. COMSOL Multiphysics is used to solve this complex HTPEM fuel cell model. The conservation equations are solved sequentially (and grouped) for various variables, such as: velocity vector, pressure, mass/mole fraction of the species, electric and protonic phase potentials, solid and fluid – gas – phase temperatures.

Momentum transfer is modeled using Free and Porous Media Flow interface.

Laminar flow in the channels is described by the Navier-Stokes equations (dimensionless formulation), for the steady state in case of no external forces:

$$(v\cdot\nabla)v+\nabla p-\frac{1}{R_e}\nabla\left(\nabla v+(\nabla v)^T\right)=0;\quad \nabla\cdot v=0 \tag{1}$$

where unknown depended variables are p – pressure and v – velocity; R_e is a dimensionless Reynolds number.

Porous gas diffusion layers (GDLs) and electrode flow can be given by a similar set of differential equations, when the Brinkman formulation is used:

$$kv+\nabla p-\frac{1}{R_e}\nabla\left(\nabla v+(\nabla v)^T\right)=0 \qquad \nabla\cdot v=0 \tag{2}$$

where k is the Brinkman parameter, defined as: $k=\frac{H^2}{R_e\cdot\kappa}$, with H – channel height and κ - permeability.

Conservation of species was solved for the flow channels, GDLs and porous electrode using the Maxwell Stefan equations in two different Transport of Concentrated Species interfaces.

It solves for the fluxes of each species (H_2 and H_2O in the anode compartment, O_2 and H_2O in the cathode compartment) in terms of mass fraction. The Maxwell-Stefan equation, defined for each component of the mixture of the three gases, is presented below [11]:

$$\nabla\cdot\left[-\rho\omega_i\sum_{j=1}^{3}D_{ij}\left\{\frac{M}{M_j}\left(\nabla\omega_j+\omega_j\frac{\nabla M}{M}\right)+\left(x_j-\omega_j\right)\frac{\nabla p}{p}\right\}+\omega_i\rho v\right]=0 \tag{3}$$

In Eq. (3), x_j is the molar fraction of each gas j, parameters ω_i and ω_j are the mass fractions of gases i and j, respectively, parameter ρ is the overall mass density of the air mixture obtained from the ideal gas law, D_{ij} is the binary diffusion coefficient, M is the total molar mass of the mixture and M_j is the molecular weight of gas j.

Conservation of the electric charge is based on two currents: an ionic current formed by the protons travelling through the membrane and an electronic current formed by the electrons passing through the solid matrix of electrodes. The current continuity equations are obtained by using Ohm's law [8].

3 Results and discussions

Fig. 2 shows the effect of changing the cathode gas flow velocity at values under 0.42 m/s on the fuel cell performance. When the gas flow velocity is increased from 0.02 to 0.12 m/s, the fuel cell performance is clearly enhanced, especially at lower operating fuel cell voltages.

The rate of the electrochemical reaction is increased due to the increase in oxygen gas through the gas diffusion layer to reaction sites. Due to the low membrane humidification, this enhancing performance effect is minimal at high operating voltages.

So, the air stream is able to supply oxygen with the required rate for gas flow velocity values over 0.1 m/s in the case of the cell model with gas channel dimensions from Table 1.

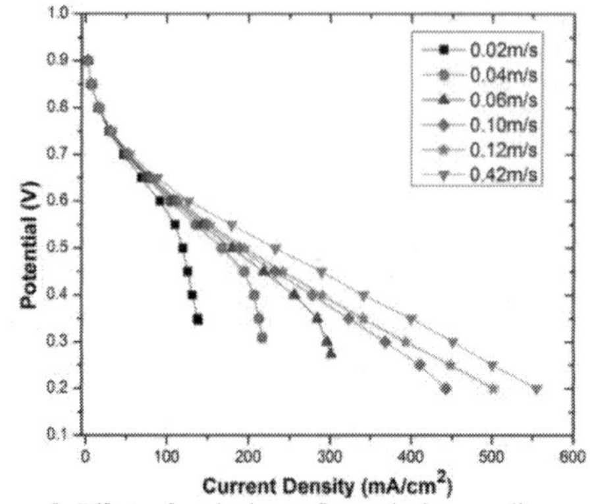

Figure 2. Effect of cathode gas flow velocity on cell performance at base conditions for different cathode inlet flow velocities (between 0.02 and 0.42 m/s).

Fig. 3 shows the effect of channel depth and width on the cell polarization curves for a cathode inlet flow velocity of 0.06 m/s. Cell performance was reduced to lower current densities at a channel depth of 1mm, with a limiting value of only 301 mA/cm^2 for 0.8x1 mm^2 geometry. So, the channel with the lowest depth presented a resistance highest enough to not be overcome by the buoyant force exerted by the fluid. With a channel depth of 2 and 3mm, fluid flow can take place easier through the channel and hence the cell operates at higher current densities.

For a given channel depth, the limiting current density was improved to increase in channel width. Maximum limiting current density of 484 mA/cm^2 at a cell voltage of 0.2V was obtained with a channel width of 1.6 mm for a channel depth of 3 mm.

The channel having the biggest width and lower flow resistance should induce air flowing into the channel more easily. They demonstrated that if the channel width is too big then the flow and the mass/heat transfer are also limited by the thermal boundary layer effect and the 3 mm gas channel with rib size and height fixed can achieve the best performance.

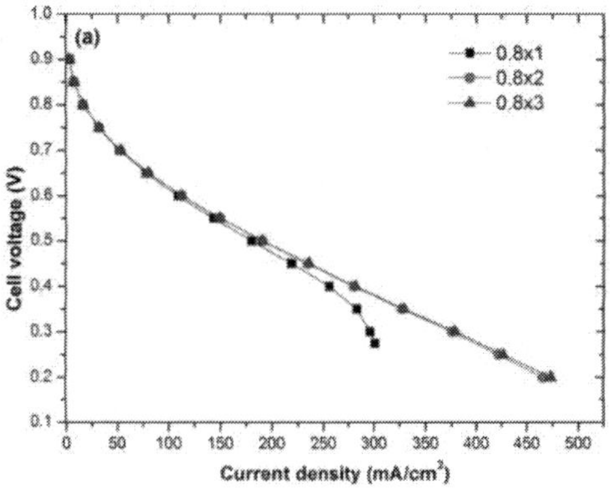

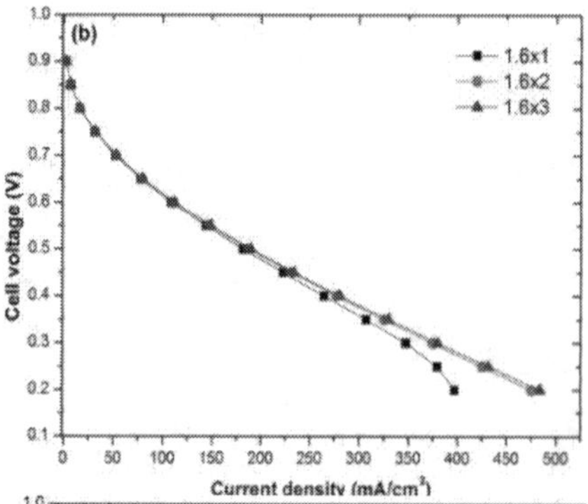

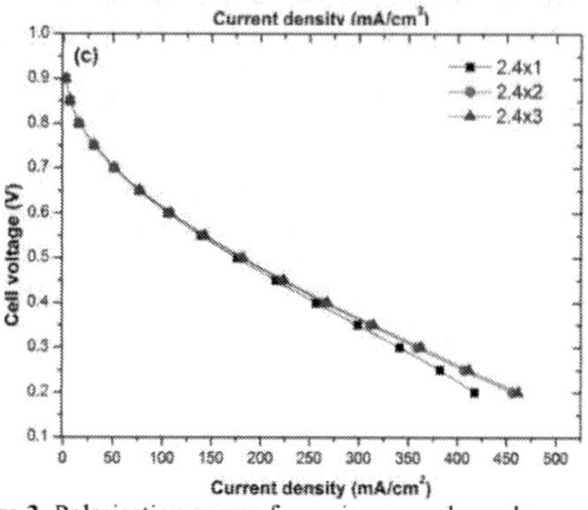

Figure 3. Polarisation curves for various gas channels geometries: a) 0.8x1 mm^2, 0.8x2 mm^2, 0.8x3 mm^2; b) 1.6x1 m^2, 1.6x2 mm^2, 1.6x3 mm^2; c) 2.4x1 mm^2; 2.4x2 mm^2, 2.4x3 mm^2.

Fig. 4 presented the variation of oxygen molar concentration in the cathode part of the cell model, for a channel with 1.6 mm width and 1 mm depth (see Fig. 4a) and with 1.6 mm width at 3 mm depth, respectively (see Fig. 4b). The point of interest in this figure is related to the concentration distribution in the cathode catalyst layer – GDL channel interface. Simulation was performed at a cell voltage of 0.3V. In Fig.4a oxygen molar concentration in the gas channel and at the catalyst layer - GDL channel interface is almost identical, whereas in Fig. 4b, we noticed that the molar concentration in the GDL – gas channel interface is higher than in the catalyst layer. The cell performance is limited in both cases for various reasons: in the case (a) channel presented a higher resistance to buoyancy flow and in the case (b) porous media offered a significant diffusion resistance.

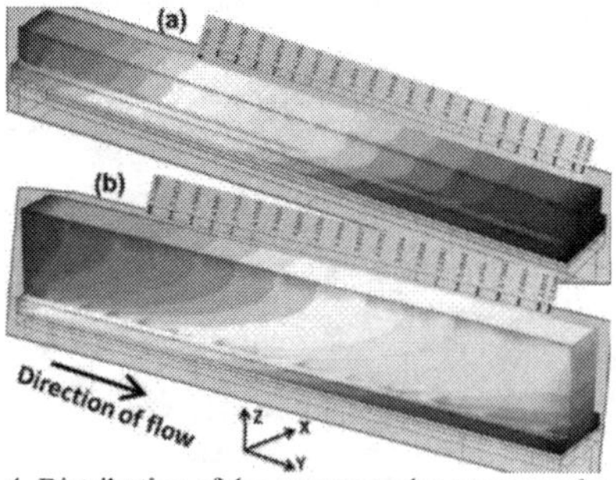

Figure 4. Distribution of the oxygen molar concentration (Mol/m^3) along the cathode side of the cell for: (a) 1.6x1 mm^2 and (b) 1.6x3 mm^2 gas channel geometries at a cell voltage of 0.3V.

Fig. 5a showed the local current density distribution along the cell height for a channel with 1.6 mm width and 1 mm depth at three polarization voltages. The local current production is very constant along the channel height at high voltages. At low voltages (0.25V, for example) the current was generated in higher amount at the beginning of the channel, decreasing across the channel height. Due to the buoyancy effect, the oxygen concentration is more at the channel inlet and for this reason current density is high here.

Fig. 5b presented the local current density variation across the cell height at 0.3 V for different channel depths: 1, 2, 3 mm. The channel width was preserved 1 mm for this study. It can be seen that the current production has the biggest variation for a channel depth of 1 mm. At higher channel depths, a superior amount of oxygen enters the cathode channel.

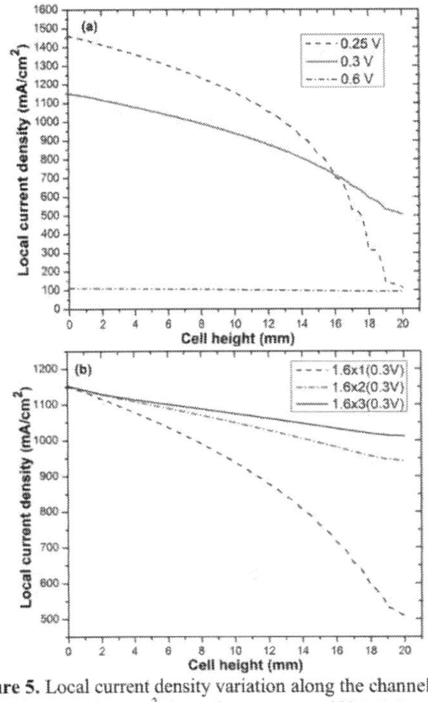

Figure 5. Local current density variation along the channel height for: a) 1.6x1 mm^2 channel geometry and b) 1.6x1 mm^2, 1.6x2 mm^2, 1.6x3 mm^2 channel geometries.

4 Conclusions

In order the study the performance of a HT-PEM fuel cell under varying gas channel parameters, a steady-state 3D computational model was implemented here using Comsol Multiphysics software.

The results showed that the cathode gas flow velocity affects cell performance at values under 0.1 m/s. Optimized gas channel geometry was established from the polarization curves of the cell, at a channel width of 1.6 mm and channel depth of 3 mm.

The high flow resistance given by the gas channel depth of 1 mm reduced the cell performance and diffusion resistance presented by the porous media in the case of 3 mm depth channel limited also the cell efficacy.

Local current density presented a uniform variation along the cell having an optimized gas channel geometry due to more amount of oxygen entering the channel.

References

1. F. Segura, J. M. Andujar, E. Duran, IEEE Trans. Ind. Electron. **58**, 1171, (2011).
2. J. Jia, G. Wang, Y. T. Cham, Y. Wang, M. Han, IEEE Trans. Ind. Electron. **57** 1945, (2010).
3. T. V. Nguyen, *ECS Trans*. **3**, 1171, (2006).
4. H. Abtahi, A. Zilouchian, A. Saengrung, 2005 IEEE International Conference on Systems, Man and Cybernetics **4**, 3486, (2005).
5. 5. G. Petrone and G. Cammarata, *Modelling and Simulation*, In Teach Education and Publishing, Croatia, **677**, (2008).
6. J. Zhang, X. Xie, Y. Tang, C. Song, T. Navessin, Z. Shi, D. Song, H. Wang, D.P. Wilkinson, Z.S. Liu, S. Holdcroft, J. Power Sources **160**, 872, (2006).
7. T. Berning, N. Djilali, J Electrochem Soc, **150**, 1589, (2003).
8. E.U. Ubong, Z. Shi, X. Wang, J Electrochem Soc **156**, 1276, (2009).
9. D.F. Cheddie, N.D.H. Munroe, J. Power Sources **160**, 215, (2006).
10. H.L. Lin, T.L. Yu, W.K. Chang, C.P. Cheng, C.R. Hu, G.B. Jung, . Power Sources **164**, 481, (2007).
11. T. Henriques, B. César, P.J. Costa Branco, Applied Energy **87**, 1400, (2010).

MATEC Web of Conferences 77, 12002 (2016)
ICMMR 2016 DOI: 10.1051/matecconf/20167712002

Simulation of the Battery Pack Exploitation Based on Real Measurement Data of Applied Chemical Cells

Marcin Koniak , Andrzej Czerepicki, Piotr Tomczuk[1], and Burak Şamşul[2]

[1]*Warsaw University of Technology, Faculty of Transportation, 00-662, Koszykowa 75, Warsaw, Poland*
[2]*Milper Propeller Technologies Inc., Teknopark Bulvarı - 1/2A Z08 34906, Pendik – İstanbul*

Abstract. Authors of following article present the method of battery work simulation based on real measurement data of applied chemical cells and workload data. Description of method as its assumptions are presented. Next authors present workplace designed and constructed for acquiring operating characteristics according to assumed research plan. Article introduces computer system responsible for running simulations. At the end of publication results of selected runs are presented. Also described are types of tested cells and means of acquiring data concerning battery workload. Article ends with summarization and approximate plans of method development.

1 Introduction

Development of renewable energy sources and electric cars is inherent with use of batteries. Batteries applied in energetics have the task of storing energy that comes from low usage and overproduction from renewable sources and release it in consumption peak or during unfavourable weather, that causes lowered generation by wind turbines and solar panels. They can also function as a power peak check in nodes, breaking capacity and regulator of: frequency, voltage and power factor.

In case of electric vehicles battery is the only one, in other solutions main or as in hybrids supporting energy source. But in every single application its cost is substantial, and choice must be specifically made based on data closest to actual. Such choice is possible based on analysis of battery work in real operating conditions. In following article authors propose battery work simulation method based on characteristics of storage and actual measurement data used in chemical cells storage.

Described method was developed and utilized in international project "Renewable Energy Powered Hybrid Innovative Sailing Yacht" goal of which is development of yacht with electric engine powered from battery and characterized by possibility of charging from renewable source such as sun and wind.

2 Method description

There are many models describing battery work, among them: electric, electrochemical, analytical and stochastic. But a barrier for industrial application are simplifying assumptions and complexity of calculating procedure. Many of them require experimental data input, which can be only acquired with help of highly specialized lab tests.

Literature analysis shows, lack of universal methods that determine battery operating characteristics based on unified series of data for different technologies. Existing methods were not tested for newest technologies such as LFP, or lithium-titanate cells.

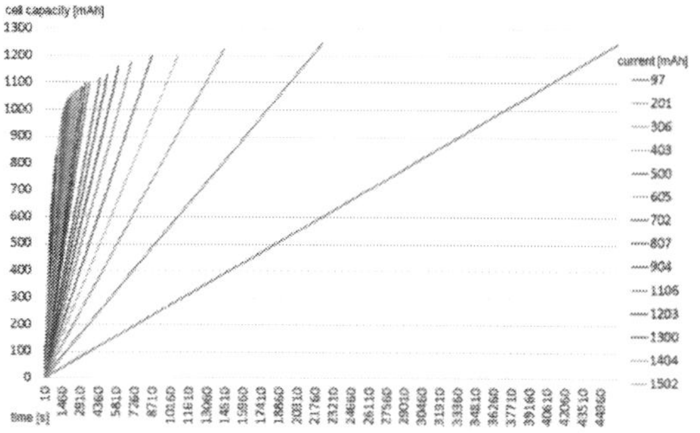

Figure 1. The dependence of available capacity, depending on the charging current as a function of the charging time.

The need for creation of a model projecting work of any battery was stated, for which input data can be acquired simply with help of universal research plan. In proposed solution fitting characteristic from a family of characteristics is plotted in response to instantaneous current of a cell, example of which can be found in Fig. 1. For each change in a current occurs a change of a characteristic that a process follows. Method assumes using real measurement data. So solution was developed and proposed that digitally models operating characteristic of a battery. Method assumes possibility of simulating whole battery by graduating a result of work simulation for a single cell. Possibility of such graduation was confirmed by a literature research [1], [2] and is a result of a previous experience team shares in that matter. That angle takes into account such characteristics as:

- Available cell capacity related to charge/discharge current,
- Actual energy losses related to cell heating up,
- Change in a cell temperature related to operating current, this parameter is substantial for safety issues and allows to decide about addition of a cooling system to a battery.

For cell operation rating and proper selection of cell technology for a battery following data are essential:

- Operating temperature range,
- Allowed operating currents,
- Safety requirements,
- Service requirements,
- Assumed tray cost
- Influence of charging and discharging current on available capacity,
- Influence of temperature on available capacity,
- Correlation between lifetime and number and depth of ran work cycles [3].

Not all of the parameters above are available in manufacturers' materials. Only way of acquiring those missing data is running our own tests.

Running simulations of cells work would not be possible based only on manufacturers' catalogue data. So developing research plan was necessary, during which all necessary characteristics would be acquired. Such plan was made, also following criteria that must be met by measurements:

- Time constant of consecutive measurements for each cell must be as close as possible,
- Each measurement must consist of following data:
- Timestamp,
- Cell temperature,
- Work current,
- Energy sent/received,
- Cell voltage.

Key recorded parameters are described below:

2.1 Current

Operating current is a basic component influencing utilization of battery. Cell operating temperature, capacity and lifetime depend on its value. Because of that, range and step of measurements must be accordant to values in which cell will be eventually applied.

The first step is to find current from which measurements will start and it can be set at 0,1 A.

Next, the step between consecutive measurements must be set and it can also be 0,1 A.

Tests will be conducted until reaching: Maximal charging/discharging current permissible by manufacturer, or maximal assumed operating current in a pack for a given cell.

2.2 Time

Measuring time must be registered from a moment of its start until the end of the process. 5 s can be assumed as a time step between consecutive measurements. That value seems enough because changes occurring in cells have a slowly changing character. This parameter is necessary to control the progress of measurement data acquirement and duration of measurement, it also allows to verify completeness of collected results.

2.3 Voltage

Each cell must be tested only in the range given as safe in data provided by manufacturer. It will be confirmed in what range cell reaches capacity given by manufacturer and in that range it will be tested. Essential is also the way of voltage measurement, because measurement on the output of the device can be affected by error connected to voltage drop on power cords and all occurring connections eg. installed fuse or connectors.

2.4 Number of work cycles

Each type of cells will undergo tests that are intended to define actual number of work cycles. Those tests will consist of running series of charge/discharge measurements with given set operating currents, until reaching capacity loss equal to 30% of starting capacity. Those tests are planned for specified discharging depths, reflecting operating at full capacity, half and so called micro-cycles.

2.5 Measurements results

The measurements result for each cell is a catalog containing full measurement dataset acquired for each current. Tests will be run on workplace consisting of 4 independent measurement sets. Each of them will be running tests according to its own plan, adapted to operating parameters given by manufacturer. Tests monitoring will be possible from the computer.

3 Research workplace

It is necessary to run series of measurements in order to acquire data, that will be used in creation of individual cell types models. Those measurements must meet described earlier criteria.

Because of time-consuming research and necessary changes in operating currents in charging – discharging cycles, workplace must be controlled by computer with possibility of remote preview of results and operating conditions via Internet.

Two kinds of research are planned to take place at workplace:

- Accumulating work characteristics, which will become database for battery work simulation,
- Running series of charging and discharging measurements with specified operating currents, until reaching 30% loss of maximum capacity – cell lifetime tests.

Because of quality of measurement data it is necessary that environment in which cells are keeps constant temperature. Also temperature of cells must be constantly registered and monitored. In case of breaching assumed thresholds test must be cancelled, because of safety reasons.

Based on market analysis and specifications, integrated charger with specially modified for us firmware was selected as system that carries charging/discharging process. It can also be controlled via the serial port. It combines executive (programmable power supply, programmable workload) and measuring (currents, voltage, energy sent/received, temperature) systems. It assures processes to be executed in required range of currents and voltages for selected cells.

To ensure continuation of process in case of power blackout and to enable energy return from discharging cell, additional batteries are used in the power supply system. Each of four chargers is supplied from separate set: lab power supply and 12 V lead-acid battery as an energy buffer.

Keeping constant temperature of the process is electronically controlled refrigerator with Peltier cells.

With the controlling computer case is (as with chargers) keeping in mind supporting its work during blackouts, so laptop was selected because of built-in high capacity battery.

Finished measurement workplace is shown in Fig. 2. From the left: four programmable chargers, device keeping process temperature, controlling computer and batteries with power supplies.

Figure 2. Laboratory test bench to collect the operating characteristics of chemical cells

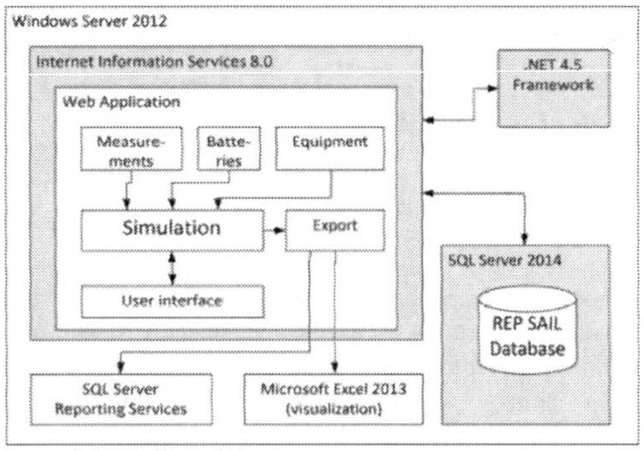

Figure 3. REP-SAIL IT system components

4 Computer system

REP SAIL computer system was designed with Microsoft technology. System platform is Microsoft Windows Server operating system. In order to store and process data relational database was used.

Application is available for user via WWW interface. Also there is provided possibility of sharing this application in Internet, what allows for remote access to research results for other members of consortium.

The WWW application is made of few interlocking modules: measurement supervision (import, validation, saving to database), battery supervision (defining properties and battery configuration), request supervision (defining workload and battery power supply), simulation of battery operation process in defined conditions, export to Excel to visualize data in table or pivot chart.

Microsoft Excel is optional component and can be installed on workstation to use data exported from system. Application is hosted on a WWW internet service server.

To run battery operation process simulation, simulation module input must contain following data:

- Configuration of battery for which simulation is run,
- Results of measurements for kind of cell on which battery is build,
- Data regarding energy flows (summed requirements for energy and delivered energy from power supplies),
- Starting battery charge level, in % of maximum capacity.

All of the above should be saved in computer system database. Modelling algorithm involves an iterative calculation of battery energy changes according to formula 1.

$$E(t, I) = E_0 + \sum_{n=1}^{k} \Delta E_k \quad (1)$$

where:

E_0 - starting battery charge level,

ΔE_k - battery energy changes in time length of t (t_i, t_{i+1}) , for which current has constant value, calculated as:

$$\Delta E_k = E(t_{i+1}, I(t_{i+1})) - E(t_i, I(t_i)) \quad (2)$$

5 Result examples

For research in REP-SAIL project lithium technologies were selected because of their high volu- and gravimetric energy densities and possible high number of work cycles [4]. After market analysis following cells were selected:

- Li-ion: Panasonic NCR 18650B
- Li-ion: Samsung INR 18650-20r
- LiFePo4: A123 ANR 26650
- Li-poly: REDOX battery

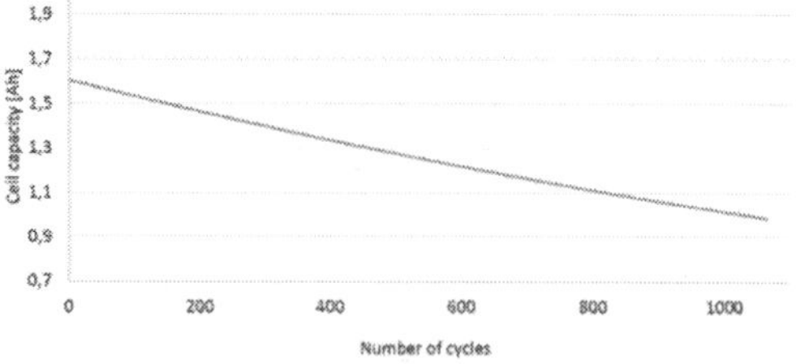

Figure 4. The dependence of the available capacity versus the number of cycles for the depth of discharge of 50% of total capacity.

For them operating characteristics were acquired in current ranges permitted by cells manufacturers. Additionally they were tested for collected currents at cycling life time angle. Fig. 4 shows example aging characteristic based on the tests results.

This kind of tests allow to determine if aging process is linear or if at some moment characteristic breaks and there is need to include it at selection and exploitation of such battery [5].

To make operational characteristics for batteries in assumed applications it is necessary to acquire progress of workload with which they will be loaded. There are many ways of acquiring that type of data, including:

- Measuring real objects. It requires dedicated gear and service time. Results are most accurate and their research value - highest.
- Simulations using existing work models based on physical dependences. In this case results require additional validation. Acquired progress may differ from actual values, the cause – simplifications occurring in used models.
- Simulations using real measurement data. Example can be characteristics of bus ride time based on schedule and traffic modelled with help of actual operational runs. That way of simulation is flexible and can be used in scenarios in which real measures are impossible. Acquired results satisfactory reflect real conditions.
- Demand prognosing with help of summing power of receivers used in vehicle in function of time and assumed drive operating characteristic. This method is inaccurate, but allows for easy calculation of heaviest possible exploitation scenarios.

Created simulation method of battery work is excellent in cooperation with any of the mentioned above models. In energetic applications most popular data sources are measurements acquired in actual objects. Example set of data is shown in Fig. 5. This characteristic consists of measurements of such devices as: fridge, basic load, washing machine, dishwasher, kettle and household solar power plant. The total load current is converted to one that corresponds to the work of single cell.

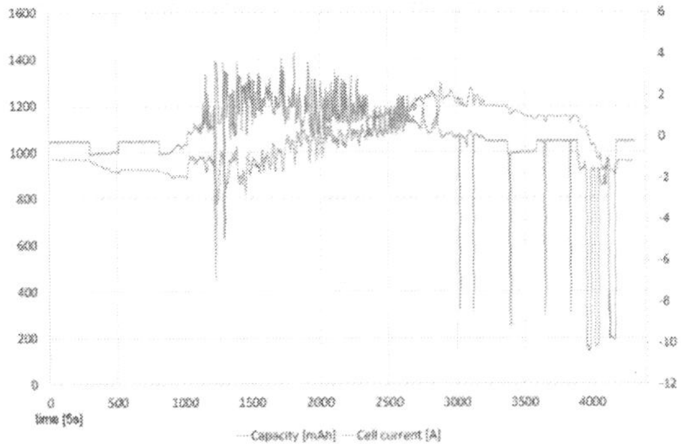

Figure 5. The dependence of available capacity as a function of time for the real data municipal type load.

One cell of the battery energy storage response to load above, generated during simulation is shown in Fig. 5.

REP-SAIL project implements workload model based on power summing of particular energy receivers in a function of time. Workload characteristic consists of following device groups: cabin equipment, navigation and yacht positioning systems, lighting, drive and its controls, yacht controls. Additionally assumed was energy production from photovoltaic panels with total power of 1,5 kW. Result workload plot with a battery response is shown in Fig. 6.

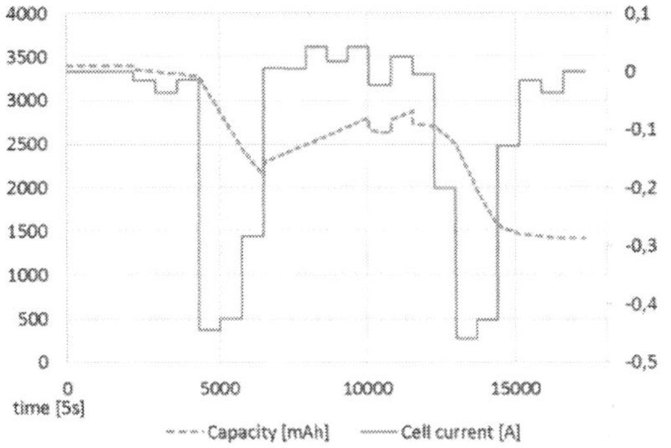

Figure 6. The dependence of available capacity as a function of time for the load created on the basis of projected energy demand for the work of REP-SAIL yacht in the motor cruise mode.

6 Summary

Finished analysis and acquired runs allow for observation of characteristics of battery work in given operating conditions. It's possible to simulate work of multiple types of cells based on measurement data acquired according to universal research plan. Thanks to aging process analysis for given conditions results closely reflect battery lifetime that is possible to achieve in particular application. Assembled laboratory stand will be serve to collect cells operational data in the ongoing project "Renewable Energy Powered Hybrid Innovative Sailing Yacht" implemented at the Faculty of Transport, Warsaw University of Technology. The collected data will be used to build computer models of chemical cells for the defined cycles cruise operational analysis and the types of cells selection for the battery pack cells selection.

Authors plan to develop computer model responsible for cell and battery work simulations, as well as the workplace so the test of whole battery modules may be possible.

Acknowledgement

The project is funded by the National Center for Research and Development under the ERA - NET TRANSPORT programme

References

1. M. Dubarry, N. Vuillaume, B. Y. Liaw, *From Li-ion single cell model to battery pack simulation,* 17th IEEE CCA, pp. 708-713 (2008).

2. M. Dubarry, N. Vuillaume, B. Y. Liaw. *From single cell model to battery pack simulation for Li-ion batteries*. J. Power Sources, **186**, 500-507 (2009).
3. T. GUENA, P. LEBLANC, *How Depth of Discharge Affects the Cycle Life of Lithium-Metal-Polymer Batteries,* INTELEC (2006).
4. J.M. Tarascon, M. Armand, *Issues and challenges facing rechargeable lithium batteries*, Nature **414** 359-367 (2001).
5. H. Popp, J. Attia, F. Delcorso, A. Trifonova, *Lifetime analysis of four different lithium ion batteries for (plug – in) electric vehicle,* TRA Proceedings. (2014).

MATEC Web of Conferences 77, 12003 (2016) DOI: 10.1051/matecconf/20167712003
ICMMR 2016

Automatic System of Monitoring and Diagnostics of Sparking in Contact

Sergey Solyonyj [1], Oksana Solenaya[1] and Andrey Roznhin[1,2]

[1]*SUAI, 67, Bolshaya Morskaia str., St. Petersburg, 190000, Russia*
[2]*SPIIRAS 39, 14 line, St. Petersburg, 199178, Russia*

Abstract. Most common occurrence of a short circuit is a consequence of a combined impact of various ignition sources – electrical sparking or monotonic alteration of resistance in contact joints – on the low–voltage network. An automatic system of monitoring and diagnostics of electrical wiring in case of an electrical sparking or monotonic alteration of the electrical contact resistance was developed. We suggest two ways of applying the elaborated system: the stationary and the mobile one. The system has two ways of applying: the stationary and the mobile one. It estimates permanently the level of fire safety of the power supply system and promptly form a signal for the abnormal electrical wiring power cut and a signal of its level of fire risk. In addition, the introduced system can serve as a subsystem for providing a controlled level of fire safety in the "Smart house" or "Smart Grid" systems.

1 Introduction

Up to 1.5 million fires occur yearly worldwide in facilities related to human vital activity [1], meanwhile nearly 30 per cent of them take place due to emergency implications of electric current in electric wiring [2]. There are several main emergency modes of electrical wiring marked out: the short circuit – 80%, electrical sparking, warming or monotonic alteration of electrical contact resistance – 11%, overloads – 3,5%, earth leakage – 1,6%, overvoltage – 2,5%, other causes hard to classify – 1,4%. Thus, the emergency events of short circuit prevail, but it is correct not in all cases. As nearly all the electrical wiring is insulated, the short circuit appearance requires preliminary insulation breakdown, which can be caused by action of electrical sparking, warming or monotonic alteration of electrical contact resistance. [3] Today technologies aimed at detection and further analysis of emergency modes of electrical wiring described above are developed and tested. Almost all of them suggest that physically the sparking alarm would be received by means of current sensor, put on the phase wire [4]. Besides the methods using indirect approach to determine the sparking point or warming point of the electrical wiring, the most widespread is the infrared temperature measurement by pyrometers [5]. However, the methods of pyrometry cannot always be applied practically, as not all of the wiring is available for analysis. Moreover, some false conclusions can be made while using these methods, as the strongly loaded phase in the contact joint point without sparking can have the same temperature as with sparking.

A separate part of researches are scientific works on modeling emergency sparking processes and arc formation in contact joints, which allow one to test without any fire risk consequences the materials used for creating wire and cable, contacts and protective and commutative appliances as well as analyze the alteration of basic physical quantities (current, voltage, frequency) in case of sparking [6].

In works [7]-[9] the methods based on Wavelet transform, allowing one to convert the original sparking alarm from time representation to time-and-frequency representation are reviewed. Thus, the physical quantities of the original sparking alarm become more estimable, which allows one to reduce the initial data set for the generalized spectral analysis. These methods allow one to identify the sparking wiring; however, they require high computational and hardware costs.

The methods of determining the impedance of the wiring, suggested in works [10], [11], enable one to estimate the ohmic resistance value of the wiring in terms of nominal and emergency exploitation (due to the alteration of the electrical contact resistance) with a subsequent comparison of the obtained results. Though these methods are attractive, they do not always give adequate convergence of the results of experiments and simulations, as modern electrical loads can change their input impedance in operation.

There are also methods of identifying the sparking alarm, which are based on using morphological filtering [12]. However, they are not suitable for use with alternating current, as they a priori imply that the power system is based on a direct current and no noise is present.

The most promising are the methods that enable one to free from periodic and random interference. They are based on the use of spectral analysis of signals before and after the appearance of sparks in the wiring. This approach is similar to methods of wavelet analysis, but it involves the assessment and work with several physical quantities (current, voltage) at a time [13], [14].

Since any electrical wiring is a set of wires and cables, interconnected by different types of contacts and the protection and commutation devices, there is always a chance of accidental damage of the insulation, poor adjustment of contacts, etc. during the installation. Consequently, the initial quality of wiring installation is the key to its future operational and fire safety.

According to the current regulations for the structure of electrical installations [15], the wiring should meet the norms of acceptance tests, due to which the wires and cables, switchgears, panels and wireways of voltage up to 1 kV from distribution points to current collectors shall be tested only for the measurement of the insulation resistance. From this it follows that the current Regulations for the structure of electrical installations do not provide adequate methods and tools for monitoring and diagnostics of electrical wiring on the quality of installation, ageing or carrying out acceptance tests. The principles and approaches outlined in mentioned works allow one to develop an automatic system of monitoring and diagnostics of sparking in contact joints of electrical wiring.

2 The principle of detection of electrical sparking and contact heating

The basis of operation of the developed automatic system of monitoring and diagnostics of wiring is the following principle: any electric sparking or monotonic alteration of the electrical contact resistance is accompanied not only by heating, but also by the appearance of load in the operational current, connected to the wiring with a frequency of 50 Hz, high-frequency harmonic components (HHC) (Fig. 1) with a frequency of 500 Hz and higher [16].

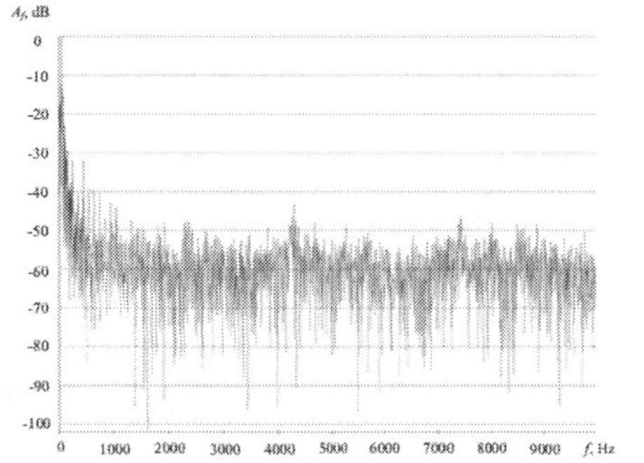

Figure 1. Spectral densities of a signal with a frequency of 50 Hz with high-frequency harmonic components with a frequency of 500 Hz and higher

Thus, knowing the duration of the existence of HHC (contact heating) and the duration of pauses between them (contact cooling) it is possible to determine the degree of heating of contact joints, using the averaged experimental curves of heating and cooling of the contact with given load power (Fig. 2). This will allow one to form signals corresponding to the levels of fire risk of existing wiring or that put into operation.

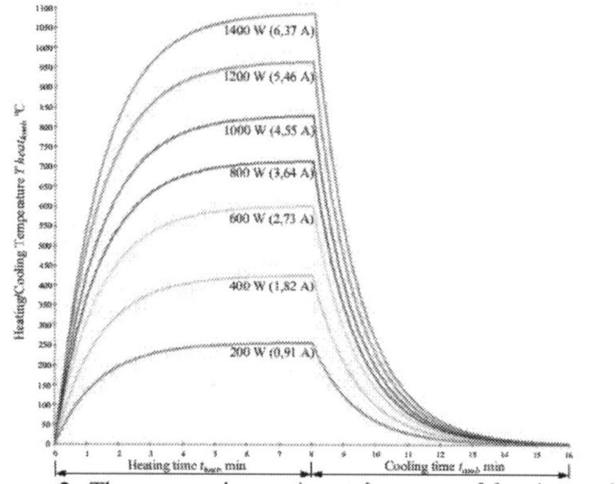

Figure 2. The averaged experimental curves of heating and cooling of the contact with a given load power

It should also be noted that in case of electric sparking, contact resistance (R_c) randomly varies from nearly zero value (best contact) to infinity (no contact). The cooling of the contact occurs in both of these opposite modes; however, with a better contact the joint can cool down maximally to a temperature conditioned by nominal load current and heat maximally when arcing (contact breaking and closing).

Using a digital oscilloscope – a measuring multi-channel registering device "RECON-08MC" with a matching unit (shunt SU 10/50), oscillograms of load current with power of 100 W were recorded while electrical sparking or monotonic alteration of contact resistance. For processing the signals a specialized software was used – a registering archiver support program WinRec-MC, version 2.8.3.0. Fig. 3 represents an oscillogram of load current with power of 100 W at sparking or monotonic alteration of contact resistance, through which it is connected to the wiring: 1 – section of the best contact ($R_c \approx 0$, rated current flowing through the load); 2 – section of no contact ($R_c \approx \infty$, current not flowing through the load); 3 – section of resumption of the current flow through the load, accompanied by the transition process (with current increasing more than rated, due to inductive and capacitive nature of the load); 4 – section of electrical sparking or monotonic alteration of contact resistance.

As it can be seen from Fig. 3, heating and cooling of the contact are similar to the heating and cooling of electrical devices in intermittent duty [17], [18]. This mode is characterized by a duty cycle of the electric device, in our case let it be called the length of Contact Sparking (CS):

$$CS_{CJ} = \frac{t_{heat}}{t_{heat} + t_{cool}} \cdot 100,\% \qquad (1)$$

where t_{heat} is the heating time of the contact (with electrical sparking or monotonic alteration of contact resistance and $R_c \approx 0$); t_{cool} – the cooling time (when R_c

≈ ∞ and also when $R_c \approx 0$ if section 3 is replaced by section 1 (Fig. 3)).

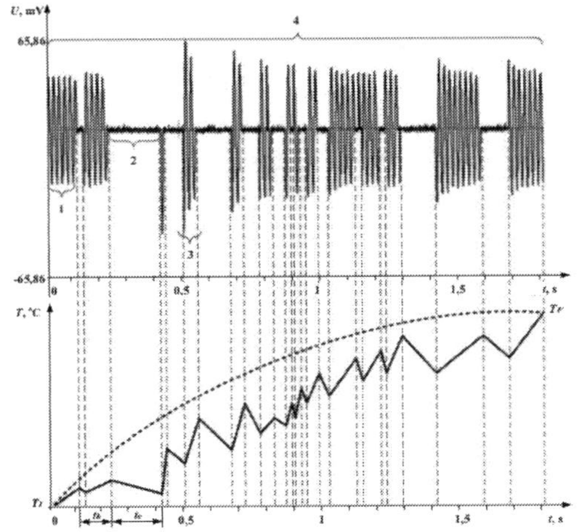

Figure 3. An oscillogram of load current with power of 100 W at electrical sparking or monotonic alteration of contact resistance and also a schematic graph of its heating/cooling

Thus, in case the electrical sparking or monotonic alteration of the contact resistance occur, the power contact connection begins to heat up from the initial temperature T_I (permissible operating temperature of the insulation depending on its class), to a final temperature T_F, which in some cases may be above the permissible operating temperature of the insulation, i.e. become the ignition temperature.

3 Flowchart of the automatic system of monitoring and diagnostics of electrical wiring

The proposed principle of detection of electrical sparking and contact heating was used for development of the automatic system of monitoring and diagnostics of wiring, which is presented in Fig. 4. To start diagnosing the internal electrical wiring must be disconnected from the external through the automatic circuit breaker QF1, whereas the circuit breakers QF2 and QF3 stay on. To the lighting part of the internal wiring, which consists of a junction box 8 and a load-break switch 9, the glow lamps 10 are connected ($cos\varphi$ = 1 and P_L = 100 W) by the number of electrical sockets. The active loads 12 ($cos\varphi$ = 1 and P_L = 100 W) are connected to the female connector of the internal wiring, consisting of a junction box 8 and an electrical receptacle 11, by the number of receptacles (glow lamps can be used). In case there are multiple varieties of electrical wiring (heating, water heating and other), the active loads are connected to them due to the number of places for electrical sockets and receptacles. A current load sensor 1 of the automatic system of monitoring and diagnostics of electrical wiring 7 is put on the phase wire. After performing the operations mentioned above the automatic circuit breaker QF1 is turned on and the internal wiring is ready for diagnostics.

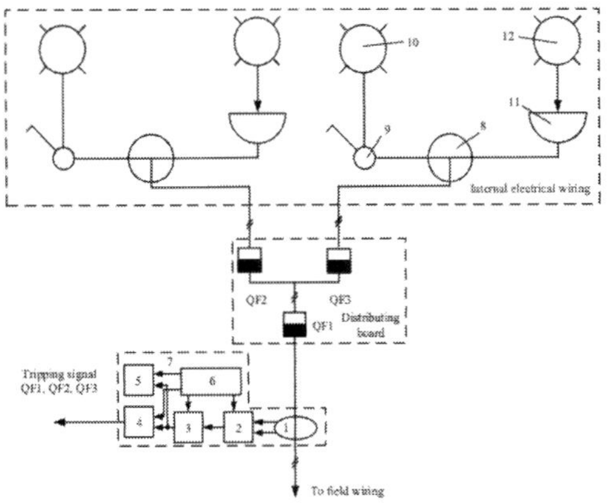

Figure 4. The flowchart of the automatic system of monitoring and diagnostics of electrical wiring

If the installation of internal wiring (wires, cables and contacts) had been conducted poorly (mechanical damage to wires and cables takes place, lose or faulty contacts are present), electrical sparking or monotonic change of contact resistance would appear in these places – the electrical commutation (Fig. 1, 3).

The signal (Fig. 1, 3) is taken by means of the current load sensor 1. Next, the signal is sent to HHCs selection box 2. In block 2 the primary signal enhancement, high-frequency filtration, secondary signal enhancement, automatic offset from periodic and stochastic interference, conditioned by the work of electrical equipment, and detection of the received signal take place.

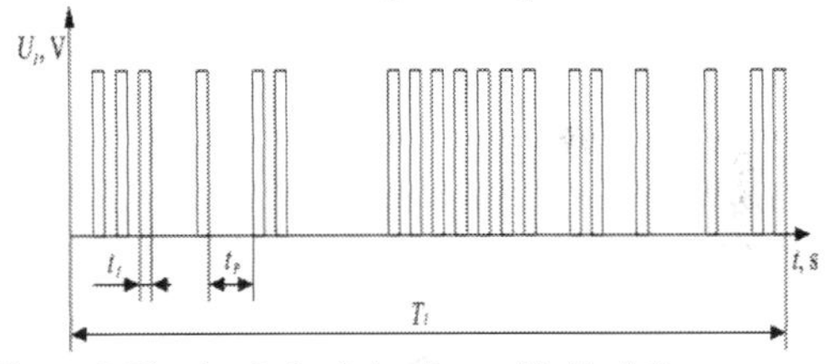

Figure 5. The signal after being detected in block 2

The signal after the detector is presented in Fig. 5. It consists of a random number of impulses having the same amplitude (voltage U_l corresponds to the logical unit and depends on the system power - block 6) and the duration (t_l is the occurrence time of the impulse).The duration between impulses (t_P – pause time between impulses) is different, since electrical sparking or monotonic alteration of contact resistance is accidental. (Fig. 3).

From block 2 the signal is sent to the block of shaping the internal wiring fire risk alarm 3, in which the following logic of work is incorporated. Block 2 integrates all impulses over the measuring time (Tm). This time can also be called the diagnostics time. The quantity T_l depends on the contact time constant τ, which is a function of contact mass:

$$\tau \sim f(m_c),$$

where m_c is the contact mass.

The analysis of the curves presented in Fig. 2 shows that $T_I \approx (5 \div 8) \cdot \tau$ (as over the time τ under continuous electric arcing or monotonic alteration of contact resistance the contact will be warmed up by 63 %), which corresponds to the calculated value of τ according to GOST 12.1.004-91 "Fire safety. General requirements" [19].

Therefore, if during the diagnostics T_I the pause time between the impulses $t_P = 0$, then the signal has a duty ratio of $Q = 1$ (the maximum level of fire hazard of internal wiring). If during the diagnostics T_I the pause time between the impulses $t_P \neq 0$, then the average duty ratio (Q_{avrg}) of such a signal can be determined using the formula:

$$Q_{avrg} = \frac{\sum_{1}^{n} t_I}{T_I} \tag{3}$$

where n is the quantity of impulses during time T_I, according to Fig. 5.

This value indicates the level of fire hazard of internal wiring.

To realize such a work logic of the block 2, the following assumptions should be considered:

1) in the internal wiring, which is operated in real conditions, the amplitude of the impulses (Fig. 5) may be different due to different power and type of loads (active, active-inductive, active-capacitive, etc.). Therefore, the method of monitoring and diagnostics of wiring by means of the automatic system involves connecting predetermined resistive loads ($cos\varphi = 1$ and $P_L = 100$ W), as that will allow one to detect a signal with the same amplitude of impulses;

2) the detection of signal in block 2 (Fig. 4) is unipolar and considers only the positive half-waves of the load current sine wave, so the signal needs to be doubled (intensified) to account the energy of the negative half-waves, i.e. to reduce Q_{avrg} by half;

3) the adequacy of the device work is logically implemented provided that one spot of electrical sparking or monotonic alteration of contact resistance is present, whereas when there are two or more simultaneous emergency spots, the parallel superposition of signals from different sources should be taken into account to initiate high-frequency harmonic components.

Based on the foregoing, 5 levels of fire danger of internal wiring are embedded in block 2. The characteristics of the levels for total load P_L= 1400 W:

– Level 1 – electrical sparking or monotonic alteration of contact resistance is absent (technical inspection of internal wiring is not required);

– Level 2 – electrical sparking or monotonic alteration of contact resistance exists in the contact for 5 seconds with a pause (t_P), which allows one to dissipate the produced heat into the environment (a technical inspection of the internal wiring and correction of the malfunction is necessary within a week);

– Level 3 – electrical sparking or monotonic alteration of contact resistance exists in the contact for 10 seconds with a pause (t_P), which does not allow one to fully dissipate the produced heat into the environment (a technical inspection of the internal wiring and correction of the malfunction is necessary within a day);

– Level 4 - electrical sparking or monotonic alteration of contact resistance exists in the contact for 15 seconds with a pause (t_P), which does not allow one to dissipate the produced heat into the environment (a technical inspection of the internal wiring and correction of the malfunction is necessary within 12 hours);

– Level 5 - electrical sparking or monotonic alteration of contact resistance exists in the contact for 20 seconds without any pause (t_P) which would have allowed one to dissipate the produced heat into the environment (an immediate shutdown of the internal wiring of the building is required till the malfunction correction). In this case, the system is able to generate signals to disconnect automatic circuit breakers QF1-QF3 sent to the additional releases (not shown in Fig. 4).

Further, the signal of a fire hazard of internal wiring, formed in block 3, is shown in the digital display of the levels of fire danger 4, which includes liquid-crystal screen to display the levels of fire hazard of the internal wiring. Starting from level 3 (high level), the block of sound indication of the levels of fire danger 5 turns on, and an audio signal takes place.

The entire automatic system of monitoring and diagnostics of the wiring 7 is powered by a power supply unit 6, which is a portable lithium-ion battery with a possibility of recharge. Should the system 7 be used continuously on site, its powering can be done stationary.

To determine the emergency spots in the internal wiring one should disconnect consistently and systematically the active loads 10 and 12. If the level of fire hazard is reduced or becomes acceptable after having disabled one of the active loads, this section of internal wiring must be subjected to preventive inspection for mechanical damage of wires and cables as well as defective or aged contacts.

If the installation of wires, cables and contacts was done thoroughly, then automatic system 7 shows the first level (low level) of fire hazard. Such wiring can be operated after the measurement of the resistance of its insulation in accordance with the rules for electrical installation [15].

The proposed automatic system of monitoring and diagnostics of electrical wiring is considered for implementation in applications of smart houses, smart grids other case studies of cyberphysical systems [20]-[25].

4 Conclusion

The suggested principle and scheme of functioning of the automatic system of monitoring and diagnostics of wiring allows one to assess the quality of the installation and the aging of wires, cables and power contact connections, which provide an acceptable level of wiring fire safety before putting it into service. Moreover, this eliminates the future possibility of formation of sources of ignition for insulation and other inflammable materials (dust, rags, etc.) due to the occurrence of electrical sparking or monotonic alteration of contact resistance in the wiring. This approach of monitoring and diagnostics does not require large expenditures of man-hours and can be performed by one installer, thereby reducing its cost and determining the economic efficiency of its use in real operating systems in smart environments and distributed robotic systems

This work is partially supported by the Russian Foundation for Basic Research (grant № 16-29-04101-ofi_m).

References

1. N. Brushlinsky, M. Ahrens, S. Sokolov, P. Wagner. World Fire Statistics, Center of Fire Statistics of CTIF, **20**: 63, (2015).
2. G.I. Smelkov. Fire safety of the electrical wiring, M: Kabel: **328**, (2009).
3. S.V. Soleniy, O.Ya. Solenaya, A.P. Kovalev, G.V. Demchenko. Formation of algorithms of functioning of the automatic system of monitoring and diagnostics of electrical wiring, Information management systems, **2/75**: 30-36, (2015).
4. A. Kovalyov, S. Soleniy, O. Solenaya. Development of electric current sensor for the device protecting electrical wiring against sparks, Material science forum, **856**: 337-342, (2016).
5. Md. Manjur Ahmed, A.S.N. Huda, Nor Ashidi Mat Isa. Recursive construction of output-context fuzzy systems for the condition monitoring of electrical hotspots based on infrared thermography, Engineering Applications of Artificial Intelligence, **39**: 120-131, (2015).
6. Qi Zi-bo, Gao Wei, Zhang Ying-cong. The Development of Electric Arc Fault Simulation Test Device, Procedia Engineering, **52**: 297-301, (2013).
7. Qiongfang Yu, Dezhong Zheng, Yi Yang, Aihua Dong. An Arc Fault Detection Method Based on Wavelet Feature Extraction and the Design and Realization by LabWindows, CVI. Journal of computers, **8/2**: 417-424, (2013).
8. Chi-Jui Wu, Yu-Wei Liu. Smart detection technology of serial arc fault on low-voltage indoor power lines, Electrical Power and Energy Systems, **69**: 391-398, (2015).
9. Chen Nan, Tang Yixing, Zhang Qian. Experimental research about the current characteristic of the fault arc in series based on wavelet analysis, Procedia Engineering, **84**: 736-741, (2014).
10. V. I. Gudim, U. I. Rudik, P. G. Stoliarchuk, V. M. Van'ko. Analyzer of resistance of two-wire electrical networks up to 1000 V, Patent of invention 80325 Ukraine, № a200508664, **14**: 8, (2007).
11. V. Torres, J.L. Guardado, H.F. Ruiz, S. Maximov. Modeling and detection of high impedance faults, Electrical Power and Energy Systems, **61**: 163-172, (2014).
12. Chen Luping, Wang Peng, Xu Liangjun. Novel detection method for DC series arc faults by using morphological filtering, Electric Power Systems Research, **119**: 91-99, (2015).
13. E. Tisserand, J. Lezama, P. Schweitzer, Y. Berviller. Series arcing detection by algebraic derivative of the current, The Journal of China Universities of Posts and Telecommunications, **22/5**: 84-91, (2015).
14. J. Lezamaa, P. Schweitzera, E. Tisseranda, J.B. Humberta, S. Webera, P. Joyeux. An embedded system for AC series arc detection by inter-period correlations of current, Electric Power Systems Research, **129**: 227-234, (2015).
15. Rules of installation for electrical devices (M.: publ «Knorus», Moskow), **488**, 2014.
16. S.V. Soleniy. Increasing the fire safety of the facilities with eletrical networks of a current with voltage up to 1 kW (Autoref. dis. for applying for the degree Candidate of Engineering Sciences / UkrNIIPB. Kiev): **20**, (2011).
17. O.Ya. Solenaya. Forecasting, monitoring and prevention of appearance of sources of ignition of combustible material in electrified areas (Autoref. dis. for applying for the degree Candidate of Engineering Sciences / DonNTU. Donetsk): 22, (2014).
18. V.A. Kazakov. Electrical appliances. A textbook for High schools (M.: publ. "RadioSoft"): **372**, (2010).
19. GOST 12.1.004-91. Fire safety. General requirements (M.: Publ. of standards): **77**, (1992).
20. R.M. Yusupov, A.L. Ronzhin. From Smart Devices to Smart Space, Herald of the Russian Academy of Sciences, **80/1**: 45-51, (2010).
21. L.I. Chubraeva, A.L. Ronzhin, A.V. Shyshlakov, Al.L. Ronzhin, V.F. Shyshlakov. The Concept of Building Intelligent Systems Management for Protected Objects Decentralized Energy, SPIIRAS Proceedings, **33**: 207-226, (2014).
22. V.P. Dashevsky, M.M. Bizin. Overview of onboard computers based SMARC-modules for robotic systems, Bulletin of TUSUR, **3/37**: 91-97, (2015).
23. A.A. Karpov, A. Lale, A.L. Ronzhin. Multimodal assistive systems for a smart living environment. SPIIRAS Proceedings, **19**: 48-64, (2011).
24. A.I. Motienko, A.L. Ronzhin, N.A. Pavlyuk. Modern developments of rescue robots: possibilities and principles of their application, Science Bulletin of the NSTU, **3/60**: 147-165, (2015).
25. A.L. Ronzhin, V.Yu. Budkov. Multimodal Interaction with Intelligent Meeting Room Facilities from Inside and Outside, NEW2AN/ruSMART 2009, LNCS 5764: 77-88, (2009).

MATEC Web of Conferences 77, 13001 (2016) DOI: 10.1051/matecconf/20167713001
ICMMR 2016

The Research on Lung Cancer Significant Detection Combined with Shape Feature of Target

Guilai Han, Yuan Jiao and Xiaoling Li

Faculty of Information Technology Hainan Medical College Haikou, China

Abstract. At present, the research on detection of lung cancer includes the sample image segmentation, extracting visual features of lung cancer and generating the classification model by training learning, then according to the classification model generated to classify the inspected images. But this kind of method usually needs a large amount of calculation and speed is slow. In order to find the region of interest as soon as possible and improve the detection speed, this paper attempts to introduce the current popular Itti visual attention model into the lung cancer detection. However, because medical images usually have low contrast, the Itti method is not directly applied to extract the region of interest in medical image. Therefore the selective visual attention mechanism combined with shape feature of target is proposed. Firstly some primary features are chosen, such as gray scale, direction, corner point and edge to generate saliency map, and then the significant regions are segmented and judged. Compared to popular lung cancer detection method, this method can improve the detection rate of suspected lung cancer and has great significance for the early detection, early diagnosis and treatment of lung cancer.

1 Introduction

Malignancies which are commonly known as cancers bring great harm to patients and their families and consume too much valuable social medical resources. But the effect of treatment is very little. Lung cancer is one of the largest cancers which hazards human life and health. The mortality rate of the lung cancer is much higher than other cancers, and is increasing year by year. The statistics shows that lung cancer has replaced liver cancer to become the highest death rate of cancer, accounting for 22.7% of all cancer deaths, and the incidence and mortality rates continue to rise rapidly. In 2014, the incidence rate of lung cancer in China grew 26.9%.If we do not take effective measures, Chinese lung cancer patients will reach 100 million in 2025, and China will become the world's first lung cancer country. At average, one person died from lung cancer every 30 seconds in the world. In fact Lung cancer has been the first deputy killer. Early detection and early diagnosis is an important measure to reduce lung cancer mortality. If lung cancer can be diagnosed and treated at an early stage, the 5-year survival rate will increase from 14% to 49% [1].

The important feature of lung cancer early is pulmonary nodules, especially malignant pulmonary nodules have burr and leaf features. Using medical image, doctors can find cancer in time, and provide a Diagnosis opinion. Computed tomography (CT) is the most common chest radiographic image which is widely used in lung tumors and lung nodules detection. Due to the huge number of images, the analysis of staff chest CT image is a boring and tedious process. Especially for a high-resolution scan image, a number of images scanned only a patient can reach tens or even hundreds of layers, which bring the enormous workload to doctors, thereby increase the chances of misdiagnosis and missed diagnosis. Computer-aided detection (CAD) for CT images can be automatically prompted to perform the analysis of the image of staff suspicious pulmonary nodules. It can effectively improve diagnostic accuracy and reduce the rate of misdiagnosis. Using computer-aided detection to detect a variety of early cancers has become a focus of attention.

2 Computer-aided Detection and Visual Attention Mechanisms

The current research on computer-aided detection of early lung cancer includes the sample image segmentation, extracting visual features of lung cancer, generating the classification model by training and learning, and according to the classification model to classify the inspected images. However, this kind of method usually needs a large amount of calculation and the speed is slow which severely restricted the use of such software in the actual clinic. In order to find the region of interest as soon as possible and improve the detection speed, the visual attention model is introduced into lung cancer detection.

Human can quickly search for the region of interest from complex scenes. The ability to choose is called visual attention mechanism. The role of visual attention mechanism lies in extracting visual information, sorting things by priority, then dealing with higher priority information and ignoring lower priority information at the fastest speed. The traditional image processing technology processes all of the pixels unitary, regardless of priority. But in many image analysis tasks, such as image retrieval, scene analysis, target identification and target tracking, they have priorities. So the regardless of priority image processing not only wastes computing resources, but also increases the complexity of the analysis process. So the study of visual attention mechanisms has more important scientific value and practical significance.

Now that there is not yet a unified mechanism of theoretical and computational model of visual attention, the visual attention model Itti et al [2] proposed is more recognized academic, which using Gaussian pyramid and considering three visual features of brightness, color, direction to obtain multi-scale significant measure by Center-surround operator, and the final saliency map obtained through fusion and normalization, finally using saliency map to evaluate a given area differences in the surrounding area.

The Itti visual attention model has achieved good effects in the natural image target detection, but the effect is not ideal when it is used in medical images which are mainly gray images. Especially in the early stages of lung cancer, the focus on lung cancer is not the most significant region in medical images, and sometimes it is very difficult to be found. Although Itti visual attention model can obtain the position of a significant target, it does not reflect the size and shape of significant goal. For this reason, the selective visual attention mechanism combined with the shape features of lung cancer, was proposed to improve Itti visual attention model in this paper. The specific method extracts the features of gray, directions, corners, edges and local entropy to generate feature saliency map.

3 Lung Cancer Significant Detection Combined with Shape Feature of Target

Because lung cancer are usually small in the whole chest CT images, the Itti method is used to detect the most significant area which is not the area of lung cancer in most cases. So this paper attempts to combine the shape features of lung cancer to detect saliency. Lung cancer usually has features of burr and leaf. In order to simplify calculations, the primary visual features which are simple and easy to be got, such as gray, direction, corner, edge and local entropy, are used to generate saliency map of specific target area. The flow diagram of visual attention model combined with shape feature of target was shown in Fig. 1.

Human visual system based on different types of receptive field to extract the image information step by step. In recent years, neuroscientists have acquired a lot of experimental results about the characteristics and types of receptive field through continuous research. In 1965, Rodieck pointed out that the receptive field of human eye can be expressed by two Gaussian functions difference [3]. The receptive field was expressed by polar coordinates as eq.1.

$$DOG(r) = A\exp(-r^2 / \sigma_A^2) - B\exp(-r^2 / \sigma_B^2) \quad (1)$$

Among them, r expresses the distance between any point and the central point in receptive field.

Neurobiology found through further study, the difference of Gaussian function (DOG) and center surrounding receptive field are very similar. Due to the difference between receptive field and DOG function is very small, DOG filter is usually adopted to simulate the human eye receptive fields in the visual design attention mechanism mode [4] l. In the visual attention mechanism model, the two concepts of multi scale and center-surround difference are put forward based on receptive field.

In order to obtain multi scale sampling, the three-tier pyramid model is adopted in fig.1. The original image is the first layer. The second layer and the third layer are sequentially sampled on the basis of the upper layer, and their size is 1/2 of the upper layer. The different features of the image were filtered by using center-surround operation. By comparing the differences between the candidate region and the surrounding area in the brightness, direction, corner, edge and local entropy of the underlying features, the saliency maps of respective features were obtained.

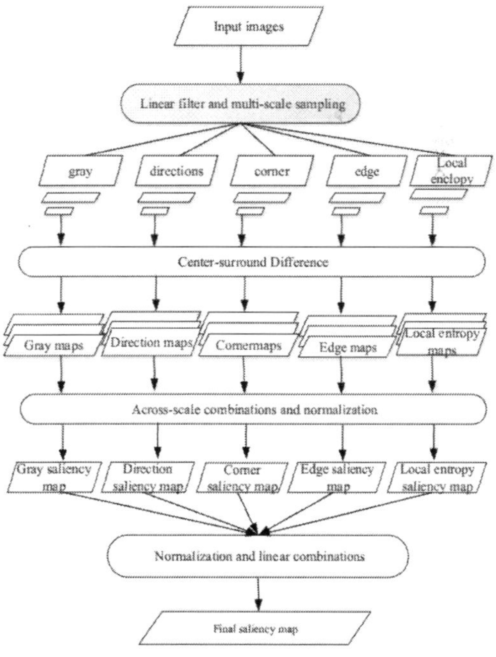

Figure 1. The flow diagram of visual attention model combined with shape feature of target

In the multi feature map integration stage, the feature maps which coming from the same feature and different layers were integrated by across scale combination and normalization. For the saliency maps of different feature s, according to the feature integration theory, different feature information of significant measure is not the same, so it is necessary to integrate the different feature channels to obtain the final saliency map by normalization and linear or nonlinear method.

In order to get better results, median filter was used firstly to eliminate the influence of noise, and the lung parenchyma image shown in Fig.2 was got from chest CT image through image segmentation.

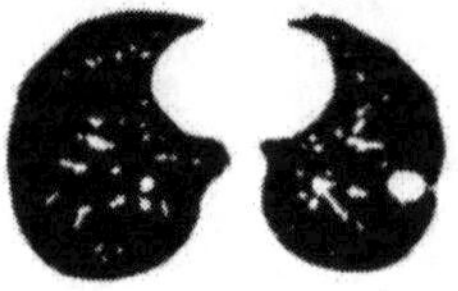

Figure 2. Lung parenchyma image

3.1 Gray feature

Gray is one of the most important features of medical image, and it is the most direct expression of the difference between the lesion and normal human body. After a series of operations, such as multi scale sampling, center-surround difference and normalization can get gray saliency map, as shown in fig.3.

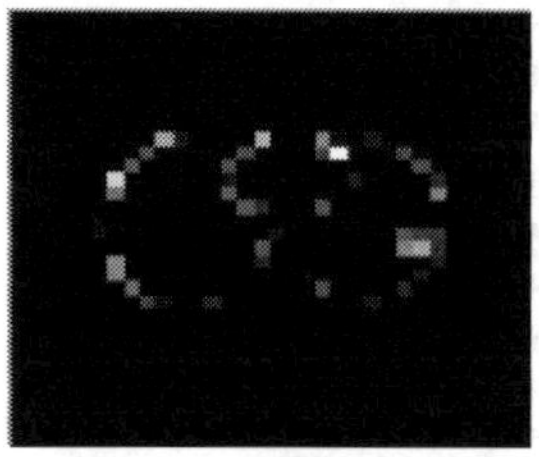

Figure 3. Gray feature saliency map

3.2 Direction

Using Gabor filter on the brightness image to filter in the four direction of $0^{\circ},45^{\circ},90^{\circ},135^{\circ}$, to acquire the direction images. After multi scale sampling, center-surround difference and normalization can get direction saliency map, as shown in fig.4.

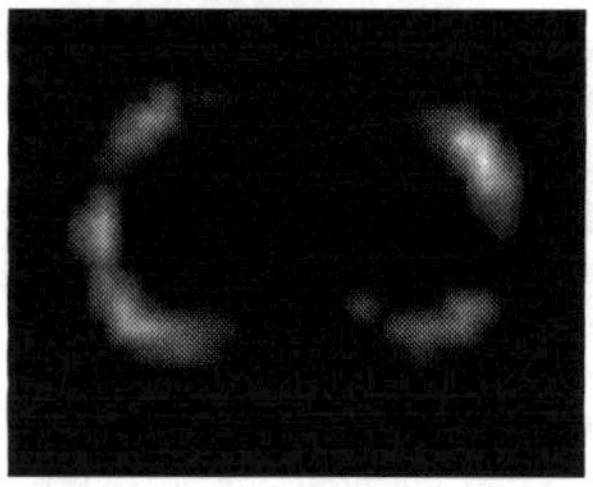

Figure 4. Direction feature saliency map

3.3 Corner

Corner is the point where the image brightness changes sharply or the edge of the image has maximum curvature. While retaining the important characteristics of image, the corner can reduce the amount of information and improve the calculation speed effectively. The identification of corner points is usually used in a local small window to slide in the image, through the observation of the gray value changes in the window to achieve. Harris corner detection operator and SUSAN detection operator are used commonly corner detection operators[5].Harris corner operator is chosen as an example in this paper. Harris corner feature and its saliency map are shown in Fig. 5.

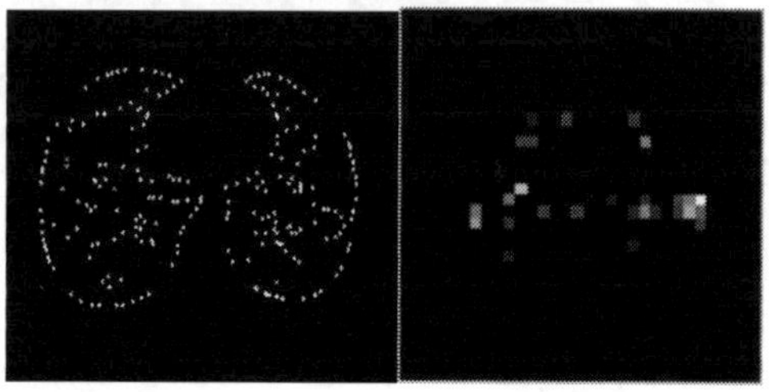

Figure 5. Harris corner feature and its saliency map

3.4 Edge

An edge is a collection of pixels whose gray value changes sharply. As the most important basis of image segmentation, edge often exists between target and background region. Edge is the sign of the location and it is not sensitive to the change of the gray value. Therefore, the edge is an important feature of image matching. sobel operator, log operator, canny operator and so on are Commonly used edge detection operators. Canny operator which has the most excellent performance in all of the operators is widely used by everyone [6].Canny edge feature and its saliency map are shown in Fig.6.

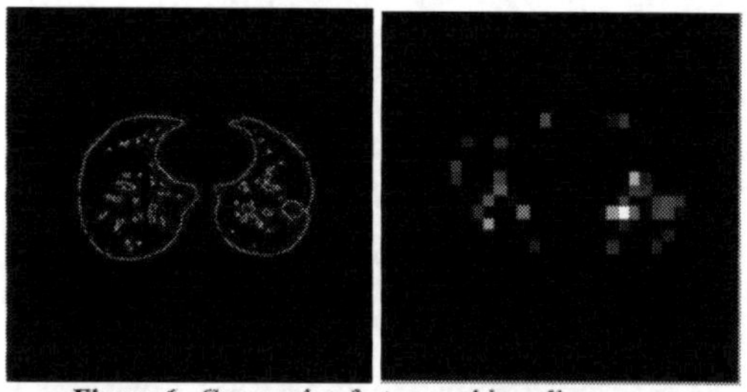

Figure 6. Canny edge feature and its saliency map

3.5 Entropy

Entropy is the concept used to measure the amount of information in information theory. If the system is more orderly, the information entropy is lower; on the contrary, if the system is more chaotic, the information entropy is higher. By the concept of entropy in information theory, the entropy of the image is a measure of the amount of information contained in the image. The bigger the entropy of the image is, the more the image information is. Then the region is more likely to be the region of interest (ROI)in the image [7].

Assume that the probability of the pixel whose gray level is m appears as in the image. According to the definition of the information entropy, the information entropy of the gray level can be defined as eq. 2.

(2)

Where in . For the gray image, the information entropy of the image can be defined as eq. 3.

(3)

It can be seen from the definition of the entropy of the image that the image information entropy only considers the global statistics of the image, and the spatial distribution information of the image is discarded. Therefore, in order to represent local features of the image, the concept of local entropy is introduced, whose idea is that the image is divided into a number of the same size sub images, and then the entropy of each sub image is calculated. Local entropy feature and its saliency map are shown in Fig.7.

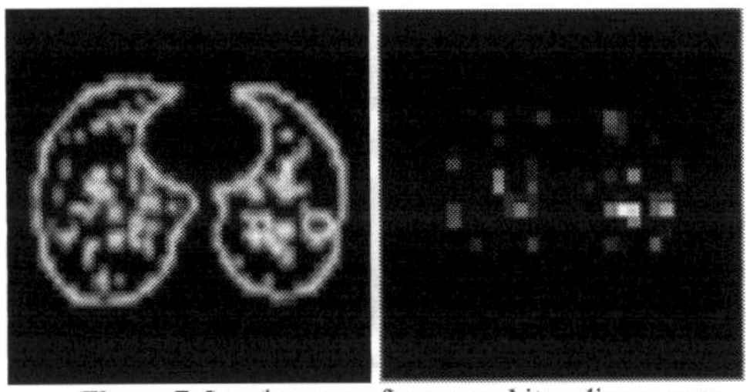

Figure 7. Local entropy feature and its saliency map

4 Conclusion

Because the medical image is usually a gray image, its color features can't be got, The saliency map shown in Fig.8 is obtained by the two features of brightness and direction based on Itti visual attention mechanism. The final saliency map shown in Fig.9 was acquired by the visual attention model combined with shape feature of target. Fig.8 and Fig.9 can be seen in contrast, the saliency map obtained by combining with the shape features of lung cancer can better highlight the region of interest which the lung cancer corresponds to than Itti feature saliency map.

In order to more quickly find lung cancer region, some primary visual features such as corners, edges and local entropy are introduced to improve the detection of lung cancer in this paper. Experiments show that the method can be faster and more accurate to find significant area compared to the Itti method. Using vision attention mechanism to find the significant region is only rough location. In order to more accurately detect lung cancer and decrease the false positive rate, segmenting the significant areas in saliency map is needed. And it is need to extract more detailed features and to classify according to classification model established on the basis of sample image.

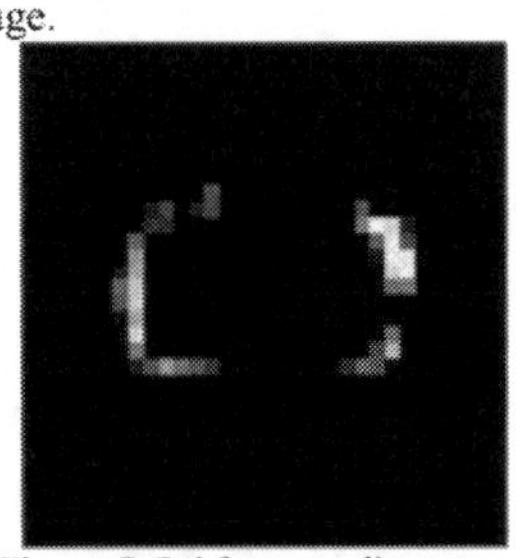

Figure 8. Itti feature saliency map

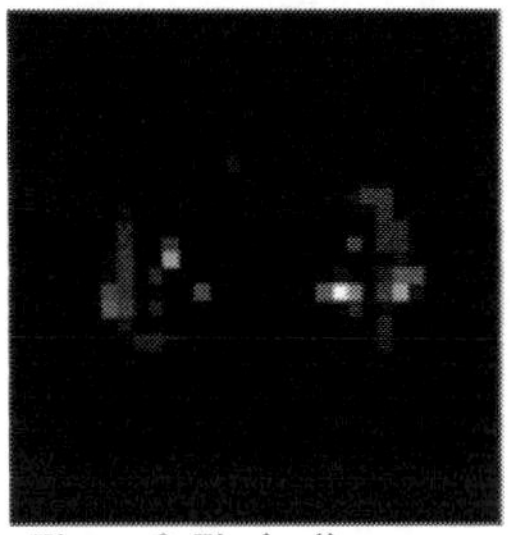

Figure 9. Final saliency map

Acknowledgment

This research was supported by the Hainan Province Nature Science Fund Item (Grant No:20156224).

References

1. H. Jie, C. H. Wanqing, *Annual report on status of cancer in China, 2014,* (TsingHua University Pres, 2014)
2. L. Itti, C. Koch , E. Niebur, A model of saliency-based visual attention for rapid scene analysis, TPAMI, **20**,6(1998).
3. Rodieck R W, Stone J J. Analysis of receptive fields of cat retina ganglion cells. JON, **28**,17(1965)
4. Z. Ming, L. Youfu, M. Qinghao, Y. Ting , L. Jian, H. Tiemao, Finding Regions of Interest Based on Scale-Space Keypoint Detection. CCIS, **202**, 8(2011).
5. ZH. Congpeng, W. Xueguang, Rectangle detection based on Harris corner. OPE, **89**, 7(2014).
6. T. Lu-lu, ZH. Qi-can, H. Song, An Improved Algorithm for Canny Edge Detection with Adaptive Threshold. OEE, **38**, 6 (2011).
7. SH. Yuhao, H. Weiwang, SH. Rui, Y. Yangchao, O. Congjie, Novel image segmentation algorithm based on entropy and clustering theory, ARC, **33**,6 (2016).

MATEC Web of Conferences 77, 13002 (2016) DOI: 10.1051/matecconf/20167713002
ICMMR 2016

A research on the reliability of light-emitting diode based on analyzing of chip image

Jian Ruan and Ke-Yuan Qian

Key Laboratory of Information Science and Technology, Graduate School of Tsinghua University at Shenzhen, Shenzhen 518055, China

Abstract. This paper proposes a new method for studying the reliability of high-power light-emitting diode (LED) by analyzing chip images taken from a batch of LEDs which are selected to conduct the accelerated aging test lasting for 1400 hours. To exclude the disturbance of electrode in these images, an image processing algorithm based on projection is used to extract the interested section. An index called "dark point" which is related to non-radiative combination to describe the reliability of LED is proposed. The method is based on analyzing the variation trends of dark points in the extracted section of these images. The results show that the proportion of the dark point increases with aging time increasing. The lifetime of LED with higher increasing rate is shorter. Based on the increasing rate of dark point, the lifetime of the chip can be predicted in an easier way.

1 Introduction

Compared to traditional incandescent lamps, GaN-based light-emitting diode (LED) has lots of obvious advantages, such as their high efficiency, long lifetime, fast modulation speed, and their linear behaviour under pulsedwidth modulation dimming [1]. Therefore it is believed that LED will replace the widely used incandescent lamps. However, the reliability of LED is still a shortcoming, and the reliability of the high-power LED chip has significant influence on expansion of solid state lighting applications. Therefore the research of the reliability of LED becomes a hot direction.

So far most of research focus on the property of the electrical characteristics and optical characteristics of LED by measuring current-voltage(I-V),capacitance-voltage, output power versus input current(L-I), electroluminescence(EL) and so on. However, these researches still emphasize on measuring the discrete curve of the LED's characteristic curves [2]. They don't essentially reveal the inner mechanisms of the aging of LED. Although there are some researches pointing out that continuous increasement of non-radiative combination centres are the main reason lead to the aging of LED [3-6], they don't give a concrete relation between the density of non-radiative combination centres and aging time. In this paper, a model is used to describe the variation of non-radiative combination centres with aging time.

Also, some researches try to find the relationship between microstructure and luminescence efficiency [7-9]. And they study the point defects, extended structural defects and other defects on the surface of the chip by atomic force microscopy (AFM), cathode luminescence measurements and transmission electron microscopy to observe the variation of pits on a microchip. However, these methods are complex, and they can only observe the micro area of the chip. In this paper, the structure of the chip in a macro perspective will be studied. The correlation between the dark area that led by the threading dislocations and the non-radiative combination centres will be directly characterized. This method is not only simple, but also very useful. The increasing rate can be used to describe the aging rate of LED. And this index can help us to examine the reliability of LED in a more simple way.

2 Experiment

2.1 Measuring instruments

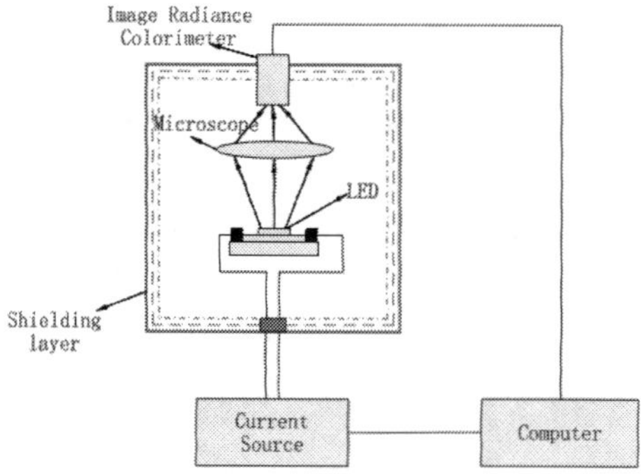

Figure 1. Experimental apparatus for taking images of LED chip.

 DOI: 10.1051/matecconf/20167713002

Figure 1 shows the experimental apparatus of getting the image of the LED chip. To prevent the chip images from being saturated, the injection current should be as week as μA. The current supply source can provide current for the LED whose range is from 10^{-7}A to 10^{-3}A. The LED works at 30μA in this experiment. A shielding layer is used to shield the ambient light. The LED is put on the stage that can rotate 180° about the shaft in steps of 0.5°. Through this system, the chip image is taken and imported into the computer, then it can be analyzed.

2.2 Aging test

A group of commercial blue LEDs are selected to conduct aging test. The parameters of these samples are shown in Table 1.

Table 1. Parameters of samples.

Parameters	Value
Power rating	1W
Surface Size	1mm×1mm
Active Layer Thickness	100nm
Substrate Temperature	60°C
Aging Current	500mA
Aging Time	400h,700h,1000h,1400h
Samples Amount	5

2.3 Data process

This aim of our study was to analyze the reliability of LED through the variation trend of the images taken from a batch of chips. However, with the LED chip surface covered with complex structures of the electrode, it is hard to find the rule that needed. To exclude the interference of electrode, it is necessary to get rid of the part of electrode in the picture by image processing techniques. Also, these images that got are in large number. It is necessary to use some techniques to deal with these images by computer automagically.

2.3.1 Image tilt correction

It is necessary to rotate the image to an appropriate angle before it is segmented. The algorithm of projection is used to get the skew angle. The original image is set in the coordinate system. The method is based on the principle that the active length that the number of pixel value projected on X-axis is the longer when the image is not upright. Just like the Figure 2 (a) and Figure 2(b), the upright image that projected on X-axis is a length of 2a while another is longer than 2a. According to the geometric features, the oblique image can be easily rotated to an appropriate angle. The result of processed image as Figure 2 (c):

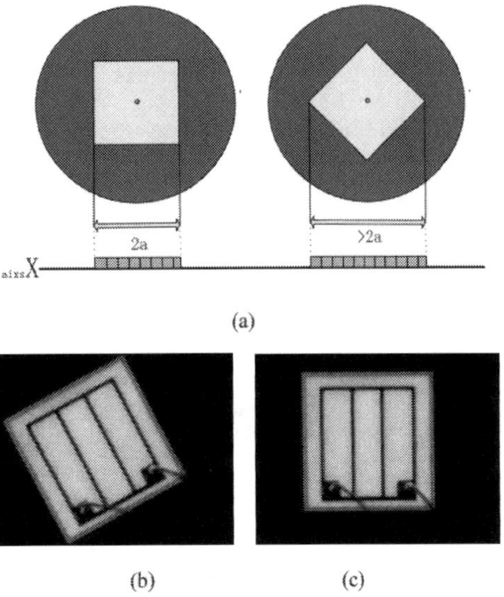

Figure 2. (a)Sketch map of rotating (b) Original image (c) Image after rotating.

2.3.2 Segmentation of image

It is need to identify the electrode that is linear shape. It is easy to find the center of the image by the value that projected in X-axis and Y-axis first. Then the boundary of the image can be found by drifting the center point. The Figure 3 (a) shows the procedure of the algorithm. And the information of the structure can be got by other measurement method .Then the images can be segmented by the information of the chip structure and the position of the boundary. Figure 3 (b) is the image got from the procedure of image rotating. At last the image processed is like Figure 3 (c), the part segmented is signed by red line.

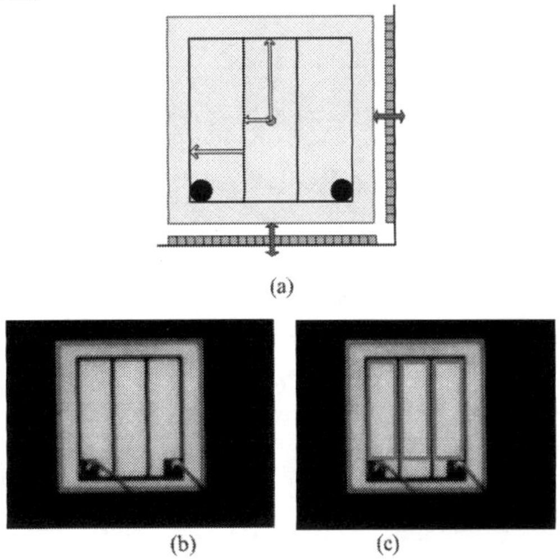

Figure 3. (a)Sketched image (b) Image before segmenting (c) Image after segmenting.

3 Theory

It is confirmed that non-radiative recombination centers are produced by the ionization of compound of $Mg-H_2$ [10]. When the energy of carriers in the chip is larger than the chemical energy of $Mg-H_2$, the compound of $Mg-H_2$ can be ionized and become defect points. Assume the density of the carriers that can be ionized is D_h, the initial density of the non-radiative combination center is D_o, D_c is the density of the defect, and then it will be:

$$D_c \propto D_0 + \int_0^t D_h(t)\,dt \tag{1}$$

Due to Fermi distribution function, then integrating on the amount of carriers, the ionization energy of $Mg-H_2$ complexes is E_c, the Fermi energy is E_f, the D_g is effective density of states on the basics of solid state physics and then integrate on it, there will be:

$$D_h = D_g \exp(-\frac{E_c - E_f}{kT}) \tag{2}$$

Due to the directional kinetic energy of carriers in the chip, the energy of carrier must be changed to E_c -E, E is the electron kinetic energy, then:

$$D_h = D_g \exp\{[E_c - E_f - \frac{1}{2}m(\frac{I}{nqs})^2]/(kT)\} \tag{3}$$

Then, as D_h is an invariant constant, the density of the non-radiative combination centers can be:

$$D_c = D_0 + D_h t \tag{4}$$

Under the condition of constant current, define a short time process Δt, suppose during Δt the carriers needed to transport is n, and suppose the active length is L, under the effect of non-radiative combinations centers with a density of D_c, the loss of carriers when passing through a length of dL is :

$$dn \propto -nD_c \times dL \tag{5}$$

And the formulation can be expressed as:

$$\int_{n1}^{n2} \frac{dn}{n} \propto \int_0^L Dc \times dL \tag{6}$$

Then it can be expressed as:

$$n_1 - n_0 \propto -\exp(-D_c L) \tag{7}$$

During a short period Δt, the amount of carriers compounded is Δn, and $\Delta n \propto \exp(-D_c L)$. Under constant current I, during a unit time, the carriers compounded should be proportional to I, then it will be:

$$I_0 \propto I \exp(-DcL) \tag{8}$$

Then the efficiency of light output will be:

$$\eta = \frac{I_0}{I} = a \exp(-DcL) \tag{9}$$

4 Result and Discussion

4.1 Definition of dark points

It is widely recognized that high density of dislocations, which act as non-radiative recombination centers, are introduced in epitaxial layers grown by using of the heteroepitaxial [11]. And it has been proved that the failure of an LED is closely related to the increasing of non-radiative recombination centers during its lifetime. With aging time increasing, the non-radiative recombination defects increases in the active region of the LED. G.Meneghesso found [7-9], when the defects of LED aggregated to an extent, it will change the surface topography of the chip. Under the atomic force microscopy, it can be observed that the dark areas increase in the surface of the chip after aging. If the images of the chip are taken, it can be found some areas in the images that the pixels value is low which can be seem equivalent to dark areas. Then it can be observed the dark points by counting dark pixels in these images. Before the definition of dark point, Figure 4 is given to describe it in an intuitive way.

Figure 4. Sketch map of definition of dark points.

Image of the chip is segmented by image process technology referred before. Part of electrode is cut off and the uniform part that occupy the most part of the image in the surface of chip is got. Extracting the interested part from images of LED, like the part signed by red line in Figure 4. Then the means of the whole pixel value in the segmented image is used to normalize it. Then a normalized part of segmented image can be obtained. The dark point in the image is defined as below:

In these segmented and normalized images, the pixel value less than a threshold which is set as 0.25 in this experiment is defined as dark point shown in Figure 4.

4.2 The change of dark points with aging time increasing

The images of aging for 300h, 700h, 1000h and 1400h is got. The interested part is extracted and the proportion of dark point that occupy the segmented part is calculated. A broken line of these data is draw as Figure5:

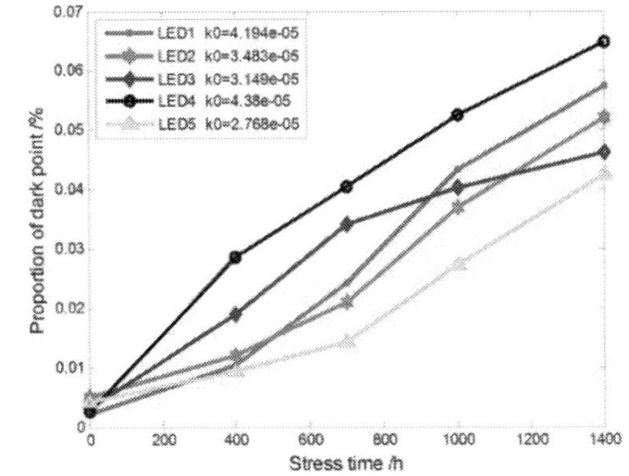

Figure 5. Proportion of dark points with stress time.

In Figure 5, the value of vertical coordinate is the proportion of dark points that occupy the segmented part of LED image. From the broken line graph, it is concluded that the density of the dark point is almost in direct proportion to aging time. The rate of the line is fitted out. For each curve, the slope to the horizontal axis, which is equal to the increasing rate of dark points, is defined as an index k_0 shown in Figure 5. From the Figure 5, it can be concluded that each chip has its own k_0, and k_0 decides the increase rate of the dark points in the chip.

Based on the previous study [12-14], it is known that the darkening of the current-injection region is caused by the diffusion of point defects, and the density of the point defects have a relationship with the injected current. And the density of point defects was increased with the increase of non-radiative recombination centers in the active area. It is inferred that the dark area is caused by the increasing of non-radiative recombination centers. And from formulation (4) and the variation trend of Figure 4, the relation between D_c and k_0 can be concluded:

$$D_c = D_0 + ck_0 t \tag{10}$$

In formulation (10), k_0 is the increasing rate of the defect points referred before. And c is a constant coefficient, which can be got from Figure 4 by using the average of the five LEDs' initial value.

4.3 Lifetime of LEDs with increasing rate of defect points

As we known, the luminous flux satisfy the formulation below:

$$p = p_0 e^{-\beta t} \tag{11}$$

In the formulation, the p_0 is the initial value of luminous flux, ß is the attenuation coeffcient. Under rated operating current of LED, we measure the change of luminous flux with time variation, when $p=0.7p_0$, the time is the lifetime of LED:

$$\tau = \frac{-In0.7}{\beta} \tag{12}$$

From formulation (9) and (10), the relation between k_0 and the lifetime of LED is concluded as below:

$$\tau = \frac{\frac{1}{L} In\frac{a}{0.7} - D_0}{ck_0} \tag{13}$$

As formulation (13) has only one variable parameter, then it can be simplified as below:

$$\tau = \frac{m}{k_0} \tag{14}$$

By measuring the luminous flux of these LEDs at 400h,700h,1000h and 1400h, the lifetime of the LEDs can be calculated based on formulation (11). According to the lifetime of LEDs and k_0, Figure 6 below is the fitted curve.

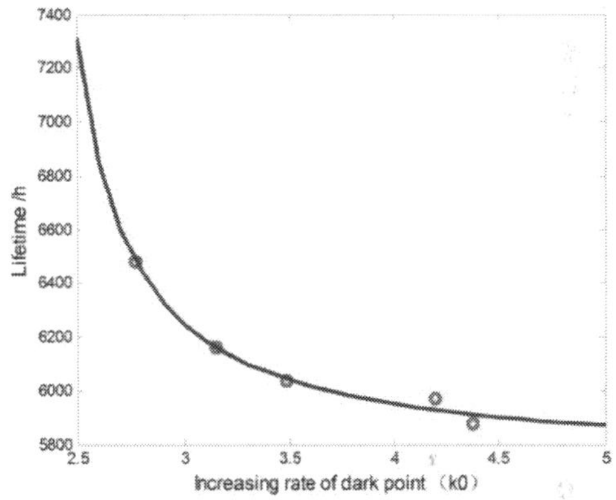

Figure 6. Proportion of dark points with stress time.

From this curve, it can be inferred the lifetime of a LED according to its diffusion rate of the defect points-- k_0. The lifetime of the same kind of LED can be estimated through the fitting formulation. And this method is not only effective but also simple to calculate the lifetime of LED.

5 Conclusions

A method of analyzing the reliability of high-power light-emitting diode (LED) by analyzing the chip image is proposed. This method can be used to all GaN-based LEDs. With the self-developed measuring system, chip image is taken and analyzed. The interested part of these images are extracted by using the algorithm of projection and the feature of LED's structure. An index called "dark point" to describe the reliability of LED is proposed. By counting up the variation trend of "dark point", it is found that the increasing rate of dark points is strongly associated with the reliability of LED as the LED with higher increasing rate has shorter lifetime. Through the relation between increasing rate of dark points and the lifetime of LED shown as formulation (14), the lifetime of LED can be calculated in a simpler way.

References

1. G. Zhang, Z. Chen , Phys, **33**: 833-842(2004)
2. M. Meneghini, L. Trevisanell, G. Meneghesso, IEEE T Device Mat Re, **8**:323–330(2008)
3. A. Hangleiter, F. Hitzel, C. Netzel, Phys. Rev. B, **95**: 1-4(2005)
4. S.D. Lester, F.A. Ponce, M.G. Craford, Appl. Phys. Lett, **66**: 1249-1251(1995)
5. X.A. Cao, P.M. Sandvik, S.F. LeBoeuf, Microelectron Reliab, **43**: 1987-1991(2003)
6. X.A. Cao, K. Topol, F. Shahedipour-Sandvik, Proc. SPIE, 4776: 105-113(2002)
7. G. Meneghesso, S. Levada, R.Pierobon. Ieice T Electorn, E86-C: 2032-2038(2003)
8. G. Meneghesso, S. Levada, R. Pierobon, IEDM, 103-106(2002)
9. G. Meneghesso, S. Levada, E. Zanoni, IRPS, 474-478(2004)
10. F. Rossi, M. Pavesi, M. Meneghini, J. Appl. Phys, 99: 053104(2006)
11. T. Egawa, H. Ishikawa, T. Jimbo, Appl. Phys. Lett, 69, 830 (1996)
12. Tomiya, O. Goto, and M. Ikeda, Proc. IEEE **98**(7), 1208–1213 (2010)
13. H. W. Shim, R. J. Choi, S. M. Jeong, L. Van Vinh, C.-H. Hong, E.-K. Suh, H. J. Lee, Y.-W. Kim, and Y. G. Hwang, Appl. Phys. Lett, **81**:3552–3554(2002)
14. X. H. Wu, C. R. Elsass, A. Abare, M. Mack, S. Keller, P. M. Petroff, S. P. DenBaars, J. S. Speck, and S. J. Rosner, B Appl. Phys. Lett., **72**:692-694(1998).

The Behaviour of Gas Bubble during Rest Period of Pulse-Activated Electrolysis Hydrogen Production

Thanet Vilasmongkolchai, Roongrojana Songprakorp, Kaokanya Sudaprasert

Division of Energy Technology, School of Energy, Environment and Materials, King Mongkut's University of Technology Thonburi, Bangkok, Thailand

Abstract. The pulse-activated electrolyzer has been developed and used for several years. With the capability of enhancing the efficiency of an electrolytic process and easy operation, this technique becomes an interesting process for hydrogen production. Unfortunately during electrolytic reaction, the creation of bubbles becomes a reaction inhibitor and consumes energy. This paper aims to study the proper rest period that gives the bubble free rise-off the solution without additional bubble created. The mathematical method and acoustic emission method were used for investigation of bubble's rising velocity. The result shows that the variation of rest period on pulse-activated makes production efficiency enhanced. For the practicality of use and set control parameters, duty cycle and frequency were demonstrated instead of rest period.

1 Introduction

Due to energy crisis and global warming awareness, the study of renewable energy also up-to-date challenged research problem. Hydrogen is one of alternative clean energy with a possibly use in several (e.g. direct combustion, fuel cell). Nowadays, the most economical hydrogen production process needs to use fossil fuels. The steam reforming process is not only consuming hydrocarbon products, but also releasing carbon monoxide. The electrolytic process is the use of electrical field activated with electrodes so, the water split into hydrogen and oxygen. This method is an advantage of producing extremely pure hydrogen and oxygen without any carbon emission. The pulse-activated has been implemented to use with electrolysis process for two decades with the good result in productivity and efficiency enhancement. According to the previous study, the electrical circuit has been used to describe electrolysis system [1], [2] and the results were revealed that the natural frequency make system's efficiency improved [2], [3].

Unlike this study, the research was focused on the increasing of Ohmic's loss due to the bubble created. Based on the main idea that has been proved in several literatures, during the electrolytic hydrogen generation process, gaseous products make the resistivity of the system increases and hence lower system efficiency [1], [4]-[6], so the degasing is the effective method to achieve higher efficiency. The rest period of the pulse-activated is the advantage to allow the bubbles rising-off the solution with no more bubble created and make system neutralized before getting activated in next cycle. It seems nonsense when getting longer rest period could reduce loss and less productivity, then the shortest rest period should be where there is no bubble in electrolyte.

2 Approach and experiment procedure

When current supply to the electrolysis cell, reaction occurred and gas bubble created. The bubble form on electrode surface then detach and rise-off electrolyte surface. Before the bubble rise-off the solution, this stage caused more energy consuming and augment higher resistivity. In addition, the bubble coverage reduce active area of electrode and make reaction prohibited. These unavoidable problems could basically solve by making bubble rise-off the solution quicken to get higher efficiency. So, the parameters that effect bubble behavior (e.g. bubble size, bubble rising velocity) should be considered.

This research uses a combination method to study a role of a rest period in hydrogen electrolysis efficiency improvement. Fig. 1 shows an overall structure of the system, the arrows connecting the block diagrams demonstrate the electrical flow (solid line) and signal flow (dashed line). The electrolyzer used 20%wt KOH solution as electrolyte under atmospheric pressure and the temperature was controlled at 353K. Both electrodes were made of stainless steel connected to the adjustable current source power supply of 10-50A in range which used to control activation parameters via a pulse generator. The gases obtained from the cathode and the anode were collected separately in gas tubes. The volume of gas produced was measured using gas mass flow meter, and re-checked with acoustic emission technique.

Therefore, broadband acoustic emission sensors were installed beside both electrodes in the chamber.

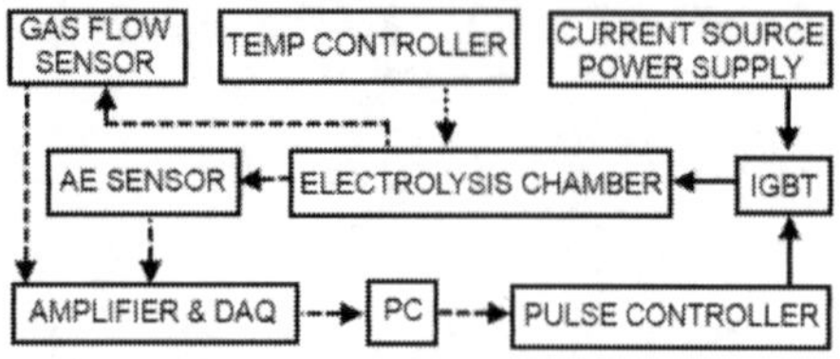

Figure 1. Overall structure of the system.

2.1 Bubble

The gaseous product of electrolyzer is generally known as a reaction inhibitor [4]-[6]. The coverage bubble on electrodes is directly affected with electrolytic resistivity, so the rest time in pulse-activated achieved the system's parameter reset, with no current input and the existing bubbles rise off the electrolyte solution. From the simple law of motion, the rest period (t_{rst}) can be calculated by

$$t_{rst} = y / v_t \tag{1}$$

According to equation (1), the average distance of bubbles traveling in y-axis (y) has been assumed as no effect in bubble diffusion. The bubble's terminal velocity (v_t) lies on the bubble size. The bigger can reach the top surface quicker although drag force would be greater. It is assumed that all bubbles are spherical (with constant shape), inviscid and irrotational flow outside bubbles, no boundary layer separation, and no internal flow circulation inside bubbles. From the previous study [7], the rise velocity of the single bubble (v_b) was calculated and the free stream velocity or swarm velocity (v_s) has been derived as a function of the voidage (ε). Hence, the mathematical estimation of rest period indicated as equation (2) with the following parameters: bubble's radius (r), difference of density between inside and outside bubble ($\Delta\rho$), gravitational acceleration (g), electrolyte's dynamic viscosity (μ).

$$t_{rst} = 9y\cdot\mu / (r^2\cdot\Delta\rho\cdot g)(1 - \varepsilon)^3 \tag{2}$$

So, the minimum rest period mainly depends on the bubble size and the void fraction. In other words, the activation current was a significant parameter because it has a direct effect on the void fraction. Although the void fraction usually become a lower value due to rest period, but the greater activation current always make greater value of void fraction.

2.2 Pulse activation

The past literature on the pulse-activated hydrogen electrolysis presented the advantage of activation frequency that led more efficiency [8]. So, the later researches explained the frequency response behavior of the pulse-activated to the influence of natural frequency using the electric circuit transfer model and set the activation frequency in range of kilohertz and over [2], [3]. According to earlier experiments, all results showed the efficiency enhancement on the natural frequency activation. However, the frequency selected in modeling should be migrated during activation because of eigenvalue migration effect, and the experimental parameter and environment were sensitive. In practice, the activation frequency cannot be estimated as a constant. Following this idea to measurement and tracking natural frequency, the control unit and sensor with high precision and high sensitivity are needed. Consequently, it was very difficult to maintenance the system depending on the sophistication of the system without high cost. Moreover, the measurement of some parameters cannot be taken accurately during activation. Hence, the low frequency should be used in this research to study the effect of bubble to rest period of a pulse activated without natural frequency awareness.

2.3 Bubble observation

At present, the study of bubble's dynamic using high speed camera is popular, with the advantage in a good accuracy estimating bubble's size in a high volume of bubble and/or in depth dimension.

Acoustic Emission (AE) is one of the non-destructive testing methods that uses the phenomenon of instantaneous acoustic waves propagating in media when a physical change occurs in the media's structure. So, in electrochemical reaction, AE has been used widespread for electrode monitoring and corrosion [9]-[11]. According to the experiments, natural frequency of the breaking hydrogen bubble is about 120-150 kHz, and the parameter "AE count" is varied depending on gas production rate.

The measurement of bubble breaking's natural frequency is an indirect measurement of bubble size. The determination of the breaking acoustic frequency (f_b) corresponding to the bubble radius (r_0) can be formulated as equation (3) [11] with the following parameters; static pressure in the fluid (P_0), specific heats for the gas (γ), Newtonian fluid of constant density (ρ_0).

$$f_b \approx sqrt((3\gamma\cdot P_0)/\rho_0)/(2\pi\cdot r_0) \tag{3}$$

3 Results and discussion

3.1 Bubble and rising velocity

In the experimental setup, the measured parameters include activation peak current, averaged activation current, voltage of electrolyzer, gas production rate, and acoustic signal from AE sensor. Those were captured and recorded in time domain.

With the proposed method, the rest period will be recognized as a priority. According to a determination of the rest period required when no bubble remained in electrolytic solution, equations (1) and (2) were used with some calculation details: 10A activation current, 200 micron bubble diameter, 5 centimetres of averaged distance when bubble travels to the top of electrolyte surface, and the gas bubble set as oxygen (higher density than hydrogen; rising time would be dominant). It is found that at least 455 milliseconds of rest period is needed to meet the requirement. Nevertheless, the actual rest period measured with the AE sensor was about 150 millisecond (red border in Fig. 2), the difference in the results depends on the bubble size in the major bubbles.

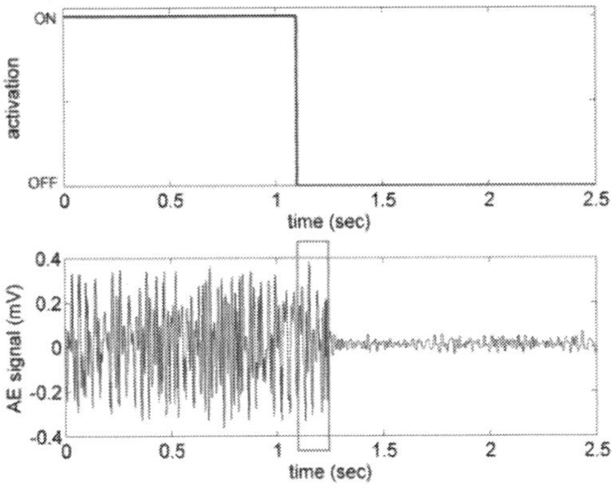

Figure 2. Time-domain acoustic emission of bubble breaking

Having considered on the bubble size using the acoustic signal measurement during bubble rising on the rest period, the frequency response was shown in Fig. 3. There are two controversial frequency bandwidths at 120 kHz and 260 kHz. By the equation (3), the two represented frequencies figure that the bubbles consist of two major bubble sizes. Oxygen bubbles reflect lower bandwidth and hydrogen bubbles attribute higher frequency.

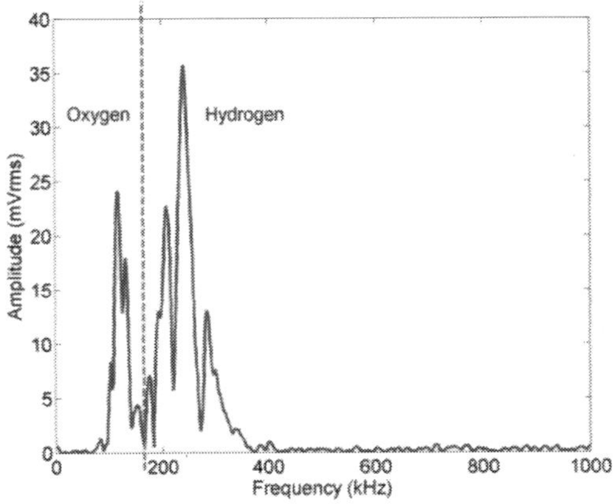

Figure 3. Acoustic frequency responses of bubble breaking.

3.2 Production efficiency

From the experimental results, only hydrogen production efficiency was considered to determine heating power as the output with the electrical power input. As shown in Fig. 4, the production efficiency presented in terms of rest period varying from 2 milliseconds to 800 milliseconds with the activations of 10, 20, 30, 40, and 50A peak current and the frequencies of 1, 5, 10, 50, and 100Hz. Due to the rest period and the activation frequency has algebraic relationship and the large variation of rest period, the x axis also represent as logarithmic scale. According to rest period variation, the results can be grouped by activation frequency ranges and shown as almost linear relationship to the production efficiency. In the other words, each of activation frequency has individual of proper rest period. Focusing on the activation current $30A_{peak}$ at the 160 millisecond rest period, this point was the highest efficiency and interesting that the rest period was approximate to the rest period requirement obtained from AE measurement. However, the variation of rest period cannot simply clarify in terms of the efficiency explanation, so the rest period was represented by duty cycle (rational of activation period and cyclic period) for easy interpretation and practical setup in the future.

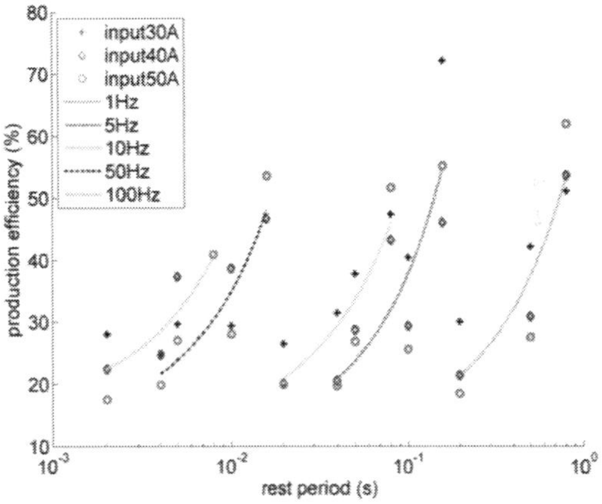

Figure 4. Production efficiency in terms of rest period in various frequencies.

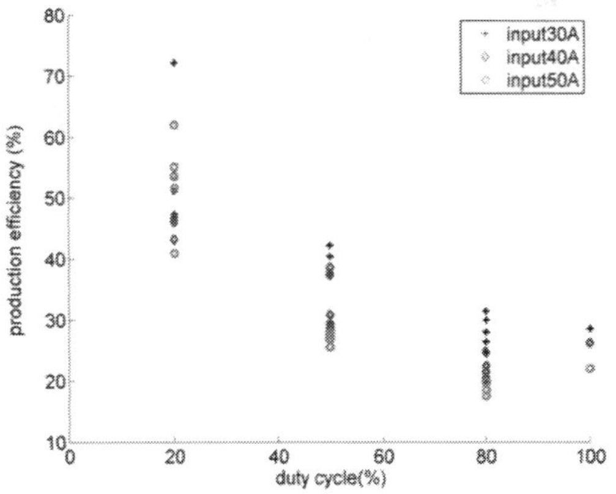

Figure 5. Production efficiency in term of duty cycle

The relationship between production efficiency and duty cycle was shown on Fig. 5. At the same result of rest period variation, duty cycle was represented instead. The pulse-activated was set on 20%, 50%, 80%, and 100% duty cycle, with frequency 1, 5, 10, 50, and 100Hz. The markers represent as peak current amplitude scattered on production efficiency and duty cycle plane. Similar to Fig.

4, duty cycle cannot describe the optimum operating condition of the pulse activated, but the activation with lower duty cycle tends to higher efficiency especially in lower activation peak current. Moreover, some of pulse-activated lack of efficiency improvement ability than the 100% duty cycle (continuous activation), that meant this is not always a suitable choice for improving efficiency of electrolysis hydrogen production. However, the operation with lower duty cycle must be concerned.

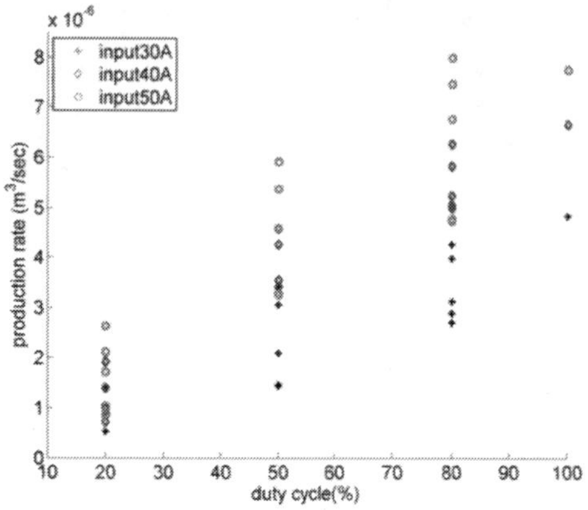

Figure 6. Hydrogen production rate in terms of duty cycle

3.3 Production rate and resistivity

Taking into account on the hydrogen production rate, it is known that the bubbles created reduce the active area of the electrodes and increase resistivity of the system, the result also presented in difference to the efficiency. Fig. 6 presents the relationship of hydrogen production rate as a function of activation duty cycle, by the increase of duty cycle, the production rate also has a tendency to increase according to Faraday's law of electrolysis. The high activation current amplitude and/or high duty cycle gives high production rate which is proportional to the averaged input current. In Fig. 6 there are some scatters of production rate at the same current and same duty cycle values. This is because of the effect of activation frequency. The lower frequency tends to achieve higher production rate. In comparison with continuous activation, the production rate of a low frequency pulse-activated was preferable because the lower frequency has longer rest period and makes bubble coverage clear from the electrodes to get ready for activation. The advantage on production rate of the lower frequency activation scheme was consistent with the use of low frequency on the view of efficiency. Corresponding to Fig. 5, the low frequency provides higher efficiency at the same duty cycle. Our conclusion is that the selected activation regime as low frequency and low duty cycle has been proved to be suitable for the electrolyzer setup.

In the previous study, increasing of activation current resulted in more fine bubble [12]. Agree with the considering on resistivity and bubble size relationship [13], too small bubble size trend to higher resistivity but the big bubble was consistent. Therefore, the increase of activation current was relative to higher resistivity due to bubble size effect. Fig. 7 demonstrates the resistivity of electrolytic solution during pulse-activated rest period as a function of average input current. The small increase rate of resistivity meant that there is a little change of bubble size. In accordance of AE bubble's bandwidth measuring on the experiment, the breaking frequency described that there was a small variation of bubble size due to the activation current increase. Hence, the smaller bubble due to increase of activation current got slower rising velocity and the resistivity also increase. Likewise the result in efficiency, the low activation frequency gives better advantage.

Consequently, the use of pulse-activated in industrial application should consider both production rate and efficiency. The use of pulse-activated for efficiency enhancement but remain production rate as continuous activation needs to activate multiple cells in sequence. However, in the industrial application hydrogen production efficiency was not only determine as production efficiency, but the efficiency in the rational of installation space was necessary.

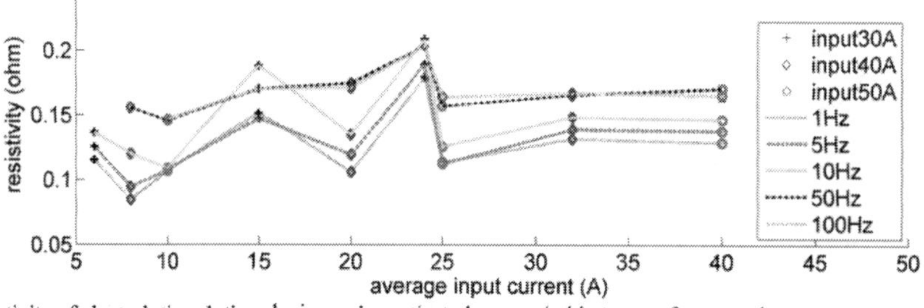

Figure 7. Resistivity of electrolytic solution during pulse-activated rest period in terms of average input current

Though, the pulse activated electrolysis hydrogen production takes more installation space than the conventional, but it gains higher efficiency and gives electrode's life time longer due to less crystal deposited on electrode surfaces [6].

4 Conclusion

The experimental results confirm the strong effect of bubble interacting with production efficiency. The lower frequency is advantageous in long rest period according

to bubble rising time while the longer rest period makes productivity depreciated. So, the minimum rest period for free rising bubble, until empty, needs to be determined.
In the study, the activation scheme comprising 30A activation current at 5Hz activation frequency and the rest period of 160 millisecond gives the best result and agreed with the least rest period that achieved from AE records. Duty cycle has been proposed to use as a practical setup parameter for simple analysis. Similar to the result emerging on the rest period, the production efficiency corrupts when increasing duty cycle. According to the rest period of 160 millisecond, the same activation scheme of pulse-activated shows higher efficiency than others, in contrast, simultaneously the other lower activation currents have a tendency to gain higher efficiency.

Moreover, the lower frequency tends to achieve higher production rate because the pulse-activated help the bubbles detaching from the electrode easier and increase active area. The experiment was given the effective criteria to use pulse-activated on electrolysis hydrogen production by controlling activation scheme with low frequency and low duty cycle. With the considering on the resistivity, the higher resistivity caused higher in Ohmic's loss and efficiency reduction. The use of lower activation frequency affected the lower resistivity similar to the production efficiency.

According to Faraday's law of electrolysis, the gas production is proportional to the averaged input current. So, the increase of duty cycle leads the production rate has a tendency to increase and vice versa. Definitely, the volumetric gaseous produced by electrolysis was strictly conformed to the Faraday's law, so the bubble size was the dominant parameter that indicated remaining voidage within specific time.

To manifest the bubble interact with electrolysis reaction, the overpotential regarding to bubble size requires a further study for the most effective activation procedure on bubble size controlling.

Acknowledgement

This materials based on acoustic emission testing is supported by Maintenance Technology Center (MTC), King Mongkut's University of Technology Thonburi.

References

1. K. Mazloomi, N.B. Sulaiman, and H. Moayedi, Int. J. Electrochem. Sci., **7**, 3314-3326, (2012).
2. K. Mazloomi, N.B. Sulaiman, and H. Moayedi, Int. J. Electrochem. Sci., **7**, 3466-3481, (2012).
3. B. Mandal, A. Sirkar, A. Shau, et al., Int. J. Renew. Energy Res., **2**, 99-102, (2012).
4. K. Zeng, and D. Zhang, Prog Energy Combust Sci., **36**, 307-326, (2010).
5. L.J.J. Jansseen, and E. Barendrecht, Electrochem. Acta, **28**, 341-346, (1983).
6. D. Hristova, I. Betova, and T. Tzvetkoff, *EMCR-2012*, (2012).
7. A.A. Kendoush, Chem. Eng. Sci., **56,** 235-238, (2001).
8. N. Shimizu, S. Hotta, T. Sekiya, et al., J. Appl. Electrochem., **36**, 419-423, (2005).
9. N. Kircheva, S. Tant, B. Legros, S. Genies, et al., *30th EWGAE and 7th ICAE,* (2012).
10. K. Wu, S. Lee, W. Lee, et. al., *TMAM 2015*, 33-37, (2015).
11. C. Jirarungsatian, A. Prateepasen, Corros. Sci., **52**, 187-197, (2010).
12. Khosla, N.K., and Venkatachalam, S., Pulsed Electrogeneration of Bubbles for Electroflotation, Journal of Applied Electrochemistry, vol. 21, pp. 986-990, (1991).
13. K. Tangphant, K. Sudaprasert, S. Channarong, Russ J Electrochem, **50**, No. 3, 253-259, (2014).

Study on Absorption Signal Interference of Gas Concentration Measurement Using Laser Absorption Spectroscopy

Saerom Maeng[1,2], Sunghyun So[1,2], Miyeon Yoo[1], Jungho Hwang[2] and Changyeop Lee[1]

[1]*Korea Institute of Industrial Technology, Thermochemical Energy System R&D Group, 31056 Seobuk-gu Cheonan-si Chungcheongnam-do, Republic of Korea*
[2]*Yeonsei University, Mechanical Engineering, 03722 Seodaemun-gu Seoul, Republic of Korea*

Abstract. In order to optimize the combustion condition of the combustion system, it is important to know the information of the physical properties which vary during combustion. Gas concentration and temperature are the major target properties but it is difficult to measure exactly at combustion system. In this paper, a distributed feedback diode laser which wavelength is tunable in accordance with a function generator's output wave is applied to realize the laser absorption spectroscopy measurement. A concentration measuring test for 99% CO_2 gas was performed as basic experiment and major experiments were conducted on separation of interfered absorption signals at CO_2 and CO mixed condition.

1 Introduction

In order to optimize the combustion condition of a large combustion system such as a powerplant boiler, it is important to obtain the informations about the temperature, the gas concentrations and other properties at various combustion conditions. [1] Generally, a gas analyzer is used to measure the combustion gas concentrations. And a semiconductor sensor, a thermal sensor and an electrochemical sensor are the gas measuring sensors in most gas analyzers. However, these sensors are sensitive to the temperature and humidity changes, sensor replacement is required frequently to secure the reliability.

Almost gas concentration measuring techniques have been limited as sampling type until now, but technical development of the laser absorption spectroscopy enables to measure the internal temperature and the gas concentration of a huge combustion system in real time. [2], [3] It is very important to select a wavelength in this technique because the absorption signals of other species are easily interfered in a wavelength region as shown in Fig. 1. Moreover, there are lots of interference between near absorption signals in a single gas component, it is very difficult to find an independent absorption signal. As shown in Fig. 2, it is good for measuring combustion product gases such as CO and CO_2 at 6250~6550cm^{-1} region, becase there is no H_2O absorption signal. However, considerable number of signals are interfered by themselves.

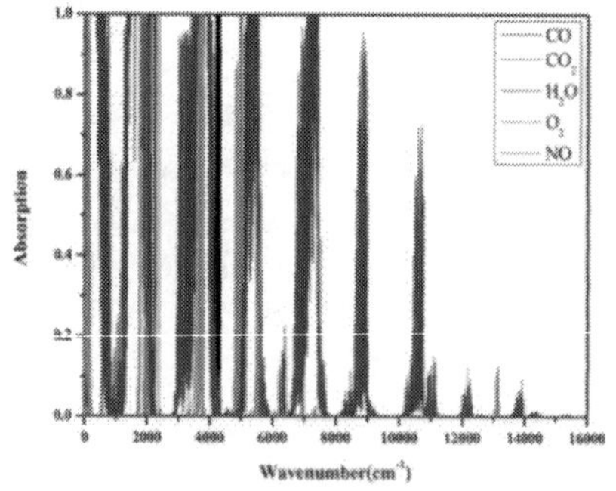

Figure. 1. Absorption spectrum of various chemical species at 296K [4].

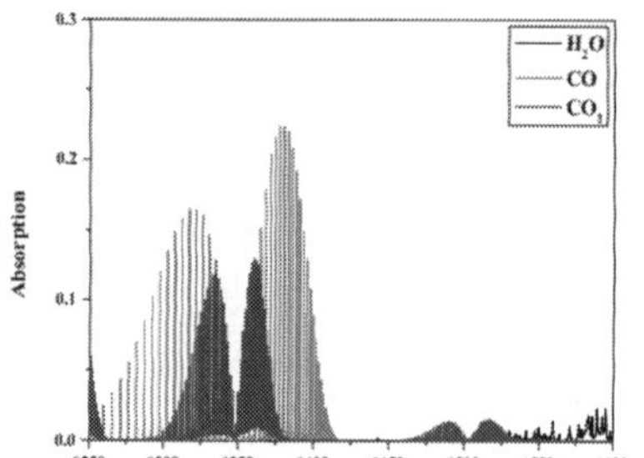

Figure. 2. Absorption spectrum of CO, CO_2 and H_2O at 296K [4].

 DOI: 10.1051/matecconf/20167714002

In this paper, a separation method of the interfered absorption signals is suggested to measure the respective gas concentrations of CO_2 & CO mixture gas and experimental results are discussed with focusing on the measurement error.

2 Basic principle

Beer-Lambert law is the basic principle of laser absorption spectroscopy that relates the transmitted intensity I through a medium of length L to the incident intensity I_0 as

$$T_v = \frac{I}{I_0} = \exp(-k_v \times L) \qquad (1)$$

where T_v is the transmissivity and k_v is the spectral absorption coefficient. k_v is expressed as follows.

$$k_v = PXS(T)\varphi_v \qquad (2)$$

where P is the pressure, X is the mole fraction of the absorbing species, $S(T)$ the linestrength of the transitions and φ_v the lineshape function. The $k_v L$ is known as the absorbance $\alpha(v)$

$$\alpha(v) = k_v L = -\ln\frac{I}{I_0} = PXS(T)\varphi_v L \qquad (3)$$

Since the lineshape function $\alpha(v)$ is normalized such that 1, the integrated absorbance can be expressed as

$$A = \int \alpha(v)dv = PXS(T)L \qquad (4)$$

Therefore, the mole fraction of absorbing species can be obtained:

$$X = \frac{A}{PS(T)L} \qquad (5)$$

3 Experimental setup

Fig. 3 shows the optical experimental system layout to measure the gas concentration using laser absorption spectroscopy. We used a distributed feedback laser(DFB laser) as a light source and wavenumber of the light is about 6371 cm^{-1} where the CO_2 and CO gas absorption signals including overlapping signals are variously distributed. And the wavelength was modulated slightly using a function generator. And an aluminum oxide(Al_2O_3) tube was used to trap the target gas. CO_2 & CO mixture gas was used to evaluate the utility of proposed signal separation method.

The modulated laser light was divided by two direction through a 1x2 coupler, the first one employed an incident light passing a solid etalon. The solid etalon was used for wavelength conversion. The second light was changed into parallel light through the collimating lens(CL1), passed the gas medium inside the aluminum oxide tube and then arrived at the photo detector(PD2). Then optical absorption signals were obtained and calculated by analysis algorithms.

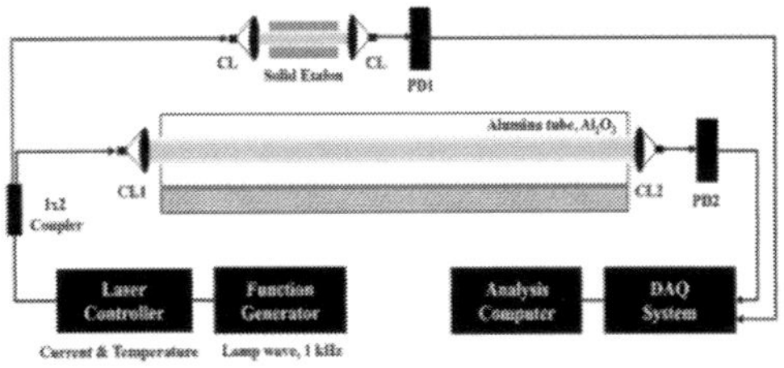

Figure. 3. Optical measurement system layout (CL: collimateng lens, PD: photo detector)

4 Results

Fig. 4 shows the absorption signals of the CO_2 gas 99% and CO_2 85% & CO 15% mixture gas in the same wavelength region. When single gas (CO_2 99%) is distributed in the tube, Only one absorption signal was observed. And two absorption signals were observed at mixed gas (CO_2 85% & CO 15%) condition. As shown in Fig.4 (b) graph, it is difficult to decide the absorption area that is essential for calculating concentration because of CO_2 and CO signal overlapping.

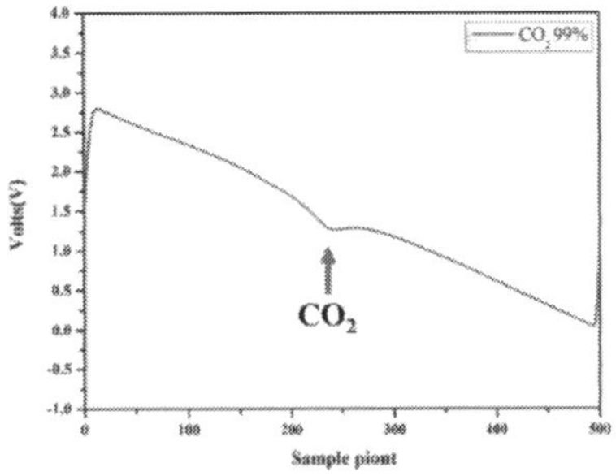

(a) Absorption signal CO_2 gas

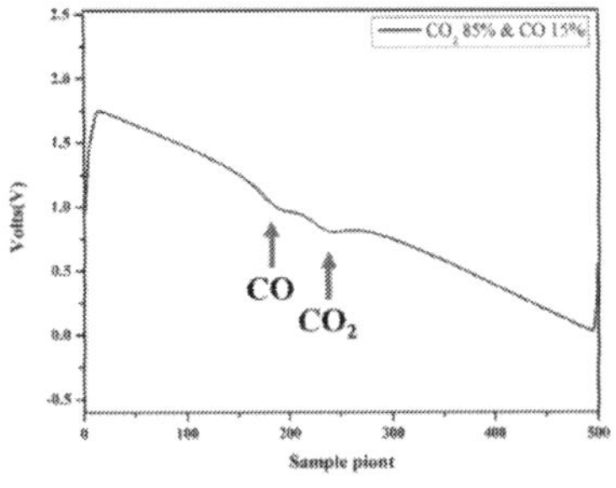

(b) Absorption signal CO_2 & CO mixed gas

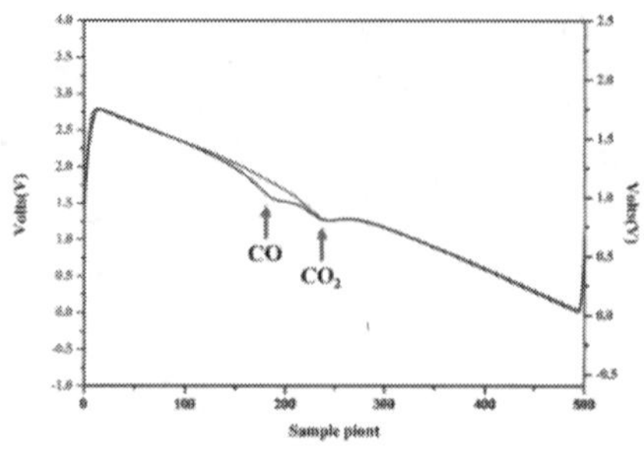

(c) Comparison of absorption signals

Figure. 4. Absorption signal measurement data of CO_2 and CO_2 & CO gas

However the single absorption lineshape of CO_2 gas is symmetric as shown in Fig.5 (a). Then, respective lineshapes can be inferred from the interfering signal as considering the symmetric characteristic of absorption signal. It is possible only when signals are not totally overlapped, in other words, each signal's full maximum height point and more than half absorption lineshape should be exposed.

Figure 5 (b) Graph shows the inferred absorption lineshapes, the dotted lineshapes are inferred. Table 1 shows the analysis results of absorption signals. Precise concentration analysis had been achieved in case 1, when the absorption signal was isolated alone. Case 2 shows the result of each inferred absorption lineshapes which are inferred by above mentioned manner. Measurement gas concentrations of CO_2 and CO were calculated by 93.76% and 18.12% each, however, the real gas concentrations were 85% and 15%. The calculate values were bigger than real values. It results from the problem of inference method. When the overlapping occurs between two absorption signals, it is expected that overlapped parts of the combined absorption signal contribute to make higher full maximum heights of the signal. Because respective absorption lineshapes were inferred from already higher height signals, absorption area values were calculated larger as well and they induced bigger predicted concentration values. The problem was also found to check the absorbance area. The area sum of inferred absorption lineshapes was 15.07% bigger than real absorbance area of combined absorption signal.

Case 3 of Table 1 shows the result of compensating analysis method. In this case, inferred absorption lineshapes were used to find the area ratio information only between CO_2 and CO gas absorption signals. And quantitative values were obtained from integral calculus of combined absorption signal. As the area ratio of two inferred absorption lineshapes were 0.56 versus 0.44, real absorbance was divided by that ratio. And calculated gas concentrations of CO_2 and CO using this manner were 81.12% and 15.69% respectively. Error rates of case 3 were reduced considerably than those of case 2.

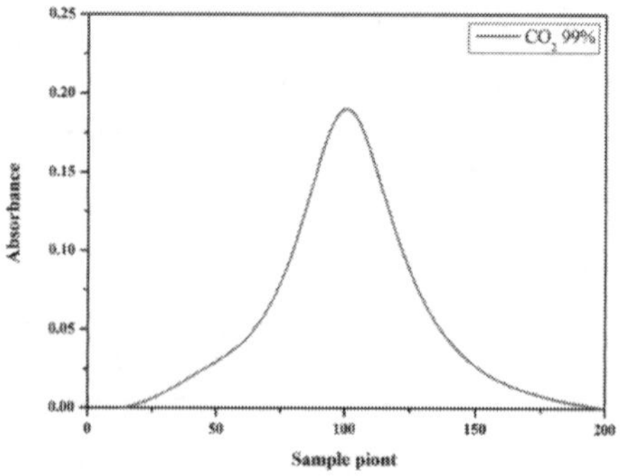

(a) Absorption lineshape of CO_2 gas

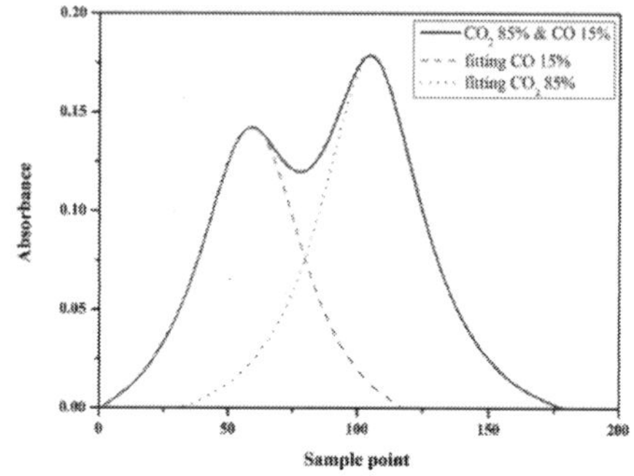

(b) Absorption lineshape of CO_2 & CO mixed gas

Fig. 5. Absorbance measurements

Table 1. Setting Word's margins.

(a) Case 1

Gas species	**CO_2 (99%)**
Measured concentration	99.04%
Error rate	0.04%

(b) Case 2

Gas species	**CO (15%)**	**CO_2 (85%)**
Measured concentration	18.14%	93.76%
Error rate	20.9%	10.3%

(c) Case 3

Gas species	**CO (15%)**	**CO_2 (99%)**
Measured concentration	15.69%	81.124%
Error rate	4.6%	4.56%

5 Conclusion

The optical absorptions of CO_2 and CO gas are occurred in the infrared region, and many of them are overlapped signals that precise informations are not revealed. In this paper, lineshapes were inferred from the interfering signal as considering the symmetric characteristic of absorption signal at partially overlapped condition. As a results, respective gas concentrations were successfully calculated using absorbance area ratio of inferred lineshape and real integral area value of combined absorption signal. The possibility that the interfered absorption signals can be used to analyze the gas concentration and temperature has been confirmed through this study. It will useful to analyze the physical properties at high temperature and multi-species mixed condition.

Acknowledgement

This work was supported by the Korea Institute of Energy Technology Evaluation and Planning(KETEP) and the Ministry of Trade, Industry & Energy(MOTIE) of the Republic of Korea (No. 20152010103630).

References

1. M. A. Allen, E. R. Furlong, R. K. Hanson, Applied Combustion Diagnostics, edited by K. Kohse-Hoinghaus and J.B Jeffries, Taylor and Francis, New York (2002)
2. R. M. Mihalcea, M. E. Webber, D. S. Baer, R. K. Hanson, G. S. Feller, a W. B. Chapman, Appl. Phys. B, 67 (1998)
3. A. Farooq, H. Li, Jay B. Jeffries, and R. K. Hanson, 43rd AIAA/ASME/SAE/ASEE Joint Propulsion Conference & Exhibit Cincinnate, AIAA
4. HITRAN DATABASE infermation on http://hitran.iao.ru/

MATEC Web of Conferences 77, 15001 (2016) DOI: 10.1051/matecconf/20167715001
ICMMR 2016

Aircraft Path Planning under Adverse Weather Conditions

Xie Z. and Zhong Z. W.

School of Mechanical and Aerospace Engineering, Nanyang Technological University, Republic of Singapore

Abstract. In recent years, flight safety is still one of the main issues for all airlines. En route civil airplanes may encounter adverse weather conditions. Some fatal airplane accidents happened because of the weather disturbance. Moreover, we should also design path to avoid the prohibited area. Therefore a good path planning algorithm plays an increasingly important role in air traffic management. An efficient path planning algorithm can help the plane to avoid severe weather conditions, restricted areas and moving obstacles to ensure the safety of the cabin crews and passengers. Here, we build our algorithm based on the A* search algorithm. Moreover, our algorithm can also find the path with least energy costs. As a result, our algorithm can improve the safety operation of the airplanes and reduce the workload of pilots and air traffic controllers.

1 Introduction

In recently years, some airplane accidents happened around the world. How to ensure the air traffic safety operation has been a main topic in Air Traffic Management (ATM). Researchers around the globe try their best to think of ways to not only increase the safety but also improve the efficiency of the airplane. As we can see in Fig. 1, adverse weather is one of the main factors which can cause the fatal aircraft incidents. How to avoid the adverse weather and create a new efficient path is our research goal. Adverse weather will also have effect on the visibility of the aircraft. [1] We also take into consideration of the effect from other aircraft in the same airspace and limitations of airspace restriction areas. The prohibited areas can also cause danger to the civil airplanes. For instance, civil airplanes can interrupt military drills or can be hit by missiles and fighters if they enter military prohibited areas or war zones.

Figure 1. Adverse Weather condition

Free flight concepts [2] was first introduced by Radio Technical Commission for Aeronautics (RCTA) in 1995. As shown in Fig. 2, free flight means the en route aircraft pilots are more responsible for the separation of aircraft while reducing the workload of air traffic controllers (ATCs), but in case of emergency, ATCs should intervene if pilots fail to execute separations. This concepts can free the ATCs so that they can focus more on the users' request.

Figure 2. Pilots should have more freedom and responsibility to separate and maneuver

Christina [3] introduced a novel 4D trajectory optimization method that creates a path based on adverse weather data. This application can successfully reduce the detour and fuel cost. They created a reference route for comparison of the detour deduction. But the author only considered storm cell related circumnavigation, other weather and traffic conflicts. Airspace restrictions were still not taken into consideration yet.

Wang [4] developed an air route network node optimization method in order to avoid the tree areas, including prohibited area, restricted area and danger area. A cellular automata model was deployed to find the shortest path for optimization problems. The authors used a case study to test the effectiveness of the model. In the end, the methods can reduce the conflict risks of cross network and decrease flight costs and the nonlinear coefficient.

The FAA rules [5] state that the minimum separation of en route flights is 5 nmi; 3 nmi for lower level terminal space. When aircraft are detected in the same airspace, the minimum separation must be satisfied in case of emergency. In our simulation case study, we use single dot to represent aircraft which share the same airspace. We take the minimum separation into consideration when optimizing the route under specific conditions.

A* search algorithm was first introduced by Hart and Nilsson [6] in 1968. It can transfer heuristic information to mathematical theory of graph searching. Also A* search algorithm is proved to be effective in finding the optimal route while examining the minimum number of nodes. The cost function contains two part, one is the cost from the initial position to current position; another one is the estimate cost from the preferred goal to the current position. We build our algorithm based on A* search algorithm and use our own cost function to get the most efficient solution.

Soltani [7] compared Dijkstra, A* and Genetic Algorithm (GA) for path planning in constructive sites. The objective function is combined of shortest path, high visibility and lowest risks. The three path planning methods are compared in time consumption, accuracy and complexity. He found that A* search algorithm will be admissible if the distance from the current position to the preferred goal is always available by heuristic function h. This is the restriction of A* search algorithm that can cause to overestimate the cost to the goal.

Garau [8] utilized A* search algorithm on path planning of unmanned underwater vehicles in a real Mediterranean Sea situation. The aim of their research is to find the optimal energy saving path based on a current field model and satellite altimetry. Moreover, the vehicle is capable of obstacle avoidance. The results show that the path planner is robust and efficient for the autonomous underwater vehicle to replan a path to the destination with shorter time and less energy cost.

Adem [9] compared A* search algorithm with GA in dynamic mobile robot path planning. He found that GA performed better than conventional A* search algorithm both in travelling distance and computation time. Also, modified A* search algorithm has better computation time but it is easy to reach a local minimum. GA can find the global optimal solution efficiently.

Chee and Zhong [10] used a nonlinear complementary filter to estimate the attitude of unmanned aerial vehicle (UAV) and implemented a reactive quad-directional algorithm to avoid obstacles. The results show the UAV can achieve attitude control and navigate open environment autonomously with obstacle avoidance capability.

Xie and Zhong [11] developed methods for the unmanned surface vehicle (USV) to create path automatically to avoid the obstacles on the sea surface. The simulation results in MATLAB show that the USV can find the most efficient path and lead the vehicle to the destination.

2 Methodology

As we can see in Fig. 3, a commercial airplane encounters an adverse weather condition. It may threaten the on-board equipment and also it may lead to the operation failure of the pilots. This is a dangerous situation which may cause a fatal disaster. We build our own algorithm based on the civil flight circumstance and simulate the situation using MATLAB when an airplane is under an adverse weather condition and airspace limitations. Paul [12] of Mathworks also created a MATLAB tutorial to compute the shortest path to the goal location using the A* search algorithm to help users better understand the function of the A* search algorithm.

Figure 3. Civil airplane under an adverse weather condition

Air flight path planning is different from ground vehicle path planning. We trim the A* algorithm to make it more fit to the en route environment. We change the optimization function to consider the deviation and turning angle as well.

2.1 Rules

Flight routes must also obey the following rules for the safety of aircraft and restriction of the airspace.

- Restricted or prohibited area cannot be violated.
- Weather conditions are forecasted and updated.
- The minimum distance between two aircraft en route should be 5 nmi [5].
- Adverse weather should be avoided for safety reasons [3].
- Find the least cost to reach the minimum fuel burning goal.
- The airplane has the ability of obstacles avoidance.

2.2 Algorithm

We use a grid to describe a 60x60 nmi area. One unit means 2 nmi. The plane starts from the start gate and ends at the end gate. The start gate and end gate can be chosen randomly before the simulation. The obstacles including the restricted areas, adverse weather areas and other aircraft are randomly distributed as well. The function of the algorithm is to avoid all the obstacles and satisfy the rules at the same time. The optimal path will be shown in black line and the obstacles are shown as red blocks.

The basic A* search algorithm objective function is as follows [6].

$$f(n) = h(n) + g(n) \quad (1)$$

$f(n)$ stands for the optimization function.

DOI: 10.1051/matecconf/20167715001

$h(n)$ stands for the estimate cost from the current position to the final position.

$g(n)$ stands for the cost for reaching the current position from the initial position.

2.2.1 Optimization function

Based on the flight route requirement, we build our own optimization function. The function can be seen below in Eq. 2.

$$F(n) = H(n) + G(n) + A(n) + D(n) \tag{2}$$

$F(n)$ stands for the objective function.

$H(n)$ stands for the estimate cost from current position to the final position.

$G(n)$ stands for the cost for reaching the current position from the start position.

$A(n)$ is the turning angle of the aircraft.

$D(n)$ is the route deviation from the reference route.

Our goal is to get the smallest value of F(n), but at the same time, the optimised route must also follow the restrictions and limitations to ensure the safe operation of the flight route.

2.2.2 Restrictions and limitations

As shown in Eq. 3, the turning angle for the airplane should not exceed the maximum turning angle [13].

$$A(n) < A_{\max} \tag{3}$$

As shown in Eq. 4, the route deviation should also be within a limit. It means the airplane should not deviate from its original path too much.

$$D(n) < D_{\max} \tag{4}$$

As can be seen in Eq. 5, the distance between two aircraft which share the same airspace should be at least 5 nmi [5].

$$Dist(x, y) > 5nmi \tag{5}$$

As can be seen in Eq. 6, the cost function should achieve the least value among all the other entries.

$$F(n^*) < F(n), n \in N \tag{6}$$

3 Results analysis and discussion

3.1. Case 1

As can be seen in Fig. 4, in case 1, the aircraft starts from the initial position which is located at (0, 22.5) of the grid map and ends at the final position which is located in the lower right corner of the map. The obstacles are marked with pink circles in our simulation. The squares with the pink circles are closed for aircraft to pass. In this situation, the closed squares include prohibited area, adverse weather and moving aircraft. The black line shows the optimal path with the least energy cost and travelling distance which is generated by the algorithm. We can see that the algorithm finds the optimal path while avoiding the adverse weather, prohibited areas and other aircraft. The two aircraft which share the same airspace also are separated further than the minimum safety distance, which is 5 nmi. This autonomously generated path to avoid the moving aircraft can reduce the workloads of the ATCs and ensure the safe operation of airspace.

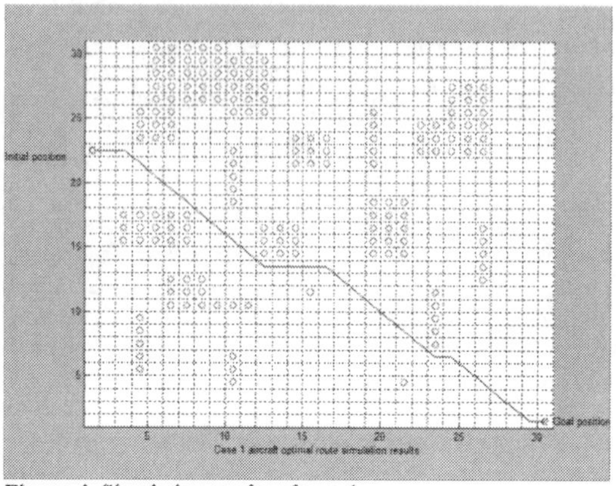

Figure 4. Simulation results of case 1

3.2 Case 2

Scenario 2 is designed for the unmanned vehicle (UV). The UV starts from the top left corner and ends at the lower right corner. This time the obstacles only include adverse weather and prohibited areas. The black line shows the optimal path in this case. We can clearly see in Fig. 5 that the path is the shortest and has the least deviation from the original path. Finally, the UV reaches the destination accurately.

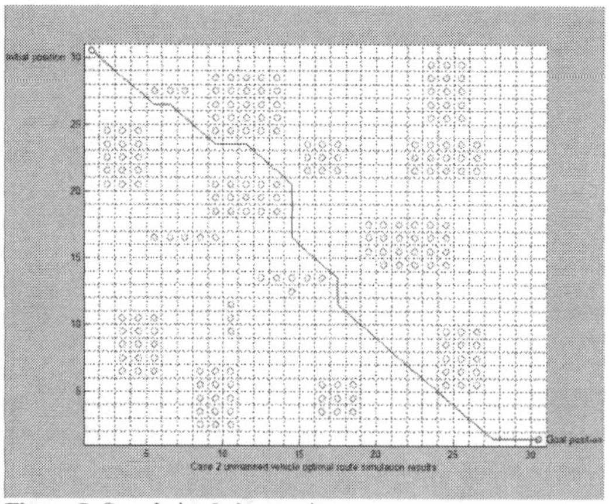

Figure 5. Case 2 simulation results

4 Conclusion

Our path planning algorithm based on the A* search algorithm can successfully find the most efficient path while avoiding the hazardous weather and airspace conditions. Also the obstacles can be randomly distributed so that we can have as many scenarios as possible to test the efficiency of our algorithm. The algorithm can find the optimal path with the least travelling distance and least energy cost while avoiding prohibited areas, adverse weather and surrounding aircraft. In conclusion, it can benefit ATM in decreasing the chance of accidents caused by bad weather and other conflicts, saving energy consumption and making the route more eco-friendly.

Acknowledgement

This research was sponsored by the ATMRI of NTU and CAAS via ATMRI Project No. 2014-D2-ZHONG for Regional Airspace Capacity Enhancement – ASEAN Pilot.

References

1. Xie, Z. and Z.W. Zhong, *Simulated Civil Airplane Visual Conspicuity Experiments During Approaching and Departure in the Airport Vicinity*, in *7th International Conference on Intelligent Systems, Modelling and Simulation*. Bangkok, (2016).
2. Endsley, M.R., et al., *Effect of free flight conditions on controller performance, workload, and situation awareness*, DTIC Document, (1997).
3. Schilke, C. and P. Hecker. *Dynamic Route Optimization Based on Adverse Weather Data*. in *Fourth SESAR Innovation Days*.
4. Wang, S.-j. and Y.-h. Gong, *Research on air route network nodes optimization with avoiding the three areas*. Safety Science. **66**: 9-18, (2014).
5. 7110.65U, F.O., *Para 5-5-4 "[Radar Separation] Minima*. FAA, (2013).
6. Hart, P.E., N.J. Nilsson, and B. Raphael, *A Formal Basis for the Heuristic Determination of Minimum Cost Paths*. Systems Science and Cybernetics, IEEE Transactions on. **4**(2): 100-107, (1968).
7. Soltani, A.R., et al., *Path planning in construction sites: performance evaluation of the Dijkstra, A*, and GA search algorithms*. Advanced Engineering Informatics, **16**(4): 291-303, (2002).
8. Garau, B., et al., *Path planning for autonomous underwater vehicles in realistic oceanic current fields: Application to gliders in the western mediterranean sea*. Journal of Maritime Research, **6**(2): 5-22, (2014).
9. Tuncer, A. and M. Yildirim, *Dynamic path planning of mobile robots with improved genetic algorithm*. Computers & Electrical Engineering,. **38**(6): 1564-1572, (2012).
10. Chee, K.Y. and Z.W. Zhong, *Control, navigation and collision avoidance for an unmanned aerial vehicle*. Sensors and Actuators A: Physical. **190**: 66-76, (2013).
11. Xie, Z. and Z.W. Zhong, *Unmanned Vehicle Path Optimization Based on Markov Chain Monte Carlo Methods*. Applied Mechanics & Materials, **829**, (2016).
12. Premakumar, P. *A* (A Star) search for path planning tutorial*. 2010; Available from: http://www.mathworks.com/matlabcentral/fileexchange/26248-a---a-star--search-for-path-planning-tutorial.
13. Phillips, W.F., *Mechanics of flight*. 2004: John Wiley & Sons.

Synthesis of Phase-locked Loop System Structure with Adaptation Based on Combined-maximum Principle

Andrey Aleksandrovich Kostoglotov [1], Sergey Valerievich Lazarenko [2], Dmitriy Sergeevich Andrashitov [3] and Igor Vladimirovich Deryabkin [1]

[1]*Rostov State Transport University, Department of Informatics, Rostov-on-Don, Russian Federation*
[2]*Don State Technical University, Department of Radio Engineering and Electronics, Rostov-on-Don, Russian Federation*
[3]*Moscow Vitte University, Department of Mathematics and Informatics, Moscow, Russian Federation*

Abstract. The synthesis of the structure of the phase-locked loop system is performed based on the generalized power function maximum condition. The system effectiveness in the lock-on mode is studied via numerical simulation. Compared to the classical one phase-locked loop system with RC-filter this solution is characterized by the higher stability with respect to external influences and by reduced phase error in the steady-state.

1 Introduction

The phase-locked loop (PLL) systems have a wide range of applications in communications technology. The design of PLL is often brought to choosing a structure and parameters of the system at a given initial data and conditions of usage so it would satisfy the required criteria in the best way. PLL is now widely used in the systems described by the nonlinear differential equations of second and higher orders. The design of such PLL needs to determine the conditions for stability of the required mode and the qualitative characteristics of the stabilization process. Many methods providing the synthesis problem solution are known. But they implement a rational but not optimal solutions. Therefore, building an optimal compensating circuit for the phase-locked loop system is actual problem.

In this paper the problem of synthesis of a compensating circuit is formulated as an optimal control problem [1]. As initial system we consider PLL based on single-section RC-filter [2].

The aim of this paper is synthesis and study of the phase-locked loop system with adaptation to external influences on the basis of the combined-maximum principle.

2 Problem formulation

The basic assumptions of dynamics, the Newton's axioms and d'Alembert's principle, allow to formulate the laws of motion in the form of differential equations, which are an integral part of the formulation of optimization problems. But the equal possibilities are provided by the variational principles which determine the stationary properties of some quantities depending on the system parameters.

Let us consider the holonomic controlled dynamical system. Its action integral has the form [3]

$$R = \int_0^{t_1} (T + A)dt, \tag{1}$$

where $T = \frac{1}{2}\sum_{s,k=1}^{n} a_{sk}\dot{q}_s\dot{q}_k$ is the kinetic energy; $\mathbf{q} = [q_1,\ldots q_n]^{\mathrm{T}}$ is the vector of generalized coordinates; a_{sk} is the inertia factors; $A = \int_{\mathbf{q}(0)}^{\mathbf{q}(t_1)} \sum_{s=1}^{n} Q_s dq_s$ is work of the generalized external forces; $\mathbf{Q} = [Q_1(\mathbf{q},\dot{\mathbf{q}},\mathbf{u}),\ldots Q_1(\mathbf{q},\dot{\mathbf{q}},\mathbf{u})]^{\mathrm{T}}$ is the vector of generalized forces which is continuous with respect to all its variables; $\mathbf{u} = [u_1,\ldots u_m]^{\mathrm{T}}$ is the control vector; $\mathbf{q}(0)$, $\mathbf{q}(t)$ are the initial and current values of the generalized coordinates vector; $\mathrm{n} = \dim \mathbf{q} \geq \mathrm{m} = \dim \mathbf{u}$; ^{T} is the symbol of matrix transposition, and the dot denotes time derivative.

Now we require that when the system moves from initial state

$$t = 0, \quad \mathbf{q}(0) = [q_{10},\ldots q_{n0}]^{\mathrm{T}}, \dot{\mathbf{q}}(0) = [\dot{q}_{10},\ldots \dot{q}_{n0}]^{\mathrm{T}} \tag{2}$$

to the final state

$$t = t_1, \quad \mathbf{q}(t_1) = [q_{11},\ldots q_{n1}]^{\mathrm{T}}, \dot{\mathbf{q}}(0) = [\dot{q}_{11},\ldots \dot{q}_{n1}]^{\mathrm{T}} \tag{3}$$

under control $\mathbf{u}$ the relation similar to the Hamilton-Ostrogradski principle for action integral (1) is true:

$$\delta R = \int_0^{t_1} (\delta T + \delta' A)dt = 0, \tag{4}$$

where δT is variation of T, δ' is elementary increment [4], and

$$\delta' A = \sum_{s=1}^{n} Q_s \delta q_s. \tag{5}$$

Note that this requirement lets us state that the Lagrange equations of the second kind are true. For a holonomic system they can be written in the form [3]:

$$\frac{d}{dt}\left(\frac{\partial T}{\partial \dot{q}_s}\right) - \frac{\partial T}{\partial q_s} = Q_s, \; s = \overline{1,n}. \tag{6}$$

Let a constant sign, scalar, continuous with its partial derivatives on the the entire domain function $F(\mathbf{q})$ is given.

Formulation of the optimal control problem is as follows [5]: we need to find the vector-functions $\mathbf{Q}(\mathbf{q},\dot{\mathbf{q}})$ and $\mathbf{q}$ which minimize the functional

$$J_1 = \int_0^{t_1} F(\mathbf{q})dt \to \min \tag{7}$$

with condition (4) and the restriction

$$\mathbf{Q} \in \overline{\mathrm{G}}_Q \tag{8}$$

where $\overline{\mathrm{G}}_Q$ is closed set of valid controls in the space of functions which are sectionally continuous in the finite time interval $[0, t_1]$.

Let us consider the extended functional of the following form:

$$J = \int_0^{t_1} \big(F(\mathbf{q}) + \lambda(T + A)\big)dt, \tag{9}$$

where λ is the Lagrange factor.

Let $\hat{\mathbf{Q}} \in \overline{\mathrm{G}}_Q$ be an arbitrary possible generalized force. Then if $\tilde{\mathbf{Q}}$ minimizes functional (9) it is necessary that its variation is nonnegative $\delta J \geq 0$ for any possible variations $\delta\mathbf{Q}$.

Let us take into consideration the function $\Phi(\mathbf{q},\dot{\mathbf{q}},\mathbf{Q},\lambda)$

$$\Phi = \sum_{s=1}^{n} \big[\lambda Q_s(\mathbf{q},\dot{\mathbf{q}},\mathbf{u}) + V_s(\mathbf{q})\big]\dot{q}_s, \tag{10}$$

where $V_s = \delta F / \delta q_s$. The physical meaning of Φ is a power of actual generalized forces $Q_s = Q_s(\mathbf{q},\dot{\mathbf{q}},\mathbf{u})$ and fictitious generalized forces $V_s(\mathbf{q},\dot{\mathbf{q}}) = \delta F / \delta q_s$ which are defined by the form of the given functional. The function $\Phi(\mathbf{q},\dot{\mathbf{q}},\mathbf{Q},\lambda)$, $\lambda \in \mathrm{R}$, is continuous with respect to the set of variables and is defined in the interval

$$(\mathbf{q},\dot{\mathbf{q}},\mathbf{Q},\lambda) \in \Omega \times \overline{\mathrm{G}}_Q \times \mathrm{R}, \tag{11}$$

where Ω is some area which is bounded in the space of functions and continuous within the time interval $[0, t_1]$. The area represents the set of solutions to equation (5). Boundedness of the area Ω comes from a continuous dependence of solutions to equation (5) with respect to the generalized force.

Theorem [1, 4]. If the generalized force $\tilde{\mathbf{Q}}(t)$ minimizes functional (7) under conditions (2), (3), (4), (6), then there exists such constant Lagrange factor λ, that for any $t \in [0, t_1]$ function Φ of the variables $(\mathbf{q},\dot{\mathbf{q}},\mathbf{Q}) \in \Omega \times \overline{\mathrm{G}}_Q$ reaches at the point $(\tilde{\mathbf{Q}}(t), \tilde{\mathbf{q}}(t), \dot{\tilde{\mathbf{q}}}(t))$ its maximum

$$\Phi(\tilde{\mathbf{Q}},\tilde{\mathbf{q}},\dot{\tilde{\mathbf{q}}}) = \max_{\mathbf{Q}(\mathbf{q},\dot{\mathbf{q}}) \in \overline{\mathrm{G}}_Q} \Phi(\mathbf{q},\dot{\mathbf{q}},\mathbf{Q},\lambda). \tag{12}$$

3 The synthesis of the phase-locked loop structure

Choosing the single-section RC-filter as a loop filter we get the nonlinear differential equation describing the phase-locked loop (PLL) system [3]:

$$T\frac{d^2\varphi}{dt^2} + \frac{d\varphi}{dt} + \Omega_{\mathrm{y}} F(\varphi) = \omega_{\mathrm{H}}. \tag{13}$$

where $\varphi(t) = \varphi_{\mathrm{c}}(t) - \varphi_0(t)$ is difference between phases of signal and adjustable oscillator; $F(\varphi)$ is output of the phase discriminator, $\omega_{\mathrm{H}} = \dot{\varphi}_{\mathrm{c}}(t_0) - \dot{\varphi}_0(t_0)$ is difference between cyclic frequency of signal and adjustable oscillator at the initial time t_0, $(\varphi(t_0), \dot{\varphi}(t_0))$ are initial conditions. This equation can be represented in the form of the Lagrange equation of the second kind.

Therefore, we can formulate the problem to find the adaptive control for PLL system under conditions of external influences based on the combined-maximum principle method.

We now consider the problem to find $U(\varphi,\dot{\varphi}) < |U|$ providing the transition of the dynamic system

$$T\frac{d^2\varphi}{dt^2} + \frac{d\varphi}{dt} + \Omega_{\mathrm{y}} F(\varphi) = \omega_{\mathrm{H}} + U \tag{14}$$

from any initial arbitrary point of the phase space $(\varphi(t_0), \dot{\varphi}(t_0))$ to the final point (0, 0), so that

$$J = \frac{1}{2}\int_{t_0}^{t_1} \varphi^2 dt \to \min. \qquad (15)$$

In accordance with eq. (12) from the condition of maximum of the function $\Phi = [\lambda U(\varphi, \dot{\varphi}) + \varphi]\dot{\varphi}$ with restrictions $U(\varphi, \dot{\varphi}) < |U|$, we get the following PLL [5]:

$$U = -\Omega_y \lambda^{-1} |U| \operatorname{sign}\left(\frac{D|\dot{\varphi}|\dot{\varphi}}{L|\varphi| + \varepsilon} \frac{T}{\Omega_y} + \varphi \right), \qquad (16)$$

where λ^{-1} is undefined Lagrange factor, L is a constant depending on the form of the switching line, ε, D are constants which define correspondingly the shift of switching lines and diameter of the ellipse in the phase space of states of the PPL system.

The PLL structure is defined by the following equation:

$$T\frac{d^2\varphi}{dt^2} + \frac{d\varphi}{dt} + \Omega_y F(\varphi) = \omega_н - \\ -\Omega_y \lambda^{-1} |U| \operatorname{sign}\left(\frac{D|\dot{\varphi}|\dot{\varphi}}{L|\varphi| + \varepsilon} \frac{T}{\Omega_y} + \varphi \right). \qquad (17)$$

During numerical experiment we have obtained the solutions to the differential equations describing the PLL system with the RC-filter (13) and the PLL system with the autoregulation loop (16) on the basis of the combined-maximum principle. The phase discriminator is:

$$F(\varphi) = \sin(\varphi). \qquad (18)$$

The simulation has been performed for the following values: T = 0,014 с; Ωy = 21 рад/с; |U| = 5; λ–1 = 1; ε = 0,001; D = 1; L = 4.

The Fig. 1 shows the lock-on process on the phase plane.

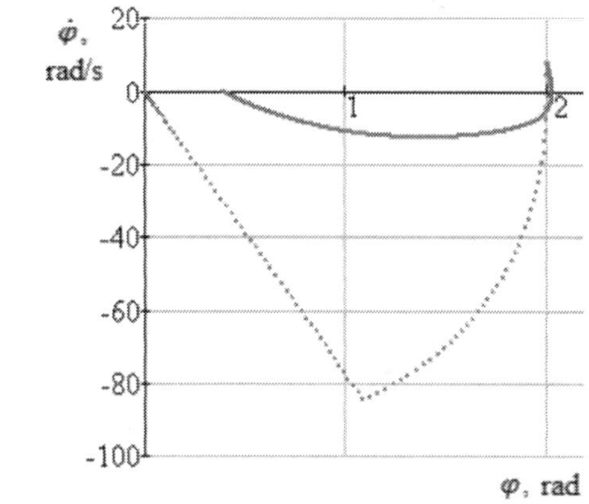

Figure 1. The phase plane of the PLL under consideration.

The Fig. 2 shows the dependence of the lock-on time $t_з$ for the considered PLL systems on the initial frequency mismatch $\dot{\varphi}(t_0)$ for three initial phases mismatches $\varphi(t_0)$ = 0 radian, $\varphi(t_0)$ = 1 radian, and $\varphi(t_0)$ = 2 radian.

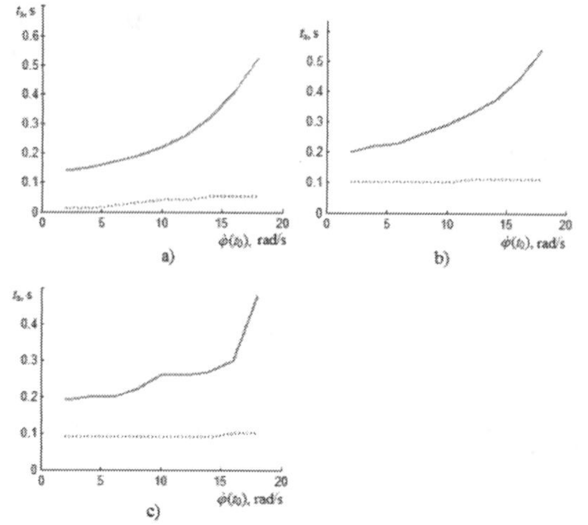

Figure 2. The dependence of the lock-on time on the the initial frequency mismatch: a) at φ(t_0) = 0 radian; b) at φ(t_0) = 1 radian; c) при φ(t_0) = 2 radian.

These results shows that the lock-on time of the PLL-controlled system using the combined-maximum principle method is approximately 2–6 times less and is almost independent on the initial conditions. For this PLL in the holding mode the steady phase difference $\varphi_y = 0$, unlike PLL with RC-filter for which φ_y = arcsin(ωн/Ωy) + 2πk. In addition for the PLL using the combined-maximum principle we see the lock-on possibility for the initial phases mismatches greater than for PLL with RC-filter (which demonstrates beats, see Fig. 3).

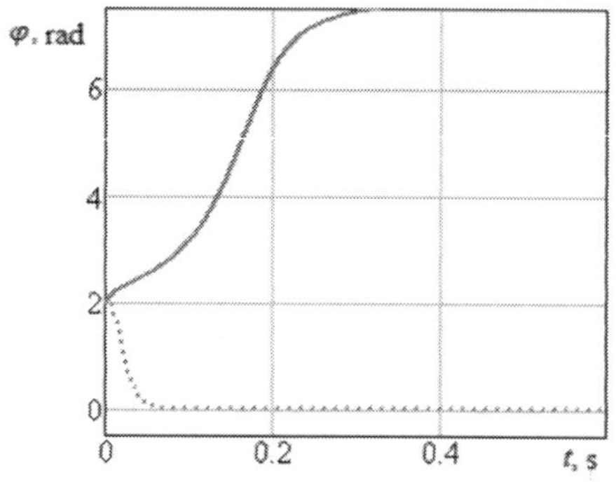

Figure 3. The dependence $\varphi(t)$ when the initial frequency mismatch with respect to the lock-on frequency band of the PLL system with RC-filter ($\varphi(t_0)$ = 2 radian, $\dot{\varphi}(t_0)$ = 22 radian/sec)

The simulation has been also performed with additional model external random interference with $D_\varphi = 1 \cdot 10^{-4}$ radian^2. The results of the simulation are shown on Fig. 4. The numbers mean: 1 – the phase trajectory (16), 2 – the phase trajectory (12). The analysis of the phase portraits lets us declare the higher degree of resistance of the system (16) to external influences compared to (12) and the decrease of the phase errors in steady mode.

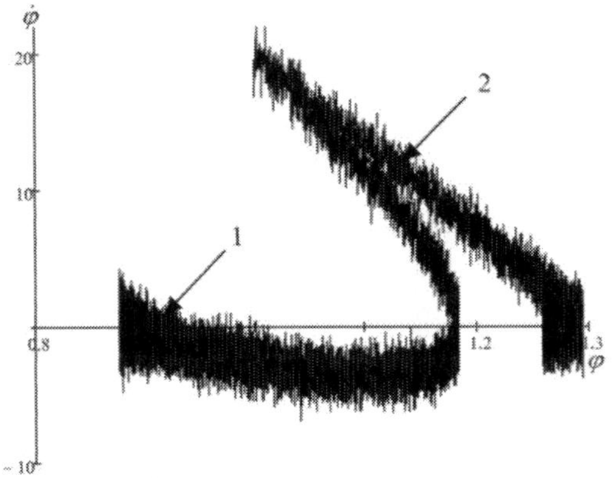

Figure 4. The phase portrait

We can getting alternative equation of PLL based on combined maximum principle

$$T\frac{d^2\varphi}{dt^2}+\frac{d\varphi}{dt}+\Omega_y F(\varphi)=\omega_н-$$
$$-\lambda^{-1}\left(\frac{TD|\dot{\varphi}|\dot{\varphi}}{L|\varphi|+\varepsilon}+\Omega_y\varphi\right). \quad (19)$$

The results of modeling are illustrated on Fig. 5. Its demonstrate are capture process on a phase plane. There are the dotted line shows the solution corresponding to (13), and the solid (19). In this case, the system phase-locked loop with loop autoregulatory advantageously has a high resistance to accidental external interference in comparison with the system phase-locked loop with a simple RC-filter.

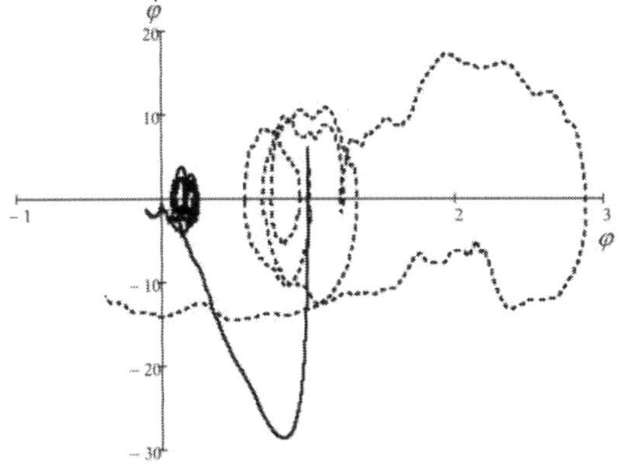

Figure 5. The phase portrait

The analysis of the functioning of the phase-locked loop system with adaptive feedback based on the combined maximum principle shows that the obtained solution can improve the efficiency of the classical scheme with respect to the criterion of minimal lock-on time and minimal synchronization errors. It has found out that this system allows to decrease the lock-on time 2–6 times approximately and to increase the PLL holding range 1.5 times compared to the single-section RC-filter case. In the absence of observation noise the zero phase error is provided in the synchronization mode. This allows to use the obtained solution for constructing precision instruments to measure range and velocity.

4 Conclusion

Analysis of the results of mathematical simulation leads to the following conclusions:

1. The system reduces the lock-on time approximately 2–6 times compared to the PLL with the single-section RC-filter.
2. The system increases about 1.5 times the holding range compared to the PLL with the single-section RC-filter.

In addition, the study of operation of the synthesized phase-locked loop system under external influences has been carried out. The use of the system provides a higher stability when solving the phase lock-on problem and reduce the phase error in the steady mode compared to the classical phase-locked system.

Acknowledgment

The paper has been accomplished with the support of Russian Federal Property Fund grants No. 15-08-03798, 15-38-20835, 16-37-60034, 16-38-00665.

References

1. A.A. Kostoglotov, "Solution of Fuller's problem on the basis of the joint Pontryagin-Hamilton-Ostrogradskii principle," Automatic Control and Computer Sciences, **41**, pp. 179 – 187 (2007).
2. V.V. Shahgil'djan, and A.A. Ljahovskij, Frequency phase-locked systems. Moscow: Svjaz' (1972).
3. A.I. Lur'e, Analytical mechanics. Moscow: Gos. Izd. Fiz.-Matem. Liter. (1961).
4. A.A. Kostoglotov, A.I. Kostoglotov, and S.V. Lazarenko, "Joint maximum principle in the problem of synthesizing an optimal control of nonlinear systems," Automatic Control and Computer Sciences, **41**, pp. 274 – 281 (2007).
5. A.A. Kostoglotov, A.I. Kostoglotov, S.V. Lazarenko, and B.M. Tsennykh, "Method of Structural-Parametric Identification of Lagrangian Dynamic Systems in the Processing of Measurement Information," Measurement Techniques, **57**, pp. 153 – 159 (2014)

MATEC Web of Conferences 77, 15003 (2016) DOI: 10.1051/matecconf/20167715003
ICMMR 2016

Method of Quasi-Optimal Synthesis Using Invariants

Andrey Aleksandrovich Kostoglotov[1], Sergey Valerievich Lazarenko[2], Anton Aleksandrovich Kuznetcov[3] and Zoya Vladimirovna Lyashchenko[1]

[1] *Rostov State Transport University, Department of Informatics, Rostov-on-Don, Russian Federation*
[2] *Don State Technical University, Department of Radio Engineering and Electronics, Rostov-on-Don, Russian Federation*
[3] *Air Force Military Training and Research Center "Air Force Academy named after Professor N.E. Zhukovsky and Yu.A. Gagarin", Department of Metrology and Metrological support, Voronezh, Russian Federation*

Abstract. This paper studies the problem of synthesis of terminal control of a dynamic system. It is shown that to solve the problem the convolution product of objective functional and Gaussian constraint can be used as expanded functional. The use of needle variation allows to get a required condition of the objective functional minimum and to reduce the optimization problem to the boundary problem, which can be solved in closed form. The quasi-optimal solution to the problem of optimal speed of operation is obtained. The modes of operation with choosing the control law parameters were studied.

1 Introduction

The research results show [1]-[5] that the convolution product of the objective functional and the action integral gives the control structure accurate to a synthesis function. The function can be built using the stationarity conditions of the energy invariants which can be used to determine the switching surface. This allows to find that solutions to the extreme problem which satisfy to variational principle (the dynamics base) and provide the stability of the controlled motion in accordance with A.M. Lyapunov's theorem.

In this work it is shown that the application of combined-maximum principle methodology for finding a required condition for the expanded functional minimum as a convolution product of the objective criterion and the Gaussian constraint [4] provides a synthesis of quasi-optimal controls. In contrast to known results this leads to a solvable boundary problem and doesn't require to build the synthesis function.

The aim of this investigation is synthesis of terminal control using criterion of operation speed using Gaussian constraint.

2 Definition of the synthesis problem

According to the Gauss principle at every time t the dynamic system moves in such a way that constraint [5]

$$Z=\sum_{s=1}^{n}\frac{1}{2}m_s\left(\ddot{q}_s-\frac{Q_s}{m_s}\right)^2, s=\overline{1,n} \quad (1)$$

corresponding to actual way is minimized by the accelerations $\ddot{q}_s$, so that

$$\delta Z=0, \quad (2)$$

where m_s is the material point mass; q_s – is the coordinate of the material point relative to a static Cartesian coordinate system; Q_s is the resultant force applied to the material point; n is a number of degrees of freedom of the dynamic system.

From (2) the Appell equations [5] follow in the form:

$$\frac{\partial G}{\partial \ddot{q}_s}=Q_s, \quad (3)$$

where G is the Gibbs function, and the time derivative is denoted by two dots.

Let the dynamics of the studied system satisfies (1) and is described by equations (3).

It is required to find the restricted possible forces $\mathbf{Q}\in\overline{G_Q}$, which transfer system (3) from the initial state $t=t_0$ to the final state $t=t_1$, under the condition of minimum of the objective functional

$$J=\int_{t_0}^{t_1}F(\mathbf{q})dt\to\min, \quad (4)$$

where a convex function $F(\mathbf{q})$ is of constant sign and is continuous along with its partial derivatives in the whole domain, and t_0, t_1 is the time of start and finish of the control accordingly [2, 6].

3 Method of synthesis using invariants

The searching for the required condition for the objective functional minimum (4) is performed by the Lagrange undetermined multipliers method. Let us consider the expanded functional [4]

$$J_1 = J + \int_{t_0}^{t_1} \lambda Z dt \to \mathbf{min}, \tag{5}$$

where λ – is the Lagrange undetermined multiplier.

Let the arbitrary generalized force is determined by expression:

$$\mathbf{Q} = \hat{\mathbf{Q}} + \delta\mathbf{Q}, \tag{6}$$

where $\hat{\mathbf{Q}}$ – is the generalized force minimizing the objective functional, $\delta\mathbf{Q} = 0$ when $t \overline{\in} [\tau, \tau + \Delta t]$, $\tau \in (t_0, t_1)$ is the given continuity point of the function $\hat{\mathbf{Q}}(t)$, $\Delta t \in [\tau, t_1]$ is the given small finite time interval; $\Delta t > 0$.

Then the full variation of the functional has the following form:

$$\Delta J_1 = [\lambda Z + F]\Delta t\Big|_{t_0}^{t_1} + \sum_{s=1}^{n} \int_{t_0}^{t_1} [\lambda \delta Z + \delta F] dt = \\ = [\lambda Z + F]\Delta t\Big|_{t_0}^{t_1} + \sum_{s=1}^{n} \int_{t_0}^{t_1} \left[\lambda \left(\frac{\partial Z}{\partial \hat{\ddot{q}}_s} \delta \ddot{q}_s \right) + V_s \delta q_s \right] dt \geq 0, \tag{7}$$

where $V_s = \dfrac{\partial F}{\partial \hat{q}_s}$ is the fictious generalized force.

The relations on the trajectory ends are the transversality conditions:

$$\lambda Z + F = 0, \tag{8}$$

if the interval $[t_1 - t_0]$ is fixed, or

$$\Delta t = 0 \tag{9}$$

if the interval $[t_1 - t_0]$ is not fixed.

When $t \in [t_0, \tau]$ the varied force and the minimizing (4) force coincide, so $\Delta J_1 = 0$.

When $t \in [\tau, \tau + \Delta t]$ $\Delta J_1 \neq 0$ and

$$\delta Z_{t \in [\tau, \tau + \Delta t]} = \frac{\partial \sum_{s=1}^{n} \frac{1}{2} m_s \left(\ddot{q}_s - \frac{Q_s}{m_s} \right)^2}{\partial \hat{\ddot{q}}_s} \delta \ddot{q}_s = \\ = \sum_{s=1}^{n} (m_s \ddot{q}_s - Q_s) \delta \ddot{q}_s. \tag{10}$$

When $t \in [\tau + \Delta t, t_1]$ $\Delta J_1 \neq 0$, but the arbitrary force Q_s and the minimizing (4) force $\hat{Q}_s$ coincide, then

$$\delta Z_{t \notin [\tau, \tau + \Delta t]} = \frac{\partial \sum_{s=1}^{n} \frac{1}{2} m_s \left(\ddot{q}_s - \frac{\hat{Q}_s}{m_s} \right)^2}{\partial \hat{\ddot{q}}_s} \delta \ddot{q}_s = \\ = \sum_{s=1}^{n} (m_s \ddot{q}_s - \hat{Q}_s) \delta \ddot{q}_s. \tag{11}$$

An increment of the integrand of the objective functional F is calculated in the following way:

$$\delta F_{t \notin [\tau, \tau + \Delta t]} = \sum_{s=1}^{n} \frac{\partial F}{\partial \hat{q}_s} \delta q_s = \sum_{s=1}^{n} \hat{V}_s (\hat{\mathbf{Q}}) \delta q_s. \tag{12}$$

Then (7) takes the form:

$$\Delta J_1 = \sum_{s=1}^{n} \int_{\tau}^{\tau + \Delta t} [V_s \delta q_s + \lambda (m_s \ddot{q}_s - Q_s) \delta \ddot{q}_s] dt + \\ + \sum_{s=1}^{n} \int_{\tau + \Delta t}^{t_1} [\lambda (m_s \ddot{q}_s - \hat{Q}_s) \delta \ddot{q}_s + \hat{V}_s \delta q_s] dt. \tag{13}$$

When $t \in [\tau, \tau + \Delta t]$ the following equation is true:

$$\sum_{s=1}^{n} \int_{\tau}^{\tau + \Delta t} [V_s \delta q_s + \lambda (m_s \ddot{q}_s - Q_s) \delta \ddot{q}_s] dt = \\ = \sum_{s=1}^{n} \int_{\tau}^{\tau + \Delta t} [\delta V_s \delta q_s + \lambda (m_s \delta \ddot{q}_s - \delta Q_s) \delta \ddot{q}_s] dt + \\ + \sum_{s=1}^{n} \int_{\tau}^{\tau + \Delta t} [\hat{V}_s \delta q_s + \lambda (m_s \ddot{q}_s - \hat{Q}_s) \delta \ddot{q}_s] dt, \tag{14}$$

where $\delta V_s = V_s - \hat{V}_s$. Because by (1) $m_s \hat{\ddot{q}}_s - \hat{Q}_s = 0$, we get

$$\Delta J_1 = \sum_{s=1}^{n} \int_{\tau}^{\tau + \Delta t} [\delta V_s \delta q_s - \lambda \delta Q_s \delta \ddot{q}_s] dt + \\ + \sum_{s=1}^{n} \int_{\tau}^{\tau + \Delta t} \lambda m_s \delta \ddot{q}_s^2 dt + \sum_{s=1}^{n} \int_{\tau}^{t_1} \hat{V}_s \delta q_s dt + \\ + \sum_{s=1}^{n} \int_{\tau + \Delta t}^{t_1} [\lambda (m_s \ddot{q}_s - \hat{Q}_s) \delta \ddot{q}_s] dt. \tag{15}$$

Then since $\lambda \left(m_s \left(\hat{\ddot{q}}_s + \delta \ddot{q}_s \right) - \hat{Q}_s \right) \delta \ddot{q}_s = \lambda m_s \delta \ddot{q}_s^2 \geq 0$, $\hat{V}_s \delta q_s \geq 0$, and the integrand F is a positively defined function, to fulfill (7) is enough that

$$\sum_{s=1}^{n} \int_{\tau}^{\tau + \Delta t} \left[(V_s - \hat{V}_s) \delta q_s - \lambda (Q_s - \hat{Q}_s) \delta \ddot{q}_s \right] dt \geq 0. \tag{16}$$

Because according to the Gauss principle the trajectories and velocities don't be varied, then the integration by parts leads to the following expression:

$$\sum_{s=1}^{n}\int_{\tau}^{\tau+\Delta t}\left[\frac{d^2\hat{Q}_s}{dt^2}-\lambda^{-1}\hat{V}_s\right]\delta q_s dt \ge$$
$$\ge \sum_{s=1}^{n}\int_{\tau}^{\tau+\Delta t}\left[\frac{d^2 Q_s}{dt^2}-\lambda^{-1}V_s\right]\delta q_s dt. \qquad (17)$$

So

$$\sum_{s=1}^{n}\frac{d^2\left(\hat{Q}_s\left(\hat{\mathbf{q}},\dot{\hat{\mathbf{q}}}\right)\right)}{dt^2}=\lambda^{-1}\sum_{s=1}^{n}\hat{V}_s\left(\hat{\mathbf{q}}\right),$$
$$t=\tau,\ \hat{\mathbf{q}}=\hat{\mathbf{q}}(\tau),\ \dot{\hat{\mathbf{q}}}=\dot{\hat{\mathbf{q}}}(\tau), \qquad (18)$$
$$t=\tau+\Delta t,\ \hat{\mathbf{q}}=\hat{\mathbf{q}}(\tau+\Delta t),\ \dot{\hat{\mathbf{q}}}=\dot{\hat{\mathbf{q}}}(\tau+\Delta t).$$

The developing of this equation should be performed separately for each specific case of synthesis problem.

4 The results of mathematical simulation

Let dynamic system (3) has the form:

$$\ddot{q}=Q,$$
$$t_0=0, q(t_0)=0,\ \dot{q}(t_0)=0. \qquad (19)$$

It is required to synthesize the law of optimal control of the dynamic system (19), transferring it from initial state to the phase space point $(0,0)$ fulfilling the condition of minimum of the objective functional

$$J=\int_{t_0}^{t_1}dt\rightarrow\min; \qquad (20)$$

We take $t_1=3$ s.

In accordance with (18)

$$q^{IV}=0, \qquad (21)$$

where from

$$Q=\frac{6A^2(q-D)+\left(6ABC-2B^3\right)}{2A(\dot{q}-C)-\left(2B^2-6AC\right)}, \qquad (22)$$

where

$$A=-t_1^{-3}\left[12\left(q(t_1)-q(0)-\dot{q}(0)t_1\right)-6t_1\left(\dot{q}(t_1)-\dot{q}(0)\right)\right];$$
$$B=t_1^{-2}\left[6\left(q(t_1)-q(0)-\dot{q}(0)t_1\right)-2t_1\left(\dot{q}(t_1)-\dot{q}(0)\right)\right]; \qquad (23)$$
$$C=\dot{q}(0);\ D=q(0).$$

The estimation of efficiency of the suggested solution is performed on the basis of comparison with the quasi-optimal law of "soft" terminal control [7, 8]:

$$Q=\frac{12\left(q(t_1)-q(t)\right)}{\left(t_1-t\right)^2}-\frac{6\dot{q}(t_1)-6\dot{q}(t)}{t_1-t}. \qquad (24)$$

The results of the mathematical simulation are shown in the Fig, 1, where the number 1 denotes the phase trajectory of the system (19) with the right part (22), and the number 2 denotes the phase trajectory of the system (19) with the right part (24).

Fig, 2 presents the structure of the controlling generalized forces. There are the following notations: 1 – the control (22), 2 – the control (24). It can be seen that singularity (24) at the end time leads to a sharp increase of control force in opposite to (22).

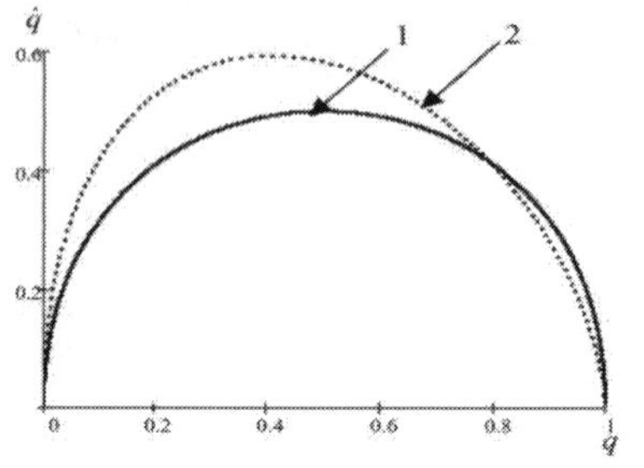

Figure 1. Phase portrait.

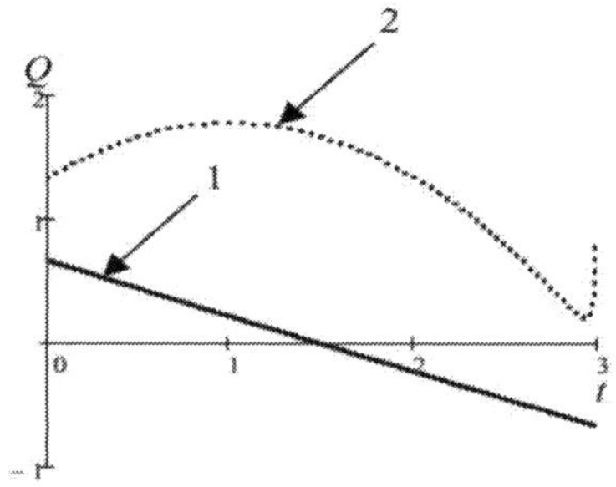

Figure 2. Structure of control.

Considered the other terminal control problem. Let a dynamical system (19) expression (18) takes the form

$$\frac{d^2}{\partial t^2}\frac{\partial^2 G}{\partial\ddot{q}}-\lambda^{-1}q=0,$$
$$t=0,\ q=-1,\ \dot{q}=-3,$$
$$t=t_1,\ q=2,\ \dot{q}=0; \qquad (25)$$
$$J=\int_0^{t_1}q^2\rightarrow\min.$$

Then

$$q^{IV}+\lambda^{-1}q=0. \qquad (26)$$

The solution of this problem can be prove by function of A.N. Krylov

$$q=\sum_{i=1}^{4}A_iK_i(\beta t)=A_1K_1(\beta t)+A_2K_2(\beta t)+$$
$$+A_3K_3(\beta t)+A_4K_4(\beta t), \qquad (27)$$

where $\beta=\sqrt[4]{\dfrac{\lambda^{-1}}{4}}$, a

$$
\begin{aligned}
&K_1(\beta t) = ch\beta t \cos \beta t, \\
&K_2(\beta t) = \frac{1}{2}\left(ch\beta t \sin \beta t + sh\beta t \cos \beta t\right), \\
&K_3(\beta t) = \frac{1}{2} sh\beta t \sin \beta t, \\
&K_4(\beta t) = \frac{1}{4}\left(ch\beta t \sin \beta t - sh\beta t \cos \beta t\right),
\end{aligned} \qquad (28)
$$

are Krylov's functions.

The control equation is

$$
\begin{aligned}
&Q = \sum_{i=1}^{4} A_i \ddot{K}_i(\beta t) = -4\beta_s^2 A_1 K_3(\beta_s t) - \\
&-4\beta_s A_2 K_4(\beta_s t)\beta_s^2 A_3 K_1(\beta_s t) + \beta_s^2 A_4 K_2(\beta_s t),
\end{aligned} \qquad (29)
$$

where

$$
\begin{aligned}
&\ddot{K}_1(\beta t) = -4\beta^2 K_3(\beta t); \quad \ddot{K}_2 = -4\beta^2 K_4(\beta t); \\
&\ddot{K}_3(\beta t) = \beta^2 K_1(\beta t); \quad \ddot{K}_4 = \beta^2 K_2(\beta t).
\end{aligned} \qquad (30)
$$

The arbitrary constant determinate by boundary conditions :

$$
\begin{aligned}
&t = 0, \quad q^0 = A_1, \quad \dot{q}^0 = A_2\beta; \\
&t = t_1, \\
&q^* - A_1 K_1(\beta t_1) - A_2 K_2(\beta t_1) = \\
&= A_3 K_3(\beta t_1) + A_4 K_4(\beta t_1), \\
&\dot{q}^* + A_1 4\beta K_4(\beta t_1) - A_2 \beta K_1(\beta t_1) = \\
&= A_3 \beta K_2(\beta t_1) + A_4 \beta K_3(\beta t_1).
\end{aligned} \qquad (31)
$$

The results of modeling are illustrated on Fig. 3. The solution based on the Pontryagin's maximum principle allow getting these results. This confirms the validity of the developed method.

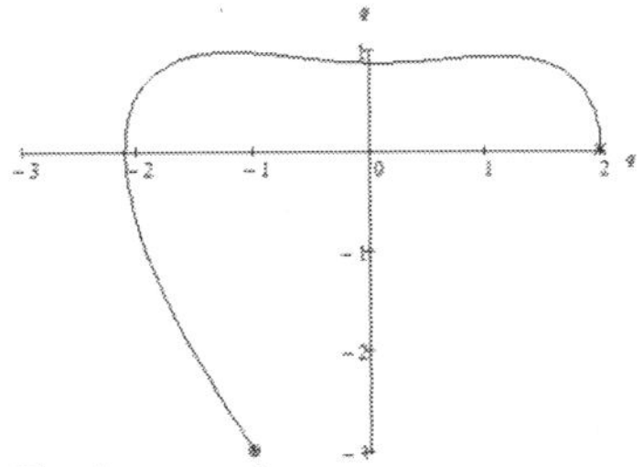

Figure 3. The phase portrait

5 Conclusion

The solution (22) unlike the known solution (24) [7] did not contain singularities at the finite time. As a result, the discontinuity of the control generalized force at the final time is absent. This allows to use it in practice without additional transformation.

Acknowledgement

The paper has been accomplished with the support of Russian Federal Property Fund grants No. 15-08-03798, 15-38-20835, 16-37-60034, 16-38-00665.

References

1. A.A. Kostoglotov, "Application of invariant sets for synthesis of optimum systems," Avtomatika i Vychislitel'naya Tekhnika, **5**, pp. 26 – 35 (2002).
2. A.A. Kostoglotov, A.I. Kostoglotov, and S.V. Lazarenko, "Joint maximum principle in the problem of synthesizing an optimal control of nonlinear systems," Automatic Control and Computer Sciences, **41**, pp. 274 – 281 (2007).
3. A.A. Kostoglotov, A.I. Kostoglotov, S.V. Lazarenko, and L.A. Shevcova, «Synthesis of optimal control based on the combined-maximum principle,» University News. North-Caucasian region. Technical Sciences Series, **2**, pp. 31–37 (2010).
4. S.V. Lazarenko, "Method of synthesis of optimum control with use of the principle of Gauss," Informacionno-izmeritel'nye i upravljajushhie sistemy, **12**, pp. 37 – 43 (2013).
5. A.I. Lur'e, Analytical mechanics. Moscow: Gos. Izd. Fiz.-Matem. Liter. (1961).
6. B.N. Petrov and P.D. Krut'ko, "Design of algorithms for flight control based on solving inverse dynamics problems," Izv. AN SSSR Tekhn. Kibernetika, **3**, 161–172 (1981).
7. G.N. Razorenov, "A method for synthesis of "soft" and "super-soft" control laws for final states of dynamic systems," Journal of Computer and Systems Sciences International, **52**, pp. 1–15 (2013).
8. G.N. Razorenov and A. A. Samarin, Optimal Control Systems and Theory. Moscow: MO. (2007).

Fault level prediction for distribution network using fuzzy logic identifier

Fang Shi[1] and Bin Wang[2]

[1]*Key Laboratory of Power System Intelligent Dispatch and Control of the Ministry of Education, Jinan 250061, China*
[2]*Shanghai Normal University, Shanghai 200234, China*

Abstract. With the increasing penetration of the renewable power energy sources, the potential fault current of the distribution power systems changes more frequently as the connection structure of the distribution power system varies from time to time. Traditionally, the fault level can be estimated through short circuit analysis which is time consuming and sometimes difficult as it needs to know the parameters of the transmission line and transformers as well as the structure of the power system. In this paper, an online-used fault level prediction method is proposed via monitoring the phasor value of the local positive-sequence voltage and current. The ratio of the voltage change and current change are used to distinguish the natural deviation of the load from the switching operations or disturbances on the grid side. Several continuous changes of the voltage and current caused by load fluctuations are recorded and used to parameterize the equivalent circuit of the power system and to estimate the fault current level. A fuzzy logic identifier is used for adaptively selecting and recording the satisfactory changes by defining an index of confidence level. The implementation of the proposed scheme is demonstrated in a relay after introducing a low-voltage blocking function. A typical distribution power system with renewable generators is established in PSCAD/EMTDC and is used to verify the effectiveness and accuracy of the proposed method under various load changing conditions.

1 Introduction

Circuit breakers are of great importance to the safety of the transmission and distribution power system, which will disconnect the fault component or the electricity supply affected sections of the grid in case of a fault. These circuit breakers must be chosen with a high enough capacity to break the possible flow of current even under the worst fault conditions [1]. The network infrastructure also needs to be able to withstand the current flow during the period before the circuit breakers isolate the affected network to prevent destructive failure. Therefore, the fault level prediction (FLP) is needed both for the planning and operation of a power system. Then the satisfying circuit breaker is selected from the available product list. Normally, the fault level is calculated by investigating the parameters of the transmission line and transformers in the power system and there will be a margin for the breaking capacity so the breakers can have the ability to cope with the increase of the fault level in a near future [2].

However, with the increasing penetration of distribution generators (DG), such as the wind power plants and the photovoltaic (PV) power plants, the fault level or fault current of the distribution power grid changes dramatically more often than ever before [3]. Sometimes the extreme fault current will exceeds the capacity of the breakers and lead to switching off failure which will threat the safe operation of the power system. Nevertheless, the settings of the protection relays also need to be changed according to the new operation conditions [4]. For example, if a big proportion of the supplying power source changes from traditional generators into PV plants, the fault current will decrease as the convertors' contributions to the fault current is negligible. Under this circumstance, the previous setting value of the protection relays could be invalid for disconnecting short circuit fault, which may lead to overheating problems for the devices, such as transmission lines and transformers. Therefore, an online used FLP is required to predict the real-time fault level of the power system.

Traditionally, the fault current can be calculated by utilities in two ways, or a combination of both: (1) Using empirical principles to calculate the fault level by summing the known contributions from the major elements in the power grid, i.e., transformers or generators; (2) Creating an analytical parameterized model of the power grid section that is closely related to the fault level and compute the result mathematically. Both the methods are valid for the FLP during the power system planning period. The rough model can be used and a conservative value can be acquired using the first method, while the latter method can provide an accurate result but is complex and time consuming, so it has been used mainly at higher network voltages (for example at 33kV and above)[5, 6].

In the last two decades, some model free methods have been proposed to predict the fault level of a radial network or a radialized section of an interconnected network using the measurements of the local voltage and current [7-9]. The impedance of the power-supply network is evaluated from the naturally occurring disturbances in [10], then the fault level is calculated. The results are encouraging and the method can be used online for continuous assessment of the fault level. Some portable device can be expediently used for calculating the fault level by connecting to the terminal of the breaker and measuring the voltage and current change induced by the deviation of the loads or by artificial disturbances [11]. A predictive fault level algorithm is proposed using the locally measured voltage and current in a protection relays by distinguishing the disturbance on the load side from on the grid side in [12]. However, only the sudden load change is used and a fixed threshold value is set for identifying the change of the voltage or the current. For the slow load changing scenarios, the method could be invalid for the power grid.

There are also some other methods can be used for calculating the equivalent impedance of the power system as well as the fault level in a more active way. In [5], the artificial disturbances are applied to the power system and the change of the voltage and current are used for calculating the impedance. In [13], the harmonic current with a frequency band around the fundamental frequency is injected to the power system and then the impedance and fault current are estimated. Some hybrid techniques use both the artificial (or controlled) disturbances and natural disturbances to measure the voltage and current change and to predict the fault level [14]. The disadvantages for these methods are obvious as extra hardware investment is needed.

In this paper, an adaptive fuzzy logic identifier is designed to detect the feasible load fluctuations for estimating the fault level. The change of the voltage and current are recorded and monitored simultaneously for a specified time duration. Three quantized variables, ie., the change of the voltage, the change of the current and the calculated equivalent impedance, are analyzed and the weighted summation of the three states is used as the index to identify the desired fluctuation occurs on the load side. Then the fault level is calculated when the fluctuation is obvious enough and a confidential level of the result is given at the same time.

2 Basic model and problem formulation

Thevenin Equivalent circuit can be used for calculating the fault current level when only local voltage and current is available. To make the paper self-contained, the basic theory is illustrated as following. The details can also be found in [12].

Take one distribution power network with DGs (in Fig. 1) as one example. Suppose the fault level before the breaker CB2 needs to be determined. The power system upstream the breaker can be represented using the equivalent circuit shown in Fig. 2 by Thevenin's theory, where E_{eq} and Z_{eq} are the inner voltage and the impedance of the equivalent voltage source, respectively. If E_{eq} and Z_{eq} are known, then the fault level can be simply predicted as:

$$I_{FL} = E_{eq}/Z_{eq} \tag{1}$$

So the key point is to calculate E_{eq} and Z_{eq}.

Suppose E_{eq} and Z_{eq} are constant and Z_L changes from Z_{L1} to Z_{L2}, which is based on the following two assumptions: 1) There is no major changes on the structure of the power grid upstream of CB2; 2) Only the load downstream of CB2 changes significantly so as the voltage and current at CB2 change from $U_1$and I_1 to U_2 and I_2accordingly. Normally, there will be some large capacity synchronous load connected to the feeder, such as production line of factories or large electric motors. If these loads trip on/off or fluctuate in a short time during which there is no changes or only trivial alteration on the structure and operation states on the grid side, then the above assumptions are justifiable.

Consequently, the following equations will hold (the directions of the current and voltage are as indicated in Fig. 2):

$$\dot{E}_{eq} = \dot{U}_1 + \dot{I}_1 Z_{eq} \tag{2}$$

$$\dot{E}_{eq} = \dot{U}_2 + \dot{I}_2 Z_{eq} \tag{3}$$

This equation set is solvable if $U_1$and I_1 are different from U_2 and I_2. Essentially, it is a parameter identification problem for the linear circuit model. Two different states are enough to estimate E_{eq} and Z_{eq} but more different states would yield an accurate result.But if the difference is trivial, the errors could be quite large because of the inexactitude in the measurement, analog-digital conversion transition as well as the numerical calculation.

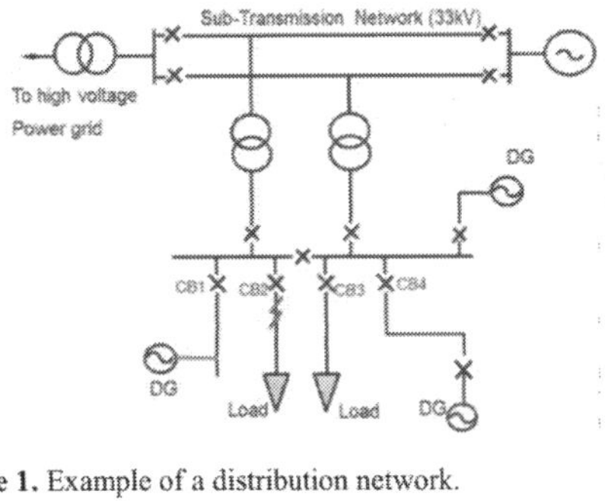

Figure 1. Example of a distribution network.

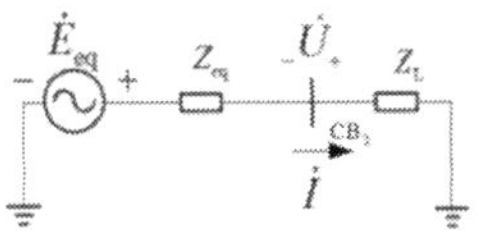

Figure 2. The equivalent circuit for fault level caculation.

However, in the practical power systems, the change of the voltage and current can also be caused by the disturbance occurs on the grid side (before CB_2 in Fig. 2). Under this circumstance, Z_{eq} will change in the above two equations. Consequently, we have to wait until the next load fluctuation occurs to form the above two equations with a new value of Z_{eq}. Therefore, it is critical to identify the disturbance occur on the grid side or on the load side.

When the disturbance occurs on the load side, from the above two equations, it can be written as:

$$Z_{eq} = -\frac{\dot{U}_2-\dot{U}_1}{\dot{I}_2-\dot{I}_1} = -\frac{\Delta\dot{U}}{\Delta\dot{I}} \tag{4}$$

On the contrary, when the disturbance occurs on the grid side, ie., Z_L is constant and E_{eq} and Z_{eq} are changeable, we can have the following two equations using the Kirchhoff's Voltage Law (KVL) on the circuit from the breaker to the downstream load:

$$0 = \dot{U}_1 - \dot{I}_1 Z_L \tag{5}$$

$$0 = \dot{U}_2 - \dot{I}_2 Z_L \tag{6}$$

So we have:

$$Z_L = \frac{\dot{U}_2-\dot{U}_1}{\dot{I}_2-\dot{I}_1} = \frac{\Delta\dot{U}}{\Delta\dot{I}} \tag{7}$$

For the practical power system, the real part of Z_{eq} and Z_L will be greater than zero, which implies that the transfer conductance of the power grid is not ignored and there exits active load on the feeder, respectively. Then for a set of directional unknown voltage and current change, the following index can be used to distinguish whether the disturbance occurs on the grid side or on the load side:

$$Dir = -real\left(\frac{\Delta\dot{U}}{\Delta\dot{I}}\right) \tag{8}$$

Specifically, when $Dir > 0$, it means the disturbance or fluctuation occur on the load side; when $Dir < 0$, the disturbance or fluctuation will be seen as occurring on the grid side.

The above conculstion is based on the assumption that the disturbance never occur simultaneously both on grid side and on load side. It could happen (practically and legally) but the chances are slim as the monitoring devices can distinguish the two consistive changes in a very short time, for example, two cycles (40ms). Even they did happen exactly at the same time, this kind of disturbance can be ignored and only the desired flucuation on the load side is used to calculate the fault level.

So there are three requirements for an accurate FLP using the above mentioned method. Firstly, the location or the direction of the disturbance need to be clearly identified, only the fluctuation on the load side can be taken advantage of. Secondly, and thirdly, non-tirval changes of the voltage and current at the breaker are needed to calculate the parameters of the equivalent ciucuit with acceptable errors. The threshold value are needed to quantitively define the non-trivial change of the voltage or current. However, it is hard to find a once-for-all settings as different power gird with a variaty of load type will have a diversity of characteristics. A fuzzy logic identifier is proposed to assess the disturbance in a more adaptive manner, which will be illustrated in the sequel.

3 Design of fuzzy logic identifier

The design of a fuzzy logic identifier usually includes three parts: (1) The rules or principles of the problem; (2) Selecting proper linguistic terms which can describe the input and output variables properly; (3) Defining fuzzy sets for all fuzzy variables [16-18].

3.1 The Rules for Disturbance Identification

As mentioned above, there are three variables need to be considered in fault level calculating process. Although the disturbance direction identification principle is clear as it directly compares *Dir* and zero, in practical application a margin need to be given. If *Dir* is very close to zero, it could lead to a mis-judgement as there exists errors both in PT and CT sensor as well as in the computation process. So the practical criterion is specified as:

$$\begin{cases} Dir < -\varepsilon, \ grid\ side\ disturbance \\ -\varepsilon <= Dir <= \varepsilon, ignored \\ Dir > \varepsilon, load\ side\ disturbance \end{cases} \tag{9}$$

where ε is a predefined small positive value. Using this criterion, the unclear disturbance conditions, such as the disturbances simultaneously occurred on both sides, are ignored directly.

For the identification of the non-trivial voltage/current change, it is obvious that a relative value (percentage) will give a more reasonable result as for the same variation the one with larger voltage/current amplitude will be affected more slightly and vice versa. So the percentage of the voltage and current change relative to their average values are used for finding a non-trivial disturbance which are noted as follows:

$$\Delta\hat{U} = \frac{|\Delta\dot{U}|}{ave(|\dot{U}_1|,|\dot{U}_2|)} = \frac{2|\dot{U}_2-\dot{U}_1|}{|\dot{U}_1|+|\dot{U}_2|} \tag{10}$$

$$\Delta\hat{I} = \frac{|\Delta\dot{I}|}{ave(|\dot{I}_1|,|\dot{I}_2|)} = \frac{2|\dot{I}_2-\dot{I}_1|}{|\dot{I}_1|+|\dot{I}_2|} \tag{11}$$

It is obvious that these two positive parameters, range from 0 to 2, can be used for evaluating severity of the disturbance or the extent of the load fluctuation.

3.2 Linguistic terms of the input-output variables

It is an important step to represent the problem in fuzzy terms when fuzzy logic is used for problem-solving procedure. Four linguistic terms, "Large", "Medium", "Small" and "Zero" are picked out to describe the input and output variables; appending positive or negative sign to nonzero terms, totally 4 terms are used to describe input variables *Dir*, $\Delta U$and ΔI in the fuzzy system [16]:

(O, PS, PM, PL)

O – Zero;
PS – Positive Small;
PM – Positive Medium;
PL – Positive Large.

Confidential level is the output variable, note as C_L, which represents the reliability of the disturbance and the final fault level result, seven terms of none negative value is required:

(O, PVS, PS, PLM, PM, PML, PL)

O – Zero;
PVS – Positive Very Small;

DOI: 10.1051/matecconf/20167716001

PS – Positive Small;
PLM – Positive Light Medium;
PM – Positive Medium;
PML Positive Medium Large;
PL – Positive Large.

The input and output variables of a fuzzy logic identifier are usually specified within a range. But most of the time, the variation range of a variable can't be confirmed accurately. So a proper closed interval is needed, which the actual variation range can be projected to, for facilitating the calculation.

According to the feasible operation states and experiments, *Dir* can vary from $-\infty$ to $+\infty$, but will be mostly in the range of $[-10, 10]$ and ε can be defined as 0.001; For ΔU, it normally varies in the range of $[0, 0.3]$ as the voltage will not beyond the operating limitation; for ΔI, it can be varied in a wide range of [0,2] supposing the load can change in a wide range. In practical usage, only the conditions with positive *Dir* value are concerned, because the disturbance with a negative *Dir* will be ignored as they are identified to occur on the grid side.

The number of the elements in the basic universe should be more than twice of the total number of fuzzy linguistic terms to ensure that the fuzzy sets can cover the whole domain, and to avoid losing important information. Here, the universes of *Dir*, ΔU, ΔI,are chosen as:

(0, 1, 2, 3, 4, 5, 6, 7)

The universe of the output variable C_L are:

(0, 1, 2, 3, 4, 5, 6, 7, 8, 9, 10, 11, 12, 13)

3.3 Membership functions

The values of the membership function are all in the interval [0, 1]. 1 means that the object entirely belongs to the member of the set and 0 means it is totally irrelevant. When designing the fuzzy logic identifier, trapezoidal membership is an effective choice. The linguistic terms of input variables *Dir*, ΔU,ΔI, the universes of them are all the same. Therefore, they have the same membership function (in Fig. 3).The membership function of the output C_L is shown in Fig. 4.

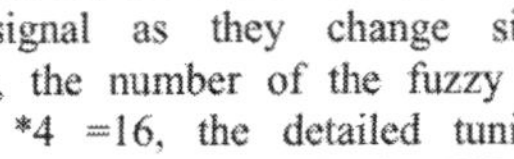

Figure 3. Membership functions of input variables.

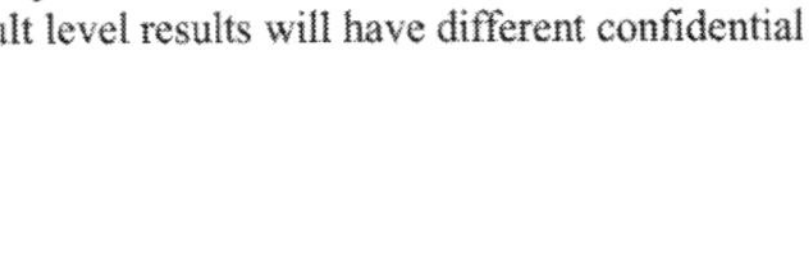

Figure 4. Membership function of output variables.

Taking advantage of the close relation between ΔU and ΔI, we combine these two variables as one input signal as they change simultaneously. Consequently, the number of the fuzzy rules can be reduced to4 *4 =16, the detailed tuning rules are explicitly illustrated in Table I. So for different scenarios, the fault level results will have different confidential level. Those of significant load change and clear positive *Dir* will be of high reliability.

Table 1. The fuzzy logic rules for forming theC_L

ΔUI \ *Dir*	O	PS	PM	PL
O	*O*	*PVS*	*PVS*	*PLM*
PS	*O*	*PVS*	*PLM*	*PM*
PM	*O*	*PVS*	*PM*	*PML*
PL	*O*	*PLM*	*PML*	*PL*

4 Simulation results

The proposed scheme can be used in the protection relays without extra investment. If there is a fault on the feeder line, the relays will output trip signal in 100ms and the FLP will be blocked at the same time because the fault condition will lead to serious errors as voltage distortion and the saturation of PT and CT. The practical implementation can be found in Fig. 5.The FLP also has a low voltage/current block function. If the voltage of the feeder is under 0.7pu with respect to the rating voltage, the FLP will be blocked to avoid giving a wrong result.

The phasor value of the positive voltage and current at the studied breaker is recorded and the *Dir*, ΔU andΔI are calculated. The confidential level will be analyzed using the fuzzy rules in the fuzzy logic identifier. When the disturbance on the load side is detected, the fault level current value will be calculated. The output of the FLP module consists of the predicted fault level and the confidential level of the result. Only the local information is used for FLP and it can be directly realized in the relays without extra hardware investment. Though the fault current obtained is non-real time and is not exactly accurate. It is still helpful for the operators to know the change of the fault level as well as the potential danger so as to update the breakers or re-set the protection settings timely.

A distribution power grid (shown in Fig. 6) is used to test the effectiveness of the propose method. The topology of the model is established in PSCAD/EMTDC by referring to the typical complex distribution network with the integration of DGs. The fault level at breaker CB2 is studied and different types of load are connected to the feeder, including motors, electric arc furnace, constant resistance load and the composition of them.

The disturbance applied to the test system is as follows: 1) At 0s, all the main-sources are disconnected (Bi1 ~Bi6 are open); 2) At 6s, half of the main-sources are connected (Bi1~Bi3 closed); 3) 10s, all the other main-sources are connected (Bi4~Bi6 closed). The composition load changes randomly, the voltage and current at CB2

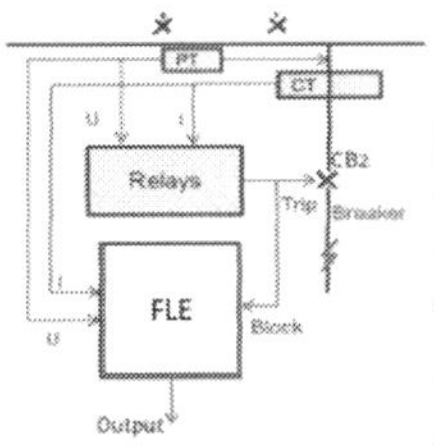

Figure 5. The practical implementation of the proposed scheme.

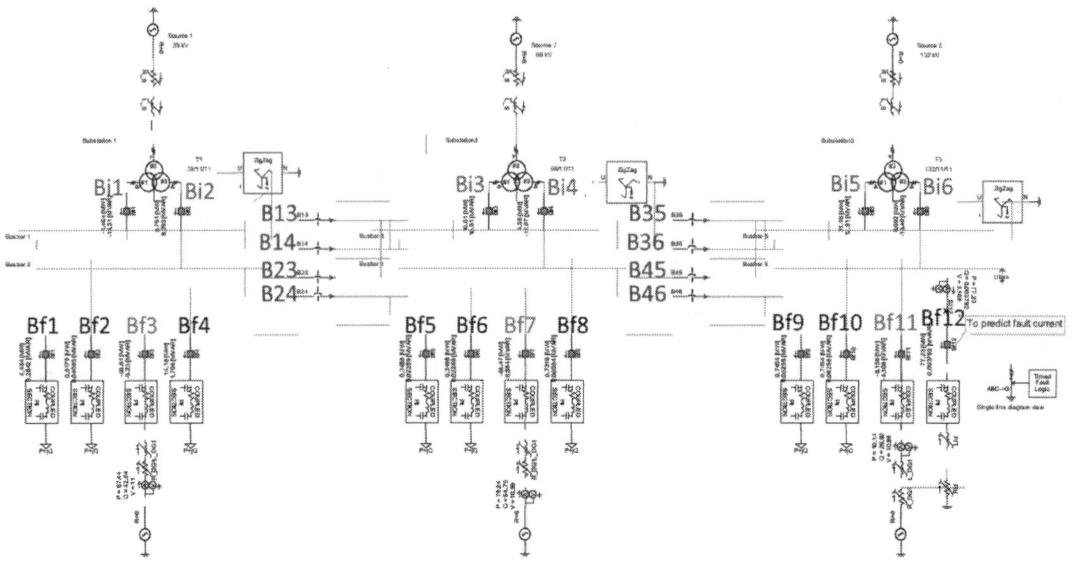

Figure 6. The test power system model in PSCAD/EMTDC.

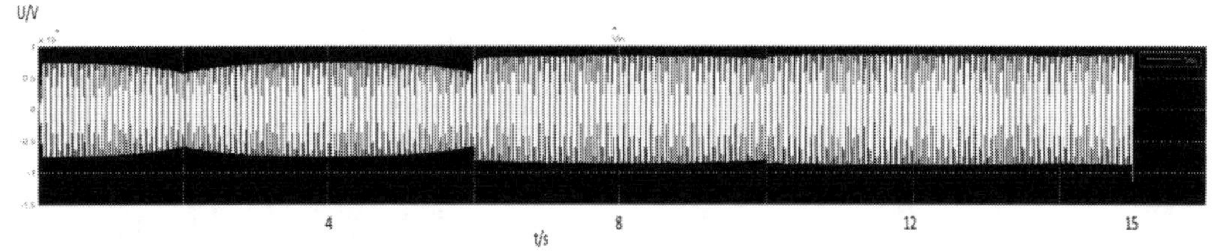

Figure 7. The instantaneous voltage at Bf12.

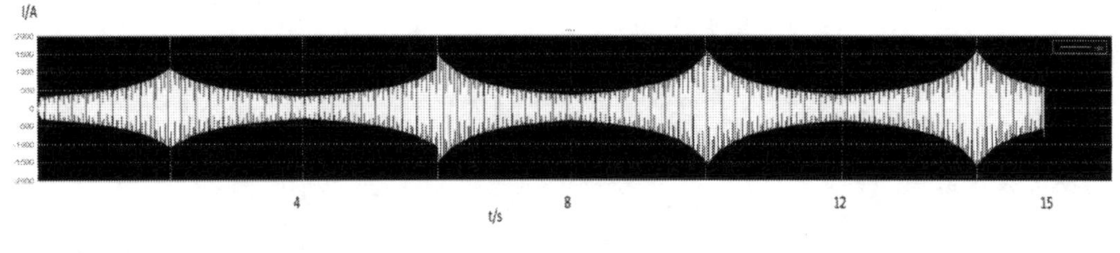

Figure 8. The instantaneous current at Bf12.

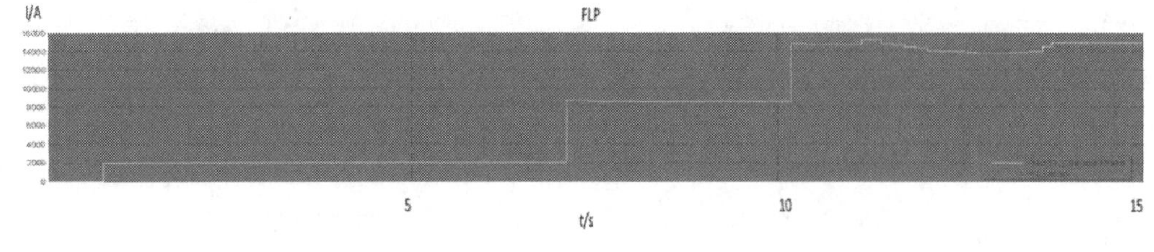

Figure 9. The actual fault current and the predicted fault current.

are shown in Fig. 7 and Fig. 8 respectively. The actual fault current (simulated using PSCAD) and the predicted fault current are shown in Fig. 9 with dashed line and solid line, respectively. Because the switching operation on the grid side, the actual fault current will change dramatically at 6s and 10s, which is clearly shown with the purple dashed line. For the predicted fault current, because it needs to detect a non-trivial load fluctuation before the FLP working, there is a delay for its updating. But as soon as the fluctuation on the load side is detected, the predicted fault level will be given (as the solid line in

Fig. 9) and it has a high similarity with the actual fault current value.

Eight other disturbances caused by motor power adjusting and electric arc furnace operating on the feeder are also tested using the proposed scheme. The maximum change of the voltage and current are different for each case. The dynamics of the voltage and current are not plotted here because of page limitation. The final statistic results are given in Table 2, in which the predicted fault current and the corresponding confidence level are given. It can be seen that the errors between the predicted fault current and the real value is no more than 5%, which is accurate enough for the planning and for reminding the operators to upgrade the breaker or reset the protection relays settings. The case with more obvious voltage and current change will give out a more accurate result with a high confidence level.

Table 2. Test results for different scenarios.

Load Type	$\max(\Delta UI)$ (%)	Errors between actual and predicted fault current (%)	Confidential level
Motor load	2e-2	4.46%	4
	3e-2	4.0%	3
	5e-2	3.07%	5
	10e-2	2.82%	7
electric arc furnace	2e-2	5.0%	3
	3e-2	4.6%	5
	5e-2	3.66%	6
	10e-2	2.58%	8

5 Conclusion

In this paper, a fault level predicting method is proposed using local measurement of the voltage and current. A fuzzy logic identifier is designed to adaptively rank the severity of the disturbance and to distinguish the direction of the disturbance in a distribution power system. The member function of the fuzzy logic identifier, which is essential to generate the confidential level of the final result, is also detailed illustrated. The implementation of the proposed scheme into a protective relay is introduced and a distribution power system is used to test the proposed method. The results show the effectiveness and accuracy of the presented scheme.

Acknowledgement

This work is supported by National Natural Science Foundation of China (No: 51407109), China Postdoctoral Science Foundation (2014M551904), and the Fundamental Research Funds of Shandong University (No. 2082014GN034).

References

1. G. Williamson, N. Jenkins, and G. Cornfield.Use of naturally occurring system disturbances to estimate the fault current contribution of induction motors.IEE Proceedings in Generation, Transmission and Distribution, **143**, 3 (1996)
2. C. Mortley. The contribution to distribution network fault levels from the connection of distributed generation, KEMA Conference, London, UK, (2005)
3. S. Conner and A. Cruden.An automatic transient detection system which can be incorporated into an algorithm to accurately determine the fault level in networks with DG, International Universities Power Engineering Conference, London, UK, (2012)
4. H. Yazdanpanahi, Y. W. Li, and W. Xu. A new control strategy to mitigate the impact of inverter-based dgs on protection system. IEEE Transactions on Smart Grid, **3**, 3(2012)
5. J. Berry, S. Jupe, M. Meisinger, and J. Outram. Implementation of an active fault level monitoring system for distributed generation integration.International Conference and Exhibition on Electricity Distribution , Stockholm,Sweden (2013)
6. P. Vovos and S. Hwachang. Analytical approach to large-scale system splitting for the regulation of fault levels. IEEE PowerTech Conference, Eindhoven, Holland, (2015)
7. M. S. Yahaya, M. F. Basar, Z. Ibrahim, M. N. N. Nasir, M. Y. Lada, and W. M. Bukhari. A Review on the Impact of Embedded Generation to Network Fault Level. International Conference on Mathematics Engineering and Industrial Applications,Penang, Malaysia (2014)
8. K. S. Ratnadeep, Y. N. Bhosale, and S. Kulkarni.Fault level analysis of power distribution system.Energy Efficient Technologies for Sustainability Conference (ICEETS), Nagercoil, India (2013)
9. E. S. Ibrahim and M. Idres. On-line fault level measurement. Electric Machines and Power Systems.**26**, 2(1998)
10. G. C. Cornfield.Estimating system fault level from naturally occurring disturbances. IEE Colloquium onFault Level Assessment Guessing with Greater Precision, London, UK (1996)
11. O. R. Ltd. Fault Level Prediction. [Online]. Available at: http://www.outramresearch.co.uk/flm/pages/product_fault_level_overview.shtml
12. H. H, S. Subramanian, and K. Venkataraman.A novel technique for fault level prediction in active distribution networks. Advanced Power System Automation and Protection Conference, Nanjing, China(2015)
13. E. S. IBRAHIM.Experimental Measurements of Fault Level For Sceco West, Kingdom of Saudi Arabia. Engineering Sciences, **15**, 1 (2004)
14. M. R. Rylander, W. M. Grady, A. Mansoor, and F. Gorgette.Passive agent system impedance monitoring station and method.US Patents (2013)
15. G. Gurrala and I. Sen.Power system stabilizers design for interconnected power systems.IEEE Transactions on Power Systems, **25**, 2 (2010)
16. H. Zhang, F. Shi, and Y. Liu.Enhancing optimal excitation control by adaptive fuzzy logic rules, International Journal of Electrical Power & Energy Systems, **63**, 1 (2014)

17. J. M. Mendel, R. I. John, and F. Liu. Interval type-2 fuzzy logic systems made simple. IEEE Transactions on Fuzzy Systems, **14**, 6(2006)

18. P. S. Bhati and R. Gupta. Robust fuzzy logic power system stabilizer based on evolution and learning. International Journal of Electrical Power & Energy Systems, **53**, 1 (2013)

MATEC Web of Conferences 77, 16002 (2016) DOI: 10.1051/matecconf/20167716002
ICMMR 2016

A comparative study of modelling AC electric arc by one-dimensional interpolation for power system harmonics analysis

Yu-Jen Liu, Jen-Pan Hung, Shang-I Chen and Cheng-Wei Lin

National Chung Cheng University, Department of Electrical Engineering, 62102 Chia-Yi, Taiwan (R.O.C.)

Abstract. Electric arc is a discharge phenomenon caused by particular electrical events and arc produced facilities in power system, for example the occurrence of short-circuit fault in feeders and the use of electric arc furnace for steel-making. All of these electric arcs have a highly nonlinear nature and can be considered as a significant source of power quality problems. To investigate the impacts of the electric arcs on power quality studies the development of the electric arc models for simulation analysis is became necessary. This paper thus presents the use of different one-dimensional interpolation methods for modelling the alternating current electric arc. The performances of different methods are compared by simulation results that are derived from the modelling of an alternating current electric arc furnace load and a low-frequency fluorescent tube.

1 Introduction

It is known that computer-aided simulation analysis is one of the most effective ways to assess the power quality (PQ) problems in power system. Therefore, to assist the simulation of the operation of an electric power network controlling an electric arc load, for instance an arc-discharge fluorescence lamp/tube (FL/FT) or an electric arc furnace (EAF), an appropriate model for the electric arc load is necessary. The usefulness of simulation performance highly relies on the applicable models of the electric components that are used to describe the operation of the system and a well understanding of the sensitivities of the models with respect to the data available for model construction. Due to the complex electric arc property, the v-i characteristic of the arc voltage and current is always nonlinear and it is thus difficult to develop a deterministic electric arc model for study.

In past efforts, many approaches are provided for electric arc modelling. For example, references [1-3] model the electric arc by a resistance and nonlinear resistance but they cannot suitably present the physics behaviour of the electric arc. Electric arc conductance models are proposed by [4, 5] that are classic differential equations for representation of an arc with high or low current conditions, separately. Harmonic voltage model and harmonic current injection model are the two common used methods for electric arc modelling in power quality study that are based on the field measurements of arc voltage and current [6]. In addition, some intelligent methods such as artificial neural network and wavelet transform have been proposed for specified studies and the performance of them are proved to be effective; however, the structures of these models are more complicated and thus is not easily be carried out on simulation. To provide an efficient and easy work modelling approach for the electric arc on PQ study, especially for the harmonic problem analysis, this paper presents the comparative study of using one-dimensional interpolation methods for electric arc modelling. Measured electric arc data that are collected from an alternating current (AC) EAF and a traditional T9 low-frequency FT are used for simulations and model validations. The performance of different interpolation models are then compared by simulation results for the assessments of harmonic distortions on the AC EAF and FT. The organization of the paper is as follows. In Section II, a brief description of the characteristic of electric arc is presented. The concept of modeling AC electric arc with different interpolation methods is addressed in Section III. Experiments and simulation results are illustrated in Section IV to show the performance of the methods for modeling the AC electric arc, and then the conclusion follows.

2 Characteristics of the AC electric arc

This paper mainly discusses the modelling issue of AC electric arc which is a gas ionization phenomenon that produces continued electrical discharge between two electrodes. An AC electric arc may re-strike on each half cycle of the fundamental frequency current period. The common features of the AC electric arc can be characterized by time-domain waveforms and the distributions of frequency spectra in frequency-domain [7]. Figure 1 shows the measured AC electric arc signal; from the view of observing time-domain waveform, all

AC electric arcs consist of three regions that named cathode drop zone, anode drop zone and arc column no matter what the total arc length is. The major arc voltage is developed within the arc column, and the cathode drop (Vc) and anode drop (Va) of a few volts are presented within a little bit millimetre of each electrode. The arc voltage will reach to peak value (reignited voltage), after the zero-crossing point, when the current lies in near zero duration (shoulders phenomenon), [8], and the shape of the arc voltage waveform is closed to square-wave. To compare with a pure sinusoidal current waveform, the peak value of the arc current may less than it and has a steep current raise rate when the current increase. Furthermore, the arc voltage and arc current are sometimes in phase, which hence presents the resistive/conductance properties on its v-i characteristic.

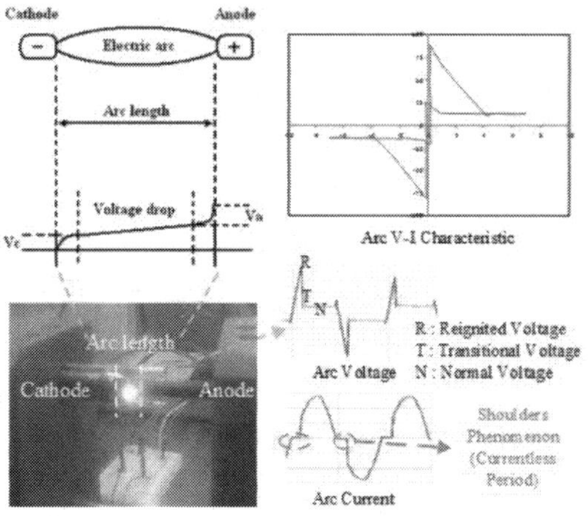

Figure 1. Characteristics of the AC electric arc.

3 Mathematical modelling

In practice, the field measurement of AC electric arc voltages and currents is the most effective way to obtain the required electrical parameters for modelling purpose. The modelling concept in this study is to develop the nonlinear conductance model based on different interpolation techniques; meanwhile, the parameters of AC electric arc conductance are collected from the above-mentioned field measurement. Interpolation is used to estimate new arc conductance values between consecutive known data points from measurements. Figure 2 shows the diagram of the presented models in this paper and the descriptions of each modelling method are illustrated as following:

3.1 Nearest neighbor interpolation

Nearest neighbor is the most basic interpolation technique and requires the least processing time of all the interpolation algorithms. This method sets the value of an interpolated point to the value of the nearest data point. Therefore, this method does not generate any new data points. Equation (1) shows the arc conductance interpolant function of the method; meanwhile, x is the original point closest to x'. Figure 3 presents the concept of this method.

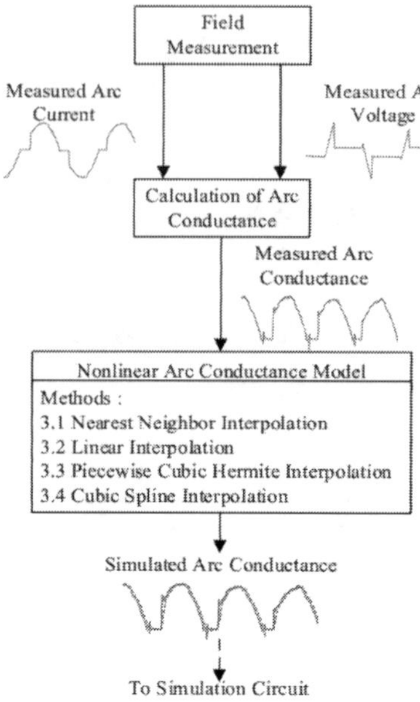

Figure 2. The concept of modelling AC electric arc by interpolation methods.

$$G(x') = f(x) \tag{1}$$

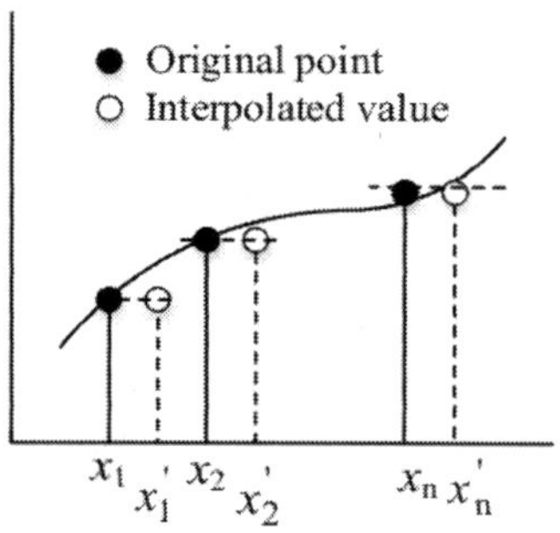

Figure 3. Nearest neighbor interpolation.

3.2 Linear interpolation

The linear interpolation is piecewise straight lines connecting two consecutive data points. The arc conductance linear interpolant function in the interval $[x_i, x_{i+1}]$ is given

$$G(x) = y_i + s \cdot \delta_i \tag{2}$$

It must pass through (x_i, y_i) and (x_{i+1}, y_{i+1}) and a continuity condition of $G_i(x_{i+1}) = G_{i+1}(x_{i+1})$ must be satisfied. In (2), y is arc conductance at original data point i, $\delta_i = (y_{i+1} - y_i) / (x_{i+1} - x_i)$ presents the slope of the line and the local variable $s = x - x_i$.

3.3 Piecewise cubic Hermite interpolation (PCHIP)

This interpolation method is to develop an interpolant cubic polynomial of the arc conductance function at known data points among a whole interval $[x_0,x_n]$. To this end, the general cubic polynomial at i[th] interval is defined by

$$P_i(x)=a_i+b_i(x-x_i)+c_i(x-x_i)^2+d_i(x-x_i)^3 \quad (3)$$

where P is arc conductance; a, b, c, and d are coefficients to be found; $i=0,1,2,\ldots,n$; and the completed arc conductance model by PCHIP is given by

$$G(x)=\begin{cases} P_0(x) & x_0\le x\le x_1 \\ P_1(x) & x_1\le x\le x_2 \\ \vdots & \vdots \\ P_n(x) & x_{n-1}\le x\le x_n \end{cases} \quad (4)$$

For a function P in (3), in order to find the coefficients of a to d it must satisfies four interpolation conditions, two on function values and two on the possibly unknown derivative values, which are given in (5) to (8).

$$P_i(x_i)=f(x_i) \quad (5)$$

$$P_i(x_{i+1})=f(x_{i+1}) \quad (6)$$

$$P_{i+1}(x_{i+1})=P_i(x_{i+1}) \quad (7)$$

$$P'_{i+1}(x_{i+1})=P'_i(x_{i+1}) \quad (8)$$

If both function values and first derivative values at a set of data points can be known then PCHIP can reproduce those data.

3.4 Cubic spline interpolation

The method is also used to create a similar cubic polynomial as in (3) but that is continuous in both the first and second derivatives. Therefore, besides (5) to (8) an additional condition in (9) must be satisfied as well.

$$P''_{i+1}(x_{i+1})=P''_i(x_{i+1}) \quad (9)$$

For a unique solution, it needs to specify further end (or boundary) conditions at x_0 and x_n, and the most common used are

- Natural end condition - assume the second derivative at the end points are zero

$$P''(x_0)=P''(x_n)=0 \quad (10)$$

- Clamped end condition - assume the first derivatives at the first and last points are known

$$P'(x_0)=f'(x_0);P'(x_n)=f'(x_n) \quad (11)$$

- Not-a-knot end condition - force continuity of the third derivative at the second and penultimate points, which result in the first two intervals having the same spline function and the last two intervals having the same spline function. In this paper, only this end condition is considered for modelling.

$$P'''(x_0)=f'''(x_1);P'''(x_n)=f'''(x_{n-1}) \quad (12)$$

4 Experiments and simulations

Two case studies are presented to estimate the performance of the different models:

4.1 AC electric arc furnace modelling

Figure 4 shows the one-line diagram of the measured steel plant power system which includes a 50 ton AC EAF load. For harmonic distortion analysis, the EAF voltages and currents are recorded at the duration when the EAF load operates at refine stage (static-state condition). Measured EAF characteristics are given in Fig. 5 and simulated results for each interpolation method are shown in Fig. 6 to Fig. 9. To observe the time-domain waveforms in these results, some of the features are found: (i) higher odd harmonic components in voltage and current are provided by the AC EAF load; (ii) the non-smooth serrated waveforms are clearly obtained by Nearest; (iii) arc voltage waveform is distorted in Linear; and (iv) the results from PCHIP and Spline are similar but the differences are presented in data boundary. Figure 10 shows the harmonic spectra of EAF voltage and current, and the total harmonic voltage/current distortions (THD_V/THD_I) relate with these spectra are listed in Table 1. It is found the better performance can be presented by using Spline which provides about 3.98% and 3.09 errors on THD_V and THD_I, separately.

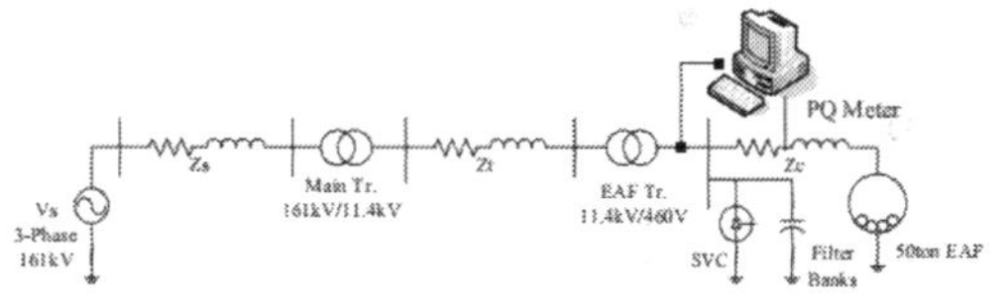

Figure 4. One-line diagram of the measured steel plant power system.

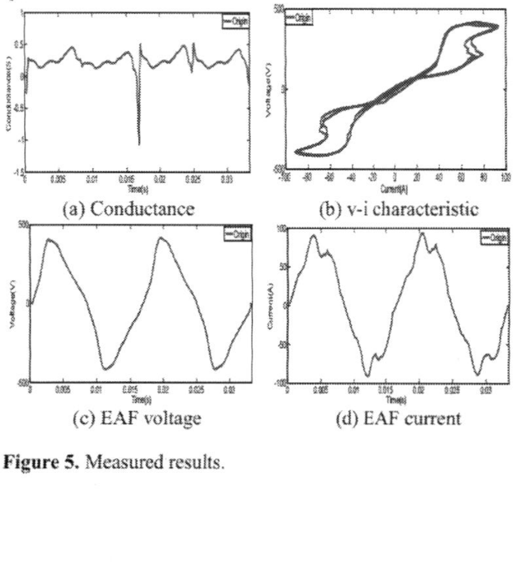

(a) Conductance (b) v-i characteristic

(c) EAF voltage (d) EAF current

Figure 5. Measured results.

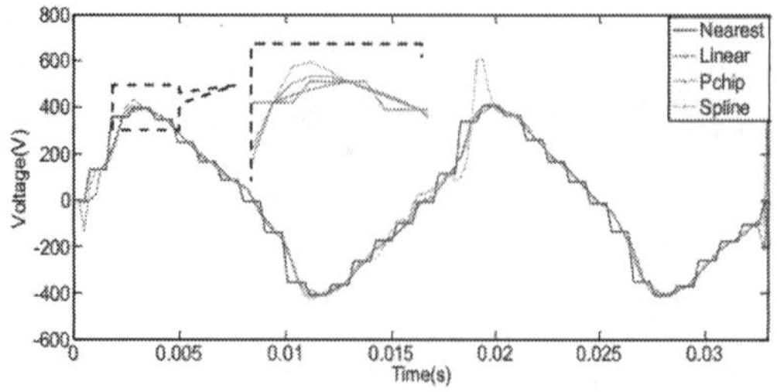

Figure 6. Simulated conductance.

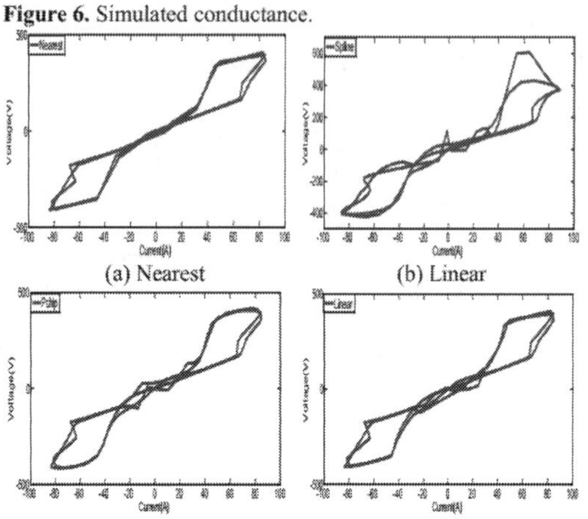

(a) Nearest (b) Linear

(c) Pchip (d) Spline

Figure 7. Simulated v-i characteristic.

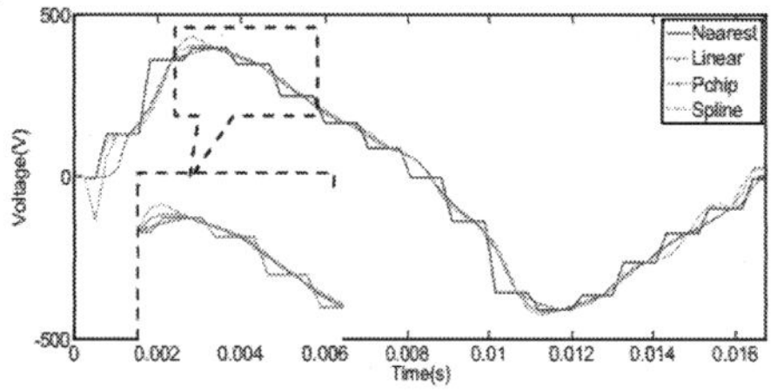

Figure 8. Simulated EAF voltage.

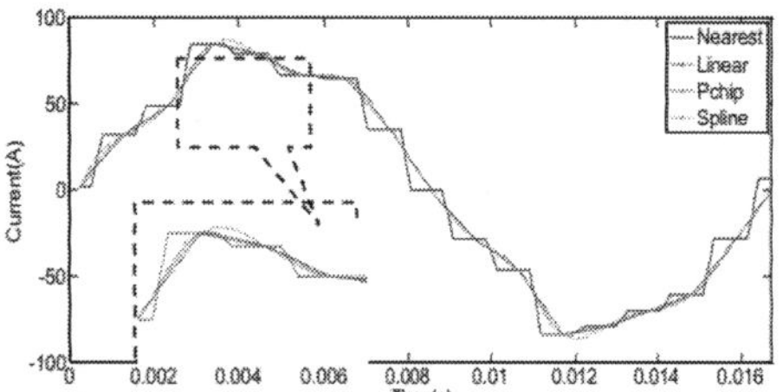

Figure 9. Simulated EAF current.

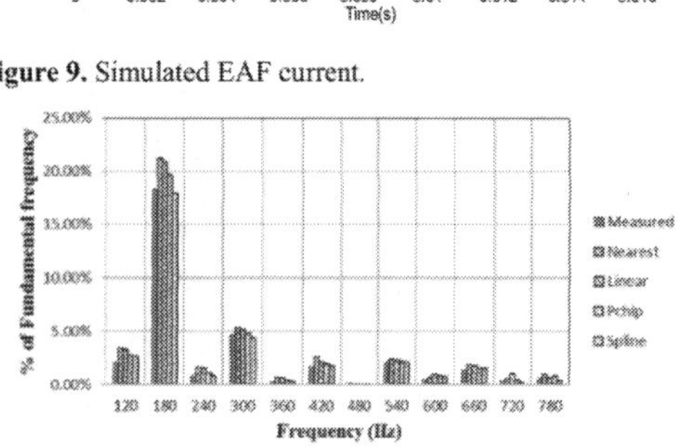

(a) Voltage

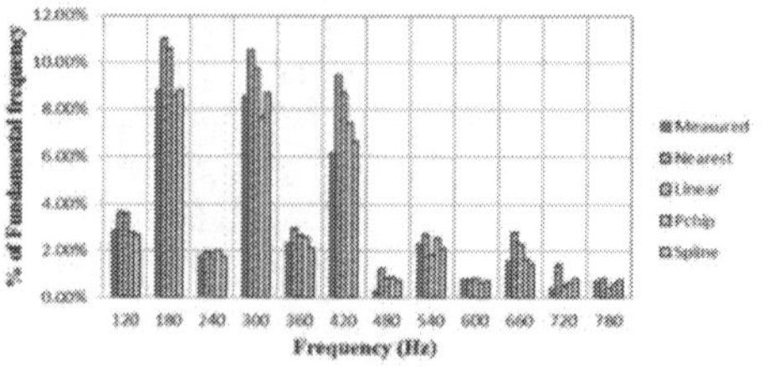

(b) Current

Figure 10. Harmonic spectra of EAF.

Table 1. Total harmonic distortion of EAF voltages and currents.

Index / Method	THD_V (%)	THD_I (%)
Measured	19.25	13.44
Nearest	29.44	23.92
Linear	27.72	20.28
Pchip	24.47	18.46
Spline	23.21	16.53

4.2 Low frequency fluorescent tube modelling

Fluorescent lighting generally requires a relatively long evacuated tube to develop sufficient ionized gas for operation. It is possible to have a nonlinear nature in the v-i characteristic and thus the tube current for an applied sinusoidal voltage is rich in odd harmonics. This can be observed from the test circuit in Fig. 11 and measured results in Fig. 12. To observe the simulation results on Fig. 13 to Fig. 17 it is found that the arc conductance on fluorescent tube presents a regular variation than EAF load, and harmonics distortions shown in Table 2 present smaller differences among each method when compared to measured results.

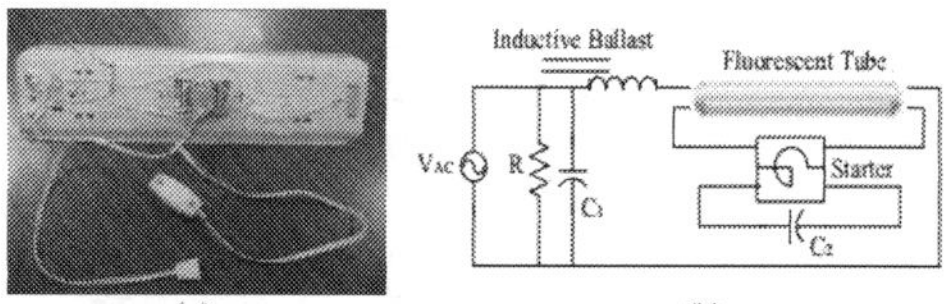

(a) (b)

Figure 11. Fluorescent tube circuit.

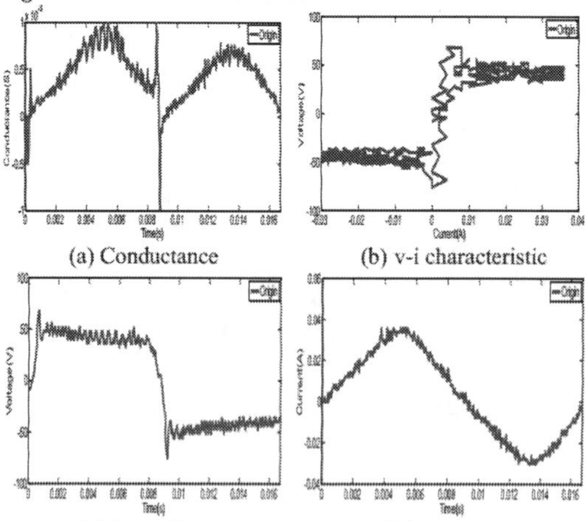

(a) Conductance (b) v-i characteristic

(c) Arc voltage (d) Arc current

Figure 12. Measured results.

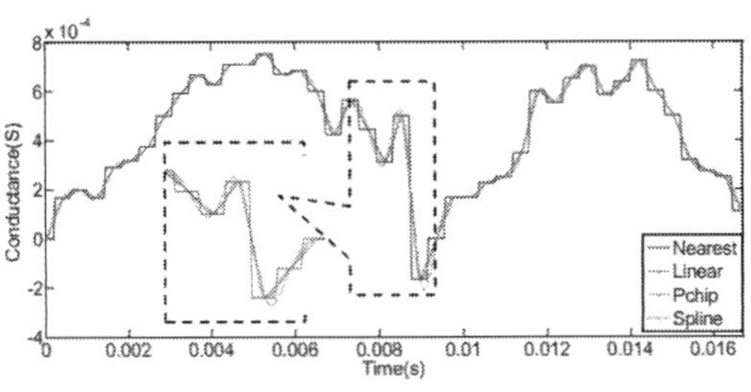

Figure 13. Simulated conductance.

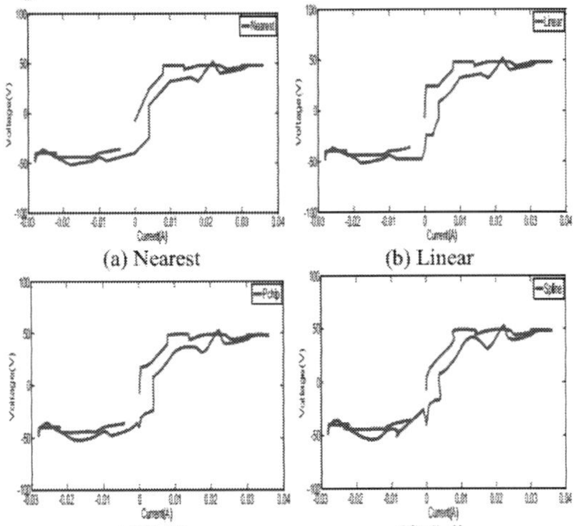

(a) Nearest (b) Linear

(c) Pchip (d) Spline

Figure 14. Simulated v-i characteristic.

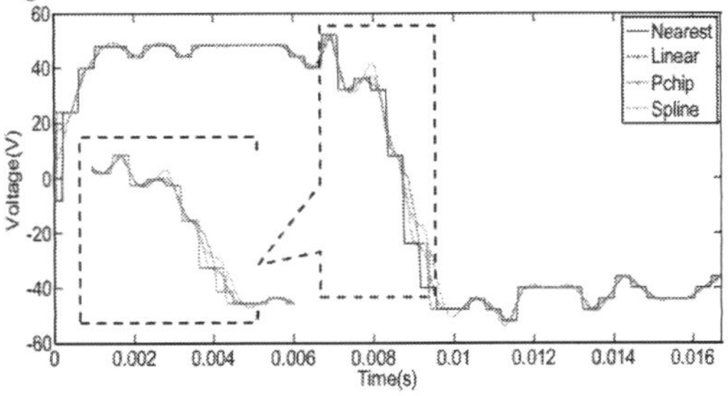

Figure 15. Simulated FT voltage.

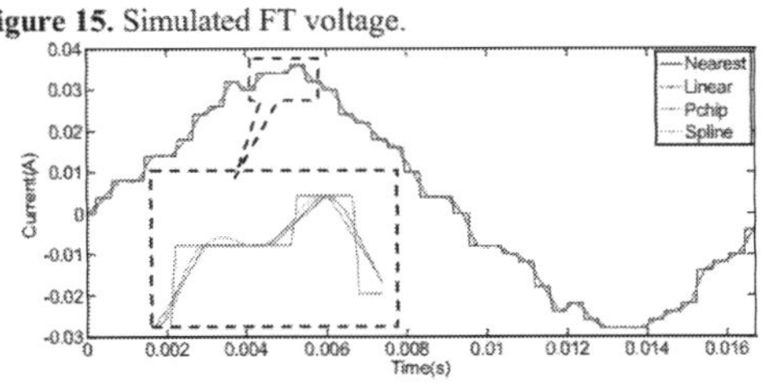

Figure 16. Simulated FT current.

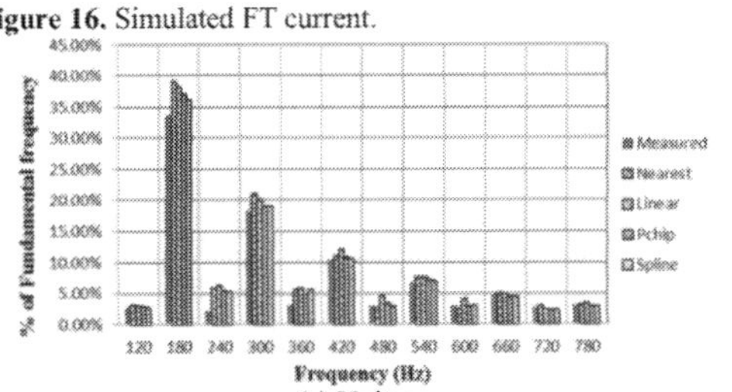

(a) Voltage

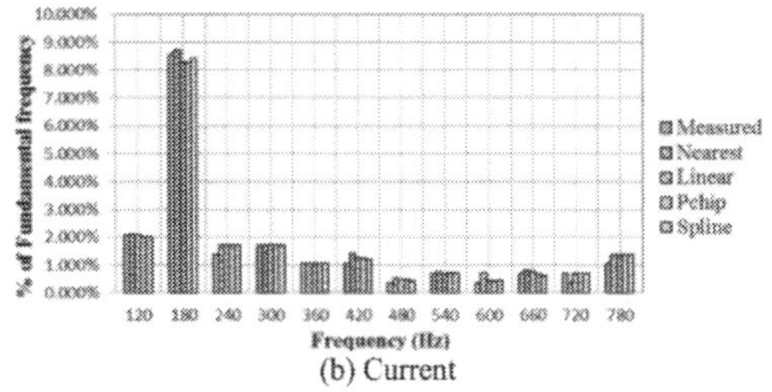

(b) Current

Figure 17. Harmonic Spectra of FT.

Table 2. Total harmonic distortion of lamp voltages and currents

Method \ Index	THD_V (%)	THD_I (%)
Measured	40.16	11.71
Nearest	36.61	12.44
Linear	36.36	10.20
Pchip	35.14	10.88
Spline	34.81	10.94

5 Conclusions

In this paper, different interpolation methods for modelling AC electric arc are presented. Experimental results taken from a steel plant power system and a fluorescence tube are used to show the accuracy of the compared models through computer simulations.

Results have shown that:

- Nearest is fastest in computation time but with a less accuracy;
- Linear requires more computation time than Nearest and the accuracy is a little bit better than that;
- Both PCHIP and Cubic Spline can obtain results with higher accuracy, but for PCHIP the derivative needs to be specified and the derivative of Cubic Spline is not specified but enforced;
- Polynomial interpolation may be good at low orders; however, higher order polynomials may "over-fit" the data and do not predict the curve well in between interpolation points;
- Cubic Spline is much good in smoothly interpolating data.

Acknowledgment

The paper is financially supported by the Ministry of Science and Technology of Taiwan (R. O. C.), under Grant no. MOST 104-2218-E-194-006.

References

1. GIGRE Working Group, Electra **36**, 5, 35 - 54 (1981).
2. S. Varadan, E.B. Makram, A.A. Girgis, IEEE Trans. Power Deliv. **11**, 3, 1685-1691 (1996)
3. M.A.P. Alonso, M.P. Donsion, IEEE Trans. Power Deliv. **19**, 1, 367-373 (2004)
4. K. J. Tseng, Y. Wang, D.M. Vilathgamuwa, IEEE Trans. Power Electron. **12**, 3, 429-436 (1997)

5. H. Mokhtari, M. Hejri, Proc. of the IEEE Asia-Pacific T & D Conf. and Exhi., **3**, 2078-2083 (2002)
6. G. W. Chang, Y. J. Liu, and C. I. Chen, Proc. of IEEE PES General Meeting, 1-6 (2008)
7. S. C. Wang, C. J. Wu, and Y. J. Wang, International Journal of Circuits, Systems, and Signal Processing, **6**, 5, 324-331 (2012)
8. G. D. Gregory, K. Wong, and R. F. Dvorak, IEEE Transactions on Industry Appl., **40**, 1006-1011 (2004)

A novel synthetic test system for thyristor level in the converter valve of HVDC power transmission

Longchen Liu[1], Ke Yue[2], Lei Pang[2], Xinghai Zhang[1], Yawei Li[1] and Qiaogen Zhang[2]

[1]*State Grid Sichuan Electric Power Research Institute, 610072 Chengdu, China*
[2]*School of Electrical Engineering, Xi'an Jiaotong University, 710049 Xi'an, China*

Abstract. The converter valve is the core equipment in the HVDC power transmission system, a+-nd its performance has a direct effect on the reliability, stability and efficiency of the whole power system. As the basic unit of HVDC converter valve, the thyristor level needs to be test routinely in order to grasp the state of the converter valve equipment. Therefore, it is urgent to develop a novel synthetic test system for the thyristor level with thyristor control unit (TCU). However, currently there is no specific test scheme for the thyristor level of HVDC converter valve. In this paper, the synthetic test principle, content and methods for the thyristor level with TCU are presented based on the analysis of the thyristor reverse recovery characteristic and the IEC technology standard. And a transient high-voltage pulse is applied to the thyristor level during its reverse recovery period in order to test the characteristics of thyristor level. Then, the synthetic test system for the thyristor level is applied to the converter valve test of ±800 kV HVDC power transmission project, and the practical test result verifies the reasonability and validity of the proposed synthetic test system.

1 Introduction

As the core element of HVDC converter valve, thyristor is massively adopted to transmit electric power energy because of its high power capacity. The converter valve usually consists of a large number of thyristor levels, which are connected in series. Each thyristor level needs to be test routinely during the maintenance period of electrical equipment in order to make sure that their electrical properties and functions are in normal state. Considering the large number of thyristors, there is a heavy workload of thyristor level test. Therefore, it is important to develop a synthetic test system for thyristor level in HVDC converter valve.

As to electrical test of HVDC converter valve, IEC 60700-1-2008 and GB/T20992-2007 technology standard provide some relevant provisions [1, 2]. However, these provisions mainly focus on the type test, such as dielectric test and operation test. Although these standards provide some general descriptions for routine test (for instance, the inspections of wiring, equalizer circuit parameters and withstand voltage), corresponding scheme and specific content of judgement methods in the routine test are scarcely mentioned. To date, some researchers introduced and studied type test and operation test of the converter valve. Ding [3] and Luo et al. [4] studied type test of converter valve and provided specific experiment protocol, requirement and improvement plan [5]. Zha et al. [6, 7] and Hu et al. [8] studied operation test under various actual operating conditions. However, studies on in-situ routine test for thyristor level of converter valve are rare. Zha et al. [9] studied routine test and proposed a thyristor level unit routine test program, but failed to investigate the TCU testing. Some international companies developed comprehensive testing devices for in-situ synthetic test of thyristor level, for example ABB, Siemens and Alstom. Yet, the domestic standard of relevant routine test for converter valve products is still in preliminary stage [10-12]. Therefore, it is necessary to establish a routine synthetic test system for thyristor level, which can ensure the performance of thyristor level and provide technical support for the operation and maintenance of HVDC converter valve.

Based on the analysis of the electrical principle and failure mode of thyristor level unit, this paper proposed a synthetic test scheme for in-situ routine test, and the synthetic test system is applied to the converter valve test of a practical ±800 kV HVDC power project.

2 Electrical structure and principles of thyristor level

The converter valve is arranged in modules, which contains a large number of thyristor level units connected in series. As shown in Figure 1, thyristor level unit with TCU is the basic structural unit in converter valve, which consists of thyristor, damping circuit ($R_{1\text{-}6}$, C_{c1}, C_{c2}), equalizing circuit (R_{41}, R_{42}), thyristor control unit (TCU) and energy storage circuit for TCU (C_3, R_3).

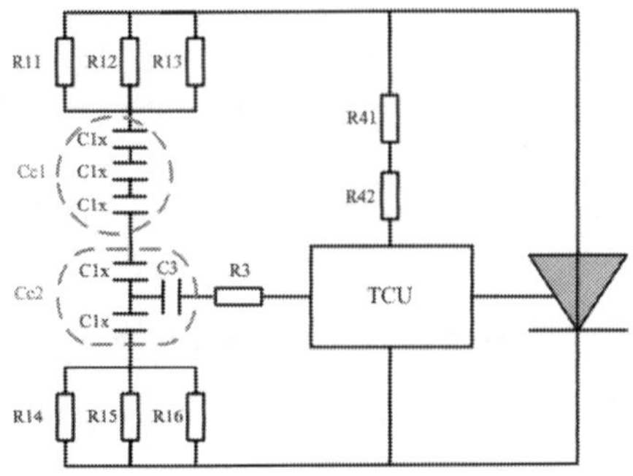

Figure 1. Electrical diagram of thyristor level with TCU.

Under the condition of inductive loads, reverse recovery overvoltage would be generated across the thyristor during the turn off process, which even exceeds the reverse repeat peak voltage of the thyristor and breaks the thyristor. Therefore, a parallel damping circuit is usually connected to the thyristor to absorb the reverse recovery charge and prevent the voltage overshot during the turn off process, which is shown in Fig. 1.

TCU is the core component in thyristor level unit. Thyristor level unit is at a high potential, whereas the control module of the converter valve is at ground potential. Therefore, an electric isolation is required in the transmission of trigger signal. Optoelectronic isolation has been widely applied in electric isolation. Valve control unit (VCU) at ground potential converts the electrical trigger pulse, which is produced by the control module of the converter valve, into optical pulse and transmits this optical pulse to TCU by the optical fiber. TCU reconverts the optical pulse into electrical pulse, and transmits this electrical pulse signal to the gate pole of the thyristor. Then the thyristor is triggered. Except the normal trigger function described above, TCU includes two protective trigger circuits: the protective firing (PF) circuit and the recovery protection (RP) circuit. The working voltage of TCU is obtained by coupling the voltage of thyristor level unit.

If the thyristor cannot be normally triggered, the forward voltage across the thyristor would continue to increase. In case that the forward voltage may exceed the rated value, PF circuit would send an emergency trigger pulse automatically in order to force the thyristor to be conducted, which avoids the thyristor being damaged. During the turn off process, TCU would detect the terminal voltage of the thyristor, and send an emergency trigger pulse to trigger the thyristor and prevent breaking if the disturbance voltage exceeds the rated value.

3 Research and development of synthetic test system for thyristor level

3.1 Electrical structure of synthetic test system

Figure 2 shows the structure and principle of the synthetic test system for thyristor level of HVDC converter valve. The synthetic test system is composed of the main circuit, control unit, communication module, peripheral equipment interface circuit and the corresponding power distribution protection module. As shown in Figure 2, the main circuit of synthetic test system includes three circuit modules: fundamental frequency (50Hz) power supply module, high voltage impulse power supply module and high frequency sinusoidal wave signal generator, these three power modules can be switched by the time control of high-voltage relay. The electromagnetic interference (EMI) module in the figure is a filter, which is used to prevent electromagnetic interference.

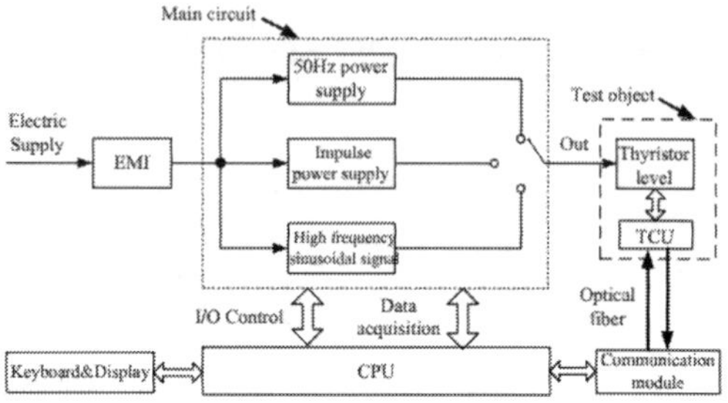

Figure 2. Diagram of test system for thyristor level of HVDC converter valve.

The electrical structure of high voltage impulse power supply module is shown in Figure 3. This power supply module consists of seven parts: a bidirectional thyristor V_1, a boost transformer T_2, a controllable rectifier V_2, a RLC discharge circuit, a power electronic switch (V_3~V_6), a internal protection circuit (V_7~V_{10}) and a driving circuit (Q_3, Q_4). As shown in Figure 3(a), the switch S_2 is mainly used for switching power to different power modules; R_1 and R_2 is the protection resistance; the energy storage capacitor C_1 is charged through the boost transformer T_2 in order to get a high voltage impulse; the output voltage can be adjusted through changing the AC positive and reverse conduction angle of the bidirectional thyristor V_1, and the thyristor V_1 must be triggered on zero by control unit Q_2 to ensure the output waveform is a standard sine wave. In Figure 3(b), C_1 is the energy storage capacity; L is the inductance of RLC oscillation circuit; L_1 is used to adjust the impulse voltage waveform; L_2 is a key part of the discharge circuit; V_3~V_6 is used as a forward switch; and V_7~V_{10} is the switch of discharge circuit. And the energy of high voltage impulse can be released through the loop V_7~V_{10}-R_3-R_4-V_3~V_6.

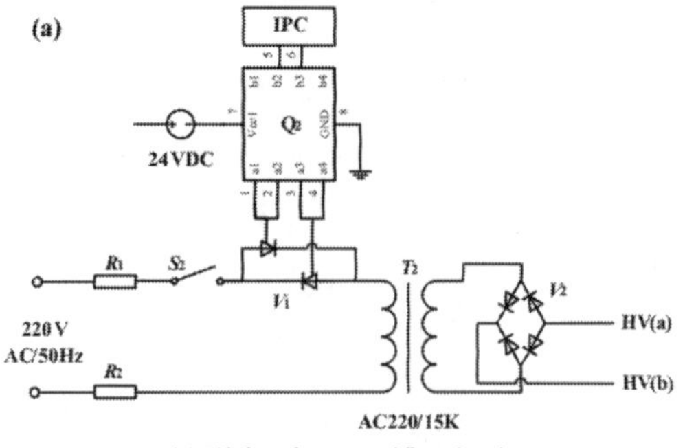

(a) High voltage rectifier circuit

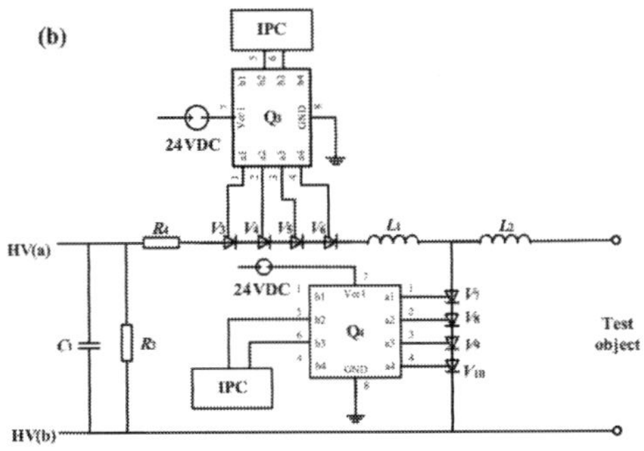

(b) High voltage impulse circuit

Figure 3. Electrical diagram of high voltage impulse power supply module.

3.2 Synthetic test contents of thyristor level

By analyzing common faults of thyristor level unit, a test scheme of in-situ synthetic test is developed on the basis of its failure modes. Combing the standard IEC 60700-1 and the working principle, the test projects for the in-situ synthetic test of the thyristor level unit are developed. Items of the routine synthetic test are shown in Table 1.

Table 1. Synthetic test items of the thyristor level.

Test items with low voltage	Test items with high voltage
Impedance test	TCU protective function test
Firing test	Reverse test

3.2.1 Impedance test

Impedance test detects the short circuit or the open circuit within the voltage equalizing circuit and damping circuit. It is a low voltage test, the voltage of which does not reach thyristor turn on voltage, therefore the corresponding impedance is mainly refer to the impedance of capacitors and resistors within the equalizing and damping circuits.

The circuit diagram of impedance test is shown in Figure 4(a), where the voltages with different frequencies generated by the sine wave generator are imposed on thyristor level unit. The peak voltage and current across the thyristor are measured during the test, and the impedances of the circuit under different frequencies are calculated. The obtained impedance is compared with the threshold, therefore the conditions of resistors and capacitors in the thyristor level unit can be evaluated.

3.2.2 Fire test

Fire test detects whether TCU is able to respond properly to optical pulse signal and send a trigger pulse to the thyristor. The test circuit of fire test is shown in Figure 4(b), where AC circuit with power frequency applies a forward voltage to the thyristor, and the electronic power switch controls the phase angle.

3.2.3 TCU protective function test

Ensure the performance of thyristor under the following conditions: (I) the thyristor suffers from an overvoltage exceeding the designed voltage threshold; (II) the thyristor is subjected to a transient forward impulse during its turn off process, the TCU should be tested by the PF and RF tests. In this study, RP test includes PRL tests and PRH tests. The parameters of impulse voltage used for RP test are shown in Table 2. PRL test is performed during the reverse recovery period of the thyristor. After the thyristor regain its reverse blocking ability, PRH test is performed.

The test circuit of TCU protective function is shown in Figure 4(c). The AC circuit provides a normal forward voltage for the thyristor, and the impulse circuit generates a forward impulse voltage. The power electronic switch is able to forcedly cut off the on state current and control trigger phase angle. These three modules are putting into operation according to a predefined controlling sequence.

Table 2. Parameters of the RP test.

Test project	Amplitude	Time
RPL	2 kV	$(t_{rr})_{max}$+ 50 μs
RPH	10 kV	$1.2(t_{rr})_{max}$

3.2.4 Reverse test

The blocking ability of the thyristor to reverse voltage can be evaluated through reverse test. The test circuit of reverse test is shown in Figure 4(d), where the impulse circuit provides a reverse voltage impulse, and the voltage divider is used to measure the impulse voltage.

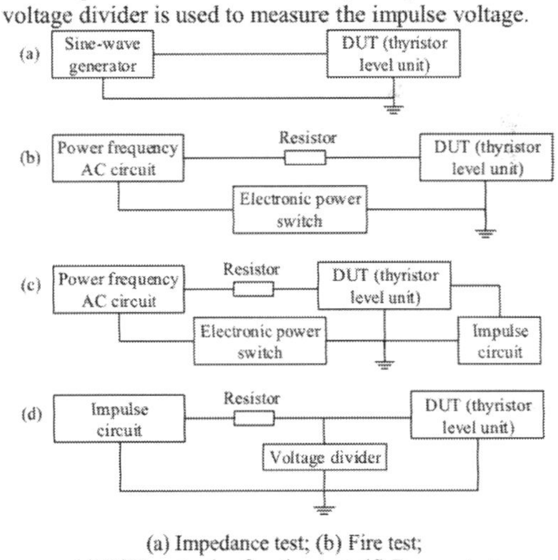

(a) Impedance test; (b) Fire test;
(c) TCU protective function test; (d) Reverse test

Figure 4. The synthetic test diagram of thyristor level.

4 Application of synthetic test system for thyristor level

The synthetic test system was applied to the in-situ test of the KGWF03-5000/±800 thyristor level in the converter value of HVDC power transmission. The acceptance standard and the interrelated technical regulations of the in-situ synthetic test are listed in Table 3, and the test results and analyses are presented as follows.

Table 3. Technical specifications of ±800kV HVDC power transmission.

Test items	Acceptance criteria
Impedance test	110Hz: 664~1727Ω 6000Hz: 26~36Ω
Firing test	U_{th}: 32±10V
TCU protective function test	U_{PF}: 8050~8500V U_{RPL}: 1475~1775V U_{RPH}: 8050~8500V
Reverse test	U_{REV}: 8040~8340V

Figure 5 shows the impedance test waveform. The PRL and RPH test worked on frequencies of 110 Hz and 6 kHz respectively. The peak voltage and current of the thyristor were measured during the tests, and then the impedance under each corresponding frequency was calculated. The impedances in PRL and PRH test were 685Ω and 29Ω respectively, which met the requirements of the technical specifications.

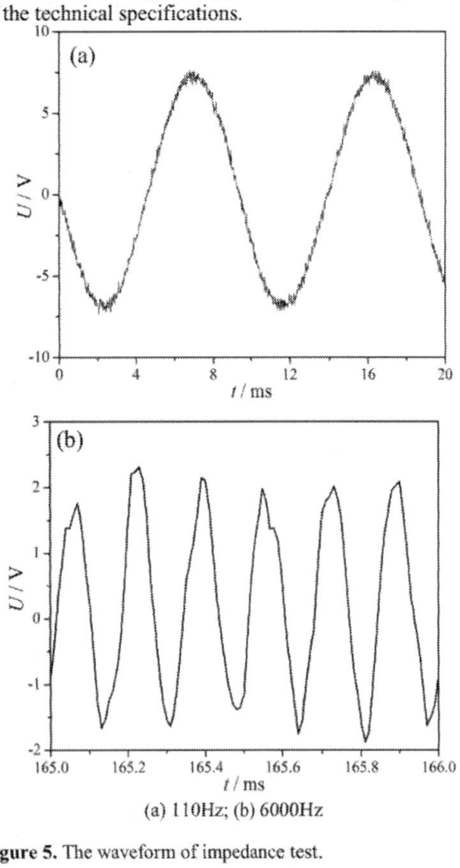

(a) 110Hz; (b) 6000Hz

Figure 5. The waveform of impedance test.

Figure 6 shows the waveform of the fire test. The thyristor level unit was imposed by a power frequency voltage. It can be found that TCU took the power and then triggered the thyristor. The measured voltage of the triggered thyristor was 35 V, which met the demand of technical specifications.

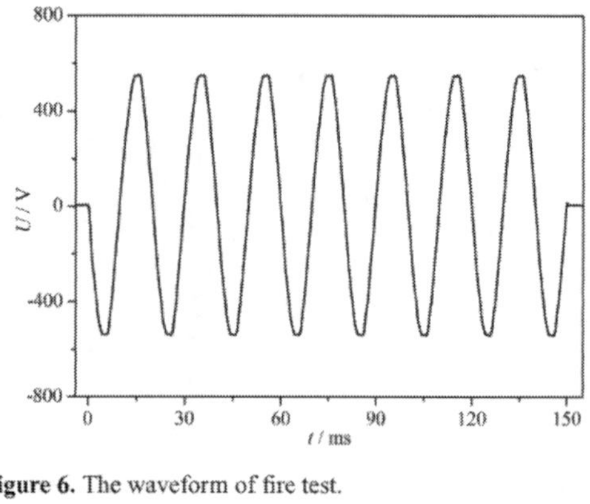

Figure 6. The waveform of fire test.

Figure 7(a) shows the waveform of the PF test. When the impulse voltage reaches the protection threshold of TCU, TCU will send an emergency pulse to trigger the thyristor. The peak value of forward voltage was measured to be 8067 V when the thyristor was triggered by TCU. Figure 7(b) and 7(c) show the waveforms of the RPL and RPH test respectively. A power frequency voltage was applied to the thyristor, and then the reverse voltage turned off the thyristor. During the reverse recovery process, a low voltage impulse was applied through the impulse circuit at 1050 μs later than the moment when the thyristor voltage began to reverse. This action threshold of TCU was 1475~1775 V, and TCU operated when the impulse voltage was 1660 V. Figure 7(c) shows the thyristor voltage waveform of the RPH test. A high impulse voltage was applied by the impulse circuit at 1200 μs later than the moment when the thyristor voltage began to reverse. This action threshold of TCU is 8050~8500 V, and TCU operated when the impulse voltage was 8099 V. The TCU protective function met the technical requirements as expected.

Figure 8 shows the waveform of the reverse test. Reverse voltage impulse was imposed on the thyristor, and ensure that thyristor breakdown would not take place. The protective circuit released the energy of high voltage impulse and protected the thyristor. The voltage peak value of reverse impulse on the thyristor was measured to be 8290 V. It was indicated that the reverse test passed.

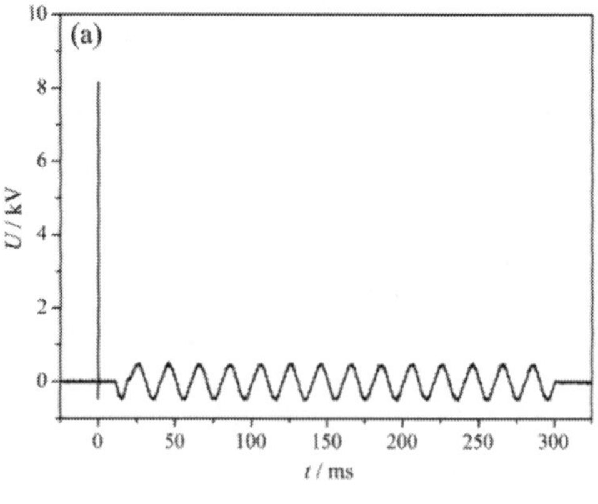

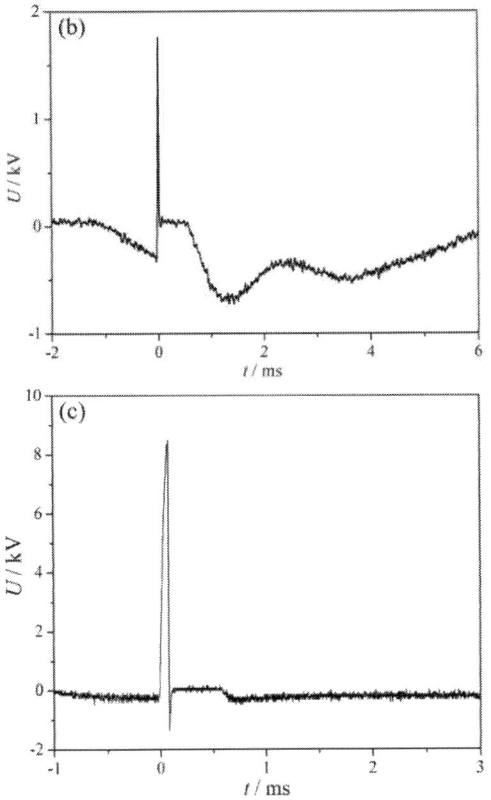

(a) PF test; (b) RPL test; (c) RPH test

Figure 7. The waveforms of TCU protective function test.

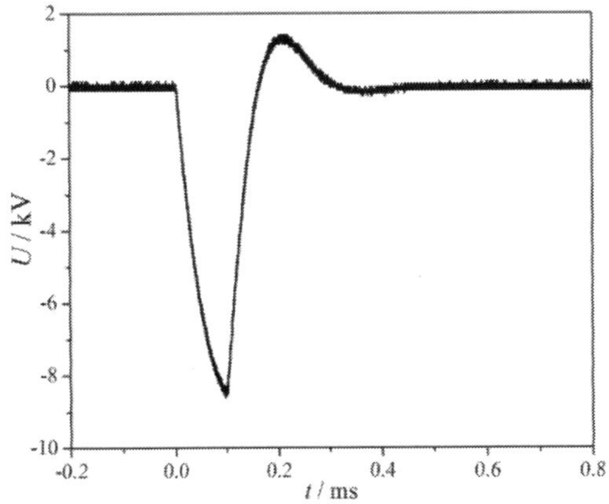

Figure 8. The waveform of reverse test.

By analyzing the test waveforms above, the thyristor level unit of the HVDC converter value met the technical requirements and all tests passed, which proved and verified the reasonability and validity of the proposed in-situ synthetic test system and method.

5 Conclusions

In this paper, the in-situ synthetic test technology of thyristor level in HVDC converter valve was investigated. Through analyzing the structure and working principle of thyristor level in the converter valve, and combining with the technical standards IEC 60700-1-2008, the synthetic test content and method for the thyristor level with TCU was proposed. And the in-situ test of thyristor level included impedance test, fire test, TCU protective function test and reverse test. The in-situ synthetic test system was developed and applied to the thyristor level test of ±800kV HVDC converter valve in order to verify the rationality and validity of this test system and method. The test waveforms indicated the performance of thyristor level in the converter valve met the technical requirements of HVDC power transmission. The results showed that this test system and method was reasonable, and satisfied the demand of engineering applications.

References

1. IEC60700-1. Geneva: International Electrotechnical Commission (2008)
2. GB/T 20992-2007. Chinese standard SN for books. Beijing: China Standard Press (2007)
3. Y. G. Ding, Electric Power Construction, **29**, 2 (2008)
4. X. Luo, G. F. Tang, Power System Technology, **34**, 5 (2010)
5. E. Nho, B. Han, Y. Chung, IEEE Transaction on power delivery, **27**, 2 (2012)
6. K. P. Zha, J. L. Wen, Power System Technology, **34**, 12 (2010)
7. K. P. Zha, C. Gao, Power System Technology, **36**, 2 (2000)
8. Z. L. Hu, F. Xu, High Voltage Apparatus, **49**, 4 (2013)
9. K. P. Zha, G. Y. Wang, High Voltage Engineering, **38**, 11 (2012)
10. T. Bauer, H. P. Lips, IEEE Transaction on Power Delivery, **12**, 3 (1997)
11. E. Nho, B. Han, Y. Chung, IEEE Transaction on Power Electronics, **29**, 7 (2014)
12. G. F. Tang, K. P. Zha, IEEE Transaction on power delivery, **28**, 3 (2013)

Root cause analysis of the folding deformation for the expansion joint in the nuclear power station

Li Nan, Che Yin hui, Li Guo dong, Zu Shuai and Guan Jian jun

SuZhou Nuclear Power research Institute, ShenZhen, 518038, China

Abstract. In the nuclear power station, the expansion joint of moisture separator re-heater (MSR for short) was founded locally folding deformation under the outage condition. Folding deformation increasing will lead expansion joint leak, it would cause emergency stop when leaking seriously. To solve these problems, this paper analysis mainly on three aspects: the material organization performance, the expansion joint installation and the stress characteristic of expansion joint. Both the verticality deviation in the process of installation and the material hardness value exceed standard are the root cause of these problems. At the end of the paper, the improvements have been put forward for this problem.

1 Introduction

The moisture separator re-heater (MSR for short) is set up between high pressure cylinder and low pressure cylinder in the Pressurized Water Reactor (PWR for short) nuclear power station. Its role is to separate new saturated steam from high pressure cylinder whose humidity is increasing after working. Then the new saturated steam go through two levels of re-heater, making the exhaust steam of high pressure cylinder become superheated steam after working and before entering the medium and low pressure cylinder. After through the separator, the water in steam has been wiped out, and it can reduce the water droplets on the turbine blade erosion, improve the efficiency of thermal system [1].

The first and second level heating steam which is coming from the main steam system mostly condenses into water after finish heat-exchange in the re-heater tube, then going through the first and second level hydrophobic pipeline to the drain tank. The hydrophobic pipeline should be through the MSR shell, when the plant is running, the pipe material expanding or the stress appearing causes the horizontal displacement increases and the expansion amount of shell not synchronized. Therefore, set the expansion joints at the connection of drain pipe and shell to compensate thermal expanding displacement in every direction [2].

This paper through metallographic analysis, mechanical property analysis and other analysis methods, on the basis of displacement measurement of expansion joint, establishes the expansion joint failure models of moisture separator re-heater, analyses the expansion joint stress, and explores failure causes of the MSR expansion joint in the nuclear power plant under the engineering application.

2 Metallography analysis

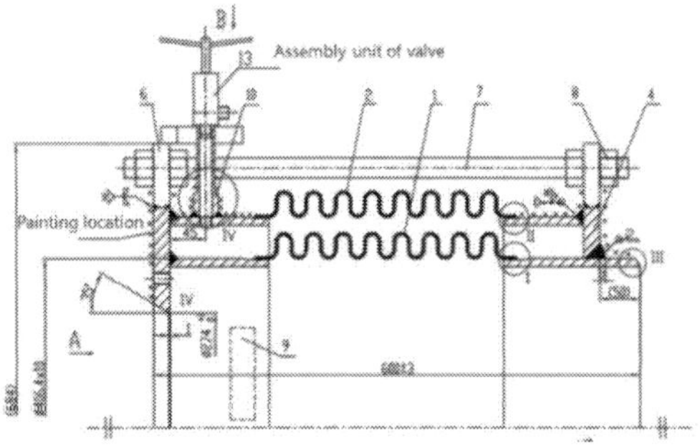

Figure 1. First and second level hydrophobic expansion joint.

Table 1. The equipment parameters of expansion joint.

item	parameter
Design pressure	1.4Mpa
Design temperature	311℃
compensation dosage	axial direction is 10mm, radial direction is 4mm
stiffness	axial direction is 1087N/mm, radial direction is 441N/mm
Design fatigue life	＞1000
installation characteristics	Welding

The folding deformation expansion joint is secondary hydrophobic expansion joint of MSR in this nuclear power station. Expansion joint has inside and outside part; each part divided into 5 layers and each layer is 0.6 mm.

DOI: 10.1051/matecconf/20167716004

Through hydraulic forming making the internal and external layers fit closely, as shown in figure 1. According to maintenance manual operation, the equipment parameters of first and second level hydrophobic expansion joint are as in Table 1.

2.1 Macroscopic observation analysis

Three corrugated segments of expansion joint has folding deformation on the side where is taking over the end surface area. The maximum deformation position has about 20 mm length crevasse shape defect, it has obviously plastic deformation on both ends of the defect, as shown in figure 2. When cut the both ends of the expansion joint, found that deformation is only in the innermost layer 1 where is taking over the end surface area near the bottom part of the expansion joint, no deformation appears in the other four layers. The outside part expansion joint does not have deformation, the ends of the first layer and the rest four layers exists gap, which was not totally fit.

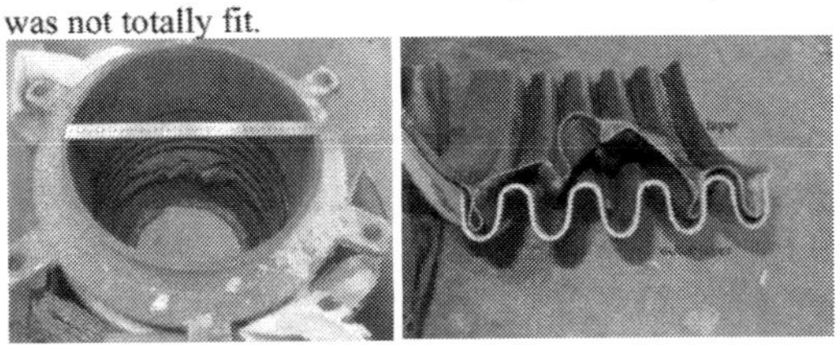

Figure 2. The deformation morphology of expansion joint.

Peeling off the innermost layer 1 where has folding deformation, found that the rest four layers of corrugated pipe have no crack, and no traces of corrosion around the crevasse. Using asana microscope observed inside and outside section morphology of the expansion joint, wave height of inside part is 29.6 mm by measured, pitch of wave is 41.6 mm. The wave height of outside part is 29.1 mm by measured, pitch of wave is 36.2 mm. The design of wave height is 29 mm, pitch of wave is 36 mm. Measurement data shows that the inside part of expansion joint has been out of shape.

2.2 Chemical composition analysis

Sampling for the innermost layer of expansion joint, using the PerkinElmer Optima 2100DV inductively coupled plasma emission spectrometer and element C、S analyzer for chemical composition analysis; the results are shown in table 2.

Chemical composition results show that the chemical composition of innermost layer 1 has big differences with the product quality book which is provided by upstream suppliers. The content of C element beyond the specified value in the ASTM A240, other elements contents are in accordance with the ASTM A240 standard requirement for 304L stainless steel chemical composition [3].

2.3 Hardness analysis

According to GB/T 4340.1-2009 'metal materials Vickers hardness test Part 1: test method'. Vickers hardness test was carried out on the sample; the results are shown in table 3.

Vickers hardness test results show that the hardness values of sampling location are higher than ASTM A240 standard of 304L stainless steel. Hardness values exceeds the standard would make the material brittleness, and prone to cracking.

2.4 Metallographic analysis

The inner layer with and without deformation has been metallographic analysis, sampling inspection results show that the sampling position of microstructure are twin austenitic, grain size is class 7-8. Organization is relatively uniform, and microstructure has no problem.

3 Displacement detection analysis

The calculation instruction shows that in order to meet the requirements of the fatigue life, the maximum horizontal displacement of the expansion joint when it is running should satisfy the following conditions: the horizontal displacement of operating equipment (≤4mm) + horizontal displacement of installation (≤5mm) = total horizontal displacement (≤9mm), as shown in figure 3. According to the calculation results, after the radial displacement changes into 9 mm, the fatigue life significantly will reduce to 1115 times (> 1000) which is about 1/10 of the initial design value (4 mm) [4].

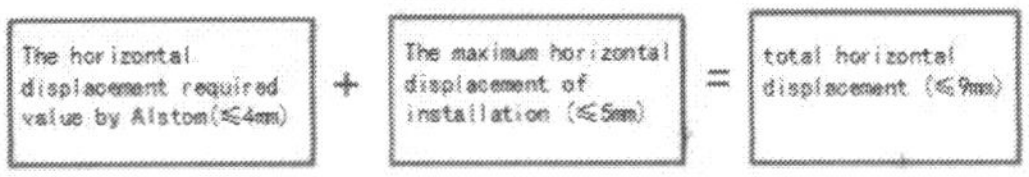

Figure 3. The displacement design requirements of expansion joint.

Analysed the reason of excessive displacement, finding the possible reasons mainly include the following two aspects:

(1) The assembling influence before installation: welding the expansion joint with MSR by poor control verticality, may cause the expansion joint deflect [2].

Query unit 1 # and 2 # MSR's End of Manufacturing Report (short for EOMR), the EOMR shows that, after the assembling complete, the maximum horizontal displacement is 4 mm. There is little amount of permissible error for on-site installation and adjustment is difficult when installing.

(2) The influence of installation: welding the expansion joint with the MSR's bottom drain pipe by poor control verticality, may cause the expansion joint deflect.

4 The displacement measurement

The expansion joint of unit 1 plant was measured under the thermal state. The measurement scheme: using the plumb line to measure, if the plumb line close to the top period of the expansion joint first. The bottom section of

Table 2. Chemical composition test results.

item	chemical composition wt%						
	C	Mn	P	S	Si	Cr	Ni
inner layer	0.038	1.55	0.009	0.007	0.36	18.06	8.12
the quality of the product specification	0.0154	1.09	0.023	0.001	0.55	18.4	9.2
ASTM A240	≤ 0.030	≤ 2.00	≤ 0.045	≤ 0.030	≤ 0.75	18.0~20.0	8.0~12.0

Table 3. Vickers hardness (HV1) test results.

NO.	location	Vickers hardness /HV1				average value
The inner layer without deformation	cross section	237.6	242.3	240.5	247.0	242
	longitudinal section	231.9	226.1	234.3	238.3	233
The inner layer with deformation	cross section	254.4	251.0	254.3	248.6	252
	longitudinal section	245.8	256.8	242.4	225.9	243
ASTM A240 GB/T 1172		≤ 201HB (≤ 206HV)				

the expansion joints deviate from the plumb line is negative. If near the bottom part first, the upper part deviation value is positive.

The unit condition is: the nuclear power is 80.7% Pn, electric power is 850 MWe, the measurement results are shown in table 4. Measurement results show that, when the nuclear power is 80.7% Pn, the expansion joint of the horizontal displacement has been reached or exceed the maximum requirements of design. Nuclear power can reach 100% Pn under normal conditions, and the horizontal displacement those are measured in the table 4 will be further increased.

Table 4. The horizontal displacement of unit 1.

NO.	Location	horizontal displacement top	horizontal displacement bottom	Horizontal displacement offset
1#	1	69	72	-3
	2	66	71	-6
	3	49	40	-5
	4	38	29	5
	5	41	36	9
	6	56	62	9

5 Stress simulation of displacement

According to the material parameters of expansion joint and combining with the conditions of operation and installation of expansion joint, the innermost layer corrugated pipe caused by displacement stress simulation has been carried out on the finite element stress analysis. The original designed displacement value is 4 mm and the maximum allowable value is 9 mm. These two situations are simulated to analysis the expansion joint's stress. The finite element calculation results are as follows:

(1) When the horizontal displacement is 4 mm, the stress nephogram is shown in figure 4

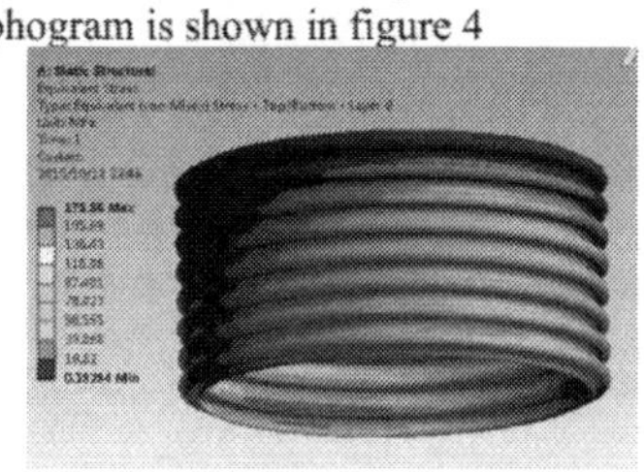

Figure 4. The stress nephogram of expansion joint.

(2) When the horizontal displacement is 9 mm, the stress nephogram is shown in figure 5.

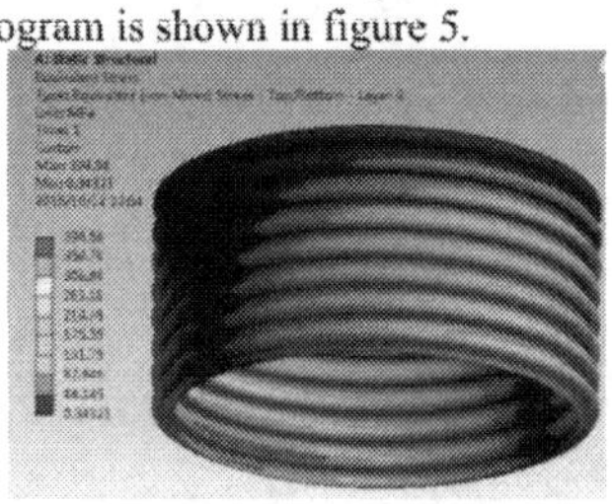

Figure 5. The stress nephogram of expansion joint.

Stress simulation results show that when the expansion joint stress deformed, the highest stress position is on the end of corrugated pipe where is connecting with the pipe. The simulation result is consistent with the deformed position. When the horizontal displacement is 4 mm, the maximum stress is 175.56 Mpa [5]; When the horizontal displacement is 9 mm, the maximum stress is up to 394.56 Mpa, almost excesses material yield limit [6].

6 Analysis and discussion

1) The metallography analysis of the MSR expansion joint shows that the expansion joint is made of 304L stainless steel, C element content of the expansion joint is beyond the specified value in the ASTM A240[7]. The hardness values of inside and outside parts of expansion joint are also significantly higher than the specified value in ASTM A240. High hardness makes the material's brittleness increase, toughness decrease. When materials under external force beyond yield limit, it will prone to cracking.

2)When the expansion joint is under the force, the highest stress position is on the end of the corrugated pipe where is connecting with the pipe. The simulation results are in conformity with the deformation position. Expansion joint folding deformation position locates in the maximum horizontal displacement. Macro and micro analysis shows, the welding ends and corrugated pipe has no abnormalities, those exclude the possibility that gas first came into the expansion joint interlayer. When the interlayer of expansion joint has water, the expansion joint has condition to deform. Based on the above evidence, this paper thinks that the innermost layer of expansion joint in maximize deformation first occurs to crack.

7 Conclusions

For the MSR expansion joint folding deformation problem, this paper analysis through the on-site survey, displacement measurement analysis, laboratory analysis and the stress calculation analysis, etc. The analysis results show the horizontal displacement allowance is small leading the hydrophobic expansion joint has poor adaptation. When the expansion joint is installed, the horizontal displacement is beyond standard. After the unit running, the horizontal displacement beyond design standards, and the expansion joint material hardness values exceed standard, finally leading to the expansion joint fatigue life greatly reduced.

After operation, the expansion joint of MSR appearing thermal expanding displacement makes expansion joint under the force. The expansion joint innermost layer 1 where connecting with pipe has forced the maximum stress began to cracking. When the unit start-up, the working medium at low temperature coming into the interlayer. As the unit power rising, the working medium which come into layers became vaporization with temperature rising, affecting internal pressure of the expansion joint interlayer, making the inside and outside expansion joint has different pressure, and finally leading to folding deformation [8].

Through comprehensive analysis, determined the root cause of the hydrophobic expansion joint deformation is: the expansion joint adaption is poor which is easy to lead the installation has big deviation, and the material hardness values exceed the standard. The contributing cause of the problem is: the material C element content beyond the standard.

8 Improvement measures

According to the nuclear power plant hydrophobic expansion joint deformation problem, making the following improvements:

1) Given the expansion joint has quality defects, whose displacement has small allowance and installation precision is difficult to control, suggesting that choosing bigger displacement allowance expansion joint, and waiting right time to replace it.

2) Before the expansion joint replacement has been finished, evaluating the expansion joint of MSR system. According to the evaluation results, arrange the reasonable replacement order.

3) Update version of installation procedures, the expansion joint of the original horizontal displacement and radial displacement measurement, ensuring that the expansion joint displacement allowance is enough.

References

1. Guangdong nuclear power training center. Devices and Systems of 900MW PWR [M]. Beijing: Atomic Energy Press.
2. Chen Lisu and so on. The application and design manufacturing characteristics of nuclear level's expansion joint. Pipeline technique and equipment. 2012(1)
3. Dong Tingkai. The failure reasons of expansion joint [J]. Shandong metallurgical. 2006, 28(1): 67- 68.
4. Hou Jinghong. Handling of over-proof horizontal displacement of the expansion joint for the re-heater in nuclear power station[J]. Power Station Auxiliary Equipment. 2012, 33(2)
5. Zhao Cheng long .Engineering thermodynamics [M], Nanjing aviation professional teaching materials published editorial group,1984, 2.
6. Zhao Jinxing and so on. External pressure type corrugated expansion joint failure case analysis[J], Pipeline technology and equipment, 2002(8)
7. Stainless steel and heat resistant steel general requirements for rolling steel plate, sheet and strip ASTM SA-480/SA-480M.
8. General technical requirements of metal bellows expansion joint. GB/T 12777- 2008.Bejing: National Quality Technical Supervise Department. 2008, 6

DOI: 10.1051/matecconf/20167716005

Fracture analysis of the planet support used in a nuclear power plant

Shuai Zu[1], Xingang Guo[2], Guangliang Xu[2], Liangjia Chen[2], Yinhui Che[1] and Jianjun Guan[1]

[1]*Suzhou Nuclear Power Research Institute, Equipment Management Department, 518000 Shenzhen, PR China*
[2]*HongYanHe Nuclear Power Plant, System Equipment Branch, 116000 DaLian, PR China*

Abstract. In this paper, the fracture analysis of the planet support, which was made of aluminium alloy and used in an electric actuator, was investigated. The laboratory analysis was conducted to find out whether there are defects generated during manufacturing process. Besides, the finite element analysis was carried out to calculate the maximum tensile stress and compared it with the material strength. The mechanism of the fracture was found to be low-cycle fatigue.

1 Introduction

With a constant consumption of the fossil energy, nuclear energy is now playing an important role all over the world. During the safety management of the nuclear power plant, root cause analysis of the equipment failure is quite important to avoid the incident happening again. Electric actuators are widely used in nuclear power plants to drive valve, many of them are directly related to the nuclear safety.

Lots of analysis of the equipment failures in nuclear power plants was conducted. Failures of the different valves were investigated, such as slide valve, safety valve and forged valve [1, 2, 3]. But in this study, the fracture analysis of the planet support used in an electric actuator was conducted. The planet support was used to rotate the output shaft of the actuator and was made of aluminium alloy. The electric actuator used to work under an abnormal condition in which the motor starts frequently in a short time. SEM (scanning electron microscope) examination and metallography examinations were conducted to see micro-images of the material structure. Hardness and tensile strength tests were conducted to evaluate the mechanical properties of the aluminium alloy. Also, the stress calculation through finite element method was carried out to see the maximum tensile stress.

2 Materials and methods

2.1 Materials

The material of the planet support was aluminium alloy and the grade was EN AC-43100 (AlSi10Mg) without any artificial age-hardening treatment and solution treatment. According to the standard EN_1706 made by European Committee for Standardization, the element contents and mechanical properties, such as yield strength, tensile strength and fatigue strength of the AC-43100 had been clearly stipulated.

2.2 Methods

The sample of the aluminium alloy was taken out to do chemical composition analysis by infrared absorption method with inductive coupled high frequency plasma. The SEM images were observed at different scale of 100um and 500um using a scanning electron microscope of TESCAN VEGA TS5136XM. The metallographic examination was carried out using a metallographic microscope of ZEISSAXIOVERT 200.

Two different areas were chosen to do hardness test with the load of 10kg. A cylinder sample was taken out to do the tensile test.

3 Results and discussion

3.1 Visual analysis

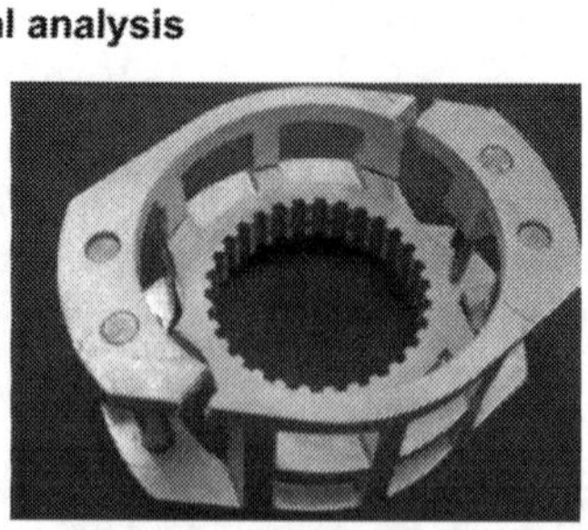

Figure 1. The fracture image of the planet support.

The planet support was broken into three parts. Through restoring these three parts, the two fractures were obvious and could not be recovered to the original position as shown in Fig. 1. This phenomenon was an

DOI: 10.1051/matecconf/20167716005

indication that the two fractures were generated firstly. Besides, no obvious deformation was found on all fractures and all crack originations were found located at the corners in the structure.

3.2 Chemical composition analysis

The result of the chemical composition analysis was shown in Table 1. As we can see, all element contents were qualified to the standard EN_1706. No abnormality was found during the chemical composition analysis.

Table 1. Chemical composition analysis of the planet support.

	Element contents (wt%)	Standard EN_1706
Si	9.77	9.0-11.0
Mg	0.22	0.20-0.45
Mn	0.12	≤0.45
Fe	0.38	≤0.55
Cu	0.011	≤0.10
Zn	0.014	≤0.10
Ti	0.11	≤0.15

3.3 SEM images and Metallographic examination

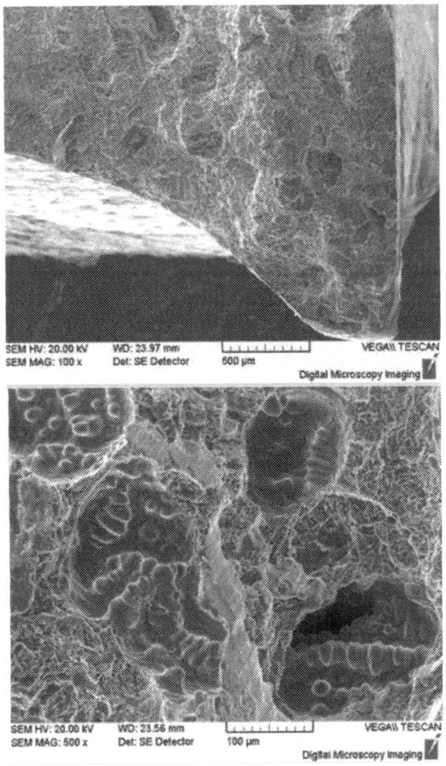

Figure 2. The SEM images of the fracture surface.

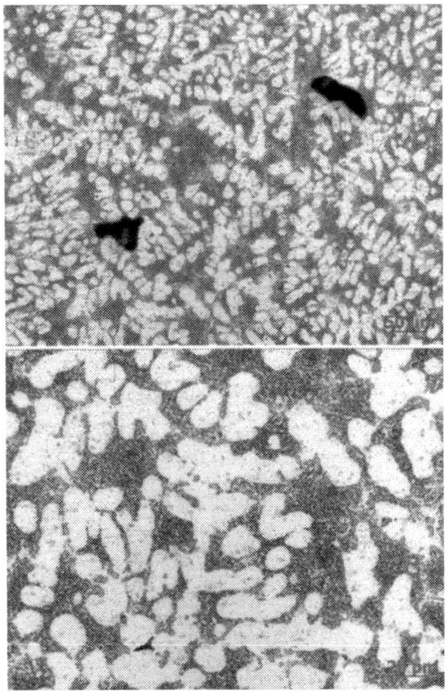

Figure 3. The metallographic images of the structure.

The micro-images of the fracture surface were shown in Fig. 2. The crack origination was observed under the scale of 500um, no plastic deformation was found. And through Energy Dispersive Spectrometer (EDS) measurement, no corrosion element was found near the crack origination.

Some holes were found in the structure as shown in the image of 100um scale, and the maximum diameter of the hole was 200um. The holes were generated during manufacturing, and could affect the mechanical properties of the structure. The size of the holes will affect the mechanical properties as shown in the previous study [4].

The metallographic examination images were shown in Fig. 3. The primary crystal phase and eutectic phase could be observed through the picture. Also, some holes were found in metallographic images. No abnormal phenomenon was found through metallographic examination.

3.4 Hardness test and tensile test result

Table 2. Hardness test result of the planet support.

No	Hardness value (HB)			Average value
1	56.4	54.5	56.0	55.6
2	63.4	63.7	62.5	63.2
	HB≥55 (According to EN_1706)			

DOI: 10.1051/matecconf/20167716005

Table 3. Tensile test result of the planet support.

	Yield strength (MPa)	Tensile strength (MPa)	Ductility(%)
Results	136	191	3.5
EN_1706	90	180	2.5

The result of hardness test was shown in Table 2. All values are qualified to the standard except the single value 54.5. But the average hardness values of these two different areas are qualified to the standard.

The tensile test result was shown in Table 3. As we can see, the mechanical properties such as yield strength, tensile strength and ductility value are all qualified to the standard.

3.5 Finite element analysis

A 3D mode of the planet support was created to show the stress distribution on the structure (as shown in Fig.4). The elastic-plastic mode was adopted in order to consider the non-linear character of the material.

Figure 4. The 3D mode of the planet support.

A typical curve of out-put torque of the asynchronous motor was shown in Fig. 5. As we can see, the out-put torque is high when the motor starts. This start torque is nearly equal to the stall torque the motor. So the applied load on the planet support was nearly 840N•M which corresponds to the maximum out-put torque of the electric actuator (stall torque).

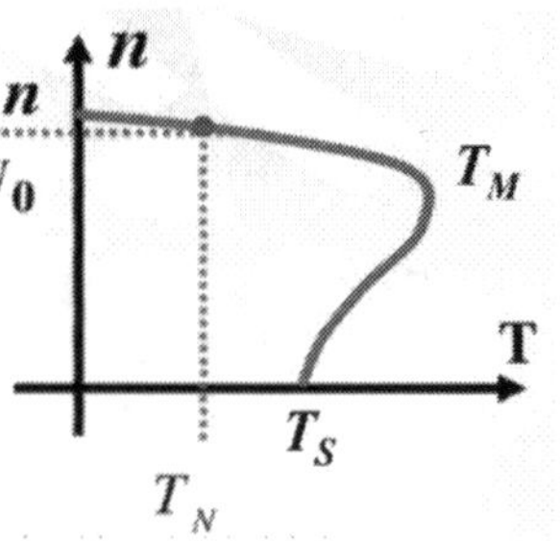

Figure 5. The out-put torque curve of the asynchronous motor.

The maximum tensile stress distribution was shown in Fig. 6. The location of the maximum tensile stress was corresponding to the crack origination in the two obvious fractures. Besides, the maximum tensile stress under the load 840N•M was 132MPa. According to the standard EN_1706, the yield strength and tensile strength were 90MPa and 180MPa respectively. The planet support should not be broken even if the motor was stalled. But, the planet support could also be broken if the motor started frequently in a short time, which corresponds to the fatigue fracture.

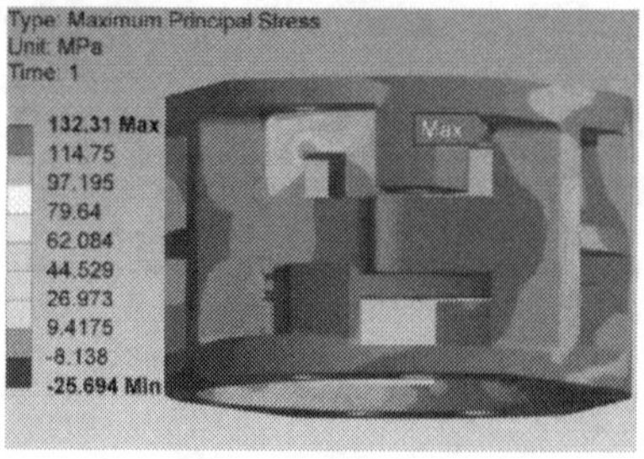

Figure 6. The tensile stress distribution.

4 Discussion

According to the laboratory analysis, no obvious manufacturing defects were found. Some holes were observed, while the mechanical properties were qualified to the standard through the tensile test. But, with the increase of the size of the holes, the fatigue life would be decreased [4]. It is also important to improve the casting process and enhance the quality control.

Through finite element analysis, the maximum tensile stress was less than the tensile strength of the material. But the planet support could also be broken due to the fatigue. According to the standard EN_1706, the fatigue strength of the AC-43100 was 80-110MPa which corresponds to the fatigue life of 10^6 times.

As we know, there is an inverse ratio relationship between the fatigue life and the stress or strain level. Fig. 7 shows a typical curve between the fatigue life and the strain level of the aluminium alloy AlSi7Mg with T6 treatment [5]. The maximum tensile stress of 132 MPa which nearly corresponds to the strain 0.5%, is excessively beyond the fatigue strength. If the cyclic strain is nearly 0.5% of the planet support, the fatigue life should be less than 1000 times as refer to Fig. 7.

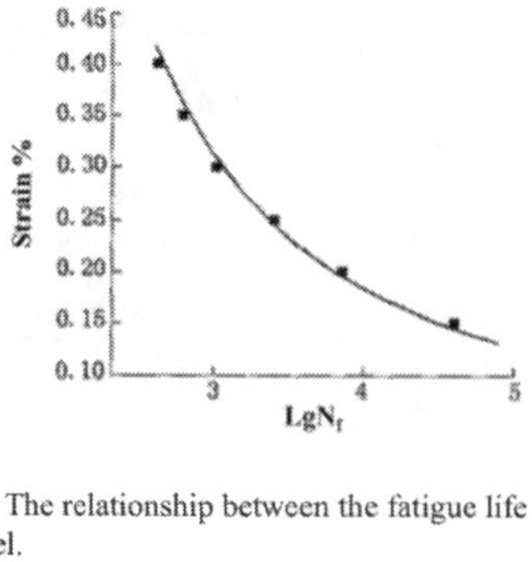

Figure 7. The relationship between the fatigue life and the strain level.

DOI: 10.1051/matecconf/20167716005

The motor of the electric actuator used to start frequently due to the failure of the self-locker which was used to keep the rotation irreversible. The motor started every 0.2s according to the current curve as shown in Fig. 8. The total time was nearly 144s which corresponds to 720 cycles. The 720 cycles also corresponds to the low-cycle fatigue fracture as shown in Fig. 7.

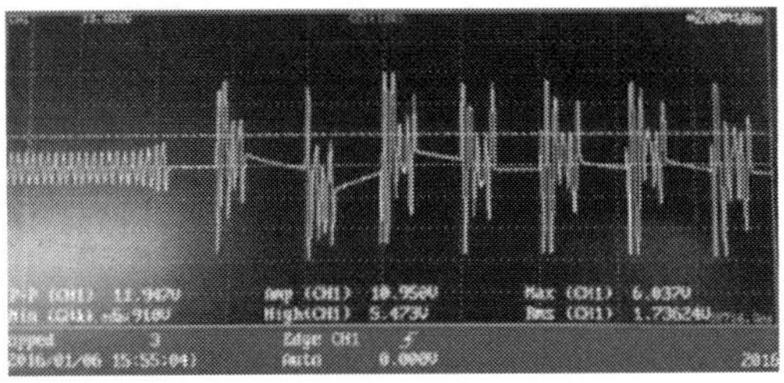

Figure 8. The current curve corresponds to the motor starts frequently.

5 Conclusion

Through chemical composition analysis, hardness test and tensile strength test, no abnormal phenomenon was found. Some holes were observed in SEM images and metallographic examination. The holes should be taken into consideration seriously during manufacturing in order to improve the mechanical properties.

The mechanism of the fracture of the planet support was proved to be low-cycle fatigue. Firstly, the abnormal working condition must be eliminated. On the other hand, as for design aspect, large radius is benefit for the structure due to the decrease of stress concentration.

Further studies should be focused on the effect on the mechanical strength caused by the defects generated during manufacturing process. Meanwhile, the manufacturing craft should also be improved to avoid the generation of such defects.

References

1. C.B Liu. S.L Jiang. Y.G Zheng. Experimental and computational failure analysis of a valve in anuclear power plant. Eng. Fail. Anal **22**, 1 (2012)
2. B Panda, M Sujata, M Madan, S.K Bhaumik. Stress corrosion cracking in 316L stainless steel bellows of a pressure safety valve. Eng. Fail. Anal **36**, 379 (2014)
3. D Gallina. Finite element prediction of crack formation induced by quenching in a forged valve. Eng. Fail. Anal **18**, 2250 (2011)
4. Xueyuan Xiang, Guoqiu He, Defeng Mo, Zhengfei Hu, Bing Liu. Effect of microstructure on Al-Si casting alloy in fatigue lifetime progress. Shanghai Materials **31**, 49 (2009)
5. Zhengyu Zhu, Guoqiu He, Chengshu Chen, Xia ngqun Ding, Xiaoshan Liu. Low Cycle Fatigue Behavior of ZL101 Aluminum Alloy. Special Casting & Nonferrous Alloys **27**, 256 (2007)

MATEC Web of Conferences 77, 16006 (2016) DOI: 10.1051/matecconf/20167716006
ICMMR 2016

Pressure fluctuation analysis for charging pump of chemical and volume control system of nuclear power plant

Qiang Chen , Yin-hui Che and Jian-jun Guan

SuZhou Nuclear Power Research Institute Co. Ltd., Shenzhen, Guang dong, 518038, China

Abstract. Equipment Failure Root Cause Analysis (ERCA) methodology is employed in this paper to investigate the root cause for charging pump's pressure fluctuation of chemical and volume control system (RCV) in pressurized water reactor (PWR) nuclear power plant. RCA project task group has been set up at the beginning of the analysis process. The possible failure modes are listed according to the characteristics of charging pump's actual pressure fluctuation and maintenance experience during the analysis process. And the failure modes are analysed in proper sequence by the evidence-collecting. It suggests that the gradually untightened and loosed shaft nut in service should be the root cause. And corresponding corrective actions are put forward in details.

1 Introduction

Chemical and Volume Control System (RCV) is one of the most important systems of primary circuit in pressurized water reactor (PWR) nuclear power plant to provide the reactor coolant system (RCP) with following services as it is designed. On one hand, RCV is able to maintain the pressurizer a programmed water level by utilizing its charging and let-down function. On the other hand, RCV in combination with the make-up system (REA) can tune the boron concentration so as to reduce the reactivity variations induced by temperature change, fuel burning and xenon transients. Through measurement of removing corrosion and demineralization of fission products in ionic and particles, the reactor coolant activity level can be effectively controlled. There are three charging pumps in RCV system capable of supplying charging flow with sufficient pressure. To be specifically, the RCV002PO and RCV003PO can be switched in routine service, while the RCV001PO is functioned as a standby pump when RCV002PO and RCV003PO are being switched. In addition, the charging pumps in RCV system is considered as high head safety injection pumps in case small breaks occur in the primary circuit pipeline. Therefore, the charging pump is much accounted of nuclear power station, RCC-M safety Class 2 as one of the most important devices.

Recently, the RCV002PO has been restarted about ten times in a power nuclear power plant from June to September in 2014, with charging pressure decreasing 0.5 bar between neighbouring setups. The charging pressure has dropped from the initial 177 bar to the final 173 bar until September. However, the charging pressure remained stable while the pump was operating continuously (as shown in Fig. 1(a)). In October, the RCV system pushed up the flow, resulting in the charging pressure decreasing from 173 bar to 168.5 bar and apparent low pressure alarm triggered (alarm value is 169 bar). In spite of consecutively dramatic decrease, the charging pressure recovered to 175 bar after a periodic test RPB010 (Reactor protection 010 Test, B column) being carried out on 24th October. However, on 27th October it redropped to 168.5 bar again when performing another periodic test RPA011 (Reactor protection 011 Test, A column).

Under the circumstance that charging pump pressure drops ceaselessly, this phenomenon may make RCV system be isolated when water level becomes out controlled by the primary circuit. Degrading function of charging pump, thereby affects high pressure safety injection function. Hence from we are motivated to unveil this phenomena by virtue of a systematically investigation.

Root Cause Analysis (RCA) is a method that is used to address a problem or non-conformance, in order to get to the "root cause" of the problem. It is used so we can correct or eliminate the root cause, and prevent the problem from recurring. RCA method arose in the 1950s as a formal study by the National Aeronautics and Space Administration (NASA) in America. But it didn't become a technical until that was used in three island nuclear accident. There are several commercial RCA methods available in the world now. Listed are examples of some popular methods, TapRooT® system, Apollo Root Cause Analysis (ARCA), REASON® and PROACT®. These methods have different characteristics and analysis steps. They are used in a wide variety of industries, such as resources, petro-chemical, manufacturing, aerospace, telecommunications and healthcare, to guide problem analysis. In a nuclear power station, Equipment Failure

 DOI: 10.1051/matecconf/20167716006

Root Cause Analysis method (ERCA) was usually used. It utilizes a 6-step sequential process to find the root cause of an event: a) Define the problem of an event, b) Collecting equipment data information, c) Listed the equipment failure modes and analysis their possibility, d) Identify the root cause, e) Formulate the corresponding corrective actions, f) Tracking and assessment the effects of the corrective actions, as shown in figure 2.

This study introduces the ERCA's process of this event to find out the root cause for pressure fluctuation and come up with corresponding corrective actions.

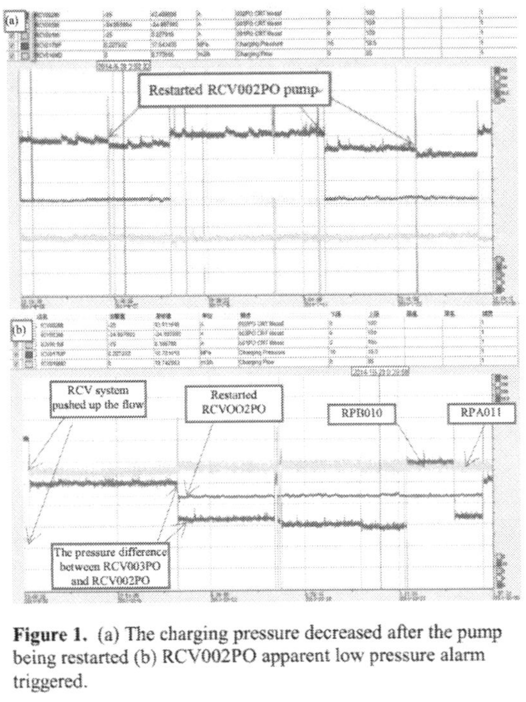

Figure 1. (a) The charging pressure decreased after the pump being restarted (b) RCV002PO apparent low pressure alarm triggered.

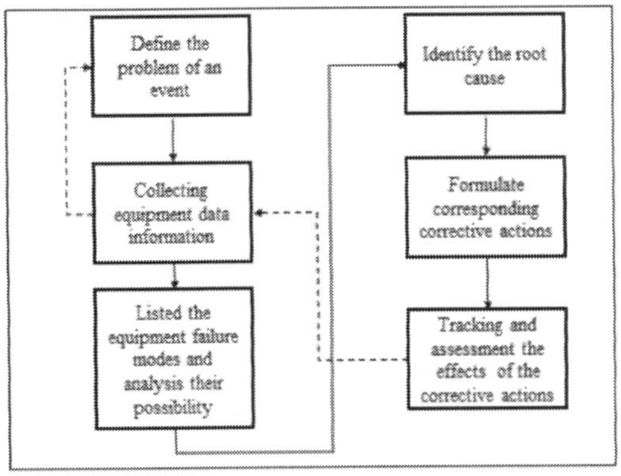

Figure 2. The 6-step of Equipment Failure Root Cause Analysis method (ERCA).

2 Root cause analysis

2.1 Process analysis

The investigation and analysis are performed in accordance with following steps:

1) Setting up RCA project task group, to collect the event background and equipment information, and experience feedback from similar equipment.
2) Surveying on the charging pump running records, to confirm the problem and the scope of the fault, such as real-time monitoring records, regular test records and charging pump's weekly running state records, etc.
3) Surveying and collecting equipment data information by methods of personnel-interviewing and so on, including equipment design, maintenance manuals, drawings, maintenance of the original data and relevant papers, to analyse relevant data and evidence.
4) Enumerating the possible failure modes according to the charging pump's actual pressure fluctuation in the nuclear power plant, to formulate action plans.
5) Analysing, demonstrating and concluding all failure modes step by step after evidence-collecting. The subsequent forensic analysis on characteristics of each failure model can collect additional evidences.
6) Probing into the most possible causes and related influencing factors of the pump pressure fluctuation through comprehensive analysis concerning the survey and data of the failure modes.
7) Constructing the logical process of the event to explain the mechanism of the pump pressure fluctuation. Address the root cause and formulate corresponding corrective actions.
8) Tracking and assessment the effects of the corrective actions and feedback the experience.

2.2 Overview of the charging pump equipment

The charging pump group is composed of 6.6 kV driving motor, speed increasing gear box, pump and other components. The schematic illustration of pump group is shown in Fig. 3. The motor is manufactured by Jiamusi Electric Machine CO., ltd. (China). The charging pump is horizontal multistage centrifugal pump produced by KSB AG (Germany).

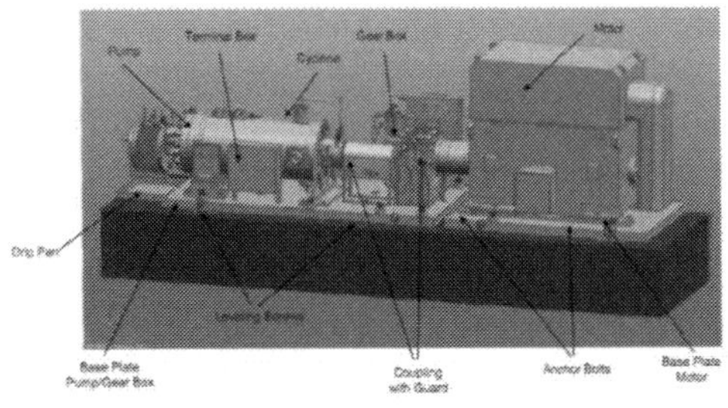

Figure 3. Charging pump group.

The pump is mainly made from impeller, guide vane, pump shaft, mechanical seal, bearing, balance drum, pump casing and pump cover. There are a radial bearing in driving end of the pump and a thrust bearing in non-driving end, as shown in Fig. 4. The pump is designed with full core pulling which is very convenient during the overhaul.

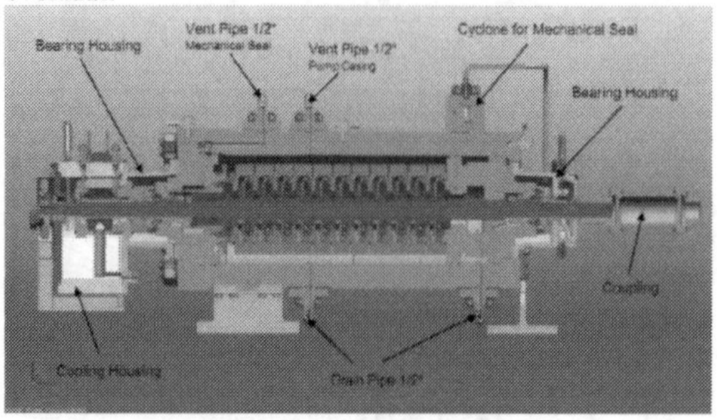

Figure 4. Charging pump body structure.

- Pump Type/Model: horizontal ring-section with 12 stages RHM 100-205.12
- Pump Capacity Maximum: 160m3/h
- Pump Capacity Minimum: 13.6m3/h
- Head /Speed: max. 1810m/ 4,657r/min
- Pump Bearing: single roller bearing (DE) double tapered roller bearing (NDE)
- Lubrication type: oil bath (DE)/oil injection (NDE)
- Motor Capacity: 630kW
- Safety Class: Class 2, acc. to RCC-M Edition 2000, Addendum June 2002

2.3 Failure mode analysis

Analysing and troubleshooting the charging pressure decrease is of extreme importance to have the reasons for charging pump's pressure fluctuation be uncovered quickly and accurately. And the possible failure modes are listed ad below:

1) Valves leakage
2) Small flow anomaly
3) Foreign materials in pipeline or pump body
4) Pipeline gas accumulation or gas entrance
5) Wear of internal pump parts
6) Centre offsetting between impeller and diffuser

2.3.1 Valves leakage

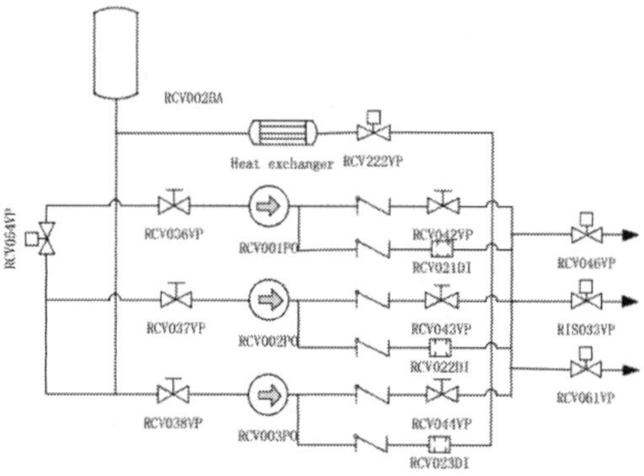

Figure 5. Local flow chart of RCV system.

In order to facilitate the inter-switching among three charging pumps, the inlet and outlet pipeline isolation valves are not allowed to close as usual. The check valves could prevent backflow from the pump outlet and small flow line (as showed in Fig. 5). If the check valve leaked, the pump outlet flow will be diverted, thereby giving rise to the charging pressure decrease.

However, there is no abnormal phenomenon occurring in the case of disintegration of the outlet check valves during the overhaul. The result shows that the possibility of valves leakage is low.

2.3.2 Small flow anomaly

If the flow rate of small discharge pipe of the pump outlet increased, the charging pressure will decrease. Once that scenario was found in an incident, investigators may measure the flow data using ultrasonic flow meter, as it can be seen in table 1.

Table 1. Pressure and flow data of the pumps.

	Inlet pressure (bar)	Charging pressure (bar)	Outlet flow (m^3/h)	Small flow pipeline flow (m^3/h)
RCV002PO single pump operation(2)	1.66	168.5	39.11	14.77
RCV001PO single pump operation	1.68	175.3	38.40	12.89
RCV003PO single pump operation	1.80	172	38.60	13.90

As shown in Fig. 6, when the flow rate of charging pump increases 1m3/h, the outlet pressure will drop by 0.6 bar. In the case of the outlet pressure of the pump arriving at 169 bar, namely the alarmed value, the bypass flow will accordingly increase at least 10 m^3/h. But the flow of small flow line cannot approach to 23.6 m^3/h (as it illustrated in table 1).

The aforementioned result suggests that the possibility of the flow rate increase of the small pipe is low.

 DOI: 10.1051/matecconf/20167716006

2.3.3 Foreign materials in pipeline or pump body

Foreign materials in pipeline or pump body is of extreme importance that it may result in the pressure of pump. However, the results investigated in our study have shown that the very small possibility for foreign materials blocking the pipeline and pump body.

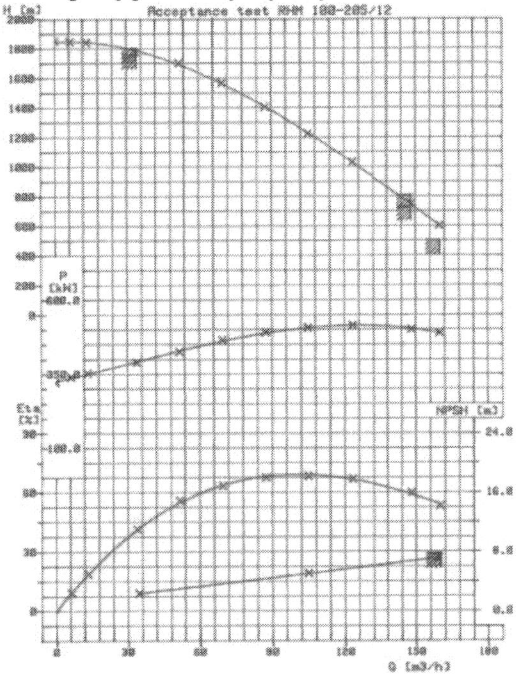

Figure 6. Performance curves of RHM 100-205.12 pump.

2.3.4 Pipeline gas accumulation or gas entrance

If the gas in pump body or pipelines accumulated, it will accordingly make the pump pressure be decreased. And continuous intake gas will also affect the pump efficiency, resulting in reduced pump output.

In line with process requirements, the charging pump was filled with water and exhausted gas after the charging pressure low alarm had been reported, but the outlet pressure of the pump was still low. The vent and drain pipelines are showed in Fig. 4. The possibility of gas accumulation can be ruled out.

If the gas entered the pump body continually during operating, the gas-liquid two-phase flow will lead to the pump vibration, and the motor current will be reduced. But the pump group vibration was normal and the motor current maintained a smooth during the event. Therefore, the possibility of continuous intake is low.

2.3.5 Wear of internal pump parts

If the rotation parts of pump were substandard or O-rings seal were broken, they will increase the risk of l internal flow leakage, resulting in reduced efficiency to the pump.

The inspection results concluded during overhaul are listed as following:

- Inlet clearance of the impellers meets the requirements.
- Outlet clearance of impellers meets the requirements.
- The clearance between the throttle sleeve and the balance drum meet the requirements.
- All kinds of O-ring seals meet requirements.

There were no significant anomalies of the rotation parts of pump during overhaul inspection (as shown in Fig. 7). The impeller and the balance drum wore lightly where the sizes of the balance drum met the standard. Therefore, the possibility of the rotation parts of pump wear or O-rings seal broken can be excluded.

Figure 7. Balance drum and the impeller of pump.

2.3.6 Centre offsetting between impeller and diffuser

Due to the existence of various capacity losses, the actual efficiency of the pump is lower than the theoretical value (pump shaft power) in the actual operation of the pump. The pressure head becomes lower while the efficiency of the pump is decreasing. The actual pump efficiency (η) depends on the volume efficiency, mechanical efficiency and hydraulic efficiency.

$$\eta = \eta_h \eta_v \eta_m \qquad (1)$$

η_v: The volume efficiency of the pump

η_m: The mechanical efficiency

η_h: The hydraulic efficiency

The hydraulic efficiency of the charging pump refers to the capacity loss induced by the change of the flow resistance, the area and the direction of the flow, and the circulation and vortex in the impeller passage. When the deviation between the centre of the impeller and the centre of the diffuser reaches its lowest, the hydraulic efficiency of the pump will be the highest [1]. While the hydraulic centre changes, the flow resistance increases, the hydraulic efficiency of the pump will be decreased (as shown in Fig. 8).

The radial cylindrical roller bearing (NU type) in charging pump's driven end (DE) tunes the inner ring movement with the shaft and there is a thrust bearing in pump's non-driven end (NDE) shafting expansion dead point. The shaft nut can be utilized to tight the thrust bearing, but it can be expanded only towards the driven end in the case that shaft is heated via flow.

When the pump operated, the axial forces of the impeller, balance drum and thrust bearing achieved a balance. If the impeller axial force was greater than the balance drum axial force, the temperature of DE roller of thrust bearing will be higher than the NDE roller, vice-versa (as shown in Fig. 9).

However, an inferior situation may occur. To be specifically, if the shaft nut loosed, and the impeller axial force was greater than the balance drum axial force, the actual efficiency of the pump will be decreased simultaneously. Since the multistage impellers moves towards the drive end with the shaft shoulder, it has led to the deviation between centres of impeller and the diffuser. In fact, it was the loosed shaft nut actually found during the overhaul.

In this event, charging pressure decreased when starting up the pump in comparison with the last run in each time. However, the charging pressure remained stable when the pump was operating continuously, because there was an axial impact during axial force balance process once the charging pump had been started. If the tighten force of the shaft nut was not strong enough, it will be loosed slightly after starting the pump in each time [2-5]. But the shaft nut will not get loosed when the pump keeps operating, because the impeller will take the shaft with the nut abutting the thrust bearing.

When the balance drum axial force is greater than the impeller axial force, the balance drum will bring the shaft moving towards to the NDE through snap ring. In this situation, the centres of impeller and the diffuser will not be deviated even if the shaft nut was loosed.

The charging pressure was reverted to 175 bar when a periodic test RPB010 had been carried out on 24 October, in which it was used to open a valve downstream of the pump. After that, the charging pressure re-dropped to 168.5 bar again when performing another periodic test RPA011 on 27 October and the RPB011 test was approaching to a valve in small follow pipeline of the pump. It is worthy to notice that the temperature of the thrust bearing changed when executed RPB010 and RPA011 test (as showed in Fig. 10), meaning that the axial forces of impeller and balance drum changed. This is the RHM 100-205.12 pumps' characteristic. When the inter parts wear or the flow changes, the axial force will change [6].

After completing RPB010 test, the balance drum axial force is larger than the impeller axial force. The channel centre of the suction stage impeller and the diffuser coincides again. The outlet pressure of the pump recovers. But the RPA011 test make the axial force change again, the efficiency of the pump decreases once more.

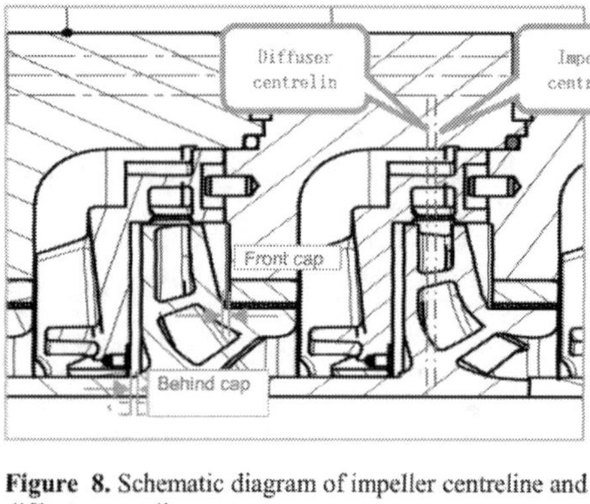

Figure 8. Schematic diagram of impeller centreline and diffuser centreline.

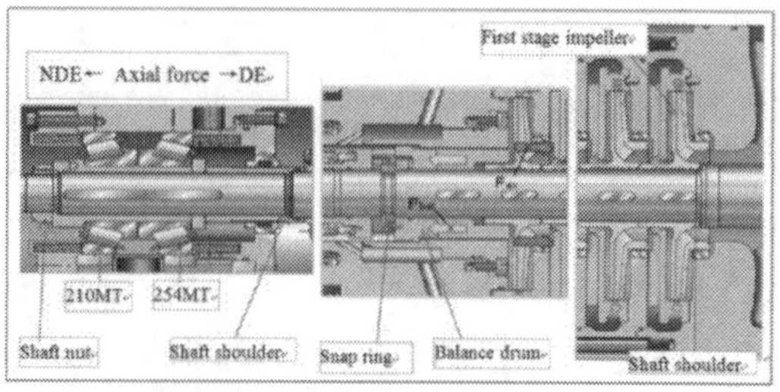

Figure 9. Schematic diagram of charging pump's axial force.

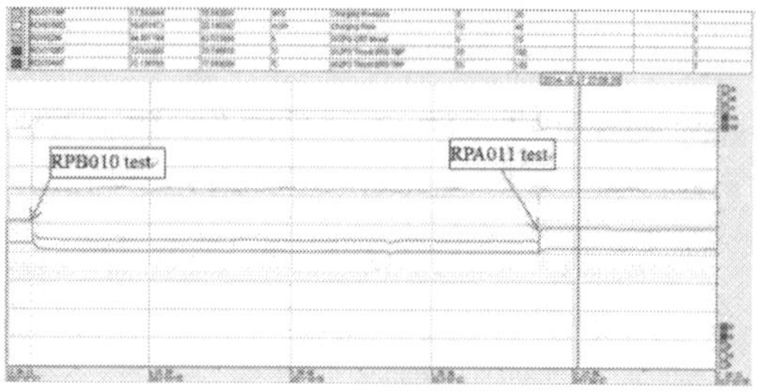

Figure 10. The RPB010 test and RPA011 test.

In summary, the failure mode of centre offsetting between impeller and diffuser of the pump will influence the hydraulic efficiency, and it can be accounted for the entire failure phenomenon. Its possibility is highest than other factors.

3 The result and corrective action

3.1 Cause and effect diagram

The cause and effect diagram was illustrated as below for better understanding of this event and finding out the root cause (as showed in figure 11). The possible worst consequence is that the charging pump may be damaged if the shaft nut continuously loosed.

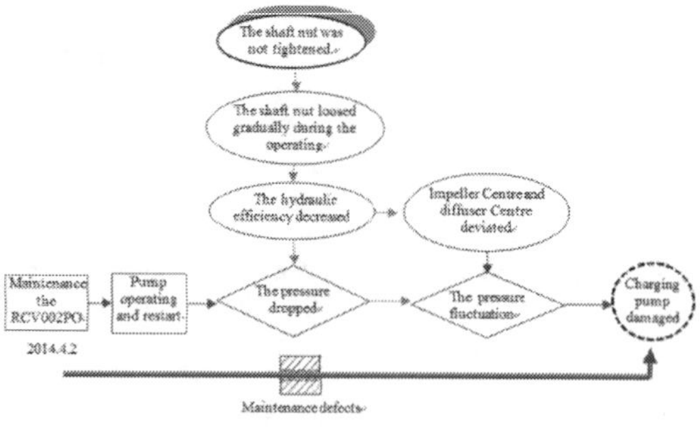

Figure 11. Cause and effect diagram.

3.2 The root cause

The root cause of the charging pump's pressure fluctuation is that the shaft nut fails to be tightened. It has been loosing gradually during the operating, resulting in changed position of impeller centre and diffuser centre.

And the reduced hydraulic efficiency of the pump led to outlet pressure fluctuations.

3.3 Corrective action

The shaft nut should be twisted and the glue should be used to prevent the nut from being loosed. Updating the maintenance program to ensure the similar incident will not occur again. Learn the feedback of this event in the whole nuclear power station.

4 Conclusion

In this paper, Equipment Failure Root Cause Analysis method is introduced to analysis the charging pump's pressure fluctuation event. The 6-steps sequential process of ERCA are introduced: a) Define the problem of an event, b) Collecting equipment data information, c) Listed the equipment failure modes and analysis their possibility, d) Identify the root cause, e) Formulate the corresponding corrective actions, f) Tracking and assessment the effects of the corrective actions. ERCA' detail analysis process has been showed to find the root cause of this event. And each of 6-steps had been extended analysis.

The root cause of pressure fluctuation of charging pump is the untightened shaft nut, which has been loosing gradually during the operating. The corresponding corrective actions have been formulated. The outlet pressure of the pump has been recovered now.

References

1. J. Zhu Theory of pump clearance [J]. Journal of Daya bay Nuclear Power, 2002(1):
2. H.Z. He Investigation on screw thread connection loosing problem and the anti-loosing measures [J]. Mechanical development (Chinese), 1993, 3: 43-47.
3. S.Y. Hou Liao R.D. Research status and development trend of screw thread connection loosing process [J]. Structure and Environment Engineering (Chinese), 2014, 41(2): 39-52.
4. H. Tang Loosening and Damage Mechanism of Thread Joined Structures in Nuclear Power Equipment [J]. Nuclear Power Engineering, 1999, 20(2): 111-116.
5. Y. Zhang Zeng G.Y. Zhao D.F. Experimental study on the threaded connection looseness in vibration environment [J]. China Heavy Equipment, 2010 (3): 23-25.
6. KSB, Calculation of axial forces Caused by pump impellers and the suitable axial thrust balancing device, 2014

 DOI: 10.1051/matecconf/20167716007

Bottlenecks reduction using superconductors in high voltage transmission lines

Labib Daloub[1] , Abdulatif Khrwat[1], Rajeb Ibsaim[2] and Mohamed Regeai[3]

[1]*University of Aljabel Algharbi, Gharian, Libya*
[2]*University of Zawia, Zawia, Libya*
[3]*College of Electronic Technology, Tripoli, Libya*

Abstract. Energy flow bottlenecks in high voltage transmission lines known as congestions are one of the challenges facing power utilities in fast developing countries. Bottlenecks occur in selected power lines when transmission systems are operated at or beyond their transfer limits. In these cases, congestions result in preventing new power supply contracts, infeasibility in existing contracts, price spike and market power abuse. The "Superconductor Technology" in electric power transmission cables has been used as a solution to solve the problem of bottlenecks in energy transmission at high voltage underground cables and overhead lines. The increase in demand on power generation and transmission happening due to fast development and linked to the intensive usage of transmission network in certain points, which in turn, lead to often frequent congestion in getting the required power across to where it is needed. In this paper, a bottleneck in high voltage double overhead transmission line with Aluminum Conductor Steel Reinforced was modeled using conductor parameters and replaced by Gap-Type Superconductor to assess the benefit of upgrading to higher temperature superconductor and obtain higher current carrying capacity. This proved to reduce the high loading of traditional aluminum conductors and allow more power transfer over the line using superconductor within the same existing right-of-way, steel towers, insulators and fittings, thus reducing the upgrade cost of building new lines.

1 Introduction

Congestions has been exists in both new and traditional power systems but when it comes to managements, it is more complex in case of competitive power markets due to the unbundled nature which calls for more coordination. In the system under study "GECOL System" (General Electric Company of Libya), the case is different with its integrated structure and due to non-availability of competitors and other utilities trading in power industry where GECOL dominating the market, in which congestion would relatively be an easier issue to handle. The issue is mainly related to the conductor temperature rise as the transmission lines are loaded which expand its length, thus increasing the line sag. Since in high transmission lines, the sag is calculated based on the clearance to earth surface, the line conductor loading is restricted to a certain degree of temperature to maintain the sag within the limits and therefore saving the line conductors from being melted [1]. In this research, a study was made to a heavily congestive overhead transmission line linking one of the most important peaking capacity plant of 4x165MW known as TSPS (Tripoli South Power Station), with focal substation feeding the Greater Capital City of Tripoli from the south bond. The Feeding S/S situated at HKSS (Hadba Khadra) district sum 18 km away from the generating plant [2].

The transmission line is 220kV double conductor Red-Wing ACSR, constructed sum 35 years ago when the line route was clear from any obstacles.

2 Problem tackling

A few limited scenarios can be recommended to tackle the issue of bottlenecks for power transmission to this high voltage line, however, most of them have advantages and disadvantages. These are as follows:

- By installing a simple cycle peaking capacity generating plant, which cannot be visible due to noise, fuel storage, fuel supply, etc.
- By installing a new overhead transmission line with similar specifications that goes all the way parallel with the existing line keeping the appropriate right of way which is no longer possible with route space unavailability.
- By upgrading to a bundle conductor which will require replacement of all fittings and accessories.
- By replacement of the traditional ACSR conductors with high temerature superconductor, in which is the most appropriate scenario as it extended in the same tower structures keeping the same insulations, fittings and accessories.

3 Congested line

One of the bottlenecks experienced by Libyan power grid represented in 220kV double transmission line linking a main power injection points to the Libya Capital of Tripoli from the south bond, which is TSPS (Tripoli South Power Station) with HKD (Hadba Khadra) S/S, where the suffering of these 2 line from drastic load increase become very unacceptable, especially when disconnecting of other lines for conducting regular maintenance or routine check. In this paper, data on line constants, power loading and congestions were accumulated from the main utility GECOL (General Electricity Company of Libya) to model the 2 overhead lines of TSPS-HKD being the worst in overloading and one of the most important injection line to Greater Tripoli [3].

4 Conventional conductors

Most of power transmission lines in service utilize aluminum (AAC), aluminum-alloy (AAAC) or aluminum conductors steel-reinforced (ACSR) due to high conductivity of aluminum and low weight density, while steel is added in ACSR for extra strength to resist sags. The extruded aluminum strands used in conventional conductors carries all or most of the tension in the cable to provide adequate strength, however, heating a cold-worked conductor may cause it to anneal. When aluminum anneals, the dislocations of its crystal structure begin to release, reducing the material's strength, thus, conductors aren't operated at temperatures that cause them to anneal. This is the basis of the operating temperature of most conductors. Moreover, aluminum conductors exhibit significant sag due to low elastic modulus.

Table 1. ACSR conductor parameters.

Conductor type:	ACSR "Redwing"
Sectional Area (A):	445.1 mm^2 (AL 30/3.92, ST 19/2.35)
Diameter (D):	2.746 cm
Weight (W):	1.686 kg/m
Ultimate strength (UTL)	15697 kg
Modules Of Elasticity (E):	8000 kg/mm^2 (78453 N/mm^2)
Coefficient of Linear Expansion (α)	18 x 10^{-6}
DC Resistance (R_{DC})	0.07978 ohm/km at 20 ^oC
Temperature Coefficient of Resistance	0.004/ ^oC

AAC, AAAC, and ACSR cables are limited to operating temperatures of 90-100°C. Above that limit, the alloyed metals which changed the nature of the metallic crystal increased its annealing temperature. Aluminum conductor will begin to anneal and lose strength, therefore, transmission lines with such type of cables are designed to operate below 60-75°C to limit their sag. The parameters of the existing conductor under study are shown in Table 1.

5 Superconducting power lines

High-temperature superconducting cables promise to revolutionize power transmission and distribution by providing lossless transmission of electrical energy. The development of superconductors with transition temperatures higher than the boiling point of liquid nitrogen has made the concept of superconducting power lines commercially feasible, at least for high-load applications [4]. One of the properties of a superconductor is that it will exclude magnetic fields, a phenomenon known as Meissner Effect [5]. It's all became after an observation made by H. Kammerlingh Onnes, who states that If mercury is cooled below 4.2 K, it loses all electric resistance and becomes a superconductor [6], (0 degrees Kelvin = -273 degrees Celsius). The observations were then made that other metals also exhibit zero resistivity below a certain critical temperature. The highest critical temperature was about 23 K until the discovery in 1986 of some high temperature superconductors. The 'high' temperatures are still low, the "125K is considered very high". With widespread use of superconducting technology in grid 'highways' energy consumption can be reduced by 5-7%, which means an equivalent reduction in Co2 emissions from electricity generation. Alloys with Zr materials often listed as Thermal-resistant and Super-thermal-resistant aluminum alloys (TAl and ZTAl). These alloys are designed to operate at 150°C and 210°C respectively. (Super) Thermal-resistant Aluminum Alloy Cables, Steel Reinforced (Z) TACSR are formed similarly to ACSR, but utilize TAl or ZTAl rather than the typical work-hardened aluminum conductor stranding. TAl and ZTAl materials are also used in GTASCR and ACCR cables [7].

6 GAP-Type superconductor

Gap-type superconductor is a stranded core surrounded by hollow cylinder of trap-wire, forming a gap between the core and the conductor which is filled with thermal-resistant grease. G (Z) TACSR utilizes a thermal-resistant aluminum alloy conductor, much like ACCR. The core is generally made of galvanized steel, the same as would be found in ACSS or ACSR. In the current study, the scenario of high temperature superconductor was considered for three types of materials (GTACSR, GZTACSR and ZTACIR). The current carrying capacity of various material types were compared and the GTACSR conductor was chosen as the best option for its high current carrying and was modeled in this study.

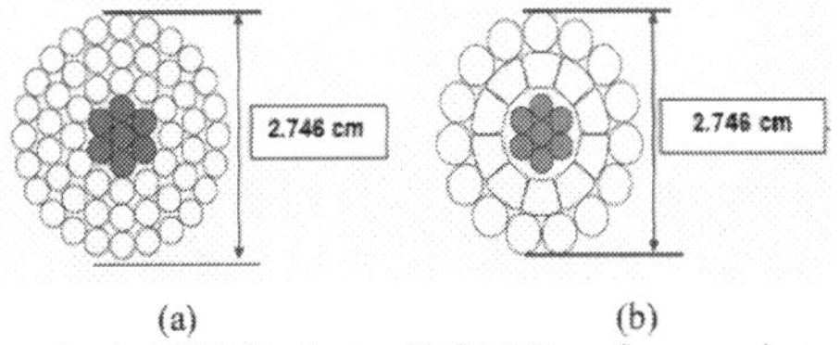

(a) (b)

Figure 1. a) ACSR Conductor. b) GAP-Type Superconductor.

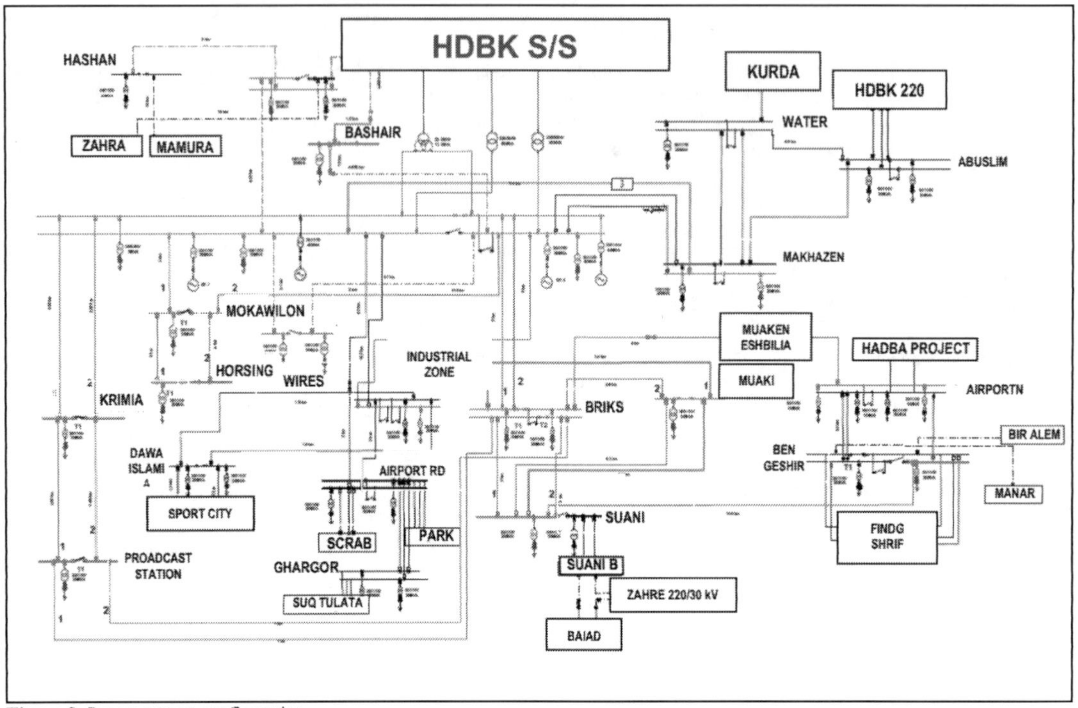

Figure 2. Power system configuration.

7 The simulation model

The simulation model was carried out using NEPLAN 260 Software to represent the actual situation of lines loading during daily peak hours and the solutions recommended to provide the most appropriate solution that can be used to overcome the problem of congestions.

The data used in the simulation is a real valid data collected through direct contact with Libyan power utility GECOL.

The congested transmission line was simulated with its existing situation and the results obtained clearly indicate the presence of congestion in the existing line. The line was again modeled by replacing the traditional ACSR by GTACSR superconductor and the obtained results proved and upgraded the transmission capacity of the line and cleared the bottleneck. The load flow result by the modeled transmission line is illustrated in the tables and figures above.

Table 2. Result of load flow of existing ACSR line.

	From	To	P Loss	Q Loss	P Imp	Q Imp	P Gen	Q Load	P Load	Q Load	Gen Cost	Qc Sunt	Ql Shunt	Q Comp	Iron Losses
	Area/Zone	Area/Zone	MW	MVar	MW	MVar	MW	MVar	MW	MVar	Curr. Units	MVar	MVar	MVar	MW
1	Network		222.4	-1149.4	642.3	-15.3	5599.3	-3079.3	5376.9	-3408.6	0	0	366.5	0	0
2	Area 1		222.4	-1149.4	0	0	5599.3	-3079.3	5376.9	-3408.6	0	0	366.5	1112.3	0
3	Zone 1		38.52	-6.95	-1165.3	723.07	2013	-807.31	809.13	-557.9	0	0	366.5	114.15	0
4	Zone 2		24.06	-367.8	1747.5	-1294.4	413	-546.6	2136.4	-1473.2	0	0	0	0	0
5	Zone 3		91.52	-157.3	-908.6	982.3	2461	-1466.6	1460.6	-1007.2	0	0		680.19	0
6	Zone 4		51.45	-531.2	952.2	-508.6	70	-243.4	970.7	-370.2	0	0	0	149.3	0
7	Zone 5		16.81	-86.19	-625.5	97.73	642.3	-15.35	0	0	0	0	0	168.6	0
8											0	0			0
9	Un		P Loss L	Q Loss L	P Loss T	Q Loss									
1	kV		MW	MVar	MW	MVar									
1	66		0.001	0	0.202	4.35									
1	220		143.97	-1425.2	29.26	704.1									
1	400		34.089	-880.47	14.847	447.79									
1															
1	From	To			P Tie	Q Tie	P Tie Sc								
1	Area/Zone	Area/Zone			MW	MVar	MW								
1	Zone 1	Zone 2			1165.3	-723.07									
1	Zone 2	Zone 3			-582.16	571.4									
1	Zone 3	Zone 4			952.19	-508.64									
2	Zone 3	Zone 5			-625.5	97.73									
2															

DOI: 10.1051/matecconf/20167716007

Table 3. Result of load flow simulation using GTACSR.

	From	To	P Loss	Q Loss	P Imp	Q Imp	P Gen	Q Load	P Load	Q Load	Gen Cost	Qc Sunt	Ql Shunt	Q Comp	Iron Losses
	Area/Zone	Area/Zone	MW	MVar	MW	MVar	MW	MVar	MW	MVar	Curr. Units	MVar	MVar	MVar	MW
1	Network		225.7	-1142.4	642.3	-15.35	5599.3	-3066.3	5373.5	-3402.7	0	0	366.4	0	0
2	Area 1		225.7	-1142.4	0	0	5599.3	-3066.3	5373.6	-3402.8	0	0	366.4	1112.4	0
3	Zone 1		38.44	-7.852	-1165.9	713.17	2013	-797.47	808.62	-556.9	0	0	366.4	114.12	0
4	Zone 2		27.46	-360.7	1749.5	-1292.8	413	-538.6	2136.1	-1470.7	0	0	0	0	0
5	Zone 3		91.59	-156.02	-909.7	990.2	2461	-1471.4	1459.7	-1005.5	0	0		680.35	0
6	Zone 4		51.41	-531.5	951.5	-508.3	70	-243.4	970.1	-369.6	0	0	0	149.4	0
7	Zone 5		16.81	-86.19	-625.5	97.72	642.3	-15.35	0	0	0	0	0	168.5	0
8											0	0			0
9	Un		P Loss L	Q Loss L	P Loss T	Q Loss									
1	kV		MW	MVar	MW	MVar									
1	66		0.001	0	0.201	4.34									
1	220		147.32	-1416.8	29.19	702.31									
1	400		34.16	-879.77	14.843	447.55									
1															
1	From	To			P Tie	Q Tie	P Tie Sc								
1	Area/Zone	Area/Zone			MW	MVar	MW								
1	Zone 1	Zone 2			1165.9	-713.17									
1	Zone 2	Zone 3			-583.64	579.6									
1	Zone 3	Zone 4			951.49	-508.31									
2	Zone 3	Zone 5			-625.48	97.72									
2															

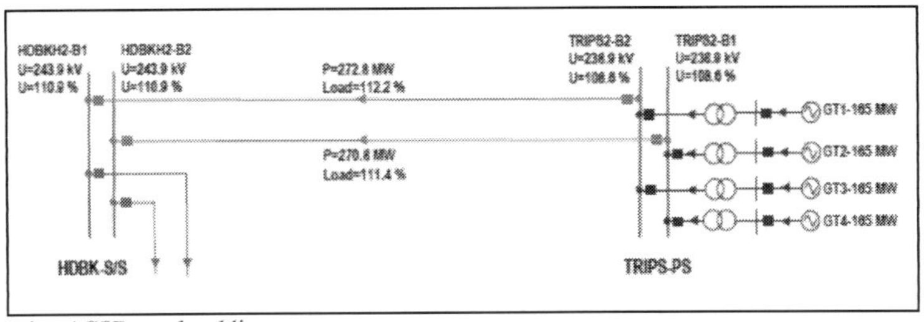

Figure 3. Load flow using ACSR overhead line.

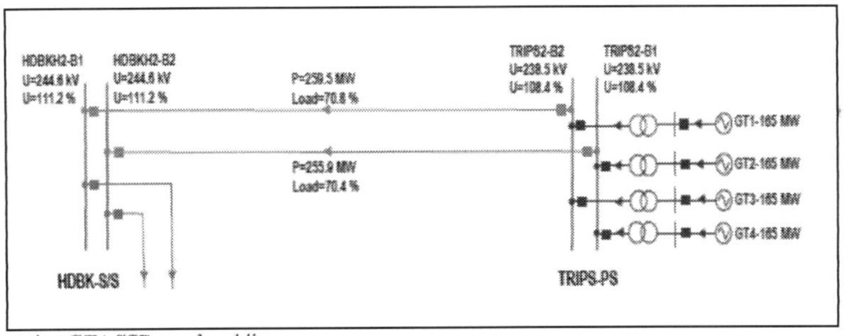

Figure 4. Load flow using GTACSR overhead line.

By considering the Temperature Coefficient of Resistance of 0.004/ ^{o}C with an average span between towers of 420m, the assumed current carrying capacity is calculated based on the following weather conditions:

- Ambient Temp: 40 ^{o}C
- Wind Velocity (cross wind): 0.5 m/sec
- Solar radiation: 0.1 W/cm^2
- Emissivity constant: 0.9

The following formula is used:

$$I\ (amp) = \sqrt{\frac{\pi . D . k . \theta}{R\ ac\ (T+\theta) . 10^{-6}}}$$

where:

- D = *Diameter of conductor*
- k = *Coefficient of Solar radiation*
- T = *Ambient Temp*
- θ = *Rate of Temp Rise*
- R_{dc} = *DC Resistance at 20* ^{o}C
- $R_{ac} = k \times R_{dc}$

8 Simulation results

- Line Losses:

When modeling the congested transmission line by generating the amount of capacity required at "TWPS" to feed the load center of "HKSS" via the congested line, the results obtained a high losses in the line with its traditional ACSR conductor. When the line conductor is remodeled using GTACSR conductor, the losses on the line was decreased and to very low impedance.

- Line Loading:

By modeling the transmission line with ACSR, to carry a power of power rated at 272.8 MW, the simulation resulted in 112.2% of loading, whereas, when the line modeled using GTACSR conductor, with the same MW capacity, the percentage load obtained was only 70.8%.

9 Other concerns

With the ongoing situation in Libya, concerns have been raised about electricity blackouts which is seen as vital to the country's stability and economy. Several incidents have contributed to these worries, including a power outage in the western and eastern regions and a blackout across the entire country.

With widespread use of superconducting technology grid highway energy consumption can be reduced by 5-7%, which means an equivalent reduction in CO2 emissions from electricity generation. But superconductors can also become part of a simplification of the electrical infrastructure. This is because they can transmit massive currents, something which can further be utilized to reduce the number of voltage levels (fewer transformer substations). This has also been feasible in underground transmission in warm and cold environment by utilizing power cables using liquid nitrogen cooling system operating at (-200 ^{o}C) [8]. A 500m and 1000m DC superconducting power cables design projects were started in Chubu University in Japan during 2013 to connect a data centre with DC from large scale array photovoltaic plant. During the design and the construction in the field, a test benches were constructed to investigate the performance of the individual components of the cable system. The design value of the heat leak from the current lead was 35 W/kA and the experimental value was 30.5 W/kA at the rated current [9]. Thus control of the thermal contraction and expansion are rather important to maintain safety of power cables. Such designs will enable cheaper electricity highways and superconductivity materials may dominate the electric energy transmission in the near future.

10 Conclusion

A simulation was made to a congested 220kV double conductor transmission line to tackle the problem of bottleneck. The obtained results when modeling a superconductor material as the main transmission proved to be the most appropriate option. The solution is recommended to the utility to be implemented as to clear the line congestion from the point-of-view of cost effective, time consuming and using all existing fittings. By using such high temp gap-type superconductors in areas of significant growth in power demand, up to double current carrying capacity can be obtained which will reducing of congested bottlenecks during peak loads and prevent outages of heavy loaded lines. While similar cross section area of the gap-type superconductor to that of original installed ACSR conductors and apart from the compressed connectors and dead-end clamps, all existing insulator strings or posts and transmission tower arrangements such as clevis, extension clevis and yokes can still be used and therefore reduce the overall replacement investment.

References

1. Mehran Keshavarzian, Charles H. Priebe, "Sag and Tension Calculation for Overhead Transmission Lines at High Temperatures-Modified Ruling Span Method" IEEE Transaction on Power Delivery, **Vol. 15, No. 2**, (2000)
2. GECOL (General Electric Company of Libya), Updated Power Transmission System Layout, (2013)
3. GECOL (General Electric Company of Libya), Overhead Transmission Line Design Standards, **Version 1.0**, 2007
4. Superconducting power line to shore up New York grid, " Technical Report" by New Scientist Tech, New York Power Grid Authority, May (2007)
5. J. E Hirsch, "The Origin of the Meissner Effect in New and Old Superconductors", Physical Scripta, 85: 035704, (2012)
6. Rudolf de Bruyn Ouboter, "Heike Kamerlingh Onnes's Discovery of Superconductivity", Scientific American p 98, March (1997)
7. Brian Wareing, "Types and Uses of High Temperature Conductors", Cigré WG48, AG06 Seminar, Bangkok, Thailand, (2006)
8. Mbunwe Muncho Josephine, Development Trends of Superconducting Cables Technology for Power Transmission Applications, World Congress of Engineering & Computer Science (WCECS 2015), Vol I, San Francisco, USA, (2015) .
9. S. Yamaguchi, H. Koshizuka, K. Hayashi and T. Sawamura, "Concept and Design of 500m & 1000m DC Superconducting Power Cables in Ishikari, Japan", IEEE Transactions on Applied Superconductivit, Volume 25, Issue 3, (2015).

MATEC Web of Conferences 77, 16008 (2016) DOI: 10.1051/matecconf/20167716008
ICMMR 2016

Fault Transient Analysis and Protection Performance Evaluation within a Large-scale PV Power Plant

Jinghua Wen[1], Meiling Luo[1], Chenjie Gu[2], Hongsheng Wei[2] and Bohan Liu[3]

[1]*State Grid Ningxia Electric Power Company, Ningxia, China*
[2]*State Key Lab of Alternate Electrical Power System with Renewable Energy Sources, North China Electric Power University, Beijing, 102206, China*
[3]*State Grid AC Project Construction Co., Ltd, Beijing, China*

Abstract. In this paper, a short-circuit test within a large-scale PV power plant with a total capacity of 850MWp is discussed. The fault currents supplied by the PV generation units are presented and analysed. According to the fault behaviour, the existing protection coordination principles within the plant are considered and their performances are evaluated. Moreover, these protections are examined in simulation platform under different operating situations. A simple measure with communication system is proposed to deal with the foreseeable problem about the current protection scheme in the PV power plant.

1 Introduction

The recent progress in renewable-energy (RE) technologies has led to the erection of RE power plants up to the order of several hundred megawatts [1]. According to the Global New Energy Development Report 2014, the world's photovoltaic market is shifting from Europe to Asia, with China becoming the largest market [2]. In China, the installed capacity of photovoltaic (PV) power plants currently accounts for over 80% of the total PV installed capacity. Such large PV power plants are naturally part of the transmission networks and hold non-negligible proportions of the generation. Consequently, the fault ride-through requirement (FRT) has become an essential part of modern grid codes [3].

As a result of the FRT schemes, the PV generators' impacts on protective relaying become challenging. Some researchers have studied the effect of fault on PV power plants, specially the inverters and have investigated measures in protecting the PV inverters [4,5]. However, an overall protection scheme has yet to come to solve the protection crisis in PV power plants.

Like distribution networks, where overcurrent protection based on the assumption of a single path for the fault current is predominant, the collection system in a PV power plant has a similar radial topology with overcurrent protection, but the terminals are PV generators instead of loads. The relay coordination might be affected by the additional current paths which are introduced by the PV generators, or the protection system selectivity may be at risk because of bidirectional fault currents. Various researches on distribution networks relaying have be done and some effective remedies are offered. In [6], the issues are identified such as PV generators feeding faults after utility protection opens, PV generators feeding the faults on adjacent feeders, and reclosure coordination. [7] focused on the coordination loss. [8] investigated the loss of speed in protective devices, sympathetic tripping, and unintentional islanding. They have suggested the use of a communication system in order to address these issues. The adaptive relaying algorithm is adopted to automatically amend the protection settings in the presence of inverter-based distributed generations in [9,10].

Different with distribution networks penetrated with DGs, the power from the PV generators flows to the substation under the normal condition. The fault behavior within the plant is necessary to be revealed and an overall protection scheme is in great need to help reduce the protection crisis in PV power plants.

The rest of this paper is arranged as follow. Section 2 presents a detailed description about the PV power plant. In Section 3, the fault currents recorded in the short-circuit test are analyzed. Based on the exposed fault behavior, Section 4 presents the existing protection configuration and the performance is evaluated. Section 5 summarizes the derived conclusion.

This work was supported in part by the National Basic Research Program (973 Program) (2012CB215206), by the National Natural Science Foundation of China (NSFC) project under Grant 51407067, by Fundamental Research Funds for the Central Universities- FRFCU (2016ZZD01)

 DOI: 10.1051/matecconf/20167716008

2 Field test scenario

The PV power plant has a capacity of 850MWp, with a 330kV substation comprising four main transformers. The partial topology in Fig. 1 involves two main transformers. Six overhead lines are connected into the 35kV side of each main transformer and the other end of each line is connected with a corresponding collection station bus. This sort of collection station has two buses and collects 64 PV generation units generally. A single PV generation unit has a capacity of 1MWp and includes two 500kW voltage source converters (VSC), and a step-up transformer boosting the voltage up to the medium level.

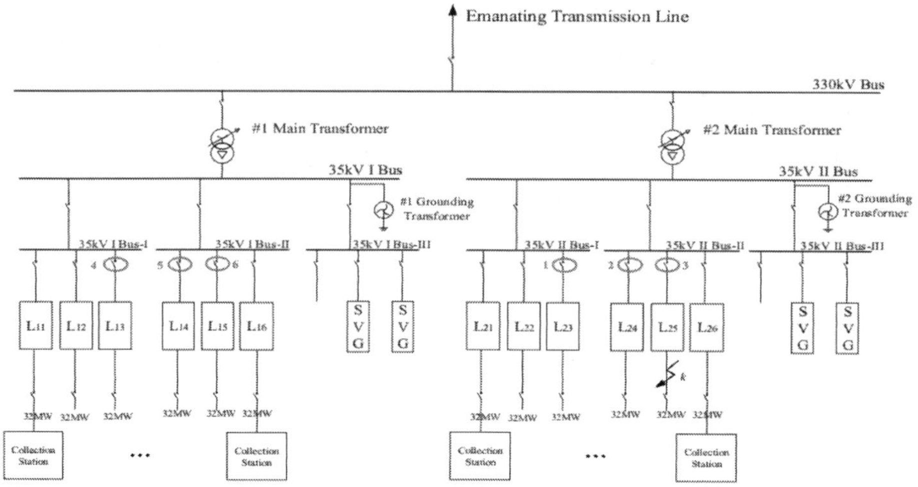

Figure 1. The partial topology of the PV power plant.

The details about the collection station and the PV generation unit will be described in Section 4. The remaining two main transformers are connected to the 330kV bus in parallel with #1 and #2 main transformers.

In the test, the fault was applied on the overhead line L_{25} close to the exit of the collection station, *k* in Fig. 1. At 8.86s, the phase B to ground (BG) fault occurred, and developed into BC to ground (BCG) at 9.09s, and was isolated at 9.17s. It should be noted that #2 grounding transformer and the SVGs did not work during the fault.

The PV power plant was not operating under the rated conditions for the local non-optimum weather. The power factor (PF) was almost unity one. Table 1 presents the prefault power delivered through certain overhead lines.

Table 1. Powers delivered through the overhead lines

Overhead line	P(MW)	Q(MVar)	PF
L_{13}	22.9	-0.83	0.999
L_{14}	20.4	-1.2	0.998
L_{15}	18.8	-1.0	0.998
L_{23}	23.1	-2.0	0.996
L_{24}	18.3	-0.1	1.000
L_{25}	26.2	-1.3	0.999

3. Fault currents analysis

Measurement points are marked with the ovals in Fig. 1 and can be divided into two groups according to the electrical distance from the fault point. The measurement points on L_{23}, L_{24}, L_{25} near to the fault point are marked as 1, 2, 3 sequentially, while the ones on L_{13}, L_{14}, L_{15} comparatively farther from the fault point are marked as 4, 5, 6 sequentially. In this section, the fault currents of typical measurement points are analyzed. In the frequency domain, the currents are analyzed with the method of moving window fast Fourier transformation (FFT). The window length is three cycles on a 50Hz base and the step is one cycle. The first window is between 8.80s and 8.86s.

3.1 Measurement points 2

During the BG fault, for the step-up transformers have ungrounded star primary and delta secondary, voltages at the secondary side do not detect the primary fault presence. So the current supplied by the PV systems are not affected by such primary single line-to-ground (SLG) fault as shown in Fig. 2.(b).

During the BCG fault, due to the overhead line impedance, the fault-phase voltages on 35kV Bus II are not equal to zero in Fig. 2.(a). The currents of L_{24} are symmetrical in Fig. 2.(b). The fundamental current of each phase increases to 0.68pu by about 17.2%. The harmonics perform as odd ones and the third-order one accounts for approximately 4% of the fundamental, in Fig. 3.

3.2 Measurement point 5

During the BG fault, a SLG fault at primary side has no impacts on the voltages at the secondary side, for the wingding connection type of main transformers (YNd11). Therefore, the currents from corresponding PV systems are same to the prefault ones, as shown in Fig. 4.

During the BCG fault, the amplitude of the fault-phase voltages on 35kV Bus I decreases into about

0.85pu in Fig. 4.(a). At the fault inception, the instantaneous values of I_b and I_c increase to the maximum 0.784pu and 0.815pu, respectively. During the steady state, I_a reaches the maximum 0.816pu and three-phase currents are substantially symmetrical, as shown in Fig. 4.(b). The fundamental current value of each phase is about 0.69pu except that I_a is 0.72pu, about 12% larger than the prefault value. The harmonics mainly perform between 100Hz and 150Hz in Fig. 5.

Through the analysis above, one can find that the fault currents contributed by PV systems are at low level and the instantaneous maximum value could even not reach the rated value, because the fault behavior is governed by the control system that determines the switching pulses of the grid-connected converter [11] and the operating point of PV arrays may deviate from maximum power point (MPP) during faults. The odd harmonic component is relevant to the inverter modulation under unbalanced faults [12]. The harmonics values are too low to have effects.

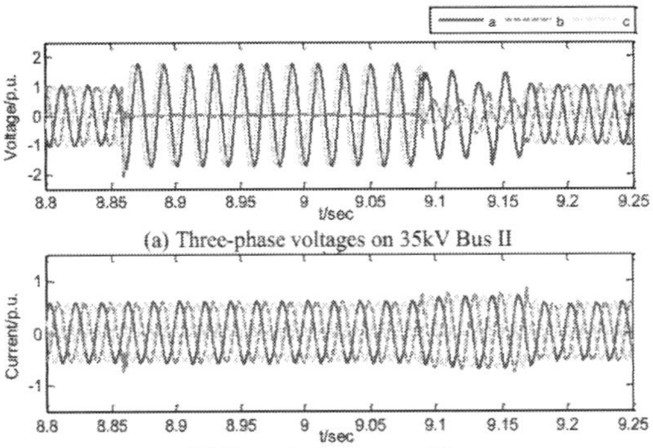

(a) Three-phase voltages on 35kV Bus II

(b) Three-phase currents of L_{24}

Figure 2. Voltages and currents recorded in the field test.

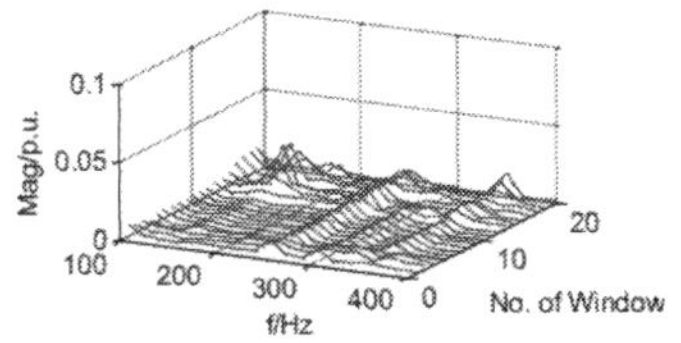

Figure 3. The current spectra in high band of the current of L_{24}.

4 Protection Configuration and Evaluation

A collection station has two 35kV buses and each bus collects 4 branch cables and an overhead line is connected with each bus. 8 PV generation units converge via a branch cable. An overhead line is configured with three sets of protection, the pilot current protection as the primary one, the overcurrent protection as the backup and the zero-sequence current protection for the grounded faults. The branch cable is equipped with two sets of protection, the overcurrent protection as the primary one and the zero-sequence current protection for the grounded faults. The specific coordination principles are displayed in Table 2.

4.1 Pilot current protection

The analysis in Section 3 indicates that the currents contributed by PV generation units increase not too much during faults. Considering that the system's short-circuit capacity is much greater, the short-circuit current from upstream will be much greater than the one from downstream. The ratio between the differential current and the braking current will be small so that the sensitivity will be reduced. Moreover, the PV power plant does not operate under the rated conditions for most time of the day, the sensitivity will be lower.

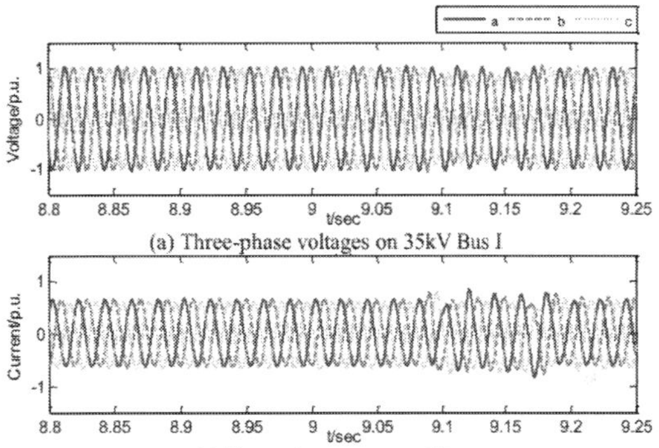

(a) Three-phase voltages on 35kV Bus I

(b) Three-phase currents of L_{15}

Figure 4. Voltages and currents recorded in the field test.

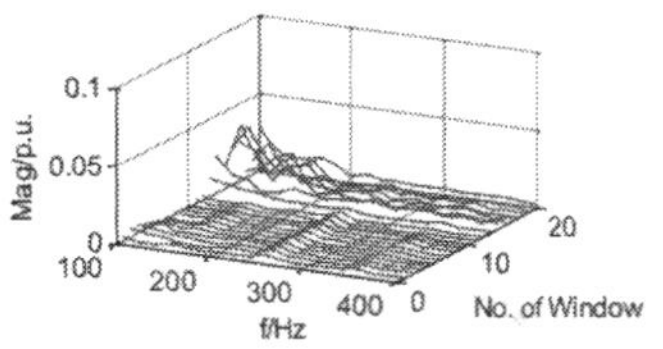

Figure 5. The current spectra in high band of the current of L_{14}.

4.2 Overcurrent protection

Suppose that a fault occurs on a branch cable, the current through the relay on the faulty cable is supplied by the system, the adjacent overhead lines and the neighboring branches. The current value must be larger than the setting values of the overcurrent protection on a branch cable. At this time, the coordination among the stages of the branch cable and the overhead line is same to a traditional radical distribution network and then accurate operation can be assured. In addition, the relays of neighboring branches and adjacent overhead lines will not malfunction for the low-level fault currents supplied by PV systems.

Suppose that a fault occurs on an overhead line, the current from the upstream is supplied by the system and the adjacent overhead lines. For the first stage assuring the sensitivity of 2 in case faults at the collection station bus, the upstream short-circuit current will be much greater than setting values of the upstream overcurrent protection. On the other side, the current from the downstream is contributed by the corresponding PV generation units. The first-stage setting value is the upstream counterpart divided by 1.1. Obviously, the short-circuit current from downstream cannot reach this level. The second-stage setting value is $1.2 \times 6I_e$ and higher than the maximum fault current supplied by 4 branch cables and the third stage will not pick up because the output current of typical commercial PV inverter products are designed for below 1.2pu [13]. Similarly, the overcurrent of a branch cable will not pick up. So, this will result in the rejection of the downstream overcurrent protection.

Therefore, a fault on an overhead line will be isolated by the upstream protection with a time delay when the pilot protection is out of service. Unfortunately, the PV generation units being cut off will continue to contribute currents to the failure point. This could damage the components in the circuit and bring security risk to the maintenance personnel. Most seriously, this may result in asynchronous reclosure and cause stoke currents, and reclosure failure for the fault still supplied by the PV generation units. The stability and security of the system and the plant will be influenced.

4.3 Zero-Sequence current protection

For a SLG fault on a 35kV line has almost no effect on both the system and the PV generation units. But a SLG fault may lead to a worse fault, a SLG fault within the collection system must be isolated immediately in a PV power plant [14]. To detect a SLG fault and isolate the failure, the zero-sequence current protection is established with a grounding transformer.

The first and second stage at different locations have same setting values indicated by I_a and I_b, respectively. The coordination is based on time intervals. Given step-up transformers with ungrounded primary and main transformers with delta secondary, in the event of earth faults, the zero-sequence circuit is composed of the distributed capacitance of overhead lines and branch cables, the zero-sequence impedance from the fault point to the grounding transformer, and that of the grounding transformer and its grounded resistance.

Suppose now that a SLG fault occurs on a branch cable, the time grading can ensure the shortest section should be disconnected.

Suppose now that a SLG fault occurs on an overhead line, the zero-sequence current flowing towards the downstream is only the capacitive current and it may not start the first stage. While that flowing towards the upstream is high enough to start the upstream first stage. The time intervals for the first stages at both ends are identical so that the upstream first stage will operate prior to the downstream one. Then, this case may result in the same risks as discussed in overcurrent protection.

Table 2. Existing protection coordination

Location	Protection Type		Coordination Principle	Setting Value	Delay
Substation Side (**Upstream**)	Overcurrent	I	Sensitivity of 2 when a fault occurs at the end	$0.866I_{f(3)}/2$	0.1s
		II	Coordinate with next stage I	$1.2*6I_e$	0.6s
		III	Maximum load current	$1.3nI_e$	0.9s
	Zero-sequence Current	I	Grounding resistance limit current	I_a	0.3s
		II	Line capacitive current	I_b	0.9s
Collection Station Side (**Downstream**)	Overcurrent	I	Coordinating with upstream stage I	$0.866I_{f(3)}/2/1.1$	0s
		II	Coordinating with next stage I	$1.2\times 6I_e$	0.3s
		III	Maximum load current	$1.3nI_e$	0.6s
	Zero-sequence Current	I	Grounding resistance limit current	I_a	0.3s
		II	Line capacitive current	I_b	0.6s
Collection Cable	Overcurrent	I	Step-up transformer inrush currents	$6I_e$	0s
		II	Maximum load current	$1.3I_e$	0.3s
	Zero-sequence Current	I	Grounding resistance limit current	I_a	0s
		II	Line capacitive current	I_b	0.3s

$I_{f(3)}$ means the short-circuit current in case of a three-phase fault at the 35kV bus in a collection station.
I_e is the current of a branch cable at the rated conditions.
n means the number of branches connected into a bus in a collection station.
I_a and I_b are the setting values.

4.4 Simulation results

To explain the problem aforementioned better, some simulations in case faults on the overhead line are conducted. The results in Fig. 6 and Fig. 7 are derived when the PV generation units operate at 0.6pu in unity PF, In Fig. 6, the subscript *down* and *up* mean the downstream and the upstream. The phase differences of fault-phase currents are shown in Fig. 7.

In Fig. 6 and Fig. 7, an AB to ground fault (ABG) occurs within the zone of the pilot relay on the overhead line. The during-fault currents from the PV system are symmetrical and the value increases not great much compared with the prefault one. The upstream current is about 20 times the downstream current. The prefault currents from the downstream and those from the upstream are in inverse phase. During the fault, I_{downA} lags I_{sysA} by 21°, while I_{downB} leads I_{upB} by 40°. Though there is a phase shift between the currents from two ends, the operating current of the pilot protection is almost equal to the braking current for faulty phases. Obviously, the ratio between the operating current and the braking current is close to one so that the sensitivity of the pilot element is reduced.

With the same fault condition but operating at 0.2pu in unity PF, the upstream current is about 40 times the value of the downstream current during the fault. The ratio of the operating current and the braking current is closer to 1.

The fault-phase currents from the upstream are high enough to ensure the upstream overcurrent protection to pick up. However, the current of each phase from the downstream could not reach the zones of the downstream overcurrent protection.

As for the zero-sequence current protection, the upstream current in the zero-sequence circuit will be many times higher than the downstream capacitive current and the same problem exists. To avoid redundancy, this is not displayed in this paper.

In order to improve the existing protection configuration, a set of transmitting and receiving device could be installed along the protection of overhead lines. Once the upstream relay operates, the downstream relay will receive an operating signal from upstream and operate immediately. This method is simple but its drawback is that extra communication apparatuses will increase the cost.

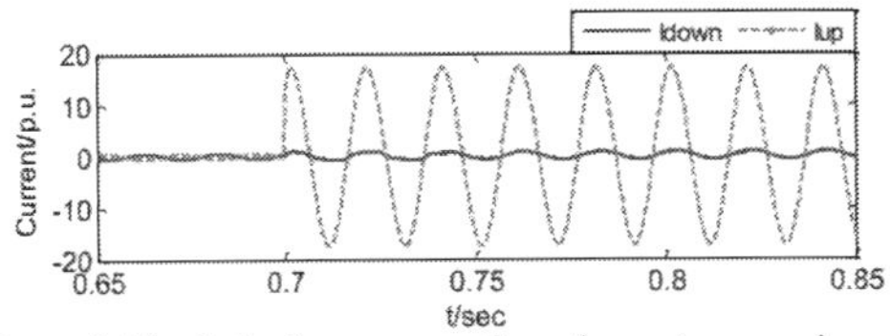

Figure 6. The fault-phase current from the upstream and downstream side under ABG with the operation of 0.6pu in unity PF

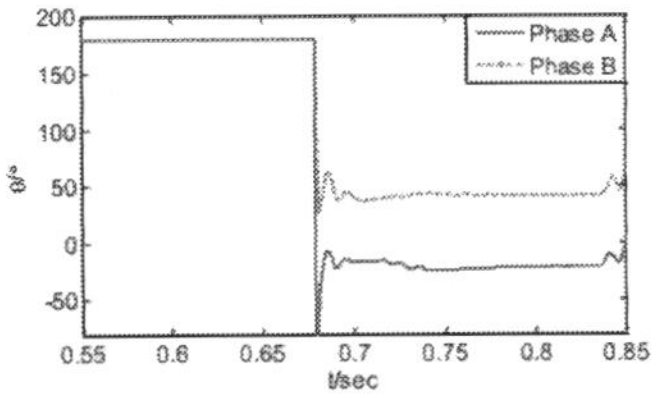

Figure 7. Phase difference between the downstream current and the upstream current of faulty phases.

5 Conclusion

The analysis of the fault currents recorded in a field test shows that the short-circuit currents supplied by the PV system increase not much and the postfault current level cannot be discriminated from the prefault one. Faults on the overhead line will not be detected by the overcurrent protection downstream. And the sensitivity of the pilot protection will be reduced for the same reason. This may result in the upstream protection operating prior to the downstream one with security risks. However, possible false trips of the overcurrent protections at adjacent lines can be eliminated, which is a significant problem in Wind Farms. The same problem exists for the zero-sequence current protection, as the zero-sequence circuit do not include PV generation units. The problem occurs when the pilot current protection fails, and this could be avoided via a simple method. The present protection configuration can response correctly to most requirements and situations and this protection scheme could be referred for PV power plants with similar topology.

References

1. A. Hooshyar , M.A. Azzouz, and E.F. El-Saadany, "Distance Protection of Lines Emanating from Full-Scale Converter-Interfaced Renewable Energy Power Plants—Part I: Problem Statement." IEEE Trans. Power Deliv., **30**, 11 (2015)
2. M. Ding, Z.C. Xu, W.S. Wang, X.L. Wang, Y.T. Song and D.Z. Chen, "A review on China' s large-scale PV integration: Progress, challenges and recommendations," Renew Sustain Energy Rev, **53**, 14 (2016)
3. A. Marinopoulos, F. Papandrea, M. Reza, S. Norrga, F. Spertino, and R. Napoli, "Grid integration aspects of large solar PV installations: LVRT capability and reactive power/voltage support requirements," *Proc. IEEE Trondheim PowerTech*, 8 (2011)
4. M. Mirhosseini, J. Pou, and V. Agelidis, "Single and two-stage inverter based grid-connected photovoltaic power plants with ride-through capability under grid faults," IEEE Trans. Sustain. Energy, 10 (2014)
5. M. Mirhosseini, J. Pou, and V. Agelidis, "Individual phase current Control with the capability to avoid overvoltage in grid-connected photovoltaic power plants under unbalanced voltage sags." IEEE Trans. Power Electron., **30**, 6 (2015)
6. R. A. Walling, R. Saint, R. C. Dugan, J. Burke, and L. A. Kojovic, "Summary of distributed resources impact on power delivery systems," IEEE Trans. Power Del., **23** 9 (2008)
7. Y. Lu, J. Du, X. Lin, and J. Ma, "An asymmetrical fault line selection based on I2 scalar product research in distribution system with DGs," *Proc. IEEE PES Gen. Meet.*, 6 (2008)
8. J. Driesen and R. Belmans, "Distributed generation: Challenges and possible solutions," *Proc. IEEE Power Eng. Soc. General Meeting*, 8 (2006)
9. M. A. Haj-Ahmed and M. S. Illindala, "The Influence of Inverter-Based DGs and Their Controllers on Distribution Network Protection," IEEE Trans. on Industry Applications, **50**, 10 (2014)
10. F. Coffele, C. Booth, and A. Dysko, "An adaptive overcurrent protection scheme for distribution networks," IEEE Trans. Power Del., **30**, 8 (2015)
11. J. Rocabert, A. Luna, F. Blaabjerg and P. Rodriguez, "Control of power converters in AC microgrids." IEEE Trans. Power Deliv., **27** 16(2012)
12. S.H. Sun, H.Q. Ben, Y.G. Li, L.Y. Wang and S. Yu, "Theory analysis of harmonic current generation and harmonic current extraction of three-phase grid-connected inverter." *9th International Conference on Power Electronics and ECCE Asia*, **6** (2015)
13. E. Ebrahimi, J.S. Mohammad, and G.B. Gevork, "Control of three-phase inverter-based DG system during fault condition without changing protection coordination." Int J of Electr Power Energy Syst, **63**, 8 (2014)
14. China Electric Power Research Institute. GB/T 19964-2012. Technical requirements for connecting photovoltaic power station to power system. Beijing: Standards Press of China; (2013)

AUTHOR INDEX

AUTHOR INDEX

AUTHOR INDEX

AUTHOR INDEX

AUTHOR INDEX